Carbonate Rock Depositional Models

Prentice Hall Advanced Reference Series

Physical and Life Sciences

BINKLEY *The Pineal: Endocrine and Nonendocrine Function*
CAROZZI *Carbonate Rock Depositional Models: A Microfacies Approach*
EISEN *Mathematical Methods and Models in the Biological Sciences:
    Linear and One-Dimensional Theory*
EISEN *Mathematical Methods and Models in the Biological Sciences:
    Nonlinear and Multidimensional Theory*
FRASER *Stratigraphic Evolution of Clastic Depositional Sequences*
JEGER, ED. *Spatial Components of Plant Disease Epidemics*
McLENNAN *Introduction to Nonequilibrium Statistical Mechanics*
PLISCHKE AND BERGERSEN *Equilibrium Statistical Physics*
VALENZUELA AND MYERS *Adsorption Equilibrium Data Handbook*
WARREN *Evaporite Sedimentology: Importance in Hydrocarbon
    Accumulation*

CARBONATE ROCK DEPOSITIONAL MODELS

A Microfacies Approach

ALBERT V. CAROZZI

University of Illinois at Urbana-Champaign

PRENTICE HALL
Englewood Cliffs, New Jersey 07632

Library of Congress Cataloging-in-Publication Data

Carozzi, Albert V.
   Carbonate rock depositional models : a microfacies approach /
 Albert V. Carozzi.
       p.   cm.—(Prentice-Hall advanced reference series)
   Bibliography: p.
   Includes index.
   ISBN 0-13-114398-0
   1. Rocks, Carbonate.   2. Sedimentation and deposition.   I. Title.
 II. Series.
 QE471.15.C3C38 1989   88–2536
 552′.5—dc19   CIP

Editorial/production supervision
  and interior design: Kathryn Gollin Marshak
Cover design: Photo Plus Art
Manufacturing buyer: Mary Ann Gloriande

Cover illustration: Spastolites in Hamburg Oolite (Lower Mississippian),
  Hamburg, southwest Illinois, U.S.A.

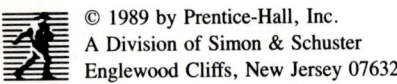 © 1989 by Prentice-Hall, Inc.
A Division of Simon & Schuster
Englewood Cliffs, New Jersey 07632

All rights reserved. No part of this book may be
reproduced, in any form or by any means,
without permission in writing from the publisher.

Printed in the United States of America
10 9 8 7 6 5 4 3 2 1

ISBN 0-13-114398-0

Prentice-Hall International (UK) Limited, *London*
Prentice-Hall of Australia Pty. Limited, *Sydney*
Prentice-Hall Canada Inc., *Toronto*
Prentice-Hall Hispanoamericana, S.A., *Mexico*
Prentice-Hall of India Private Limited, *New Delhi*
Prentice-Hall of Japan, Inc., *Tokyo*
Simon & Schuster Asia Pte. Ltd., *Singapore*
Editora Prentice-Hall do Brasil, Ltda., *Rio de Janeiro*

To Marguerite, once more

Contents

Preface   xix

Introduction   1

PART I PRINCIPLES

1 Carbonate Rocks   4

2 Microfacies Techniques   24

Introduction, 24 / Field Sampling, 25 / Preparation of Thin Sections, 25 / Inventory of Constituents, Matrix, Cement, and Textures, 27 / Qualitative Classification of Thin Sections into Preliminary Microfacies, 27 / Classification of Microfacies into a Preliminary Shallowing-Upward Sequence, 27 / Quantitative Analysis of Thin Sections, 28 / Statistical Evaluation and Final Characterization of Microfacies, 30 / Correlation Coefficients of Microfacies and Final Classification in an Ideal Shallowing-Upward Sequence, 31 / Graphic Representation of Microfacies, 32 / The Ideal Depositional Model, 33

3 Carbonate Ramp-Bar-Platform Evolution   34

Method of Study, 34 / Description of Microfacies, 35 / Depositional Models, 39 / The Ideal Depositional Model 1, 39 / The Ideal Depositional Model 2, 41 / The Ideal Depositional Model 3, 43 / Stratigraphic Sections and Vertical Succession of Microfacies, 43 / Vertical Succession of Depositional Models and Basinwide Evolution, 44 / Diagenesis, 45

PART II CARBONATE RAMPS

4 Carbonate Ramps in Open Marine Environment — 57

4.1 CHARACTERISTIC FEATURES: EFFECTS OF BIOTURBATION PROCESSES (Platteville Group, Middle Ordovician, Northern Illinois) 57

Description of Microfacies, *59* / The Ideal Depositional Models, *61*

4.2 CHARACTERISTIC FEATURES: EFFECTS OF STORM PROCESSES (Galena Group, Middle Ordovician, Upper Mississippi Valley) 61

Description of Microfacies, *61* / The Ideal Fair-Weather–Storm Shallowing-Upward Sequence, *64* / The Ideal Fair-Weather–Storm Depositional Model, *66* / General Evolution of the Depositional Environments, *76* / Diagenetic Evolution, *77*

4.3 CHARACTERISTIC FEATURES: EFFECTS OF CURRENT REWORKING PROCESSES (Menard Formation, Upper Mississippian, Southwest Illinois) 81

Description of Microfacies, *83* / The Ideal Shallowing-Upward Sequence, *86* / The Ideal Depositional Model, *90* / Relationship between Degree of Neomorphism and Environmental Energy Level, *90*

4.4 CHARACTERISTIC FEATURES: EFFECTS OF DETRITAL QUARTZ SEDIMENT YIELD (Bird Spring Group, Lower-Middle Pennsylvanian, Southeast Nevada) 90

Description of Microfacies, *92* / The Ideal Shallowing-Upward Sequences of the Morrowan, *92* / The Ideal Shallowing-Upward Sequence of the Atokan–Desmoinesian–Lower Missourian, *96* / The Ideal Depositional Model, *96*

5 Carbonate Ramps in Semirestricted Marine Environment — 101

5.1 CHARACTERISTIC CONSTITUENTS: AGGLUTINATED BENTHIC FORAMINIFERS (Itaituba–Nova Olinda Formations, Carboniferous-Permian, Amazon Basin, Brazil) 101

Description of Microfacies, *101* / The Ideal Depositional Model, *106*

6 Carbonate Ramps in Restricted Marine Environment — 108

6.1 CHARACTERISTIC CONSTITUENTS: OOIDS AND ONCOIDS (Barra Nova Formation, Albian-Cenomanian, Espirito Santo Basin, Brazil) 108

Description of Microfacies of Lower Regencia Member, *108* / The Ideal Depositional Model, *110* / Description of Microfacies of Middle Regencia Member, *110* / The Ideal Depositional Model, *115* / Diagenesis, *115*

7 Carbonate Ramps with Hydrodynamic Buildups — 119

7.1 CHARACTERISTIC CONSTITUENTS: CRINOIDS, SMALLER BENTHIC FORAMINIFERS, AND OOIDS (Salem Limestone, Middle Mississippian, Southeast Illinois) 119

Description of Microfacies, *119* / The Ideal Shallowing-Upward Sequences, *123* / The Ideal Depositional Models, *124* / Diagenetic Sequence, *125* / Reservoir Generation, *128*

8 Carbonate Ramps with Bioaccumulated to Hydrodynamic Buildups — 131

8.1 CHARACTERISTIC CONSTITUENTS: BRYOZOANS, CRINOIDS, AND OOIDS (Kinkaid Formation, Upper Mississippian, Southern Illinois and Western Kentucky) 131

Description of Microfacies of Model 1, *131* / The Ideal Depositional Model 1, *132* / Description of Microfacies of Models 2 and 2A, *135* / The Ideal Depositional Models 2 and 2A, *143* / Description of Microfacies of Model 3, *143* / The Ideal Depositional Model 3, *144* / The General Evolution of the Depositional Environments, *145*

8.2 CHARACTERISTIC CONSTITUENTS: CRINOIDS (Rundle Group, Mississippian, Alberta, Canada) 145

Description of Microfacies, *151* / The Ideal Shallowing-Upward Sequence, *153* / The Ideal Depositional Model, *154* / The General Evolution of the Depositional Environments, *155*

9 Carbonate Ramps with Bioconstructed Buildups — 156

9.1 CHARACTERISTIC CONSTITUENTS: STROMATOLITES (Upper Silurian, Cayugan, Northern Ohio) 156

Description of Microfacies, *156* / The Ideal Depositional Model, *159*

9.2 CHARACTERISTIC CONSTITUENTS: *AMPHIPORA* AND STROMATOPOROIDS (Arrow Canyon Formation and Crystal Pass Limestone, Upper Devonian, Southeast Nevada) 159

Description of Microfacies, *159* / The Ideal Shallowing-Upward Sequence, *161* / The Ideal Depositional Models, *161*

9.3 CHARACTERISTIC CONSTITUENTS: PHYLLOID AND RED ALGAE (Iola Formation, Pennsylvanian, Southeast Kansas) 165

Description of Microfacies, *165* / The Ideal Depositional Models, *171* / Diagenesis, *171*

9.4 CHARACTERISTIC CONSTITUENTS: PHYLLOID ALGAE (Brereton Limestone, Middle Pennsylvanian, Southwest Illinois) 173

Description of Microfacies, *175* / The Ideal Depositional Models, *176*

PART III CARBONATE PLATFORMS

10 Carbonate Platforms with Frontal Bioaccumulated Buildups — 191

10.1 CHARACTERISTIC CONSTITUENTS: MICRITE, BRYOZOANS (STROMATACTIS), AND CRINOIDS (Middle Silurian, Niagaran, Central Indiana) 191

Description of Microfacies, *191* / The Ideal Shallowing-Upward Sequence, *193* / The Ideal Depositional Model, *193*

10.2 CHARACTERISTIC CONSTITUENTS: MICRITE AND PELECYPODS (Chachao Formation, Valanginian, Neuquén Basin, Argentina) 193

Description of Microfacies, *195* / The Ideal Vertical Sequence of Microfacies, *201* / The Ideal Depositional Model, *202*

xii   Contents

11 Carbonate Platforms with Frontal Bioaccumulated to Hydrodynamic Buildups 209

11.1 CHARACTERISTIC CONSTITUENTS: PELLETS, *GIRVANELLA* ONCOIDS, AND *NUIA* (Pogonip Group, Lower Ordovician, Southeast Nevada) 209

Description of Microfacies, 209 / The Ideal Shallowing-Upward Sequence of the *Nuia* Model, 213 / The Ideal *Nuia* Depositional Model, 213 / The Ideal Shallowing-Upward Sequence of the Oncoid Model, 213 / The Ideal Oncoid Depositional Model, 213 / General Evolution of the Depositional Environments and Storm Deposits, 216

11.2 CHARACTERISTIC CONSTITUENTS: CRINOIDS AND BRYOZOANS (Burlington Limestone, Middle Mississippian, Upper Mississippi Valley) 218

Description of Microfacies, 220 / The Ideal Vertical Sequence, 226 / The Ideal Depositional Model, 227 / The Vertical Environmental Evolution, 227 / Synsedimentary Chert Breccia Interpreted as a Tempestite, 229

11.3 CHARACTERISTIC CONSTITUENTS: CRINOIDS (Edwardsville Formation, Middle Mississippian, Southern Indiana) 230

Description of Microfacies, 231 / The Ideal Vertical Sequence, 233 / The Ideal Depositional Model, 233

11.4 CHARACTERISTIC CONSTITUENTS: PELLETS AND CRINOIDS (Monte Cristo Group, Mississippian, Southeast Nevada) 235

Description of Microfacies, 238 / The Ideal Shallowing-Upward Sequences, 244 / The Ideal Depositional Model, 244 / General Evolution of the Depositional Environments, 249

11.5 CHARACTERISTIC CONSTITUENTS: CRINOIDS AND OOIDS (Bird Spring Group, Upper Pennsylvanian-Lower Permian, Southeast Nevada) 251

Description of Microfacies, 251 / The Ideal Shallowing-Upward Sequence, 253 / The Ideal Depositional Model, 255 / General Evolution of the Depositional Environments, 255

11.6 CHARACTERISTIC CONSTITUENTS: *GIRVANELLA* ONCOIDS (Macaé Formation, Albian-Cenomanian, Campos Basin, Offshore Brazil) 256

Description of Microfacies, 257 / The Ideal Shallowing-Upward Sequence, 264 / The Ideal Depositional Models, 268 / Diagenesis, 269

11.7 CHARACTERISTIC CONSTITUENTS: RED ALGAE AND LARGER BENTHIC FORAMINIFERS (Amapá Formation, Paleocene-Middle Miocene, Foz do Amazonas Basin, Offshore Brazil) 273

Description of Microfacies of Model 1, 275 / The Ideal Depositional Model 1, 278 / Diagenesis, 278 / Porosity Evolution, 279 / Description of Microfacies of Model 2, 280 / The Ideal Depositional Model 2, 287 / Diagenesis, 287 / Porosity Evolution, 287 / General Evolution of the Depositional Environments, 296

12 Carbonate Platforms with Frontal Hydrodynamic Buildups 303

12.1 CHARACTERISTIC CONSTITUENTS: SMALLER BENTHIC FORAMINIFERS, CRINOIDS, AND OOIDS (Salem Limestone, Middle Mississippian, Southwest Illinois) 303

Description of Microfacies, 303 / The Ideal Shallowing-Upward Sequence, 305 / The Ideal Depositional Model, 305 / General Evolution of the Depositional Environments, 307

12.2 CHARACTERISTIC CONSTITUENTS: PELLETS, CRINOIDS, AND OOIDS (Ste. Genevieve Limestone, Middle Mississippian, Southern Illinois) 307

Description of Microfacies, 307 / Relationships of Microfacies, 313 / The Ideal Shallowing-Upward Sequences, 313 / The Ideal Depositional Model, 313 / Analysis of Autochthonous and Allochthonous Oolitic Environments, 314

12.3 CHARACTERISTIC CONSTITUENTS: SMALLER BENTHIC FORAMINIFERS, CRINOIDS, BRYOZOANS, OOIDS, AND ONCOIDS (Glen Dean Formation, Middle Mississippian, Southern Illinois-Indiana-Kentucky) 318

Description of Carbonate Microfacies, 319 / Description of Siliciclastic Microfacies, 321 / The Ideal Carbonate Shallowing-Upward Sequence, 323 / The Ideal Carbonate Depositional Model, 323 / The Ideal Siliciclastic Depositional Model, 327 / Carbonate Diagenesis, 327

12.4 CHARACTERISTIC CONSTITUENTS: SMALLER BENTHIC FORAMINIFERS, CRINOIDS, AND BRYOZOANS (Glen Dean Formation, Middle Mississippian, Southwest Illinois) 329

Description of Microfacies, 329 / The Ideal Shallowing-Upward Sequence, 331 / The Ideal Depositional Model, 332 / The General Evolution of the Depositional Environments, 334

12.5 CHARACTERISTIC CONSTITUENTS: OOIDS (Quintuco–Loma Montosa Formation, Lower Cretaceous, Neuquén Basin, Argentina) 334

Description of Microfacies, 334 / The Vertical Depositional Sequences, 340 / The Ideal Depositional Models, 340

13 Carbonate Platforms with Frontal Bioconstructed to Hydrodynamic Buildups 352

13.1 CHARACTERISTIC CONSTITUENTS: STROMATOPOROIDS, CORALS, AND *AMPHIPORA* (Jeffersonville Limestone, Middle Devonian, Southeast Indiana) 352

Description of Microfacies, 352 / The Ideal Shallowing-Upward Sequence, 354 / The Ideal Depositional Model, 355

13.2 CHARACTERISTIC CONSTITUENTS: STROMATOPOROIDS AND CORALS (Jeffersonville Limestone, Middle Devonian, Southeast Indiana) 355

Description of Microfacies, 358 / The Ideal Vertical Sequence, 359 / The Ideal Depositional Model, 359

13.3 CHARACTERISTIC CONSTITUENTS: STROMATOPOROIDS AND *AMPHIPORA* (Beaverhill Lake Formation, Upper Devonian, Alberta, Canada) 359

Description of Microfacies, 362 / The Ideal Shallowing-Upward Sequence, 363 / The Ideal Depositional Model, 364 / The General Evolution of the Depositional Environments, 364

13.4 CHARACTERISTIC CONSTITUENTS: CORALS, STROMATOPOROIDS, AND STROMATOLITES (Traverse Group, Givétian, Southern Peninsula of Michigan 364

Description of Microfacies, *365* / Description of Sections and Ideal Depositional Models, *373*

13.5 CHARACTERISTIC CONSTITUENTS: *DONEZELLA* AND CRINOIDS (Atokan Limestone, Middle Pennsylvanian, Delaware Basin, Texas) 377

Description of Microfacies, *377* / The Ideal Shallowing-Upward Sequence, *381* / The Ideal Depositional Model, *381* / Diagenesis, *386* / The Evolution of the Depositional Environments, *390*

13.6 CHARACTERISTIC CONSTITUENTS: SPONGES, CORALS, STROMATOPOROIDS, AND PELECYPODS (Upper Jurassic, Salève, Haute-Savoie, France) 398

Description of Microfacies, *398* / The Ideal Shallowing-Upward Sequence, *405* / The Ideal Depositional Model, *405*

13.7 CHARACTERISTIC CONSTITUENTS: RED AND GREEN ALGAE (Bonfim Formation, Cenomanian, Barreirinhas Basin, Brazil) 405

Description of Microfacies, *407* / The Ideal Depositional Models, *417*

14 Carbonate Platforms with Frontal Bioconstructed Buildups 419

14.1 CHARACTERISTIC CONSTITUENTS: STROMATOLITES UNDER STORM PROCESSES (Allentown Dolomite, Upper Cambrian, New Jersey) 419

Description of Microfacies, *419* / The Ideal Shallowing-Upward Sequence, *421* / The Ideal Depositional Model, *421* / General Evolution of the Depositional Environments, *422*

14.2 CHARACTERISTIC CONSTITUENTS: STROMATOLITES (Shakopee Dolomite, Lower Ordovician, Southwest Wisconsin) 423

Description of Microfacies, *423* / The Ideal Shallowing-Upward Sequence, *425* / The Ideal Depositional Model, *425*

14.3 CHARACTERISTIC CONSTITUENTS: STROMATOLITES AND *NUIA* (Joachim Dolomite, Middle Ordovician, Upper Mississippi Valley) 425

Description of Microfacies, *428* / The Ideal Shallowing-Upward Sequence, *431* / The Ideal Depositional Model, *437*

14.4 CHARACTERISTIC CONSTITUENTS: STROMATOLITES (*SPONGIOSTROMATA*) (Burnt Bluff Group, Middle Silurian, Southeast Wisconsin) 441

Description of Microfacies, *441* / Ideal Shallowing-Upward Sequence of Nasbro Buildup, *446* / The Ideal Depositional Model of Nasbro Buildup, *446* / The Ideal Shallowing-Upward Sequence of Sturgeon Bay Buildup, *448* / The Ideal Depositional Model of Sturgeon Bay Buildup, *448* / Organically Encrusted Hydrodynamic Buildup of Chilton, *448*

14.5 CHARACTERISTIC CONSTITUENTS: CORALS, RED ALGAE, AND ENCRUSTING FORAMINIFERS (Miocene, Visayan Islands, Central Philippines) 448

Description of Microfacies, *451* / Diagenesis, *457* / The Ideal Depositional Model, *460*

PART IV CARBONATE SLOPES AND BASINS

15 Carbonate Turbidites in Cratonic and Orogenic Settings — 465

15.1 CHARACTERISTIC CONSTITUENTS: CRINOIDS (Middle Silurian, Niagaran, Central Indiana) 465

Description of Microfacies, 465 / The Vertical Sequences, 467 / The Ideal Depositional Model, 467

15.2 CHARACTERISTIC CONSTITUENTS: CRINOIDS AND PELLETS (Bailey Limestone, Lower Devonian, Southern Illinois) 468

Description of Microfacies, 469 / The Ideal Depositional Sequence, 469 / The Origin of Components, 474 / The Vertical Evolution of the Investigated Section, 474

15.3 CHARACTERISTIC CONSTITUENTS: BENTHIC BIOCLASTS, PELLETS, OOIDS, AND QUARTZ (Upper Jurassic, Morcles Nappe, Haute-Savoie, France) 475

Description of Microfacies, 475 / Horizontal Distribution and Correlation of the Turbidites, 478 / Horizontal Grading, 478

16 Carbonate Slopes in Cratonic Settings — 482

16.1 CHARACTERISTIC FEATURES: PHOSPHATIZATION PROCESSES (Galena-Maquoketa Contact, Middle-Upper Ordovician, Upper Mississippi Valley) 482

Description of Microfacies, 483 / The Ideal Shallowing-Upward Sequence, 487 / The Ideal Depositional Model, 491 / Diagenesis and Regional Paleogeography, 495

17 Lacustrine Carbonates with Bioaccumulated to Hydrodynamic Buildups — 496

17.1 CHARACTERISTIC CONSTITUENTS: PELECYPODS, OSTRACODS, AND BASIC HYALOCLASTITES (Lagoa Feia Formation, Lower Cretaceous, Campos Basin, Offshore Brazil) 496

Description of Microfacies, 496 / Vertical Sequences of Microfacies, 500 / Ideal Vertical Sequences of Microfacies, 500 / The Ideal Playa Lake Depositional Model, 502 / The Ideal Pluvial Lake Depositional Model, 502 / The Ideal Subaqueous Basic Volcanism Model, 503 / Diagenesis and Porosity Evolution, 506

PART V TOWARD AN EXPLANATION

18 Synthesis — 516

18.1 INTRODUCTION TO THE GENETIC CLASSIFICATION OF CARBONATE ROCK DEPOSITIONAL MODELS 516

18.2 GENETIC CLASSIFICATION OF SIMPLE RAMPS 519

Type 1. Platteville Group, Middle Ordovician, Northern Illinois, 519 / Type 2. Galena Group, Middle Ordovician, Upper Mississippi Valley, 519 / Type 3. Menard Formation, Upper Mississippian, Southwest Illinois, 519 / Type 4. Bird Spring Group, Lower-Middle Pennsylvanian, Southeast Nevada, and Arrow Canyon Formation and Crystal Pass

Limestone, Upper Devonian, Southeast Nevada, *519* / Type 5. Kinkaid Formation, Upper Mississippian, Southern Illinois, and St. Louis Limestone, Middle Mississippian, Illinois Basin, *521* / Type 6. Itaituba–Nova Olinda Formations, Carboniferous-Permian, Amazon Basin, Brazil, *523* / Distribution of Potential Primary to Early Diagenetic Reservoirs in Simple Ramps, *523*

18.3 GENETIC CLASSIFICATION OF RAMPS WITH BIOACCUMULATED TO HYDRODYNAMIC BUILDUPS *523*

Type 1. Brereton Limestone, Middle Pennsylvanian, Southwest Illinois, *524* / Type 2. Rundle Group, Mississippian, Alberta, Canada, *524* / Type 3. St. Louis Limestone, Middle Mississippian, Illinois Basin, *525* / Type 4. Salem Limestone, Middle Mississippian, Southeast Illinois, *525* / Type 5. Kinkaid Formation, Upper Mississippian, Southern Illinois, *525* / Distribution of Potential Primary to Early Diagenetic Reservoirs in Ramps with Bioaccumulated to Hydrodynamic Buildups, *528*

18.4 GENETIC CLASSIFICATION OF RAMPS WITH BIOCONSTRUCTED BUILDUPS *528*

Type 1. Brereton Limestone, Middle Pennsylvanian, Southwest Illinois, *528* / Type 2. Arrow Canyon Formation and Crystal Pass Limestone, Upper Devonian, Southeast Nevada, *528* / Type 3. Upper Silurian, Cayugan, Northern Ohio, *528* / Distribution of Potential Primary to Early Diagenetic Reservoirs in Ramps with Bioconstructed Buildups, *530*

18.5 GENERAL SYNTHESIS OF RAMPS *530*

Fair-Weather Wave Base, Tidal Currents, and Submarine Morphology, *530* / Coexistence of *in Situ* and Transported Organic and Inorganic Constituents, *531* / Distribution Pattern of Oolitic Microfacies, *532* / Distribution Pattern of Oncoids, *532* / Distribution Pattern of Phosphates and Glauconite, *532* / Distribution Pattern of Extrabasinal Materials, *532* / Distribution Pattern of Dolomitization, *533* / Distribution Pattern of Storm Deposits, *533* / Distribution Pattern of Potential Primary to Early Diagenetic Reservoirs, *534*

18.6 GENETIC CLASSIFICATION OF PLATFORMS WITH FRONTAL BIOACCUMULATED BUILDUPS *534*

Type 1. Chachao Formation, Valanginian, Neuquén Basin, Argentina, *534*

18.7 GENETIC CLASSIFICATION OF PLATFORMS WITH FRONTAL BIOACCUMULATED TO HYDRODYNAMIC BUILDUPS *534*

Type 1. Amapá Formation, Paleocene-Middle Miocene, Foz do Amazonas Basin, Offshore Brazil, *534* / Type 2. Pogonip Group, Lower Ordovician, Southeast Nevada, *536* / Type 3. Monte Cristo Group, Mississippian, Southeast Nevada, and Bird Spring Group, Upper Pennsylvanian–Lower Permian, Southeast Nevada, *538* / Type 4. St. Louis Limestone, Middle Mississippian, Illinois Basin, and Ste. Genevieve Limestone, Middle Mississippian, Southern Illinois, *539* / Type 5. Salem Limestone, Middle Mississippian, Southwest Illinois, Glen Dean Formation, Middle Mississippian, Western Shelf of Illinois Basin, and Glen Dean Formation, Middle Mississippian, Eastern Shelf of Illinois Basin, *541* / Type 6. Macaé Formation, Albian-Cenomanian, Campos Basin, Offshore Brazil, *543*

18.8 GENERAL SYNTHESIS OF PLATFORMS WITH FRONTAL BIOACCUMULATED TO HYDRODYNAMIC BUILDUPS *544*

Fair-Weather Wave Base, Tidal Currents, and Submarine Morphology, *544* / Coexistence of *in Situ* and Transported Organic and Inorganic Constituents, *545* / Distribution Pattern of Oolitic Microfacies, *545* / Distribution Pattern of Oncoids, *546* / Distribution Pattern of Phosphates and Glauconite, *546* / Distribution Pattern of Extrabasinal Materials, *546* / Distribution Pattern of Dolomitization, *546* / Distribution Pattern of Storm Deposits, *547* / Distribution Pattern of Potential Primary to Early Diagenetic Reservoirs, *547*

18.9 GENETIC CLASSIFICATION OF PLATFORMS WITH FRONTAL BIOCONSTRUCTED TO HYDRODYNAMIC BUILDUPS 547

Type 1. Beaverhill Lake Formation, Upper Devonian, Alberta, Canada, and Atokan Limestone, Middle Pennsylvanian, Delaware Basin, Texas, *547* / Type 2. Joachim Dolomite, Middle Ordovician, Upper Mississippi Valley, *549* / Type 3. Allentown Dolomite, Upper Cambrian, New Jersey, and Shakopee Dolomite, Lower Ordovician, Southwest Wisconsin, and Miocene, Visayan Islands, Central Philippines, *550* / Type 4. Bonfim Formation, Cenomanian, Barreirinhas Basin, Brazil, and Jeffersonville Limestone, Middle Devonian, Southeast Indiana, and Traverse Group, Givétian, Southern Peninsula of Michigan, *552*

18.10 GENERAL SYNTHESIS OF PLATFORMS WITH FRONTAL BIOCONSTRUCTED TO HYDRODYNAMIC BUILDUPS 554

Fair-Weather Wave Base, Tidal Currents, and Submarine Morphology, *554* / Coexistence of *in Situ* and Transported Organic and Inorganic Constituents, *555* / Distribution Pattern of Oolitic Microfacies, *555* / Distribution Pattern of Oncoids, *555* / Distribution Pattern of Phosphates and Glauconite, *555* / Distribution Pattern of Extrabasinal Materials, *555* / Distribution Pattern of Dolomitization, *555* / Distribution Pattern of Storm Deposits, *555* / Distribution Pattern of Potential Primary to Early Diagenetic Reservoirs, *555*

18.11 GENETIC CLASSIFICATION OF LACUSTRINE MODELS 555

Type 1. Lagoa Feia Formation, Playa Lake Stage, Lower Cretaceous, Campos Basin, Offshore Brazil, *556* / Type 2. Lagoa Feia Formation, Pluvial Lake Stage, Lower Cretaceous, Campos Basin, Offshore Brazil, *557*

18.12 GENERAL SYNTHESIS OF LACUSTRINE MODELS 557

Fair-Weather Wave Base, Seiches, and Sublacustrine Morphology, *557* / Coexistence of *in Situ* and Transported Organic and Inorganic Constituents, *558* / Distribution Pattern of Oolitic Microfacies, *558* / Distribution Pattern of Oncoids, *558* / Distribution Pattern of Extrabasinal Materials, *558* / Distribution Pattern of Dolomitization, *558* / Distribution Pattern of Storm Deposits, *558* / Distribution Pattern of Potential Primary to Early Diagenetic Reservoirs, *558*

18.13 PALEOCLIMATIC IMPLICATIONS OF CARBONATE ROCK DEPOSITIONAL MODELS 558

Paleogeographic Interpretations, *559* / Interpretation of Storm Deposits, *561*

18.14 CYCLICITY OF CARBONATE SEDIMENTATION 562

18.15 A NEW EUSTATIC MODEL 568

References   571

Index   579

Preface

Every method starts with a few simple concepts, proceeds through a time of growth with increasing sophistication, and finally reaches a time of synthesis when case histories can be brought together. This book summarizes more than forty-five years of personal research on the petrography and depositional environments of carbonate rocks. Although it naturally rests on initial concepts and later interpretations and contributions in the field, it differs from other approaches in carbonate petrography by representing the work of successive generations of graduate students guided by the same author, by the painstaking petrographic measurements completed on more than 50,000 thin sections, and by the result of many challenging discussions and new ideas afforded by the interactions between student and adviser and between coworker and consultant, which all led to the perfecting of the method.

ACKNOWLEDGMENTS

In the course of this work I have greatly benefited from the constant suggestions by all my graduate students and coworkers whose names are recorded in joint publications, particularly William C. Dawson, and in final analysis, from the critical readers of these papers. I am indebted also to Nancy R. Black and Sadat Feiznia for allowing me to use materials extensively from their unpublished dissertations.

Sincere thanks are extended to George deVries Klein for his thorough critical reading of the manuscript and his numerous provocative suggestions, which have greatly enhanced the presentation, and to Daniel A. Textoris for his review and pertinent comments.

I greatly appreciate the constant support given by David E. Anderson, Head of the Department of Geology at the University of Illinois, particularly his understanding that a blending of academic duties and consulting activities is a source of constant challenge and therefore of benefit to all involved.

Many thanks are extended to Jacques Deferne of the Natural History Museum of Geneva for his generous technical help, to Jan L. Reichelderfer for her preparation of most of the plates of photomicrographs, to David R. Phillips, Jessie Knox, and Gilles Roth for their outstanding drafting of numerous figures, and to Karolyn Roberts for a perfect final typing.

I deeply appreciate the keen interest of Michael Hays, Senior Editor and Assistant Vice-President, College Editorial Division, and the help of Kathryn G. Marshak, Senior Production Editor (Freelance).

Finally, I must repeat that I shall never be able to acknowledge the debt of gratitude I owe to my wife Marguerite for her constant encouragement and help through the sunny and cloudy days of these many years.

Albert V. Carozzi
Urbana and Geneva

Introduction

The concept of index of clasticity pertaining to the maximum apparent grain size of detrital minerals (quartz, glauconite, muscovite) and of an index of frequency (abundance) of benthic organisms (*Inoceramus* prisms and *Lagena*) and planktonic organisms (*Gümbelina*), as determined in thin sections, were first presented and illustrated graphically by means of vertical variation curves in a study of two stratigraphic sections of Upper Cretaceous limestones of the High Calcareous Alps (Paréjas and Lillie, 1935a,b). The index of clasticity of detrital minerals was interpreted as a direct expression of the power of transporting agents, often inversely related to distance from shoreline, and depth of water. The frequency of benthic organisms displayed a general relationship to quartz clasticity, whereas an opposite behavior was observed between the frequencies of benthic and planktonic organisms (Fig. I.1). From these observations, one of the sections investigated appeared shallower than the other as shown by its larger clasticity and greater abundance of detrital minerals, although both sections showed a rapid deepening in time revealed by the vertical decrease in clasticity of detrital minerals and the change of a benthic fauna into a planktonic fauna. In this pioneer study, microscopic parameters were used as criteria for environmental interpretation and correlation in time and space.

This technique was applied subsequently to the Purbeckian freshwater limestones of the Jura Mountains (Carozzi, 1948) and to various sections of the Lower Cretaceous of the Alps (Carozzi, 1949). The environmental interpretation was represented graphically for the first time in both works by a "curve of relative bathymetry" expressing the existence of numerous shallowing-upward sequences designated at the time as "asymmetric sedimentary rhythms." More precise definitions of the index of clasticity and frequency, as well as their application to a variety of detrital, authigenic, and organic constituents of carbonate sequences in the Alps and in Canada, were proposed the following year (Carozzi, 1950). This more refined approach was tested on the entire Cretaceous System of the High Calcareous Alps (Carozzi, 1951a,b, 1952a) and used as a correlation technique (Carozzi, 1951c).

Under the influence of Philip H. Kuenen, the hypothesis of turbidity currents was introduced in Alpine sedimentation (Carozzi, 1952b), and among carbonate rocks, it was first applied to the so-called "reworked zones" of the Upper Jurassic deep-water calcilutites of the High Calcareous Alps (Carozzi, 1952c,d), and later extended to several other tectonic units (Kuenen and Carozzi, 1953).

The association of the author with the University of Illinois spurred a collaboration with Harold R. Wanless and the application of what had just become known as "microfacies techniques" to the Ordovician carbonates of the Middle West (Carozzi, 1956; Wanless *et al.*, 1956).

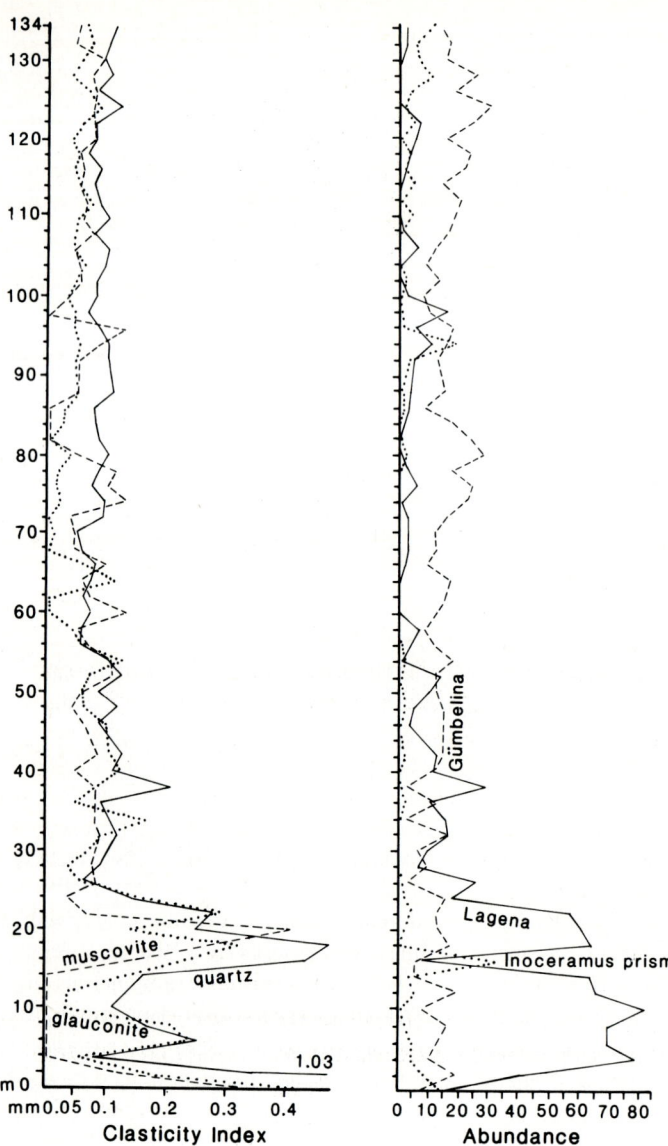

FIGURE I.1 Example of typical vertical variation of clasticity index and abundance of microscopic parameters in the Upper Cretaceous limestone sequence of Châtelard-en-Bauges, High Calcareous Alps, Savoy, France. Redrawn from Paréjas and Lillie (1935a).

These first studies were followed by a long and fruitful relationship with graduate students with whom he studied the Paleozoic carbonates of the Appalachians, the Middle West, and Nevada and, under the sponsorship of major oil companies, carbonates of many parts of the world, including recently those of the People's Republic of China (Yang and Carozzi, 1987).

Case histories in this book are not organized chronologically, although it is easy to find them in such an order and assess the increasing degree of sophistication of the approach as time went by. Indeed, initial studies which showed a few variation curves of parameters in simple shallowing-upward sequences, with little concern for diagenesis, were followed by computerized curves which illustrate complex depositional-diagenetic models in which diagenesis takes an increasing importance. This trend of increased sophistication is traced to the involvement of oil companies from Shell Oil Canada in 1957 to Texaco U.S.A. in 1984. His long association with Petróleo Brasileiro S.A. PETROBRÁS since 1968, mainly under the influence of its director, Carlos Walter M. Campos, provided the author with the triple opportunity to undertake microfacies research directly related to exploration, to teach its principles in training programs, and to direct doctoral dissertations at the University of Illinois on the microfacies of major offshore carbonate units of Brazil (Falkenhein et al., 1981; Bertani ànd Carozzi, 1984; Wolff and Carozzi, 1984).

The Société des Pétroles d'Aquitaine, now Elf-Aquitaine, sponsored a volume on the carbonate microfacies of the Jurassic of the Aquitaine Basin which was a first attempt at integrating petrography, geochemistry of trace elements, clay mineralogy, paleosalinity, and petrophysics into the concept of microfacies (Carozzi *et al.*, 1972). This encompassing definition was also used in a microfacies synthesis of the Miocene reefs of the Visayan Sea in the Philippines published by Philippine Oil Development Company (Carozzi *et al.*, 1976).

In 1978, the author began a productive association with Yacimientos Petrolíferos Fiscales of Argentina (Y.P.F.) which introduced carbonate petrography and microfacies techniques to that country and led, in collaboration with the Geological Association of Argentina, to the first symposium on carbonate rocks at the National Congress of 1981 in San Luis. The president of the Association, Gualter A. Chebli, was the enthusiastic supporter of the publication of a book on carbonate depositional models (Carozzi, 1983) which appeared in spite of tremendous financial problems. In the author's first synthesis of carbonate depositional models, examples he had investigated by microfacies methods were combined with those of other researchers who had used conventional techniques. Written in Spanish, the book provided the Latin American geologists with a much needed introduction to carbonate rocks in their own language.

From the work noted above evolved the present volume, which is organized as follows. The first part, "Principles," includes a review of the study of carbonate rocks, a full account of the technique of description and interpretation of microfacies by means of sets of variation curves of their microscopic parameters, and a completely described Paleozoic example of the application of microfacies techniques to the natural evolution of a carbonate depositional model which starts from a simple ramp, proceeds through a subtidal bar, and terminates in a complex platform.

Part II, "Carbonate Ramps," Part III, "Carbonate Platforms," and Part IV, "Carbonate Slopes and Basins," describe the various types of environments and/or of buildups which are used as criteria for a second-order subdivision. This subdivision includes specific case histories in stratigraphic order for the purpose of providing data on the Phanerozoic evolution of benthic communities, which are the major controlling factors of carbonate sedimentation. The description of previously published case histories has been summarized and streamlined in terms of petrographic and environmental terminology. The major part of the original graphic representation, with its scale in meters or feet, has been reproduced, whereas some cases were reinterpreted and redrafted to ensure a uniform treatment. Complete sets of photomicrographs were prepared to illustrate both previous and new work. For the sake of brevity, variation curves of parameters were not described as exhaustively as in the original papers, which should also be consulted for matters pertaining to regional stratigraphy and detailed geographic location of investigated sections. At first glance, the reader may therefore feel confronted by a vast and perhaps bewildering array of curves whose significance necessarily may not be grasped instantly. He or she will find it necessary to analyze the graphic representations with care to see the associations and disassociations which illustrate the pattern of physical and biological constituents. These patterns characterize microfacies which resulted from processes peculiar to distinct environments genetically assembled into depositional models.

Part V, "Toward an Explanation," is an attempt at a genetic classification of carbonate ramps, platforms, and lacustrine models based on the most characteristic case histories selected from the previous descriptive parts of the book. This genetic classification is process oriented and emphasizes the role of the major depositional and diagenetic factors as a function of the increase in energy and in complexity of the models. The interest of some of these aspects for reservoir generation of hydrocarbons and their exploration is also stressed. A discussion of problems related to the nature and possible causes of cyclicity of carbonates in general, and of shallowing-upward sequences in particular, precedes the presentation of a new eustatic model.

The author is aware that further analysis of these complex interrelationships between microscopic parameters and certainly other interpretations may be warranted. He hopes that he has provided material for future studies to other carbonate sedimentologists, particularly in view of the acute need for more data on both shallowing-upward sequences and depositional models derived from them.

1

Carbonate Rocks

Investigation of carbonate microfacies relies necessarily on previous contributions, among which a choice has to be made with respect to definitions, concepts, and classifications. The method presented here was therefore influenced by a series of choices, as described below.

The modern systematic petrographic study of carbonates was established by Cayeux (1935). It was further developed by Carozzi (1953, 1960, 1972), and Bathurst (1975) made it a fully recognized independent field of investigation. At the present time, specialization in this field has reached such proportions that a separate volume is required for a satisfactory treatment of each of its numerous subdivisions, ranging from coated grains (Peryt, 1983) to carbonate cements (Schneidermann and Harris, 1985).

Any attempt at describing carbonate rocks in detail under the microscope requires a classification (Fig. 1.1). The scheme used in this book (Carozzi, 1983) has been tested for many years and calls for a full sentence for each thin section which names all major constituents and their textural relationships. Such a description may be, for example: coarse grain-supported crinoid and bryozoan calcarenite with a cavity-filling mosaic of equant sparite cement. This apparently rigid approach is indispensable for microfacies characterization and comparison, whereas use of the term "biosparite" for the above-described rock is inadequate. This classification has borrowed many aspects from earlier ones (Folk, 1962; Dunham, 1962), and like them, it is genetic and based on the concept of energy (bedshear) of the environment of deposition (grain-supported versus mud-supported, cement/matrix ratio).

The environmental interpretation of carbonate rocks has been attempted by countless authors with the realization of the important role played by discrete masses constructed or accumulated by sedentary organisms (Longman, 1981; Toomey, 1981; Harris, 1983). These "reefs" have been for many years the object of a widespread controversy and their designation as "buildups" (Heckel, 1974), with qualifying adjectives, has been found well adapted to most circumstances and is used in this book (Fig. 1.2).

The almost complete control of carbonate sedimentation by benthic communities is a well-known fact (Heckel, 1974; James, 1983). It implies that regardless of the existence of buildups, depositional models proposed for carbonate rocks of different geological ages display important differences originating from the paleoecological community prevalent at a particular time, a situation further complicated by extinction (Fig. 1.3). Therefore, among ancient carbonate environments, with the exception of a few mechanical processes, the present is hardly the key to the past, as shown by the Bahama syndrome. Consequently, ancient carbonate environ-

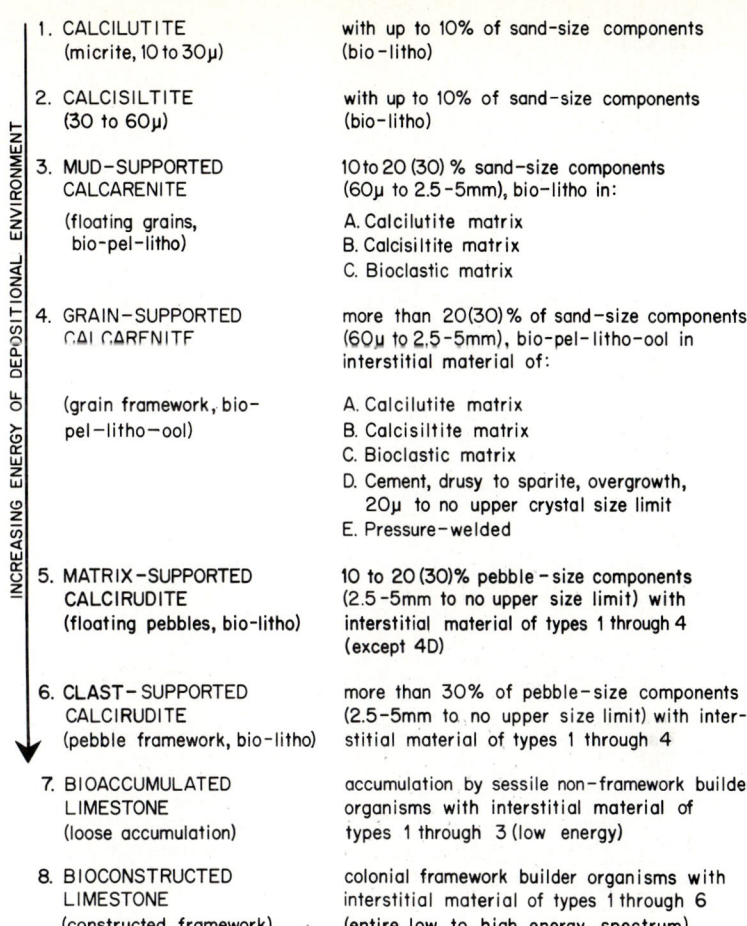

FIGURE 1.1 Practical classification of carbonate rocks. Modified from Carozzi (1983).

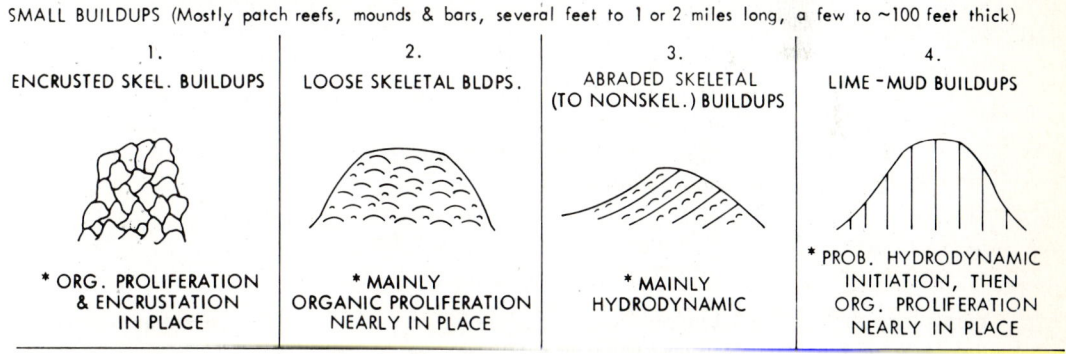

FIGURE 1.2 Diagram showing major compositional types of carbonate buildups grouped by common size range. Asterisks mark most probable origins. From Heckel (1974). Reprinted by permission of the Society of Economic Paleontologists and Mineralogists.

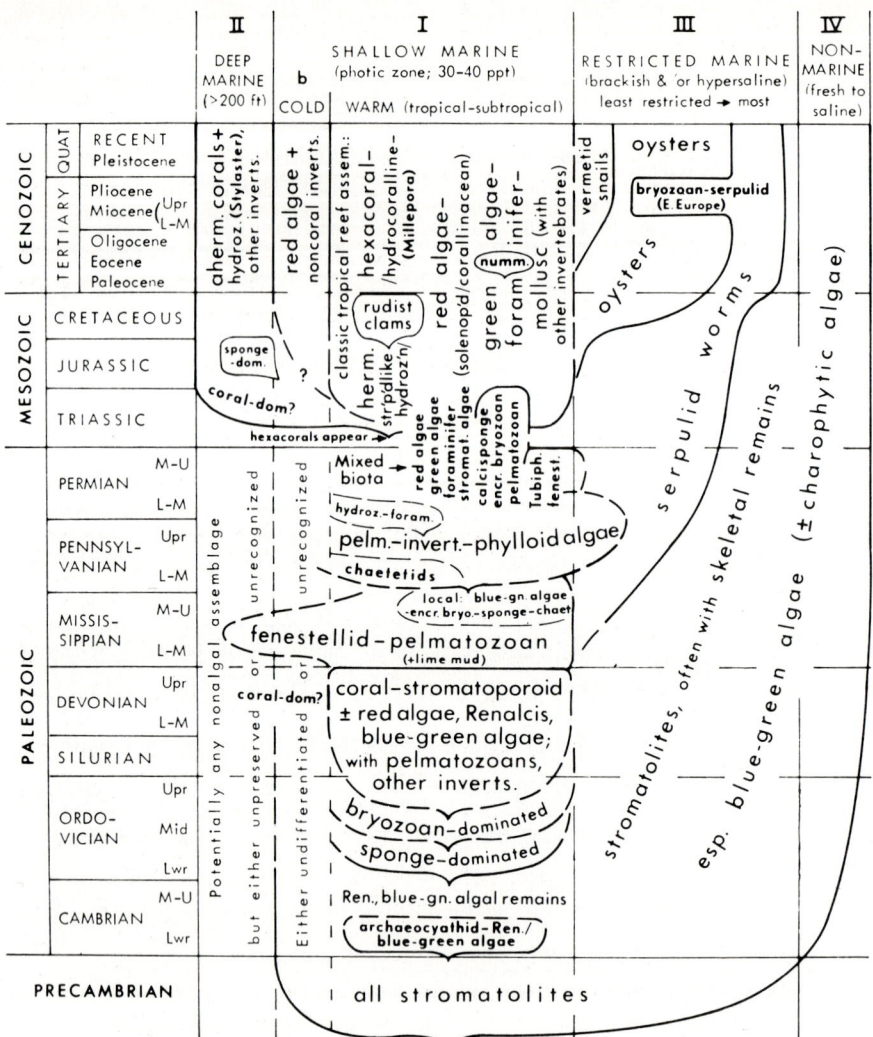

FIGURE 1.3 Geological history of major skeletal buildup assemblages in major environmental regimes. From Heckel (1974). Reprinted by permission of the Society of Economic Paleontologists and Mineralogists.

ments are challenging situations to be interpreted on the basis of their own intrinsic characteristics and by means of the hypothesis of simplicity.

The most characteristic feature of carbonate rocks is the extraordinary variety of shallowing-upward sequences (Fig. 1.4), both at small and large scale. The causes of this arrangement with its predictive capability remain obscure. Nevertheless, the shallowing-upward sequence finds its best expression when represented, according to Walther's law (Middleton, 1973), as the carbonate ramp depositional model. The latter was proposed by Irwin (1965) in two fundamental diagrams (Figs. 1.5, 1.6), further developed by Ahr (1973), Read (1982, 1985), and summarized by Nelson (1978).

The carbonate ramp model (Fig. 1.7), although initiated as a gentle seaward slope (perhaps inherited from an underlying siliciclastic environment), rarely maintains such a simple geometry for very long because at the location where the wave base intersects the submarine slope, conditions are generated which favor the development of subtidal bioclastic bars (hydrodynamic buildups) and incipient constructed buildups. The prolific carbonate production at that particular location, associated with related processes of sediment trapping and baffling possesses the capacity to change a carbonate ramp with its incipient buildups into a carbonate platform. In this new depositional model, buildups become the oceanward edge of a carbonate system (rimmed carbonate

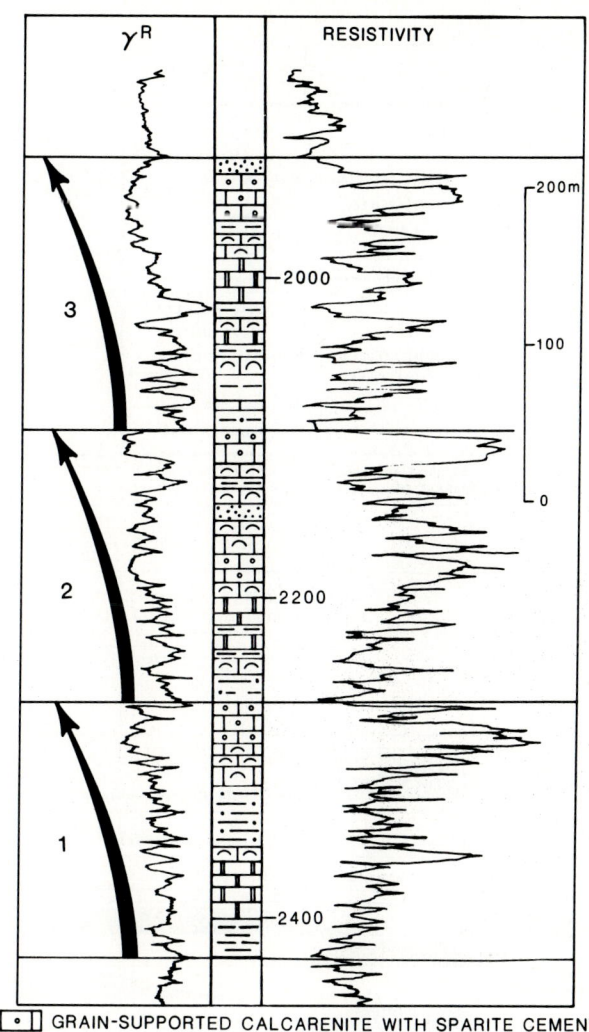

FIGURE 1.4 Examples of shallowing-upward sequences (arrows) in the Upper Member of the Quintuco Formation, Lower Cretaceous, Loma La Lata field, Neuquén Basin, Argentina. Redrawn from Carozzi et al. (1982).

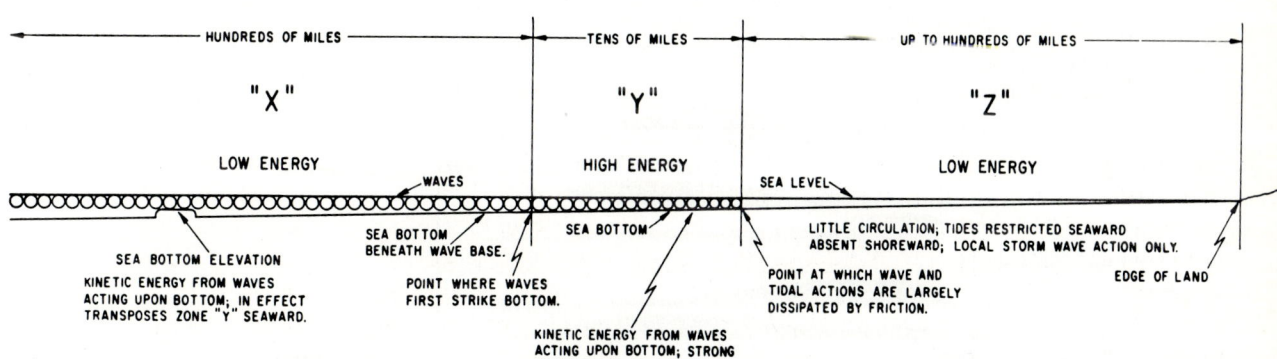

FIGURE 1.5 Section showing energy zones in epeiric seas, not to scale. From Irwin (1965). Reprinted by permission of the American Association of Petroleum Geologists.

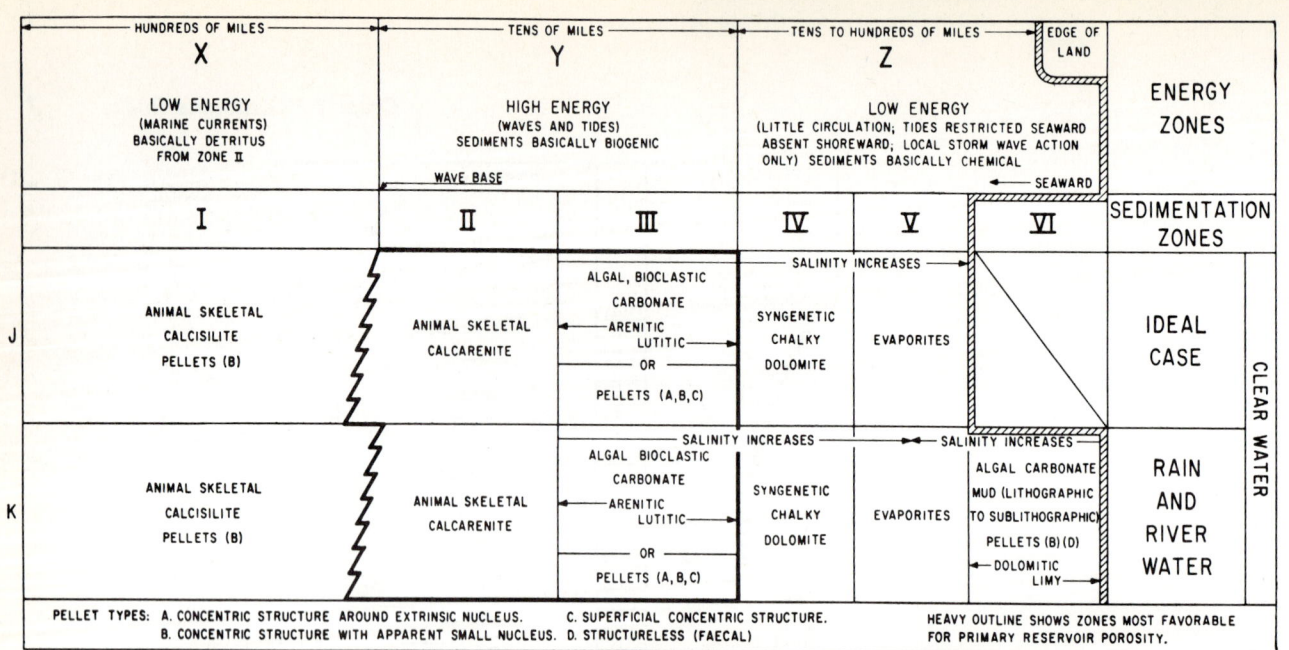

FIGURE 1.6 Relationships between clear water energy and sedimentation zones. From Irwin (1965). Reprinted by permission of the American Association of Petroleum Geologists.

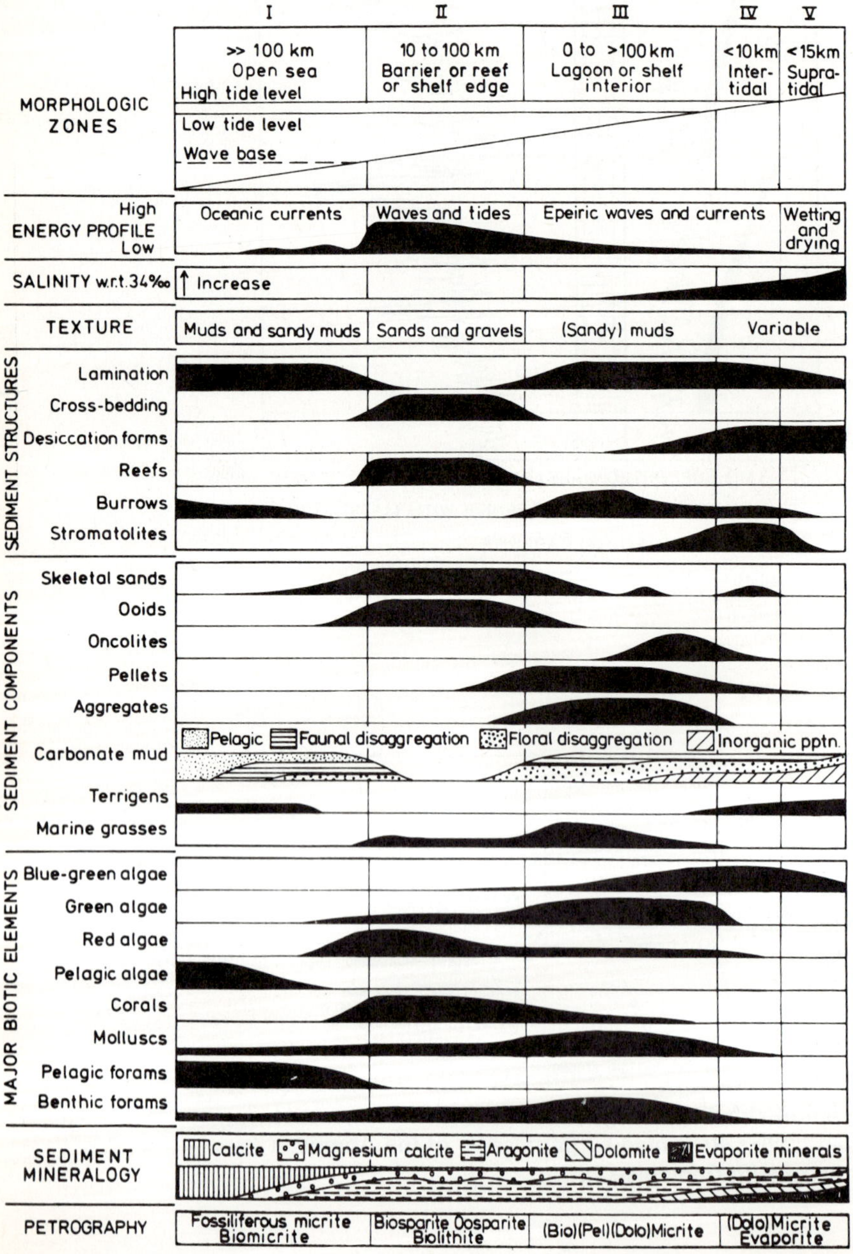

FIGURE 1.7 Theoretical model of shallow marine carbonate ramp sedimentation. From Nelson (1978). Reprinted by permission of the International Association of Sedimentologists.

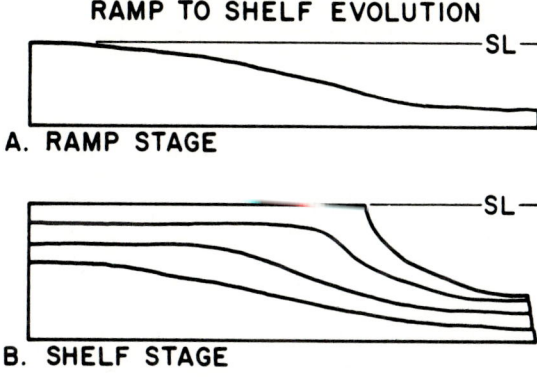
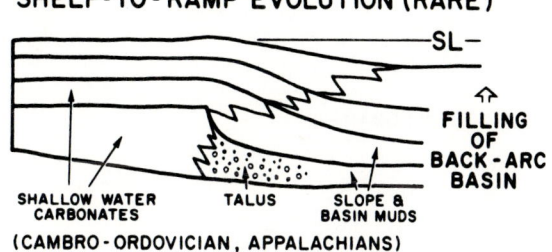

FIGURE 1.8 Evolution of a carbonate ramp to a rimmed shelf and evolution of the latter into a ramp with filling of marginal basin. From Read (1982). Reprinted by permission of Elsevier Publishing Company.

shelf) whose internal complexity increases through time (Harris, 1984), whereas the transition to basinal carbonates occurs by means of debris flows, turbidites, and submarine fans identical to siliciclastic ones (Cook *et al.*, 1972, 1983; Cook and Enos, 1977; James and Mountjoy, 1983; Mountjoy *et al.*, 1972; Crevello and Harris, 1985). Simultaneously, basin margins display continuous sequences of lagoonal to supratidal evaporitic sediments of sabkha type (Fisher, 1977; Zenger *et al.*, 1980).

The ramp to platform evolution in time is characteristic of the margins of most carbonate basins (Fig. 1.8) and generally involves a slow process of sedimentation. It remains incomplete often and may end at any stage by the effects of subsidence, active tectonism, eustatism, or influx of siliciclastic sediments. Platform growth displays a stationary margin when the rate of carbonate accretion equals essentially the rate of relative sea level rise, or displays an offlap or prograding margin when carbonate accretion is faster than the rate of relative sea level rise (Fig. 1.9). An important factor in the final shape of carbonate platforms is the intensity of waves impinging on their margins (Wilson, 1974, 1975), which leads to the recognition of three major types of fossil platform edges (Fig. 1.10). The reverse evolution, platform or shelf to ramp, is rare and results from drowning of the carbonate system due to rise of sea level or through burial by prograding clastics (Fig. 1.8).

In the final analysis, the nature of carbonate models depends on numerous factors, among which the paleobiology of the organisms capable of forming buildups and their environmental setting are the most critical. A temporal sequence can be established (Fig. 1.11) showing times when only carbonate ramps existed, others characterized by low-energy bioaccumulated mounds, and still others when mounds and well-developed buildups were associated. In the latter situation, complex morphologies are common (Figs. 1.12, 1.13) and require precise terminologies.

Examples of carbonate depositional models have been described in numerous publications. Excellent introductions may be found in Davis (1983), Walker (1984), and Reading (1986). A spectacular treatment with lavish color illustrations has been presented by Scholle *et al.* (1983). More specific discussions of carbonate depositional models in terms of exploration for hydrocarbon reservoirs have been written by the sedimentologists of Elf-Aquitaine (1975, 1977), emphasizing the principles of analysis and interpretation, with a condensed version in English (Elf-Aquitaine *et al.*, 1982), and also by Halley and Loucks (1980).

Recent advances in carbonate petrography and geochemistry have amply demonstrated the importance of diagenetic processes and of the related diagenetic environments they characterize. Carbonate sediments undergo these processes immediately after deposition, and during gradual burial they pass through various diagenetic environments, following a great number of pathways that can frequently lead to partial or total destruction of previously acquired diagenetic textures. At present, the complexity of carbonate diagenesis under shallow or deep conditions of burial is a field open for speculation

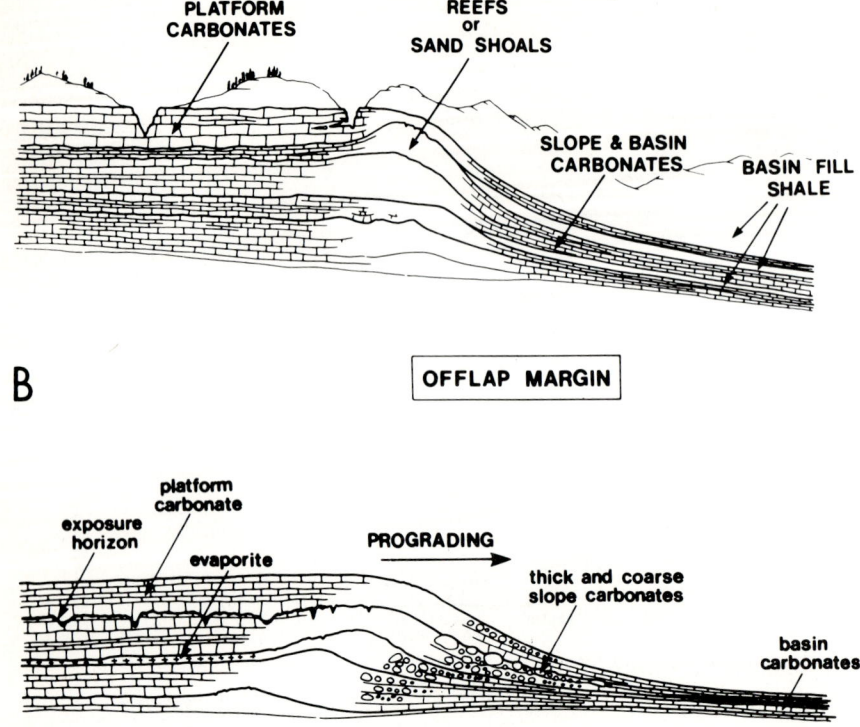

FIGURE 1.9 A. Sketch illustrating the main elements of a fossil carbonate platform margin. In this example the shelf-slope break is in a *stationary* mode, remaining more or less in the same position as the platform grew. B. Diagram of a carbonate platform in which the rate of accretion has exceeded the relative rate of sea level rise and the shelf slope is in the *offlap* mode, prograding over older slope deposits. From James and Mountjoy (1983). Reprinted by permission of the Society of Economic Paleontologists and Mineralogists.

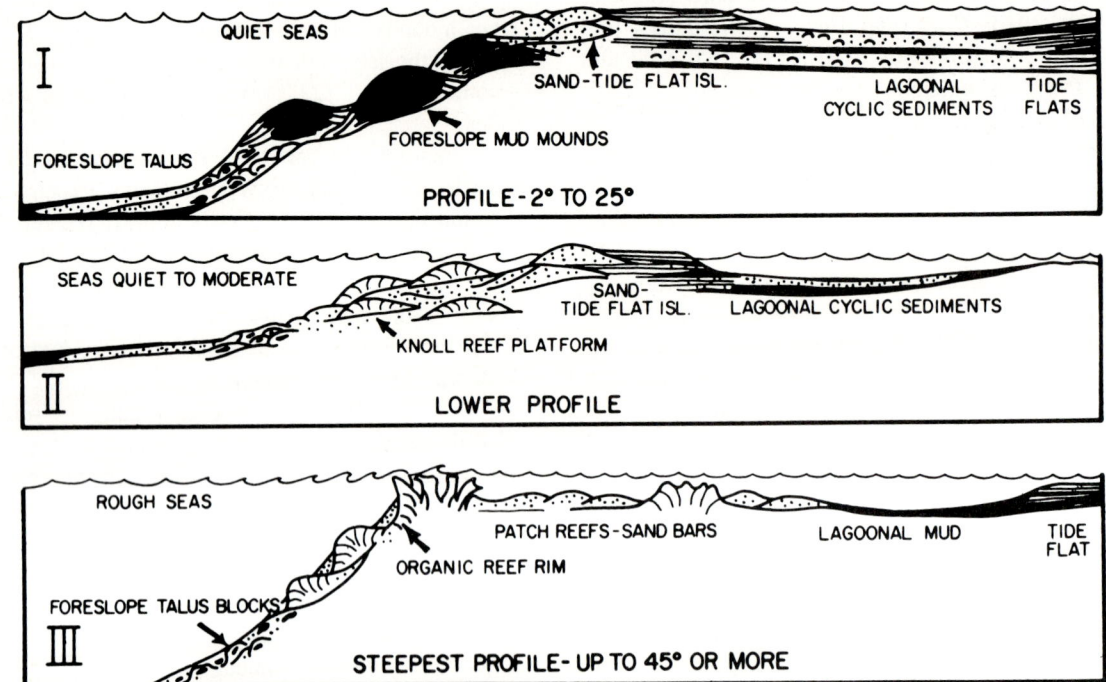

FIGURE 1.10 Formation of margins of shelf areas by carbonate sedimentation. I. Mud mounds on the shelf slope; sediment binding organisms build quiet water bioherms. II. Knoll reef platform with a clear predominance of bioclastic debris in the interreef areas. III. Framework reef rim in the turbulent zone adjoining the reef complex. From Wilson (1974). Reprinted by permission of the American Association of Petroleum Geologists.

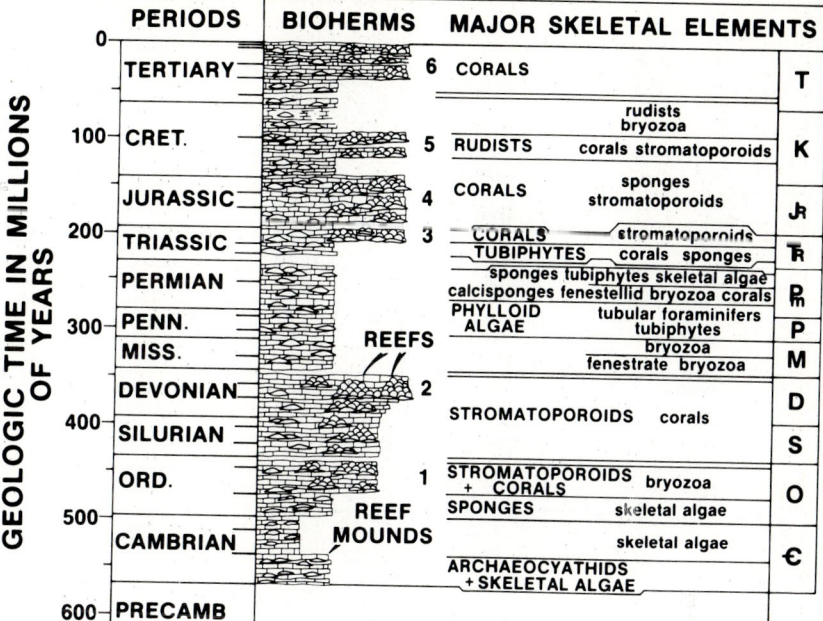

FIGURE 1.11 Idealized stratigraphic column representing the Phanerozoic and illustrating times lacking reefs or bioherms (gaps), times of reef mounds only, and times of both reefs and reef mounds, and the organisms that built them. From James (1983). Reprinted by permission of the American Association of Petroleum Geologists.

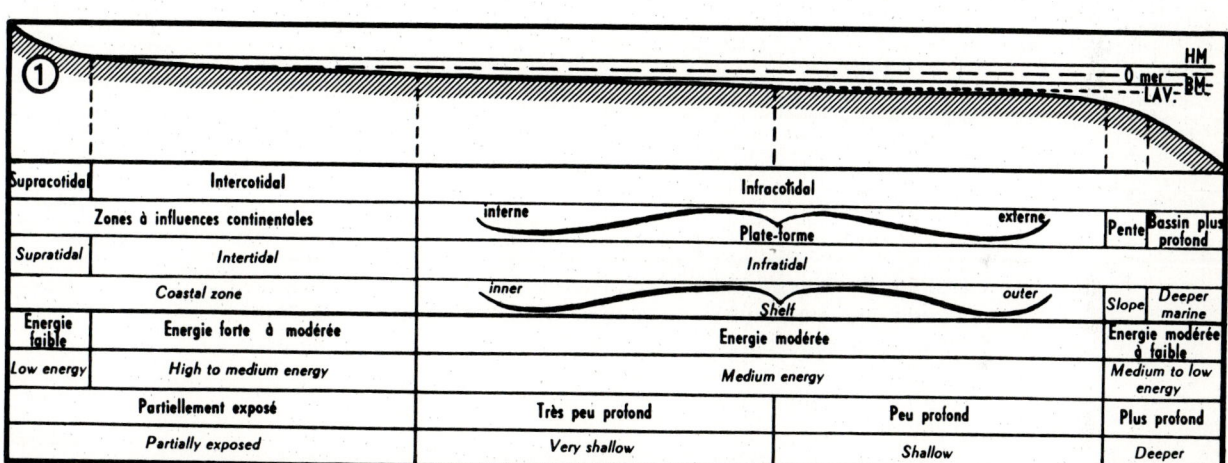

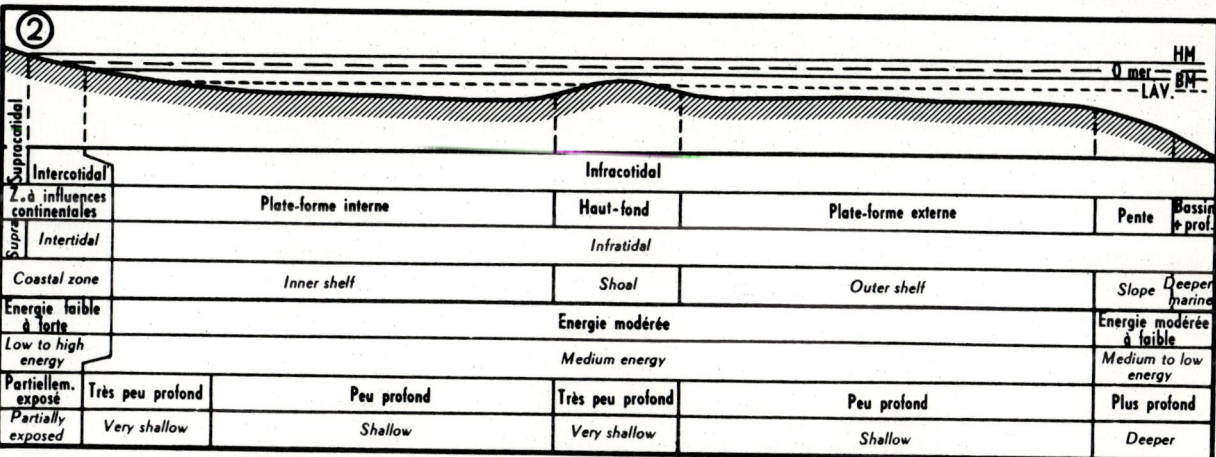

FIGURE 1.12 Theoretical models of distribution of carbonate marine environments. HM, high tide; BM, low tide; LAV, wave base. 1. Ramp to simple platform, 2. Platform with infratidal shoal. From Carozzi et al. (1972). Reprinted by permission of Société Nationale des Pétroles d'Aquitaine.

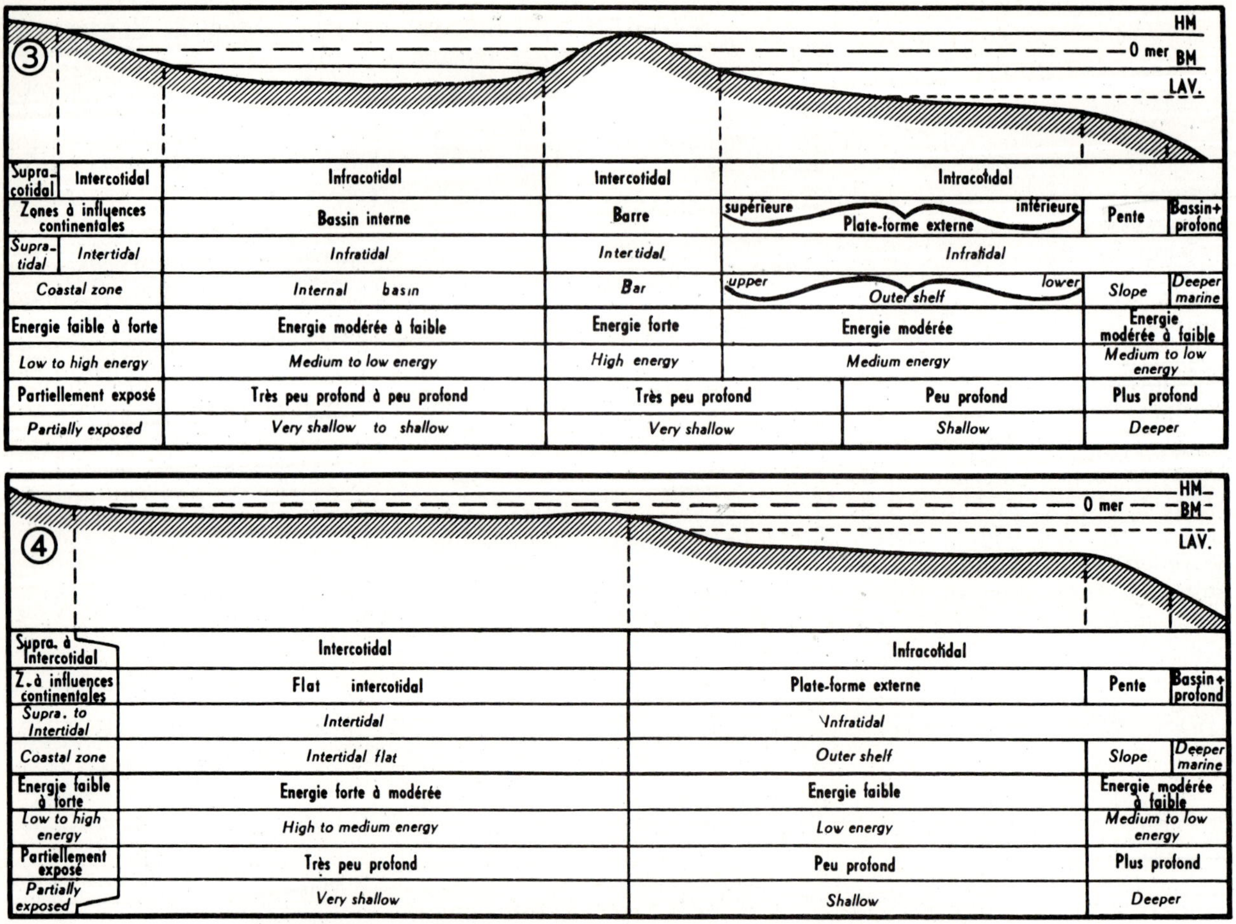

12 Part I / Principles

Supra-cotidal	Intercotidal	Infracotidal	Intercotidal	Intracotidal		
Zones à influences continentales		Bassin interne	Barre	supérieure inférieure Plate-forme externe	Pente	Bassin + profond
Supra-tidal	Intertidal	Infratidal	Intertidal	Infratidal		
Coastal zone		Internal basin	Bar	upper lower Outer shelf	Slope	Deeper marine
Energie faible à forte		Energie modérée à faible	Energie forte	Energie modérée	Energie modérée à faible	
Low to high energy		Medium to low energy	High energy	Medium energy	Medium to low energy	
Partiellement exposé		Très peu profond à peu profond	Très peu profond	Peu profond	Plus profond	
Partially exposed		Very shallow to shallow	Very shallow	Shallow	Deeper	

Supra. à Intercotidal	Intercotidal	Infracotidal		
Z. à influences continentales	Flat intercotidal	Plate-forme externe	Pente	Bassin + profond
Supra. to Intertidal	Intertidal	Infratidal		
Coastal zone	Intertidal flat	Outer shelf	Slope	Deeper marine
Energie faible à forte	Energie forte à modérée	Energie faible	Energie modérée à faible	
Low to high energy	High to medium energy	Low energy	Medium to low energy	
Partiellement exposé	Très peu profond	Peu profond	Plus profond	
Partially exposed	Very shallow	Shallow	Deeper	

FIGURE 1.13 Theoretical models of distribution of carbonate marine environments. HM, high tide; M, low tide; LAV, wave base. 3. Platform with intertidal bar, 4. Platform with intertidal flat. From Carozzi et al. (1972). Reprinted by permission of Société Nationale des Pétroles d'Aquitaine.

in which this book does not indulge. At this stage, oversimplifications are required (Longman, 1980) and on the basis of textural characteristics, three diagenetic environments can be distinguished in the shallow subsurface involving an ideal carbonate body (Fig. 1.14): marine phreatic (Figs. 1.15, 1.16), freshwater vadose (Fig. 1.17), and freshwater phreatic (Fig. 1.18). Each environment has its own set of petrographic characteristics. Under particular circumstances, a fourth diagenetic environment may be added: mixed freshwater-seawater phreatic (Figs. 1.19, 1.20). It is the location of the dorag model of secondary dolomitization (Badiozamani, 1973; Zenger et al., 1980). Further subsidence takes the carbonate rocks into deep phreatic burial conditions, an extremely complex and still little understood diagenetic environment where the slow circulation of many brines causes far-reaching effects, among which are burial dolomitization (Mattes and Mountjoy, 1980) and generation of secondary porosity.

The description and understanding of porosity in carbonate rocks requires a systematic nomenclature (Fig. 1.21) which has proven extremely useful (Choquette and Pray, 1970). The strict application of this terminology is critical for the classification of petroleum reservoirs (Roehl and Choquette, 1985).

To establish the diagenetic pathway of a carbonate sediment on its way to subsurface burial and to unravel its potential to become a reservoir is one of the major tasks of microfacies studies. The case history of such a depositional-diagenetic sequence in the Quintuco Formation, Lower Cretaceous of Argentina (Rodriguez Schelotto et al. 1981; Carozzi et al., 1982) is presented here as an example. It consists of the following stages (Fig. 1.22): marine phreatic (micritization, cementation, and

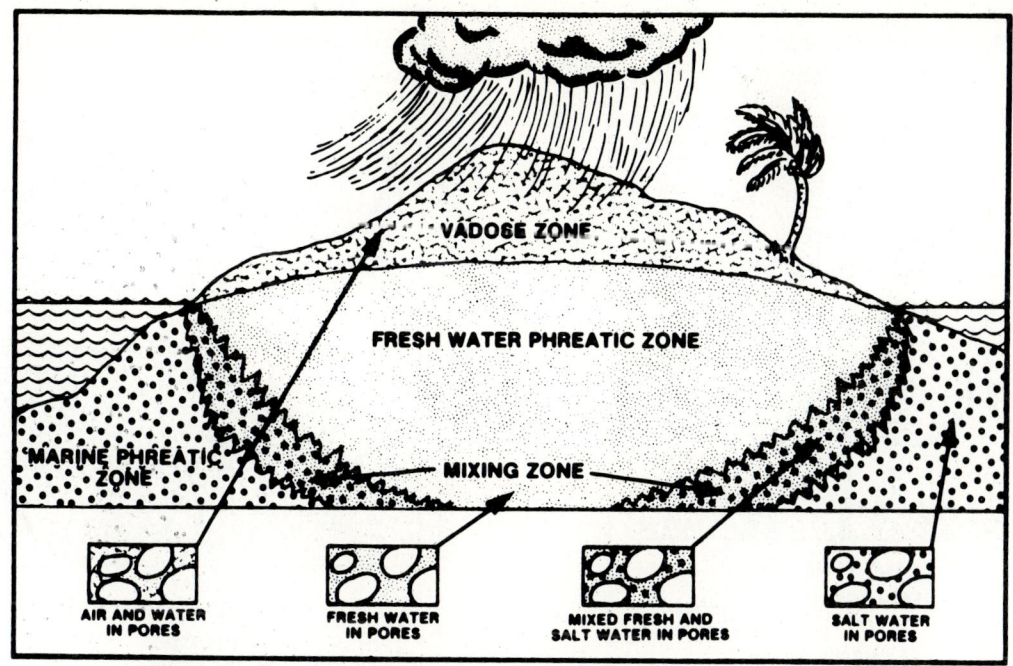

FIGURE 1.14 Cross section showing distribution and relations of major diagenetic environments in shallow subsurface in ideal permeable carbonate-sand island. Not to scale, but vertical distance would typically represent tens of meters and horizontal distance would be a few kilometers. From Longman (1980). Reprinted by permission of the American Association of Petroleum Geologists.

early compaction); freshwater vadose (dissolution of aragonite tests); mixing freshwater-marine phreatic (early dolomitization); freshwater phreatic (sparite cementation and neomorphism), and burial (late compaction, late dolomitization, and late dissolution generating secondary porosity). The petrographic illustration of these processes (Figs. 1.23 to 1.27) represents but one pathway of carbonate diagenesis; many others are described in subsequent chapters.

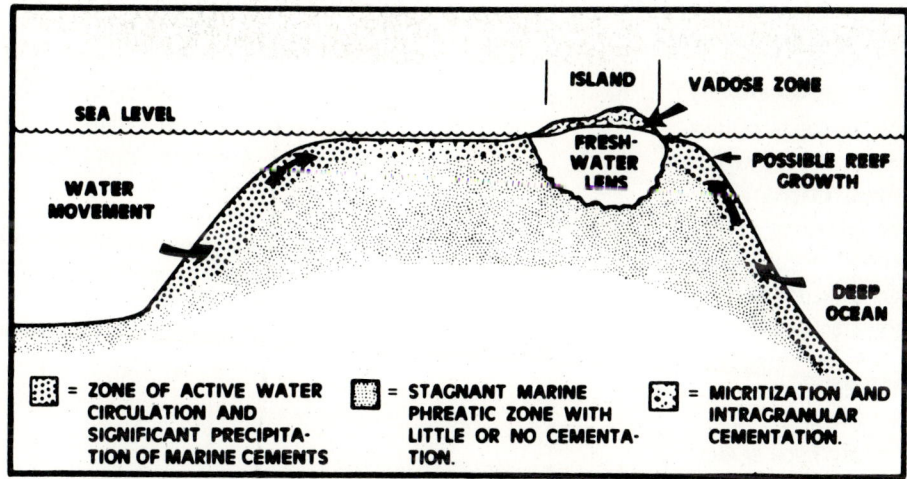

FIGURE 1.15 Cross section of carbonate bank with marine phreatic zone divided into active circulation area (cementation) and stagnant (little cementation). Not to scale. From Longman (1980). Reprinted by permission of the American Association of Petroleum Geologists.

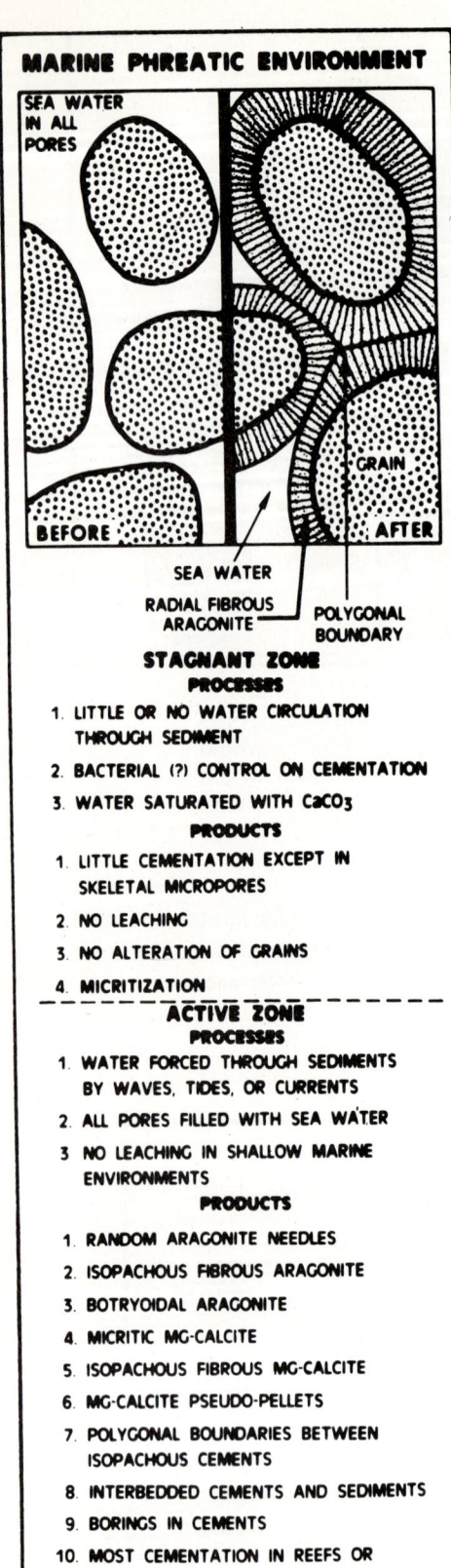

FIGURE 1.16 Characteristics of marine phreatic diagenetic environment. From Longman (1980). Reprinted by permission of the American Association of Petroleum Geologists.

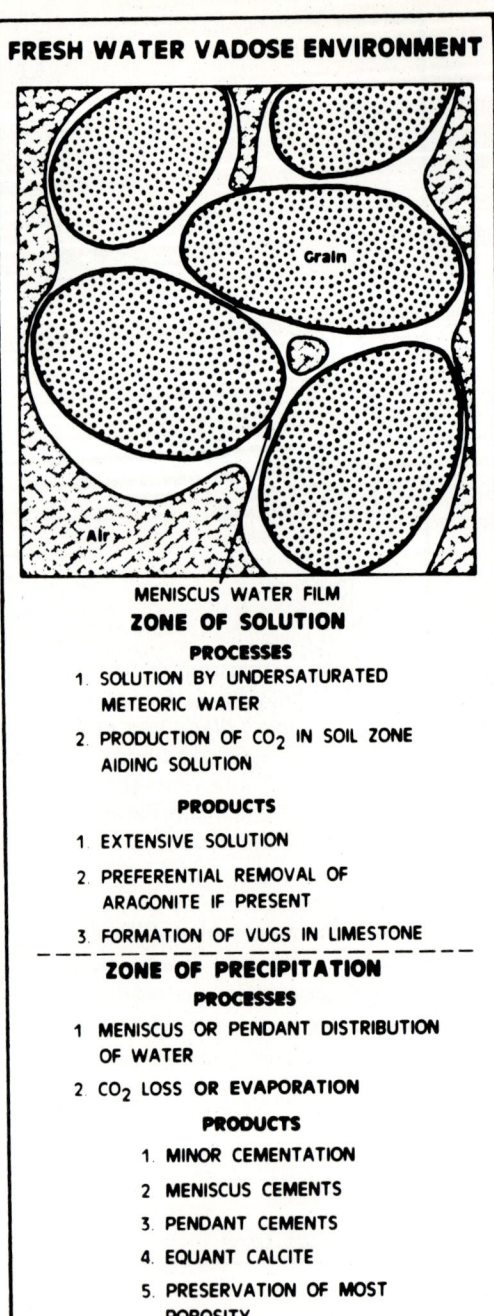

FIGURE 1.17 Characteristics of freshwater vadose diagenetic environment. From Longman (1980). Reprinted by permission of the American Association of Petroleum Geologists.

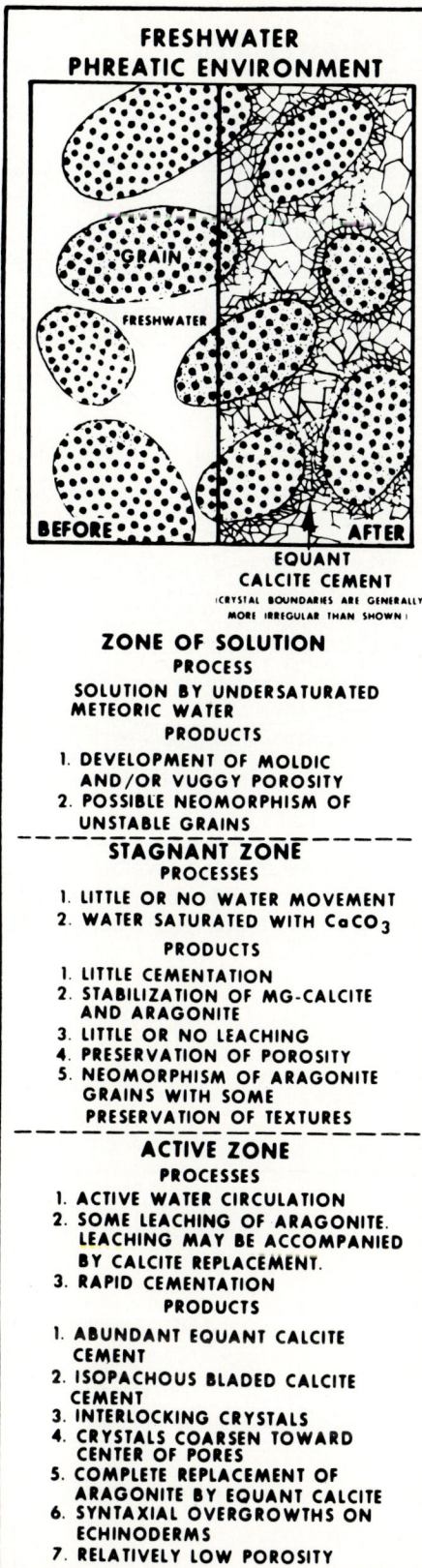

FIGURE 1.18 Characteristics of freshwater phreatic diagenetic environment. From Longman (1980). Reprinted by permission of the American Association of Petroleum Geologists.

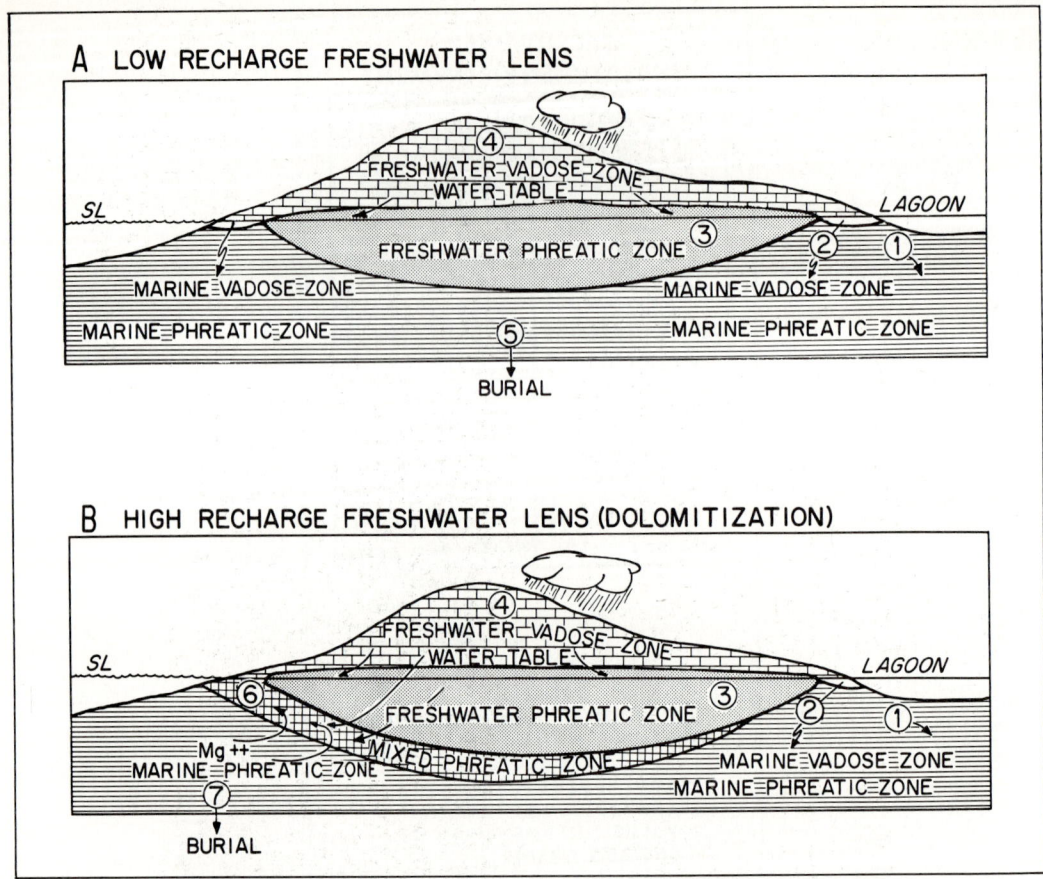

FIGURE 1.19 Cross sections showing development of a mixed freshwater-seawater phreatic zone under conditions of large recharge of the freshwater lens. From Carozzi (1981). Reprinted by permission of Scientific Press Ltd.

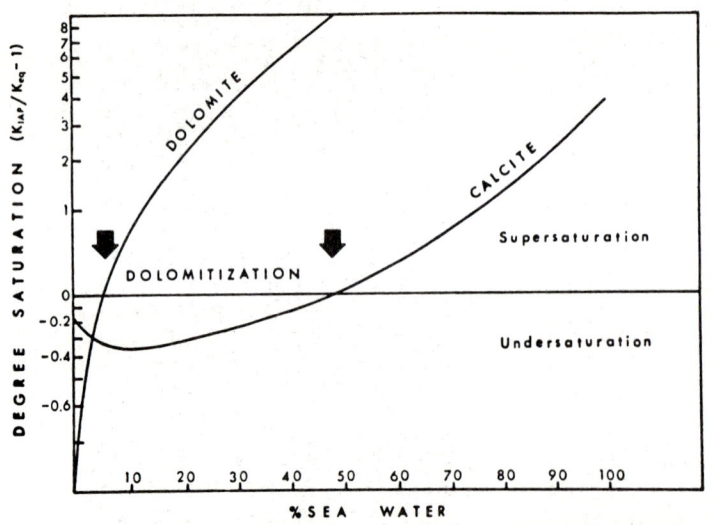

FIGURE 1.20 Effect of mixing of Yucatan ground water with sea water on saturation degrees of calcite and dolomite. From Badiozamani (1973). Reprinted by permission of the Society of Economic Paleontologists and Mineralogists.

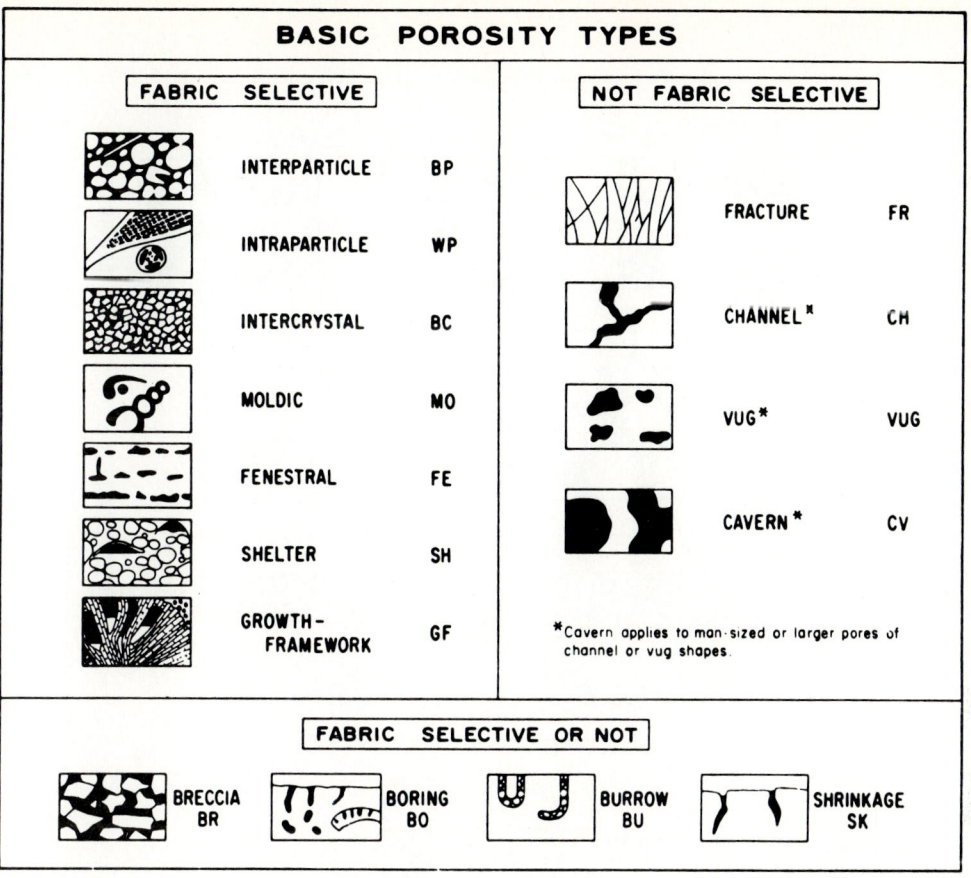

FIGURE 1.21 Geologic classification of pores and pore systems in carbonate rocks. From Choquette and Pray (1970). Reprinted by permission of the American Association of Petroleum Geologists.

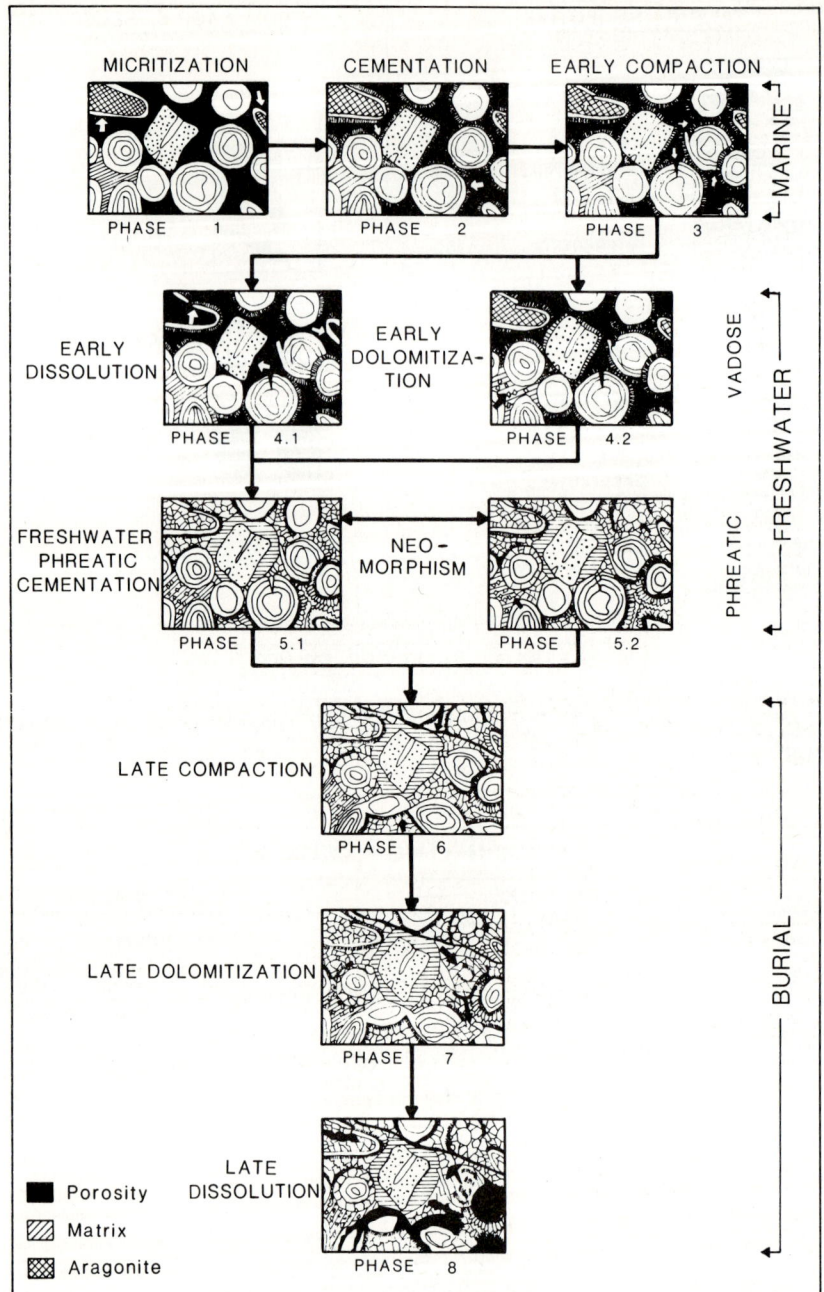

FIGURE 1.22 Depositional-diagenetic sequence of the Upper Member of the Quintuco Formation, Lower Cretaceous, Loma La Lata field, Neuquén Basin, Argentina. Redrawn from Carozzi *et al*. (1982).

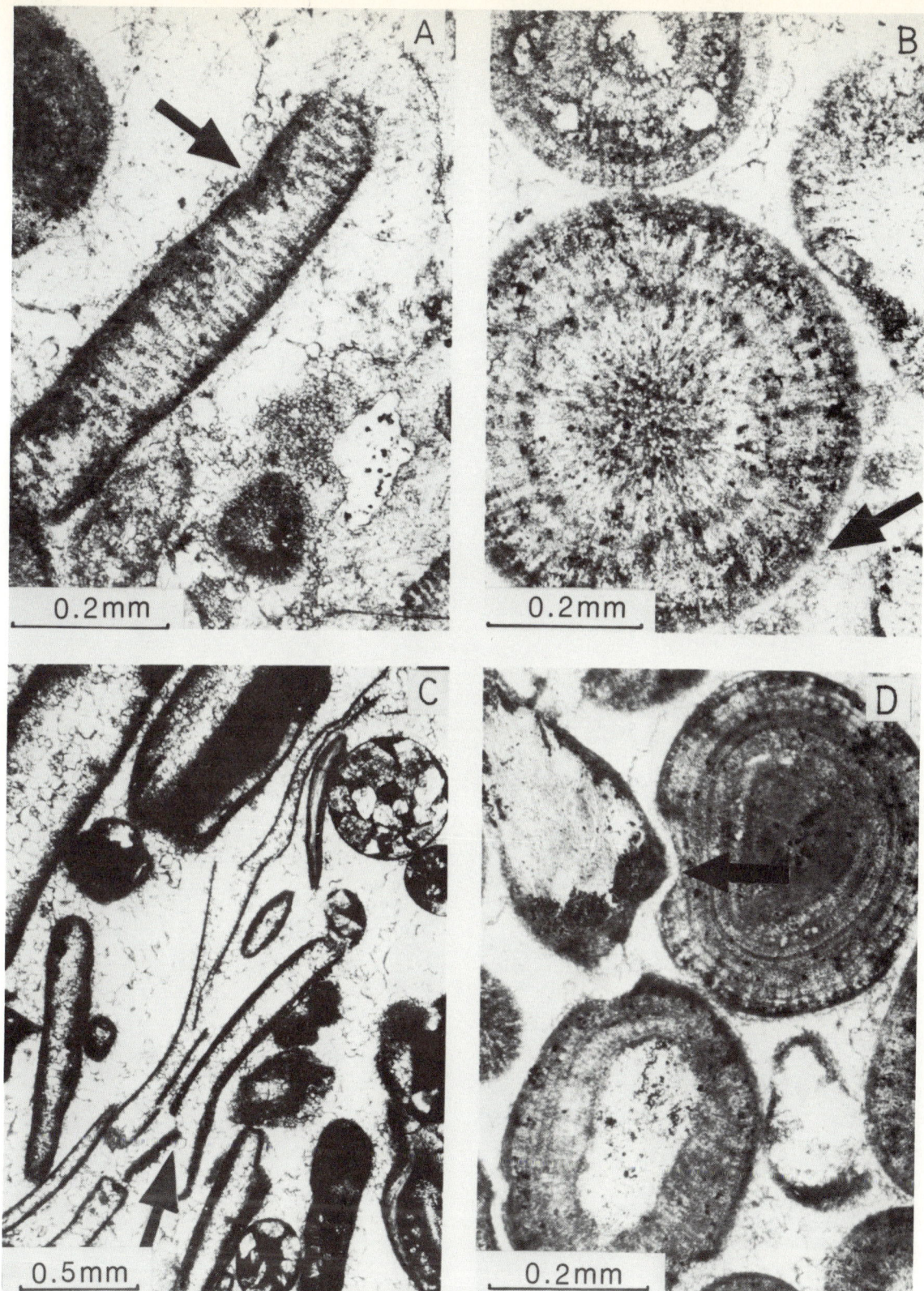

FIGURE 1.23 Upper Member, Quintuco Formation, Lower Cretaceous, Loma La Lata field, Neuquén Basin, Argentina. *Marine phreatic diagenesis.* A. Echinoderm fragment with thick micrite envelope (arrow). B. Fibrous isopachous rim cement (arrow) around ooids. C. Micrite envelopes (arrow) collapsed by early compaction. D. Pressure-solution due to early compaction after deposition of isopachous rim cement. All photomicrographs: plane-polarized light. From Carozzi *et al.* (1982).

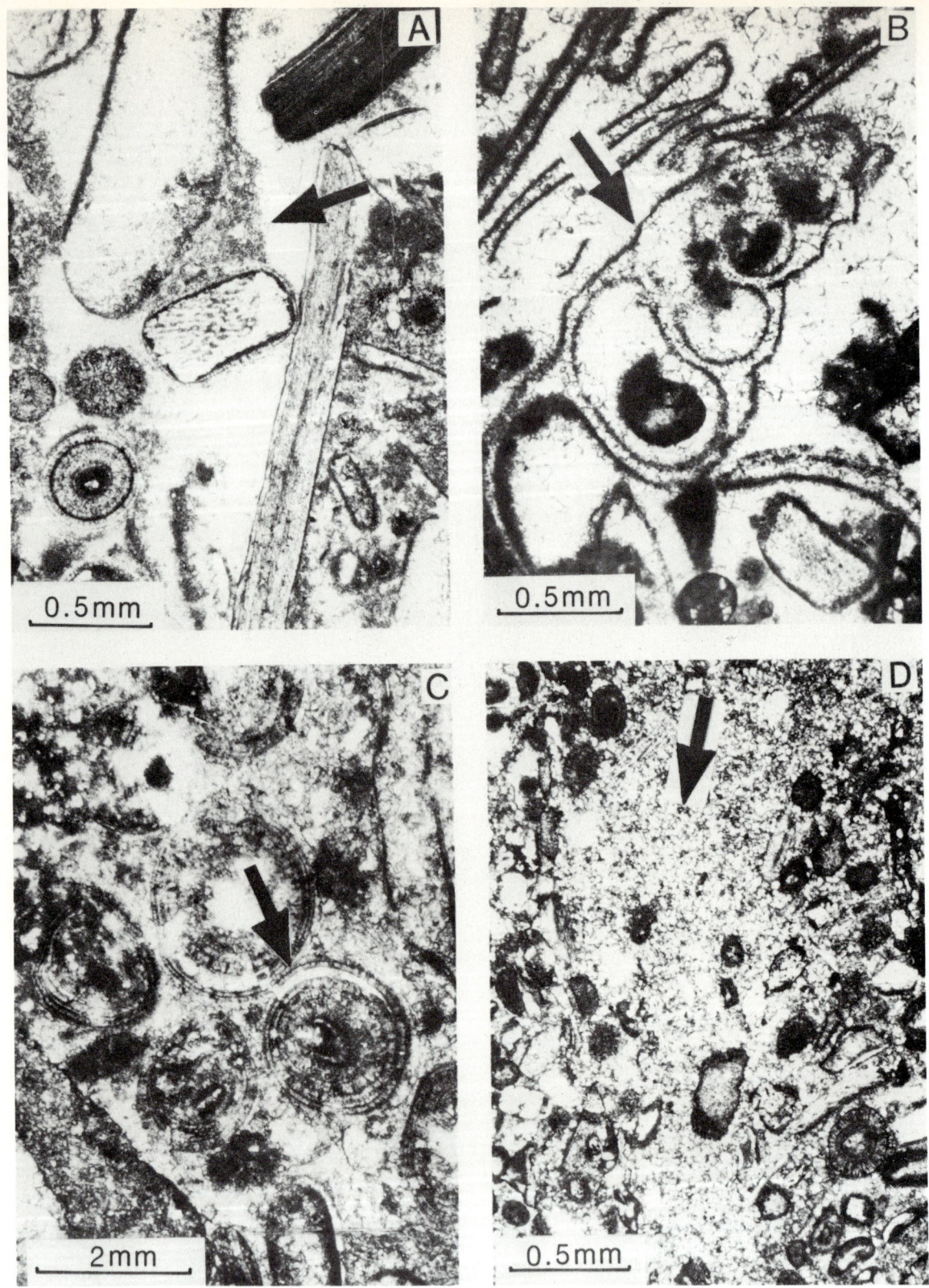

FIGURE 1.24 Upper Member, Quintuco Formation, Lower Cretaceous, Loma La Lata field, Neuquén Basin, Argentina. *Freshwater vadose, freshwater phreatic, and mixing freshwater-seawater phreatic diagenesis.* A. Vadose silt (arrow) overlain by phreatic sparite cement. B. Aragonite gastropod shell with micrite envelope (arrow), dissolved and refilled by phreatic sparite cement. C. Phreatic calcite cement-filling space (arrow) between oolitic concentric layers spalled by early compaction. D. Partial dolomitization by mosaic of fine anhedral dolomite crystals (arrow). All photomicrographs: plane-polarized light. From Carozzi *et al.* (1982).

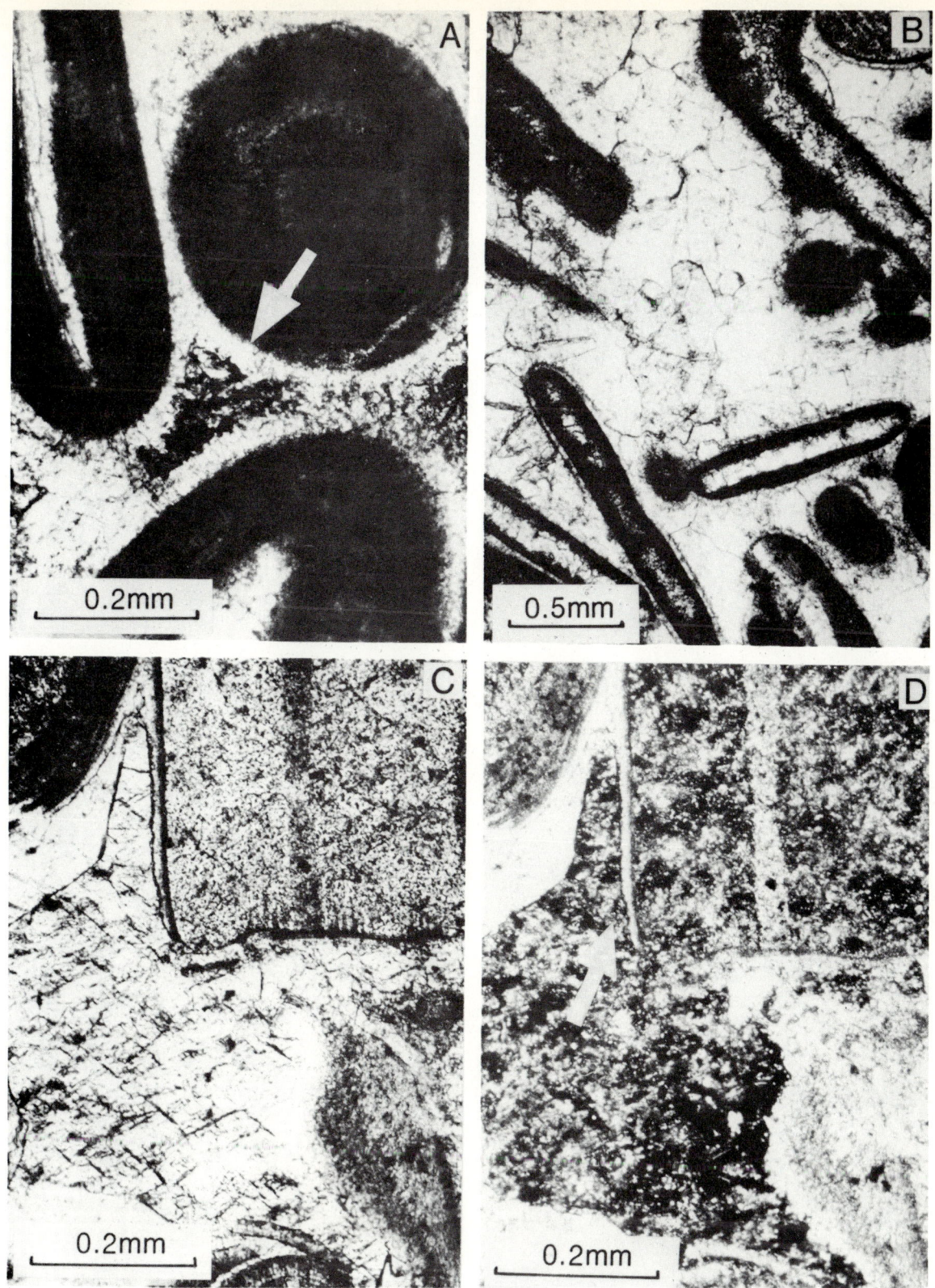

FIGURE 1.25 Upper Member, Quintuco Formation, Lower Cretaceous, Loma La Lata field, Neuquén Basin, Argentina. *Freshwater phreatic diagenesis*. A. Bladed calcite cement (arrow) nucleated on isopachous fibrous rim cement. B. Equant sparite mosaic over partially dissolved isopachous fibrous rim cement. C. Syntaxial overgrowth around echinoderm bioclast and resting on isopachous fibrous rim cement. D. Same photomicrograph as C but in crossed nicols. All photomicrographs except D: plane-polarized light. From Carozzi *et al.* (1982).

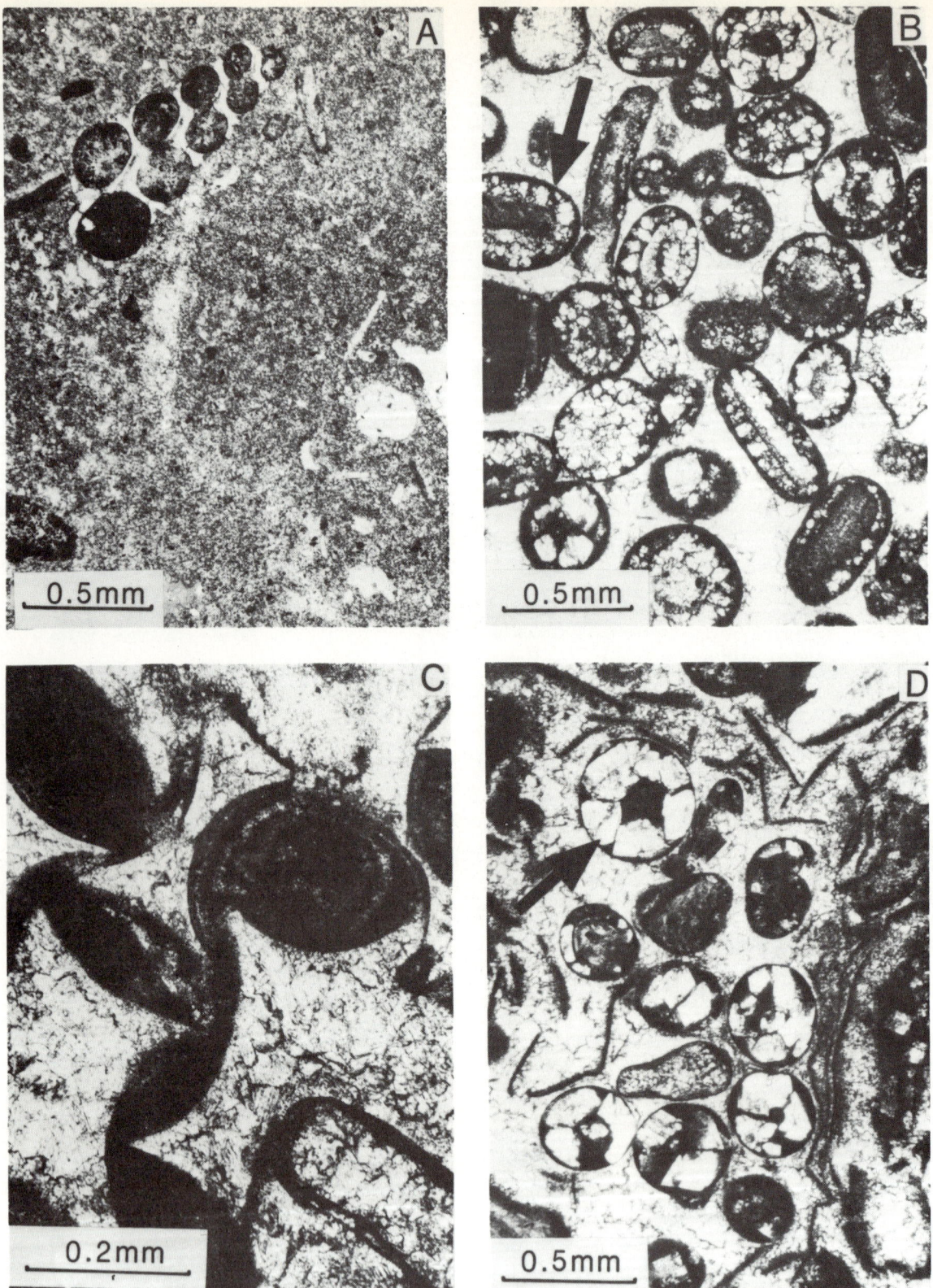

FIGURE 1.26 Upper Member, Quintuco Formation, Lower Cretaceous, Loma La Lata field, Neuquén Basin, Argentina. *Freshwater phreatic and burial diagenesis.* A. Aggrading neomorphism of micrite matrix to pseudomicrosparite. Gastropod chambers filled by collophanite. B. Aggrading neomorphism of ooids by aggregates of pseudosparite crystals (arrow). C. Spastolites generated by late compaction during burial. D. Differential burial dolomitization of ooids by crystals of dolosparite (arrow). All photomicrographs: plane-polarized light. From Carozzi *et al*. (1982).

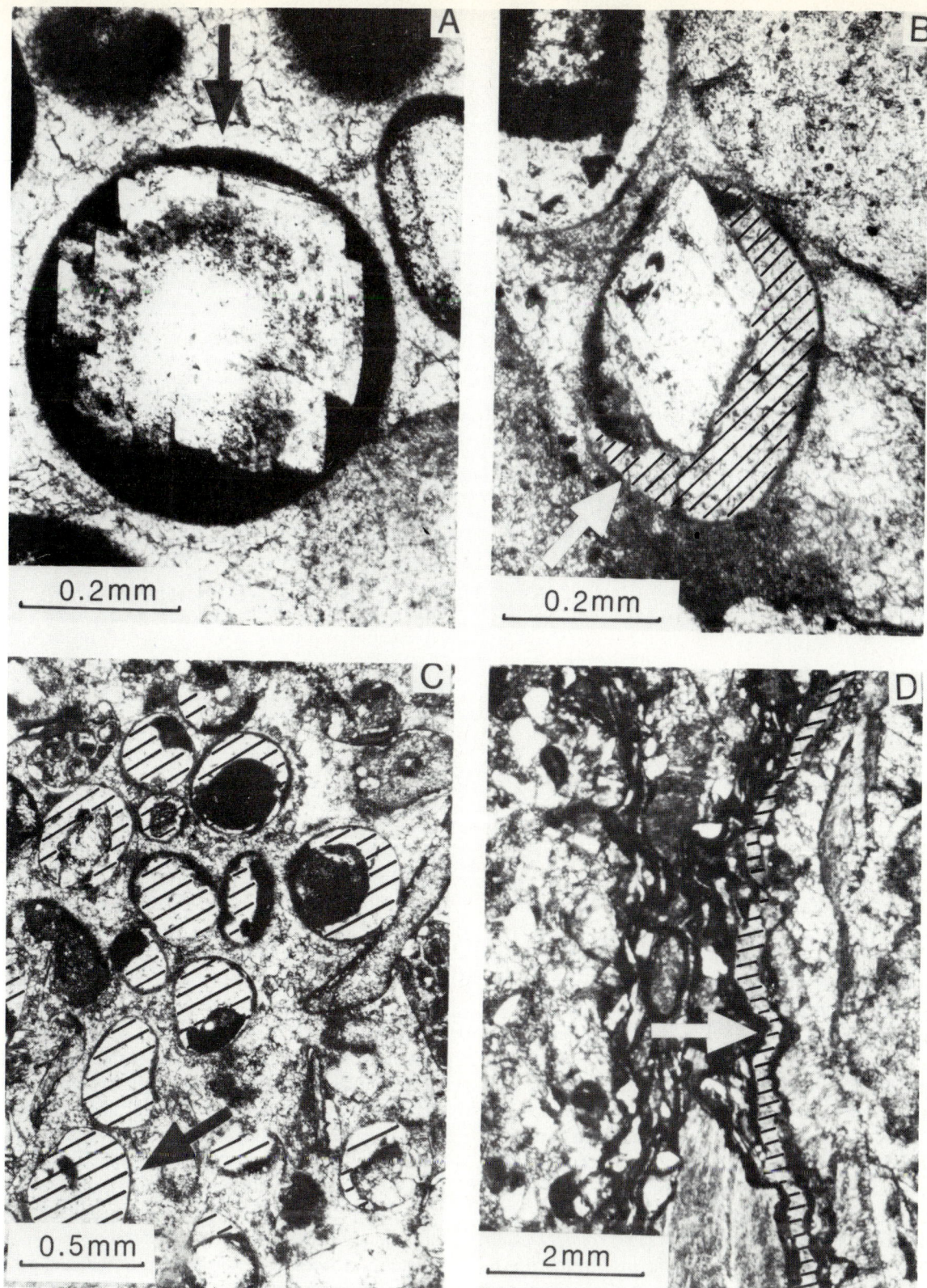

FIGURE 1.27 Upper Member, Quintuco Formation, Lower Cretaceous, Loma La Lata field, Neuquén Basin, Argentina. *Burial diagenesis.* A. Large crystals of saddle dolomite (arrow) inside ooid display "ghosts" of its original concentric structure. B. Partial dissolution of dolomitized ooid with partially corroded dolomite crystal. Porosity shown by oblique ruling (arrow). C. Oomoldic porosity, the latter is shown by oblique ruling (arrow). Notice partially geopetal dropped cores. D. Stylolite porosity (arrow) is shown by oblique ruling. All photomicrographs: plane-polarized light. From Carozzi *et al.* (1982).

2

Microfacies Techniques

Introduction

The term "microfacies," which represents the application of Gressly's classic facies concept (1838) on a microscopic scale, was suggested by Brown (1943), namely that rocks consist of microfacies defined by criteria visible in thin sections under the microscope. Early minor objections by different authors (Calkins, 1943; Campbell, 1944; Alling, 1945) to the original definition did not prevent its acceptance. The latter gained worldwide acceptance with work by Cuvillier (1952, 1958, 1961), who independently introduced the name of microfacies to characterize micropaleontological and petrographic criteria in thin sections which he described as "sedimentary landscapes" (Cuvillier and Sacal, 1956, p. 10), infinitely variable as a whole but usable for local and regional correlations within given basins. In the sense of Cuvillier, a microfacies is the sum of all the micropaleontologic and petrographic properties recognized in a given thin section, and hence allows its characterization and classification. Attempts at splitting this encompassing definition into microlithofacies and microbiofacies were unsuccessful (Fairbridge, 1954). The tendency was, in fact, just the opposite, particularly after the publication of many volumes describing carbonate microfacies in various basins of the world (see, for instance, Carozzi and Textoris, 1967), preceded by the pioneer work of Hovelacque and Kilian (1900).

It became apparent that for an even better characterization of the microfacies and its environment of deposition, the original definition should be enlarged to include criteria not visible under the microscope. This was attempted in a study of the Jurassic microfacies of the Aquitaine Basin (Carozzi *et al.*, 1972) in which the definition of microfacies includes the following data: identification of minerals by X-ray diffraction, composition of clay mineral assemblages, geochemically measured contents of organic carbon, sulfur, trace metals, and boron, measurements of porosity and permeability, and capillary pressure curves. With the possible exception in some cases of petrophysical data, all the parameters above are primary depositional features. Moreover, it is appropriate to proceed even further and to include any early or late diagenetic features which add further discrimination to the microfacies, a common practice in the study of dolomitized limestones, for instance. Consequently, the following definition of microfacies was proposed (Carozzi *et al.*, 1976): the total of the mineralogic, paleontologic, textural, diagenetic, geochemical, and petrophysical features of a carbonate rock. In summary, one can now examine Cuvillier's "sedimentary landscapes" with more penetrating eyes. An examination of the volume by Flügel (1982) shows that microfacies analysis of limestones has indeed become a major discipline by itself in the study and interpretation of sedimentary rocks.

To improve the qualitative and partially subjective nature of the current definitions of microfacies just reviewed, the author developed since 1948, with his students and coworkers, a technique of quantification and graphic representation of microfacies. The aim was to record by absolute and relative measurements under the microscope, the organic, inorganic, and textural parameters of a given microfacies. These parameters provide a more precise definition of a microfacies based on numbers that can be treated statistically and hence will give a more adequate classification within any suitable scheme and a more sophisticated environmental and diagenetic interpretation.

If a stratigraphic superposition of microfacies is considered in field sections or cores, the vertical variations of all these parameters expressed as curves drawn alongside the stratigraphic column, in a way similar to electric logs, display in a clear fashion the cyclic nature of the superposition of microfacies which corresponds generally to a succession of shallowing-upward sequences of variable completeness.

The microscopic analysis of these sequences allows one to prepare *an ideal shallowing-upward sequence* for a given stratigraphic interval which represents the statistically most complete sequence of microfacies that could have been deposited within a certain time interval. The environmental interpretation of this ideal sequence is expressed by a variety of curves which depend on the complexity of the conditions of deposition. They may express relative bathymetry, relative energy levels, or the succession of distinct depositional environments through time.

The environmental interpretation of microfacies can be improved further when the ideal shallowing-upward sequence is represented horizontally following Walther's law (see Middleton, 1973) as an *ideal depositional model*. In it, the various microfacies are distributed in adjacent environments and along a subaqueous topographic profile parallel to the depositional dip. The variations of all the measured parameters can be analyzed in detail and genetically interpreted as a function of this more elaborate graphic representation of the conditions of deposition.

The ideal shallowing-upward sequence, the ideal depositional model, and observation data represented by the stratigraphic variation curves of the various parameters can be treated by a variety of statistical techniques for the purpose of comparison, correlation, prediction, detection of trends, and so on, within any aspect of basin analysis for either scientific or applied investigations.

In practice, the microfacies technique consists of a sequence of distinct operations which are summarized in the following sections. The subsequent examination of individual cases illustrates the modalities of application of the technique to a variety of carbonate environments.

Field Sampling

Selection of field sections that are exposed as continuously as possible, or of cores, is indispensable for a microfacies study. A debate exists about whether the sampling should be uniformly spaced vertically or whether local megascopic properties require an increase or decrease in spacing. In practice, textures and structures of carbonate rocks, with a few exceptions such as cross-bedding or graded bedding, are difficult to observe in the field. Consequently, the various types of bedding are the most commonly used criteria to separate field units. Ideally, a sample should be taken from the lower, middle, and upper part of each field unit while attempting to keep a vertical interval between successive samples of 15 to 30 cm. Tests have shown (Nowak and Carozzi, 1972) that an average 20 cm vertical interval of sampling is sufficient to account for microfacies variations in most carbonate environments (Fig. 2.1). Vertical intervals of as much as 1 to 2 m may be sufficient, in preliminary studies, to unravel the main depositional variations. However, it is always useful to start with a test section and to establish the scale of the vertical variation of the microscopic parameters with respect to field units. The two scales of observation do not coincide generally, and it is easy to choose the most appropriate vertical sampling distance.

Samples are collected usually of a size sufficient to cut several thin sections. Their tops are clearly marked, and they are numbered in stratigraphic order in agreement with the field units. Whenever several samples from a given unit are collected, they are labeled A, B, C, and so on, following the number of the unit. The sampling procedure is accompanied by a standard field or core description which allows one subsequently to locate with precision a particular specimen, or to undertake checking or additional sampling.

Preparation of Thin Sections

Standard thin sections (30 μm thickness) are cut perpendicular to bedding in order to display maximum textural variability. Initially, the thin sections are not covered, to allow various types of staining and other investigating techniques. Later, only selected thin sections are covered, in particular for the purpose of photographing representative examples of microfacies or of stages of depositional-

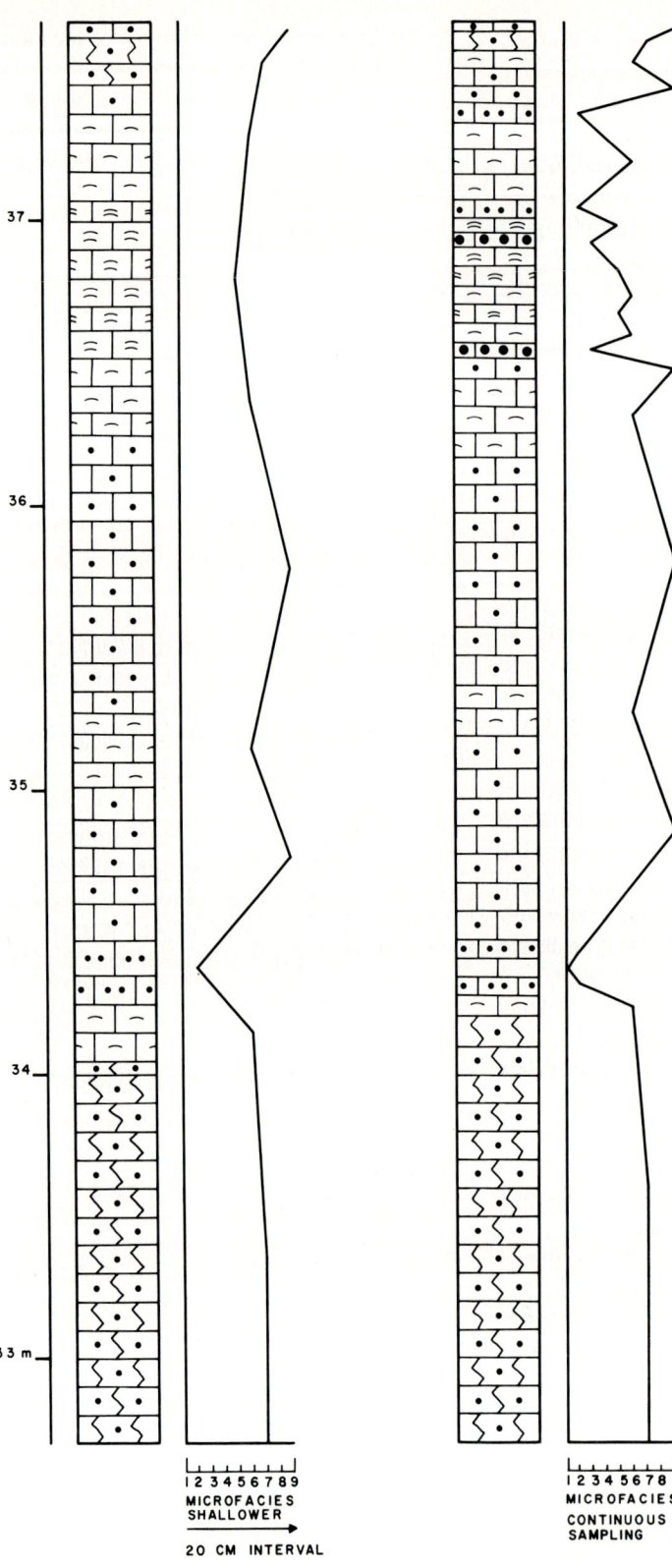

FIGURE 2.1 Comparison of results between different sampling techniques in the Upper Bird Spring Group (Pennsylvanian-Permian), Arrow Canyon Range, Clark County, Nevada. Although continuous sampling provides more information and leads to a better distinction of genetically related microfacies, a 20-cm-average vertical sampling interval is sufficient to unravel the main trend of evolution of the carbonate sedimentation. From Nowak and Carozzi (1972).

diagenetic sequences. Dolomites are cut thicker to preserve color contrasts and usually investigated under diffused light to detect the "ghost" textures of the original limestone. Contrary to common belief, many dolomites display sufficient textural features to allow a microfacies investigation.

Thin sections are also prepared routinely from well cuttings. Microfacies techniques are applied to them after their "calibration" against data obtained from cores, and with the usual reservations due to contamination from overlying units. Cuttings should be sufficiently coarse to include fragments of identifiable constituents and/or matrix, and picked by a trained technician (see Rodriguez Schelotto and Carozzi, 1981).

Inventory of Constituents, Matrix, Cement, and Textures

In the first general examination of thin sections, the emphasis, regardless of their stratigraphic order, is placed on the identification of the various types of benthic and planktonic bioclasts and of the inorganic constituents such as intraclasts, pellets, ooids, extrabasinal detrital minerals, glauconite, and phosphates. The purpose of this is to separate those constituents which can be submitted, in a later phase, to absolute measurements of clasticity and/or frequency from those constituents whose abundance can only be visually estimated in a relative fashion (in percentage, for instance), or simply recorded by presence or absence.

The limit between organic and inorganic clastic components and matrix is usually placed at 50 μm because of the difficulty of identifying and counting grains below this size by means of a conventional petrographic microscope. The various types of intergranular cements are also characterized to avoid any possible confusion with products of neomorphism or replacement (dolomite, for instance).

As a consequence of the visualization of grains, matrix, and cement, the concept of grain-supported versus mud(matrix)-supported texture is established for the suite of thin sections to be examined. This concept, which underlies all the classifications of carbonate rocks and is an expression of the energy (bedshear) of the environment of deposition, carries critical implications for the subsequent subdivision of thin sections in preliminary microfacies. At any rate, the purpose of this first examination of the appearance of thin sections is to familiarize the observer with the entire spectrum of constituents and textures that must be associated in a meaningful manner in the next phase of investigation.

Qualitative Classification of Thin Sections into Preliminary Microfacies

Thin sections are grouped into a certain number of preliminary microfacies using the previously established inventory of major and minor organic and inorganic components, their general reciprocal relationships, the ratio of matrix versus cement, and the concept of mud-supported versus grain-supported texture. A clustering of thin sections into distinct microfacies can be obtained rather rapidly. Those of intermediate character are attributed arbitrarily to the nearest similar group. The subsequent quantitative petrographic analysis provides the final answer to this particular problem.

Classification of Microfacies into a Preliminary Shallowing-Upward Sequence

The previously discussed criteria used to characterize preliminary microfacies afford sufficient data to establish a tentative shallowing-upward sequence among microfacies at this early stage of the investigation. Ancient carbonate sediments deposited in ramp or slope environments usually seem to have reached a mechanical equilibrium with their environment of deposition that reveals a direct relationship between their energy level (expressed by their texture and representing equivalent bedshear) and relative bathymetry. Hence it is common to find the following general shallowing-upward sequence: subtidal quiet water calcilutites and calcisiltites with sparse benthic or abundant planktonic fauna, followed by calcisiltites with scattered sand-size bioclasts, mud-supported biocalcarenites, grain-supported biocalcarenites with calcisiltite matrix, grain-supported biocalcarenites with associated matrix and sparite cement (usually indicating wave base), and finally high-energy subtidal grain-supported biocalcarenites with sparite cement and increasingly well-sorted and abraded bioclasts. This shallowing environment with increasing energy level is followed by even shallower subtidal to intertidal conditions represented by oolitic calcarenites with sparite cement, frequently associated with bioconstructed limestones representing various types of greater energy buildups. This environment, often representing a shelf edge, is in general the end term of the direct relationship between shallowing-upward conditions and energy level. In shallow platform or lagoonal conditions, located behind the protection afforded by oolitic banks and buildups, finer-grained carbonate sediments reappear and eventually grade into well-characterized carbonate mud flats and supratidal microfacies. The relationship between shallowing and energy level is reversed or becomes extremely complex under these conditions. However, paleoecological criteria and sedimentary structures typical of extremely shallow water allow their identification and prevent possible confusion with fine-grained deep-water carbonates of the other end of the spectrum.

In summary, at this stage it is possible to organize microfacies into a preliminary order of relative shallowing-upward environments and to designate (by letters or numbers) microfacies accordingly. This classification is susceptible to later modifications and is finalized by means of quantitative analysis of microfacies and related statistical treatments.

Quantitative Analysis of Thin Sections

A data sheet (Fig. 2.2) is prepared in order to record in a systematic fashion measurements made under the microscope for each thin section. Indices of clasticity and frequency are measured for organic and inorganic detrital components, whereas only frequency is measured for nondetrital benthic and planktonic organic constituents.

The clasticity index of a given detrital component is defined as the mean apparent diameter (expressed in millimeters) of the 10 largest grains of that mineral present in the slide (Carozzi, 1948, 1950, 1958). In general, measurements are made over six adjacent low-power microscope fields, arranged in two rows of three fields which cover the entire surface of an average standard thin section. The determination of the index of clasticity is easy and rapid, and variations around the mean maximum apparent diameter are relatively small. Only in rare cases of poorly sorted or irregularly textured microfacies, measurement of more than 10 grains may be required to obtain a representative value of clasticity. In the case of bimodal sediments, both modes are measured. Because an apparent diameter is determined from the average of random point intersections, the index of clasticity is systematically slightly smaller (by a constant value of 0.7) than the real maximum diameter which would theoretically be obtained by sieving (Textoris, 1971).

The index of clasticity is a measure of the energy (bedshear) level of the environment of deposition because it is restricted to detrital particles transported by traction or suspension processes. Besides detrital minerals such as grains of quartz or glauconite with specific densities, or intrabasinal detrital particles such as intraclasts, pellets, ooids, or oncoids, the majority of the transported grains in carbonate environments are bioclasts derived from mechanical and biological fragmentation or dissociation of benthic organisms, or the organisms themselves, such as smaller foraminifers when behaving as individual grains.

The shape and size of a bioclast is unquestionably very much influenced by the internal structure of the type of test from which it is derived. Furthermore, its density varies as a function of intraparticle porosity. This situation restricts the choice of bioclasts for the purpose of clasticity measurements to fragments of echinoderms, in particular to crinoid columnals. Indeed, the latter show an original subcylindrical shape which by abrasion is easily converted into a subrounded grain. They consist, furthermore, of a relatively homogeneous microporous calcite network. The index of clasticity of crinoidal debris, which are ubiquitous in Paleozoic and Mesozoic carbonates, is therefore the most widely used parameter to establish the energy level of carbonates. The reliability of this parameter has been demonstrated repeatedly by comparisons with the behavior of intrabasinal detrital components such as intraclasts, pellets, and ooids and with extrabasinal detrital minerals such as quartz whenever present.

The same technique of measuring the mean apparent maximum size is also applied to the size of crystals developed in interparticle calcite cements or during diagenetic processes such as neomorphism, dolomitization, anhydritization, or silicification and is called crystallinity index. It is an expression of the crystallinity of these various components, and in the case of dolomites, the following subdivisions can be used for classification and description (Okhravi and Carozzi, 1983):

Extremely coarsely crystalline	>4 mm
Very coarsely crystalline	1–4 mm
Coarsely crystalline	0.25–1 mm
Medium crystalline	0.062–0.25 mm
Finely crystalline	0.016–0.062 mm
Very finely crystalline	0.004–0.016 mm
Aphanocrystalline	<0.004 mm

In many dolomites (see Zadnik and Carozzi, 1963), crystallinity is related closely and directly to clasticity of the original limestone and thus provides an important clue for microfacies interpretation of rocks which have been replaced totally.

The frequency index of a given constituent is defined as the number of grains, fragments, or complete individuals of that component within a standard area of the thin section (Carozzi, 1948, 1950, 1958). In the case of grain-supported fabric, it is preferable to use an area which is at least 10 times larger than the average grain size of the constituent so as to minimize the possible effect of an increase of grain size which inherently decreases the frequency. In practice, this problem is essentially eliminated because most of the constituents of cal-

Slide #   Microfacies #

COMPONENTS	FREQUENCIES				TOTAL
	1	2	3	4	
Bryozoans					
Brachiopods					
Pelecypods					
Gastropods					
Trilobites					
Ostracods					
Endothyridae					
Earlandidae					
Echinoid spines					
Brachiopod spines					
Sponge spicules					
Calcispheres					

		FREQUENCIES IN 4 FIELDS CLASTICITIES IN 6 FIELDS				TOTAL	AVERAGE
Crinoids	f						
	c						
Ooids	f						
	c						
Oncolites	f						
	c						
Intraclasts/Lithoclasts	f						
	c						
Detrital quartz	f						
	c						
Cement in 100% of interst. material							

Minor fossil
  fragments
Coral (Rugose)
Archaediscidae
Stacheiinae
Girvanella
Asphaltina
Rectangulina
Dasycladaceae

Clastic particles
Alkali-feldspar
Plagioclase feldspar
Muscovite
Phosphate
Pyrite
Limonite
Hematite

Pellets
Bioturbation
Worm tube
Geopetal feature
Lamination

Pressure-solution
Stylolite
Fracture
Distortion and
  deformation
Compaction

Dolomitization
  (ankerite)
Silicification
Hematitization
Limonitization
Pyritization
Glauconitization

Additional remarks: Types of ooids: well-developed, superficial, chain ooid,
                                         cerebroid ooid

                      Cores of ooids: crinoid, bryozoan, brachiopod, mollusk

                      Types of bryozoans: fenestrate, ramose (hollow, solid)
                                             encrusting forms

FIGURE 2.2 Example of data sheet for the Glen Dean Formation (Upper Mississippian) of the Illinois Basin. From Feiznia (1983).

carenites rarely reach above the coarse-sand size range so that one can use, as for the index of clasticity, a surface of 300 to 400 mm² which corresponds to six adjacent lower-power microscope fields arranged in two rows of three fields. In this fashion, practically the entire field of an average standard thin section is covered.

The frequency index of a given constituent is the sum of the number of particles recorded over the six adjacent fields. In practice, the counting of grains and fragments (a rather tedious operation) is made over one-quarter of a field at a time and the 24 values are then added. However, when certain constituents are extremely small and uniformly abundant, a smaller area may be selected for the measurement of their frequency to avoid the time-consuming task of counting an extremely large number of grains or individuals. These changes of reference surface should be clearly stated at the beginning of each particular study and kept constant throughout. This procedure is used commonly for such abundant and small components as planktonic foraminifers, calcispheres, or sponge spicules in contrast to crinoidal bioclasts, oncoids, or ooids.

Contrary to measurements of the index of clasticity which are absolute figures in millimeters, the values of frequency are necessarily a function of the surface over which they are measured. Consequently, comparisons between frequencies of various constituents are undertaken more in terms of reciprocal positions of peaks and lows rather than on the basis of their real values.

The index of frequency for detrital minerals, bioclasts, or transported entire individuals is an indication of the traction or suspension load of transporting agents in the environment of deposition. An adequate yield for a given constituent is shown by the parallel behavior of its curves of clasticity and frequency. As a result of insufficient yield or winnowing processes, the two curves tend to vary in opposite direction, thus expressing the case of a small amount of grains of relatively large size, or vice versa (Carozzi, 1950, 1958).

For nondetrital benthic and planktonic organic components, the index of frequency expresses the abundance of a given population and is of critical importance for expressing the paleoecological characteristics of the various microfacies.

The determination of the index of frequency is also applied to individual crystals developed in place by diagenetic processes such as authigenic quartz and feldspars. Its relationship with the above-mentioned index of crystallinity gives a complete picture of these diagenetic products.

In order to determine the frequency of particulate components which are not abundant enough to be computed statistically, such as a wide range of nonparticulate constituents, including algal mats, matrix, and cement, or of various depositional (bioturbation) and diagenetic textures (neomorphism, dolomitization, porosity), visual estimation in percentage, using standard charts, is adopted (see Flügel, 1982). This estimation is computed as the average percentage over the previously described six adjacent low-power microscope fields. The results are graphically represented by variation curves of the estimated percentages or by lines of variable thickness which represent subdivisions such as rare, present, common, and abundant. The ranges of percentage corresponding to these subdivisions vary from one environment to another and are left to the discretion of the operator. Another representation of certain constituents is done by symbols which indicate presence or absence. The purpose of visual estimations is to provide a graphic expression for the greatest possible number of additional characteristics of microfacies which may be used for the final environmental interpretation.

When the measurements of clasticity and frequency of all the available constituents under the microscope are finished, a complete data sheet becomes available for each thin section. The preliminary microfacies classified in order of relative shallowing-upward environments are now as fully characterized as possible with respect to all the parameters of environmental and diagenetic significance.

Statistical Evaluation and Final Characterization of Microfacies

All the measured parameters for each thin section, together with their subdivision into preliminary microfacies, are submitted to a variety of computer programs which can evaluate and improve classifications so that the original subdivision is tested for statistical validity. Over the years, several techniques were successively used, modified, and abandoned in favor of more improved ones; they are described in publications pertaining to specific models.

At present, for routine purposes, a modified Iterim subroutine of the iterative classification program of Demirmen (1969), which can accommodate up to 3000 variables is used by means of an IBM 360/75 computer. Demirmen's program operates on the basis of statistical criteria which measure the degree of "compactness" or quality of a classification by means of an analysis of variance of a matrix of n items (in this case thin sections), m groups (microfacies), and p variables (petrographic parameters). From the analysis of variance, the three following criteria are measured: "the within-group sum of squares," "the between-group sum of squares," and "the total sum of squares." This leads to an orthogonal comparison of groups (microfacies) and evaluation of the classification by means of iterations. After the iterations are stabilized, a percentage of thin sections (core ratio) will be retained in their original groups, whereas others are "rejected" to groups other than the ones to which they belonged originally. In most studies, the core ratio is 90% or higher, which means that about 10% of the thin sections were "rejected" by the program because they showed particular compositional or textural variations which were pronounced enough to affect the parameters of the constituents and cause reclassification of these thin sections in microfacies other than the ones to which they were originally attributed.

The "rejected" thin sections are reexamined petrographically in detail to determine the reasons of their reclassification. Two possibilities may occur: (1) the placement by the computer program of some thin sections in another microfacies expresses a deficiency of observation which led to an incorrect original subdivision into microfacies, hence the changes introduced by the computer program are accepted; (2) the placement of some thin sections in another microfacies is statistically valid but sedimentologically inacceptable. In this case these thin sections are returned to their original microfacies. This discrepancy results from the fact that the computer program is based entirely on quantitative data and thus attributes equal weight to all variables, a situation which may not express sedimentological reality. Indeed, the computer program does not take into consideration the qualitative to semiquantitative parameters pertaining to sedimentary textures and structures which may be of critical importance to characterize certain microfacies. A typical example is represented by the distinction between a bioaccumulated brachiopod limestone with shells *in situ*, well aligned in a biocalcisiltite matrix, and a similar-appearing biocalcisiltite containing transported shells of brachiopods aligned more or less parallel to bedding. Only the occurrence of geopetal features in the former microfacies and subtle textural differences in the alignment of the shells in the latter, or differences in the species of brachiopods, may be the critical features for the distinction of the two microfacies, not the data fed into the computer program.

A statistical treatment that is supposed to evaluate and improve classifications can only be applied to a suite of microfacies which have at least a minimum amount of common parameters. Other microfacies may occur in the same suite, as for instance a stromatolite-constructed limestone or a calcisiltite devoid of bioclasts, which display such a unique character or lack sufficient constituents that a meaningful comparison with the other associated microfacies is statistically impossible. These microfacies should therefore be excluded from the computer evaluation, and their characterization must remain the one obtained from the preliminary subdivision.

A final classification of all thin sections into distinct microfacies is reached at the end of the computer operations. By using another appropriate program, the SOUPAC (1976) statistical package program, the average values of all measured parameters for each microfacies are calculated, and each microfacies is now fully characterized. For microfacies not included in the statistical evaluation, the average values of their parameters are calculated independently.

Correlation Coefficients of Microfacies and Final Classification in an Ideal Shallowing-Upward Sequence

Although each microfacies of a given sequence is now characterized fully by a petrographic description and a set of average values of its major and minor microscopic parameters, the classification of these microfacies in order of relative shallowing-upward environments, as established during the early stages of the investigation, must also be evaluated statistically. This evaluation can be done by any appropriate computer program such as SOUPAC (1976), which calculates Pearson multivariate correlation coefficients between microfacies on the basis of all the average values of their parameters.

The table of correlation coefficients between all microfacies indicates their statistically most probable order of superposition within an ideal relative shallowing-upward sequence. This order is generally an improvement over the one established earlier before the computer operations. However, because certain microfacies with unique properties have been excluded from the statistical computation, it is necessary now to determine their position within the ideal shallowing-upward sequence, and also to test the latter against the real sequences displayed by field sections or cores. Indeed, these consist of cyclic repetitions of shallowing-upward sequences displaying variable degrees of completeness. Shallowing-upward sequences are recognized in the field by a variety of sedimentary features indicating a gradual change upward from smaller to greater energy environments within the range of subtidal to supratidal. Cases often display calcilutites grading into calcarenites, coarsening-upward calcarenites, or the appearance of constructed buildups in the upper part. These sequences generally terminate with a sharp erosional contact separating them from the base of the overlying ones. In cases of extensive environmental changes, deposits of tidal flats, of sabkhas with evaporites, and of subaerial exposure indicate the upper part of sequences. The megascopic aspect of shallowing-upward sequences has been extensively documented (James, 1984).

Comparison between the ideal shallowing-upward sequence derived from microfacies relationships and its real megascopic field expression establishes the *final* ideal shallowing-upward sequence, and tables are prepared that show microfacies classified and numbered in a shallowing order along with the average values of all their respective measured parameters.

In most situations, a single sequential numbering of microfacies in shallowing-upward order such, as 1, 2, 3, . . . , is sufficient to account for the observed facts. When the shallowing-upward sequence is more complex and consists of groups of microfacies, such as a group of calcisiltites followed by a group of calcarenites, the most adequate numbering is the use of series. For instance, series 20 could be used for the various calcarenites, usually with the number increasing with the energy of the environment: microfacies 20 is calcarenite with micrite matrix, 21 is calcarenite with micrite matrix associated with sparite cement, and 22 is calcarenite with sparite cement. The series system is useful in studies related to oil exploration, where interpretations have to be updated constantly to incorporate newly discovered microfacies. The series system is also more appropriate to computer treatment.

In complex environments, it is convenient to number microfacies by means of series corresponding to the shallowing-upward sequence of subenvironments. For instance, series 10 represents basinal carbonates; series 20, slope; series 30, platform, and series 40, lagoon. Under certain circumstances, and in order to stress genetic relationships, series 30 may be used, for instance, for various types of red algae constructed microfacies, and series 300 for the calcarenites resulting from their reworking.

The final choice of the use of numbers or letters or a combination of both is left to the discretion of the operator. The case histories presented in this book show a wide spectrum of possibilities. The ideal shallowing-upward sequence of microfacies is a tool with predictive capabilities for the interpretation of environments of deposition when sequences are incomplete, truncated, or differ in any way from the ideal one.

Graphic Representation of Microfacies

Field sections or cores are drafted as stratigraphic columns (for example, Fig. 3.13) of microfacies (not of megascopic field units) with appropriate vertical scale in feet or meters, and precise location of the control points (samples and corresponding thin sections). The choice of graphic symbols for microfacies (for example, Fig. 3.5) is as systematic as possible and expresses, for instance, their micritic, bioclastic, intraclastic, or oolitic character or any observed combination, as well as any associations or groups of microfacies with common characters which may occur. However, the relative large number of microfacies encountered in most carbonate environments and their petrographic complexity does not allow the use of conventional lithologic representations; new sets of graphic symbols, as distinct from each other as possible, have been created (for example, Fig. 4.15). These symbols change from one study to another, and no attempt has yet been made to reach a uniform style of graphic representation. Furthermore, any partial coincidence with conventional lithologic symbols is fortuitous, with the exception of the oblique ruling, which is also used to indicate dolomitization.

To the right of the microfacies column are drafted the variation curves of all measured parameters (for example, Fig. 3.13). As a general rule, unless stated otherwise in individual case histories, curves of clasticity are solid lines, whereas curves of frequency are dashed or dotted lines. Parameters are generally organized as follows from left to right: detrital minerals, inorganic carbonate components (intraclasts, pellets, ooids, etc.), bioclasts of benthic organisms, planktonic organisms, and textural and compositional parameters. This systematic order is, however, modified frequently to save space in diagrams, to avoid excessive lengths of empty columns when some constituents show a discontinuous distribution, and particularly to stress reciprocal relationships between parameters.

At the extreme right of the diagrams is the environmental interpretation, which, according to available data, can be expressed in four ways: (1) by a curve of relative bathymetry (in a few cases calibrated to real depth) with a scale in which microfacies are classified in decreasing depth order from left to right; (2) by a curve of relative energy level with a scale in which microfacies are classified in increasing energy order from left to right; (3) by an association of both curves; and (4) by a curve of shallowing depositional environments with a scale in which the various environments and their corresponding microfacies are arranged, for instance, from basinal to supratidal from left to right.

Any one of these modes of interpretation illustrates clearly the various patterns of cyclic repetition of carbonate sequences. Megasequences and general trends of evolution of sedimentation are shown by envelope curves.

Graphic representation of the ideal shallowing-upward sequence is identical to that of field sections or cores except that the thickness of each microfacies is either constant and arbitrarily chosen for best graphic expression, or is proportional to the average thickness in percentage of all microfacies of a given investigated column.

In complex environments, petrographic and statistical studies often reveal the existence of several synchronous ideal shallowing-upward sequences which correspond to particular locations within the area of extent of a given formation. These sequences require their own individual graphic representation, and a diagram of filiation of microfacies is generally used to illustrate their space relationships.

The Ideal Depositional Model

The ideal depositional model is obtained from the horizontal representation of the ideal shallowing-upward sequence according to Walther's law (see Textoris and Carozzi, 1964; Middleton, 1973). In this final mode of graphic expression, individual microfacies or associations of microfacies are given a more elaborate environmental interpretation expressed by a subaqueous topographic profile, parallel to the depositional dip, and with the basinal direction to the left. The horizontal extent of each microfacies is either constant and chosen arbitrarily, or proportional to its average thickness in percentage of all other microfacies as expressed previously in the ideal shallowing-upward sequence.

The profile of the ideal depositional model shows juxtaposition of the environments of deposition interpreted, according to available data, in terms of wave base, tide amplitude, current patterns, and detailed submarine morphology such as basin, slope, platform edge, buildup, bank, or lagoon. Below this profile are drafted the variation curves of all measured parameters as well as the distribution of constituents estimated visually or represented by presence or absence. A curve of relative energy level is also provided at the lower part of the illustration which allows comparison with the inferred submarine morphology.

Based on this complete graphic representation, it is possible to see all the data at a glance and to analyze in detail the reciprocal behavior of all parameters and their relationships to various environments. This analysis leads to a genetic interpretation of the various constituents in terms of depositional processes, paleoecology, diagenesis, and porosity generation. An ideal depositional model therefore shows a wide spectrum of predictive capabilities and is an extremely useful tool adaptable to any specific purpose required by basin analysis.

An examination of the case histories presented in this volume indicates the existence of an infinite variety of depositional models, as infinite as the number of possible interactions between biological and physical factors in a carbonate environment. Therefore, the concept of standard microfacies types (Wilson, 1970, 1974, 1975; Flügel, 1972, 1982), namely the attempt to combine the microfacies of various ages into major types, to associate standard microfacies into belts, which in turn build a generalized depositional model, appears as a premature oversimplification. Even Flügel (1982) agreed that the concept of standard microfacies should not be overrated. Indeed, such oversimplification even contradicts the purpose of microfacies techniques, which is merely to unravel and explain the variety rather than to standardize the results into a rigid framework before its time.

3

Carbonate Ramp-Bar-Platform Evolution

It seems appropriate to present here, in a more detailed manner than in the subsequent case histories discussed in this book, an example which shows not only all the steps involved in a microfacies investigation and the wealth of environmental information provided for interpretation, but also an almost ideal case describing the change from a carbonate ramp through a subtidal bioclastic bar into a complex oolitic platform. The St. Louis Limestone, Middle Mississippian, of the Illinois Basin (Diaby and Carozzi, 1984) shows the unique quality of displaying a succession in time of the three depositional models just mentioned which took place in the short interval of two conodont zones and hence was undisturbed by external factors.

Method of Study

This study is based on eight cores and six field sections located in Illinois, southwestern Indiana, and western Kentucky. These stratigraphic columns were measured, described, and sampled from each megascopic lithologic unit. A total of 1542 samples were collected from a total thickness of 3073 ft (921.9 m), giving an average sampling interval of 2 ft (0.61 m).

For the petrographic study of the oriented thin sections, the limit between organic and inorganic components and matrix was set at 50 μm. The most abundant organic components are crinoids and echinodermal debris (hereafter lumped as crinoids without differentiation) and bryozoans. Endothyrids, echinoid spines, brachiopod shells and spines, ostracods, and smaller benthic foraminifers are common.

Intraclasts, pellets, and ooids are the major inorganic components. Intraclasts are defined as fragments of previously indurated biocalcisiltite sediment larger than 250 μm, subangular with irregular polygonal shapes, and poorly sorted. Lithic pellets are defined as ellipsoidal grains of similar material as the intraclasts and ranging in size from 50 to 250 μm. The fecal pellets are ellipsoidal grains of calcisiltitic material that occur in association with bioturbation tracks. The fecal pellet sizes range from 50 to 250 μm.

Normal ooids show well-developed and multiple cortical layers and often display cerebroid structure. The ooid cores, in order of decreasing abundance are crinoids, intraclasts, detrital quartz, bryozoans, brachiopods, and *Endothyra*. Superficial ooids are common and associated with the ooids.

Following the inventory of the various components, the clasticity index was measured for intraclasts, lithic pellets, detrital quartz, fecal pellets, ooids, superficial ooids, and crinoids. The frequency index was measured for the following components: intraclasts, lithic pellets, fecal pellets, ooids, superficial ooids, detrital quartz, crinoids, echinoid spines, bryozoans, *Endothyra*, smaller benthic foraminifers, brachiopods, calcispheres, sponge spicules, and ostracods. All frequency counts were done within 390 mm² of a thin section area corresponding to six low-power microscope fields of view.

Intensity of dolomitization and anhydritization was determined by measuring average rhomb or crystal size, and by visual estimation of abundance in percentage. Sparite cement, calcisiltite matrix, and bioclastic matrix abundances were also estimated visually. Relative abundance was estimated for algal laminae, bioturbation, and bioaccumulation with the following adjectives: rare (less than 5%), present (5 to 15%), common (15 to 85%), and abundant (over 85%).

The statistical evaluation of the initial microfacies undertaken by the modified Demirmen program gave a core ration of 89%. The 11% of "rejected" thin sections, after another petrographic examination, were returned to their original microfacies because textural and compositional variations which had led to the computer reclassification were not found to be significant sedimentologically. Stromatolites and calcisiltites were not subjected to statistical analysis because they lacked sufficient components for meaningful quantitative comparison with other microfacies.

The SOUPAC (1976) statistical package program was used to calculate average values of all measured parameters in each microfacies (Table 3.1). This table was completed by calculation of average values of components for stromatolites and calcisiltites. The table of average values of components was in turn used to calculate Pearson multivariate correlation coefficients between the microfacies (Table 3.2) with SOUPAC (1976).

The combination of the Pearson multivariate correlation coefficients between microfacies with the study of the vertical superposition of microfacies in field sections and cores led to the recognition of three vertical shallowing-upward sequences of microfacies, which were in turn interpreted horizontally as three distinct depositional models.

Description of Microfacies

The numbering system for microfacies is as follows:

1. *Series 10*: oolitic biocalcarenites ranging from well-sorted grain-supported or pressure-welded oolitic biocalcarenite with sparite cement (10) through bioclastic oolitic sparite cemented calcarenites (11 and 12) to oolitic biocalcarenite with calcisiltite matrix (13)
2. *Series 20*: intraclastic to lithic pelletoidal biocalcarenites ranging from coarse intraclastic and sparite cemented biocalcarenite (21) through lithic pelletoidal sparite cemented biocalcarenite (22) to lithic pelletoidal biocalcarenite with calcisiltite matrix and sparite cement (23)
3. *Series 30*: coarse sparite cemented biocalcarenites (31 and 32) ranging through grain-supported biocalcarenite with bioclastic matrix (33), mud-supported biocalcarenites (34 and 36), and bioaccumulated bryozoan-brachiopod limestone (35), to bioclastic calcisiltites (37 and 38)
4. *Series 40*: calcisiltites, which may be bioturbated, or show collapse brecciation, and local beds of very bituminous calcisiltites containing abundant scattered small euhedral anhydrite crystals (numbered 50)
5. *Series 60*: stromatolites

Microfacies of the St. Louis Limestone were divided into the following three broad textural groups:

Group 1. Calcisiltites (Including Stromatolites)

This group consists of calcisiltites sometimes with small anhydrite crystals or fine-grained bioclasts of crinoids, bryozoans, and brachiopods. Bioturbation is profuse. Brecciation caused by dissolution of evaporites may be extensive. The microfacies of this group, in order of increasing energy are: 60, 50, and 40.

Microfacies 60 (Fig. 3.1A). Stromatolitic bituminous bioconstructed limestone with regularly to irregularly laminated reddish to dark brown algal mats. Intermat calcisiltite laminae contain rare minute bioclasts of crinoids and bryozoans.

Microfacies 50 (Fig. 3.1B). Laminated bituminous calcisiltite with abundant small euhedral anhydrite crystals and common pyrite pigments.

Microfacies 40. Bituminous, slightly pelletoidal, and bioturbated calcisiltite with scattered sand-size to silt-size bioclasts of crinoids, bryozoans, brachiopods, and ostracods.

Group 2. Pelletoidal Calcisiltites to Bioclastic-Intraclastic Calcarenites

These microfacies range from pelletoidal calcisiltites with scattered fine-grained bioclasts to pressure-welded sparite cemented biocalcarenites that may be intraclastic. The microfacies of this group are in order of increasing relative energy: 37, 38, 36, 35, 34, 33, 23, 32, 22, 31, and 21.

TABLE 3.1

St. Louis Limestone (Middle Mississippian), Illinois Basin, Average Values of Components per Microfacies

Microfacies Components	10	11–12	13	21	22	23	31	32	33	34	35	36	37	38	40	60
Intraclast frequency	14.20	90.02	24.41	31.06	9.66	14.67	6.36	11.67	1.66	10.95		6.81		4.84		
Intraclast size	0.50	0.67	0.38	1.27	0.06	0.25	0.10	0.10	0.03	0.09		0.12		0.05		
Pellet frequency	59.90	105.67	74.00	37.19	95.66	603.25	28.76	83.65	16.30	123.30	16.62	97.30	318.88	184.53	15.00	
Pellet size	0.04	0.05	0.05	0.03	0.14	0.12	0.02	0.05	0.01	0.07	0.01	0.07	0.07	0.06	0.12	
Ooid frequency	205.70	175.80	59.40	35.90	7.65			1.62				0.02				
Ooid size	1.00	0.61	0.69	0.64	0.03			0.01								
Superficial ooid frequency	14.40	16.22	0.52	4.87						0.09		0.07				
Superficial ooid size	0.31	0.06	0.02	0.10						0.01		0.01				
Crinoid frequency	74.95	176.88	103.15	142.44	380.66	127.59	148.50	184.15	118.70	133.40	106.16	77.70	37.22	52.23	16.00	
Crinoid size	1.78	1.20	1.15	2.42	0.74	0.69	1.74	1.38	1.57	1.45	1.53	1.11	0.37	0.66	1.10	
Quartz frequency	0.96	5.24	74.10		8.46	29.86	4.11	0.31	1.39	5.17		12.76	86.60	50.54	13.00	
Quartz size	0.10	0.09	0.06		0.06	0.05	0.01	0.02	0.01	0.02		0.02	0.02	0.02	0.10	
Percent dolomitization			1.11		0.36	2.62	0.32	0.22	0.58	1.67	1.66	2.66	1.33	3.77		
Average rhomb size			0.23		0.01	0.02	0.01	0.01	0.01	0.02	0.01	0.03	0.21	0.03		
Percent anhydritization	2.70	0.85	0.85	2.43	0.74	0.73	0.83	0.44	0.22	0.85	0.24	0.80	2.44	1.61		
Average crystal size	0.08	0.06	0.13	0.24	0.04	0.04	0.05	0.02	0.03	0.07	0.02	0.04	0.06	0.09		
Echinoid spines frequency	1.08	2.57	1.11	4.81	0.56	0.51	1.27	1.50	1.33	1.09	1.28	0.22	0.22	0.33		
Bryozoan frequency	54.13	116.71	59.07	133.20	369.92	41.43	231.03	99.91	181.20	55.99	174.69	65.00	14.66	31.84	9.20	
Endothyra frequency	10.95	21.40	9.29	9.06	14.20	10.31	12.17	165.05	19.63	23.16	2.95	4.00	3.77	5.45	3.50	
Small foraminifer frequency	2.34	2.48	0.33	0.62	2.72	0.08	0.25	0.94	1.15	0.52	0.35	0.20		0.64		
Brachiopod frequency	15.87	12.69	9.85	17.37	15.85	4.41	14.23	7.72	11.71	7.46	24.34	9.00	4.33	4.50	1.50	
Calcisphere frequency	0.35	0.93	0.11		8.13	6.86	0.17	0.44	3.74	2.13	0.12	0.88	5.55	7.16		
Sponge spicules frequency						0.65				0.16		0.84	95.77	1.08		
Ostracod frequency	15.52	6.22	16.81	4.44	12.20	28.43	4.06	12.41	9.38	6.89	9.07	10.85	40.55	25.93		
Percent sparite cement	13.48	13.03	4.48	14.75	11.82	6.99	16.79	8.60	4.07	4.60	8.10	3.21	3.00	3.61		55
Percent calcisiltite matrix	7.74	6.05	29.85	4.06	7.20	31.57	3.27	16.87	0.18	20.49	22.97	42.44	61.11	60.91	90	
Percent bioclastic matrix		0.36	1.11		0.56		1.56	0.19	25.91	0.25						
Algal laminations[a]																A
Bioturbation			P		R	P		R	R	R	R	R	P	R		R
Bioaccumulation												A		P		

[a] R, rare; A, abundant; P, present.
Source: Diaby and Carozzi (1984).

TABLE 3.2

St. Louis Limestone (Middle Mississippian), Illinois Basin, Multivariate Correlation Coefficients of Microfacies

	10	11–12	13	31	32	21	22	33	23	34	35	36	37	38
10	1.00													
11–12	0.85	1.00												
13	0.61	0.81	1.00											
31	0.31	0.62	0.60	1.00										
32	0.29	0.59	0.61	0.67	1.00									
21	0.60	0.85	0.75	0.90	0.70	1.00								
22	0.34	0.69	0.69	0.96	0.74	0.91	1.00							
33	0.92	0.82	0.58	0.53	0.34	0.71	0.52	1.00						
23	0.89	0.71	0.55	0.06	0.17	0.33	0.13	0.71	1.00					
34	0.36	0.69	0.78	0.65	0.79	0.73	0.77	0.30	0.37	1.00				
35	0.23	0.49	0.48	0.88	0.54	0.77	0.83	0.45	0.01	0.53	1.00			
36	0.34	0.64	0.79	0.69	0.68	0.70	0.75	0.31	0.37	0.94	0.60	1.00		
37	0.15	0.27	0.51	0.08	0.26	0.11	0.16	−0.04	0.47	0.63	0.03	0.69	1.00	
38	0.06	0.15	0.30	0.10	0.18	0.09	0.15	−0.02	0.24	0.40	0.06	0.45	0.50	1.00

Source: Diaby and Carozzi (1984).

Microfacies 37 (Fig. 3.1C). Pelletoidal bituminous calcisiltite with common monaxonic sponge spicules, calcispheres, ostracods, and crinoids.

Microfacies 38. Finely pelletoidal bituminous calcisiltite with common sand-size to silt-size bioclasts of crinoids, bryozoans, ostracods, calcispheres, sponge spicules, and brachiopods.

Microfacies 36 (Fig. 3.1D). Poorly sorted bioturbated mud-supported biocalcarenite. Bioclasts of crinoids, bryozoans, echinoid spines, and brachiopods occur as "floating" grains within a bituminous calcisiltite matrix.

Microfacies 35 (Fig. 3.1E). Bioaccumulated bryozoan-brachiopod limestone with pelletoidal bituminous calcisiltite matrix containing minor amounts of crinoids, coral fragments, ostracods, and pelecypods.

Microfacies 34. Mud-supported crinoid-*Endothyra*-bryozoan calcarenite. Bioclasts are "floating" in a bioturbated bituminous calcisiltite matrix containing silt-size bioclasts of crinoids, bryozoans, and ostracods.

Microfacies 33 (Fig. 3.1F). Mud-supported to grain-supported crinoid-bryozoan-*Endothyra* calcarenite with a bioturbated matrix composed of very fine-grained bioclasts of crinoids and bryozoans in a bituminous calcisiltite groundmass.

Microfacies 23 (Fig. 3.1G). Grain-supported fine-grained pelletoidal crinoid-bryozoan calcarenite with calcisiltite matrix and sparite cement. Crinoidal bioclasts show syntaxial calcite overgrowth that forms most of the sparite cement of the rock and is associated with extensive pressure-solution.

Microfacies 32. Grain-supported crinoid-*Endothyra*-bryozoan calcarenite with calcisiltite matrix and sparite cement. The cement consists of syntaxial overgrowths on crinoids and interparticle sparite mosaic. The calcisiltite matrix shows an irregular distribution due to bioturbation and is slightly more abundant than the cement.

Microfacies 22 (Fig. 3.1H). Grain-supported intraclast-crinoid-bryozoan calcarenite with sparite cement as syntaxial overgrowths on crinoids, and common pressure-solution.

Microfacies 31 (Fig. 3.2A). Coarse grain-supported to pressure-welded crinoid-bryozoan calcarenite with sparite cement and local pressure-solution. Micritization of *Endothyra* and bryozoan bioclasts is common. Cement consists of syntaxial overgrowths on crinoids grading into cavity-filling sparite cement.

Microfacies 21 (Fig. 3.2B). Coarse, well-rounded, grain-supported crinoid-intraclast-bryozoan biocalcarenite with intergranular cement of sparite mosaic and syntaxial overgrowths on crinoids.

Group 3. Grain-Supported Oolitic Biocalcarenites

This group consists of ooid-intraclastic calcarenites with calcisiltite matrix to grain-supported sparite cemented oolitic calcarenites. The microfacies of this group are in order of increasing relative energy: 13, 12, 11, and 10.

Microfacies 13. Mud-supported to grain-supported crinoid-ooid intraclast calcarenite with calcisiltite matrix. Ooids are commonly abraded and broken.

Microfacies 12. Grain-supported crinoid-intraclast-ooid calcarenite with sparite cement and pressure-welding. Crinoidal fragments often possess syntaxial

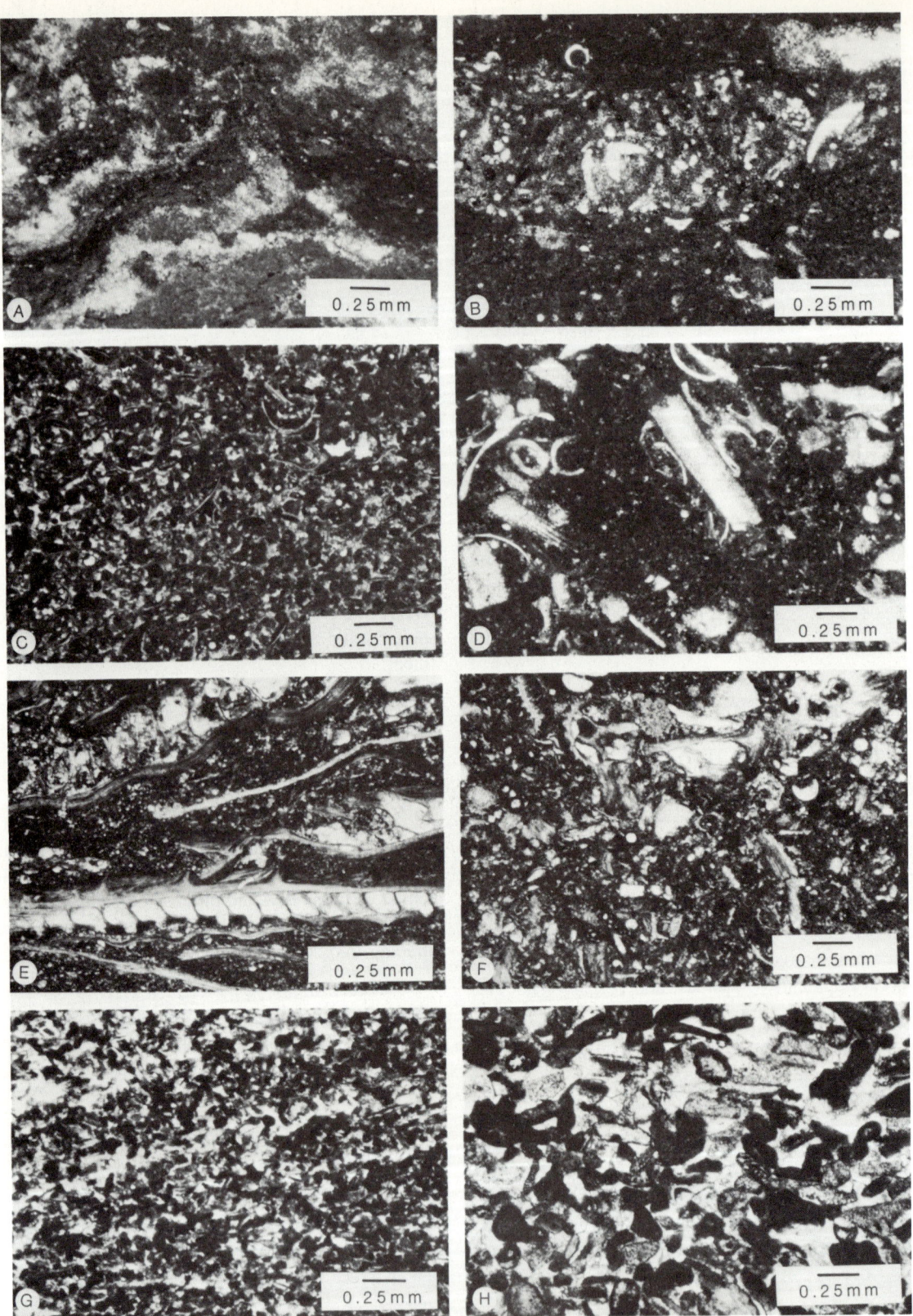

FIGURE 3.1 St. Louis Limestone (Middle Mississippian), Illinois Basin. Typical microfacies. A. Microfacies 60. B. Microfacies 50. C. Microfacies 37. D. Microfacies 36. E. Microfacies 35. F. Microfacies 33. G. Microfacies 23. H. Microfacies 22. See text for detailed descriptions. All photomicrographs: plane-polarized light. From Diaby and Carozzi (1984).

overgrowths which are associated with interparticle mosaic cement. Ooids and superficial ooids show evidence of reworking such as breakage, abrasion, or truncation. Bioclasts of ostracods, echinoid spines, and *Stachyodes* are rare accessory components.

Microfacies 11 (Fig. 3.2C). Grain-supported ooid-crinoid-intraclast calcarenite with sparite cement and local pressure-solution. The minor components are bryozoans, brachiopods, *Endothyra*, and gastropods. Pressure-solution is common and is shown mostly by straight ooid contacts. Cement is mostly intergranular sparite with rare calcite overgrowths on crinoids.

Microfacies 10 (Fig. 3.2D). Grain-supported and pressure-welded oolitic calcarenite with sparite cement. Minor bioclasts consist of *Endothyra*, brachiopods, ostracods, echinoid spines, and bryozoans. Pressure-solution is shown by straight ooid contacts, chain ooids, and spalling of ooid outer cortical layers. Cement is intergranular void filling sparite mosaic.

Depositional Models

Microfacies Associations

Examination of the vertical succession and reciprocal relationships of microfacies shows that three associations of microfacies followed each other vertically in cores and sections expressing an increasing complexity of the carbonate environments during St. Louis time. These microfacies associations are:

1. A smaller energy group consisting of microfacies 60, 40, 38, 37, 36, 35, 34, 33, and 23.
2. A smaller and moderate energy group consisting of microfacies 60, 40, 38, 37, 36, 35, 34, 33, 23, 32, 22, 31, and 21.
3. A smaller, medium, and greater energy group consisting of microfacies 60, 40, 38, 37, 36, 35, 34, 33, 23, 32, 22, 31, 12, 11, and 10.

The core from Caldwell County, Kentucky, exemplifies the succession of the three associations of microfacies (Fig. 3.4). Such a distinct evolution requires preparation of three depositional models which followed each other in time. For graphical reasons, the three models are shown as juxtaposed from left to right. The apparent abrupt change from a given model was located arbitrarily at the first recognized appearance, in a given section or core, of the typical microfacies association of the next model. In reality, the change between successive models is gradual. Each depositional model is illustrated first by an ideal vertical shallowing-upward sequence with average values of components and a relative depth variation curve at the extreme right (Figs. 3.5, 3.6, 3.8, 3.10). Each vertical ideal sequence is then transposed into a horizontal depositional model (Figs. 3.7, 3.9, 3.11). The description will be limited to the graphic representation of the latter, which consists of curves of average values of component frequencies and clasticities, of average matrix and cement percentages, of relative abundance of certain minor biotic components, textural features, evaporites and dolomitization, and of a relative energy variation curve.

The Ideal Depositional Model 1

It consists of a gentle carbonate ramp (Fig. 3.7) which displays a complete sequence of microfacies from supratidal evaporitic flats to basinal conditions. The upper ramp is characterized by abnormal smaller energy, a sparse fauna, and pervasive dolomitization represented by small dolomite rhombs scattered in the calcisiltites. This unusual association of features occurs only in model 1. It indicates a restricted hypersaline environment which can be attributed to the combination of baffling effects of crinoidal buildups of the lower ramp with the seepage reflux of heavy brines generated by large evaporation rates in the supratidal flats.

Component Variation

Maxima of crinoid frequency and size occur in the agitated moderate energy environment of the lower ramp (microfacies 23, 33, and 34) in coincidence with the wave base and representing the peak of energy of this model. From this area of favorable growth, which will subsequently evolve into a bioclastic bar, the crinoids decrease in importance both seaward and landward. Distribution of crinoids by currents was more toward the basin than toward the upper ramp, probably as an effect of the baffle zone created by the crinoids themselves in association with bryozoans and brachiopods. At this early stage, the crinoid-bryozoan brachiopod baffle community affects to a large extent the distribution of all biogenic and inorganic components, the matrix, and the cement.

The peaks of frequency and size of lithic pellets and intraclasts are both centered on the moderate energy lower ramp environment in association with the baffling crinoid-bryozoan-brachiopod community and indicate active intraformational reworking of sediments in that particular zone with intraclast generation extending farther into the upper ramp. Fecal pellet distribution corresponds to smaller energy conditions in both basinal and protected upper ramp environments which are more amenable to extensive biogenic activity and preservation of fecal pellets. The greatest frequency of fecal pellets is in the basinal environment.

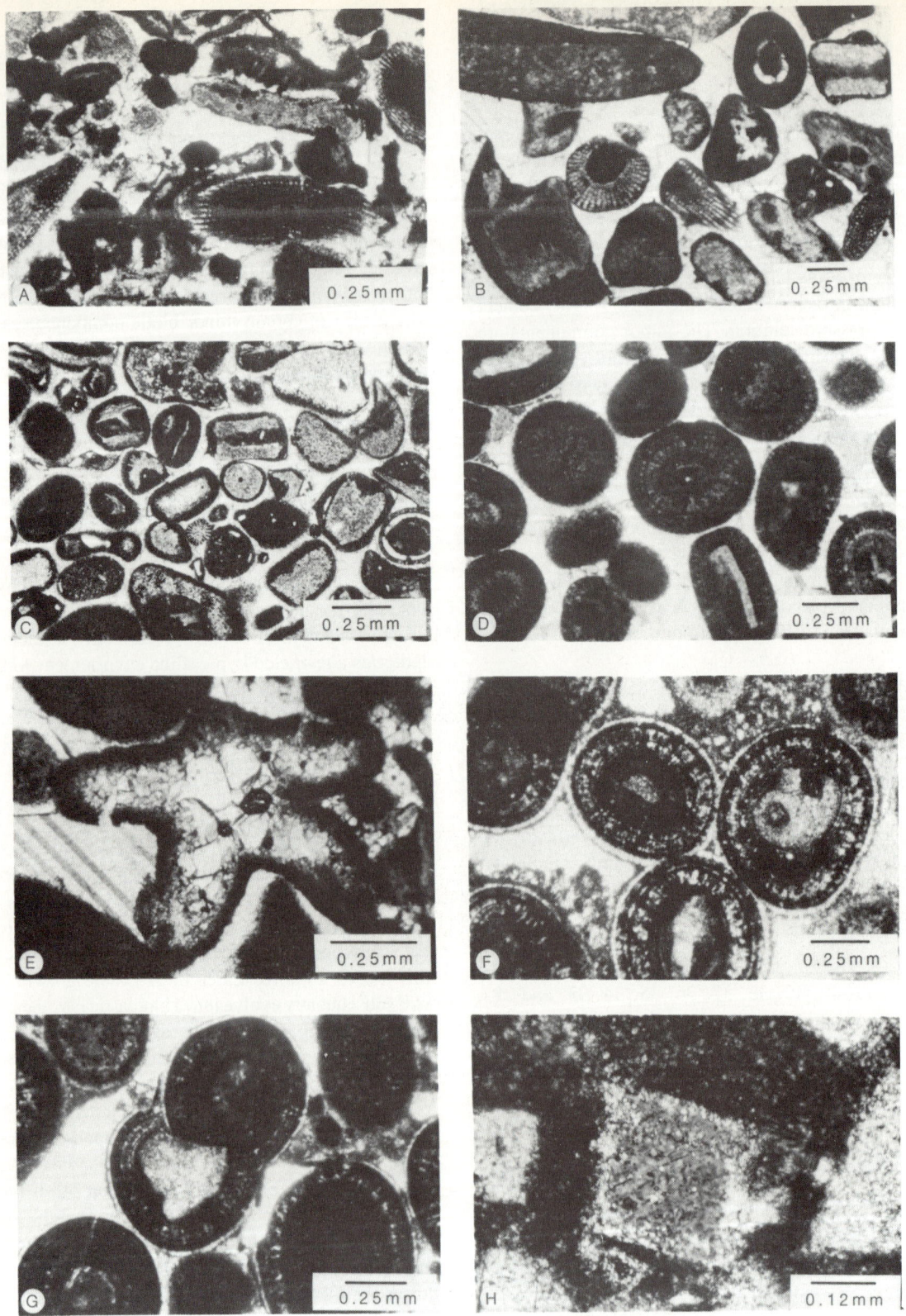

FIGURE 3.2 St. Louis Limestone (Middle Mississippian), Illinois Basin. Typical microfacies and diagenesis. A. Microfacies 31. B. Microfacies 21. C. Microfacies 11. D. Microfacies 10. See text for detailed descriptions. *Marine phreatic environment.* E. Micritization of margins of a green algal fragment. The micritized envelope is fairly thick, broken in places, and is now infilled with microsparitic calcite. F. Early compaction II of well-formed ooids causing local solution of thin isopachous rim cement. Geopetal pelletoidal internal sediment is present between ooids and is overlain by freshwater phreatic

Frequency and clasticity of detrital quartz show peaks in two environments: at the foot of the lower ramp (average grain size of 0.05 mm) and in the upper ramp calcisiltites (average grain size of 0.02 mm). The upper ramp fine-grained quartz appears land derived and wind blown, in contrast to the former coarse-grained quartz, which was probably transported by rivers into the basin, distributed laterally by currents, and eventually trapped in the crinoidal baffle zone of the lower ramp.

Bryozoans, brachiopods, and echinoid spines show maximum frequencies in the lower ramp and their distribution curves are slightly offset upslope from that of crinoids, indicating that these organisms lived behind the protection of the crinoid baffle systems and participated significantly in constructing the incipient bioclastic bar.

Sparite cement and micrite matrix show an expected inverse relationship. More matrix occurs on the upper ramp than in the basin because of the smaller energy of currents in the upper ramp, a consequence of the baffling effect of the crinoidal buildups of the lower ramp. More cement occurs in the basinal microfacies than in the upper ramp microfacies because of more current winnowing out of micritic matrix in the basinal environment. The relative energy curve for the environments is derived from the cement/matrix ratios as well as from grain sizes, rounding and sorting in the various microfacies.

Ostracods, calcispheres, and sponge spicules are common in the basinal environment, with an apparent trapping of calcispheres in the crinoidal baffle zone of the lower ramp, indicating general basinward dispersion.

The Ideal Depositional Model 2

In this model (Fig. 3.9) a subtidal to low intertidal crinoidal-bryozoan bioclastic bar (hydrodynamic buildup) has developed which separates a lagoonal environment from slope and basinal conditions. Compared to model 1, the extension of the evaporitic flat is appreciably reduced and the related dolomitization by seepage reflux therefore becomes a moderately active process.

Component Variation

Crinoid frequency and clasticity curves reach their peaks in coincidence with the bioclastic bar, which corresponds to the maximum energy of this model. As in the previous model, the behavior of crinoids is a major controlling factor.

Intraclast peaks of frequency and clasticity in the bioclastic bar coincide with crinoids, whereas lithic pellets reach their peak for both frequency and clasticity behind the bioclastic bar as a result of abrasion because of slight lagoonward transport. Generation of intraclasts confirms the location of the maximum energy at the crinoidal bar. Fecal pellets show their frequency peak in the basin and decrease rapidly up the slope and toward the bioclastic bar. Another occurrence of fecal pellets is located in the lagoon. Fecal pellets are thus abundant in smaller-energy environments where extensive infaunal activity can occur and preserve its products.

Detrital quartz behaves as in model 1. A minor occurrence of ooids in the bioclastic bar forecasts its evolution into the oolitic bar-to-bank environment characterizing model 3.

Bryozoan and brachiopod frequency curves show parallel behavior. A first peak is in the bioclastic bar, indicating a community associated with crinoids and greater energy conditions. Bioclasts of this community are transported basinward. A second peak is located in the lagoon, which represents a community living in much smaller energy. Preservation is enhanced and fragments tend to be larger.

A peak of *Endothyra* frequency occurs in front of the bioclastic bar and a rapid decrease occurs in both directions away from the bioclastic bar. Some *Endothyra* appear reworked; they were transported across the bar into the lagoon.

Sparite cement and micrite matrix show an expected inverse relationship with the cement reaching its peak in the bioclastic bar, and again, as in model 1, more micrite matrix in the lagoonal than in basinal conditions.

Ostracods, calcispheres, and sponge spicules have settled in the quiet basinal environment. Together with

FIGURE 3.2 (*continued*) sparite cement. G. Early compaction II also causes deformation and strong interpenetration of ooids with thin isopachous rim cement. Minor spalling of ooid cortical layers is present. Intergranular spaces are filled by geopetal pelletoidal internal sediment overlain by freshwater phreatic calcite cement. *Marine vadose environment.* H. Development of euhedral crystals of anhydrite (anhydrite I) by replacement of bituminous calcisiltite, with inclusions of unreplaced matrix. Subsequent marginal calcitization (freshwater phreatic) is visible in crystal extinction in center of photomicrograph (nicols half-crossed). All other anhydrite crystals are entirely calcitized. All photomicrographs: plane-polarized light, except D and H, half-crossed nicols. From Diaby and Carozzi (1984).

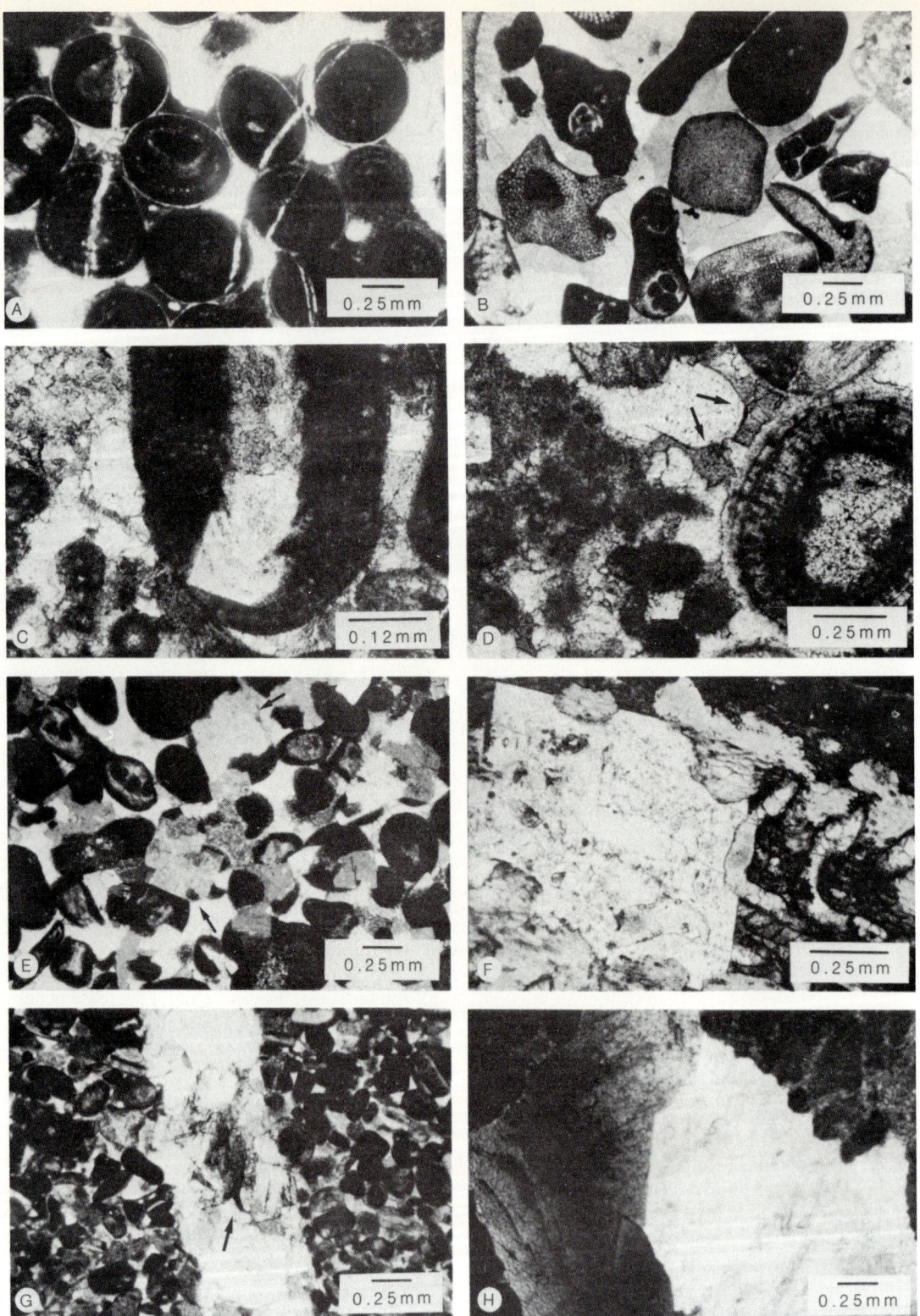

FIGURE 3.3 St. Louis Limestone (Middle Mississippian), Illinois Basin. Diagenesis. *Undersaturated freshwater phreatic environment.* A. Early fracturation of ooids after deposition of geopetal internal sediment. The thin fractures cut across ooids, isopachous rim cement, and geopetal internal sediment. The fractures are filled with microsparitic cement from the saturated freshwater phreatic environment. The remaining intergranular spaces are filled with coarser freshwater phreatic cement. *Saturated freshwater phreatic environment.* B. Syntaxial calcite overgrowths enclosing bryozoans developed around crinoids and formed an interlocking mosaic. *Mixing*

other constituents, they indicate dispersion across the bioclastic bar seaward, landward, or both under moderate action of currents and tides.

The Ideal Depositional Model 3

It represents the final stage of deposition in St. Louis time. The well-developed bioclastic bar of model 2 has now evolved into a system of oolitic bars and banks (Fig. 3.11) separating lagoonal from basinal environment. Compared to model 2, the extension of the evaporitic flat is further reduced and the related dolomitization by seepage reflux therefore becomes a minor process.

Component Variation

Crinoids are the major contributors to the bar-to-bank system. Widespread distribution of crinoids expresses greater energy conditions under which the bar-to-bank system developed. Crinoids were distributed basinward mechanically from the bar-to-bank area and a second population of larger (average size of columnals over 3 mm) crinoids occurs in the lagoon associated with peaks of frequency of bryozoans and brachiopods. As in the previous models, the position of the oolitic-crinoidal bar-to-bank system affects the distribution of almost all components.

Both lithic pellets and intraclasts are most abundant in the oolitic bar-to-bank environment, indicating active intraformational reworking and distribution. As previously, fecal pellets show two peaks of abundance, first in the basin with decrease to zero as one progresses up the slope toward the bar-to-bank environment, and second, in the lagoon. Both fecal pellet peaks occur under conditions of smaller energy with active infauna and good preservation of pellets.

Fine sand-size detrital quartz occurs throughout the oolitic bar-to-bank environment, whereas fine-grained wind blown concentrations occur along the landward edge of the lagoon.

Ooids are a major component of the bar-to-bank environment with corresponding maxima of their frequency and clasticity. Clasticity is largest near the frontal area of this environment where maximum energy would be expected as well as best conditions for oolitization associated with currents rising up the slope.

Frequency curves of bryozoans and brachiopods are roughly parallel. A broad peak centers on the oolitic bar-to-bank environment, suggesting a distinct population living in greater-energy conditions but partially dispersed by currents. A second but narrower peak of greater frequency indicates another population living in the quieter lagoon, which was apparently still well connected with open sea conditions.

The inverse relationship between sparite cement and micrite matrix continues in this model, and the matrix remains more abundant in lagoonal than in basinal conditions. As in the previous model, ostracods, calcispheres, and sponge spicules settled in the lower-energy basinal environment after a dispersion seaward by currents and tides. A similar redistribution both seaward and landward affects crinoid, bryozoan, and brachiopod bioclasts. Only stromatolites live under the influence of the seepage reflux processes.

Stratigraphic Sections and Vertical Succession of Microfacies

Petrographic data derived from the study of thin sections were plotted for all stratigraphic sections. The section of the Caldwell County core (Fig. 3.4) is a perfect example of vertical superposition of deposits of the three depositional models and can be used to introduce the type of graphic representation in this study (Figs. 3.12, 3.13).

FIGURE 3.3 (continued) *marine freshwater phreatic environment.* C. Pervasive silicification resulting in growth of secondary euhedral quartz crystals within an ooid core and inner cortical layers, with "ghosts" of the cortical layers inside the crystal. *Burial environment.* D. Burial dissolution of intergranular sparite cement creates secondary porosity (3.6%) filled with blue epoxy (appearing gray, arrows). E. Partial anhydrite replacement of bioclasts (crinoids and bryozoans), ooids, and sparite cement (arrows). The anhydrite is subhedral to euhedral (anhydrite II). F. Euhedral crystal of anhydrite replacing a partially silicified bryozoan showing that the second phase of anhydritization follows silicification. G. Late burial fracturation and sparite cementation of fractures, with also partial anhydritization (arrow) of fracture-filling calcite cement by bladed anhydrite III. H. Late dolomitization by saddle dolomite in cavities (fractures) in a pelletoidal calcisiltite. Upper right and upper left show margins of the fracture. All photomicrographs: crossed nicols except A, D, and F, plane-polarized light. From Diaby and Carozzi (1984).

44   Part I / Principles

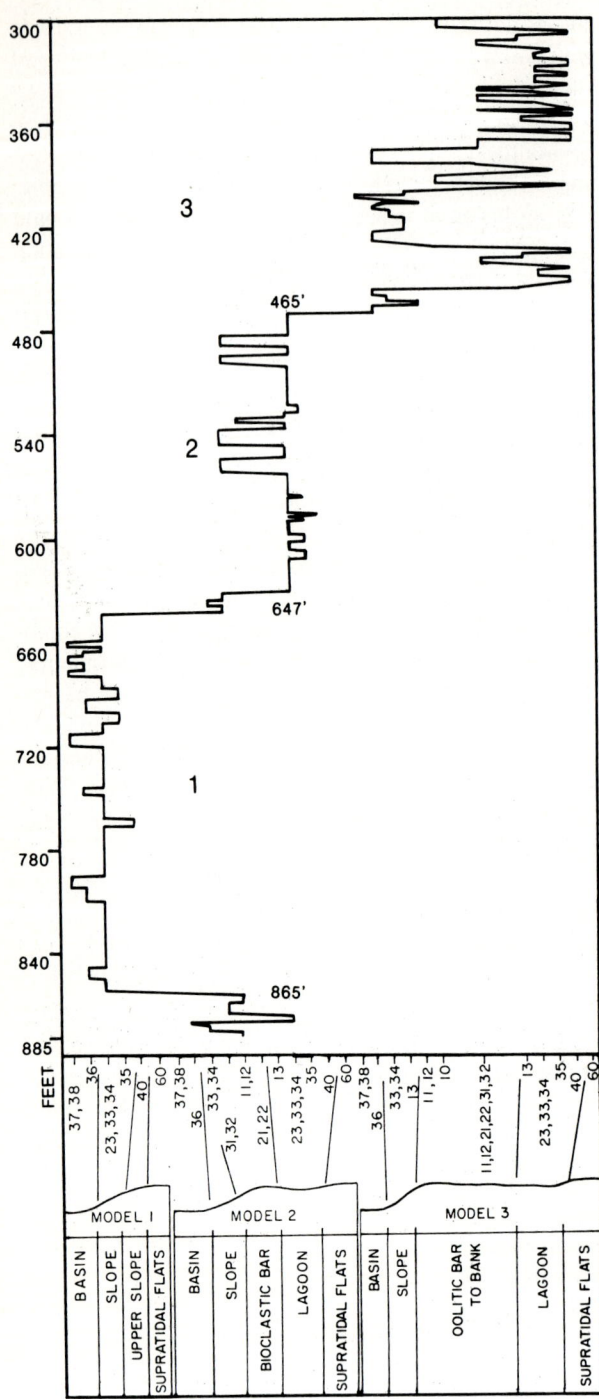

FIGURE 3.4 St. Louis Limestone (Middle Mississippian), Illinois Basin. Environmental variation curve for Caldwell County core, Kentucky. From Diaby and Carozzi (1984).

This graphic representation includes the following (from left to right):

1. Column of microfacies identified in the sections with control points and depths for the cores and distance above base for field sections in feet.
2. Set of four vertical curves showing clasticity and frequency for selected components as well as percentage of cement and matrix for the whole rock.
3. On the right, a simplified sketch of the three depositional models is used as a horizontal scale for the curve of evolution of the environments through time.

Models 1, 2, and 3 are juxtaposed from left to right in Figures 3.12 and 3.13. All three share the microfacies of depositional model 1 (microfacies 60, 40, 34, 35, 33, 23, 36, 38, and 37). The environmental variation curve is moved from model 1 to model 2 (Fig. 3.12) when the sparite-cemented biocalcarenites to pelletoidal biocalcarenites (microfacies 32, 22, and 21), characteristic of the bioclastic bar (hydrodynamic buildup) of model 2, are encountered in sections or cores. Similarly, the environmental variation curve is moved from model 2 to model 3 (Fig. 3.13) when grain-supported to pressure-welded oolitic biocalcarenites (microfacies 10, 11, and 12) of the oolitic bar-to-bank environment of model 3 appear in the sections.

Vertical Succession of Depositional Models and Basinwide Evolution

The St. Louis Limestone in the Illinois Basin (Fig. 3.14) represents a vast embayment open to the south with a present-day northern erosional limit beneath the Pennsylvanian and a portion of the depocenter detectable in southeastern Illinois and western Kentucky. The distribution of the evaporite-bearing zone of the lower part of the St. Louis Limestone corresponding to model 1 indicates the great extent during that particular time interval of the supratidal evaporitic flats (coastal sabkha) grading basinward into a carbonate ramp.

A simplified basin cross section of the St. Louis Limestone, oriented southwest-northeast, was prepared from the following cores: Crawford County, Indiana; Warren County, Kentucky; Christian County, Kentucky; and Caldwell County, Kentucky (Fig. 3.15) to show the time and space relationships of the three depositional models from basin margins in the northeast to the depocenter in the southwest.

Environmental variation curves representing the interpretation of microfacies were plotted in each vertical section and correlated across the basin using their breaks (changes of depositional model), thus revealing the limits of the three successive depositional models. The Crawford County, Indiana, core is on the margin of the basin and shows only model 1 deposits throughout St. Louis

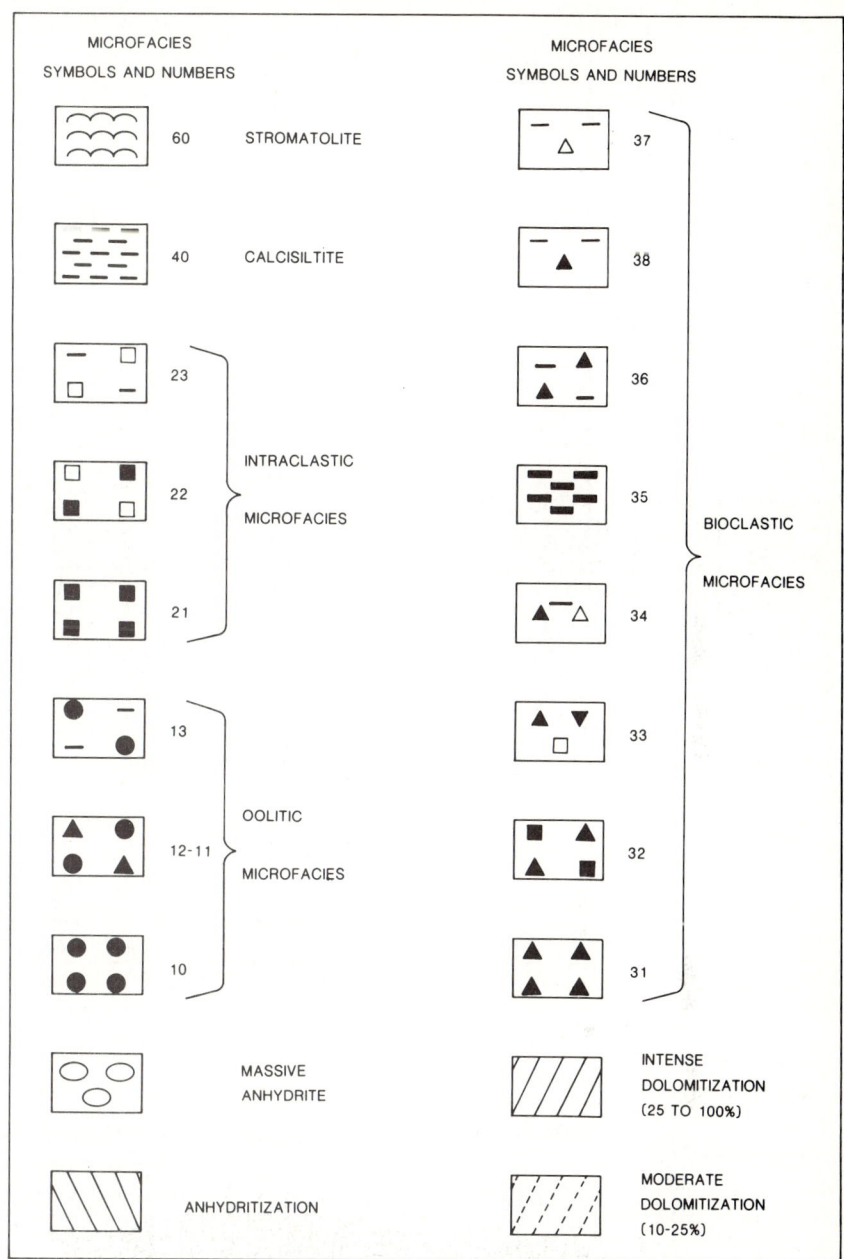

FIGURE 3.5 St. Louis Limestone (Middle Mississippian), Illinois Basin. Symbols for microfacies and diagenesis. From Diaby and Carozzi (1984).

time. The Warren County, Kentucky, core, farther away from the basin margin, shows depositional model 1 microfacies over most of its length except for the top third, where microfacies representing depositional model 2 are encountered. More toward the center of the basin, the Christian County, Kentucky, core shows a superposition of deposits from models 1, 2, and 3 and the Caldwell County, Kentucky, core an ideal succession with full development of deposits from models 1, 2, and 3. The depocenter probably corresponds to continuous deposition of basinal bioclastic calcisiltites.

Diagenesis

Diagenetic features observed in the St. Louis Limestone (Fig. 3.16) are complex but can be interpreted as representing depositional and postdepositional changes that took place in the following environments: (1) marine phreatic, (2) marine vadose, (3) undersaturated freshwater phreatic, (4) saturated freshwater phreatic, (5) mixing marine freshwater phreatic, and (6) burial. Diagenetic features observed will be described and interpreted in the order of environments mentioned above.

46 Part I / Principles

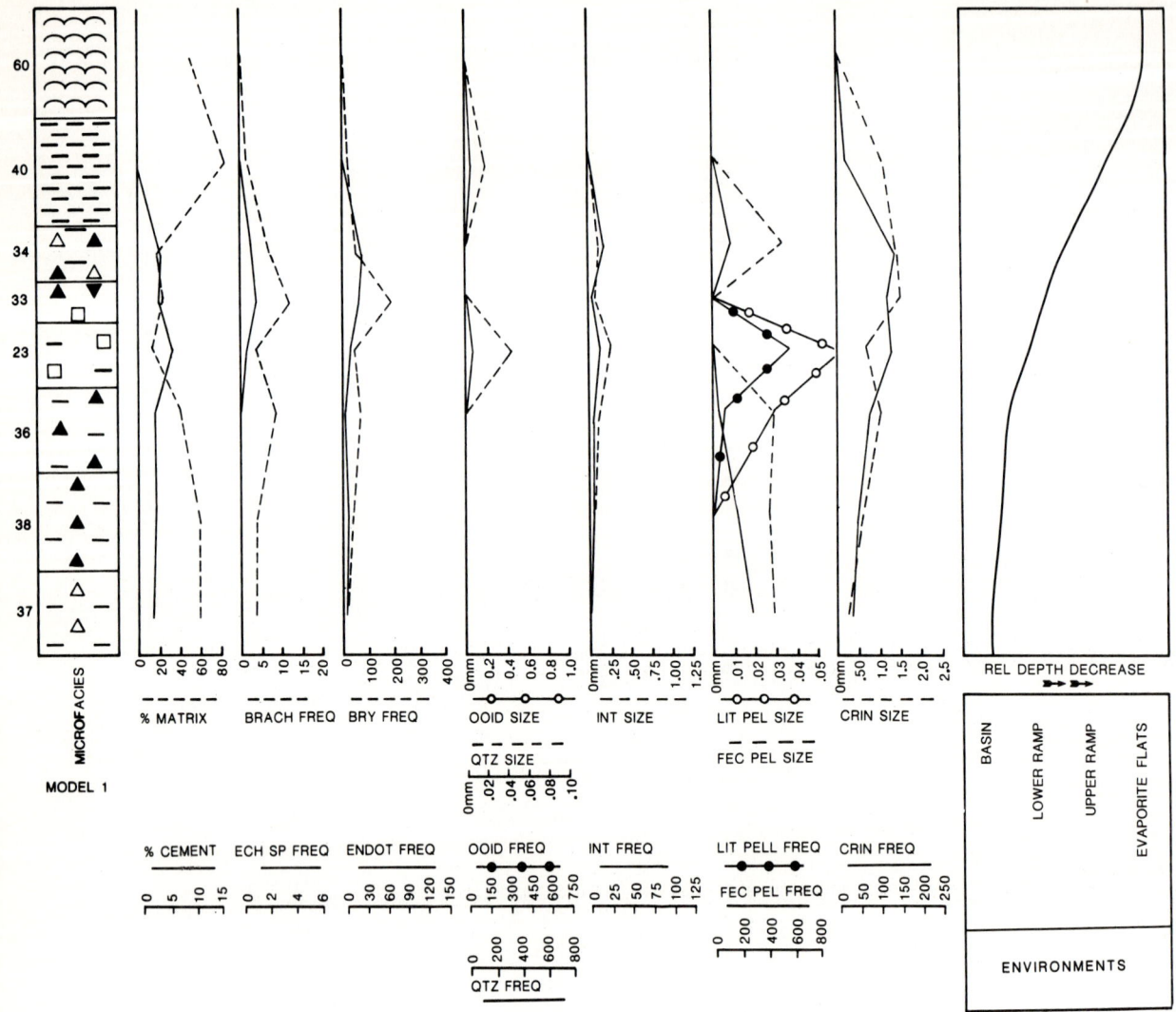

FIGURE 3.6 St. Louis Limestone (Middle Mississippian), Illinois Basin. Ideal shallowing-upward sequence for model 1. From Diaby and Carozzi (1984).

Marine Phreatic Environment

Micritization of all bioclasts (Fig. 3.2E) is as widespread as generation of intraclasts in high-energy conditions, which demonstrates early submarine lithification. Early compaction I is expressed by pressure-solution and reciprocal deformation of bioclasts, deformation at ooid contacts, minor spalling of outer cortical layers, and formation of chain ooids. The early nature of this episode of compaction is demonstrated by the absence of submarine rim cement between deformed components, which is elsewhere very well developed (Fig. 3.2F).

Early compaction II follows precipitation of the isopachous rim cement. The latter shows partial dissolution at contact of ooids as well as cases of reciprocal interpenetration of two ooids with isopachous rim cement and minor spalling (Fig. 3.2G).

Geopetal pelletoidal internal sediment infills the interooid cavities, overlies the isopachous fibrous rim cement, and is in turn overlain by freshwater phreatic calcite cement, which fills the remainder of the intergranular voids (Figs. 3.2F,G, 3.3A).

Marine Vadose Environment

Pervasive dolomitization takes place in the calcisiltites (microfacies 40) of the upper ramp (model 1) or edge of the lagoon (models 1 and 2). Dolomite occurs

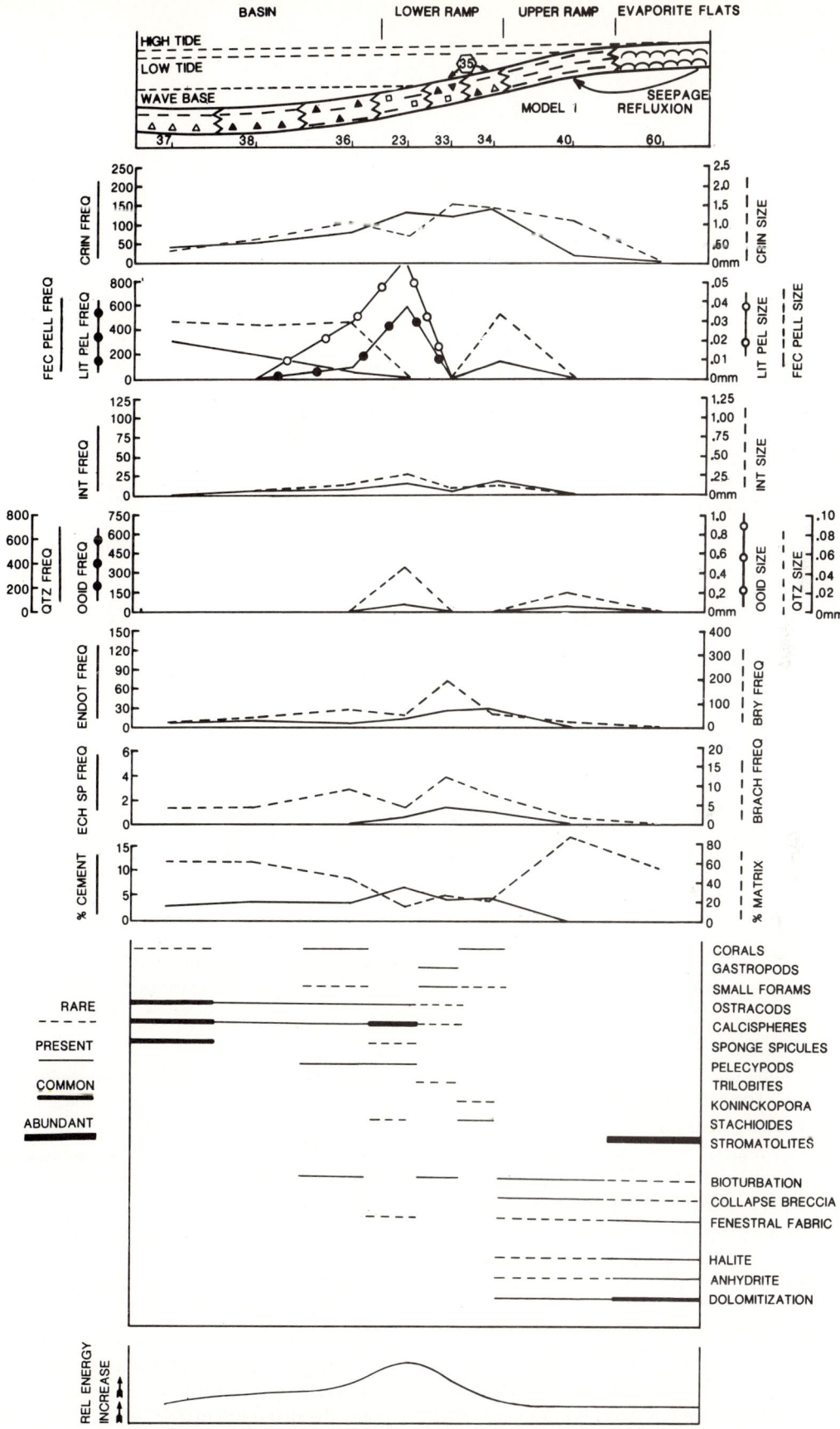

FIGURE 3.7 St. Louis Limestone (Middle Mississippian), Illinois Basin. Horizontal depositional model 1. From Diaby and Carozzi (1984).

48 Part I / Principles

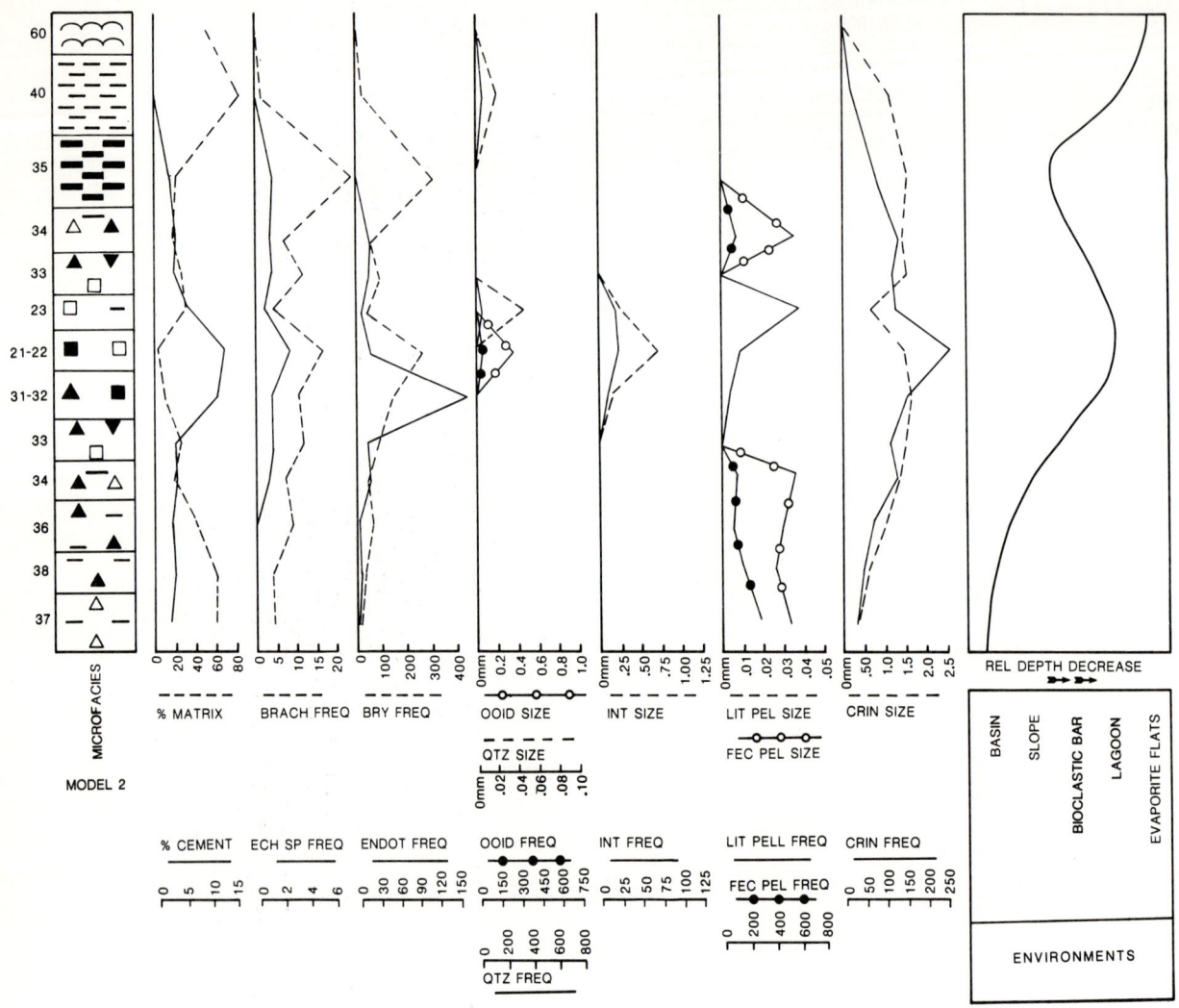

FIGURE 3.8 St. Louis Limestone (Middle Mississippian), Illinois Basin. Ideal shallowing-upward sequence for model 2. From Diaby and Carozzi (1984).

as scattered rhombs (25 to 50 μm) representing 40 to 60% of the rock. The supersaturated brines are derived from large evaporation rates in the supratidal evaporite flats (sabkha environment); they migrated by seepage refluxion toward the calcisiltites of the upper ramp or edge of the lagoon, where dolomitization took place. In association with the pervasive dolomitization, small scattered euhedral anhydrite crystals (anhydrite I) also form from the sabkha-derived brines (Fig. 3.2H).

Undersaturated Freshwater Phreatic Environment

In this diagenetic environment, leaching of thus far preserved aragonitic shells takes place. Micritized molds of molluscan shells are infilled subsequently with bladed to mosaic sparite cement. Some dissolution of anhydrite crystals and of small dolomite rhombs may also occur, creating crystal moldic porosity. Very characteristic are subparallel early fractures perpendicular to bedding, which cut across the ooids, the fibrous rim cement, and the geopetal internal sediment. Fractures are filled with calcite cement which is in optical continuity with the cavity-filling cement only at the extremities of fractures and when the latter are wide (Fig. 3.3A).

The criteria above indicate that this early fracturing phase occurred in the undersaturated freshwater phreatic environment. Then, either late in that environment or very early in the saturated freshwater phreatic environment (before most of the cavity-filling sparite cement formed), fractures were infilled with calcite cement. Fracturing is probably due to dissolution of subjacent evapo-

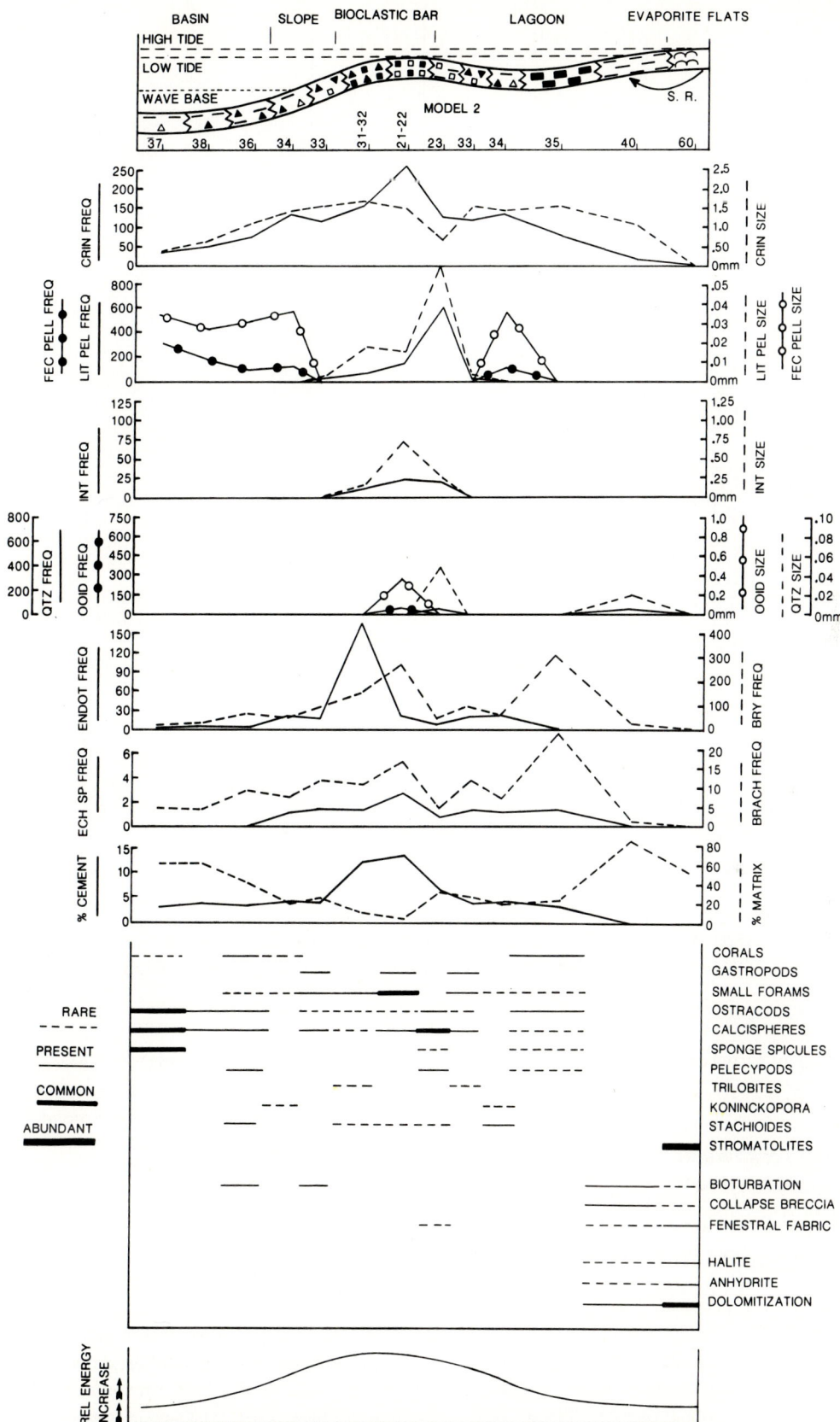

FIGURE 3.9 St. Louis Limestone (Middle Mississippian), Illinois Basin. Horizontal depositional model 2. From Diaby and Carozzi (1984).

50 Part I / Principles

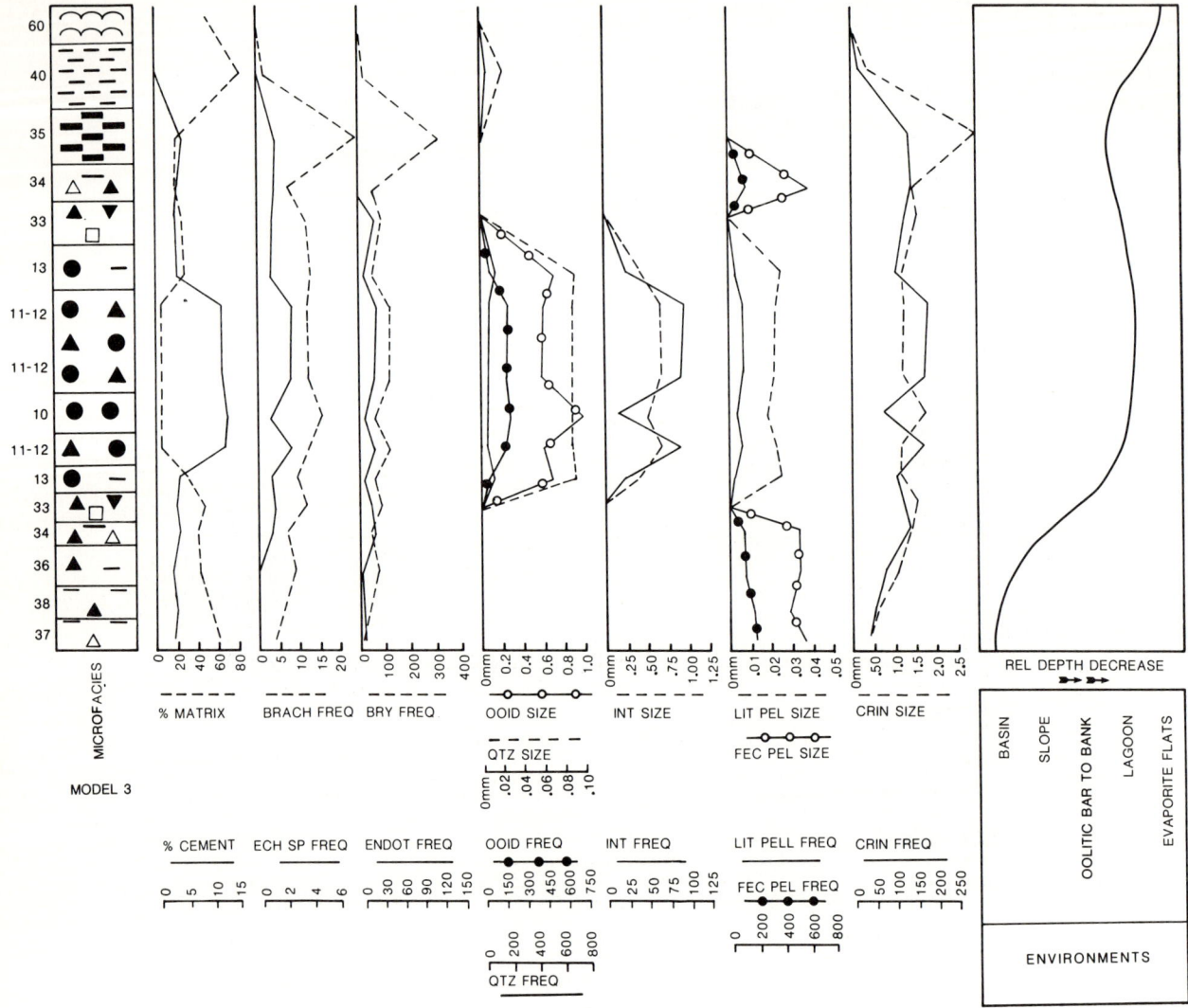

FIGURE 3.10 St. Louis Limestone (Middle Mississippian), Illinois Basin. Ideal shallowing-upward sequence for model 3. From Diaby and Carozzi (1984).

rites with consequent partial collapse of the oolitic biocalcarenite.

Saturated Freshwater Phreatic Environment

In this diagenetic environment, partial to complete pseudomorphic calcitization of euhedral anhydrite takes place (Fig. 3.2H). Disseminated small dolomite rhombs from the marine vadose pervasive dolomitization phase may remain unaffected. The most common process is the formation of clear syntaxial calcite overgrowths on the nonmicritized crinoidal bioclasts which engulf the other constituents. The syntaxial overgrowths are juxtaposed with intergranular cavity-filling sparite mosaic (Fig. 3.3B). The latter, at places, grades into fracture-filling (early fracturing phase) microcrystalline calcite (Fig. 3.3A). In this diagenetic environment, neomorphism is widespread among gastropods and bioclasts of green algae (dasyclads?).

Mixing Marine Freshwater Phreatic Environment

Silicification occurs as euhedral quartz crystals replacing cores and outer cortical layers of ooids (Fig. 3.3C) and as pervasive fibroradiated microcrystalline quartz occurring inside anhydrite nodules. This episode of silicification follows the freshwater phreatic cementation because cases exist where sparite cemented biocalcarenites (microfacies 32) are entirely silicified. The attribution of silicification to the mixing marine freshwater

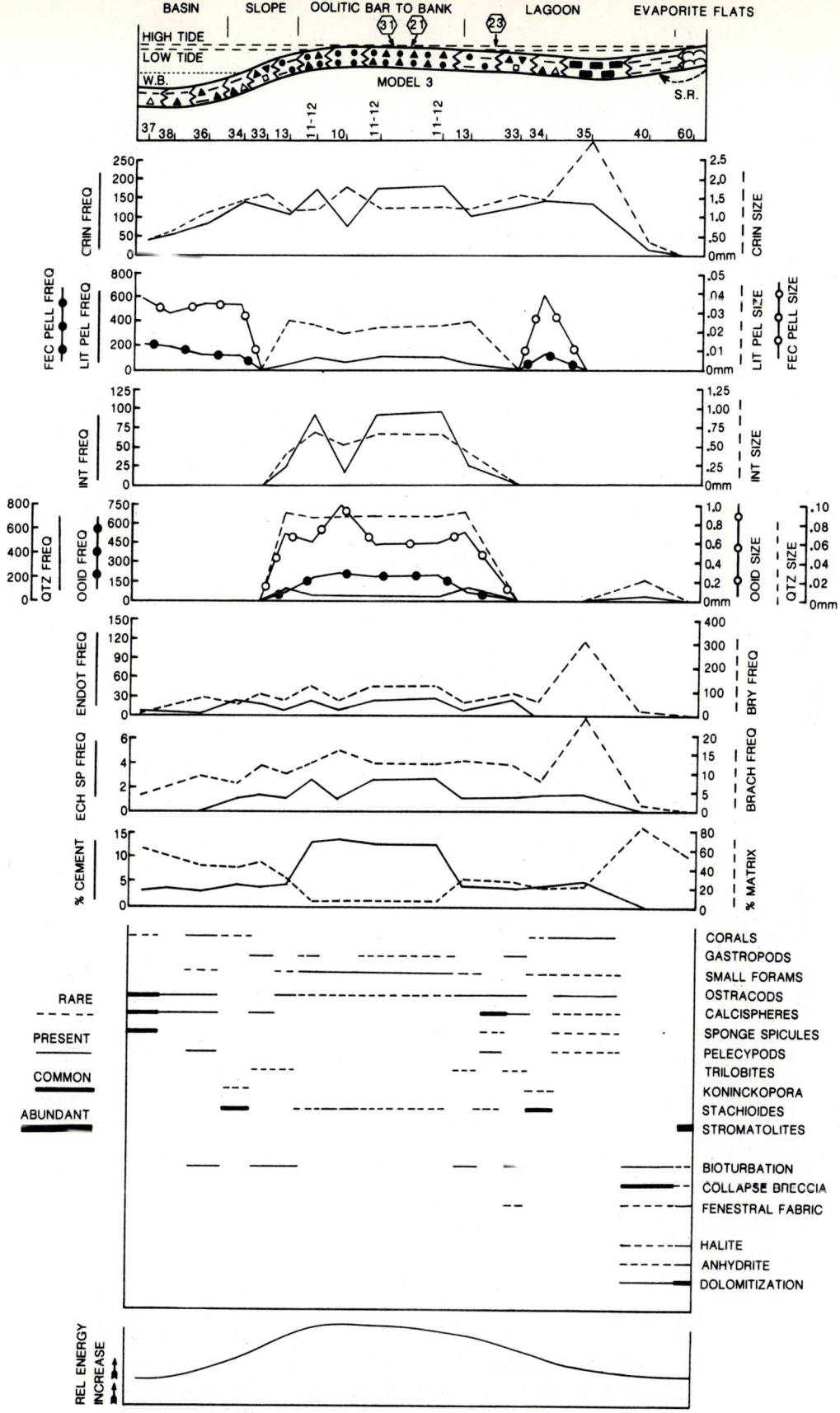

FIGURE 3.11 St. Louis Limestone (Middle Mississippian), Illinois Basin. Horizontal depositional model 3. From Diaby and Carozzi (1984).

52 Part I / Principles

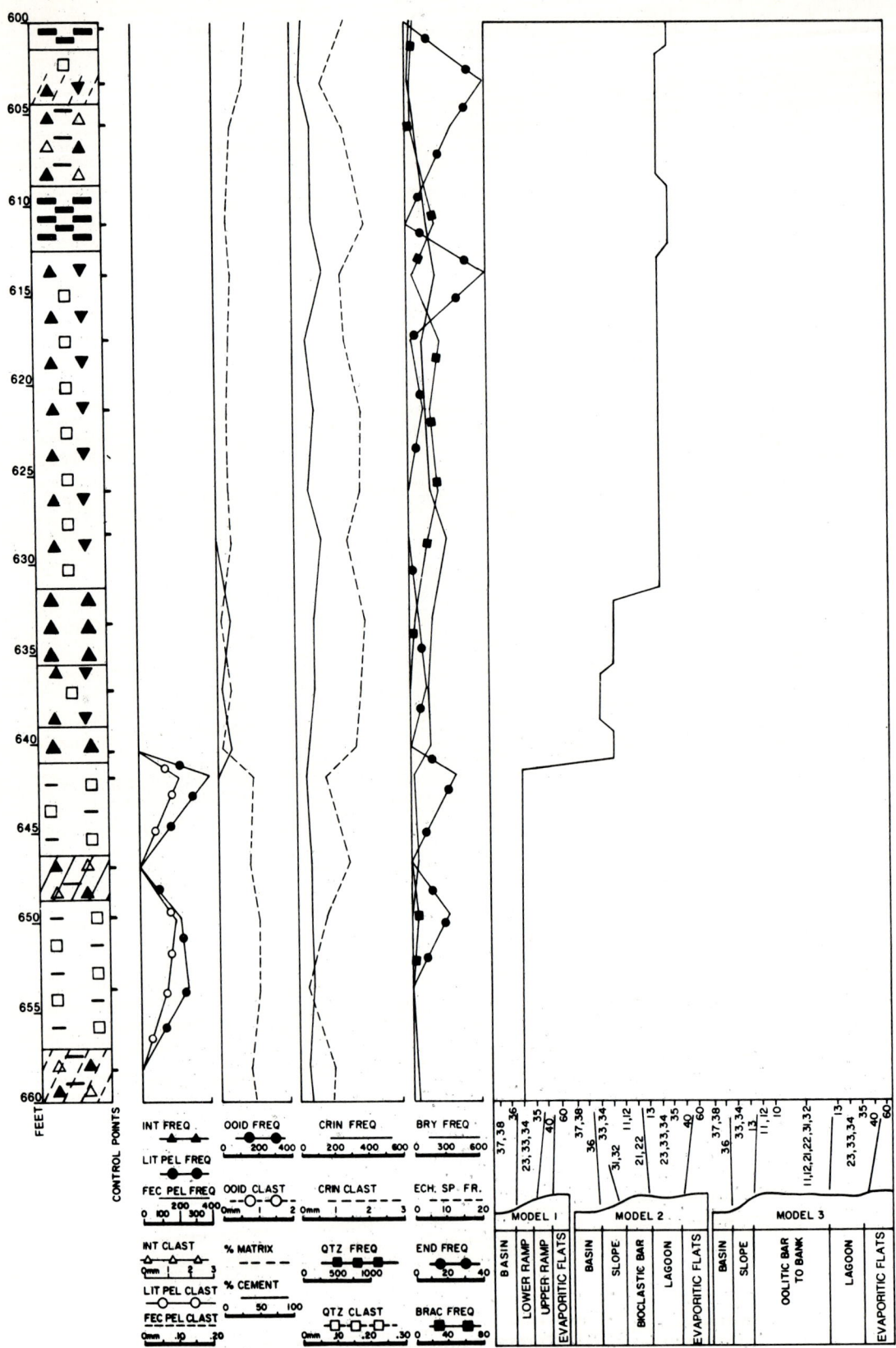

FIGURE 3.12 St. Louis Limestone (Middle Mississippian), Illinois Basin. Shallowing-upward sequence in Caldwell County core, Kentucky (600 to 660 ft). From Diaby and Carozzi (1984).

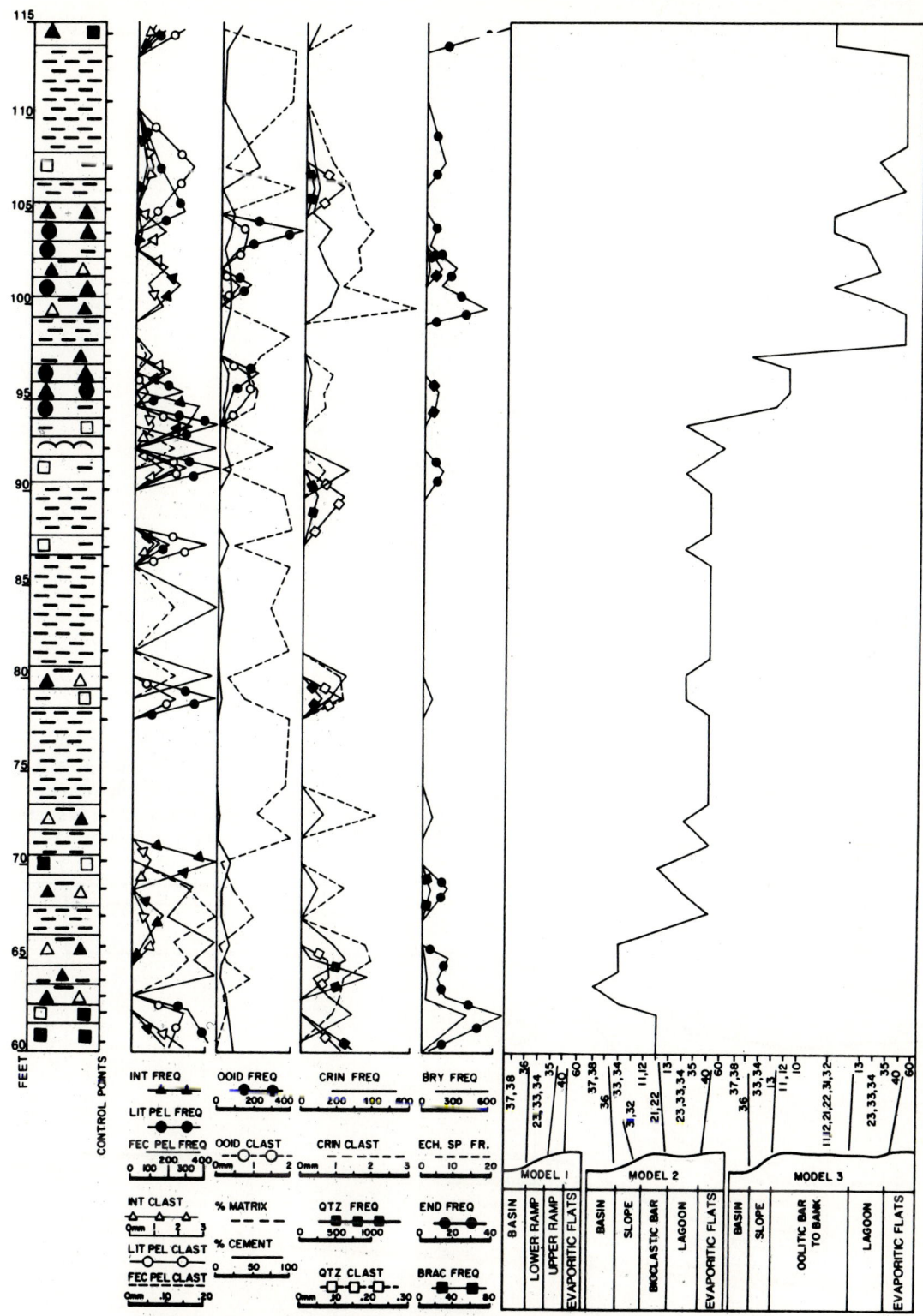

FIGURE 3.13 St. Louis Limestone (Middle Mississippian), Illinois Basin. Shallowing-upward sequence in Olin quarry, Alton, Illinois (60 to 115 ft). From Diaby and Carozzi (1984).

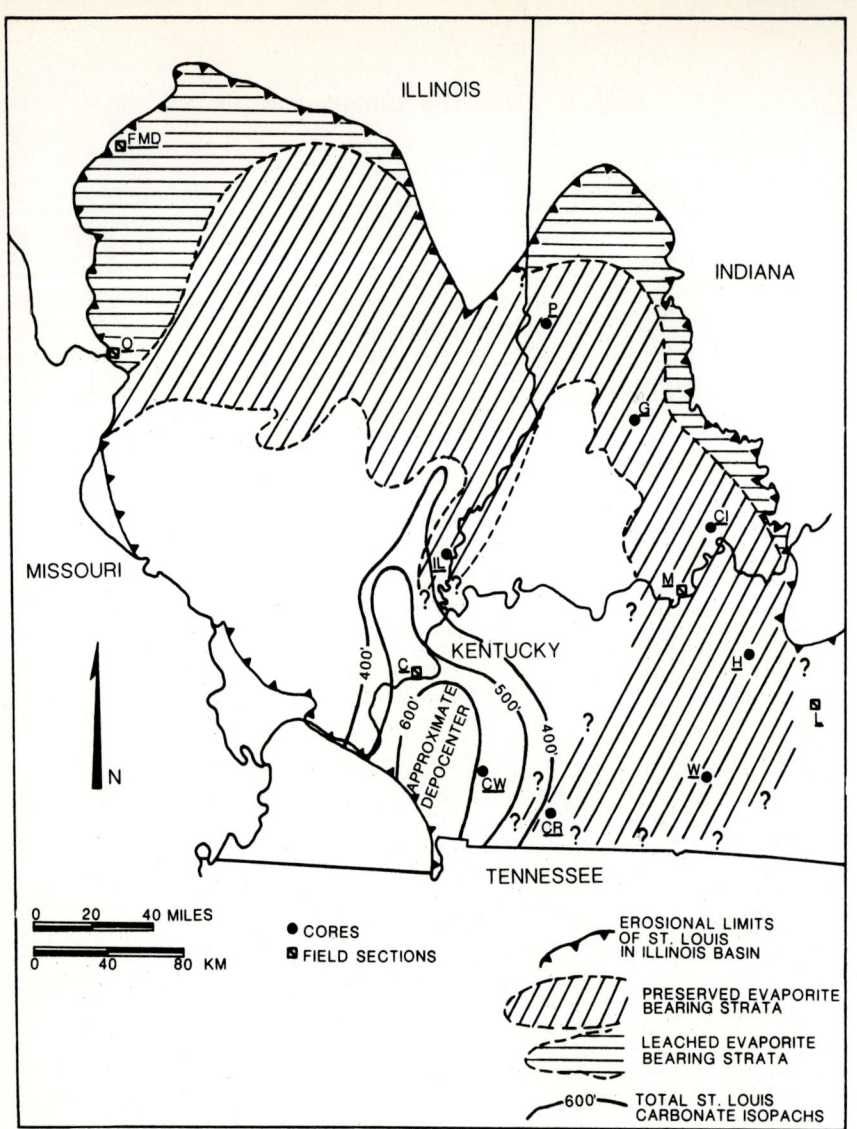

FIGURE 3.14 St. Louis Limestone (Middle Mississippian), Illinois Basin. Map of distribution of lithofacies and location of sections. From Diaby and Carozzi (1984).

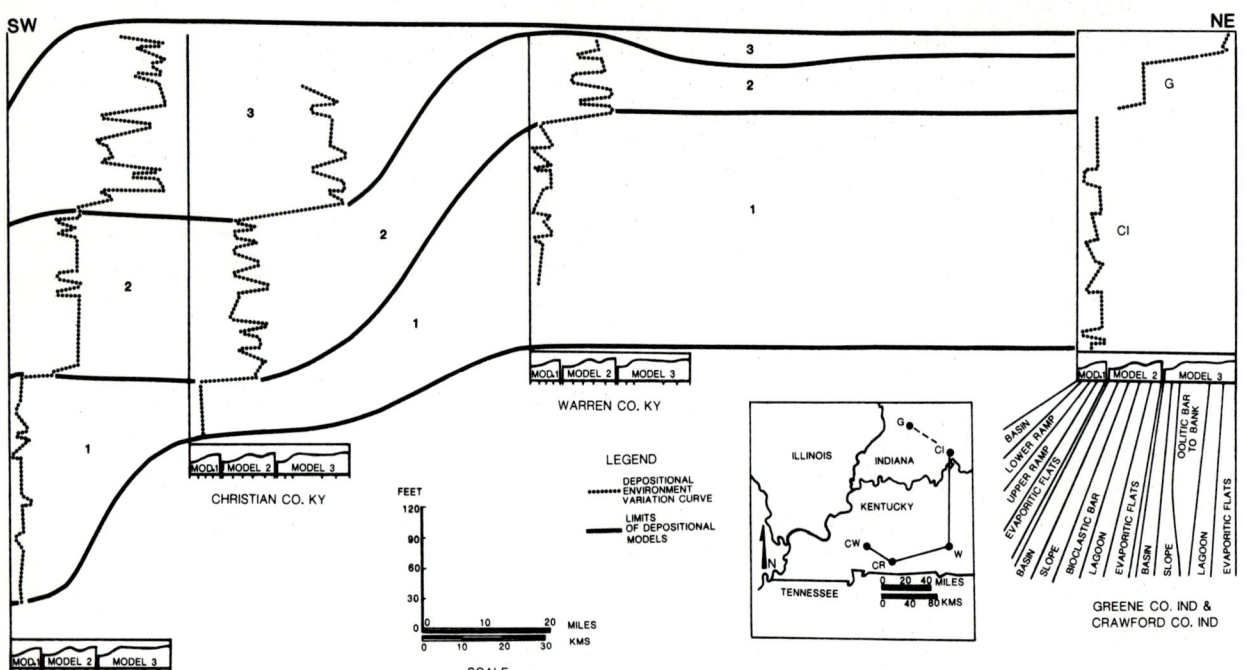

FIGURE 3.15 St. Louis Limestone (Middle Mississippian), Illinois Basin. Cross section with basinwide evolution in time from ramp to bar to platform. From Diaby and Carozzi (1984).

DIAGENETIC FEATURES \ DIAGENETIC ENVIRONMENTS	MARINE PHREATIC	MARINE VADOSE (SABKHA BRINES)	UNDERSATURATED FRESHWATER PHREATIC	SATURATED FRESHWATER PHREATIC	MIXING MARINE FRESHWATER PHREATIC	BURIAL
MICRITIZATION	▬▬					
INTRACLAST CEMENTATION OXIDATION & REWORKING	▬▬					
EARLY COMPACTION (I) (PRESSURE-SOLUTION)	▬▬					
THIN ISOPACHOUS RIM CEMENT	▬▬					
EARLY COMPACTION (II)	▬▬					
GEOPETAL ORGANIZATION OF INTERNAL SEDIMENT	▬▬					
PERVASIVE DOLOMITIZATION		▬▬				
ANHYDRITIZATION (I) EUHEDRAL CRYSTALS & NODULES IN MICRITE		▬▬				
LEACHING OF ARAGONITIC SHELLS & DISSOLUTION OF EVAPORITES & DOLOMITE			▬▬			
EARLY FRACTURATION			▬▬			
FRACTURE FILLING CALCITE				▬▬		
CALCITIZATION OF EUHEDRAL ANHYDRITE				▬▬		
SYNTAXIAL RIM CEMENT				▬▬		
SPARITE MOSAIC CEMENT				▬▬		
NEOMORPHISM (STABILIZATION OF RIM CEMENT)				▬▬		
PERVASIVE SILICIFICATION					▬▬	
LATE COMPACTION (STYLOLITES)						▬▬
BURIAL DISSOLUTION OF CALCITE, DOLOMITE? ANHYDRITE?						▬▬
ANHYDRITE (II) REPLACEMENT OF BIOCLASTS, OOIDS & SPARITE CEMENT						▬▬
LATE BURIAL FRACTURATION						▬▬
LATE FRACTURE FILLING CALCITE CEMENT						▬▬
BLADED ANHYDRITE (III) REPLACING LATE FRACTURE FILLING CALCITE						▬▬
LATE DOLOMITIZATION (SADDLE DOLOMITE)						▬▬
	▸	▸	▸ TIME	▸	▸	▸

FIGURE 3.16 St. Louis Limestone (Middle Mississippian), Illinois Basin. Diagenetic features and environments. From Diaby and Carozzi (1984).

phreatic environment is based on its association in other instances with the dorag-type dolomitization.

Burial Environment

Late compaction, expressed by stylolites with thick insoluble residue, occurs both in pelletoidal biocalcarenites with calcisiltite matrix and in biocalcarenites, intersecting all constituents and cement. Partial dissolution of cavity-filling calcite cement occurs at this stage, creating secondary porosity in oolitic biocalcarenites (Fig. 3.3D). Possibly, dolomite and anhydrite, which may have survived the leaching of the undersaturated freshwater phreatic environment, are eventually dissolved in this environment.

Pervasive anhydrite (anhydrite II) replacement of bioclasts, ooids, and sparite cement in the oolitic biocalcarenites by connate waters also occurs at this stage (Fig. 3.3E). Anhydrite is coarse, euhedral to subhedral, sometimes lath shaped. When replacing cavity-filling sparite cement, anhydrite is often poikilotopic. The timing of this second phase of anhydritization is shown by the fact that it replaces silicified bryozoans (Fig. 3.3F). It is possible that the source of anhydrite-rich burial fluids may be a recycling of anhydrite of the lower part of the St. Louis (Fig. 3.14).

Oolitic biocalcarenites display late burial fractures which cut across the partially anhydritized sparite-cemented oolitic biocalcarenites of the previous diagenetic event and are filled by sparite calcite cement (Fig. 3.3G). This cement is in turn partially replaced by bladed anhydrite III (Fig. 3.3G).

In large fractures of some pelletoidal calcisiltites, late dolomitic cement occurs with undulose extinction (Fig. 3.3H). None of the samples on hand permitted exact determination of the relative timing of the saddle dolomitization with respect to other burial events. However, the fact that saddle dolomite is known to form in burial environments at temperatures on the order of 100 to 150°C would allow one to consider it as one of the latest, if not the last burial diagenesis event in the samples studied.

4

Carbonate Ramps in Open Marine Environment

Ramps with their typical gentle seaward slope represent the first morphologic expression of carbonate sedimentation. As long as they keep this simple geometry, they tend to display extensive and rather uniform lesser energy subtidal microfacies. These fine-grained carbonates can often be distinguished only by means of minute textural features among which bioturbation patterns and trace fossils can be very specific.

4.1 CHARACTERISTIC FEATURES: EFFECTS OF BIOTURBATION PROCESSES

The Platteville Group (Middle Ordovician) of northern Illinois was investigated by means of six stratigraphic field sections (Kuhnhenn and Carozzi, 1977). The average thickness of the Platteville Group is 35 m and 951 samples were investigated petrographically with an average vertical interval of 7.4 cm. These carbonates consist of a repetition of eight distinct microfacies. They have been grouped into a noncomminuted series (N-1, N-2, and N-3) where biogenic debris have not undergone extensive breakage by bioturbation processes (Fig. 4.1A), and a comminuted series (C-1, C-2, and C-3) characterized by extensive breakage (Fig. 4.1B). Microfacies 4 and 5 are considered noncomminuted but their allochthonous nature requires to put them in a distinct category.

A detailed petrographic study was undertaken of burrow fillings and tabulated with the other parameters in the detailed columns. Six varieties were recognized: (1) medium crystalline anhedral dolomite often zoned (Fig. 4.1C); (2) coarsely crystalline calcite with "cauliflower-like" crystal aggregates (Fig. 4.1D); (3) finely crystalline pseudomicrosparite with scattered minute zoned dolomite rhombs (Fig. 4.1E); (4) medium crystalline sparite cement often with geopetal texture over internal sediment (Fig. 4.1F); (5) biogenic debris or pellets in geopetal position overlain by sparite cement (Fig. 4.2A); and (6) comminuted debris parallel to burrow walls and forming spiral arrangements (Fig. 4.2B). These types of burrows are nonmicrofacies selective, except type 6, which is closely related to the comminuted series.

Certain Platteville sections of the investigated area were variably dolomitized by a large-scale and late dorag-type process which occurred when these carbonates, together with the overlying Galena Group, were exposed around the Wisconsin Arch during or at the end of Maquoketa time (late Ordovician), an episode of worldwide lowering of sea level (see Fig. 4.27). In the Platteville, complete dolomitization appears as even-textured, fine to medium crystalline, anhedral dolomite, often with burrows containing concentrations of St. Peter-type quartz grains (Fig. 4.2C); medium to coarse crystalline subhedral zoned dolomite rhombs (Fig. 4.2D); and coarse crystalline subhedral to euhedral zoned rhombs often

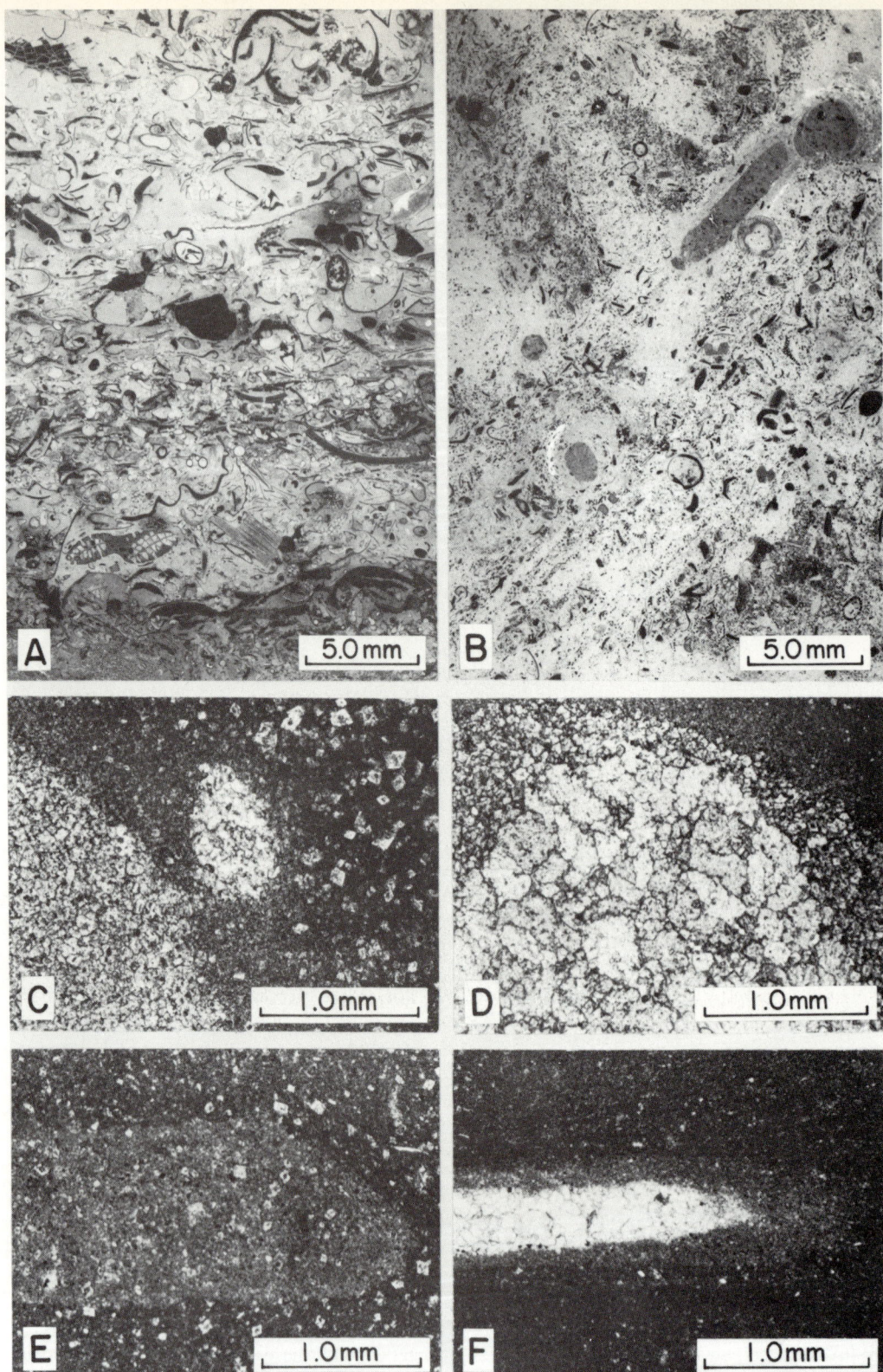

FIGURE 4.1 Platteville Group (Middle Ordovician), north-central Illinois. Bioturbation types. A. Negative print from thin section showing noncomminuted type of bioturbation. Burrowing has not caused extensive breakage of bioclasts. B. Negative print from thin section showing comminuted type of bioturbation where browsing organisms have produced extensive breakage of bioclasts. Much of the debris is comminuted *Vermiporella*. Petrographic varieties of burrow fillings: C. Burrows filled with medium crystalline anhedral dolomite that is sometimes zoned (variety 1). D. Large burrow filled with

with moldic to vuggy porosity (Fig. 4.2E). Incomplete dolomitization appears as scattered zoned rhombs in calcisiltites (Fig. 4.2F), and as fine to medium crystalline anhedral dolomite partially replacing laminated calcisiltites (Fig. 4.2G). However, instances of synsedimentary to early postdepositional dolomitization occur in limestone sections of the Platteville Group. They result from a mixed seawater-freshwater system related to temporary and local exposure of the carbonates. This origin is demonstrated by intraclasts in litho-biocalcarenites which contain euhedral dolomite rhombs truncated by intraclast boundaries (Fig. 4.2H).

Description of Microfacies

Noncomminuted Series

Its microfacies are described in order of decreasing relative depth.

Microfacies N-1 (Fig. 4.3A). Fine, evenly textured calcisiltite with up to 10% scattered bioclasts of pelmatozoans, brachiopods, and ostracods.

Microfacies N-2 (Fig. 4.3B). Mud-supported biocalcarenite with fine calcisiltite matrix. The bioclasts which range from 10 to 30% are similar to those of the preceding microfacies but are associated with more abundant fragments of trilobites, gastropods, and pelecypods.

Microfacies N-3 (Fig. 4.3C). Grain-supported biocalcarenite with fine calcisiltite matrix. The bioclasts are identical to those of the preceding microfacies and form a framework after reaching more than 30% frequency.

Comminuted Series

Its microfacies are described in order of decreasing relative depth.

Microfacies C-1 (Fig. 4.3D). Fine to medium, bioclastic to pelletoidal, calcisiltite with scattered sand-size bioclasts. Identifiable bioclasts range up to 10% and are dominated by brachiopods, bryozoans, gastropods, and pelecypods.

Microfacies C-2 (Fig. 4.3E). Mud-supported biocalcarenite with fine to medium bioclastic to pelletoidal calcisiltite matrix. Dominant bioclasts (10 to 30%) are similar to those of the preceding microfacies with the addition of frequent *Vermiporella*.

Microfacies C-3 (Fig. 4.3F). Grain-supported biocalcarenite with fine to medium bioclastic to pelletoidal calcisiltite matrix. The major bioclasts, ranging up to more than 30%, are *Vermiporella*, pelmatozoans, brachiopods, ostracods, gastropods, and trilobites.

Allochthonous Series

These microfacies represent temporary short-lived high-energy deposits which interrupted the previous two series of microfacies at irregular time intervals. They are interpreted as storm deposits, as shown by their sharp and irregular basal contacts, internal imbrication, and incipient graded bedding of the bioclasts, as well as their extensive and thin nature (average 7 cm thickness).

Microfacies 4 (Fig. 4.3G). Grain-supported litho-biocalcarenite with intraclasts up to 1 cm diameter consisting of partially dolomitized or normal calcisiltites with scattered bioclasts. They are associated with numerous bioclasts of pelmatozoans, brachiopods, and ostracods. Quartz grains of St. Peter type are abundant. Interstitial material is cavity-filling sparite in the lower part which grades, with the fining-upward texture, into a calcisiltite matrix often neomorphosed into pseudomicrosparite.

Microfacies 5 (Fig. 4.3H). Medium to coarse, grain-supported pelletoidal biocalcarenite with association of sparite cement and calcisiltite matrix neomorphosed into pseudomicrosparite. Predominant bioclasts are pelmatozoans, brachiopods, and ostracods.

The range of the average microscopic parameter values for the two series of microfacies is represented graphically by standardized Z values which show a mean of zero and a standard deviation of 1. This facilitates

FIGURE 4.1 (*continued*) medium to very coarsely crystalline calcite displaying distinctive "cauliflower-like" crystal aggregate, with rim of variety 1 burrow filling (variety 2). E. Longitudinal section of burrow showing somewhat clearer, finely crystalline pseudosparite. Zoned euhedral dolomite rhombs are scattered throughout (variety 3). F. Longitudinal section of burrow filled with medium crystalline sparite. Burrow also contains some variety 3 pseudosparite (variety 4). All photomicrographs: plane-polarized light. From Kuhnhenn and Carozzi (1977).

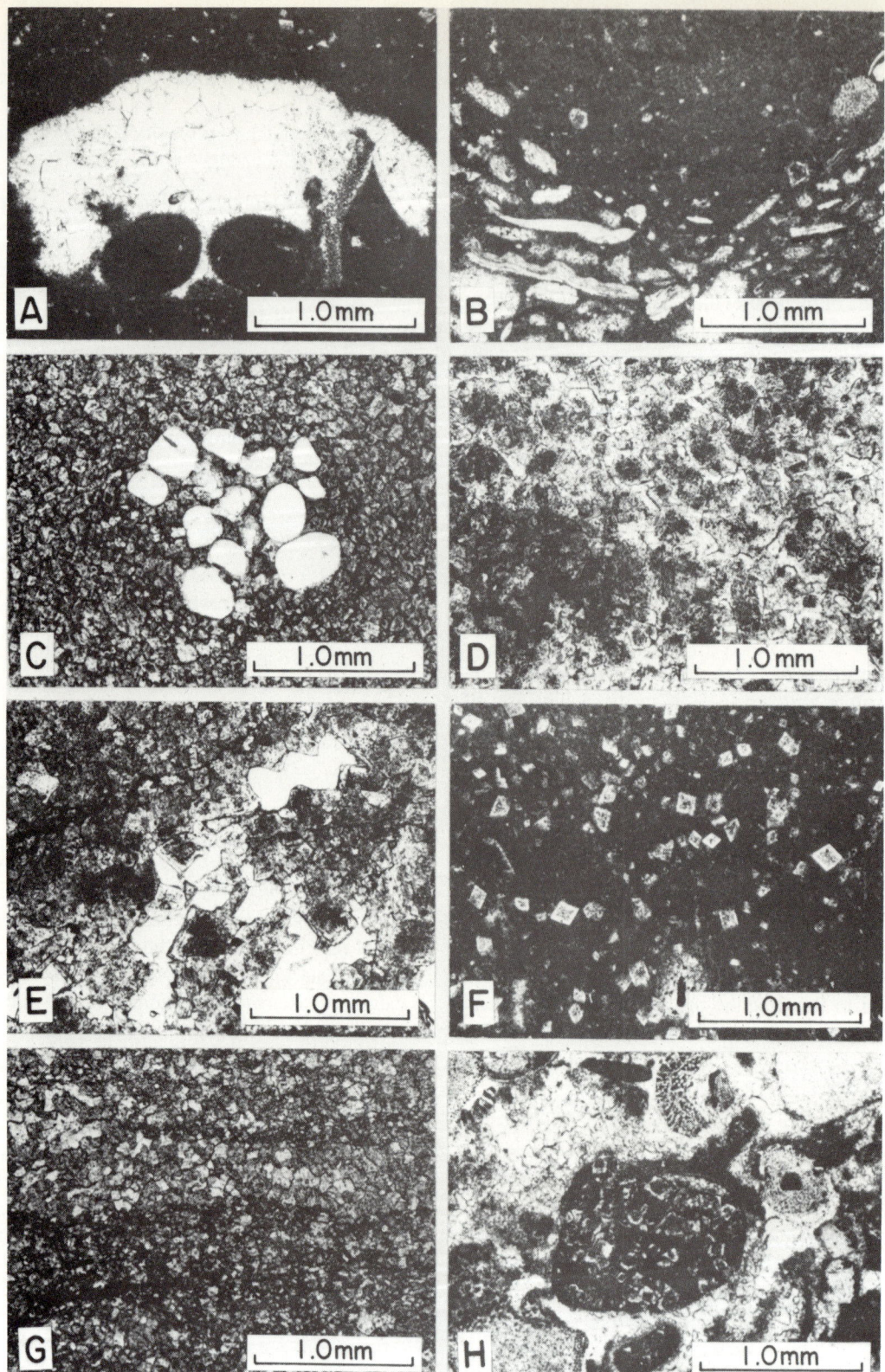

FIGURE 4.2 Platteville Group (Middle Ordovician), north-central Illinois. Petrographic varieties of burrows. A. Burrow containing biogenic debris (pelmatozoan fragments) and pellets with associated void-filling sparite. The burrow is geopetal and contains recrystallized calcisiltite (variety 5). B. Burrow wall (boundary) is marked by parallel alignment of bioclasts with interior of burrow relatively free of debris (variety 6). Dolomitization: C. Even-textured, fine to medium crystalline, anhedral dolomite. Well-rounded St. Peter sandstone-type quartz grains are contained in a burrow delineated subtly by a slight

graphic representation and allows easy comparison between average values even though uncoded averages may vary greatly (Fig. 4.4). A typical section (Figs. 4.5, 4.6) shows detailed stratigraphic variation of the microscopic parameters.

The Ideal Depositional Models

The deposition of the Platteville Group carbonates can be interpreted as the result of the stratigraphic superposition of three models. These models are similar, generally, and consist of variations of gently sloping infratidal ramps in open marine conditions with an inferred shoreline (Fig. 4.7). The first model displays noncomminuted microfacies and a break in slope on the open marine side due to the effect of a submerged preexisting bar. The second model shows comminuted microfacies and a general shallower position in the infratidal environment. It displays a gentle break in slope as the final influence of the submerged bar. The third model shows a return to noncomminuted microfacies, a simple ramp morphology, and deeper infratidal conditions.

The two noncomminuted models display rather stable depositional conditions with minor fluctuations, in contrast to the comminuted model, which is characterized by well-developed symmetric oscillations (Fig. 4.8). The relative depth evolution of the three models represents an overall deepening which was temporarily interrupted in the middle by shallower conditions.

4.2 CHARACTERISTIC FEATURES: EFFECTS OF STORM PROCESSES

The predominant subtidal nature of many fine-grained ramp carbonates did not prevent them from undergoing extensive reworking by storms. The latter interfered with shallowing-upward processes, but at the same time, provided further means for subdivision and correlation of these sequences.

The Galena Group (Middle Ordovician) was investigated petrographically along the Mississippi River in Iowa, Wisconsin, Illinois, and Missouri by means of six stratigraphic sections and four cores (Bakush and Carozzi, 1986). A total of 1773 thin sections were prepared from samples collected at an average vertical interval of 15 cm for limestones and 30 to 60 cm for dolomites. Although the aggregated thickness of investigated rocks was 415 m, the microfacies study was limited to 187 m of limestones.

Description of Microfacies

The petrographic study led to the identification of seven microfacies which are described in order of increasing relative energy and general shallowing.

Microfacies 1 (Fig. 4.9A). Moderately bioturbated and slightly pelletoidal calcisiltite with up to 10% scattered sand-size bioclasts of crinoids, brachiopods, ostracods, trilobites, gastropods, pelecypods, bryozoans, dasyclads, and *Receptaculites*. Calcified monaxonic and tetraxonic sponge spicules as well as fecal pellets are common. Bioturbation consists mostly of horizontal burrows often represented by concentrations of "flowery" pseudosparite or aggregates of dolomite rhombs.

Microfacies 2 (Fig. 4.9B). Moderately bioturbated mud-supported to grain-supported brachiopod-pelecypod bioaccumulated limestone with calcisiltite matrix. Most of the shells are aligned parallel to bedding with common umbrella effects of sparite cement beneath them. However, local bioturbation has disrupted the original texture. Frequent whole gastropods are associated with bioclasts of bryozoans, crinoids, trilobites, and ostracods.

Microfacies 3 (Fig. 4.9C). Moderately to heavily bioturbated mud-supported biocalcarenite with pelletoidal calcisiltite matrix. Bioclasts of crinoids and brachiopods predominate over gastropods, ostracods,

FIGURE 4.2 (*continued*) decrease in crystal size and increase in iron oxide and clay content. D. Medium to coarse crystalline, subhedral dolomite. Cloudy centers of some crystals were probably pelmatozoan fragments. The clearer areas may have been syntaxial overgrowths of pseudosparite. E. Coarse crystalline, subhedral to euhedral dolomite forming in secondary voids (moldic porosity). Crystals display cloudy centers changing to clear (limpid) dolomite toward margins. F. Zoned euhedral dolomite rhombs found commonly in "clean" calcisiltites. G. Well-laminated, fine to medium crystalline, anhedral dolomite. H. Lithoclast, in a litho-biocalcarenite, showing zoned euhedral dolomite rhombs truncated by the fragment's boundary indicating early postdepositional dolomitization preceding formation of lithoclast. All photomicrographs: plane-polarized light. From Kuhnhenn and Carozzi (1977).

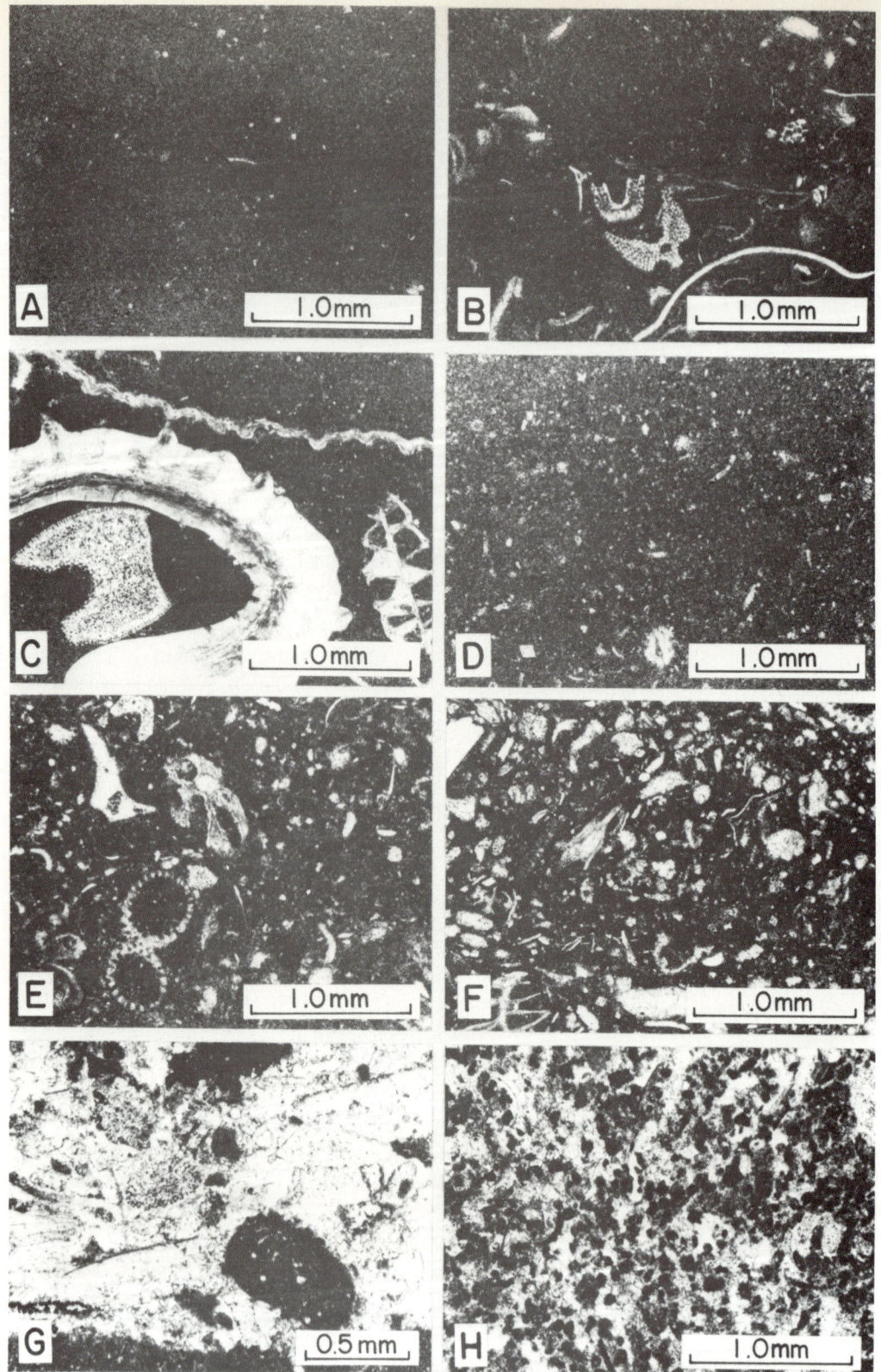

FIGURE 4.3 Platteville Group (Middle Ordovician), north-central Illinois. Noncomminuted microfacies: A. Microfacies N-1. B. Microfacies N-2. C. Microfacies N-3. Comminuted microfacies: D. Microfacies C-1. E. Microfacies C-2. F. Microfacies C-3. Allochthonous microfacies: G. Microfacies 4. H. Microfacies 5. See text for detailed descriptions. All photomicrographs: plane-polarized light. From Kuhnhenn and Carozzi (1977).

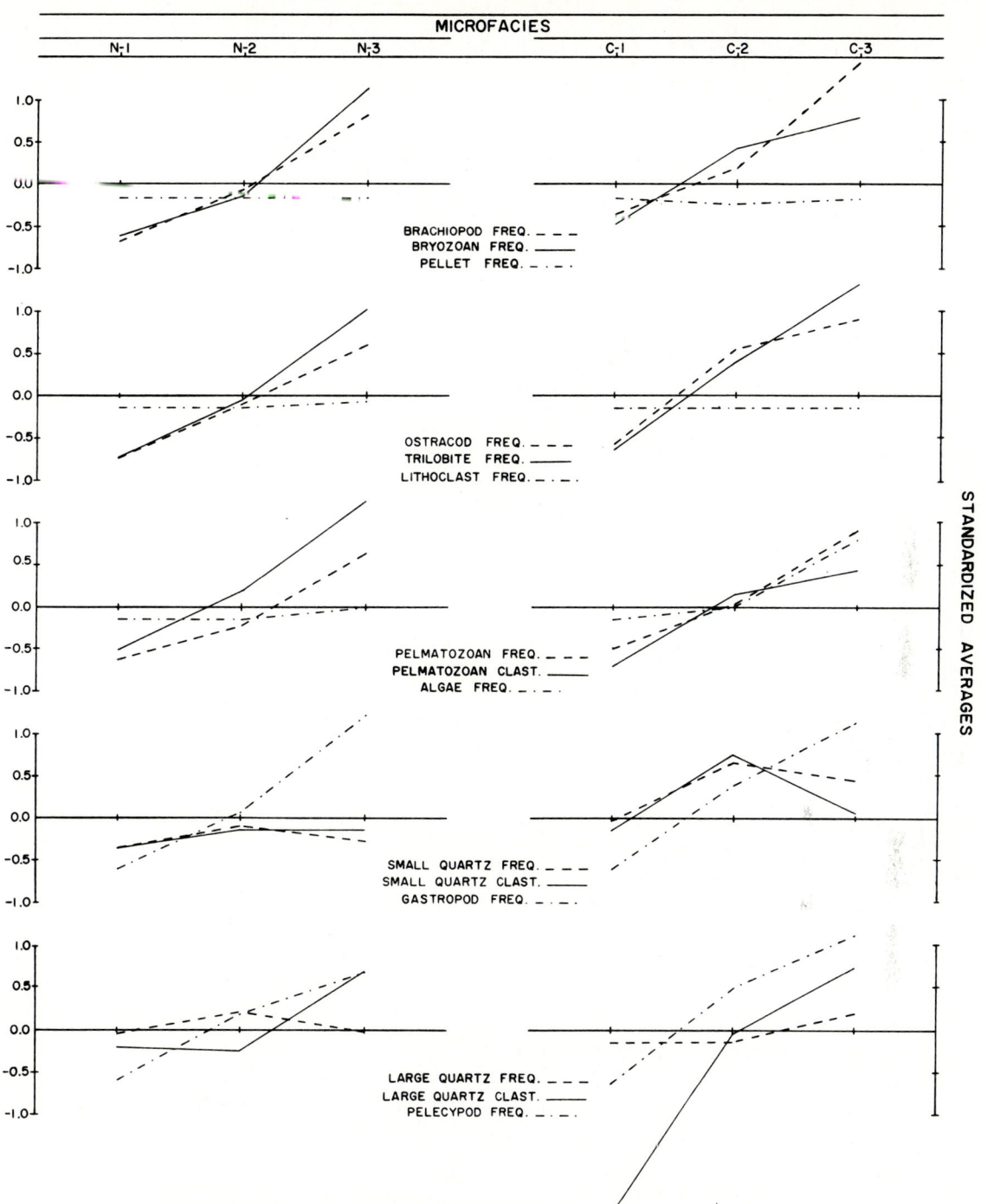

FIGURE 4.4 Platteville Group (Middle Ordovician), north-central Illinois. Illustration of standardized component variation of microfacies. From Kuhnhenn and Carozzi (1977).

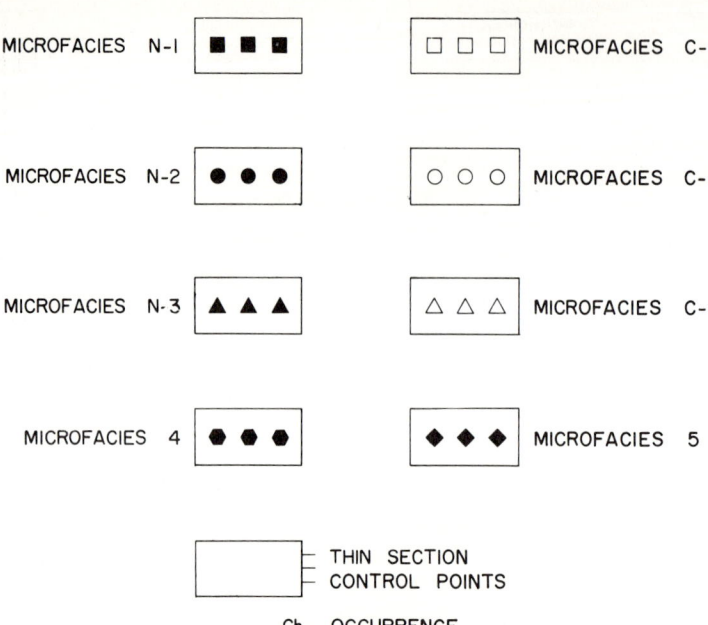

FIGURE 4.5 Platteville Group (Middle Ordovician), north-central Illinois. Symbols for microfacies. From Kuhnhenn and Carozzi (1977).

trilobites, *Receptaculites*, dasyclads, bryozoans, and monaxonic sponge spicules. Bioturbation is expressed by abundant scattered patches of pseudosparite within the matrix.

Microfacies 4 (Fig. 4.9D). Heavily bioturbated grain-supported biocalcarenite with pelletoidal calcisiltite matrix and rare sparite cement. Bioclasts of crinoids and brachiopods predominate over gastropods, *Vermiporella*, trilobites, ostracods, bryozoans, pelecypods, and *Receptaculites*. Calcisiltite intraclasts and fine grains of detrital quartz are conspicuous. Bioturbation consists of numerous burrows often filled by pseudosparite or aggregates of fine euhedral dolomite crystals.

Microfacies 5 (Figs. 4.9E,F, 4.10, 4.11). Bioturbated and disorganized, grain-supported biocalcirudite with sparite cement and rare pelletoidal calcisiltite matrix. Bioclasts of brachiopods, pelecypods, and crinoids dominate over common bryozoans, gastropods, ostracods, trilobites, dasyclads, and *Receptaculites*. The larger bioclasts are oriented in various directions, often imbricated and even standing perpendicular to bedding, expressing the effects of extensive reworking. Single valves of brachiopods and pelecypods occur in a horizontal or weakly inclined position with associated umbrella effects. In many instances, the finer bioclasts display an incipient graded bedding with interstitial material changing upward from sparite cement into pelletoidal calcisiltite matrix. The lower contacts of the beds of microfacies 5 are often erosional with scouring and clasts from the underlying lithology, whereas the upper contacts are always gradational. The combination of these features indicates a storm origin. Bioturbation may introduce locally further disruption of the depositional textures.

Microfacies 6 (Figs. 4.9G, 4.12, 4.13). Moderately bioturbated coarse grain-supported to pressure-welded crinoid-bryozoan calcarenite with sparite cement and rare calcisiltite matrix. Associated with the predominant bioclasts are brachiopods, trilobites, ostracods, gastropods, pelecypods, and *Receptaculites*. This microfacies is finer grained than the preceding one and less disorganized, although entire valves and bioclasts display variable orientations, imbrication, and common incipient graded bedding. All other sedimentary textures are similar to those of microfacies 5 and indicate a storm origin of lesser intensity also modified by subsequent bioturbation.

Microfacies 7 (Fig. 4.9H). Moderately bioturbated grain-supported to pressure-welded crinoid-brachiopod-bryozoan calcarenite with hematitic sparite cement. Crinoid columnals are well sorted and abraded strongly, whereas bryozoans are highly fragmented. They are associated with bioclasts of trilobites, ostracods, gastropods, pelecypods, dasyclads, *Receptaculites*, corals, and an appreciable amount of calcisiltite intraclasts. Cementation combines syntaxial overgrowths on crinoid columnals and interparticle equant sparite mosaic.

The Ideal Fair-Weather–Storm Shallowing-Upward Sequence

The analysis of the stratigraphic columns and the correlation coefficients between the various microfacies indicate that microfacies 1, 2, 3, 4, and 7 are fair-weather deposits

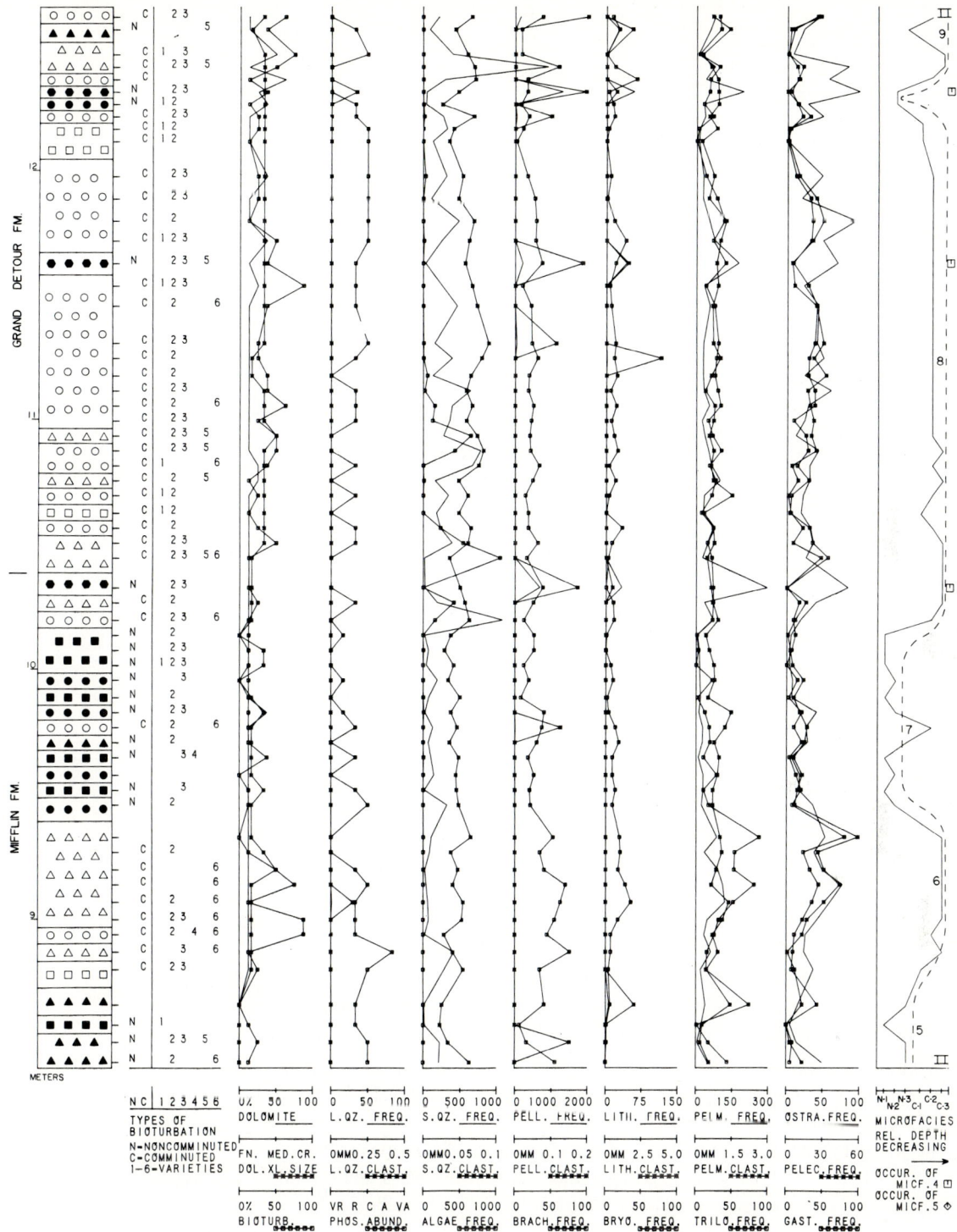

FIGURE 4.6 Platteville Group (Middle Ordovician), north-central Illinois. Typical example of vertical variation of microscopic parameters. From Kuhnhenn and Carozzi (1977).

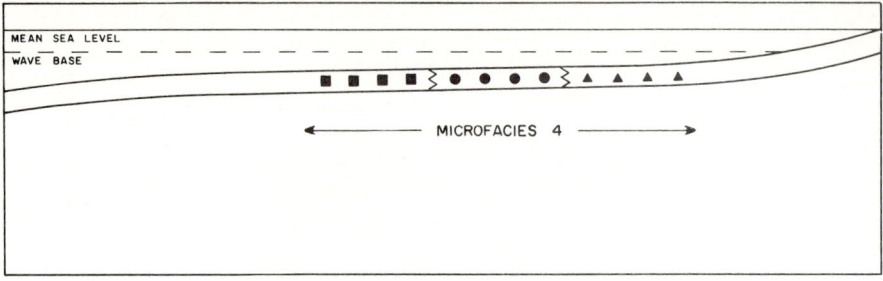

C: NONCOMMINUTED MODEL (UPPER)

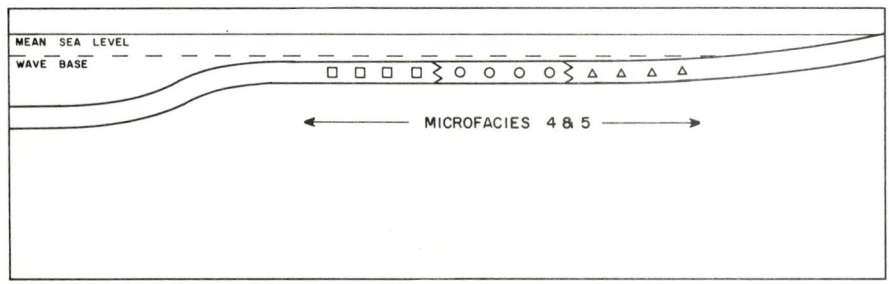

B: COMMINUTED MODEL (MIDDLE)

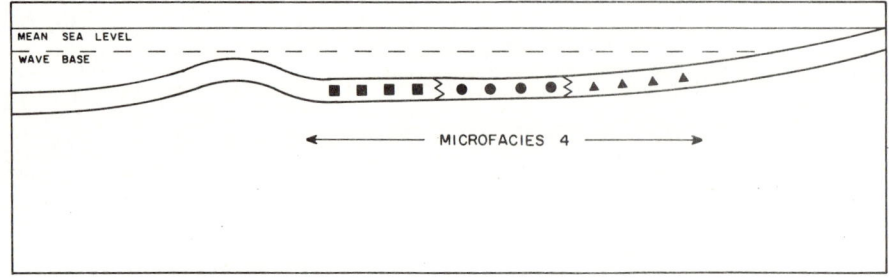

A: NONCOMMINUTED MODEL (LOWER)

FIGURE 4.7 Platteville Group (Middle Ordovician), north-central Illinois. Ideal depositional models. From Kuhnhenn and Carozzi (1977).

which build an ideal shallowing-upward sequence (Fig. 4.14) with the statistically most common intercalation of microfacies 5 (greater energy tempestite) between 3 and 4 and the intercalation of microfacies 6 (medium-energy tempestite) between 4 and 7. The variation of the microscopic parameters are shown as continuous curves for the ideal shallowing-upward sequence of fair-weather microfacies, whereas the corresponding values for the intercalated storm microfacies are indicated by individual peaks (Fig. 4.14). This mode of representation allows one to compare the effects of storm processes on the fair-weather sedimentation. The detailed variation of microscopic parameters are shown in two instances of superposed shallowing-upward sequences (Figs. 4.15, 4.16, 4.17).

The Ideal Fair-Weather–Storm Depositional Model

The proposed depositional model (Figs. 4.18, 4.19) represents a shallow infratidal and open marine carbonate ramp submitted to episodic storm action. Morphologically, it can be divided from north to south into three environments: a slope, an extensive outer platform with incipient bioaccumulation at the distal end, and a shoal. The expected more shoreward environments were not detected in the studied area apparently because of the great extent of the sea in which the Galena was deposited over the American Midcontinent, which led to the shorelines being far to the west-southwest and north.

The behavior of most of the benthic constituents

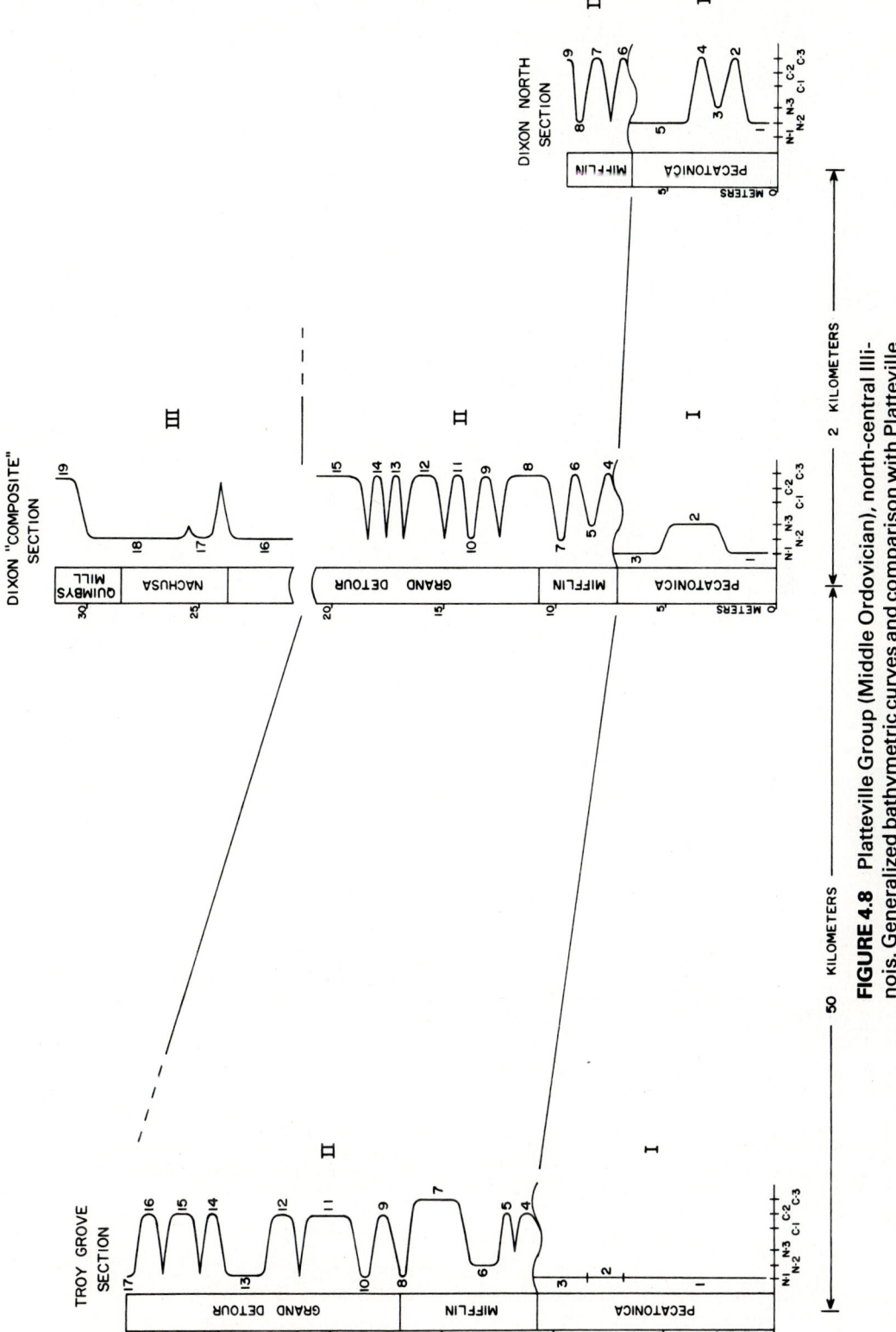

FIGURE 4.8 Platteville Group (Middle Ordovician), north-central Illinois. Generalized bathymetric curves and comparison with Platteville stratigraphy. From Kuhnhenn and Carozzi (1977).

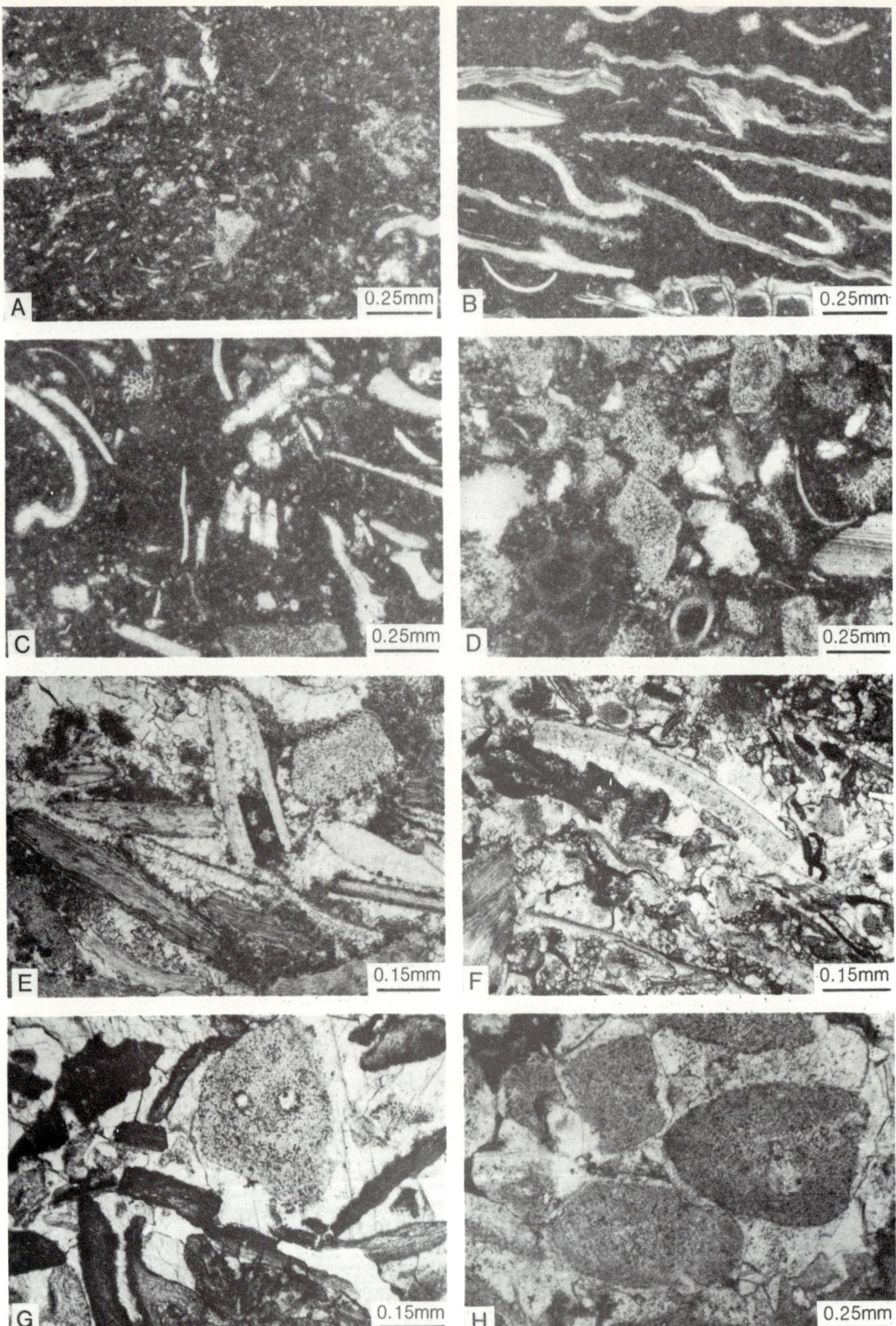

FIGURE 4.9 Galena Group (Middle Ordovician), Upper Mississippi Valley. Typical microfacies. A. Microfacies 1. B. Microfacies 2. C. Microfacies 3. D. Microfacies 4. E and F. Microfacies 5. G. Microfacies 6. H. Microfacies 7. See text for detailed descriptions. All photomicrographs: plane-polarized light. From Bakush and Carozzi (1986).

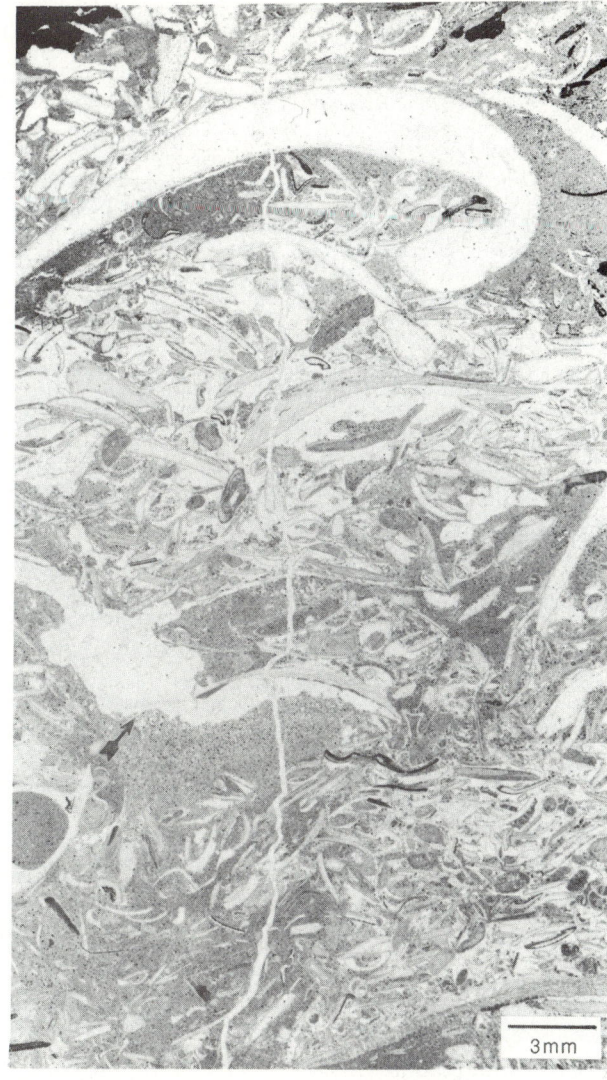

FIGURE 4.10 Galena Group (Middle Ordovician), Upper Mississippi Valley. Typical tempestite (microfacies 5). Moderately bioturbated and disorganized grain-supported biocalcirudite with sparite cement and calcisiltite matrix. Brachiopods and pelecypods predominate over gastropods, bryozoans, crinoids, and trilobites. The bioclasts display an irregular texture where the fine to medium components are concentrated in lower half of photomicrograph with calcite cement (at lower right corner) and calcisiltite matrix (at lower left corner). Larger shells of brachiopods and pelecypods are in equilibrium position at top of photograph with calcite cement and rare calcisiltite matrix. Dissolution feature in large cavity with typical geopetal internal sediment overlain by sparite cement (arrow). Plane-polarized light. From Bakush and Carozzi (1986).

of the fair-weather microfacies such as crinoids, bryozoans, gastropods, brachiopods, pelecypods, and *Receptaculites* shows *in situ* assemblages. This situation indicates that the carbonate ramp possessed a very gentle slope, and paleoecologic data, mainly based on the living conditions of *Receptaculites*, lead to an estimation of maximum water depth of the proposed model at about 60 ft (18.3 m).

The frequency and size (not clasticity) of the debris released in place after the death of the crinoids are at a peak in microfacies 7 and decrease gradually toward deeper and lower-energy microfacies. The behavior of the two curves (Fig. 4.18) indicates that abundant and large species of crinoids were living in the shallow and high-energy shoal environment and that with increasing depth, crinoids became smaller, more delicate, and less abundant. Other evidence of the autochthonous character of the fauna are as follows: two distinct assemblages of bryozoans in the shoal area (microfacies 7) and in

the distal part of the outer platform (microfacies 2), the latter in coincidence with the bioaccumulation of the brachiopods; the peak of gastropods in the proximal part of the outer platform (microfacies 4); and finally the maximum frequency of the pelecypods in microfacies 3 with a curve following a trend entirely different from that of the other constituents.

The weak currents flowing downslope and decreasing in intensity in that direction are responsible for the local intraformational generation of intraclasts and lithic pellets from freshly deposited, semi-indurated muds. This situation is expressed by the gradual and parallel decrease downslope of the clasticity and frequency of the intraclasts. The trend is slightly different for the pellets, where a frequency peak in microfacies 4 coinciding with the frequency peak of both the gastropods and the intensity of bioturbation may indicate an addition of fecal pellets.

The same weak currents distribute silt-size grains of detrital quartz with a gradual downslope decrease of

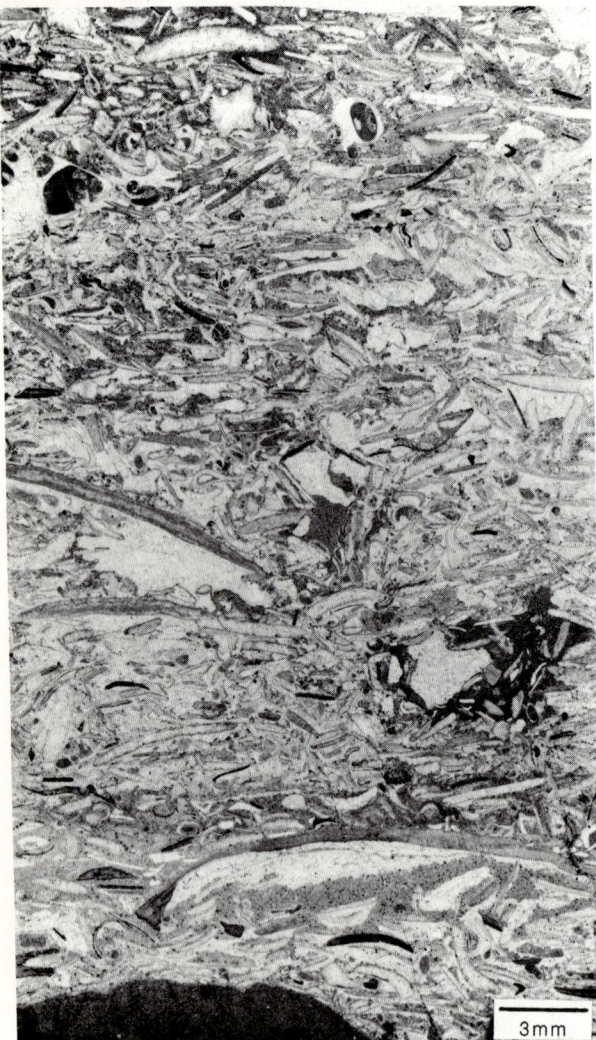

FIGURE 4.11 Galena Group (Middle Ordovician), Upper Mississippi Valley. Typical tempestite (microfacies 5). Moderately bioturbated and disorganized grain-supported biocalcirudite with sparite cement and peletoidal calcisiltite matrix. Brachiopods and pelecypods predominate over gastropods, crinoids, phosphatic trilobites, and bryozoans. Bioclasts show various orientations either perpendicular, inclined at various angles, or parallel to bedding. Both surface of mud at bottom and geopetal internal sediment are overlain by sparite cement under brachiopod shells (middle left and bottom). Burrows are filled partially with sparite cement and calcisiltite matrix. Plane-polarized light. From Bakush and Carozzi (1986).

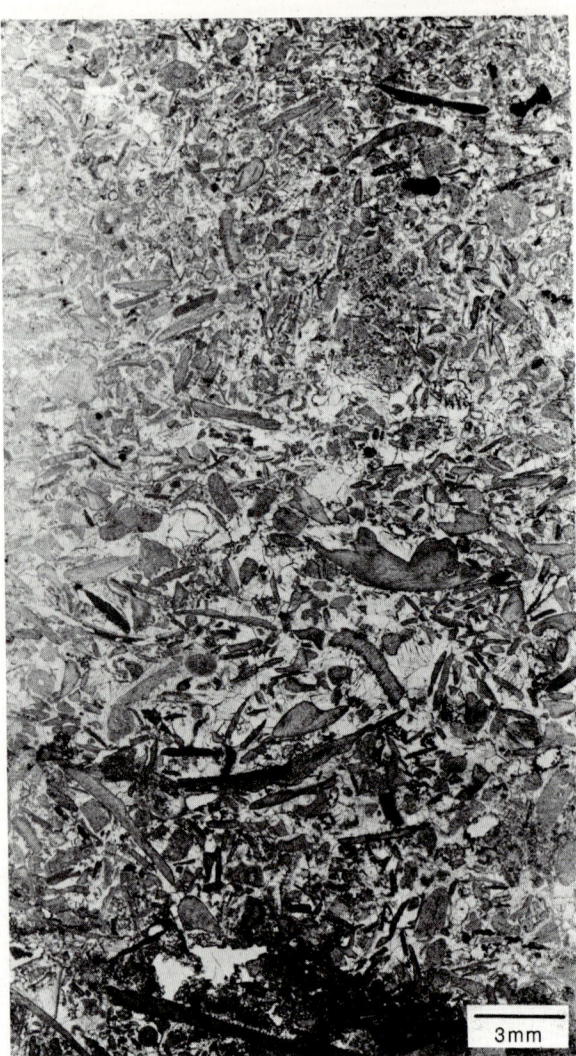

FIGURE 4.12 Galena Group (Middle Ordovician), Upper Mississippi Valley. Typical tempestite (microfacies 6). Bioturbated grain-supported biocalcarenite with sparite cement and rare calcisiltite matrix. Brachiopod and crinoid fragments predominate over trilobites, pelecypods, bryozoans, and ostracods. Reworking features include breakage, imbrication, random orientation of large shell fragments, and incipient graded bedding of bioclasts. The calcisiltite matrix is mostly at the bottom and the sparite cement at middle and top. Plane-polarized light. From Bakush and Carozzi (1986).

Chap. 4 / Carbonate Ramps in Open Marine Environment

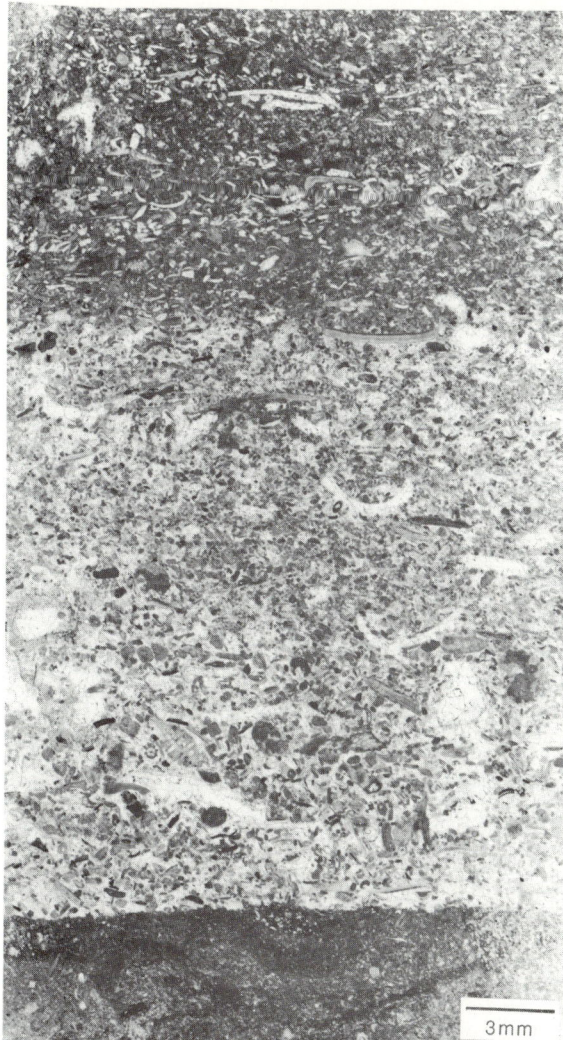

FIGURE 4.13 Galena Group (Middle Ordovician), Upper Mississippi Valley. Typical tempestite (microfacies 6). Moderately bioturbated grain-supported biocalcarenite with sparite cement and calcisiltite matrix. Crinoids and gastropods predominate over brachiopods, ostracods, trilobites, bryozoans, and neomorphosed pelecypod fragments. Other components are lithic pellets and calcisiltite intraclasts. Observe incipient graded bedding, concave upward fragments of pelecypod shells, and sharp erosional contact with underlying microfacies 1. Calcisiltite matrix is concentrated at top and sparite cement is at middle and bottom. Plane-polarized light. From Bakush and Carozzi (1986).

clasticity and frequency preceded by lows in microfacies 7, indicating that these small particles could not remain in high-energy conditions and were winnowed away. In a similar manner, clay minerals increase in volume downslope and settle eventually in the deepest environment.

Storm deposits of microfacies 5 and 6 consist of materials originating from nearby fair-weather microfacies. This local derivation is shown by the estimate in percentage of the number of times that storm microfacies 5 and 6 rest, in the field sections and cores, on the fair-weather microfacies (1, 2, 3, 4, and 7). Microfacies 5 rests 15%, 15.2%, 20.3%, 46%, and 3.5% on fair-weather microfacies 1, 2, 3, 4, and 7, respectively, whereas microfacies 6 rests 36% and 64% on fair-weather microfacies 4 and 7, respectively. Therefore, in the ideal depositional model, the location of the storm microfacies is based on the highest percentage. Thus microfacies 5 is located between 3 and 4, corresponding to the greatest intensity storm action accompanied by the maximum lowering of storm wave base, whereas microfacies 6 is located between 4 and 7, corresponding to lesser-intensity storm action. Both types of tempestites display textures indicating processes which put the constituents of the fair-weather microfacies into suspension and redeposited them essentially in place, or within a short distance. Apparently, storm-induced turbidites due to backflow currents were not generated, perhaps due to the very gentle slope of the carbonate ramp, or the lack of development of sufficient density contrast in water masses. If they had been generated, the foregoing relations of percentages would not occur.

Comparison of the behavior of the various parameters between fair-weather conditions and storm conditions indicates that in the case of the greatest-intensity tempestite (microfacies 5), most of the constituents increase in frequency by mechanical concentration except the sturdy or heavy organisms (gastropods and pelecypods), which were apparently not affected appreciably by the process. In the tempestite of lesser intensity (microfacies 6), most of the constituents decrease significantly in frequency or remain unchanged, indicating winnowing or a simple reorganization.

Storms were reaching first the deeper areas of the Galena sea in the north and then the shallower areas to the south (Figs. 4.18, 4.19). The strongest effect on the sediment-water interface occurred, therefore, in the proximal part of the outer platform where the storm wave base first intersected the seafloor, generating the disorganized calcirudite of microfacies 5. As storms moved to the southern transition outer platform shoal, they decreased in intensity and generated the graded-bedded calcarenites of microfacies 6.

Examination of the field sections and cores shows

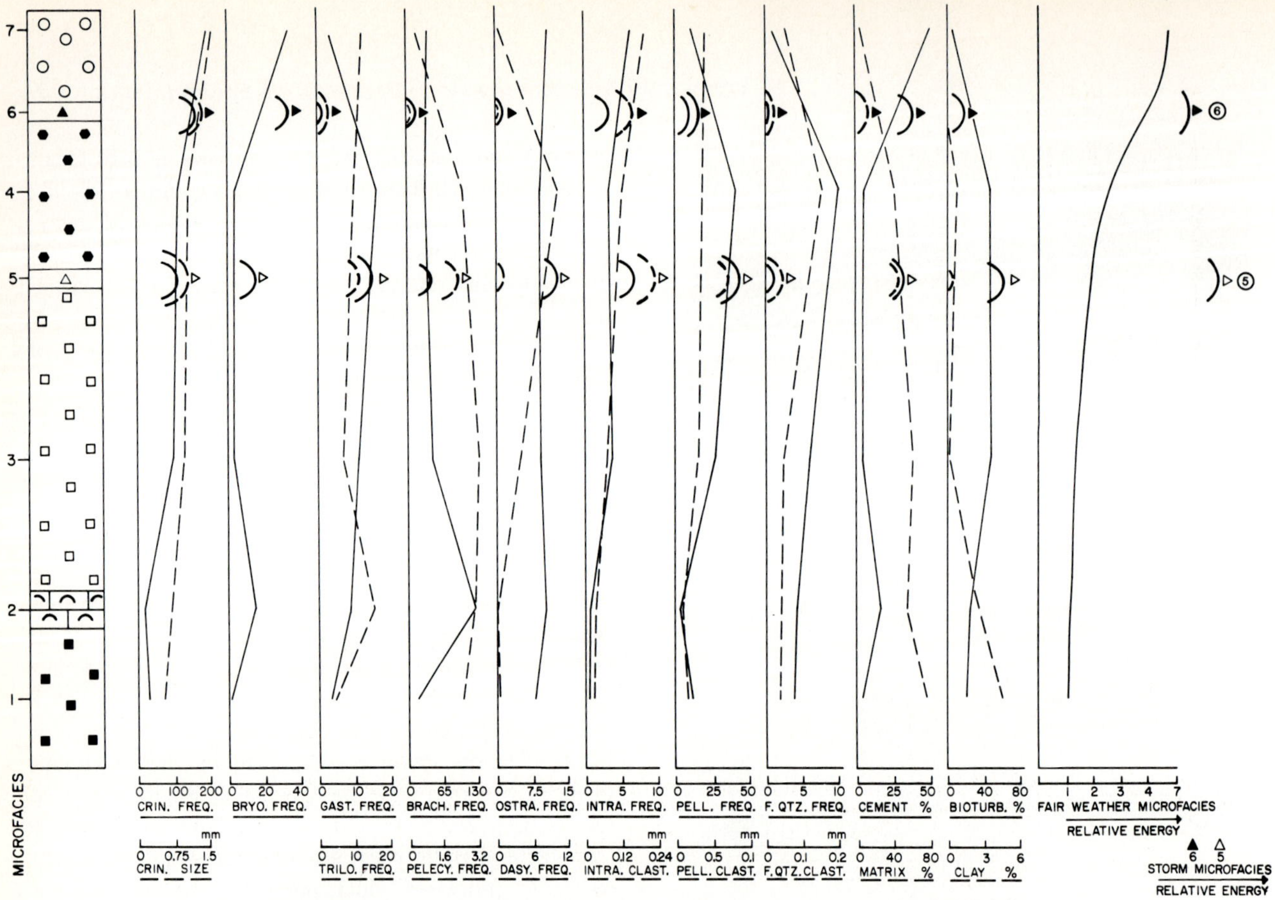

FIGURE 4.14 Galena Group (Middle Ordovician), Upper Mississippi Valley. Ideal shallowing-upward sequence. From Bakush and Carozzi (1986).

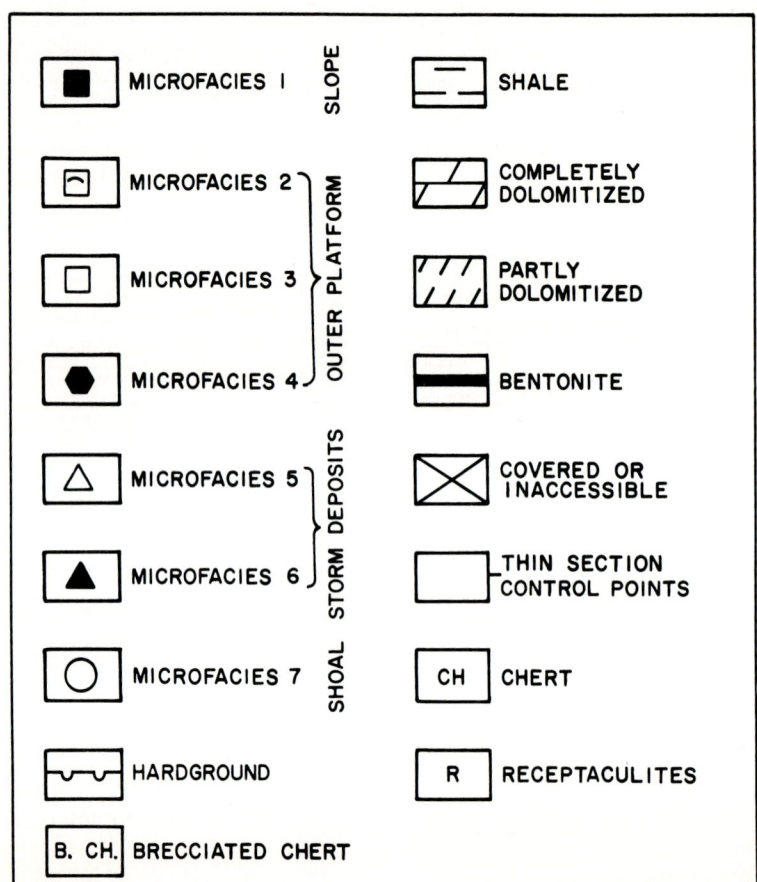

FIGURE 4.15 Galena Group (Middle Ordovician), Upper Mississippi Valley. Symbols for microfacies. From Bakush and Carozzi (1986).

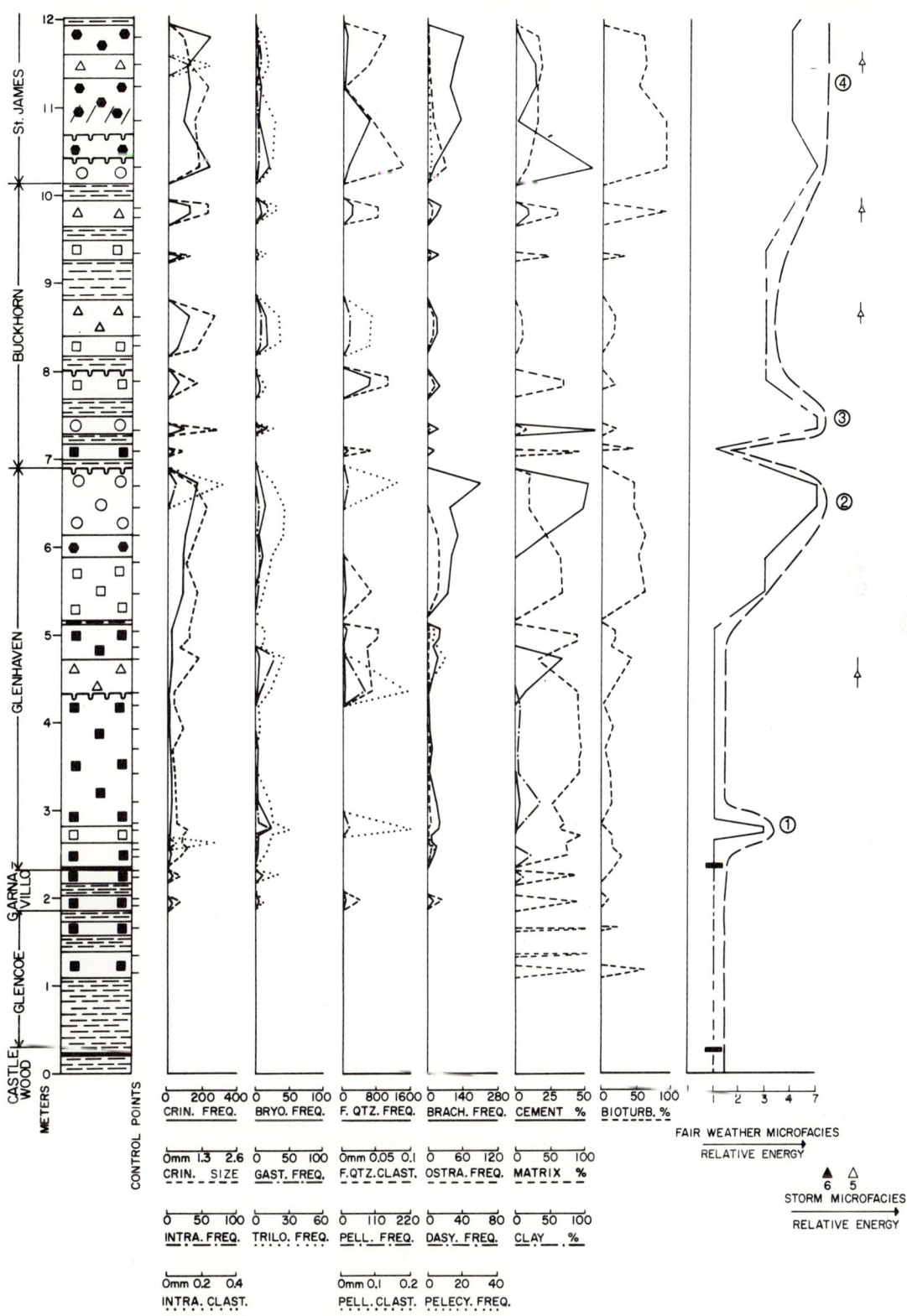

FIGURE 4.16 Galena Group (Middle Ordovician), Upper Mississippi Valley. Typical example of vertical variation of microscopic parameters during shallowing-upward sequence, field section C near Guttenberg, Iowa. From Bakush and Carozzi (1986).

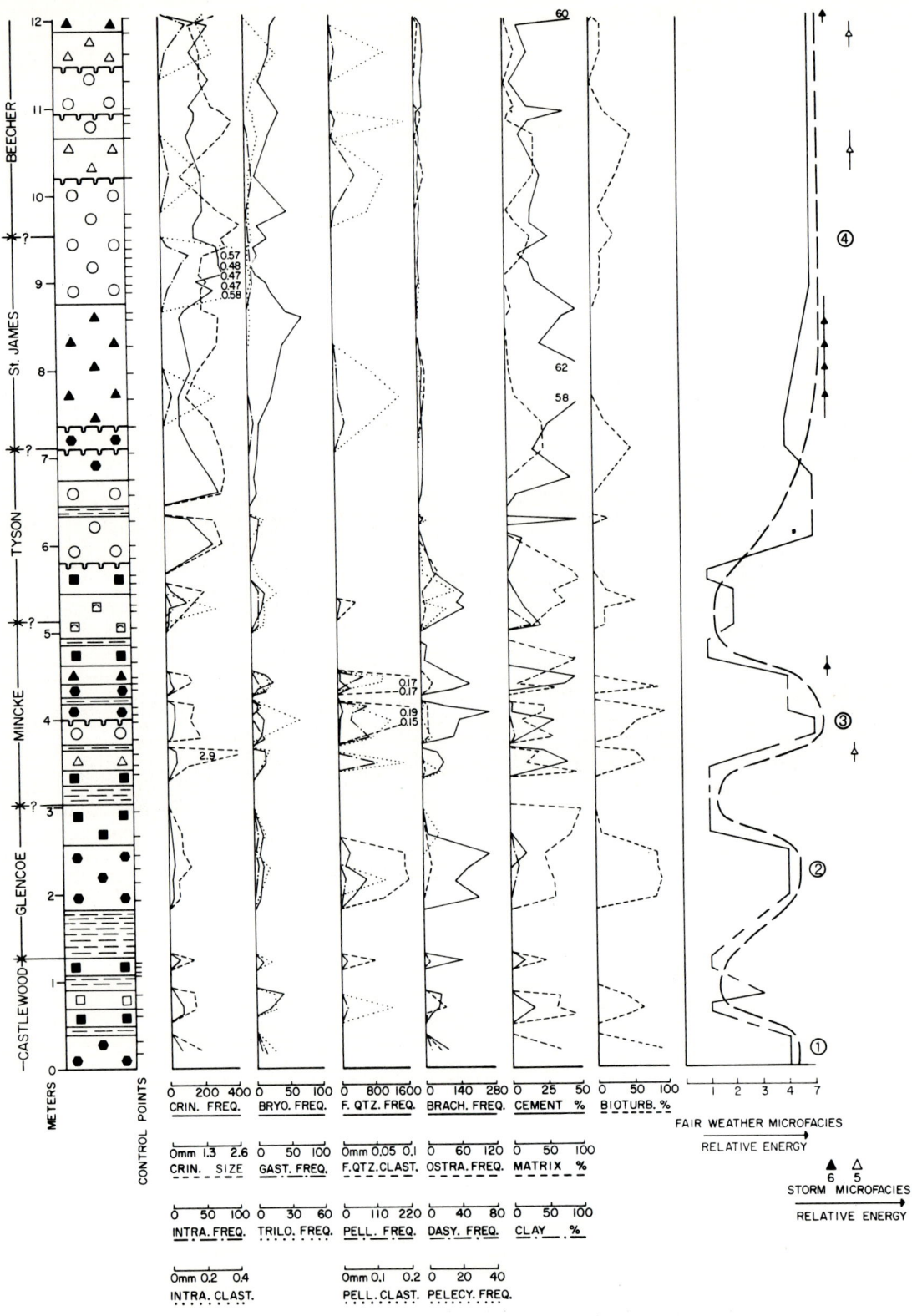

FIGURE 4.17 Galena Group (Middle Ordovician), Upper Mississippi Valley. Typical example of vertical variation of microscopic parameters during shallowing-upward sequence, core section L near Bloomsdale, Missouri. From Bakush and Carozzi (1986).

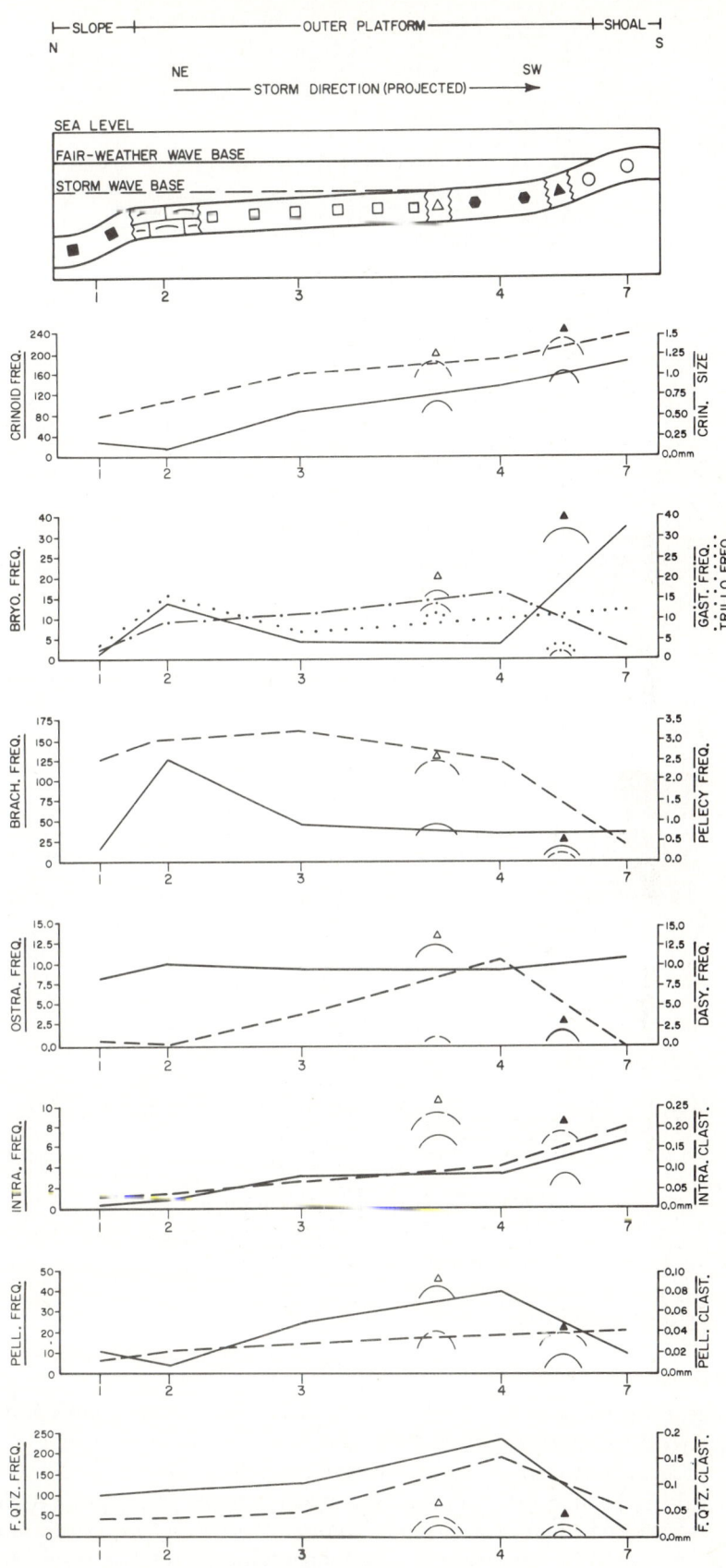

FIGURE 4.18 Galena Group (Middle Ordovician), Upper Mississippi Valley. Ideal fair-weather–storm depositional model. From Bakush and Carozzi (1986).

76 Part II / Carbonate Ramps

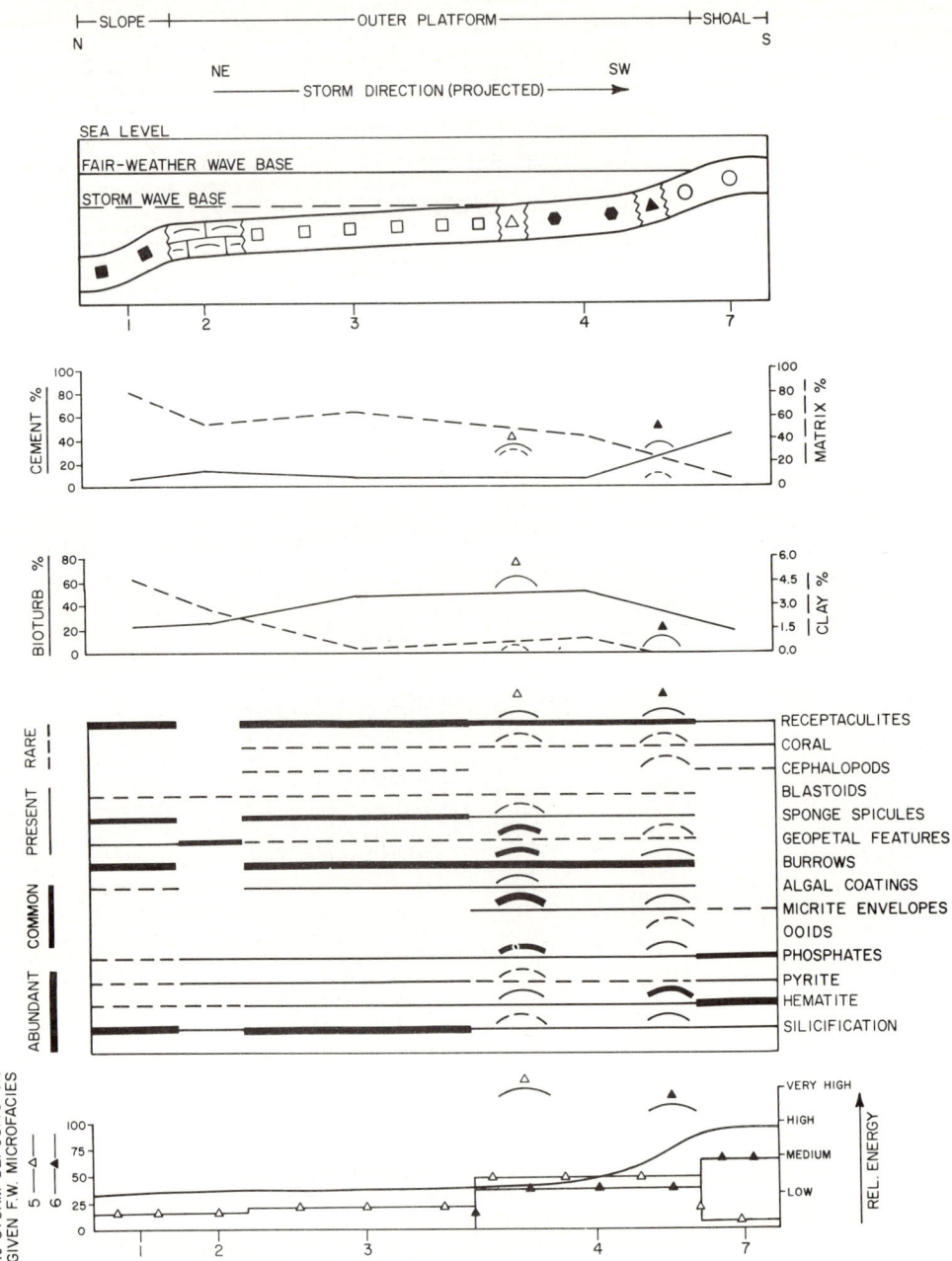

FIGURE 4.19 Galena Group (Middle Ordovician), Upper Mississippi Valley. Ideal fair-weather–storm depositional model (continued). From Bakush and Carozzi (1986).

that storm deposits increase in number from north to south toward the shallower environment. The seafloor in the deeper northern area is inferred to have been affected only by the stronger storms, whereas the shallower southern area would have been affected also by weaker storms, thus displaying a greater recorded number of storm deposits than in the north.

General Evolution of the Depositional Environments

A composite section of the investigated limestone portion of the Galena Group (Fig. 4.20) shows that it consists of 25 environmental oscillations displaying a broad range of amplitude and an association of asymmetric (shallow-

Chap. 4 / Carbonate Ramps in Open Marine Environment 77

ing-upward) and symmetric types (shallowing- and deepening-upward). It is observed generally that in carbonates, the shallower environments tend to record a greater number of oscillations than the deeper ones during a given time interval. In the Galena, just the opposite occurs (Fig. 4.21). This is due probably to the fact that the relatively deep infratidal environment in the north consists of different microfacies, whereas the relatively shallower infratidal environment in the south consists only of crinoidal-bryozoan microfacies.

Because the Galena Group corresponds to a fair-weather–storm depositional model, a cross section from north to south can be drawn to show the relationships between the environmental oscillations and the pattern of distribution of storms (Fig. 4.21). At least five successive alternating predominant storm and fair-weather episodes can be outlined. The boundaries of these episodes were established by the presence or absence of single and/or combined storm events in different sections and independently from any lithostratigraphic correlations. An examination of episodes I, II, and V, which are storm dominated, illustrates clearly that the type of predominant tempestite changes from north to south from microfacies 5 to 6 and that simultaneously, the overall number of recorded events increases. Detailed correlations have shown tempestites (microfacies 5), 10 to 12 cm thick, continuously traceable for more than 48 miles (77 km).

Diagenetic Evolution

The Galena Group was dolomitized extensively in northern Illinois and eastern Iowa by a large-scale dorag mechanism. Therefore, the description of the diagenetic evolution concerns first the limestones and then the dolomites.

Diagenetic Evolution of the Limestones

Observed features and their time relationships (Figs. 4.22, 4.23) reveal a fairly complex evolution with some phases deserving additional comments. Extensively bioturbated calcisiltites display burrows with concentrations of fine to medium euhedral rhombs of dolomite (Fig. 4.23A). This early submarine dolomitization was controlled, apparently, by permeability inhomogeneities within the borrows, and the needed magnesium may have been provided by organic constituents such as minute crinoid bioclasts, and subsequently concentrated in solutions.

An unusual combination of early submarine silicification and of storm effects is revealed by ellipsoidal chert nodules rotated and embedded in crinoidal calcisiltites which themselves display a soft pebble monogenic brecciated texture. Some of the chert nodules are either

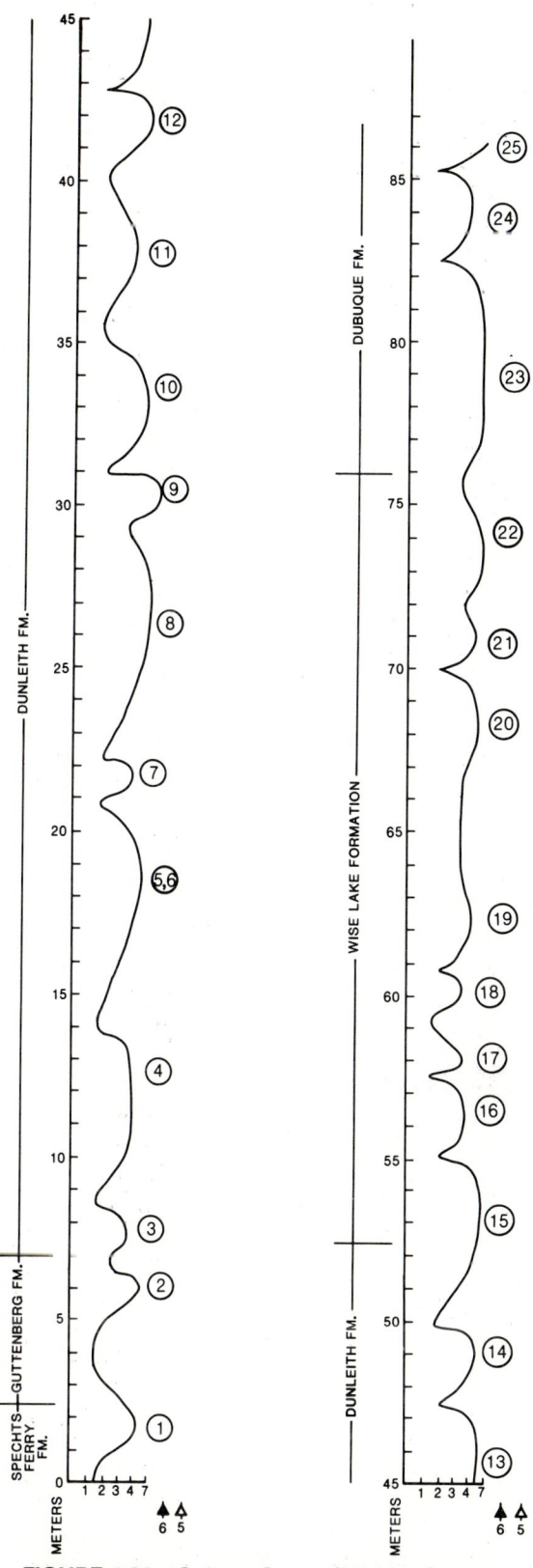

FIGURE 4.20 Galena Group (Middle Ordovician), Upper Mississippi Valley. Composite section showing superposition of depositional cycles. From Bakush and Carozzi (1986).

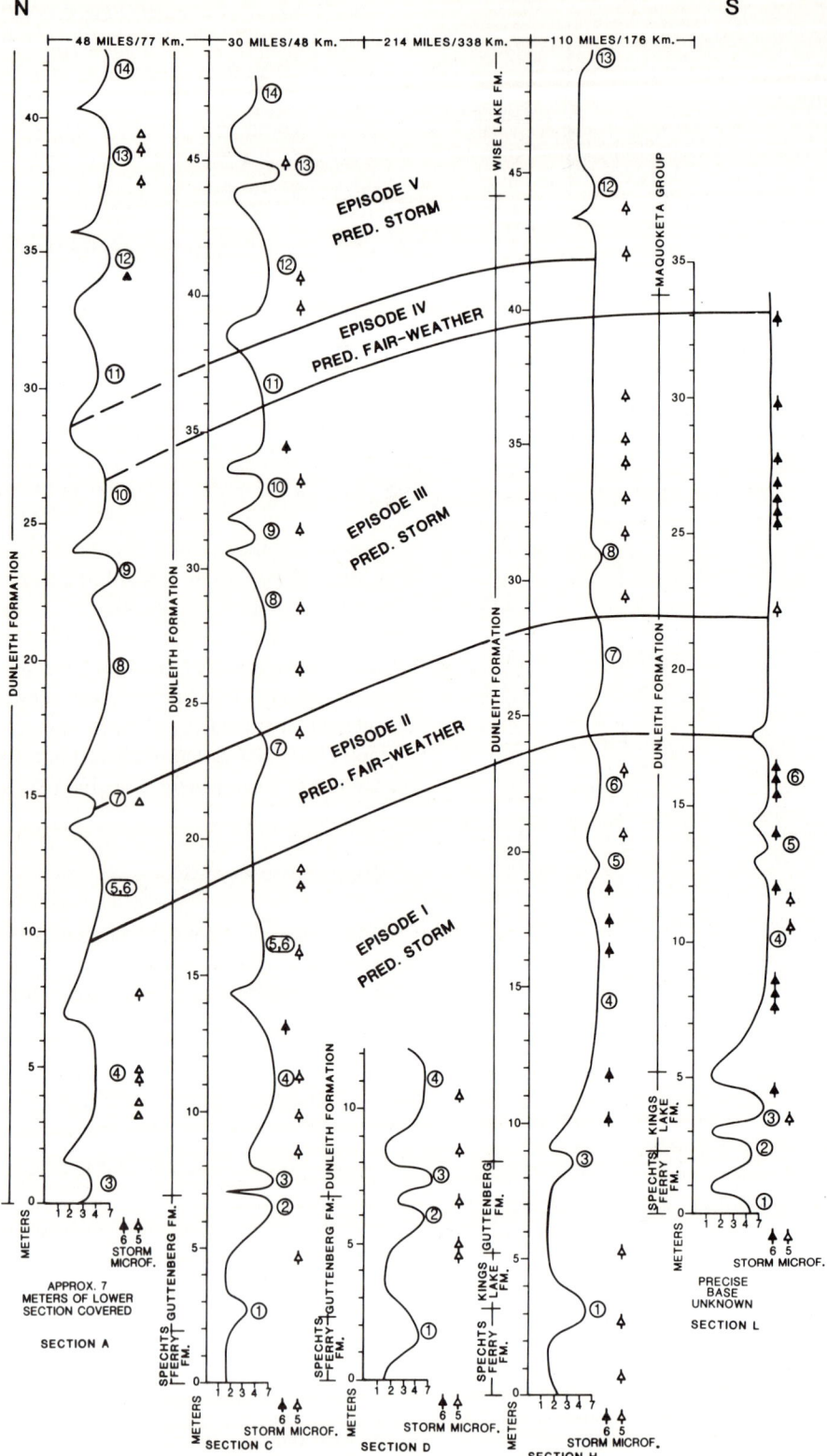

FIGURE 4.21 Galena Group (Middle Ordovician), Upper Mississippi Valley. North–south tentative environmental correlations by cycles and fair-weather–storm episodes. The cycles decrease in number from north to south. From Bakush and Carozzi (1986).

DIAGENETIC FEATURES \ DIAGENETIC ENVIRONMENTS	MARINE PHREATIC	UNDERSATURATED FRESHWATER PHREATIC	SATURATED FRESHWATER PHREATIC	MIXING MARINE FRESHWATER PHREATIC	BURIAL
BIOTURBATION (BURROWS AND BORING OF BIOCLASTS)	——				
EARLY SUBMARINE DOLOMITIZATION	——				
SILICIFICATION (I) (EARLY SUBMARINE)	——				
HARDGROUNDS WITH IRON IMPREGNATION	——				
SYNERESIS	——				
MICRITIZATION (ALGAL COATINGS AND MICRITE ENVELOPES)	——				
INTERNAL SEDIMENTS (WITH GEOPETAL FEATURES)	——				
ISOPACHOUS RIM CEMENT	——				
EARLY COMPACTION	——				
LEACHING OF ARAGONITIC SKELETAL DEBRIS		——			
CAVITY FILLING SPARITE CEMENT			——		
SYNTAXIAL OVERGROWTH			——		
AGGRADING NEOMORPHISM (STABILIZATION OF RIM CEMENT)			——		
SILICIFICATION II				——	
LATE COMPACTION (STYLOLITIZATION)					——
MINERALIZATION (POST EARLY PERMIAN-PRE LATE CRETACEOUS)					——
LATE FRACTURATION					——
LATE FRACTURE FILLING CALCITE CEMENT					——

TIME ⟶

FIGURE 4.22 Galena Group (Middle Ordovician), Upper Mississippi Valley. Diagenetic features and environments of limestone portion. From Bakush and Carozzi (1986).

partly or completely broken in place, with the country rock filling the space between the fragments, although others appear shattered into numerous minute polyhedral fragments (Fig. 4.24). These unusual chert nodules demonstrate the following succession of events: (1) deposition of crinoidal calcisiltite bed; (2) syngenetic submarine silicification generating completely indurated but easily fracturable chert nodules by replacement of the host sediment; (3) high-energy storm event which uplifted, rotated, and transported the chert nodules for a short distance while breaking them in various ways, while simultaneously reworking the semiconsolidated host sediment into a soft pebble monogenic breccia; and (4) final deposition by gravity of all the constituents. A very similar

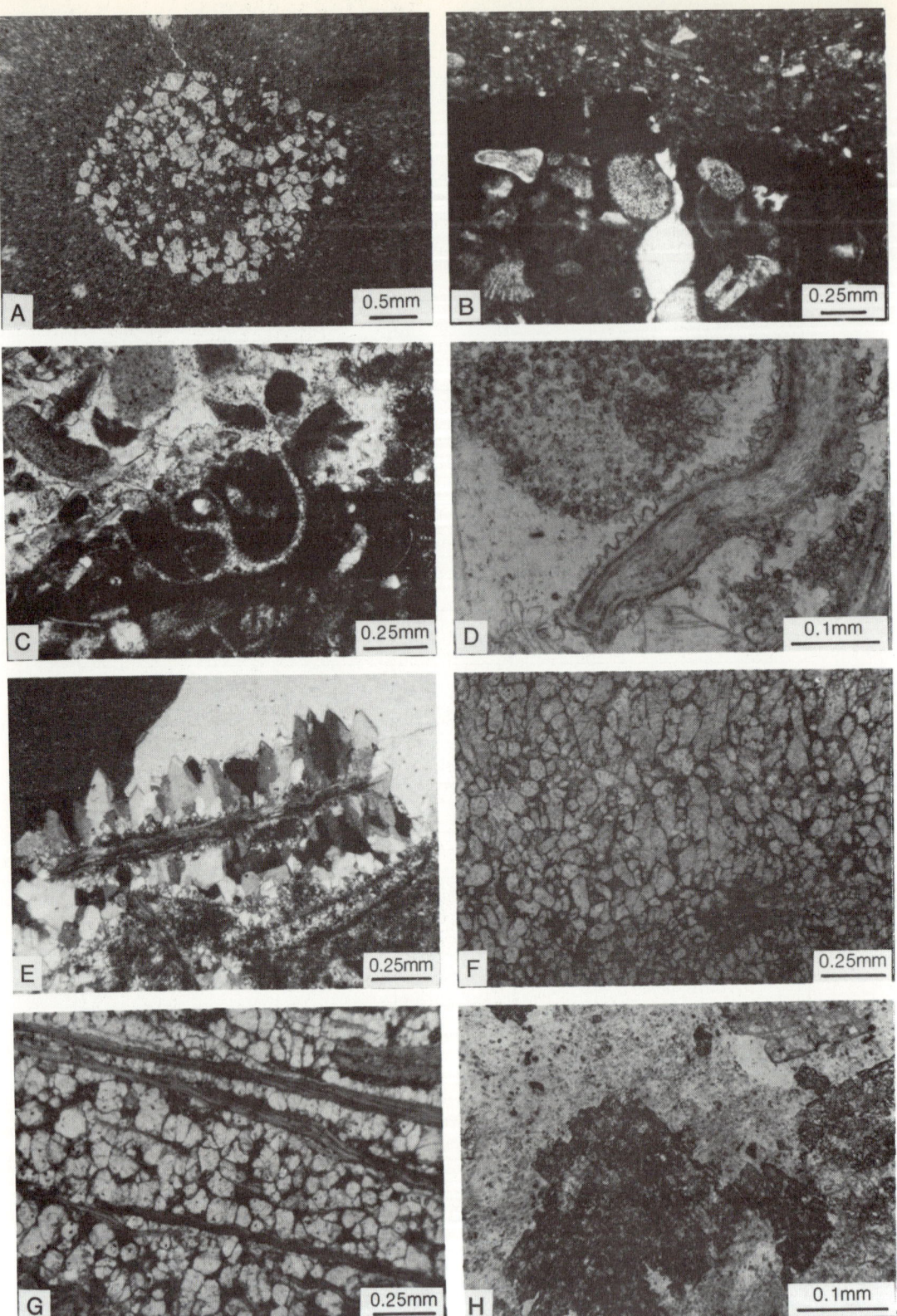

FIGURE 4.23 Galena Group (Middle Ordovician), Upper Mississippi Valley. Diagenetic sequence. *Marine phreatic environment.* A. Horizontal burrow in calcisiltite partially and selectively replaced by dolomite rhombs. B. Smooth hardground between grain-supported biocalcarenite and overlying calcisiltite. Iron oxides are concentrated in crust of hardground and decreasing in importance downward, and microfault infilled with sparite cement. C. Irregular hardground surface between grain-supported biocalcarenite with calcisiltite matrix and overlying grain-supported to pressure-welded biocalcarenite with sparite cement. Gastropod shell is truncated by hardground

case of synsedimentary chert breccia in the Burlington Limestone was interpreted as a Mississippian tempestite (Carozzi and Gerber, 1978) and is described in Section 11.2 of this book.

Hardgrounds (Fig. 4.23B,C) are extremely common in the investigated rocks and may represent a hiatus of 15 to 30% of Galena time. They are smooth to irregular surfaces of discontinuity between two microfacies or inside the same microfacies due to chemical dissolution of early submarine cemented sediment accompanied by various types of mineralization (hematite, pyrite, phosphates), or nondeposition, or submarine mechanical erosion.

A second phase of silicification expressing a mixed marine freshwater phreatic environment is represented by a megaquartz mosaic filling cavities formed by dissolution of an equant calcite mosaic of previous freshwater phreatic origin.

Diagenetic Evolution of the Dolomites

The observed features and their time relationships (Figs. 4.25, 4.26) indicate that the limestones became extensively dolomitized after the second phase of silicification in the mixed marine freshwater phreatic environment. The isopachs of the dolomites formed by this regional dorag model follow a southwest-northeast trend (Fig. 4.27). The contours of the thickest dolomite occurrence correspond to the emergent part of the Galena Group between the Wisconsin Arch and the Mississippi River Arch, which was exposed during or at the end of Maquoketa time (Late Ordovician), an episode of worldwide lowering of sea level. A large active freshwater lens would have existed in this emergent area, leading in turn to the development of an extensive mixed marine-freshwater system.

Petrographically, dolomitization displays three habits: fine anhedral crystals (0.01 to 0.12 mm) in calcisiltites; interlocking subhedral rhombs (0.03 to 0.24 mm) replacing calcisiltites and mud-supported biocalcarenites; zoned euhedral medium to coarse rhombs (0.22 to 0.24 mm) replacing grain-supported biocalcarenites. Occurrence of these habits reflects the grain size of the original carbonates (Zadnik and Carozzi, 1963; see also Chapter 14 in this volume). A third phase of silicification follows dolomitization and consists of fibroradiated microcrystalline quartz replacing entirely or marginally dolomite rhombs, or euhedral quartz crystals replacing rhombs.

Uplifting of the major positive structures of the investigated area was reactivated during Champlainian times and may have continued later. Consequently, the dolomitized Galena Group was elevated and underwent a second diagenetic evolution before final burial (Fig. 4.25). It consists of dedolomitization and of an interesting phase of pyrite mineralization which took place after late compaction, dissolving and replacing both coarse equant sparite cement and dolomite rhombs. The latter display severe marginal dissolution and show a "floating" texture within pyrite. This mineralization is an aspect of the Mississippi-type ore deposits for which a basinal brine hypothesis is favored at present. The age of the mineralization is debated and considered to range from post Early Permian to pre Late Cretaceous; at any rate, the late vertical fracturation observed in the diagenetic sequence of the dolomites cuts pyrite seams.

4.3 CHARACTERISTIC FEATURES: EFFECTS OF CURRENT REWORKING PROCESSES

The rate of subtidal sedimentation of many fine-grained ramp carbonates was relatively small. These conditions allowed freshly deposited and unconsolidated muds to be reworked and redistributed by the action of well-developed systems of gentle bottom currents.

FIGURE 4.23 (*continued*)   surface. D. Thin isopachous rim cement around trilobite surrounded by equant sparite cement. Notice small scalenohedral calcite crystals of the rim cement due to stabilization in the freshwater phreatic environment. *Saturated freshwater phreatic environment.* E. Second generation of cavity-filling sparite cement nucleated on thin isopachous rim cement itself stabilized in the form of scalenohedral bladed calcite crystals. The cement grades from small to large bladed to large equant mosaic. F. Aggrading neomorphism from pseudomicrosparite to pseudosparite with elongate loafish crystals in mud-supported biocalcarenite with argillaceous micrite groundmass between crystals. G. Pseudosparite in bioaccumulated limestone. Observe relatively uniform crystal size and argillaceous micrite groundmass between crystals. *Mixing marine freshwater phreatic environment.* H. Silicification after dolomitization where silica replaces dolomite crystals. Note marginal replacement of dark dolomite rhombs. All photomicrographs: plane-polarized light, except E and H, crossed nicols. From Bakush and Carozzi (1986).

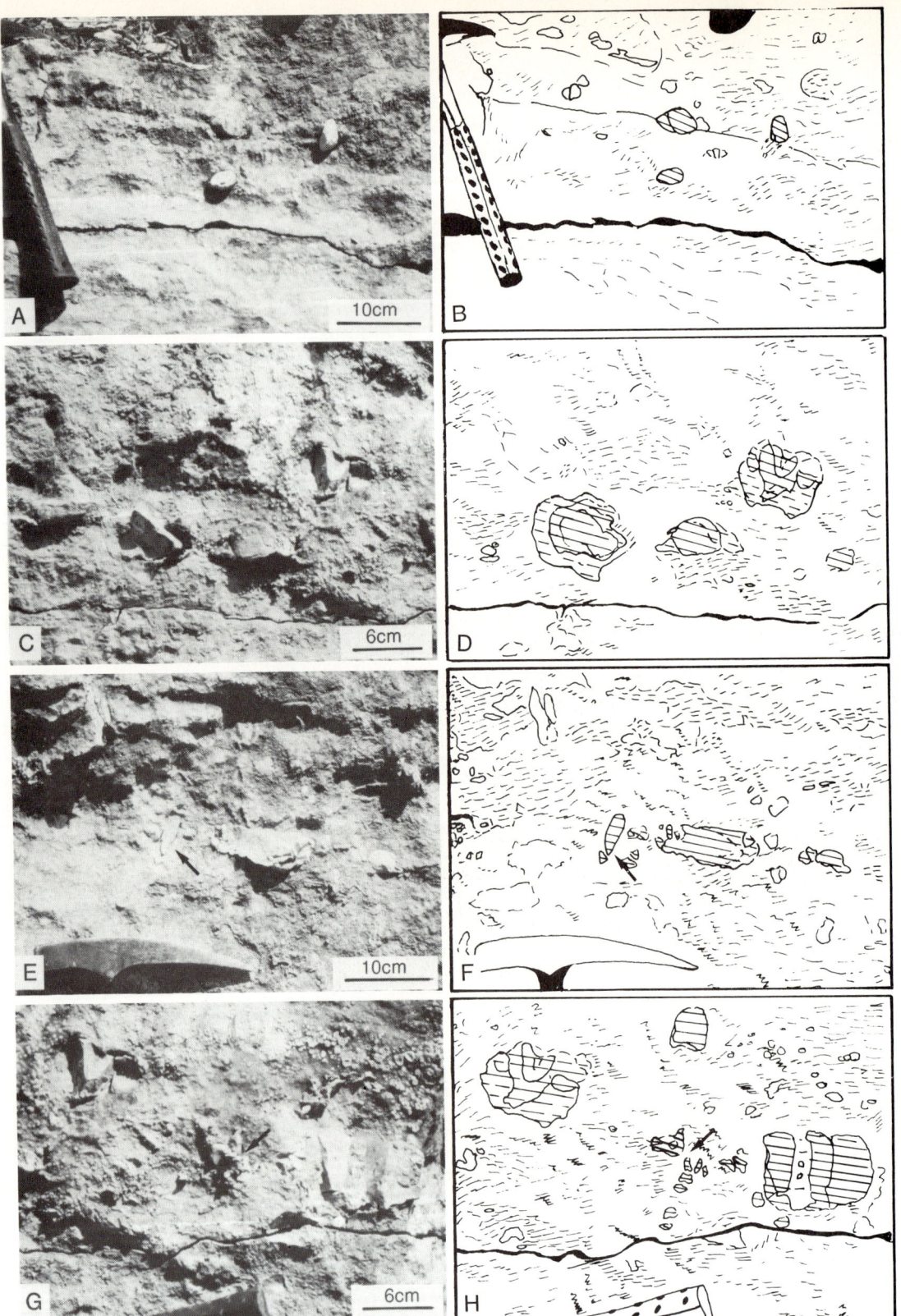

FIGURE 4.24 Galena Group (Middle Ordovician), Upper Mississippi Valley. Diagenetic sequence (continued). *Marine phreatic environment and storm effects.* A and B. Field photograph and sketch showing three ellipsoidal chert nodules with different orientations in brecciated crinoidal calcisiltite. C and D. Field photograph and sketch showing three large chert nodules with different orientation in brecciated crinoidal calcisiltite. E and F. Field photograph and sketch showing partially fragmented elongated chert nodule (center right) and

DIAGENETIC FEATURES \ DIAGENETIC ENVIRONMENTS	MARINE PHREATIC	UNDERSATURATED FRESHWATER PHREATIC I	SATURATED FRESHWATER PHREATIC I	MIXED MARINE FRESHWATER PHREATIC	UNDERSATURATED FRESHWATER PHREATIC II	SATURATED FRESHWATER PHREATIC II	BURIAL
BIOTURBATION (BURROWS AND BORING OF BIOCLASTS)	—						
EARLY SUBMARINE DOLOMITIZATION	—						
SILICIFICATION (I) (EARLY SUBMARINE)	—						
HARDGROUNDS WITH IRON IMPREGNATION	—						
SYNERESIS	—						
MICRITIZATION (ALGAL COATINGS AND MICRITE ENVELOPES)	—						
INTERNAL SEDIMENTS (WITH GEOPETAL FEATURES)	—						
ISOPACHOUS RIM CEMENT	—						
EARLY COMPACTION	—						
LEACHING OF ARAGONITIC SKELETAL DEBRIS		—					
CAVITY FILLING SPARITE CEMENT			—				
SYNTAXIAL OVERGROWTH			—				
AGGRADING NEOMORPHISM (STABILIZATION OF RIM CEMENT)			—				
SILICIFICATION II				—			
DOLOMITIZATION (DORAG MODEL) (U. ORD.-MAQUOKETA)				—			
SILICIFICATION III				—			
UPLIFT							
DEDOLOMITIZATION					—		
POIKILOTOPIC CALCITE CEMENT						—	
LATE COMPACTION (STYLOLITIZATION)							—
MINERALIZATION (POST EARLY PERMIAN-PRE LATE CRETACEOUS)							—
LATE FRACTURATION							—
LATE FRACTURE FILLING CALCITE CEMENT							—
TIME →							

FIGURE 4.25 Galena Group (Middle Ordovician), Upper Mississippi Valley. Diagenetic features and environments of dolomite portion. From Bakush and Carozzi (1986).

The Menard Formation, Upper Mississippian of the southwest margin of the Illinois Basin (Carozzi and Roche, 1968), is part of the Upper Chester Series, in which carbonate units are separated or truncated by channel sandstones deposited in estuarine conditions (Fig. 4.28). The petrographic investigation consisted of nearly 400 thin sections with an average interval of 15 cm.

Description of Microfacies

Six distinct microfacies were recognized, but one of them, a pressure-welded biocalcarenite (microfacies 3d), was found to cut across the microfacies boundaries of all the grain-supported rocks regardless of matrix type, and it was therefore not used as a potential indicator of environment in the otherwise shallowing-upward sequence because of its obvious generation through diagenetic processes.

Microfacies 1 (Fig. 4.29A). Bioturbated biocalcisiltite with scattered fine sand-size bioclasts of crinoids, ostracods, and calcified monaxonic sponge spicules. Calcispheres, dasyclads, bryozoans, and arenaceous foraminifers are rare. Occasional brachiopod and pelecypod shells and debris of trilobites occur as large, nontransported components, indicating an *in situ* community. Reworked ooids and silt-size detrital quartz are sparse.

Microfacies 2 (Fig. 4.29B). Bioturbated mud-supported biocalcarenite with calcisiltite matrix. "Floating" bioclasts are mainly crinoids and echinoids, calci-

FIGURE 4.24 (*continued*) elongated chert nodule fragment (arrow) in vertical position. G and H. Field photograph and sketch showing chert nodule shattered into minute polyhedral fragments (arrow) and large chert nodule (right) broken vertically into two halves with penetration of country rock between fragments. All illustrations from outcrop at Highway 9, Decorah, Iowa. From Bakush and Carozzi (1986).

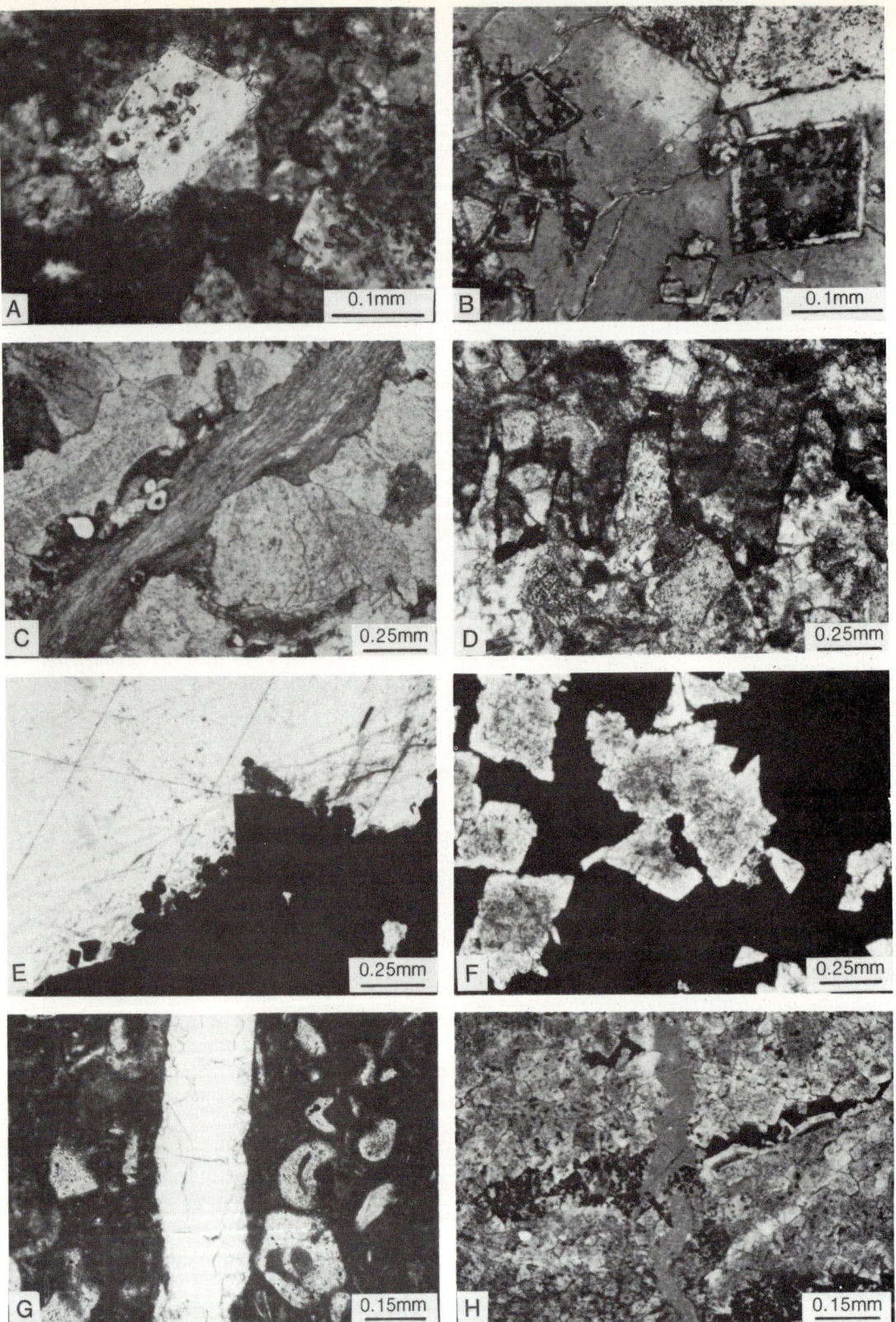

FIGURE 4.26 Galena Group (Middle Ordovician), Upper Mississippi Valley. Diagenetic sequence (continued). *Mixing marine-freshwater phreatic environment.* A. Diagenetic quartz (clear crystal) replacing dolomite rhombs. Iron oxide inclusions within the dolomite survived silicification. *Saturated freshwater phreatic environment II.* B. Poikilotopic calcite cement following in time fabric selective dedolomitization in which dolomite rhombs "float" in calcite and are infilled with dark iron-rich calcite. *Burial environment.* C. Late compaction

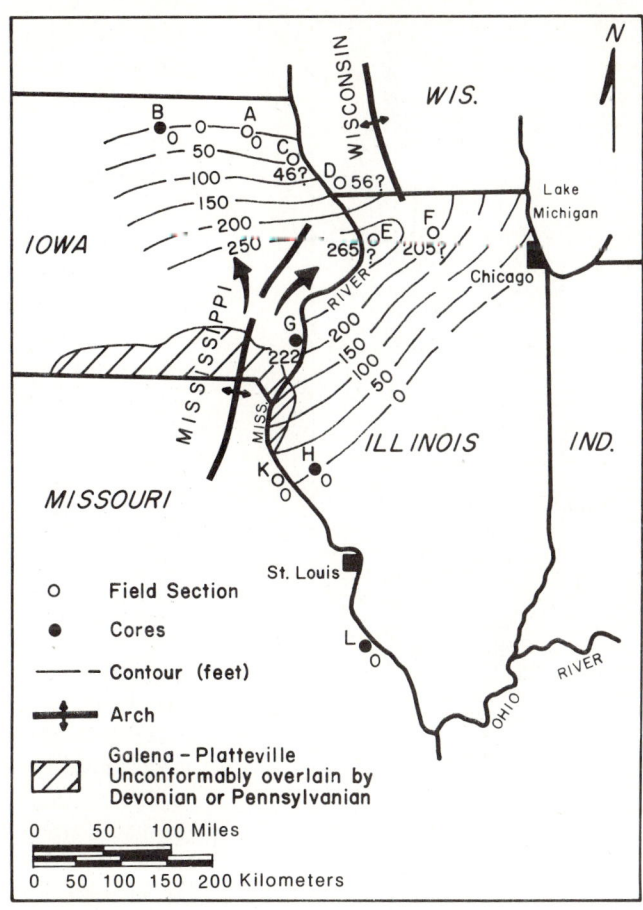

FIGURE 4.27 Galena Group (Middle Ordovician), Upper Mississippi Valley. Isopach map of dolomite. Arrows around Mississippi River Arch are the possible pathways of circulation of mixed marine-freshwater solutions. From Bakush and Carozzi (1986).

fied monaxonic sponge spicules and bryozoans. Accessory components are calcispheres, ostracods, small arenaceous foraminifers, and smaller bioclasts of brachiopods, pelecypods, gastropods, and trilobites. Fine sand-size detrital quartz is accompanied by a few reworked ooids.

Microfacies 3a (Fig. 4.29C). Bioturbated grain-supported biocalcarenite with a calcisiltite matrix in which neomorphism into pseudomicrosparite or pseudosparite does not exceed 20%. The predominant bioclasts are crinoids, fenestrate bryozoans (*Archimedes*) associated with brachiopods, pelecypods, gastropods, dasyclads, and trilobites. Detrital quartz grains are fairly abundant and reach 0.1 mm clasticity.

Microfacies 3b (Fig. 4.29D,E). Grain-supported biocalcarenite with a calcisiltite matrix in which neomorphism into pseudomicrosparite and pseudosparite ranges from 20 to 80%. The major bioclasts, which

FIGURE 4.26 (*continued*) causing crushing and interpenetration of brachiopod shell fragment by crinoid columnals. D. Sutured stylolite with bituminous residue. E. Pyrite replacing equant sparite cement filling large void in mud-supported biocalcarenite. F. Pyrite infilling secondary porosity after dissolution of dolomite rhombs. Observe severe marginal dissolution of dolomite rhombs and their "floating" texture within pyrite. G. Late burial fracturation in mud-supported biocalcarenite. The vertical fracture is filled with a first generation of scalenohedral calcite directly on the walls and a second generation of coarser sparite in the center. H. Late burial vertical fracture filled with calcite cement (appearing gray after staining) in dolomitized calcisiltite to mud-supported biocalcarenite. The fracture cuts pyrite seams which replaced dolomite. All photomicrographs: plane-polarized light. From Bakush and Carozzi (1986).

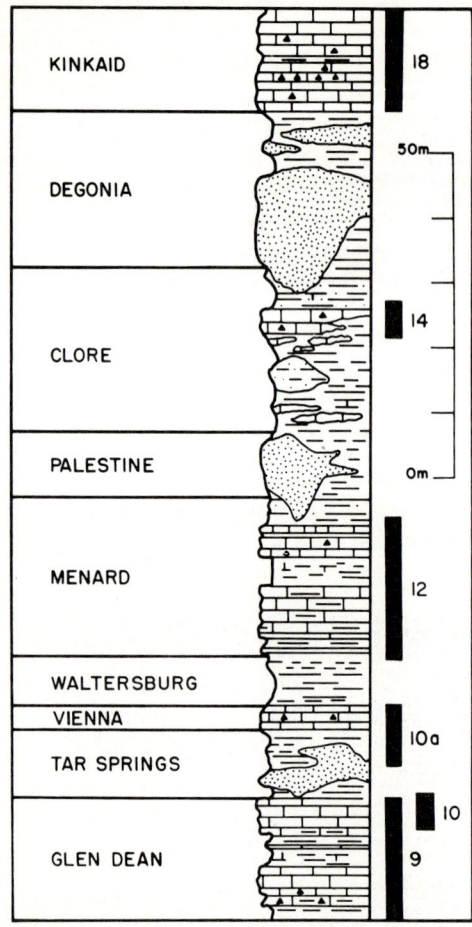

FIGURE 4.28 Menard Formation (Middle Mississippian), southwest margin of Illinois Basin. Generalized geologic column of the Upper Chester Series in the type area. Numbered black boxes indicate investigated stratigraphic sections. From Carozzi and Roche (1968). Reprinted by permission of the Illinois State Academy of Science.

show local pressure-solution, are crinoids and fenestrate bryozoans which predominate over brachiopods, pelecypods, gastropods, dasyclads, and trilobites. Reworked ooids are fairly common with nuclei of any of the bioclasts above and occur in local concentrations with pressure-solution contacts and marginal replacement by the neomorphic matrix.

Microfacies 3c (Fig. 4.29F,G). Grain-supported biocalcarenite with more than 80% of the original calcisiltite matrix changed by neomorphism into an irregular association of pseudomicrosparite and patches of pseudosparite which locally grades into a fine to coarse uniform mosaic of pseudosparite. Among the major bioclasts, fenestrate bryozoans predominate over crinoids, and reworked ooids are abundant. The remainder of the components consists of bioclasts of brachiopods, pelecy-

pods, gastropods, trilobites with rare ostracods, calcispheres, and arenaceous foraminifers.

Microfacies 3d (Fig. 4.29H). Pressure-welded crinoidal biocalcarenite in which pressure-solution ranges in intensity from incipient stages where patches of the original calcisiltite matrix are still preserved to terminal stages with deep microstylolitic interpenetration and total elimination of the matrix. The bioclasts consist essentially of crinoids and bryozoans oriented parallel to bedding and associated with frequent spalled and distorted ooids. Most of the other types of bioclasts observed in microfacies 3a, 3b, and 3c, with the exception of brachiopods and pelecypods, have largely been eliminated by pressure-solution.

The Ideal Shallowing-Upward Sequence

Microfacies 1 to 3c can be organized into a shallowing-upward sequence (Figs. 4.30, 4.31), and although the superposition of 3a–3b–3c corresponds to an increasing degree of neomorphism of the calcisiltite matrix, these three microfacies are also different with respect to their depositional parameters. The extensive pressure solution which occurred in microfacies 3d prevents its environmental interpretation.

Crinoids, which are the major components, display an increase in clasticity related to increasing energy, although the frequency reaches its peak at the first grain-supported microfacies and then decreases upward as the result of the removal of smaller debris in more agitated conditions. Crinoid parameters express the growth of an almost autochthonous population and the sorting of its skeletal debris. The frequency of bryozoans increases upward, indicating increasing fragmentation of their fronds with higher energy, whereas under the same conditions, the more resistant crinoid columnals were only sorted according to size but not broken. Ostracods and sponge spicules tend to concentrate in low-energy conditions, whereas arenaceous benthic foraminifers and calcispheres appear to be ubiquitous with only a slight tendency toward a similar concentration. Detrital quartz shows a parallelism between its clasticity and frequency curves, an indication of a regular but not abundant clastic sediment yield to the basin. The coarse-silt to fine-sand particles could not remain in the high-energy environment. They were winnowed and concentrated in microfacies 3a, where their grain size was in equilibrium with the energy of the environment; only smaller and fewer particles reached deeper waters. The relatively large ooids were transported from shallower to deeper conditions but remained unaffected by winnowing processes because of their size.

A typical detailed section (Fig. 4.32) shows the

Chap. 4 / Carbonate Ramps in Open Marine Environment 87

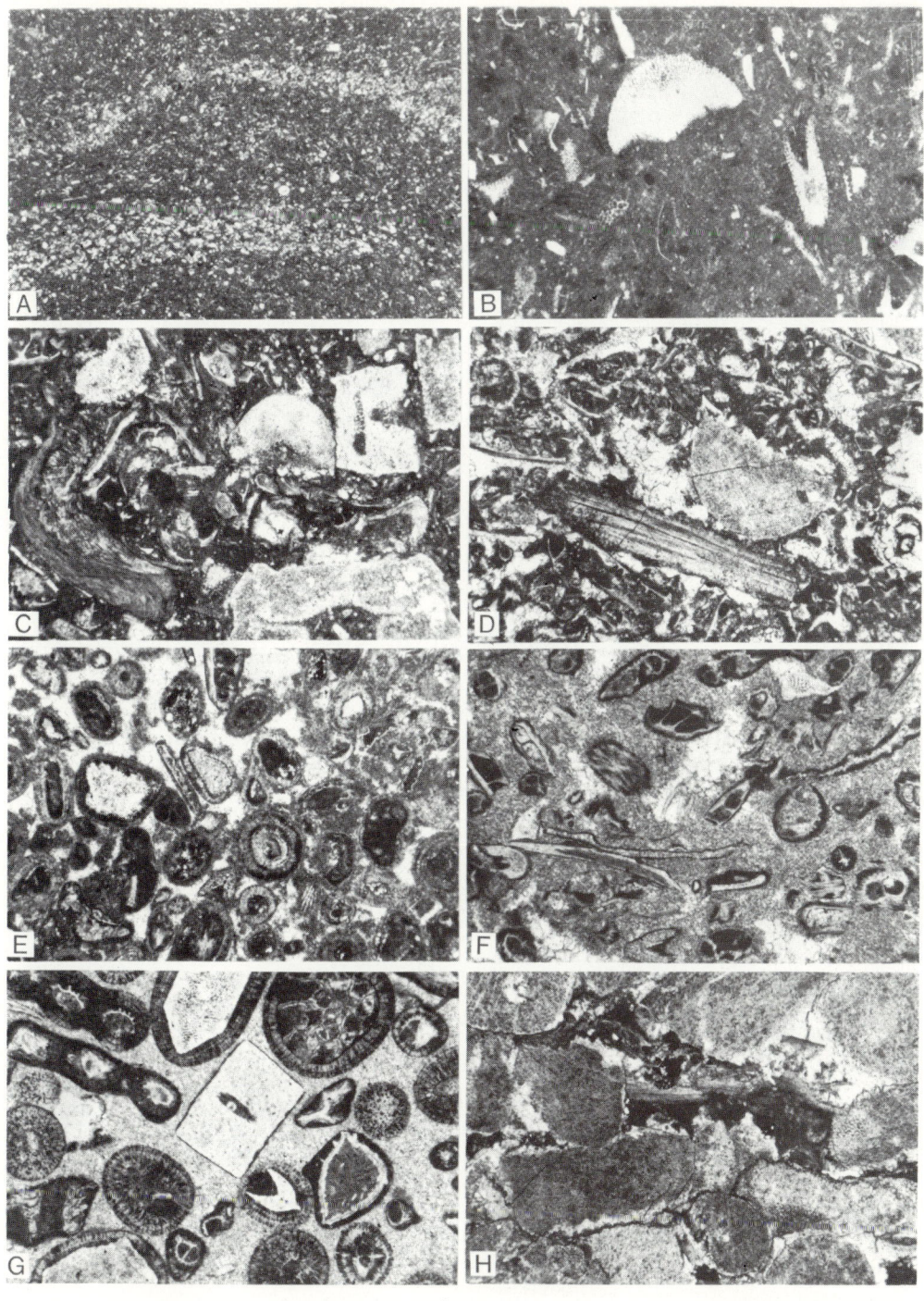

1mm

FIGURE 4.29 Menard Formation (Middle Mississippian), southwest margin of Illinois Basin. Typical microfacies. A. Microfacies 1. B. Microfacies 2. C. Microfacies 3a. D and E. Microfacies 3b. F and G. Microfacies 3c. H. Microfacies 3d. See text for detailed descriptions. All photomicrographs: plane-polarized light. From Carozzi and Roche (1968). Reprinted by permission of the Illinois State Academy of Science.

88   Part II / Carbonate Ramps

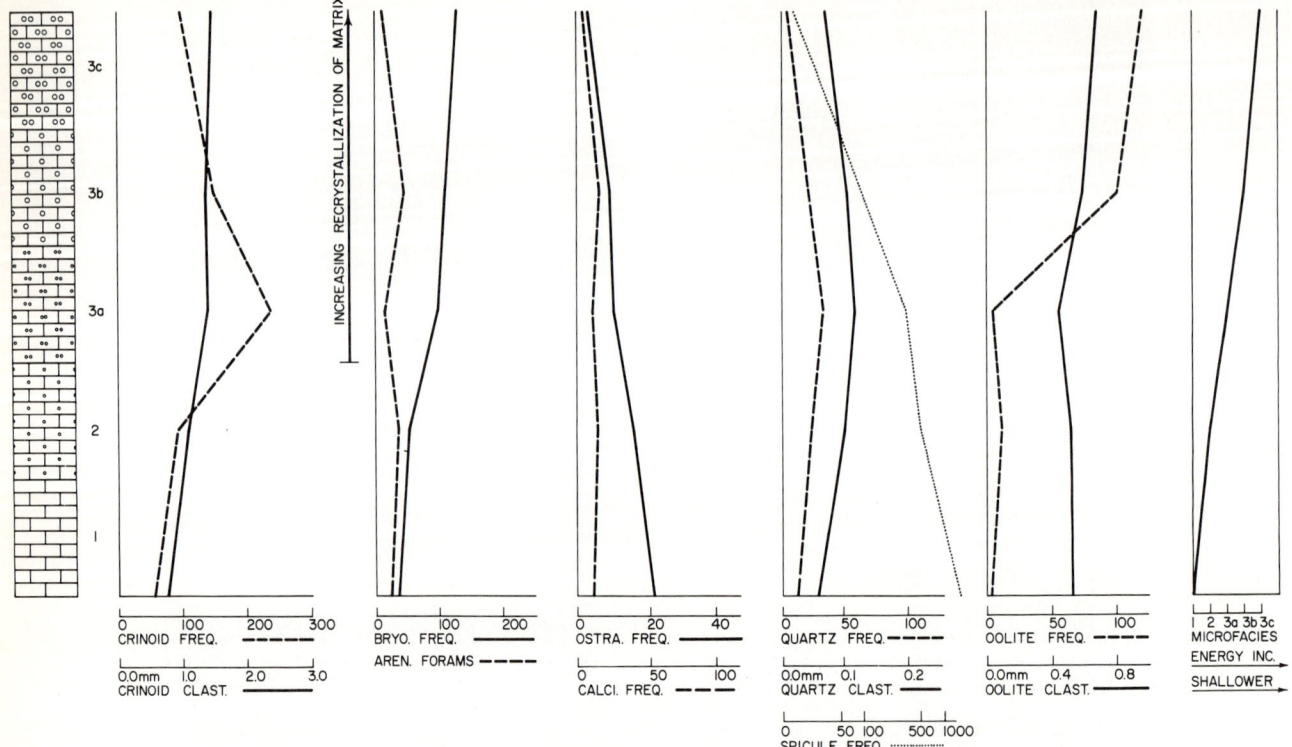

FIGURE 4.30 Menard Formation (Middle Mississippian), southwest margin of Illinois Basin. Ideal shallowing-upward sequence. From Carozzi and Roche (1968). Reprinted by permission of the Illinois State Academy of Science.

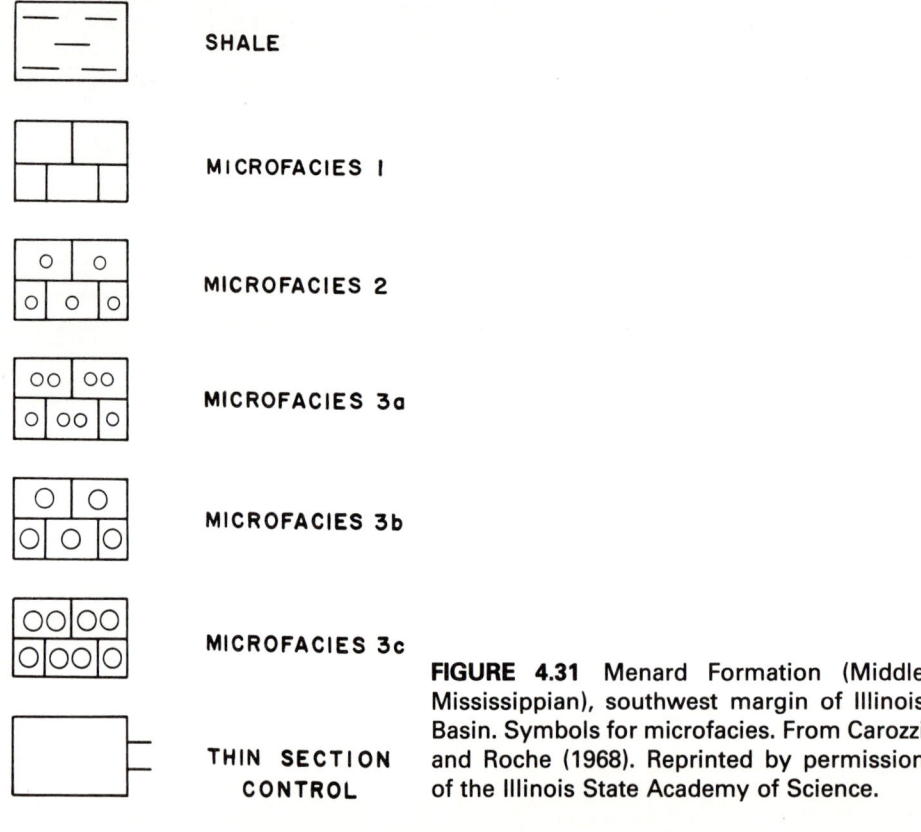

FIGURE 4.31 Menard Formation (Middle Mississippian), southwest margin of Illinois Basin. Symbols for microfacies. From Carozzi and Roche (1968). Reprinted by permission of the Illinois State Academy of Science.

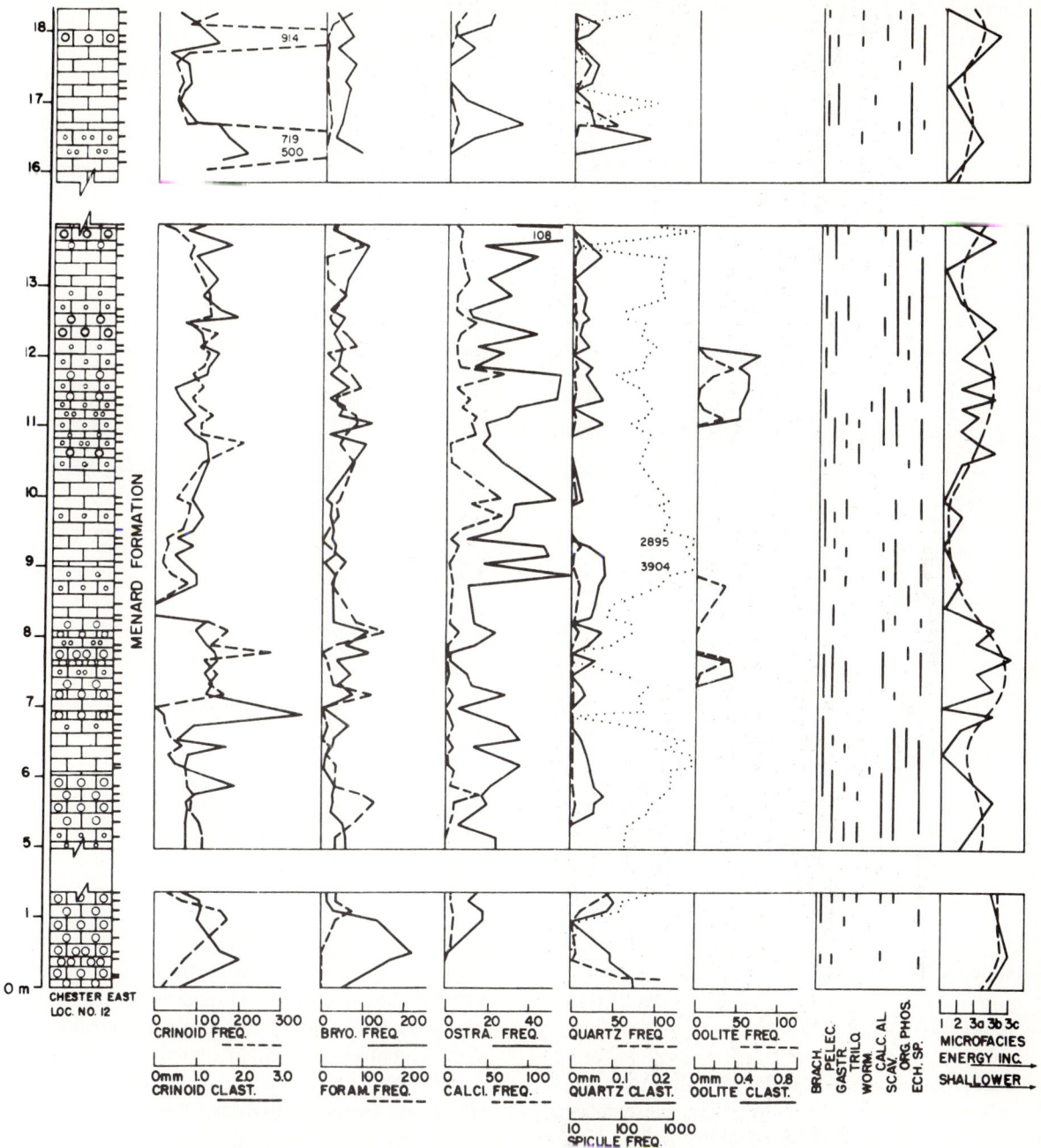

FIGURE 4.32 Menard Formation (Middle Mississippian), southwest margin of Illinois Basin. Typical example of vertical variation of microscopic parameters in section of Chester East, No. 12. See Figure 4.28 for location. From Carozzi and Roche (1968). Reprinted by permission of the Illinois State Academy of Science.

90 Part II / Carbonate Ramps

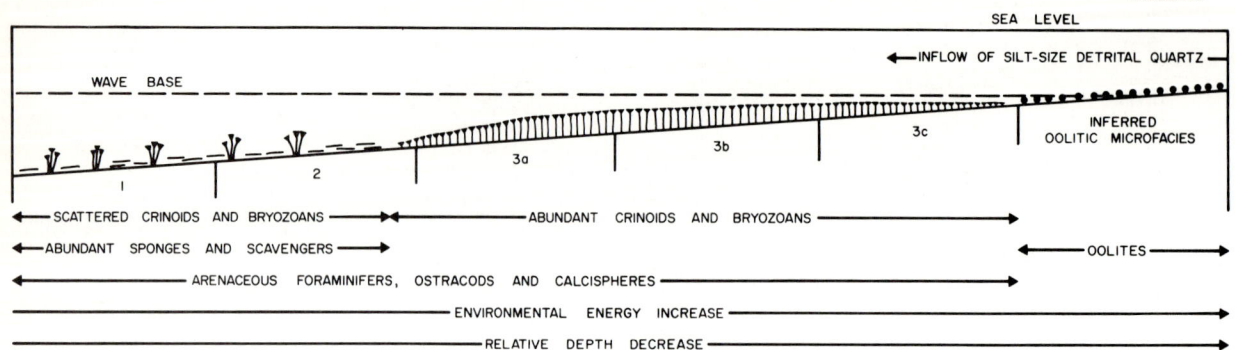

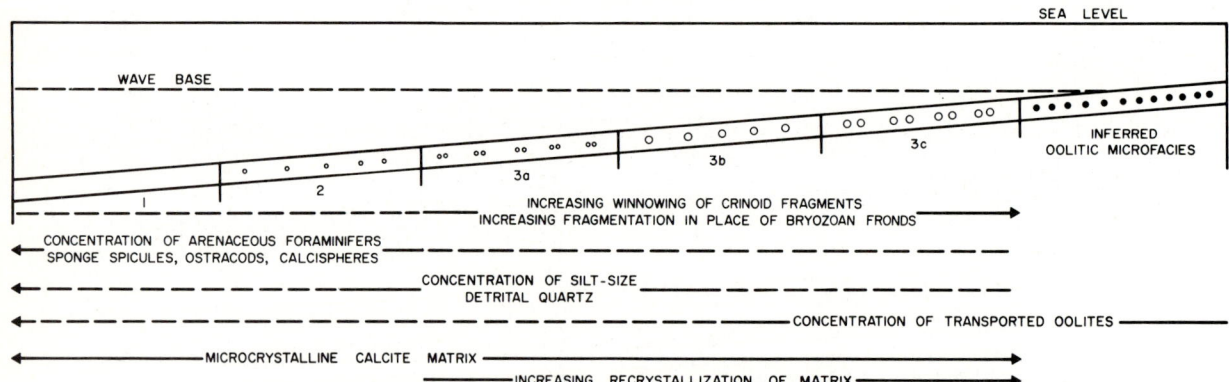

FIGURE 4.33 Menard Formation (Middle Mississippian), southwest margin of Illinois Basin. Ideal depositional model. From Carozzi and Roche (1968). Reprinted by permission of the Illinois State Academy of Science.

reciprocal behavior of microscopic parameters during several depositional sequences.

The Ideal Depositional Model

The ideal depositional model (Fig. 4.33) shows the original ecological conditions and the effects of submarine currents which have modified the crinoid-bryozoan community, concentrated silt-size detrital quartz in microfacies 3a and a variety of small-size constituents in quiet deeper waters, and distributed ooids from an inferred landward facies belt across the entire carbonate system.

Relationship Between Degree of Neomorphism and Environmental Energy Level

This relationship (Carozzi, 1971a), which ranges from 20 to 100% development of pseudomicrosparite and pseudosparite at the expense of the original calcisiltite matrix (Fig. 4.34), appears to characterize only the calcarenites

overlain by channel sandstones (see Fig. 4.28). It is suggested that the brackish or freshwater environment corresponding to the channel sandstones was responsible for the early change of the original calcisiltite matrix of the carbonate substratum into a neomorphic calcite cement. These fine-grained sandstones range from mature pure quartz arenites with calcite-hematite cement to calcite-cemented feldspathic lithic arenites with plant debris. They are forerunners of the Pennsylvanian coal-bearing cyclothems.

4.4 CHARACTERISTIC FEATURES: EFFECTS OF DETRITAL QUARTZ SEDIMENT YIELD

Some carbonate ramps were juxtaposed to siliciclastic environments, and hence changed temporarily to mixed carbonate-clastic depositional models. Under such circumstances, detrital quartz sediment yield can cause profound effects on carbonate sedimentation.

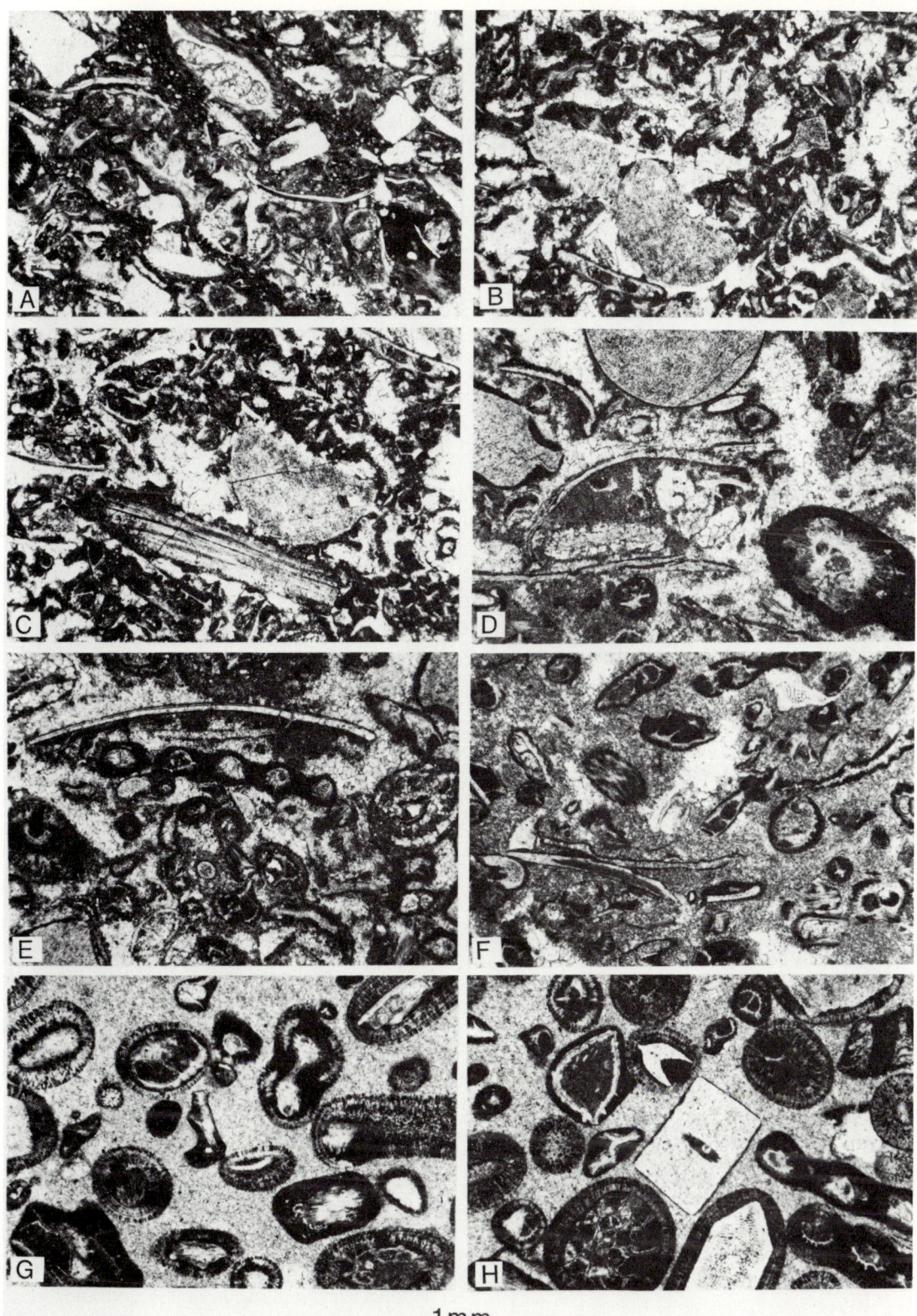

FIGURE 4.34 Menard Formation (Middle Mississippian), southwest margin of Illinois Basin. Gradual increase of neomorphic pseudomicrosparite and sparite from microfacies 3b (A to D) to microfacies 3c (E to H). From Carozzi (1971). Reprinted by permission of the Johns Hopkins Press.

A stratigraphic interval of 1846 ft (553.8 m) belonging to the Lower and Middle Bird Spring Group (Middle Morrowan to Lower Missourian) of the Arrow Canyon Range, Clark County, southeast Nevada (Heath *et al.*, 1967), was investigated petrographically by means of more than 2095 thin sections from samples collected at an average vertical interval of 10 in. (25.4 cm).

Two shallowing-upward sequences have been recognized. The first one characterizes the Morrowan and displays normal and quartz-rich carbonates; the second one covers the Atokan–Desmoinesian–Lower Missourian interval and consists again of pure carbonates.

Description of Microfacies

Microfacies within each group are described in order of decreasing relative depth.

Normal Microfacies Group

Microfacies 0 (Fig. 4.35A). Calcisiltite with less than 3% scattered minute bioclasts of crinoids and brachiopods, and calcified monaxonic sponge spicules.

Microfacies 1 (Fig. 4.35B). Bioclastic, pelletoidal calcisiltite with 3 to 30% scattered sand-size bioclasts among which crinoids, brachiopods, and bryozoans predominate over ostracods, pelecypods, and arenaceous benthic foraminifers.

Microfacies 2 (Fig. 4.35C). Bioturbated pelletoidal calcisiltite with abundant scattered sand-size bioclasts (30 to 40%) among which crinoids, brachiopods, encrusting bryozoans, and ostracods predominate.

Microfacies 3 (Fig. 4.35D,E,F). Grain-supported biocalcarenite with pelletoidal calcisiltite matrix. The larger bioclasts are crinoids, brachiopods, and bryozoans. Brachiopod spines and trilobites are frequent, and pressure-solution between all components is common. This microfacies shows the first appearance of oncoids, irregularly shaped and frequently developed around nuclei of gastropod shells.

Microfacies 4 (Figs. 4.35G,H, 4.36A). Grain-supported pelletoidal biocalcarenite with sparite cement. The well-rounded bioclasts consist predominantly of crinoids, brachiopod shells and spines, and bryozoans. The majority of the pellets are lithic in origin and may grade locally into intraclasts containing small bioclasts. The cavity-filling cement consists of sparite mosaic and syntaxial overgrowths around crinoidal bioclasts. Locally, well-developed oncoids are present with cores consisting of bioclasts of gastropods, crinoids, and brachiopods.

Microfacies 5 (Fig. 4.36B,C). Grain-supported oolitic biocalcarenite with sparite cement. Normal, superficial, and half-moon ooids are associated, but the latter show random geopetal orientation indicating reworking. The cores of the ooids are identical to the noncoated bioclasts which are predominantly crinoids, brachiopod shells and spines, bryozoans, ostracods, and arenaceous benthic foraminifers. Cavity-filling sparite cement predominates over local patches of calcisiltite matrix changed by neomorphism into pseudomicrosparite.

Quartz-Rich Microfacies Group

Microfacies 0a (Fig. 4.36D). Bioturbated quartz-rich calcisiltite to calcite-cemented quartz siltstone. The quartz grains are angular and devoid of overgrowths; the patches of calcisiltite matrix are often pelletoidal and pyrite rich. The rare small bioclasts are crinoids, brachiopod shells, and ostracods.

Microfacies 1a (Fig. 4.36E). Bioturbated quartz-rich pelletoidal calcisiltite with scattered sand-size bioclasts. Among the latter, brachiopods predominate over crinoids, bryozoans, and ostracods.

Microfacies 2a (Fig. 4.36F). Bioturbated quartz-rich pelletoidal calcisiltite with abundant scattered sand-size bioclasts. The latter consist mainly of crinoids, brachiopod spines, arenaceous benthic foraminifers, and ostracods.

Microfacies 3a (Fig. 4.36G). Bioturbated quartz-rich, grain-supported biocalcarenite with calcisiltite matrix. The bioclasts are limited to crinoids and brachiopod shells. Lithic pellets appear more abundant than fecal ones.

Microfacies 4a (Fig. 4.36H). Quartz-rich, grain-supported biocalcarenite with sparite cement. The bioclasts consist of echinoid spines, crinoid columnals, brachiopod shells and spines, and bryozoans. An appreciable portion of the bioclasts was altered into pseudosparite by neomorphism. Quartz grains are much coarser than in any other microfacies and subangular to rounded.

The Ideal Shallowing-Upward Sequences of the Morrowan

For the normal microfacies group (Fig. 4.37), the microscopic parameter variations appear as follows: microfacies 0: very quiet conditions with concentration of very fine quartz grains and a rarity of bioclasts; microfacies 1: quiet conditions with largest frequency of ostracods and fecal pellets; microfacies 2: weakly agitated conditions with largest frequency of crinoids and brachiopods together with development of lithic pellets; microfacies 3: moderately agitated conditions with the greatest variety of bioclasts and maximum frequency of arenaceous ben-

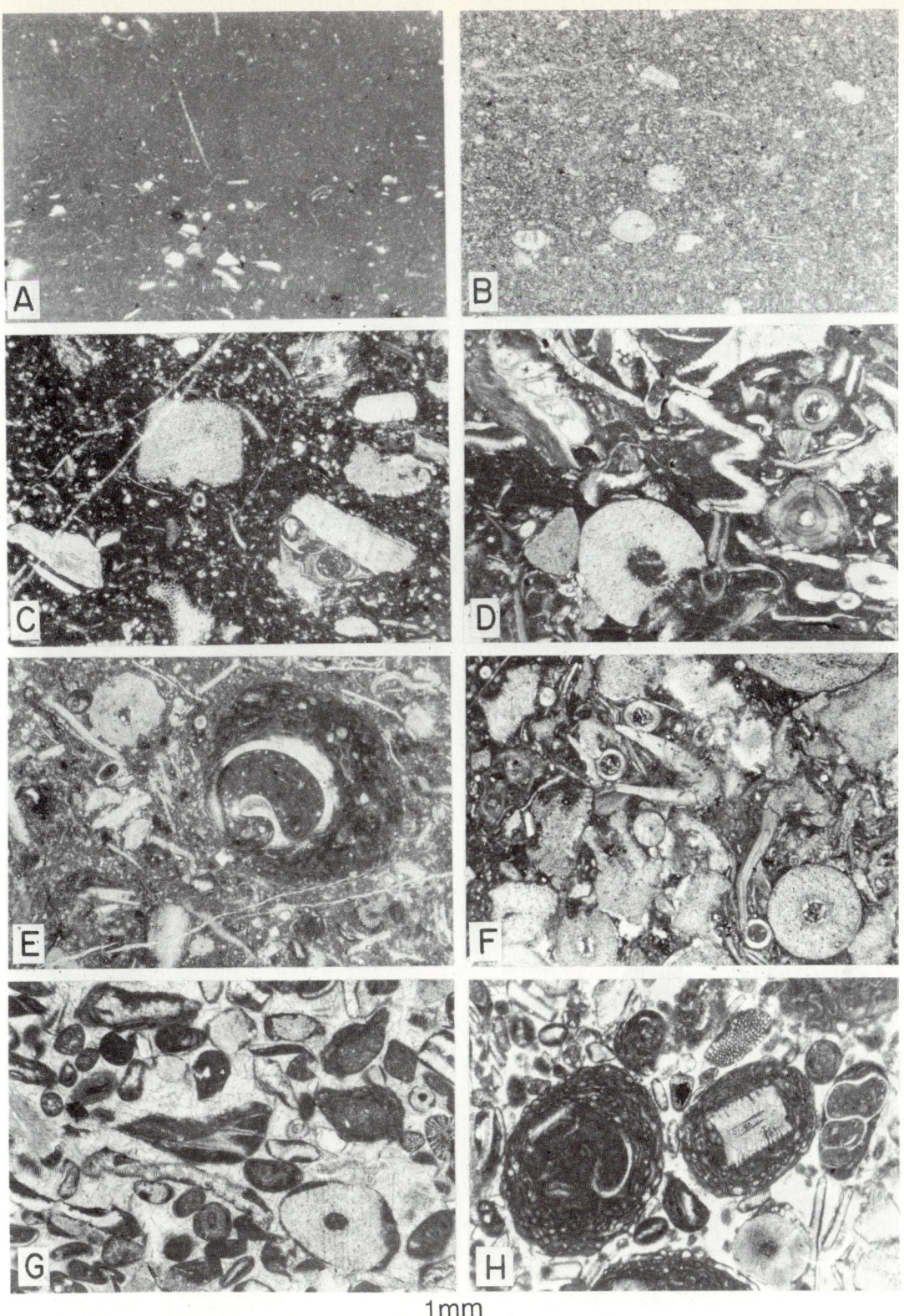

FIGURE 4.35 Bird Spring Group (Middle Morrowan to Lower Missourian), Arrow Canyon Range, Clark County, Nevada. Typical microfacies. A. Microfacies 0. B. Microfacies 1. C. Microfacies 2. D to F. Microfacies 3. G and H. Microfacies 4. See text for detailed descriptions. All photomicrographs: plane-polarized light. From Heath *et al.* (1967). Reprinted by permission of the Society of Economic Paleontologists and Mineralogists.

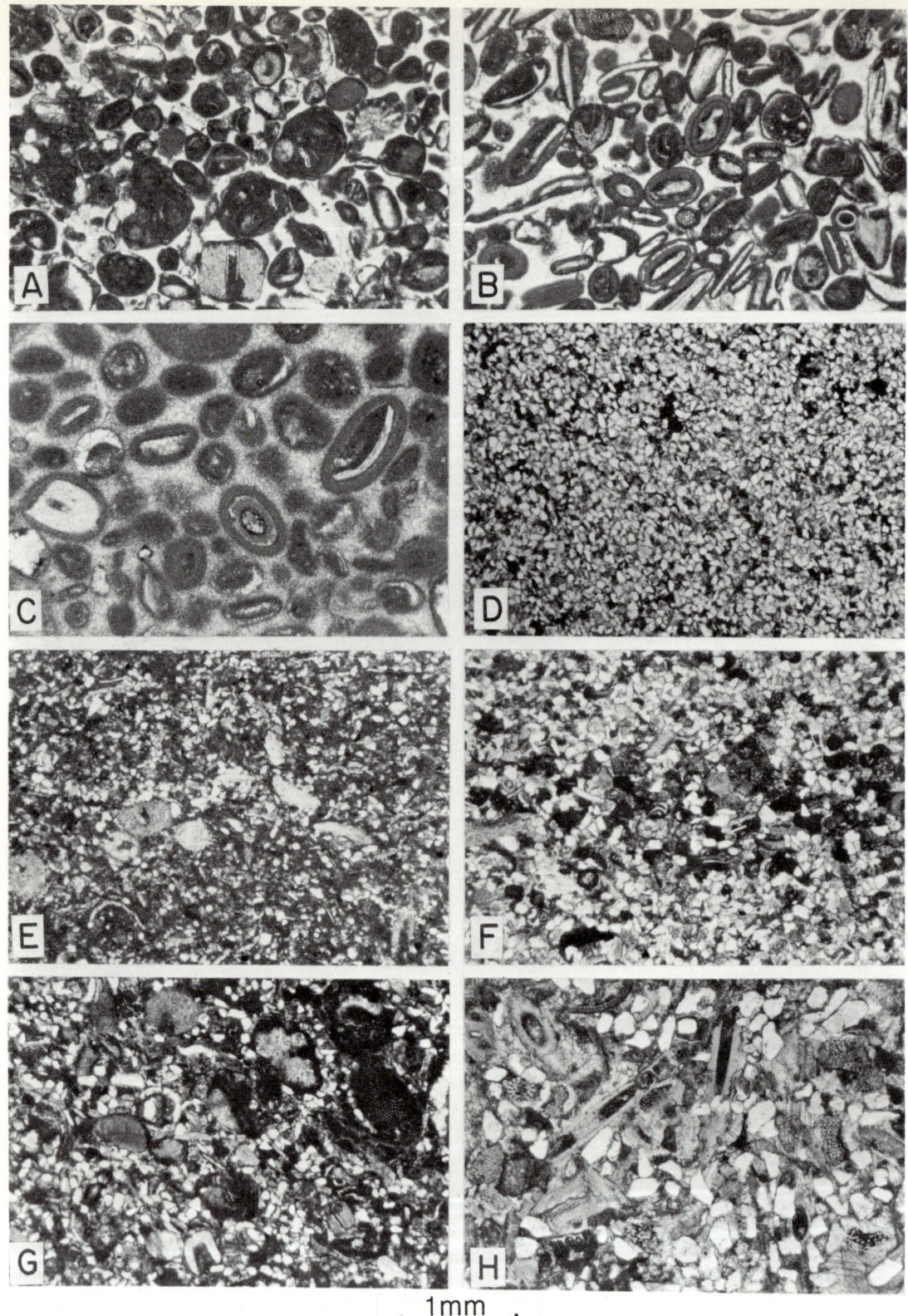

FIGURE 4.36 Bird Spring Group (Middle Morrowan to Lower Missourian), Arrow Canyon Range, Clark County, Nevada. Typical microfacies (continued). A. Microfacies 4. B and C. Microfacies 5. D. Microfacies 0a. E. Microfacies 1a. F. Microfacies 2a. G. Microfacies 3a. H. Microfacies 4a. See text for detailed descriptions. All photomicrographs: plane-polarized light. From Heath et al. (1967). Reprinted by permission of the Society of Economic Paleontologists and Mineralogists.

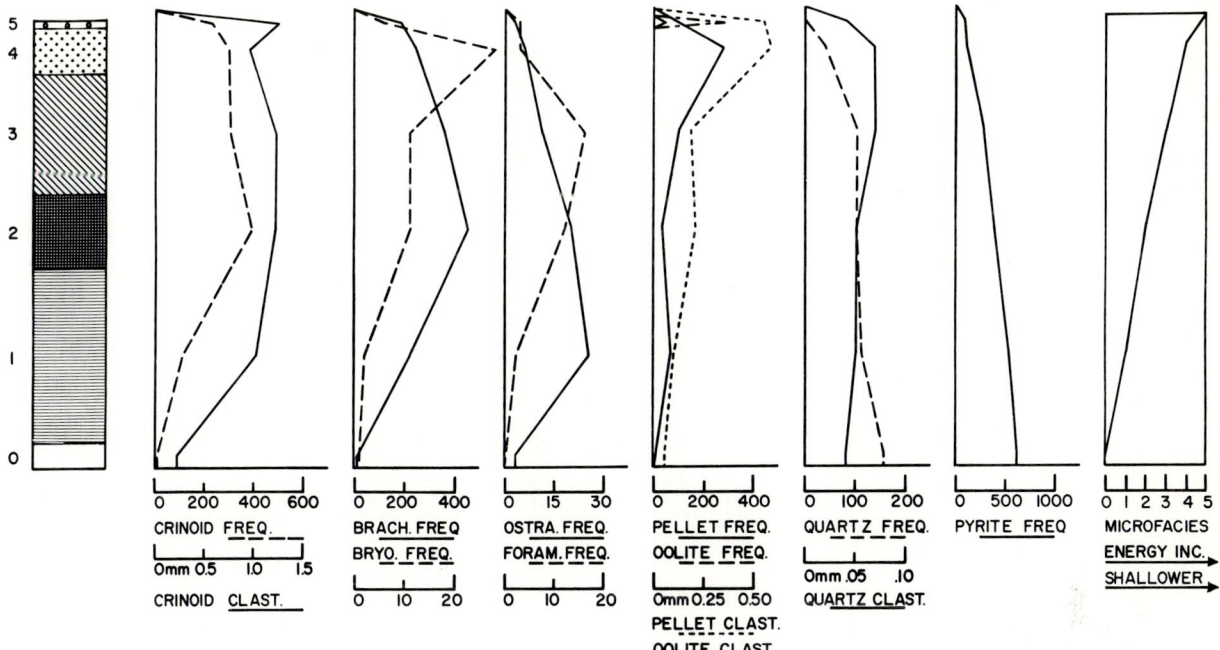

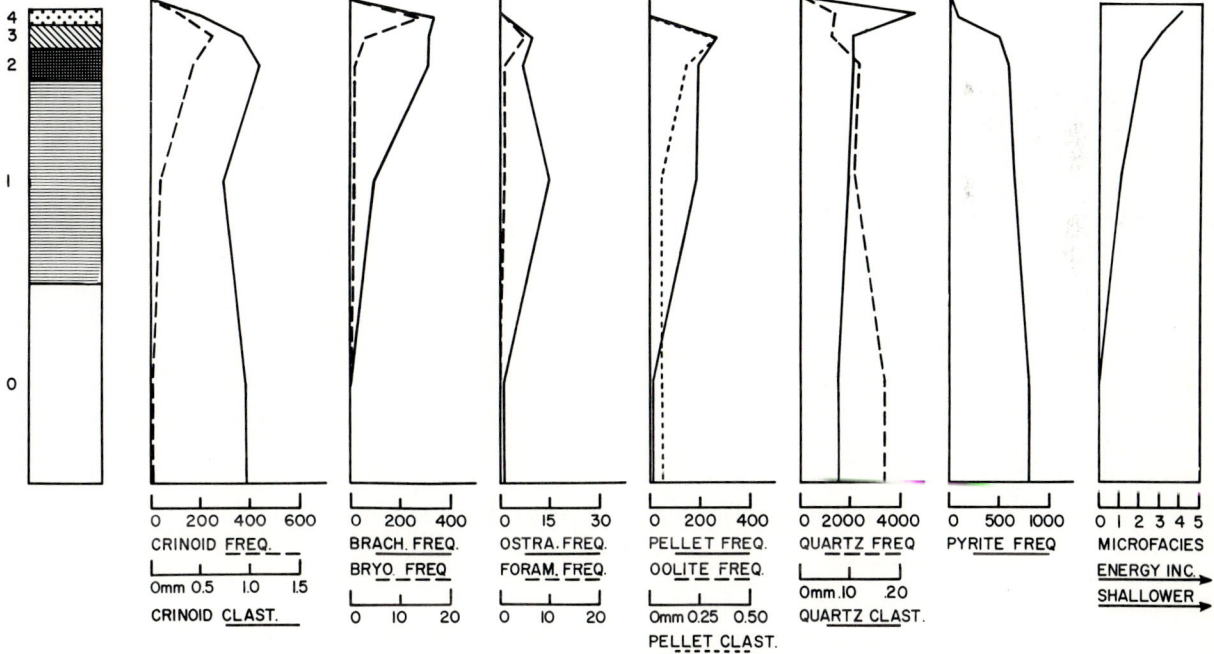

FIGURE 4.37 Bird Spring Group (Middle Morrowan to Lower Missourian), Arrow Canyon Range, Clark County, Nevada. Ideal shallowing-upward sequences of the Morrowan. From Heath *et al.* (1967). Reprinted by permission of the Society of Economic Paleontologists and Mineralogists.

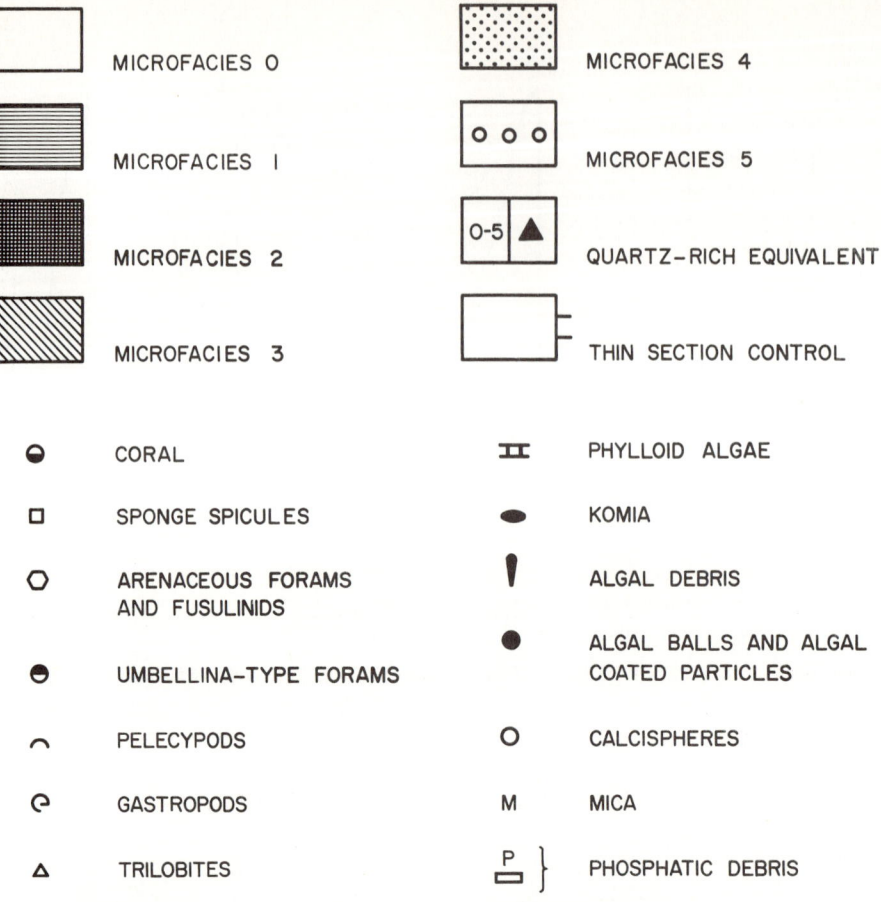

FIGURE 4.38 Bird Spring Group (Middle Morrowan to Lower Missourian), Arrow Canyon Range, Clark County, Nevada. Symbols for microfacies.

thic foraminifers; microfacies 4: agitated conditions with largest frequency of bryozoans and lithic pellets; microfacies 5: well-agitated conditions, maximum size of crinoids and oolitization.

For the quartz-rich microfacies group (Fig. 4.37) the abundance of detrital quartz reduced the space available for bioclasts and created a well-marked destructive action on them. The general effect was to shift some of the frequency peaks in shallower conditions and to eliminate the oolitic environment. A typical section (Figs. 4.38, 4.39) shows stratigraphic variation of microscopic parameters.

The Ideal Shallowing-Upward Sequence of the Atokan–Desmoinesian–Lower Missourian

Microscopic parameter variation appears as follows (Fig. 4.40) in the absence of microfacies 0, microfacies 1: quiet conditions with peaks of the frequency of detrital quartz and pyrite; microfacies 2: weakly agitated conditions with frequency peak of ostracod valves and predominance of fecal pellets; microfacies 3: moderately agitated conditions with frequency peaks of brachiopods, bryozoans, arenaceous benthic foraminifers, and calcispheres as well as maximum clasticity of lithic pellets: microfacies 4: agitated conditions with frequency peaks of crinoids and complete ostracods, maximum clasticity of detrital quartz; microfacies 5: well-agitated conditions with predominance of ooids.

The Ideal Depositional Model

Reciprocal relations of microfacies and behavior of the various components (Fig. 4.41) indicate a typical carbonate ramp with wave base located in microfacies 3 and 4 corresponding to the gradual disappearance of the calcisiltite matrix and its replacement by cavity-filling sparite.

A total of 78 complete oscillations were recognized in the interval investigated. The Morrowan with 46 oscillations (Fig. 4.42) corresponds to a gradual deepening with a massive influx of detrital quartz coinciding with

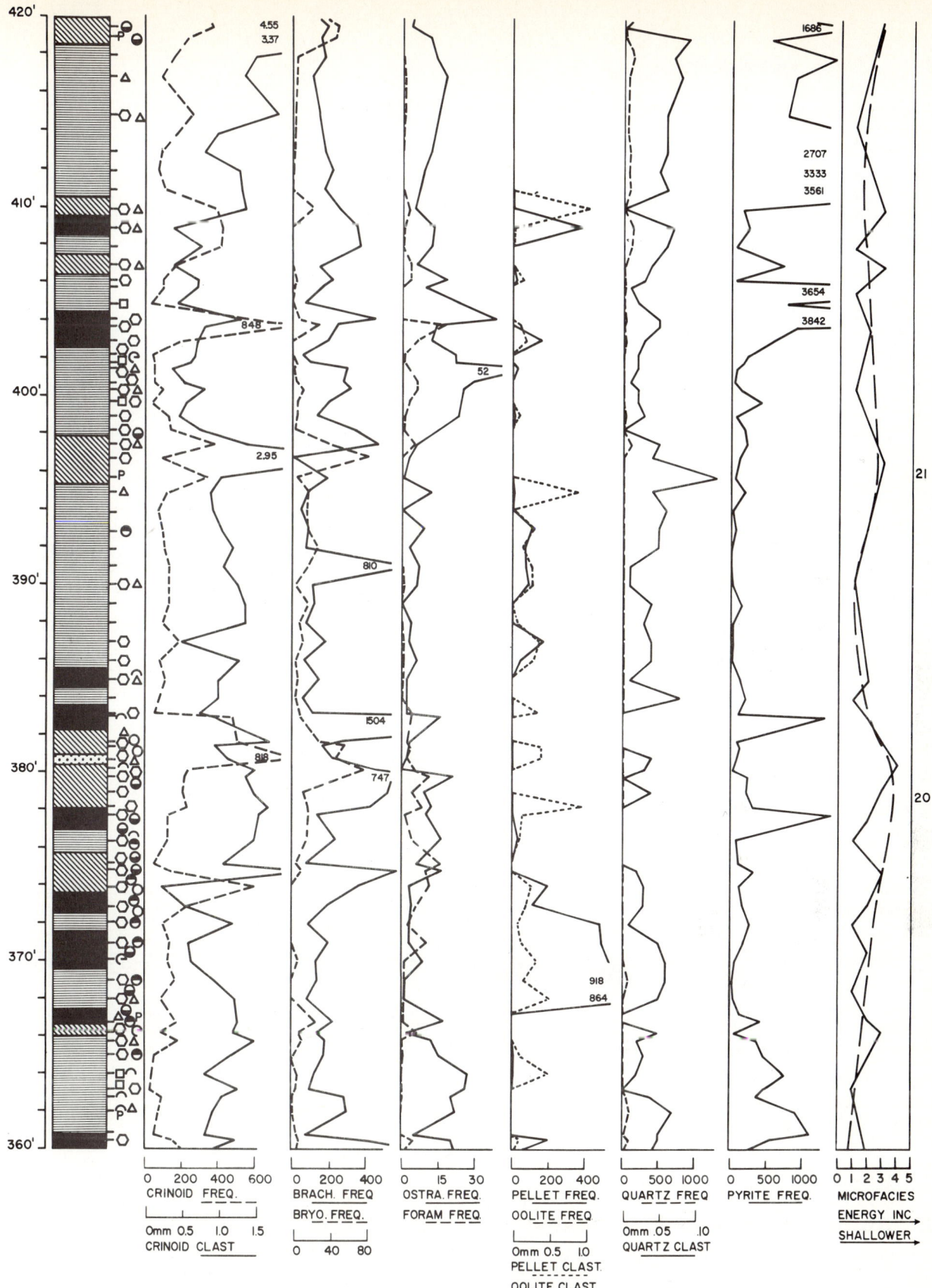

FIGURE 4.39 Bird Spring Group (Middle Morrowan to Lower Missourian), Arrow Canyon Range, Clark County, Nevada. Typical example of vertical variation of microscopic parameters for a portion of the Morrowan. From Heath *et al.* (1967). Reprinted by permission of the Society of Economic Paleontologists and Mineralogists.

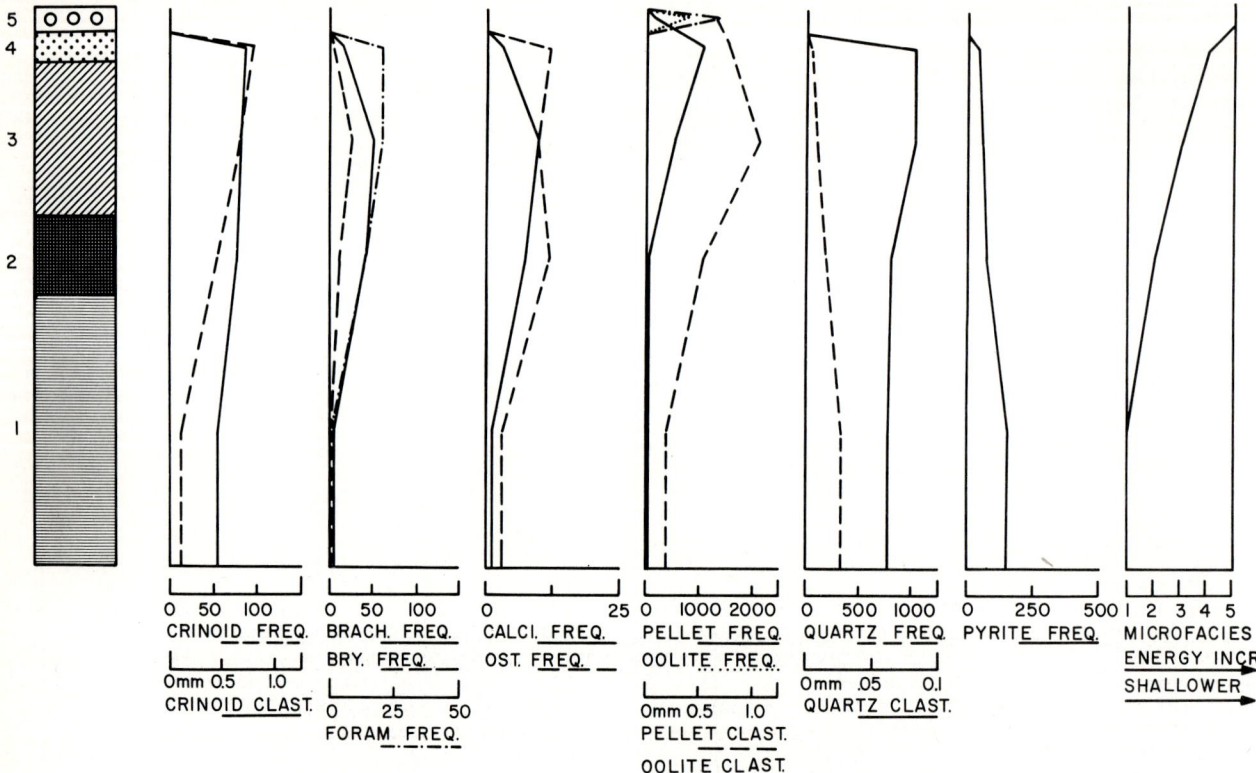

FIGURE 4.40 Bird Spring Group (Middle Morrowan to Lower Missourian), Arrow Canyon Range, Clark County, Nevada. Ideal shallowing-upward sequence of the Atokan–Desmoinesian–Lower Missourian. From Heath *et al.* (1967). Reprinted by permission of the Society of Economic Paleontologists and Mineralogists.

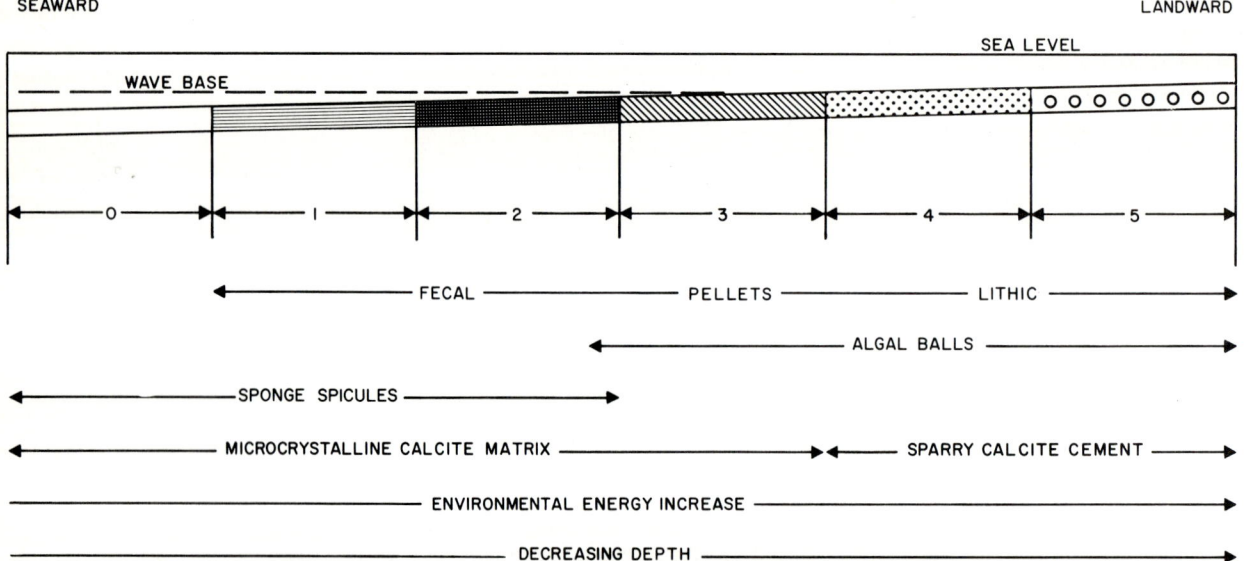

FIGURE 4.41 Bird Spring Group (Middle Morrowan to Lower Missourian), Arrow Canyon Range, Clark County, Nevada. Ideal depositional model. From Heath *et al.* (1967). Reprinted by permission of the Society of Economic Paleontologists and Mineralogists.

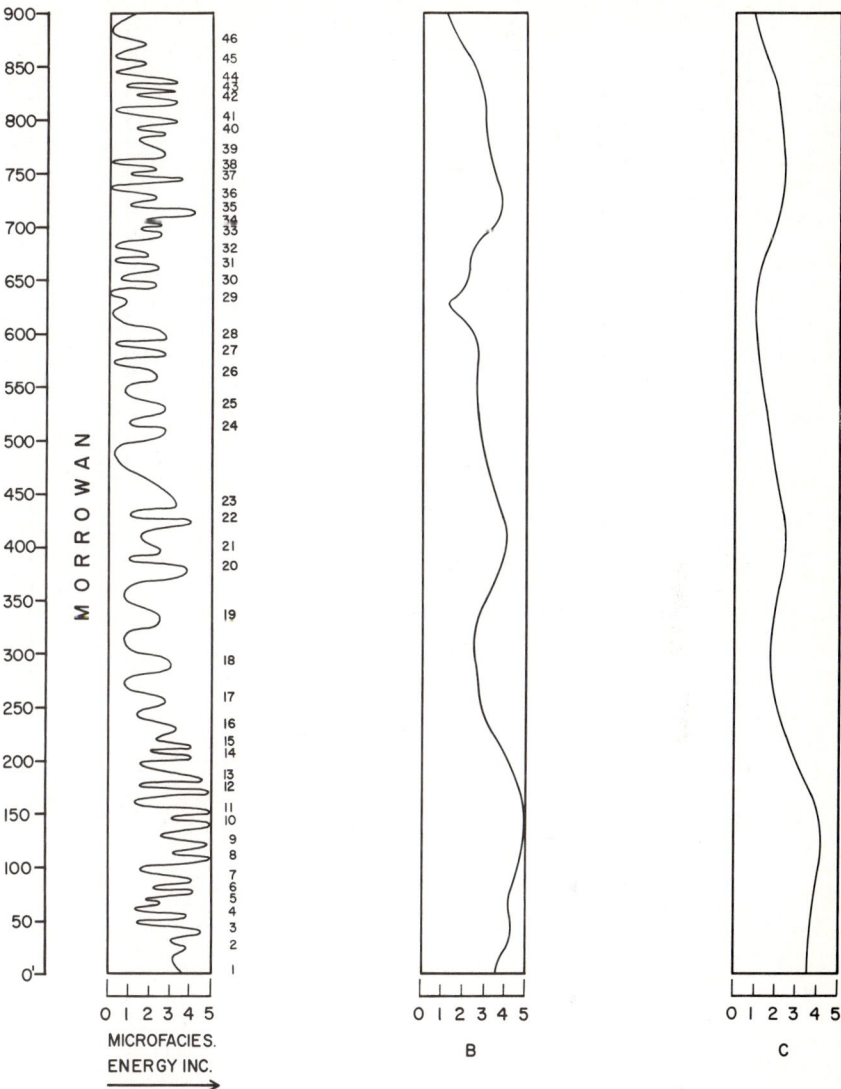

FIGURE 4.42 Bird Spring Group (Middle Morrowan to Lower Missourian), Arrow Canyon Range, Clark County, Nevada. Generalized relative bathymetric curves of the Morrowan. From Heath *et al.* (1967). Reprinted by permission of the Society of Economic Paleontologists and Mineralogists.

its maximum. The rest of the sequence (Fig. 4.43) consists of 32 oscillations, indicating a gradual shallowing. It is interesting to note that in the same time interval the corresponding sequence (Abbott, Spoon, Carbondale, and Modesto Formations) in the Illinois Basin contains 33 named cyclothems. This similarity in environmental oscillations indicates that the cyclic sedimentation of both regions may have been controlled by the same set of factors, namely, glacio-eustatic changes of sea level (Lumsden, 1965).

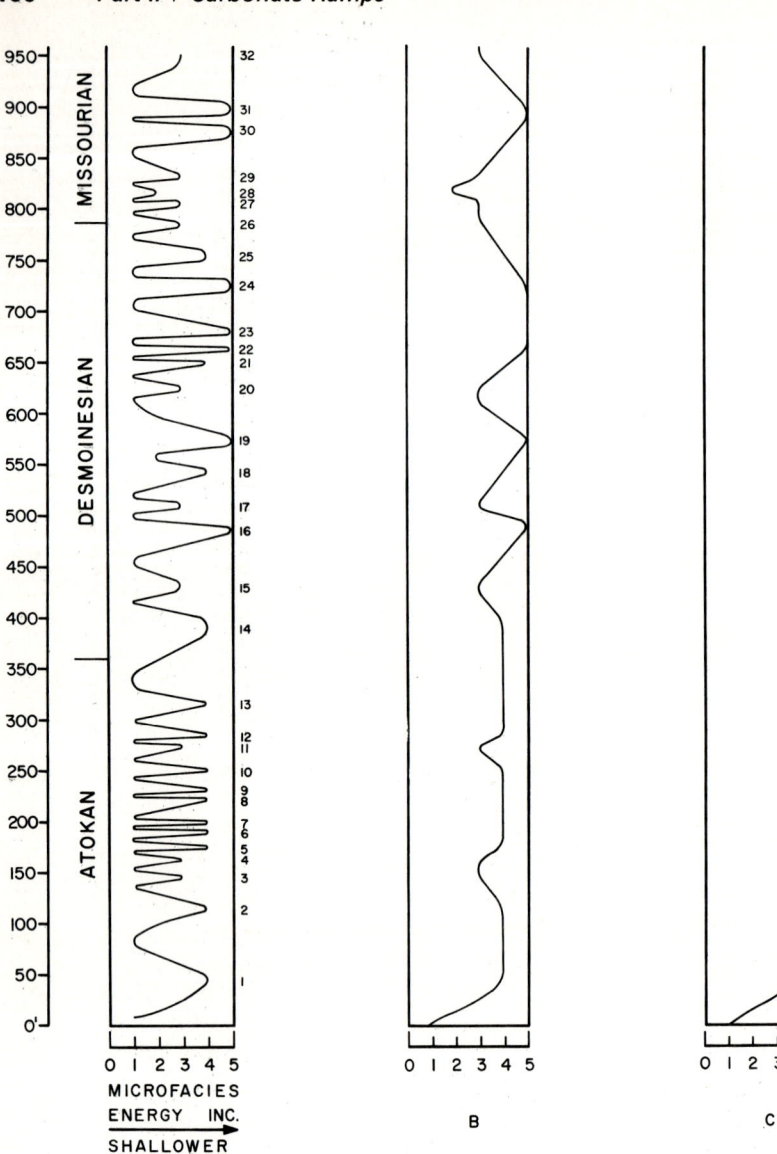

FIGURE 4.43 Bird Spring Group (Middle Morrowan to Lower Missourian), Arrow Canyon Range, Clark County, Nevada. Generalized relative bathymetric curves for the Atokan–Desmoinesian–Lower Missourian interval. From Heath *et al.* (1967). Reprinted by permission of the Society of Economic Paleontologists and Mineralogists.

5

Carbonate Ramps in Semirestricted Marine Environment

Semirestricted conditions tend to generate ramps characterized by very long geologic duration, slow shallowing-upward processes, extensive lesser-energy microfacies, and unusual but diversified ecological conditions.

5.1 CHARACTERISTIC CONSTITUENTS: AGGLUTINATED BENTHIC FORAMINIFERS

The Itaituba and Nova Olinda Formations (Carboniferous-Permian) in the Amazon Basin, northern Brazil (Carozzi et al., 1974) are 1700 m thick and were studied in cores from 65 exploratory wells distributed over 750,000 km^2. A total of 2500 thin sections representing the major megascopically recognizable carbonate facies were investigated petrographically.

During this time interval, the Amazon Basin was an enormous and relatively shallow embayment with a narrow connection to the open sea in a west to southwest direction. Consequently, carbonate sedimentation, although influenced by tides of moderate amplitude, consists mainly of extremely widespread associations of relatively low energy microfacies. Among them, and particularly in the intertidal environment, are irregularly scattered concentrations of medium- to high-energy textural types represented by oolitic and intraclastic calcarenites with sparite cement. The latter appear to have formed large shoals and bars shifting probably under the action of wind-generated currents.

If the distribution of carbonate microfacies is relatively uniform, the organic constituents, on the other hand, mostly *in situ*, are abundant, diversified, and extremely sensitive to changes in bathymetry and salinity. They allow a classification of microfacies in four distinct environments: infratidal, low intertidal, high intertidal, and supratidal.

Description of Microfacies

The different microfacies are described within each environment in order of decreasing importance.

Infratidal Environment

Microfacies 1 (Fig. 5.1A). Grain-supported biocalcarenite with abundant bioclastic matrix. Predominant bioclasts are brachiopods, crinoids, bryozoans, echinoids, followed by calcareous benthic foraminifers, trilobites, and thick pelecypods. The bioclastic matrix consists of smaller debris of the above-mentioned organisms. Elongated bioclasts are oriented parallel to bedding due to compaction associated with frequent pressure-solution.

Microfacies 2 (Fig. 5.1B). Grain-supported biocalcarenite with calcisiltite matrix. Major bioclasts

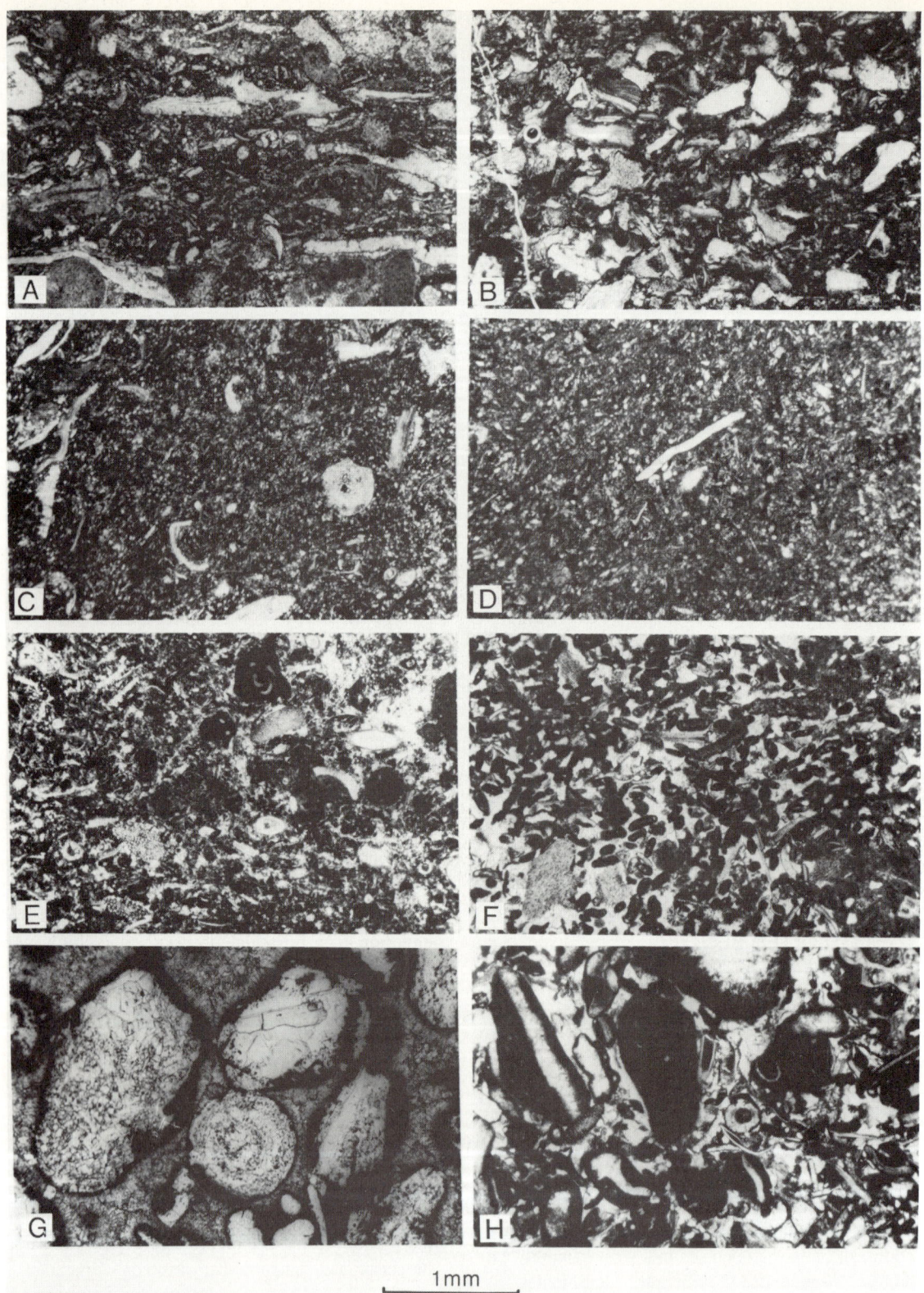

FIGURE 5.1 Itaituba–Nova Olinda Formations (Carboniferous-Permian), Amazon Basin, northern Brazil. Typical microfacies. A. Microfacies 1. B. Microfacies 2. C. Microfacies 3. D. Microfacies 4. E. Microfacies 5. F. Microfacies 6. G. Microfacies 7. H. Microfacies 8. See text for detailed descriptions. All photomicrographs: plane-polarized light. From Carozzi et al. (1974).

include brachiopods, crinoids, thick pelecypods, and ostracods. Bioturbation and strong pressure-solution of randomly oriented bioclasts are widespread.

Microfacies 3 (Fig. 5.1C). Biocalcisiltite with abundant sand-size bioclasts of brachiopods, crinoids, pelecypods, ostracods, trilobites, echinoids, calcareous benthic foraminifers, and bryozoans. Very fine detrital quartz and phosphatic debris are rare.

Microfacies 4 (Fig. 5.1D). Biocalcisiltite with rare scattered sand-size bioclasts of brachiopods, echinoids, ostracods, calcareous and agglutinated benthic foraminifers, crinoids, and monaxonic sponge spicules.

Low Intertidal Environment

Microfacies 5 (Fig. 5.1E). Grain-supported biocalcarenite with calcisiltite matrix. Bioclasts consist mainly of agglutinated benthic foraminifers, brachiopods, and echinoids. Plant remains and detrital quartz are rare. Bioturbation is moderate.

Microfacies 6 (Fig. 5.1F). Grain-supported biocalcarenite with association of cavity-filling sparite cement and calcisiltite matrix. The bioclasts consist mainly of agglutinated benthic foraminifers with frequent brachiopods and echinoids. Sorting of the bioclasts is moderate.

Microfacies 7 (Fig. 5.1G). Grain-supported intraclastic calcarenite with cavity-filling sparite cement. The intraclasts consist of reworked calcarenites and calcisiltites with agglutinated benthic foraminifers and brachiopods. They are replaced often by coarse crystalline anhydrite, leaving dark unreplaced rims.

Microfacies 8 (Fig. 5.1H). Grain-supported oncoidal calcarenite. The oncoids display envelopes consisting of blue-green algal mats alternating with encrusting foraminifers developed around cores which are either bioclasts or intraclasts. Other constituents are bioclasts of agglutinated benthic foraminifers, brachiopods, echinoids, and bryozoans. Interstitial material is an association of calcisiltite matrix and cavity-filling sparite.

High Intertidal Environment

Microfacies 9 (Fig. 5.2A). Grain-supported biocalcarenite with calcisiltite matrix often altered into patches of pseudosparite by neomorphism. Main bioclasts are agglutinated benthic foraminifers, abundant ostracods associated with thin pelecypods, and calcispheres. Secondary dolomitization and anhydritization are widespread.

Microfacies 10 (Fig. 5.2B). Pelecypod-bioaccumulated limestone with calcisiltite matrix containing ostracods and agglutinated benthic foraminifers. Pelecypod tests are entirely altered into pseudosparite by neomorphism.

Microfacies 11 (Fig. 5.2C). Spiculitic calcisiltite with irregular concentrations of monaxonic sponge spicules due to bioturbation and associated agglutinated benthic foraminifers and ostracods. Secondary dolomitization is common.

Microfacies 12 (Fig. 5.2D). Grain-supported pelletoidal calcarenite with associated calcisiltite matrix and cavity-filling sparite cement. The fecal pellets together with gastropods frequently appear deformed and merged by early compaction. Secondary dolomitization is widespread.

Microfacies 13 (Fig. 5.2E). Calcisiltite with abundant ostracods and rare agglutinated benthic foraminifers and calcispheres. Plant debris aligned parallel to bedding are common. Bioturbation is abundant. Secondary dolomitization and patchy anhydritization are widespread.

Microfacies 14 (Fig. 5.2F). Banded bituminous calcilutite interbedded with biocalcisiltite rich in echinoids, agglutinated benthic foraminifers, and ostracods. Frequent fluidal texture is due to compaction.

Microfacies 15 (Fig. 5.2G). Arenaceous and heavily bioturbated calcisiltite with scattered debris of ostracods and agglutinated benthic foraminifers. Burrows are commonly filled by darker micrite and differential dolomitization of the calcisiltitic matrix is widespread.

Microfacies 16 (Fig. 5.2H). Grain-supported oolitic calcarenite with cavity-filling sparite cement. Concentric rings of ooids are very fine and dark, nuclei consist of crinoidal columnals and agglutinated benthic foraminifers.

Supratidal Environment

Microfacies 17 (Fig. 5.3A). Stromatolite-constructed limestone consisting of blue-green algal mats, often pelleted by desiccation and trapping large grains of angular detrital quartz. Intermat thin layers of fecal pelletoidal calcarenite with cavity-filling sparite cement. Secondary dolomitization and anhydritization are frequent.

Microfacies 18 (Fig. 5.3B). Intraclastic desiccation calcirudite with angular clasts of calcilutite occasionally coated with iron oxides. The original interstitial cement of sparite is often replaced by secondary dolomite and megaquartz mosaic.

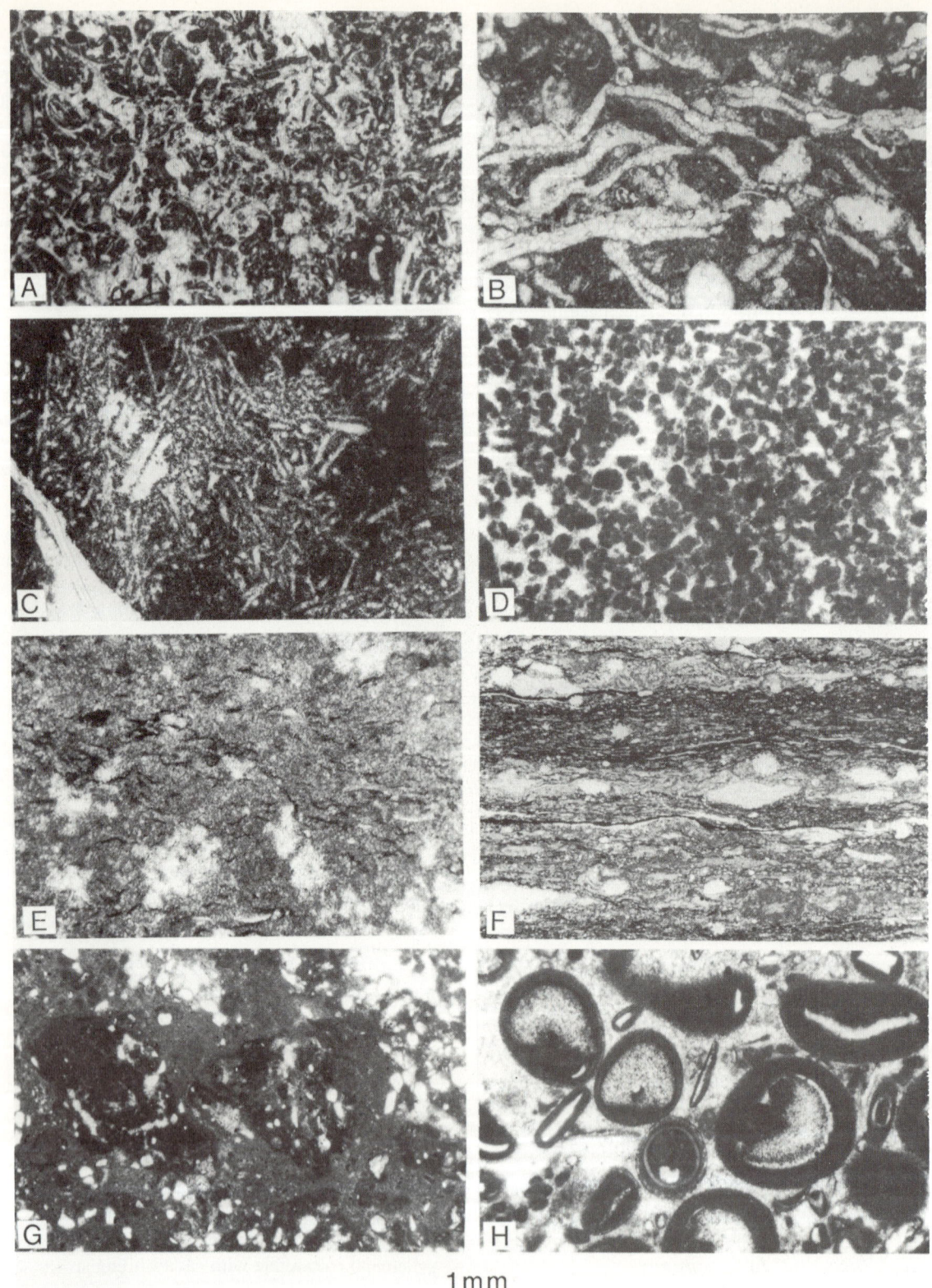

FIGURE 5.2 Itaituba–Nova Olinda Formations (Carboniferous-Permian), Amazon Basin, northern Brazil. Typical microfacies (continued). A. Microfacies 9. B. Microfacies 10. C. Microfacies 11. D. Microfacies 12. E. Microfacies 13. F. Microfacies 14. G. Microfacies 15. H. Microfacies 16. See text for detailed descriptions. All photomicrographs: plane-polarized light. From Carozzi *et al.* (1974).

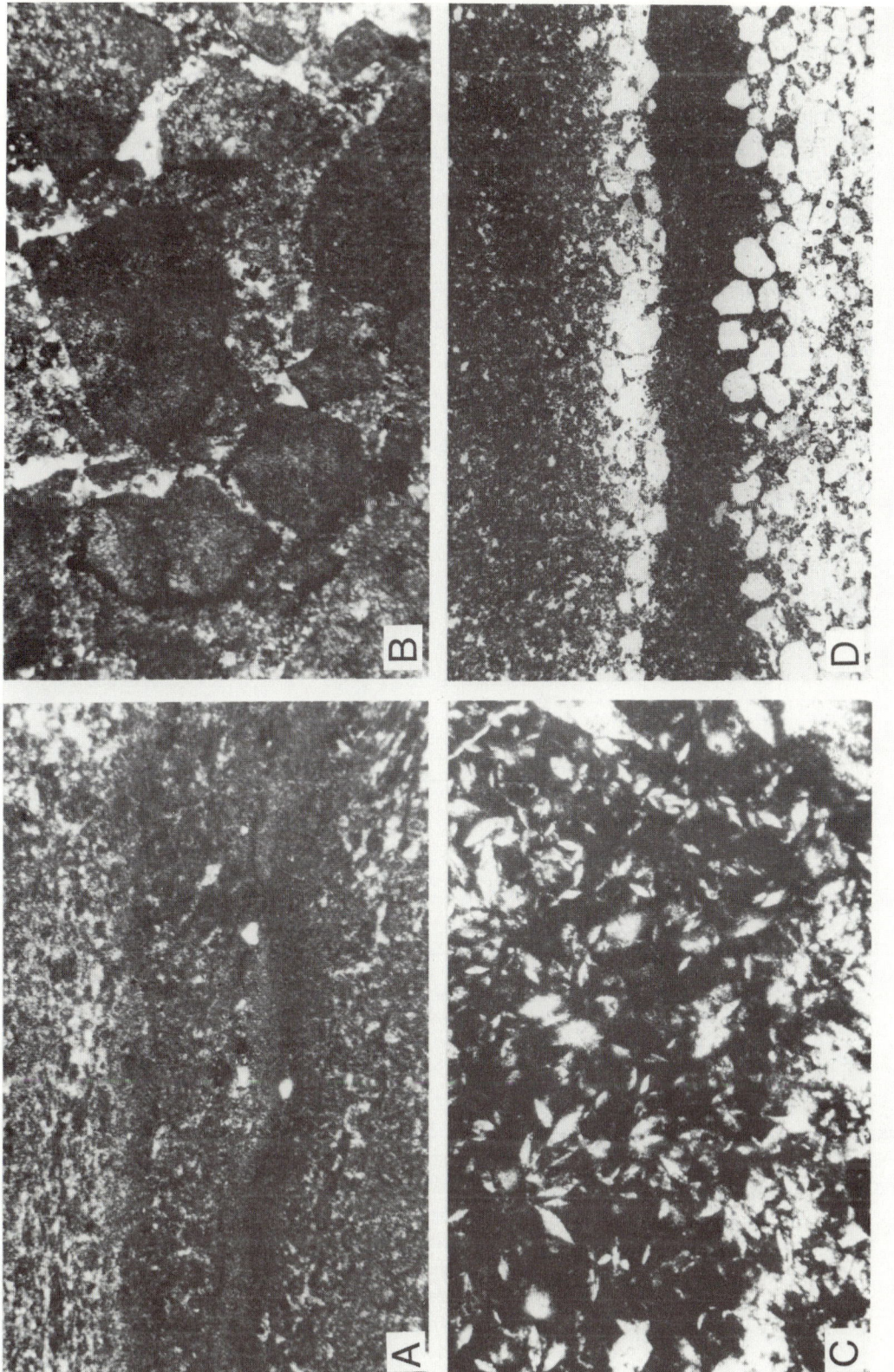

FIGURE 5.3 Itaituba–Nova Olinda Formations (Carboniferous–Permian), Amazon Basin, northern Brazil. Typical microfacies (continued). A. Microfacies 17. B. Microfacies 18. C. Microfacies 19. D. Microfacies 20. See text for detailed descriptions. All photomicrographs: plane-polarized light. From Carozzi *et al.* (1974).

106 *Part II / Carbonate Ramps*

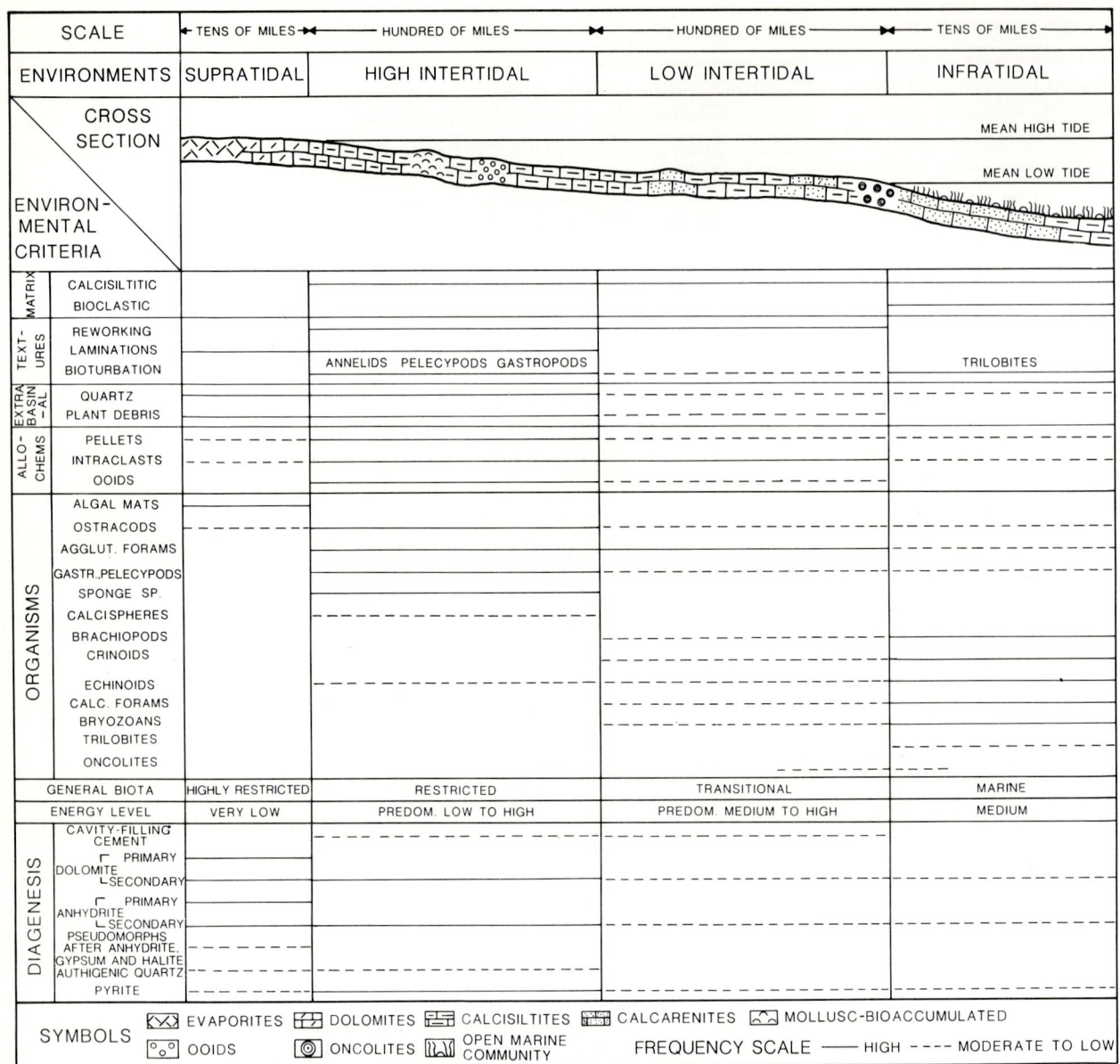

FIGURE 5.4 Itaituba–Nova Olinda Formations (Carboniferous-Permian), Amazon Basin, northern Brazil. Ideal depositional model. From Carozzi *et al.* (1974).

Microfacies 19 (Fig. 5.3C). Bituminous calcisiltite with patches of secondary anhydrite and pseudomorphs of anhydrite after gypsum. Secondary dolomitization is widespread.

Microfacies 20 (Fig. 5.3D). Alternating laminae of dark dolomicrosparite and pure quartz arenite with dolomicrosparite cement corroding quartz grains.

The Ideal Depositional Model

As a consequence of the ubiquitous distribution of many of the microfacies above, the proposed model of carbonate ramp relies heavily on paleoecological criteria (Fig. 5.4). The infratidal normal marine environment is confirmed by a rich brachiopod assemblage (predominance of productids, spiriferids, and rhynchonellids), a characteristic crinoid-echinoid fauna, the presence of fragile, branching forms of bryozoans (*Fenestella*, *Polypora*), and the rich community of calcareous benthic foraminifers (*Millerella*, *Paleotextularia*, *Endothyra*, *Plectogyra*, and *Tetrataxis*). The rarity of fusulinids and of solitary corals confirms that a fully open marine environment did not exist in Carboniferous-Permian times in the Amazon Basin. The low intertidal environment is characterized by a predominance of agglutinated benthic foraminifers associated with frequent brachiopods and echinoderms. The high intertidal environment is typified by an association of agglutinated benthic foraminifers, ostracods, thin pelecypods, gastropods, sponge spicules, and calcispheres. The supratidal environment is represented typically by stromatolitic algal mats, rare ostracods, and interbedding of the carbonates with sabkha-type evaporites. The seepage reflux of the sabkha brines is responsible for the widespread dolomitization and anhydritization of the intertidal microfacies.

6

Carbonate Ramps in Restricted Marine Environment

The main effect of restricted marine conditions on carbonate ramps is generally threefold under increasing energy environments: micritic pelletoidal microfacies with common contamination by influx of detrital quartz and clay minerals, stromatolitic mats and oncoids, and oolitic microfacies.

6.1 CHARACTERISTIC CONSTITUENTS: OOIDS AND ONCOIDS

In the northeast onland portion of the Basin of Espirito Santo, eastern Brazil, the Barra Nova Formation of Albian-Cenomanian age occurs in the subsurface and forms the platform of São Mateus of that particular basin (Carvalho et al., 1982a,b). It was investigated in the northern area of the platform over a surface of 500 km^2 by means of cores from 13 exploratory wells. A total of 344 samples of carbonate rocks were studied petrographically with a vertical spacing of about 50 cm.

The Barra Nova Formation, which is equivalent to the Macaé Formation, of the Campos Basin, offshore Rio de Janeiro, represents the internal quiet lagoonal portion of a widespread embayment protected by an unknown platform located offshore. It is a relatively thin unit which has been divided into the Lower Regencia Member (about 18 m thick), the Middle Regencia Member (about 22 m thick), and the Upper Regencia Member, which is restricted to the southeastern part of the platform and not included in this study. The first two members are shallowing-upward sequences which grade landward into coeval fluviodeltaic siliciclastic sediments of the São Mateus Member.

Description of Microfacies of Lower Regencia Member

Microfacies are described from the distal to the proximal portion of the lagoon.

Distal Lagoon

Microfacies 1 (Fig. 6.1A). Argillaceous and laminated micrite, locally bioturbated with no microfossils except rare plant debris. Minute grains of detrital quartz and biotite flakes are rare, whereas pyrite pigments are abundant. Neomorphism into pseudomicrosparite is extensive and distributed in irregular patches.

Microfacies 2 (Fig. 6.1B). Argillaceous and laminated micrite with ostracods, rare miliolids, and plant debris. Neomorphism into pseudomicrosparite is common, and irregular concentrations of sparite cement fill bioturbation tunnels.

Microfacies 3 (Fig. 6.1C). Stromatolitic micrite with irregularly distributed dark algal mats, slightly

Chap. 6 / *Carbonate Ramps in Restricted Marine Environment* **109**

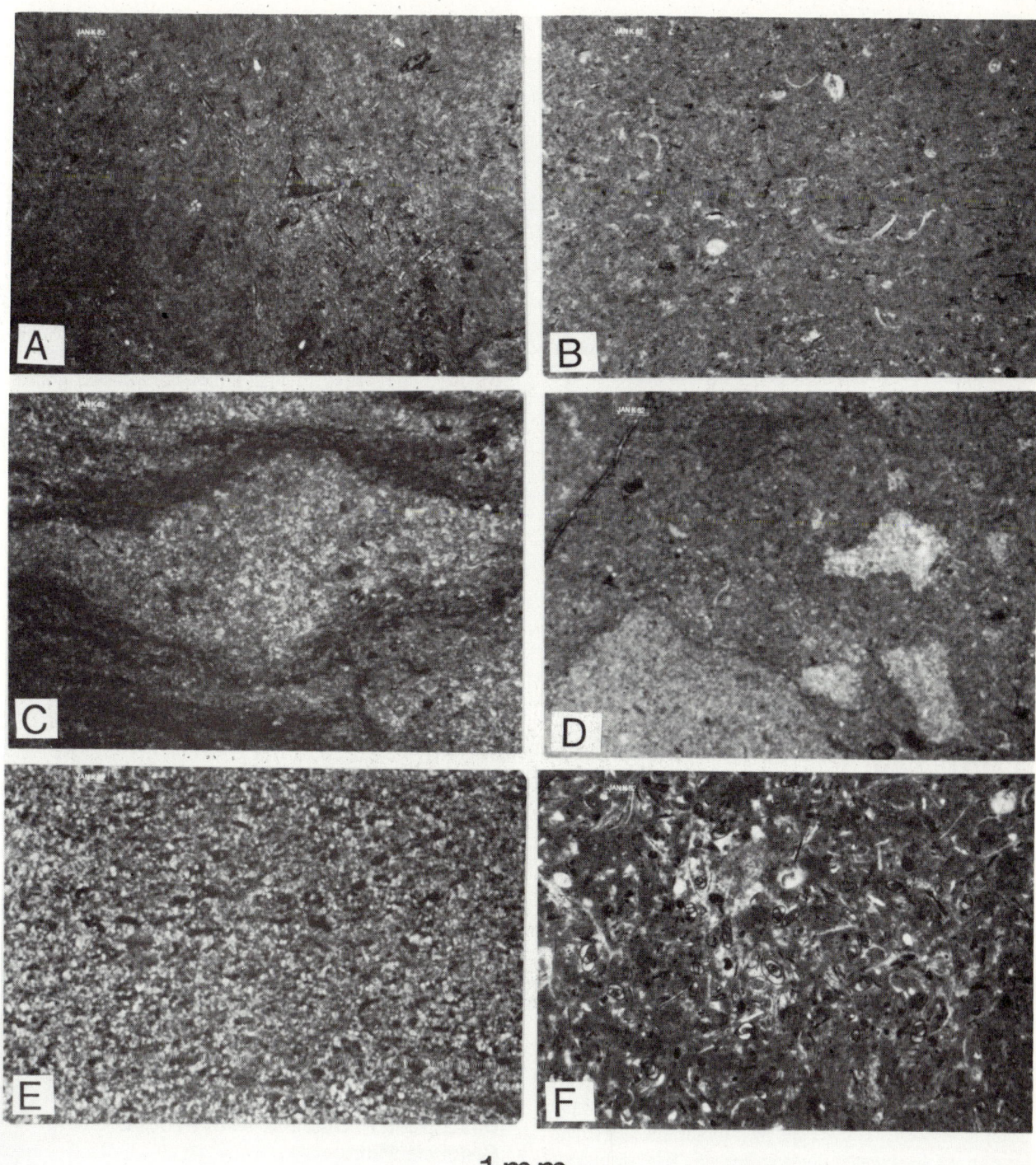

FIGURE 6.1 Lower Regencia Member, Barra Nova Formation (Albian-Cenomanian), Espirito Santo Basin, eastern Brazil. Typical microfacies. A. Microfacies 1. B. Microfacies 2. C. Microfacies 3. D. Microfacies 4. E. Microfacies 5. F. Microfacies 6. See text for detailed descriptions. All photomicrographs: plane-polarized light. From Carvalho *et al.* (1982a). Reprinted by permission of Petróleo Brasileiro S.A. PETROBRÁS.

argillaceous, devoid of bioclasts but showing silt-size grains of detrital quartz, biotite flakes, and pyrite pigments.

Proximal Lagoon

Microfacies 4 (Fig. 6.1D). Intraclastic-pelletoidal micrite in which clasts of lighter pelletoidal micrite are scattered within a similar matrix of dark color. The only bioclasts are rare ostracods.

Microfacies 5 (Fig. 6.1E). Argillaceous, laminated, pelletoidal micrite, locally with a clotted texture. Pellets appear predominantly fecal, although lithic ones can be observed. Bioclasts of pelecypods and plant debris occur scattered with silt-size grains of detrital quartz, biotite flakes, and pyrite pigments. Quartz grains appear marginally replaced by the matrix when the latter is changed into pseudomicrosparite by neomorphism.

Microfacies 6 (Figs. 6.1F, 6.2A). Laminated micrite with abundant ostracods and miliolids associated with plant debris. Rare silt-size grains of detrital quartz are associated with biotite flakes and pyrite pigments. This microfacies is locally argillaceous and bioturbated. Quartz grains are marginally replaced by the matrix when the latter is altered into pseudomicrosparite by neomorphism.

Oolitic Bar

Microfacies 7 (Fig. 6.2B). Grain-supported pelletoidal calcarenite with micrite matrix grading locally into a pelletoidal micrite. Fecal pellets predominate over lithic ones. Bioclasts consist of ostracods, pelecypods, and miliolids associated with reworked ooids. Silt-size grains of detrital quartz and flakes of muscovite and biotite occur together with pyrite pigments.

Microfacies 8 (Fig. 6.2C). Grain-supported oolitic biocalcarenite with micrite matrix. Normal ooids are generally 0.3 mm in size, with rare larger ones reaching 2 mm. Their cores consist of bioclasts of pelecypods and ostracods, grains of detrital quartz, and pellets. Many ooids have been dissolved and their molds filled by single crystals of freshwater phreatic sparite. Rare ostracods and miliolids occur in the interstitial micrite matrix. Miliolids and pelecypod shells are occasionally neomorphosed into pseudomicrosparite and pseudosparite, respectively.

Microfacies 9 (Fig. 6.2D,E,F). Grain-supported oolitic biocalcarenite with associated micrite matrix and sparite cement. Ooids, with an average size of 0.5 mm, contain cores of bioclasts of pelecypods and ostracods, pellets, and grains of detrital quartz. Bioclasts of pelecypods are frequent in the matrix, whereas grains of detrital quartz, muscovite flakes, and pyrite are rare. In this microfacies, ooids have been dissolved often and refilled by single crystals of freshwater phreatic calcite (Fig. 6.2F). In the burial environment, this calcite has been dissolved locally creating an incomplete (Fig. 6.2D) to complete (Fig. 6.2E) oomoldic porosity, whereas the matrix shows simultaneously an intercrystalline microporosity.

Restricted Lagoon

Microfacies 10 (Fig. 6.2G). Argillaceous laminated micrite with finely clotted to pelletoidal texture and abundant desiccation fenestrae and microfractures. Bioclasts are ostracods and miliolids. Silt-size grains of detrital quartz, biotite flakes, and plant debris are rare. Local neomorphism is into pseudomicrosparite.

Microfacies 11 (Fig. 6.2H). Pelecypod bioaccumulated limestone with valves broken by compaction and bioturbation. They display thick micrite envelopes and perforations by endolithic algae, and are greatly altered by neomorphism into pseudosparite. Interstitial material is a siltitic micrite with scattered grains of detrital quartz and biotite flakes. A typical section (Fig. 6.3) shows variation of microscopic parameters and the shallowing-upward sequence of the microfacies.

The Ideal Depositional Model

The general lagoonal character of all microfacies is demonstrated by the smaller diversity of benthic fauna which consists of ostracods, miliolids, thin pelecypods, and gastropods. Only some pelecypods of microfacies 11 display shells of normal thickness. The depositional model (Fig. 6.4) shows general smaller-energy conditions with a slight increase when the wave base intersects the submarine profile and generates the intraclastic microfacies 4. The oolitic bar under medium-energy conditions shows a symmetrical distribution of its pelletoidal microfacies on both sides of oolitic ones, a relationship derived from the microfacies succession in vertical columns.

Description of Microfacies of Middle Regencia Member

Microfacies are described from distal to proximal portion of the lagoon.

Distal Lagoon

Microfacies 1 (Fig. 6.5A). Argillaceous and laminated micrite devoid of microfossils except plant debris. Rare minute grains of detrital quartz are associated with flakes of biotite and abundant pyrite pigments. Frequent neomorphism is into pseudomicrosparite, and scat-

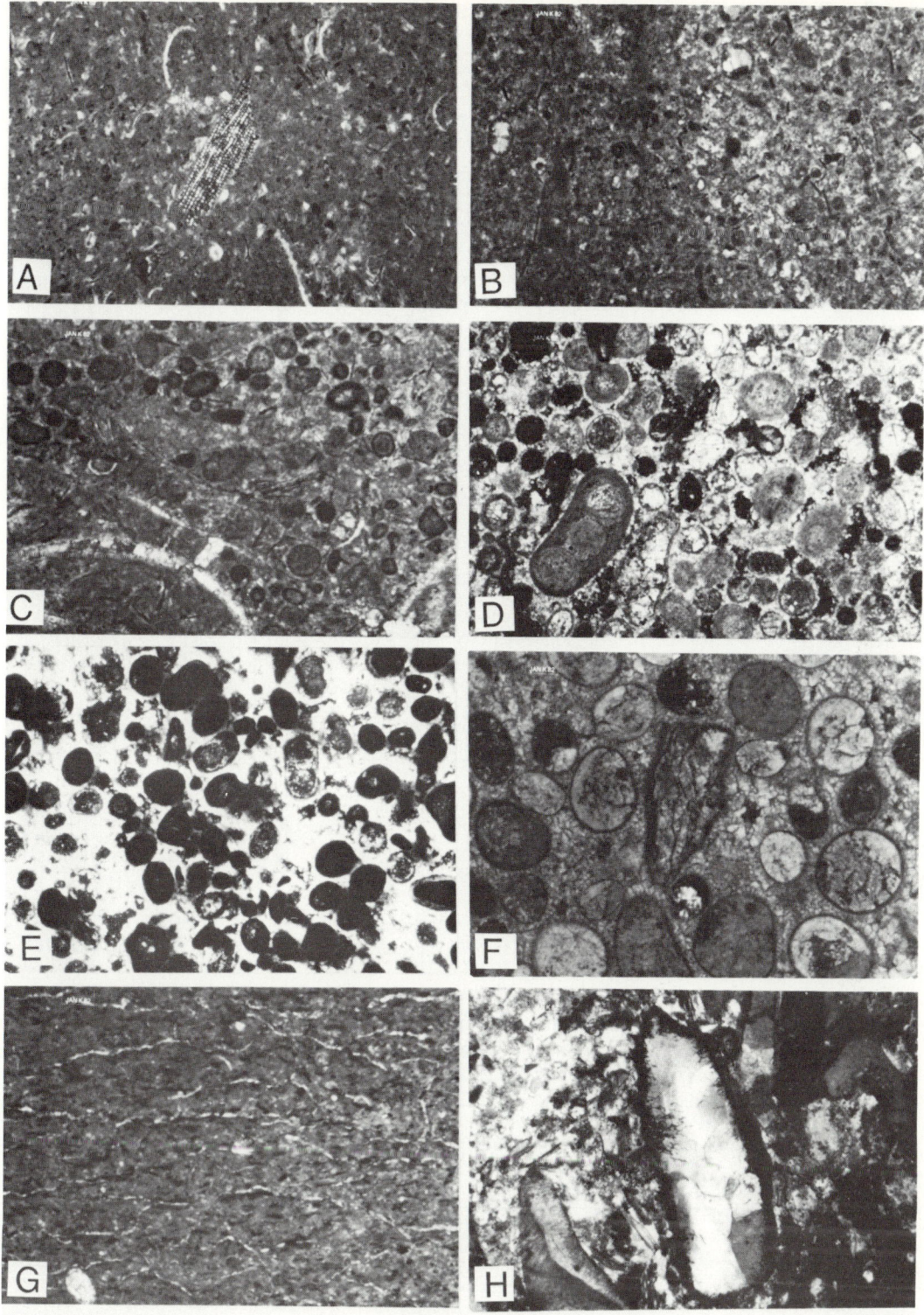

FIGURE 6.2 Lower Regencia Member, Barra Nova Formation (Albian-Cenomanian), Espirito Santo Basin, eastern Brazil. Typical microfacies (continued). A. Microfacies 6. B. Microfacies 7. C. Microfacies 8. D to F. Microfacies 9. G. Microfacies 10. H. Microfacies 11. See text for detailed descriptions. All photomicrographs: plane polarized light, except D, E, and F, crossed nicols. From Carvalho *et al.* (1982a). Reprinted by permission of Petróleo Brasileiro S.A. PETROBRÁS.

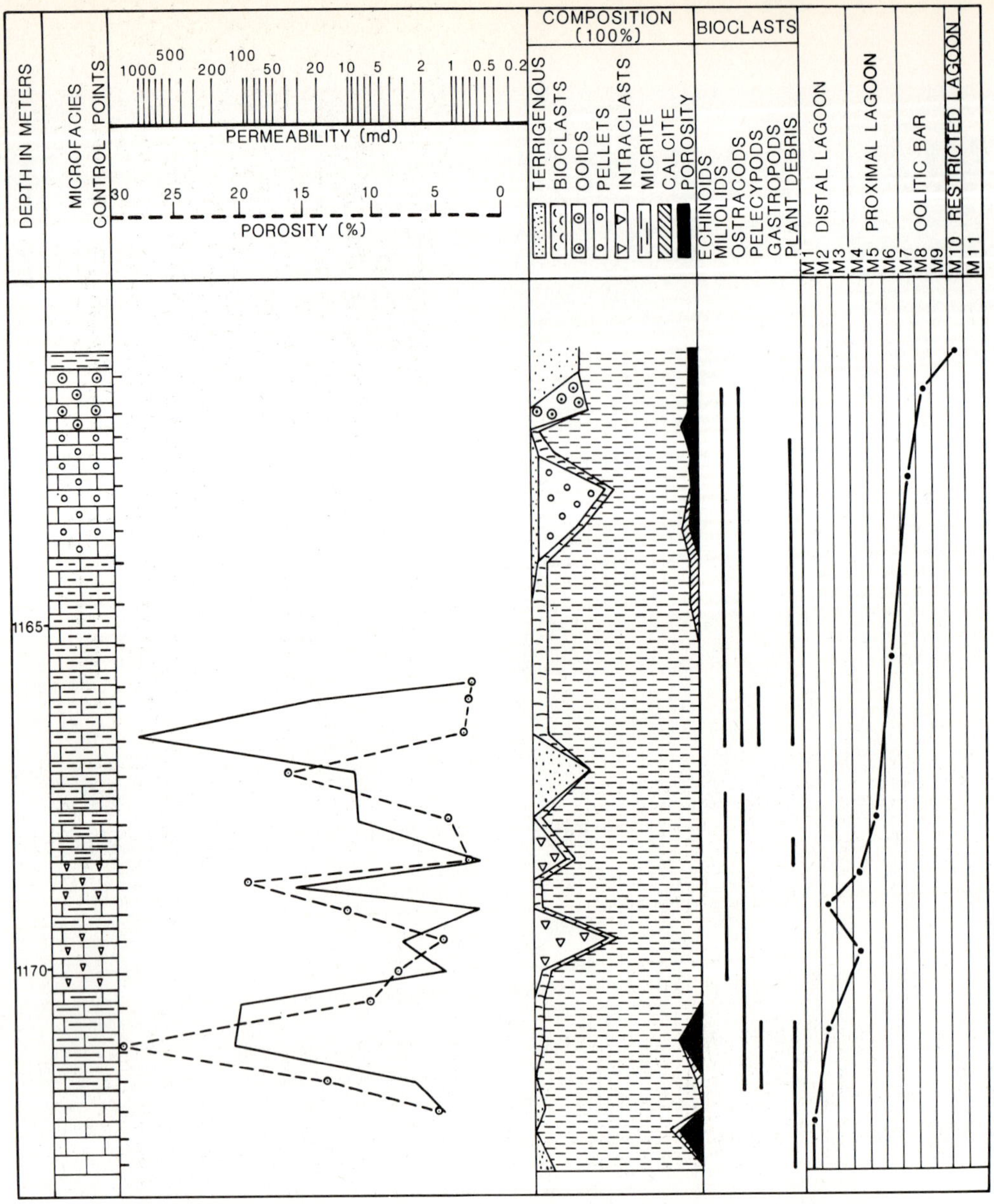

FIGURE 6.3 Lower Regencia Member, Barra Nova Formation (Albian-Cenomanian), Espirito Santo Basin, eastern Brazil. Typical vertical variation of microscopic parameters during shallowing-upward sequence. From Carvalho *et al.* (1982a). Reprinted by permission of Petróleo Brasileiro S.A. PETROBRÁS.

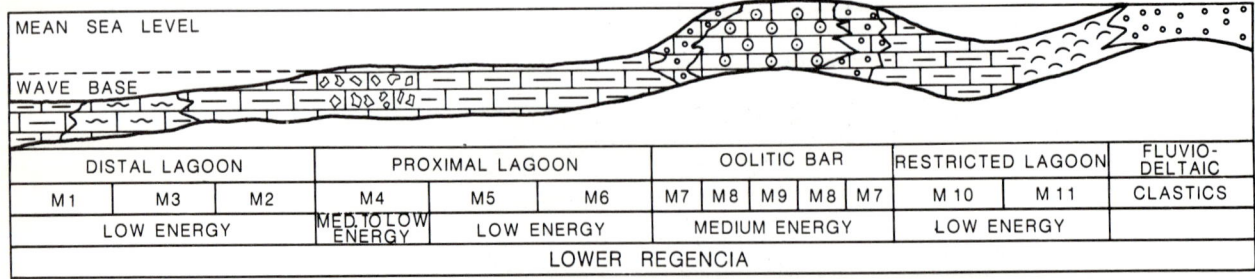

FIGURE 6.4 Lower Regencia Member, Barra Nova Formation (Albian-Cenomanian), Espirito Santo Basin, eastern Brazil. Ideal depositional model. From Carvalho *et al.* (1982a). Reprinted by permission of Petróleo Brasileiro S.A. PETROBRÁS.

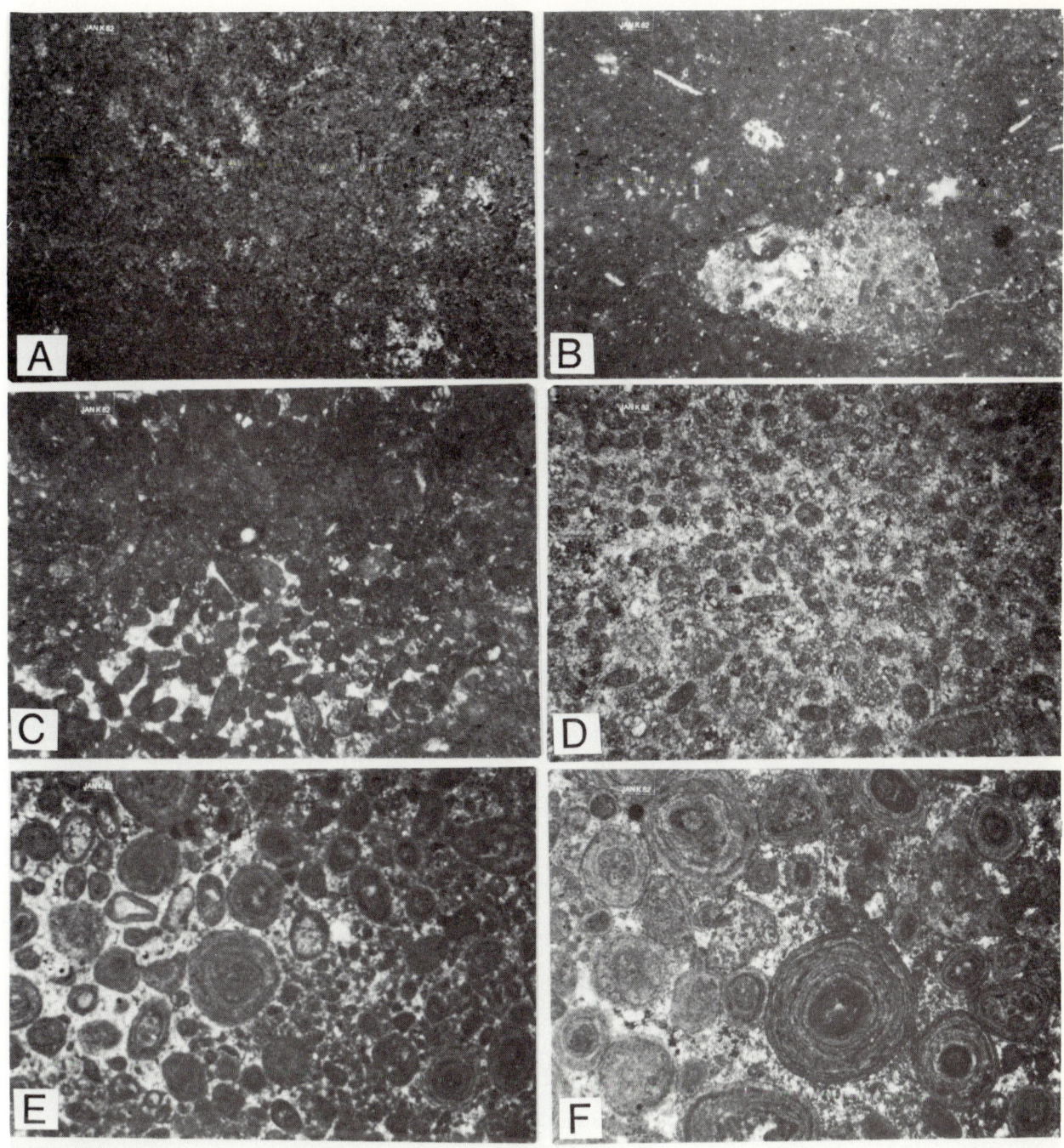

FIGURE 6.5 Middle Regencia Member, Barra Nova Formation (Albian-Cenomanian), Espirito Santo Basin, eastern Brazil. Typical microfacies. A. Microfacies 1. B. Microfacies 2. C and D. Microfacies 3. E. Microfacies 4. F. Microfacies 5. See text for detailed descriptions. All photomicrographs: plane-polarized light except D, E, and F, crossed nicols. From Carvalho *et al.* (1982a). Reprinted by permission of Petróleo Brasileiro S.A. PETROBRÁS.

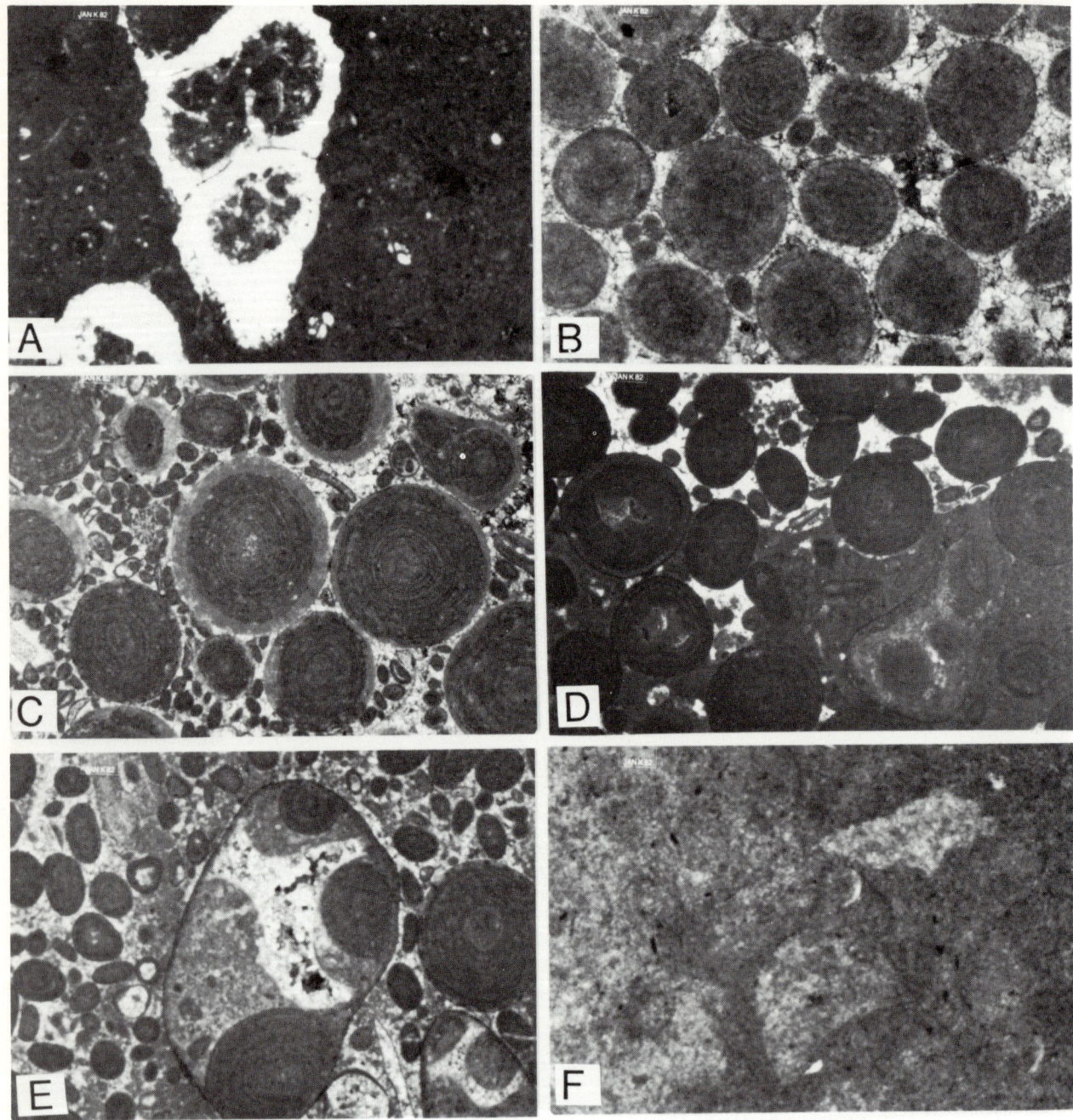

FIGURE 6.6 Middle Regencia Member, Barra Nova Formation (Albian-Cenomanian), Espirito Santo Basin, eastern Brazil. Typical microfacies (continued). A. Microfacies 6. B and C. Microfacies 7. D and E. Microfacies 8. F. Microfacies 9. See text for detailed descriptions. All photomicrographs: plane-polarized light, except B and E, crossed nicols. From Carvalho et al. (1982a). Reprinted by permission of Petróleo Brasileiro S.A. PETROBRÁS.

tered irregular patches of sparite cement are related to bioturbation.

Microfacies 2 (Fig. 6.5B). Laminated micrite, locally heavily bioturbated into concentrations of fecal pellets with interstitial sparite cement. The matrix displays common bioclasts of thin pelecypod shells, ostracods, miliolids, gastropods, and plant debris. Rare silt-size grains of detrital quartz are associated with biotite flakes and pyrite pigments. Some neomorphism of the micrite is into pseudomicrosparite.

Oncolitic Bar

Microfacies 3 (Fig. 6.5C,D). Grain-supported pelletoidal biocalcarenite with micrite to pelletoidal micrite matrix. Pellets are predominantly fecal and often are coprolites of crustaceans with typical internal structure, a few are lithic. With the pellets are associated bioclasts of pelecypods, gastropods, ostracods, and miliolids. This microfacies is argillaceous locally and bioturbated with related patches of sparite cement.

Microfacies 4 (Fig. 6.5E). Grain-supported pelletoidal and oncoidal calcarenite with associated micrite matrix and sparite cement. Pellets are either fecal in origin, or represent small oncoids with poorly defined internal structure, or are the products of intraformational reworking of larger oncoids. The latter, which show a maximum size of 1 cm, display cores of bioclasts of pelecypods, gastropods, ostracods, miliolids, pellets, and quartz grains. The matrix contains a small amount of the same components.

Microfacies 5 (Fig. 6.5F). Grain-supported oncoidal calcarenite with associated micrite matrix and sparite cement related to bioturbation. Oncoids are bimodal with a mode at 0.3 mm and another at 1 cm. Cores consisting of bioclasts of gastropods, pelecypods, ostracods, pellets, and quartz grains are surrounded by well-developed and irregular blue-green algal concentric mats separated by layers of lighter micrite. The same bioclasts, as above, are relatively rare within the interstitial matrix.

Proximal Lagoon

Microfacies 6 (Fig. 6.6A). Argillaceous and laminated micrite, locally pelletoidal and bioturbated with ostracods and miliolids. Entire pelecypods and gastropods are altered to pseudosparite by neomorphism. Silt-size grains of detrital quartz and pyrite are rare.

Oolitic Bar

Microfacies 7 (Fig. 6.6B,C). Grain-supported oolitic calcarenite with submarine isopachous rim cement followed by interstitial mosaic of freshwater sparite. Normal ooids are strongly bimodal, display micritized margins, and developed around cores consisting of bioclasts of pelecypods, gastropods, echinoid spines, ostracods, miliolids, pellets, and grains of detrital quartz. All these components, together with small intraclasts of the same oolitic calcarenite, occur in interstitial spaces. Patches of micrite matrix occur locally.

Microfacies 8 (Fig. 6.6D,E). Grain-supported oolitic calcarenite with association of micrite matrix and sparite cement. Ooids are similar to those of microfacies 7 but associated with large intraclasts of that microfacies, indicating early cementation and reworking from beachrock conditions. Other constituents are similar to those of microfacies 7.

Restricted Lagoonal

Microfacies 9 (Fig. 6.6F). Argillaceous and laminated micrite, locally pelletoidal and heavily bioturbated. Scattered bioclasts of ostracods, miliolids, and pelecypods are associated with rare grains of detrital quartz and biotite flakes.

Tidal Flat

Microfacies 10. Light tan, siltitic, bioturbated shales with minute calcareous concretions.

A typical section (Fig. 6.7) shows variation of microscopic parameters and a shallowing-upward sequence of the microfacies.

The Ideal Depositional Model

Diversity of the benthic fauna is greater than that of the Lower Regencia Member. Ostracods, miliolids, pelecypods, gastropods, echinoids, and rare agglutinated foraminifers are associated in the Middle Regencia with blue-green algal oncoids, recalling those of the Macaé Formation (see Chapter 11). Although this paleoecological association is still lagoonal, the environment is more open to marine influence. The depositional model (Fig. 6.8) shows a slightly greater general energy level, the oncoidal bar displays a symmetrical distribution of its pelletoidal microfacies on both sides of the oncoidal ones, and the oolitic bar is devoid of such a symmetry. All these relationships were derived from microfacies succession in vertical columns.

Diagenesis

Burial secondary porosity by dissolution through the action of CO_2-rich brines is encountered in a variety of microfacies of the Middle Regencia Member. It is most

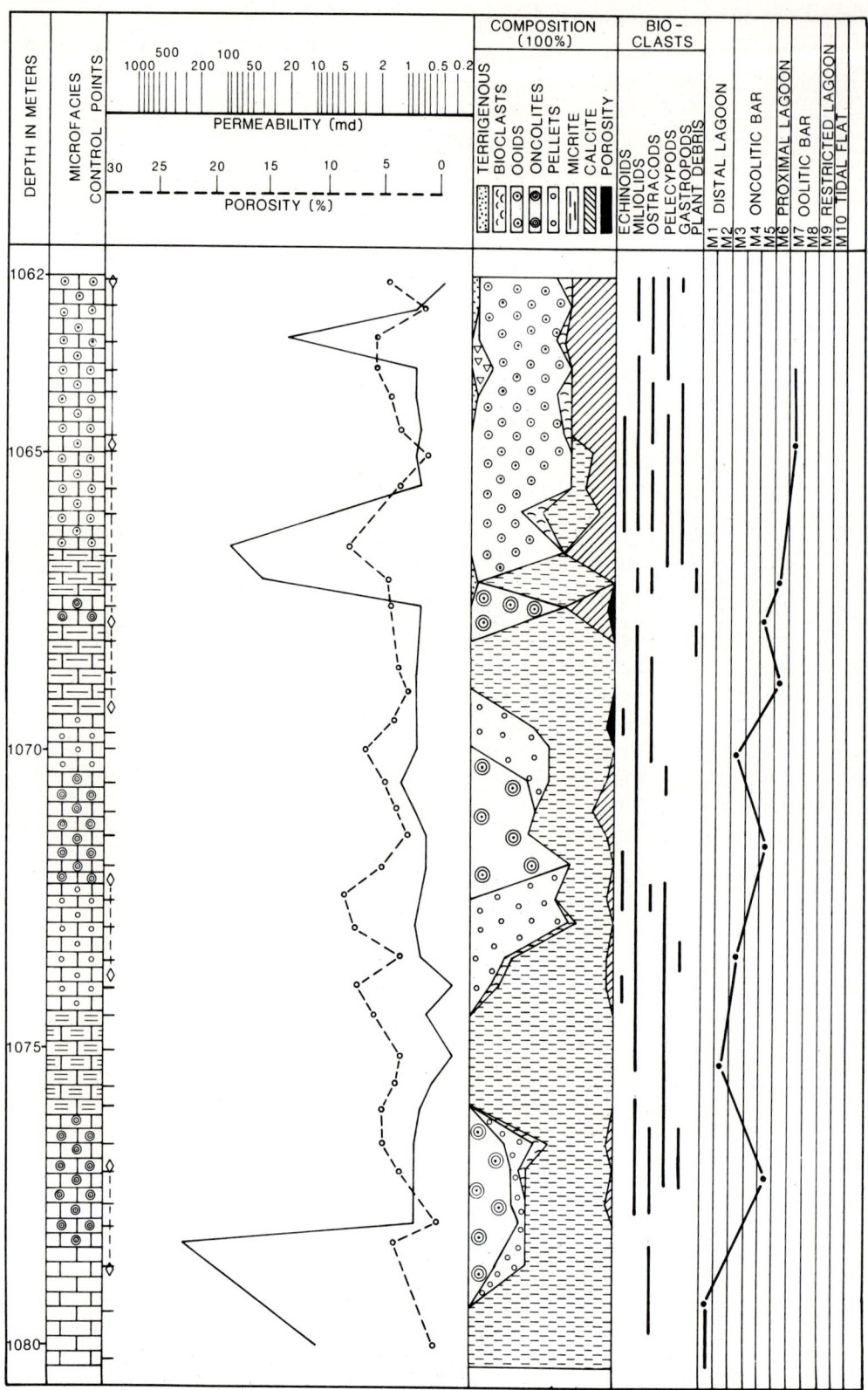

FIGURE 6.7 Middle Regencia Member, Barra Nova Formation (Albian-Cenomanian), Espirito Santo Basin, eastern Brazil. Typical vertical variation of microscopic parameters during shallowing-upward sequence. From Carvalho *et al.* (1982a). Reprinted by permission of Petróleo Brasileiro S.A. PETROBRÁS.

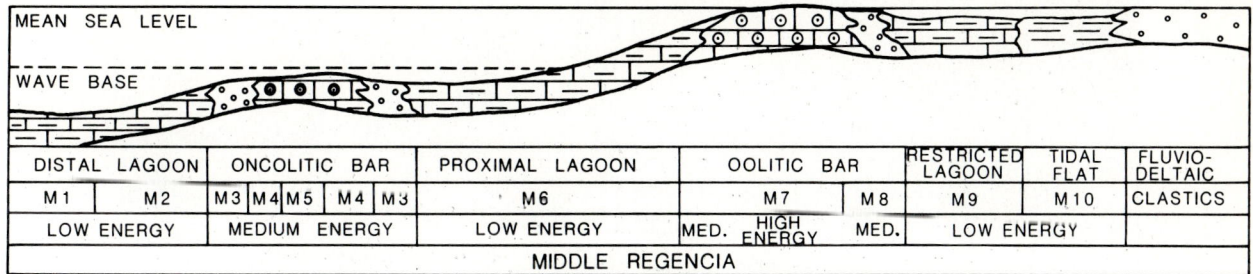

FIGURE 6.8 Middle Regencia Member, Barra Nova Formation (Albian-Cenomanian), Espirito Santo Basin, eastern Brazil. Ideal depositional model. From Carvalho et al. (1982a). Reprinted by permission of Petróleo Brasileiro S.A. PETROBRÁS.

important in the oolitic microfacies 8 (Fig. 6.9B to H), where it is interparticle, often respecting the isopachous rim cement, interparticle enlarged to vuggy, oomoldic, and moldic in pelecypod and gastropod bioclasts; it also corresponds to enlarged fracture systems. In the pelletoidal biocalcarenite microfacies 3, open stylolitic porosity was observed (Fig. 6.9A). In the micritic microfacies, secondary porosity is represented by molds of pelecypods and gastropods, micropores in the matrix, and enlarged fractures.

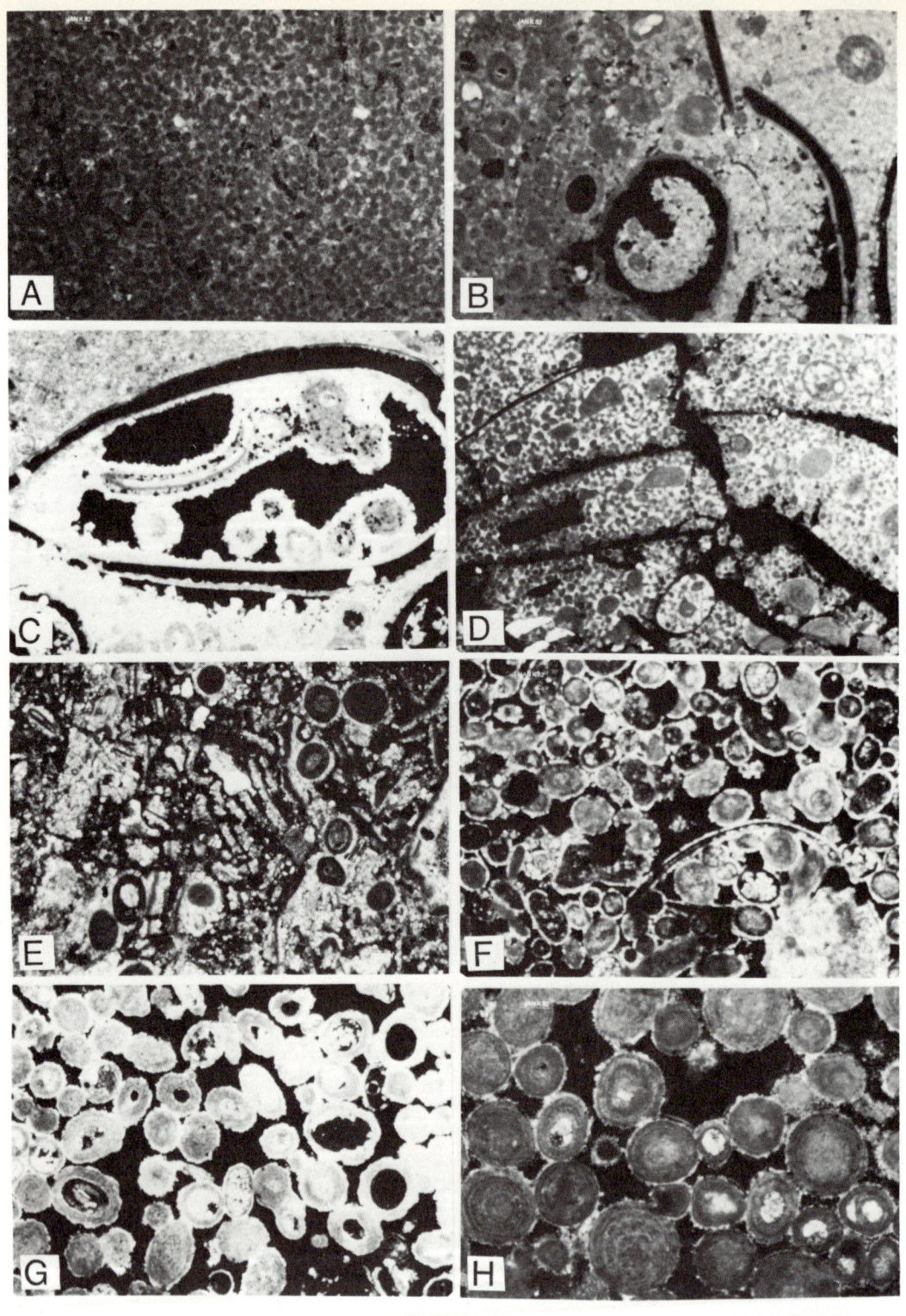

FIGURE 6.9 Middle Regencia Member, Barra Nova Formation (Albian-Cenomanian), Espirito Santo Basin, eastern Brazil. Types of porosity. A. Stylolite porosity. B. Moldic porosity in pelecypods and gastropods. C. Moldic and vuggy porosity in closed pelecypod. D. Enlarged fracture porosity. E. Enlarged interparticle porosity. F. Interparticle porosity enlarged to vuggy. G. Interparticle porosity enlarged to vuggy. H. Interparticle porosity. All photomicrographs: plane-polarized light, porosity in black due to impregnation by dead oil. From Carvalho *et al.* (1982a). Reprinted by permission of Petróleo Brasileiro S.A. PETROBRÁS.

7

Carbonate Ramps with Hydrodynamic Buildups

In general, carbonate ramps rarely maintain their typical geometry for long periods of geologic time. Indeed, at the location where the wave base intersects the gentle submarine slope, subtidal hydrodynamic buildups develop. They introduce new bioclastic and oolitic microfacies that appreciably modify the original depositional model.

7.1 CHARACTERISTIC CONSTITUENTS: CRINOIDS, SMALLER BENTHIC FORAMINIFERS, AND OOIDS

The Salem Limestone (Middle Mississippian) of the southeastern margin of the Illinois Basin (Reichelderfer, 1985; Carozzi and Reichelderfer, 1987) was investigated by means of five cores taken from oil- and gas-producing areas. More than 332 thin sections were studied petrographically from samples taken at an average vertical interval of 4.5 in. (11.43 cm) for a total thickness of 158 ft (47.4 m). This study emphasizes the relationships between microfacies, diagenetic evolution, and generation of porosity and permeability of commercial interest.

Description of Microfacies

Nine distinct microfacies were recognized in oolitic to bioclastic calcarenite bars prograding across a gently sloping carbonate ramp. This ramp was located on the southeastern margin of the Illinois Basin opposite the carbonate platform displaying frontal hydrodynamic buildups, which represents the Salem Limestone in southwestern Illinois (Carozzi and Diaby, 1981) analyzed in Chapter 12. Microfacies are described in a general and composite shallowing-upward order and across an ideal hydrodynamic bar.

Slope

Microfacies A (Fig. 7.1A). Dolomitized, fine-grained biocalcarenite with bioclastic matrix. The components in decreasing order of importance are: crinoids, *Endothyra*, smaller benthic foraminifers, bryozoans, brachiopods, ostracods, and pelecypods. Large solitary corals and syringoporid corals also occur in this microfacies. Matrix is gradational in size to the components. Pervasive formation of dolomite rhombs often has obliterated depositional textures. Cements are predominantly dolosparite, anhydrite, quartz, and saddle dolomite.

Measured porosity averages 8% and ranges from 4 to 12%. Primary porosity predominates as unfilled chambers in foraminiferal tests, interseptal spaces in corals, and zooecia of bryozoans. Secondary porosity combines biomolds of unstable constituents (pelecypods and gastropods) held by micrite envelopes, and fractures with intercrystalline type in dolomitized matrix. Because of the small size and poorly connected nature of pore spaces,

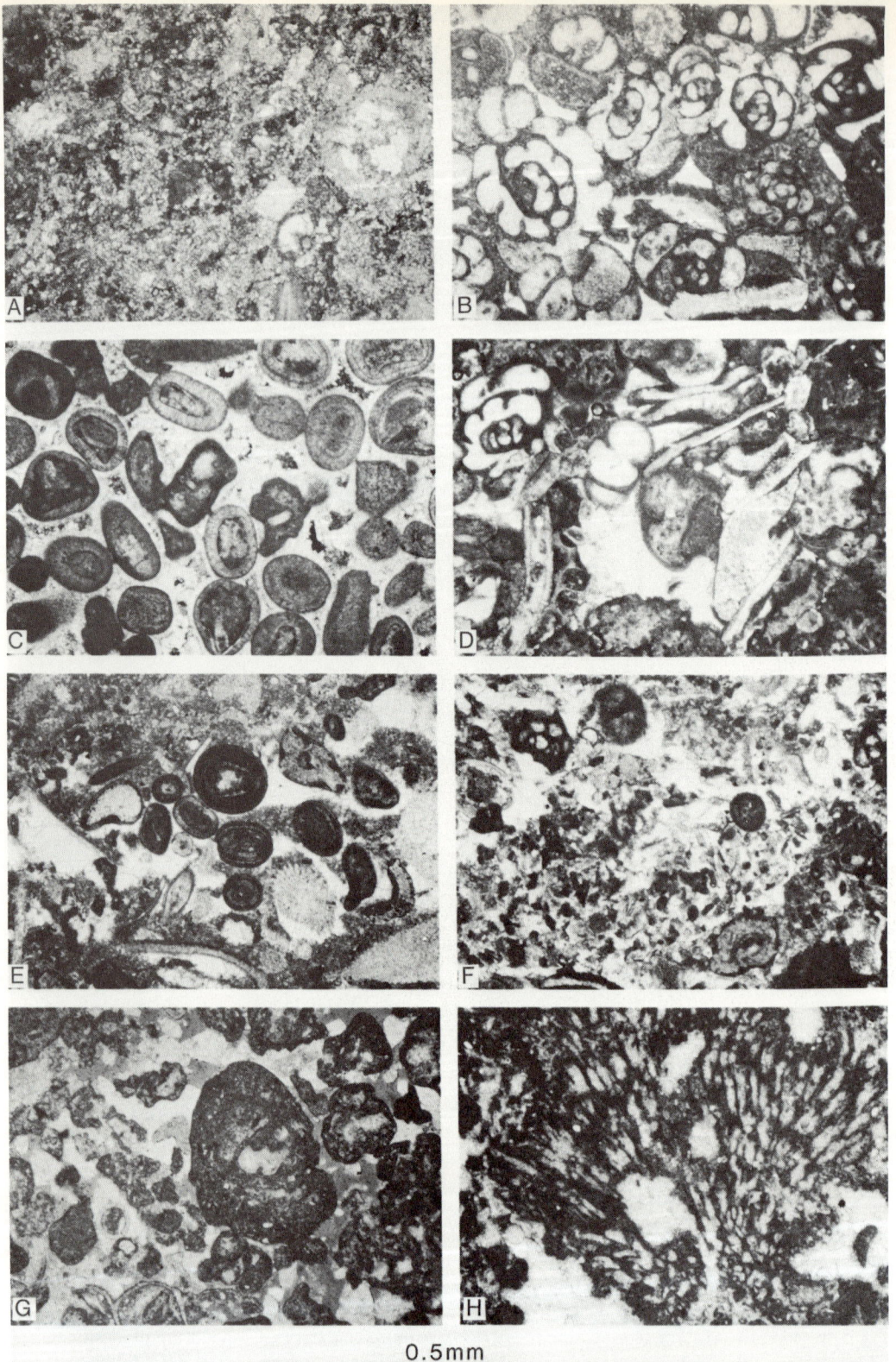

FIGURE 7.1 Salem Limestone (Middle Mississippian) of the southeastern margin of Illinois Basin. Typical microfacies. A. Microfacies A. B. Microfacies B. C. Microfacies C. D. Microfacies D. E. Microfacies E. F. Microfacies F. G. Microfacies G. H. Microfacies H. See text for detailed descriptions. All photomicrographs: plane-polarized light, except C, crossed nicols. From Carozzi and Reichelderfer (1987). Reprinted by permission of Illinois State Academy of Science.

permeability of this microfacies is 0 md. For all practical purposes this microfacies acts as a seal.

Foraminiferal Fore Bar

Microfacies B (Fig. 7.1B). Pressure-welded *Endothyra*-crinoid biocalcarenite with common syntaxial overgrowths and interparticle sparite cement. The major organic components in this microfacies are, in decreasing order of importance, *Endothyra*, crinoids, blue-green algal fragments, *Koninckopora*, smaller benthic foraminifers, brachiopods, ostracods, and mollusks, superficial ooids, and ooids. Calcite cement fills the chambers of the *Endothyra*; replacive and pore-filling saddle dolomite and anhydrite cement also occur.

Measured porosity averages 9% and ranges from 2 to 18%. Permeability averages 31 md and ranges from 0 to 142 md. Primary porosity consists of intraparticle unfilled *Endothyra* chambers, and interparticle reduced by sparite cement and syntaxial overgrowths on crinoid fragments. Abundant overgrowths create a framework that enables pore spaces left by incomplete cementation to remain open. Secondary porosity consists of biomolds of unstable constituents (mollusks) held by micrite envelopes, and fractures. Advanced stages of pressure solution tend to destroy both types of porosity in association with saddle dolomite and anhydrite cements.

Oolitic Bar

Microfacies C (Fig. 7.1C). Well-sorted compacted oolitic-crinoidal biocalcarenite with isopachous rim cement and interparticle sparite. This microfacies frequently is cross-bedded and locally bioturbated. Constituents in order of decreasing abundance are: ooids, superficial ooids, crinoids, algal clasts, *Endothyra*, smaller benthic foraminifers, pellets, ostracods, and mollusks. Rare pelletoidal geopetal internal sediment is associated with truncation surfaces. Saddle dolomite occurs as an additional cement.

Measured porosity averages 13% and ranges from 3 to 23%. Primary porosity is important as interparticle reduced by isopachous rim and sparite cements, and unfilled chambers of foraminifers. However, in numerous cases porosity is enhanced by secondary types such as biomolds of mollusks and occasional dissolved ooid concentric laminae (Fig. 7.2B,C). In samples with maximum porosity of 23%, open vugs of an unusual type due to the dissolution of pelletoidal burrow fillings with microsparite cement (Fig. 7.2D,E,F) are responsible for most of the porosity because interparticle spaces are highly reduced by calcite cementation. Permeability is often minor (average 22 md, range 0 to 80 md) even when porosity is high because vugs are poorly interconnected.

Occlusion of the secondary vug porosity is by sparite, anhydrite, and saddle dolomite (Fig. 7.2G).

Bioclastic Bar

Microfacies D (Fig. 7.1D). Pressure-welded crinoidal-algal clast-*Endothyra* biocalcarenite with syntaxial overgrowth cement. Constituents in decreasing order of importance are: crinoids, blue-green algal clasts, *Endothyra*, smaller benthic foraminifers, superficial ooids, *Koninckopora*, ooids, pellets, brachiopods, ostracods, mollusks, and bryozoans. Pressure-welding and overgrowths participate in the cementation in variable proportions.

Measured porosity averages 10% and ranges from 1 to 15%. Permeability averages 29 md and ranges from 0 to 142 md. Primary porosity is important as interparticle reduced either by cementation or pressure solution, and as intraparticle in tests of various organisms. Samples with maximum porosity display a combination of large interparticle unfilled pore spaces, and secondary porosity represented by abundant biomolds of mollusks and oversized pores where several fragments of unstable constituents were adjacent to each other and dissolved. In some instances, stylolitic and fracture porosities (Fig. 7.2H) added a small amount of secondary pore space. Advanced stages of pressure-solution together with calcite cementation tend to destroy both types of porosity.

Oolitic Back Bar

Microfacies E (Fig. 7.1E). Grain-supported to mud-supported oolitic-algal clast-crinoid-*Endothyra* biocalcarenite with sparite cement, pelletoidal geopetal internal sediment, and massive matrix. Constituents are in decreasing order of importance: ooids, algal clasts, crinoids, *Endothyra*, superficial ooids, smaller benthic foraminifers, brachiopods, mollusks, large intraclasts, and detrital quartz. Matrix is locally bioturbated and isopachous rim cement is also present. Minor saddle dolomite and anhydrite occur as cement and as a replacement phase.

Measured porosity averages 8% and ranges from 3 to 11%. Permeability averages 14 md and ranges from 0 to 51 md. Primary porosity generally predominates as reduced type by incomplete calcite cementation. Secondary porosity is again responsible for the highest value of 11% as the unusual type of open vugs due to the dissolution of pelletoidal burrow fillings (as in microfacies C) together with biomolds of mollusks, dissolved concentric layers of ooids, and stylolitic porosity. Samples with lesser porosity are either pressure welded or finer grained, and contain abundant crinoids with syntaxial overgrowths.

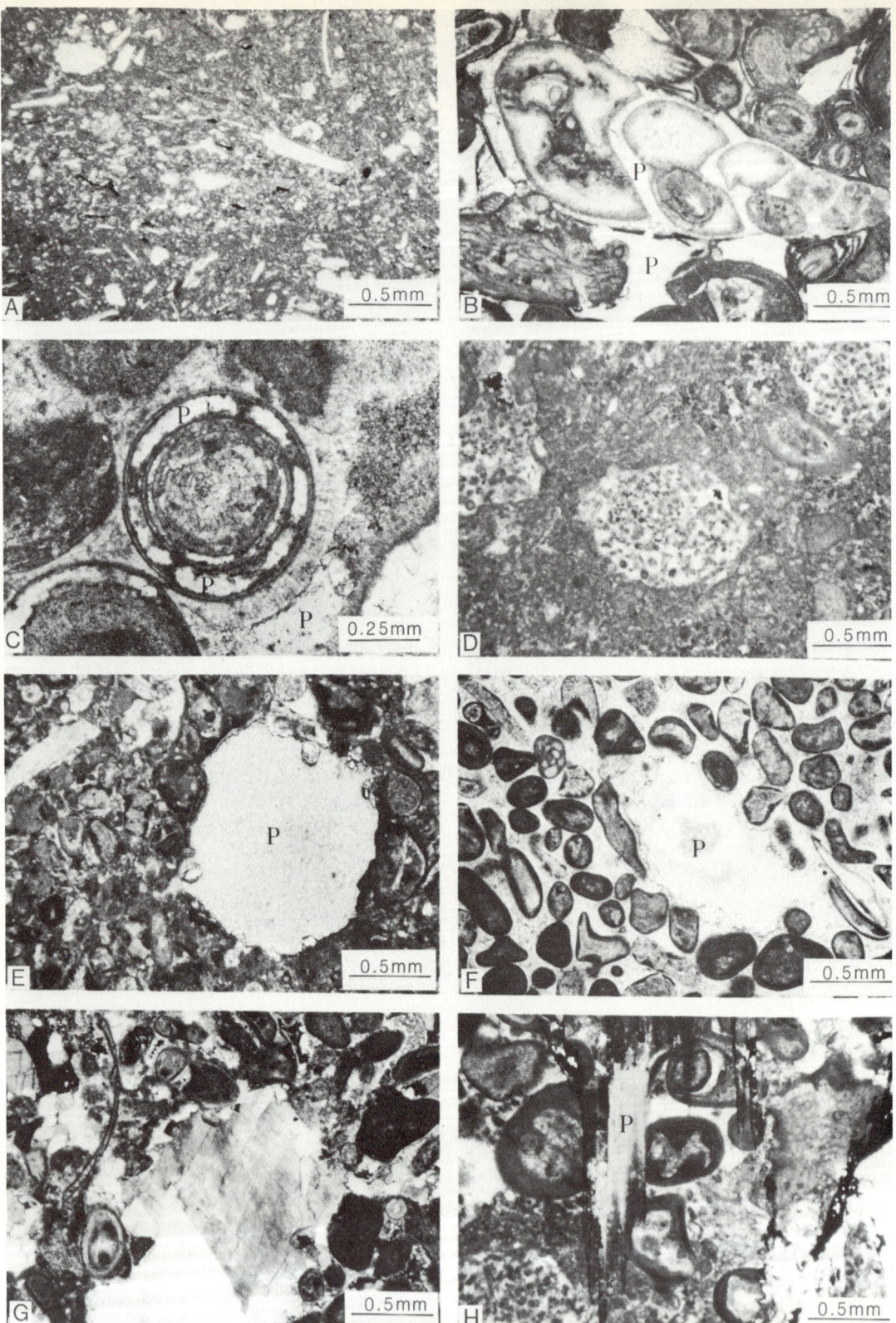

FIGURE 7.2 Salem Limestone (Middle Mississippian) of the southeastern margin of Illinois Basin. Typical microfacies (continued) and porosity types (designated by letter P). A. Microfacies I. B. Partially filled gastropod biomoldic porosity and reduced interparticle porosity. C. Selective cortical dissolution in ooid, with compacted submarine fibrous rim cement. D. Burrows filled with pellets and microsparite cement. Vugs illustrated in E, F, and G similarly are inferred to have been filled before dissolution. E. Unfilled vug, with lack of interstitial porosity in biocalcarenite. F. Unfilled vug, with well-pre-

Bioclastic Back Bar

Microfacies F (Fig. 7.1F). Fine-grained crinoid-*Endothyra* biocalcarenite with pelletoidal matrix and interparticle sparite cement. Constituents are in decreasing order of importance: crinoids, *Endothyra*, smaller benthic foraminifers, ostracods, pellets, *Koninckopora*, algal clasts, ooids, pelecypods, and grains of detrital quartz. Matrix is pelletoidal, geopetal, and frequently gradational in size with sand-size components.

Measured porosity is 3% and entirely primary reduced by incomplete interparticle cementation. Permeability is 0 md.

Reworked Algal Flat

Microfacies G (Fig. 7.1G). Grain-supported algal-crinoidal biocalcarenite with isopachous rim cement and interparticle sparite. This microfacies consists predominantly of fragments of *Stachyodes*, *Girvanella*, and *Orthonella*, associated with ooids, pellets, crinoids, mollusks, *Endothyra*, large oncoids, ostracods, brachiopods, *Koninckopora*, and smaller benthic foraminifers.

Measured porosity averages 7% and ranges from 1 to 12%. Permeability averages 13 md and ranges from 0 to 53 md. Primary porosity predominates generally as reduced type by a combination of submarine dissolution of isopachous rim cement, and incomplete interparticle sparite cementation. Samples with lesser porosity display fibrous rim cement which occluded most of the primary porosity. Minor secondary porosity consists of pelecypod biomolds. Destruction of porosity results from a sparite mosaic cement, sometimes associated with saddle dolomite filling any remaining pore space except for an occasional incompletely cemented pelecypod biomold.

Algal Flat

Microfacies H (Fig. 7.1H). Algal-constructed to grain-supported algal clast-oolitic-crinoidal biocalcarenite with isopachous rim cement, geopetal pelletoidal internal sediment, and sparite cement.

In situ algal growths of *Girvanella*, *Orthonella*, and *Stachyodes* characterize this microfacies. Other components include, in decreasing order of abundance, algal clasts, ooids, crinoids, pellets, *Endothyra*, pelecypods, ostracods, bryozoans, brachiopods, and smaller benthic foraminifers. Rare large intraclasts are algal coated. Geopetal internal sediment rests on partially dissolved isopachous rim cement and components. Interparticle sparite and saddle dolomite are common. As expected, this microfacies is greatly variable. For example, half of a thin section will show an algal-constructed framework with an interstitial algal bioclast-oolitic-crinoidal biocalcarenite and the other half will be entirely a grain-supported biocalcarenite.

Measured porosity averages 8% and ranges from 2 to 12%. Permeability is negligible (average 0 md, range 0 to 2 md). Primary porosity predominates as reduced type by isopachous rim and sparite cement. Secondary porosity consists of partially dissolved normal ooids and large vugs due to the dissolution of pelletoidal matrix filling of burrows. Samples with lesser porosity result from a combination of abundant pelletoidal matrix, thick isopachous rim cement, and well-developed sparite cement. The latter, together with saddle dolomite, fill interparticle voids and dissolution vugs.

Lagoon

Microfacies I (Fig. 7.2A). Bioturbated, mud-supported biocalcarenite with argillaceous calcisiltite and bioclastic matrix. Fine bioclasts occur as streaks of crinoids, brachiopods, bryozoans, ostracods, *Endothyra*, gastropods, trilobites, calcispheres, and smaller benthic foraminifers. The bioclastic matrix is gradational in size with sand-size components, and argillaceous calcisiltite matrix is locally pelletoidal. Bioturbation and pyrite streaks are common.

Measured porosity is 1%, but no porosity is visible in thin section. Permeability is 0 md, and for all practical purposes this microfacies acts as a seal. Primary porosity represented by minute interparticle pore spaces, and intraparticle spaces in tests as well as secondary porosity corresponding to biomolds have been occluded by microsparite cement.

The Ideal Shallowing-Upward Sequences

The Salem Limestone in the investigated area was deposited as a series of hydrodynamic bars prograding across a carbonate ramp environment. It is conceivable that certain bars were entirely oolitic and others entirely bioclastic, but it seems reasonable on the basis of the availa-

FIGURE 7.2 (*continued*) served rim cement surrounding ooids and bioclasts near vug and dissolution of mollusk fragment to right of vug. G. Vug filled with saddle dolomite and minor sparite. H. Stylolite porosity. All photomicrographs: plane-polarized light, except G, crossed nicols. From Carozzi and Reichelderfer (1987). Reprinted by permission of Illinois State Academy of Science.

ble data that an ideal bar could be visualized as possessing an oolitic central portion and bioclastic margins (Figs. 7.3, 7.4). Although the various microfacies were previously described in a general shallowing-upward order, we are dealing in reality with two such sequences, an oolitic one (Fig. 7.5A) and a bioclastic one (Fig. 7.5B), which are lateral hydrodynamic equivalents. Graphic representation of the petrographic data corresponds to percentage visual estimation of components (biogenic and lithic), matrix and porosity to which are juxtaposed measured values of porosity in percent using the mercury injection method, and of permeability in millidarcies using nitrogen gas.

The Ideal Depositional Models

The ideal models corresponding to the two shallowing-upward sequences (Figs. 7.6, 7.7) deserve a few comments. Microfacies A, which represents the slope environment, is a fine-grained, dolomitized biocalcarenite. Because the slope is the only dolomitic microfacies, it was dolomitized, probably, soon after deposition by a dorag-type model due to freshwater and marine water mixing. However, no evidence of exposure necessary to provide freshwater for dolomitization has been encountered in the sections investigated. It is known in other cases that minor subaerial exposure of carbonate bodies could provide temporary and localized conditions of mixing. Therefore, the assumption is made that some of the bars could have been exposed occasionally.

Microfacies B, the foraminiferal fore bar, consists predominantly of *Endothyra*, which are also found throughout the other microfacies. The oolitic microfacies C and the bioclastic microfacies D are lateral equivalents and occupy the same position in their respective shallowing-upward sequences. Microfacies E and F are the poorly sorted, finer-grained, lower-energy marginal facies of microfacies C and D, respectively, deposited on the shoreward side of the ideal bar and containing more matrix and internal sediment.

The ideal bar provided a lower-energy, quiet environment behind it for an algal flat (microfacies H). Longitudinal currents funneled between the bar and the algal flat reworked the algal material forming microfacies G. This sediment is variable, containing both algal growths and interstitial calcarenites. Microfacies I corresponds to the lowest-energy environment, the lagoon.

In the ideal bioclastic bar model (Fig. 7.6), a clearly opposite relationship is seen between the amount of matrix, porosity, and permeability. Both well-sorted calcarenites microfacies B and D with peaks of porosity and permeability display predominantly interparticle reduced primary porosity. Where matrix is present, as in the slope (A), bioclastic back bar (F), and lagoon (I) microfacies, porosity and permeability are reduced.

The relationship between porosity, permeability, and microfacies is more complex in the ideal oolitic bar model (Fig. 7.7). The oolitic bar (microfacies C), which shows the greatest porosity, contains numerous oolitically coated crinoids. This texture makes syntaxial overgrowths on crinoids rare. Although they reduce primary porosity when present, these overgrowths form a stable framework keeping residual pore space open during pressure-solution. Vugs also add to the porosity in this microfacies. The peak in permeability, however, is in the foraminiferal fore bar (microfacies B), because the chambers in the *Endothyra* provide numerous interconnected pore spaces.

Although matrix certainly influences the reduction of porosity and permeability in the slope (A) and lagoonal (I) microfacies, other matrix-rich microfacies, such as the oolitic back bar (E), still display high porosity. This

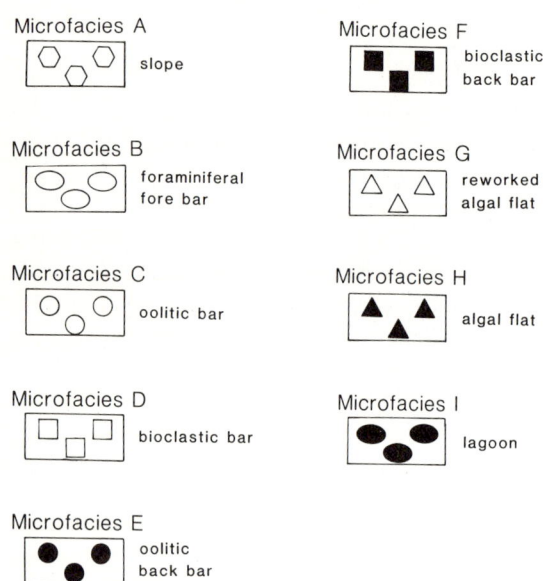

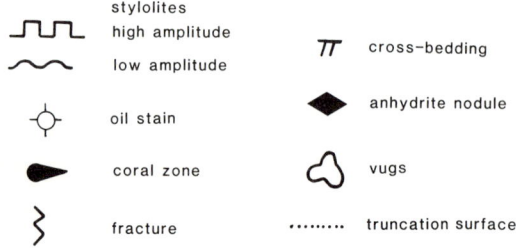

FIGURE 7.3 Salem Limestone (Middle Mississippian) of the southeastern margin of Illinois Basin. Symbols for microfacies. From Reichelderfer (1985).

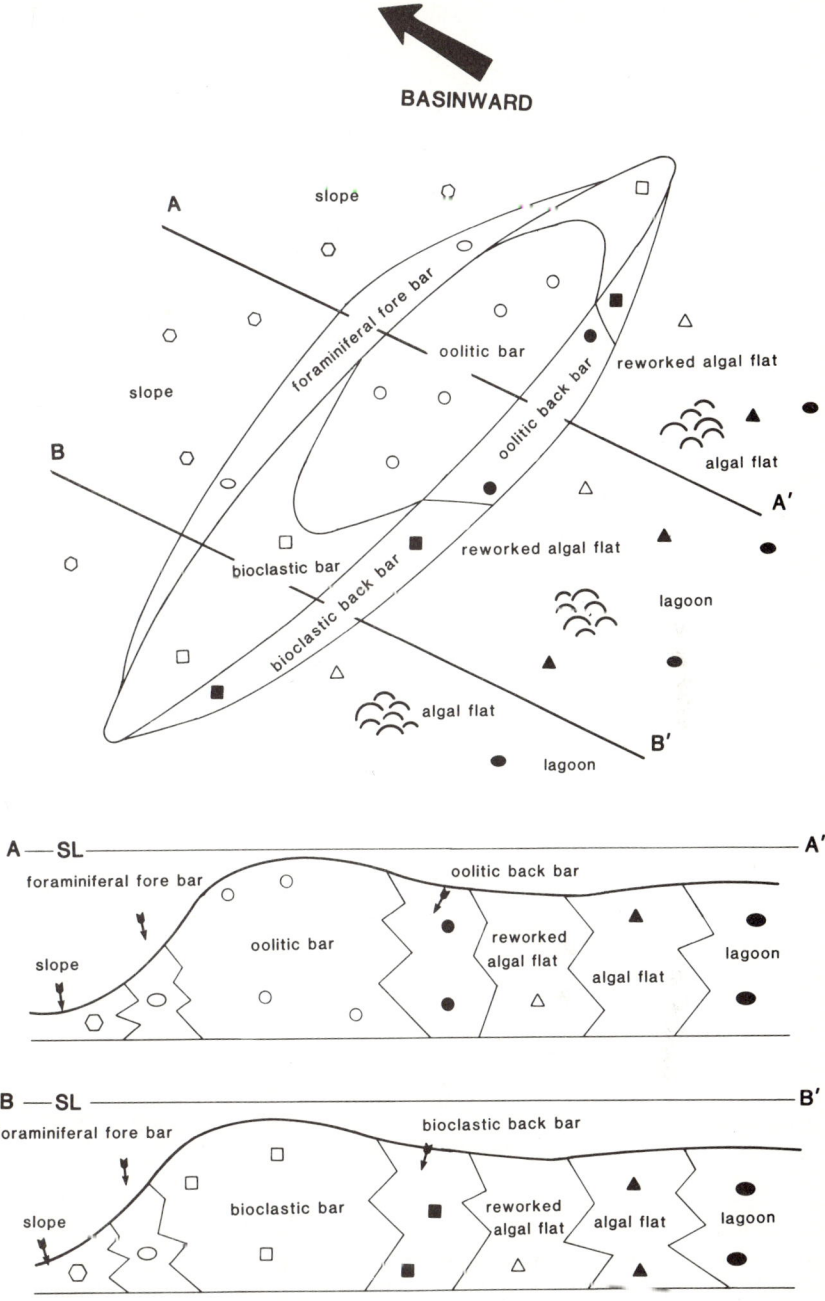

FIGURE 7.4 Salem Limestone (Middle Mississippian) of the southeastern margin of Illinois Basin. Map view and cross sections of ideal hydrodynamic bar. From Reichelderfer (1985).

porosity is related to the formation of vugs by burrow dissolution. Although algal flat microfacies (H) displays matrix, it also contains interstitial calcarenite where most of the porosity is located. Unfilled vugs add to the porosity in this microfacies, but because they are poorly interconnected, they often do not affect permeability.

A general depositional model (Fig. 7.8) shows a typical bar preceded seaward by finer-grained slope carbonates and landward by lagoonal to supratidal muds. Both environments produce sediments capable of acting as potential seals for stratigraphic traps developed in the hydrodynamic buildup.

Diagenetic Sequence

A generalized diagenetic sequence can be prepared encompassing all the investigated cores (Fig. 7.9), and explaining the origin of the porosity of the oil- and gas-producing microfacies of the bioclastic and oolitic bars.

The petrographic description of microfacies shows that the main type of porosity is primary interparticle and intraparticle, in part immediately reduced by isopachous rim cementation, which takes place in the active marine phreatic environment. Further porosity of secondary type is generated by total or partial dissolution of

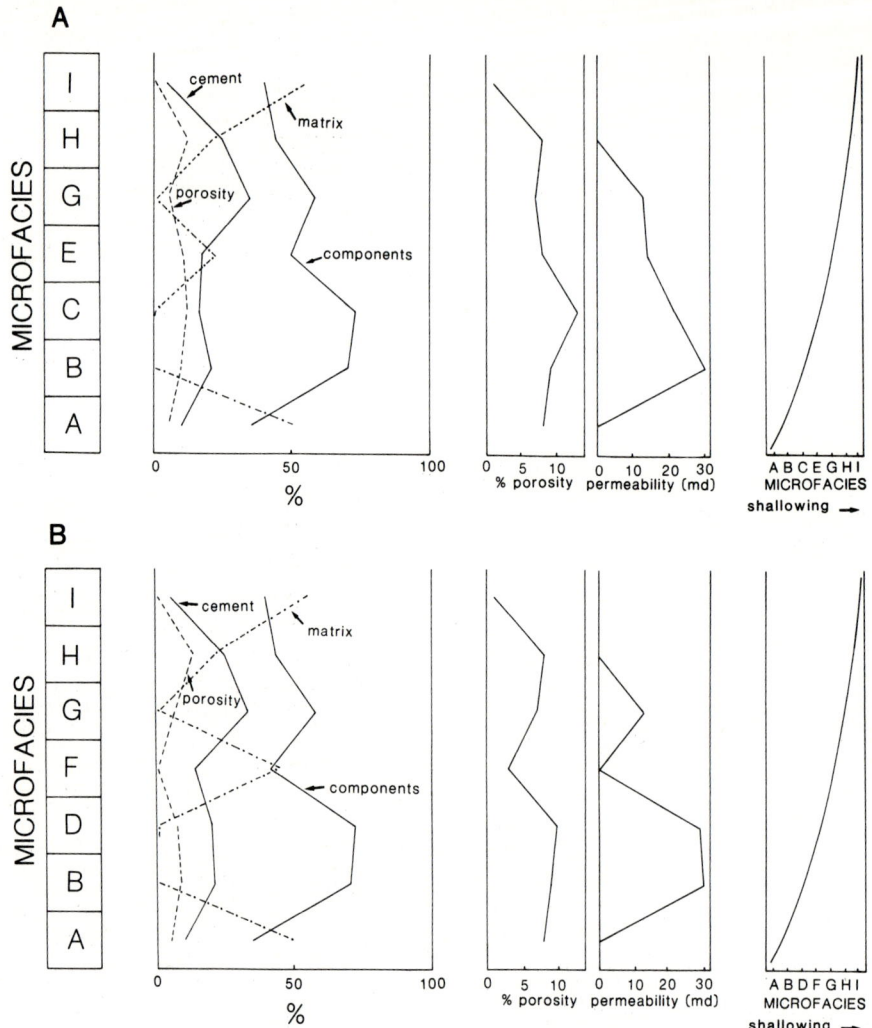

FIGURE 7.5 Salem Limestone (Middle Mississippian) of the southeastern margin of Illinois Basin. A. Ideal oolitic shallowing-upward sequence. B. Ideal bioclastic shallowing-upward sequence. From Reichelderfer (1985).

unstable aragonitic or high-magnesium constituents such as tests of pelecypods and gastropods surrounded by micrite envelopes, pelletoidal filling of burrows with submarine microsparite cement, and concentric rings of ooids. This process occurred in the undersaturated freshwater phreatic environment. A small amount of intercrystalline secondary porosity developed through a dorag process of dolomitization in the slope microfacies and was accompanied by secondary quartz generation (mixed marine and freshwater phreatic).

Early compaction caused minor fracturing, and these early fractures are difficult to distinguish from those of late burial origin. Petrographically, the cements filling the fractures do not appear different, nor could cathode luminescence distinguish between them. The fracture morphology (an even, straight fracture formed during early compaction and a fibrous, branching fracture formed during the more ductile burial stage) usually allows them to be differentiated. The relationship of stylolites to fractures also contributes to the distinction: early compaction fractures are sometimes cut by stylolites, whereas late fractures seem to be associated with the process of stylolitization. Further subsidence of the microfacies into the saturated freshwater phreatic environment led to widespread cementation of all previous types of porosity by interparticle sparite mosaic and syntaxial overgrowths (sparite cementation I).

In the burial environment cementation continued through the action of formation brines as silicification, anhydritization, and precipitation of saddle dolomite as well as some late sparite in fractures (sparite cementation II). The timing of these processes of cementation varies,

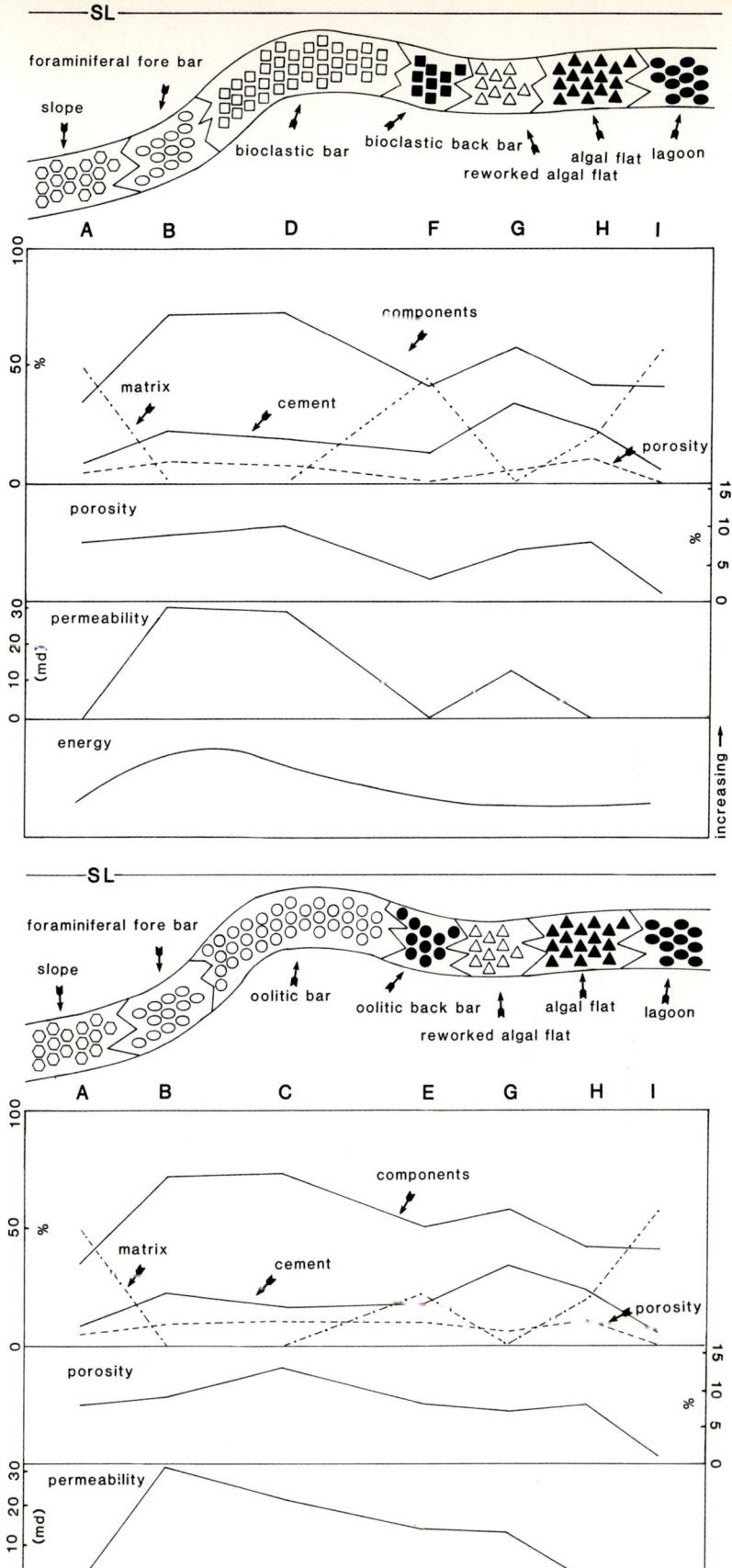

FIGURE 7.6 Salem Limestone (Middle Mississippian) of the southeastern margin of Illinois Basin. Ideal bioclastic bar model. From Carozzi and Reichelderfer (1987). Reprinted by permission of Illinois State Academy of Science.

FIGURE 7.7 Salem Limestone (Middle Mississippian) of the southeastern margin of Illinois Basin. Ideal oolitic bar model. From Carozzi and Reichelderfer (1987). Reprinted by permission of Illinois State Academy of Science.

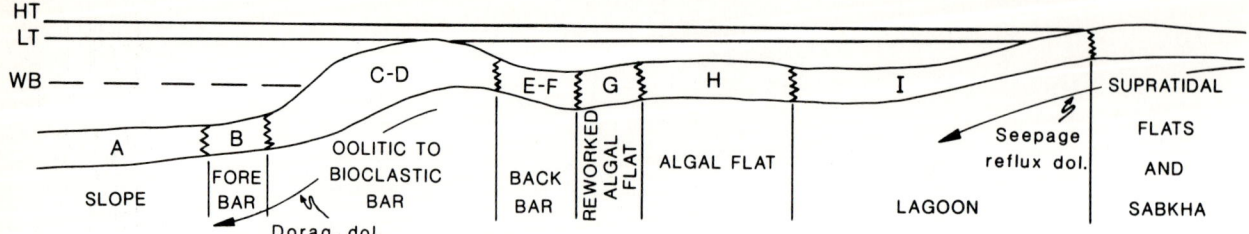

FIGURE 7.8 Salem Limestone (Middle Mississippian) of the southeast margin of the Illinois Basin. General depositional model.

but silicification seems to precede anhydritization, as shown by silicified crinoid fragments replaced by anhydrite. At least two phases of anhydritization are evident: pore filling and replacive. The second phase replaces saddle dolomite, but in another sample, a rhomb of saddle dolomite replaces massive anhydrite. Both saddle dolomite and anhydrite can replace components, matrix, and cement pervasively or act as a pore-filling cement. The reducing conditions in the burial environment probably also account for the reduction of the residual organic matter in *Endothyra* and other bioclasts to form pyrite. During burial, which reached 6000 ft (1829 m) or more during Late Pennsylvanian or Permian and is now approximately 2000 to 4000 ft (610 to 1220 m), no important dissolution processes took place with the exception of very minor generation of fracture and stylolitic secondary porosities. Present-day porosity of economic interest in the Salem Limestone consists of pores which have survived the complex diagenetic evolution described above.

Reservoir Generation

A typical example of shallowing-upward bioclastic sequence is the core from the well Texaco B-1, Francis Wente (Fig. 7.10). There, porosity is confined to the coarse-grained foraminiferal and bioclastic bars (microfacies B, D). In the coarse-grained samples from 2856 to 2848 ft, porosity can be attributed to moderate pressure-solution, especially where overgrowths on crinoid fragments show partially infilled pore space and created a stable framework that does not allow primary pore space to collapse during pressure-solution.

The largest values of porosity and permeability occur when unfilled foraminiferal tests are abundant and provide additional pore space. In samples with abundant overgrowths, components are more often sutured and no pore space is left. In the finer-grained samples, overgrowths tend to occlude all primary porosity.

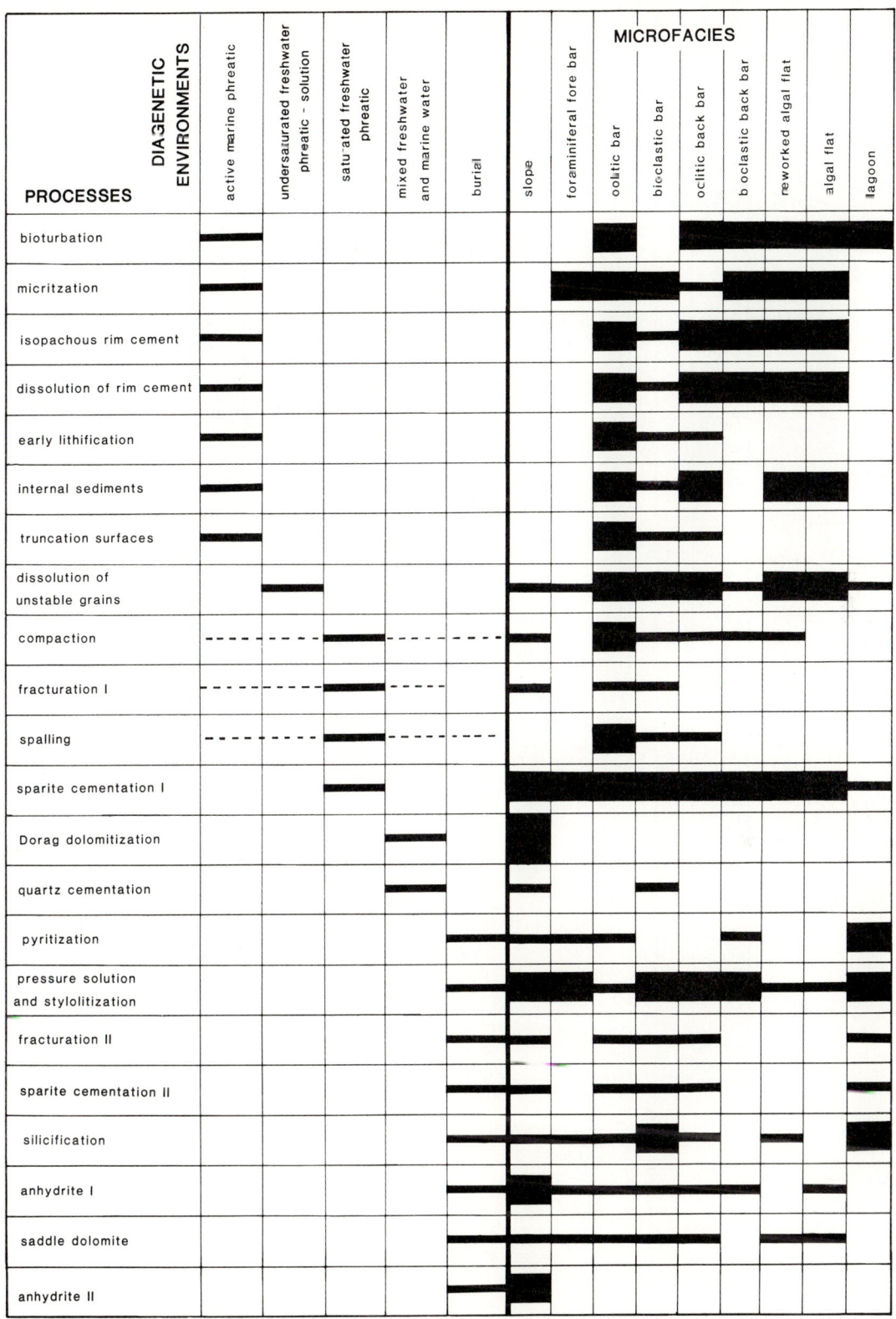

FIGURE 7.9 Salem Limestone (Middle Mississippian) of the southeastern margin of Illinois Basin. Generalized diagenetic sequence. From Carozzi and Reichelderfer (1987). Reprinted by permission of Illinois State Academy of Science.

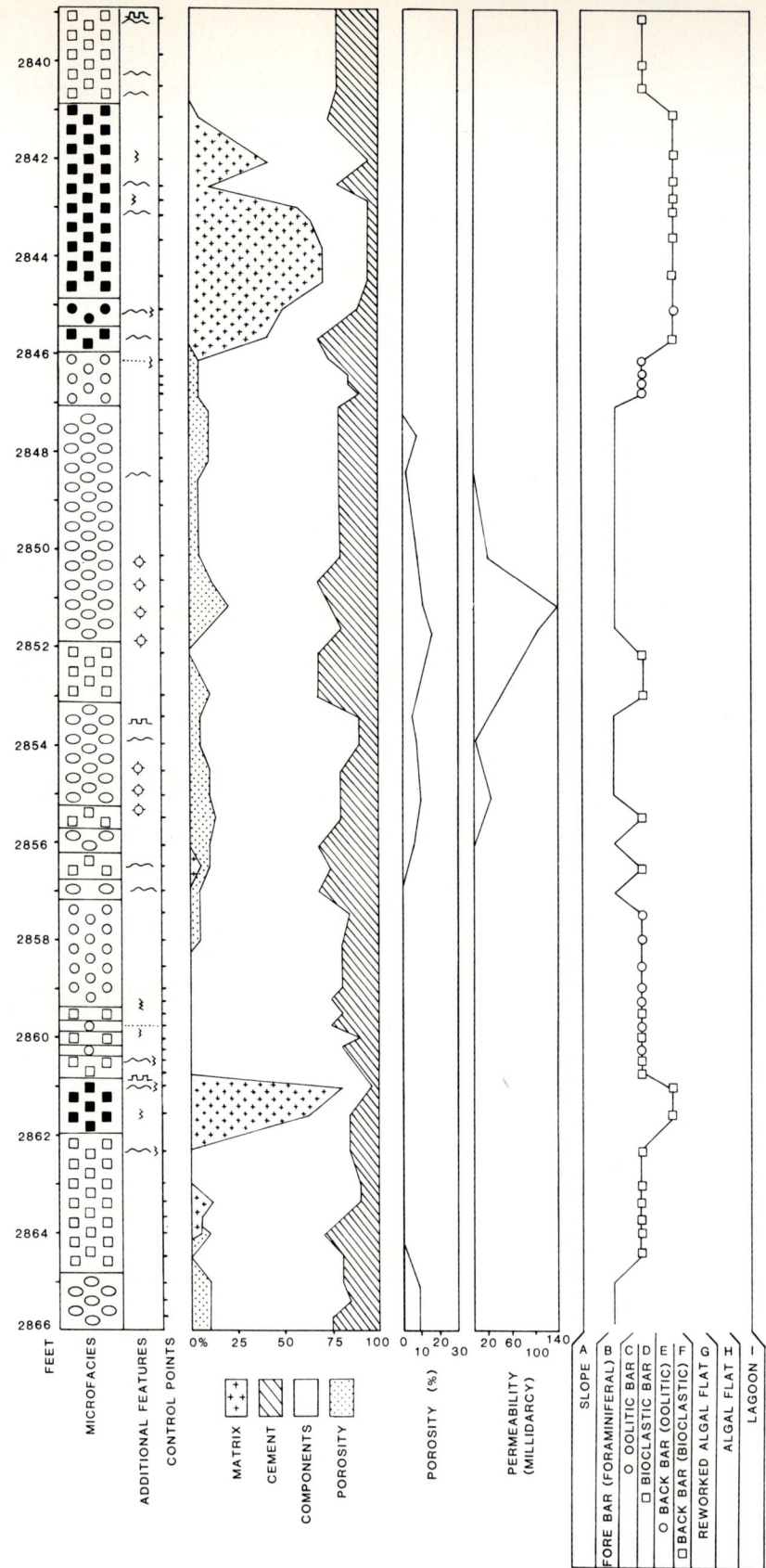

FIGURE 7.10 Salem Limestone (Middle Mississippian) of the southeastern margin of Illinois Basin. Typical vertical variation of microscopic parameters during shallowing-upward sequence, well Texaco B-1, Francis Wente. From Carozzi and Reichelderfer (1987). Reprinted by permission of Illinois State Academy of Science.

Carbonate Ramps with Bioaccumulated to Hydrodynamic Buildups

Subtidal buildups on carbonate ramps consist often of two types of sediments which express variation of energy generated where the wave base intersects the original gentle slope. Organisms encountered most frequently under these conditions are crinoids and bryozoans. They produce sediments arranged in phases of accumulation of bioclasts and phases of mechanical reworking in the immediate vicinity.

8.1 CHARACTERISTIC CONSTITUENTS: BRYOZOANS, CRINOIDS, AND OOIDS

The Kinkaid Formation, the uppermost carbonate-shale sequence of the Upper Mississippian of the southern part of the Illinois Basin (Lasemi and Carozzi, 1981), is truncated at the top by the pre-Pennsylvanian unconformity and therefore ranges in thickness from 0 to 33 m in the outcrop area and from 36 to 49 m in the less-eroded subsurface. The Kinkaid Formation was investigated by means of 12 stratigraphic sections (10 outcrops and two cores) in southern Illinois and western Kentucky, with a total of 1042 thin sections made from samples collected at an average vertical interval of 17 cm.

The depositional evolution of the Kinkaid Formation can be interpreted as resulting from the stratigraphic superposition of three models in the southwest of the Illinois Basin and of four models in the northeast and southeast where siliciclastic deltaic sediments of the Michigan River system interfered with carbonate deposition. All these models are variations of the ramp model with or without hydrodynamic buildups and are discussed in relation to the three members in which the formation is usually subdivided (Fig. 8.1).

Description of Microfacies of Model 1

Ten microfacies were recognized and are described in a shallowing-upward order.

Microfacies 1 (Fig. 8.2A). Strongly bioturbated pelletoidal calcisiltite with abundant monaxonic sponge spicules and calcispheres associated with smaller amounts of earlandiids, endothyrids, and pelmatozoan and bryozoan bioclasts. Patches of microsparite cement fill the porosity of burrows.

Microfacies 2 (Fig. 8.2B). Mud-supported and bioturbated biocalcarenite with calcisiltite matrix. Major bioclasts are gastropods altered by neomorphism, pelmatozoans, and bryozoans associated with paleotextulariids, endothyrids, and fecal pellets.

Microfacies 3 (Fig. 8.2C). Grain-supported and pressure-welded pelmatozoan-bryozoan calcarenite with calcisiltite matrix.

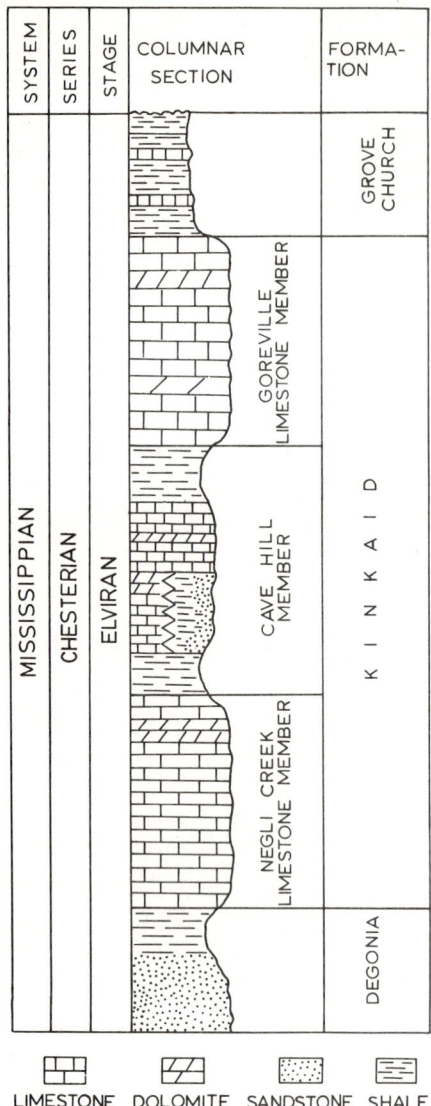

FIGURE 8.1 Kinkaid Formation (Upper Mississippian), southern part of Illinois Basin. Columnar section. From Lasemi and Carozzi (1981).

Microfacies 4 (Fig. 8.2D). Grain-supported pelmatozoan-bryozoan calcarenite with association of abundant calcisiltite matrix and patches of sparite mosaic cement and syntaxial overgrowths around crinoid bioclasts. Endothyrids and paleotextulariids are predominant among benthic foraminifers and *Archimedes* are common.

Microfacies 5 (Fig. 8.2E). Grain-supported bryozoan-pelmatozoan calcarenite with cavity-filling sparite cement and syntaxial overgrowths around crinoid bioclasts. Paleotextulariids and endothyrids are very common.

Microfacies 6 (Fig. 8.2F). Grain-supported oolitic biocalcarenite with isopachous rim cement and cavity-filling sparite cement. Ooids locally are cerebroid and also spalled by compaction. The uncoated bioclasts are the same as in microfacies 4 and 5, together with stacheiinids they form the cores of ooids. This microfacies is cross-bedded megascopically.

Microfacies 7 (Fig. 8.3A). Grain-supported algal-foraminiferal calcarenite with association of calcisiltite matrix, cavity-filling sparite, and syntaxial overgrowths around crinoid bioclasts. Stacheiinids, paleotextulariids (*Climacammina*), and endothyrids are the main constituents. Scattered incipient blue-green algal oncoids, fragments of oncoids, and algal coatings on bioclasts are frequent together with lithic and fecal pellets. Ooids are rare. This microfacies is cross-bedded megascopically.

Microfacies 8 (Fig. 8.3B,C,D). Grain-supported oncoidal biocalcarenite with calcisiltite matrix or association of the latter with sparite cement. Oncoids possess nuclei of gastropod or of prismatic brachiopod shells. They are shaped irregularly and built either by aggregation of *Girvanella* tubes (Fig. 8.3C) or by concentrically laminated stromatolitic algal mats (Fig. 8.3D). Other constituents are ammodiscids, calcareous worm tubes, abundant small bioclasts of prismatic brachiopods, calcispheres, lithic pellets, and intraclasts derived from the abrasion of oncoids.

Microfacies 9 (Fig. 8.3E). Mud-supported oncoidal biocalcarenite with abundant calcisiltite matrix. Abraded oncoids, pellets, and intraclasts derived from the former are associated with common benthic foraminifers and minute bioclasts similar to those of microfacies 8. Bioturbation is very common.

Microfacies 10 (Fig. 8.3F). Bioturbated calcisiltite with scattered sand-size bioclasts similar to those of the preceding two microfacies, but much less abundant. Common vertical burrows are filled with microsparite cement and geopetal concentrations of fecal pellets.

The Ideal Depositional Model 1

Depositional model 1 includes the Negli Creek Member and the lower part of the Cave Hill Member. Two field sections (Figs. 8.4, 8.5, 8.6) are used to show detailed variation of microscopic parameters during shallowing-upward and deepening-upward sequences.

Model 1 was divided into three major depositional environments (Figs. 8.7, 8.8) as follows. An open marine subtidal environment (microfacies 1 to 5) is characterized by a pelmatozoan-fenestellid bryozoan suite with associated pelecypods and trilobites. This faunal association

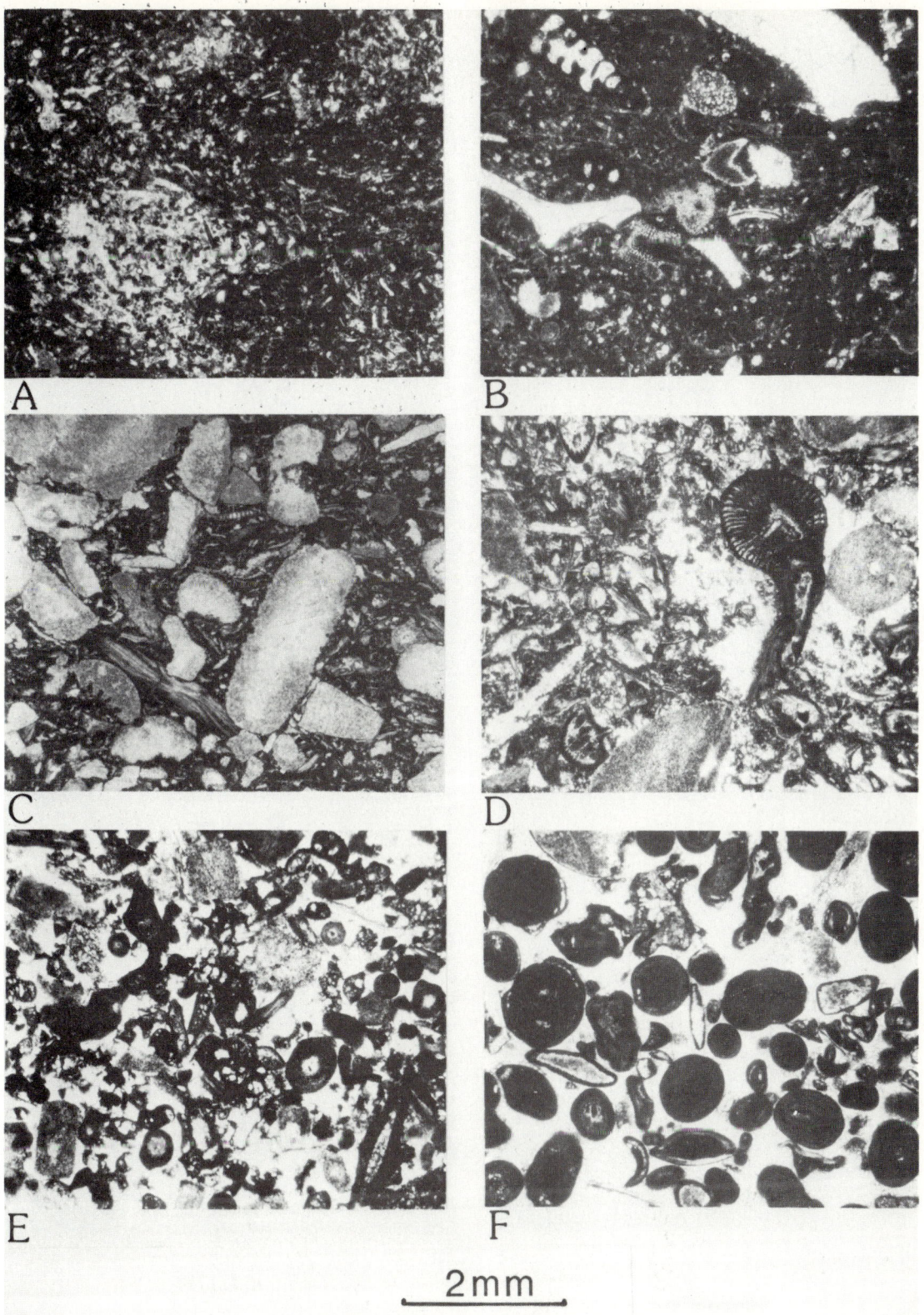

FIGURE 8.2 Kinkaid Formation (Upper Mississippian), southern part of Illinois Basin. Typical microfacies of ideal depositional model 1. A. Microfacies 1. B. Microfacies 2. C. Microfacies 3. D. Microfacies 4. E. Microfacies 5. F. Microfacies 6. See text for detailed descriptions. All photomicrographs: plane-polarized light. From Lasemi and Carozzi (1981).

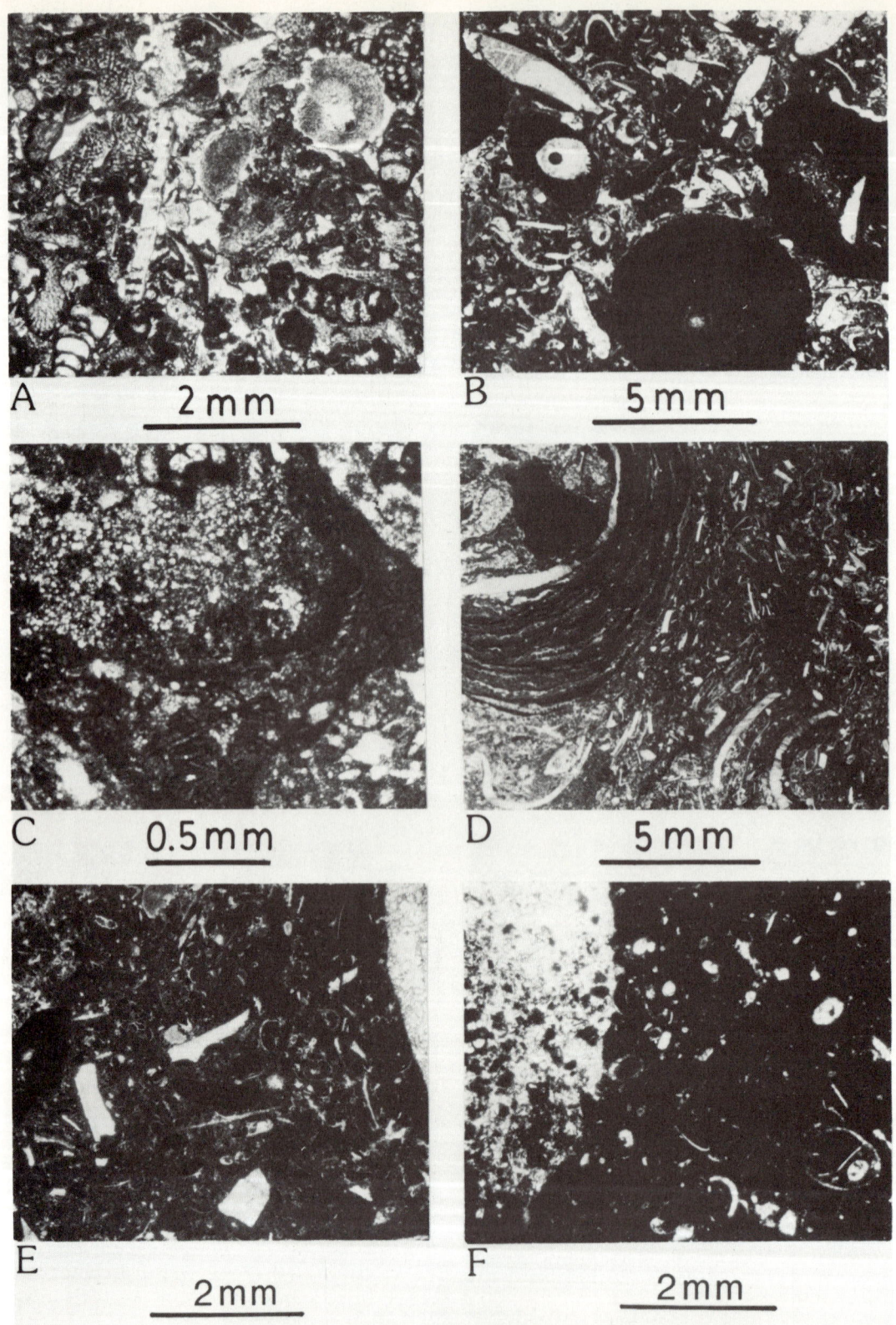

FIGURE 8.3 Kinkaid Formation (Upper Mississippian), southern part of Illinois Basin. Typical microfacies of ideal depositional model 1 (continued). A. Microfacies 7. B to D. Microfacies 8. E. Microfacies 9. F. Microfacies 10. See text for detailed descriptions. All photomicrographs: plane-polarized light, except B and D, crossed nicols. From Lasemi and Carozzi (1981).

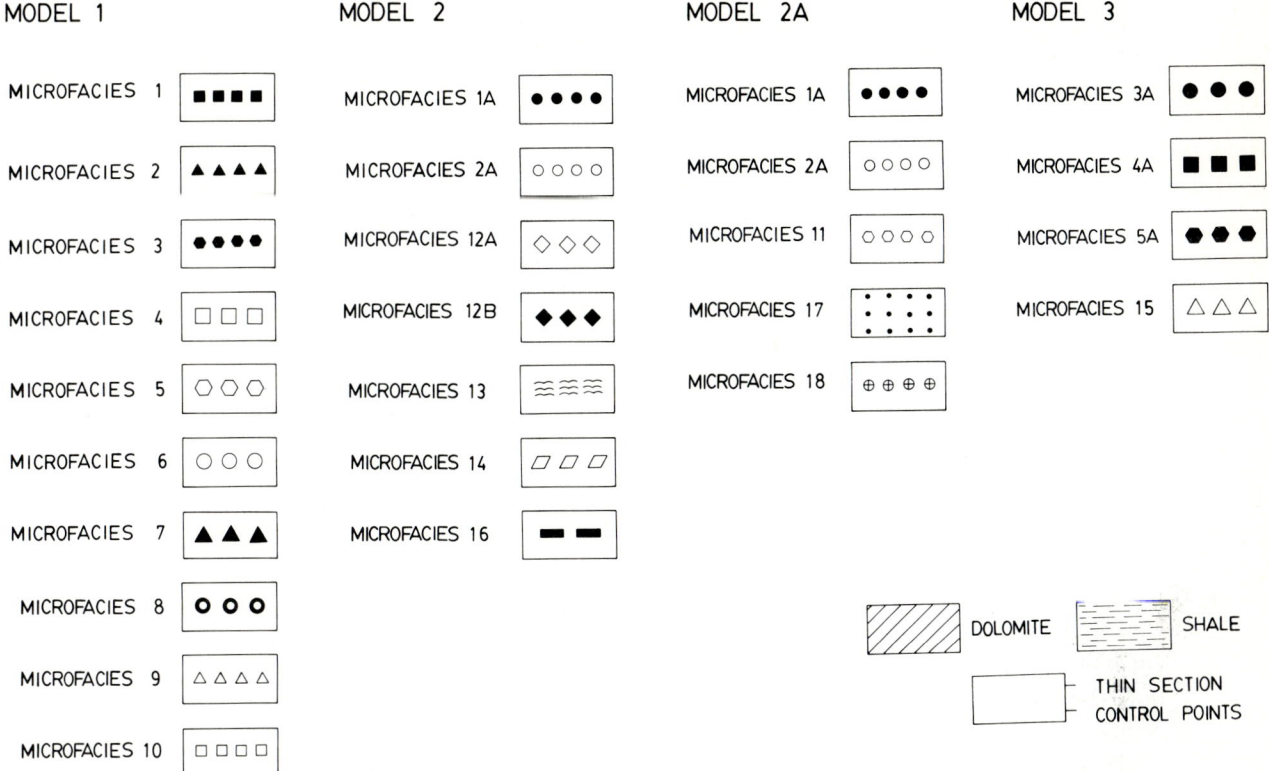

FIGURE 8.4 Kinkaid Formation (Upper Mississippian), southern part of Illinois Basin. Symbols for microfacies. From Lasemi and Carozzi (1981).

lived mostly in microfacies 4 and 5, and their debris were transported both into shallower and deeper conditions where bioturbation was also intense. Dolomitization in this environment resulted from a dorag meteoric-seawater mixing process related to a freshwater lens developed on the adjacent and temporarily exposed hydrodynamic calcarenitic to oolitic buildup. This second intertidal environment corresponds to a discontinuous hydrodynamic buildup (microfacies 6), which when not oolitic would be represented by the landward edge of microfacies 5. The hydrodynamic buildup is characterized by ooids that are reworked into adjacent microfacies, vadose silt, exposure crusts, and rare calcite pseudomorphs after anhydrite. The third environment (microfacies 7 to 10) coresponds to protected shallow subtidal to intertidal conditions developed behind the hydrodynamic buildup. Its fauna assemblage is rich in benthic foraminifers, oncoids, dasyclads, and *Chara*, and it contains frequent bioturbation as vertical burrows. The occurrence of vadose silt, exposure crusts, and moderate dolomitization by seepage reflux reflects the influence of a high intertidal to supratidal environment that has not been preserved.

Examination of the distribution of such constituents as ostracods, sponge spicules, calcispheres, intraclasts, pellets, and ooids shows dispersion seaward, landward, or both, which is indicative of the action of currents and tides.

Description of Microfacies of Models 2 and 2A

Models 2 and 2A show the general shape of ramps. They share a common open marine subtidal environment (microfacies 1A and 2A) which grades landward into an intertidal carbonate flat environment for model 2 (microfacies 12, 13, 14, and 16), and a laterally equivalent deltaic-prodeltaic environment for model 2A (microfacies 11, 17, and 18). They are geographically located in the northeast and southeast outcrop belt where the Michigan River delta system emptied into the basin.

Nine microfacies were recognized and are described in a shallowing-upward order.

Microfacies 1A (Fig. 8.9A,B). Heavily bioturbated, argillaceous biocalcisiltite with scattered monaxonic sponge spicules, sand-size bioclasts of bryozoans (*Archimedes*), pelmatozoans, brachiopods, and many types of calcareous benthic foraminifers. Fecal pellets are abundant.

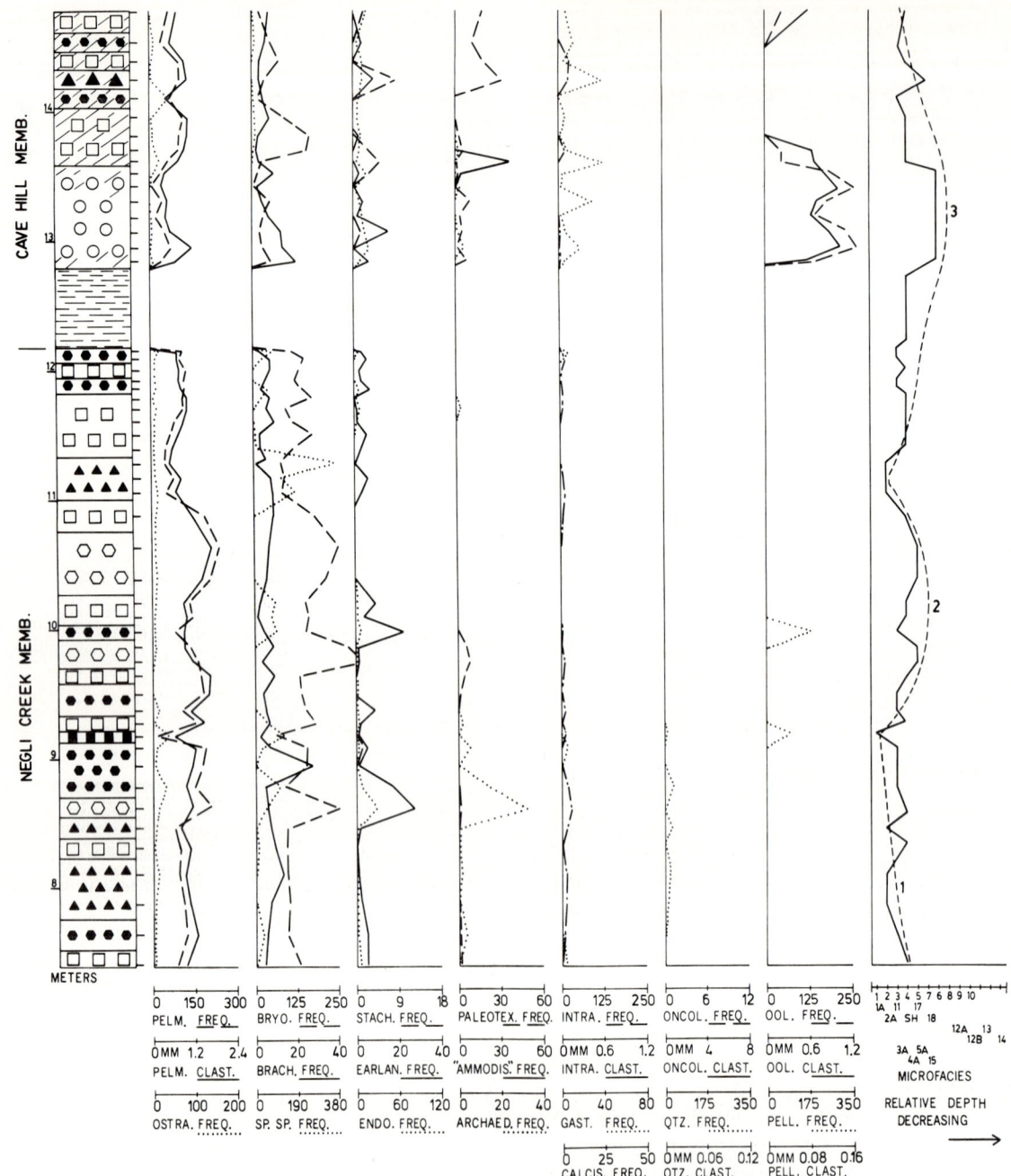

FIGURE 8.5 Kinkaid Formation (Upper Mississippian), southern part of Illinois Basin. Typical example of vertical variation of microscopic parameters. From Lasemi and Carozzi (1981).

Microfacies 2A (Fig. 8.9C). Bioturbated, argillaceous mud-supported biocalcarenite with calcisiltite matrix. Bryozoans, brachiopods, pelmatozoans, and neomorphosed pelecypods are the major components. Frequent geopetal filling of articulated shells by fecal pellets is overlain by sparite.

Microfacies 12A (Fig. 8.9D,E). Pelletoidal biocalcisiltite with scattered bioclasts among which monaxonic sponge spicules and brachiopod spines predominate over calcispheres, gastropods, ostracods, and ammodiscids. Bioturbation is very common. Locally, this microfacies grades into a grain-supported intraclastic cal-

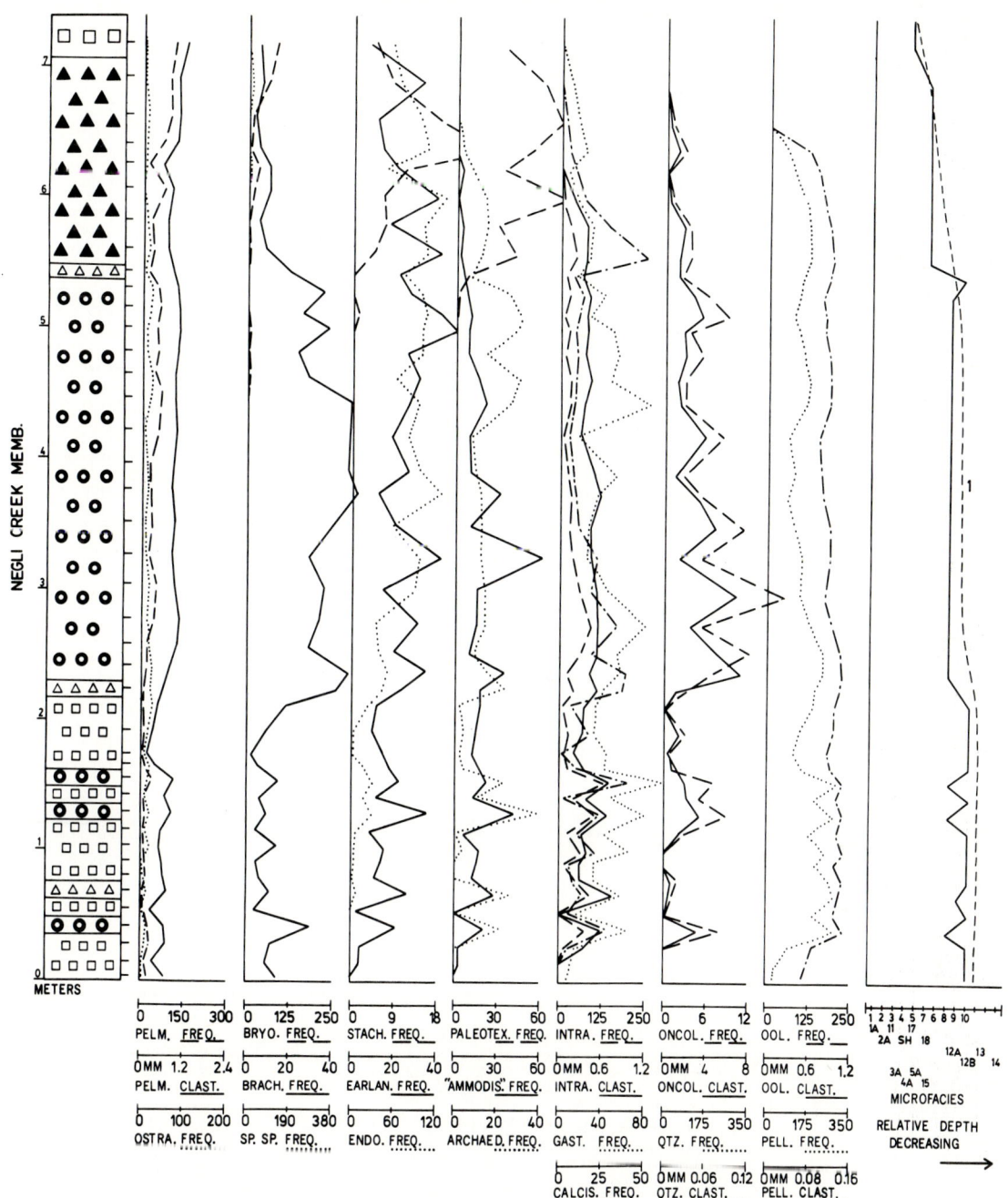

FIGURE 8.6 Kinkaid Formation (Upper Mississippian), southern part of Illinois Basin. Typical example of vertical variation of microscopic parameters. From Lasemi and Carozzi (1981).

carenite (Fig. 8.9E) with strongly bioturbated calcisiltite matrix, probably representing a tidal channel deposit.

Microfacies 12B (Figs. 8.9F, 8.10A,B). Laminated calcisiltite with alternating darker calcisiltite laminae rich in organic matter and bioclast-rich lighter laminae. Identifiable microdebris are ostracods, sponge spicules, smaller benthic foraminifers, intraclasts, pellets, and silt-size detrital quartz. Bioclast-rich laminae are often graded bedded with sharp basal contacts (Fig. 8.9F). They show desiccation cracks and fenestral fabric (Fig. 8.10A) as well as many small-scale scour-and-fill

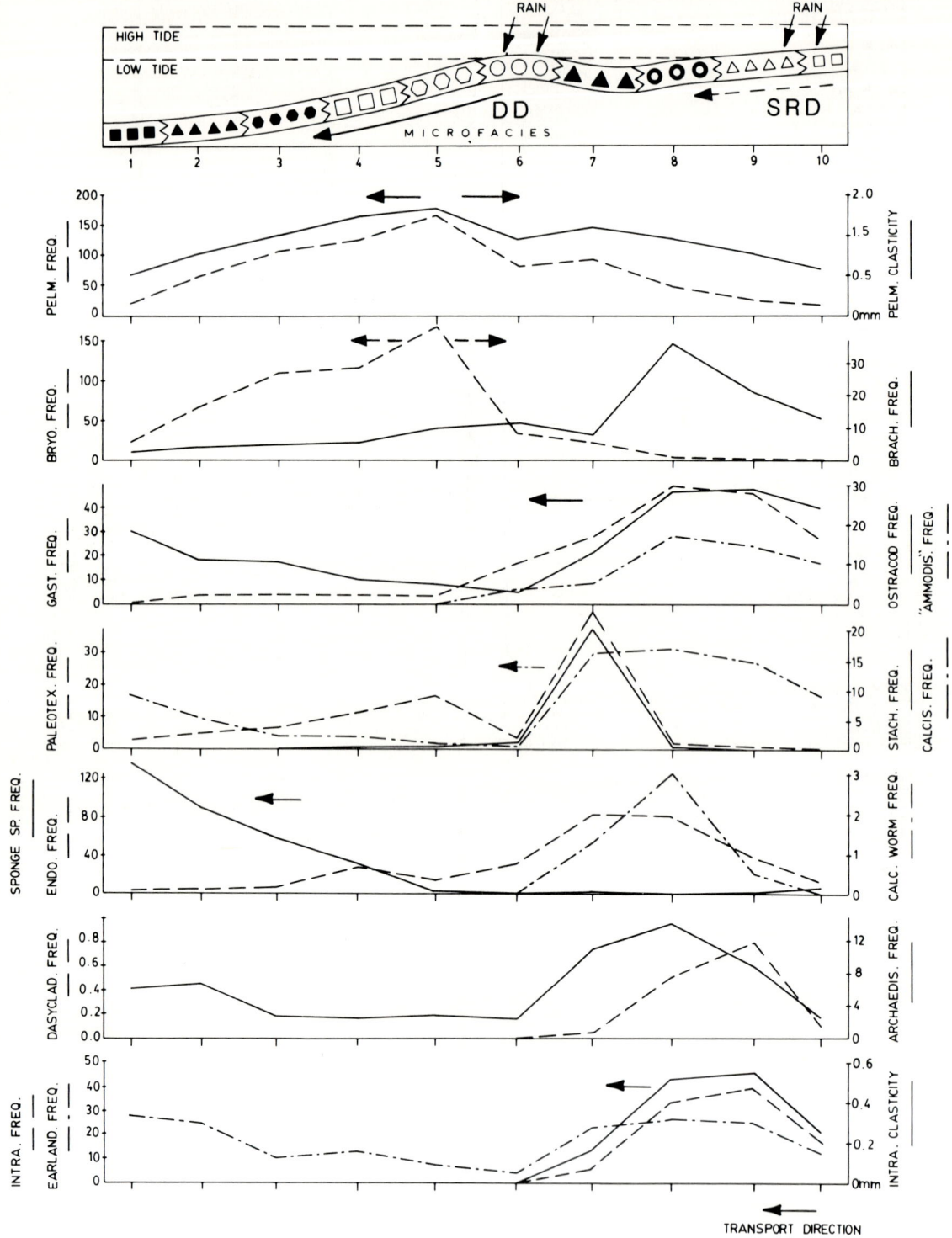

FIGURE 8.7 Kinkaid Formation (Upper Mississippian), southern part of Illinois Basin. Ideal depositional model 1. From Lasemi and Carozzi (1981).

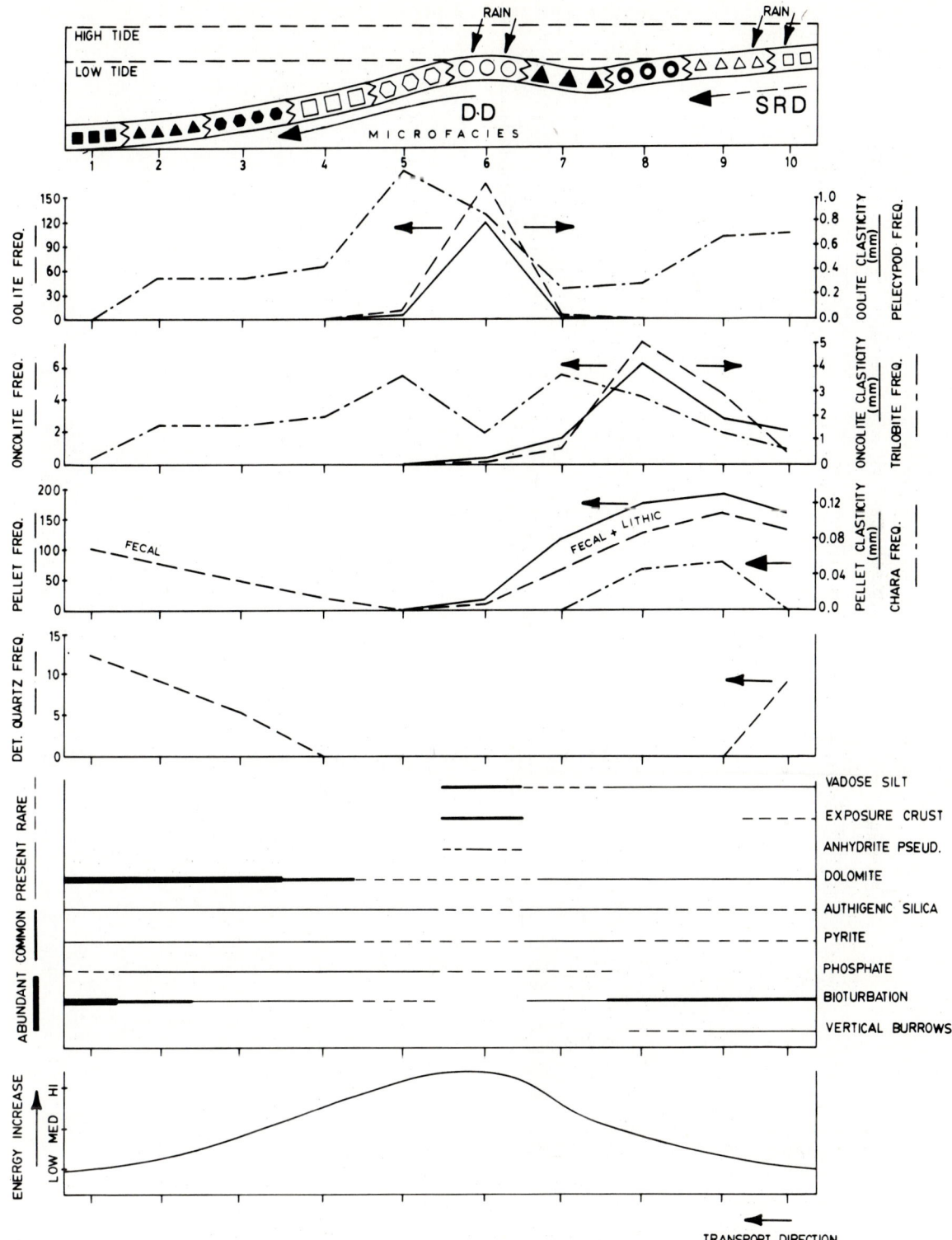

FIGURE 8.8 Kinkaid Formation (Upper Mississippian), southern part of Illinois Basin. Ideal depositional model 1 (continued). From Lasemi and Carozzi (1981).

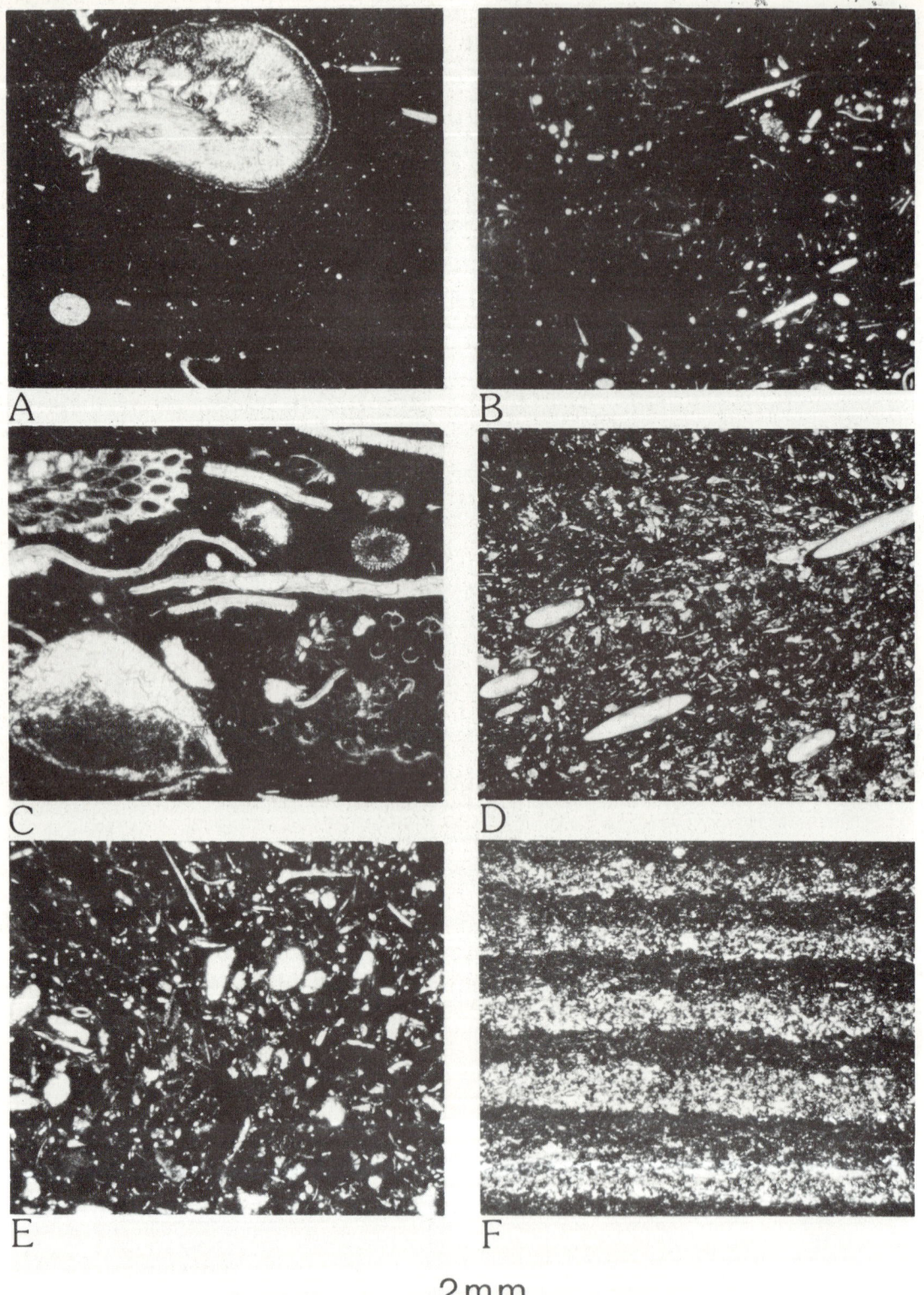

FIGURE 8.9 Kinkaid Formation (Upper Mississippian), southern part of Illinois Basin. Typical microfacies of ideal depositional models 2 and 2A. A and B. Microfacies 1A. C. Microfacies 2A. D and E. Microfacies 12A. F. Microfacies 12B. See text for detailed descriptions. All photomicrographs: plane-polarized light. From Lasemi and Carozzi (1981).

Chap. 8 / Carbonate Ramps with Bioaccumulated to Hydrodynamic Buildups   141

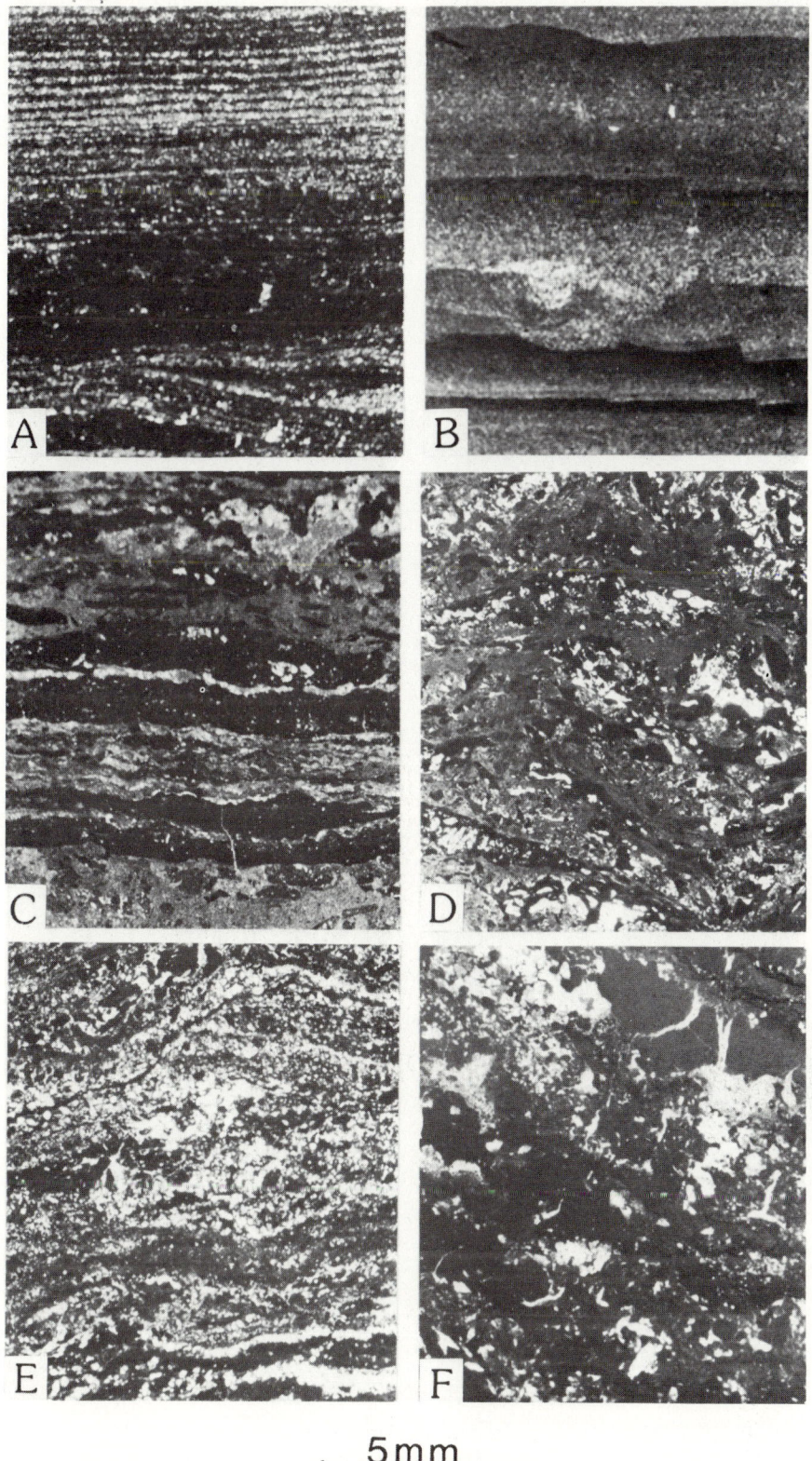

FIGURE 8.10 Kinkaid Formation (Upper Mississippian), southern part of Illinois Basin. Typical microfacies of ideal depositional models 2 and 2A (continued). A and B. Microfacies 12B. C to F. Microfacies 13. See text for detailed descriptions. All photomicrographs: plane polarized light. From Lasemi and Carozzi (1981).

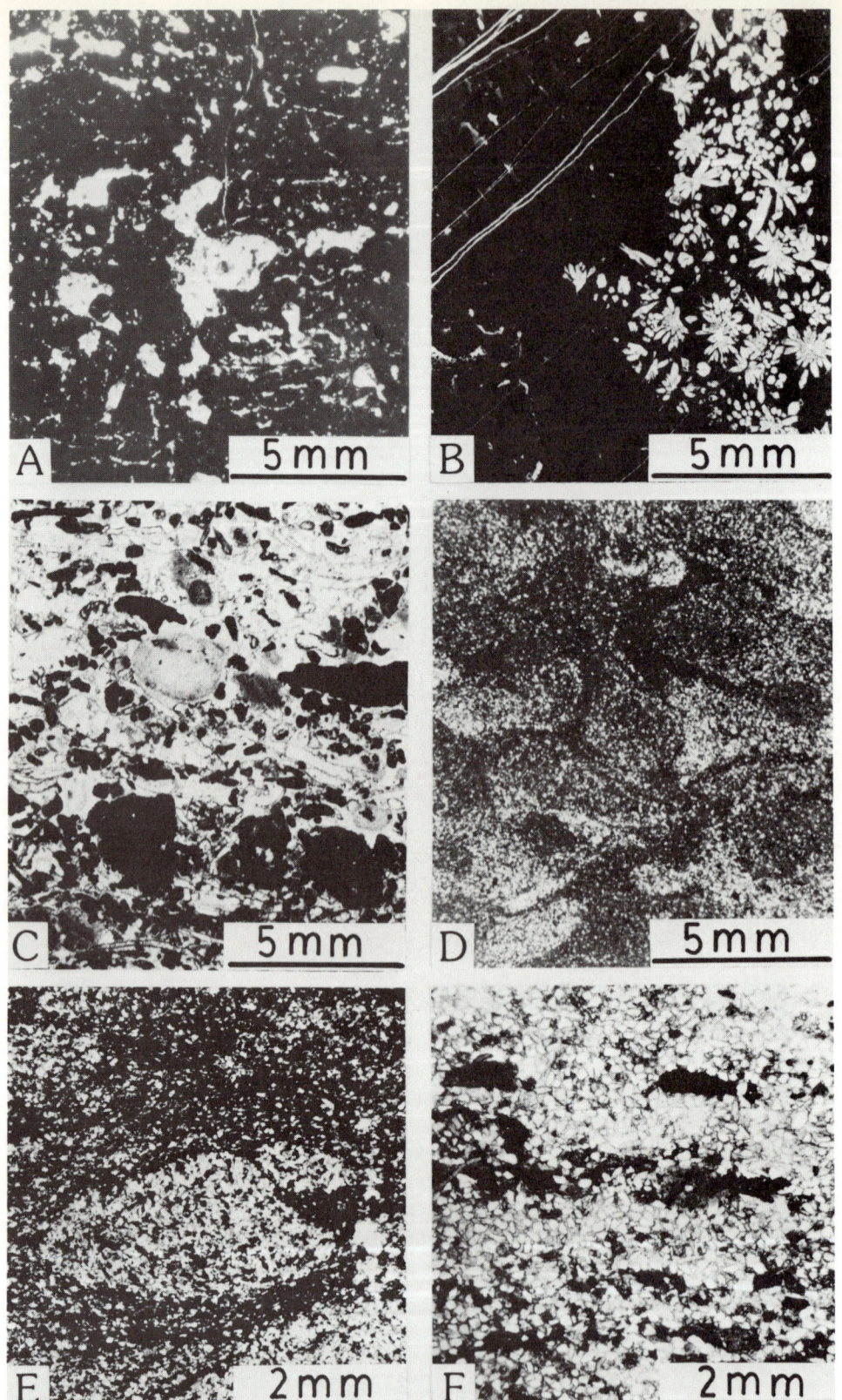

FIGURE 8.11 Kinkaid Formation (Upper Mississippian), southern part of Illinois Basin. Typical microfacies of ideal depositional models 2 and 2A (continued). A and B. Microfacies 14. C. Microfacies 16. D. Microfacies 11. E. Microfacies 17. F. Microfacies 18. See text for detailed descriptions. Photomicrographs: A, B, C, and D, crossed nicols, E, and F, plane-polarized light. From Lasemi and Carozzi (1981).

structures (Fig. 8.10B). Association of these features indicates deposition of bioclast-rich laminae by incoming tides that carried bioclasts from deeper environments over intertidal flats. Darker laminae, rich in organic matter, formed at low tide by blue-green algal activities. The sediment trapping mechanism of algae would be responsible for deposition of the bioclast-rich laminae, and upward growth of algal filaments during ebb tide would ensure their preservation.

Microfacies 13 (Fig. 8.10C to F). Flat-laminated to undulatory stromatolite constructed limestone to dolomite consisting of alternating dark pelletoidal laminae with abundant filaments of blue-green algae and of lighter laminae of trapped pellets, algal chips and intraclasts, and calcispheres. Calcite pseudomorphs after gypsum as lenticular crystals (Fig. 8.10D) and layers (Fig. 8.10E) which disrupted algal mats are widespread. Desiccation cracks (Fig. 8.10F), exposure crusts, and fenestral fabric, due to the dissolution of gypsum and anhydrite crystals, and partially filled with vadose silt, are extremely common.

Microfacies 14 (Fig. 8.11A,B). Pelletoidal calcilutite with a few ostracods and calcispheres. Tubular and irregular fenestrae filled with sparite mosaic (replacing anhydrite?) or vadose silt are widespread. Rosettes (Fig. 8.11B) and laths of calcite pseudomorphic after anhydrite are very common, as well as desiccation textures and collapse breccias resulting from evaporite dissolution.

Microfacies 16 (Fig. 8.11C). Grain-supported and graded-bedded intraclastic biocalcarenite with sparite cement. Intraclasts are derived from stromatolites and bioclasts represented by those of all the microfacies of model 2. These features and the fact that this microfacies overlies microfacies 14 with a sharp erosional bar indicates that it is a tempestite.

Microfacies 11 (Fig. 8.11D). Strongly bioturbated argillaceous-arenaceous calcisiltite with scattered intraclasts and pellets.

Microfacies 17 (Fig. 8.11E). Heavily bioturbated quartzose-argillaceous siltstone with many small-scale cross-beds and scattered pellets.

Microfacies 18 (Fig. 8.11F). Fine-grained quartz arenite, cross-bedded and bioturbated with slightly argillaceous poikilotopic sparite cement. Bioclasts of pelmatozoans, bryozoans, and brachiopods are associated with lithic pellets.

The Ideal Depositional Models 2 and 2A

Models 2 and 2A, which are lateral equivalents, extend from the lower part of the Cave Hill Member to the base of the Goreville Member. Model 2 (Figs. 8.12, 8.13) can be divided into three depositional environments as follows. One is an open marine subtidal environment (microfacies 1A and 2A) characterized by a rich fauna in which bryozoans, brachiopods, and pelmatozoans predominate. The second is a low to high intertidal environment (microfacies 12A and 12B), marked by peaks of the frequency of gastropods, ostracods, ammodiscids, endothyrids, archaeodiscids, earlandiids, and *Chara*. A third high intertidal to supratidal environment (microfacies 13 and 14) is entirely dominated by stromatolites with pellets and intraclasts, derived from their desiccation and reworking. This subevaporitic environment contains only calcispheres and displays plant roots, exposure crusts, and collapse breccias. A sabkha-type dolomitization with seepage reflux as dolomicrite to dolomicrosparite is widespread.

Examination of the distribution of some constituents, such as ostracods, sponge spicules, calcispheres, intraclasts, pellets, bryozoans, and crinoids, shows dispersion landward, seaward, or both, which is indicative of the action of currents and tides.

Model 2A (Fig. 8.14) is characterized by the deltaic-prodeltaic environment of arenaceous and bioturbated microfacies 11, 17, and 18. Its bioclasts are reworked from older microfacies, whereas the lithic-fecal pellets and intraclasts are also reworked but from contemporaneous exposure crusts of adjacent carbonate tidal flats.

Description of Microfacies of Model 3

Four microfacies have been recognized and described in a shallowing-upward order.

Microfacies 3A (Fig. 8.15A,B). Grain-supported and poorly sorted pelmatozoan-bryozoan biocalcarenite with common pressure-solution and calcisiltite matrix. Endothyrids, paleotextulariids, and other calcareous benthic foraminifers are associated with brachiopod bioclasts and calcispheres.

Microfacies 4A (Fig. 8.15C). Grain-supported pelmatozoan-bryozoan biocalcarenite with association of calcisiltite matrix and sparite mosaic cement. Endothyrids are common and associated with the same subordinate constituents as in microfacies 3A.

Microfacies 5A (Fig. 8.15D,E). Grain-supported pelmatozoan-bryozoan biocalcarenite with cavity-filling sparite cement and syntaxial rim cement around crinoid bioclasts. Brachiopods and endothyrids are common and associated with the same subordinate constituents as in microfacies 3A.

Microfacies 15A (Fig. 8.15F). Pelletoidal calcisiltite with scattered bioclasts among which ammodiscids, ostracods, and gastropods predominate. Rare fenestral fabric and vadose silt have been observed.

144 Part II / Carbonate Ramps

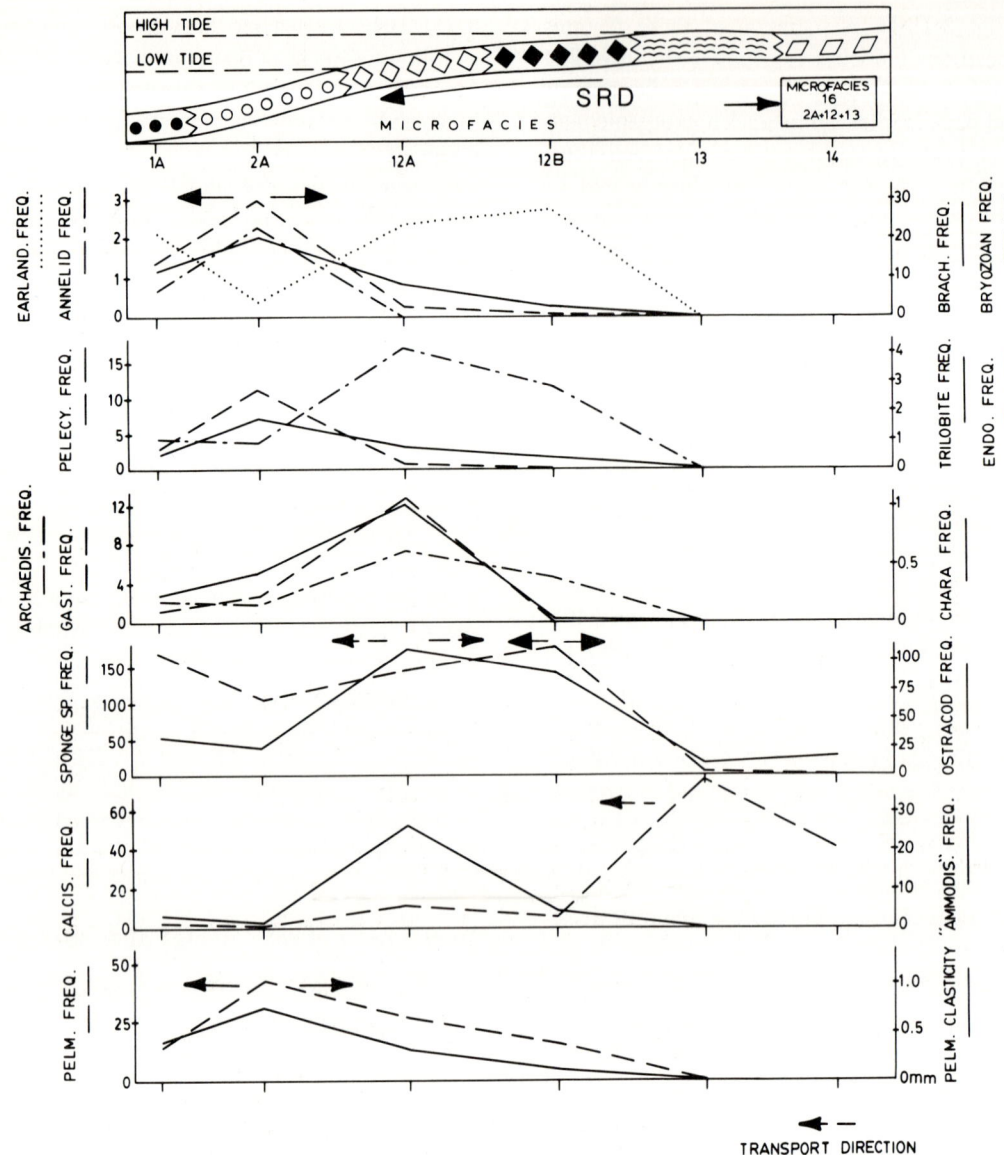

FIGURE 8.12 Kinkaid Formation (Upper Mississippian), southern part of Illinois Basin. Ideal depositional model 2. From Lasemi and Carozzi (1981).

The Ideal Depositional Model 3

This model extends from the base of the Goreville Member to the basal Pennsylvanian unconformity and was never interrupted by influx of terrigenous materials. It can be subdivided into the following two environments (Fig. 8.16). An open marine subtidal environment with microfacies 3A, 4A, and 5A which are similar essentially to the pelmatozoan-bryozoan suite of model 1. Dolomitization is interpreted as resulting from a meteoric seawater mixing (dorag model) related to a freshwater lens developed landward on the exposed portion of the model, which has subsequently been eroded. An open marine intertidal environment represented by microfacies 15 indicates some protected conditions behind the higher-energy microfacies 5A. A distinct set of components reaches here peaks of their frequency: ostracods, gastropods, sponge spicules, earlandiids, ammodiscids, and calcispheres. Bioturbation is very common, and some vadose silt is present. However, examination of the distribution of some constituents, such as crinoids, bryozoans, ostracods, sponge spicules, and pellets, shows dispersion seaward, or landward, or both, indicative of the action of currents and tides.

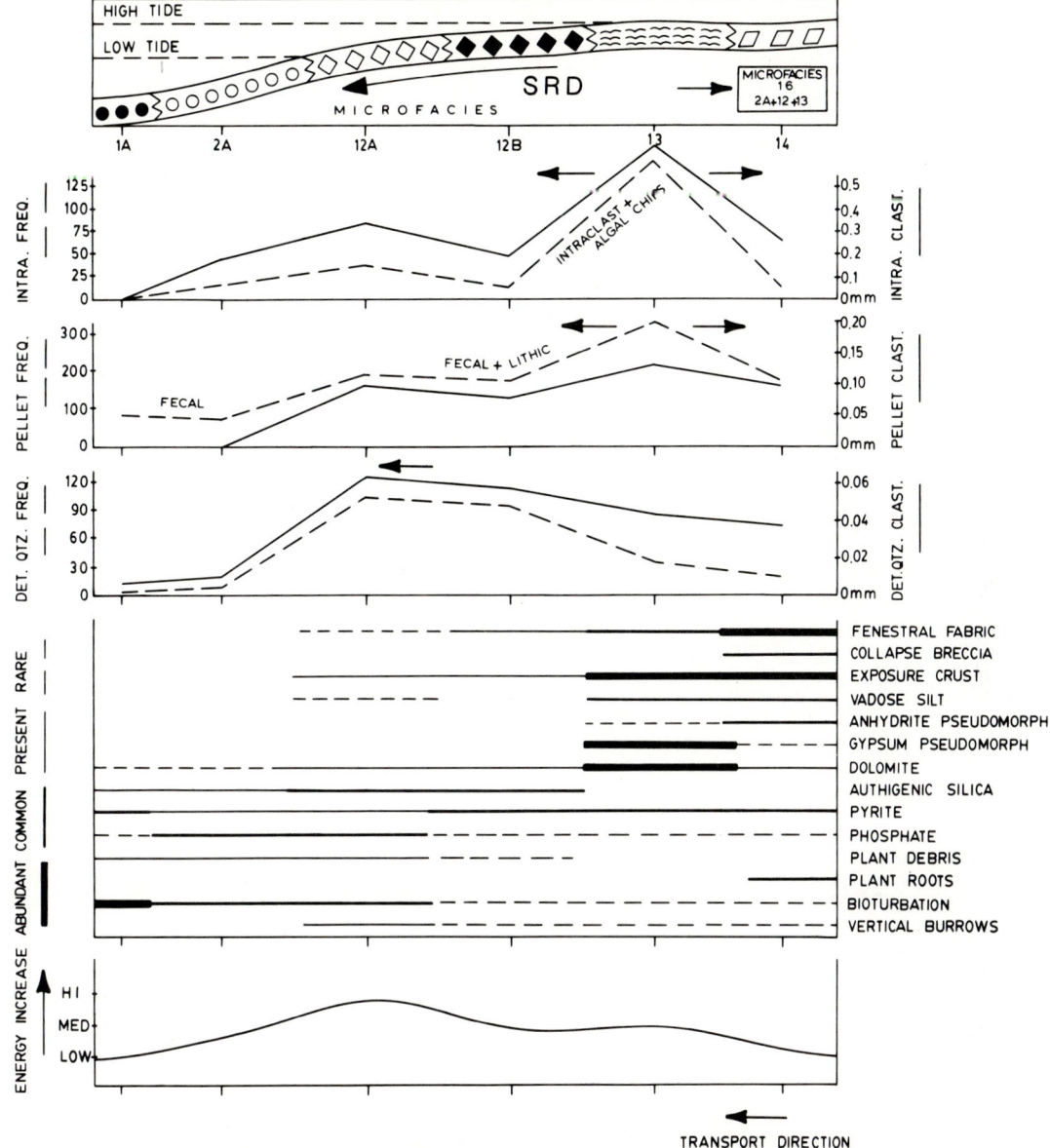

FIGURE 8.13 Kinkaid Formation (Upper Mississippian), southern part of Illinois Basin. Ideal depositional model 2 (continued). From Lasemi and Carozzi (1981).

The General Evolution of the Depositional Environments

A composite section (Fig. 8.17) consisting of the stratigraphic superposition of models 1, 2, and 3 is characteristic of the southwest of the Illinois Basin. An examination of the curve of general depth trends reveals three episodes and 15 sequences of deepening and shallowing. Episode I indicates a general deepening, episode II is a gentle shallowing, and episode III is another gentle deepening truncated by the Pennsylvanian unconformity.

8.2 CHARACTERISTIC CONSTITUENTS: CRINOIDS

In ramps submitted to greater energy conditions, crinoidal subtidal buildups may be destroyed as rapidly as they accumulate by the combined action of waves and currents. As a result, widespread dispersion of bioclasts involves the entire ramp and even reaches basinal areas.

The Rundle Group (Mississippian) of the Canadian Rocky Mountain Front Range of the Ram River area, in southwestern Alberta (Walpole and Carozzi, 1961),

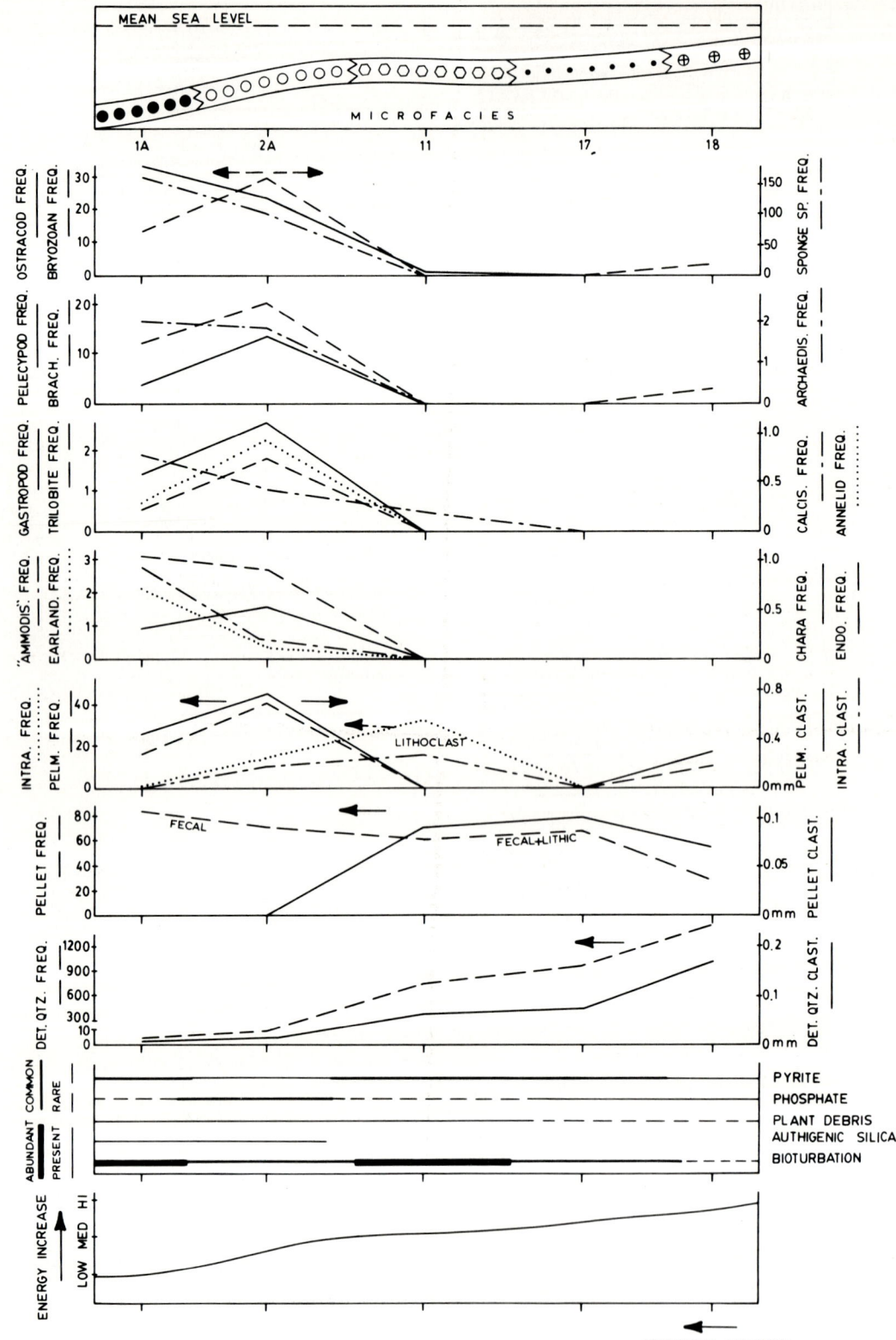

FIGURE 8.14 Kinkaid Formation (Upper Mississippian), southern part of Illinois Basin. Ideal depositional model 2A. From Lasemi and Carozzi (1981).

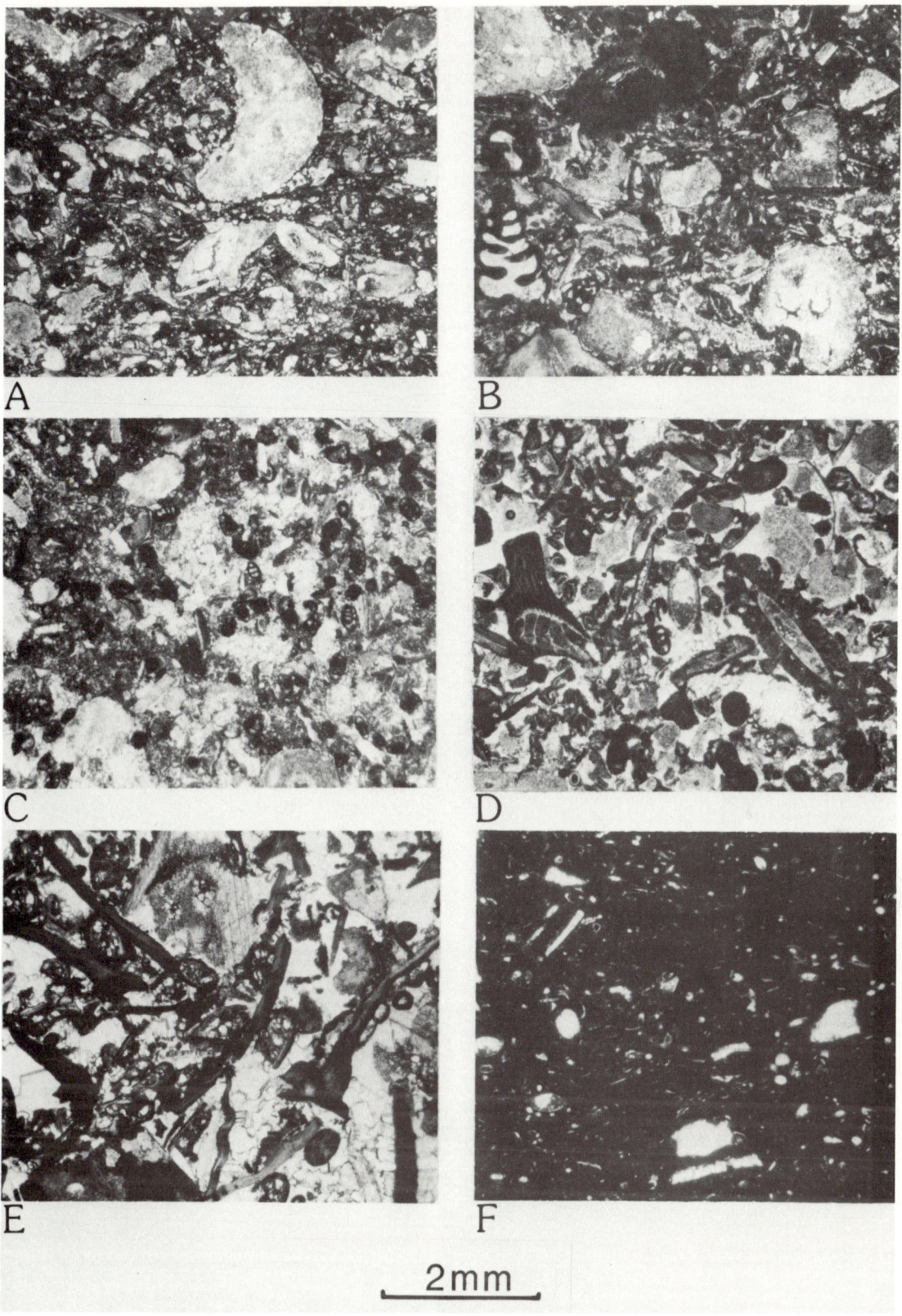

FIGURE 8.15 Kinkaid Formation (Upper Mississippian), southern part of Illinois Basin. Typical microfacies of ideal depositional model 3. A and B. Microfacies 3A. C. Microfacies 4A. D and E. Microfacies 5A. F. Microfacies 15A. See text for detailed descriptions. All photomicrographs: plane-polarized light. From Lasemi and Carozzi (1981).

148 Part II / Carbonate Ramps

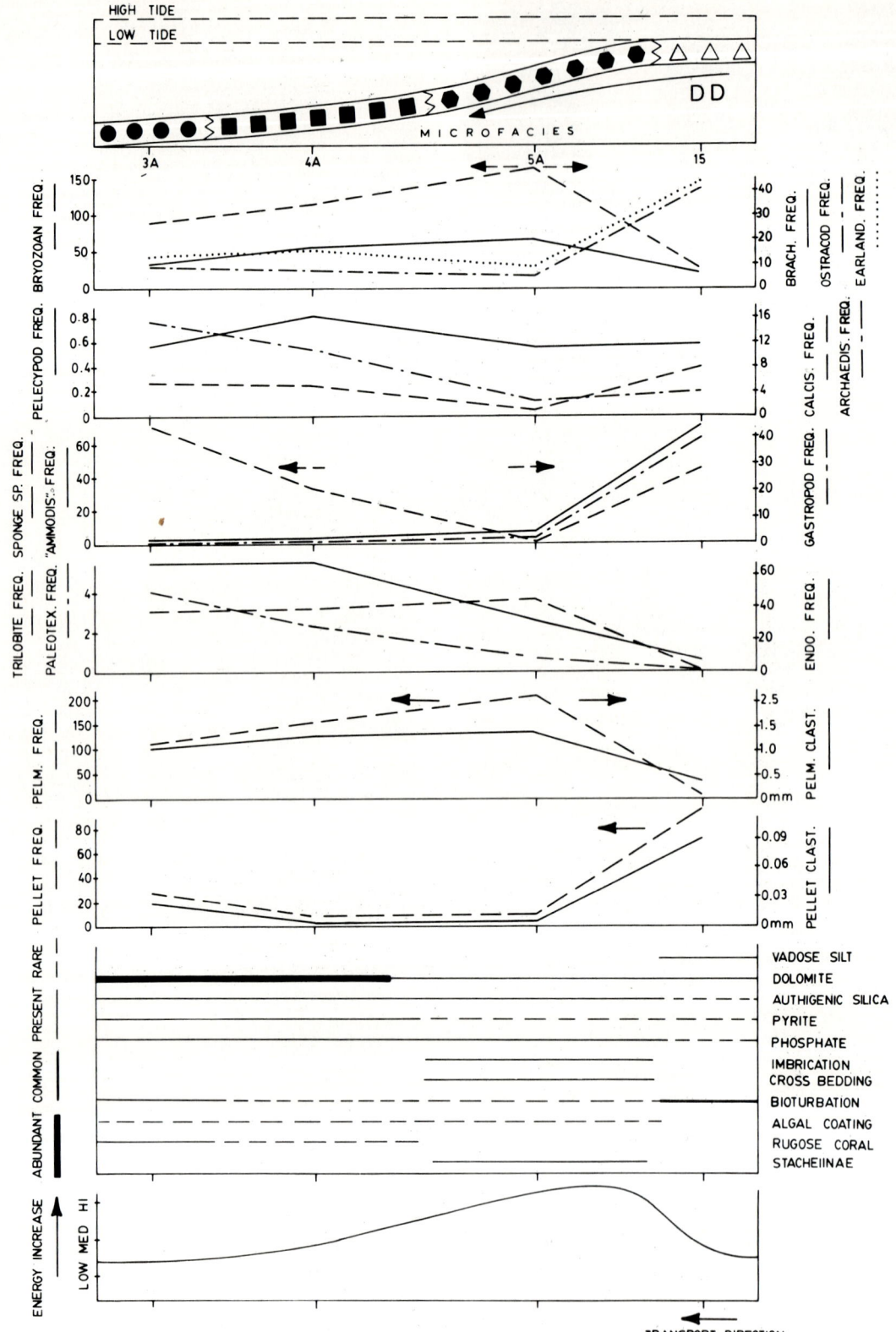

FIGURE 8.16 Kinkaid Formation (Upper Mississippian), southern part of Illinois Basin. Ideal depositional model 3. From Lasemi and Carozzi (1981).

Chap. 8 / Carbonate Ramps with Bioaccumulated to Hydrodynamic Buildups

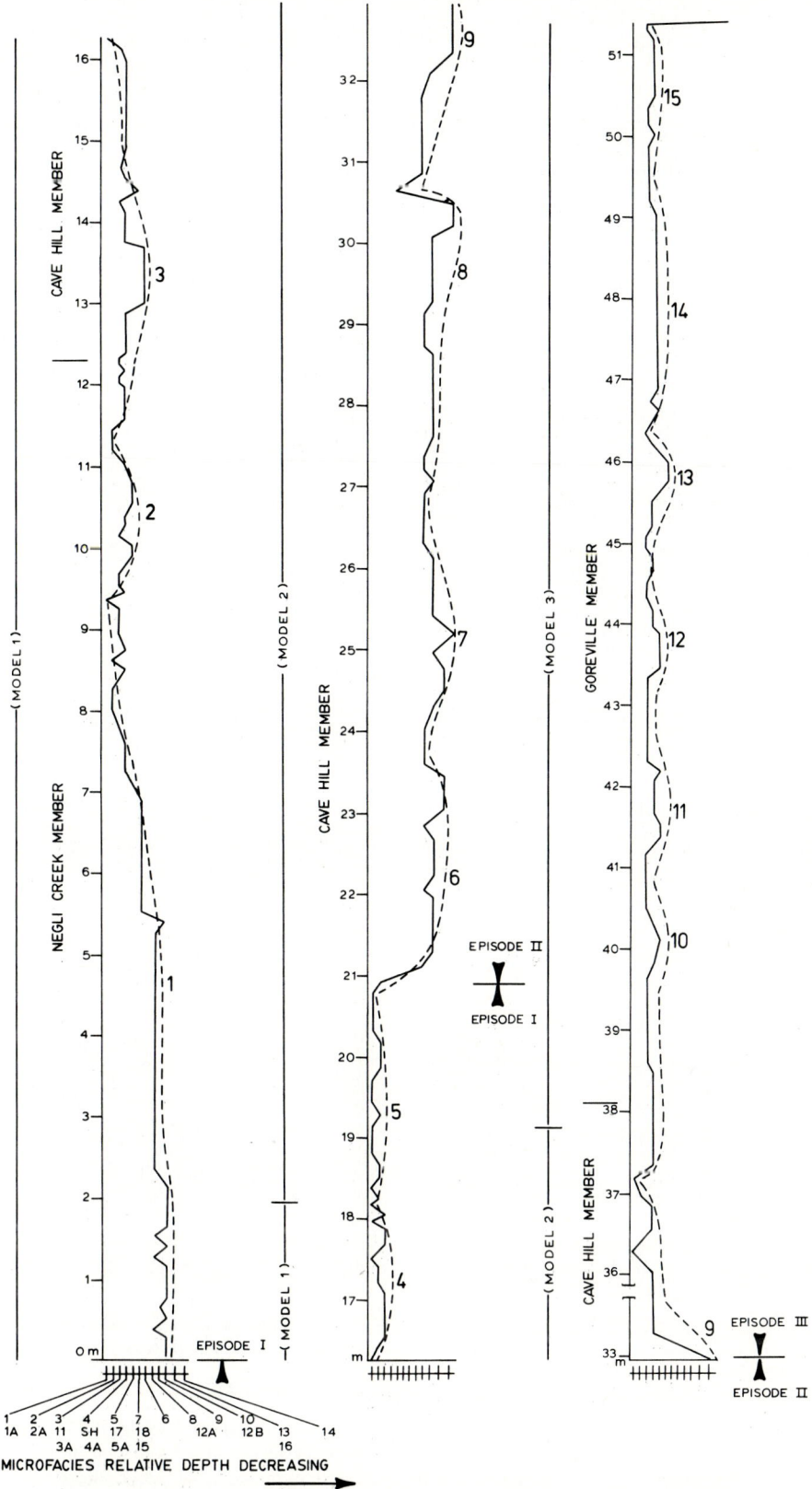

FIGURE 8.17 Kinkaid Formation (Upper Mississippian), southern part of Illinois Basin. General relative depth trends of composite section. From Lasemi and Carozzi (1981).

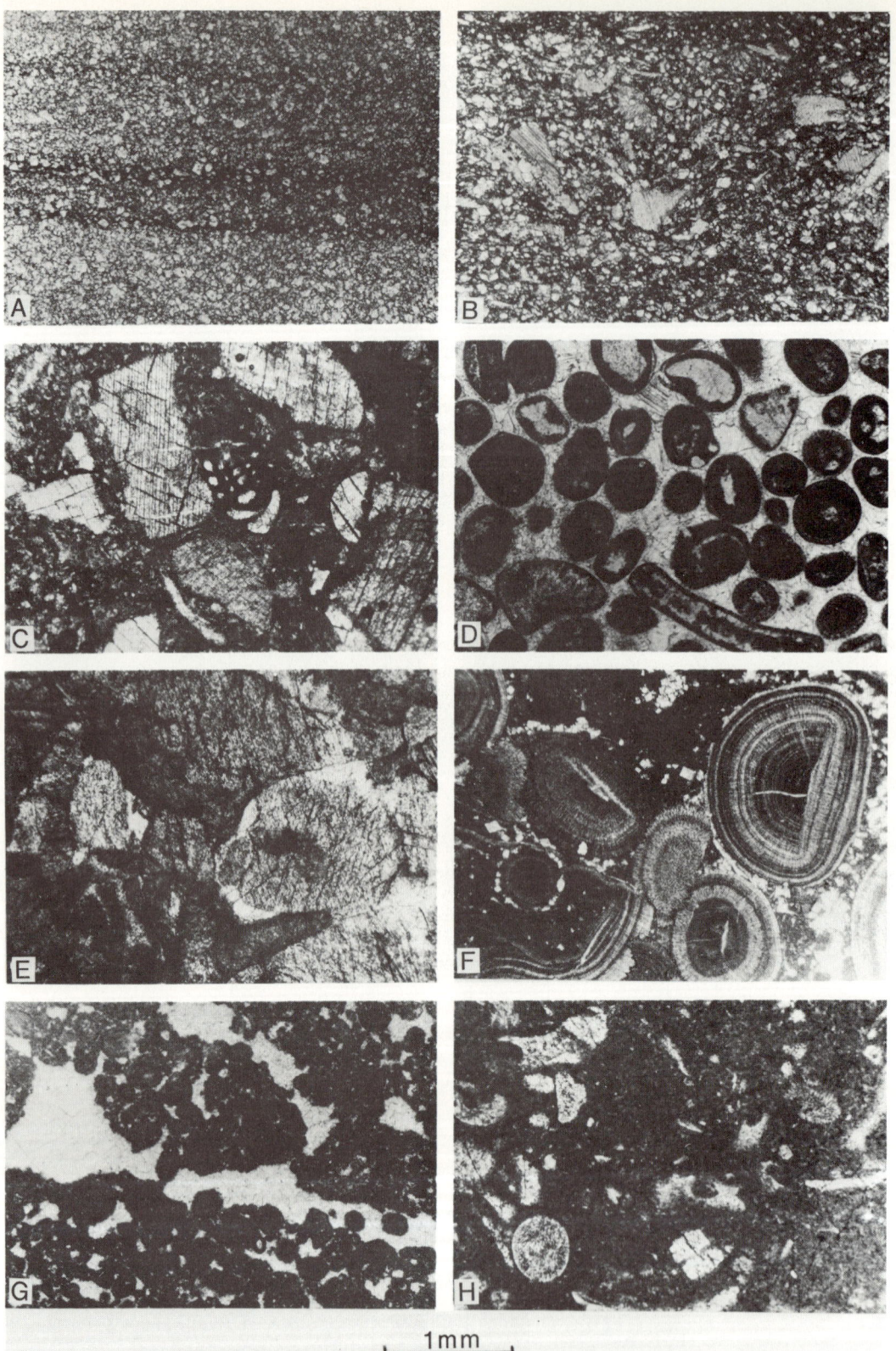

FIGURE 8.18 Rundle Group (Mississippian), Canadian Rocky Mountain Front Range, southwestern Alberta. Typical microfacies. A and B. Microfacies 1. C. Microfacies 2. D. Microfacies 3. E. Microfacies 4. F to H. Microfacies 5. See text for detailed descriptions. All photomicrographs: plane-polarized light. From Walpole and Carozzi (1961). Reprinted by permission of the American Association of Petroleum Geologists.

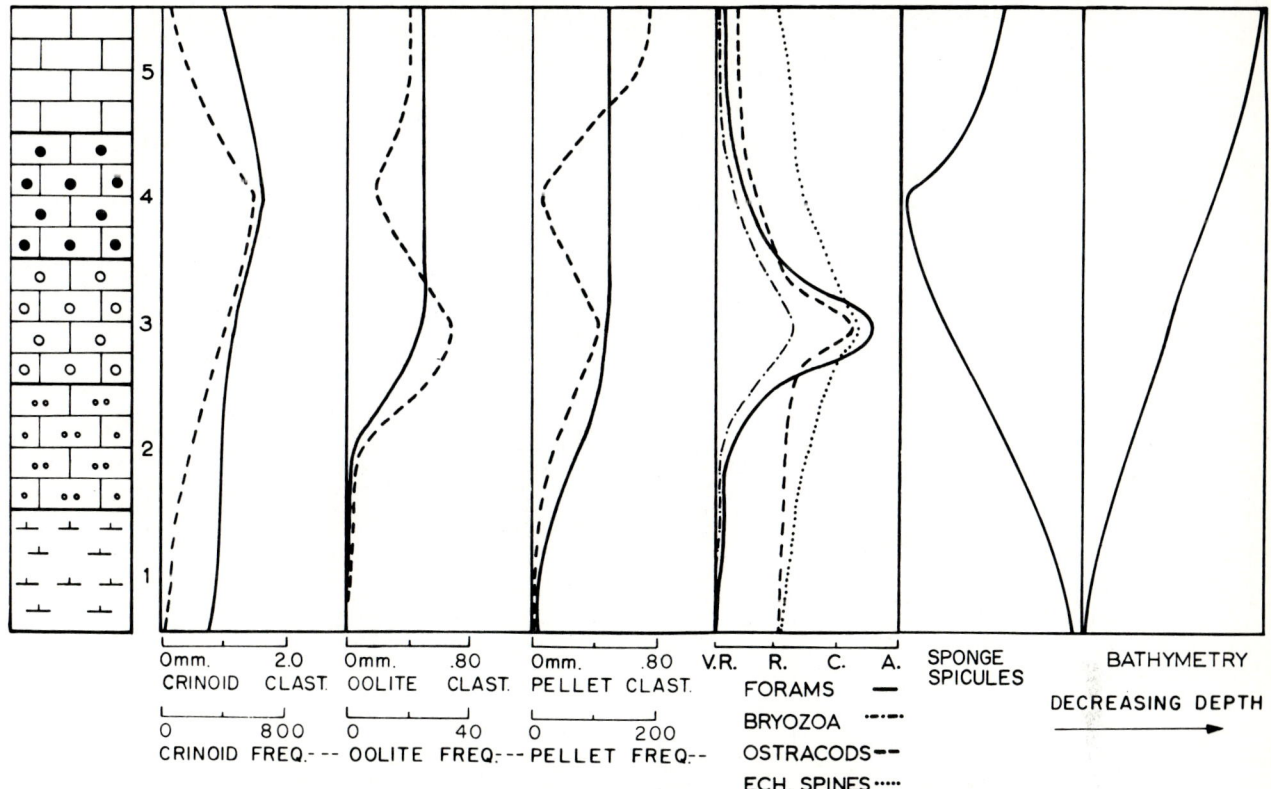

FIGURE 8.19 Rundle Group (Mississippian), Canadian Rocky Mountain Front Range, southwestern Alberta. Ideal shallowing-upward sequence. From Walpole and Carozzi (1961). Reprinted by permission of the American Association of Petroleum Geologists.

is a carbonate ramp entirely controlled by crinoid colonies and their products of mechanical destruction. It was investigated by means of three field sections as follows: North Ram River section in the First Range (995 ft, 298.5 m); South Ram River section No. 1 in the Third Range (1370 ft, 411 m); and South Ram River section No. 2 in the Fourth Range (1550 ft, 465 m). A total thickness of about 4000 ft (1200 m) of carbonates was sampled by means of 800 thin sections at an average vertical interval of 5 ft (1.5 m). Where obvious or abrupt lithologic changes occurred within this interval, the sampling distance was adjusted accordingly. By this procedure it was concluded that sufficient control was attained so as not to miss any significant sedimentological variation.

Description of Microfacies

The petrographic study led to the distinction of five microfacies which form a well-defined shallowing-upward sequence and are described in that order.

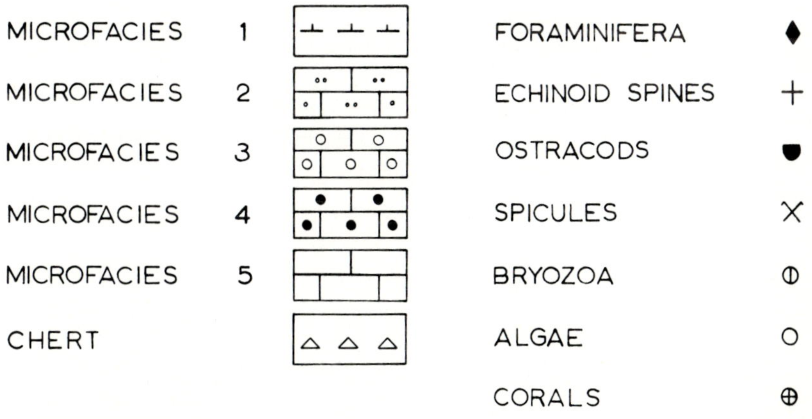

FIGURE 8.20 Rundle Group (Mississippian), Canadian Rocky Mountain Front Range, southwestern Alberta. Symbols for microfacies. From Walpole and Carozzi (1961). Reprinted by permission of the American Association of Petroleum Geologists.

152 Part II / Carbonate Ramps

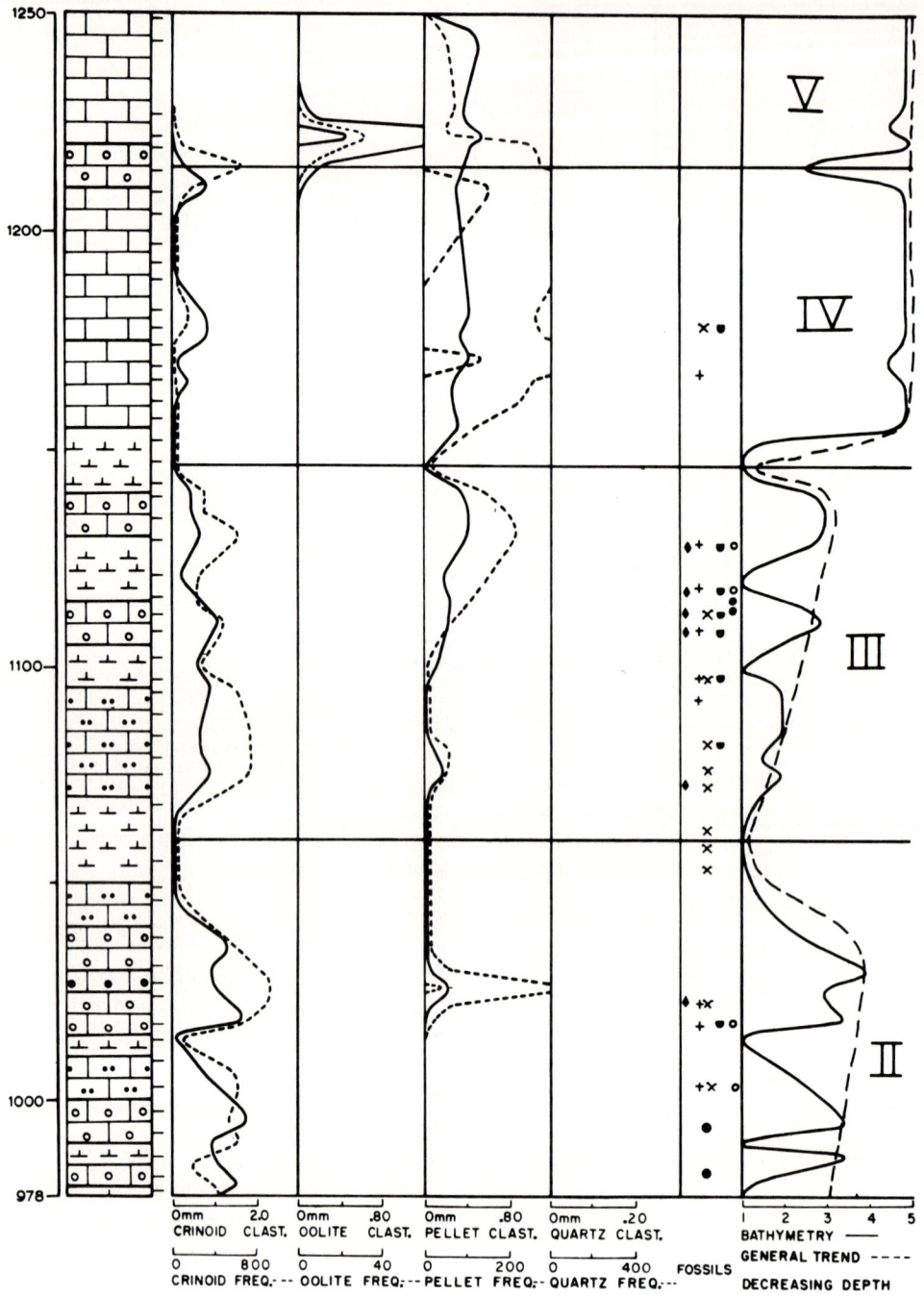

FIGURE 8.21 Rundle Group (Mississippian), Canadian Rocky Mountain Front Range, southwestern Alberta. Typical example of vertical variation of microscopic parameters. From Walpole and Carozzi (1961). Reprinted by permission of the American Association of Petroleum Geologists.

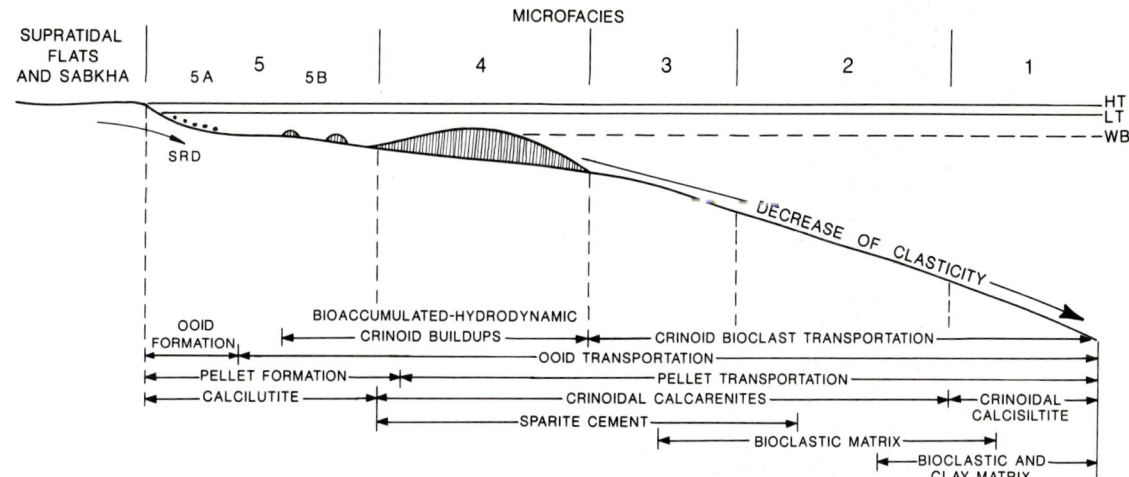

FIGURE 8.22 Rundle Group (Mississippian), Canadian Rocky Mountain Front Range, southwestern Alberta. Ideal depositional model. Redrawn from Walpole and Carozzi (1961).

Microfacies 1 (Fig. 8.18A,B). Laminated crinoidal biocalcisiltite, argillaceous and pyritic, grading into a fine-grained, mud-supported crinoidal biocalcarenite with a bioclastic matrix (Fig. 8.18B). The entire range of biogenic particles is derived from crinoids at various stages of fragmentation, associated with abundant monaxonic sponge spicules, ostracods, and rare smaller benthic foraminifers. Incipient neomorphism may result in scattered small calcite rhombs.

Microfacies 2 (Fig. 8.18C). Poorly sorted, grain-supported crinoidal biocalcarenite with bioclastic matrix. Besides the crinoid bioclasts ranging from sand size to silt size are common *Endothyra*, lithic pellets, echinoid spines, and reworked ooids.

Microfacies 3 (Fig. 8.18D). Well-sorted, grain-supported crinoidal biocalcarenite with sparite cement and small patches of fine bioclastic matrix. The well-rounded crinoidal bioclasts, which often show residual coatings of calcilutite mud, are associated with abundant reworked ooids, lithic pellets, *Endothyra*, bioclasts of bryozoans, brachiopods, pelecypods, echinoid spines, and complete ostracod valves.

Microfacies 4 (Fig. 8.18E). Coarse-grained and cross-bedded crinoidal calcarenite cemented by pressure-solution and local syntaxial overgrowths. The unsorted crinoid bioclasts display central canals filled with calcilutite. They are associated with a small number of the other constituents of microfacies 3, usually deformed by pressure-solution and overgrowths.

Microfacies 5 (Fig. 8.18 F,G, H). Microfacies 5 consists of two subtypes. Subtype 5A is a pelletoidal calcilutite with very well-developed fenestral fabric filled with sparite cement (Fig. 8.18G). Locally, it displays concentrations of large ooids to pisoids, often broken and recoated (Fig. 8.18F), indicating an early generation of the fibro-radiated structure followed by reciprocal impacts (Carozzi, 1961b). Scattered and relatively rare bioclasts belong to monaxonic sponge spicules, echinoid spines, and ostracods. Locally abundant angular grains of detrital quartz may be present. Subtype 5B (Fig. 8.18H) is a massive calcilutite with scattered sand-size bioclasts of crinoids associated in places with intact stems consisting of many columnals, and abundant monaxonic sponge spicules. Irregular distribution of many bioclasts indicates appreciable bioturbation.

The Ideal Shallowing-Upward Sequence

The five microfacies of the Rundle Group can be associated into a shallowing-upward sequence (Fig. 8.19) in which the pattern of the various parameters indicates a predominantly mechanical distribution of all the components. Crinoid bioclasts produced in microfacies 4 are distributed downslope with gradually decreasing clasticity and frequency; reworked ooids and pellets originating from the lagoonal environment of microfacies 5 indicate a similar dispersion. Smaller benthic foraminifers, bioclasts of bryozoans, ostracods, and echinoid spines are preferentially concentrated in microfacies 3 due to similarity of grain size. Monaxonic spicules of sponges living in the lagoonal conditions of microfacies 5 transited across the entire carbonate system and settled eventually in the quiet and deeper conditions of microfacies 1.

A typical detailed partial section (Figs. 8.20, 8.21) shows reciprocal behavior of the various microscopic parameters in a superposition of four shallowing-upward

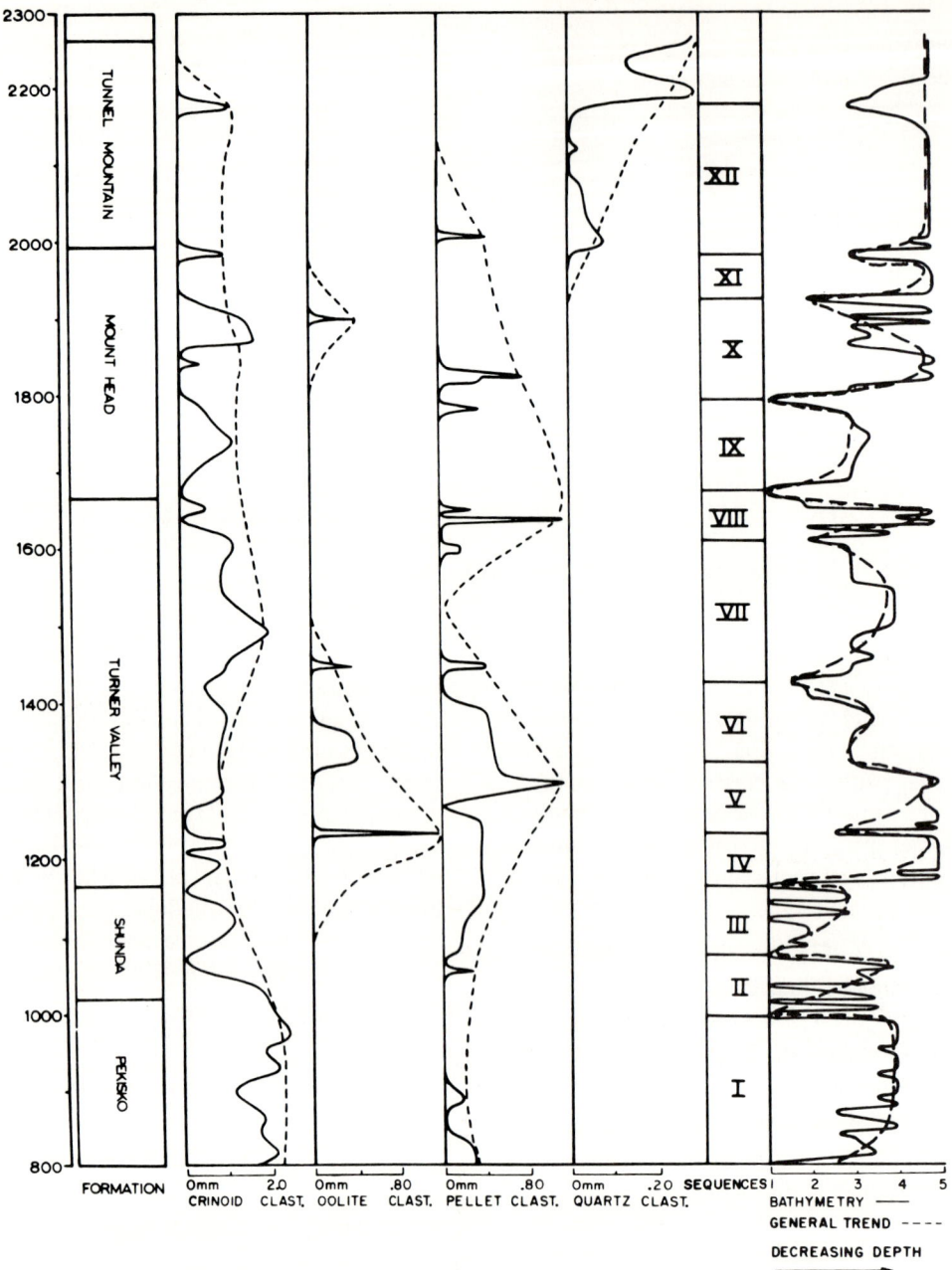

FIGURE 8.23 Rundle Group (Mississippian), Canadian Rocky Mountain Front Range, southwestern Alberta. Vertical variation of microscopic parameters and relative bathymetric evolution of the entire group. From Walpole and Carozzi (1961). Reprinted by permission of the American Association of Petroleum Geologists.

sequences which together form in turn a shallowing-upward megasequence.

The Ideal Depositional Model

In the proposed model (Fig. 8.22), the Rundle Group is interpreted as a carbonate ramp ranging from deep subtidal to intertidal lagoonal, and outside the studied area, to supratidal flats and sabkha evaporites. The lagoon is characterized by an association of quiet pelletoidal and fenestral calcilutites where sponges and scattered patches of crinoid colonies lived. Shoreward more agitated oolitic conditions prevailed. The main controlling factor was, however, a bioaccumulated to hydrodynamic buildup consisting mainly of crinoids associated with echinoids, bryozoans, pelecypods, brachiopods, and

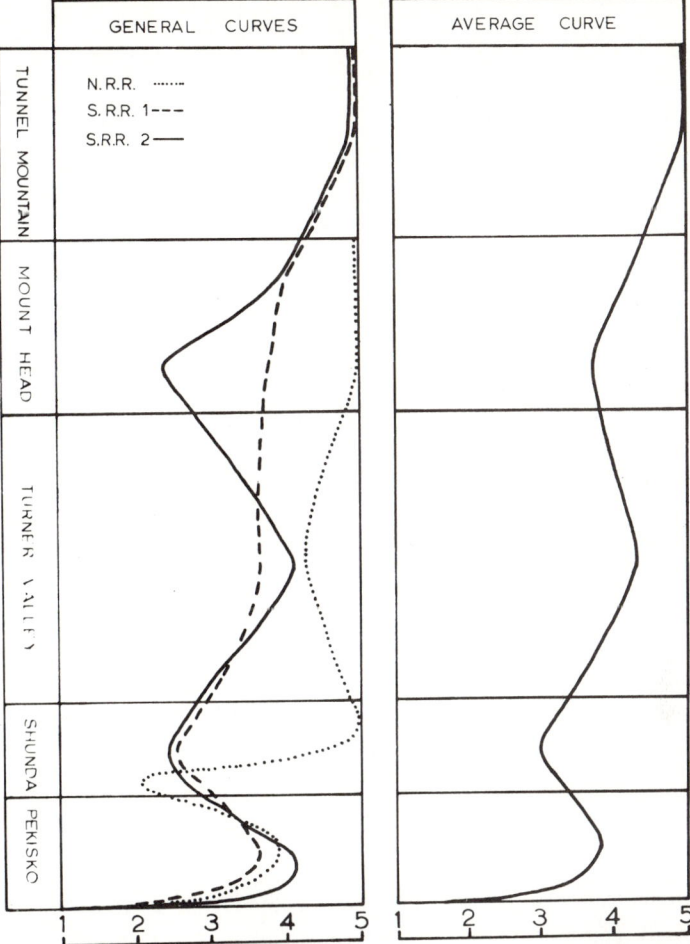

FIGURE 8.24 Rundle Group (Mississippian), Canadian Rocky Mountain Front Range, southwestern Alberta. Generalized relative bathymetric evolutions of the three investigated sections and of their average. N.R.R., North Ram River section; S.R.R.1, South Ram River section 1; S.R.R.2, South Ram River section 2. From Walpole and Carozzi (1961). Reprinted by permission of the American Association of Petroleum Geologists.

smaller benthic foraminifers. Wave action kept destroying the crinoidal colonies, leaving in place coarser bioclasts and redistributing by currents all the smaller-size constituents downslope to the deepest conditions. Similarities of grain size led to mechanical concentration of a variety of constituents in microfacies 3.

The General Evolution of the Depositional Environments

The examination of an entire section of the Rundle Group (Fig. 8.23) shows on a large scale, and with the use of envelope trend curves, that the relationship of ooids and pellets to crinoids is the same as expressed in the ideal shallowing-upward sequence (Fig. 8.19). Upon decrease of the importance of crinoids upward, the environment becomes enriched increasingly in detrital quartz preceding deposition of the overlying quartzites of the Rocky Mountain Formation. Considered as a whole, the Rundle Group shows by its superposition of shallowing-upward sequences a gradual shallowing as well. In a more generalized expression of variation of relative bathymetry (Fig. 8.24), shallowing is achieved in all three investigated sections by a succession of three positive oscillations.

9

Carbonate Ramps with Bioconstructed Buildups

Carbonate ramps reach their stage of final morphologic evolution toward platforms by means of bioconstructed buildups. The latter are generated not only under a wide range of general energy levels, but also under restricted to open marine environment. Consequently, a wide spectrum of organisms participate in the construction of these buildups, such as stromatolites, phylloid algae, *Amphipora*, and stromatoporoids.

9.1 CHARACTERISTIC CONSTITUENTS: STROMATOLITES

Total restriction of the Michigan Basin in late Silurian times (Cayugan) led to a sabkha-type evaporitic environment which was bordered by stromatolite-constructed buildups occupying the upper portion of gentle ramps (Textoris and Carozzi, 1966). The buildup discussed here (Fig. 9.1) is exposed on the southern margin of the Michigan Basin near Maumee, northern Ohio, and forms with its overlying sequence a section 85 ft (25.5 m) thick; it was investigated by means of 143 samples with an average vertical interval of 0.6 ft (5 cm).

Description of Microfacies

Entirely dolomitized microfacies are described from the main buildup to the first subevaporitic deposits. They are associated in several major units (Fig. 9.1).

Microfacies 1 (Fig. 9.2A). Hemispherical stromatolite-constructed dolomite, well laminated. Black and dark laminae of anhedral to subhedral, finely crystalline dolomite represent the algal mats. Clear intermat laminae of subhedral to euhedral medium crystalline dolomite correspond to original trapped calcilutite. Several algal mats display a pelletoidal texture due to desiccation during temporary subaerial exposure.

Microfacies 2 (Fig. 9.2B). Hemispherical stromatolite-constructed dolomite, poorly laminated of *Spongiostromata* type. The texture appears as pellet-like, gray to dark masses of finely crystalline dolomite in a clear, medium crystalline dolomite mosaic.

Microfacies 3 (Fig. 9.2C). Grain-supported worm tube dolarenite. Well-sorted terebellid tubes and mud pellets agglutinated by worms consist of dark, finely crystalline dolomite set in a clear, medium to coarsely

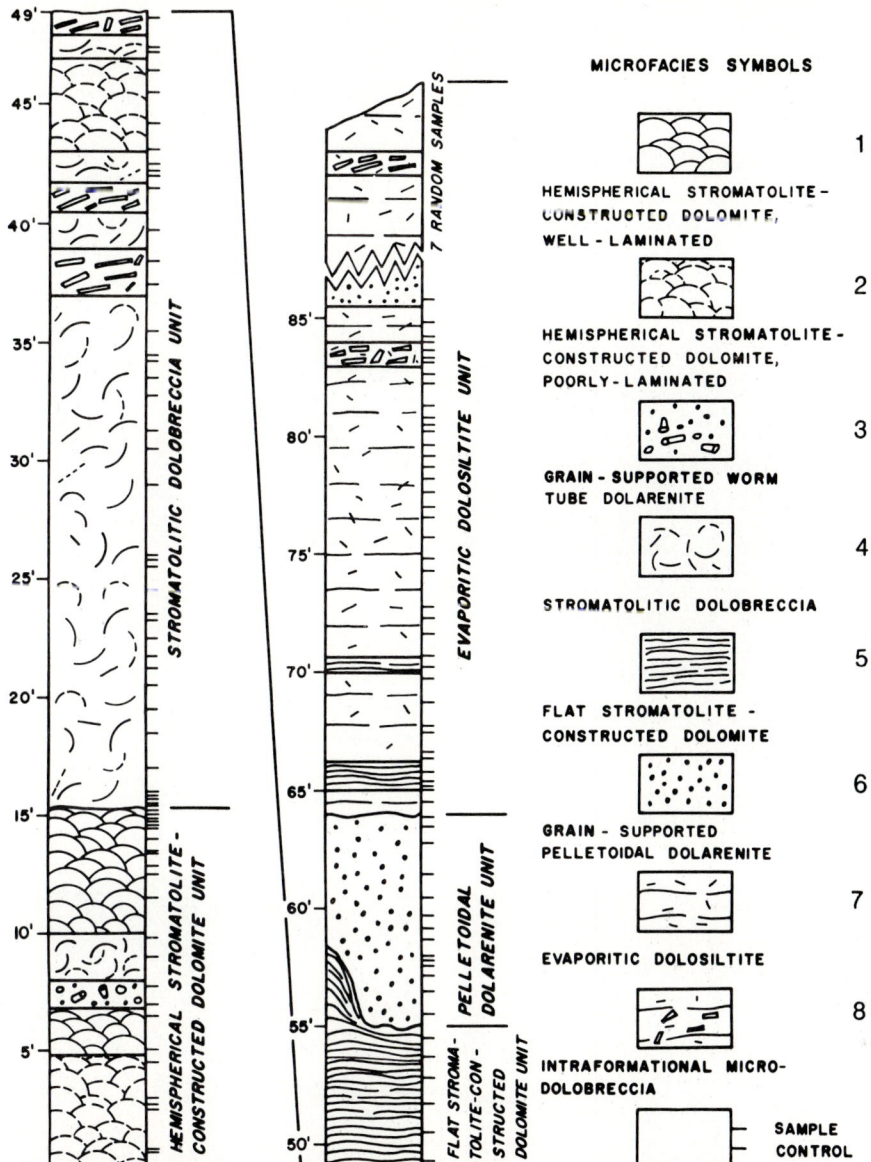

FIGURE 9.1 Cayugan (Upper Silurian), southern margin of Michigan Basin, Maumee, northern Ohio. Composite section and symbols for microfacies. From Textoris and Carozzi (1966). Reprinted by permission of the American Association of Petroleum Geologists.

crystalline dolomite mosaic representing an original sparite cement.

Microfacies 4 (Fig. 9.2D). Stromatolitic dolobreccia consisting of angular to crescentic clasts of well-laminated stromatolite colonies set in a mosaic of clear and dark, medium crystalline dolomite representing an original sparite cement. This is interpreted as a storm-induced deposit.

Microfacies 5 (Fig. 9.2E). Flat stromatolite-constructed dolomite consisting of extremely thin and dark algal mats with slight tendency toward mounding separated by intermat layers of lighter-colored, finely to medium crystalline dolomite.

Microfacies 6 (Fig. 9.2F). Grain-supported pelletoidal dolarenite consisting predominantly of rounded lithic pellets and larger intraclasts of dark, finely crystal-

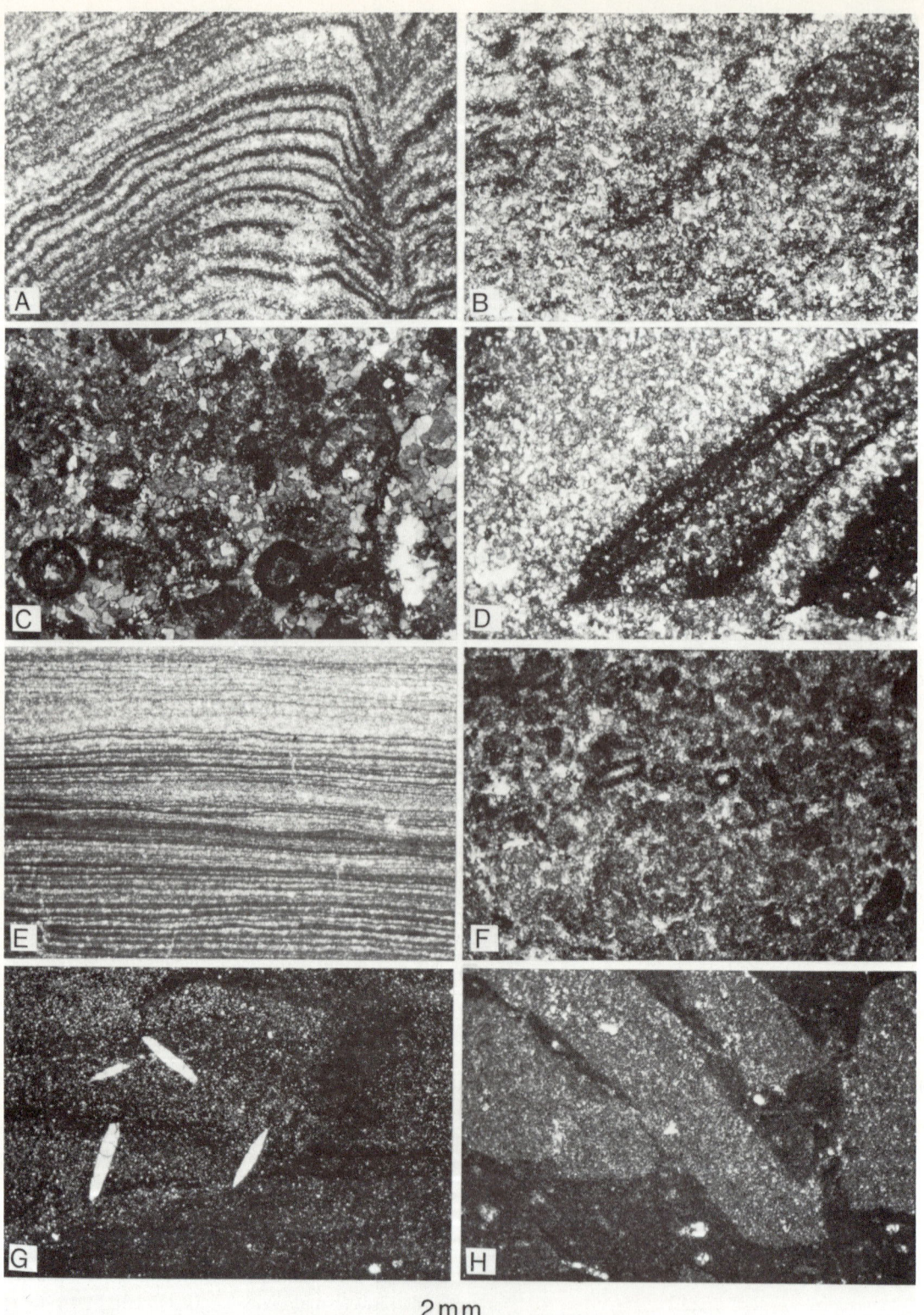

FIGURE 9.2 Cayugan (Upper Silurian), southern margin of Michigan Basin, Maumee, northern Ohio. Typical microfacies. A. Microfacies 1. B. Microfacies 2. C. Microfacies 3. D. Microfacies 4. E. Microfacies 5. F. Microfacies 6. G. Microfacies 7. H. Microfacies 8. See text for detailed descriptions. All photomicrographs: plane-polarized light. From Textoris and Carozzi (1966). Reprinted by permission of the American Association of Petroleum Geologists.

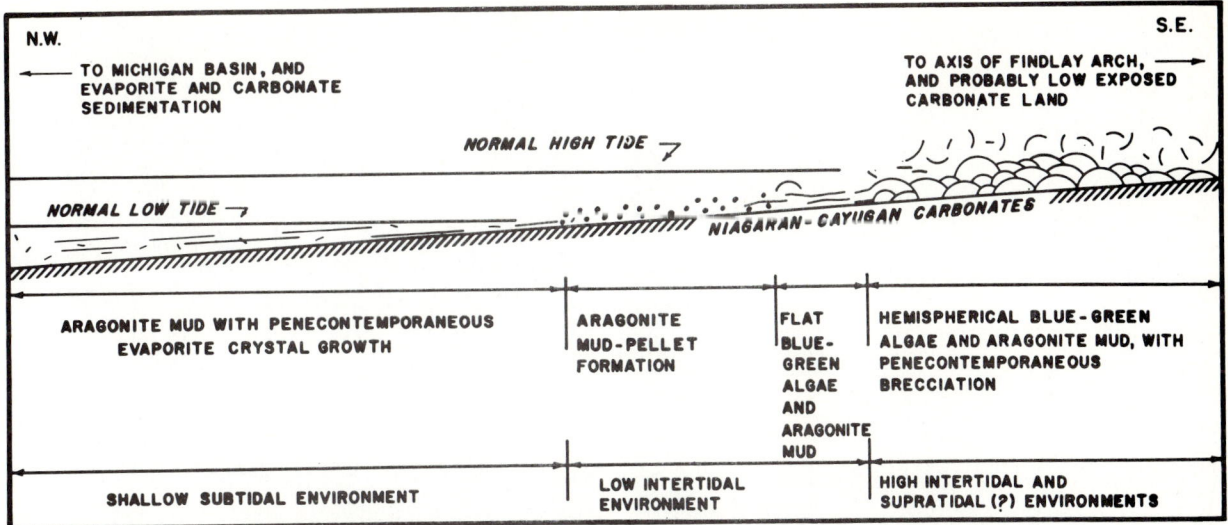

FIGURE 9.3 Cayugan (Upper Silurian), southern margin of Michigan Basin, Maumee, northern Ohio. Ideal depositional model. From Textoris and Carozzi (1966). Reprinted by permission of the American Association of Petroleum Geologists.

line dolomite associated with leperditids and rare worm tubes. The interstitial material consists of a mosaic of dark and clear, medium crystalline dolomite that represents an original association of calcisiltite matrix and patchy sparite cement.

Microfacies 7 (Fig. 9.2G). Laminated evaporitic dolosiltite with molds of penecontemporaneous gypsum crystals filled with calcite cement.

Microfacies 8 (Fig. 9.2H). Intraformational micro-dolobreccia consisting of imbricated, angular, flat pebbles of laminated evaporitic dolosiltite (microfacies 7) set in a dark matrix of finely crystalline dolomite representing an original micrite matrix. Gypsum crystal molds transect clast-matrix boundaries, indicating that they postdate the intraformational reworking processes.

The Ideal Depositional Model

The cross section of the stromatolitic buildup and associated facies along the northwestern flank of the Findlay Arch, bordering the evaporitic Michigan Basin during Early Cayugan time (Fig. 9.3), suggests a generally low energy, shallow, restricted penesaline environment with storm-induced brecciation of stromatolite colonies.

9.2 CHARACTERISTIC CONSTITUENTS: *AMPHIPORA* AND STROMATOPOROIDS

Amphipora and stromatoporoids provide the initial stage of many constructed buildups of Devonian ramps. Whether these organisms remain at an incipient stage or subsequently develop into platform edges, they effectively withstand influxes of detrital quartz and clay minerals. Therefore, they offer a very suitable substratum for more complex ecologic associations.

A section 1200 ft (360 m) thick of Upper Devonian limestone of which 920 ft (276 m) belong to the Arrow Canyon Formation and 280 ft (84 m) to the Crystal Pass Limestone, in the Arrow Canyon Range, Clark County, southeast Nevada (Carss and Carozzi, 1965), was investigated petrographically with a total of 581 samples at an average vertical interval of 2 ft (0.61 m).

Description of Microfacies

The microfacies are described in order of general shallowing. Their more complex interrelationships are discussed later.

Microfacies 1A (Fig. 9.4A). Fine and massive to laminated calcisiltite in which minute grains of detrital quartz, pyrite pigments, and unidentifiable bioclasts are either scattered at random or concentrated in streaks.

Microfacies 1B (Fig. 9.4B). Pelletoidal biocalcisiltite in which abundant minute bioclasts of bryozoans, pelmatozoans, pelecypods, calcispheres, and ostracods are scattered at random or concentrated in streaks.

Microfacies 1C (Fig. 9.4C). Strongly burrowed and bioturbated pelletoidal calcisiltite. The fecal pellets occur as irregularly shaped pocketlike concentrations displaying sharp or hazy boundaries with the surrounding matrix.

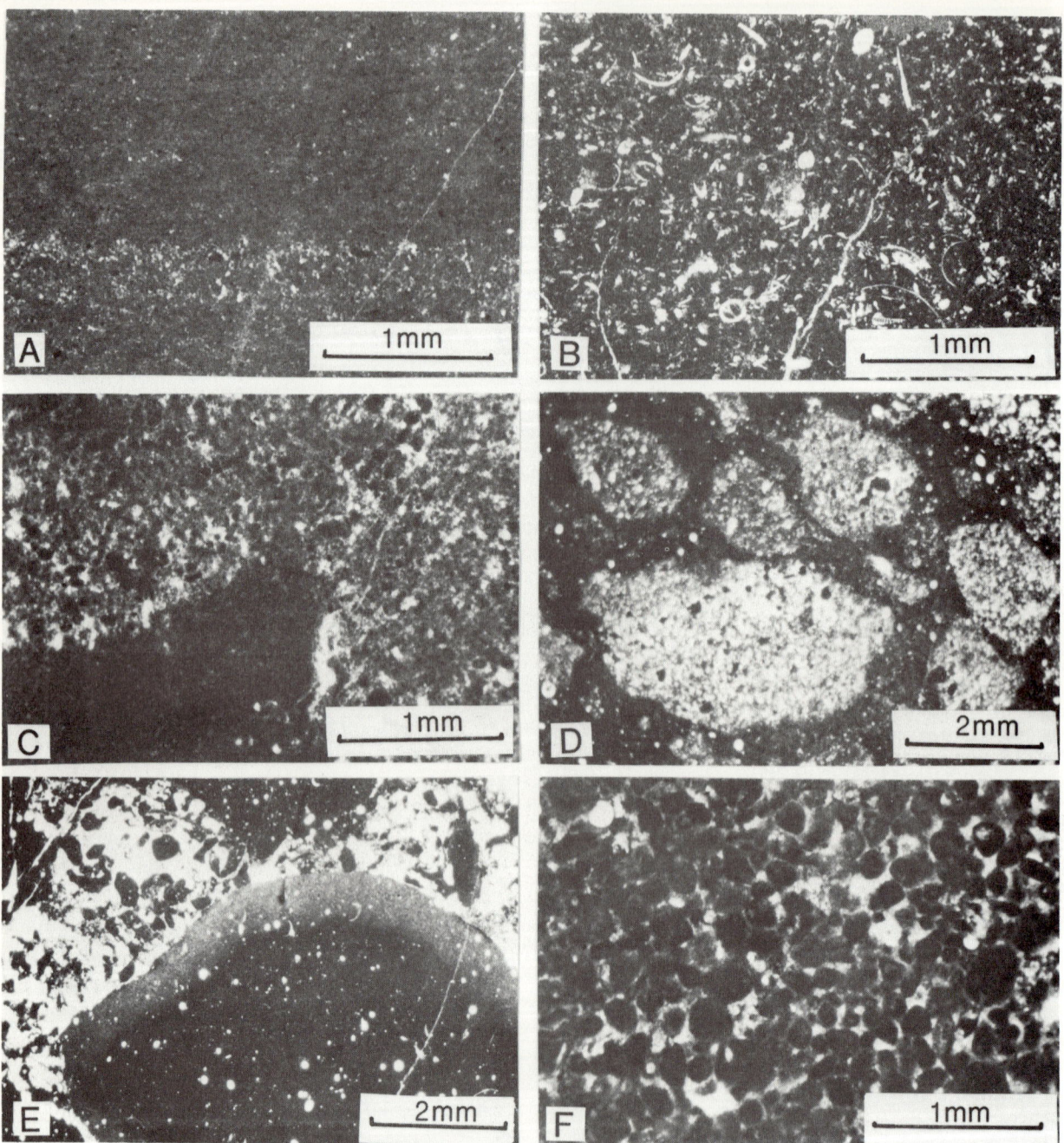

FIGURE 9.4 Arrow Canyon Formation and Crystal Pass Limestone (Upper Devonian), Arrow Canyon Range, Clark County, Nevada. Typical microfacies. A. Microfacies 1A. B. Microfacies 1B. C. Microfacies 1C. D and E. Microfacies X. F. Microfacies 3. See text for detailed descriptions. All photomicrographs: plane-polarized light. From Carss and Carozzi (1965). Reprinted by permission of Elsevier Publishing Company.

Microfacies X (Fig. 9.4D,E). The microfacies was designated "X" so that it will not be regarded as a necessary intermediate between microfacies 1A, 1C, and 3 as it would have been if designated 2. Statistical analysis shows this microfacies to be most closely related to microfacies 3.

Intraformational breccia-conglomerate derived from the reworking of any of the microfacies 1A through 6 with predominance of pelletoidal calcisiltites and neomorphosed pelletoidal calcarenites. Some of the rounded calcisiltite intraclasts display oxidized margins (Fig. 9.4E). Interstitial material ranges from pelletoidal calcisiltite to cavity-filling sparite cement with rare bioclasts of bryozoans and ostracods together with grains of detrital quartz. These features indicate storm processes which swept the entire carbonate system.

Microfacies 3 (Figs. 9.4F, 9.5A). Poorly sorted, grain-supported, pelletoidal biocalcarenite to calcirudite. In the calcarenite, fecal pellets with diffused margins are associated with well-defined dark and larger lithic pellets. Scattered bioclasts are from crinoids, bryozoans, gastropods, ostracods, and calcispheres. Interstitial material is an association of calcisiltite matrix neomorphosed into pseudomicrosparite and of patches of cavity-filling sparite cement.

The calcirudite consists of irregularly shaped intraclasts of dark calcisiltite and of imbricated flat pebbles of the same material with transverse fractures recalling desiccation cracks (Fig. 9.5A). Incipient graded bedding and interstitial sparite cement indicate that microfacies 3 as microfacies X has undergone appreciable storm effects.

Microfacies 3A (Fig. 9.5B). Poorly sorted, grain-supported, pelletoidal biocalcarenite with scattered grains of detrital quartz and a small amount of bryozoan bioclasts. Interstitial material is calcisiltite changed into pseudomicrosparite by neomorphism.

Microfacies 4 (Fig. 9.5C). Well-sorted, grain-supported, lithic pelletoidal biocalcarenite with sparite cement. Pellets range from angular to well rounded and are associated with abundant calcispheres, many of which liberated from the abrasion of pellets, and with rare bryozoans.

Microfacies 5 (Fig. 9.5D). Arenaceous, grain-supported, lithic pelletoidal calcarenite grading to a pure quartz arenite with patches of pellets; both types have an interstitial sparite cement.

Microfacies 6 (Fig. 9.5E). Pure quartz arenite consisting of well-rounded grains, interlocked by pressure-solution or by overgrowth generating a local quartzitic texture; elsewhere an interstitial mosaic of hematitic sparite predominates.

Microfacies 7 (Fig. 9.5F). *Amphipora*—stromatoporoid-constructed limestone with interstitial pelletoidal biocalcarenite with bioclasts of bryozoans, ostracods, and calcispheres set in a fine bioclastic calcisiltite matrix.

The Ideal Shallowing-Upward Sequence

In essence, this sequence (Figs. 9.6, 9.7) consists of the superposition of the following microfacies groups: calcisiltite (1A, 1B, and 1C), pelletoidal biocalcarenites to calcirudites with tempestites (X, 3, 3A, and 4), bioconstructed limestone (7), and arenaceous pelletoidal calcarenites to pure quartz arenites (5 and 6). The pattern of the detrital quartz indicates the extrabasinal sediment yield, whereas pyrite decreases upward during the change from reducing to oxidizing conditions. The clasticity peak of the pellets is related to the storm-influenced microfacies, and their frequency increases in the more agitated and shallower conditions due to increased breaking up of larger intraclasts into smaller lithic pellets. Bryozoans are well represented in the group of microfacies 1 and reach their frequency peak in association with *Amphipora* and stromatoporoids. Calcispheres and ostracods are evenly distributed with a frequency peak of the calcispheres in microfacies 4 when liberated from the abrasion of lithic pellets.

A typical section (Fig. 9.8) shows the reciprocal behavior of microscopic parameters. The horizontal ruled area on the extreme right of the diagram is enclosed between lines tangent to the maxima and minima of the bathymetric curve and expresses the range of energy fluctuations.

The Ideal Depositional Models

The study of the stratigraphic section and the statistical interrelationships between microfacies (Fig. 9.9) indicates that two types of models have followed each other in time. The first model (Fig. 9.10) applies to the lower portion of the section up to approximately 1600 ft. It corresponds to an entirely carbonate ramp with an incipient *Amphipora*—stromatoporoid-constructed buildup surrounded by a halo of fossiliferous pelletoidal biocalcisiltite (1B). The model terminates landward with an arenaceous pelletoidal biocalcarenite with extremely thin and rare intercalations of pure quartz arenites.

The second type of model and corresponding perfect ramp (Fig. 9.10) applies to the upper part of the section from 1600 ft to the top. The buildup ceases to exist and its position is taken by the well-sorted pelletoidal biocalcarenite (4). The model terminates landward by transitional arenaceous microfacies to pure quartz arenites.

162 Part II / Carbonate Ramps

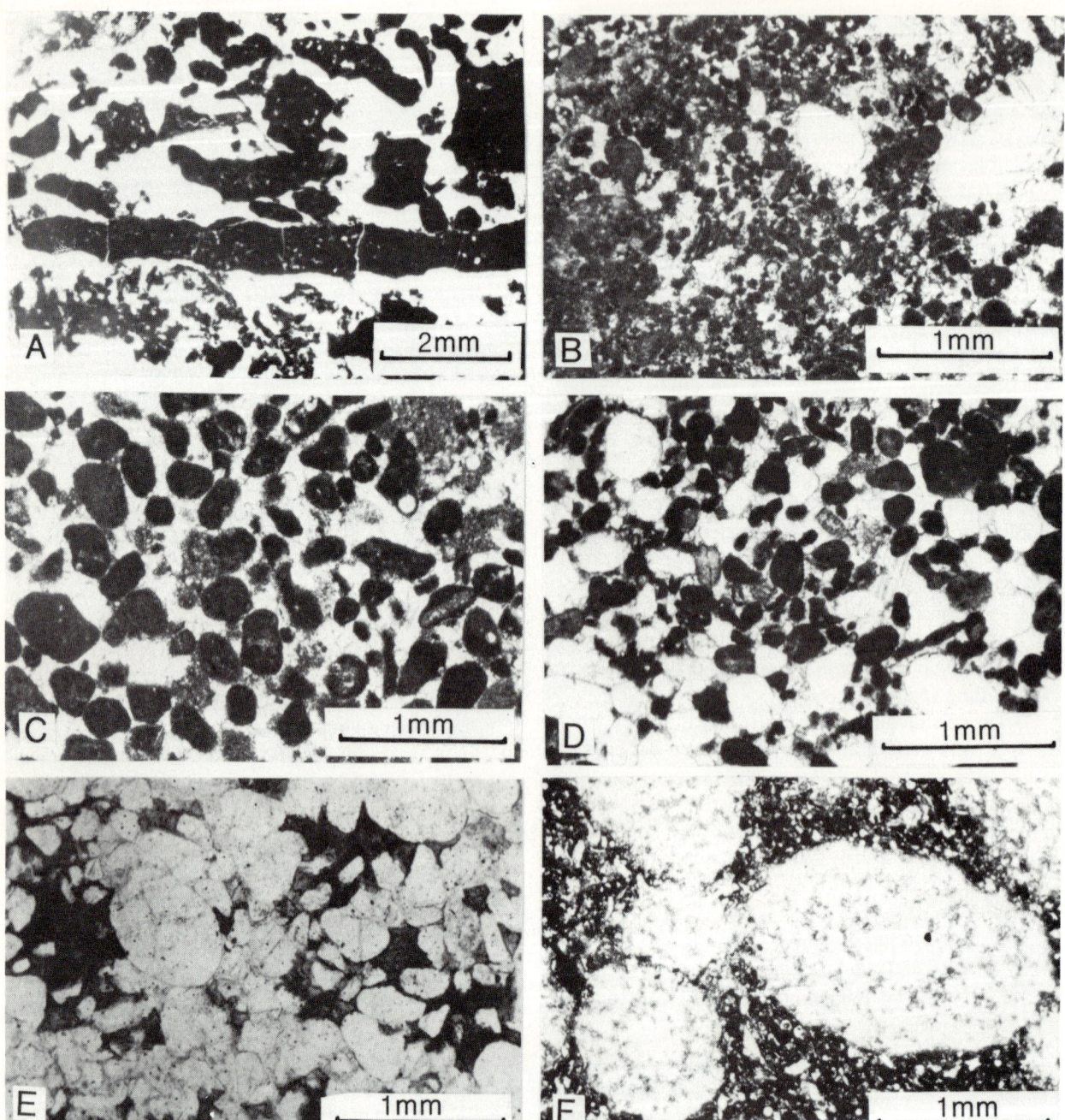

FIGURE 9.5 Arrow Canyon Formation and Crystal Pass Limestone (Upper Devonian), Arrow Canyon Range, Clark County, Nevada. Typical microfacies (continued). A. Microfacies 3. B. Microfacies 3A. C. Microfacies 4. D. Microfacies 5. E. Microfacies 6. F. Microfacies 7. See text for detailed descriptions. All photomicrographs: plane-polarized light. From Carss and Carozzi (1965). Reprinted by permission of Elsevier Publishing Company.

Chap. 9 / Carbonate Ramps with Bioconstructed Buildups

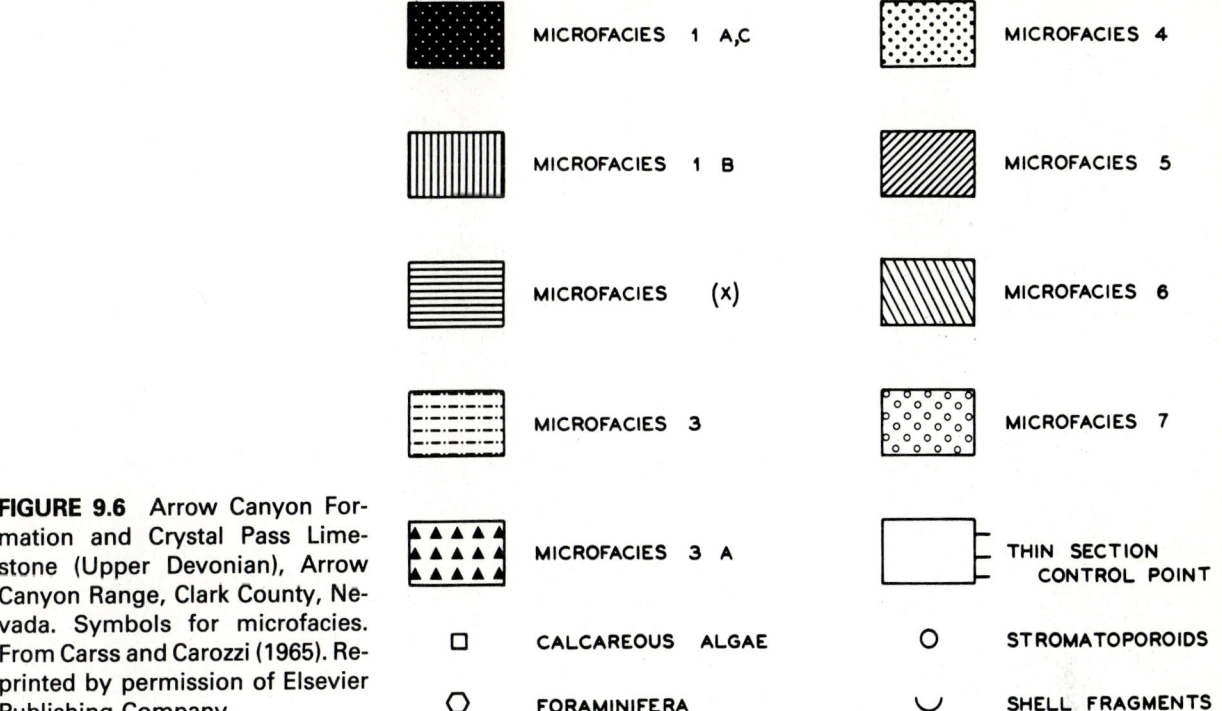

FIGURE 9.6 Arrow Canyon Formation and Crystal Pass Limestone (Upper Devonian), Arrow Canyon Range, Clark County, Nevada. Symbols for microfacies. From Carss and Carozzi (1965). Reprinted by permission of Elsevier Publishing Company.

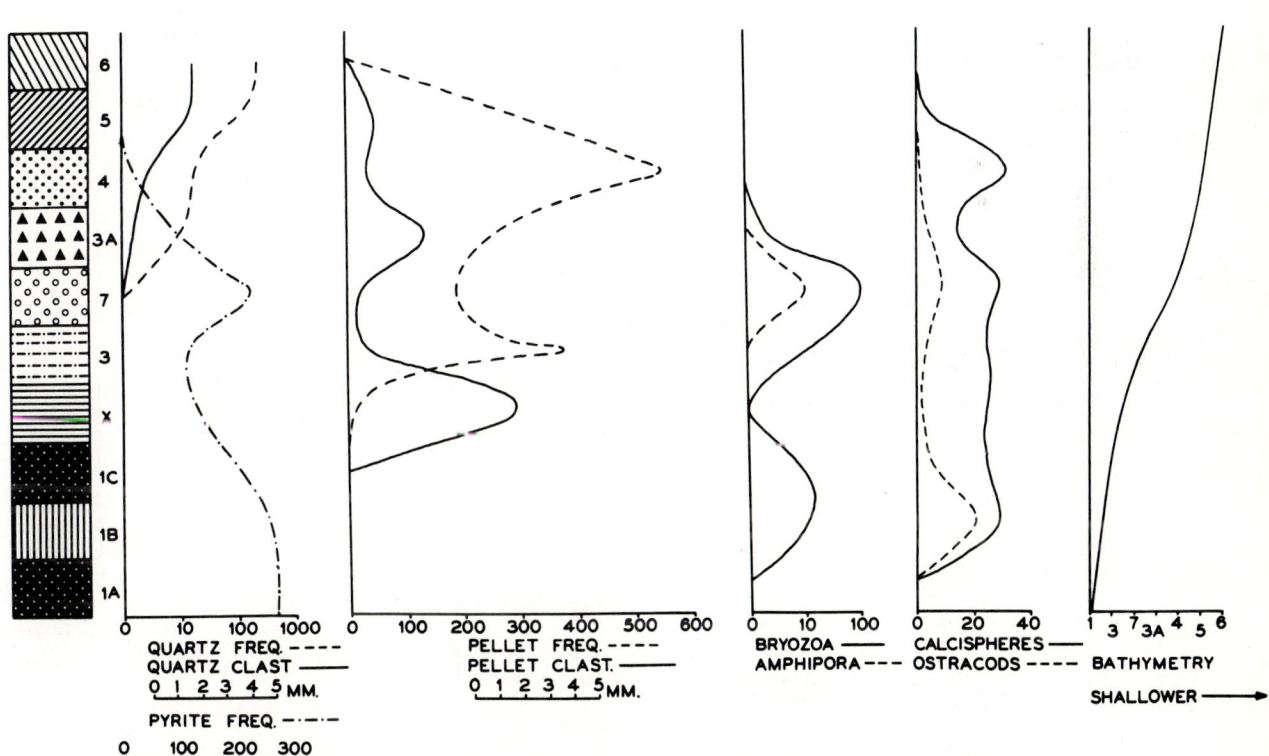

FIGURE 9.7 Arrow Canyon Formation and Crystal Pass Limestone (Upper Devonian), Arrow Canyon Range, Clark County, Nevada. Ideal shallowing-upward sequence. From Carss and Carozzi (1965). Reprinted by permission of Elsevier Publishing Company.

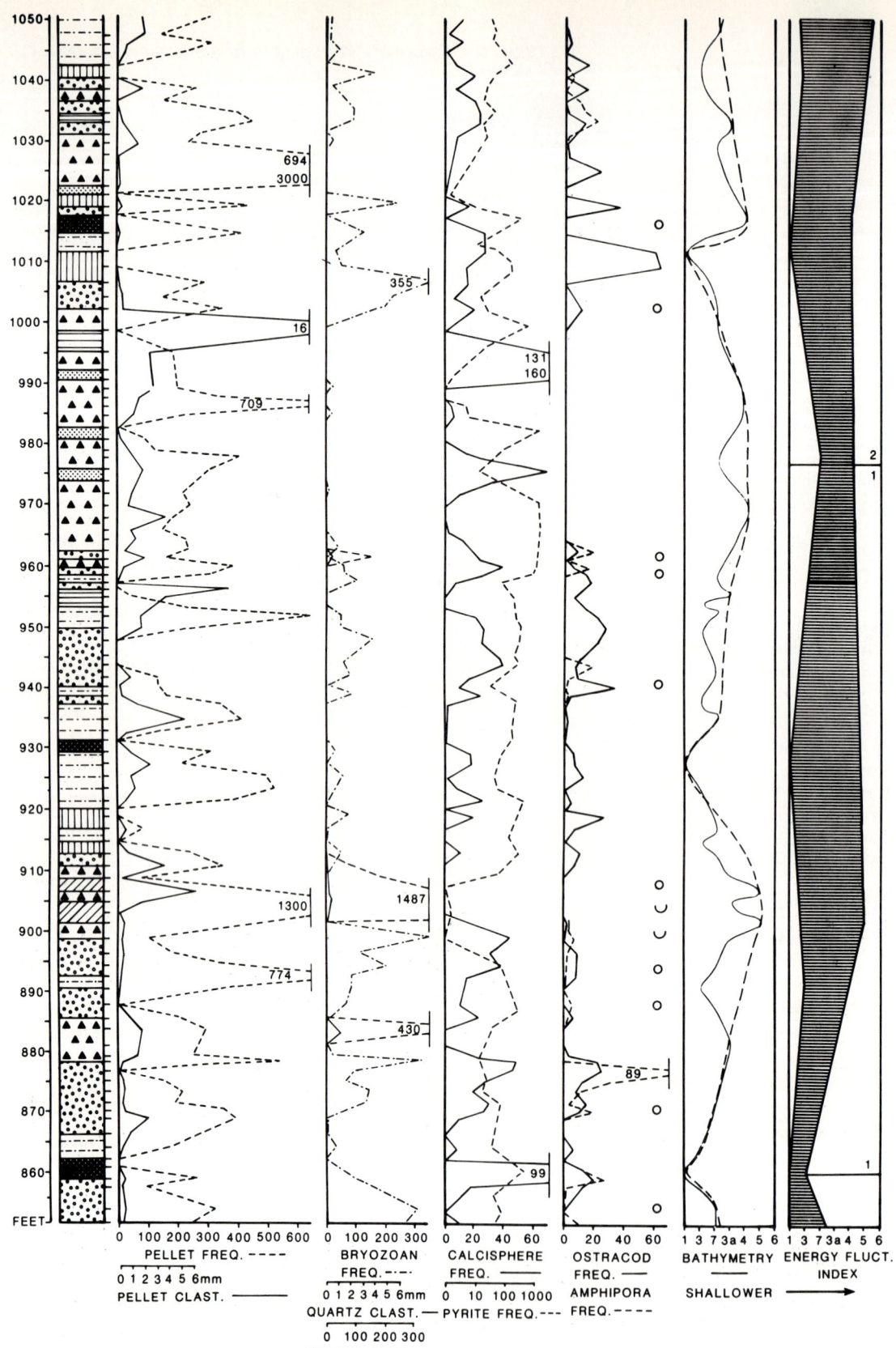

FIGURE 9.8 Arrow Canyon Formation and Crystal Pass Limestone (Upper Devonian), Arrow Canyon Range, Clark County, Nevada. Typical example of vertical variation of microscopic parameters. From Carss and Carozzi (1965). Reprinted by permission of Elsevier Publishing Company.

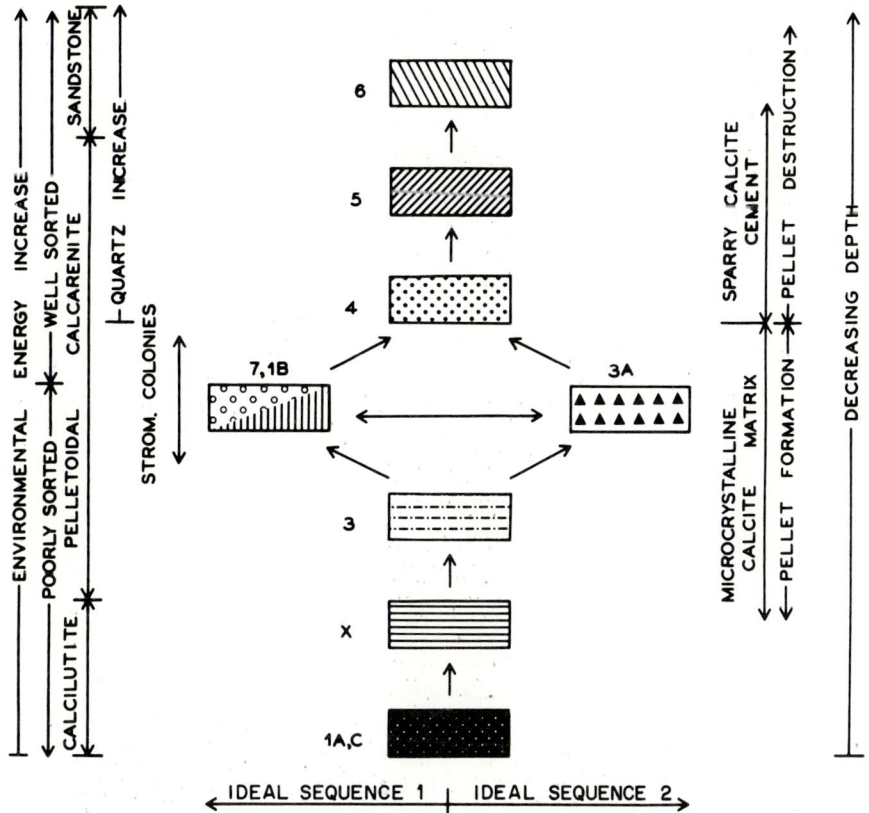

FIGURE 9.9 Arrow Canyon Formation and Crystal Pass Limestone (Upper Devonian), Arrow Canyon Range, Clark County, Nevada. Microfacies interrelationships. From Carss and Carozzi (1965). Reprinted by permission of Elsevier Publishing Company.

9.3 CHARACTERISTIC CONSTITUENTS: PHYLLOID AND RED ALGAE

Buildups in which phylloid and red algae play a major role occur in numerous entirely marine Pennsylvanian cyclothems. They formed under low-energy conditions between weakly prograding delta fronts and open marine shelves. The depositional history of many of these buildups consists of an association of hydrodynamic, bioaccumulated, and bioconstructed phases with abundant micrite contribution. Further textural and mineralogic complications are introduced by repeated diagenetic episodes ranging from subtidal to temporary exposure.

The Iola Formation (Missourian) of southeast Kansas belongs to the Pennsylvanian cyclothems of Kansas type of entirely marine nature (Fig. 9.11). It consists of two carbonate units separated by a shale. The Paola Limestone was deposited on a gentle carbonate ramp and appears to record the maximum transgression within the Iola cycle of deposition. The overlying Raytown Limestone belongs to the regressive portion of the cycle and consists of an initial bioclastic bar evolving upward into a complex phylloid algal buildup. This buildup was adjacent landward to a siliciclastic distal deltaic environment and graded basinward into lower-energy upper ramp and lower ramp carbonates.

The shallowing-upward depositional evolution of the phylloid algal buildup consisted of four distinct stages: bioclastic (hydrodynamic buildup), bioaccumulation, bioconstruction, and bioclastic (destruction in storm-dominated conditions and temporary emergence). Phylloid algal buildups are commonly very porous in subsurface and are known to form economically significant petroleum reservoirs in Oklahoma, Texas, Colorado, and Utah.

The buildup selected for this study (Dawson and Carozzi, 1986) is located in southeast Kansas, near Iola, and was investigated by means of 15 field sections sampled at a 15-cm average vertical interval. More than 700 polished slabs and 1200 thin sections were examined petrographically.

Description of Microfacies

Thin sections of both the Paola Limestone and the Raytown Limestone were divided into 11 series or textural groups using the following criteria: depositional fabric (bioclastic, bioaccumulated, or bioconstructed); fre-

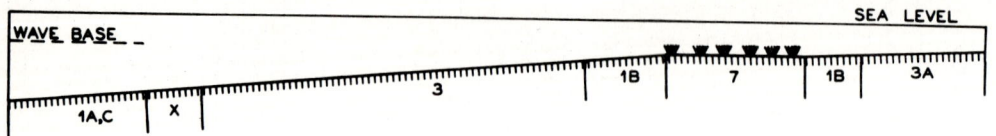

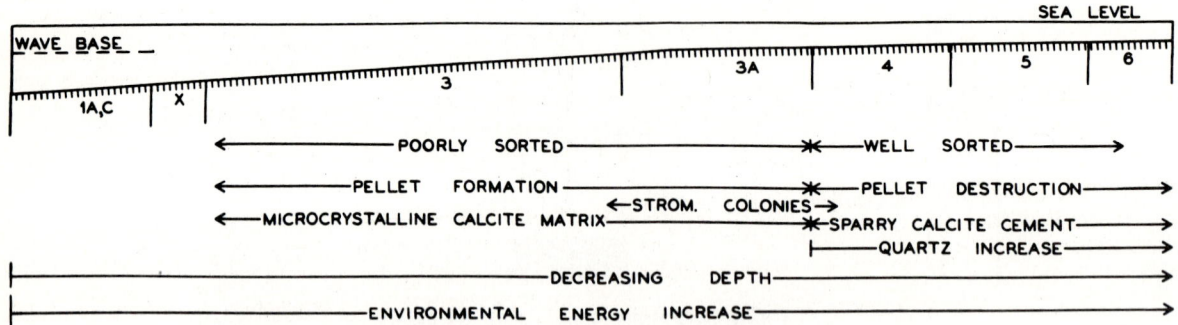

FIGURE 9.10 Arrow Canyon Formation and Crystal Pass Limestone (Upper Devonian), Arrow Canyon Range, Clark County, Nevada. Ideal depositional models. From Carss and Carozzi (1965). Reprinted by permission of Elsevier Publishing Company.

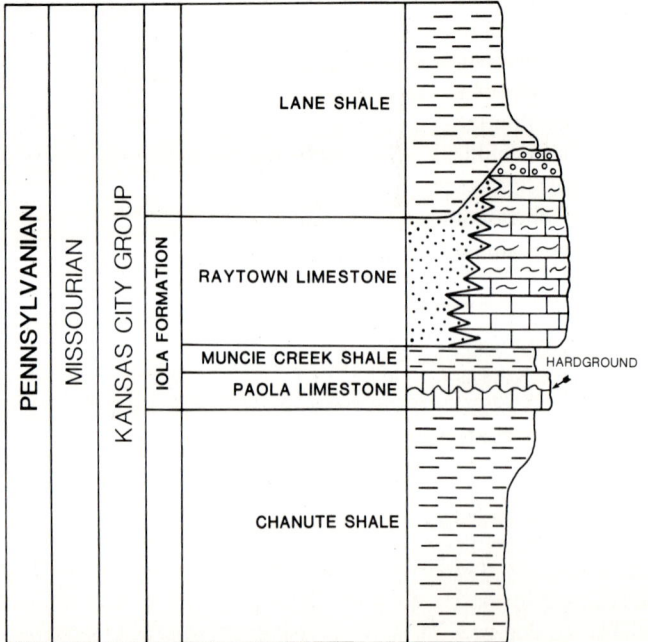

FIGURE 9.11 Iola Formation (Pennsylvanian), southeast Kansas. General stratigraphy. From Dawson and Carozzi (1986). Reprinted by permission of Elsevier Publishing Company.

quency of matrix versus frequency of bioclasts (matrix- or grain-supported); texture of matrix; percentage of cements; mean clasticity and sorting of bioclasts. The 11 textural groups were designated numerically as series 10, 20, 30, 40, 50, 60, 70, 80, 90, 100, and 200. They were subdivided further into 24 distinct microfacies on the basis of the frequency variations of major biotic components (phylloid algae, bryozoans, foraminifers, pelmatozoans, and sponge spicules).

Microfacies 11 (Fig. 9.12A,B). *Archaeolithophyllum*—bryozoan-foraminifer-encrusted biocalcisiltite in which the upper part of the bioturbated pelletoidal matrix with scattered bioclasts is scoured and encrusted by complex intergrowths of red algae and bryozoans.

Microfacies 12 (Fig. 9.12C,D,E). Matrix-supported pelmatozoan-fusulinid biocalcarenite with algaloid concretions. The bioturbated pelletoidal biocalcisiltite matrix contains "floating" bioclasts which are often pyritized and dolomitized burrows with concentrations of micronodules of collophanite.

Microfacies 21. Bioturbated, pelmatozoan-fusulinid, spiculitic biocalcisiltite.

Microfacies 22 (Fig. 9.12F). Phylloid algal, fenestrate bryozoan, matrix-supported biocalcarenite. The bioturbated matrix consists of micritic pelloids and undifferentiated comminuted skeletal debris.

Microfacies 23 (Fig. 9.12G). Strongly bioturbated spiculitic biocalcisiltite with concentration of monaxonic sponge spicules in burrows.

Microfacies 31 (Fig. 9.12H). Collapse-brecciated, pelletoidal, matrix-supported, fine-grained biocalcarenite with ostracods. The pelletoidal calcisiltitic matrix shows abundant rectangular calcite pseudomorphs after anhydrite crystals.

Microfacies 32 (Fig. 9.13A). Phylloid algal, fenestrate bryozoan, matrix-supported, fine-grained biocalcarenite. The rectangular sparite-filled fragments of phylloid algae "float" in a spiculitic pelletoidal biocalcisiltite matrix.

Microfacies 33 (Fig. 9.13B). Phylloid algal, productid brachiopod, matrix-supported, fine-grained biocalcarenite. Abundant *Composita* with intact shells and geopetal filling together with rectangular fragments of phylloid algae are scattered in the pelletoidal calcisiltite matrix.

Microfacies 34. Pelletoidal, *Archaeolithophyllum*, matrix-supported, fine-grained biocalcarenite. The pelletoidal matrix displays abundant sinuous and bifurcating red algal crusts.

Microfacies 41 (Fig. 9.13C). Phylloid algal, half-moon ooid, matrix-supported, medium-grained biocalcarenite. The major components are "floating" in a vaguely pelletoidal biocalcisiltite matrix.

Microfacies 42. Fenestrate bryozoan, pelmatozoan, matrix-supported, medium-grained biocalcarenite.

Microfacies 43 (Fig. 9.13D). Pelmatozoan, oncoids, matrix-supported, medium-grained biocalcarenite. The bioturbated pelletoidal biocalcisiltite matrix contains large oncoids formed by roughly concentric layers of red algae, blue-green algae, fistuliporid bryozoans, and encrusting foraminifers (*Tuberitina* and *Hedraites*).

Microfacies 44 (Fig. 9.13E). Foraminifer-encrusted, matrix-supported, medium-grained biocalcarenite. The vaguely pelletoidal and bioturbated biocalcisiltite matrix displays irregularly shaped, encrusted masses of *Hedraites*, *Tuberitina*, and bryozoans together with fusulinids.

Microfacies 45 (Fig. 9.13F). Spiculitic, pelmatozoan, matrix-supported, medium-grained biocalcarenite. Large pelmatozoan bioclasts are "floating" within a bioturbated spiculitic biocalcisiltite matrix.

Microfacies 51 (Fig. 9.13G). Phylloid algal, fenestrate bryozoan, bioaccumulated limestone. Unbroken and "wavy laminated" fronds of phylloid algae are oriented parallel to bedding in a bioclastic matrix and display relict cellular microstructures within the pseudosparite-replaced blades (arrow).

Microfacies 52 (Fig. 9.13H). Phylloid algal, productid brachiopod, bioaccumulated limestone. The algae and the large intact brachiopod shells are often surrounded by a groundmass of pseudosparite due to intense neomorphism of the original pelletoidal biocalcisiltite matrix.

Microfacies 53. Fenestrate bryozoan bioaccumulated limestone. Intact fronds of *Polypora* form an extensive, broadly curved to contorted framework with interstitial space filled by a fine, grain-supported, pelletoidal biocalcarenite with a calcilutite matrix.

Microfacies 61 (Fig. 9.14A,B). *Archaeolithophyllum*, fistuliporoid bryozoan, bioconstructed limestone. Red algae are often preserved as wavy to convolute micritic lamellae in a groundmass of pseudosparite with rare relicts of cellular microstructure. A pelletoidal micrite internal sediment locally occurs as a geopetal filling of algal synforms.

Microfacies 71 (Fig. 9.14C). Phylloid algal, fenestrate bryozoan, grain-supported, medium-grained biocalcarenite with pelletoidal matrix.

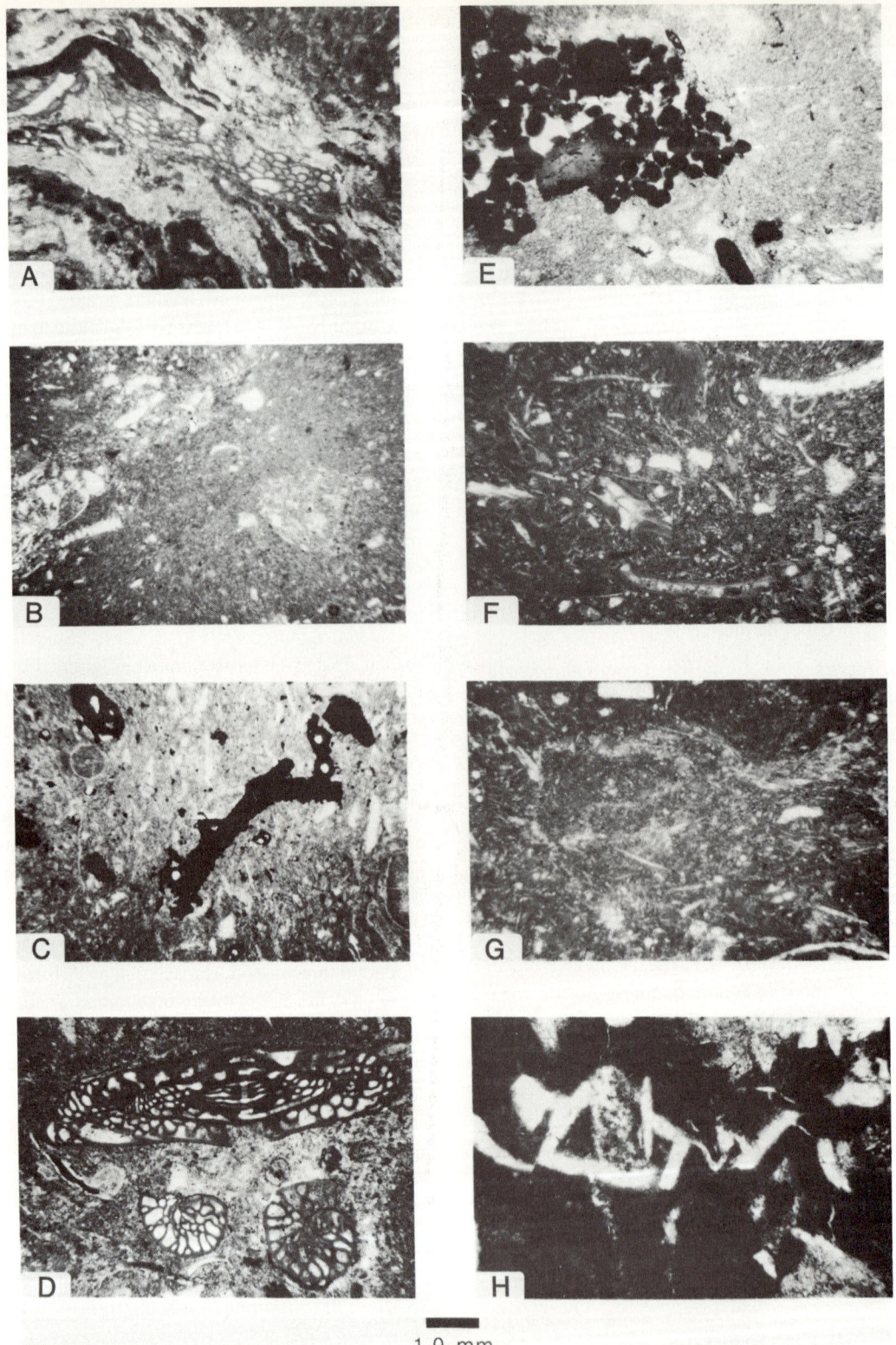

FIGURE 9.12 Iola Formation (Pennsylvanian), southeast Kansas. Typical microfacies. A and B. Microfacies 11. C to E. Microfacies 12. F. Microfacies 22. G. Microfacies 23. H. Microfacies 31. See text for detailed descriptions. All photomicrographs: plane-polarized light. From Dawson and Carozzi (1986). Reprinted by permission of Elsevier Publishing Company.

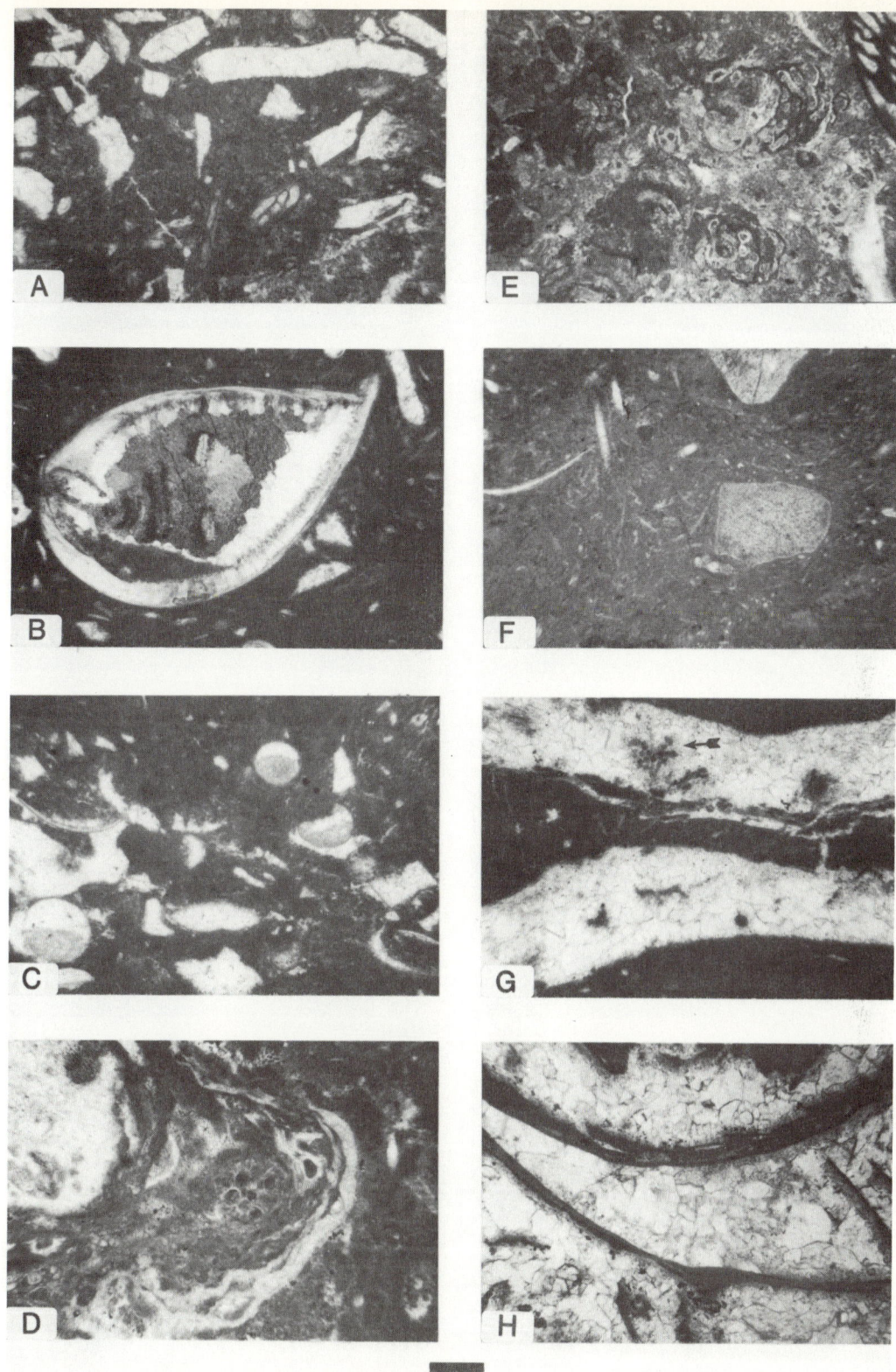

FIGURE 9.13 Iola Formation (Pennsylvanian), southeast Kansas. Typical microfacies (continued). A. Microfacies 32. B. Microfacies 33. C. Microfacies 41. D. Microfacies 43. E. Microfacies 44. F. Microfacies 45. G. Microfacies 51. H. Microfacies 52. See text for detailed descriptions. All photomicrographs: plane-polarized light, except B, crossed nicols. From Dawson and Carozzi (1986). Reprinted by permission of Elsevier Publishing Company.

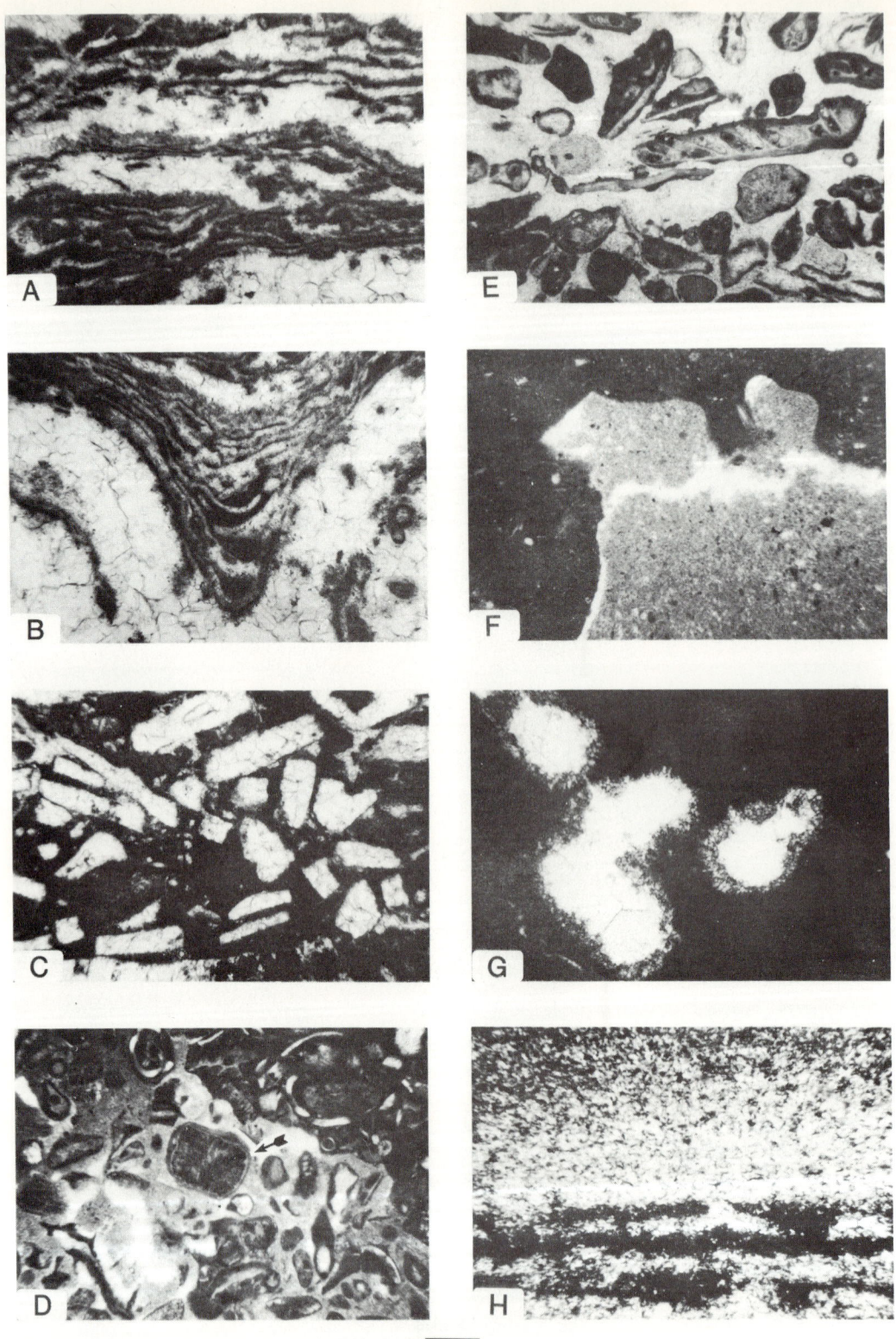

1.0 mm

FIGURE 9.14 Iola Formation (Pennsylvanian), southeast Kansas. Typical microfacies (continued). A and B. Microfacies 61. C. Microfacies 71. D. Microfacies 81. E. Microfacies 91. F and G. Microfacies 101. H. Microfacies 201. See text for detailed descriptions. All photomicrographs: plane-polarized light. From Dawson and Carozzi (1986). Reprinted by permission of Elsevier Publishing Company.

Microfacies 72. Phylloid algal, grain-supported, coarse-grained biocalcarenite with pelletoidal matrix.

Microfacies 81 (Fig. 9.14D). Pelmatozoan, fenestrate bryozoan, superficial ooid, grain-supported, coarse-grained biocalcarenite with micrite matrix (often neomorphosed into pseudomicrosparite) and sparite cement. Most superficial ooids contain cores of large pelmatozoan bioclasts (arrow).

Microfacies 91 (Fig. 9.14E). Intraclastic, pelmatozoan, fenestrate bryozoan, superficial ooid, grain-supported, coarse-grained biocalcarenite with sparite cement. The latter occurs as isopachous rims, syntaxial overgrowths, and interparticle mosaic.

Microfacies 101 (Fig. 9.14F,G). Micritic lithocalcirudite with interstitial vadose silt matrix and sparite cement. The angular to subangular, pebble-size lithoclasts of dark micrite commonly show circular to oval-shaped spar-filled fenestrae (root molds) with rims of pseudomicrosparite. Geopetal features and syneresis cracks are common in the vadose silt.

Microfacies 201 (Fig. 9.14H). Thinly laminated, ferruginous, micaceous, quartz-feldspar siltstone with streaks of carbonized plant material.

The general-shallowing upward sequence of the phylloid algal buildup (Figs. 9.15, 9.16) is shown clearly by the variation of microscopic parameters.

The Ideal Depositional Models

Model 1 (Figs. 9.17, 9.18) applies to the Paola Limestone and corresponds essentially to a carbonate ramp with microfacies 11 recording some effects of turbulence inferred as near-storm wave base conditions. The thickest part of the model (microfacies 34) represents a subwave base offshore bar.

The succession of models 2 to 4 expresses the development of the Raytown Limestone phylloid algal buildup. Model 2 (Figs. 9.19, 9.20) is controlled by microfacies 71, which is a phylloid algal-fenestrate bryozoan bioclastic bar. Model 3 (Figs. 9.21, 9.22) is the major phase of bioaccumulation and bioconstruction of the mound by phylloid and red algae, respectively. This development leads to the generation of a hypersaline lagoon separated by a bioclastic ridge (microfacies 32) from offshore carbonates. Model 4 (Figs. 9.23, 9.24) is predominantly intraclastic and bioclastic, a wave-dominated destruction phase subject to intermittent storm activity with occasional episodes of subaerial exposure.

The superposition of the four depositional models allows a cross section to be drawn (Fig. 9.25) which summarizes the sedimentological evolution of the Paola

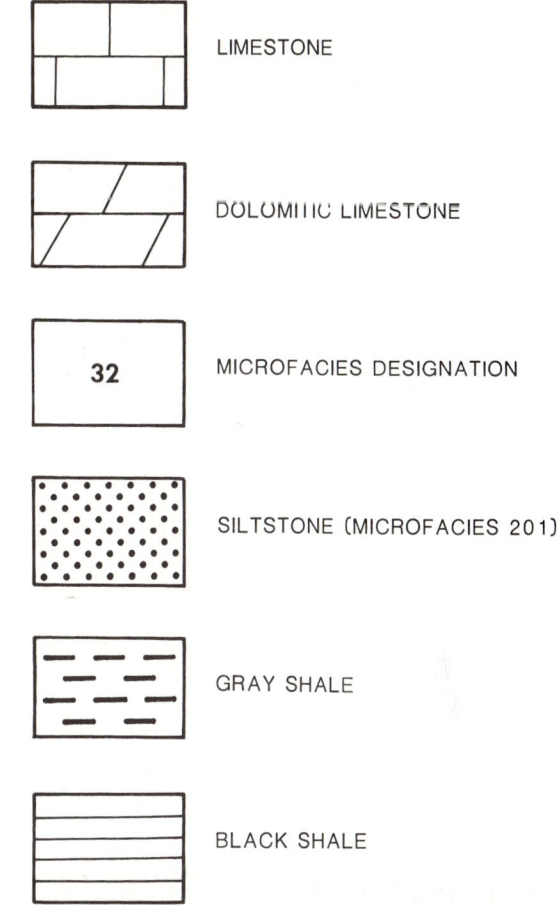

FIGURE 9.15 Iola Formation (Pennsylvanian), southeast Kansas. Symbols for microfacies. From Dawson and Carozzi (1986). Reprinted by permission of Elsevier Publishing Company.

Limestone (T1), in particular the development stages (T2 to T4) of the Raytown phylloid algal buildup.

Diagenesis

The Raytown phylloid algal buildup underwent the following stages: marine phreatic 1; freshwater vadose-phreatic undersaturated 1; marine phreatic 2; freshwater phreatic saturated 2; deep burial; and late uplift, all of which led to extensive mineralogic and textural alterations. Much of the diagenesis was microfacies specific; hence numerous diagenetic pathways are evident. Their graphic representation (Figs. 9.26, 9.27, 9.28) is self-explanatory, and only the major diagenetic products are illustrated (Fig. 9.29). A generalized (nontemporal) diagenetic sequence (Fig. 9.30) shows how relative porosity in Iola carbonate microfacies changed because of variable diagenetic processes and environments. Two major episodes of secondary porosity are evident: extensive, fabric-

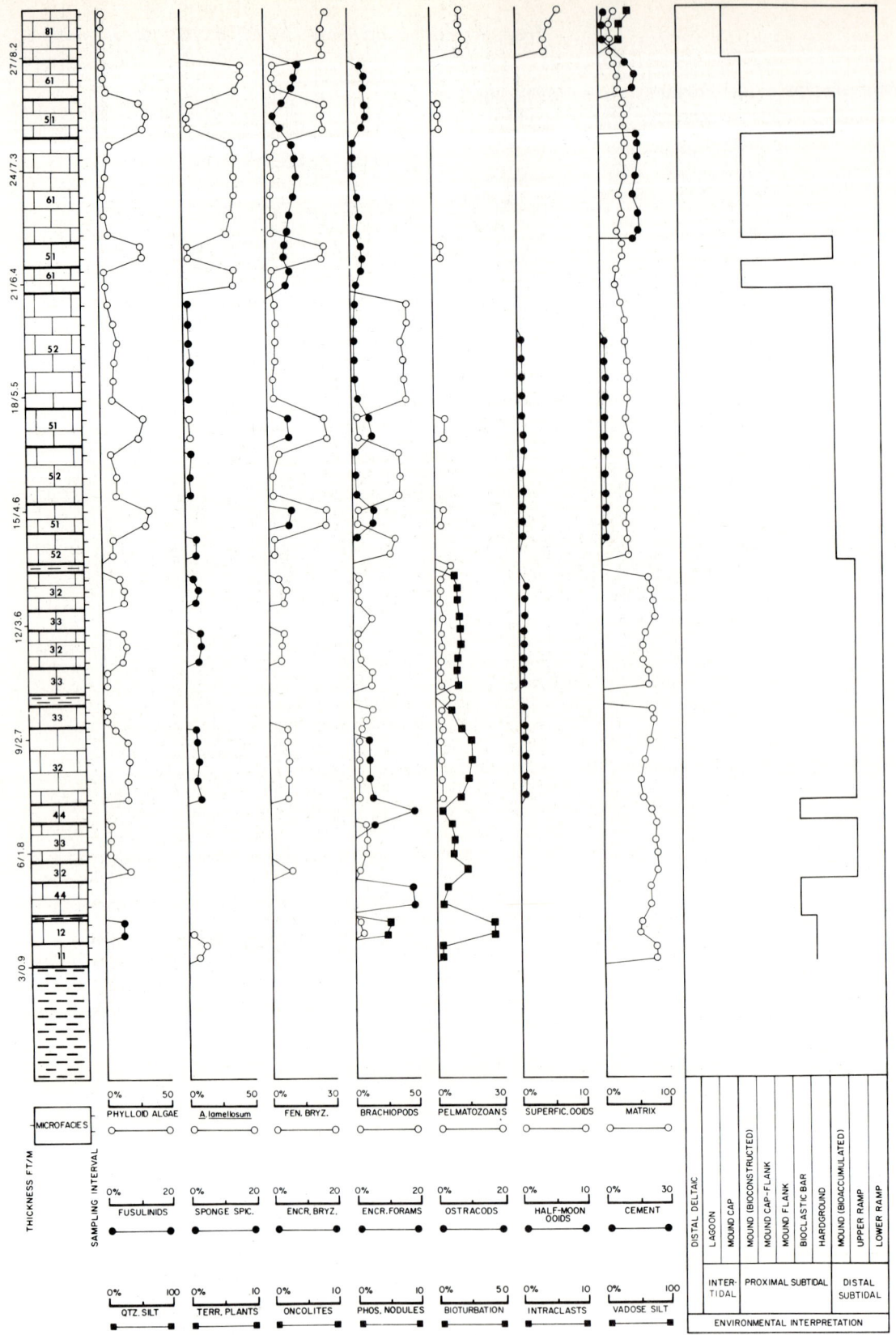

FIGURE 9.16 Iola Formation (Pennsylvanian), southeast Kansas. Typical example of vertical variation of microscopic parameters during shallowing-upward sequence of field section 3. From Dawson and Carozzi (1986). Reprinted by permission of Elsevier Publishing Company.

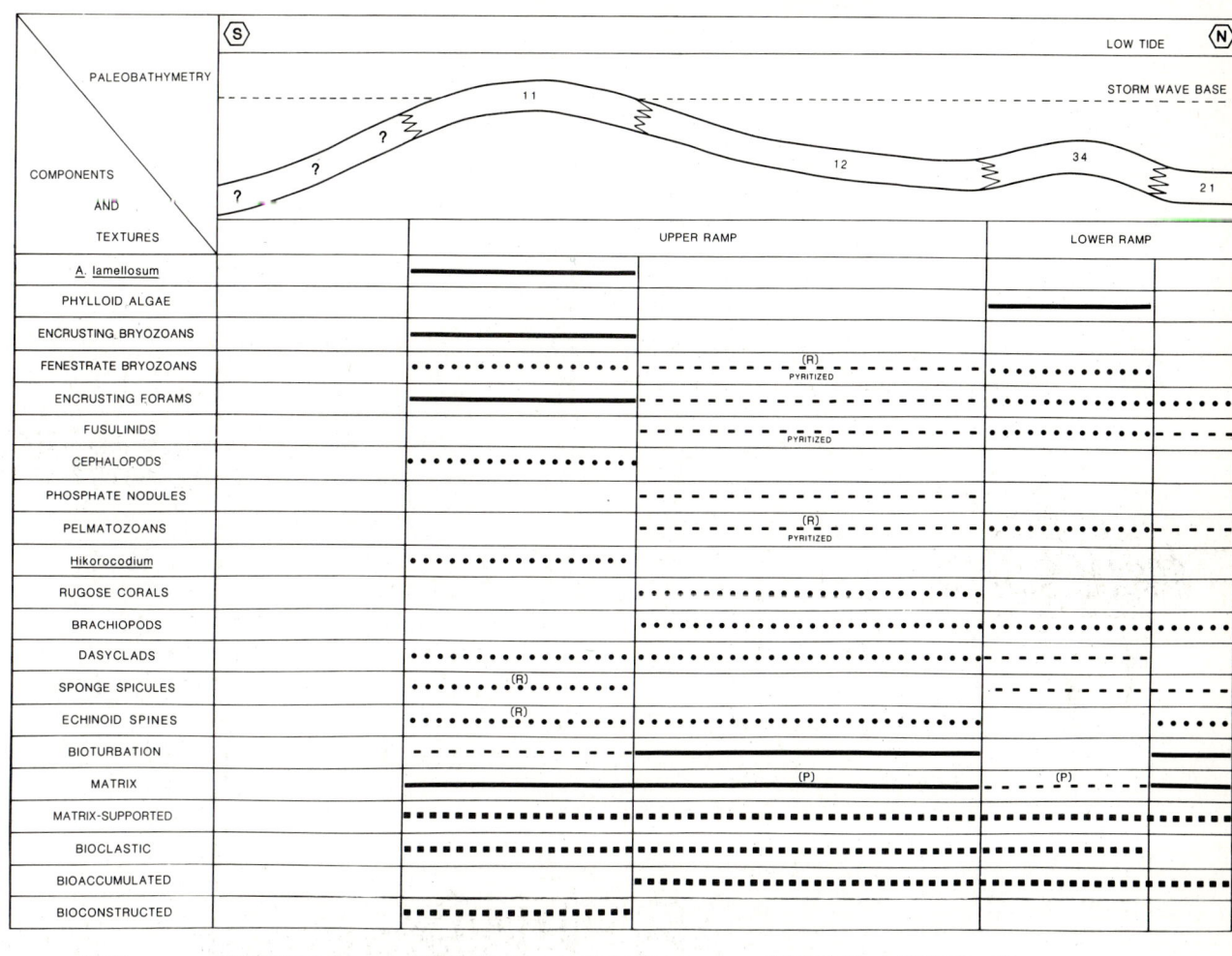

FIGURE 9.17 Iola Formation (Pennsylvanian), southeast Kansas. Ideal depositional model 1 (T1, Paola Limestone). From Dawson and Carozzi (1986). Reprinted by permission of Elsevier Publishing Company.

selective algal moldic dissolution within a freshwater phreatic environment and late pervasive fracturation resulting from uplift. Nearly all this porosity has been occluded during subsequent episodes of cement precipitation from circulating meteoric waters.

9.4 CHARACTERISTIC CONSTITUENTS: PHYLLOID ALGAE

Phylloid algae formed incipient constructed buildups within bituminous and argillaceous carbonates in the marine phase of coal-bearing Pennsylvanian cyclothems. These buildups were located in embayments between strongly prograding delta lobes and graded laterally into better developed crinoid-brachiopod hydrodynamic accumulations. These organisms apparently were able to withstand better unfavorable conditions generated by intense siliciclastic sediment yield.

The Brereton Limestone (Middle Pennsylvanian) of the southwest margin of the Illinois Basin (Fig. 9.31) is characteristic of thin marine carbonates deposited over the siliciclastic-influenced shelves of the American Mid-Continent in infratidal ramps that were transitional to prodelta conditions, and in interdeltaic troughs or embayments (Khawlie and Carozzi, 1976).

This particular limestone displays an average thickness of 2 m and was investigated in 100 cores by means of 319 samples with an average vertical spacing of 60 cm.

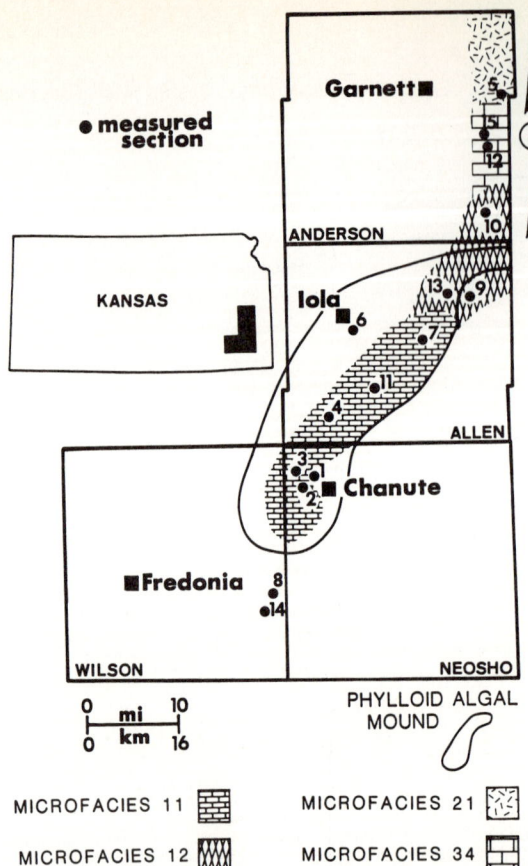

FIGURE 9.18 Iola Formation (Pennsylvanian), southeast Kansas. Microfacies map of ideal depositional model 1 (T1, Paola Limestone). From Dawson and Carozzi (1986). Reprinted by permission of Elsevier Publishing Company.

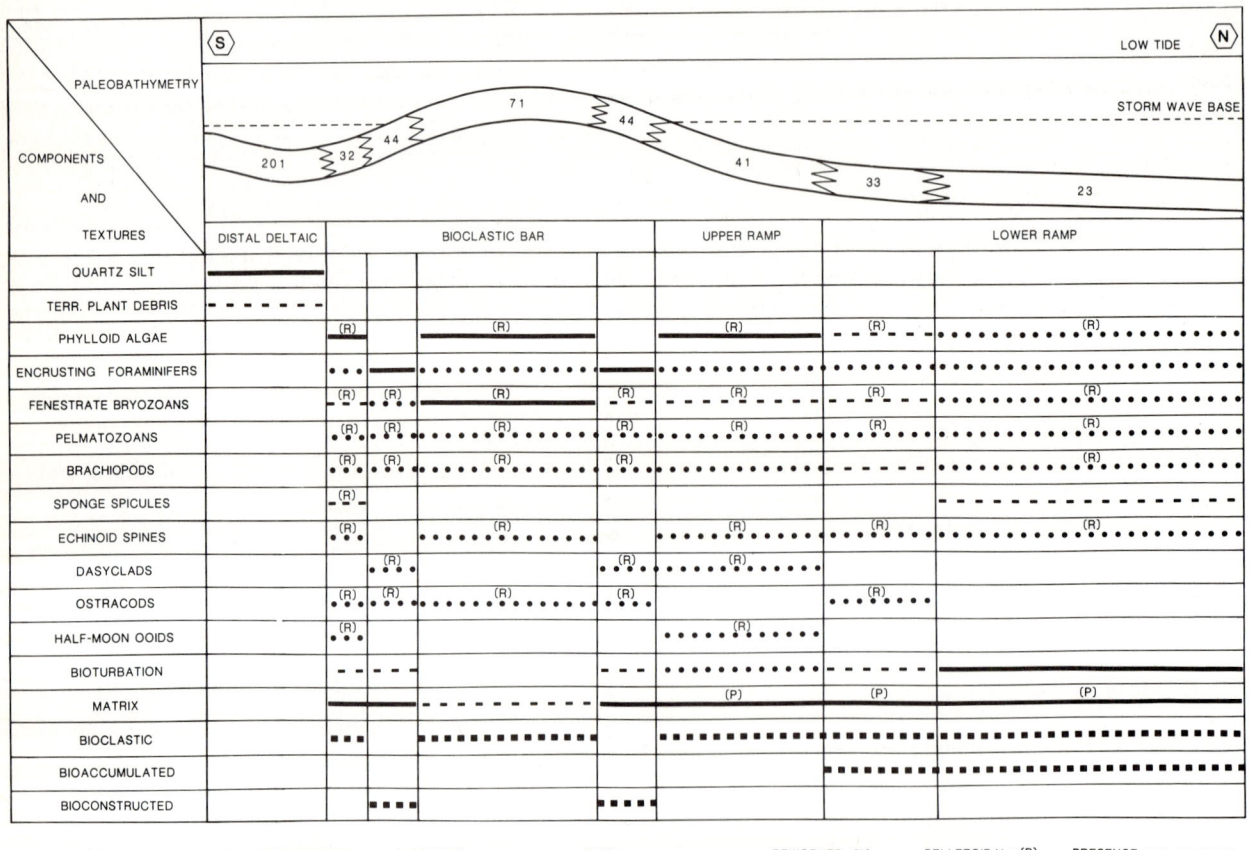

FIGURE 9.19 Iola Formation (Pennsylvanian), southeast Kansas. Ideal depositional model 2 (T2, Lower Raytown Limestone). From Dawson and Carozzi (1986). Reprinted by permission of Elsevier Publishing Company.

Chap. 9 / Carbonate Ramps with Bioconstructed Buildups **175**

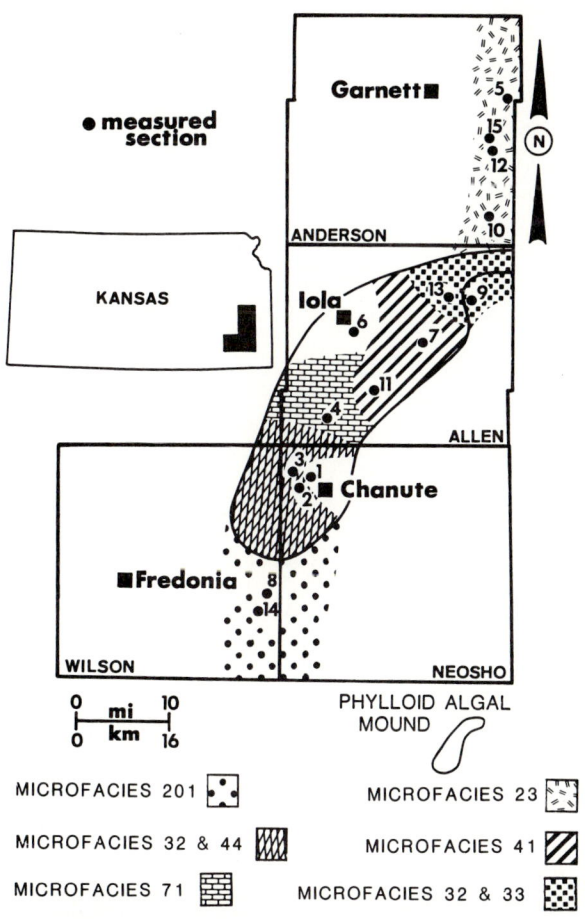

FIGURE 9.20 Iola Formation (Pennsylvanian), southeast Kansas. Microfacies map of ideal depositional model 2 (T2, Lower Raytown Limestone). From Dawson and Carozzi (1986). Reprinted by permission of Elsevier Publishing Company.

Description of Microfacies

The recognized microfacies were divided into arenaceous, argillaceous, and weakly argillaceous series (Fig. 9.32). Because of the strong deltaic influence on these carbonates, they are described in order of general relative depth increase along the gentle infratidal slope in front and between prograding delta lobes.

Arenaceous Series

Microfacies 1 (Fig. 9.33A,B). Massive to bioturbated arenaceous and argillaceous calcisiltite with scattered sand-size bioclasts of ostracods, brachiopods, and pelecypods. Detrital quartz can increase locally in clasticity and frequency to the extent of generating a quartz arenite with a calcisiltite matrix devoid of bioclasts (Fig. 9.33B).

Microfacies 2 (Fig. 9.33C). Bioturbated arenaceous and argillaceous mud-supported biocalcarenite with 20 to 40% sand-size bioclasts scattered in a calcisiltite matrix. Bioclasts of ostracods, crinoids, brachiopods, and pelecypods are associated with grains of phosphate and glauconite.

Microfacies 3 (Fig. 9.33D). Bioturbated arenaceous and argillaceous grain-supported biocalcarenite with a calcisiltite matrix. Bioclasts of ostracods, crinoids, and brachiopods predominate over bryozoans. Phosphatic pellets and glauconite grains are common.

Microfacies 4 (Fig. 9.33E). Bioturbated slightly arenaceous grain-supported biocalcarenite with sparite cement. Major bioclasts are ostracods, brachiopods, crinoids, and pelecypods associated with fecal pellets.

Microfacies 5 (Fig. 9.33F,G). This microfacies, although not arenaceous, is included in this series because it is related genetically. It is a phylloid algae-constructed limestone in which closely superposed dark and platy fronds of *Ivanovia* sp. are separated by lighter colored zones of interstitial weakly argillaceous calcisiltite which have been altered into pseudomicrosparite by neomorphism. The algal fronds are often encrusted by bryozoans (Fig. 9.33G), whereas the interstitial calcisiltite may show a few ostracods and some scattered grains of detrital quartz.

Argillaceous Series

Microfacies 6 (Fig. 9.33H). Bioturbated argillaceous, mud-supported to grain-supported biocalcarenite with a calcisiltite matrix. Bioclasts of ostracods, brachiopods, crinoids, and pelecypods predominate over trilobites, fusulinids, arenaceous foraminifers, calcispheres, and gastropods. Detrital quartz is rare, but grains of phosphate and glauconite are common.

Microfacies 7 (Fig. 9.34A). Argillaceous ostracodal calcarenite with a calcisiltite matrix. This laminated microfacies consists almost entirely of greatly compacted, distorted, and disarticulated ostracod valves. Silt-size detrital quartz is concentrated in matrix-rich laminae.

Microfacies 8 (Fig. 9.34B). Bioturbated argillaceous, mud-supported biocalcarenite with 10 to 20% sand-size bioclasts in a pyritic calcisiltite matrix. Bryozoan bioclasts predominate over ostracods and monaxonic sponge spicules. Some grains of detrital quartz are associated with rare phosphate and glauconite pellets.

Microfacies 9 (Fig. 9.34C). Bioturbated, argillaceous, mud-supported to grain-supported biocalcarenite with 20 to 40% sand-size bioclasts and a calcisiltite ma-

176 Part II / Carbonate Ramps

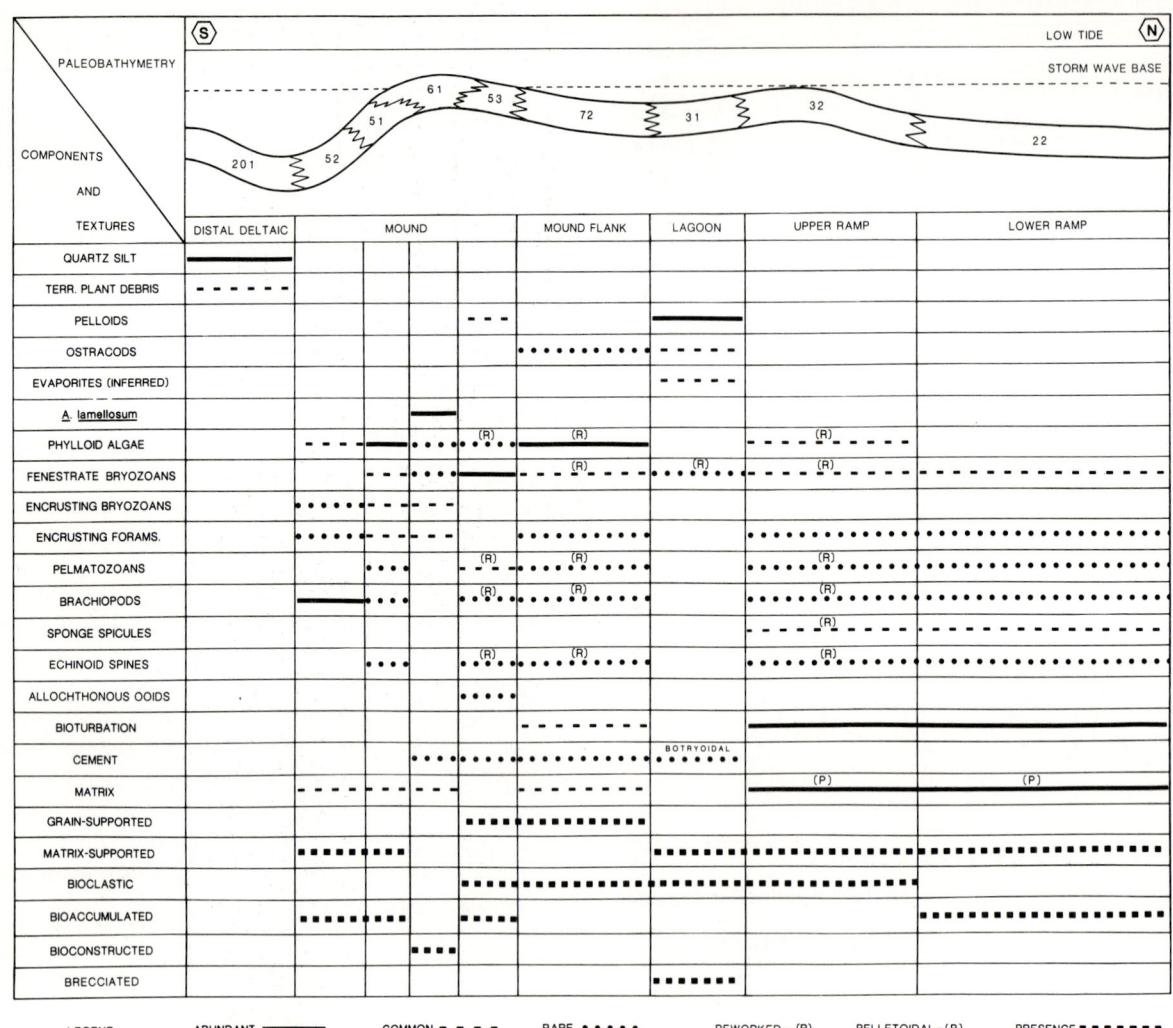

FIGURE 9.21 Iola Formation (Pennsylvanian), southeast Kansas. Ideal depositional model 3 (T3, Middle Raytown Limestone). From Dawson and Carozzi (1986). Reprinted by permission of Elsevier Publishing Company.

trix. Trilobite bioclasts predominate over fusulinids, brachiopods, crinoids, arenaceous foraminifers, and ostracods. Grains of detrital quartz, phosphate, and glauconite are rare.

Microfacies 10 (Fig. 9.34D,E). Deeply bioturbated argillaceous calcisiltite with abundant bioclasts of monaxonic sponge spicules, brachiopod spines, brachiopod shells, and ostracods. Locally, bioturbation is less developed and monaxonic sponge spicules are oriented parallel to bedding (Fig. 9.34E).

Weakly Argillaceous Series

Microfacies 11 (Fig. 9.34F). Bioturbated weakly argillaceous, mud-supported biocalcarenite with 20 to 40% sand-size bioclasts in a calcisiltite to bioclastic matrix with merged pellets. The bioclasts are monaxonic sponge spicules, ostracods, brachiopods, and pelecypods, associated with frequent paleotextulariids.

Microfacies 12 (Fig. 9.34G). Deeply bioturbated weakly argillaceous mud-supported biocalcarenite with 10 to 20% sand-size bioclasts in a calcisiltite pelletoidal matrix with abundant pyrite pigments.

Microfacies 13 (Fig. 9.34H). Deeply bioturbated spiculitic calcisiltite with rare grains of silt-size detrital quartz.

The Ideal Depositional Models

The general environment of deposition of the Brereton Limestone in the area investigated can be represented by a paleoenvironmental map (Fig. 9.35) showing the

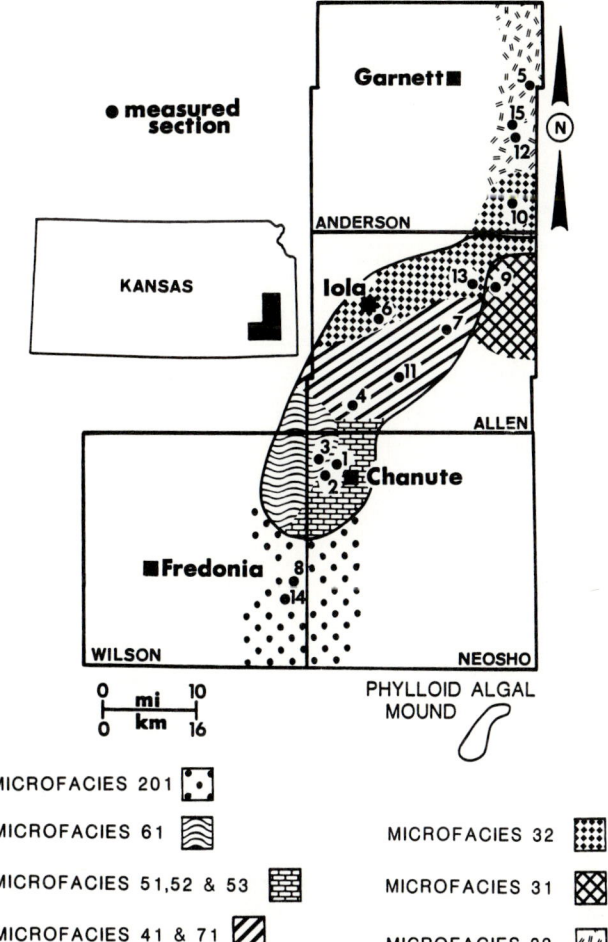

FIGURE 9.22 Iola Formation (Pennsylvanian), southeast Kansas. Microfacies map of ideal depositional model 3 (T3, Middle Raytown Limestone). From Dawson and Carozzi (1986). Reprinted by permission of Elsevier Publishing Company.

space distribution of the 13 recognized microfacies. Two sections across this map (AA' and BB') are used, respectively, to establish a depositional model for the phylloid algae-constructed buildup and for the crinoid-brachiopod hydrodynamic buildup and their associated microfacies.

The Ideal Constructed Phylloid Algae Buildup Model

In this model (Figs. 9.36, 9.37) microfacies 7 is interpreted as a depression within the carbonate ramp where detrital quartz grains and ostracod bioclasts were trapped preferentially during seaward transportation. The latter process also concentrated the lighter sponge spicules in deeper quiet conditions offshore. The association of phylloid algae and encrusting bryozoans is characteristic of the algal buildup. A partial relationship exists between the degree of bioturbation and frequency of the trilobites as if the latter were responsible for the commonly observed churning of the microfacies. Offshore, this relationship disappears, however, and the type of bioturbation changes to burrows suggesting the action of annelids in that area.

In summary, with increasing relative depth, the following ecological communities occur: arenaceous foraminifers and fusulinids; phylloid algae and encrusting bryozoans; crinoids, brachiopods, and pelecypods.

The Ideal Hydrodynamic Crinoid-Brachiopod Buildup Model

In this model (Figs. 9.38, 9.39), detrital quartz is also transported seaward across the entire carbonate system, but it is trapped mostly behind the hydrodynamic buildup while the lighter sponge spicules are finally deposited offshore. Arenaceous foraminifers and fusulinids characterize the shallower conditions behind the hydrodynamic buildup and adjacent microfacies where the association of crinoids-brachiopods-pelecypods and trapped ostracods predominates. The intensity of bioturbation increases seaward with, as in the previous model, at first a partial relationship with the frequency of trilobites

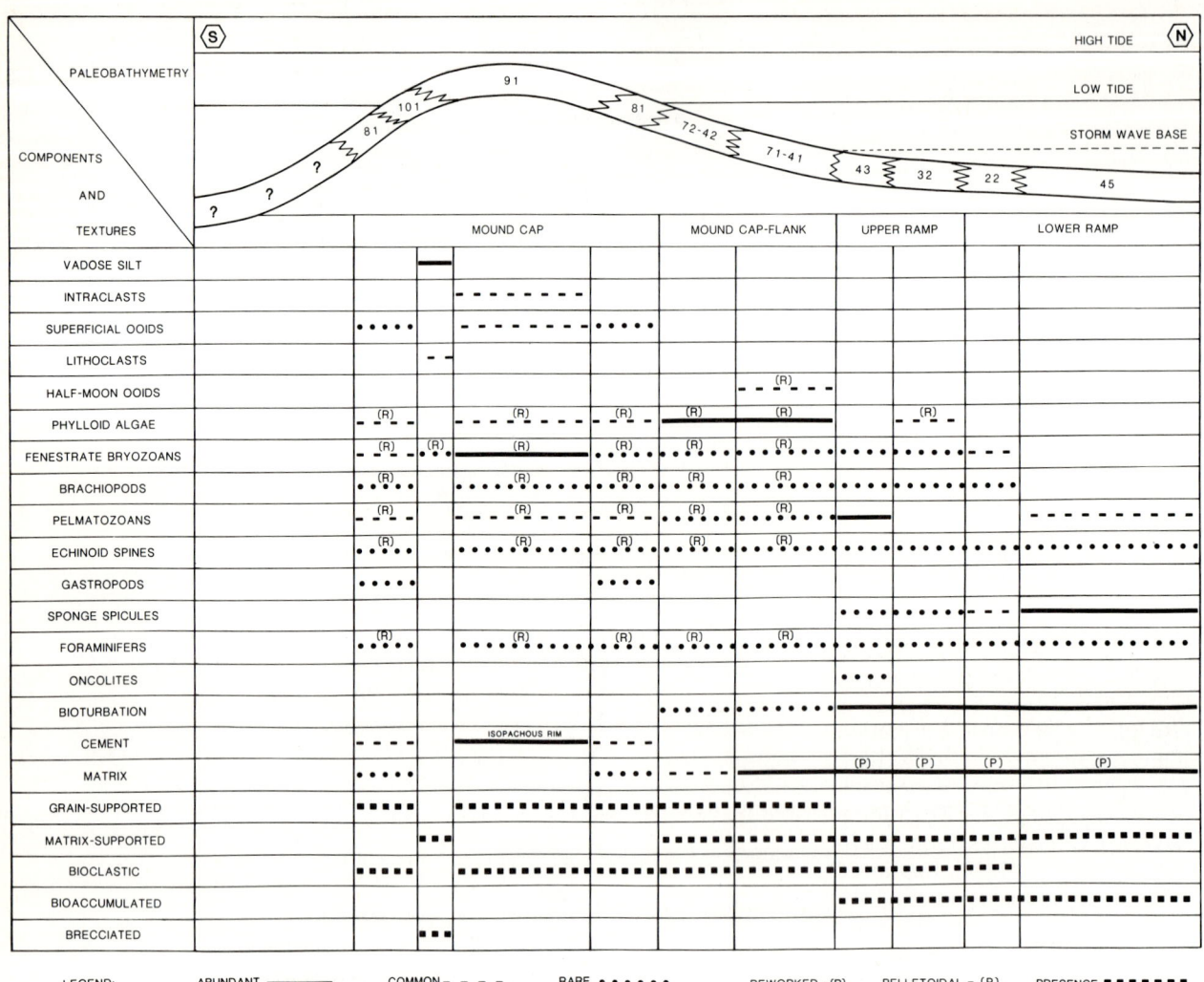

FIGURE 9.23 Iola Formation (Pennsylvanian), southeast Kansas. Ideal depositional model 4 (T4, Upper Raytown Limestone). From Dawson and Carozzi (1986). Reprinted by permission of Elsevier Publishing Company.

and then a probable activity of annelids in deeper water. Phylloid algal bioclasts are ubiquitous but minor components. They indicate nearby communities which may be responsible, as a lateral effect (Fig. 9.35), for the peak of bryozoan frequency in microfacies 8 in the lower part of the ramp.

The general distribution of components in both models indicates *in situ* benthic communities with very moderate reworking limited to hydrodynamic buildups. Only a seaward transportation of detrital quartz and of ostracods is apparent. These conditions are typical of quiet interdeltaic embayments.

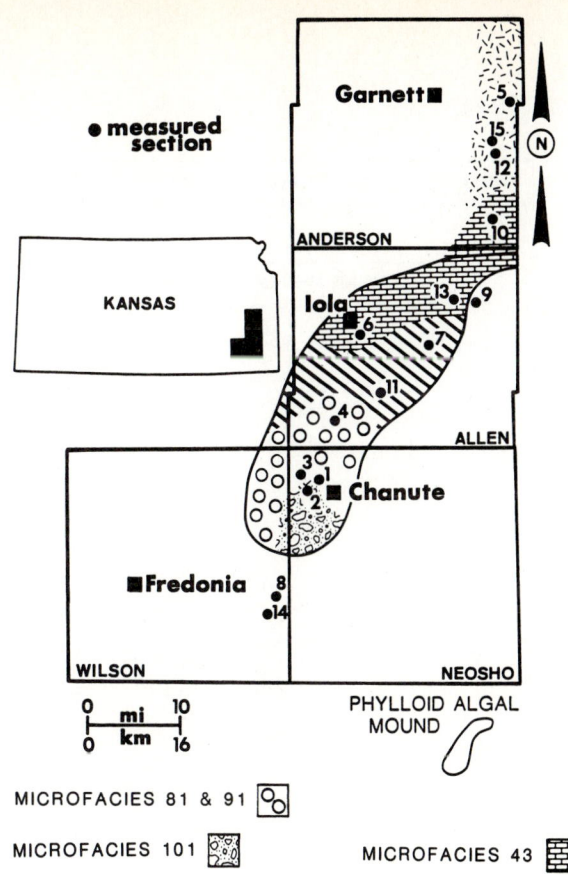

FIGURE 9.24 Iola Formation (Pennsylvanian), southeast Kansas. Microfacies map of ideal depositional model 4 (T4, Upper Raytown Limestone). From Dawson and Carozzi (1986). Reprinted by permission of Elsevier Publishing Company.

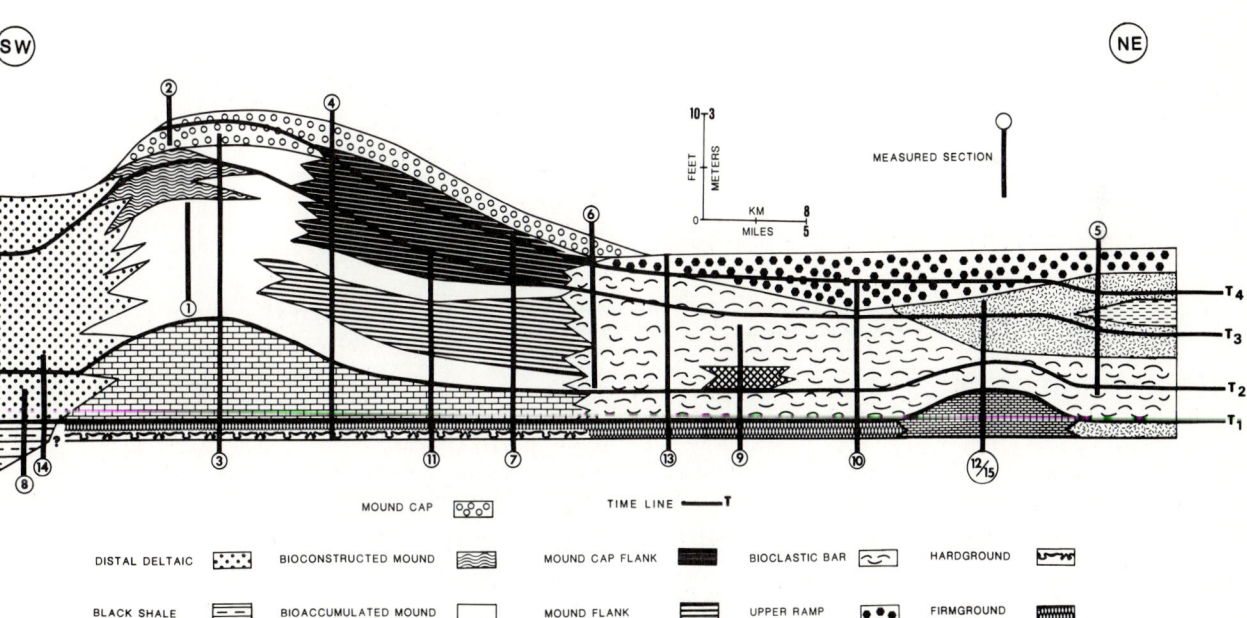

FIGURE 9.25 Iola Formation (Pennsylvanian), southeast Kansas. Cross section showing lithofacies, depositional environments, location of measured sections, and developmental stages of Paola Limestone (T1) and overlying Raytown phylloid algal buildup (T2 to T4). From Dawson and Carozzi (1986). Reprinted by permission of Elsevier Publishing Company.

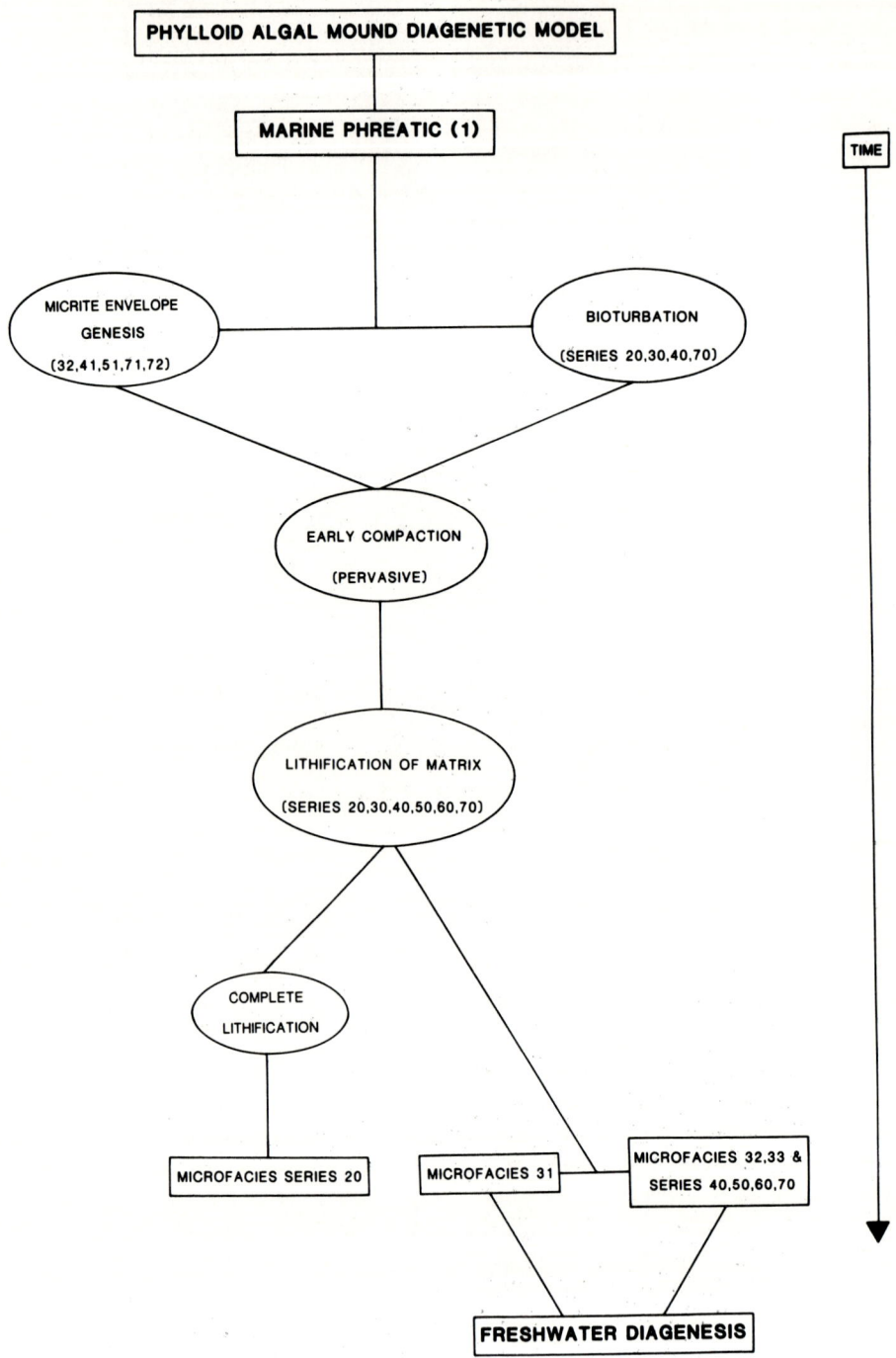

FIGURE 9.26 Iola Formation (Pennsylvanian), southeast Kansas. Diagenetic pathways of Raytown Limestone microfacies. From Dawson and Carozzi (1986). Reprinted by permission of Elsevier Publishing Company.

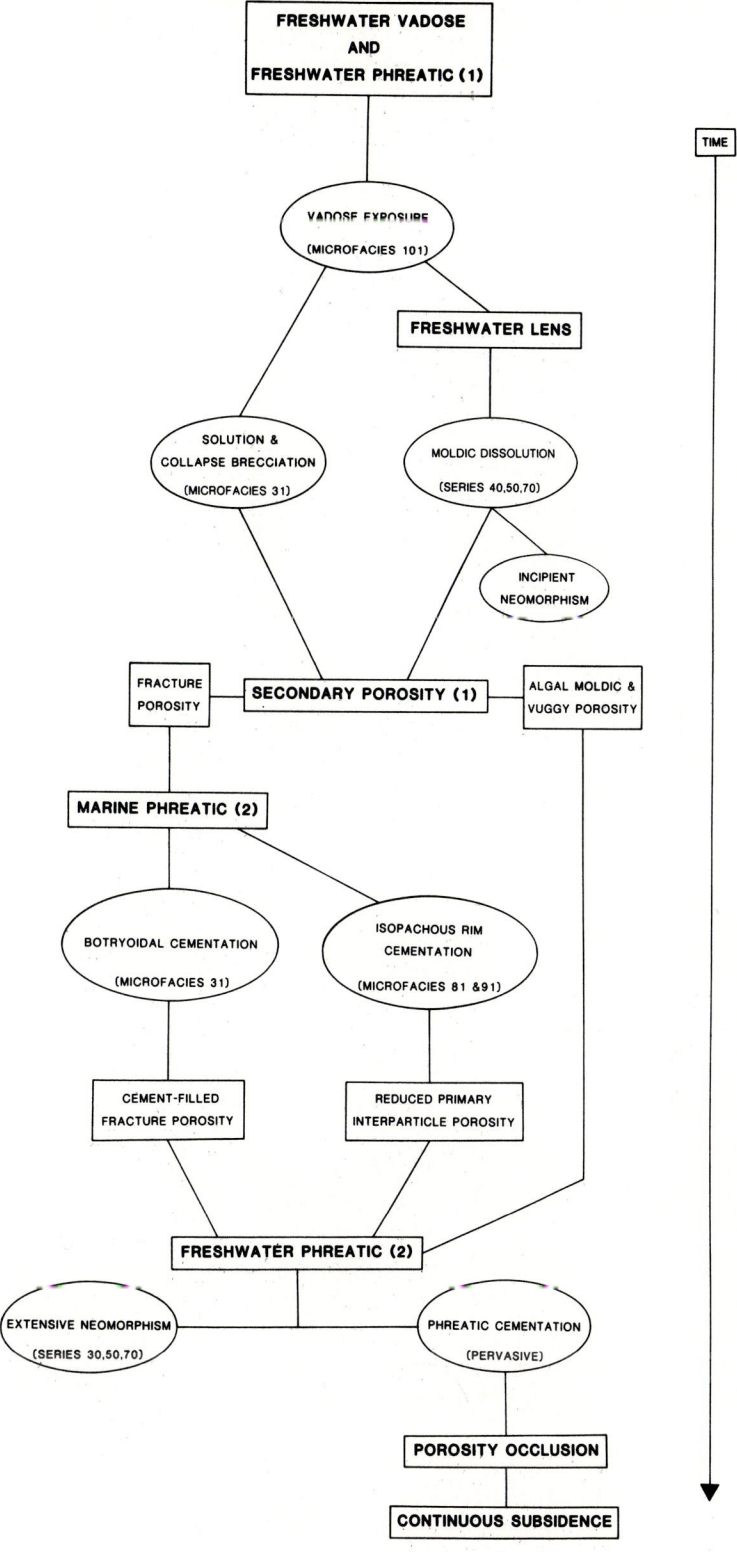

FIGURE 9.27 Iola Formation (Pennsylvanian), southeast Kansas. Diagenetic pathways of Raytown Limestone microfacies (continued). From Dawson and Carozzi (1986). Reprinted by permission of Elsevier Publishing Company.

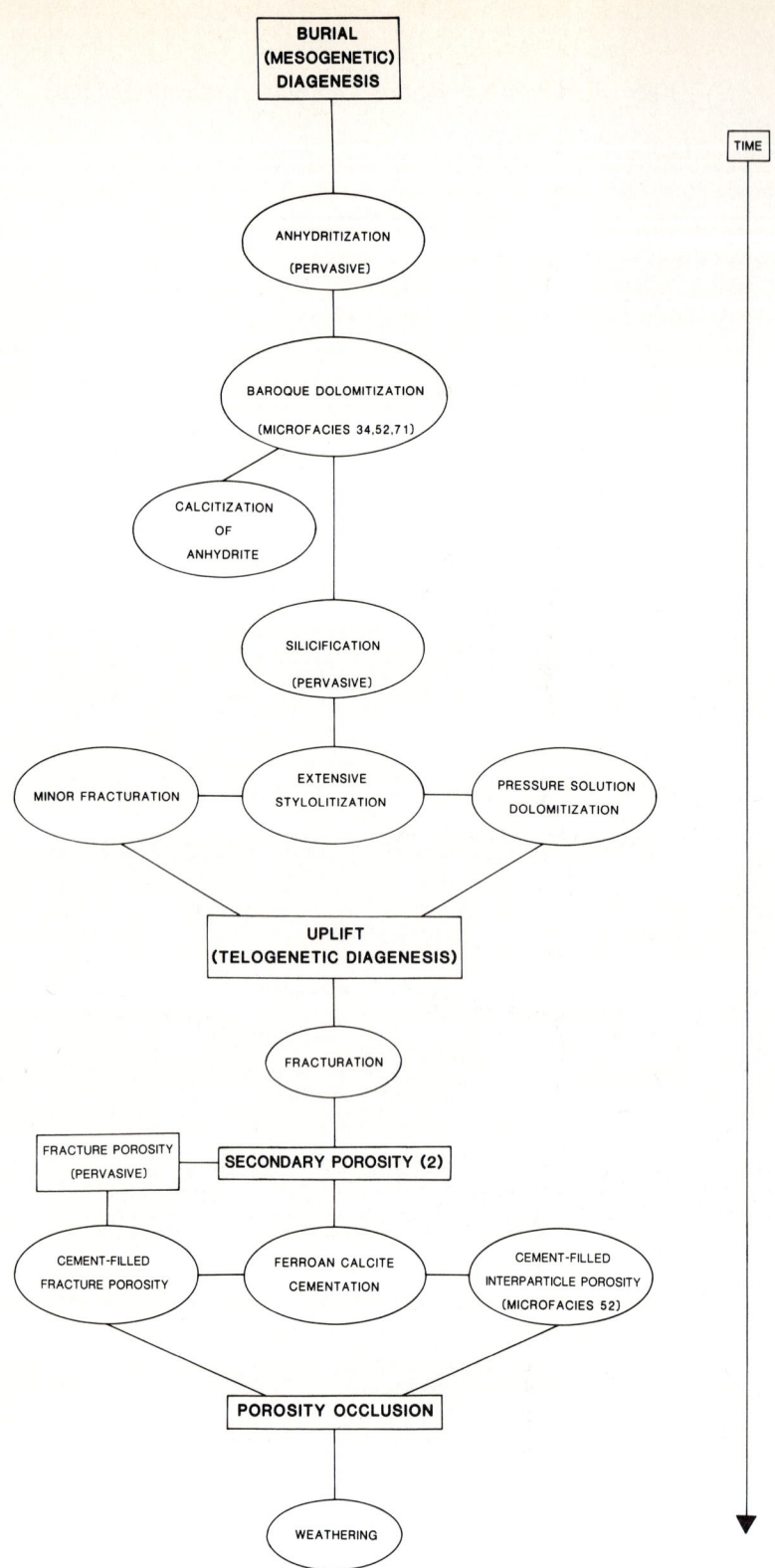

FIGURE 9.28 Iola Formation (Pennsylvanian), southeast Kansas. Diagenetic pathway of Raytown Limestone microfacies (continued). From Dawson and Carozzi (1986). Reprinted by permission of Elsevier Publishing Company.

FIGURE 9.29 Iola Formation (Pennsylvanian), southeast Kansas. Diagenetic sequence of Raytown Limestone phylloid algal buildup. *Marine phreatic (1) environment and freshwater vadose-phreatic undersaturated (1) environment.* A. Micrite envelopes outlining sparite-filled molds of phylloid algal bioclasts. Phylloid algae were dissolved during freshwater phreatic undersaturated diagenesis (1) and subsequently filled with phreatic cement. B. Syneresis cracks within vaguely pelletoidal biocalcisiltite matrix. Cracks have been filled, during freshwater diagenesis, with sparite cement. *Marine phreatic (2) environment and freshwater phreatic saturated (2) environment.* C. Relicts of several coalescing "fans" of botryoidal marine

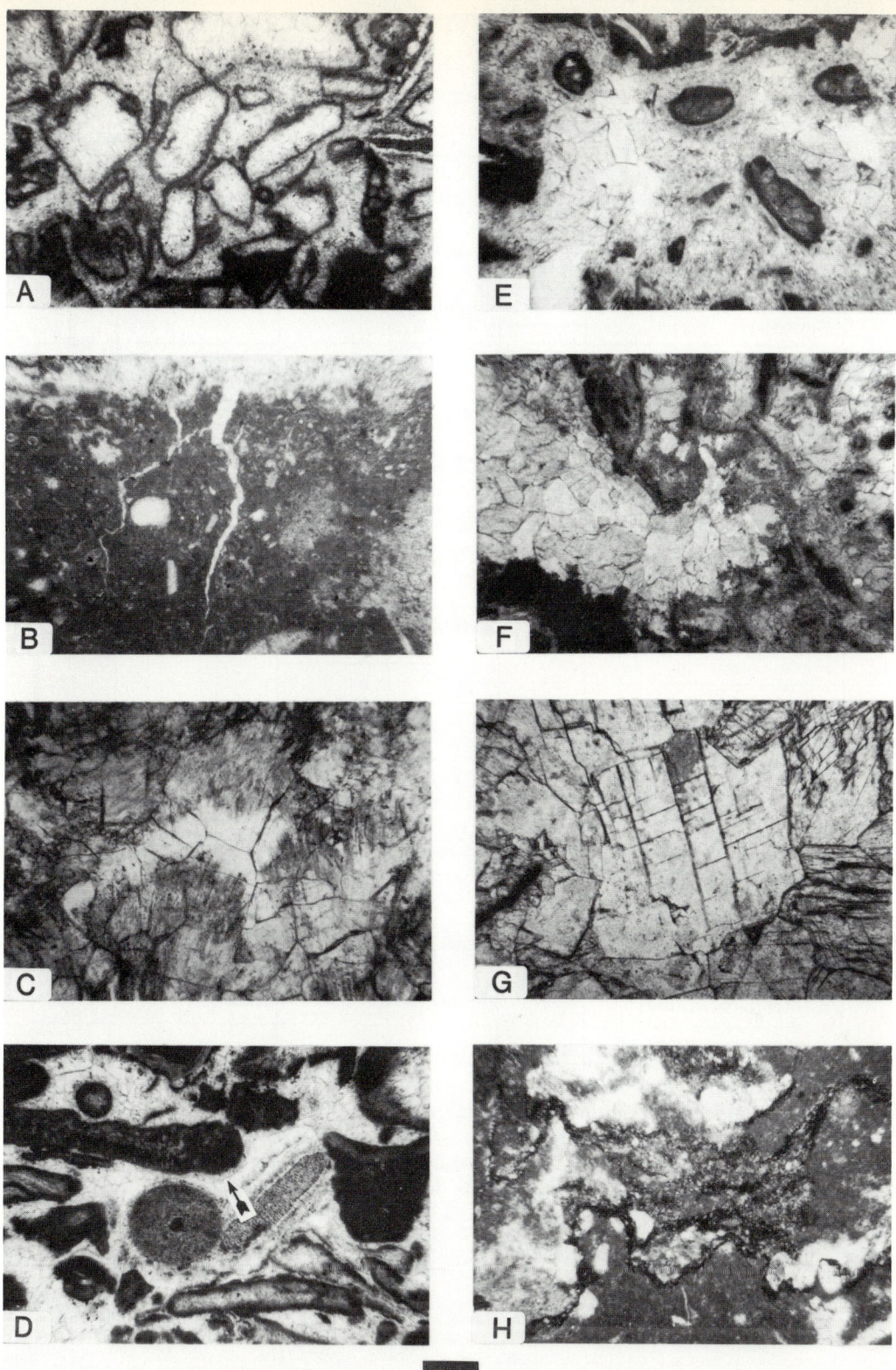

FIGURE 9.29 (*continued*) cement within pseudosparite, showing boundaries of pseudosparite crystals transecting those of original acicular crystals. D. Grain-supported biocalcarenite cemented by first generation isopachous rim cement (arrow) and second generation of coarse equant sparite. E. Pseudosparite groundmass containing numerous "floating" inclusions of fenestrate bryozoans. *Burial environment.* F. Tongue of baroque dolomite replacing inclusion-rich pseudosparite. G. Calcite pseudomorphs after anhydrite, showing preservation of original anhydrite cleavage. H. Microstylolites and associated "pressure-solution" dolomite within biocalcisiltite matrix. All photomicrographs: plane-polarized light. From Dawson and Carozzi (1986). Reprinted by permission of Elsevier Publishing Company.

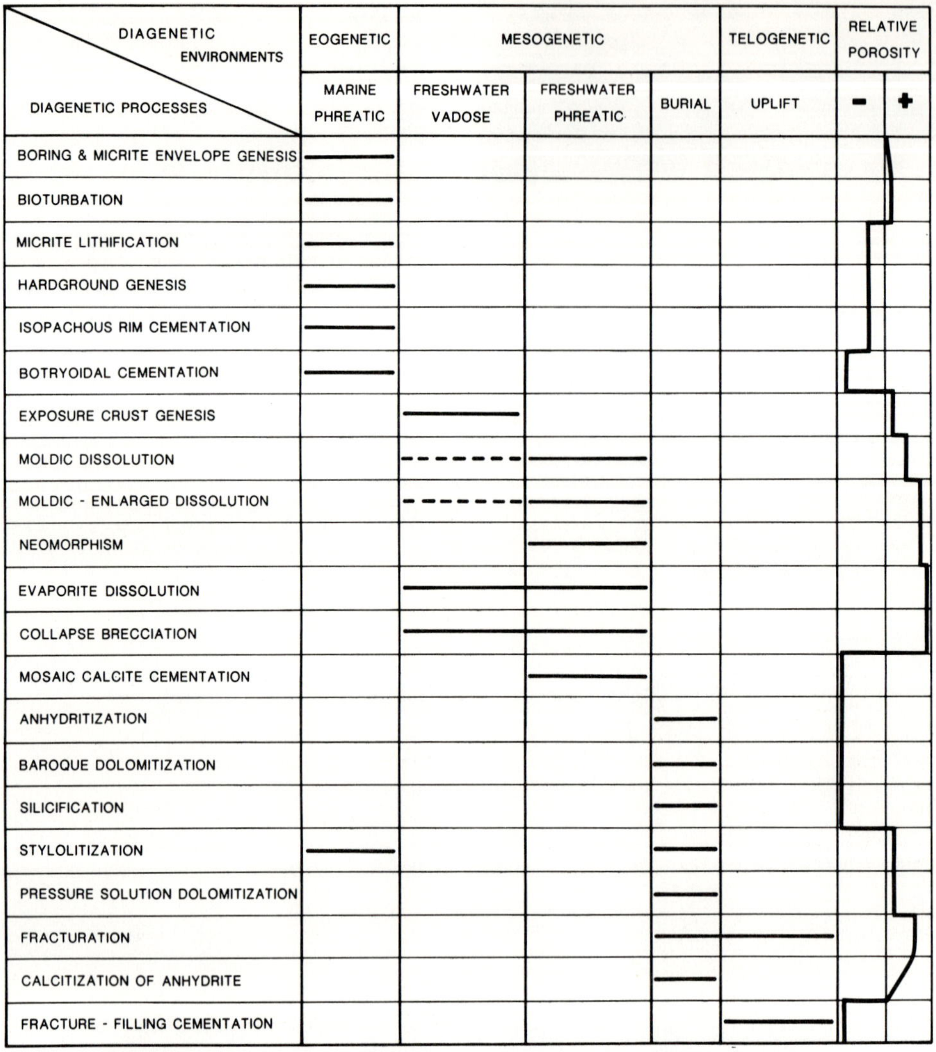

DIAGENETIC ENVIRONMENTS / DIAGENETIC PROCESSES	EOGENETIC	MESOGENETIC		TELOGENETIC	RELATIVE POROSITY
	MARINE PHREATIC	FRESHWATER VADOSE	FRESHWATER PHREATIC	BURIAL	UPLIFT	− +
BORING & MICRITE ENVELOPE GENESIS	───					
BIOTURBATION	───					
MICRITE LITHIFICATION	───					
HARDGROUND GENESIS	───					
ISOPACHOUS RIM CEMENTATION	───					
BOTRYOIDAL CEMENTATION	───					
EXPOSURE CRUST GENESIS		───				
MOLDIC DISSOLUTION		- - -	───			
MOLDIC - ENLARGED DISSOLUTION		- - -	───			
NEOMORPHISM			───			
EVAPORITE DISSOLUTION		───	───			
COLLAPSE BRECCIATION		───	───			
MOSAIC CALCITE CEMENTATION			───			
ANHYDRITIZATION				───		
BAROQUE DOLOMITIZATION				───		
SILICIFICATION				───		
STYLOLITIZATION	───			───		
PRESSURE SOLUTION DOLOMITIZATION				───		
FRACTURATION				───	───	
CALCITIZATION OF ANHYDRITE				───		
FRACTURE - FILLING CEMENTATION					───	

FIGURE 9.30 Iola Formation (Pennsylvanian), southeast Kansas. Generalized (nontemporal) diagenetic sequence of Raytown Limestone phylloid algal buildup showing inferred variations of relative porosity with changes in diagenetic environments and processes. From Dawson and Carozzi (1986). Reprinted by permission of Elsevier Publishing Company.

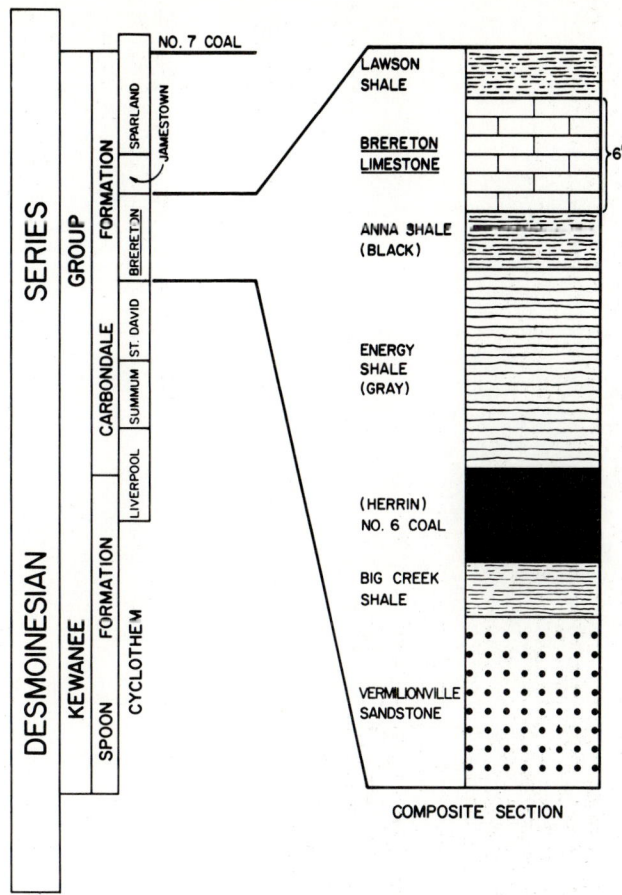

FIGURE 9.31 Brereton Limestone, Carbondale Formation (Pennsylvanian), southwest margin of Illinois Basin. Stratigraphy of Brereton cyclothem. From Khawlie and Carozzi (1976).

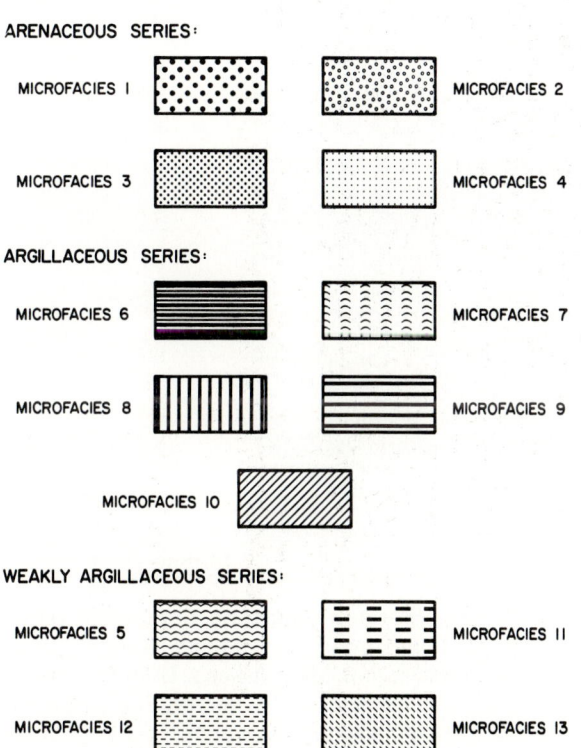

FIGURE 9.32 Brereton Limestone, Carbondale Formation (Pennsylvanian), southwest margin of Illinois Basin. Symbols for microfacies. From Khawlie and Carozzi (1976).

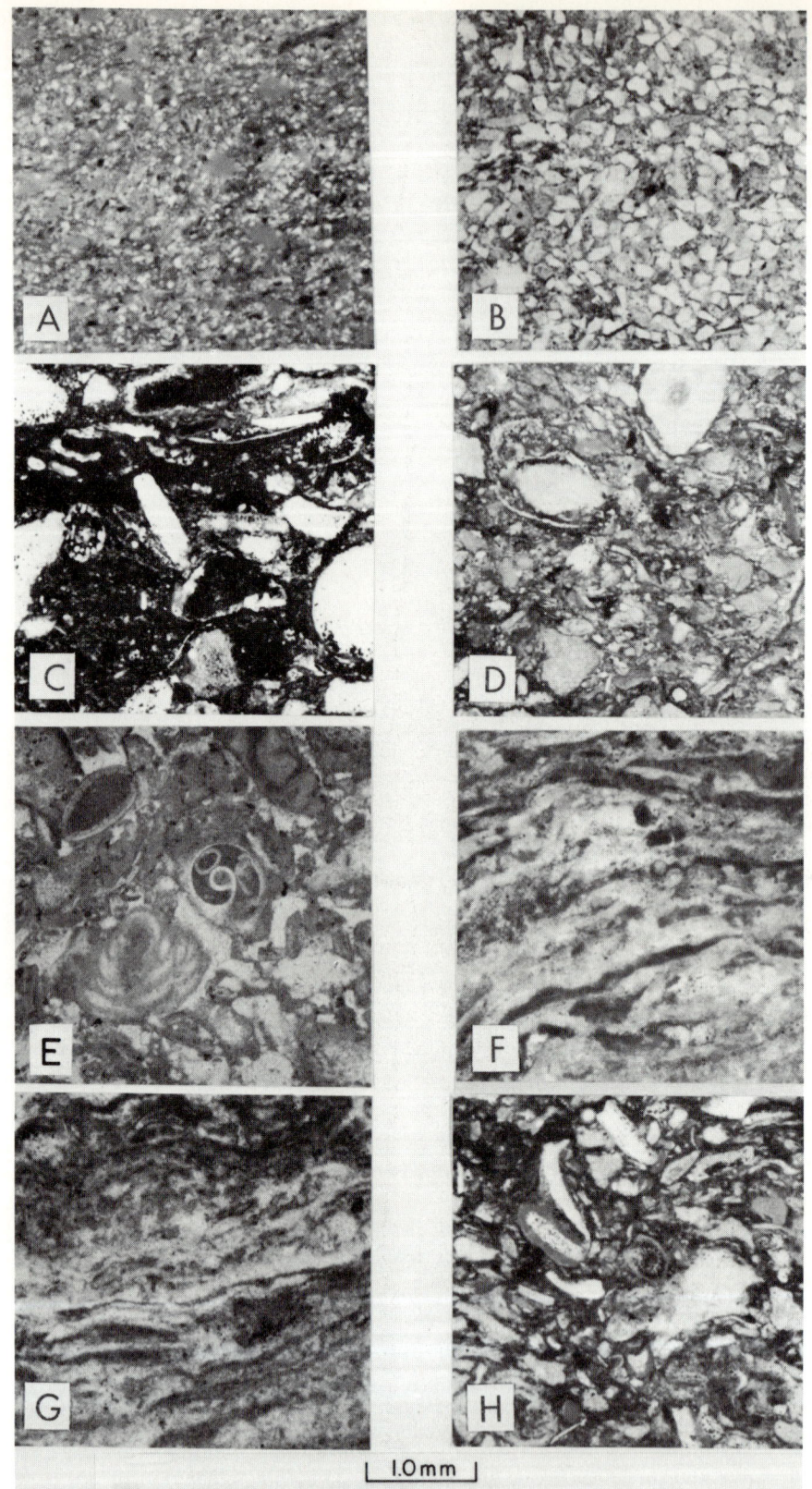

FIGURE 9.33 Brereton Limestone, Carbondale Formation (Pennsylvanian), southwest margin of Illinois Basin. Typical microfacies. A and B. Microfacies 1. C. Microfacies 2. D. Microfacies 3. E. Microfacies 4. F and G. Microfacies 5. H. Microfacies 6. See text for detailed descriptions. All photomicrographs: plane-polarized light. From Khawlie and Carozzi (1976).

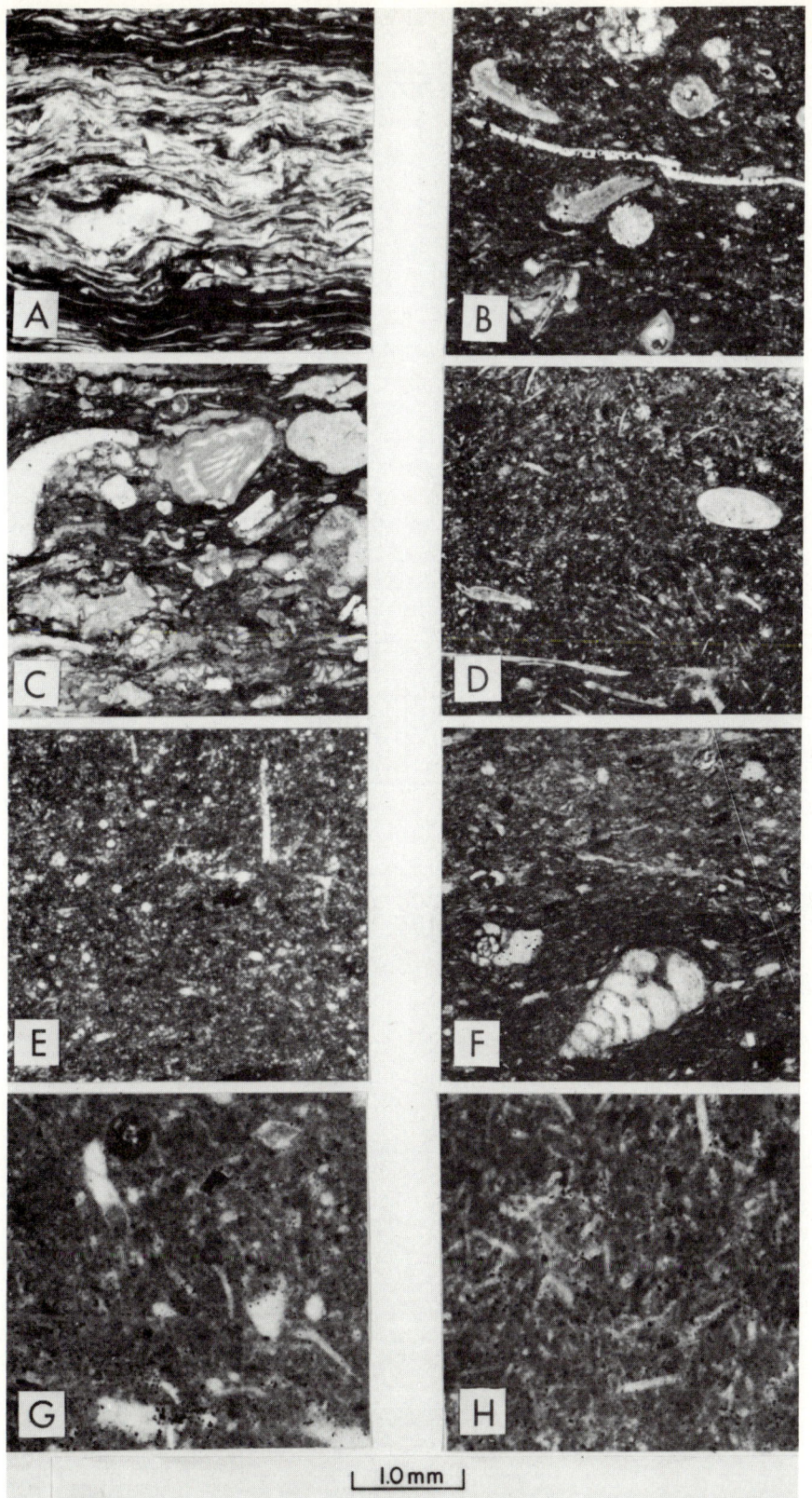

FIGURE 9.34 Brereton Limestone, Carbondale Formation (Pennsylvanian), southwest margin of Illinois Basin. Typical microfacies (continued). A. Microfacies 7. B. Microfacies 8. C. Microfacies 9. D and E. Microfacies 10. F. Microfacies 11. G. Microfacies 12. H. Microfacies 13. See text for detailed descriptions. All photomicrographs: plane-polarized light. From Khawlie and Carozzi (1976).

188 Part II / Carbonate Ramps

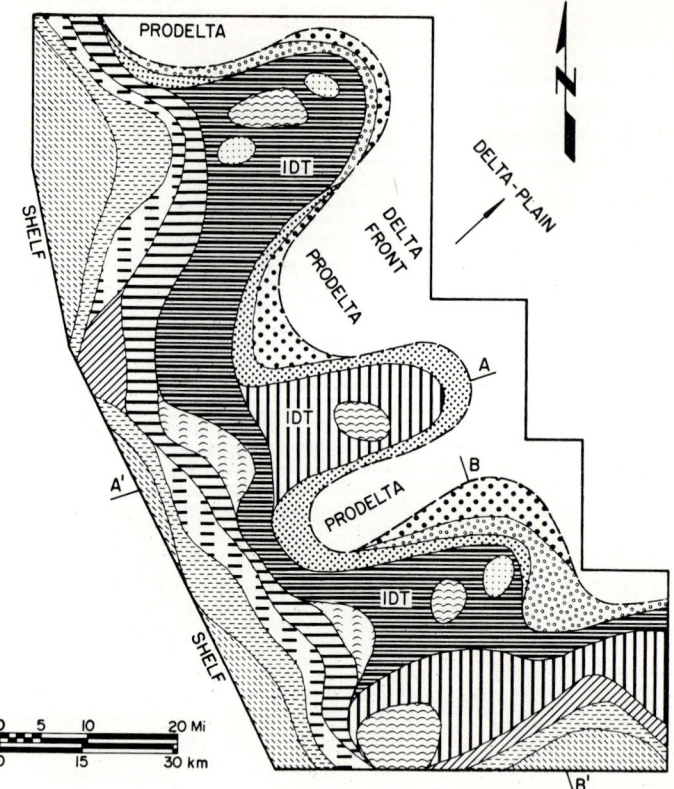

FIGURE 9.35 Brereton Limestone, Carbondale Formation (Pennsylvanian), southwest margin of Illinois Basin. Microfacies distribution map. IDT, interdeltaic embayment. Section AA′ corresponds to the constructed phylloid algae buildup model, and section BB′ to the hydrodynamic crinoid-brachiopod buildup model. From Khawlie and Carozzi (1976).

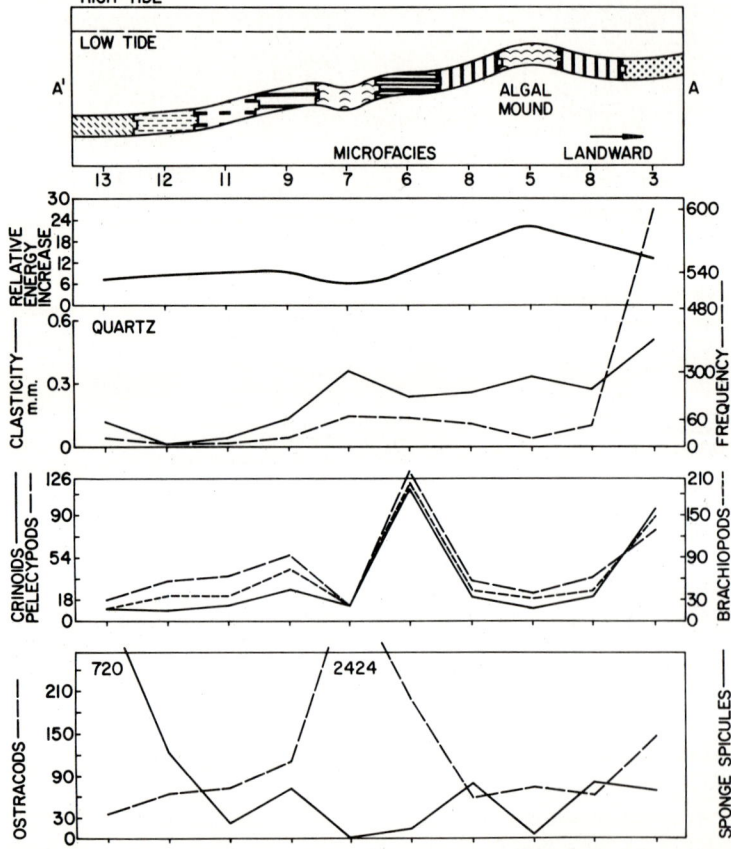

FIGURE 9.36 Brereton Limestone, Carbondale Formation (Pennsylvanian), southwest margin of Illinois Basin. Ideal constructed phylloid algae buildup model. See Figure 9.35 for location. From Khawlie and Carozzi (1976).

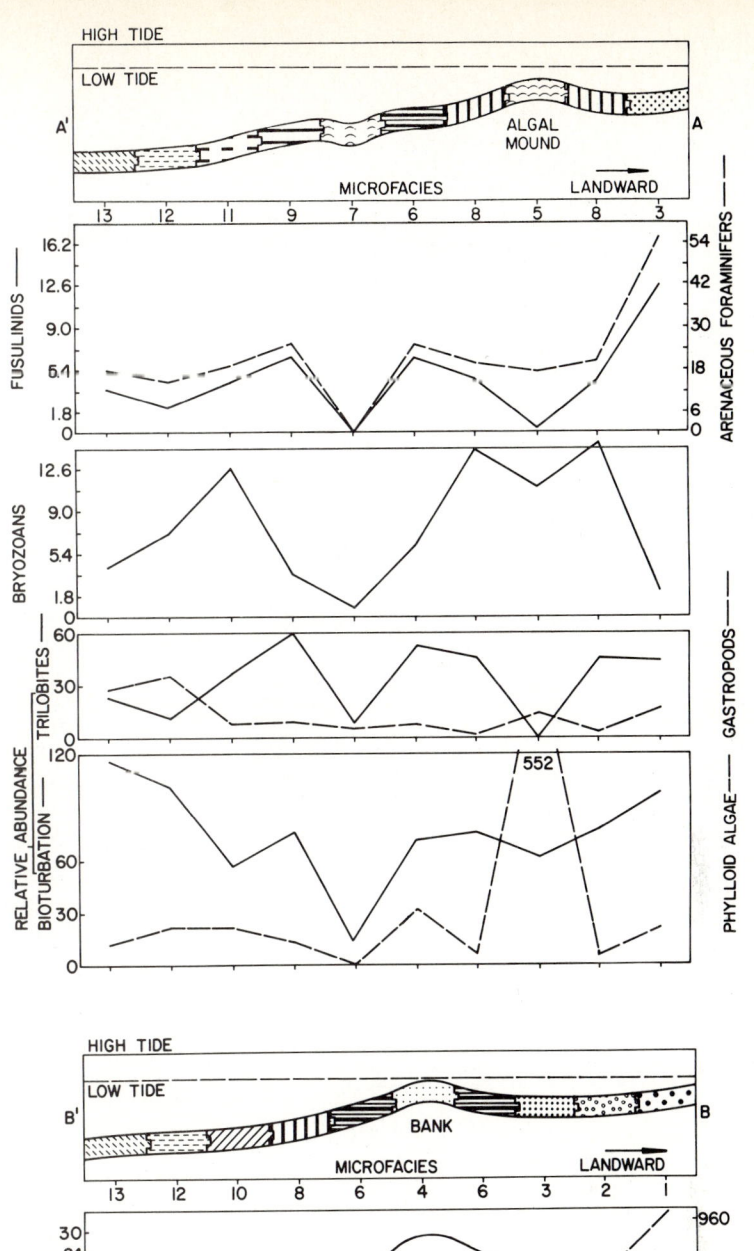

FIGURE 9.37 Brereton Limestone, Carbondale Formation (Pennsylvanian), southwest margin of Illinois Basin. Ideal constructed phylloid algae buildup model (continued). See Figure 9.35 for location. From Khawlie and Carozzi (1976).

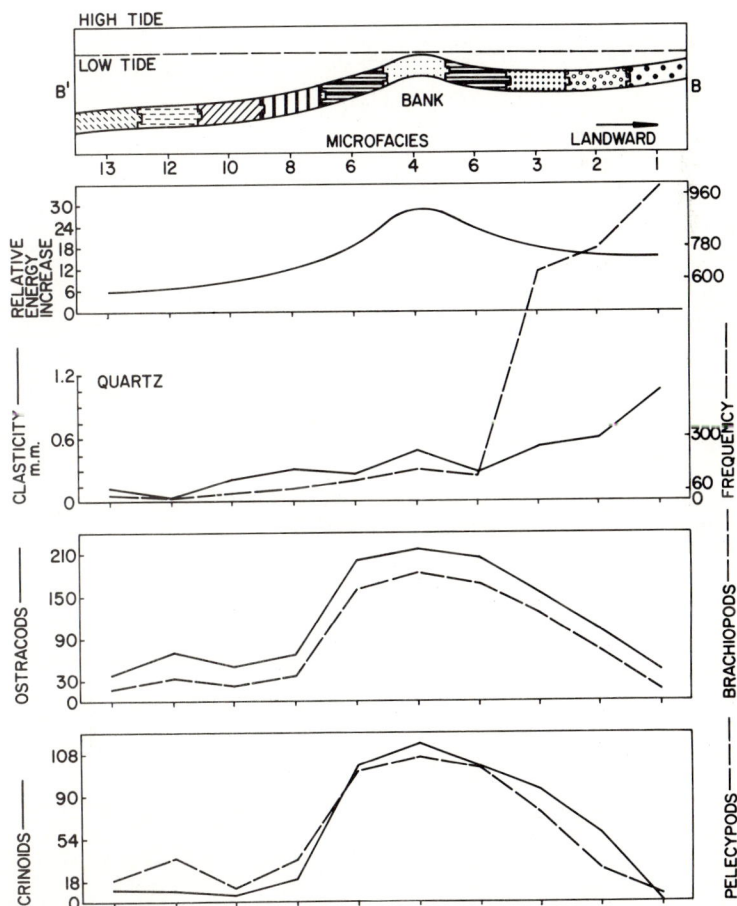

FIGURE 9.38 Brereton Limestone, Carbondale Formation (Pennsylvanian), southwest margin of Illinois Basin. Ideal hydrodynamic crinoid-brachiopod buildup model. See Figure 9.35 for location. From Khawlie and Carozzi (1976).

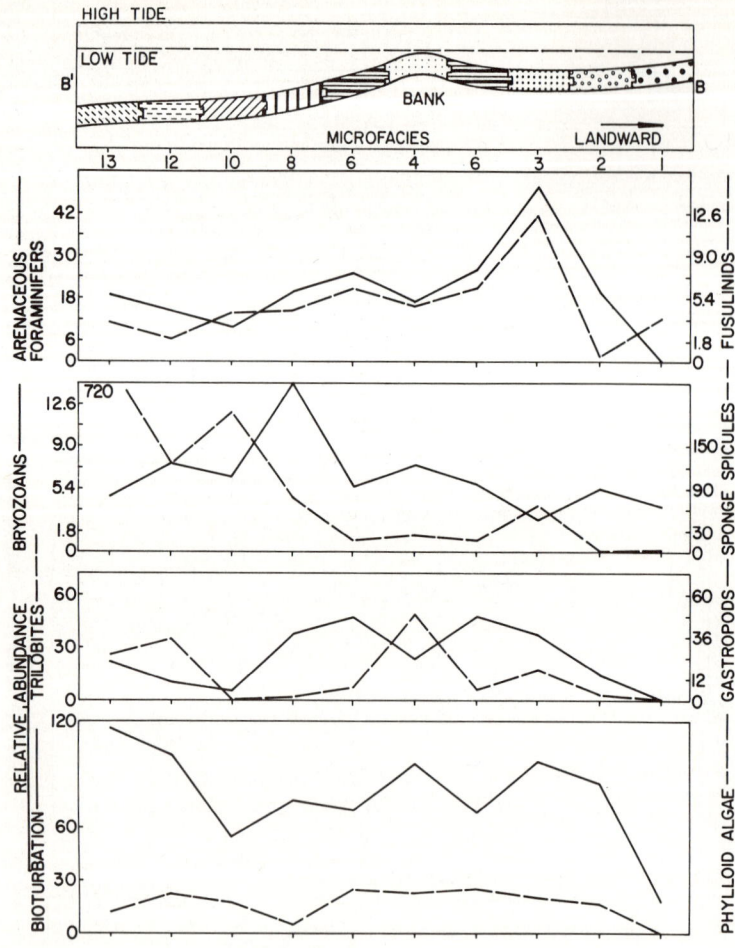

FIGURE 9.39 Brereton Limestone, Carbondale Formation (Pennsylvanian), southwest margin of Illinois Basin. Ideal hydrodynamic crinoid-brachiopod buildup model (continued). See Figure 9.35 for location. From Khawlie and Carozzi (1976).

10

Carbonate Platforms with Frontal Bioaccumulated Buildups

Along the margins of relatively low-energy intracratonic basins, micritic buildups extend from the edge of platforms down to foreslope environments. In examples from the Paleozoic, submarine stabilization of carbonate mud is often achieved by encrusting bryozoans that provide a favorable substratum for prolific crinoidal growth followed by a final bioconstructed phase. In examples from the Mesozoic, annelids and sponges play a similar stabilizing role preceding extensive bioaccumulation by numerous genera of pelecypods, followed by a final phase of coral colonies.

10.1 CHARACTERISTIC CONSTITUENTS: MICRITE, BRYOZOANS (STROMATACTIS), AND CRINOIDS

This complex evolution from stabilization through bioaccumulation to bioconstruction is characteristic of Niagaran (Middle Silurian) buildups along the southern margin of the Michigan Basin, particularly in north-central Indiana (Textoris and Carozzi, 1964). The evolution of these buildups began with small and randomly distributed bioclastic accumulations overlain by micrite mounds, which in turn acted as physical support for an extraordinary development of bryozoan and crinoid communities that eventually reached wave base. At that point, vertical growth was gradually replaced by horizontal dispersion that developed flat-topped surfaces over which a final and relatively thin phase of atoll-like bioconstruction of stromatoporoids and tabulate corals took place. It terminated with exposure features and related vadose products.

Description of Microfacies

The various microfacies are described in a general shallowing order following the above-mentioned evolution. Certain microfacies display the effects of late Silurian regional dolomitization. This study is based on more than 800 thin sections collected at an average vertical interval of 0.69 ft (21 cm).

Microfacies 1 (Fig. 10.1A,B). Argillaceous nonfossiliferous to slightly fossiliferous laminated calcisiltite (Fig. 10.1B), containing clay- and quartz-rich zones, or massive zones with scattered minute bioclasts (10%) of crinoids, bryozoans, ostracod valves, and sponge spicules. Small brachiopods and trilobite fragments are rare. Partial dolomitization is common in the form of scattered minute rhombs.

Microfacies 2 (Fig. 10.1C,D,E). Fossiliferous to pelletoidal calcisiltite with scattered bioclasts among which crinoids, bryozoans, trilobites, sponge spicules, and ostracods predominate. Locally, the encrusting bryozoan *Fistulipora* (Fig. 10.1D) is common, indicating a

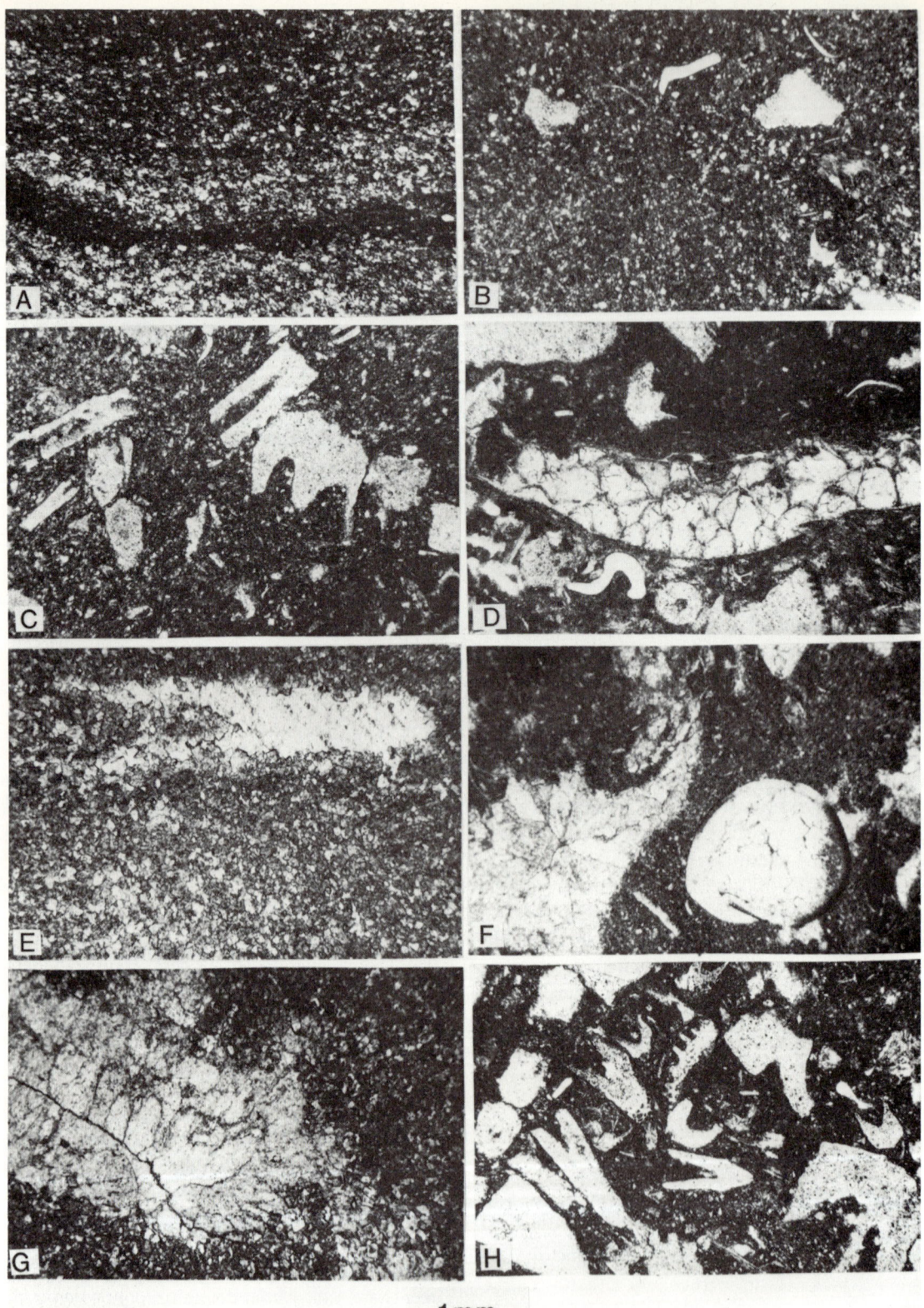

FIGURE 10.1 Niagaran (Silurian), southern margin of Michigan Basin, north-central Indiana. Typical microfacies. A and B. Microfacies 1. C to E. Microfacies 2. F and G. Microfacies 3. H. Microfacies 4A. See text for detailed descriptions. All photomicrographs: plane-polarized light. From Textoris and Carozzi (1964). Reprinted by permission of the American Association of Petroleum Geologists.

first population of the semiconsolidated micrite substratum. Dolomitization is common as a mosaic of subhedral crystals preserving only ghosts of crinoid columnals (Fig. 10.1E).

Microfacies 3 (Fig.10.1 F,G). Fossiliferous to pelletoidal calcisiltite similar to the preceding microfacies but with abundant stromatactis structures filled with coarsely bladed sparite cement. These structures, which have been the subject of widespread controversy, originate in this case (Carozzi and Textoris, 1963) from submarine dissolution of early cemented calcisiltite, apparently triggered by the presence of *Fistulipora* colonies whose remains are always found at the base of the sparite cement filling (Figs. 10.2, 10.3). The proposed mechanism of formation of these peculiar structures (Fig. 10.4) also accounts for the local presence of internal sediments often with graded bedding. Stromatactis remain calcitic within completely dolomitized matrix with only incipient marginal dolomitization at best (Fig. 10.1G).

Microfacies 4A (Figs. 10.1H, 10.5A). Mud-supported biocalcarenite with a calcisiltite matrix. Crinoid bioclasts predominate over bryozoans, ostracods, brachiopods, and trilobites. Subhedral to euhedral crystals of dolomite preserve only ghosts of crinoids and bryozoans (Fig. 10.5A).

Microfacies 4B (Fig. 10.5B,C). Grain-supported biocalcarenite consisting of crinoid bioclasts reaching their maximum frequency and clasticity associated with abundant bryozoans. Pressure-solution contacts among bioclasts are common, and the interstitial spaces range from a calcisiltite and bioclastic matrix to predominant syntaxial overgrowths around crinoids engulfing the bryozoan bioclasts and to sparite mosaic cement. Dolomitization is similar to that of microfacies 4A, but coarser crystals are generated when sparite cement was originally present (Fig. 10.5C).

Microfacies 5A (Fig. 10.5D). Mud-supported pelletoidal calcarenite with a calcisiltite to bioclastic matrix. This microfacies generally is dolomitized. Dark rounded pellets, associated with crinoid bioclasts, appear scattered in a mosaic of cloudy subhedral dolomite crystals.

Microfacies 5B (Fig. 10.5E). Grain-supported pelletoidal calcarenite with a calcisiltite to bioclastic matrix. Dark rounded pellets and associated bioclasts of crinoids in this dolomitized microfacies display interstitial cloudy subhedral dolomite crystals.

Microfacies 6 (Fig. 10.5F). Stromatoporoid constructed limestone consisting of matlike and hemispherical colonies of *Clathrodictyon* associated with rugose and tabulate corals, and *Conchydium*. This microfacies is usually porous and completely dolomitized. The associated matrix can be composed of any of the dolomitized equivalents of microfacies 2, 4A, 4B, 5A, and 5B.

Exposure Features. Following the final constructed phase of the Niagaran buildups which assumed a general atoll shape with a central pelletoidal lagoon, local development of vadose ooids and pisoids with crinoidal cores took place. They are found as filling of clastic dikes and appear now completely dolomitized (Fig. 10.5G). During the karstic phase of development of the Silurian-Devonian unconformity, dolomite-cemented quartz arenite were deposited in solution-enlarged fractures associated with green clay (Fig. 10.5H).

The Ideal Shallowing-Upward Sequence

The interrelationships between microfacies indicate that calcisiltite is the main microfacies from which two distinct series evolved, one bioclastic and the other pelletoidal (Fig. 10.6) until bioconstructing organisms such as stromatoporoids were developing gradually. The variation curves of average microscopic parameters in the various buildups (Figs. 10.7, 10.8) show clearly successive faunal associations during local variation of the shallowing-upward evolution, as well as simultaneous decrease of frequency of detrital quartz, micas, and pyrite which characterize the pre- and interbuildup environments. A typical section (Fig. 10.9) shows the detailed variation of microscopic parameters during the shallowing-upward process.

The Ideal Depositional Model

The evolution of an ideal Niagaran buildup can be summarized in six stages (Fig. 10.10), which demonstrate the succession of vertical and lateral growth. The final stage, as displayed in the Thornton quarry near Chicago (Shaver, 1977), showed a very diversified morphological and ecological zonation related to predominant wind pattern (Fig. 10.11). A generalized model (Fig. 10.12) shows how buildups were associated along the margin of the Michigan Basin, with pinnacle types on basinward slopes and landward lagoons with stromatolitic mounds.

10.2 CHARACTERISTIC CONSTITUENTS: MICRITE AND PELECYPODS

In subtidal pelecypod-dominated micritic buildups, smaller energy conditions lead to an apparent great uniformity of microfacies. Paleoecological criteria become very important in the distinction of subenvironments. These

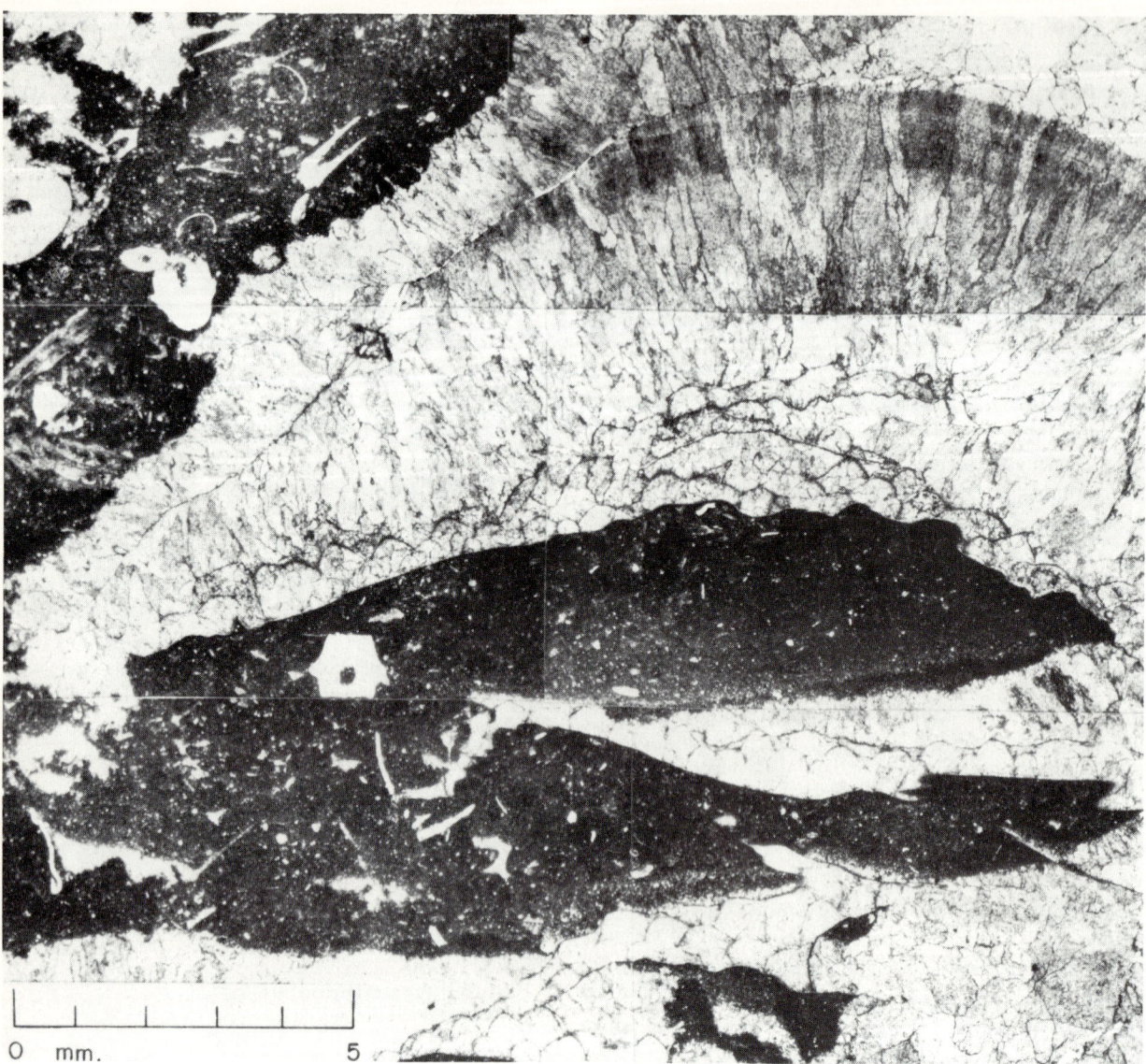

FIGURE 10.2 Niagaran (Silurian), southern margin of Michigan Basin, north-central Indiana. Fossiliferous stromatactis calcisiltite, showing part of complex stromatactis network. Upper half of photomosaic shows partially preserved fistuliporoid colony which has maintained smooth cavity base. Zooecia and zooecial interspaces that have not been destroyed are filled with fine-grained equant sparite and grade into peripheral coarser bladed sparite. Coarser equant sparite-filling is displayed in upper right corner. The only internal sediment, a fenestellid fragment on left, occurs at line juncture of blade sparite that grew from both upper and lower cavity boundaries. Lower part of photomosaic displays well-preserved fistuliporoid extremities of stromatactis. Scattered bioclasts are crinoid, bryozoan, ostracod, and trilobite. From Textoris and Carozzi (1964). Reprinted by permission of the American Association of Petroleum Geologists.

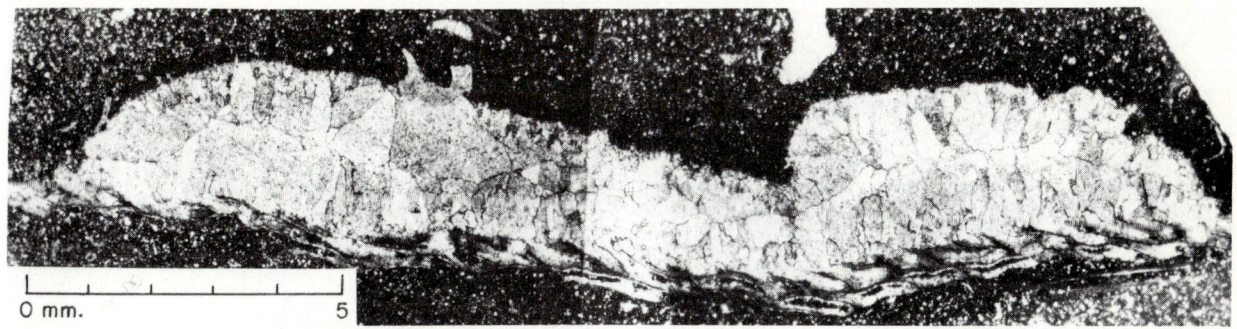

FIGURE 10.3 Niagaran (Silurian), southern margin of Michigan Basin, north-central Indiana. Fossiliferous stromatactis calcisiltite, with partly dolomitized matrix. This small, simple, individual stromatactis displays bryozoan-controlled smooth bottom, enlarged convex top, and bladed sparite-filling. At upper left, crinoid fragment forms roof pendant. From Textoris and Carozzi (1964). Reproduced by permission of the American Association of Petroleum Geologists.

include subtle microscopic features indicating gentle increases of agitation followed by short periods of subaerial exposure.

The Chachao Formation (Valanginian) of the Malargüe region, Neuquén–South Mendoza Basin, Argentina (Carozzi et al., 1981a), was studied petrographically by means of 311 samples collected at an average vertical interval of 1.16 m in 13 field sections representing an aggregated thickness of 362 m. The Chachao Formation consists essentially of an extensive pelecypod bank with abundant interstitial micrite accumulated in a low-energy subtidal environment along the hinge line of the Neuquén–South Mendoza Basin. It grades basinward into the upper part of the Vaca Muerta euxinic black shales and interfingers shoreward with a variety of inner platform clastics designated as Lomas Bayas Formation. The Chachao Formation is in turn overlain by another sequence of euxinic black shales (Agrio Formation).

Description of Microfacies

Because the Chachao carbonates do not display small-scale cyclic patterns but represent an irreversible depositional evolution, the microfacies are described in stratigraphic order. Due to their restricted range of depositional energy, the description of the microfacies depends strongly on paleoecological data.

Microfacies I (Fig. 10.13A). Moderately bioturbated grain-supported biocalcarenite with calcisiltite matrix. Pelecypod bioclasts and complete individuals consist of small oysters and *Panopea* variably altered into pseudomicrosparite and pseudosparite by neomorphism. Other constituents are fish scales, fish coprolites, and rare smaller benthic foraminifers.

Microfacies II (Fig. 10.13 B,C). Moderately bioturbated, grain-supported biocalcarenite with calcisiltite matrix. Pelecypod bioclasts and complete individuals consist of large and small oysters associated with pectinids. They reach their maximum development. Annelids (colonial serpulids) are abundant (Fig. 10.13B), as are monaxonic sponge spicules (Fig. 10.13C); echinoderms and gastropods are accessory components.

Microfacies III (Figs. 10.13D,E,F, 10.14A). An association of grain-supported biocalcarenites with calcisiltite matrix grading into interstitial cement represented by syntaxial overgrowths on echinoderm debris and intergranular microsparite (Fig. 10.13E). Bryozoan and gastropod bioclasts and complete individuals predominate over large and small oysters. Colonial corals appear in this microfacies and annelids continue to be well represented. Bioturbation is well developed and coincides with a rarity of fish remains. The appearance of interstitial cement coincides with that of glauconite in grains and as filling of organic cavities. This increase in depositional energy expresses a shallowing which leads locally to exposure features such as vadose silt (Fig. 10.13F), calcrete crusts, solution breccias with pseudomorphs of calcite after anhydrite and gypsum, and local dolomitization by a micro-dorag model due to a mixture of rainwater and seawater (Fig. 10.14A).

Microfacies III A (Fig. 10.14B,C). Coral-constructed limestone consisting of a framework of ramose and bushy colonies with an interstitial matrix which ranges from calcisiltite with calcispheres (Fig. 10.14C) and small benthic foraminifers (*Lenticulina*) to a pelecypod-calcisphere grain-supported biocalcarenite with calcisiltite matrix and patches of microsparite cement. In

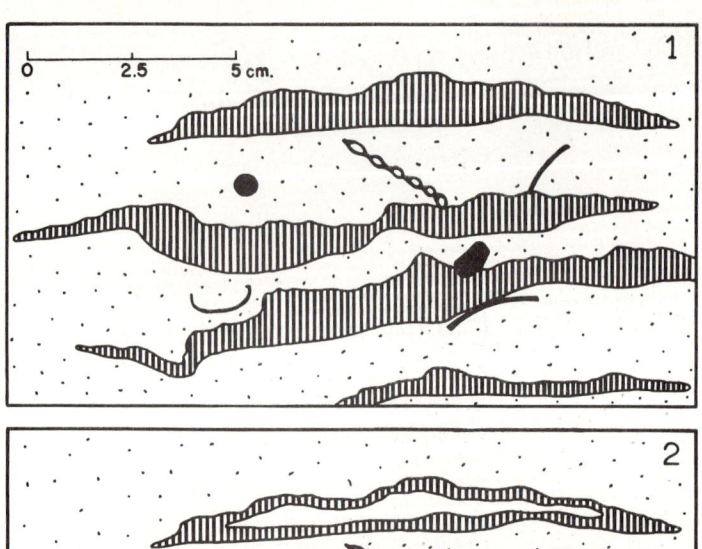
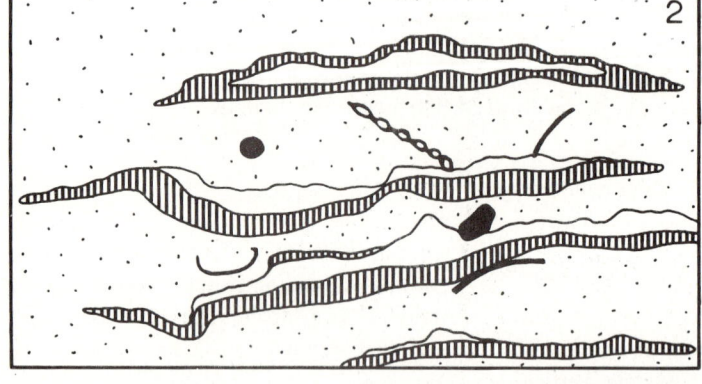
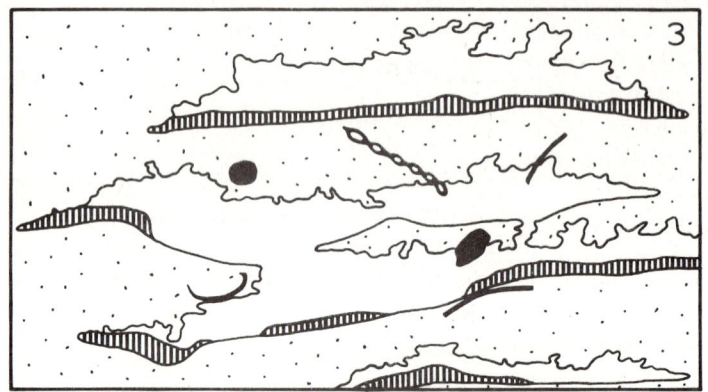
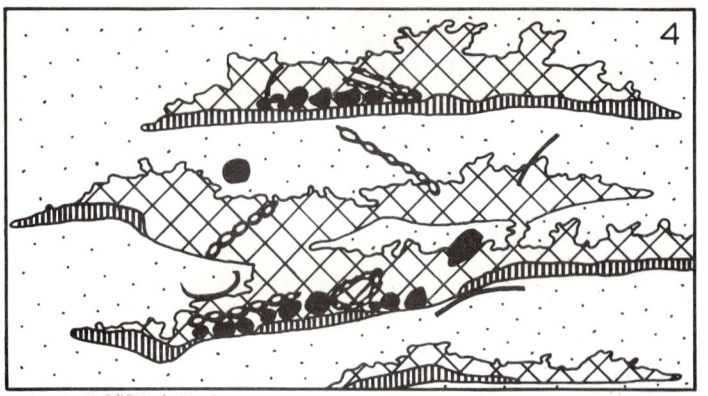

FIGURE 10.4 Niagaran (Silurian), southern margin of Michigan Basin, north-central Indiana. Stages of stromatactis development. Stippled area, calcisiltite; black areas, bioclasts; white areas, cavities; vertically lined areas, *Fistulipora*; crosshatched areas, sparite filling. From Textoris and Carozzi (1964). Reprinted by permission of the American Association of Petroleum Geologists.

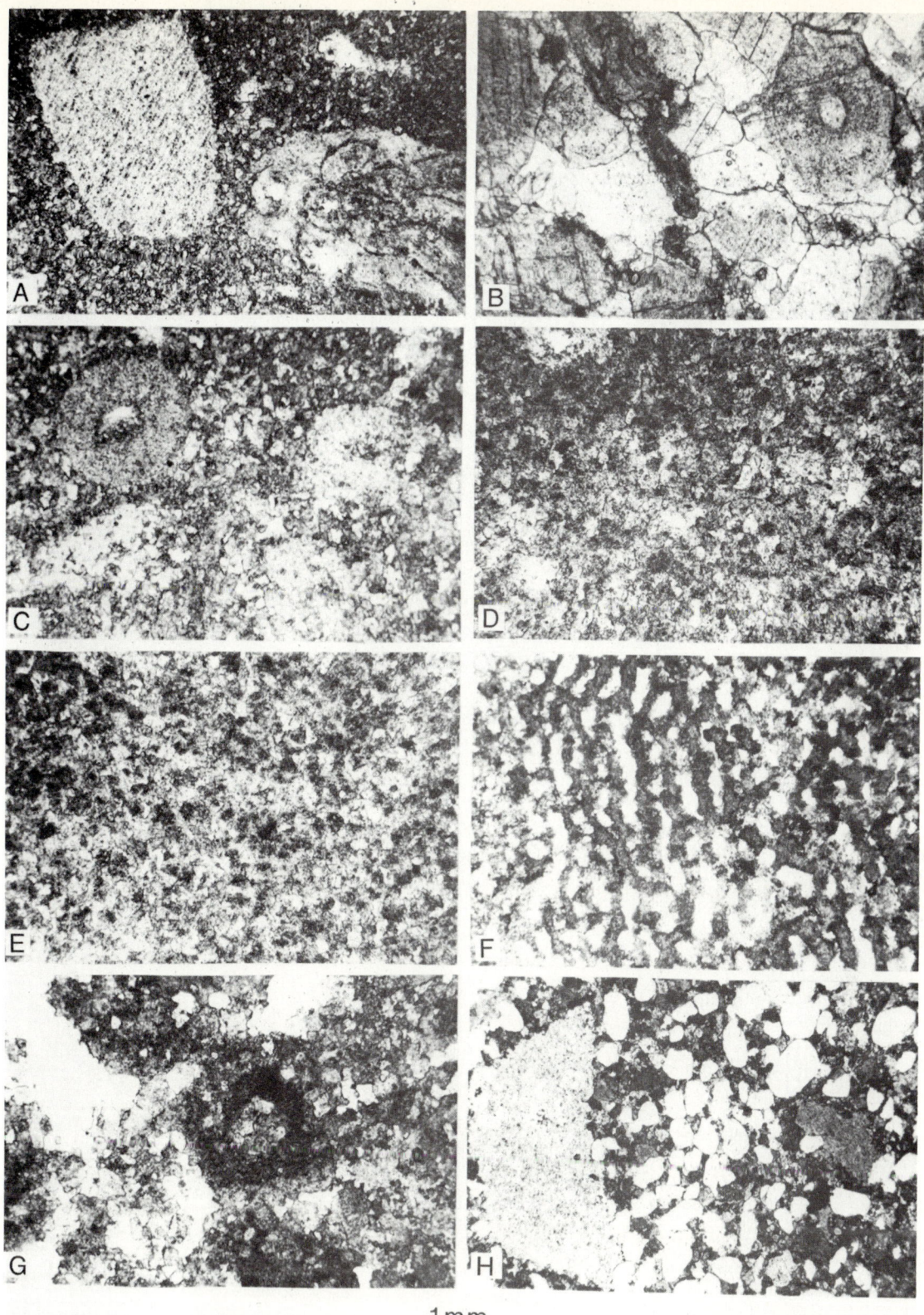

FIGURE 10.5 Niagaran (Silurian), southern margin of Michigan Basin, north-central Indiana. Typical microfacies (continued). A. Microfacies 4A. B and C. Microfacies 4B. D. Microfacies 5A. E. Microfacies 5B. F. Microfacies 6. G. Dolomitized ooid in filling of dike; large white areas are cavities. H. Dolomite-cemented quartz arenite fracture-solution filling. Well-rounded and sorted quartz grains and two large crinoid bioclasts are set in granular dolomite cement. Pressure-solution, secondary overgrowths, and marginal corrosion of quartz grains occur also. See text for detailed descriptions. All photomicrographs: plane-polarized light. From Textoris and Carozzi (1964). Reprinted by permission of the American Association of Petroleum Geologists.

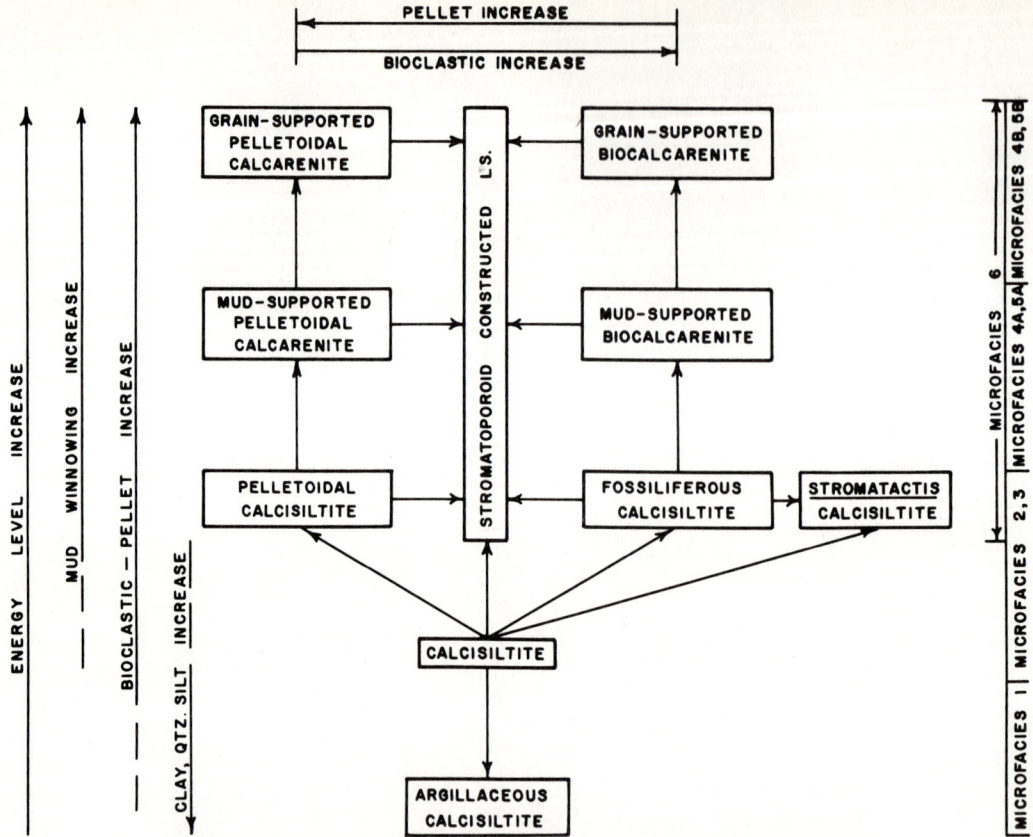

FIGURE 10.6 Niagaran (Silurian), southern margin of Michigan Basin, north-central Indiana. Relationships of microfacies. From Textoris and Carozzi (1964). Reprinted by permission of the American Association of Petroleum Geologists.

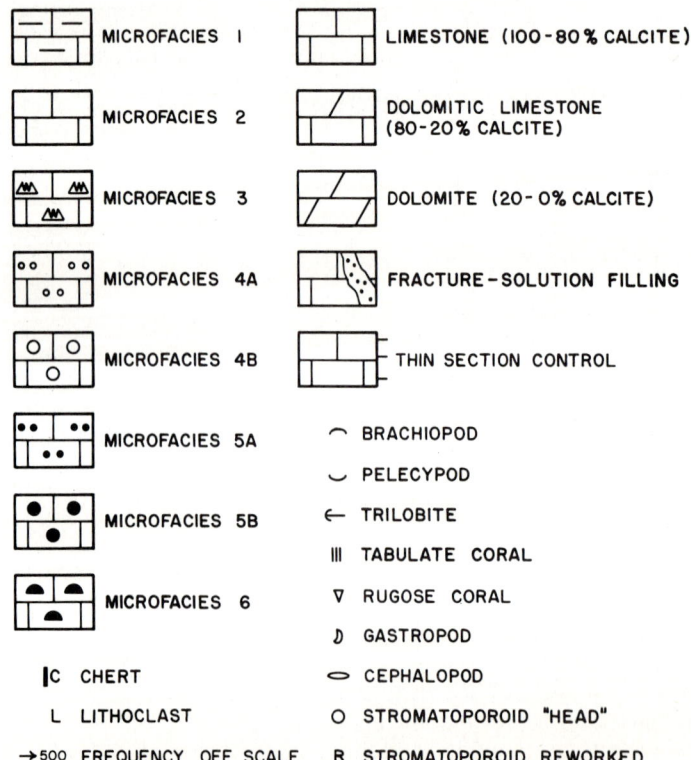

FIGURE 10.7 Niagaran (Silurian), southern margin of Michigan Basin, north-central Indiana. Symbols of microfacies. From Textoris and Carozzi (1964). Reprinted by permission of the American Association of Petroleum Geologists.

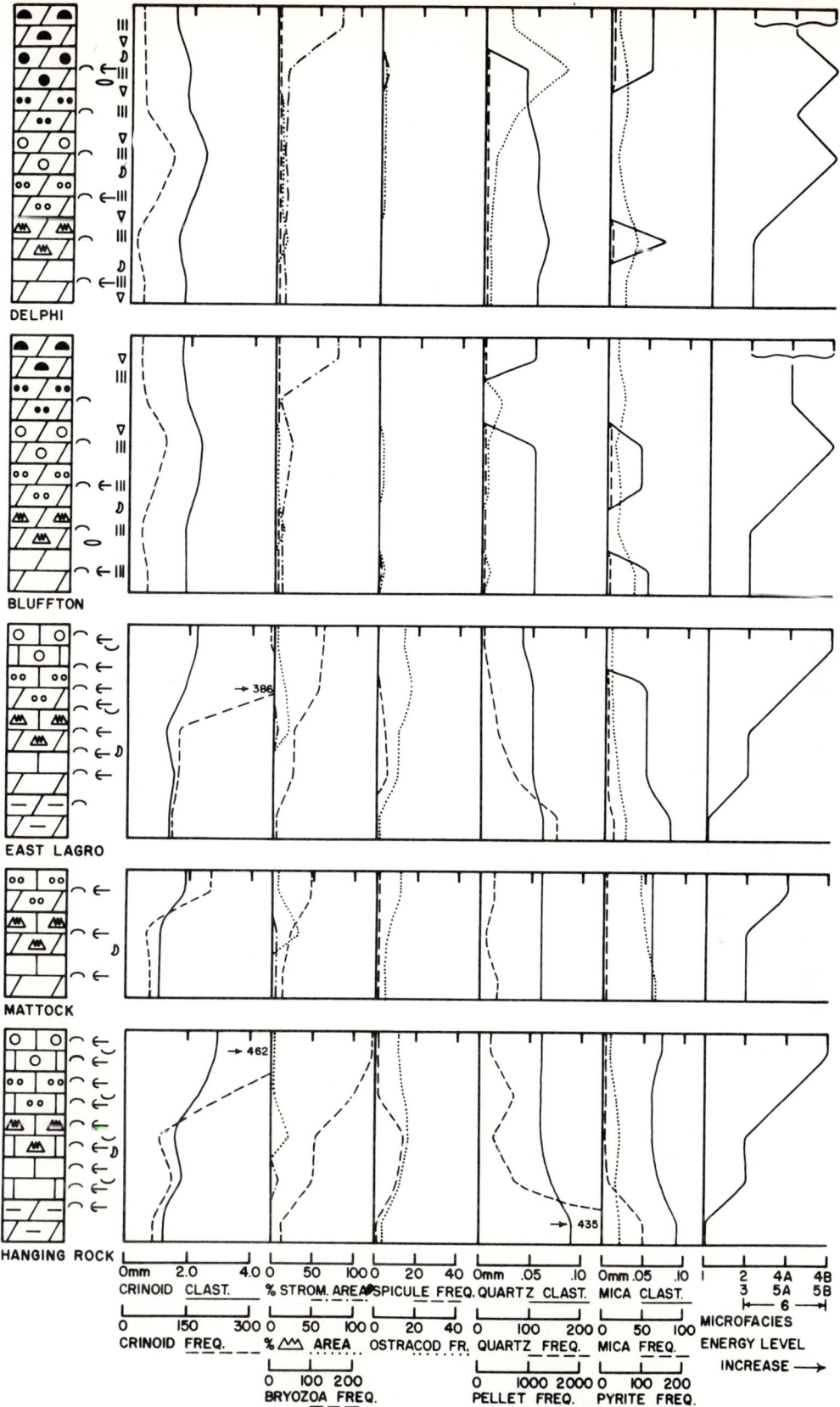

FIGURE 10.8 Niagaran (Silurian), southern margin of Michigan Basin, north-central Indiana. Vertical variation curves of average microscopic parameters during shallowing-upward evolution of various buildups. From Textoris and Carozzi (1964). Reprinted by permission of the American Association of Petroleum Geologists.

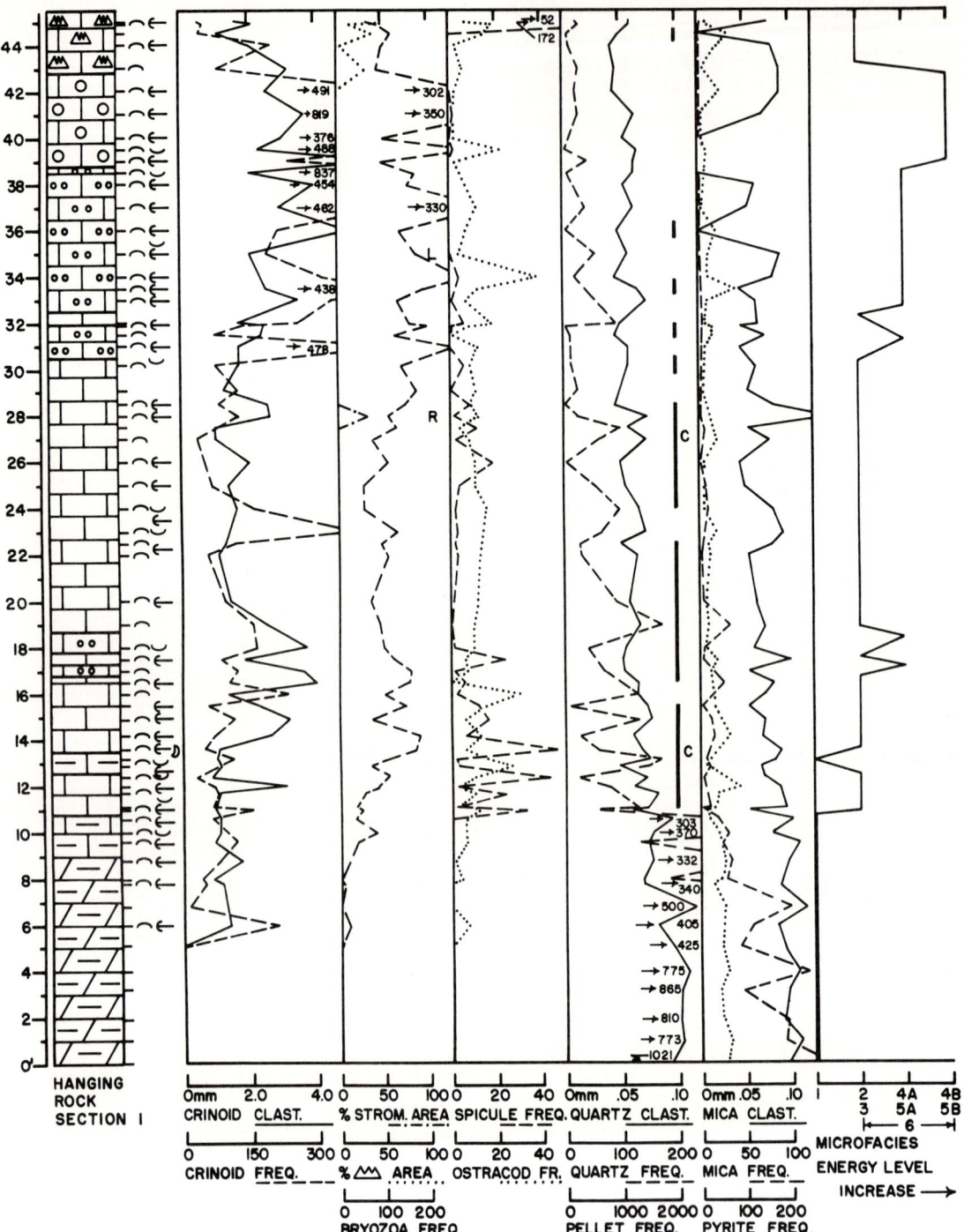

FIGURE 10.9 Niagaran (Silurian), southern margin of Michigan Basin, north-central Indiana. Typical example of vertical variation of microscopic parameters during shallowing-upward sequence. From Textoris and Carozzi (1964). Reprinted by permission of the American Association of Petroleum Geologists.

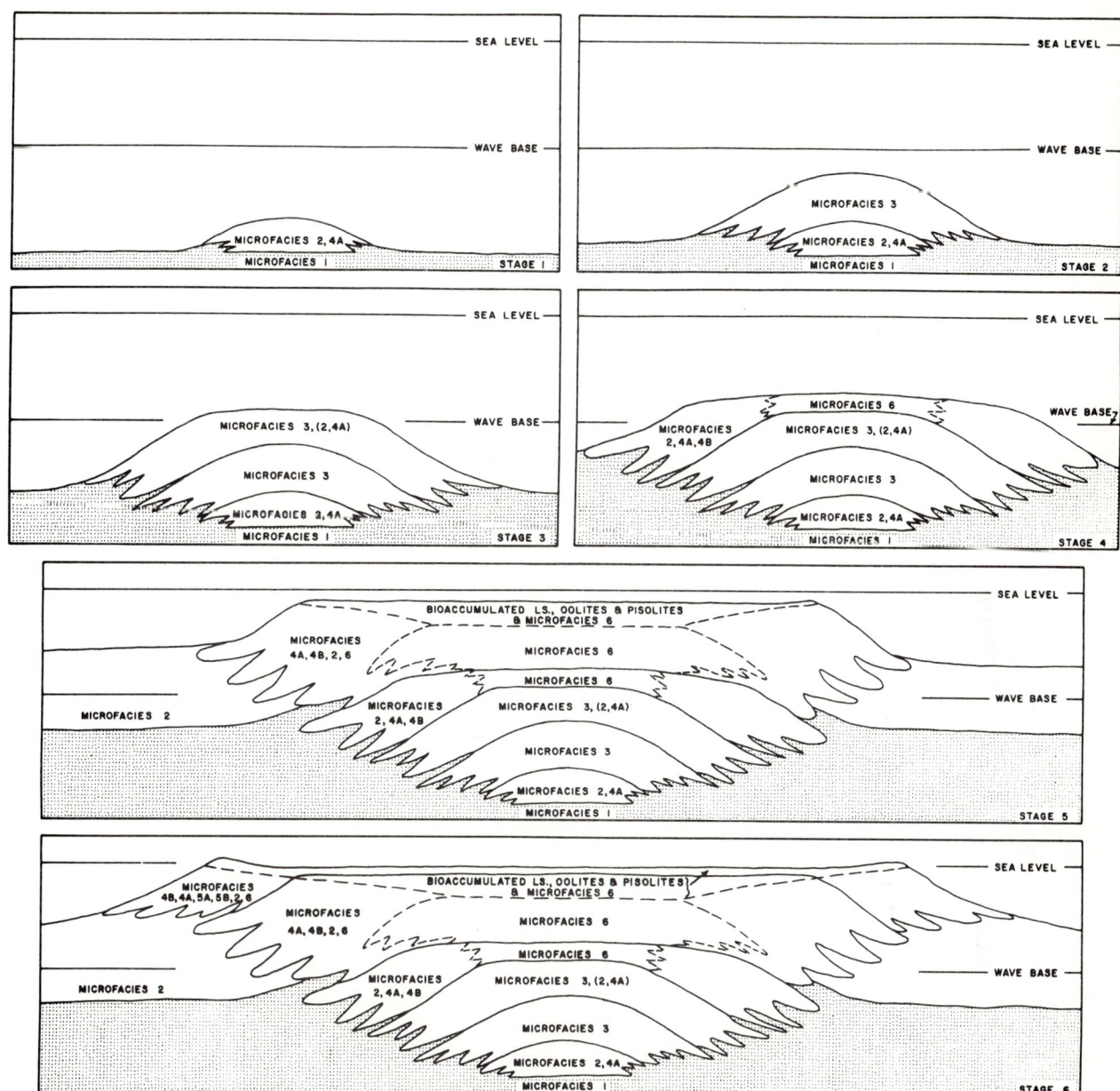

FIGURE 10.10 Niagaran (Silurian), southern margin of Michigan Basin, north-central Indiana. Stages of ideal buildup development. From Textoris and Carozzi (1964). Reprinted by permission of the American Association of Petroleum Geologists.

essence, the matrix consists of microfacies III, hence the designation of this microfacies as III A.

Microfacies IV (Fig. 10.14D,E). Deeply bioturbated grain-supported biocalcarenite with abundant calcisiltite matrix and a well-developed infauna of pelecypods (*Ptychomia*, *Eriphyla*, and *Myoconcha*) and ornamented to bulbous small gastropods, associated with abundant bioclasts of echinoids, calcispheres, smaller benthic foraminifers (*Lenticulina*), fish scales, and fish coprolites (Fig. 10.14E).

Microfacies V (Fig. 10.14F). Association of very fine grain-supported to mud-supported biocalcarenites with abundant calcisiltite matrix. Bioclasts are limited to pelecypods, gastropods, smaller benthic foraminifers (*Lenticulina*), calcispheres, and fish remains.

The Ideal Vertical Sequence of Microfacies

The Chachao Formation displays a symmetrical bathymetric evolution consisting of a shallowing-upward sequence ending with local emergence, followed by a deep-

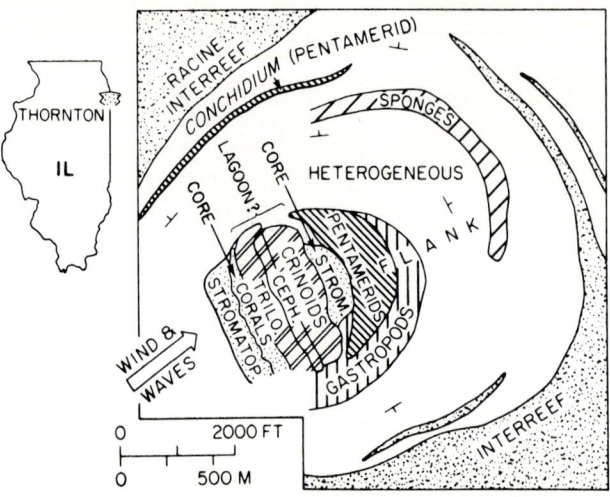

FIGURE 10.11 Niagaran (Silurian), southern margin of Michigan Basin. Paleoecological zonation of final buildup stage, Thornton, Illinois. From Shaver (1977). Reprinted by permission of the Society of Economic Paleontologists and Mineralogists.

ening-upward sequence (Fig. 10.15). The bank was initiated by pelecypods associated with fish remains (microfacies I). Gradually, the fauna became more diversified (microfacies II and III) by addition of gastropods, echinoderms, bryozoans, colonial annelids, and sponges, and it was more intensively bioturbated while the conditions were shallowing. Local colonies of corals developed eventually (microfacies III-A) and during both microfacies, local conditions of subaerial exposure accompanied by local dolomitization by a micro-dorag model took place which indicate a slight topographic relief as shown by a transverse cross section (Fig. 10.16). Deepening took place by means of microfacies IV and V, during which the pelecypod bank lost many of its previous components, which were replaced by small benthic foraminifers, calcispheres, and the return of fish remains. In summary, microfacies and paleoecological evolution were irreversible, but relative bathymetric and energy-level evolution were rather symmetrical. A typical variation of microscopic parameters is shown by the section of Puesto Moyano III (Fig. 10.17).

Because microfacies III corresponds to a time of greatest faunal diversification, a west-east cross section was drawn to illustrate the paleoecological zonation within the bank (Fig. 10.18, upper part). Inside a marginal pelecypod-gastropod ring are concentrations of sponges and echinoderms indicating more protected conditions. A similar cross section for microfacies IV, corresponding to the terminal deepening of the bank, shows a comparable relationship with central concentrations of calcispheres and small benthic foraminifers surrounding gastropods (Fig. 10.18, lower part).

The Ideal Depositional Model

The depositional model (Fig. 10.19) consists of a sequence of five time sketches showing the growth and disappearance of the bank. The Chachao pelecypod bank is a target for subsurface exploration because it grades laterally and is overlain by mature euxinic shales. It is, however, interesting as a reservoir only when it has undergone extensive tectonic fracturing which introduced secondary porosity.

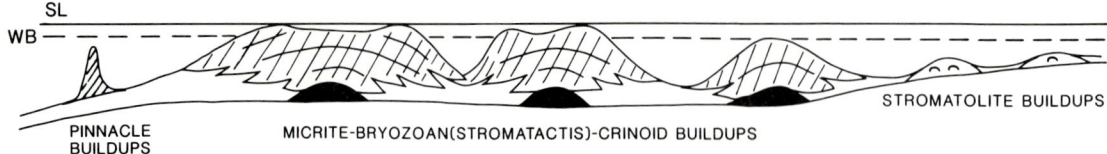

FIGURE 10.12 Niagaran (Silurian) of margins of Michigan Basin. Generalized depositional model.

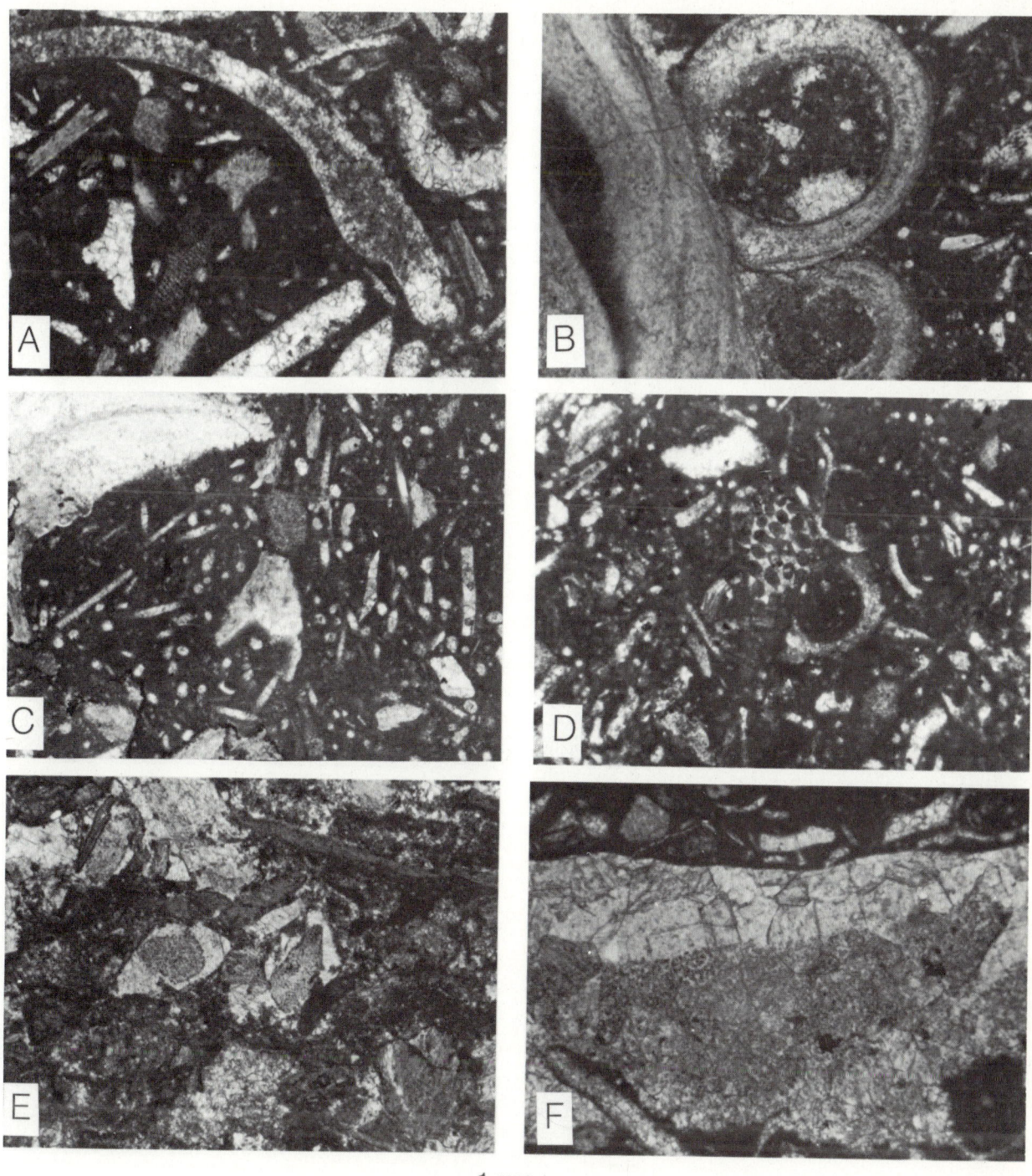

FIGURE 10.13 Chachao Formation (Valanginian), Neuquén–South Mendoza Basin, Argentina. Typical microfacies. A. Microfacies I. B and C. Microfacies II. D to F. Microfacies III. See text for detailed descriptions. All photomicrographs: plane-polarized light. From Carozzi et al. (1981a).

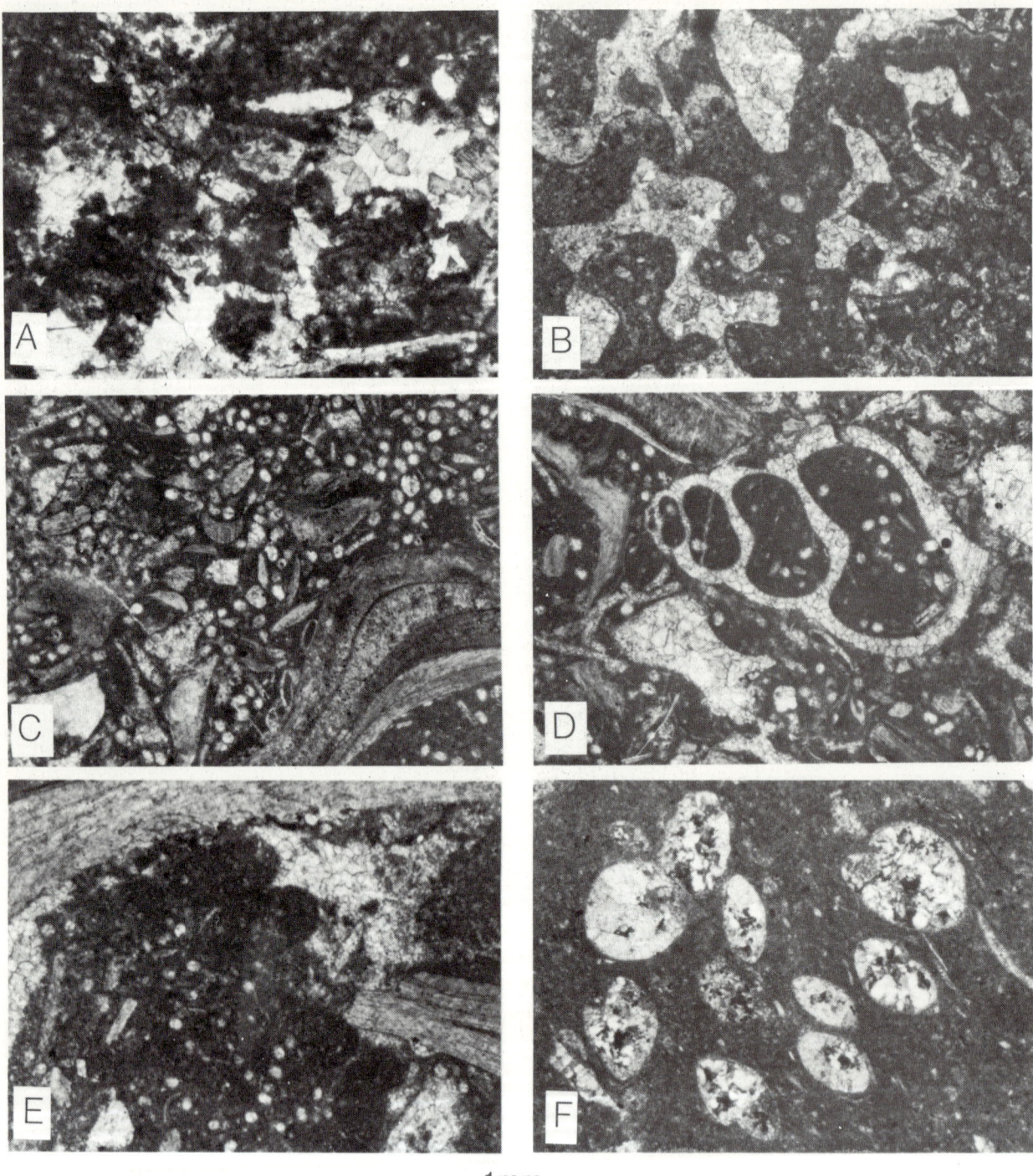

FIGURE 10.14 Chachao Formation (Valanginian), Neuquén–South Mendoza Basin, Argentina. Typical microfacies (continued). A. Microfacies III. B and C. Microfacies III-A. D and E. Microfacies IV. F. Microfacies V. See text for detailed descriptions. All photomicrographs: plane-polarized light. From Carozzi et al. (1981a).

Chap. 10 / Carbonate Platforms with Frontal Bioaccumulated Buildups   205

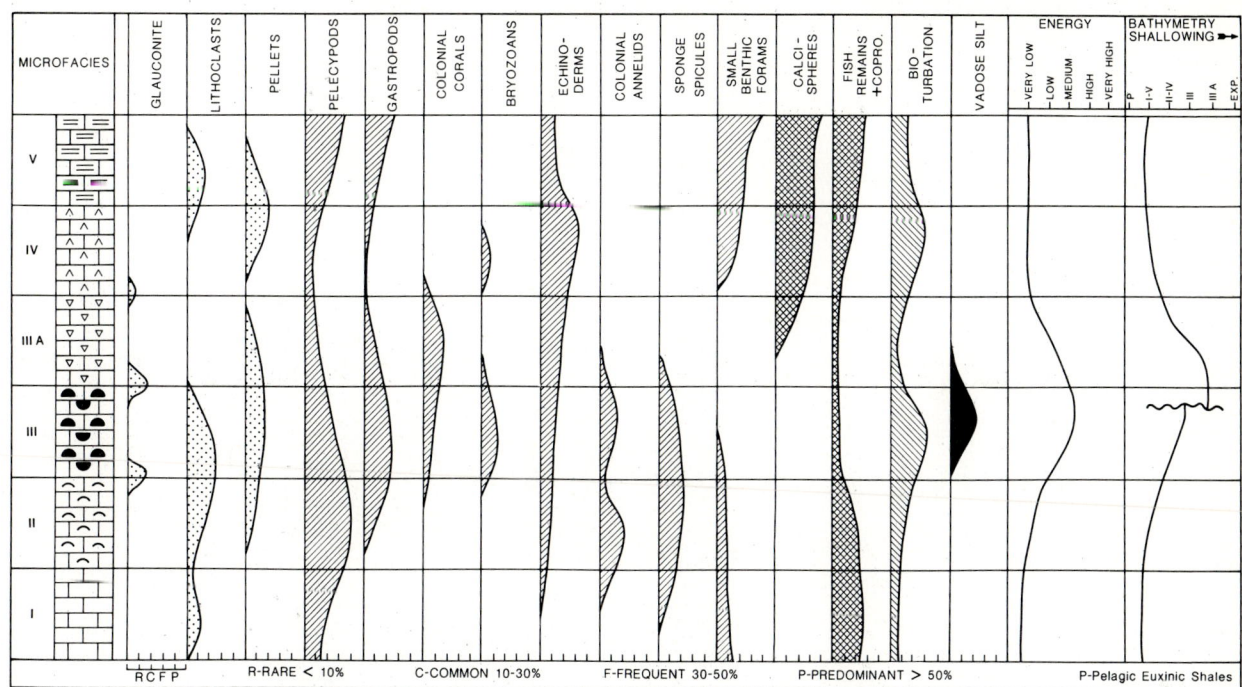

FIGURE 10.15 Chachao Formation (Valanginian), Neuquén–South Mendoza Basin, Argentina. Vertical variation of average microscopic parameters. Redrawn from Carozzi et al. (1981a).

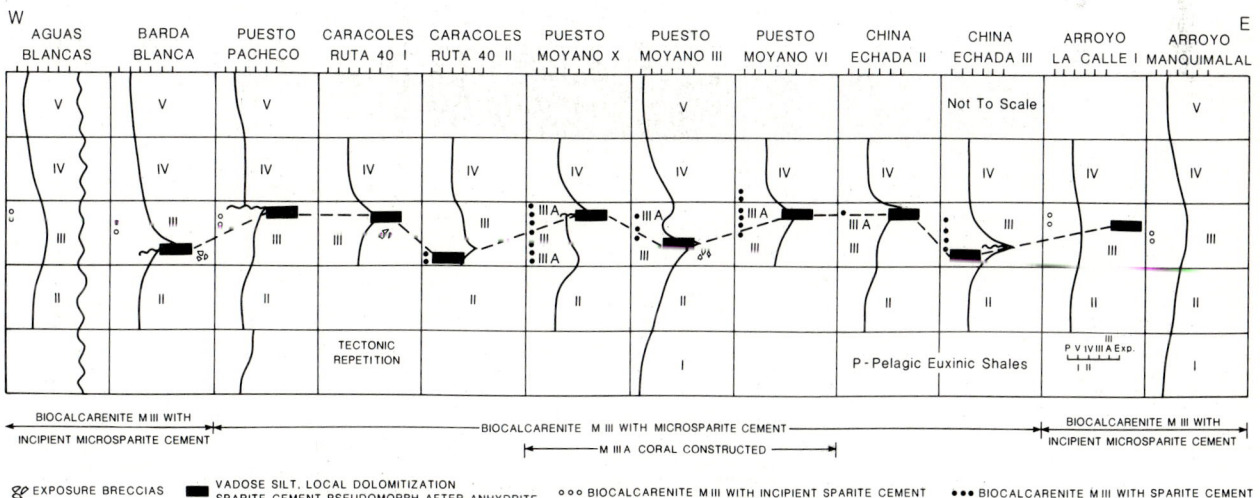

FIGURE 10.16 Chachao Formation (Valanginian), Neuquén–South Mendoza Basin, Argentina. West-east correlation of petrographic features related to temporary subaerial exposure. Redrawn from Carozzi et al. (1981a).

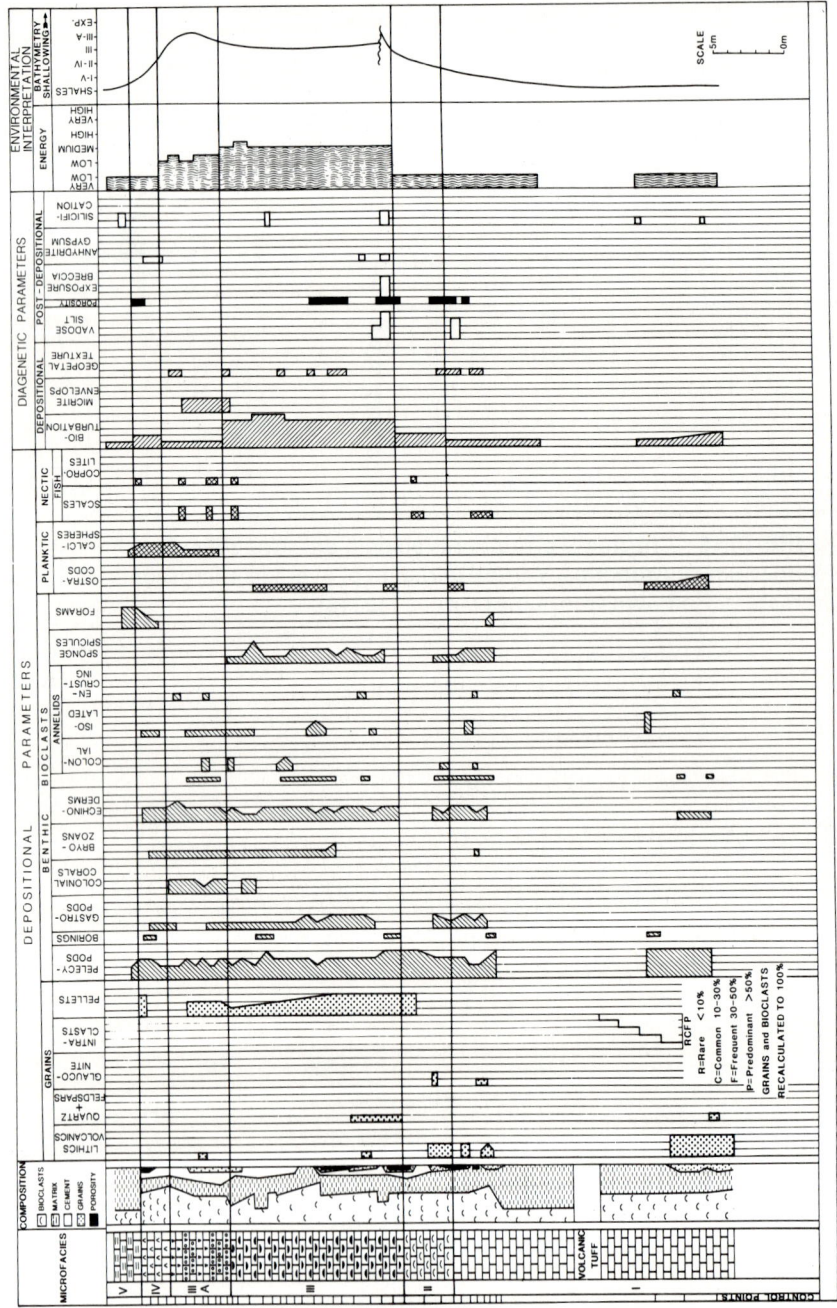

FIGURE 10.17 Chachao Formation (Valanginian), Neuquén–South Mendoza Basin, Argentina. Typical example of vertical variation of microscopic parameters. Redrawn from Carozzi *et al.* (1981a).

Chap. 10 / Carbonate Platforms with Frontal Bioaccumulated Buildups 207

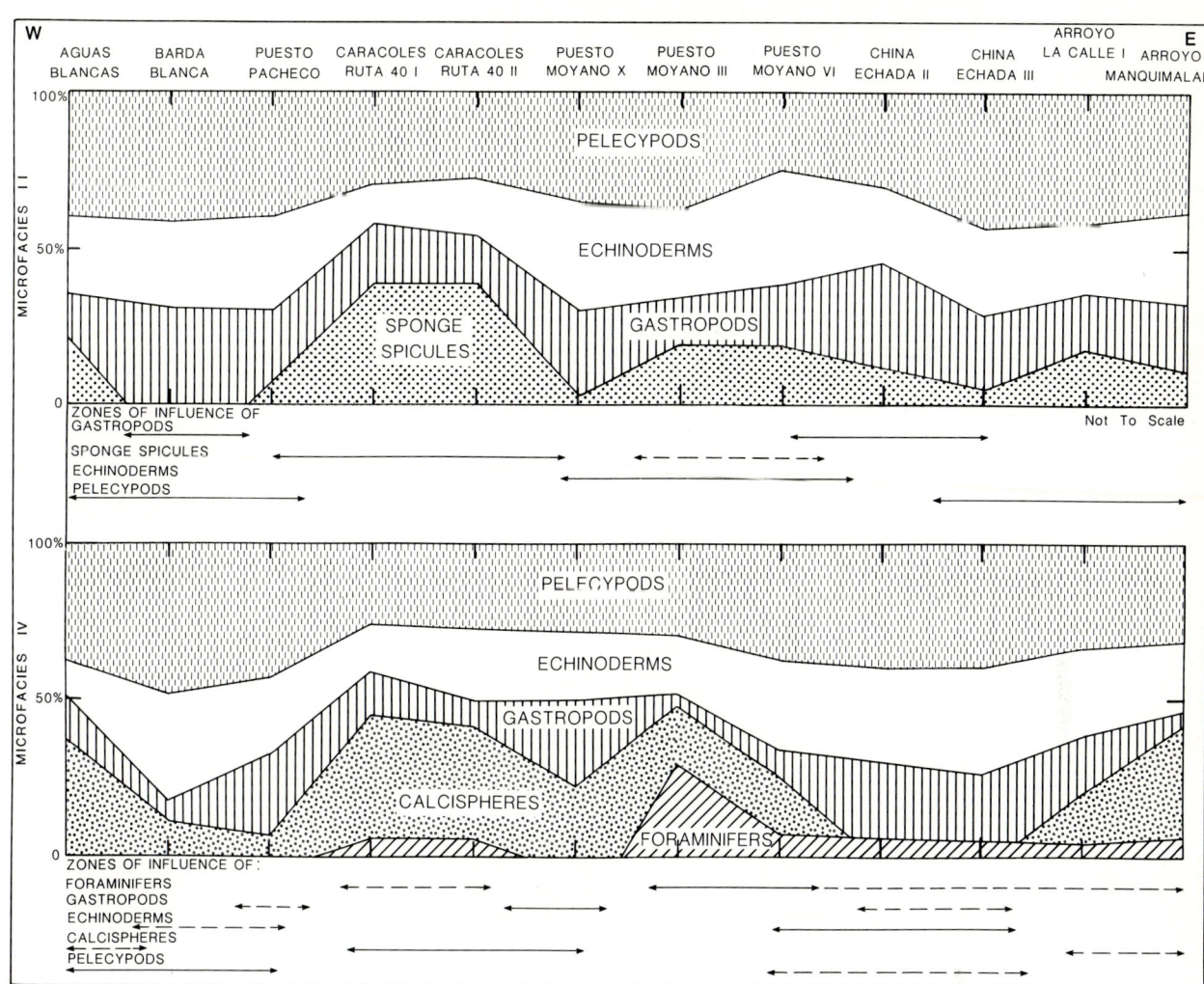

FIGURE 10.18 Chachao Formation (Valanginian), Nequén–South Mendoza Basin, Argentina. West-east cross section during deposition of microfacies III and IV showing paleoecological zonation. Redrawn from Carozzi et al. (1981a).

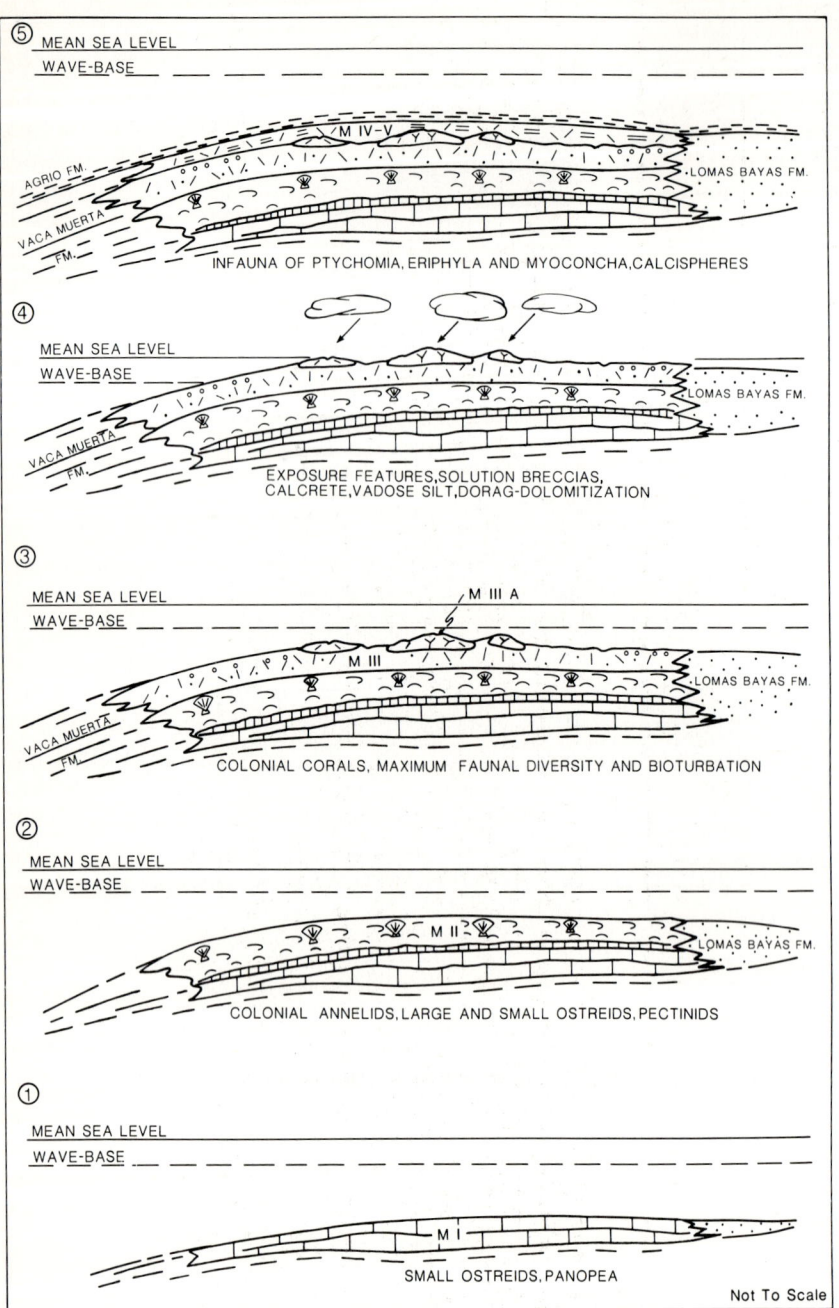

FIGURE 10.19 Chachao Formation (Valanginian), Neuquén–South Mendoza Basin, Argentina. Evolutionary stages of ideal depositional model. Redrawn from Carozzi *et al.* (1981a).

11

Carbonate Platforms with Frontal Bioaccumulated to Hydrodynamic Buildups

Carbonate platforms facing moderately agitated seas represent environments where frontal buildups undergo almost continuous reworking by the action of waves and currents. Thus abundant bioclastic and pelletoidal material is generated and redistributed locally into hydrodynamic buildups. The volume of the latter largely exceeds that of the original bioaccumulated bodies, which are often partially buried under the bioclasts produced. Organisms involved in these reworking processes throughout the Phanerozoic record include blue-green and red algae, bryozoans, crinoids, and larger benthic foraminifers.

11.1 CHARACTERISTIC CONSTITUENTS: PELLETS, *GIRVANELLA* ONCOIDS, AND *NUIA*

The Pogonip Group (Lower Ordovician) of the Arrow Canyon Range, Clark County, southeast Nevada, was investigated petrographically in a field section 407 m thick (Fig. 11.1) that consists of the superposition of two environments separated by a transition zone (Stricker and Carozzi, 1973). The first environment, characterized by the alga *Nuia sibirica,* extends from 0 to 177 m. A transition zone between 177 and 202 m leads to the second environment, characterized by *Girvanella* oncoids. This environment ranges from 202 to 407 m.

Detailed petrography on 1053 thin sections was made from samples collected at an average vertical interval of 39 cm.

Description of Microfacies

The depositional history of the Opb through Opf units was interpreted as the stratigraphic superposition of two depositional models separated by a transition zone. Both models display a similar platform geometry bordered by buildups, but they differ by the shape of their respective seaward slopes and hence share a large proportion of the recognized microfacies. Their description is in order of general deepening of the depositional environment.

Microfacies 1 (Fig. 11.2A). Finely bedded, dark, arenaceous to slightly argillaceous pelletoidal biocalcisiltite. Fine to coarse sand-size bioclasts, aligned parallel to bedding, include crinoids, trilobites, ostracods, and monaxonic sponge spicules. Locally, bioturbation destroyed the laminated texture.

Microfacies 2 (Fig. 11.2B). Massive to bioturbated calcisiltite with scattered sand-size bioclasts of crinoids, trilobites, pelecypods, brachiopods, ostracods, and monaxonic sponge spicules. Silt-size grains of detrital quartz are locally abundant. Some burrow fillings

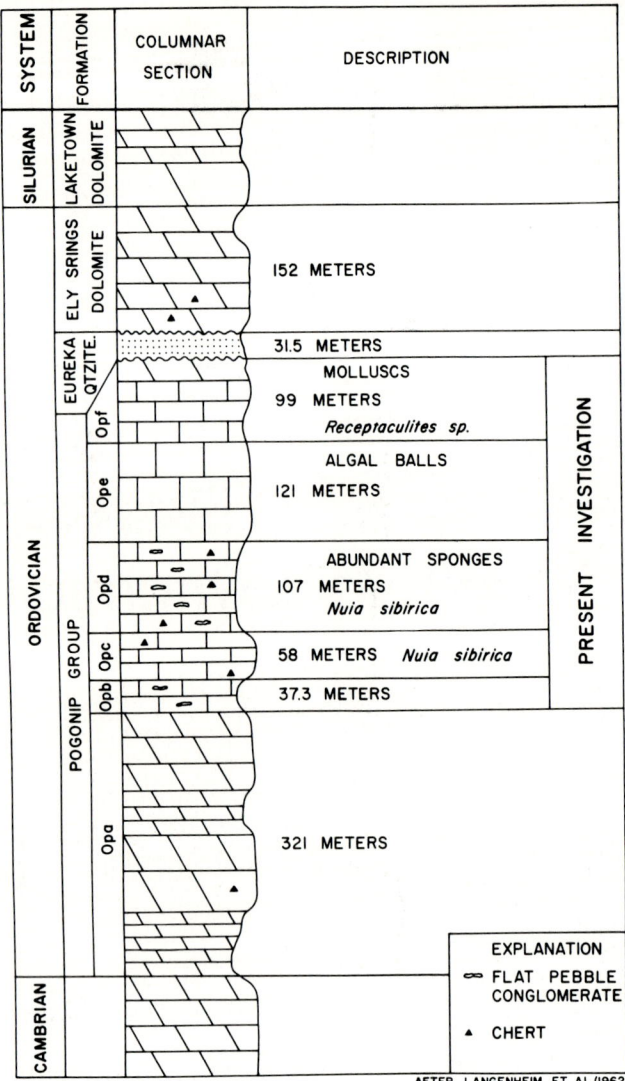

FIGURE 11.1 Pogonip Group (Lower Ordovician), Arrow Canyon Range, Clark County, Nevada. Stratigraphy of the Ordovician at Arrow Canyon Range, Nevada. From Stricker and Carozzi (1973). Reprinted by permission of Société Nationale des Pétroles d'Aquitaine.

may show a grain-supported pelletoidal texture. Neomorphism into pseudomicrosparite is widespread.

Microfacies 3 (Fig. 11.2C). Fine to coarse mud-supported biocalcarenite with calcisiltite matrix. Scattered crinoids, trilobites, large ostracods, pelecypods, brachiopods, gastropods, *Nuia*, and sponge spicules occur as bioclasts and as entire individuals or colonies. Bioturbation is similar to that of microfacies 2, but the burrows contain often concentrations of detrital quartz.

Microfacies 4A (Fig. 11.2D). Fine to coarse grain-supported pelletoidal biocalcarenite with irregular association of calcisiltite matrix and sparite cement. Lithic pellets derived from the submarine reworking of microfacies 3 are associated with clasts of oncoids derived from microfacies 6. Other biogenic constituents are bioclasts of crinoids, trilobites, brachiopods, gastropods, ostracods, and monaxonic sponge spicules. Detrital quartz is abundant.

Microfacies 4 (Fig. 11.2E). Fine to coarse grain-supported pelletoidal biocalcarenite with cavity-filling sparite cement. Fecal pellets are associated with abundant bioclasts of crinoids. Trilobites, brachiopods, *Nuia*, and ostracods are accessory constituents. Detrital quartz is abundant.

Microfacies 5 (Fig. 11.2F,G). Grain-supported *Nuia* biocalcarenite with cavity-filling sparite cement. Bioclasts and complete colonies of *Nuia sibirica* largely predominate over crinoids, which are the next most important constituents, followed by brachiopods and trilobites. Within this microfacies are intraformational clasts resulting from early submarine cementation of beachrock

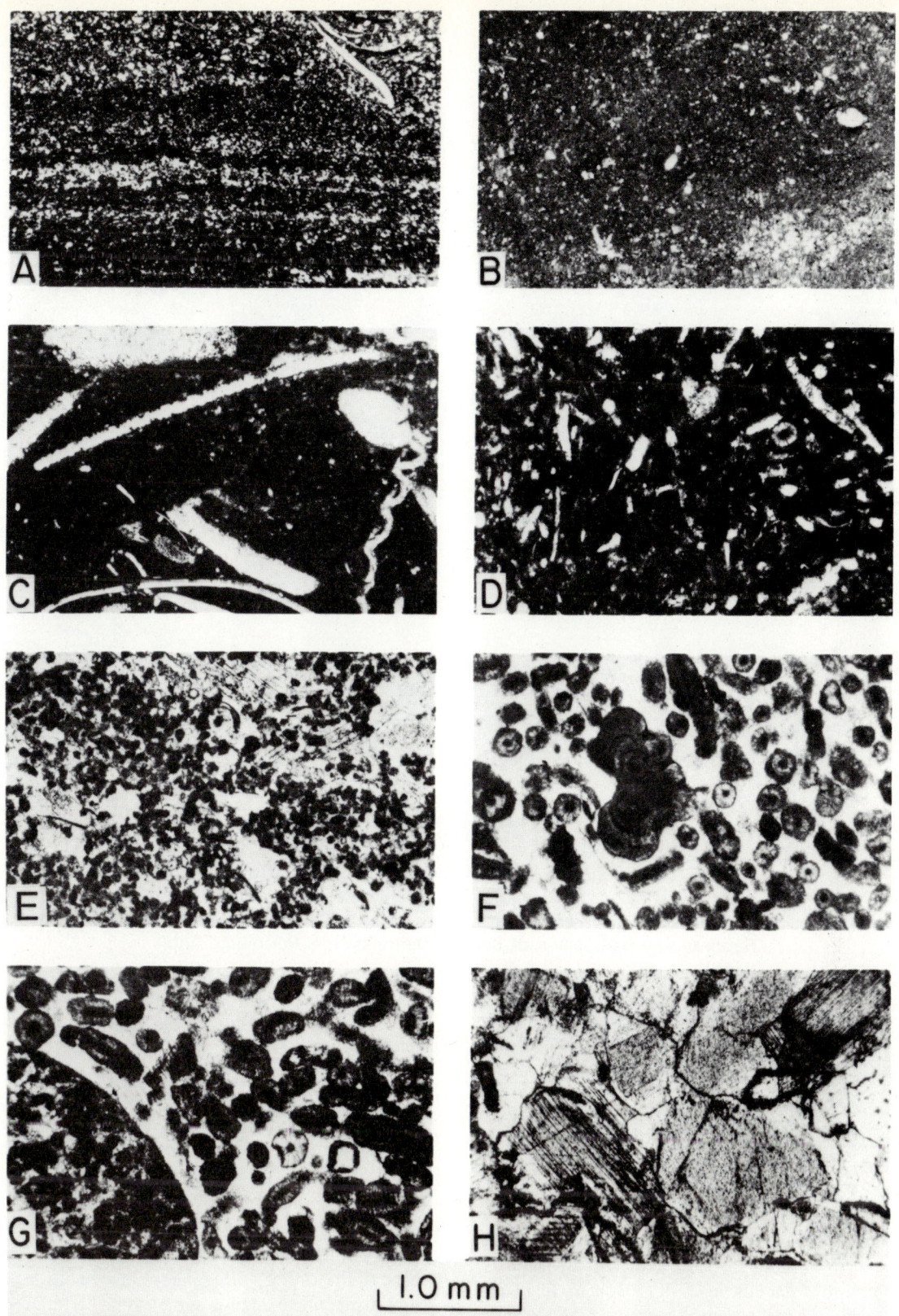

FIGURE 11.2 Pogonip Group (Lower Ordovician), Arrow Canyon Range, Clark County, Nevada. Typical microfacies. A. Microfacies 1. B. Microfacies 2. C. Microfacies 3. D. Microfacies 4A. E. Microfacies 4. F and G. Microfacies 5. H. Microfacies 7. See text for detailed descriptions. All photomicrographs: plane-polarized light. From Stricker and Carozzi (1973). Reprinted by permission of Société Nationale des Pétroles d'Aquitaine.

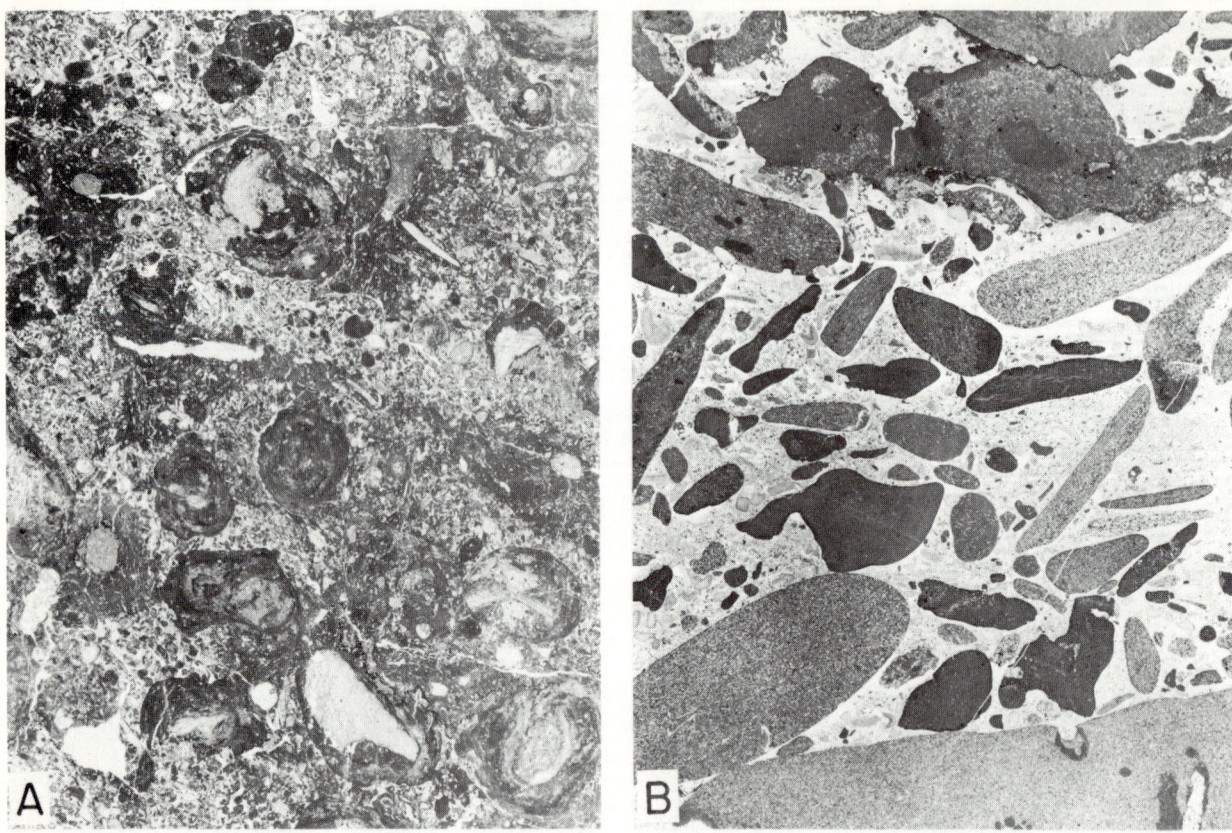

FIGURE 11.3 Pogonip Group (Lower Ordovician), Arrow Canyon Range, Clark County, Nevada. Typical microfacies (continued). A. Microfacies 6. B. Microfacies 8. From Stricker and Carozzi (1973). Reprinted by permission of Société Nationale des Pétroles d'Aquitaine.

type (Fig. 11.2G). In these clasts, the cement is a finer-grained sparite and pellets are more closely packed with a general texture recalling that of microfacies 4.

Microfacies 6 (Fig. 11.3A). Oncoidal calcirudite with an interstitial material consisting of a biocalcarenite with smaller oncoids, fragments of oncoids, bioclasts of crinoids, and pelecypods set in a cavity-filling sparite cement. Oncoids within a thin section are of approximately the same size, but in general, they range in diameter from 5 to 15 mm. They display vaguely concentric to irregular envelopes of *Girvanella* filaments or tubes with abundant trapped minute debris around cores consisting of shell fragments, crinoid bioclasts, and clasts of older algal material.

Microfacies 7 (Fig. 11.2H). Coarse crinoidal biocalcarenite consisting predominantly of columnals not displaying appreciable abrasion and cemented by a combination of syntaxial overgrowths and pressure-solution. Interstitial and deformed bioclasts of trilobites, gastropods, pelecypods, *Nuia* are associated with clasts of oncoids.

Microfacies 8 (Fig. 11.3B). Grain-supported flat pebble calcirudite with graded bedding and imbricated texture with an interstitial material consisting of smaller lithoclasts, lithic pellets derived from the abrasion of larger clasts, bioclasts of crinoids, gastropods, pelecypods, *Nuia,* and trilobites set in a cavity-filling sparite cement. Lithoclasts are large, up to 8 cm, and range in shape from extremely irregular to abraded flat desiccation chips displaying oxidation aureoles and perforations from subaerial exposure on tidal flats. Related to their imbrication is the frequent development of umbrella effects. The predominant lithologies indicate derivation from the reworking of microfacies 2, 3, and 4 by greater energy events, namely storms which are known to develop all the above-described features. The demonstration of their existence and a discussion of their periodicity is given below.

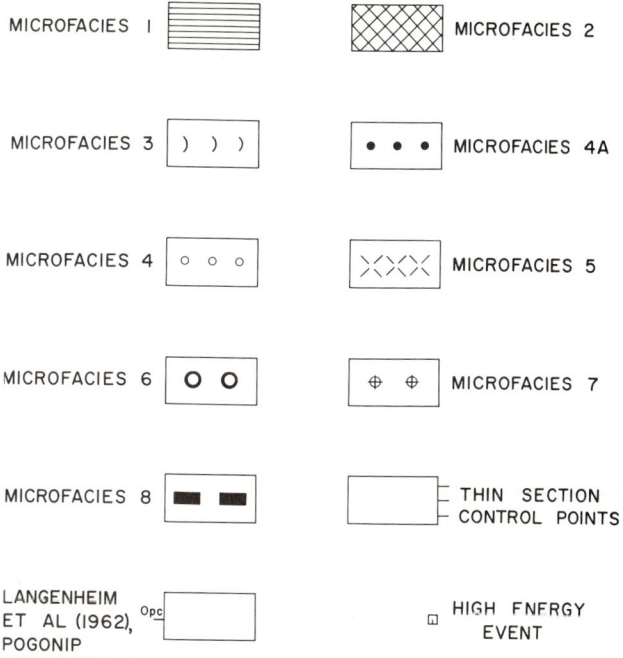

FIGURE 11.4 Pogonip Group (Lower Ordovician), Arrow Canyon Range, Clark County, Nevada. Symbols for microfacies. From Stricker and Carozzi (1973). Reprinted by permission of Société Nationale des Pétroles d'Aquitaine.

The Ideal Shallowing-Upward Sequence of the *Nuia* Model

In this portion of the investigated sequence, the shallowing-upward order of the microfacies is 7, 5, 4, 3, 2, 1. Microfacies 8, the tempestite, was not assigned a particular position because it interrupts the depositional sequence at any time. It consists of clasts derived from microfacies 5, 4, 3, and 2. A typical section (Figs. 11.4, 11.5) with abundant tempestites shows detailed variation of microscopic parameters within the superposition of four shallowing-upward sequences.

The Ideal *Nuia* Depositional Model

The proposed model (Figs. 11.6, 11.7) consists of an intertidal platform with an oceanward margin represented by a discontinuous pelletoidal buildup (microfacies 4) of medium to high energy. It was followed by an upper slope where dense colonies of *Nuia* lived, were later destroyed by wave action, and dispersed both landward and seaward. Crinoids developed in abundance in the infratidal lower slope (microfacies 7). Their bioclasts were transported landward by currents and waves across the entire carbonate system to concentrate mechanically at the edge of the platform, where, together with lithic pellets, they formed hydrodynamic buildups. Detrital quartz of landward provenance also became concentrated in the same buildups.

Lithoclasts are composed mainly of microfacies 1, 2, and 3. They originated from subaqueous erosion of hydrodynamic buildups and from reworking of desiccation chips in intertidal flats. The majority of them were moved seaward by tidal currents and deposited below wave base in microfacies 7. The figures plotted in the frequency curve indicate, in percentage, the most abundant contributing microfacies.

The behavior of many constituents is a clear indication of dispersion landward, or seaward, or both by the action of currents and tides.

The Ideal Shallowing-Upward Sequence of the Oncoid Model

In this portion of the sequence investigated, the shallowing-upward order of the microfacies is 7, 6, 4A, 3, 2, 1. Again, microfacies 8, the tempestite, was not assigned a particular position because it can interrupt the depositional sequence at any time. It consists of clasts derived from microfacies 6, 4A, 3, and 2. A typical section (Fig. 11.8) shows detailed variation of microscopic parameters within the superposition of four shallowing-upward sequences.

The Ideal Oncoid Depositional Model

The proposed model (Figs. 11.9, 11.10) consists of an intertidal platform with an oceanward margin represented by a discontinuous pelletoidal buildup (microfacies 4A) of much lower energy than in the *Nuia* model, and followed by a nearly horizontal bench below low-tide level where the oncoidal calcirudite (microfacies 6) was generated. A rather steep slope led to the deeper subtidal crinoidal environment of microfacies 7.

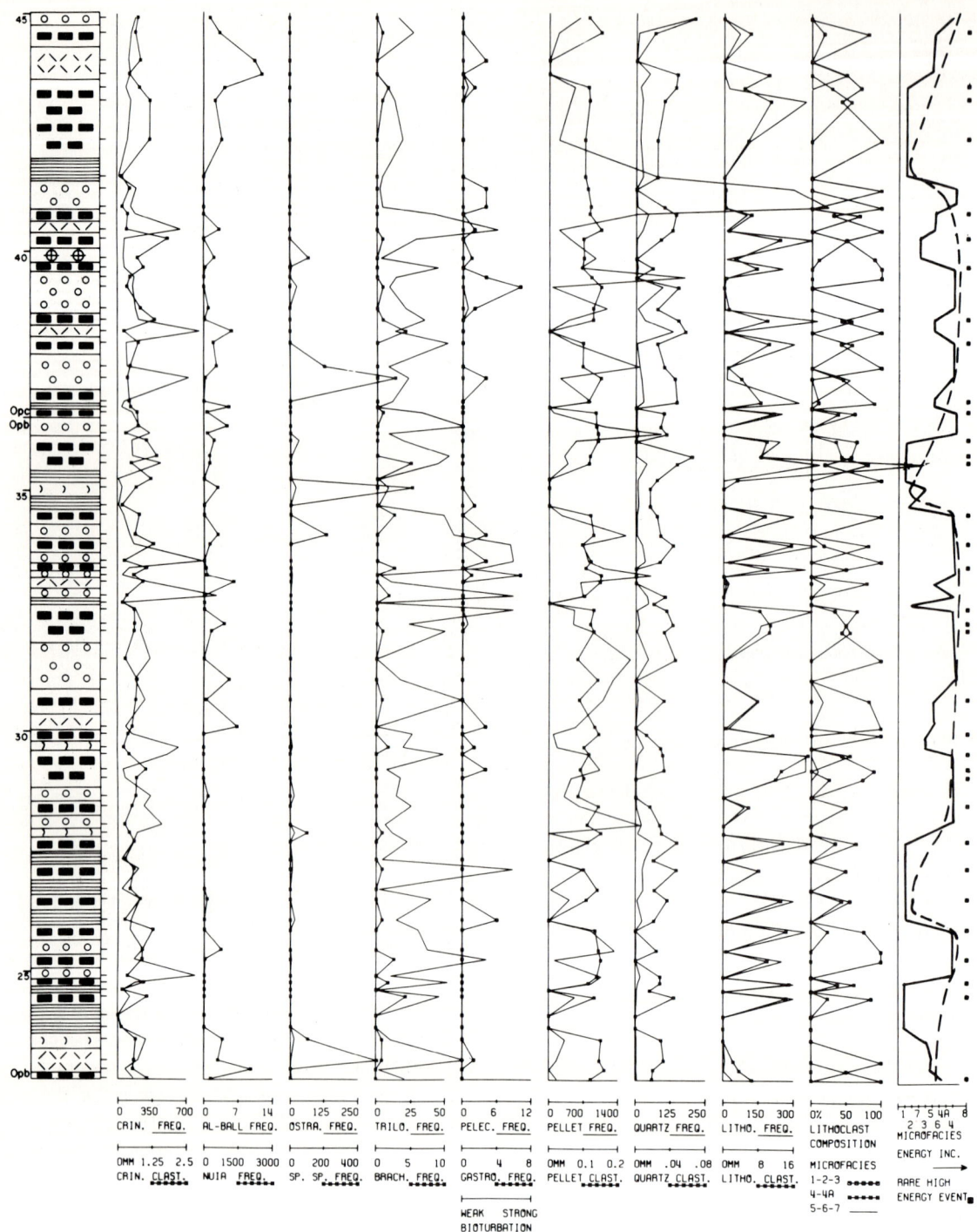

FIGURE 11.5 Pogonip Group (Lower Ordovician), Arrow Canyon Range, Clark County, Nevada. Typical example of vertical variation of microscopic parameters during four successive shallowing-upward sequences of the *Nuia* model. Modified from Stricker and Carozzi (1973).

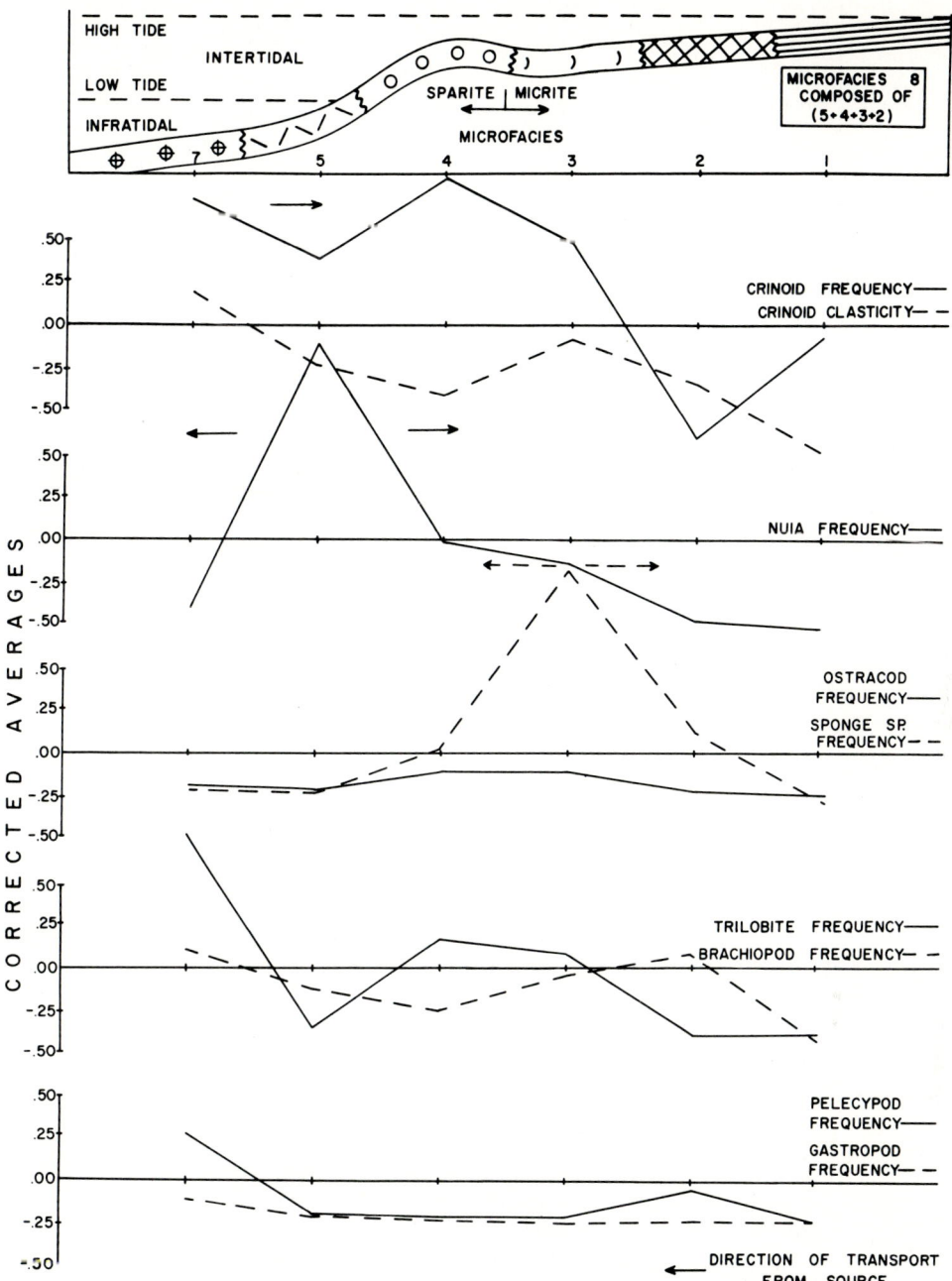

FIGURE 11.6 Pogonip Group (Lower Ordovician), Arrow Canyon Range, Clark County, Nevada. Ideal *Nuia* depositional model. From Stricker and Carozzi (1973). Reprinted by permission of Société Nationale des Pétroles d'Aquitaine.

The general energy of this model is less than that of the *Nuia* model and results mainly from the shape of the platform edge, which is broader, and hence restricted circulation occurred across it in both directions. This situation is also shown by a comparison of the behavior of some constituents in both models. For instance, the majority of the ostracods lived in the intertidal flat, with a small population in open marine conditions. An increase in their abundance in the oncoidal model indicates a more favorable environment resulting from decreasing energy and increasing restriction of the intertidal area. A similar situation combined with increasing silica sediment yield, expressed by more detrital quartz, shifted the frequency peak of the sponge spicules from microfacies 3 in the *Nuia* model to microfacies 2 in the oncoidal model. Finally, the abundance of brachio-

216 Part III / Carbonate Platforms

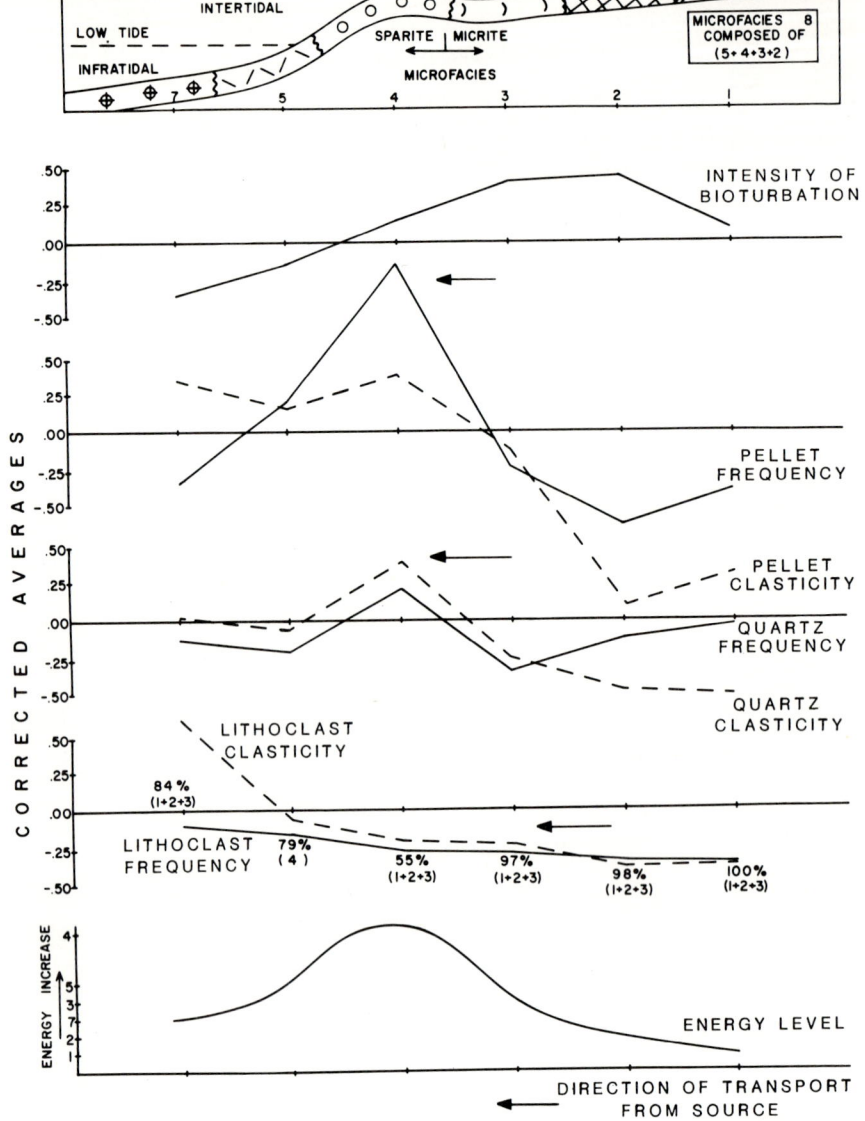

FIGURE 11.7 Pogonip Group (Lower Ordovician), Arrow Canyon Range, Clark County, Nevada. Ideal *Nuia* depositional model (continued). From Stricker and Carozzi (1973). Reprinted by permission of Société Nationale des Pétroles d'Aquitaine.

pods, pelecypods, and gastropods in the intertidal flat was also greater under the more restricted conditions of the oncoidal model.

As in the previous model, the behavior of many constituents is a clear indication of dispersion landward, or seaward, or both by the action of currents and tides.

General Evolution of the Depositional Environments and Storm Deposits

An examination of the section investigated reveals that a total of 130 intercalations of microfacies 8, interpreted as storm deposits, interrupted the successions of microfacies 1 to 7. The storm deposits are concentrated mostly in Opb and the lower part of Opc, which can be used as a typical example (Fig. 11.11).

A Fourier series analysis is an efficient tool for demonstration of the periodicity of normal sedimentation and another periodicity or its absence in the case of assumed storm deposits (Stricker and Carozzi, 1974). The smooth power spectrum of microfacies 8 only (Fig. 11.11) shows a major peak at the 14th cycle. Minor peaks occur at the 19th and 24th cycles. Other major peaks are found at approximately the 29th, 33rd, and 40th cycles. No clear cyclicity is obvious and the harmonic relationship remains undetermined. Analysis of all normal microfacies, excluding microfacies 8 (Fig. 11.11), shows a pronounced peak at the 6th cycle. A

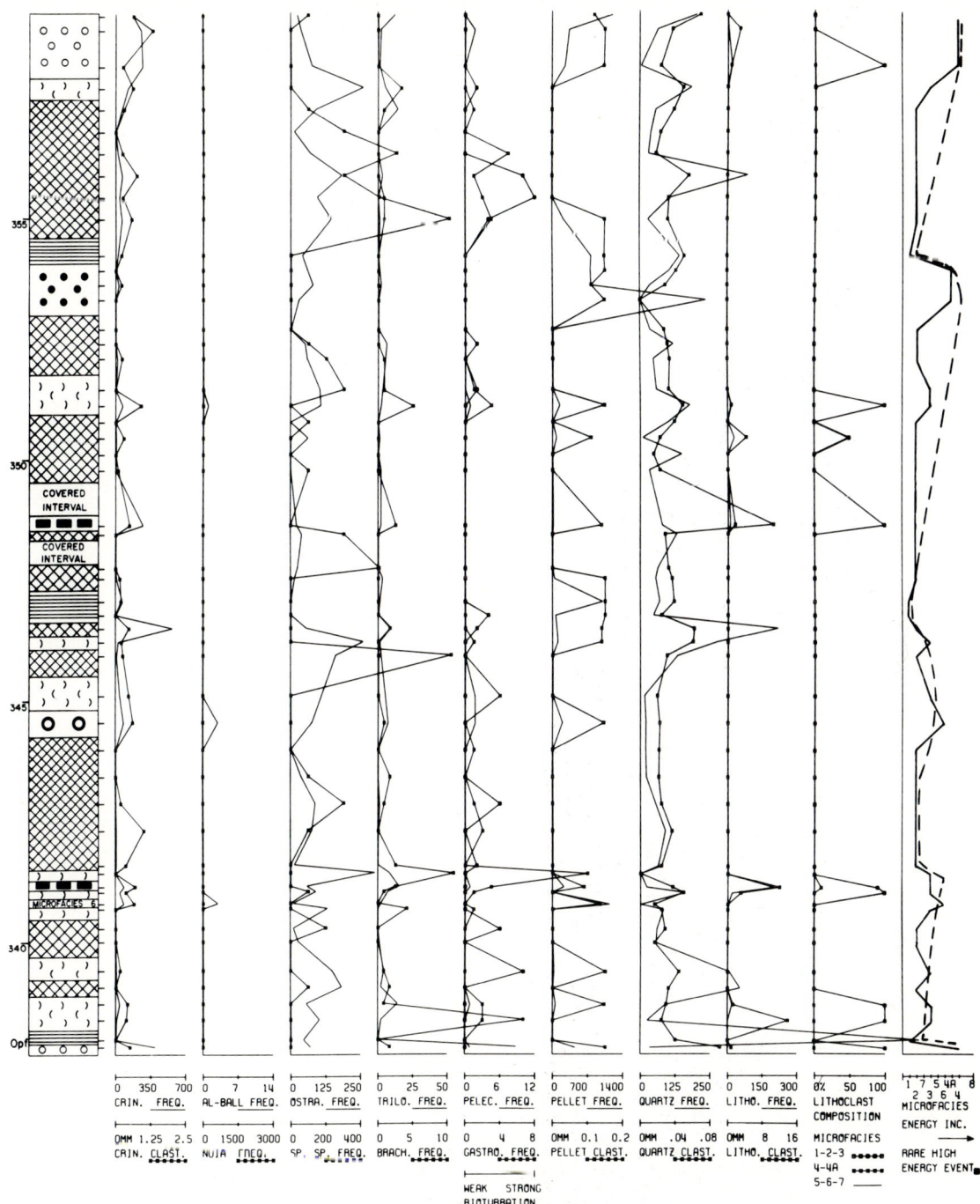

FIGURE 11.8 Pogonip Group (Lower Ordovician), Arrow Canyon Range, Clark County, Nevada. Typical example of vertical variation of microscopic parameters during four successive shallowing-upward sequences of the oncoid model. Modified from Stricker and Carozzi (1973).

second major peak is located at the 12th cycle, others at about cycles 18, 31, and 35. An apparent harmonic of 6 exists here, reinforcing the concept that normal microfacies are cyclic in arrangement. The smoothed power spectrum for all samples (Fig. 11.11) shows a peak at approximately the 6th cycle and a second major peak at the 14th cycle. Other peaks are found at the 17th cycle, an harmonic of 6, and at the 29th cycle, an harmonic of 14. Obviously, this curve shows the combined effect of normal cyclic sedimentation and random occurrence of storm deposits.

The general evolution of the depositional environ-

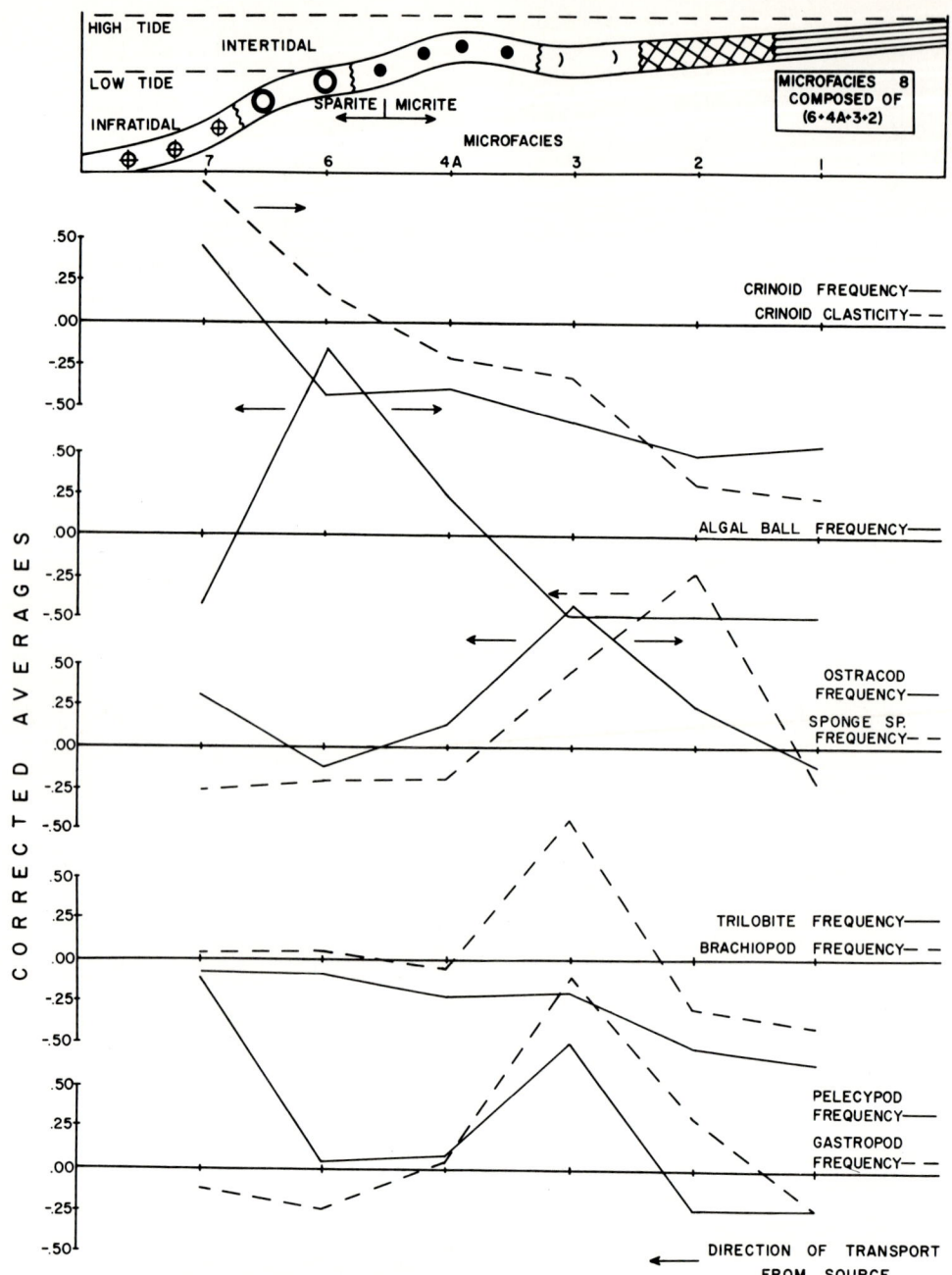

FIGURE 11.9 Pogonip Group (Lower Ordovician), Arrow Canyon Range, Clark County, Nevada. Ideal oncoid depositional model. From Stricker and Carozzi (1973). Reprinted by permission of Société Nationale des Pétroles d'Aquitaine.

ments can be expressed by the general trend of oscillation of energy, which is a dashed line tangential to the major peaks of the energy curve (Figs. 11.12, 11.13, 11.14). Six large environmental groups were found and designated by Roman numerals. Some are characterized by strong asymmetric shallowing-upward oscillations, others by gently symmetric shallowing and deepening oscillations, and still others by persistent stability interrupted by minor oscillations.

11.2 CHARACTERISTIC CONSTITUENTS: CRINOIDS AND BRYOZOANS

Carbonate platforms formed by crinoidal calcarenites with a minor contribution of bryozoans require unusual environmental conditions. In general, actively growing crinoid colonies formed numerous bioaccumulated build-ups scattered at random across platforms. Bioclasts produced in great amount by the action of waves and currents

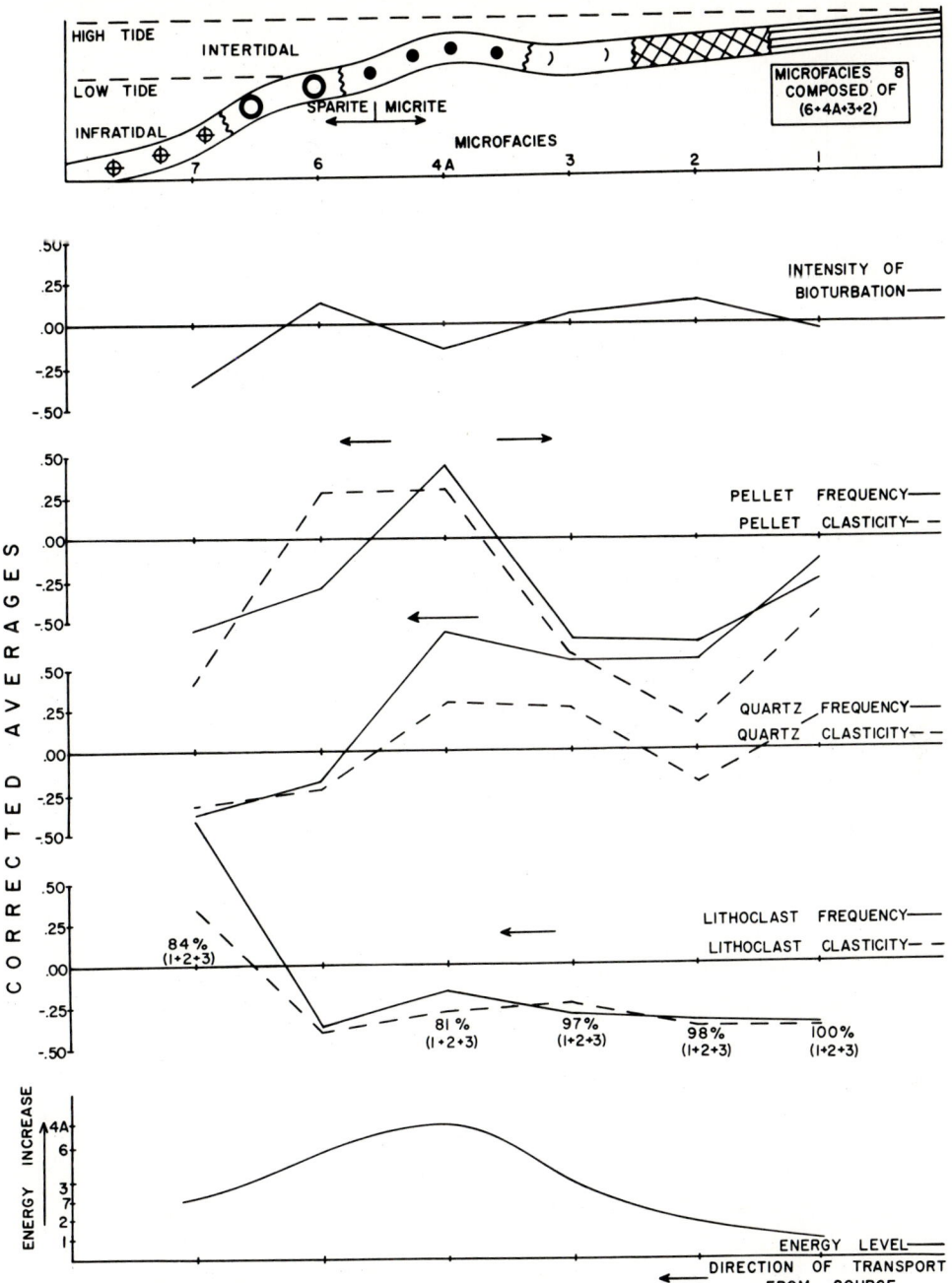

FIGURE 11.10 Pogonip Group (Lower Ordovician), Arrow Canyon Range, Clark County, Nevada. Ideal oncoid depositional model (continued). From Stricker and Carozzi (1973). Reprinted by permission of Société Nationale des Pétroles d'Aquitaine.

were dispersed radially from bioaccumulated centers of production. This combination of growth and dispersal was an effective mechanism for the generation of massive and extensive crinoidal limestones.

The Burlington Limestone (Middle Mississippian) represents an extensive carbonate platform on the western margin of the Illinois Basin which terminated eastward by a steep slope leading to basinal conditions where the distal siltstones of the coeval Borden Group turbidites were deposited (Carozzi and Gerber, 1979).

Seven field sections of the Burlington Limestone were investigated along the Mississippi River between Burlington, Iowa, and Louisiana, Missouri. It averages between 25 and 35 m in thickness. Five hundred sixteen

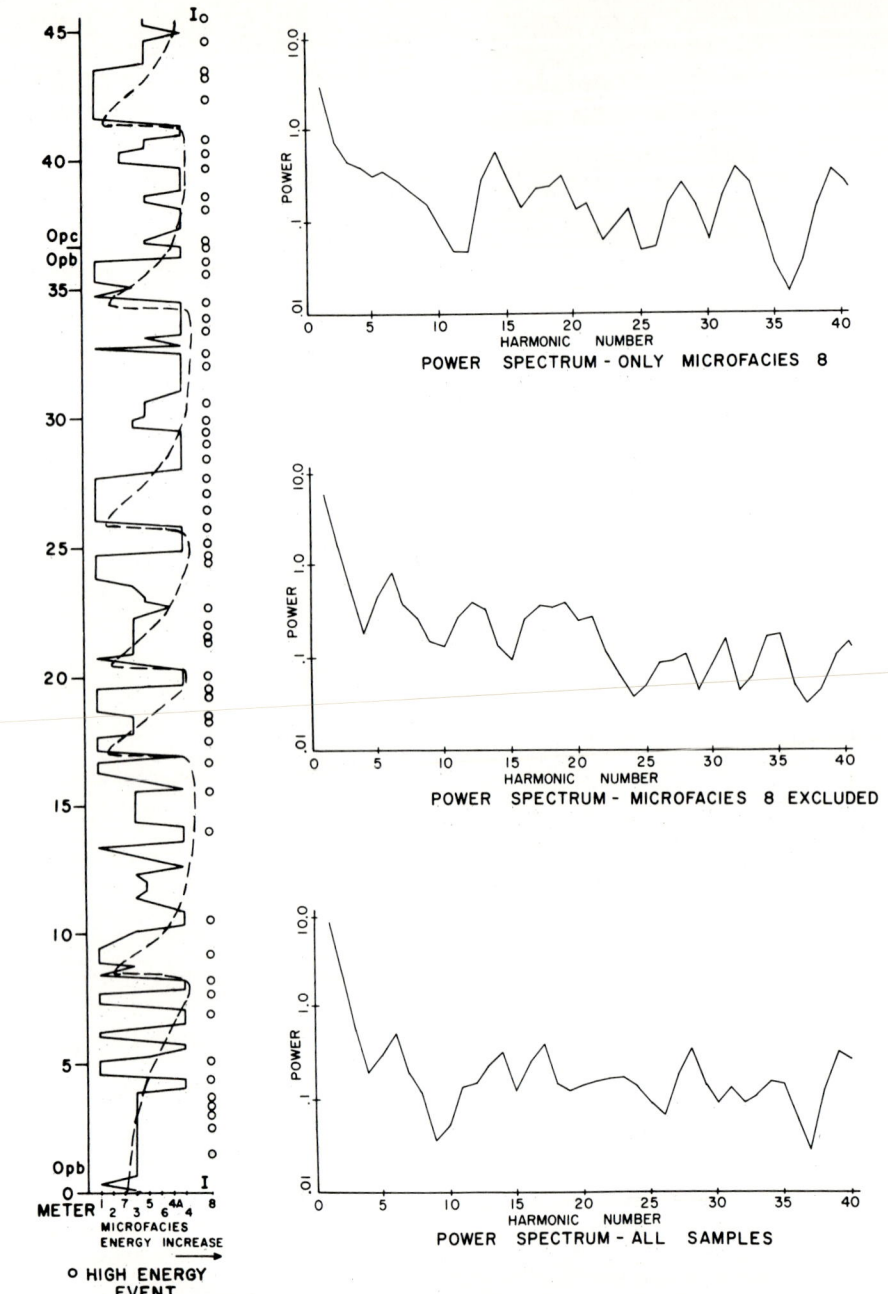

FIGURE 11.11 Pogonip Group (Lower Ordovician), Arrow Canyon Range, Clark County, Nevada. Distribution by Fourier analysis of storm deposits. From Stricker and Carozzi (1974).

samples were studied petrographically from 86 field units. The total thickness of measured section was 90.3 m and the average vertical interval of sampling was 17.3 cm.

Description of Microfacies

The major components of the Burlington Limestone are grains (crinoid and blastoid bioclasts), bioclastic matrix (less than 100 μm), and cement (syntaxial overgrowths on echinoderm grains). Hence these carbonates can be treated petrographically and interpreted like siliciclastic rocks.

Arenite Suite

The arenite suite is characterized in the field by medium to thick, even-bedded light gray to white, coarsely crystalline crinoidal limestone.

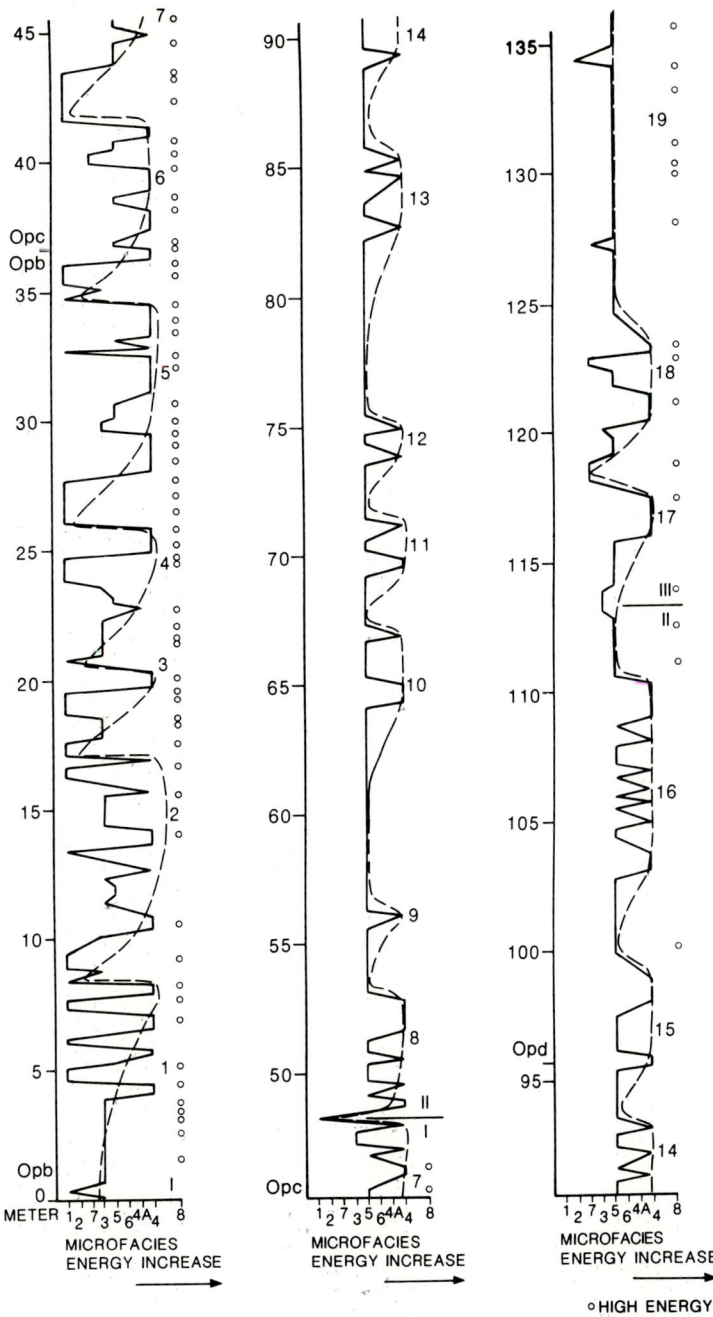

FIGURE 11.12 Pogonip Group (Lower Ordovician), Arrow Canyon Range, Clark County, Nevada. General vertical evolution of depositional environments. Modified from Stricker and Carozzi (1973).

Microfacies 1 (Fig. 11.15A,B,C). Grain-supported crinoidal calcirudite with syntaxial rim calcite cement. Granule and sand-size bioclasts (81%) consist predominantly of moderately sorted and rounded crinoids with accessory amounts of bryozoans. Less than 2% of the components are ostracod valves, bioclasts of thick-shelled brachiopods, trilobites, fish plates and scales, corals, and smaller arenaceous foraminifers. Matrix is absent. Syntaxial rim cement fills all primary intergranular pore space and engulfs the bryozoan bioclasts. It represents almost 18% of the total components. Compaction effects are minimal. This microfacies commonly displays replacement of the internal network of echinoderm grains by silica (Fig. 11.15B, letter A) and by glauconite (Fig. 11.15B, letter B) and geopetal internal sediment consisting of reorganized bioclastic matrix deposited by downward-moving seawater during temporary emergence of crinoidal sediment (Fig. 11.15E).

Microfacies 1 is characterized by small-scale trough cross-stratification (Fig. 11.15F). Foreset beds dip at

222 Part III / Carbonate Platforms

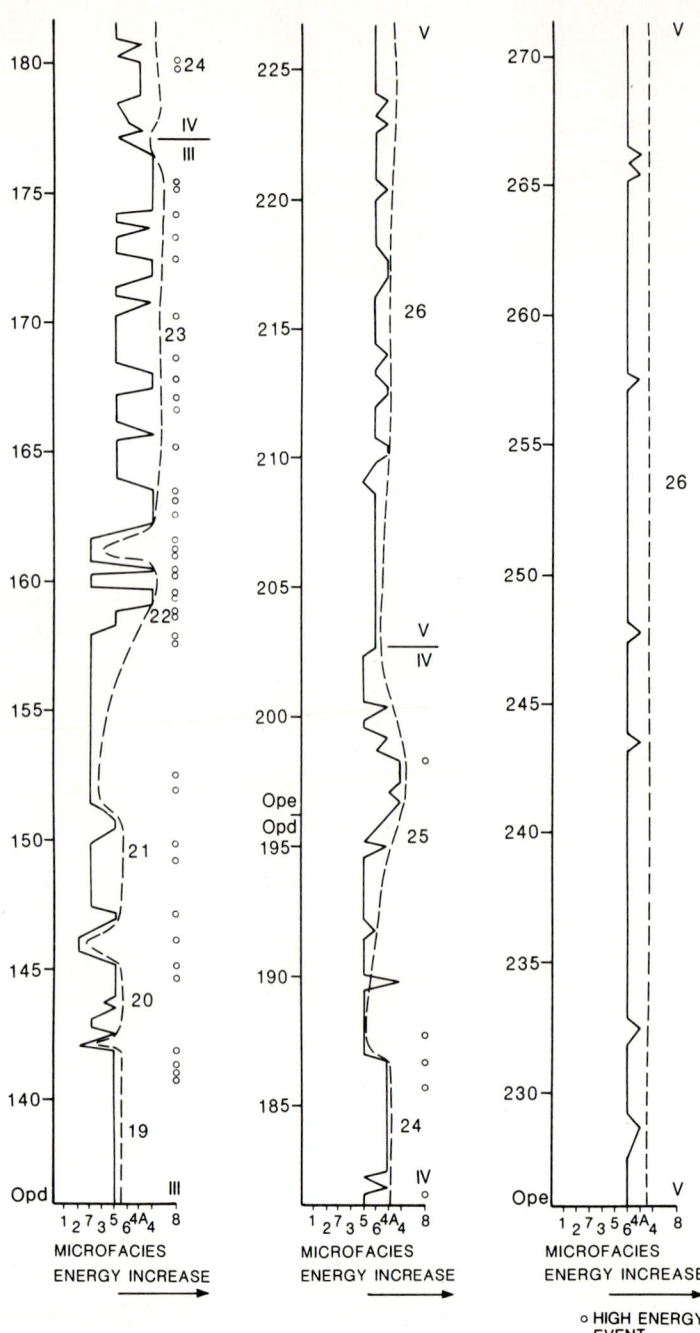

FIGURE 11.13 Pogonip Group (Lower Ordovician), Arrow Canyon Range, Clark County, Nevada. General vertical evolution of depositional environments (continued). Modified from Stricker and Carozzi (1973).

10 to 15° in many different directions, and no dominant transport direction is indicated.

Microfacies 2 (Fig. 11.15D). Grain-supported crinoid-bryozoan calcarenite with syntaxial rim calcite cement. A decrease of rim cement to 9% parallels an increase in total grains (particularly bryozoans) and matrix. Sorting of crinoid bioclasts is moderate to poor. The degree of fragmentation of the bryozoan bioclasts is reduced compared to the previous microfacies. Bryozoan zooecia are collapsed partially by compaction and are filled either with micrite or cavity-filling sparite. This microfacies occurs in massive beds devoid of crossbedding, and corresponds to slightly lower energy conditions than microfacies 1.

Environmental Interpretation. The textural features of the arenite suite suggest medium-energy traction deposition. Currents were strong enough to winnow silt-size debris from the crinoidal sediment and abrade bioclasts, but not strong enough to sort the deposit significantly. The even-bedded nature and tabular geometry

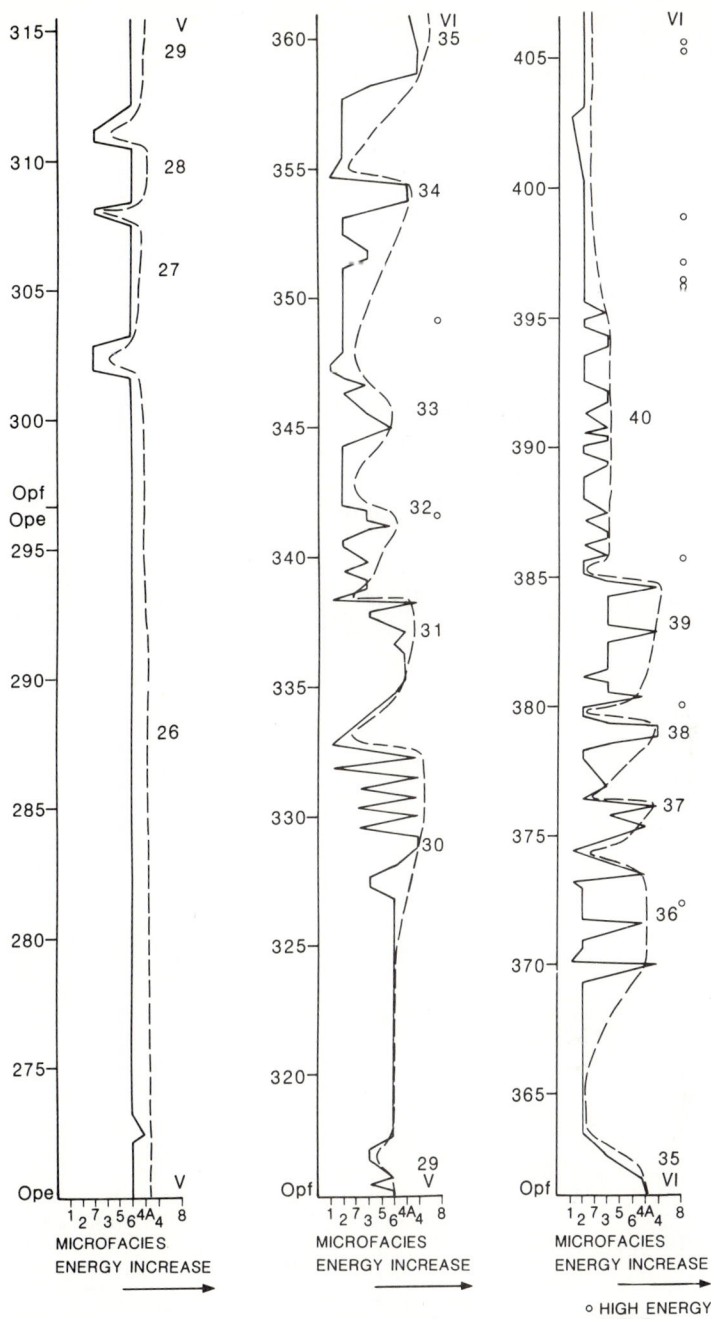

FIGURE 11.14 Pogonip Group (Lower Ordovician), Arrow Canyon Range, Clark County, Nevada. General vertical evolution of depositional environments (continued). Modified from Stricker and Carozzi (1973).

of arenite suite strata indicate shallow, subtidal crinoidal banks, slightly raised above the seafloor, at or near the wave base. Microfacies 2 with more abundant and better preserved bryozoans, and poor crinoid sorting, is inferred to represent nearly an *in situ* assemblage. It was deposited under slightly lower energy than microfacies 1, probably in the central part of crinoidal banks, where it was protected from currents by active crinoidal growth on bank margins. Microfacies 1 represents such margins where high-energy destruction of crinoid colonies led to the deposition of cross-bedded calcirudites.

Wacke Suite

The wacke suite is characterized in the field by tan to brown, medium- to thin-bedded crinoidal limestones with frequent graded bedding of crinoid debris (Fig. 11.16F). The microfacies of this suite are described in order of decreasing energy.

Microfacies 3 (Fig. 11.16A,B,E). Grain-supported to matrix-supported crinoidal calcirudite with a silt-size bioclastic matrix consisting of bryozoans and

224 Part III / Carbonate Platforms

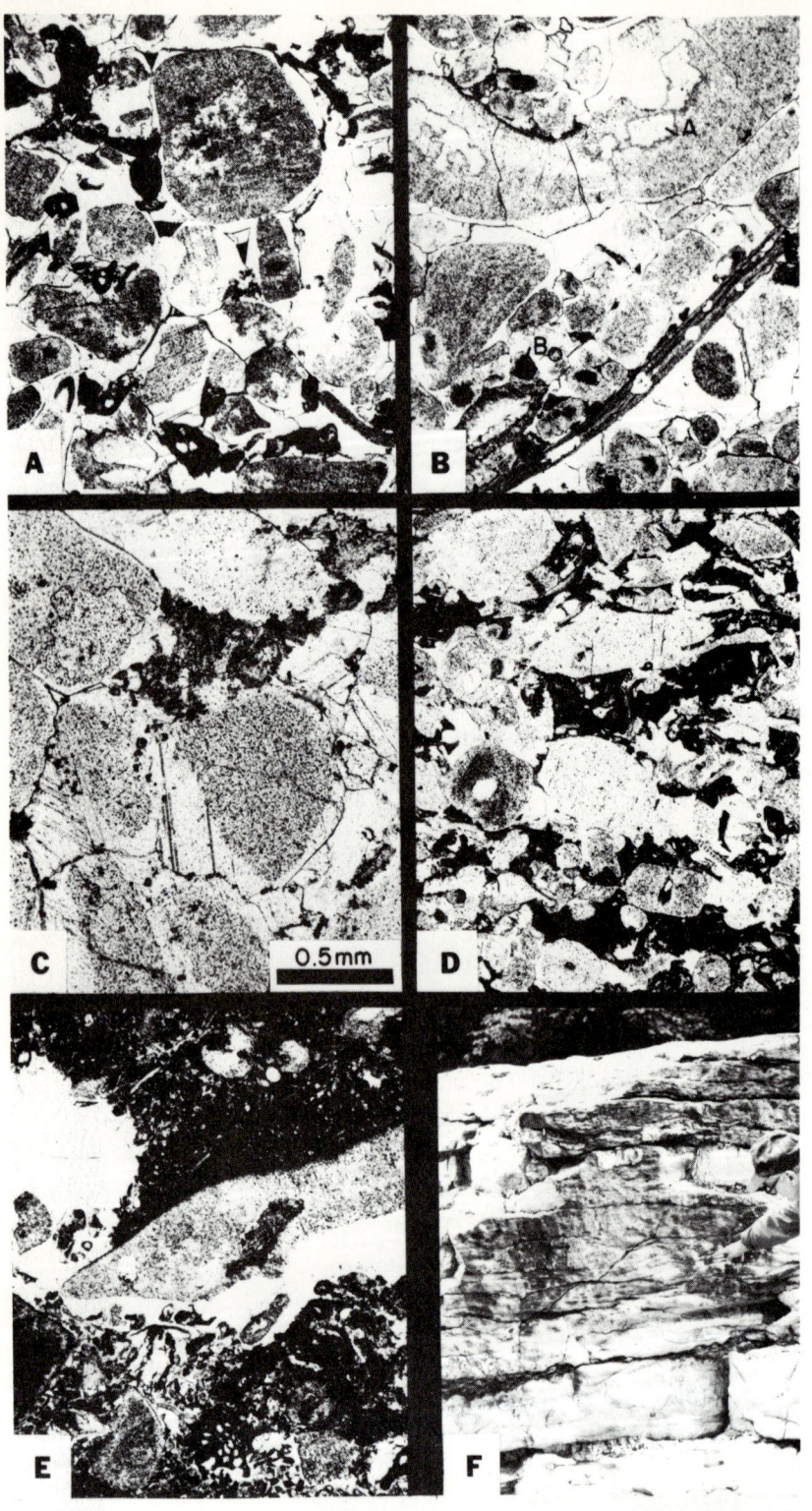

FIGURE 11.15 Burlington Limestone (Middle Mississippian), western margin of Illinois Basin. Typical microfacies. A to C. Microfacies 1. D. Microfacies 2. E. Microfacies 1. F. Field aspect of microfacies 1. See text for detailed descriptions. All photomicrographs: plane polarized light. Bar scale at lower right is for A, B, D, and E. From Carozzi and Gerber (1979). Reprinted by permission of the Southern Illinois University Press.

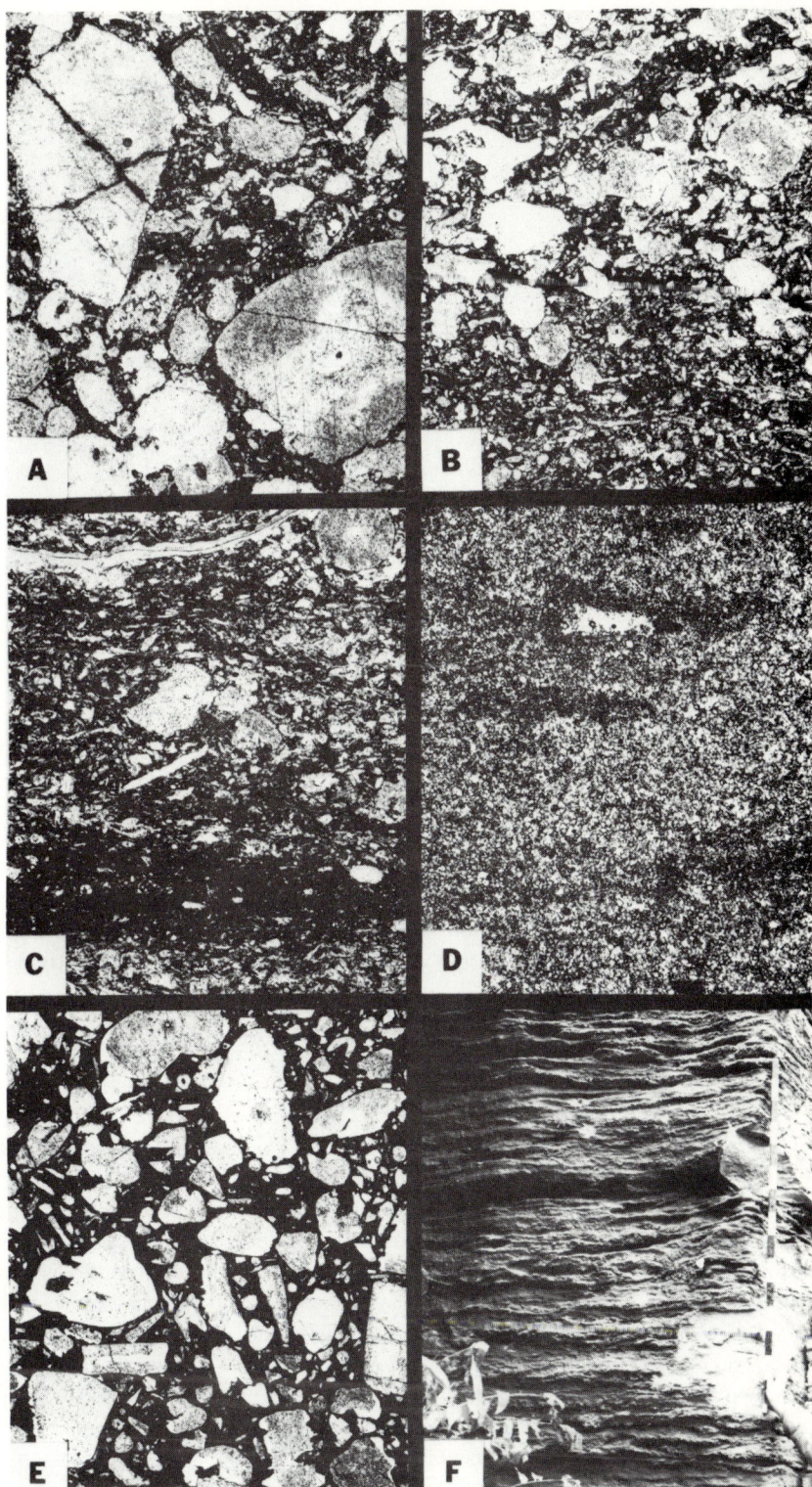

FIGURE 11.16 Burlington Limestone (Middle Mississippian), western margin of Illinois Basin. Typical microfacies (continued). A and B. Microfacies 3. C. Microfacies 4. D. Microfacies 5. E. Microfacies 3. F. Field aspect of microfacies 3. See text for detailed descriptions. All photomicrographs: plane-polarized light. Bar scale at lower right is for A to E. From Carozzi and Gerber (1979). Reprinted by permission of the Southern Illinois University Press.

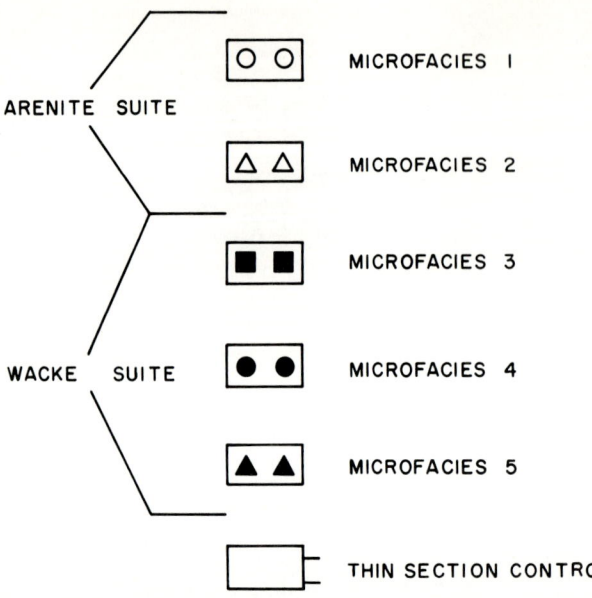

FIGURE 11.17 Burlington Limestone (Middle Mississippian), western margin of Illinois Basin. Symbols for microfacies. From Gerber (1978).

crinoids averaging 22%. Texture ranges from massive to normal and inverse coarse-tail graded bedding (Fig. 11.16B). Rim cement averages only 5% and its irregular shape indicates that it is a neomorphic replacement of the matrix (Fig. 11.16E).

Microfacies 4 (Fig. 11.16C). Crinoidal calcisiltite with scattered sand-size crinoid bioclasts associated with widely scattered brachiopod and bryozoan debris. Texture is irregularly laminated. Compaction has caused overpacking of bioclasts and grain interpenetration of fine debris, making grain boundaries difficult to distinguish.

Microfacies 5 (Fig. 11.16D). Biocalcisiltite with widely scattered sand-size crinoid bioclasts often completely dolomitized by very small rhombs. Sand-size grains average 6.1% and the rest of the rock consists of a bioclastic matrix of crinoids and bryozoans.

Environmental Interpretation. Microfacies of the wacke suite are repeatedly superposed in graded sequences ranging from several centimeters to several tenths of meters. These sequences are the following: a basal unit of microfacies 3 with sole markings, load casts, and normal or rarely inverse graded bedding, overlain by a laminated crinoidal calcisiltite (microfacies 4), grading in turn into microfacies 5. A similar horizontal grading distinguishes proximal, medial, and distal facies within the wedge-shaped geometry of the wacke suite. The latter results from the multiple repetition, below the wave base, of individual and interfering inertia flows developed along the flanks of the crinoid banks and filling the interbank areas. The inertia flows are initiated by oversteepening of the sediment pile at the margins of the banks, by overproduction of bioclasts, or by traction transport to the edge of the banks of debris derived from their internal portions.

The Ideal Vertical Sequence

The ideal vertical sequence (Figs. 11.17, 11.18) summarizes the pattern of cyclicity between crinoid bank and interbank facies. Repetitions of microfacies 1 and 2 at the base of the sequence indicate an actively growing crinoidal bank which was initiated probably on a slightly elevated portion of the seafloor under moderate current action. Crinoid bank deposition ceased abruptly and was superseded by interbank deposits of crinoid calcarenites and calcisiltites produced by inertia flows showing that the crinoid bank was buried by flows which originated from adjacent larger banks. Locally, in the middle of interbank deposits, a few thin intercalations of microfacies 1 may occur; they are interpreted as high-energy, possibly storm events which swept across nearby crinoid banks and carried coarse-grained and moderately sorted crinoid debris into interbank areas. Periods of interbank deposition were followed abruptly with renewed growth of crinoidal banks, with microfacies 1 or 2 superposed directly on microfacies 5. The interbank low areas have been built up apparently by repetition of inertia flows to bathymetric levels suitable for repopulation by crinoids. Occasional occurrences of microfacies, 3, 4, and 5 in bank sequences, result from inertia flows coming down from adjacent banks and interrupting crinoid growth.

The McCraney North field section, Illinois (Fig. 11.19), illustrates detailed variation of microscopic parameters and consists of six superposed bank-interbank episodes.

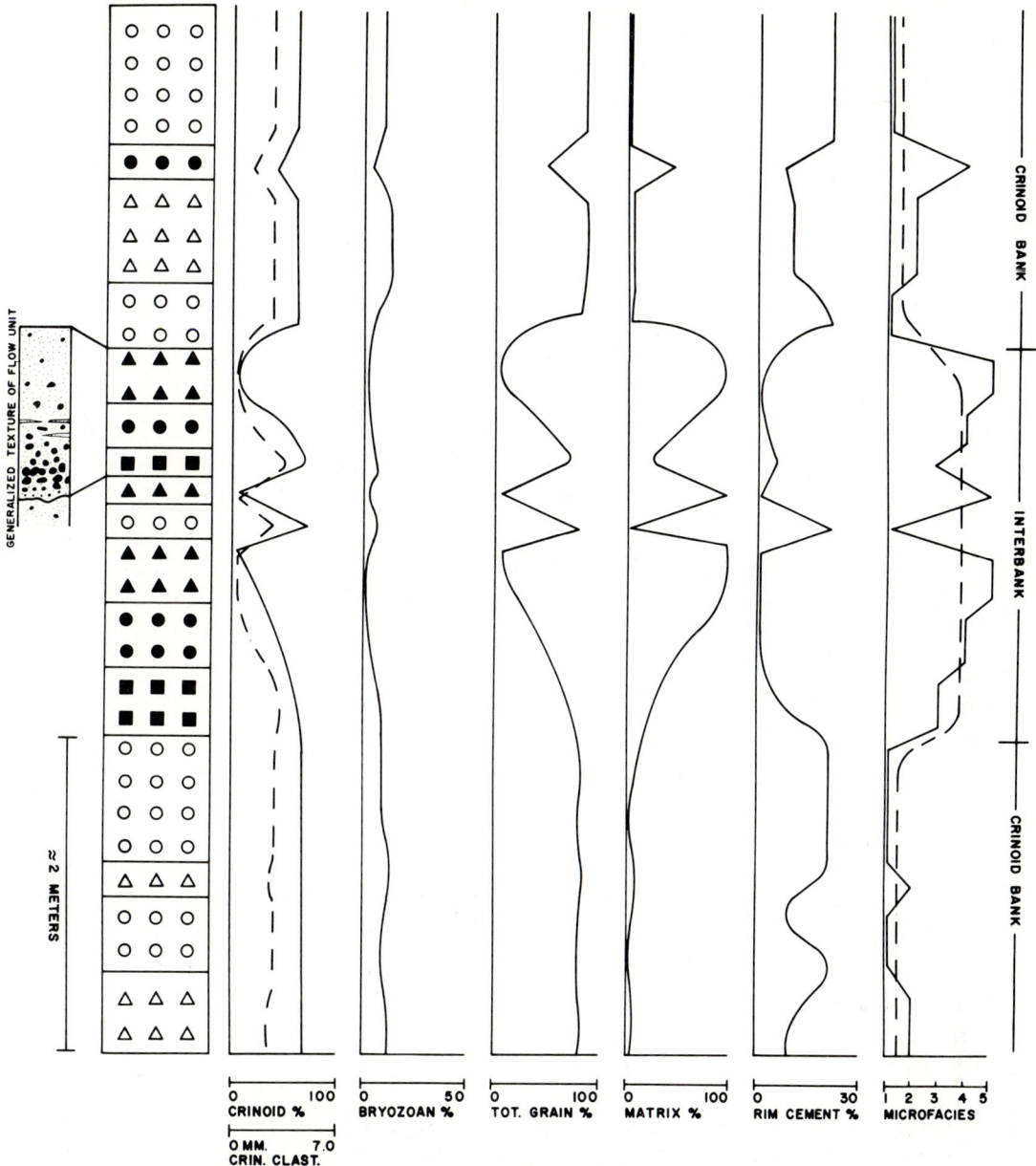

FIGURE 11.18 Burlington Limestone (Middle Mississippian), western margin of Illinois Basin. Idealized vertical variation of microfacies and average values of microscopic parameters. From Carozzi and Gerber (1979). Reprinted by permission of the Southern Illinois University Press.

The Ideal Depositional Model

The ideal depositional model (Fig. 11.20) extends from the center of a crinoid bank to the bottom of an interbank area. Variation of microscopic parameters indicates the change from traction processes on the bank to inertia flow along the margin and into the interbank area.

The Vertical Environmental Evolution

Analysis of the alternations bank-interbank in the measured sections (Fig. 11.21) shows several general trends. A systematic vertical increase in crinoid bank microfacies (1 and 2) and a corresponding decrease in interbank microfacies (3, 4, and 5) occur throughout the studied

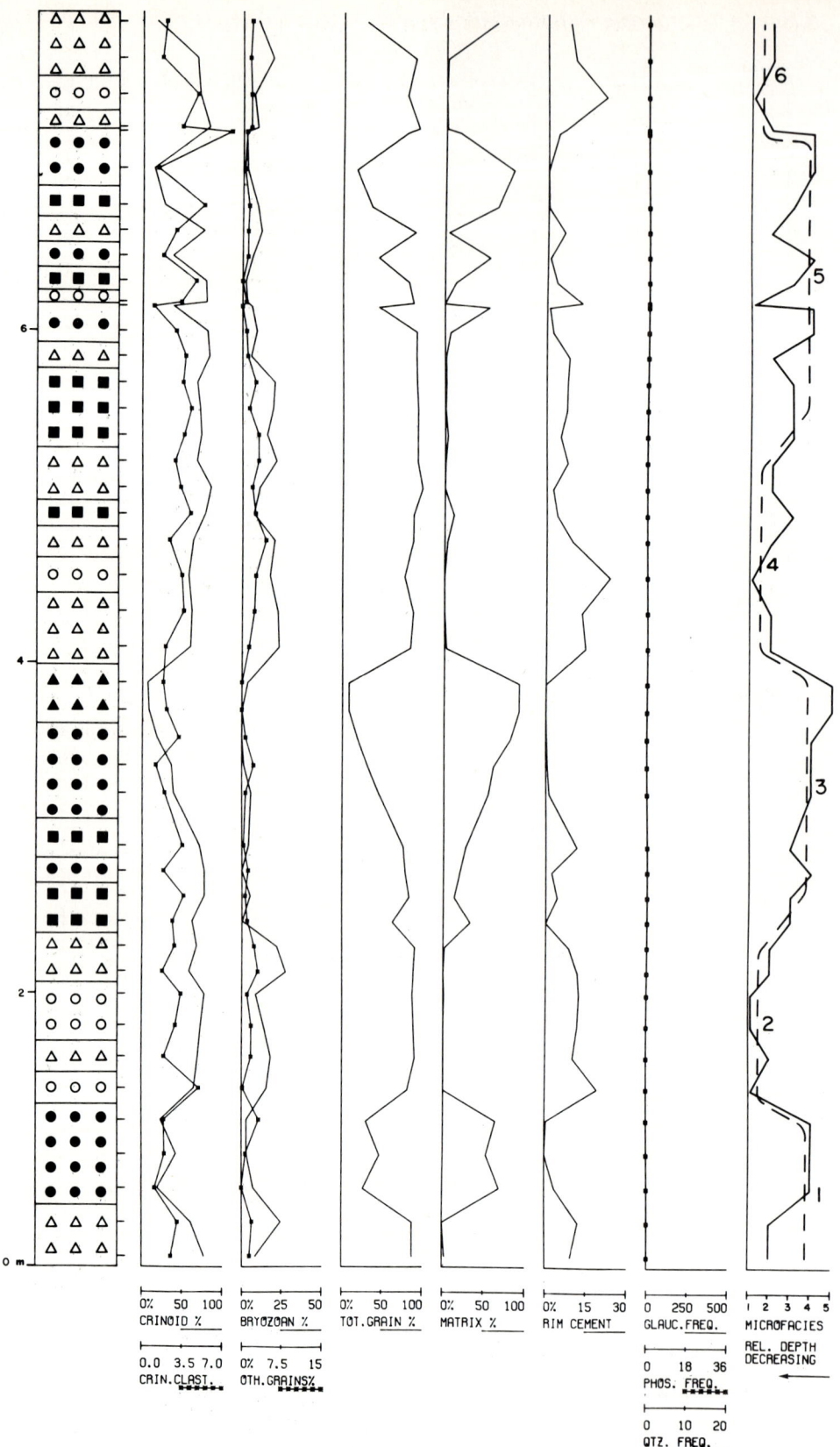

FIGURE 11.19 Burlington Limestone (Middle Mississippian), western margin of Illinois Basin. Typical example of vertical variation of microscopic parameters. From Carozzi and Gerber (1979). Reprinted by permission of the Southern Illinois University Press.

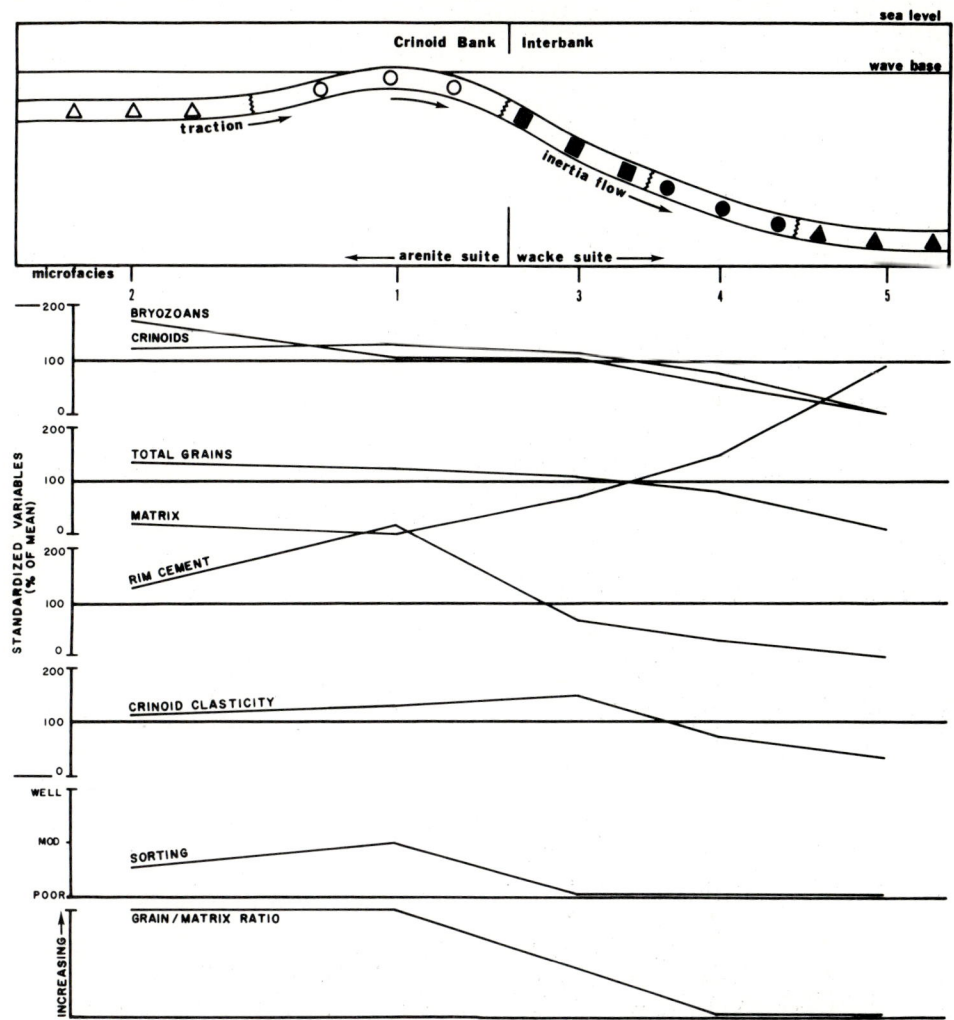

FIGURE 11.20 Burlington Limestone (Middle Mississippian), western margin of Illinois Basin. Ideal depositional model. From Carozzi and Gerber (1979). Reprinted by permission of the Southern Illinois University Press.

area. This indicates a general energy increase in time, namely a general shallowing-upward evolution. This trend is also supported by the fact that crinoid bank episodes in the upper part of the Burlington Limestone tend to consist of microfacies 1, whereas the crinoid bank episodes in the lower part of the Burlington consist primarily of microfacies 2.

Crinoidal bank and interbank episodes in the lower part of the Burlington Limestone cannot be correlated, even between closely spaced sections, which indicates that crinoidal banks were small in extent at that time and according to conodont correlation, synchronous with interbank deposits (Fig. 11.22A). During late Burlington times (Fig. 11.22B,C), vertical growth of crinoid banks increased, as did the degree of interference among their respective inertia flows until a final stage of lateral expansion of the banks was reached, which led to widespread coalescence.

Synsedimentary Chert Breccia Interpreted as a Tempestite

A section of Burlington Limestone near Hannibal, Missouri (Carozzi and Gerber, 1978), shows a bed of crinoidal calcarenite in which, over a distance of 100 ft, ellipsoidal chert nodules with a whitish patina are disturbed extensively and brecciated (Fig. 11.23A to D). This bed shows neither evidence of interruption of sedimentation nor of subaerial exposure. The only visible sedimentary structure is a general decrease in the size of chert clasts throughout the thickness of the bed and a concentration of most clasts in the lower half.

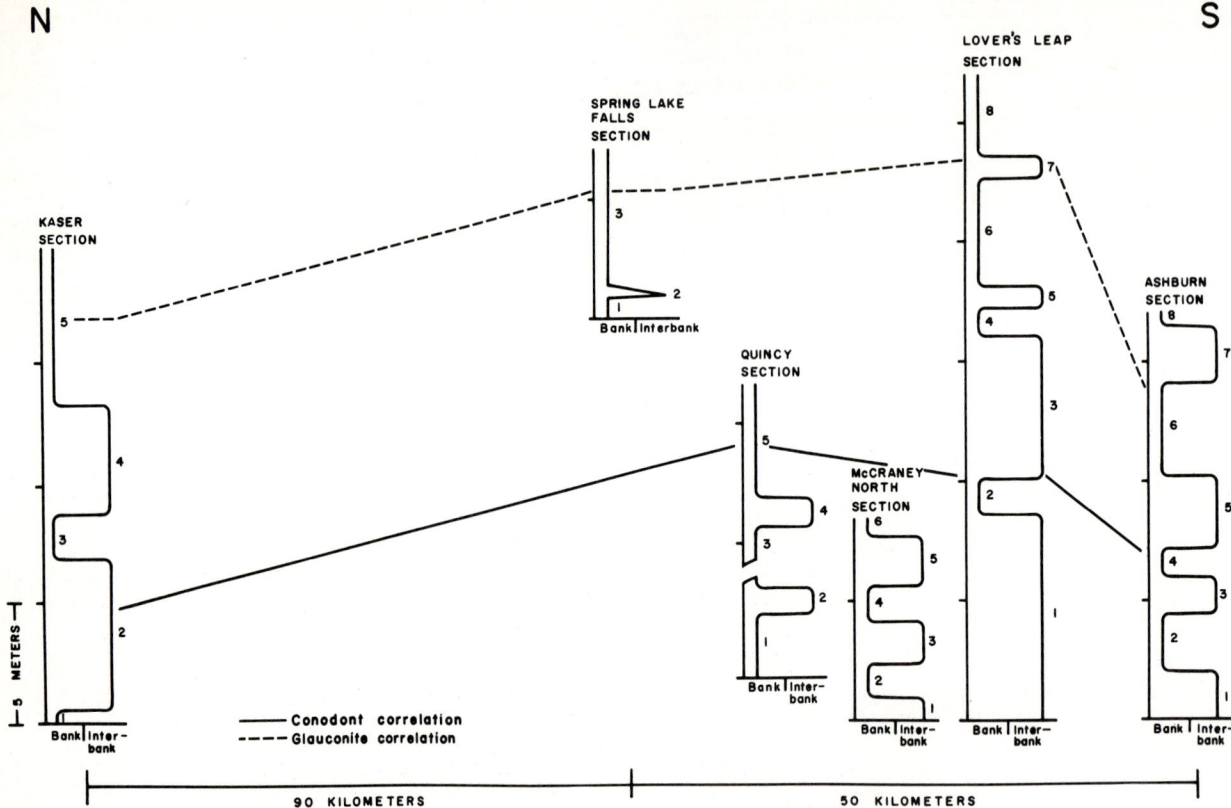

FIGURE 11.21 Burlington Limestone (Middle Mississippian), western margin of Illinois Basin. Environmental oscillations and correlation of investigated sections. From Carozzi and Gerber (1979). Reprinted by permission of the Southern Illinois University Press.

Some chert nodules were shattered *in situ* with opening of systems of fractures into which the enclosed crinoidal calcarenite was injected (Fig. 11.24A). Other nodules were broken into two parts by a major transverse fracture with subsequent rotation of the individual segments (Fig. 11.24B). Still other nodules display either truncated edges (Fig. 11.24C) or shattering, so that the nodules were reduced to numerous polyhedral fragments and slivers (Fig. 11.24A,D).

In thin section, the groundmass of the chert breccia consists of a coarse-grained, poorly sorted, crinoidal calcarenite, 70% of which are crinoidal fragments in an interstitial matrix of silt-size debris of crinoids and bryozoans. It is a variety of microfacies 3 (Fig. 11.23E) and indicates a high-energy environment of deposition. The chert fragments display the texture of a silicified calcisiltite with streaks of sand-size bioclasts of crinoids, bryozoans, and brachiopods, deposited under lesser-energy conditions and are similar to the beds underlying and overlying the chert breccia (Fig. 11.23F).

The inferred succession of events that generated this chert breccia is as follows: (1) synsedimetary generation of completely indurated chert nodules by replacement of the enclosing unconsolidated crinoidal calcisiltite (Fig. 11.25, stage 1); (2) a greater energy event which broke the nodules by reciprocal impacts and dispersed the debris a short distance (Fig. 11.25, stage 2); (3) their final deposition with crude-graded bedding within a poorly sorted crinoidal calcarenite (Fig. 11.25, stage 3); and (4) continuation of deposition of the undisturbed overlying crinoidal calcisiltite. The local nature and lack of repetition of the postulated greater energy event seems to indicate the touchdown of a tornadolike system that behaved like a funnel rather than the passage of a storm front sweeping a shallow carbonate platform.

11.3 CHARACTERISTIC CONSTITUENTS: CRINOIDS

During the Paleozoic, crinoids displayed an unusual faculty of adaptation to turbid environments of intense siliciclastic sediment yield. Isolated crinoid colonies developed microenvironments by means of baffling processes

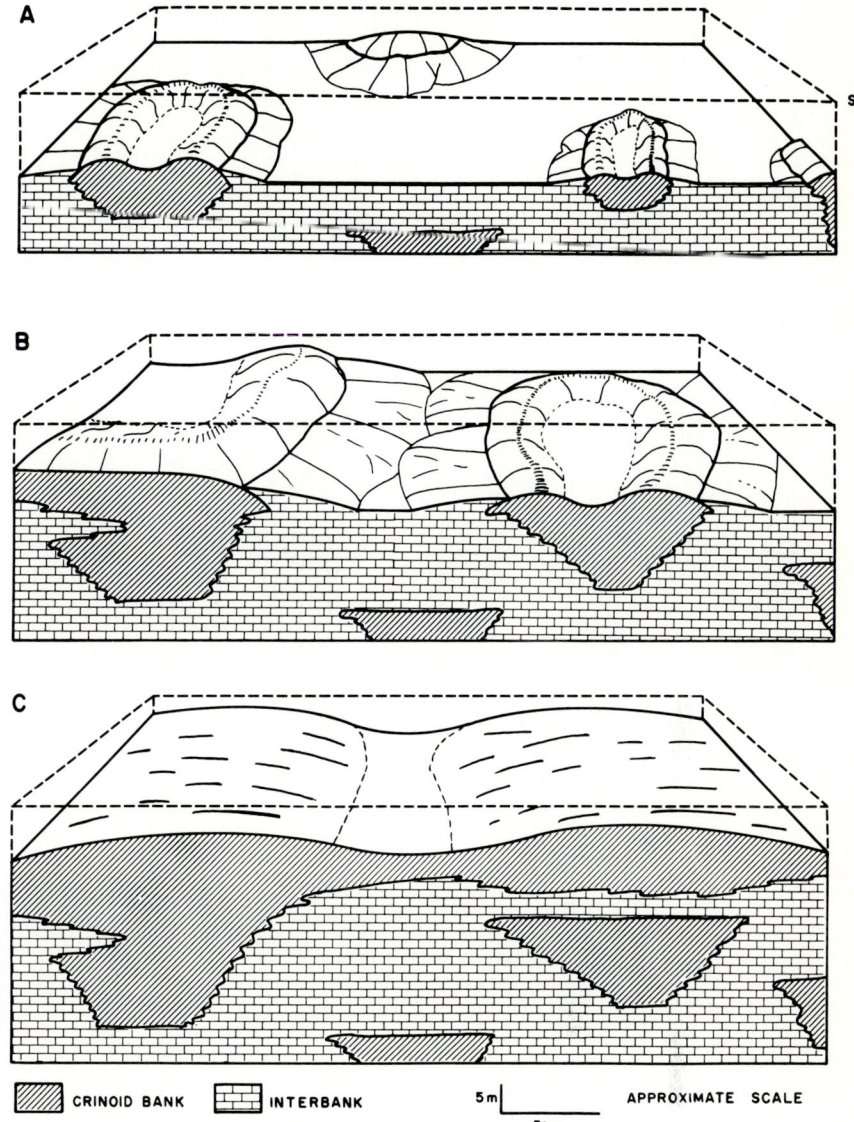

FIGURE 11.22 Burlington Limestone (Middle Mississippian), western margin of Illinois Basin. General evolution of depositional environments. See text for discussion. From Carozzi and Gerber (1979). Reprinted by permission of the Southern Illinois University Press.

that allowed them to grow on foreslopes of carbonate platforms and sometimes to reach basinal conditions.

The previously described platform of the Burlington Formation, dominated by frontal bioaccumulated to hydrodynamic crinoidal buildups, terminated abruptly eastward with a slope leading to deeper waters where the distal siltstones of contemporaneous Borden Group turbidites were deposited. Crinoidal colonies were, nevertheless, still able to survive in this turbid environment; they generated a series of tabular bioaccummulated buildups within the siltstones of the Edwardsville Formation of southern Indiana. The example investigated here (Carozzi and Soderman, 1962), located near the locality of Stobo, is 2 miles (3.22 km) long and 60 ft (18 m) thick. It was studied in seven field sections (Fig. 11.26) with a total of 235 sections at an average vertical interval of 11 in. (27.9 cm).

Description of Microfacies

Because these bioaccumulated buildups consist of numerous shallowing-upward sequences of crinoidally controlled carbonates, terminating in calcilutites and surrounded at all times by siltstones, the various microfacies are described in that order.

Microfacies 1 (Fig. 11.27A,B). Bioaccumulated crinoidal limestone, which consists mainly of complete columnals and calyx plates, only rarely broken and displaying reciprocal pressure-solution or syntaxial

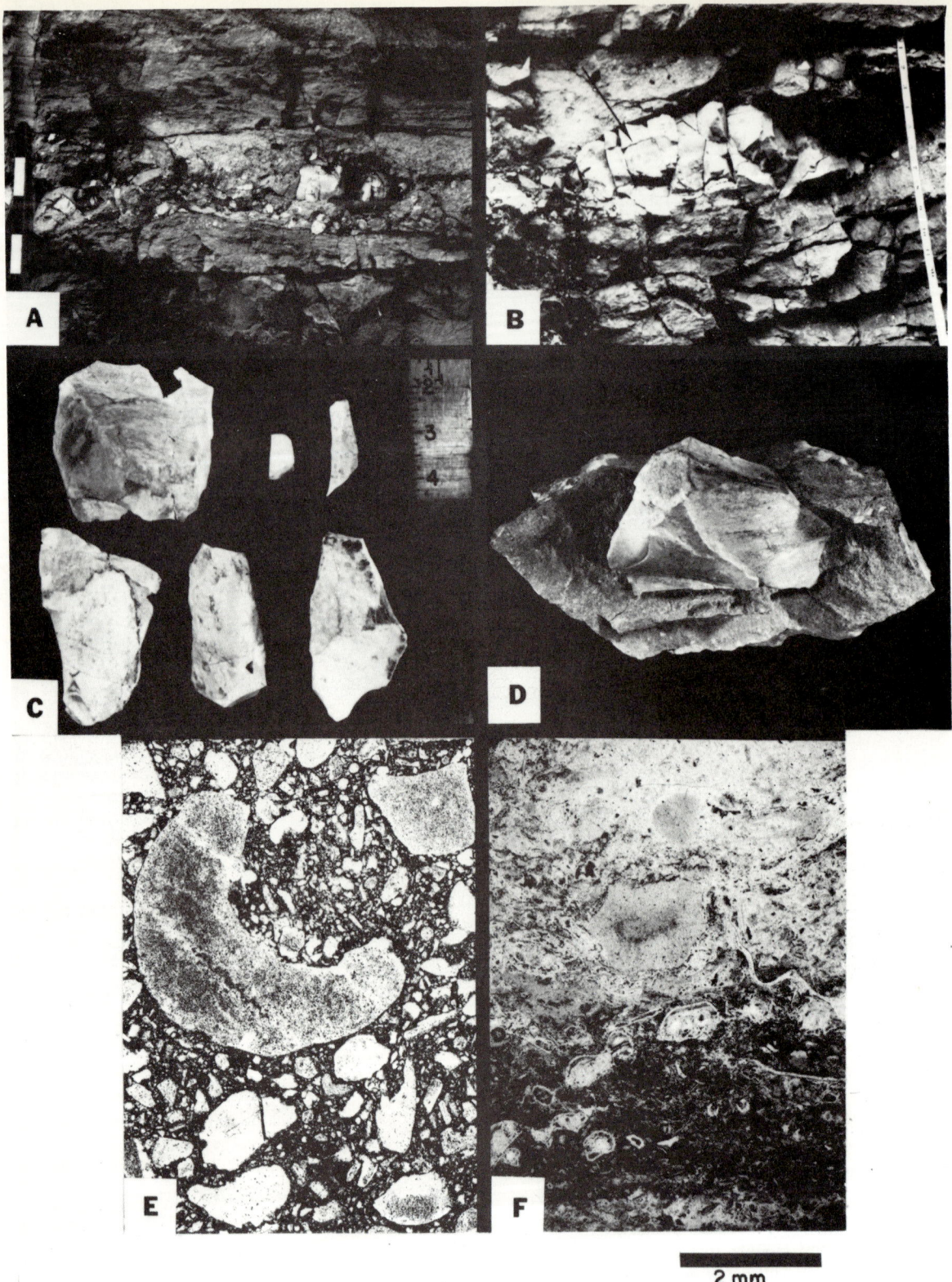

FIGURE 11.23 Burlington Limestone (Middle Mississippian), western margin of Illinois Basin. Megascopic and microscopic illustrations of tempestite chert breccia. A. General view of chert breccia. Lighter-colored chert breccia with a matrix of crinoidal calcarenite is interbedded between darker crinoidal calcisiltites. Top is even, whereas base is irregularly wavy. Concentration of angular chert clasts in lower half of bed with crude graded bedding. Scale in decimeters. B. Typical chert nodule with whitish patina (arrow) and not shattered as observed at both ends of chert breccia bed. Scale in inches. C. Examples

overgrowths. Associated bioclasts are of bryozoans, monaxonic sponge spicules, ostracods, and phosphatic fish plates (Fig. 11.27B). The remaining interstitial material consists of finer-grained crinoidal debris set in a calcilutite matrix locally neomorphosed into pseudomicrosparite.

Microfacies 2 (Fig. 11.27C). Arenaceous crinoidal biocalcarenite in which crinoid and bryozoan bioclasts are sorted incipiently and display traces of abrasion. Besides local pressure-solution, interstitial spaces are filled with fine-grained detrital quartz, iron oxide pigments, and smaller bioclastic materials.

Microfacies 3 (Fig. 11.27D). Argillaceous crinoidal biocalcarenite similar to microfacies 2 but with interstitial spaces filled by argillaceous material, iron oxide pigments, fine-grained detrital quartz, and calcilutite with minute bioclasts dominated by bryozoans.

Microfacies 4 (Fig. 11.27E). Calcilutite with scattered "floating" sand-size bioclasts of crinoids and bryozoans associated with monaxonic sponge spicules. Incipient dolomitization occurs as minute rhombs distributed at random.

Microfacies 5 (Fig. 11.27F). Calcilutite with clotted texture essentially devoid of organic debris except for rare monaxonic sponge spicules. The texture is that of "algal dust" produced by phytoplankton. Minor amounts of clay minerals detected by X-ray diffraction occur throughout the groundmass, which may locally show an incipient dolomitization by minute rhombs.

Microfacies 6 (Fig. 11.27G). Siltstone with scattered crinoidal bioclasts appreciably abraded. Abundant silt-size grains of detrital quartz are set in an argillaceous matrix that sometimes contains scattered monaxonic sponge spicules.

Microfacies 7 (Fig. 11.27H). Fine siltstone with grains of detrital quartz set in an argillaceous matrix that contains occasional large brachiopod valves and concentrations of monaxonic sponge spicules.

The Ideal Vertical Sequence

Section A (Figs. 11.28, 11.29), which is 38 ft (11.4 m) thick, is a typical example of superposition of the numerous shallowing-upward sequences consisting of crinoidal bioaccumulated limestones and overlying calcilutites and showing the detailed variations of the organic parameters during the existence of the carbonate body. The main variation trend curves (Fig. 11.30) indicate that major constituents, namely crinoids and bryozoans, show a general tendency to decrease in frequency upward until they completely disappear in the overlying siltstones. A first zone of about 13 ft with crinoids of relatively small size coincides with concentrations of fish plates and sponge spicules. Detrital quartz is completely excluded from the carbonate environment until deposition of the overlying siltstones.

An ideal vertical sequence was prepared from the investigated sections (Fig. 11.31). It shows again that crinoids and bryozoans decrease in frequency upward, indicating that conditions of development of the two major constituents were very favorable at the beginning but gradually deteriorated. The relationship between size and frequency of the crinoids shows that three distinct zones succeeded each other in time: bioaccumulation of large crinoid species (0 to 16 ft, 0 to 5 m), of small crinoid species (16 to 44 ft, 5 to 13 m), and again of large crinoid species (44 to 59 ft, 13 to 18 m). Best development of fish plates and sponge spicules coincides with the zone of small crinoid species. Peaks in the clasticity of detrital quartz also appear to be related in the same way to the zonation of size of crinoids that extends throughout the entire bioaccumulated body (Fig. 11.32).

The Ideal Depositional Model

Bioaccumulated crinoidal limestones (microfacies 1) grade laterally into crinoidal calcarenites (microfacies 2 and 3), which in turn are replaced abruptly and laterally

FIGURE 11.23 *(continued)* of variation in size and shape of chert clasts which have been extracted from enclosing crinoidal calcarenite. Scale in inches. D. Typical angular clast of chert partially embedded in crinoidal calcarenite, with conchoidal character of fracture surfaces. Scale as in C. E. Texture of groundmass of chert breccia. Coarse-grained, poorly sorted crinoidal calcarenite consisting of 70% crinoidal fragments in an interstitial matrix of silt-sized debris of crinoids and bryozoans (microfacies 3). Scale at lower right. Plane-polarized light. F. Texture of chert nodules. Silicified crinoidal calcisiltite with streaks of sand-size bioclasts of crinoids, bryozoans, and brachiopods. Observe cryptocrystalline nature of quartz in groundmass. Plane-polarized light. From Carozzi and Gerber (1978). Reproduced by permission of the Society of Economic Paleontologists and Mineralogists.

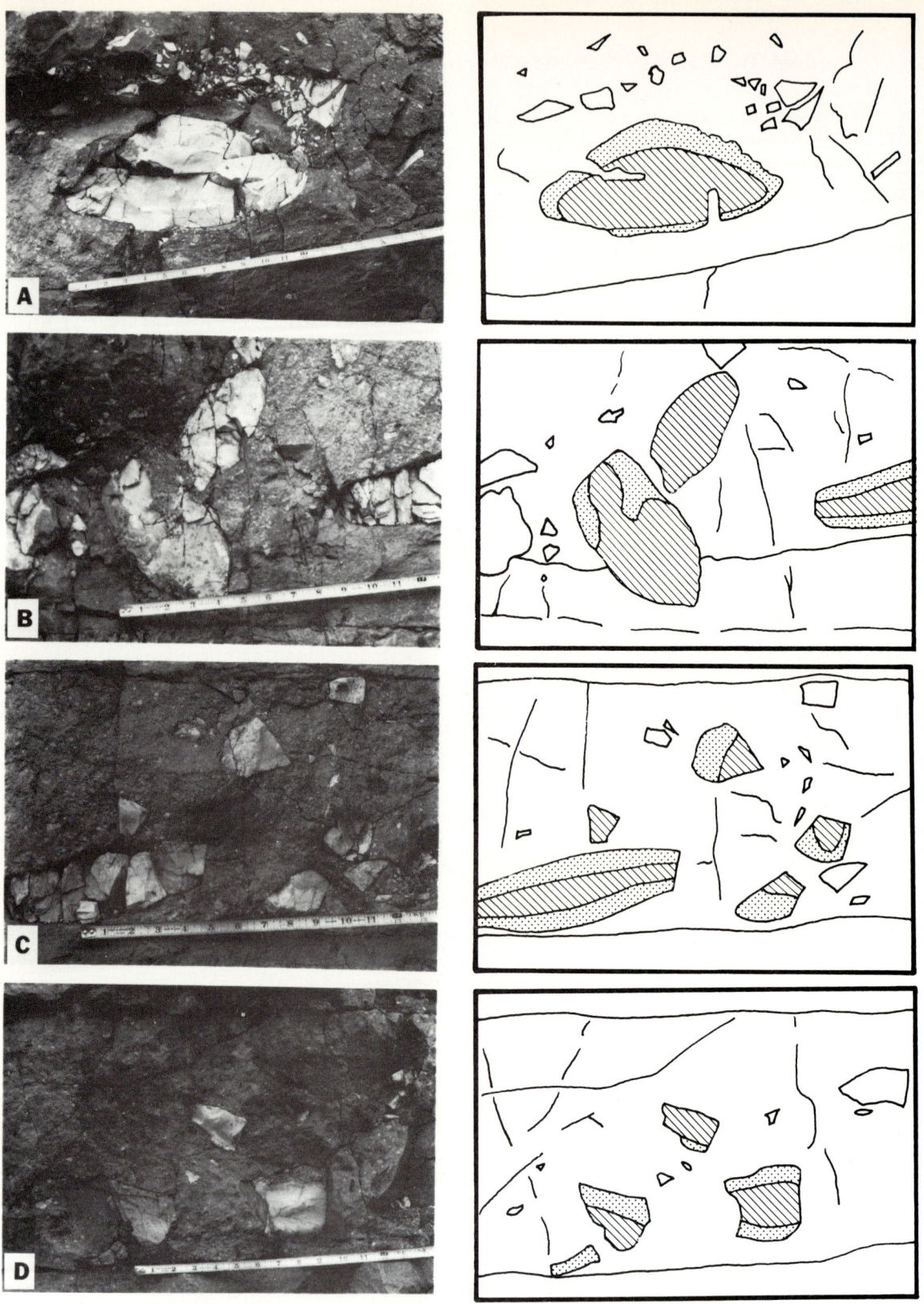

FIGURE 11.24 Burlington Limestone (Middle Mississippian), western margin of Illinois Basin. Examples of shattering of chert nodules. Central part of nodules are denoted by hachure pattern, patina is shown by a dotted pattern. Scale in inches. A. Nodule shattered *in situ* with penetration of crinoidal calcarenite in two open cracks. B. Rotated half nodules. C. Angular clasts and flat nodule with truncated edges. D. Angular clasts showing truncation of patina. From Carozzi and Gerber (1978). Reprinted by permission of the Society of Economic Paleontologists and Mineralogists.

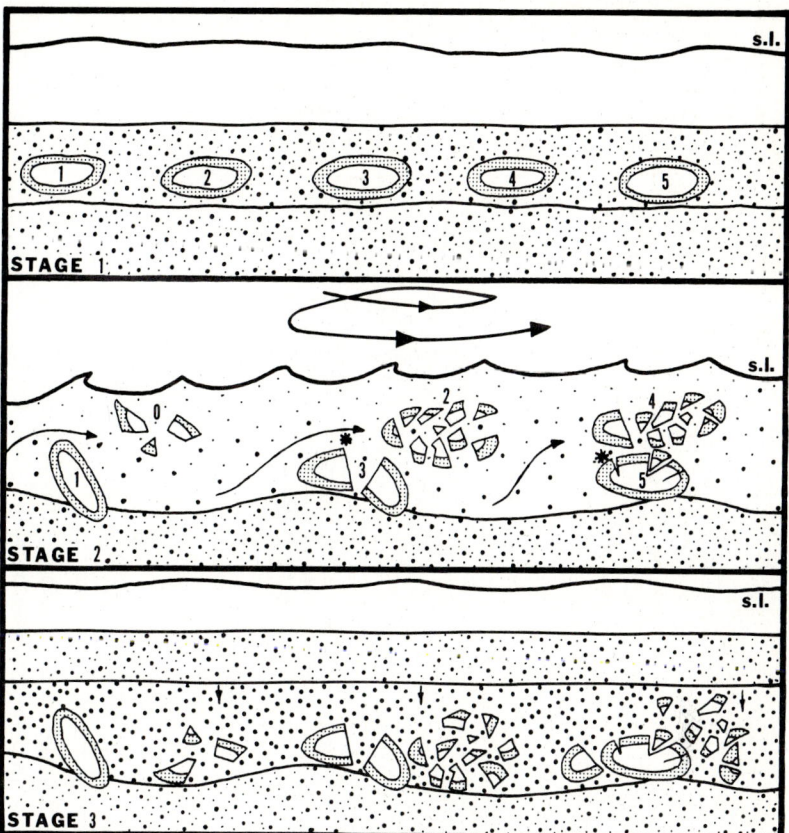

FIGURE 11.25 Burlington Limestone (Middle Mississippian), western margin of Illinois Basin. Schematic diagram illustrating process of formation of chert breccia (not to scale). Stage 1. Chert nodules with patina forming within crinoidal calcisiltite. Stage 2. Shattering of chert nodules by moderate horizontal displacement and reciprocal impact within slurry-like crinoidal calcarenite churned by high-energy tornado-like event. Stage 3. Settling of chert clasts with crude graded bedding within crinoidal calcarenite. Renewed deposition of undisturbed overlying crinoidal calcisiltite. From Carozzi and Gerber (1978). Reprinted by permission of the Society of Economic Paleontologists and Mineralogists.

by siltstones (microfacies 6 and 7). The crinoidal bioaccumulated microfacies are followed repeatedly upward by pure calcilutites (microfacies 5), sometimes preceded by the intermediate microfacies 4 (calcilutite with scattered crinoid fragments). The fine-grained carbonates are also replaced abruptly and laterally by the two types of siltstones. These relationships and the general geometry of the crinoidal buildup (Fig. 11.33) lead to the interpretation that their evolution began in a slight depression of the seafloor where crinoid colonies proliferated with a strong baffling effect which prevented major penetration of silty material. Because of the continuous production of CO_2 by metabolism of the crinoids, conditions were generated over the depression which were very favorable to the blooming of phytoplankton. Algal dust precipitated by the latter created eventually lethal conditions for the underlying crinoidal colonies. This mechanism generated a bioaccumulated crinoidal limestone overlain by a calcilutite. Upon disappearance of the baffling effect, the currents brought in clastics and made the depressions smaller while dispersing the phytoplankton. These currents introduced crinoid larvae also, which established new colonies on the remaining firm calcilutite substratum. As time progressed, the periodic influx of clastic material gradually decreased the area available for crinoid repopulation until final disappearance of the carbonate buildup by burial under the siltstones.

11.4 CHARACTERISTIC CONSTITUENTS: PELLETS AND CRINOIDS

Whenever crinoid colonies concentrate along the frontal parts of platforms, they undergo the brunt of reworking processes by the action of tides and waves. The main result is a predominant transport of bioclasts landward, generating internal rows of hydrodynamic buildups parallel to frontal bioaccumulated ones. In even more internal position, shallow lagoonal pelletoidal carbonates predominate, whereas platform slope carbonates are also fine grained and pelletoidal.

The Monte Cristo Group (Mississippian) of the Arrow Canyon Range, Clark County, southeast Nevada (Hansen and Carozzi, 1974), is 398 m thick and its petrographic investigation was based on thin sections from 799 samples with an average vertical spacing of 50 cm. This sequence of crinoid-controlled carbonates consists commonly of thick massive beds from which adequate samples were sometimes difficult to obtain. Hence the sampling distance was adapted to these local

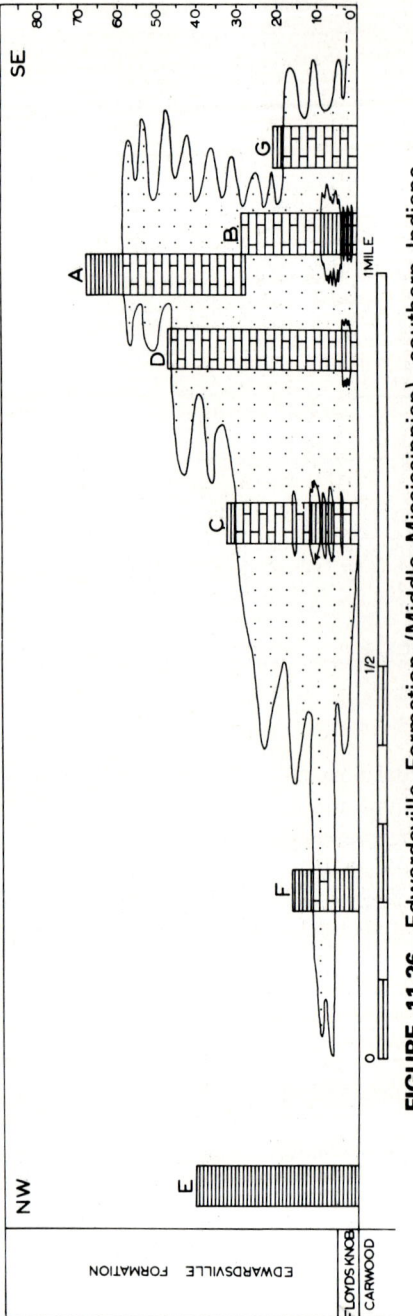

FIGURE 11.26 Edwardsville Formation (Middle Mississippian), southern Indiana. Cross section of crinoidal buildup and location of investigated sections. From Carozzi and Soderman (1962). Reprinted by permission of the Society of Economic Paleontologists and Mineralogists.

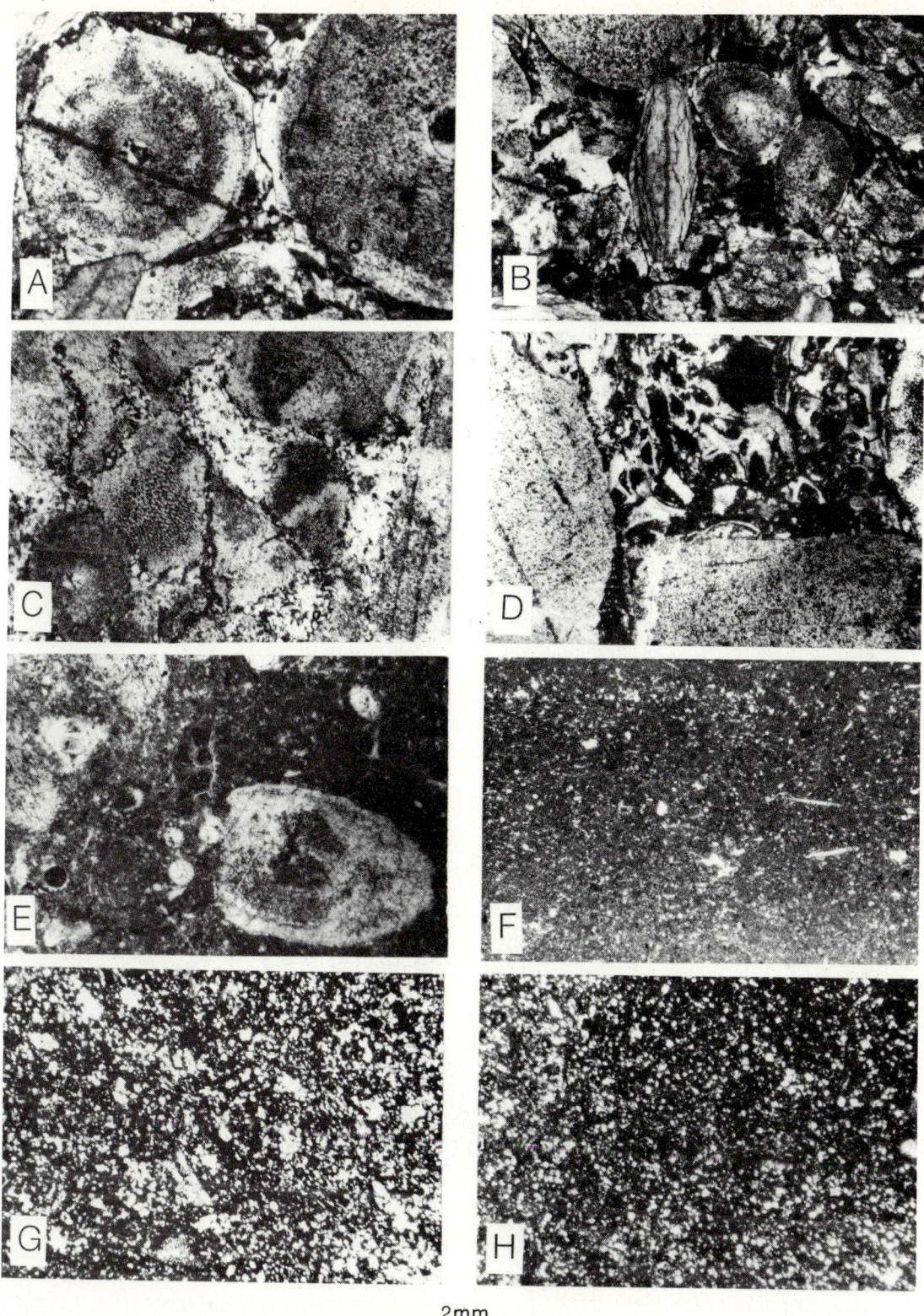

FIGURE 11.27 Edwardsville Formation (Middle Mississippian), southern Indiana. Typical microfacies. A and B. Microfacies 1. C. Microfacies 2. D. Microfacies 3. E. Microfacies 4. F. Microfacies 5. G. Microfacies 6. H. Microfacies 7. See text for detailed descriptions. All photomicrographs: plane-polarized light. From Carozzi and Soderman (1962). Reprinted by permission of the Society of Economic Paleontologists and Mineralogists.

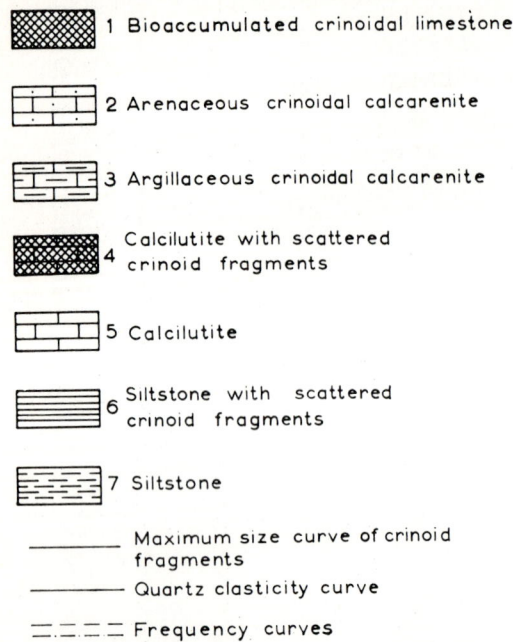

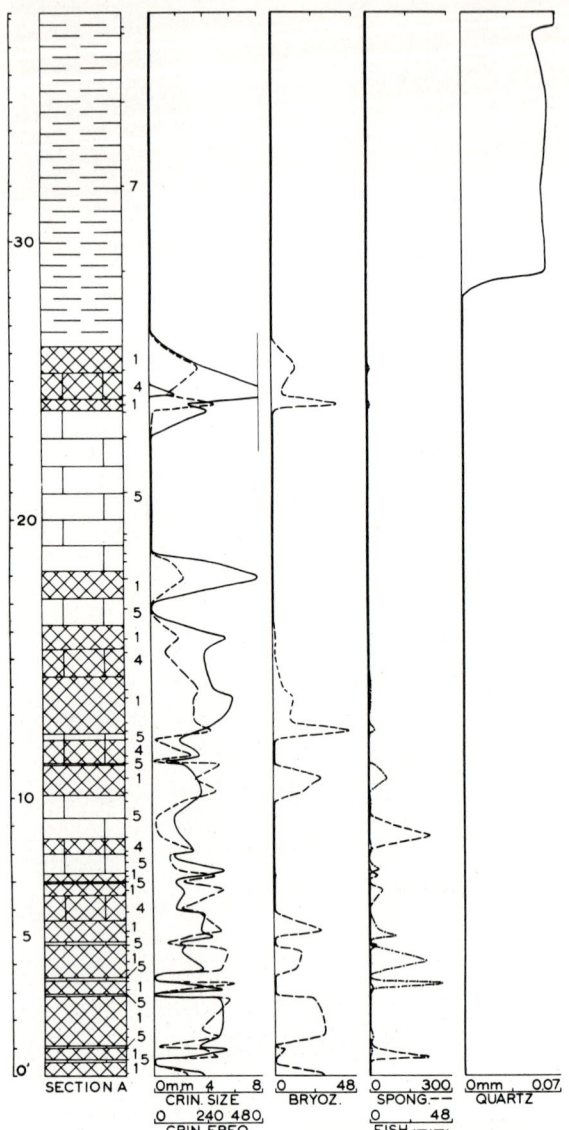

FIGURE 11.28 Edwardsville Formation (Middle Mississippian), southern Indiana. Symbols for microfacies. From Carozzi and Soderman (1962). Reprinted by permission of the Society of Economic Paleontologists and Mineralogists.

conditions, but the effect in terms of representation of conditions of sedimentation was minimal because the vertical rate of environmental variation is extremely slow.

Description of Microfacies

The entire Monte Cristo Group is a shallowing-upward megasequence which displays a total of 12 distinct microfacies divided into three suites: deeper pelletoidal, crinoidal, and shallower pelletoidal. Within each suite, microfacies are described in order of increasing percentage of sand-size bioclasts, which expresses increasing energy.

Deeper Pelletoidal Suite

Pellets are fecal in origin and associated with a significant amount of pyrite pigments, indicating lower oxidizing conditions than in the two other, shallower microfacies suites. An estimate of average original porosity of the microfacies of this suite is about 30%.

Microfacies 1 (Fig. 11.34A). Bioturbated calcisiltite with 1 to 5% of scattered sand-size components, which are mainly pellets followed in decreasing importance by calcispheres, monaxonic sponge spicules, and crinoids. Secondary dolomitization is widespread as scattered minute rhombs.

FIGURE 11.29 Edwardsville Formation (Middle Mississippian), southern Indiana. Typical example of vertical variation of microscopic parameters in section A. From Carozzi and Soderman (1962). Reprinted by permission of the Society of Economic Paleontologists and Mineralogists.

Microfacies 2 (Fig. 11.34B). Bioturbated calcisiltite with 5 to 20% scattered sand-size components which are mainly pellets, numerous monaxonic sponge spicules, calcispheres, and very few crinoid bioclasts. Secondary dolomitization is widespread as scattered minute rhombs.

Microfacies 3 (Fig. 11.34C). Bioturbated calcisiltite to mud-supported pelletoidal calcarenite with a calcisiltite to fine bioclastic matrix. Sand-size components, ranging from 20 to 40%, are predominantly pellets

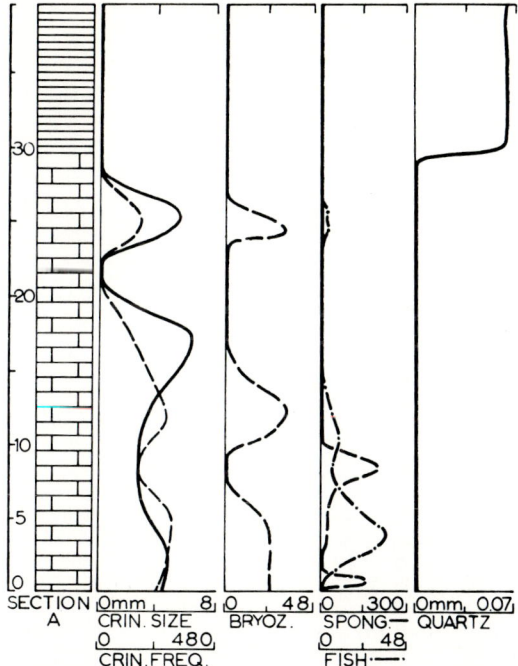

FIGURE 11.30 Edwardsville Formation (Middle Mississippian), southern Indiana. Main trend curves of microscopic parameters in section A. All the carbonate microfacies (1 to 5) are grouped with standard limestone symbol. From Carozzi and Soderman (1962). Reprinted by permission of the Society of Economic Paleontologists and Mineralogists.

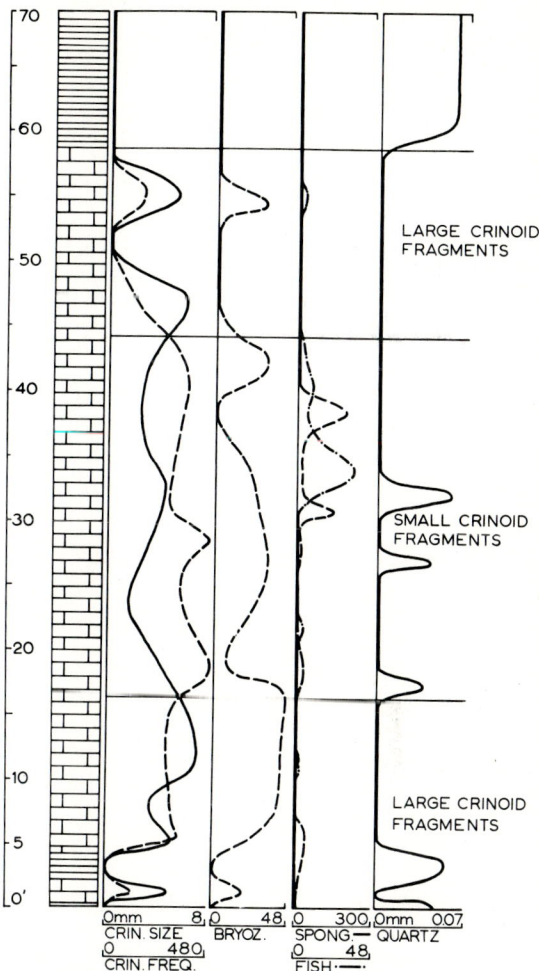

FIGURE 11.31 Edwardsville Formation (Middle Mississippian), southern Indiana. Ideal vertical sequence. From Carozzi and Soderman (1962). Reprinted by permission of the Society of Economic Paleontologists and Mineralogists.

associated with crinoid and brachiopod bioclasts. Minor constituents are ostracods, calcispheres, smaller benthic foraminifers, and monaxonic sponge spicules. Secondary dolomitization is frequent as scattered minute rhombs.

Microfacies 4 (Fig. 11.34D). Grain-supported pelletoidal calcarenite with sparite cement and more than 40% of sand-size constituents. Besides predominant pellets, smaller crinoid and brachiopod bioclasts occur. Ostracods, calcispheres, and smaller benthic foraminifers occur in minor amounts.

Crinoidal Suite

The dominant crinoids are associated with abundant bryozoans and some noteworthy concentrations of very large intact brachiopods. Sorting of these microfacies can be excellent locally, but estimated original porosity averages only 15%.

Microfacies 5 (Fig. 11.34E). Calcisiltite with 1 to 5% scattered sand-size bioclasts of crinoids, bryozoans, brachiopod shells, and spines. Minor organic constituents include monaxonic sponge spicules and ostracods. Secondary dolomitization is weak as scattered rhombs.

Microfacies 6 (Fig. 11.34F). Calcisiltite with 5 to 20% scattered sand-size bioclasts of crinoids, bryozoans, and brachiopods. Other constituents include some calcispheres, monaxonic sponge spicules, and pelecypods. Secondary dolomitization is minor as scattered rhombs.

Microfacies 7 (Fig. 11.34G). Mud-supported crinoidal biocalcarenite with 20 to 40% sand-size bioclasts of crinoids, bryozoans, brachiopods, and coral septas. The groundmass consists of an association of calcisiltite matrix and silt-size to fine sand-size crinoid bioclasts.

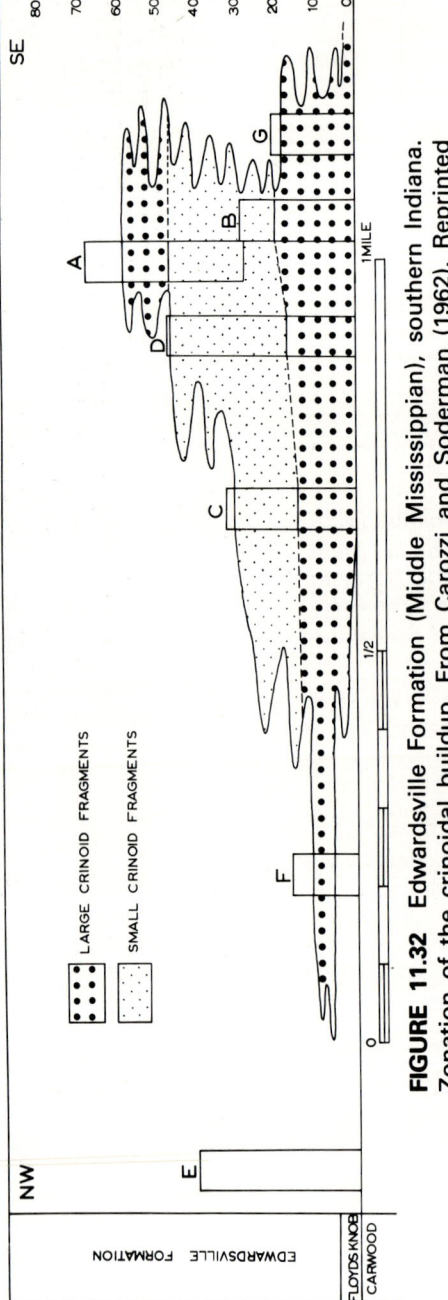

FIGURE 11.32 Edwardsville Formation (Middle Mississippian), southern Indiana. Zonation of the crinoidal buildup. From Carozzi and Soderman (1962). Reprinted by permission of the Society of Economic Paleontologists and Mineralogists.

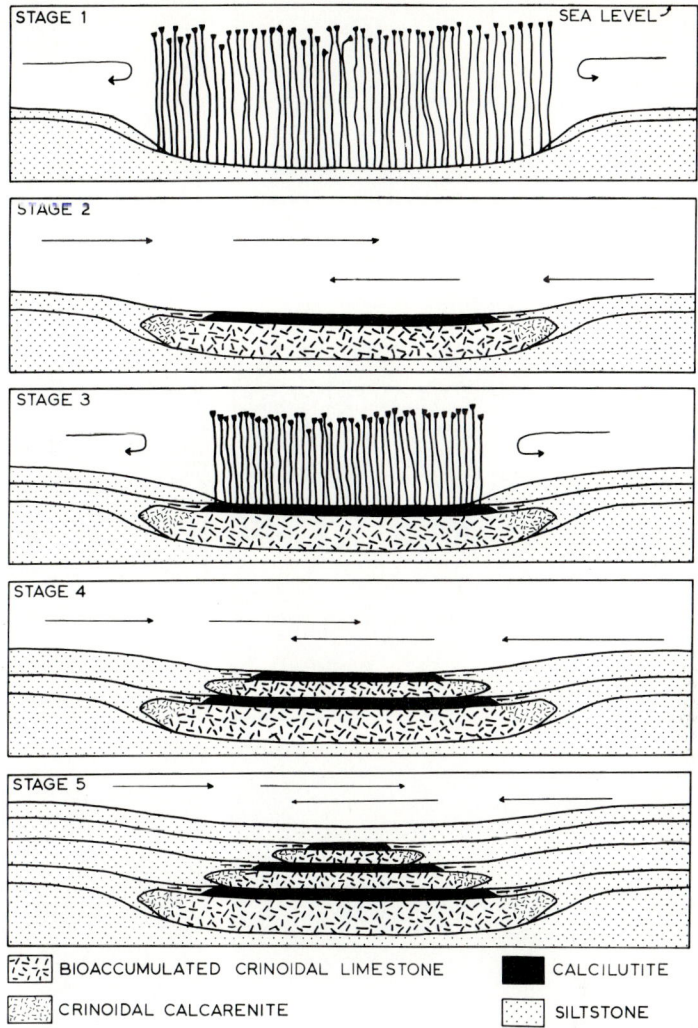

FIGURE 11.33 Edwardsville Formation (Middle Mississippian), southern Indiana. General environmental evolution of the crinoidal buildup. From Carozzi and Soderman (1962). Reprinted by permission of the Society of Economic Paleontologists and Mineralogists.

Microfacies 8 (Figs. 11.34H, 11.35A). Grain-supported crinoidal biocalcarenite with more than 40% sand-size bioclasts of crinoids, bryozoans, brachiopods, corals, and a few pellets. Interstitial material is crinoidal bioclastic associated with syntaxial overgrowths around the larger crinoid bioclasts. Some intraparticle sparite cement occurs mainly in the central canal of large ossicles. Other constituents are large spiriferid and productid brachiopods as well as tabulate corals (*Syringopora*), ostracods, and smaller benthic foraminifers (Fig. 11.35A).

This microfacies and the remaining ones of the crinoidal suite show tectonically induced multiple twinning lamellae in most crinoid bioclasts.

Microfacies 9 (Fig. 11.35B,C). Grain-supported crinoidal biocalcarenite with more than 40% sand-size bioclasts of crinoids, bryozoans, and brachiopods cemented by syntaxial overgrowths around the crinoid bioclasts. Central canals of the latter are filled generally with sparite cement. Other constituents are smaller benthic foraminifers, conodonts, and some large spiriferid brachiopods (Fig. 11.35C).

Microfacies 10 (Fig. 11.35D,E). Grain-supported and pressure-welded crinoidal biocalcarenite with more than 40% sand-size bioclasts of crinoids, bryozoans, and brachiopods. In spite of predominant pressure-welding, small patches of syntaxial overgrowth and of bioclastic matrix remain (Fig. 11.35E). Accessory constituents are concentrations of large spiriferid brachiopods.

Shallower Pelletoidal Suite

As in the deeper pelletoidal suite, pellets are fecal in origin, smaller in size (average 0.04 mm versus 0.3 mm), lighter in color, and accompanied by stronger bioturbation.

Microfacies 11 (Fig. 11.35F). Calcisiltite to mud-supported pelletoidal biocalcarenite with 1 to 40% of sand-size constituents which are mainly pellets fol-

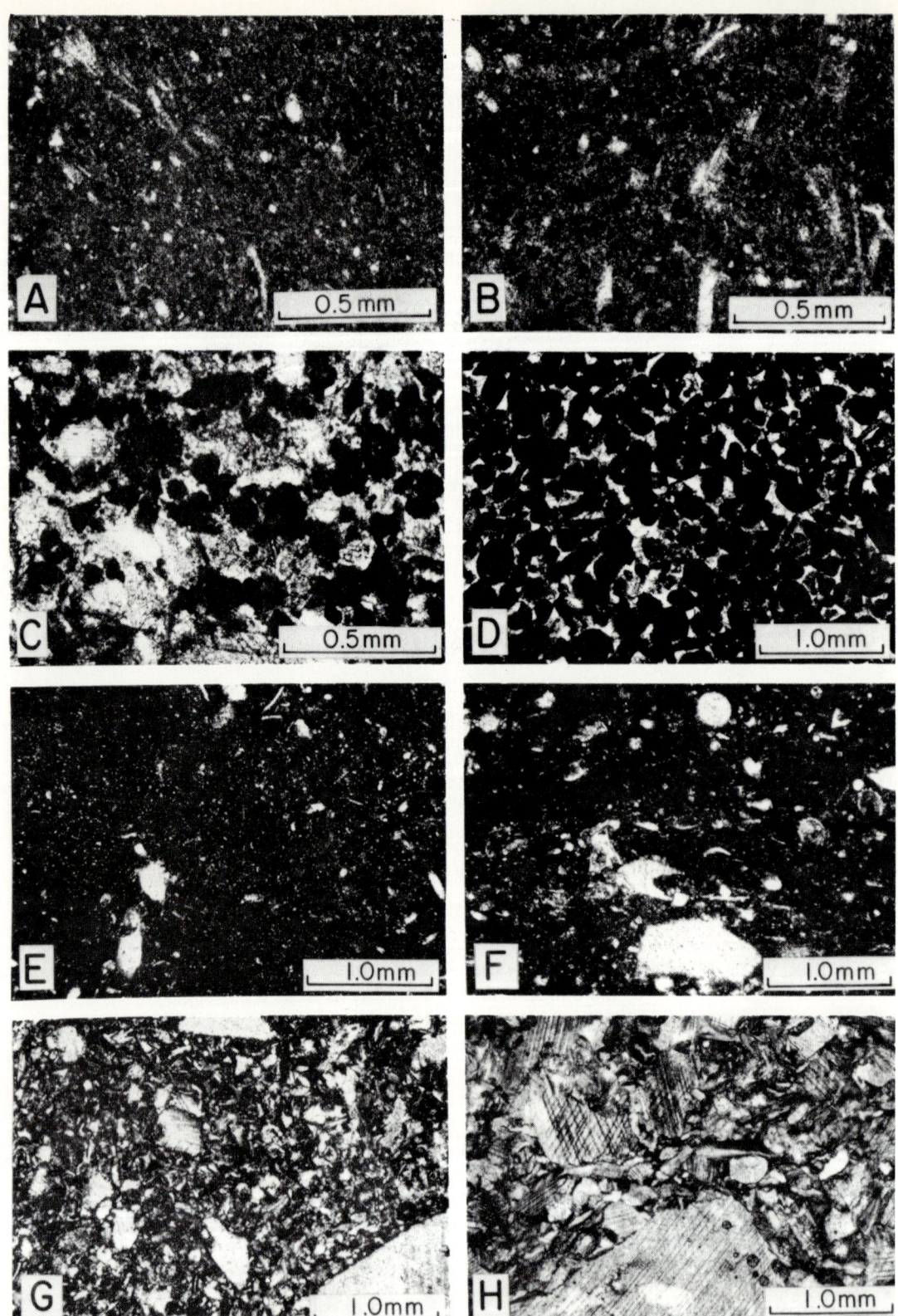

FIGURE 11.34 Monte Cristo Group (Mississippian), Arrow Canyon Range, Clark County, Nevada. Typical microfacies. A. Microfacies 1. B. Microfacies 2. C. Microfacies 3. D. Microfacies 4. E. Microfacies 5. F. Microfacies 6. G. Microfacies 7. H. Microfacies 8. See text for detailed descriptions. All photomicrographs: plane-polarized light. From Hansen and Carozzi (1974). Courtesy of the Wyoming Geological Association.

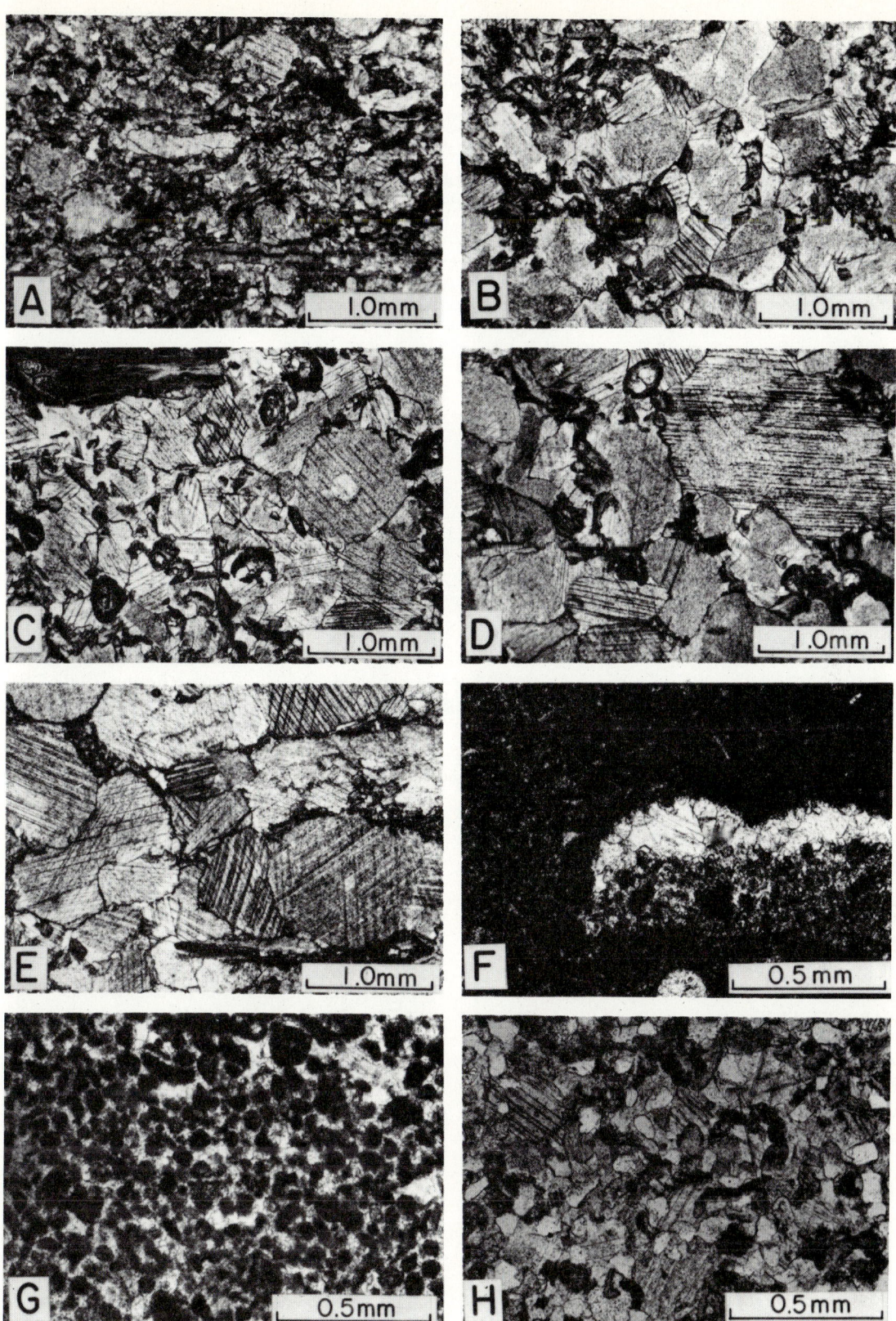

FIGURE 11.35 Monte Cristo Group (Mississippian), Arrow Canyon Range, Clark County, Nevada. Typical microfacies (continued). A. Microfacies 8. B and C. Microfacies 9. D and E. Microfacies 10. F. Microfacies 11. G and H. Microfacies 12. See text for detailed descriptions. All photomicrographs: plane-polarized light. From Hansen and Carozzi (1974). Courtesy of the Wyoming Geological Association.

244 Part III / Carbonate Platforms

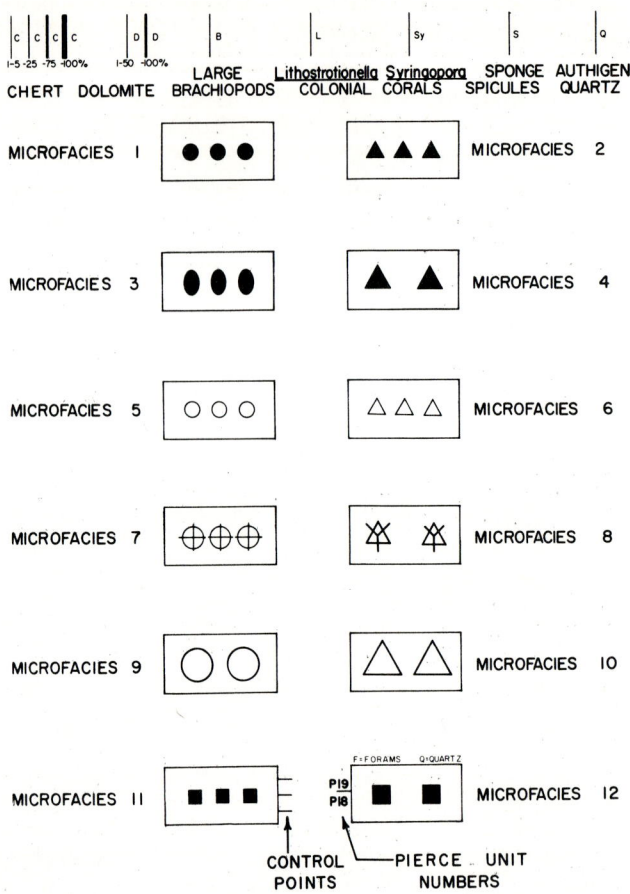

FIGURE 11.36 Monte Cristo Group (Mississippian), Arrow Canyon Range, Clark County, Nevada. Symbols for microfacies. From Hansen and Carozzi (1974). Courtesy of the Wyoming Geological Association.

lowed in decreasing importance by ostracods, bioclasts of crinoids, brachiopods, coral septas, calcispheres, smaller benthic foraminifers, and monaxonic sponge spicules. Intense bioturbation is shown by well-preserved burrows which are associated with common fenestral textures displaying geopetal pelletoidal fillings overlain by sparite cement. These textures appear related to very thin stromatolite-like laminae.

Microfacies 12 (Fig. 11.35G,H). Grain-supported pelletoidal biocalcarenite with more than 40% sand-size constituents set in sparite cement. Predominant pellets are associated with abundant arenaceous and endothyrid foraminifers, bioclasts of crinoids and brachiopods, together with ostracods and debris of rugose corals (*Lithostrotionella*). Detrital quartz grains are scattered in this microfacies and whenever it is highly arenaceous (Fig. 11.35H), the designation 12+Q is used in graphic representation.

The Ideal Shallowing-Upward Sequences

Each of the suites of microfacies just described, namely, deeper pelletoidal, crinoidal, and shallower pelletoidal, are shallowing-upward sequences, but because they consist of a small number of microfacies (4, 6, and 2, respectively) which are each relatively thick, the amplitude of the shallowing process remains small. These conditions are well displayed in typical sections with detailed microscopic parameter variation of the deeper pelletoidal suite (Figs. 11.36, 11.37), the crinoidal suite (Fig. 11.38), and the shallower pelletoidal suite (Fig. 11.39). As mentioned previously, the stratigraphic superposition of the three suites, which forms the entire Monte Cristo Group, is itself a shallowing-upward megasequence.

The Ideal Depositional Model

Field relationships and statistical behavior of microscopic parameters permit development of microfacies filiation (Fig. 11.40), which in turn leads to the preparation of the ideal depositional model illustrated first as a plan view (Fig. 11.41). The principal environment is a hydrodynamic crinoidal buildup or shoal consisting of an external portion where the crinoid colonies lived (microfacies 9) and an internal portion of accumulation of crinoidal bioclasts (microfacies 10). Microfacies 8 surrounding the buildup is characterized by brachiopods, bryozoans, and crinoid debris. A subtidal apron of mud-supported crinoidal biocalcarenites (microfacies 7) to crinoidal cal-

Chap. 11 / Carbonate Platforms with Frontal Bioaccumulated to Hydrodynamic Buildups 245

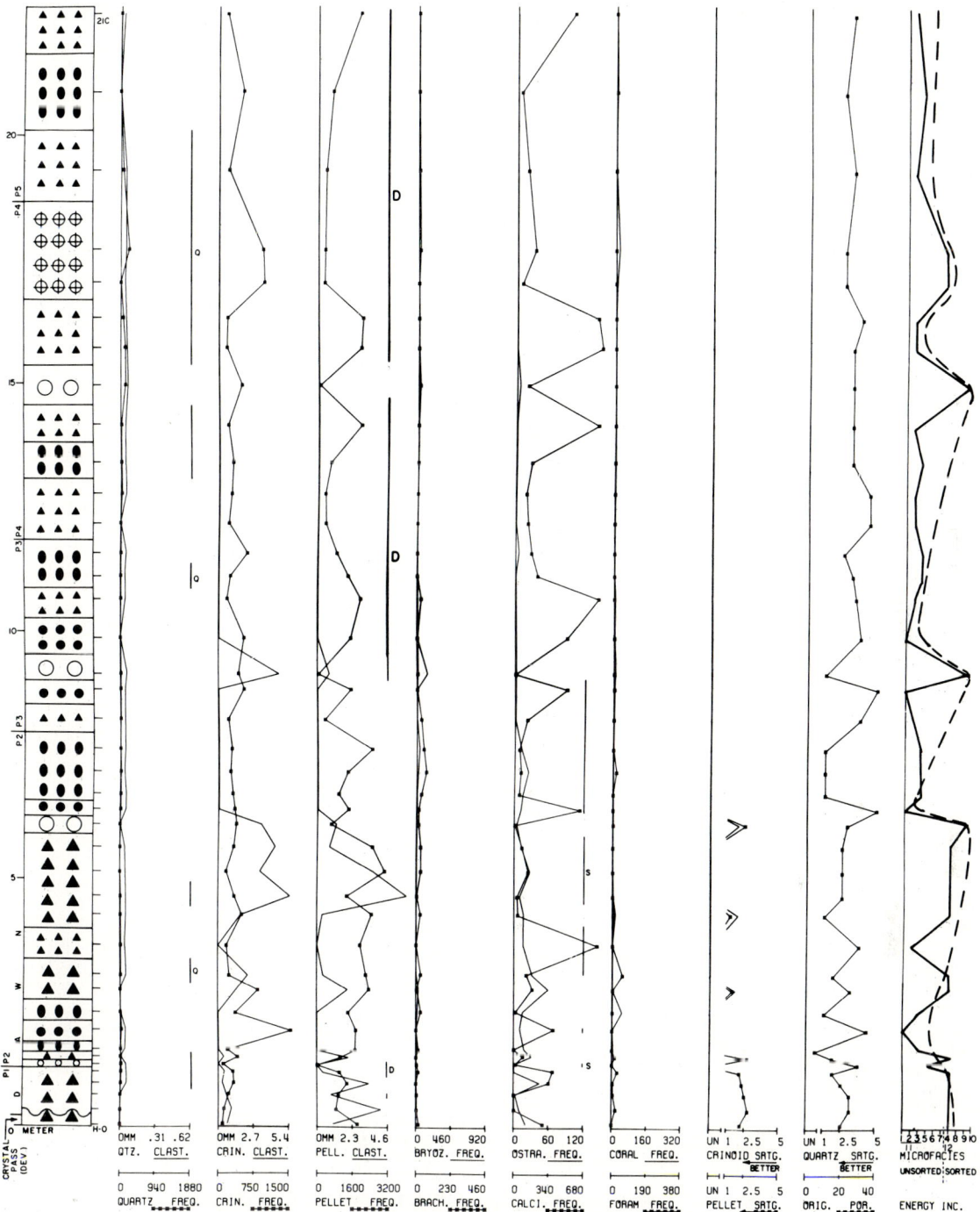

FIGURE 11.37 Monte Cristo Group (Mississippian), Arrow Canyon Range, Clark County, Nevada. Typical example of vertical variation of microscopic parameters of deeper pelletoidal suite. From Hansen and Carozzi (1974). Courtesy of the Wyoming Geological Association.

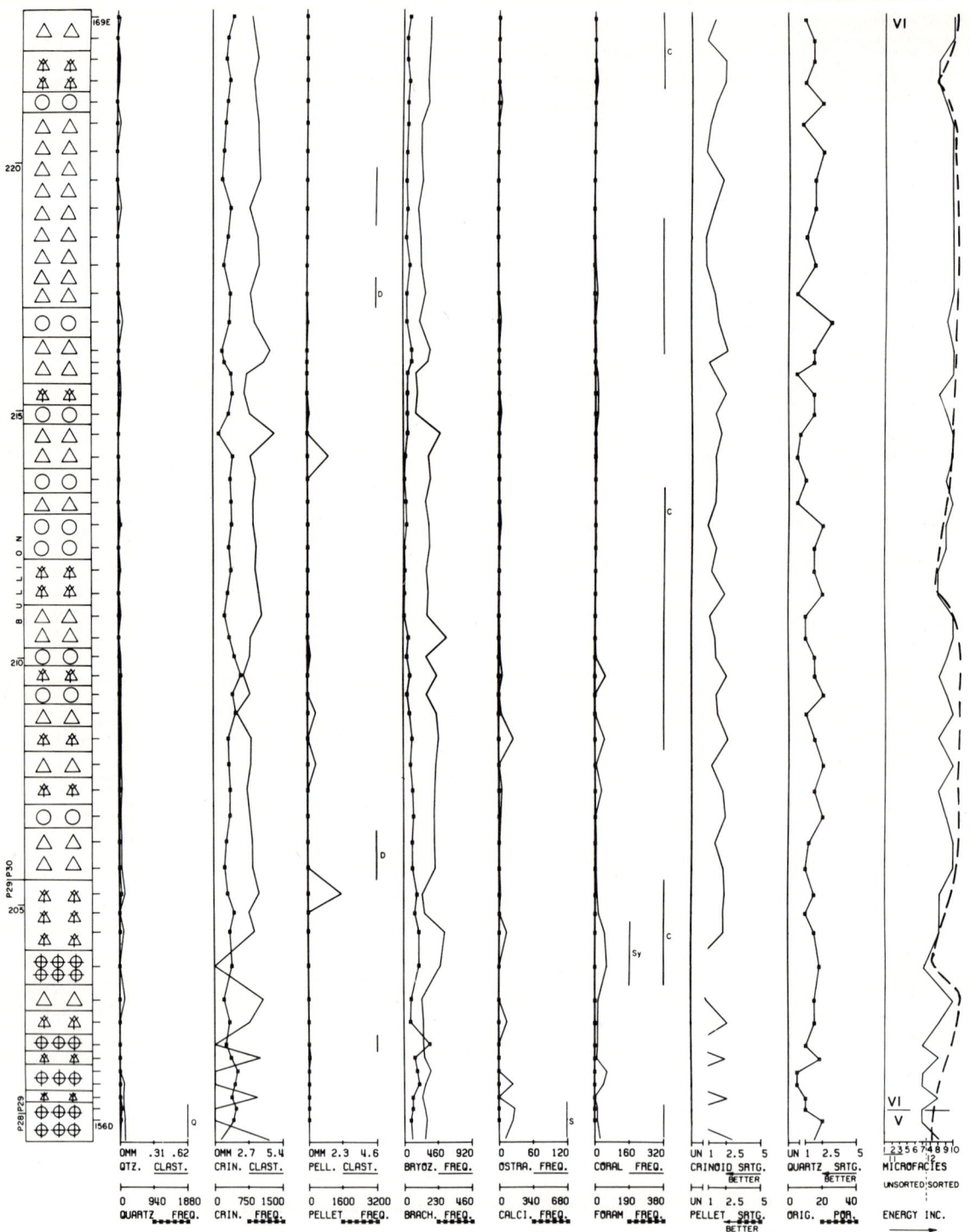

FIGURE 11.38 Monte Cristo Group (Mississippian), Arrow Canyon Range, Clark County, Nevada. Typical example of vertical variation of microscopic parameters of crinoidal suite. From Hansen and Carozzi (1974). Courtesy of the Wyoming Geological Association.

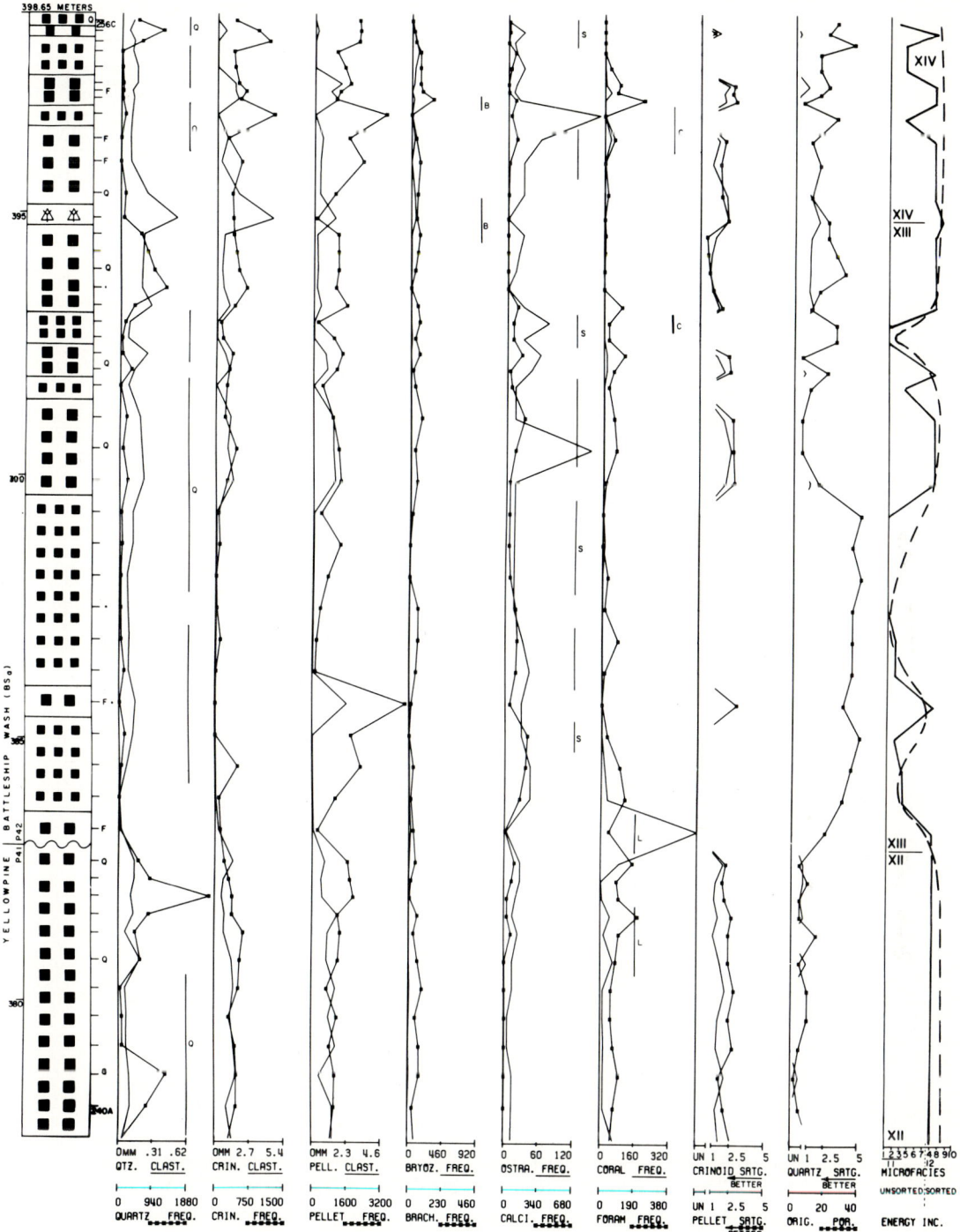

FIGURE 11.39 Monte Cristo Group (Mississippian), Arrow Canyon Range, Clark County, Nevada. Typical example of vertical variation of microscopic parameters of shallower pelletoidal suite. From Hansen and Carozzi (1974). Courtesy of the Wyoming Geological Association.

248 Part III / Carbonate Platforms

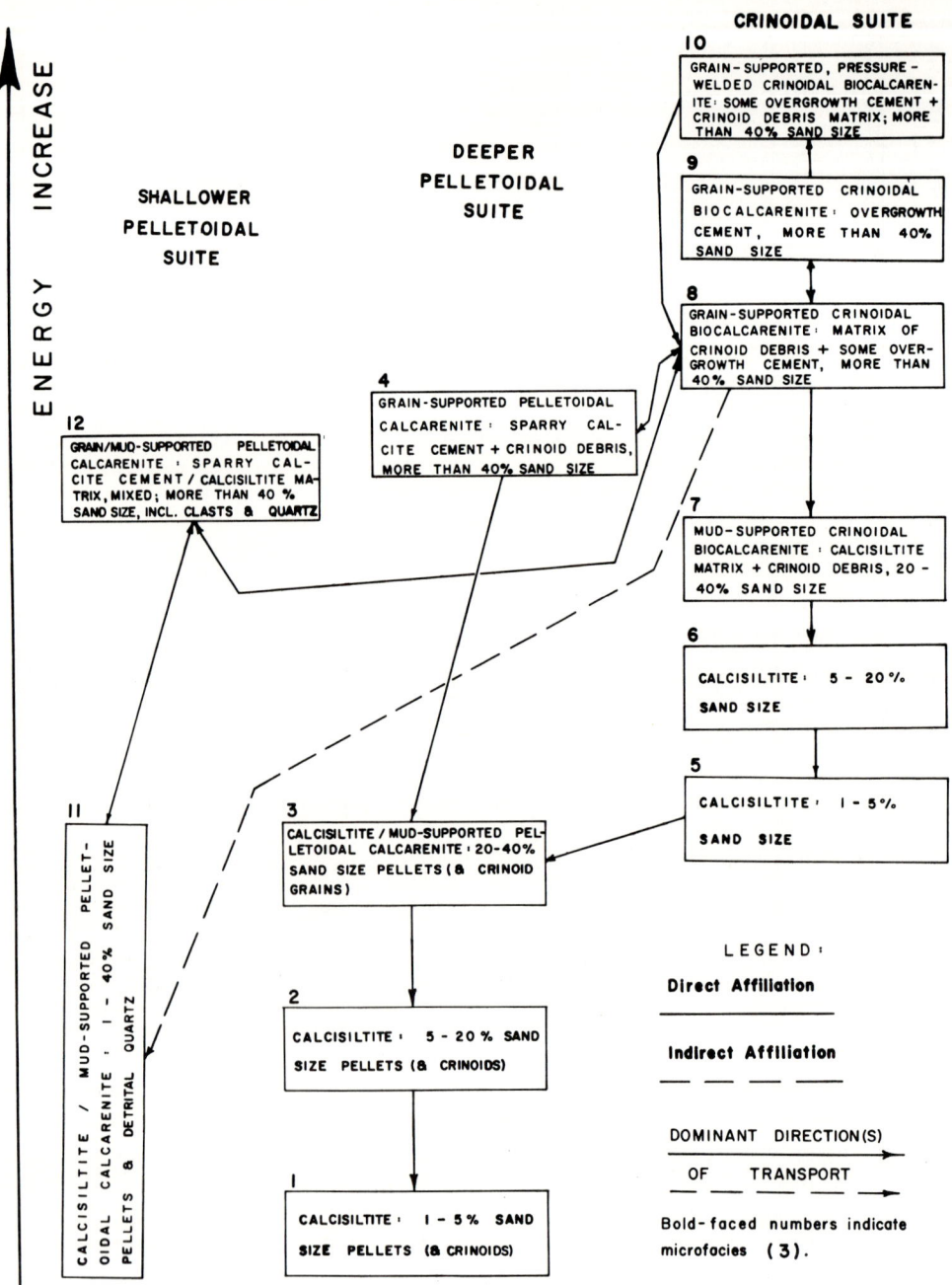

FIGURE 11.40 Monte Cristo Group (Mississippian), Arrow Canyon Range, Clark County, Nevada. Relationships of microfacies. From Hansen and Carozzi (1974). Courtesy of the Wyoming Geological Association.

cisiltites (microfacies 6 and 5) extends seaward and eventually merges with the deeper pelletoidal suite.

The hydrodynamic crinoidal buildup is considered to have been discontinuous, and this situation is shown by a subtidal channel (microfacies 4) which merges seaward with the deeper pelletoidal calcisiltites (microfacies 3, 2, and 1), and landward with microfacies 8. Another hydrodynamic buildup extends behind the crinoidal one and consists of pelletoidal biocalcarenites with abundant crinoids, brachiopods, and endothyrids stabilized by colonies of rugose corals, mainly *Lithostrotionella* (microfacies 12). This narrow bank, partially deflected lagoonward as a tidal delta when intersected by the channel, appears to have surrounded the lagoonal environment (microfacies 11), although becoming more arenaceous landward (microfacies 12+Q).

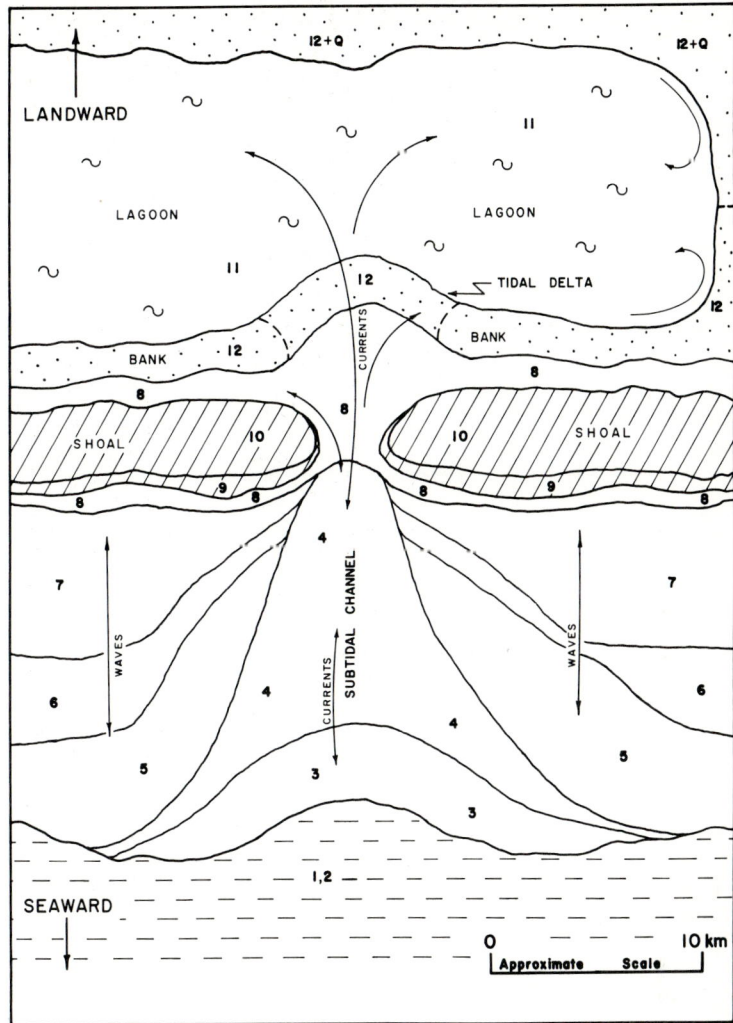

FIGURE 11.41 Monte Cristo Group (Mississippian), Arrow Canyon Range, Clark County, Nevada. Schematic plan view of ideal depositional model. From Hansen and Carozzi (1974). Courtesy of the Wyoming Geological Association.

Variations of microscopic parameters in a transverse cross section (Fig. 11.42) reveal the characteristic ecological associations of the three suites of microfacies, and the strong tidal current influence they underwent. This influence is expressed by dispersal of major bioclasts seaward, landward, or both, from their main places of original occurrence, such is the shoal with maximum development of crinoids, bryozoans, and brachiopods. Sponge spicules appear transported from shallower to deeper finer microfacies. Although fecal pellets in deeper water appear reworked from their environment into the subtidal channel, those generated in the shallower lagoon were transported seaward. Detrital quartz during its seaward distribution was trapped mainly behind the bank, and only a few larger grains mark the position of the shoal and of the subtidal channel. Calcispheres were transported seaward from the lagoon. However, ostracods, many articulated, show two distinct communities, one in front of the crinoidal shoal, the other in the lagoon. The latter community has undergone a certain amount of transportation seaward. Coral bioclasts display reduced dispersion away from their areas of development. These are sites of colonization on the frontal part of the crinoidal shoal by *Syringopora* and on the more internal pelletoidal-foraminiferal bank by *Lithostrotionella*. The smaller benthic foraminifers (endothyrids) are distributed in a manner similar to corals. They display dispersion from the bank, which might have been their original habitat, in a seaward direction. Arenaceous foraminifers are almost entirely lagoonal.

General Evolution of the Depositional Environments

A composite section of the entire Monte Cristo Group (Fig. 11.43) shows that it represents a complex shallowing-upward evolution. It consists of the following three sequences: first deepening (phases I and II) and shallowing sequence (phases III, IV, and V), second deepening (phase VI) and shallowing sequence (phases VII, VIII,

250 Part III / Carbonate Platforms

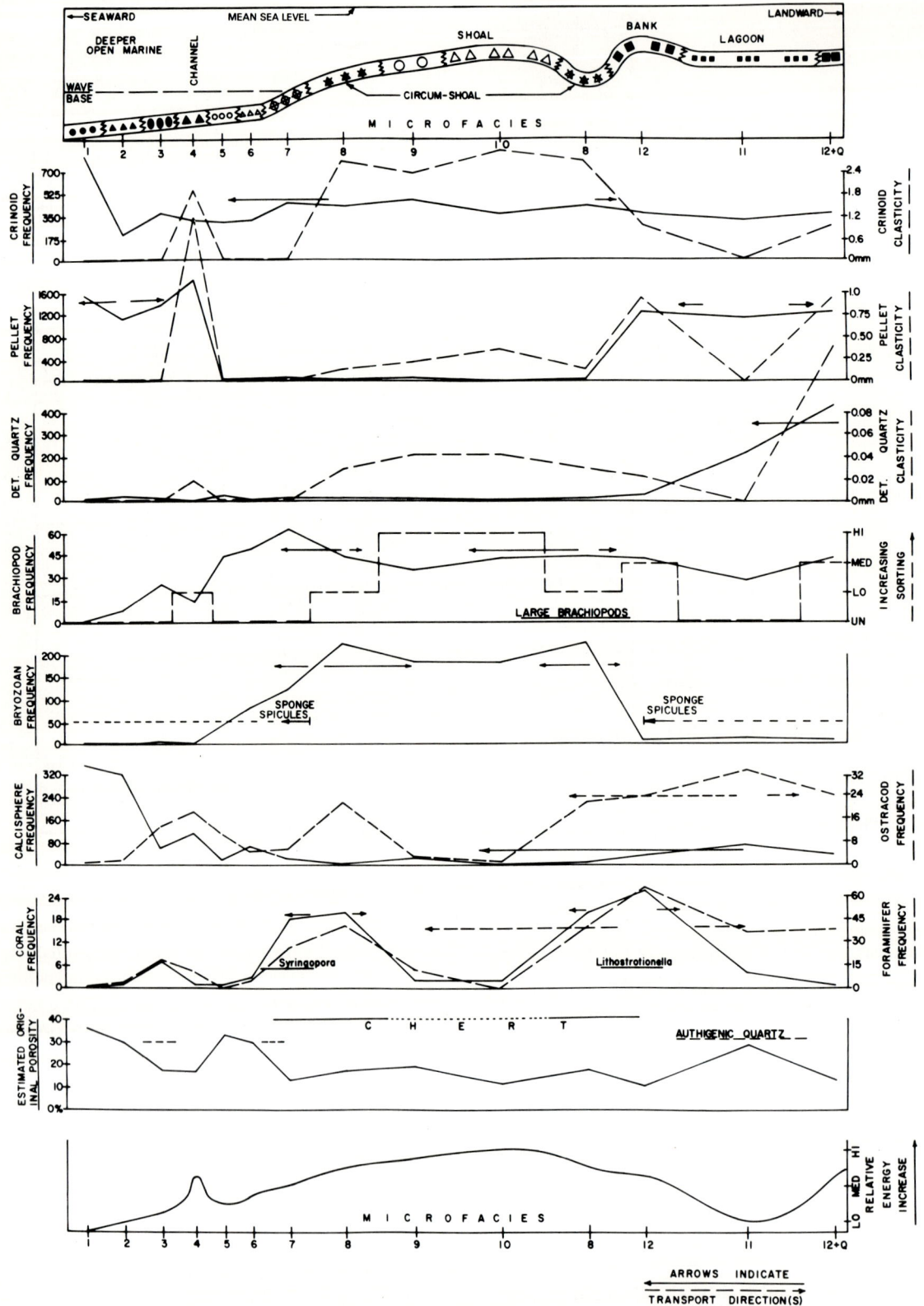

FIGURE 11.42 Monte Cristo Group (Mississippian), Arrow Canyon Range, Clark County, Nevada. Ideal depositional model. From Hansen and Carozzi (1974). Courtesy of the Wyoming Geological Association.

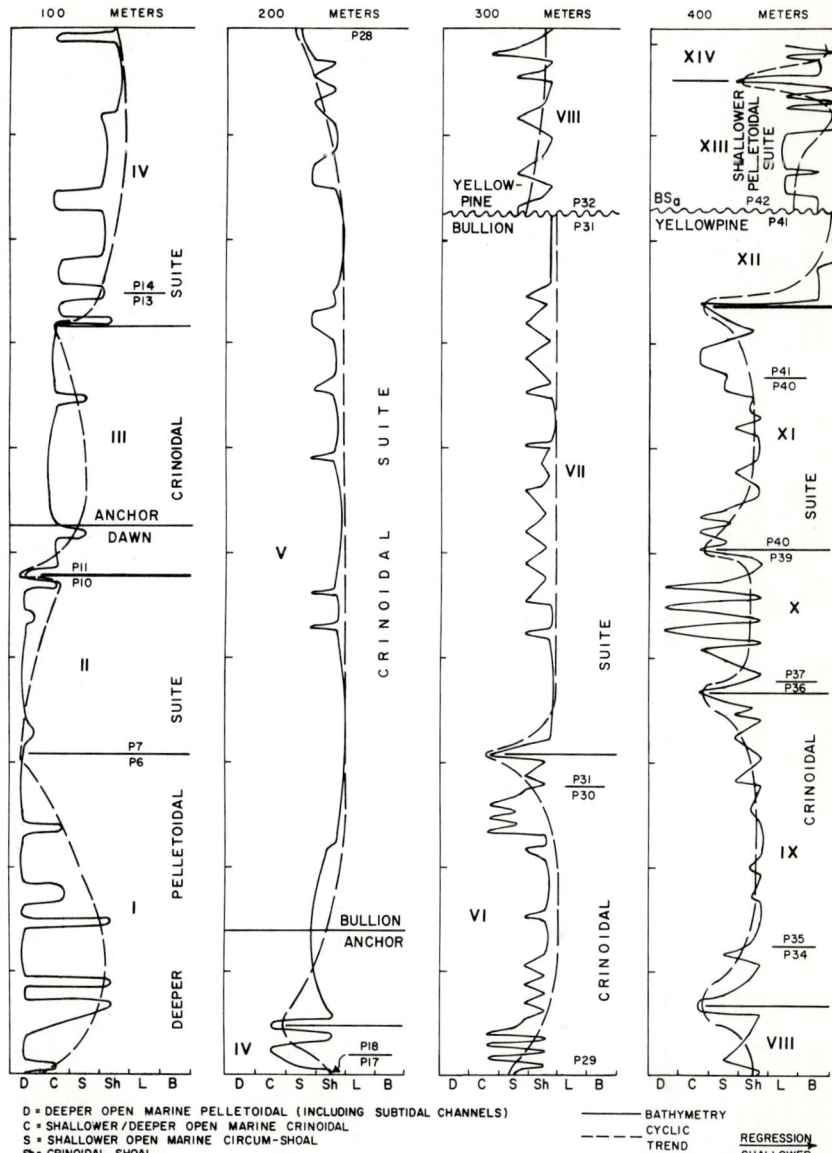

FIGURE 11.43 Monte Cristo Group (Mississippian), Arrow Canyon Range, Clark County, Nevada. General vertical evolution of depositional environments. From Hansen and Carozzi (1974). Courtesy of the Wyoming Geological Association.

and IX), and third deepening (phases X and XI) and shallowing sequence (phases XII, XIII, and XIV). In all circumstances, shallowing-upward sequences tend to be thicker than deepening ones.

11.5 CHARACTERISTIC CONSTITUENTS: CRINOIDS AND OOIDS

Carbonate platforms with frontal concentrations of bioaccumulated to hydrodynamic crinoid buildups can be submitted locally and temporarily to saturation in calcium carbonate by the effect of cold slope-ascending currents. Crinoids do not tolerate such conditions and their deposits are replaced by oolitic hydrodynamic buildups in which ooids form around crinoidal bioclasts or other available nuclei.

A section consisting of the upper 200 m of the Bird Spring Group (Upper Missourian–Wolfcampian) of the Arrow Canyon Range, Clark County, southeast Nevada, was investigated petrographically by means of 907 samples with an average vertical interval of 22 cm (Nowak and Carozzi, 1972).

Description of Microfacies

Microfacies are described in order of general shallowing.

Microfacies 1 (Fig. 11.44A). Calcisiltite with scattered sand-size bioclasts which may reach 20% and

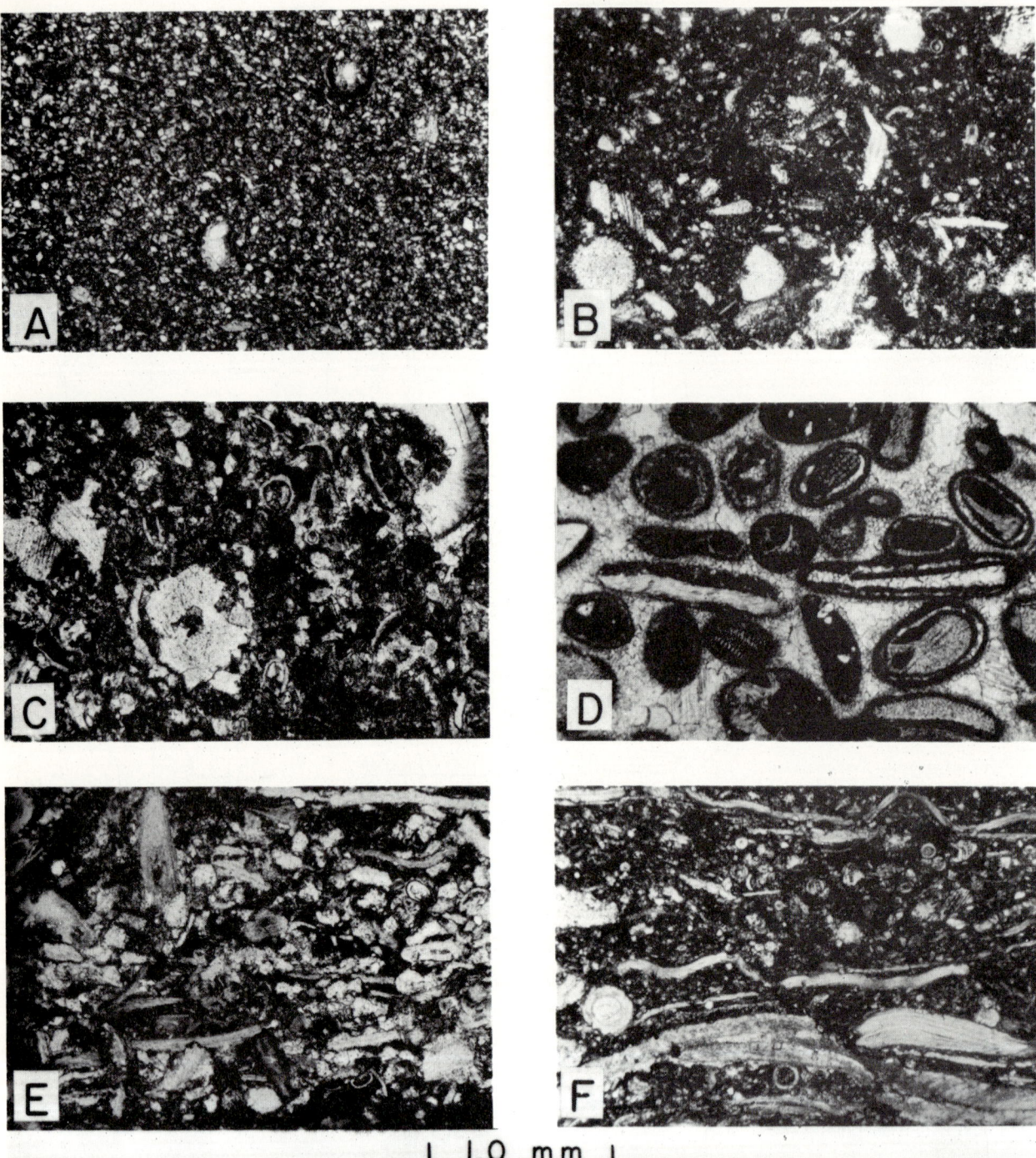

FIGURE 11.44 Bird Spring Group (Upper Missourian–Wolfcampian), Arrow Canyon Range, Clark County, Nevada. Typical microfacies. A. Microfacies 1. B. Microfacies 2. C. Microfacies 3. D. Microfacies 4. E. Microfacies 5. F. Microfacies 6. See text for detailed descriptions. All photomicrographs: plane-polarized light. From Nowak and Carozzi (1972).

consist predominantly of crinoids, brachiopods, and bryozoans associated with rare fusulinids.

Microfacies 2 (Fig. 11.44B). Mud-supported crinoidal biocalcarenite with pelletoidal calcisiltitic or bioclastic matrix. Other bioclasts are brachiopods, bryozoans, ostracods, and fusulinids. Bioturbation may concentrate bioclasts locally into patches of grain-supported texture.

Microfacies 3 (Fig. 11.44C). Grain-supported crinoidal biocalcarenite with pelletoidal or bioclastic matrix. Crinoids reach their peak of frequency and dominate all other bioclasts which also include *Endothyra* and agglutinated foraminifers. Pressure-solution contacts between bioclasts are frequent.

Microfacies 4 (Fig. 11.44D). Grain-supported oolitic biocalcarenite with sparite cement. Superficial ooids predominate over normal ones. Cores consist of bioclasts of crinoids and brachiopods which exist also as noncoated constituents and of rounded lithic pellets of calcisiltite. Rounded intraclasts of fossiliferous calcisiltite are the most common detrital components next to ooids.

Microfacies 5 (Fig. 11.44E). Grain-supported brachiopod biocalcarenite with calcisiltite matrix. Shell fragments and spines, associated with crinoids and bryozoans, are usually arranged parallel to bedding, indicating a bioaccumulated character sometimes disturbed by scavengers.

Microfacies 6 (Fig. 11.44F). Mud-supported brachiopod biocalcarenite which differs from microfacies 5 mainly by the smaller amount of sand-size bioclasts. Brachiopod shells also are aligned parallel to bedding unless disturbed by bioturbation.

Microfacies 7 (Fig. 11.45A,B). Fine-grained pelletoidal biocalcarenite with calcisiltite matrix; pellets appear of fecal origin and are merged frequently (Fig. 11.45B). Associated bioclasts belong to crinoids, brachiopods, and bryozoans with local concentrations of agglutinated benthic foraminifers.

Microfacies 8 (Fig. 11.45C,D). Fine-grained pelletoidal and foraminiferal biocalcarenite with micrite matrix often altered into pseudosparite by neomorphism. Besides fecal pellets and agglutinated benthic foraminifers, intraclasts of fine pelletoidal calcisiltite originating from microfacies 7 are abundant (Fig. 11.45D).

Microfacies 9 (Fig. 11.45E,F). Calcisiltites are of two kinds: either with irregularly scattered debris of fine sand-size bioclasts of crinoids and brachiopods, associated with flat desiccation chips of dark calcisiltite which originate from an assumed adjacent tidal flat, or displaying a laminated texture with alternating quartz-rich layers (Fig. 11.45F).

Average microscopic parameter values for each microfacies are displayed graphically by standardized Z values which show a mean of zero and a standard deviation of 1. This facilitates graphic representation and allows easy comparison between average values even though uncoded averages may vary greatly (Figs. 11.46, 11.47, 11.48).

The Ideal Shallowing-Upward Sequence

The superposition of microfacies 1 to 9 represents a general shallowing-upward evolution from a seaward slope, across a crinoidal-oolitic bar made discontinuous by tidal channels, and into a shallow lagoon with unknown landward boundary, probably adjacent to a carbonate tidal flat. The complete shallowing-upward sequence is shown without microfacies 4 (Fig. 11.49A) and also with microfacies 4 (Fig. 11.49B), which replaces microfacies 3 during temporary periods of oolitic development.

The behavior of the microscopic parameters is influenced strongly by presence of the crinoidal bar from which its bioclasts were distributed in either direction from the bar crest by tidal currents. Brachiopods preferred the flanks of the crinoidal bar, particularly the internal margin facing the lagoon where bioaccumulation took place. In the agitated conditions of the bar, lithic pellets were associated with fecal pellets, whereas fecal pellets predominated in the lower-energy conditions behind the bar. Two separate communities of agglutinated benthic foraminifers developed on the bar and behind it, with maximum frequency in the latter location. Lithic pellets and lithoclasts were formed by reworking processes both on the bar and in nearshore conditions. The pattern of distribution of the silt-size quartz grains indicates an extrabasinal source landward of the lagoon, a transit across the entire carbonate system, and some settling in deeper water.

With the development of an oolitic environment, microfacies 4 replaces microfacies 3. The general effect is a reduction in average values of all measured parameters except crinoid clasticity, bryozoan frequency, pellet frequency, and lithoclast frequency and clasticity due to energy increase in the oolitic environment. A typical section (Figs. 11.50, 11.51) shows the stratigraphic variation of microscopic parameters during the shallowing-upward evolution.

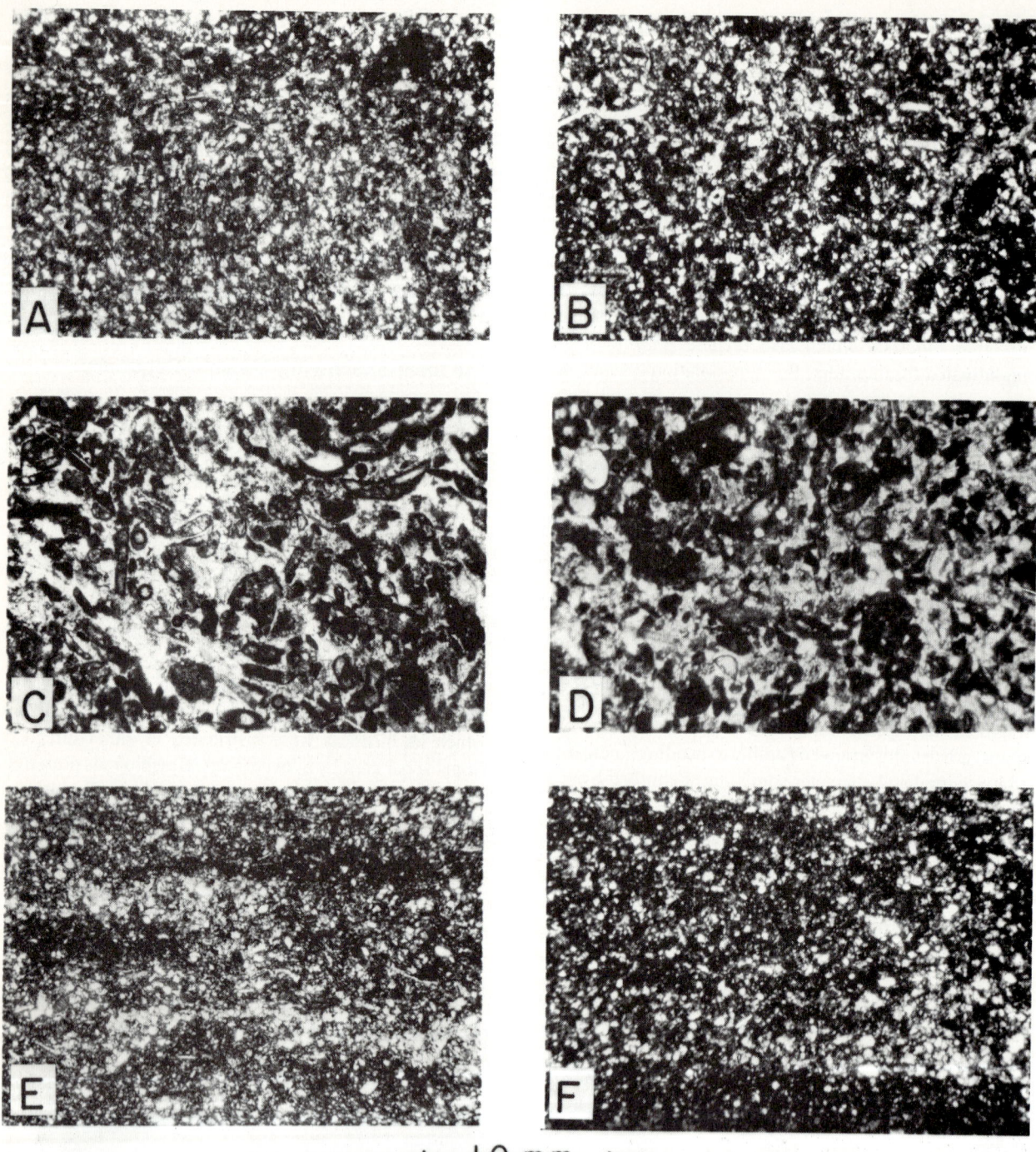

FIGURE 11.45 Bird Spring Group (Upper Missourian–Wolfcampian), Arrow Canyon Range, Clark County, Nevada. Typical microfacies (continued). A and B. Microfacies 7. C and D. Microfacies 8. E and F. Microfacies 9. See text for detailed descriptions. All photomicrographs: plane-polarized light. From Nowak and Carozzi (1972).

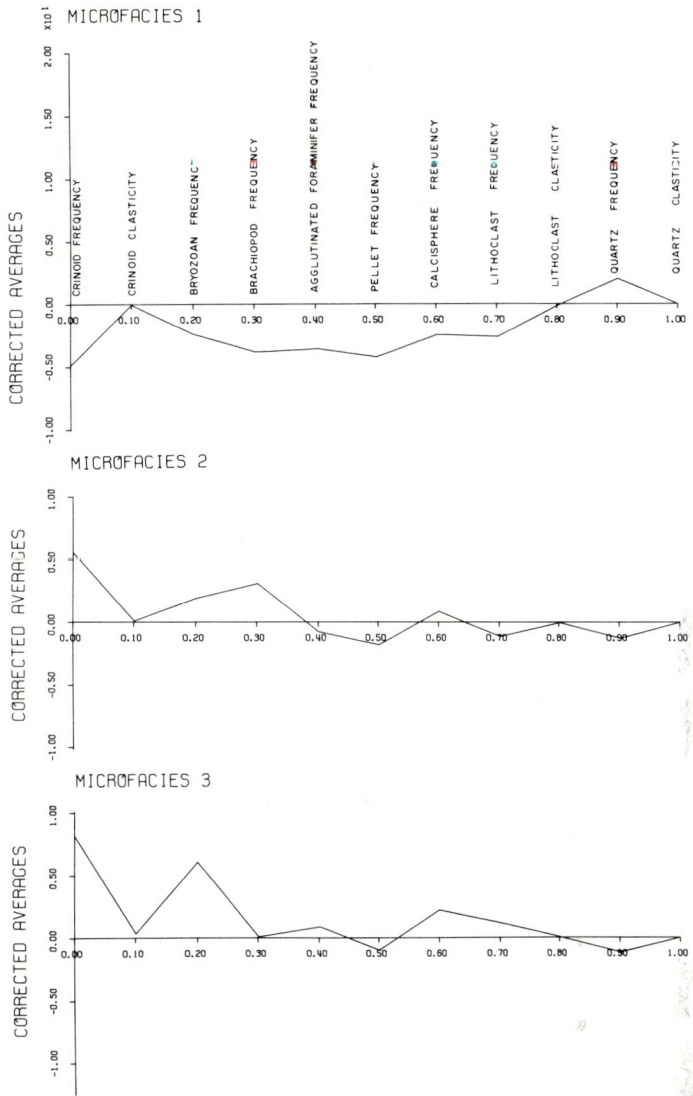

FIGURE 11.46 Bird Spring Group (Upper Missourian–Wolfcampian), Arrow Canyon Range, Clark County, Nevada. Coded average values of microscopic parameters. From Nowak and Carozzi (1972).

The Ideal Depositional Model

Variation of the coded average frequency for each component through the nine microfacies (Figs. 11.52, 11.53) allows the reconstruction of a depositional model (Fig. 11.54). It consists essentially of six subenvironments: (A) slope below wave base, (B) external flank of the crinoidal-oolitic bar, (C) crinoidal-oolitic bar, (D) internal flank of the bar, (E) main lagoon, and (F) its landward portion with temporary subaerial exposure and detrital quartz sediment yield.

In summary, the crinoidal-oolitic bar (subenvironment C) was the source of most of the bioclasts found in adjacent subenvironments as the result of dispersion seaward, or landward, or both by the action of currents and tides. This pattern also holds true for some intraclasts and lithic pellets, although fecal pellets and small micritic intraclasts were generated in place in the lagoonal subenvironment E, where agglutinated foraminifers were abundant. The internal flank of the bar (subenvironment D) acted as a trap for some of the silt-size quartz which transited across the carbonate system, and also for calcispheres.

General Evolution of the Depositional Environments

The investigated section consists of the superposition of 12 asymmetric and symmetric oscillations (Fig. 11.55) of water depth. Both the first- and second-order tangents to the maxima are straight lines indicating that there are no overall deepening or shallowing trends and that the upper portion of the Bird Spring Group was deposited in a stable, shallow environment. In comparison (Fig.

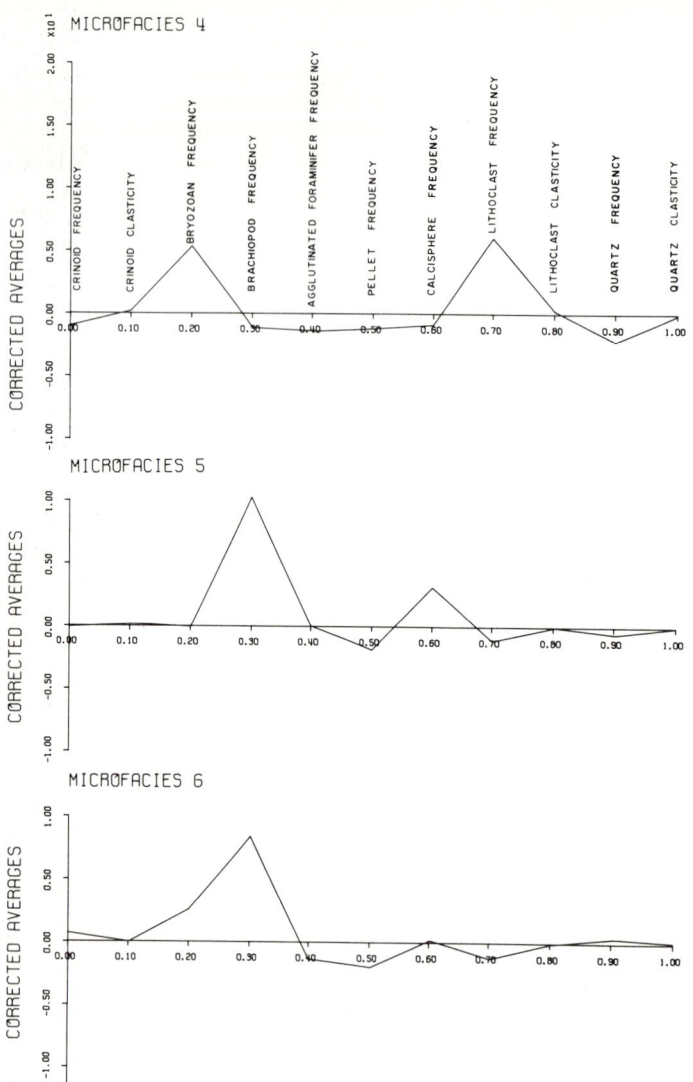

FIGURE 11.47 Bird Spring Group (Upper Missourian–Wolfcampian), Arrow Canyon Range, Clark County, Nevada. Coded average values of microscopic parameters (continued). From Nowak and Carozzi (1972).

11.56), the 562 m of the Bird Spring Group directly below the investigated section (see Chapter 4) display a deepening trend with small-amplitude oscillations during the Morrowan and a shallowing trend in two steps during Atokan through Lower Missourian. During Upper Missourian through Wolfcampian, the general bathymetry does not change.

11.6 CHARACTERISTIC CONSTITUENTS: *GIRVANELLA* ONCOIDS

Carbonate platforms which developed over passive continental margins on both sides of the proto South Atlantic display semirestricted conditions inherited from the underlying evaporitic phase. Frontal and internal bioaccumulated to hydrodynamic buildups consist entirely of blue-green algal oncoids formed under more agitated conditions within a widespread background of numerous types of pelagic calcilutites. Because of the paucity of benthic and planktonic organisms, environmental reconstructions are difficult and require the use of elaborate quantitative microscopic techniques.

The Macaé Formation (Albian-Cenomanian) of the Campos Basin, offshore Rio de Janeiro, Brazil, represents the initial event of the oceanic stage in the typical tectosedimentary evolution of Brazilian coastal basins (Falkenhein *et al.*, 1981; Carozzi and Falkenhein, 1985).

The Macaé Formation, a prolific oil producer, shows an average thickness of 1000 m in the eastern part of the Campos Basin, is truncated gradually westward, and grades into contemporaneous fluviodeltaic sandstones and shales. This extensive carbonate platform was investigated by means of 32 exploratory wells which

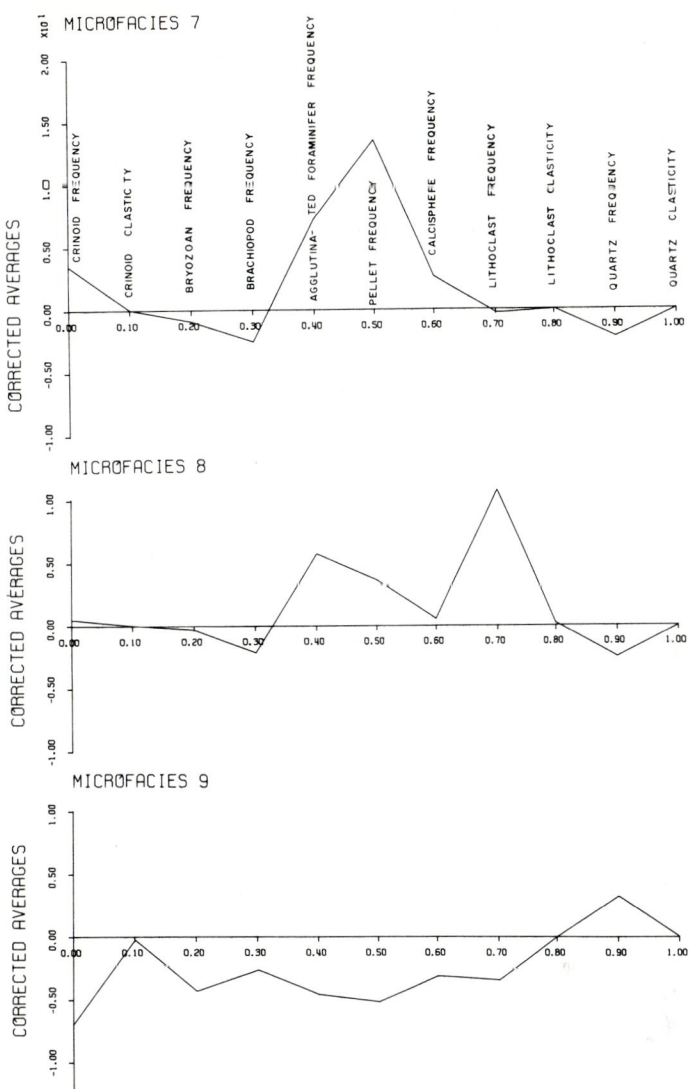

FIGURE 11.48 Bird Spring Group (Upper Missourian–Wolfcampian), Arrow Canyon Range, Clark County, Nevada. Coded average values of microscopic parameters (continued). From Nowak and Carozzi (1972).

represent an aggregate column of about 15,000 m of which 1200 m consist of cores (8%) and 13,800 m of cuttings (92%) A total of 2404 thin sections were studied petrographically, of which 828 were made from cores (35%) and 1576 from cuttings (65%). The vertical spacing of thin sections varies according to the type of samples, averaging 0.5 m in cores, 4.5 m in cuttings of calcarenites, and 12 m in cuttings of calcilutites. The average sampling control is of about 100 thin sections per well corresponding to one thin section per 5 m.

Description of Microfacies

A suite of 19 characteristic microfacies was defined. The microfacies were divided into two major groups: a suite of calcilutites and a suite of calcarenites and calcirudites. The latter suite was further subdivided on the basis of shore proximity as shoreface and offshore.

Calcilutite Suite

The seven calcilutite microfacies are described from shoreface to offshore.

Microfacies 00 (Fig. 11.57A). Calcilutite and dolomitic calcilutite with common silt-size grains of detrital quartz and patches of secondary anhydrite. Traces of plant remains are rare. The diagenetic signature consists of incomplete early dolomitization into dolomicrosparite, fenestral textures due to the decay of organic matter (probably algal), and pseudobrecciated textures related to possible evaporite dissolution. Environmental interpretation is supratidal to high intertidal.

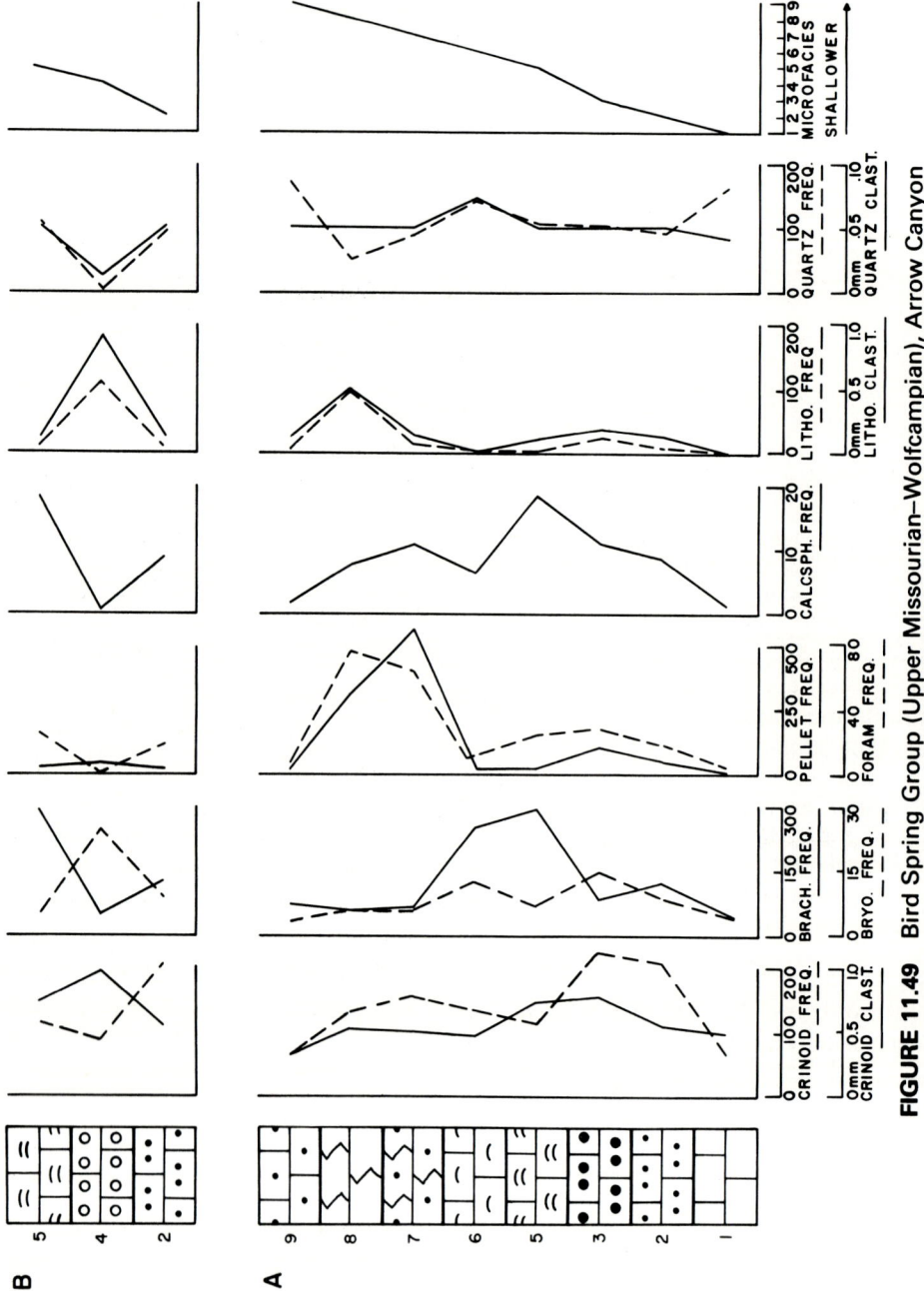

FIGURE 11.49 Bird Spring Group (Upper Missourian–Wolfcampian), Arrow Canyon Range, Clark County, Nevada. Ideal shallowing-upward sequence. From Nowak and Carozzi (1972).

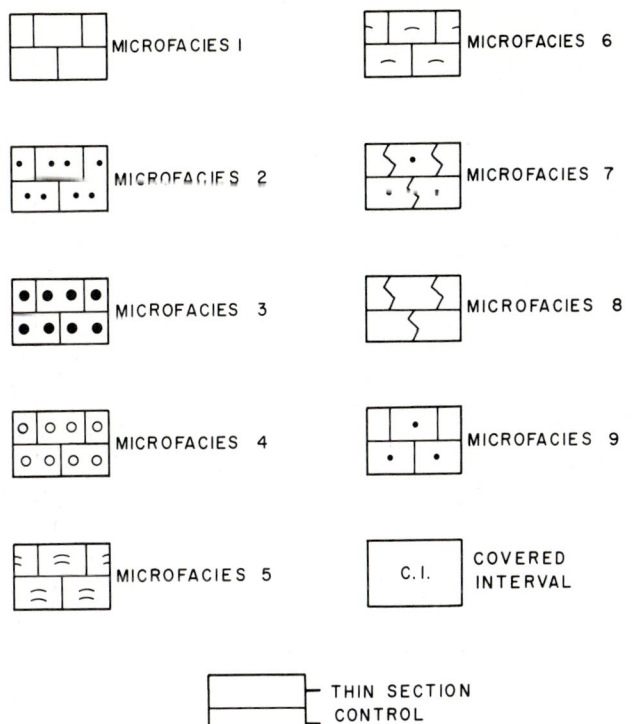

FIGURE 11.50 Bird Spring Group (Upper Missourian–Wolfcampian), Arrow Canyon Range, Clark County, Nevada. Symbols for microfacies. From Nowak and Carozzi (1972).

Microfacies 10 (Fig. 11.57B). Arenaceous calcilutite with scattered pyrite pigments and muscovite flakes with extensive neomorphism into pseudomicrosparite and dolomitization into dolomicrosparite. Bioclasts of benthic foraminifers, ostracods, and plants are rare. Environmental interpretation is nearshore intertidal to shallow subtidal.

Microfacies 20 (Fig. 11.57C). Pelletoidal calcilutite with pyrite pigments and rare minute grains of detrital quartz. Abundant microdebris of echinoids, pelecypods, calcareous worm tubes, benthic foraminifers, and ostracods predominate over planktonic foraminifers and calcispheres. Environmental interpretation is normal marine subtidal shoreface.

Microfacies 30 (Fig. 11.57D). Homogeneous to pelletoidal calcilutite with relatively abundant organic components (8 to 10%) and rare silt-size grains of detrital quartz. The benthic bioclasts predominate over planktonic organisms. Bioclasts consist of echinoderms (mostly echinoids) and pelecypods, followed by gastropods, foraminifers, worm tubes, rare ostracods, and red algae. Bioturbation is locally common. Environmental interpretation is subtidal open marine shoreface to offshore.

Microfacies 40 (Fig. 11.57E). Clotted calcilutite with scattered small grains of detrital quartz and pyrite pigments. The very small biogenic content consists mostly of planktonic foraminifers, calcispheres, and rare debris of echinoids, pelecypods, and ostracods. Environmental interpretation is subtidal open marine offshore.

Microfacies 45 (Fig. 11.57F). Clotted to homogeneous calcilutite with scattered minute grains of detrital quartz and pyrite pigments. Small amounts of planktonic foraminifers and calcispheres are associated with rare microdebris of echinoids, pelecypods, and ostracods. The biogenic content is even smaller than in microfacies 40. Environmental interpretation is subtidal open marine offshore.

Microfacies 48 (Fig. 11.57G). Hemipelagic calcilutite with abundant planktonic fauna represented by foraminifers, calcispheres, and rare radiolarians. Ostracods are very rare. Grains of fine detrital quartz, mica flakes, rare phosphate, and glauconite pellets are scattered in the groundmass. Environmental interpretation is distal offshore to shelf slope.

Diamictite and sandstone microfacies representing deep-water turbidites (designated as microfacies ST) are locally associated with microfacies 48. Diamictites (Figs. 11.57H, 11.58A) consist of clasts of oncoidal calcarenites in a groundmass of argillaceous calcilutite with local concentrations of detrital quartz, lobate glauconite pellets, and pyrite pigments. The sandstone (Fig. 11.58B) is a quartzose feldspathic arenite with an argillaceous calcilutite matrix, muscovite flakes, pyrite pigments, and subangular calcilutite clasts.

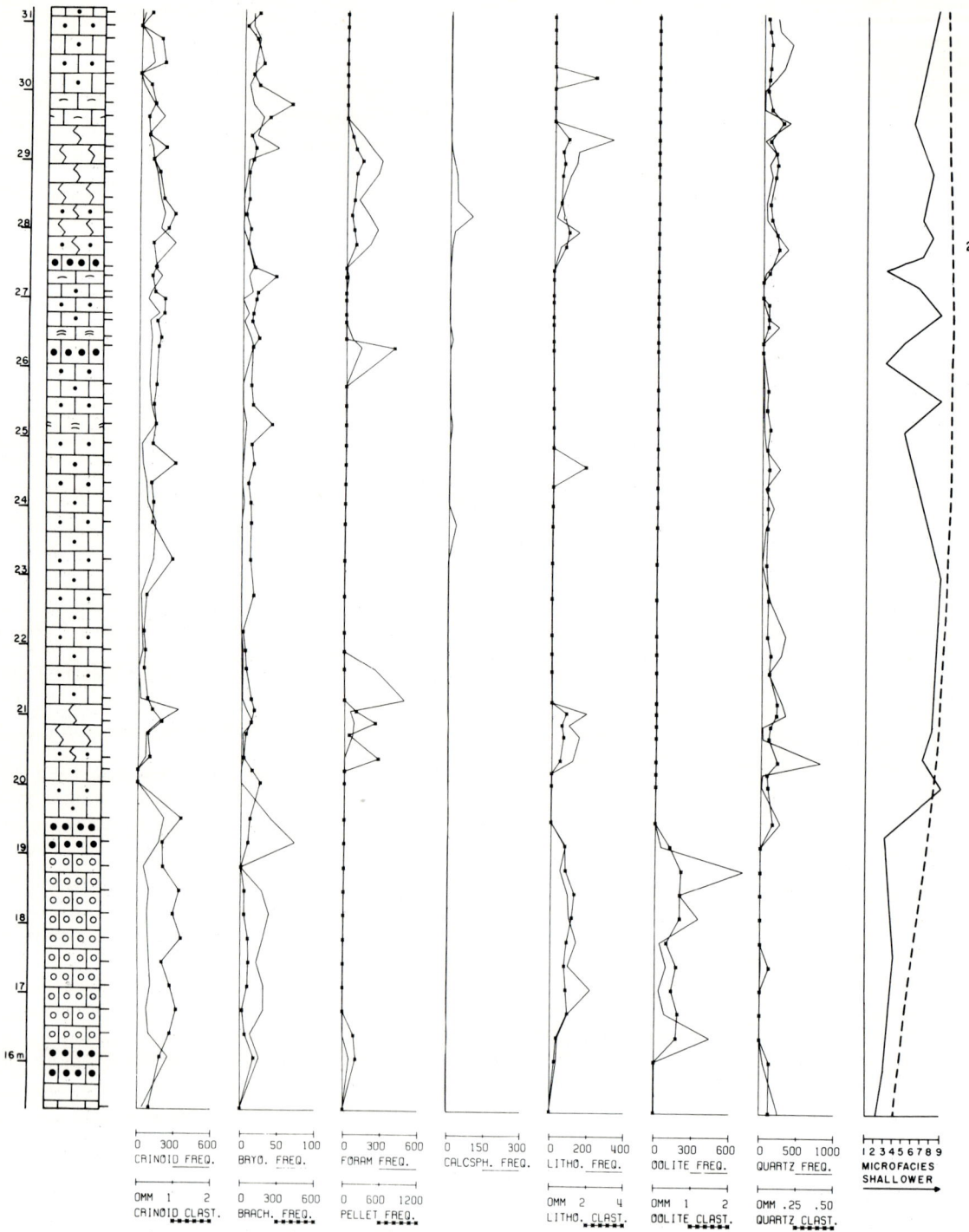

FIGURE 11.51 Bird Spring Group (Upper Missourian–Wolfcampian), Arrow Canyon Range, Clark County, Nevada. Typical example of vertical variation of microscopic parameters during shallowing-upward sequence. From Nowak and Carozzi (1972).

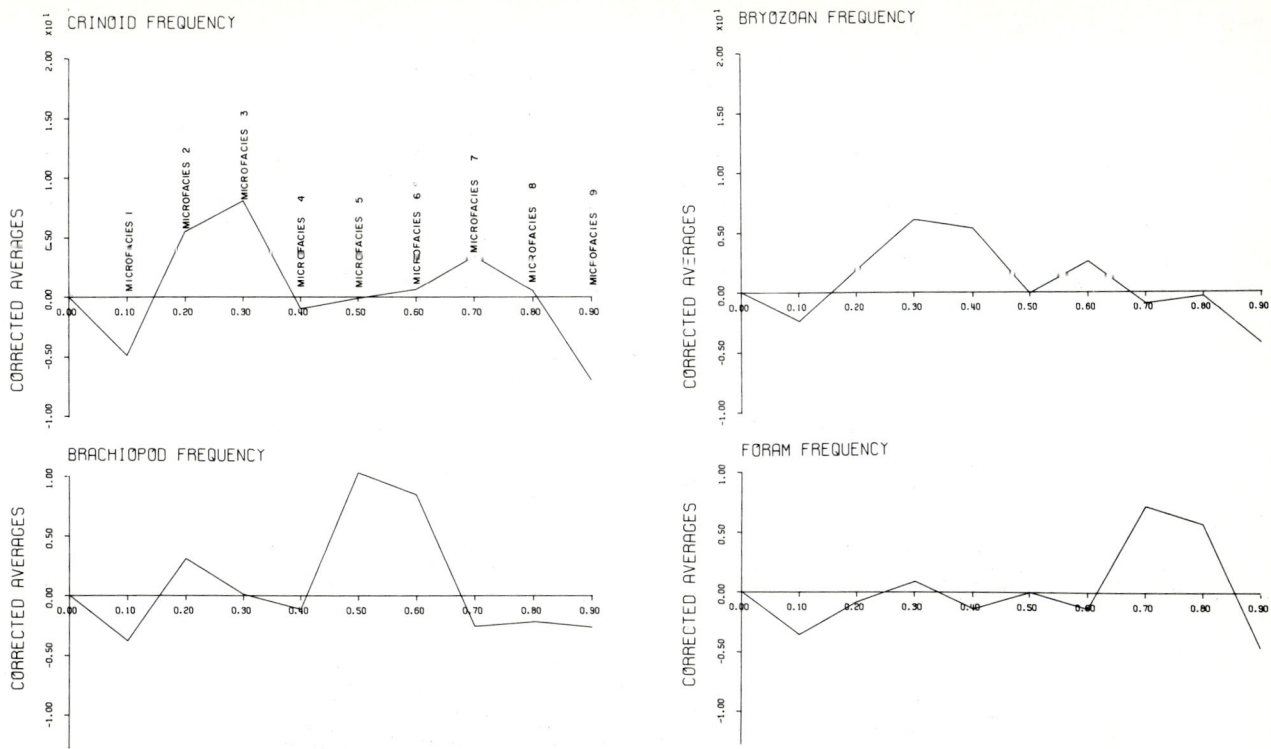

FIGURE 11.52 Bird Spring Group (Upper Missourian–Wolfcampian), Arrow Canyon Range, Clark County, Nevada. Variation of coded average values of microscopic parameters through all microfacies. From Nowak and Carozzi (1972).

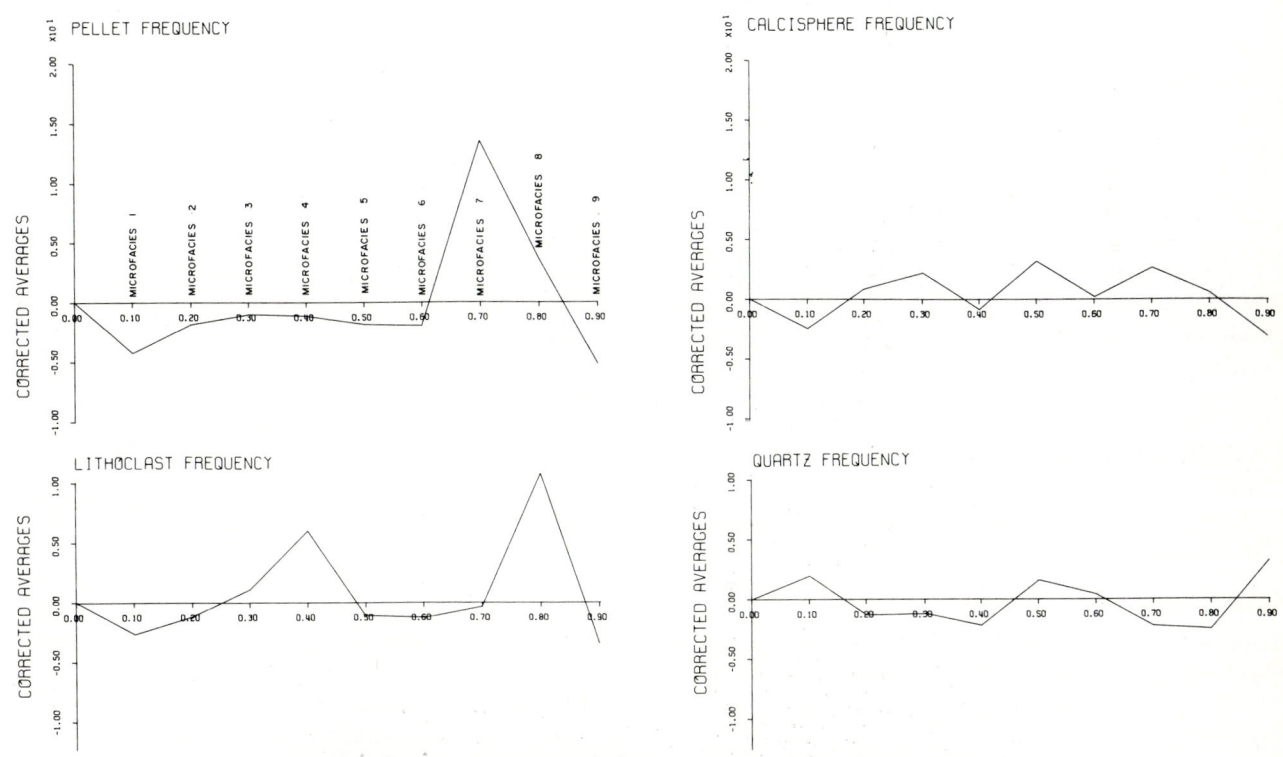

FIGURE 11.53 Bird Spring Group (Upper Missourian–Wolfcampian), Arrow Canyon Range, Clark County, Nevada. Variation of coded average values of microscopic parameters through all microfacies (continued). From Nowak and Carozzi (1972).

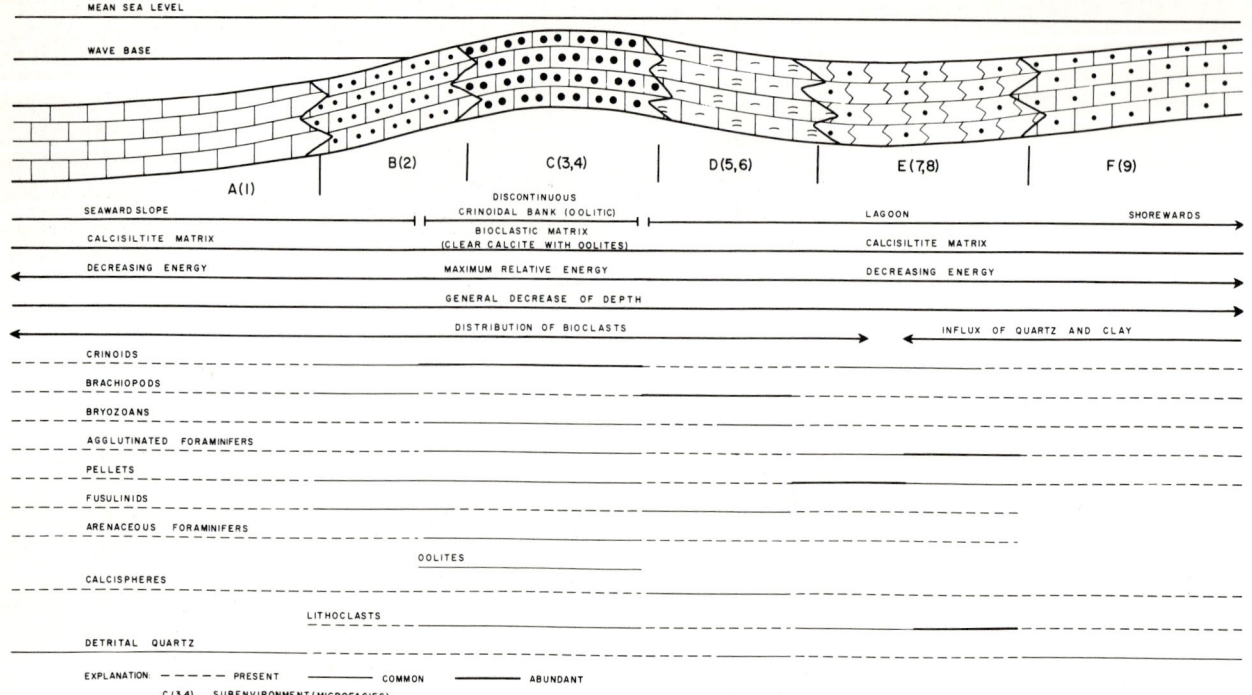

FIGURE 11.54 Bird Spring Group (Upper Missourian–Wolfcampian), Arrow Canyon Range, Clark County, Nevada. Ideal depositional model. From Nowak and Carozzi (1972).

Calcarenite and Calcirudite Suite

The 11 microfacies of the calcarenite and calcirudite suite are subdivided into shoreface and offshore groups and are described within each group in order of increasing energy. The shoreface group (microfacies 50, 60, 70, and 80) displays an association of oncoids, ooids, and terrigenous influx of isolated grains of quartz or quartz nuclei of both oncoids and ooids. The offshore group (microfacies 90, 100, 110, 120, 130, 140, and 150) contains oncoidal carbonates associated with open marine calcilutites.

Microfacies 50 (Fig. 11.58C). Grain-supported arenaceous oncoidal calcarenite to calcirudite with micrite matrix and local patches of cavity-filling sparite. Neomorphism of the interstitial material into pseudosparite or replacement by dolomicrosparite are common. Bioclasts forming the cores of the oncoids are echinoids, pelecypods, benthic foraminifers, and more rarely calcareous worm tubes, red algae, bryozoans, corals, and agglutinating foraminifers. The same bioclasts occur scattered in the interstitial material. Environmental interpretation is nearshore subtidal.

Microfacies 60 (Fig. 11.58D). Grain-supported very arenaceous oncoidal calcarenite to calcirudite with cavity-filling sparite cement and patches of calcilutite matrix. Ooids and large grains of detrital quartz are common locally. The composition and abundance of skeletal fragments are similar to those of microfacies 50 both for bioclasts and cores of oncoids with well-preserved *Girvanella* filaments (Fig. 11.58H). Environmental interpretation is shallow subtidal, shoreface.

Microfacies 70 (Fig. 11.58E). Grain-supported arenaceous oolitic-oncoidal calcarenite with rim cement of sparite and secondary vuggy porosity. Oncoids occur always associated with ooids which are locally the dominant component. Cores of oncoids and ooids consist of detrital quartz, feldspar grains, and subordinated bioclasts with a composition similar to those in microfacies 50 and 60. Environmental interpretation is shoreface beachrock shoals subjected to meteoric vadose dissolution.

Microfacies 80 (Fig. 11.58F). Grain-supported oolitic-oncoidal coarse calcarenite to calcirudite with geopetal internal sediments, sparite rim cement (cement I), and blocky sparite cavity-filling cement (cement II). This microfacies is coarser than microfacies 70 but otherwise similar to it. It has undergone a late stage of porosity occlusion by intensive freshwater phreatic cementation. Environmental interpretation is shoreface beachrock shoals with meteoric vadose dissolution followed by phreatic cementation.

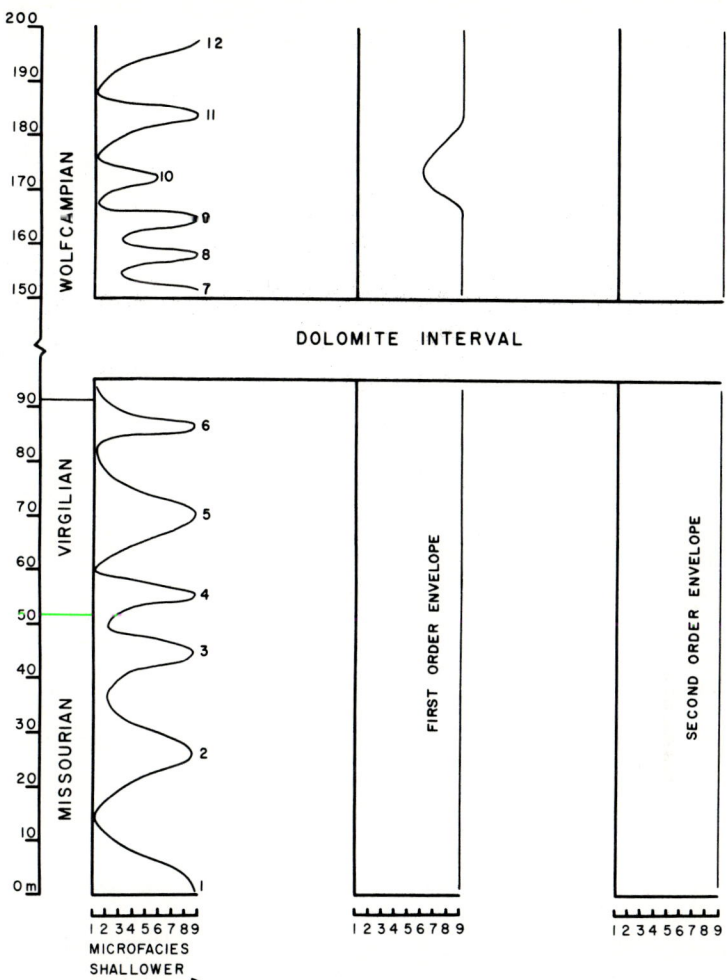

FIGURE 11.55 Bird Spring Group (Upper Missourian–Wolfcampian), Arrow Canyon Range, Clark County, Nevada. General vertical evolution of depositional environments. From Nowak and Carozzi (1972).

Microfacies 90 (Fig. 11.58G). Mud-supported oncoidal calcarenite to calcirudite with a calcilutite matrix often changed into pseudomicrosparite by neomorphism. Well-developed oncoids (with remains of *Girvanella* filaments) display nuclei consisting predominantly of echinoids, pelecypods, benthic foraminifers, rare rudistids, corals, and red algae. Scattered bioclasts in the matrix are similar to those of the nuclei, but also include worm tubes and planktonic and agglutinating foraminifers. Environmental interpretation is low energy, subtidal.

Microfacies 100 (Fig. 11.59A). Grain-supported microoncoidal calcarenite with calcilutite matrix and rare patches of sparite cement. The matrix includes algal pellets originating from abraded oncoids. Microoncoids contain cores of pelecypods and echinoid bioclasts, but generally lack visible internal structure. Rare bioclasts scattered in the matrix are mostly echinoids, pelecypods, benthic and planktonic foraminifers, and calcispheres. Environmental interpretation is subtidal offshore.

Microfacies 110 (Fig. 11.59B). Grain-supported and pressure-welded biocalcarenite with calcilutite matrix and very rare small patches of cavity-filling sparite cement. Predominant bioclasts are pelecypods, echinoids, followed by benthic foraminifers, rare red algae, worm tubes, and ostracods. Some microoncoids also occur. Environmental interpretation is subtidal, open marine.

Microfacies 120 (Fig. 11.59C). Grain-supported oncoidal calcarenite to calcirudite with calcilutite matrix and local patches of cavity-filling sparite cement. Many large oncoids possess pelecypod shells as cores. Scattered bioclasts are mostly echinoids and pelecypods followed by benthic foraminifers, rare red algae, rudistids, corals, planktonic foraminifers, and ostracods. Environmental interpretation is subtidal open marine.

Microfacies 130 (Fig. 11.59D). Grain-supported oncoidal calcarenite to calcirudite with cavity-filling sparite cement, internal sediment, and a small amount of calcilutite matrix. Abundance and composition of isolated bioclasts and those forming the nuclei of oncoids are the same as in microfacies 120. Environmental interpretation is shallow subtidal open marine.

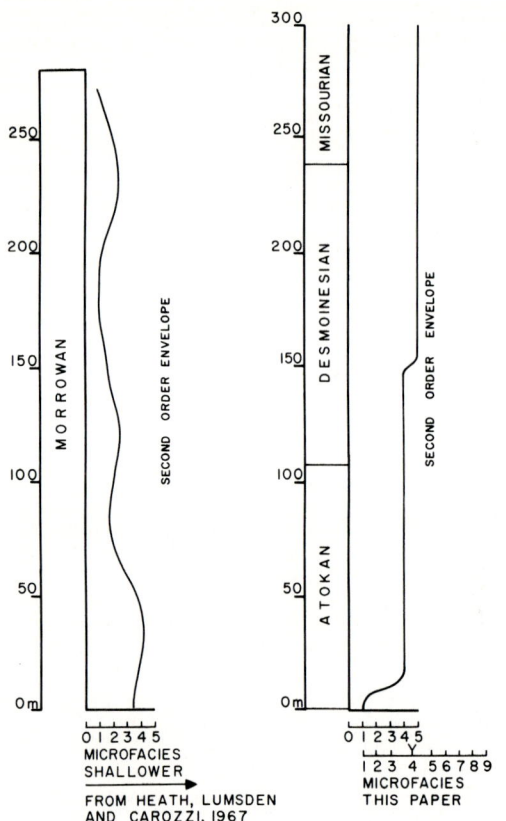
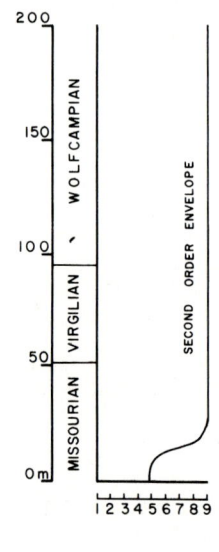

FIGURE 11.56 Bird Spring Group (Morrowan-Wolfcampian), Arrow Canyon Range, Clark County, Nevada. General vertical evolution of depositional environments. From Nowak and Carozzi (1972).

Microfacies 140 (Fig. 11.59E,F). Grain-supported oncoidal calcarenite to calcirudite with rim cement calcite and cavity-filling sparite that left primary interparticle pores. Bioclasts with a composition similar to those of microfacies 120 and 130 are rare. Type and abundance of skeletal fragments as cores of oncoids are the same as in the previous two microfacies. Locally, this microfacies (Fig. 11.59F) is very porous (23%) due to the association of primary interparticle porosity with secondary nonfabric selective vuggy porosity. Environmental interpretation is offshore beachrock shoals subjected to meteoric vadose dissolution. This is the main reservoir rock and oil producer in the Macaé Formation.

Microfacies 150 (Fig. 11.59G,H). Grain-supported oncoidal calcarenite to calcirudite with interstitial rim cement and cavity-filling sparite. Internal sediments consist of algal pellets, calcilutite, and oncoidal microdebris organized in geopetal texture (Fig. 11.59G). Interstitial bioclasts and nuclei of oncoids show composition and frequency similar to those in microfacies 120 and 130. This microfacies is similar to microfacies 140 but has undergone a late stage of porosity occlusion by intensive freshwater phreatic cementation. Environmental interpretation is offshore beachrock shoals with meteoric vadose dissolution followed by phreatic cementation.

The Ideal Shallowing-Upward Sequence

The section of well 57 (Figs. 11.60, 11.61) is representative of the type of shallowing-upward evolution of the microfacies displayed by the bars of oncoidal calcarenites of the Macaé Formation. Because standard microfacies techniques such as frequency of organic components were not able to differentiate the suite of microfacies which possessed a small and relatively constant fossil content, indexes of paleoecological nature were devised (Table 11.1). They are: an index of biotic diversity of benthic

TABLE 11.1

Index of Biotic Diversity of Benthics as Bioclasts or as Cores of Oncoids

Biotic Diversity Index	Types of Bioclasts
1	Echinoids
2	Echinoids and gastropods
3	Echinoids and pelecypods
4	Echinoids, pelecypods, and benthic foraminifers
5	Echinoids, pelecypods, benthic foraminifers, and worm tubes
6	Same as 5 plus red algae
7	Same as 5 plus rudistids
8	Same as 5 plus hexacorals

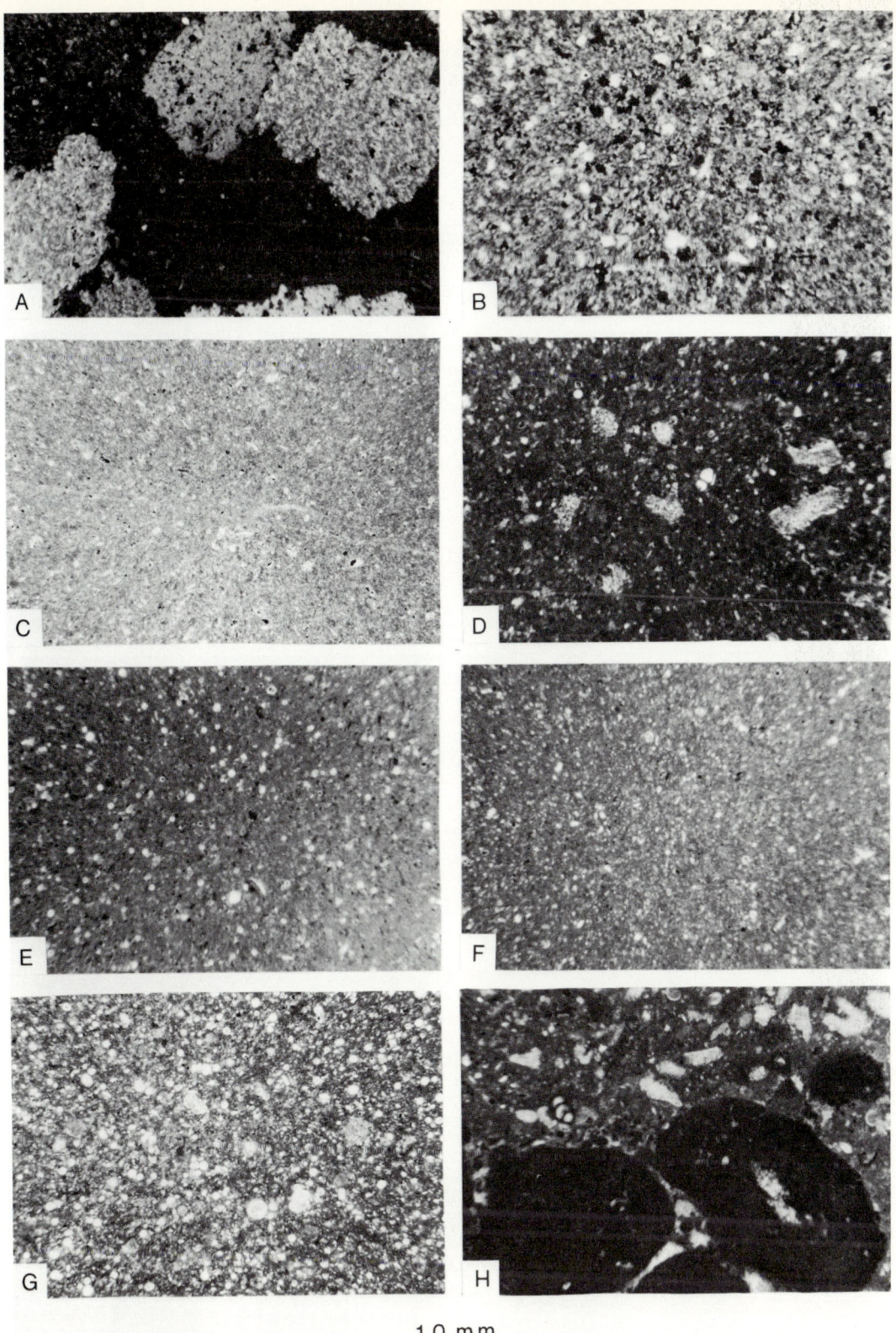

FIGURE 11.57 Macaé Formation (Albian-Cenomanian), Campos Basin, offshore Rio de Janeiro, Brazil. Typical microfacies. A. Microfacies 00. B. Microfacies 10. C. Microfacies 20. D. Microfacies 30. E. Microfacies 40. F. Microfacies 45. G. Microfacies 48. H. Microfacies ST. See text for detailed descriptions. All photomicrographs: plane-polarized light, except A and B, crossed nicols. From Falkenhein et al. (1981).

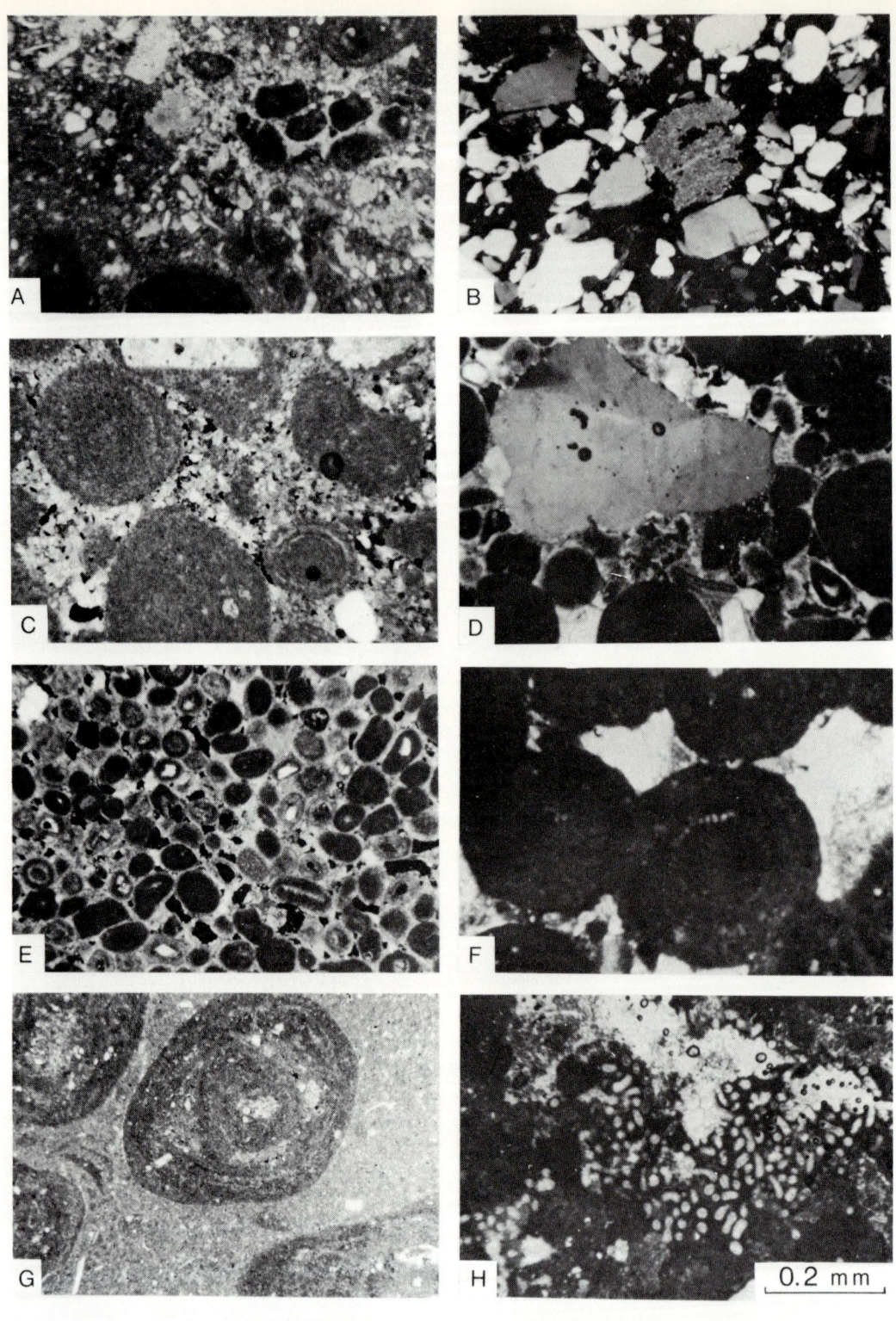

SCALE FOR A,B,C,D,E,F & G : 1.0 mm

FIGURE 11.58 Macaé Formation (Albian-Cenomanian), Campos Basin, offshore Rio de Janeiro, Brazil. Typical microfacies (continued). A and B. Microfacies ST. C. Microfacies 50. D. Microfacies 60. E. Microfacies 70. F. Microfacies 80. G. Microfacies 90. H. Microfacies 60. See text for detailed descriptions. All photomicrographs: crossed nicols, except A and H, plane-polarized light. From Falkenhein *et al.* (1981).

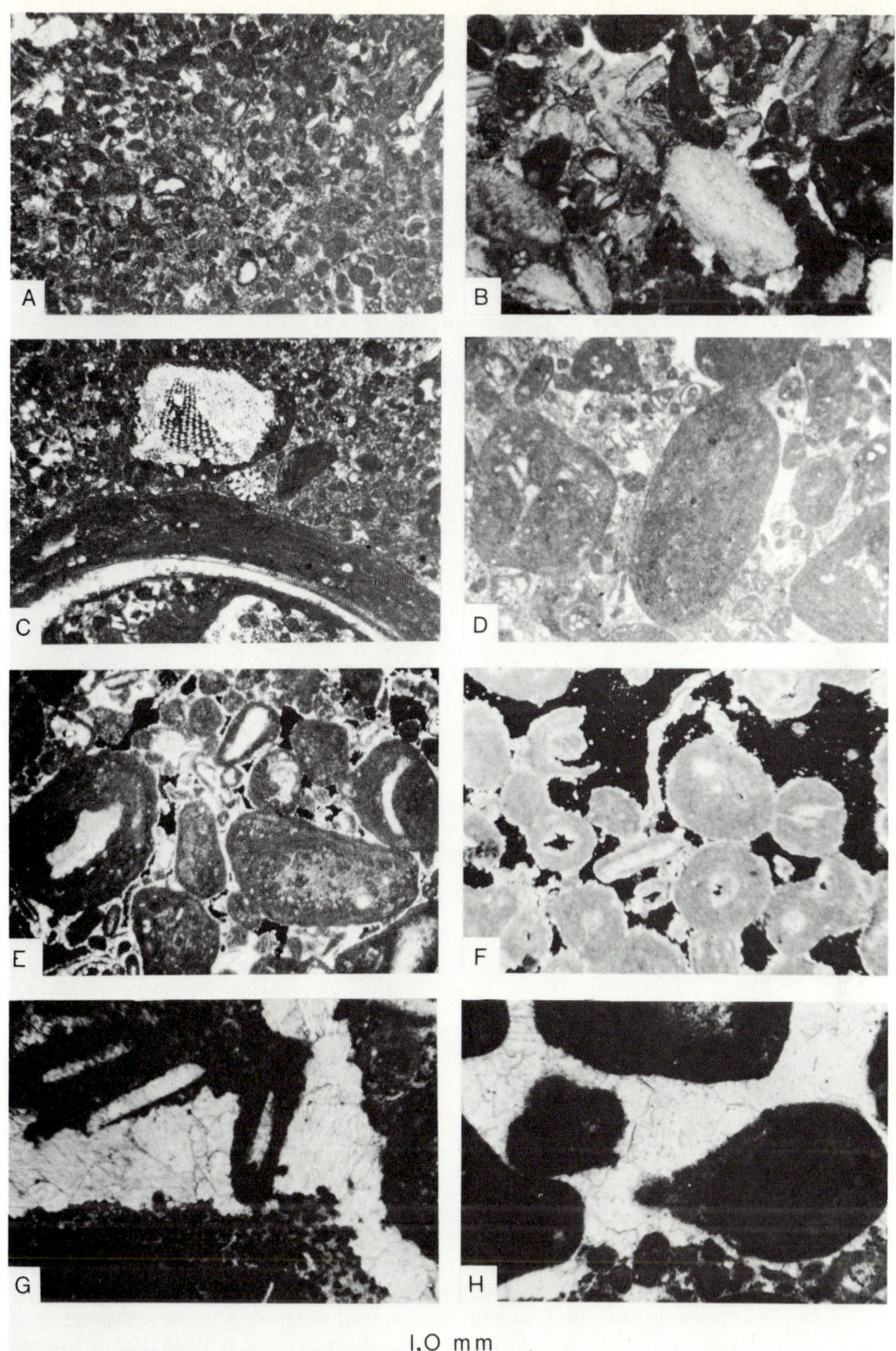

FIGURE 11.59 Macaé Formation (Albian-Cenomanian), Campos Basin, offshore Rio de Janeiro, Brazil. Typical microfacies (continued). A. Microfacies 100. B. Microfacies 110. C. Microfacies 120. D. Microfacies 130. E and F. Microfacies 140. G and H. Microfacies 150. See text for detailed descriptions. All photomicrographs: plane-polarized light, except E and F, crossed nicols. From Falkenhein *et al*. (1981).

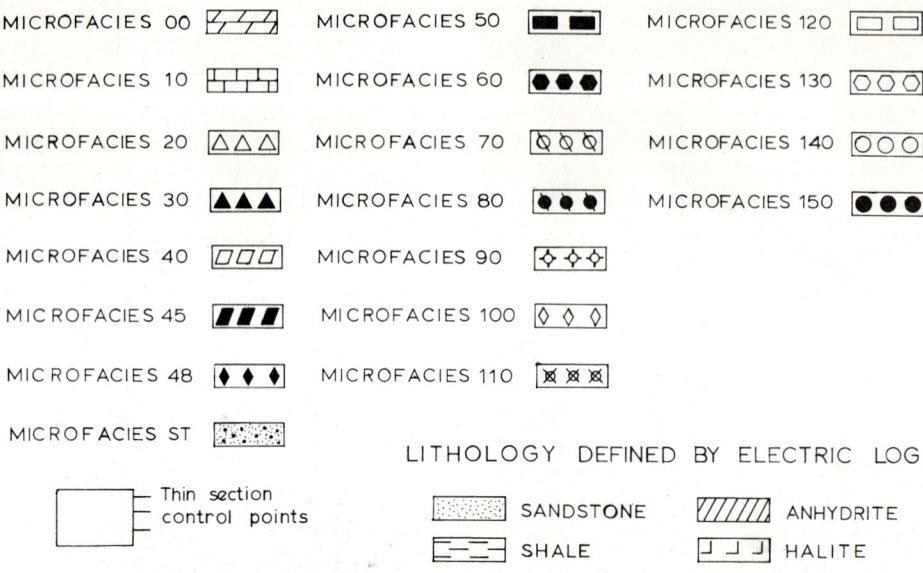

Figure 11.60 Macaé Formation (Albian-Cenomanian), Campos Basin, offshore Rio de Janeiro, Brazil. Symbols for microfacies. From Falkenhein et al. (1981).

bioclasts, an index of biotic diversity of benthic bioclasts as nuclei of oncoids or ooids, a percentage of detrital quartz as nuclei of oncoids or ooids, and a relative percentage of planktonic bioclasts among the total skeletal fragments.

The percentage of planktonic components relative to the total amount of bioclasts was used for evaluating quantitatively the paleoecologic significance of the benthic-planktonic relationship in calcilutites. The percentage of detrital quartz was used for evaluating the proximality of shore environments.

The Ideal Depositional Models

The Macaé Formation can be interpreted as forming during stratigraphic superposition of three successive depositional models. Each model shows a symmetrical occurrence of microfacies in the various shoals which is based on facies relationships between exploratory wells. This symmetry relative to a central and shallower crest of shoals reflects both landward and seaward hydraulic sorting and suggests bidirectional current, wave, and tidal action. Model 1 (Fig. 11.62) corresponds to the first 300 m of carbonate sedimentation. Shallow subtidal to high intertidal shoreface shoals were not continuous throughout the carbonate platform; thus the foreshore environment was relatively open to marine water circulation. Oncoids reached their maximum in the marginal portions of the shoals while ooids had the highest frequency in their center. The calcilutite matrix was absent on the crest of shoals and increased both landward and seaward. Internal sediments occurred only on the crest of the shoals. The index of biotic diversity of benthic bioclasts is large in the shoal deposits and correlates very well with the index of biotic diversity of bioclasts as nuclei of oncoids and ooids, indicating an adjacent stable baffle community. Besides the diagenetic parameters that relate porosity to the freshwater vadose dissolution at the crest of the shoals, moderate dolomitization by mixing marine-freshwater (dorag) occurred along the marginal portions. This process was independent from the sabkha-type dolomitization in the backshore environment. In summary, variation of components and indexes shows a suite of oncoidal calcarenites to calcirudites forming discontinuous shoreface shoals and a suite of calcilutites deposited in backshore, foreshore, and offshore environments.

Model 2 (Fig. 11.63) corresponds to a slice of carbonates between 400 and 700 m below the top of the carbonate unit. It differs from the previous model by the offshore expansion of the carbonate platform with subtidal microoncoidal shoals. On these shoals algal mats did not actively grow because baffle systems composed of immature and unstable seagrass-like communities were unable to provide protection against turbulence and could not prevent an antipathetic relationship between grazers and algal mats. Shoreface shoal deposits display the same behavior of microscopic parameters as in model 1, but deposits in the new offshore subenvironments are fossiliferous and biotic diversified calcilutites adjacent to microoncoidal shoals.

Model 3 (Fig. 11.64) represents the final episode of the depositional history of the Macaé Formation and corresponds to the upper 400 m of carbonates. This model differs from the previous one by a large development

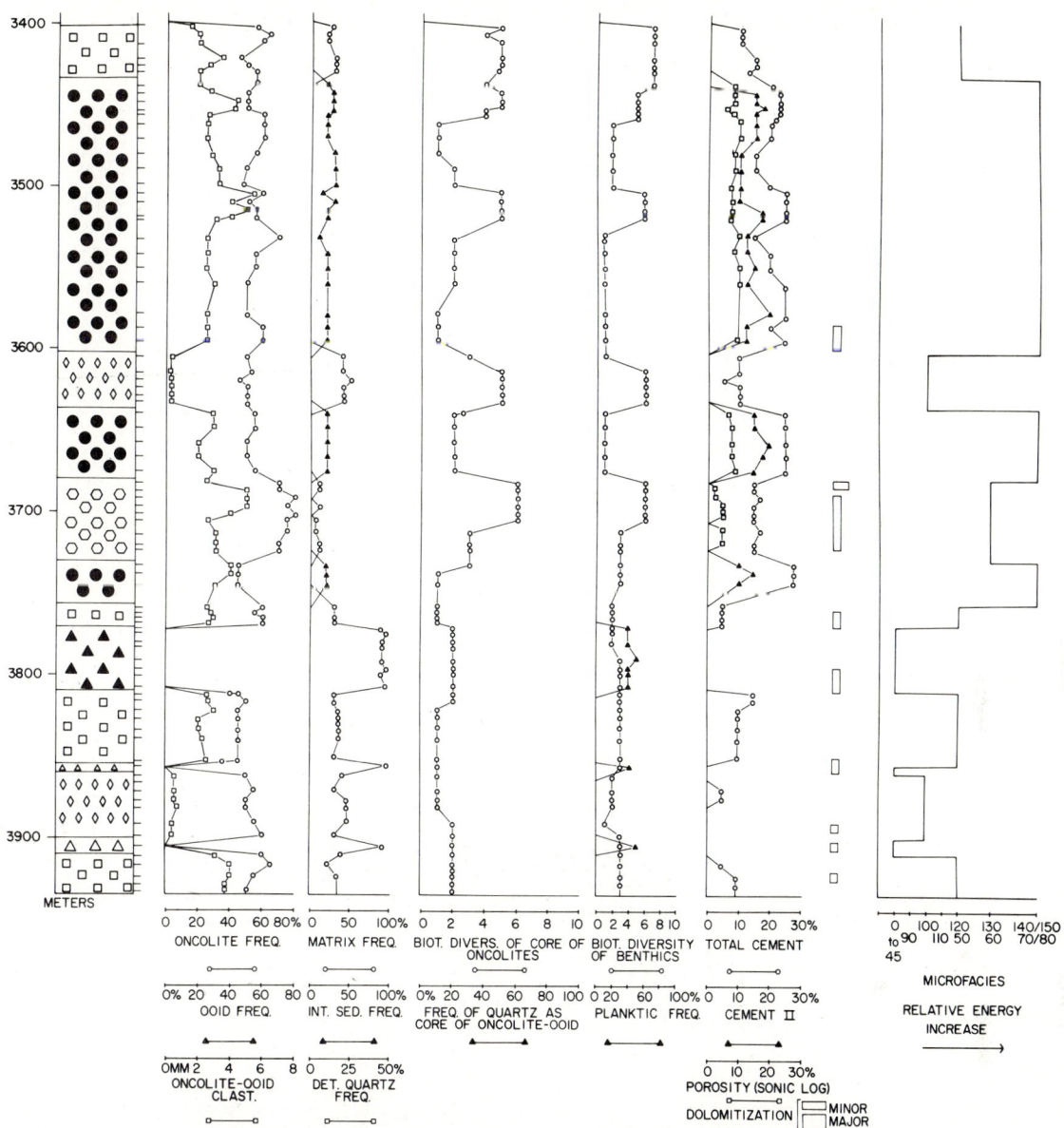

FIGURE 11.61 Macaé Formation (Albian-Cenomanian), Campos Basin, offshore Rio de Janeiro, Brazil. Typical example of vertical variation of microscopic parameters during a shallowing-upward sequence. From Falkenhein *et al.* (1981).

of offshore shallow subtidal to high intertidal oncoidal shoals which grew from the subtidal microoncoidal shoals of the previous model. This model also differs from the previous one by more differentiated offshore calcilutites. All depositional parameters and indexes show that the oncoidal offshore shoals are much better developed than their shoreface equivalents. Furthermore, they have undergone the same diagenetic processes, which led to reservoir generation of even higher quality.

Diagenesis

The oil traps in the upper part of the Macaé are of a mixed structural-stratigraphic nature, resulting from closure by doming and growth faults combined with favorable depositional-diagenetic conditions (Carozzi *et al.*, 1983; Carozzi and Falkenhein, 1985). These conditions are part of a sequence of 10 stages of diagenesis, ranging from early to burial distinguished on petrographic and cathode luminescence textures.

Low Intertidal Environment

Stage 1 (Fig. 11.65) represented deposition of initial unconsolidated sediment consisting of a framework of abraded oncoids with interstitial sand-size bioclasts and intraclasts set in a finer matrix, interpreted as "oncoid flour" and derived from the abrasion of pisooncoids. Stage 2 (Figs. 11.65, 11.66A,B) was diagenetic and consisted of precipitation of thin isopachous rim cement of fibrous calcite (high-magnesium calcite and/or aragonite?).

High Intertidal Environment

Stage 3 (Figs. 11.65, 11.66A,B,C) was depositional and involved removal and reorganization of interstitial constituents into an internal sediment with geopetal features and horizontal surfaces. Energy was high and predominant circulation of seawater downward, probably adjacent to a beachrock.

Beachrock Environment

Stage 4 (Figs. 11.65, 11.66D) was diagenetic and corresponded to lithification of internal sediment by interparticle sparite cement followed by dissolution which generated secondary vuggy to channel porosity cutting across oncoids and internal sediment. Stage 5 (Figs. 11.65, 11.66E) was diagenetic and involved precipitation of rim cement of bladed calcite (high-magnesium calcite and/or aragonite?) along boundaries of all previously generated open spaces. If preserved after burial, this stage is a reservoir.

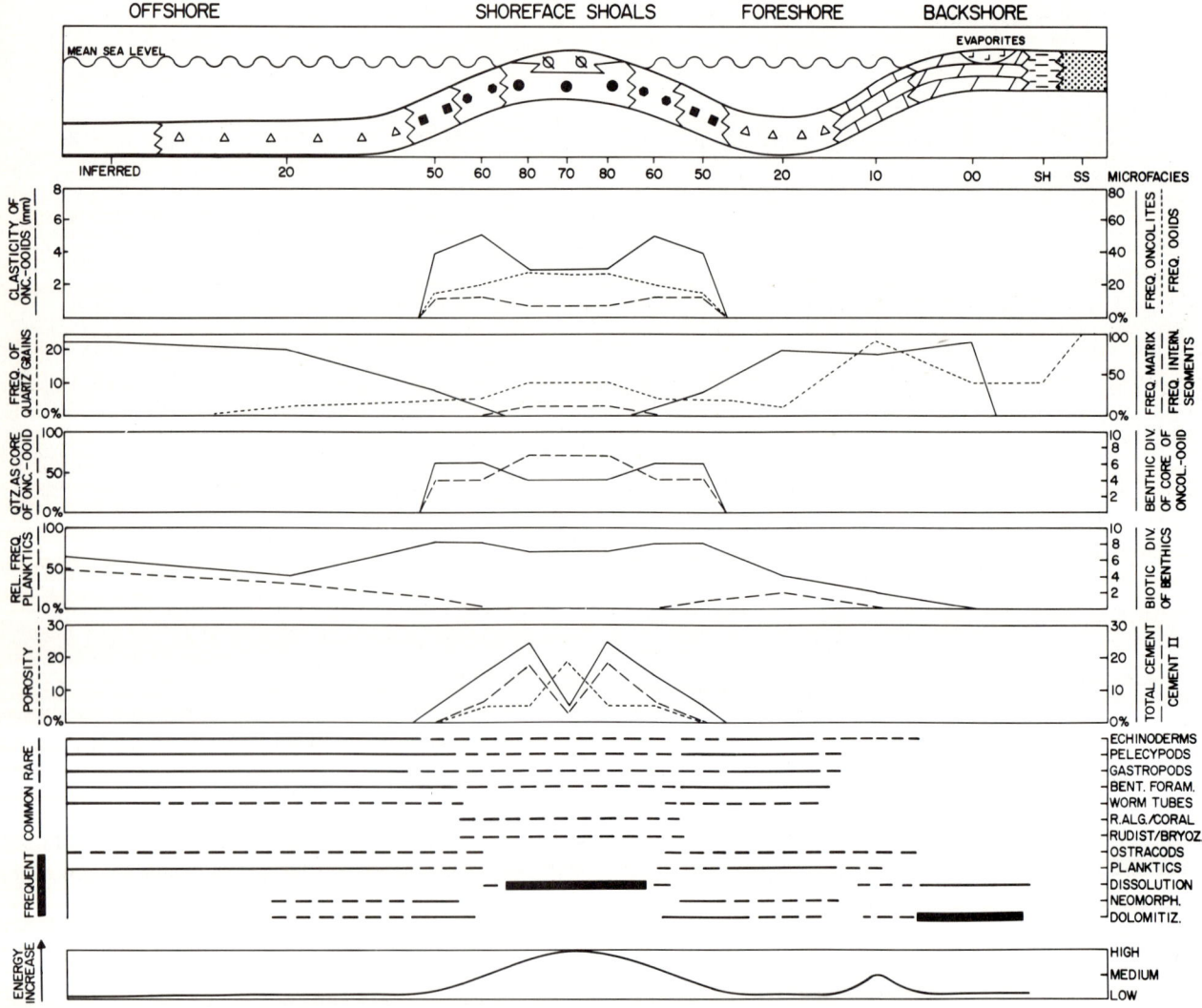

FIGURE 11.62 Macaé Formation (Albian-Cenomanian), Campos Basin, offshore Rio de Janeiro, Brazil. Ideal depositional model 1. From Falkenhein *et al.* (1981).

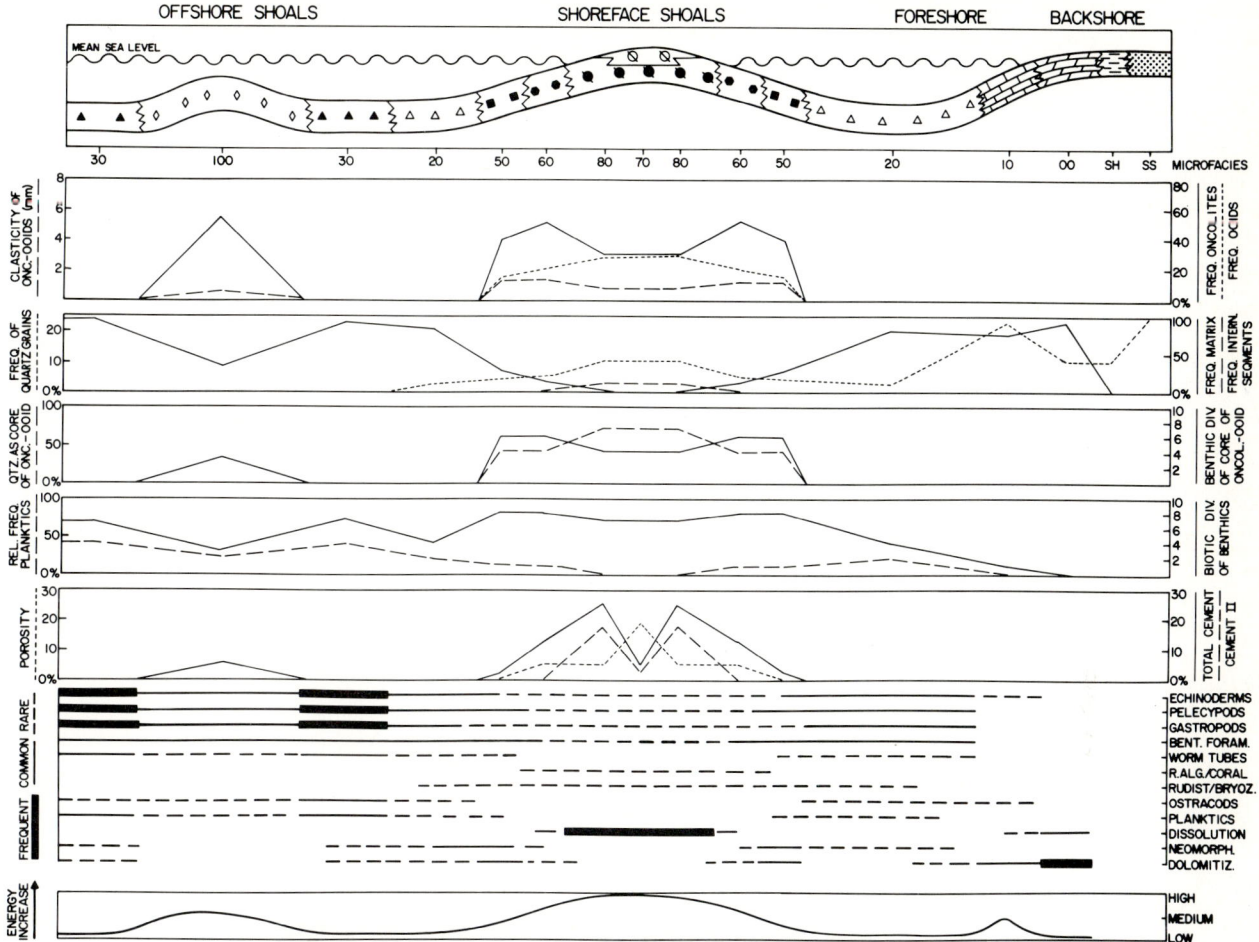

FIGURE 11.63 Macaé Formation (Albian-Cenomanian), Campos Basin, offshore Rio de Janeiro, Brazil. Ideal depositional model 2. From Falkenhein *et al.* (1981).

Beachrock Vadose Environment

Stage 6 (Figs. 11.65, 11.66F,G,H) was diagenetic and its results whenever preserved after burial correspond to the best reservoir rocks. These results are intense vadose dissolution after exposure of beachrock to subaerial conditions but without generation of a freshwater lens; removal of large amounts of bladed rim cement; corrosion of margins of oncoids; deep etching of upper surface of cemented internal sediment; and local differential solution of nuclei and concentric layers of oncoids. These mesopores provide a measured porosity of 25 to 30% and 200 to 400 md permeability.

Freshwater Meteoric Phreatic Environment

Stage 7 (Figs. 11.65, 11.67A,B) was diagenetic and corresponded to cementation in a freshwater lens generated by extensive subaerial exposure. Reservoirs formed during the previous beachrock stage, including any enhancement by subsequent meteoric vadose conditions, were occluded by calcite precipitation when transferred by subsidence into the meteoric phreatic zone prior to permanent burial. The phreatic calcite cement occurs as single crystals and coarsely crystalline irregular to highly interlocked equant mosaic. Frequent occurrence of intense and distorted twinning indicates that cementation preceded major compaction.

Burial Environment

Stage 8 (Figs. 11.65, 11.67C,D,E) involved compaction with grain breakage by radial microfractures, reciprocal grain interpenetration by pressure-solution and stylolitization, spalling of bladed rim cement with penetration in oncoids, oncoid spalling, and non-fabric-selective fracturation, at least in two generations. Early fractures were cemented and others enlarged by late dissolution. In general, pores generated during compaction remain uncemented or show a very thin rim of late calcite cement. Stage 9 (Figs. 11.65, 11.67F) included late dissolution with enlargement of all previous compaction

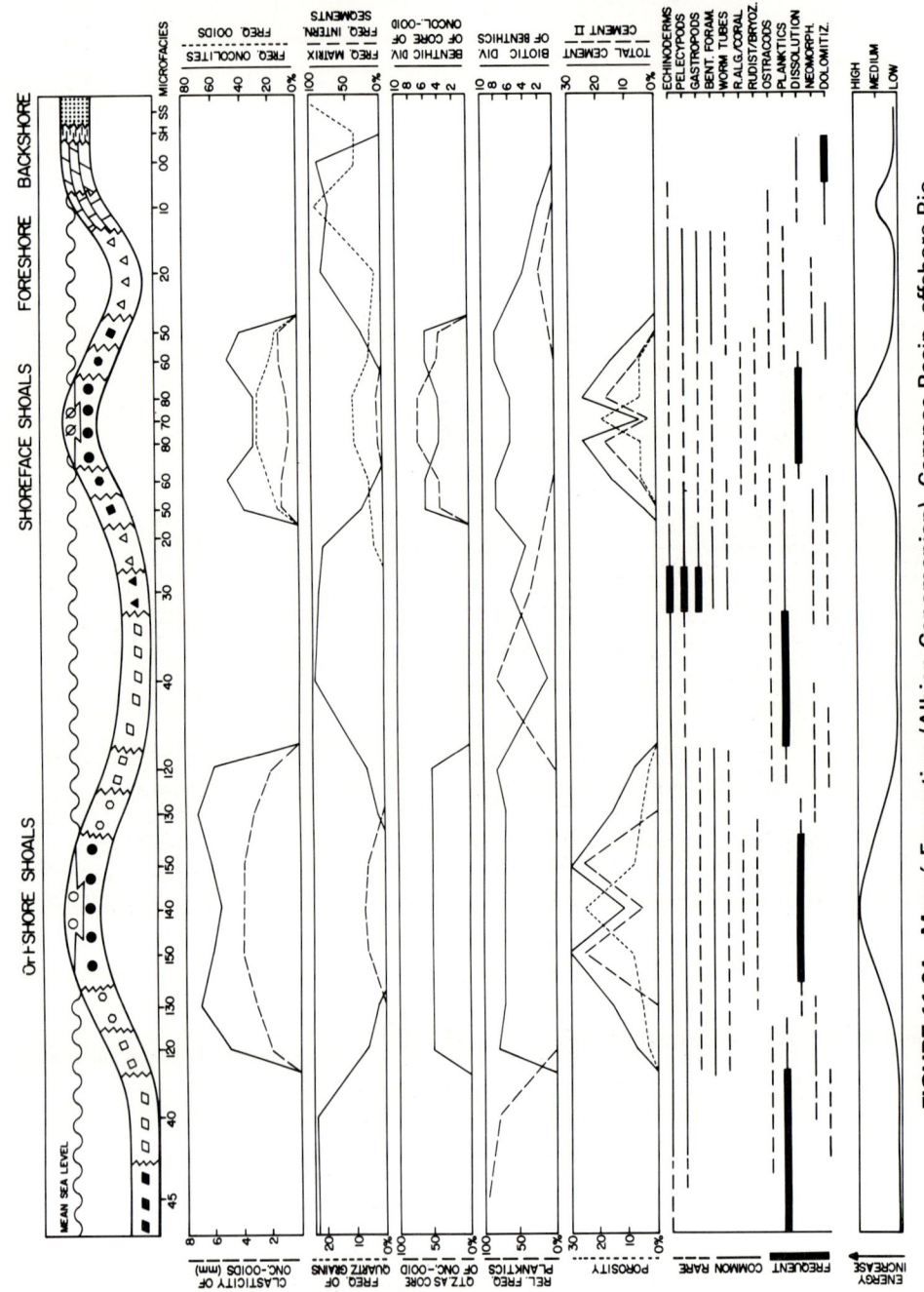

FIGURE 11.64 Macaé Formation (Albian–Cenomanian), Campos Basin, offshore Rio de Janeiro, Brazil. Ideal depositional model 3. From Falkenhein *et al.* (1981).

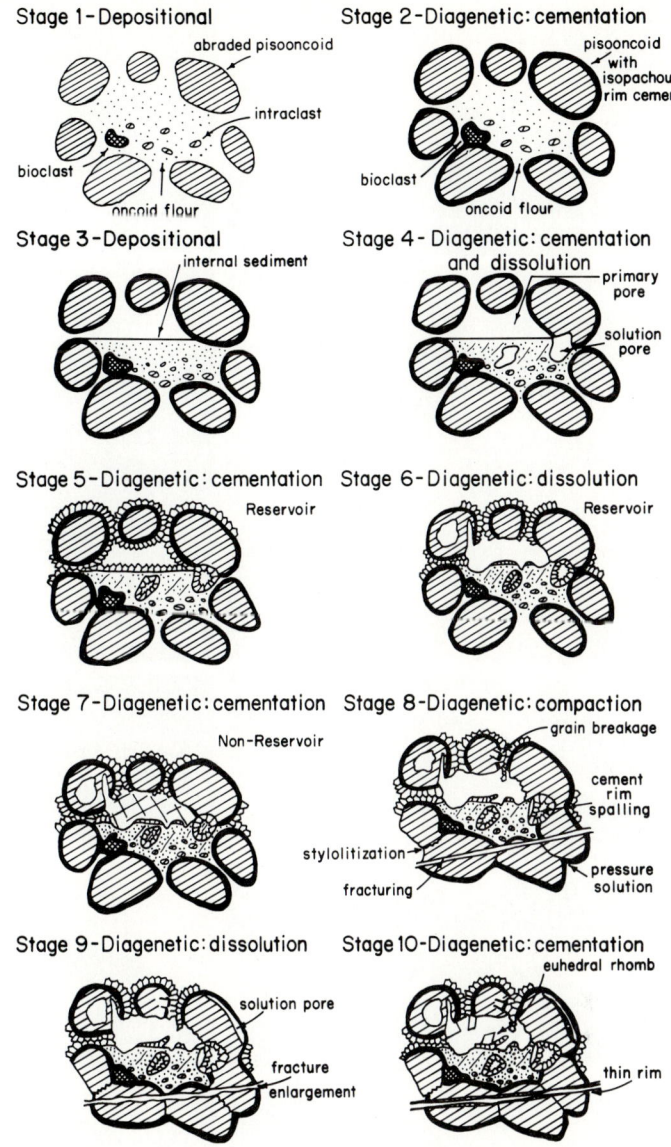

FIGURE 11.65 Macaé Formation (Albian-Cenomanian), Campos Basin, offshore Rio de Janeiro, Brazil. Depositional-diagenetic sequence. From Carozzi *et al.* (1983). Reprinted by permission of Springer-Verlag New York Inc.

features and generation of intraparticle porosity by differential solution of some concentric layers of oncoids. Late cementation was widespread in stage 10 (Figs. 11.65, 11.67F,G,H), but porosity was not greatly reduced. Cement occurs as a thin rim of calcite crystals and as perfect rhombohedral crystals with frequent growth lines and inclusions (oil?). This cement may be coeval with oil migration.

Other minor aspects of burial diagenesis include aggrading neomorphism appearing as relatively large cloudy calcite crystals with diffused boundaries and abundant inclusions of unreplaced material; rare dolomite rhombs replacing matrix and cement; extremely rare silicification as overgrowths on detrital quartz grains that postdates dolomitization; and anhydritization selectively replacing oncoidal layers and sparite cement.

11.7 CHARACTERISTIC CONSTITUENTS: RED ALGAE AND LARGER BENTHIC FORAMINIFERS

Mature platforms with bioaccumulated buildups that form on passive continental margins or along edges of intracratonic basins develop in time self-perpetuating morphologies directly influenced by the nature of framework-building organisms and the mode of dispersion of bioclasts by the action of tides and currents. A typical transverse pattern consists of various types of bioaccumulated buildups separated by active tidal channels extending landward as far as the lagoons. Tidal actions redistribute the abundant production of bioclasts into channel-like and tidal delta deposits that may terminate seaward into deep-sea fans.

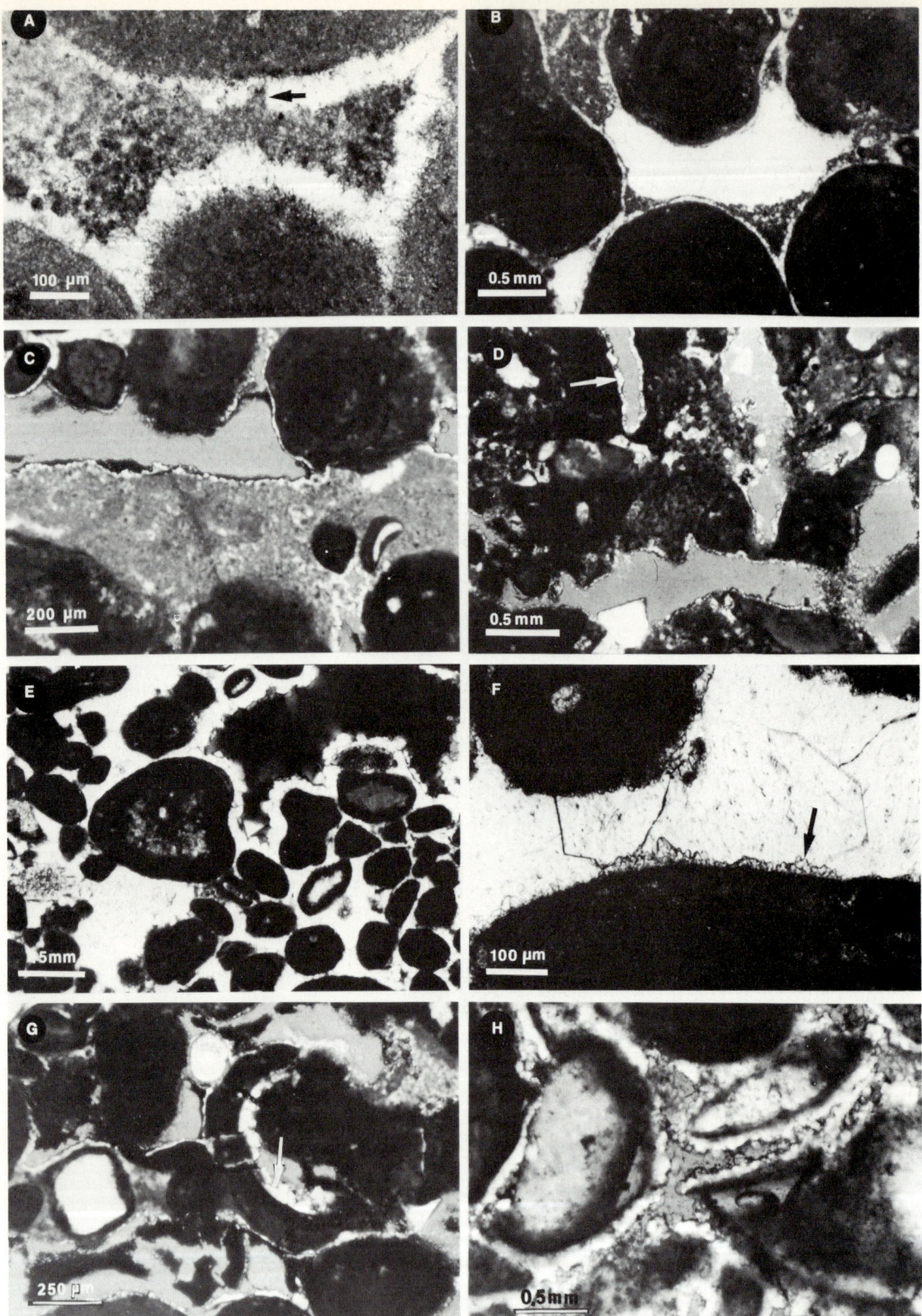

FIGURE 11.66 Macaé Formation (Albian-Cenomanian), Campos Basin, offshore Rio de Janeiro, Brazil. Illustrations of depositional-diagnetic stages. *Submarine to beachrock vadose diagenesis.* Plane-polarized pictures of thin sections impregnated with blue plastic, porosity appears in various shades of gray. A. Submarine isopachous rim cement of fibrous calcite with traces of dissolution (arrow) and surrounding graded internal sediment. B. Isopachous rim of fibrous

The Amapá Formation (Paleocene-Middle Miocene) is an extensive carbonate platform in the Foz do Amazonas Basin, reaching 4000 m thickness and located entirely offshore northeast Brazil. This platform, which probably represents one of the largest Cenozoic carbonate systems controlled by red algae and larger foraminifers, was studied petrographically by means of 2795 thin sections obtained from 18 exploratory wells with an average vertical interval of 50 cm in the cores and 6 to 18 m in cuttings (Wolff and Carozzi, 1984). The Amapá carbonate platform was adjacent to a landward fluviodeltaic system that periodically prograded over the platform. The platform became discontinuous after Oligocene time, when transverse canyons were developed that connected the deltaic complex to the open sea (Carozzi et al., 1981b; Carozzi, 1981). The deposition of the Amapá carbonates may be interpreted as resulting from the stratigraphic superposition of two distinct models separated by a petrographic discontinuity corresponding to a major global unconformity.

Description of Microfacies of Model 1

Model 1 ranges from Paleocene to Early Eocene and is interpreted as a carbonate platform consisting mainly of larger foraminifer banks (mostly *Ranikothalia* sp.) that accumulated in the outer shelf, of dasyclads, and oolitic shoals formed in the inner shelf.

Microfacies are described in order of decreasing relative water depth from the slope toward the platform.

Slope Deposits

Microfacies 102 (Fig. 11.68A). Bioturbated argillaceous calcilutite with common planktonic foraminifers and small rotaliids. Ostracods, textulariids, silt-size bioclasts of red algae, and phosphatic debris are rare. Aggrading neomorphism of micrite into pseudomicrosparite is common, and pyrite pigments are scattered in the groundmass.

Outer Shelf Deposits

Microfacies 42 (Fig. 11.68B). Bioaccumulated larger foraminifer limestone consisting predominantly of *Ranikothalia* sp. Accessory components are red algae, miliolids, dasyclads, bryozoans, and mollusks. Interstitial matrix is micrite or fine calcisiltite, frequently dolomitized.

Microfacies 420 (Fig. 11.68C). Grain-supported biocalcarenite consisting predominantly of debris of nummulitids (mostly *Ranikothalia* sp.). Other components are echinoids, mollusks, small rotaliids, red algae, dasyclads, and rare ooids. Micrite matrix is commonly altered into pseudomicrosparite by neomorphism or is dolomitized.

Inner Shelf Deposits

Microfacies 402 (Fig. 11.68D). Grain-supported oolitic biocalcarenite with sparite cement. Superficial ooids usually predominate over normal ooids characterized by poorly developed concentric layers. Ooid cores consist mainly of dasyclads and secondarily of debris of nummulitids, miliolids, echinoids, *Halimeda*, mollusks, and sand-size quartz grains. Uncoated bioclasts are composed of mollusks, green algae, nummulitids, miliolids, echinoids, and small rotaliids. A well-developed fibrous rim cement followed by a second generation of sparite mosaic characterize this microfacies.

Microfacies 500 (Fig. 11.68E,F). Grain-supported biocalcarenite with predominant dasyclads. Other common components are miliolids, codiaceans, and mollusks. Red algae, small rotaliids, echinoids, nummulitids, and ooids are minor components. These dasycladacean calcarenites display a complete range of interstitial material from pure micrite matrix to pure sparite cement (Fig. 11.68F).

FIGURE 11.66 (*continued*) calcite cement overlain by graded internal sediment with geopetal attitude, showing vadose dissolution effects preceding phreatic sparite cementation. C. Horizontal surface of geopetal internal sediment. D. Vuggy to channel secondary porosity cutting across pisooncoid (right side) and internal sediment. Pores rimmed by bladed calcite cement (arrow). E. Rim cement of bladed calcite (upper right); elsewhere, same calcite is cavity filling. F. Relict of bladed calcite rim cement on pisooncoid surface (arrow) from vadose dissolution before phreatic sparite cementation. G. Reservoir rock showing vadose dissolution effects on bladed calcite rim cement (arrow). H. Reservoir rock showing vadose dissolution of bladed calcite rim cement and of cortical layers of pisooncoid. From Carozzi et al. (1983). Reprinted by permission of Springer-Verlag New York Inc.

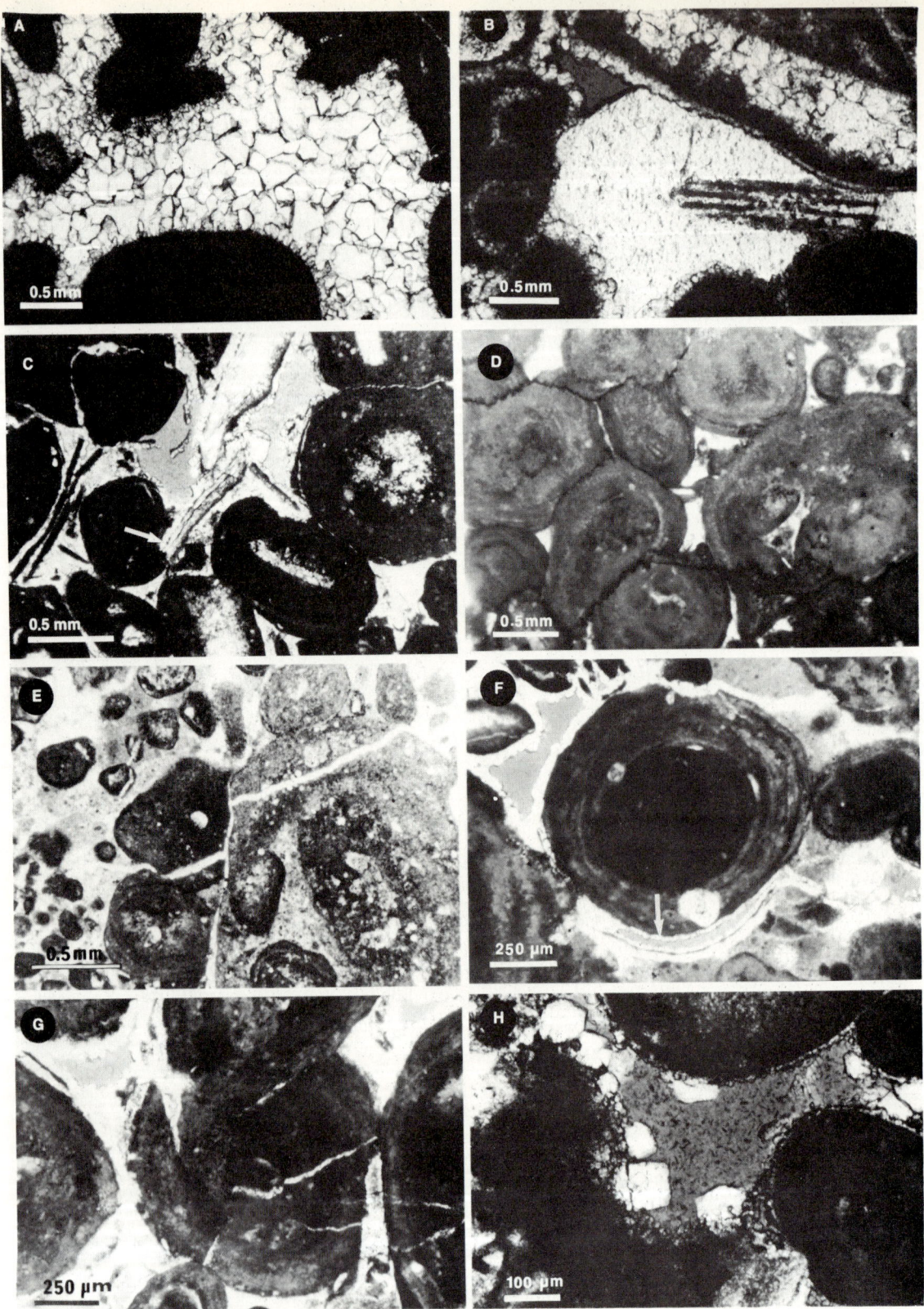

FIGURE 11.67 Macaé Formation (Albian-Cenomanian), Campos Basin, offshore Rio de Janeiro, Brazil. Illustrations of depositional-diagenetic stages (continued). *Meteoric phreatic to burial diagenesis.* Plane polarized pictures of thin sections impregnated with blue plastic, porosity appears in various shades of gray. A. Reservoir occlusion by phreatic coarsely crystalline equant mosaic of calcite. Earlier rim cement still visible. B. Almost complete occlusion of pores by single crystal of calcite overgrowth on echinoid spine. Earlier bladed rim cement still visible. C. Compaction effects: spalling of bladed calcite

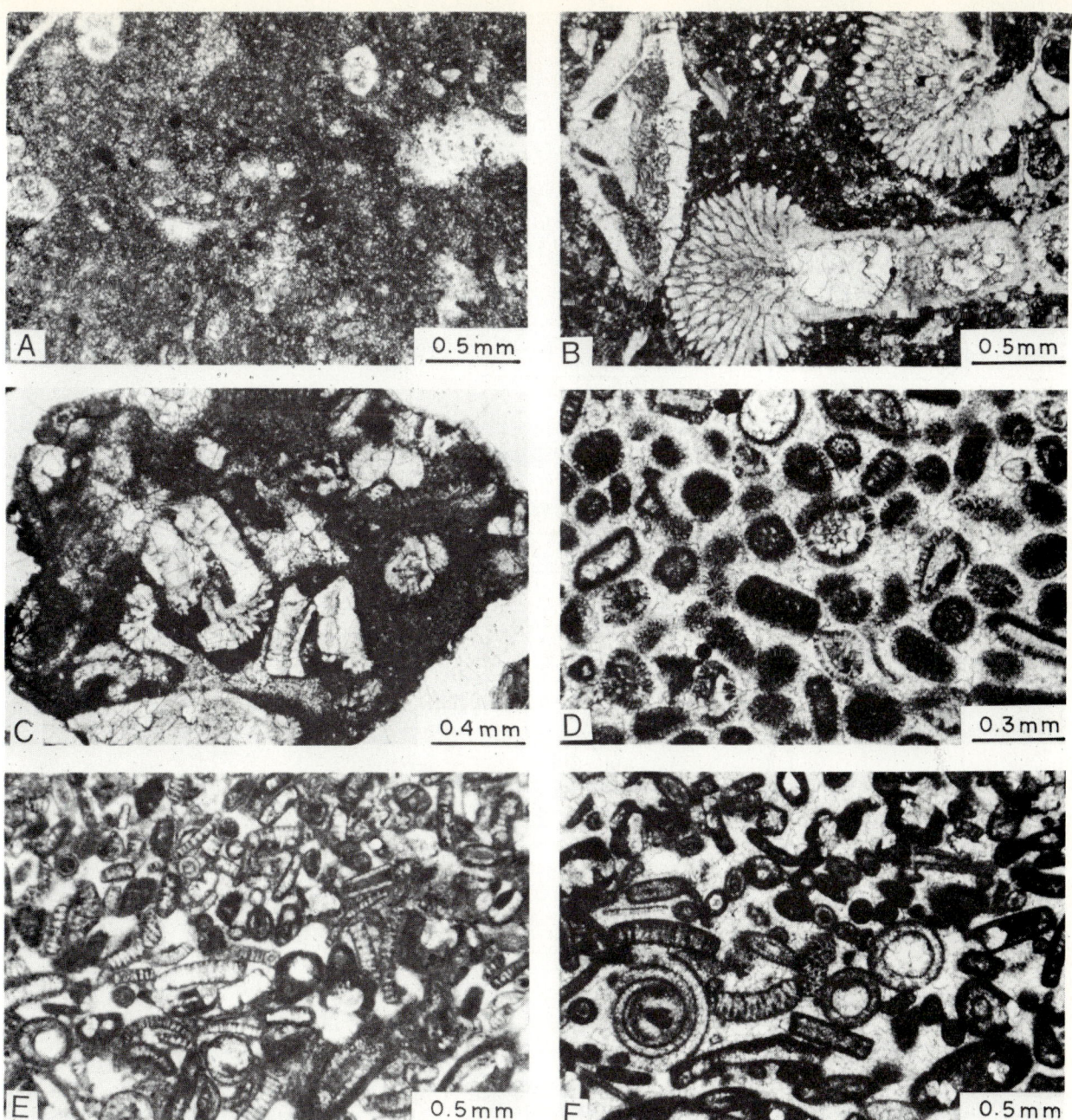

FIGURE 11.68 Amapá Formation (Paleocene–Middle Miocene), Foz do Amazonas Basin, offshore northeast Brazil. Typical microfacies of ideal depositional model 1. A. Microfacies 102. B. Microfacies 42. C. Microfacies 420. D. Microfacies 402. E and F. Microfacies 500. See text for detailed descriptions. All photomicrographs: plane-polarized light. From Wolff and Carozzi (1984).

FIGURE 11.67 (*continued*) rim cement, grain breakage by microfractures, bladed rim cement penetrating pisooncoid margin (arrow). D. Compaction effects: deformed pisooncoids, reciprocally penetrating pisooncoids and stylolitized contacts. E. Compaction effects: first generation fracture cemented and second generation fracture (trending vertical) displacing the first and open. F. Late dissolution pore (arrow) by differential solution of concentric layers of pisooncoid and showing thin rim of late calcite cement. G. Thin rim of late calcite cement filling fractures of pisooncoid deformed by compaction. H. Late cement as perfect rhombohedral crystals with oil staining growing on relicts of earlier bladed calcite rim cement. From Carozzi *et al.* (1983). Reprinted by permission of Springer-Verlag New York Inc.

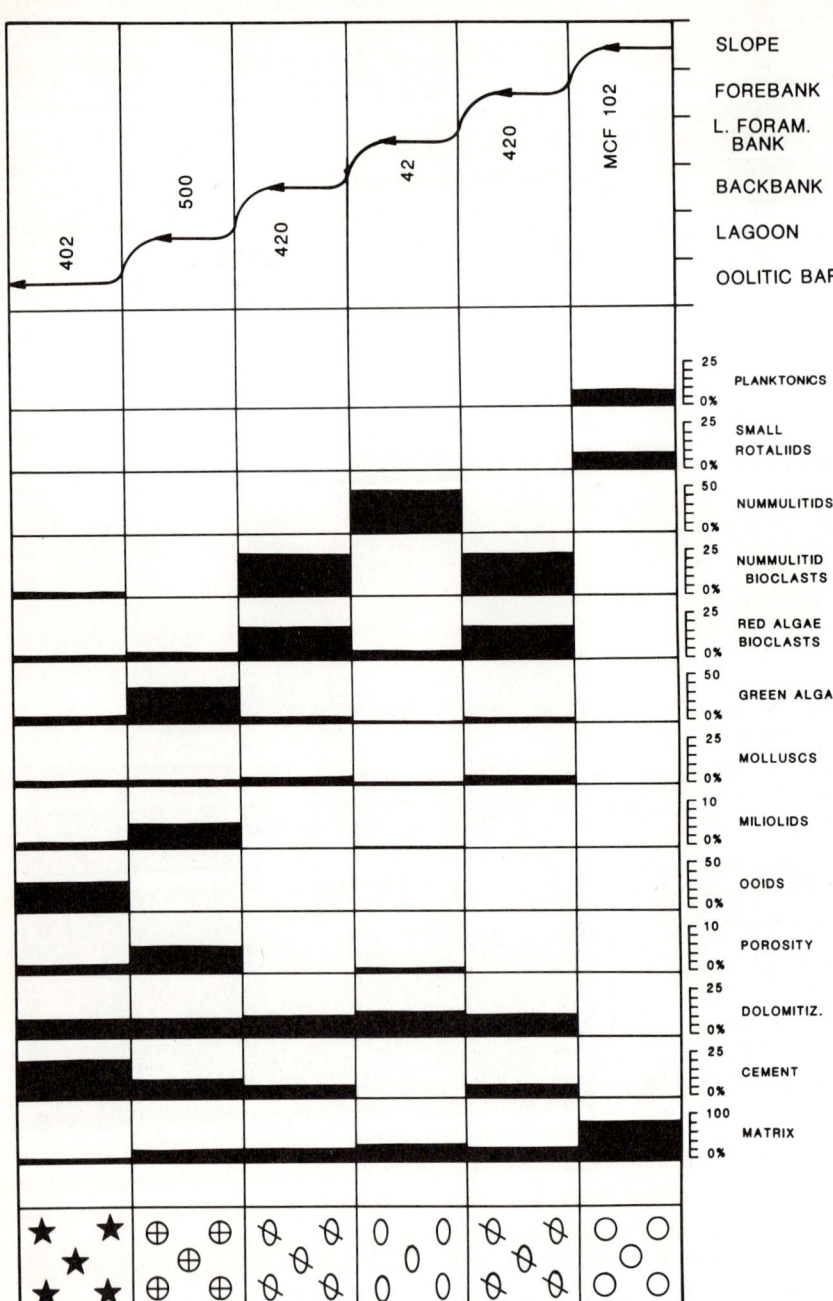

FIGURE 11.69 Amapá Formation (Paleocene–Middle Miocene), Foz do Amazonas Basin, offshore northeast Brazil. Ideal shallowing-upward sequence for model 1. From Wolff and Carozzi (1984).

The Ideal Depositional Model 1

The ideal shallowing-upward sequence of microfacies (Fig. 11.69) consists of the following environments: slope, forebank, larger foraminifer bank, backbank, lagoon, and oolitic bar. The depositional model (Fig. 11.70) shows how the larger foraminifer bank in an outer shelf position is surrounded by a destructional halo of calcarenites consisting mainly of nummulitid debris. This interpretation is supported by shallowing-upward sequences in which microfacies 42 is both preceded and followed by microfacies 420. Similarly, oolitic calcarenites are interpreted as discontinuous oolitic bars developed in a broad lagoonal environment represented by dasycladacean calcarenites (Fig. 11.71).

Diagenesis

Petrographically, a sequence of 12 diagenetic events has been established for model 1. Each individual microfacies is considered separately to illustrate best the various inferred diagenetic pathways (Fig. 11.72). These pathways represent the interpreted diagenetic processes undergone by each microfacies from original depositional site to deep burial realm and their final end products in terms of porosity. Upon deciphering individual diagenetic path-

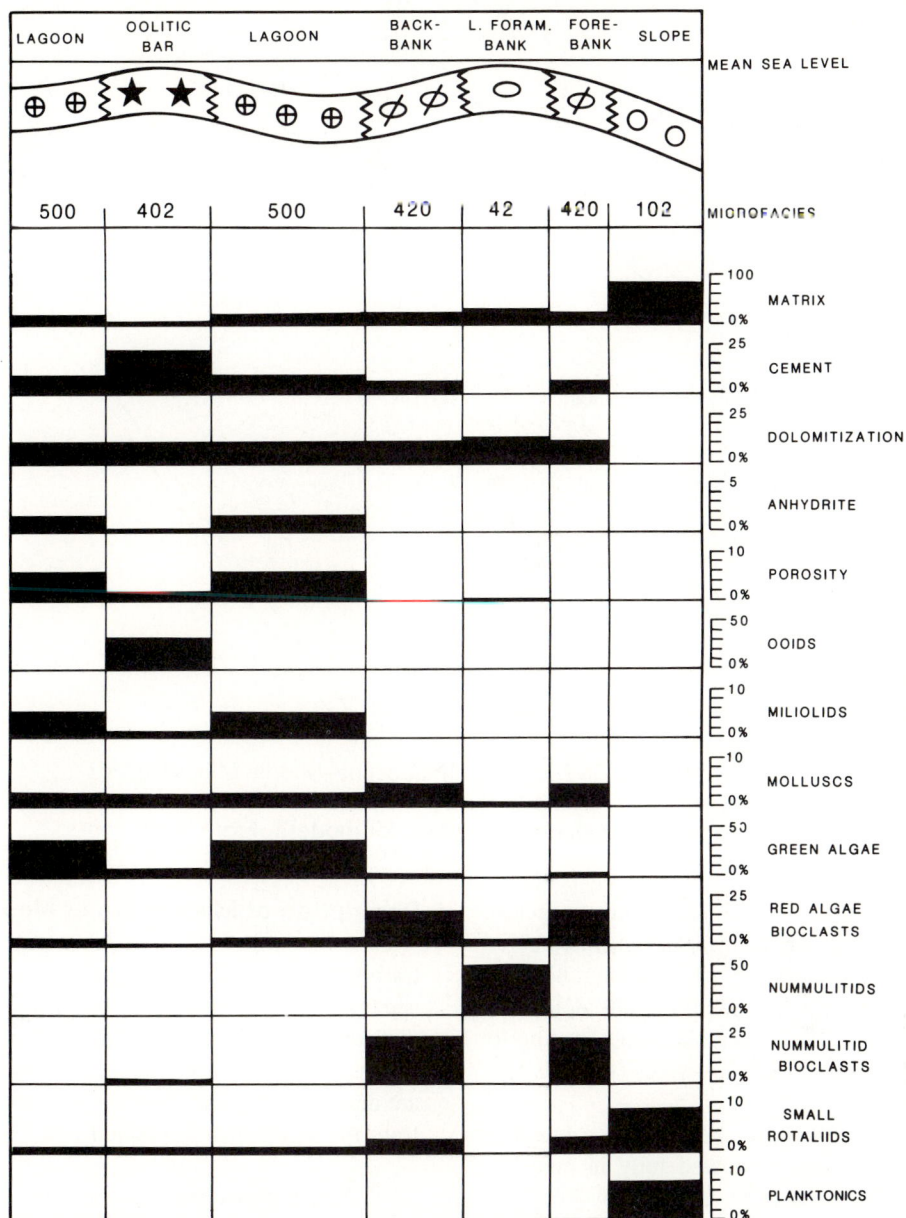

FIGURE 11.70 Amapá Formation (Paleocene–Middle Miocene), Foz do Amazonas Basin, offshore northeast Brazil. Ideal depositional model 1. From Wolff and Carozzi (1984).

ways undergone by each microfacies, the composite diagenetic sequence is reconstructed for the carbonate system as a whole (Fig. 11.73). A curve is used to represent the general loss or gain of porosity inferred for each diagenetic event. The thickness of the black bars represents the relative importance of each diagenetic event.

Porosity Evolution

Although good evidence supports a major stage of dissolution in the undersaturated zone of the freshwater phreatic environment, the porosity generated at this stage was almost completely occluded by calcite cement, subsequently precipitated in the saturated zone (Fig. 11.74A,B,C). The most important reservoirs in model 1 are due to primary porosity, locally reduced by fibrous rim cement, late euhedral calcite cement, and rare anhydrite cement (Fig. 11.74E to H). The rocks that constitute potential hydrocarbon reservoirs escaped freshwater phreatic cementation. Possible explanations to account for the interruption in the normal diagenetic sequence after marine cementation may be rapid burial and/or stagnation of diagenetic fluids.

Primary porosity is very important in some intervals of the lagoonal sediments (microfacies 500), where it reaches approximately 20%; it also represents most of

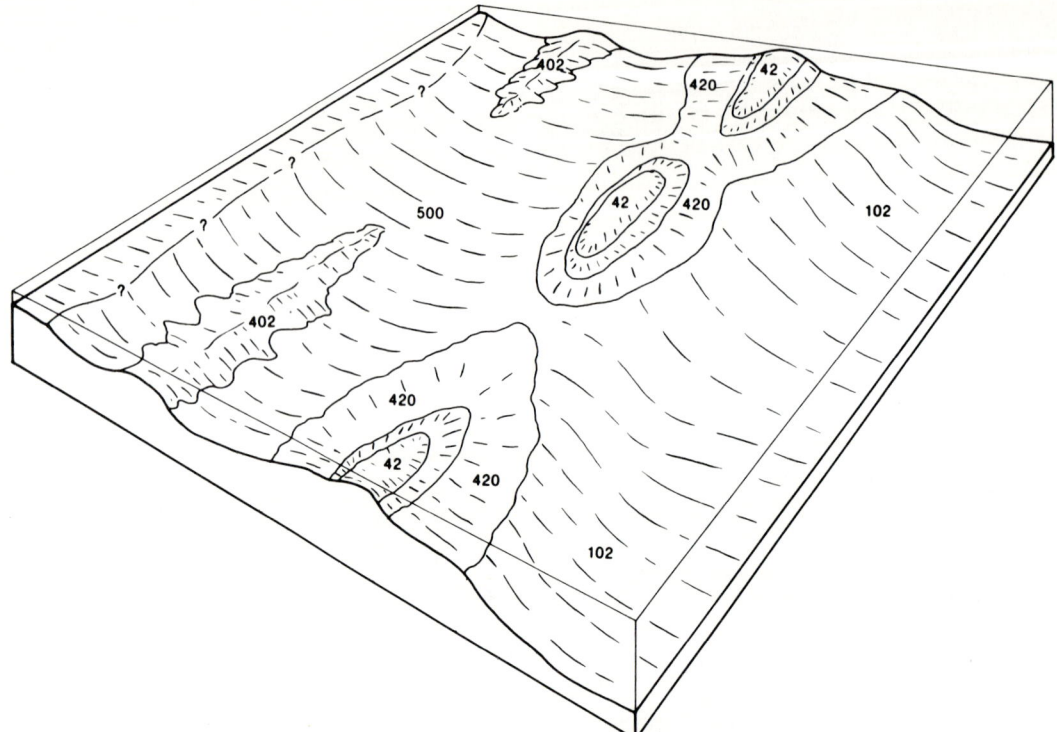

FIGURE 11.71 Amapá Formation (Paleocene–Middle Miocene), Foz do Amazonas Basin, offshore northeast Brazil. Block diagram showing paleoenvironmental reconstruction for model 1. From Wolff and Carozzi (1984).

the 6% overall porosity average for this microfacies. This type of porosity is observed locally in microfacies 402 (Fig. 11.74E) but shows no practical significance. Late burial cements (euhedral calcite, saddle dolomite, and anhydrite) are the last diagenetic events leading to total or partial occlusion of primary voids (Fig. 11.74A to D,H).

Generation of secondary intercrystalline porosity by dolomitization was not evaluated fully in model 1 because of the small number of samples available and the difficulty of estimating this type of porosity with the petrographic microscope.

The ITERIM program was used to evaluate whether or not the depositional model had controlled diagenesis. The input data matrix consisting only of diagenetic variables such as dolomitization, porosity, and cement percent was treated by computer. Samples were maintained in the same five distinct groups as defined by the microfacies technique. After stabilization of the program, a high core ratio would be considered as an indication of a large control of the depositional model over diagenesis. The program stabilized after six iterations with 62% of the thin sections left in their original groupings. The results, although statistically not very significant due to the small number of samples available for model 1, seem to agree with the petrographic observations that depositional environment did control diagenesis to some extent.

Description of Microfacies of Model 2

Carbonate sediments of this complex model range in age from Middle Eocene to Middle Miocene. It is possible to distinguish 17 juxtaposed microfacies which are classified into seven environmental subdivisions. Microfacies are described in order of decreasing relative water depth from the slope toward the platform.

Slope Deposits

Microfacies 101 (Fig. 11.75A). Bioturbated argillaceous calcilutite to calcisiltite with abundant planktonic foraminifers. Other common components are small rotaliids and locally silt-size red algae debris. Aggrading neomorphism of the micrite matrix into pseudomicrosparite is common.

Trough Deposits and Olistostromes

Microfacies 200 (Fig. 11.75B). Argillaceous mud-supported calcarenite to calcirudite with common allochthonous lithified carbonate clasts embedded in carbonate or shaly mudstone rich in planktonic foraminifers. Common components are: planktonic foraminifers, corals, red algae, larger foraminifers, and bifoliate bryozoans. Aggrading neomorphism of micrite matrix into pseudomicrosparite is frequent.

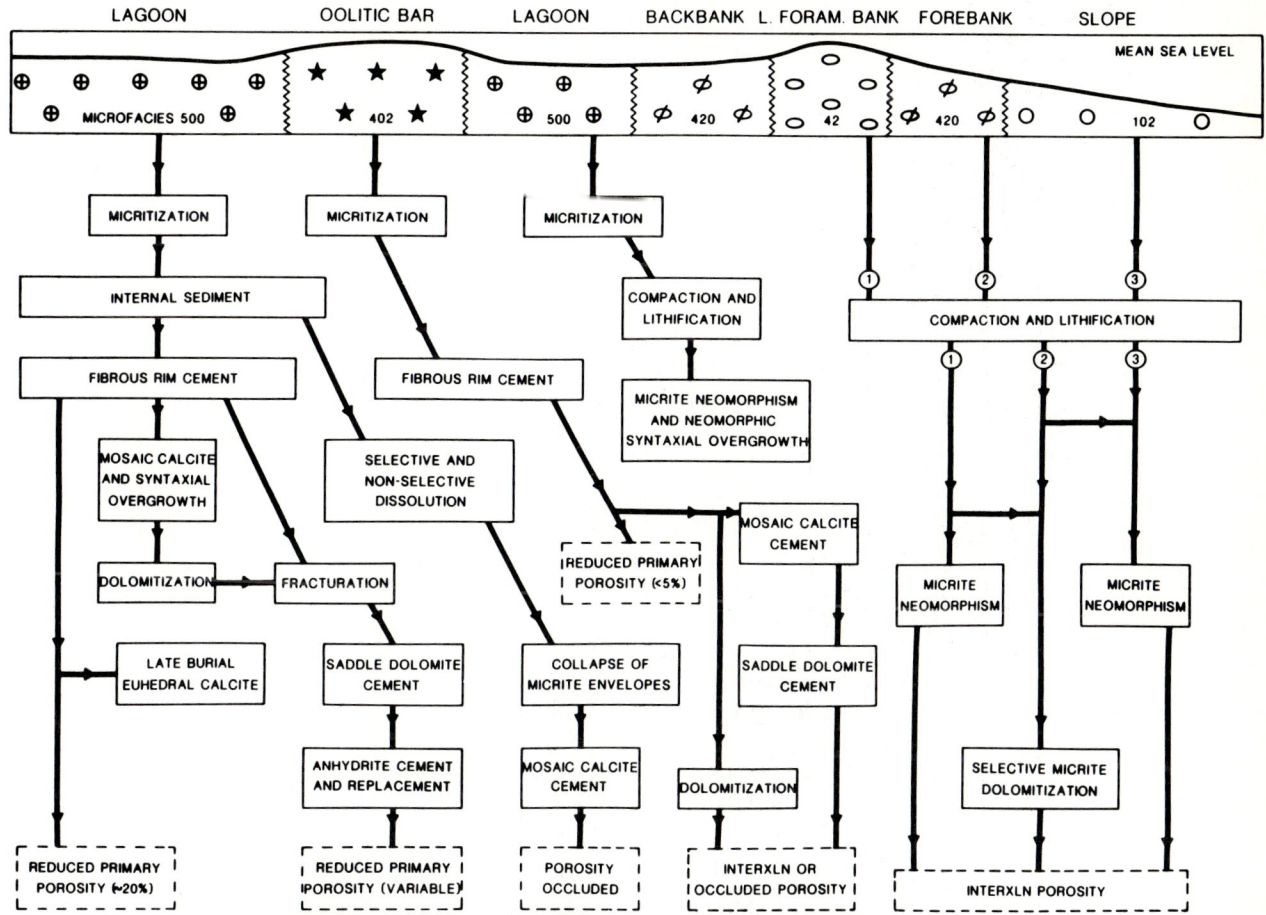

FIGURE 11.72 Amapá Formation (Paleocene–Middle Miocene), Foz do Amazonas Basin, offshore northeast Brazil. Main diagenetic pathways for each microfacies of model 1. The position of the solid boxes does not represent relative depths, and the dashed boxes indicate final porosity types and estimated amounts. From Wolff and Carozzi (1984).

Forebank Deposits

Microfacies 201 (Fig. 11.75C). Bioturbated argillaceous biocalcisiltite with abundant silt-size to very fine sand-size debris of red algae and common small rotaliids and planktonic foraminifers. Accessory components are ostracods, echinoids, mollusks, bryozoans, encrusting foraminifers, and pyrite pigments.

Microfacies 202 (Fig. 11.75D). Bioturbated argillaceous biocalcisiltite with common debris of larger foraminifers and red algae. Other components are small rotaliids, bryozoans, mollusks, echinoids, ostracods, and porcelaneous, arenaceous, and encrusting foraminifers.

Microfacies 403 (Fig. 11.75E). Grain-supported biocalcarenite consisting mainly of larger foraminifers (lepidocyclinids, discocyclinids, and nummulitids) and red algae. Small rotaliids and planktonic foraminifers are common.

Red Algae and Larger Foraminifer Bank Deposits

Microfacies 30 (Fig. 11.75F). Bioconstructed and bioaccumulated red algae limestone consisting predominantly of massive nodules, locally larger than 10 cm in diameter, associated with minor amounts of open branched colonies. Encrusting foraminifers and bryozoans are commonly intercalated between distinct types of red algae forming complex nodules. Interstitial matrix is composed of micrite and scattered bioclasts of small rotaliids, echinoids, mollusks, porcelaneous foraminifers, and corals.

Microfacies 32 (Fig. 11.75G). Bioconstructed and bioaccumulated red algae limestone consisting mainly of open branched red algae nodules in an abundant micrite matrix. Encrusting foraminifers and bryozoans are the main components associated with red algae. Other minor components scattered in the interstitial micrite

DIAGENETIC EVENTS	DIAGENETIC ENVIRONMENTS				POROSITY	
	EOGENETIC	MESOGENETIC			LOSS	GAIN
	MARINE PHREATIC	FRESHWATER PHREATIC	MIXED PHREATIC	DEEP BURIAL		
MICRITIZATION	▬					
INTERNAL SEDIMENT	▬					
FIBROUS RIM CEMENT	▬					
SELECTIVE AND NON-SELECTIVE DISSOLUTION		▬▬				
COLLAPSE OF MICRITE ENVELOPES						
MOSAIC CALCITE AND SYNTAXIAL OVERGROWTH		▬▬				
NEOMORPHISM		▬▬				
DOLOMITIZATION			▬▬			
LATE BURIAL EUHEDRAL CALCITE				▬		
FRACTURATION						
SADDLE DOLOMITE CEMENT				▬▬		
ANHYDRITE CEMENT AND REPLACEMENT				▬▬		

FIGURE 11.73 Amapá Formation (Paleocene–Middle Miocene), Foz do Amazonas Basin, offshore northeast Brazil. Composite diagenetic sequence of model 1. The thickness of the black bars represents the relative importance of each diagenetic event. From Wolff and Carozzi (1984).

matrix are benthic foraminifers, green algae, bioclasts of echinoids, mollusks, and corals.

Microfacies 301 (Fig. 11.75H). Bioturbated calcilutite to mud-supported biocalcarenite consisting mainly of red algae fragments. Encrusting foraminifers and small rotaliids are occasionally common. Accessory components include bioclasts of echinoids, mollusks, bryozoans, and corals.

Microfacies 35 (Fig. 11.76A). Bioconstructed and bioaccumulated red algae–larger foraminifer limestone consisting of an association of crustose corallines and large rotaliids. Massive red algae nodules predominate over open branched colonies. Lepidocyclinids are the most common larger foraminifers followed by discocyclinids and rare nummulitids. Other components are encrusting foraminifers and bryozoans, small rotaliids, *Amphistegina*, echinoids, and corals. The interstitial ma-

FIGURE 11.74 Amapá Formation (Paleocene–Middle Miocene), Foz do Amazonas Basin, offshore northeast Brazil. Porosity evolution in model 1. A. Fabric-selective dissolution of pelecypod shells subsequently occluded by precipitation of sparite mosaic cement (microfacies 500). B. Non-fabric-selective dissolution cutting across bioclasts and/or matrix boundaries. The porosity generated was occluded subsequently by sparite cement and syntaxial rim cement (microfacies 500). C. Collapse of pelecypod micrite envelopes and precipitation of sparite cement, the latter was preceded by deposition of some internal sediment (microfacies 500). D. Large crystal of saddle dolomite with typical wavy extinction partially occludes primary porosity inside dasycladacean cylindrical stem. It postdates internal sediment but is overlain and cross-cut by coarsely crystalline anhydrite cement, which occludes remaining void space (microfacies 500). E. Primary interparticle porosity (in black) reduced by fibrous rim cement (oolitic calcarenite, microfacies 402). F. Primary interparticle porosity (in gray) reduced by fibrous rim cement and late euhe-

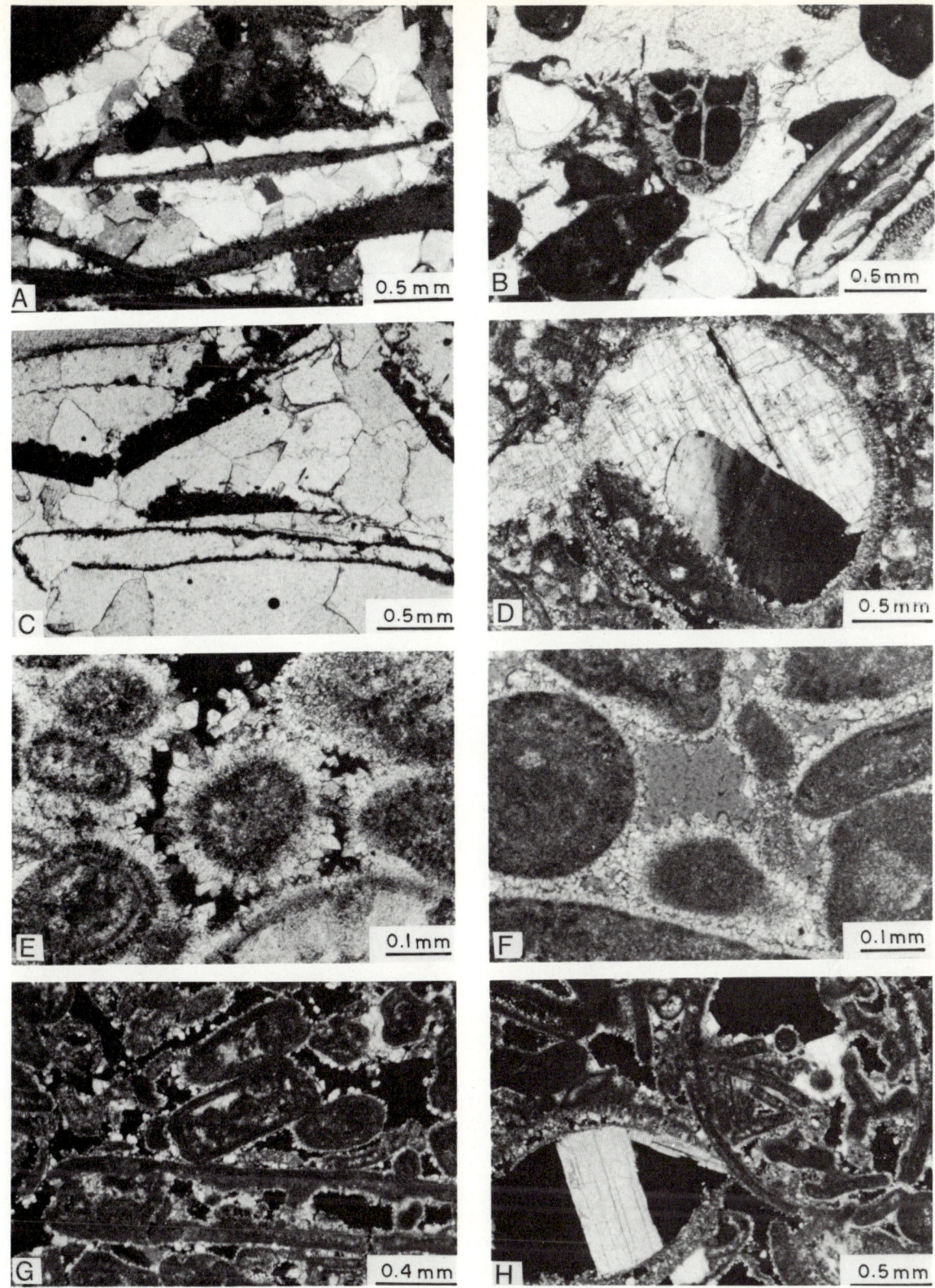

FIGURE 11.74 (*continued*) dral calcite crystals (dasycladacean calcarenite, microfacies 500). G. Primary intraparticle and interparticle porosity (in black) partially reduced by fibrous rim cement, late euhedral calcite crystals and saddle dolomite cement (microfacies 500). H. Primary intraparticle and interparticle porosity (in black) partially filled by anhydrite crystals (dasycladacean calcarenite, microfacies 500). All photomicrographs: crossed nicols, except B and C, plane-polarized light. From Wolff and Carozzi (1984).

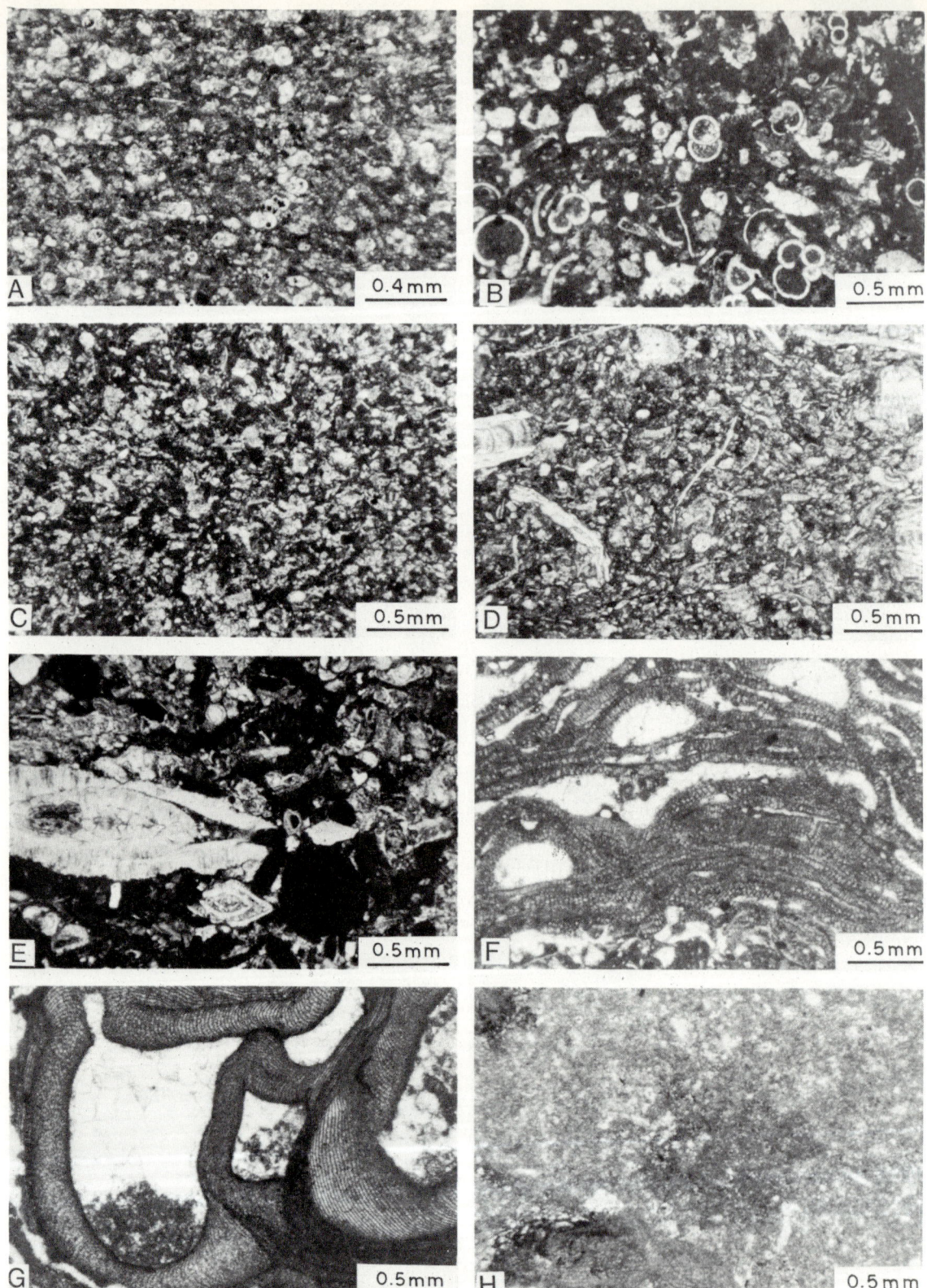

FIGURE 11.75 Amapá Formation (Paleocene–Middle Miocene), Foz do Amazonas Basin, offshore northeast Brazil. Typical microfacies of ideal depositional model 2. A. Microfacies 101. B. Microfacies 200. C. Microfacies 201. D. Microfacies 202. E. Microfacies 403. F. Microfacies 30. G. Microfacies 32. H. Microfacies 301. See text for detailed descriptions. All photomicrographs: plane-polarized light. From Wolff and Carozzi (1984).

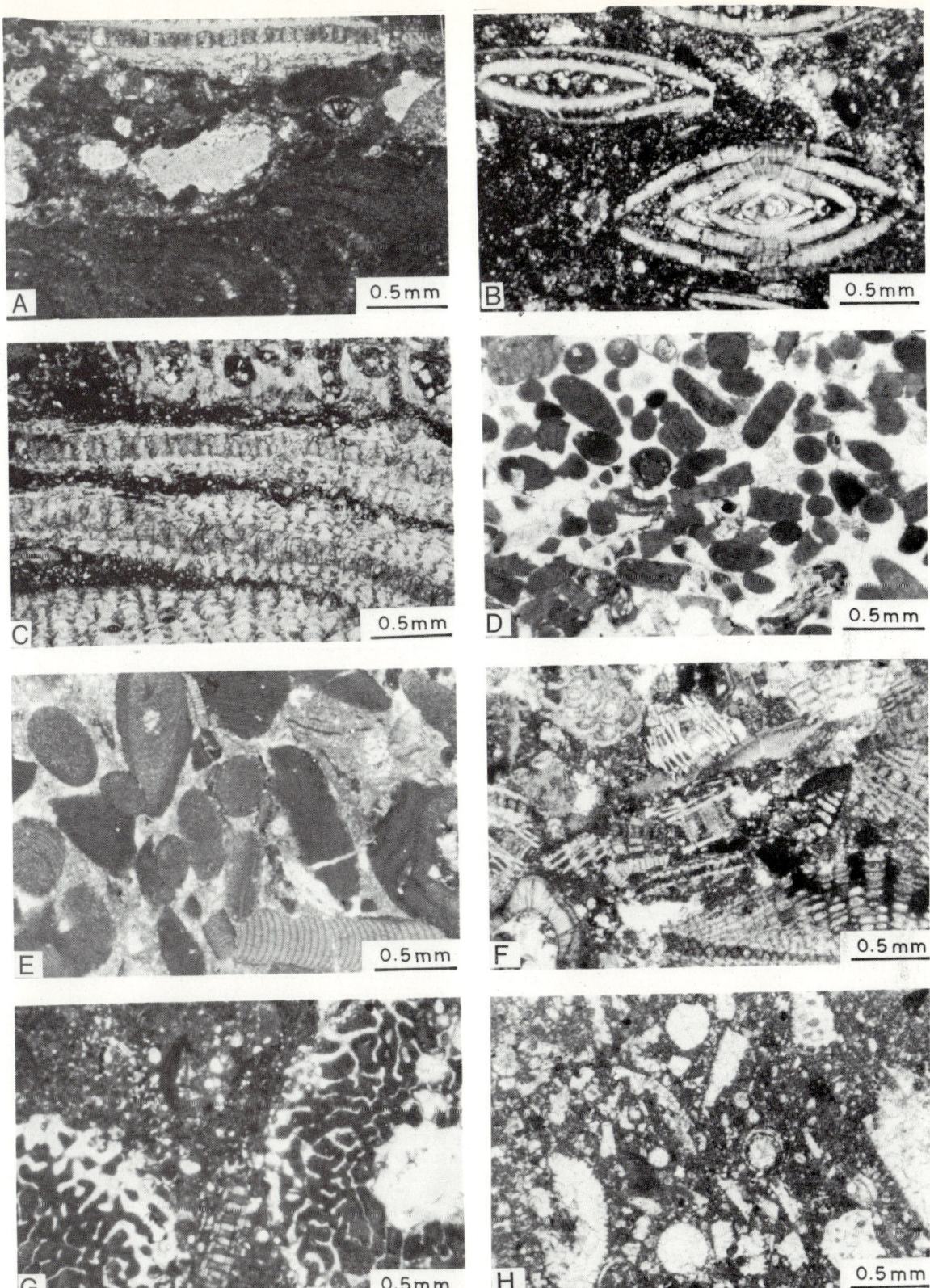

FIGURE 11.76 Amapá Formation (Paleocene–Middle Miocene), Foz do Amazonas Basin, offshore northeast Brazil. Typical microfacies of ideal depositional model 2 (continued). A. Microfacies 35. B. Microfacies 40. C. Microfacies 41. D. Microfacies 401. E. Microfacies 300. F. Microfacies 400. G. Microfacies 50. H. Microfacies 501. See text for detailed descriptions. All photomicrographs: plane-polarized light. From Wolff and Carozzi (1984).

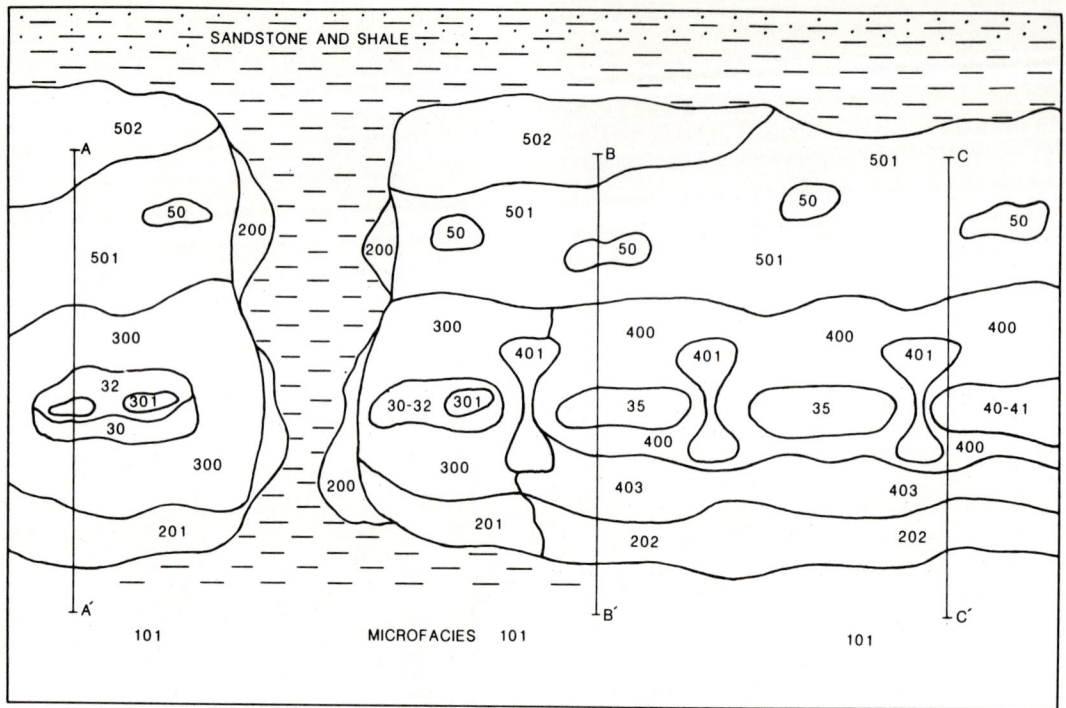

FIGURE 11.77 Amapá Formation (Paleocene–Middle Miocene), Foz do Amazonas Basin, offshore northeast Brazil. Ideal paleoenvironmental map of model 2. From Wolff and Carozzi (1984).

trix of micrite to biocalcarenite is locally altered into pseudomicrosparite by neomorphism or is dolomitized.

Microfacies 40 (Fig. 11.76B). Bioaccumulated larger foraminifer limestone consisting mainly of small species (millimeters in size) of lepidocyclinids and nummulitids locally associated with massive red algae nodules. Interstitial matrix is represented by micrite and scattered fragments of red algae, mollusks, bryozoans, and small rotaliids.

Microfacies 41 (Fig. 11.76C). Bioaccumulated larger foraminifer limestone consisting of species (centimeters in size) of large rotaliids (mainly lepidocyclinids) preferentially aligned parallel to bedding in common association with thick bifoliate bryozoans. Interstitial matrix is micrite to biocalcarenite with common red algae fragments.

Channel Deposits

Microfacies 401 (Fig. 11.76D). Cross-bedded grain-supported biocalcarenite consisting predominantly of small to medium red algae bioclasts. Accessory components are small rotaliids, *Amphistegina*, bryozoans, lepidocyclinids, encrusting foraminifers, green algae, echinoids, mollusks, and corals. Micrite matrix is completely absent or very rare. Cementation consists of a poorly developed fibrous rim cement overlain by coarse sparite mosaic and/or syntaxial rim cement.

Backbank Deposits

Microfacies 300 (Fig. 11.76E). Grain-supported biocalcarenite with mainly bioclasts of red algae. Accessory components are small rotaliids, encrusting foraminifers, bryozoans, echinoids, pelecypods, larger foraminifers, porcelaneous foraminifers, green algae, and ostracods. Interstitial matrix is predominantly micrite, commonly changed into pseudomicrosparite by neomorphism, and associated with sparite cement.

Microfacies 400 (Fig. 11.76F). Grain-supported biocalcarenite consisting predominantly of fragments of larger foraminifers (mostly lepidocyclinids), red algae, and echinoids. Accessory components are small rotaliids, *Amphistegina*, bryozoans, encrusting foraminifers, and pelecypods. Interstitial matrix of micrite is commonly changed into pseudomicrosparite by neomorphism. Cementation is by equant to mosaic sparite and syntaxial overgrowth on echinoids.

Lagoonal Deposits

Microfacies 50 (Fig. 11.76G). Bioconstructed finger coral (*Porites*) limestone with abundant clotted or bioturbated calcisiltite matrix. Accessory components are red algae, porcelaneous foraminifers, mollusks, larger foraminifers, and small rotaliids. Fabric selective moldic porosity in corals is characteristic of this microfacies, either preserved or subsequently occluded by mosaic sparite cement.

Microfacies 501 (Fig. 11.76H). Bioturbated mud-supported biocalcarenite with characteristic clotted micrite to biocalcisiltite matrix. Major components are mollusks, corals, miliolids, soritids, red algae, and green algae (mainly *Halimeda*). Accessory components are: small rotaliids, encrusting and larger foraminifers, bryozoans, and ostracods. Neomorphism of skeletal components (*Halimeda* and mollusks) into pseudosparite is common as well as leaching of aragonite skeletons generating secondary fabric selective moldic porosity.

Microfacies 502. Mud-supported to grain-supported biocalcarenite consisting predominantly of thick bifoliate and ramose bryozoans and sand-size quartz grains. Accessory components are mollusks, small rotaliids, red algae, and green algae. Micrite matrix is commonly altered into pseudomicrosparite by neomorphism. Primary intraparticle porosity is frequent in zooecia of bryozoans.

The Ideal Depositional Model 2

Because of the relative complexity of the bank deposits of this model, which is also crossed by transverse shaly channels with marginal olistostromes, it is necessary to consider three distinct transverse sections (Fig. 11.77, A-A′, B-B′, C-C′) to account for the most common observed vertical sequences. The three ideal shallowing-upward sequences of microfacies (Figs. 11.78, 11.79, 11.80) consist of the following environments: slope, forebank, red algae bank, backbank, and lagoon. Corresponding depositional models (Figs. 11.81, 11.82, 11.83) show how the various types of banks, red algae, red algae plus larger foraminifers, and larger foraminifers are surrounded by their respective destructional halos of biocalcarenites. Transverse tidal channels (microfacies 401) are well-characterized, high-energy biocalcarenites between individual banks. Lagoonal microfacies are typical both petrographically and ecologically.

Shale channels transecting the Amapá carbonates were identified by seismic profiles. They are coeval with carbonate sedimentation and contain shales alone or with olistostromes. Olistoliths consist of collapsed blocks and smaller fragments of shallow-water microfacies embedded in shale alone or in a micrite matrix with planktonic foraminifers; all these varieties are designated as microfacies 200 (Fig. 11.84).

A typical detailed section shows variation of microscopic parameters in two successive shallowing-upward sequences of model 2 (Figs. 11.85, 11.86).

Diagenesis

A petrographic sequence of 13 diagenetic events was established for model 2 and as in model 1, it is expressed by pathways followed by the individual microfacies (Fig. 11.87). Pathways lead to the composite diagenetic sequence (Fig. 11.88) in which the thickness of the black bars represents the relative importance of each diagenetic event.

Porosity Evolution

Contrary to model 1, primary porosity observed in rocks of model 2 is of no practical importance. Major reservoirs in model 2 correspond to all the microfacies from lagoonal to bank environment that experienced dissolution in the undersaturated zone of the freshwater phreatic environment and that were prevented from subsequent freshwater cementation by not having passed through the saturated zone. Main types of secondary porosity generated by dissolution are enlarged intraparticle and interparticle moldic (Fig. 11.89D,E), enlarged moldic, vuggy, and channel (Fig. 11.89F,G,H). The largest average value for porosity was 17% in microfacies 50. It consists predominantly of fabric selective moldic porosity developed within coral framework. Lagoonal calcarenites (microfacies 501) usually develop fabric selective moldic porosity in coral, pelecypod, and codiacean skeletons (Fig. 11.89D,E). Enlarged moldic porosity is also common in these rocks when aragonite skeletons are preferentially removed (moldic porosity) and dissolution spreads to the micrite matrix and/or other components. Locally, it grades to vuggy and channel porosity (Fig. 11.89F,G,H). Bank and backbank rocks (microfacies 30, 35, 41, 300, and 400) behave differently because they are composed predominantly of high Mg-calcite instead of aragonite skeletons. Consequently, micrite matrix is preferentially removed by dissolution generating intraparticle, interparticle, and vuggy porosity rather than moldic. Furthermore, these rocks represent environmental belts in which the mixing model dolomitization occurred with its highest intensity, generating an appreciable amount of intercrystalline porosity. In contrast, forebank (microfacies 403, 201, and 202) and slope deposits (microfacies 101) are usually represented by tight rocks.

The ITERIM program was used to evaluate the control of the depositional model over diagenesis. Thin sections were maintained in the same groups as defined by the microfacies technique, but only diagenetic variables (dolomitization, porosity, cement, etc.) were considered. A large core ratio (number of thin sections left in

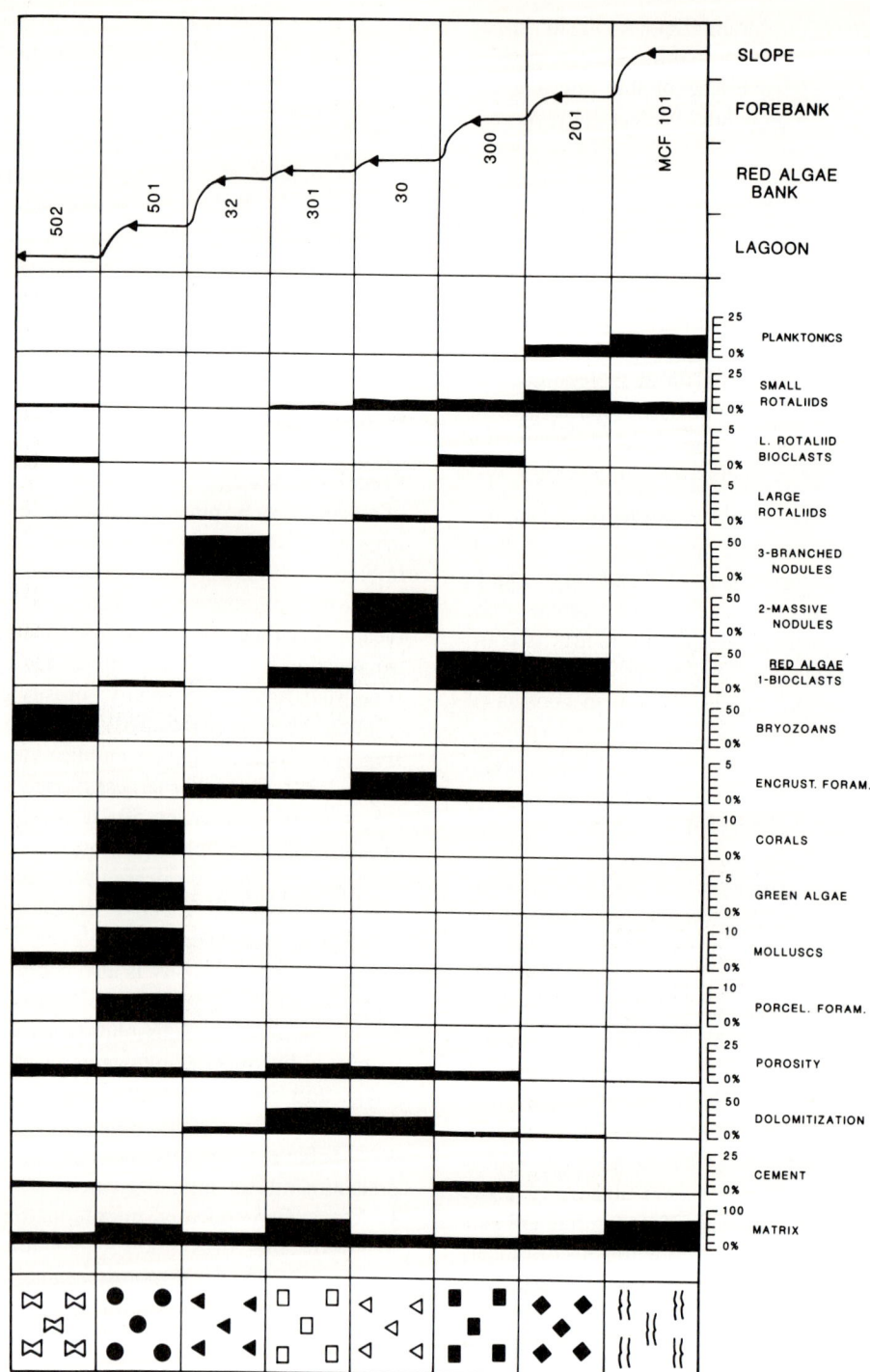

FIGURE 11.78 Amapá Formation (Paleocene–Middle Miocene), Foz do Amazonas Basin, offshore northeast Brazil. Ideal shallowing-upward sequence for model 2 (Section A-A'). See Figure 11.77 for location. From Wolff and Carozzi (1984).

Chap. 11 / Carbonate Platforms with Frontal Bioaccumulated to Hydrodynamic Buildups   289

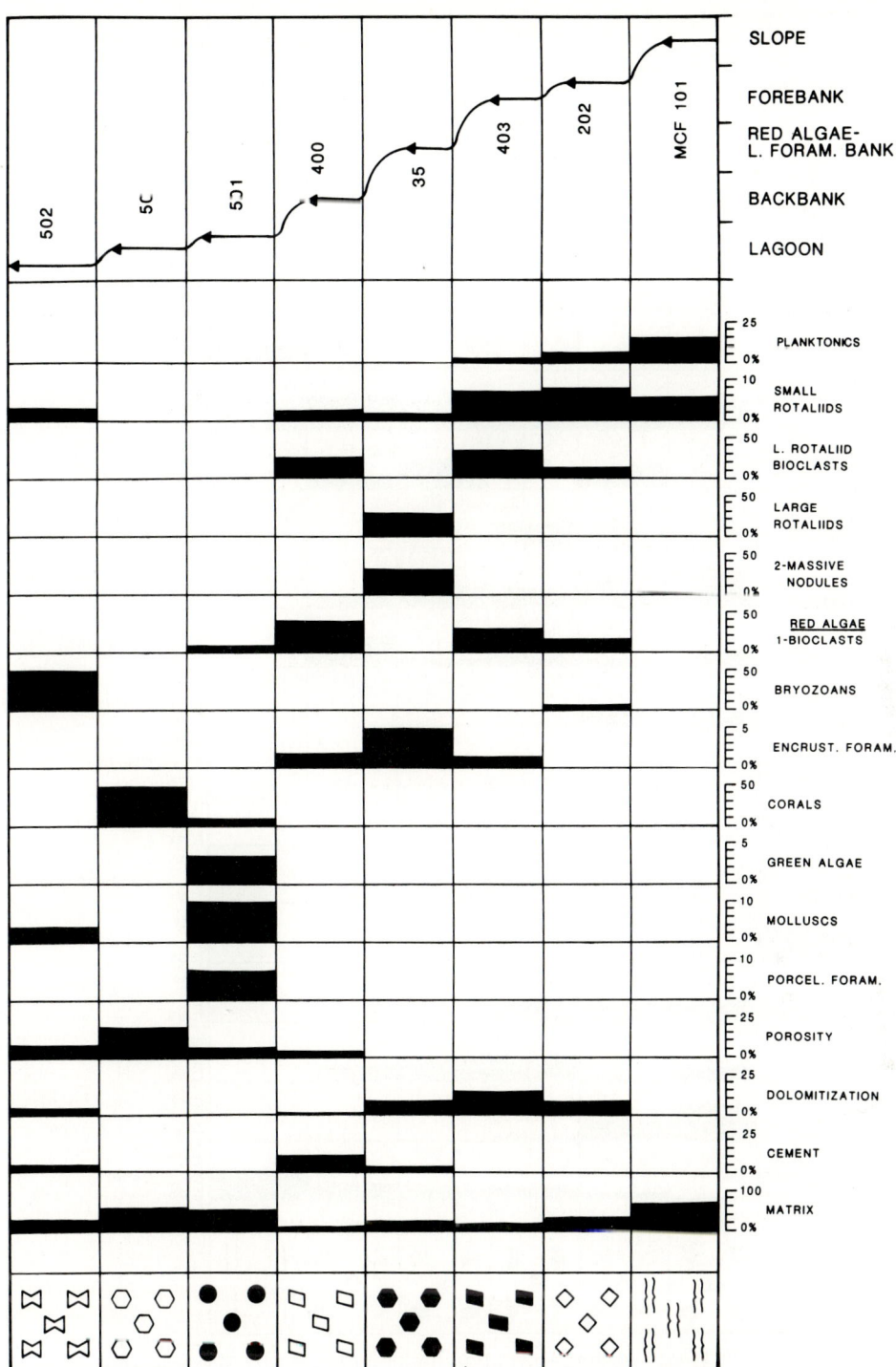

FIGURE 11.79 Amapá Formation (Paleocene–Middle Miocene), Foz do Amazonas Basin, offshore northeast Brazil. Ideal shallowing-upward sequence for model 2 (Section B-B'). See Figure 11.77 for location. From Wolff and Carozzi (1984).

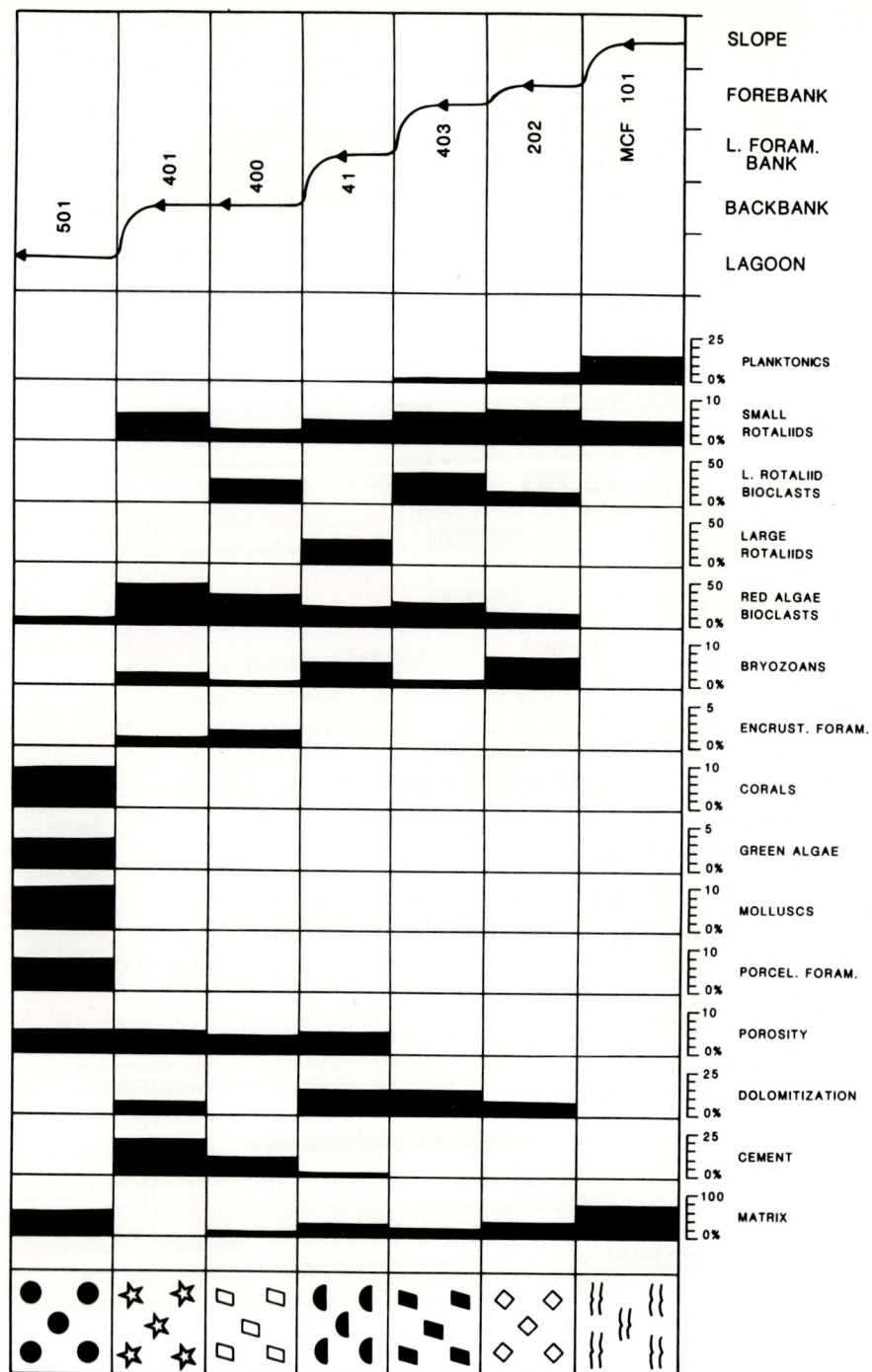

FIGURE 11.80 Amapá Formation (Paleocene–Middle Miocene), Foz do Amazonas Basin, offshore northeast Brazil. Ideal shallowing-upward sequence for model 2 (Section C-C'). See Figure 11.77 for location. From Wolff and Carozzi (1984).

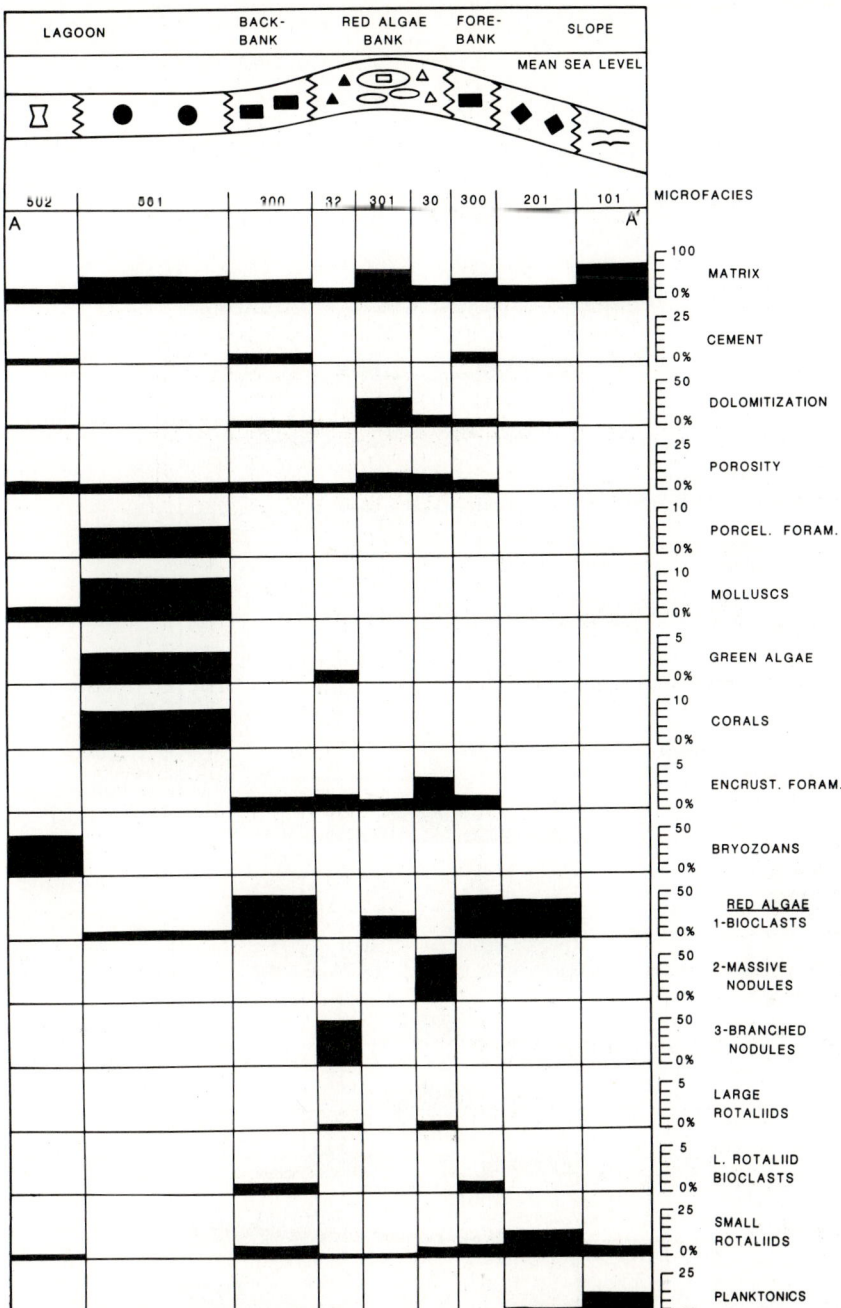

FIGURE 11.81 Amapá Formation (Paleocene–Middle Miocene), Foz do Amazonas Basin, offshore northeast Brazil. Ideal depositional model 2 (Section A-A'). See Figure 11.77 for location. From Wolff and Carozzi (1984).

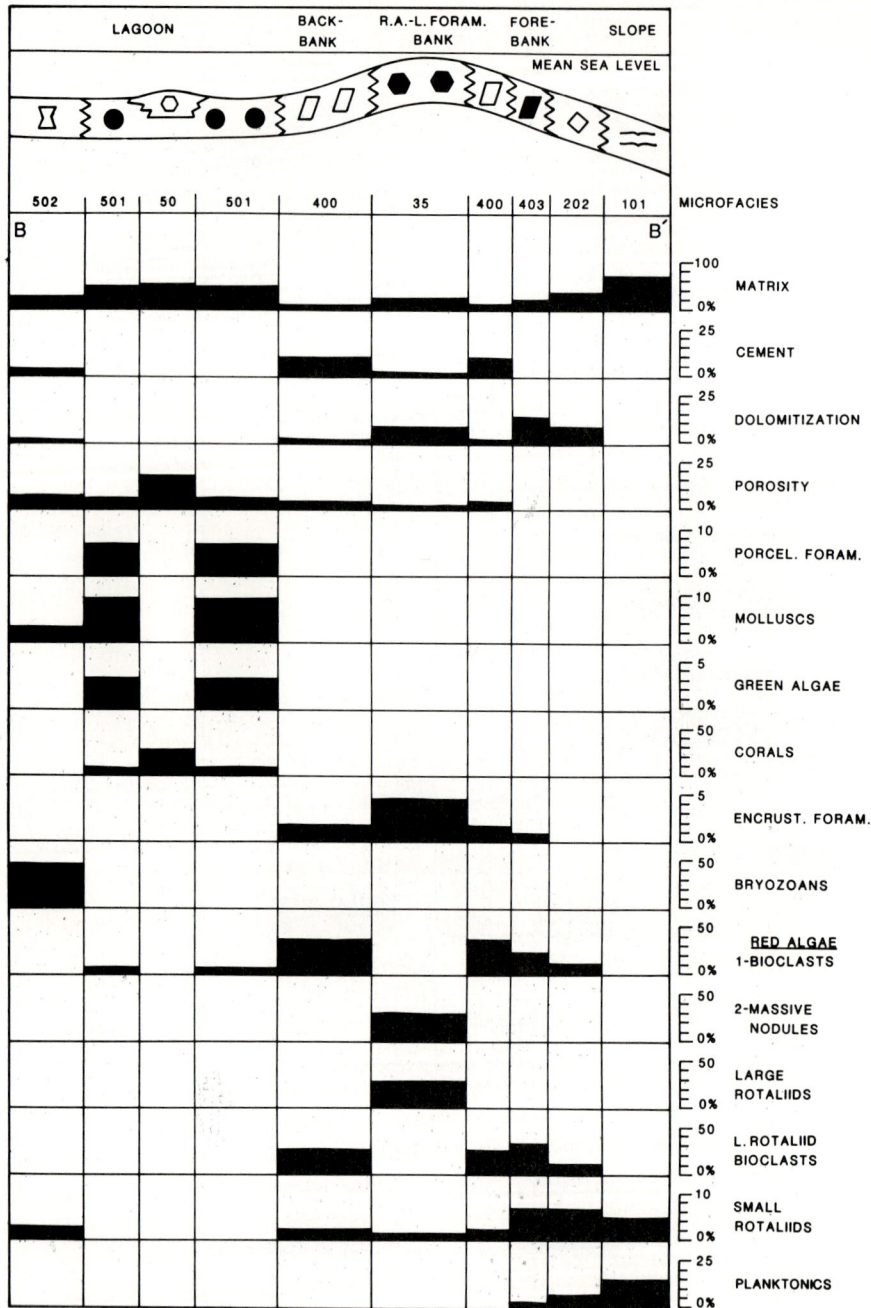

FIGURE 11.82 Amapá Formation (Paleocene–Middle Miocene), Foz do Amazonas Basin, offshore northeast Brazil. Ideal depositional model 2 (Section B-B'). See Figure 11.77 for location. From Wolff and Carozzi (1984).

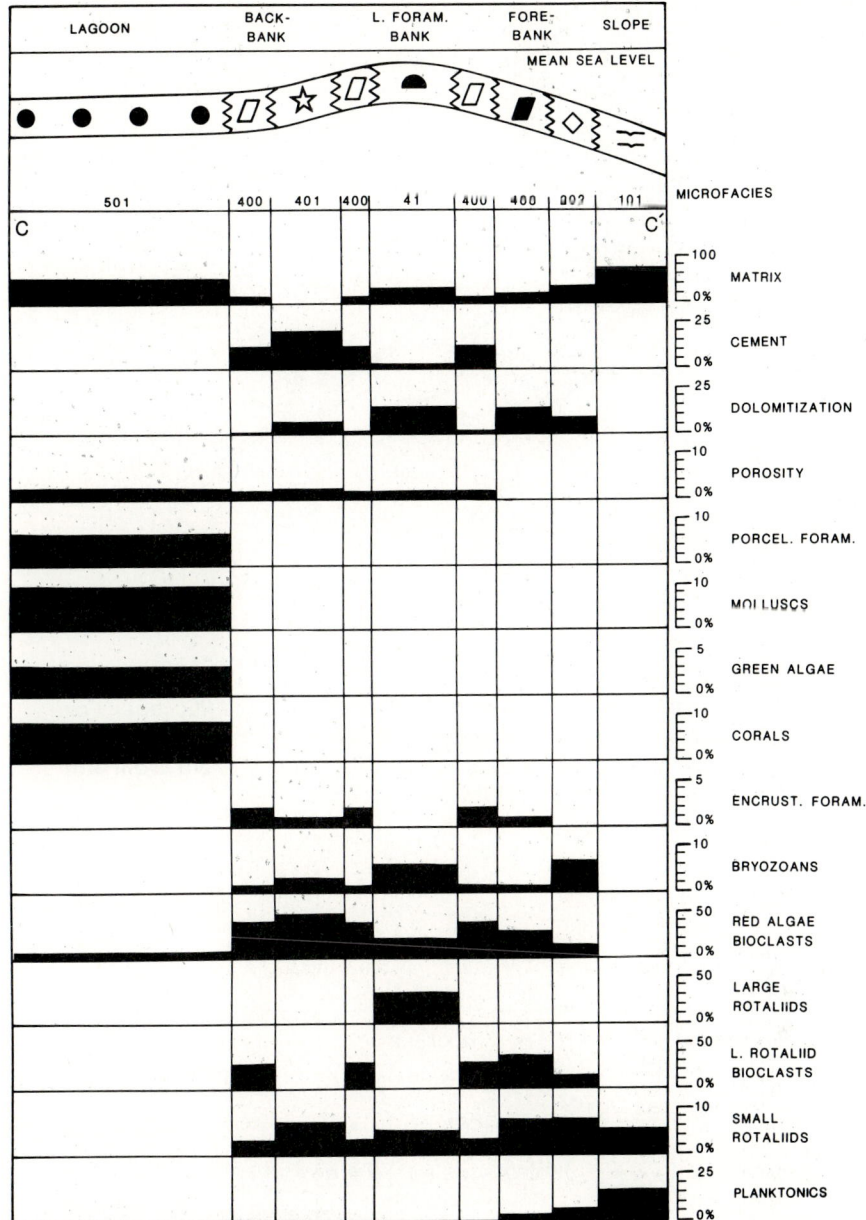

FIGURE 11.83 Amapá Formation (Paleocene–Middle Miocene), Foz do Amazonas Basin, offshore northeast Brazil. Ideal depositional model 2 (Section C-C'). See Figure 11.77 for location. From Wolff and Carozzi (1984).

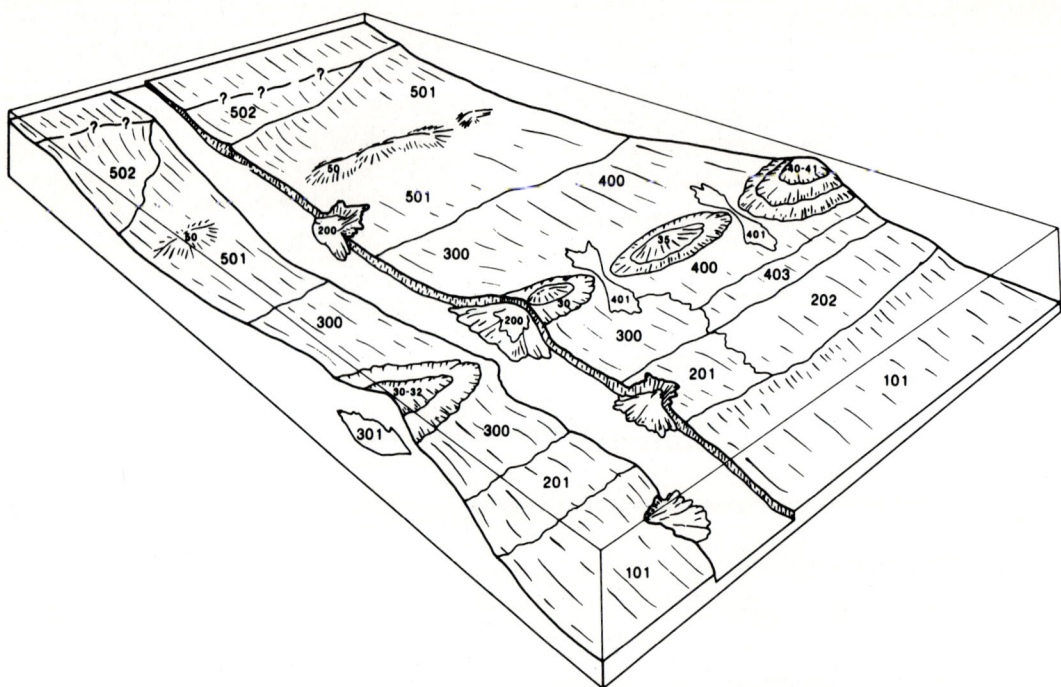

FIGURE 11.84 Amapá Formation (Paleocene–Middle Miocene), Foz do Amazonas Basin, offshore northeast Brazil. Block-diagram showing paleoenvironmental reconstruction for model 2. From Wolff and Carozzi (1984).

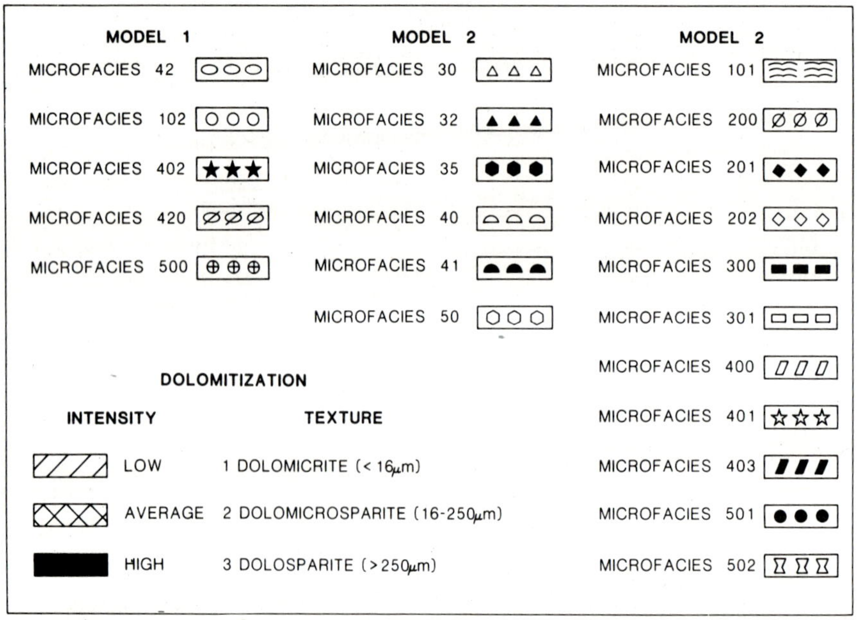

FIGURE 11.85 Amapá Formation (Paleocene–Middle Miocene), Foz do Amazonas Basin, offshore northeast Brazil. Symbols for microfacies. From Wolff and Carozzi (1984).

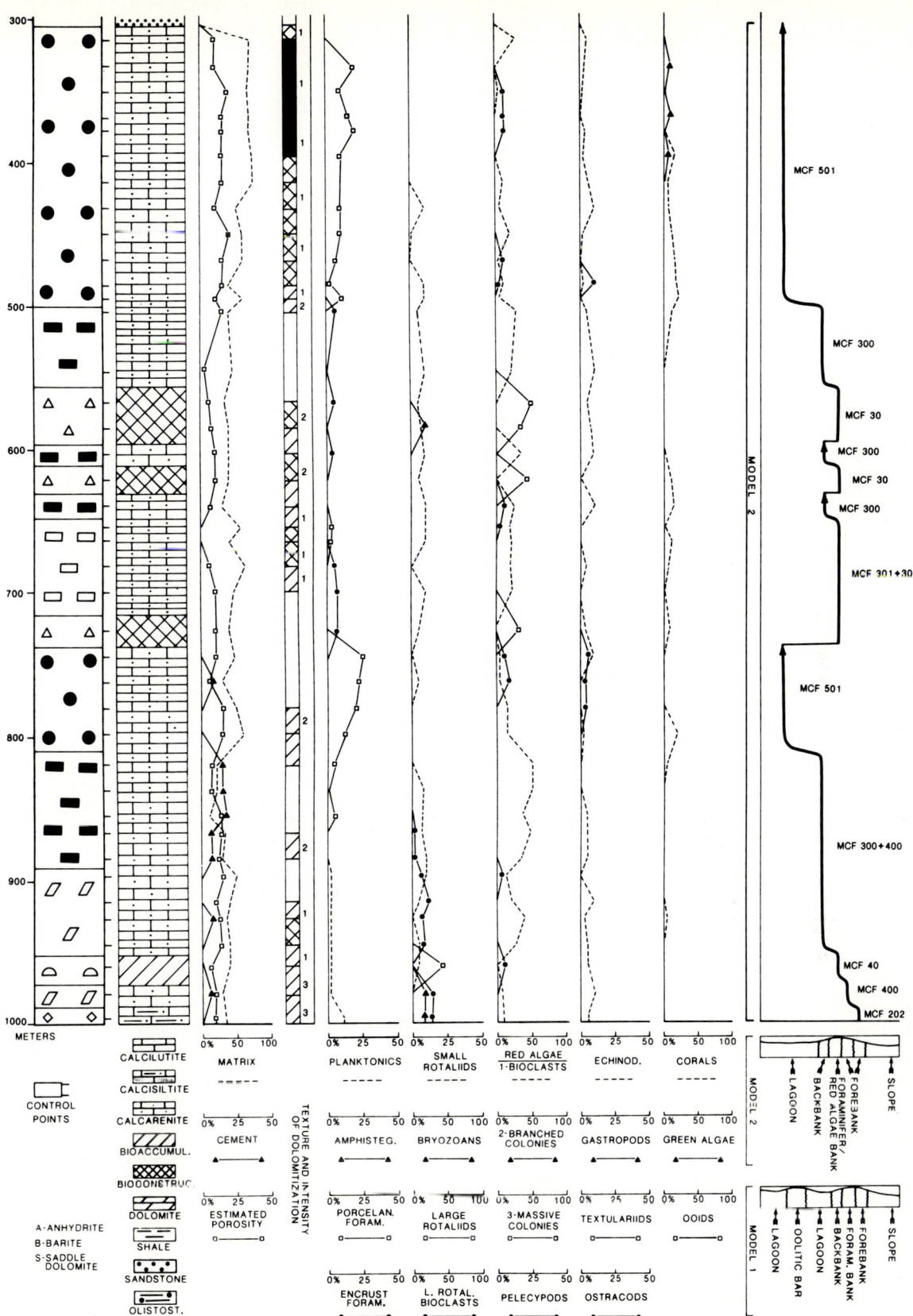

FIGURE 11.86 Amapá Formation (Paleocene–Middle Miocene), Foz do Amazonas Basin, offshore northeast Brazil. Typical example of vertical variation of microscopic parameters during two successive shallowing-upward sequences of model 2. From Wolff and Carozzi (1984).

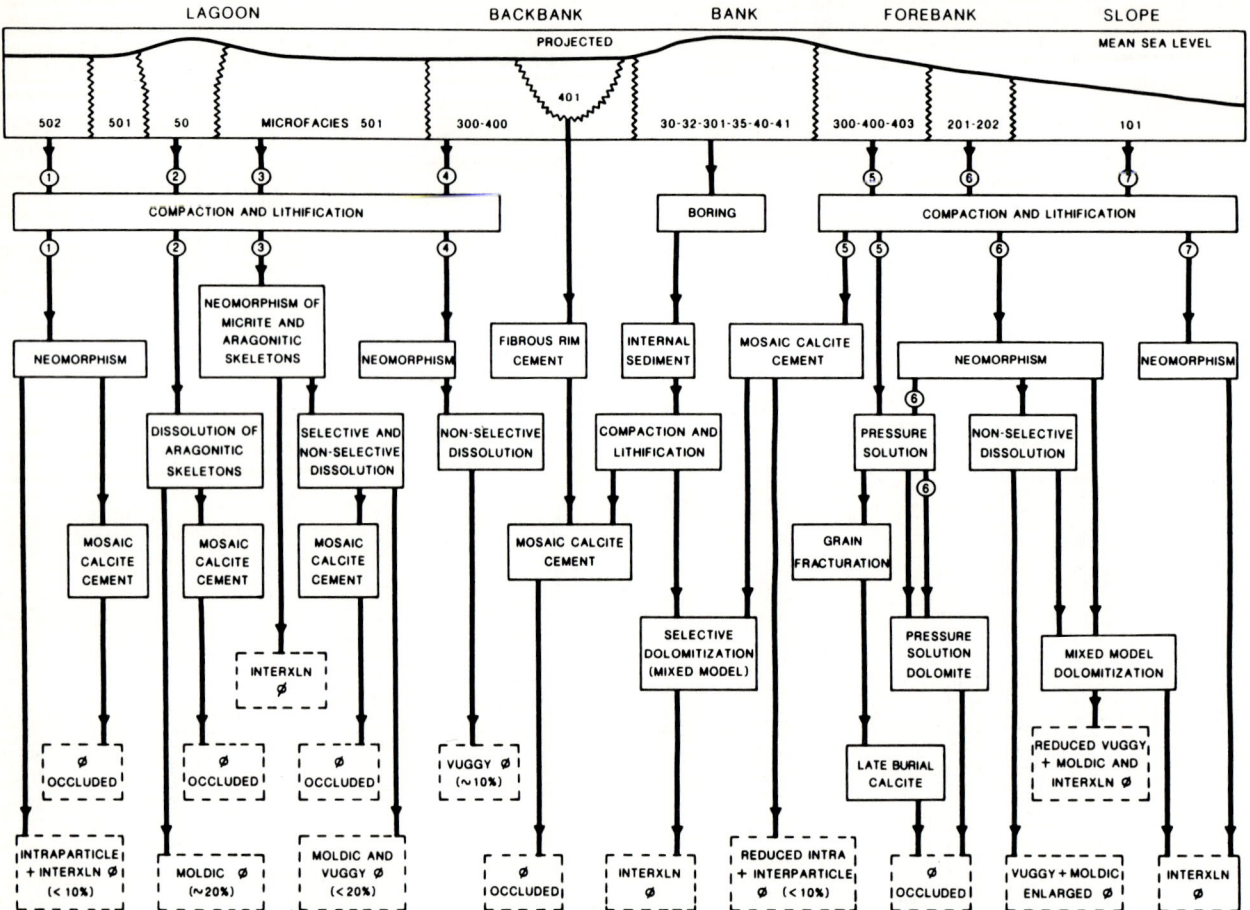

FIGURE 11.87 Amapá Formation (Paleocene–Middle Miocene), Foz do Amazonas Basin, offshore northeast Brazil. Marine diagenetic pathways for each microfacies of model 2. The position of the solid boxes does not represent relative depths, and the dashed boxes indicate final porosity types and estimated amounts. From Wolff and Carozzi (1984).

their original groups) after stabilization would be considered as indicative of a strong control of the depositional model over diagenesis. The program stabilized after 13 iterations with approximately 32% of the thin sections remaining in their original groups. These results confirm present petrographic observations, and those made by Carozzi et al. (1981b) according to which the depositional model exerted only a very minor control on diagenesis. This conclusion is illustrated by porosity distribution in these rocks: any microfacies through which undersaturated freshwater fluids percolated developed into a potential hydrocarbon reservoir.

General Evolution of the Depositional Environments

Detailed stratigraphic work shows that model 1 corresponds to a shallowing-upward sequence (I) and that model 2 results from the superposition of three others (II, III, and IV). Each sequence developed its own depositional pattern; these patterns are well expressed by paleoenvironmental maps pertaining to model 2. The map at the top of sequence II (Middle Eocene) represents the largest expansion of the carbonate platform with banks of larger foraminifers and red algae (Fig. 11.90). The map at the top of sequence III (Late Oligocene) shows the effect of the transverse canyons (Fig. 11.91), and the one at the top of sequence IV (Middle Miocene) shows the destruction of the platform reduced to red algae banks only (Fig. 11.92). To each of these paleoenvironmental maps, a diagenetic map shows the corresponding distribution of secondary porosity and dolomitization peculiar to each sequence (Figs. 11.93, 11.94, 11.95). Maximum intensities of diagenetic processes occur in sequence IV (Fig. 11.95), where vuggy to channel porosity in bank, and back-forebank deposits, moldic porosity

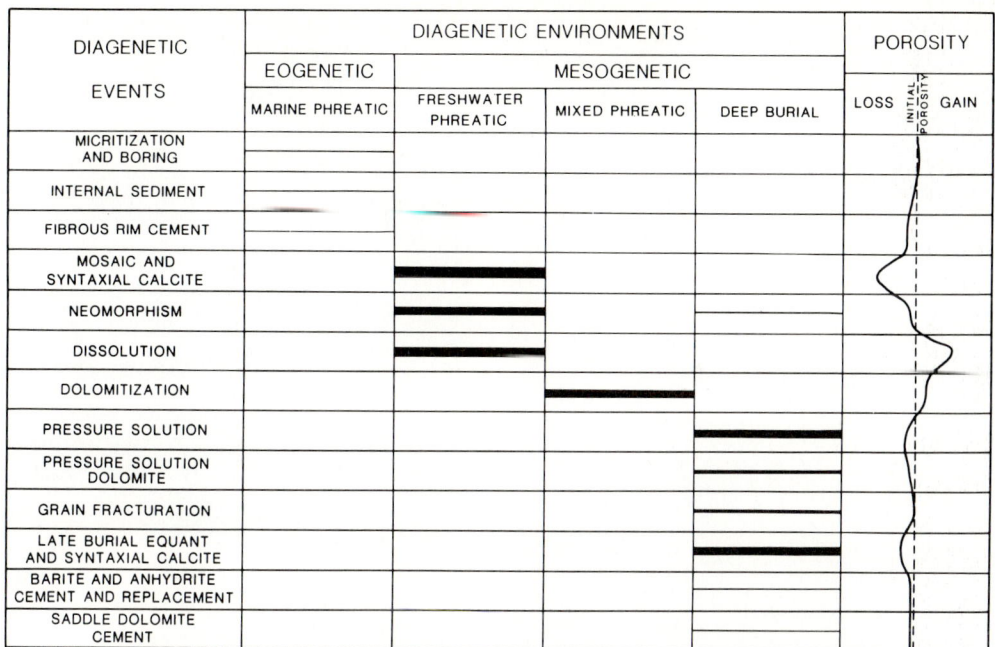

FIGURE 11.88 Amapá Formation (Paleocene–Middle Miocene), Foz do Amazonas Basin, offshore northeast Brazil. Composite diagenetic sequence of model 2. The thickness of the black bars represents the relative importance of each diagenetic event. From Wolff and Carozzi (1984).

in lagoon deposits, and intercrystalline porosity by dolomitization extend reservoir potentiality to all microfacies except slope and canyon deposits.

Diagenetic processes were active at the end of each of the four depositional sequences and coincided with low-sea-level stands. Their intensity is a direct function of the importance of sea level fall. It is suggested as a general interpretation that four episodes of sea level fall disrupted the dynamic equilibrium between rivers and ocean, and triggered distinct reactivation of groundwater systems with a magnitude proportional to that of the respective sea level fall. These disruptions would have activated extensive underground circulation of freshwater through carbonates, causing both porosity generation in the undersaturated zone and dolomitization in the mixed water zone. This interpretation does not rule out the possibility of four episodes of subaerial exposures, as suggested by Carozzi *et al.* (1981b). It is consistent only with the present petrographic study, in which no convincing evidence of subaerial exposure was found.

To investigate further the potential of the proposed mechanism to explain the observed diagenetic data, a transverse section of model 2 was drawn, with its characteristic porosity and dolomitization intensity plotted by black bars (Fig. 11.96). The flux of groundwater abnormally rich in CO_2 derived from decomposing plant material originating from the fluviodeltaic environment was greatly undersaturated with respect to calcium carbonate. It would generate presumably the largest porosity in the lagoonal facies, followed by a decreasing amount of porosity toward the slope as it reaches saturation. The mixed water zone, necessary to explain the main dolomitization process, would be located between the freshwater zone and the marine phreatic zone. Consequently, dolomitization intensity would decrease probably toward both sides of the ideal fluid mixture zone for dolomitization, in agreement with petrographic data.

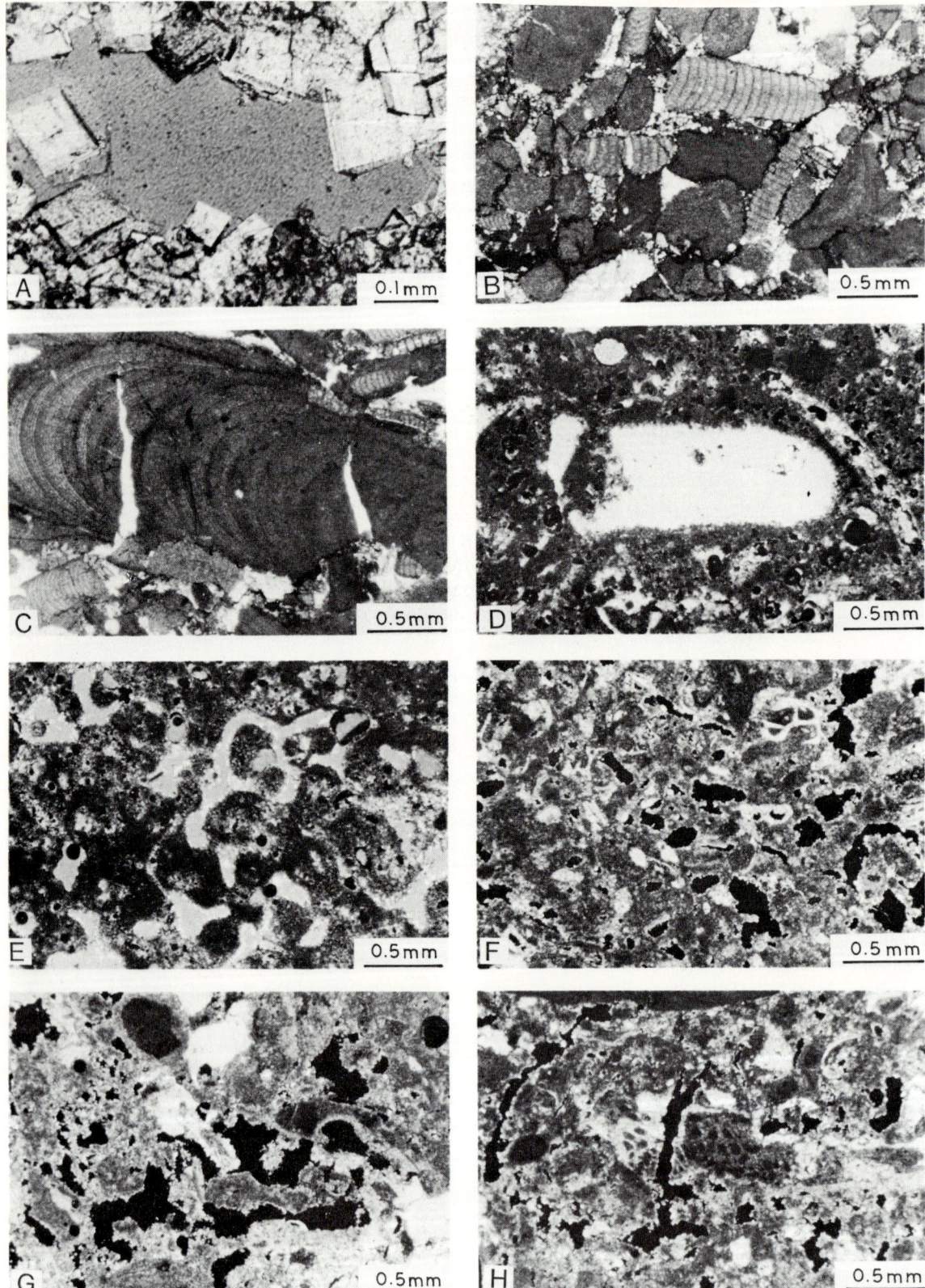

FIGURE 11.89 Amapá Formation (Paleocene–Middle Miocene), Foz do Amazonas Basin, offshore northeast Brazil. Porosity evolution in model 2. A. Larger and clearer dolomite crystals developed toward

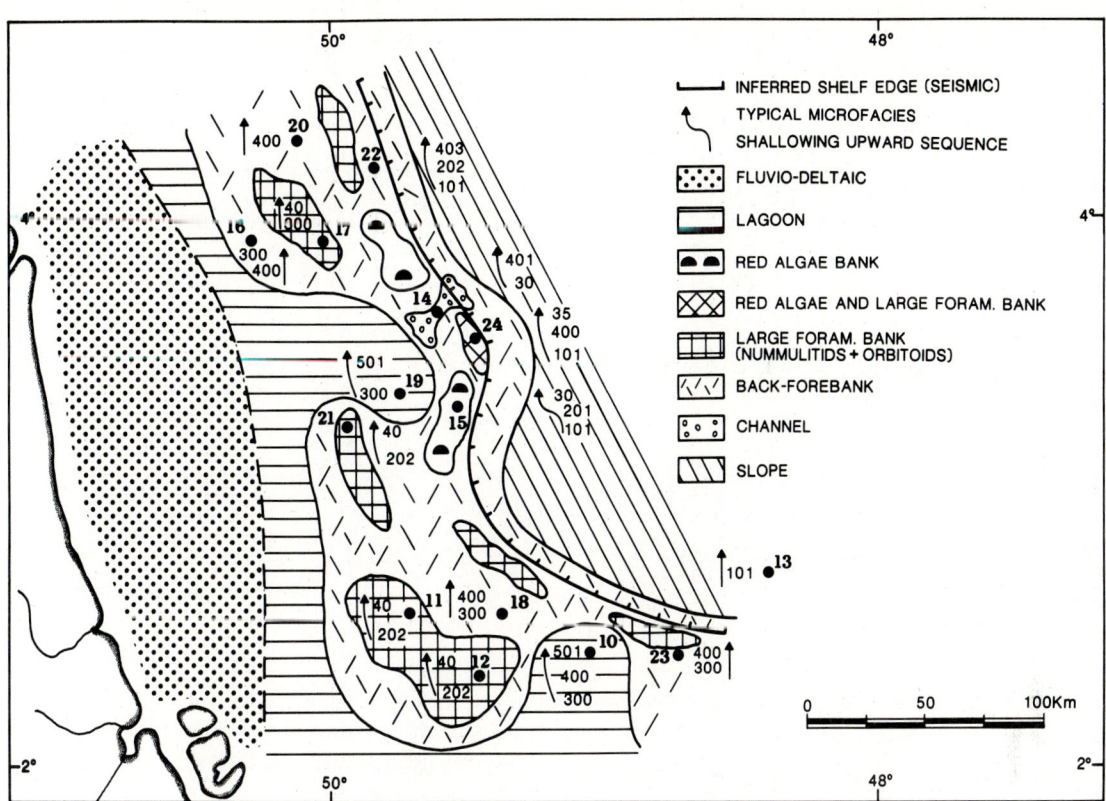

FIGURE 11.90 Amapá Formation (Paleocene–Middle Miocene), Foz do Amazonas Basin, offshore northeast Brazil. Paleoenvironmental map of top of sequence II (Middle Eocene). From Wolff and Carozzi (1984).

FIGURE 11.89 (*continued*) center of vug indicating that dissolution preceded dolomitization. Porosity in gray (microfacies 301). B. Pressure-solution in red algae calcarenite with microstylolitic concentration of insoluble residues (microfacies 300). C. Red algae colony with fractures filled by late equant to mosaic sparite cement, fractures probably generated by pressure-solution (microfacies 300). D. Fabric-selective moldic porosity (in white) in *Halimeda*, some vuggy porosity developed in micrite matrix (microfacies 501). E. Fabric-selective moldic porosity (in gray) developed by preferential dissolution of aragonite coral skeleton (microfacies 50). F. Enlarged moldic and vuggy porosity (in black) in biocalcarenite (microfacies 300). G. Nonselective vuggy porosity (in black) in biocalcarenite (microfacies 300). H. Nonselective vuggy to channel porosity (in black) in biocalcarenite (microfacies 300). All photomicrographs: crossed nicols, except B and C, plane-polarized light, and D, crossed nicols + quartz plate. From Wolff and Carozzi (1984).

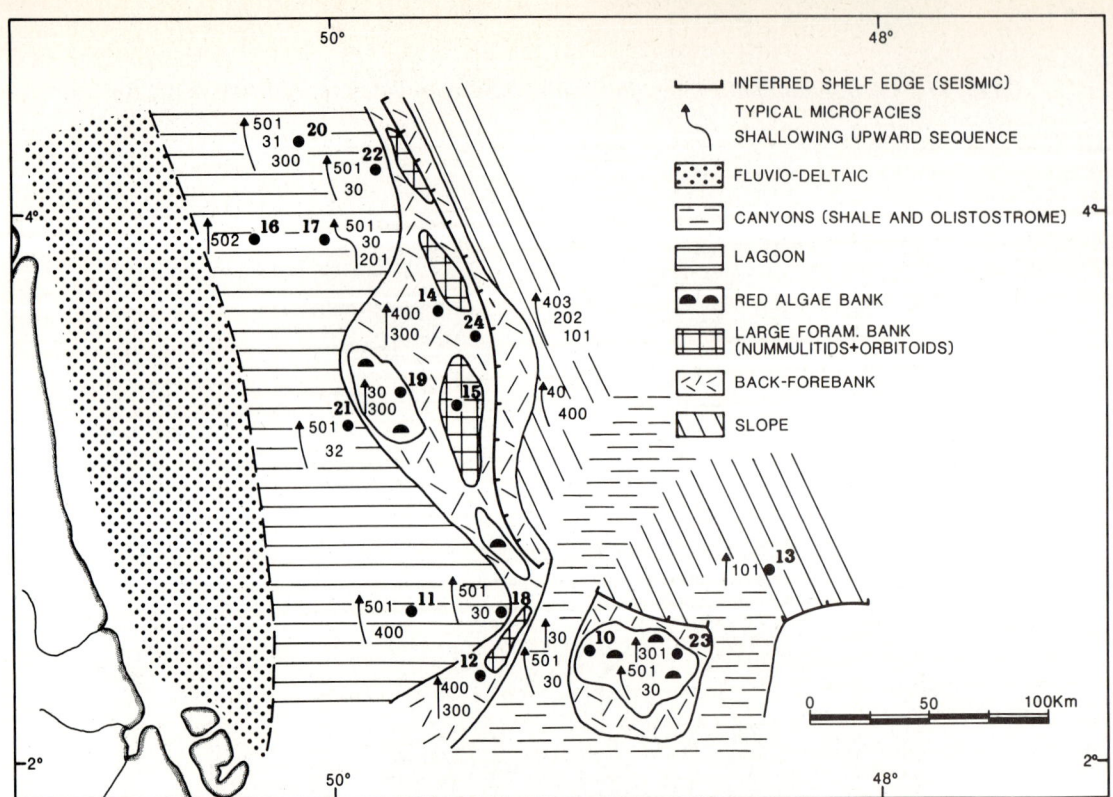

FIGURE 11.91 Amapá Formation (Paleocene–Middle Miocene), Foz do Amazonas Basin, offshore northeast Brazil. Paleoenvironmental map of top of sequence III (Late Oligocene). From Wolff and Carozzi (1984).

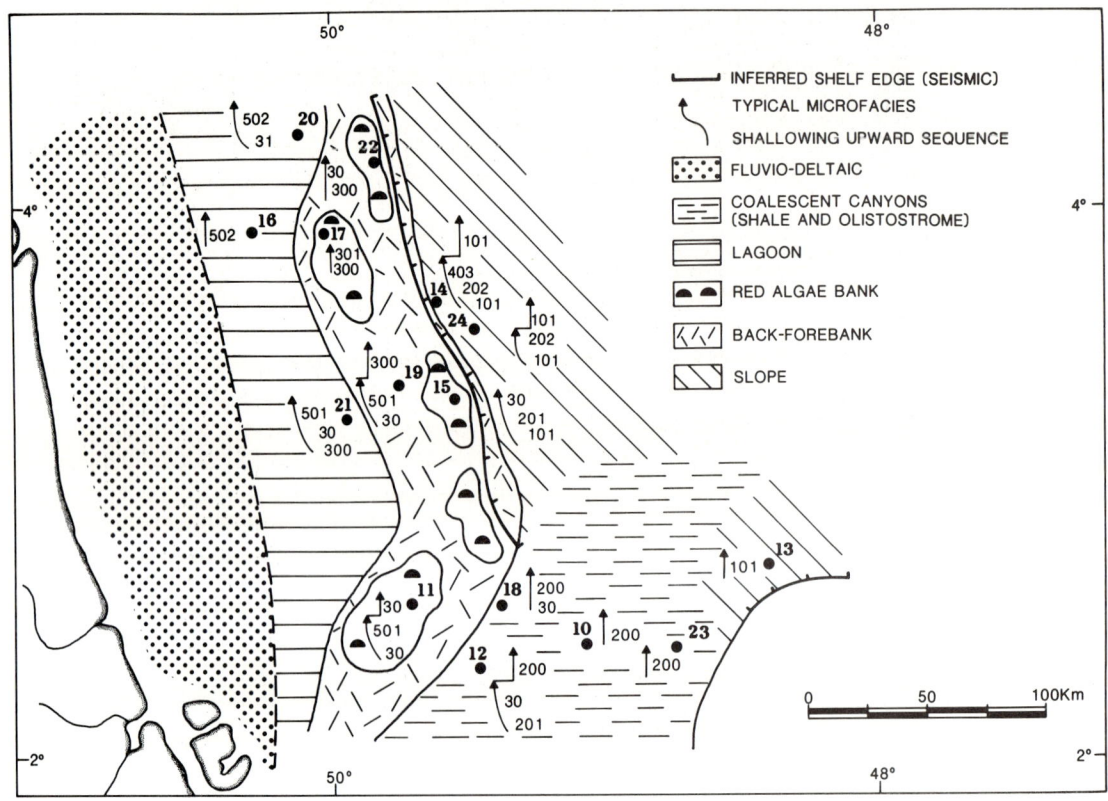

FIGURE 11.92 Amapá Formation (Paleocene–Middle Miocene), Foz do Amazonas Basin, offshore northeast Brazil. Paleoenvironmental map of top of sequence IV (Middle Miocene). From Wolff and Carozzi (1984).

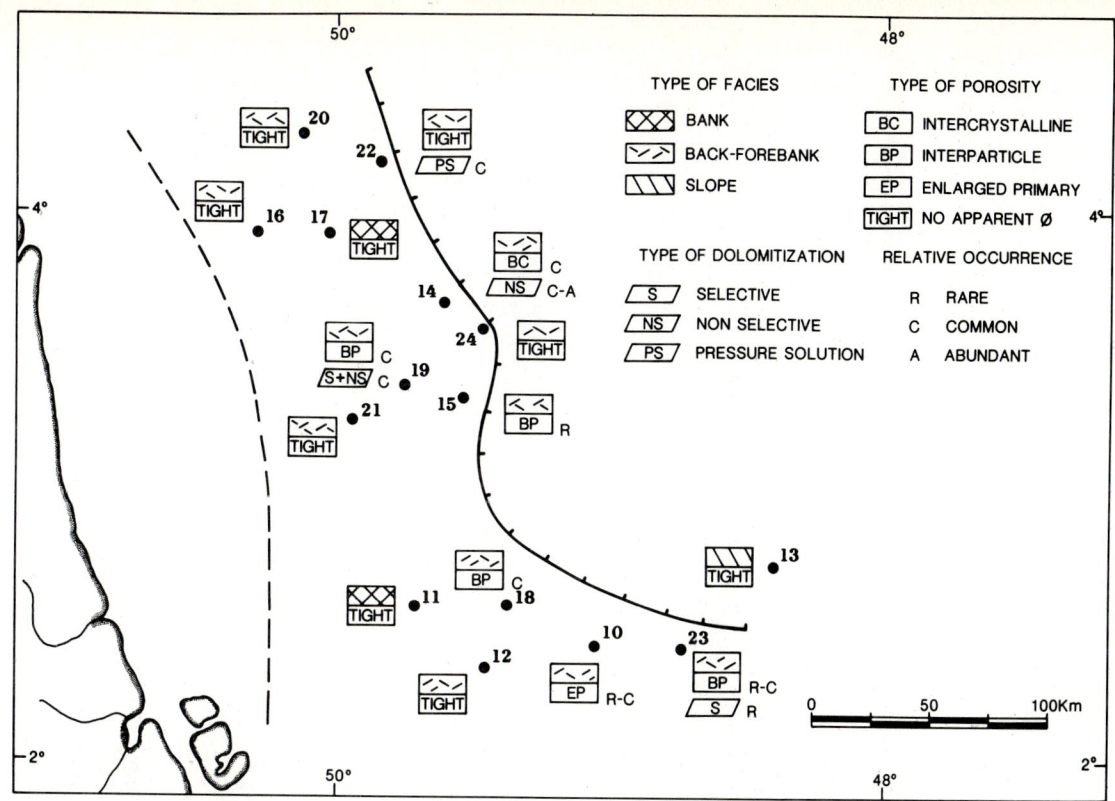

FIGURE 11.93 Amapá Formation (Paleocene–Middle Miocene), Foz do Amazonas Basin, offshore northeast Brazil. Diagenetic map of sequence II (Middle Eocene). From Wolff and Carozzi (1984).

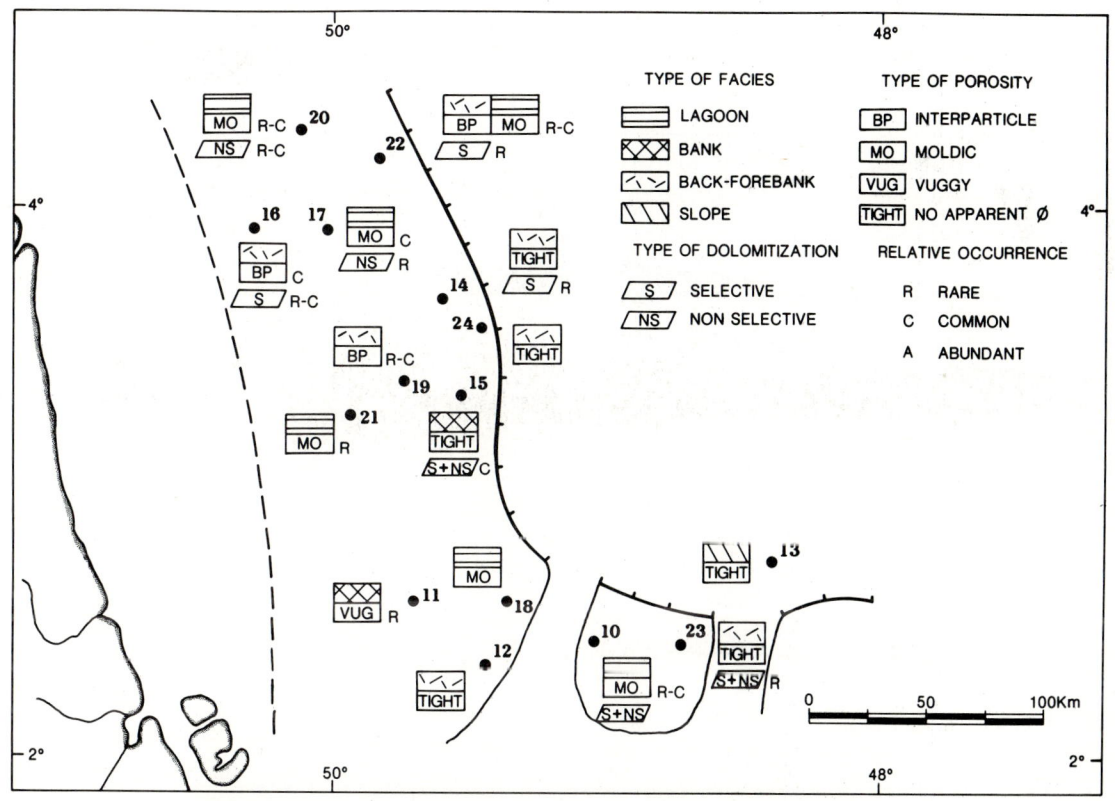

FIGURE 11.94 Amapá Formation (Paleocene–Middle Miocene), Foz do Amazonas Basin, offshore northeast Brazil. Diagenetic map of sequence III (Late Oligocene). From Wolff and Carozzi (1984).

302 Part III / Carbonate Platforms

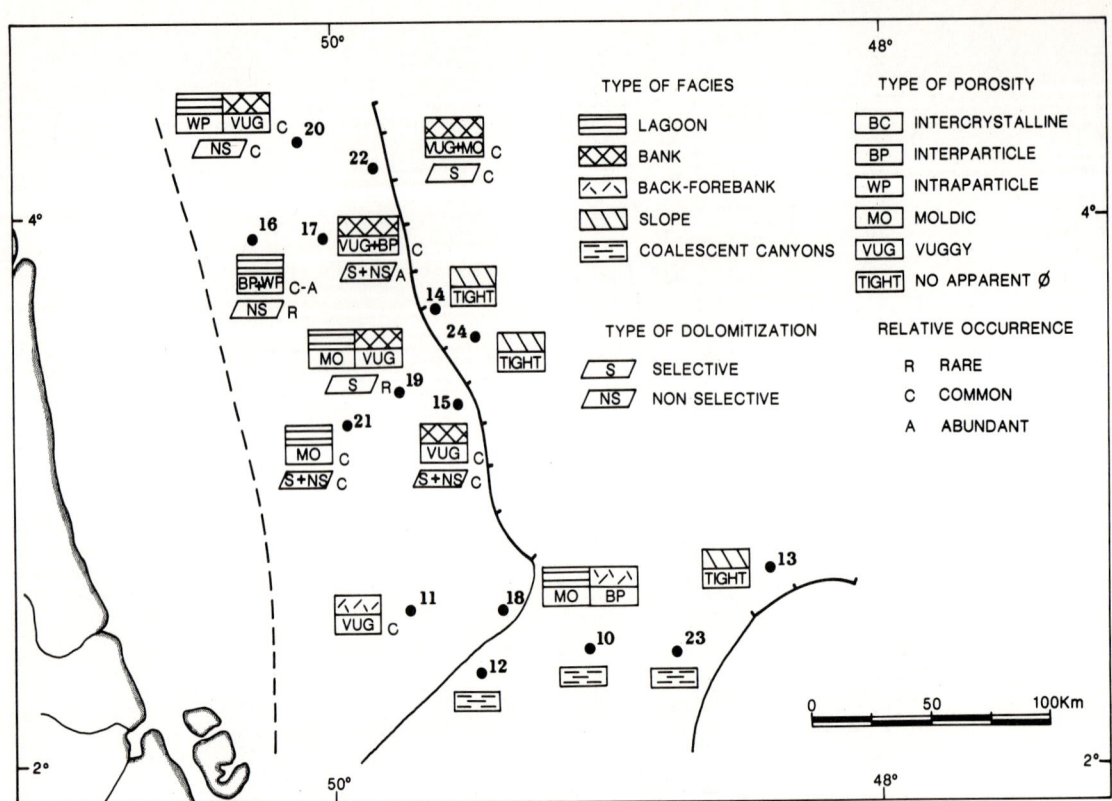

FIGURE 11.95 Amapá Formation (Paleocene–Middle Miocene), Foz do Amazonas Basin, offshore northeast Brazil. Diagenetic map of sequence IV (Middle Miocene). From Wolff and Carozzi (1984).

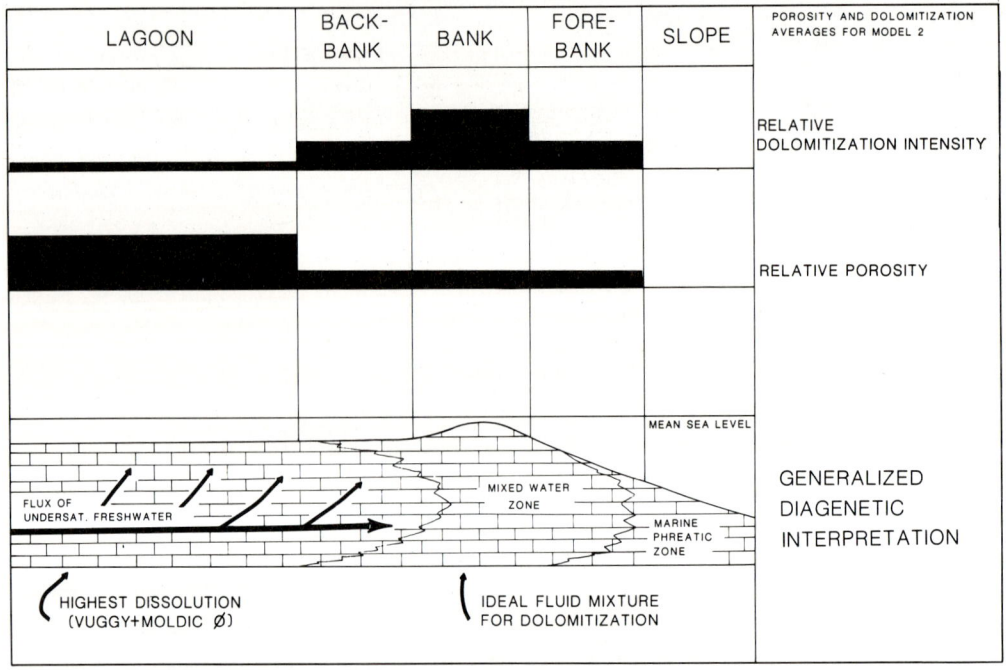

FIGURE 11.96 Amapá Formation (Paleocene–Middle Miocene), Foz do Amazonas Basin, offshore northeast Brazil. Basinwide diagenetic interpretation. From Wolff and Carozzi (1984).

12

Carbonate Platforms with Frontal Hydrodynamic Buildups

Bioaccumulated buildups of platforms undergoing the action of agitated seas are completely destroyed as they grow. Thus bioclasts generated in great quantity and rounded during transport by waves and currents are accumulated in hydrodynamic buildups. Because of their sand-size dimension, most transported bioclasts, oncoids, and pellets are mixed in variable amounts with non-abraded tests of smaller benthic foraminifers of the same original size. Some of these biogenic sands and gravels can become locally oolitized.

12.1 CHARACTERISTIC CONSTITUENTS: SMALLER BENTHIC FORAMINIFERS, CRINOIDS, AND OOIDS

The Salem Limestone (Middle Mississippian) of the southwest margin of the Illinois Basin (Carozzi and Diaby, 1981) was investigated by means of six field sections. More than 378 thin sections were studied petrographically from samples taken at an average vertical interval of 0.26 m for a total measured thickness of 97.8 m.

Description of Microfacies

Nine distinct microfacies were recognized in this carbonate platform. It displays a well-defined morphologic and environmental zonation according to which microfacies are described in a landward direction.

Lower Slope

Microfacies A (Fig. 12.1A). Very fine-grained, grain-supported, intraclastic, pelletoidal calcarenite with homogeneous calcilutite matrix and rare patches of sparite cement. Calcispheres are common and associated with entire ostracods, bryozoans, *Endothyra*, monaxonic sponge spicules, and rare oncoids.

Upper Slope

Microfacies B (Fig. 12.1B). Fine-grained, grain-supported, pelletoidal, intraclastic biocalcarenite with sparite cement and residual patches of calcilutite matrix. Common bioclasts are crinoids, calcispheres, brachiopods, pelecypods, and ostracods. *Endothyra*, red

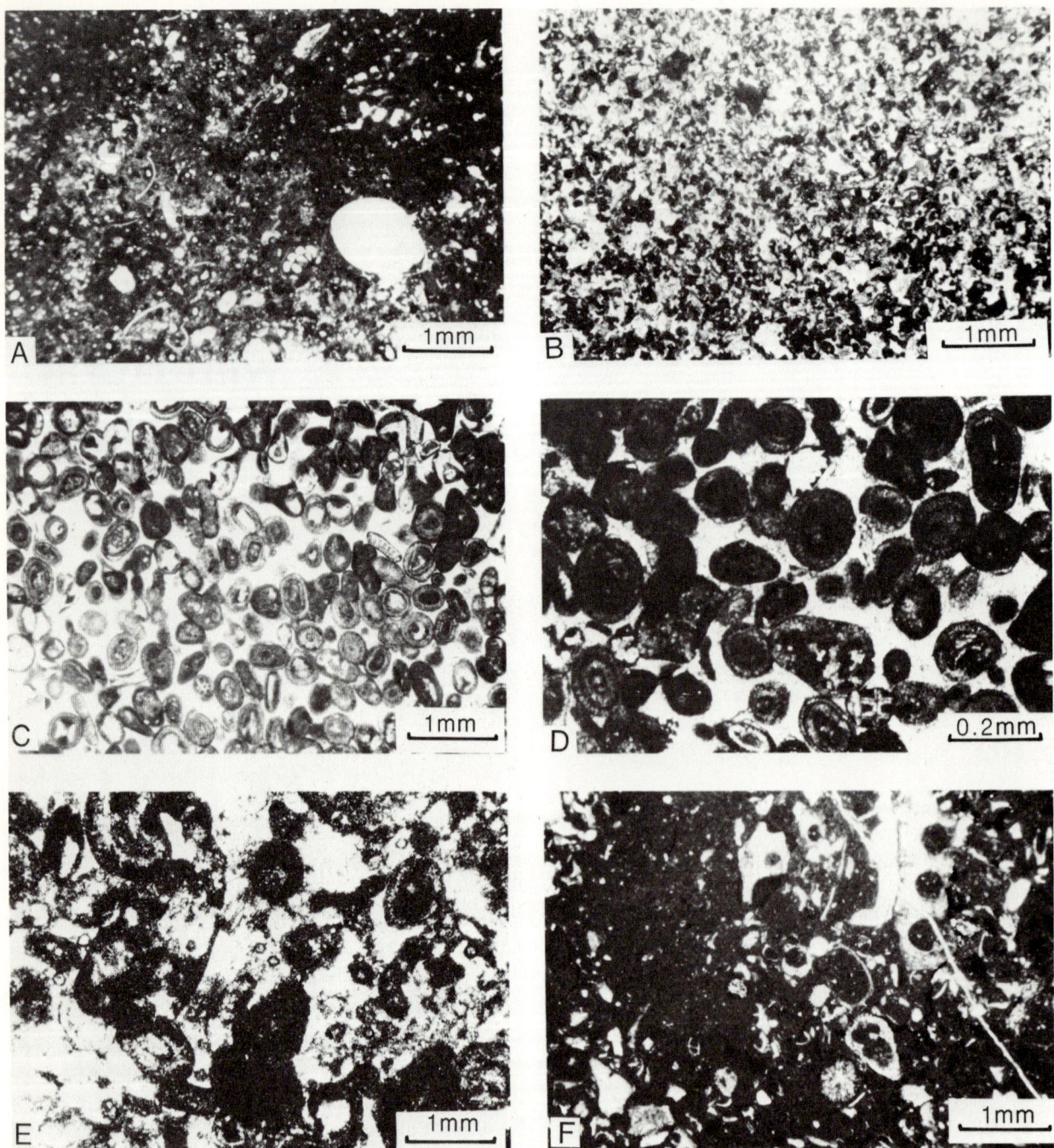

FIGURE 12.1 Salem Limestone (Middle Mississippian), southwest margin of Illinois Basin. Typical microfacies. A. Microfacies A. B. Microfacies B. C and D. Microfacies C. E. Microfacies D. F. Microfacies E. See text for detailed descriptions. All photomicrographs: plane-polarized light. From Carozzi and Diaby (1981).

algae, and reworked ooids are rare. Dolomitization of matrix and bioclasts to dolomicrosparite is common.

Oolitic Outer Bar

Microfacies C (Fig. 12.1C,D). Grain-supported, well-sorted, intraclastic, oolitic biocalcarenite with sparite cement often as syntaxial rims and with pressure-solution features. Common uncoated bioclasts are *Endothyra*, crinoids, and brachiopods. Most of these constituents also form the nuclei of the ooids (Fig. 12.1D).

Outer Open Lagoon

Microfacies D (Fig. 12.1E). Bimodal, grain-supported, intraclastic, crinoidal, *Endothyra* calcarenite with cavity-filling sparite cement. Common ooids are associated with bioclasts of *Koninckopora*, bryozoans, brachiopods, smaller benthic foraminifers, and calcispheres. Patches of interstitial calcisiltite and reorganized pelletoidal internal sediment with geopetal features are locally present.

Microfacies E (Fig. 12.1F). Grain-supported to mud-supported biocalcarenite with bioturbated calcisiltite to calcilutite matrix. Most important bioclasts are crinoids and bryozoans with subordinate amounts of brachiopods, red algae, *Endothyra*, ostracods, calcispheres, and monaxonic sponge spicules. Sparite-filled fenestral textures are locally present.

Microfacies F (Fig. 12.2A). Poorly sorted, grain-supported crinoid-bryozoan calcarenite with calcisiltite matrix and minor patches of sparite cement, mainly as syntaxial overgrowths around crinoid bioclasts. *Koninckopora* and *Endothyra* are associated with bioclasts of trilobites and gastropods. Dolomitization of the matrix into dolomicrosparite is frequent.

Bioclastic Inner Bar

Microfacies G (Fig. 12.2B). Poorly sorted, grain-supported crinoid-bryozoan calcarenite with sparite cement (mainly syntaxial overgrowths around crinoid bioclasts) and local pressure-solution. Scattered small patches of bioclastic matrix are present. Besides the large bioclasts of crinoids and fenestrate bryozoans are smaller bioclasts of brachiopods, gastropods, *Endothyra*, and ostracods. Well-rounded grains of detrital quartz are sometimes present.

Inner Restricted Lagoon

Microfacies H (Fig. 12.2C,D). Laminated to bioturbated argillaceous and pelletoidal calcisiltite with scattered sand-size bioclasts grading into a mud-supported biocalcarenite with bioclastic matrix (Fig. 12.2D). The most common bioclasts are crinoids, echinoids, brachiopods, fenestrate bryozoans, pyritized plant debris, and rare gastropods often destroyed by common dolomitization into dolomicrosparite. Silt-size angular grains of detrital quartz are present locally.

Supratidal Flats

Microfacies I (Fig. 12.2E,F). Deeply bioturbated argillaceous bioclastic calcisiltite with abundant pyritized plant debris. The minute bioclasts are from crinoids, echinoid spines, ostracods, and smaller benthic foraminifers. Angular silt-size grains of detrital quartz are common and associated with fenestral textures occluded by vadose geopetal silt and rectangular aggregates of sparite crystals with wavy extinction which are pseudomorphs after anhydrite (Fig. 12.2F). Dolomitization of the matrix into dolomicrosparite is very common.

The Ideal Shallowing-Upward Sequence

The vertical superposition of microfacies A through I represents an environmental evolution that begins in lower slope conditions, passes through a complex shallow subtidal to intertidal platform, and terminates in supratidal flats. In a general sense, the superposition thus displays a general shallowing-upward trend that is well expressed in a typical section (Figs. 12.3, 12.4), where it ranges, for instance, from upper slope to bioclastic inner bar, and displays detailed variation of major microscopic parameters.

The Ideal Depositional Model

The proposed model (Figs. 12.5, 12.6) shows clearly the relationships between constituents, microfacies, and inferred morphology of the platform. Microfacies X designates inferred offshore conditions corresponding to the basinal portion of the model not encountered in the investigated area but analyzed in Chapter 7. The association brachiopods, smaller benthic foraminifers, and *Endothyra* is characteristic of the oolitic outer bar and the related slopes, whereas the ecological community of the inner bioclastic bar consists of bryozoans, crinoids, and echinoids. The behavior of ooids illustrates the high-energy conditions of the outer bar, which is also a generator of micrite intraclasts and lithic pellets. Apparently, dispersion of the intraclasts was both landward and seaward, and seaward only for the pellets.

Two populations of detrital quartz occur, namely, a coarse-grained subangular to rounded one, and a very fine-grained angular one. The coarse grains appear to have been brought by streams that cut across the three

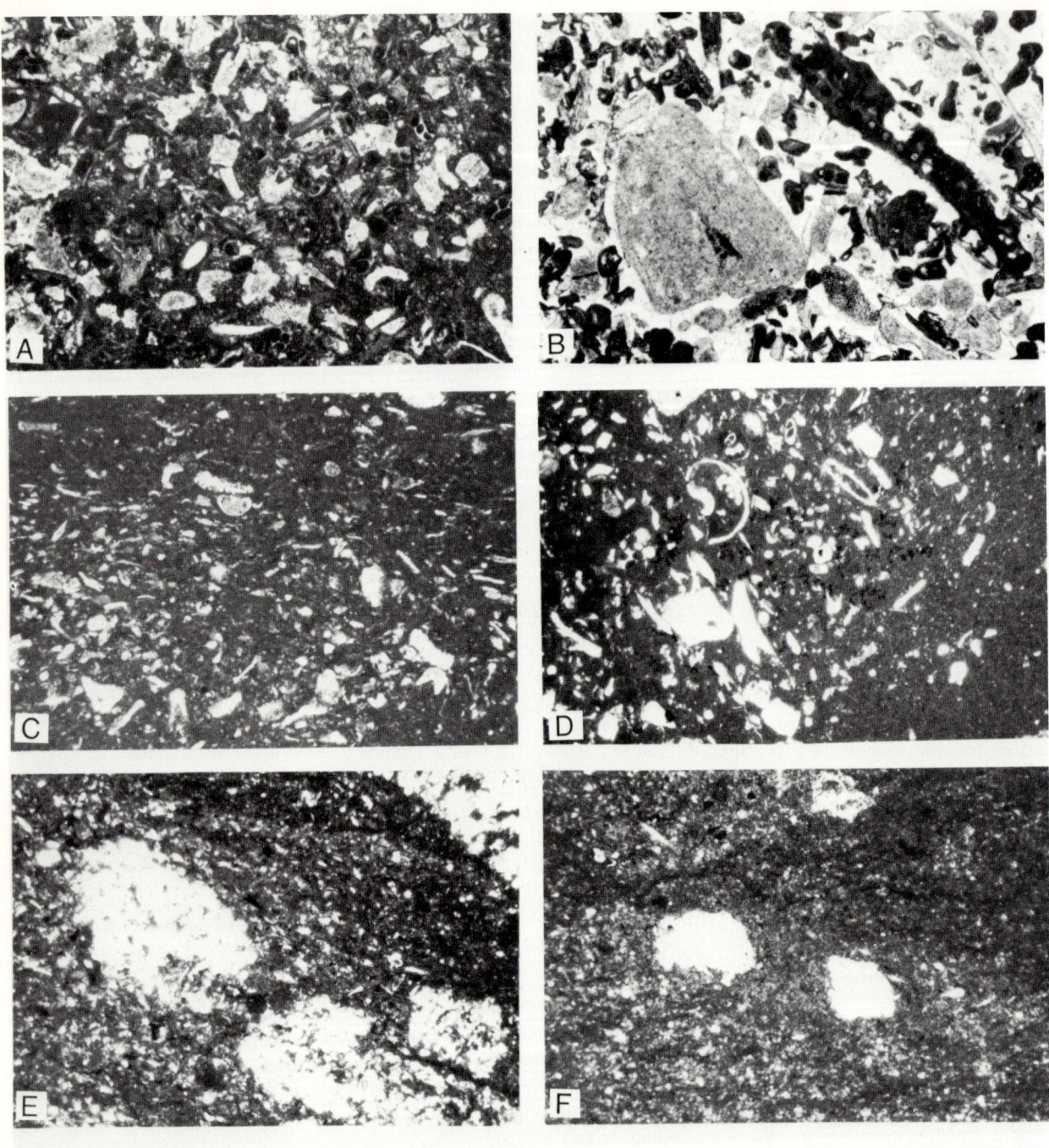

FIGURE 12.2 Salem Limestone (Middle Mississippian), southwest margin of Illinois Basin. Typical microfacies (continued). A. Microfacies F. B. Microfacies G. C and D. Microfacies H. E and F. Microfacies I. See text for detailed descriptions. All photomicrographs: plane-polarized light. From Carozzi and Diaby (1981).

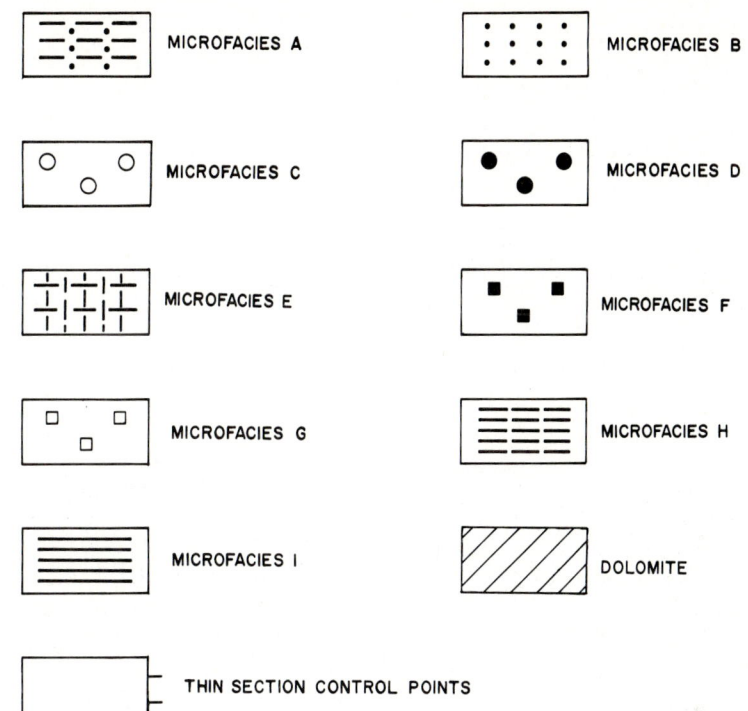

FIGURE 12.3 Salem Limestone (Middle Mississippian), southwest margin of Illinois Basin. Symbols for microfacies. From Carozzi and Diaby (1981).

internal carbonate environments and built small deltas in the outer open lagoon, from which they were further distributed by longshore currents. The finer-grained angular fraction is limited to the inner restricted lagoon and the supratidal flats. Its low frequency and clasticity as well as angularity indicates an eolian origin.

Dolomitization of sabkha type occurred in the inner restricted lagoon and the supratidal flats. Another type of dolomitization is critically located on the seaward slope of both bars, indicating that it results from a mixed seawater-freshwater process (dorag) with discharge on the seaward side when subaerial exposure of the bars provided recharge areas for their freshwater lenses. Emergence of the bioclastic inner bar is supported by the occurrence of beachrock cementation within dislocated and redeposited slabs; a similar situation is inferred for the oolitic outer bar but was not observed in this study.

General Evolution of the Depositional Environments

A composite section of the Salem Limestone (Fig. 12.7) shows that it consists of the superposition of three shallowing-upward episodes. The first episode begins in the outer open lagoon and rises gently to a termination in the supratidal flats. The second episode starts in the upper slope and rises rapidly to the bioclastic inner bar and related beachrock development. The third episode starts in the upper slope and shows an extensive development of the oolitic outer bar and back bar conditions before deepening rapidly into the lower slope environment of the overlying St. Louis Limestone.

12.2 CHARACTERISTIC CONSTITUENTS: PELLETS, CRINOIDS, AND OOIDS

Platforms with predominant pelletoidal and crinoidal constituents display intense oolitization processes along their margins when submitted to shallow agitated conditions and cold slope-ascending currents. Distribution of resulting hydrodynamic oolitic buildups is discontinuous because of preferred location of ascending currents due to slope morphology combined with limited lateral transport along platform edge.

The Ste. Genevieve Limestone (Middle Mississippian) of the southern part of the Illinois Basin (southern Illinois and eastern Missouri) was studied petrographically by means of 800 samples collected at an average vertical interval of 12 cm in four field sections (Rao and Carozzi, 1971).

Description of Microfacies

Microfacies are described in general shallowing-upward order.

Microfacies 1 (Fig. 12.8A). Calcisiltite with scattered sand-size bioclasts of crinoids and bryozoans. The groundmass displays minute angular grains of detrital

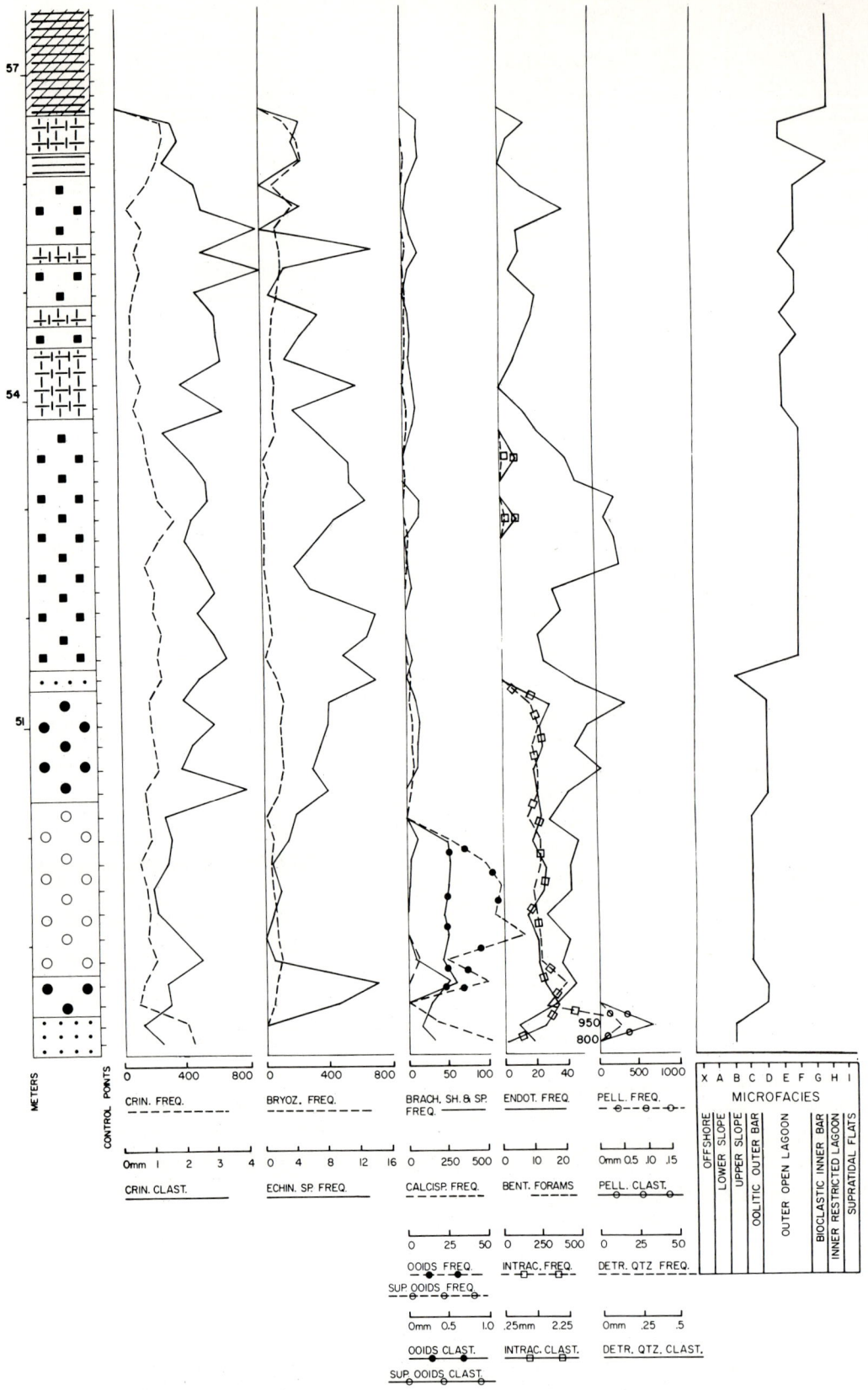

FIGURE 12.4 Salem Limestone (Middle Mississippian), southwest margin of Illinois Basin. Typical example of vertical variation of microscopic parameters during a shallowing-upward sequence. From Carozzi and Diaby (1981).

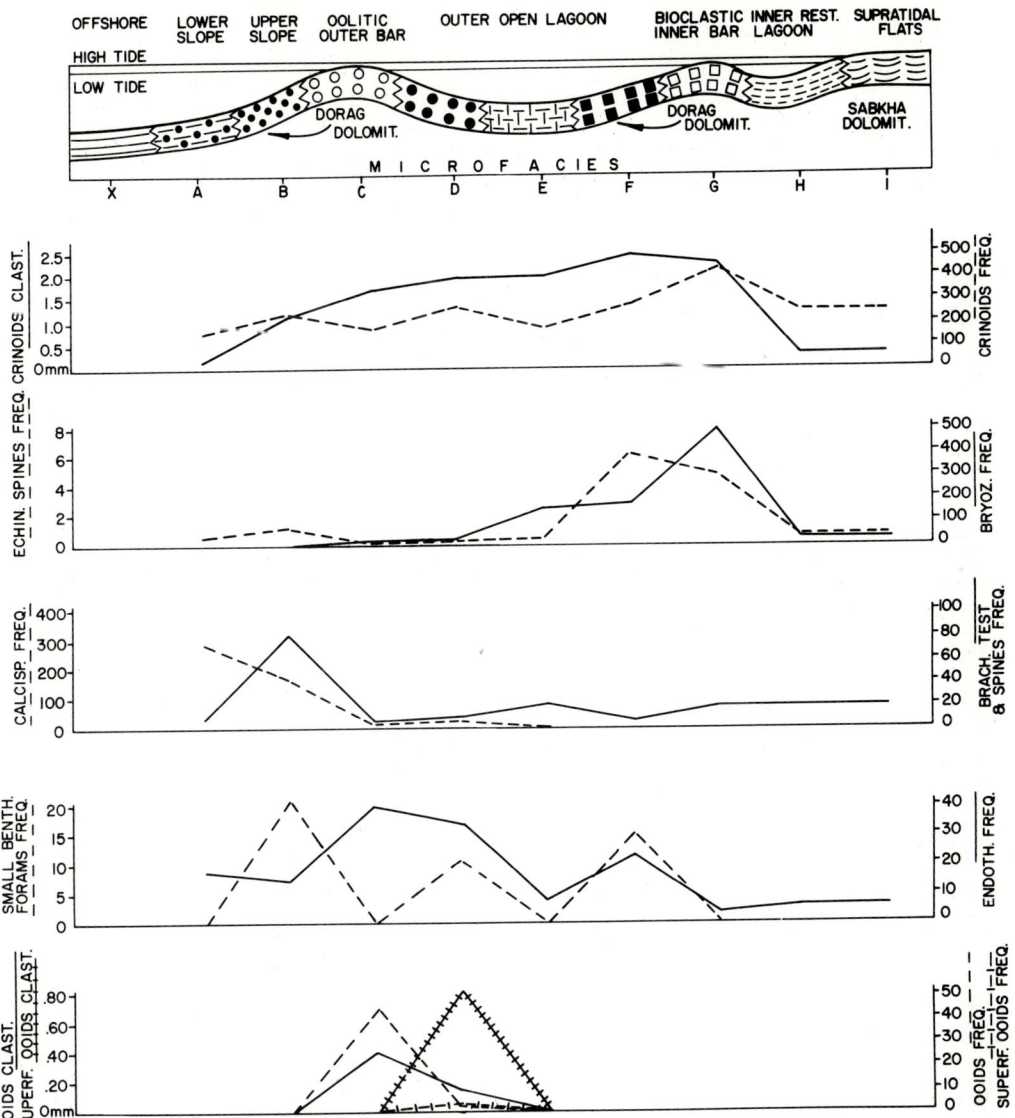

FIGURE 12.5 Salem Limestone (Middle Mississippian), southwest margin of Illinois Basin. Ideal depositional model. From Carozzi and Diaby (1981).

quartz, muscovite flakes, and a few ooids, invariably broken or abraded, which indicates transportation.

Microfacies 2 (Fig. 12.8B). Grain-supported pelletoidal biocalcarenite with calcisiltite to bioclastic matrix. Angular to subrounded intraclasts and lithic pellets predominate over smaller fecal pellets and bioclasts of crinoids, bryozoans, and ostracods.

Microfacies 3 (Fig. 12.8C). Grain-supported biocalcarenite with calcisiltite matrix. Sand-size bioclasts of crinoids and bryozoans are poorly sorted and pressure-welded locally. Rare transported ooids are present.

Microfacies 4 (Fig. 12.8D). Grain-supported, pressure-welded biocalcarenite with cavity-filling sparite cement. Well-sorted coarse sand-size bioclasts of crinoids and bryozoans display internal cavities filled with calcisiltite, which forms also irregular coatings on the bioclasts and indicates their reworking from microfacies 6. Well-rounded lithic pellets and large subrounded grains of detrital quartz are locally present. Transported, abraded, and broken ooids are common.

Microfacies 5 (Fig. 12.8E). Grain-supported pelletoidal biocalcarenite with sparite mosaic cement. Well-rounded intraclasts and lithic pellets, locally pres-

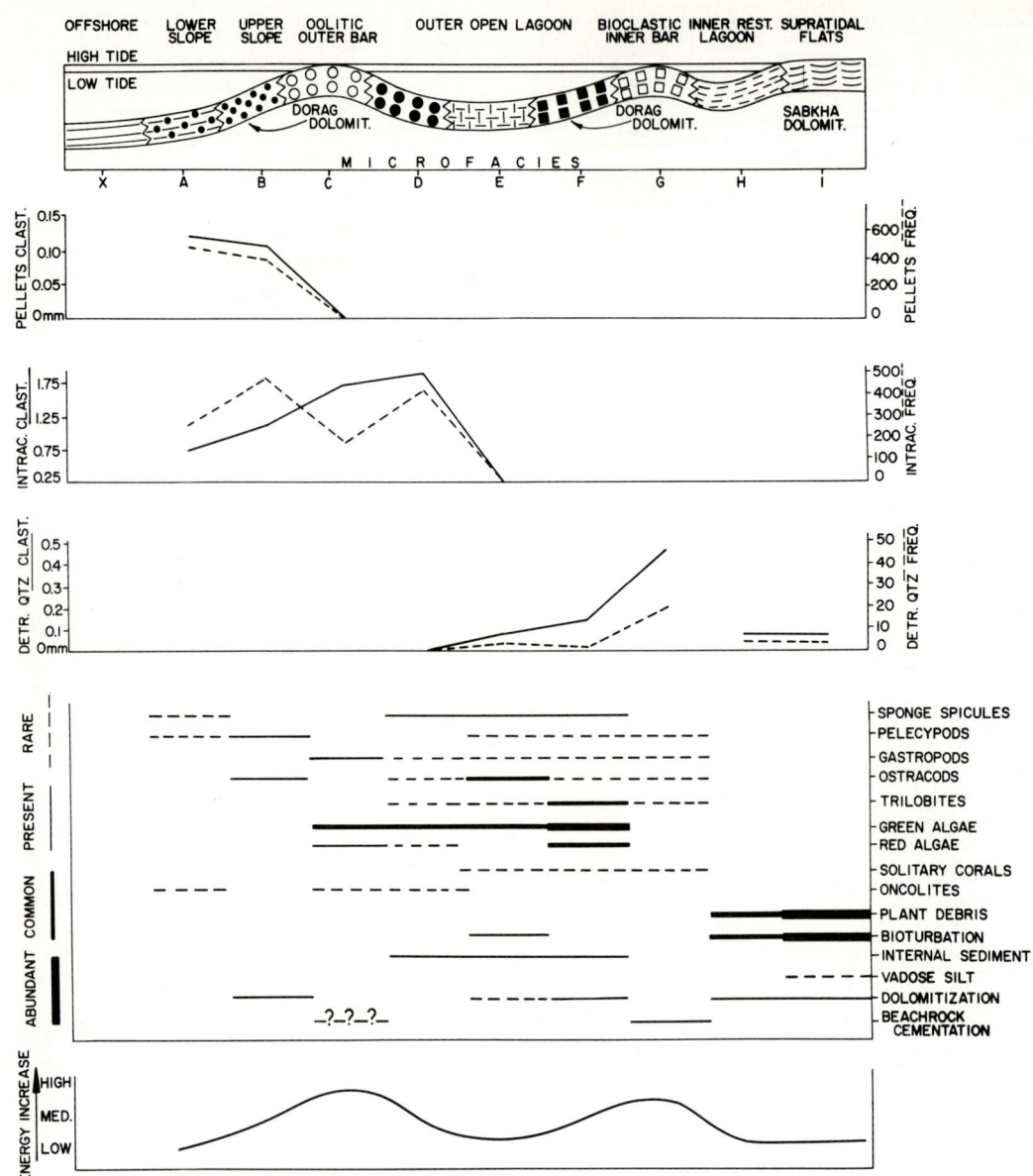

FIGURE 12.6 Salem Limestone (Middle Mississippian), southwest margin of Illinois Basin. Ideal depositional model (continued). From Carozzi and Diaby (1981).

sure-welded, predominate over bioclasts of crinoids and bryozoans, and large subrounded grains of detrital quartz. Bioclasts appear to have been reworked from microfacies 6.

Microfacies 6 (Fig. 12.8F). Grain-supported oolitic biocalcarenite with cavity-filling sparite cement. This is the typical oolitic sediment of the Ste. Genevieve Limestone. It consists of well-developed ooids (40%), some of which are cerebroid (Carozzi, 1962), well-rounded lithic pellets (23%), bioclasts of crinoids (19%), and of bryozoans (8%). Pellets and bioclasts are similar to those forming microfacies 4 and 5. Detrital quartz is fine-grained, subrounded, and found in small amounts.

Microfacies 7 (Fig. 12.8G). Grain-supported oolitic and pelletoidal biocalcarenite with calcisiltite matrix. Poorly sorted, broken, and abraded transported ooids are associated in almost equal amount with rounded lithic pellets and subordinated bioclasts of crinoids and bryozoans.

Microfacies 8 (Fig. 12.8H). Pure quartz arenite with ferruginous microsparite cement. Moderately sorted, subangular to subrounded grains of quartz with

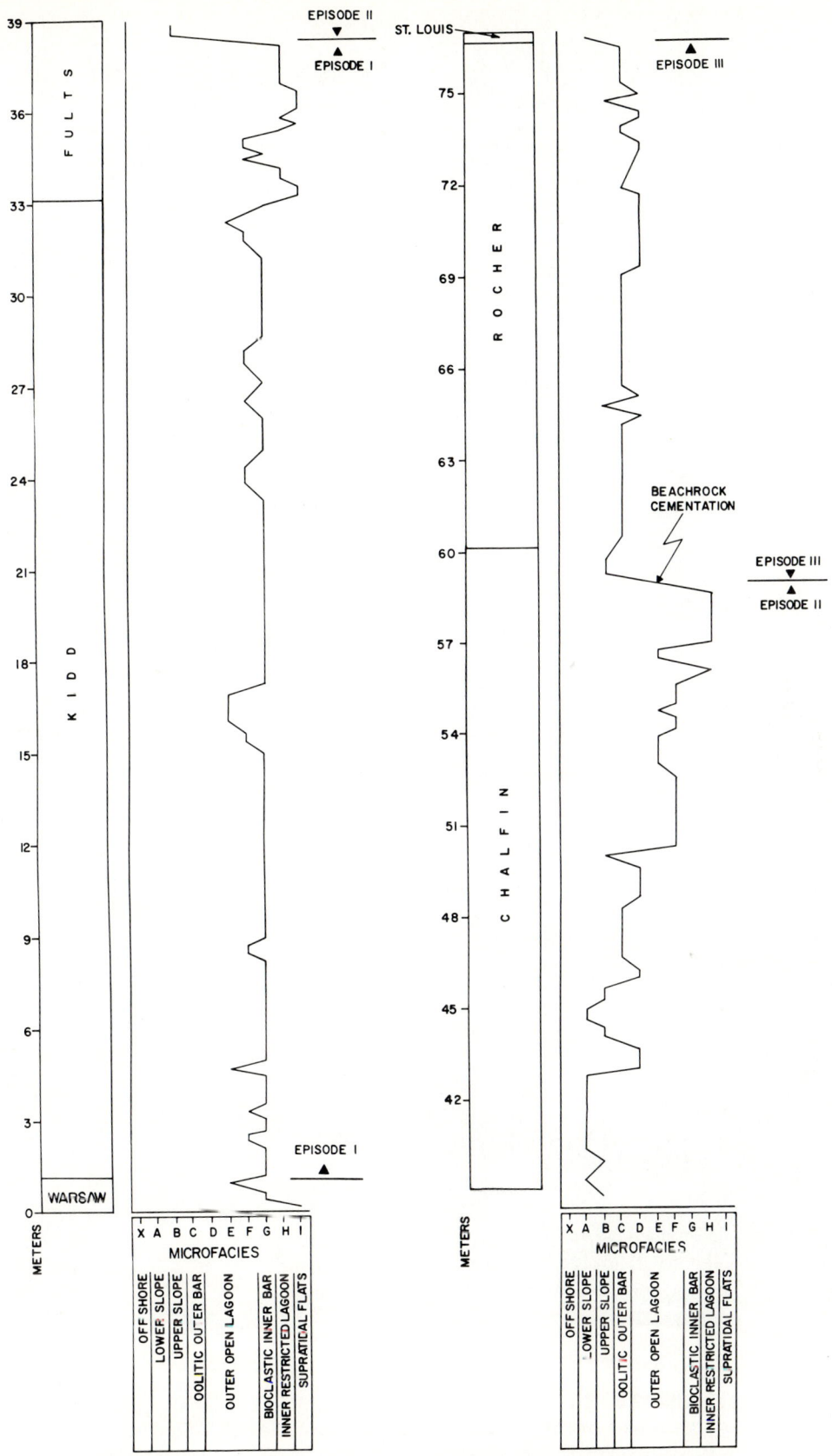

FIGURE 12.7 Salem Limestone (Middle Mississippian), southwest margin of Illinois Basin. General vertical evolution of depositional environments. From Carozzi and Diaby (1981).

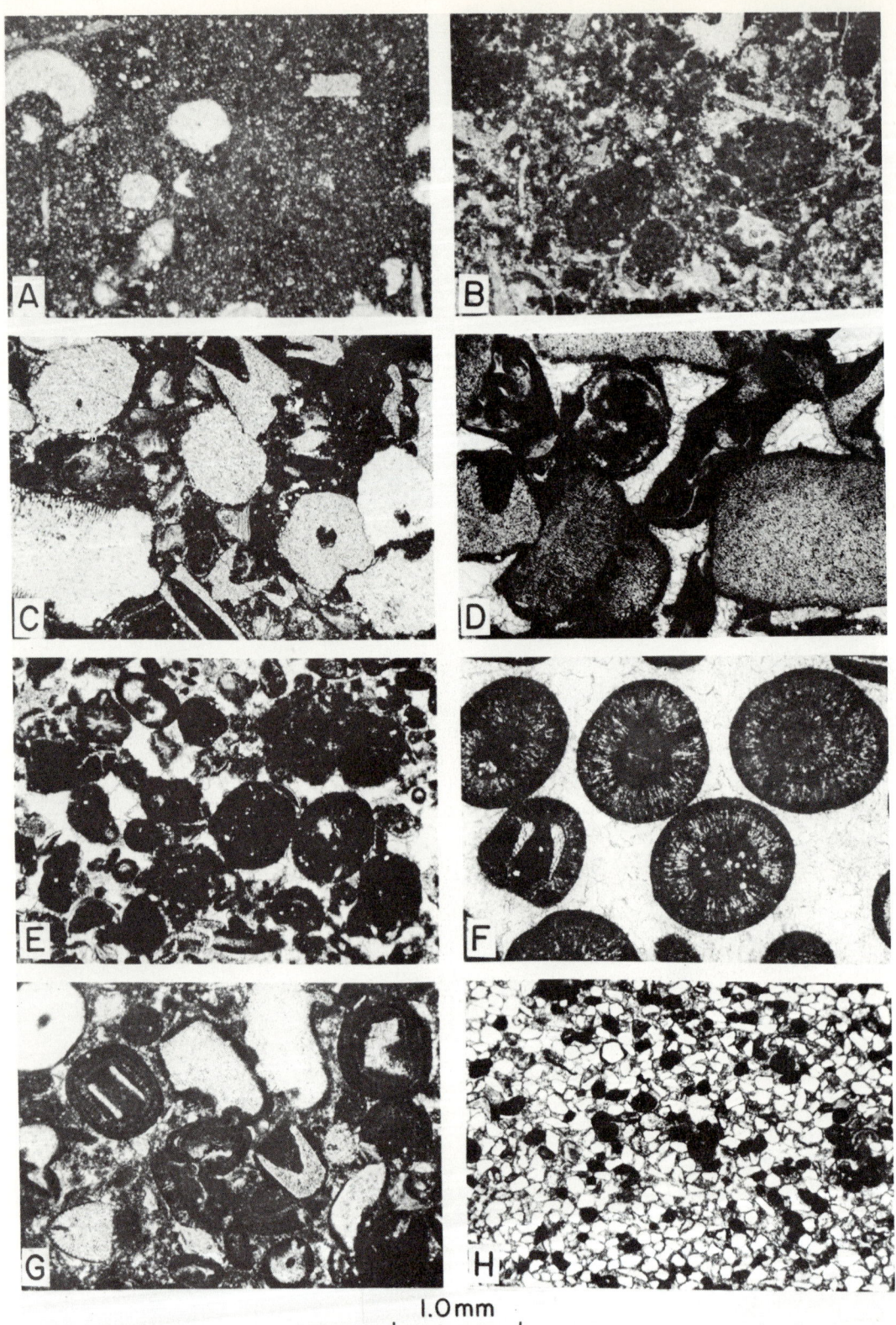

FIGURE 12.8 Ste. Genevieve Limestone (Middle Mississippian), southern part of Illinois Basin. Typical microfacies. A. Microfacies 1. B. Microfacies 2. C. Microfacies 3. D. Microfacies 4. E. Microfacies 5. F. Microfacies 6. G. Microfacies 7. H. Microfacies 8. See text for detailed descriptions. All photomicrographs: plane-polarized light. From Rao and Carozzi (1971).

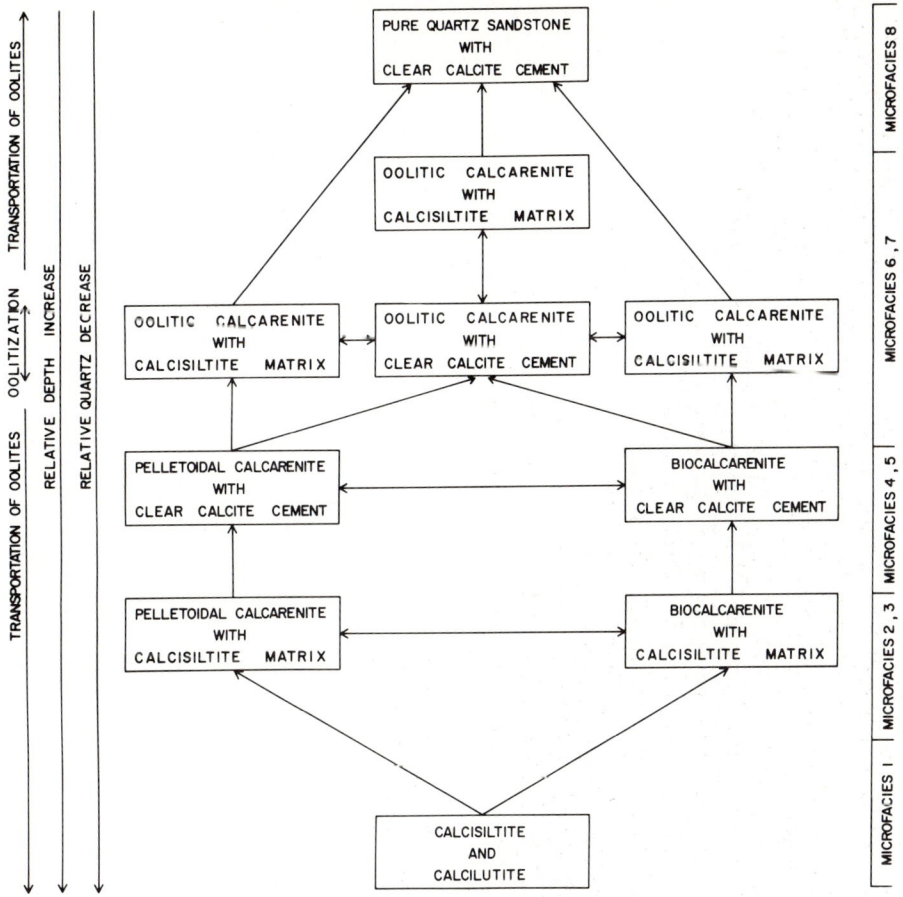

FIGURE 12.9 Ste. Genevieve Limestone (Middle Mississippian), southern part of Illinois Basin. Relationships of microfacies. From Rao and Carozzi (1971).

a few grains of feldspars and glauconite are associated with broken and abraded ooids, small lithic pellets, and rare crinoid bioclasts.

Relationships of Microfacies

The microfacies just described follow each other in the investigated sections in a shallowing-upward order according to two distinct sequences, a pelletoidal and a bioclastic sequence (Fig. 12.9). The domain of oolitization processes (microfacies 6, autochthonous ooids) is partially surrounded by oolitic calcarenites (microfacies 7) formed almost entirely of allochthonous ooids.

The Ideal Shallowing-Upward Sequences

In both the pelletoidal and bioclastic sequences (Fig. 12.10), peaks of clasticity and frequency of complete ooids in microfacies 6 indicate the environment of oolitization, whereas continuous increase of frequency of broken ooids expresses their transportation into nonoolitic microfacies as detrital particles. Curves of clasticity and frequency of lithic pellets and crinoid bioclasts are random and reflect the behavior of particles not involved in oolitization. Frequency of bryozoan bioclasts, ostracods, and endothyrids decrease in microfacies 6 because they are winnowed away as an effect of the high energy. Frequency of detrital quartz remains small with a few oscillations until its abrupt increase from microfacies 7 to 8, the latter being a pure quartz arenite. A typical detailed section (Figs. 12.11, 12.12) shows reciprocal behavior of microscopic parameters in two successive shallowing-upward sequences.

The Ideal Depositional Model

The model (Fig. 12.13) ranges from subtidal to intertidal and consists of a seaward slope, discontinuous oolitic bars and shoals, a lagoon, and finally, a delta or estuary. In the channels between oolitic bars, microfacies 4 and 5 grade through microfacies 7 into microfacies 8; because of this transition, microfacies 4 and 5 may become locally very arenaceous. A comparison was made between this model and the present Great Bahama Bank (Fig. 12.14).

314 Part III / Carbonate Platforms

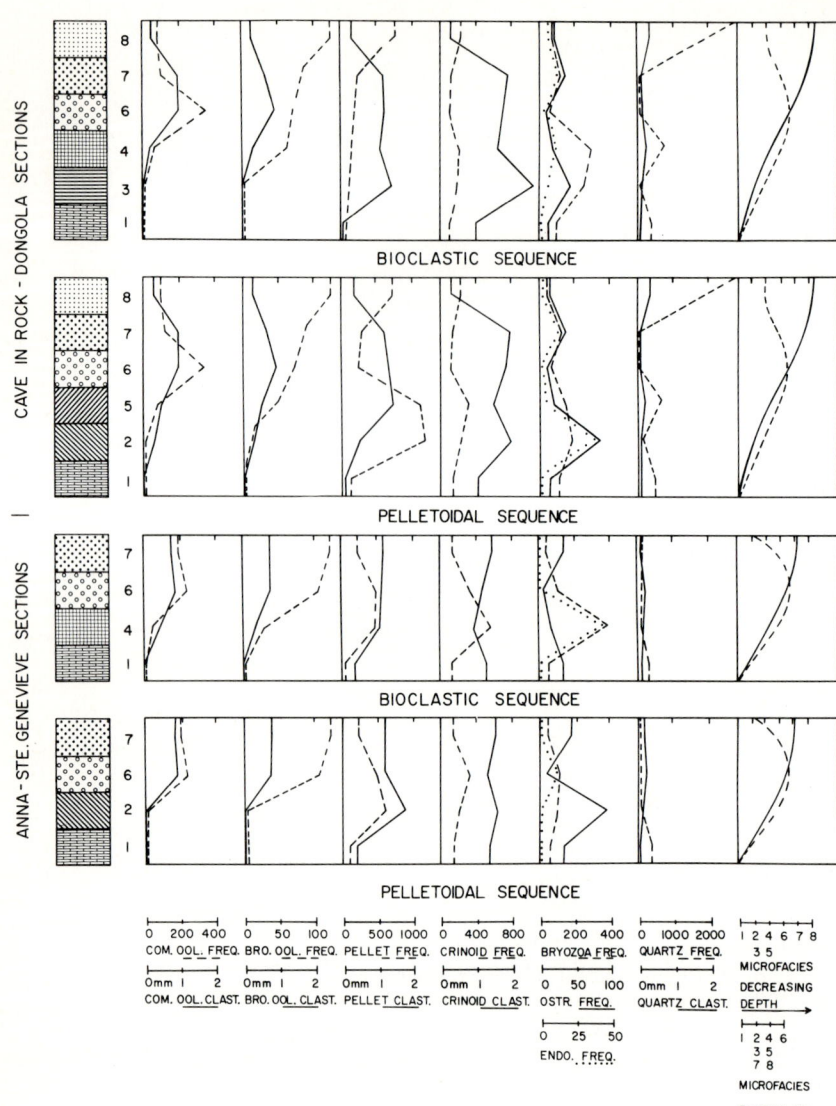

FIGURE 12.10 Ste. Genevieve Limestone (Middle Mississippian), southern part of Illinois Basin. Vertical variation curves of average microscopic parameters during shallowing-upward evolution of bioclastic and pelletoidal sequences. From Rao and Carozzi (1971).

The main difference is that allochthonous ooids are not only transported into the shelf-lagoonal microfacies 7 as in the Bahama Bank (Purdy, 1963) and in the marginal channels, but that they also occur from the shallowest to the deepest microfacies (8 to 1). Such a mechanism of distribution of transported ooids and detrital quartz in all the seaward microfacies seems to result from a system of alternating tidal currents which reached the lagoonal area and also contributed to drain the influx of freshwater from the deltaic or estuarine areas.

Analysis of Autochthonous and Allochthonous Oolitic Environments

The oolitic facies of the Ste. Genevieve Limestone is an excellent oil producer because of the combination of various types of secondary burial porosity, which ranges from oomoldic through enlarged interparticle to vuggy. In that regard it is significant that microfacies 6 shows two major textural variations which correspond to the predominance of ooids essentially in their environment of formation (autochthonous ooids) and of ooids which have undergone a certain amount of transportation (allochthonous ooids) mostly by longshore currents. By means of statistical petrography and computer techniques, it is possible to distinguish and characterize the two types of textures which correspond to different types of potential reservoirs (Lacey and Carozzi, 1967). It is well known that although calcium carbonate saturation is widespread in shallow marine environments, conditions for precipitation are very localized. Therefore, ooids are generated in relatively small areas along carbonate platform edges and distributed penecontemporaneously along depositional strike as detrital particles, from a few hundred feet to several miles.

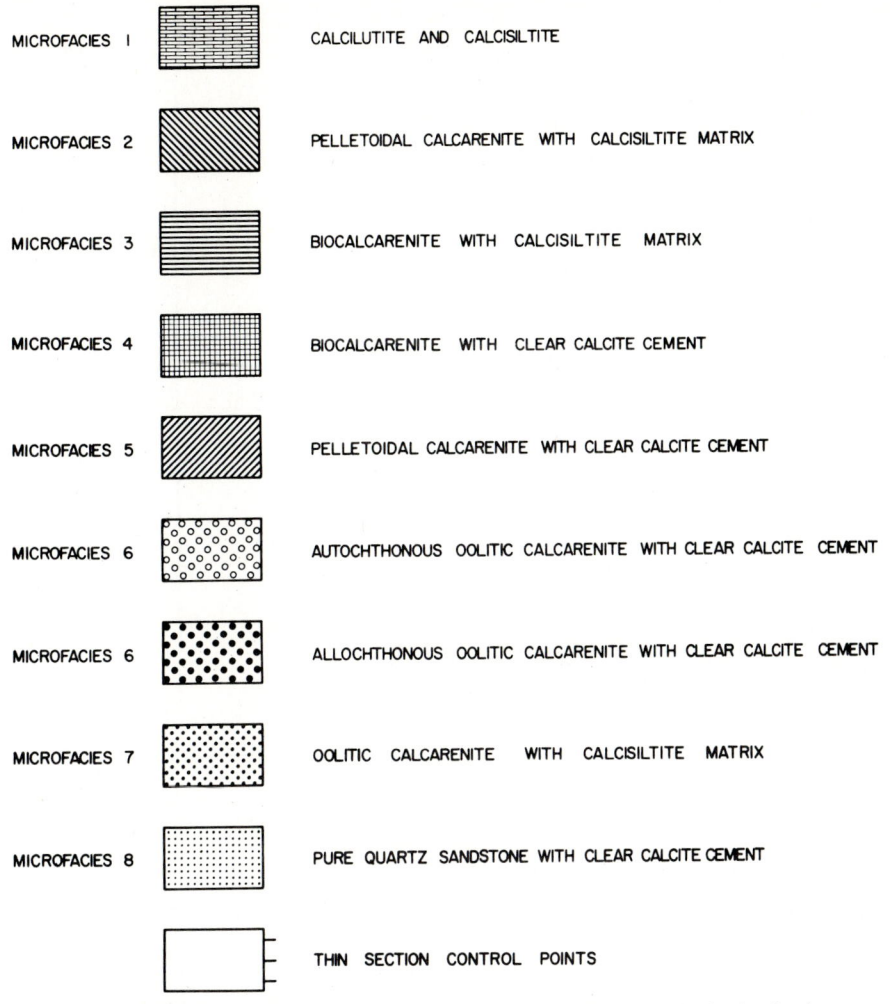

FIGURE 12.11 Ste. Genevieve Limestone (Middle Mississippian), southern part of Illinois Basin. Symbols for microfacies. From Rao and Carozzi (1971).

Parameter Variations and Their Relationships in Autochthonous Oolitic Environment

Autochthonous ooids are characterized by the general absence of abrasion of the outer concentric rings, textural and compositional similarity of concentric rings of associated ooids, and tendency of most ooids of a given sample to show same maximum size. Autochthonous ooids are often associated with recoated broken ooids which demonstrate saturated conditions.

Parameter variation shows the following relationships:

1. Maximum clasticity of ooids is independent of maximum clasticity of other "nonoolite" grains (the largest carbonate particles of any kind which were not oolitized, usually intraclasts), of the clasticity of pellets, and of the clasticity of crinoid bioclasts.

2. Maximum clasticity of "nonoolites," clasticity of pellets, and clasticity of crinoid bioclasts show a general direct relationship.

3. Pellet, crinoid, and bryozoan bioclast frequencies are generally opposite to percentage of ooids, directly related to percentage of "nonoolites," and independent of maximum clasticity of ooids.

4. Pellet, crinoid, and bryozoan bioclast frequencies are generally related.

The clasticity relationships described above are shown in theoretical model 1 (Fig. 12.15) and those of frequency in theoretical model 2 (Fig. 12.16). The textural aspect of authochthonous oolitic calcarenites (Fig.

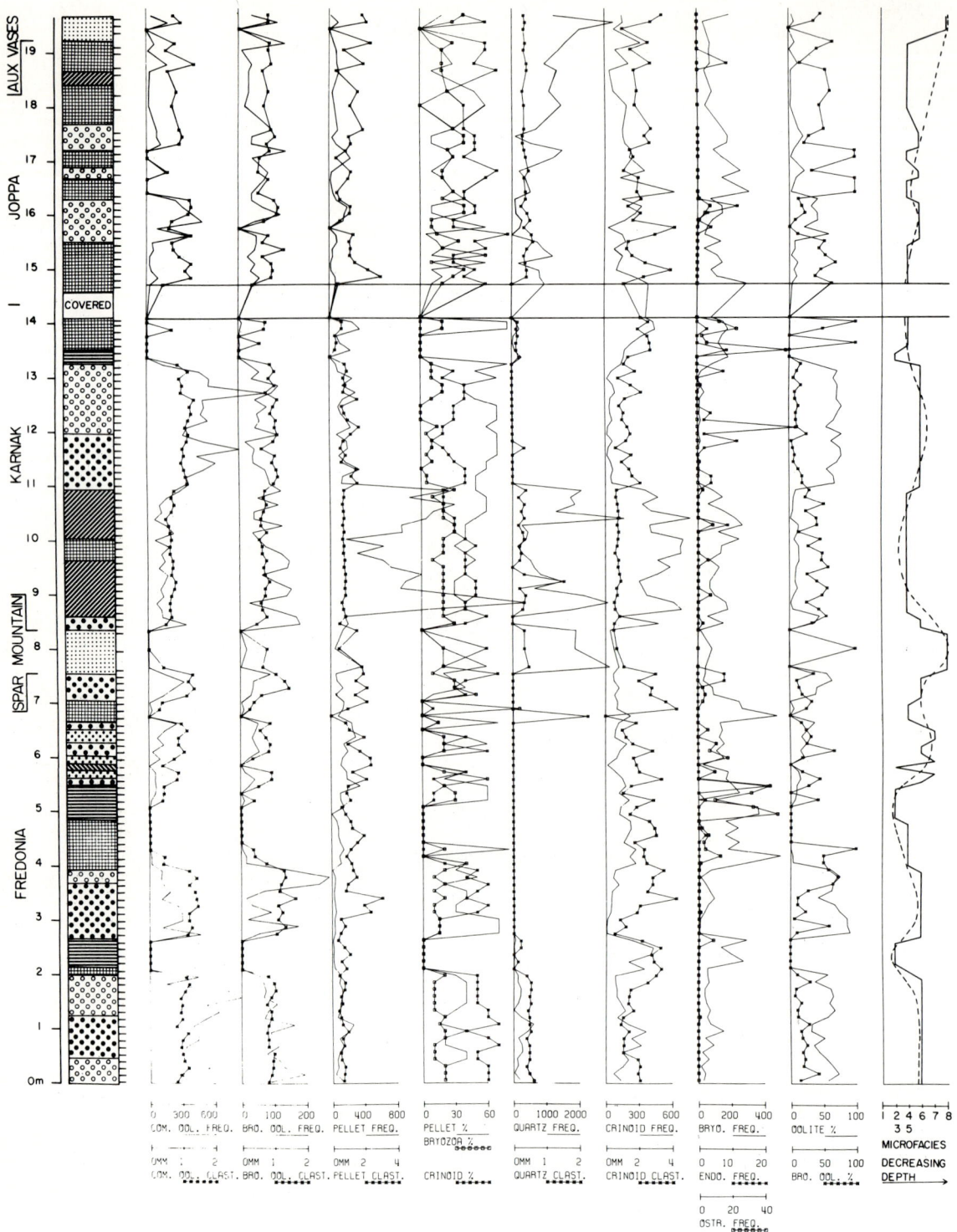

FIGURE 12.12 Ste. Genevieve Limestone (Middle Mississippian), southern part of Illinois Basin. Typical example of vertical variation of microscopic parameters. From Rao and Carozzi (1971).

Chap. 12 / *Carbonate Platforms with Frontal Hydrodynamic Buildups* 317

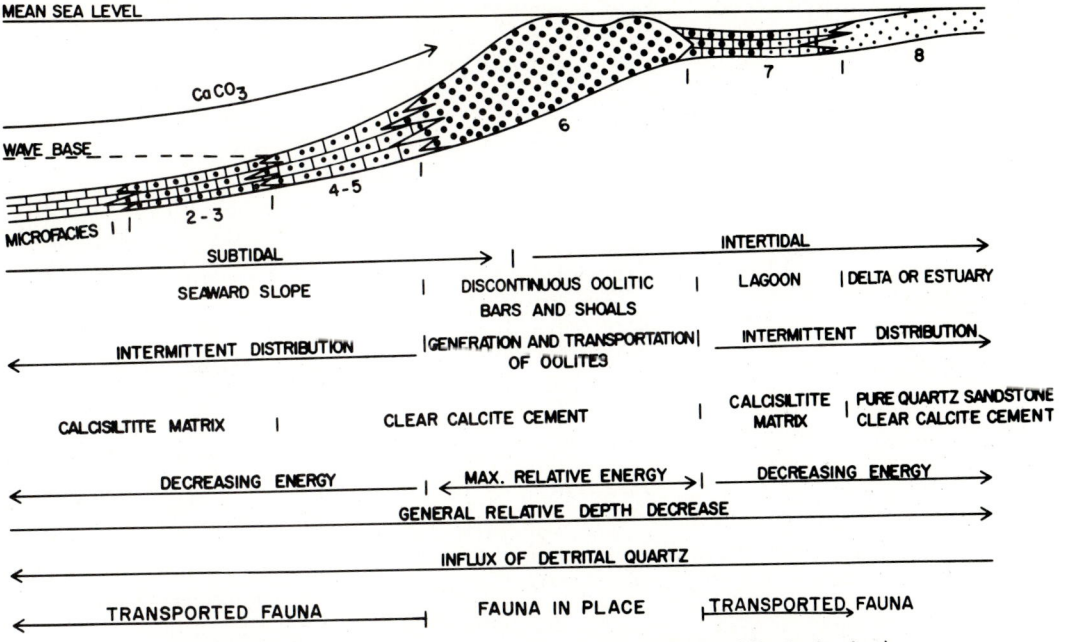

FIGURE 12.13 Ste. Genevieve Limestone (Middle Mississippian), southern part of Illinois Basin. Ideal depositional model. From Rao and Carozzi (1971).

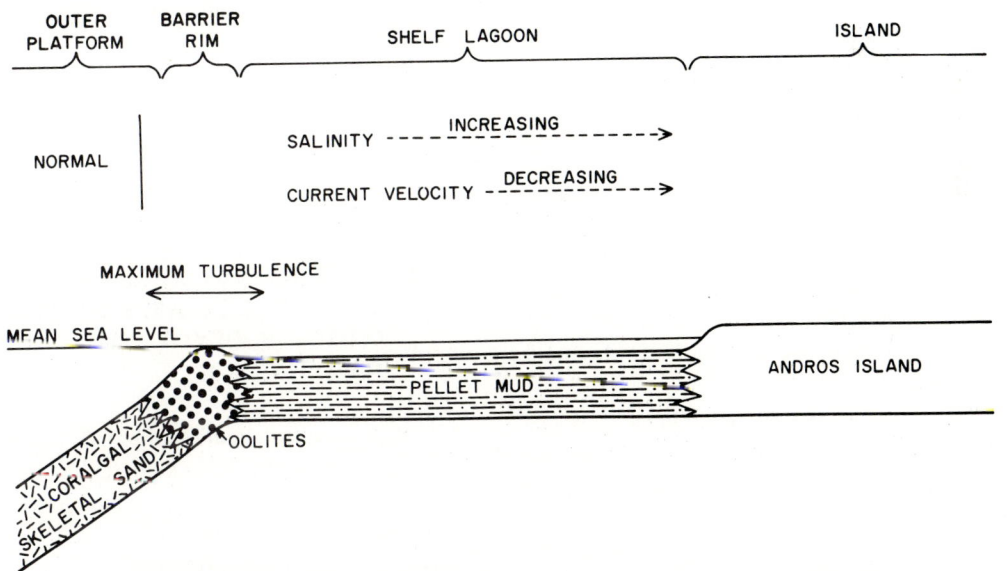

FIGURE 12.14 Ste. Genevieve Limestone (Middle Mississippian), southern part of Illinois Basin. Simplified partial model of Great Bahama Bank. From Rao and Carozzi (1971).

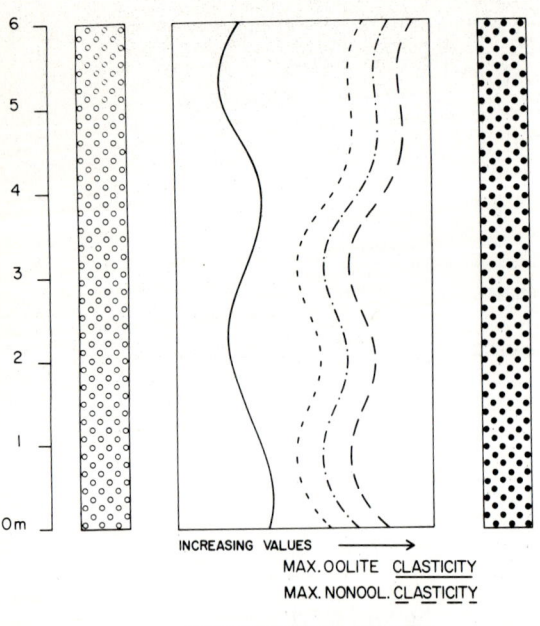

FIGURE 12.15 Ste. Genevieve Limestone (Middle Mississippian), southern part of Illinois Basin. Theoretical model 1. Clasticity relationships between autochthonous and allochthonous ooids and other constituents. See Figure 12.16 for complete explanation of symbols. From Rao and Carozzi (1971).

12.17) shows a mature carbonate sediment developed *in situ* and uncontaminated by any transportation processes or admixtures of foreign constituents.

Parameter Variations and Their Relationships in Allochthonous Oolitic Environment

Allochthonous ooids are characterized by abrasion of outer concentric rings, lack of textural and compositional similarity of concentric rings of associated ooids, and tendency of most ooids of a given sample not to show common maximum size due to mixing of ooids from different sources. Allochthonous ooids are not often associated with recoated broken ooids.

Parameter variation shows the following relationships:

1. Maximum clasticity of ooids is related directly to maximum clasticity of "nonoolites," to clasticity of pellets, and to clasticity of crinoid bioclasts.
2. Maximum clasticity of "nonoolites," clasticity of pellets, and clasticity of crinoid bioclasts show a general direct relationship.
3. Pellet, crinoid, and bryozoan bioclast frequencies are opposite to percentage of ooids, directly related to percentage of nonoolites, but opposed to maximum clasticity of ooids.
4. Pellet, crinoid, and bryozoan bioclast frequencies are mutually related.

The clasticity relationships described above are shown in theoretical model 1 (Fig. 12.15) and those of frequency in theoretical model 2 (Fig. 12.16). The textural aspect of allochthonous oolitic calcarenites (Fig. 12.18) shows an immature carbonate sediment generated by transportation processes and contaminated by a variety of foreign constituents.

Oolitic limestones display a wide range of porosity types from primary to secondary with complex time and space relationships (Rich and Carozzi, 1981). A genetic classification of these types is possible and of critical importance for reconstructing the depositional and diagenetic history of oolitic limestones and for evaluating their economic importance as potential hydrocarbon reservoirs.

12.3 CHARACTERISTIC CONSTITUENTS: SMALLER BENTHIC FORAMINIFERS, CRINOIDS, BRYOZOANS, OOIDS, AND ONCOIDS

Morphologic differentiation of broad carbonate platforms in agitated seas shows juxtaposed longitudinal rows of coarse-grained and flat-topped hydrodynamic buildups separated by finer-grained areas. Their constituents are mainly smaller benthic foraminifers associated with crinoid and bryozoan bioclasts as well as oncoids. Some restriction develops landward across these platforms leading to local generation of oolitic microfacies and eventually to lagoons almost devoid of organisms.

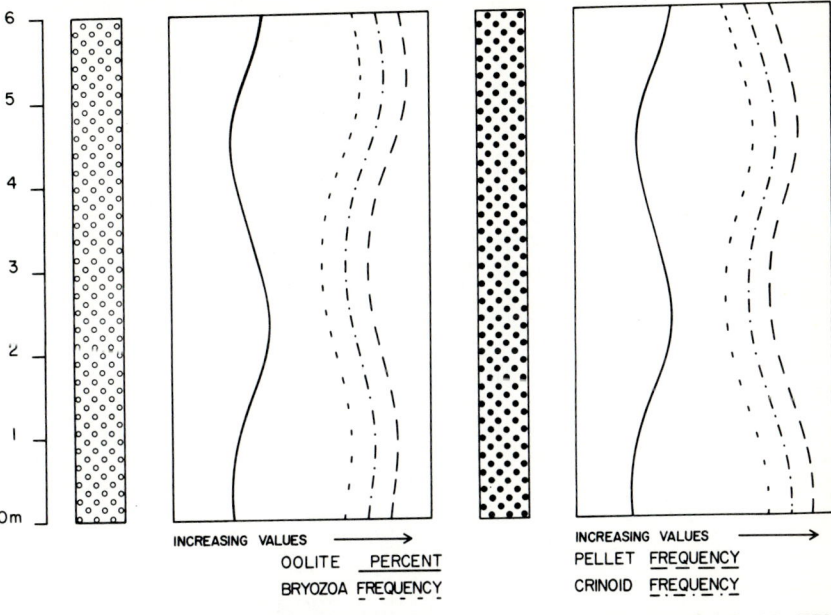

FIGURE 12.16 Ste. Genevieve Limestone (Middle Mississippian), southern part of Illinois Basin. Theoretical model 2. Frequency relationships between autochthonous ooids and allochthonous ooids and other constituents. See Figure 12.15 for complete explanation of symbols. From Rao and Carozzi (1971).

The Glen Dean Formation (Middle Mississippian) was formed by an extensive carbonate platform that represents the eastern shelf of the Illinois Basin in southern Illinois, Indiana, and Kentucky. This platform was invaded at times by siliciclastic sediments originating from prograding deltaic systems that crossed the platform by means of subtidal channels leading to their final basinal deposition (Feiznia, 1983, Feiznia and Carozzi, 1987).

The Glen Dean Formation of the eastern shelf of the Illinois Basin was investigated by means of 23 stratigraphic sections, of which 15 are field sections and 8 are cores. The petrographic study is based on 1200 samples collected with an average vertical interval of 30 cm for the sandstone and shale portions and 15 cm for the carbonate portions.

Description of Carbonate Microfacies

Carbonate microfacies are separated into groups according to their major constituents and subdivided internally on the basis of type and amount of interstitial material.

Spiculitic Group

Microfacies A (Fig. 12.19,1). Pelletoidal and bioturbated calcilutite with scattered sand-size or silt-size bioclasts of crinoids, bryozoans, and ostracods associated with rare monaxonic sponge spicules and calcispheres.

Microfacies B (Fig. 12.19,2). Bioturbated spiculitic and pelletoidal calcisiltite with calcilutite matrix. Abundant monaxonic sponge spicules and calcispheres are associated with sand-size bioclasts of crinoids, bryozoans, and brachiopods. Endothyrids, earlandiids, and archaediscids are common.

Microfacies C (Fig. 12.19,3). Heavily bioturbated, mud-supported biocalcarenite with pelletoidal calcisiltite to calcilutite matrix. Crinoid and bryozoan bioclasts are associated with monaxonic sponge spicules, ostracods, and calcispheres.

Microfacies AC. Very arenaceous, mud-supported biocalcarenite with calcisiltite to calcilutite matrix. Components similar to microfacies C but with lesser frequencies due to increased amount of detrital quartz.

Foraminiferal Group

Microfacies D1 (Fig. 12.19,4). Grain-supported foraminiferal biocalcarenite with sparite cement. Very abundant endothyrids are associated with earlandiids, crinoids, bryozoans, echinoid spines, and brachiopods.

Microfacies D2. Grain-supported foraminiferal and intraclastic biocalcarenite with association of bioturbated calcilutite matrix and sparite cement. Endothyrids, archaediscids, and earlandiids predominate over crinoids, bryozoans, and ostracods.

Microfacies D3 (Fig. 12.19,5). Grain-supported foraminiferal biocalcarenite with bioturbated calcilutite matrix. Very abundant endothyrids, earlandiids, and calcispheres are associated with echinoid spines, crinoids, bryozoans, and ostracods.

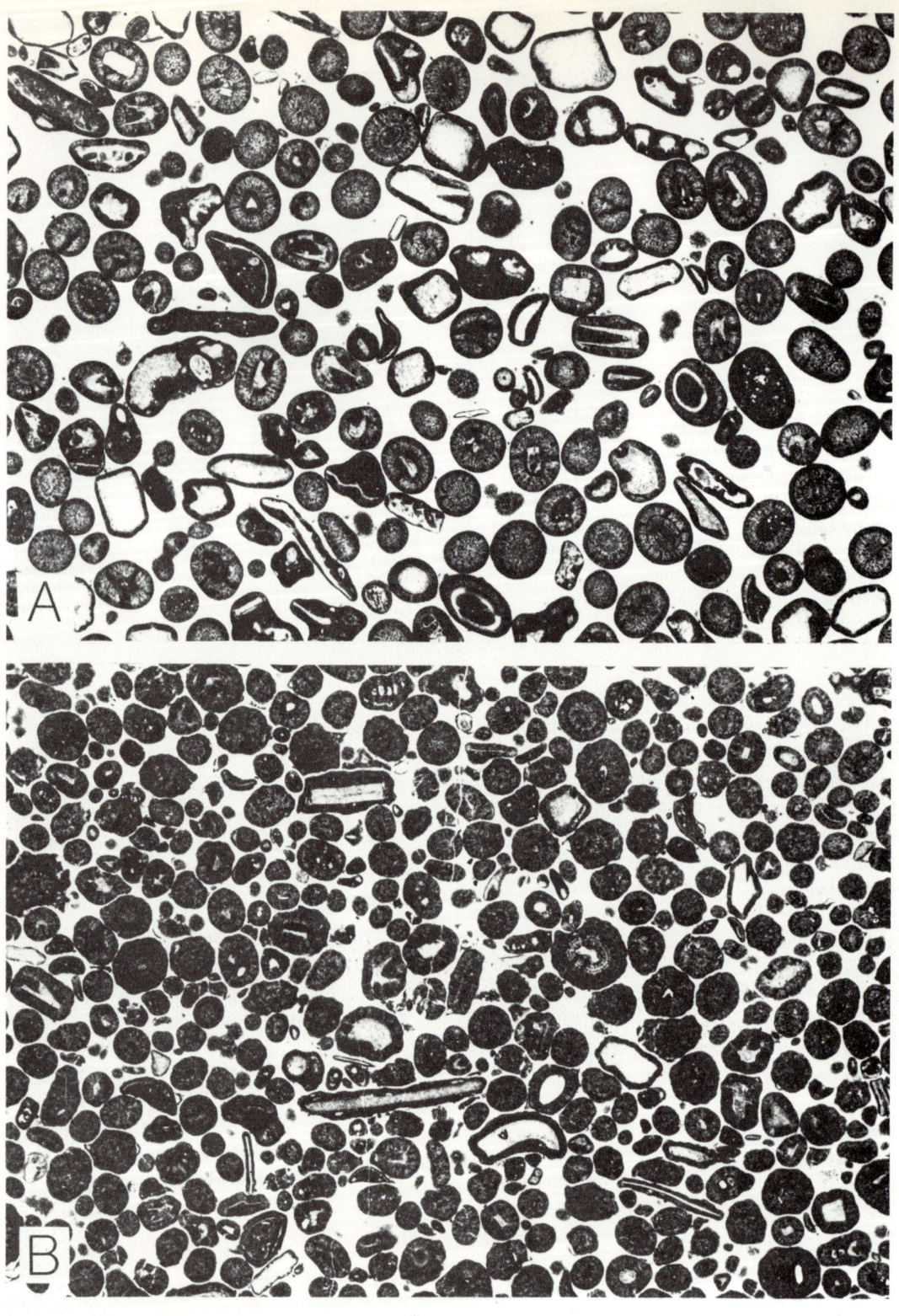

FIGURE 12.17 Ste. Genevieve Limestone (Middle Mississippian), southern part of Illinois Basin. *Autochthonous ooids*. A. Rather large ooids with cores composed primarily of crinoid and bryozoan bioclasts set in clear sparite cement. Small cores are surrounded by thick oolitic coatings and large cores by thin coatings leading to a fairly uniform external diameter. B. Most of the ooids possess small cores consisting of pellets, crinoids, and bryozoan bioclasts with thick oolitic coatings leading to a fairly uniform external diameter. The cement is clear sparite. Cerebroid ooids are present and deforma-

Microfacies AD. Very arenaceous foraminiferal biocalcarenite similar to microfacies D2, but with lower frequency of carbonate components due to increased amount of detrital quartz.

Crinoidal Group

Microfacies E1 (Fig. 12.19,6). Grain-supported to pressure-welded crinoidal calcarenite with sparite cement. Crinoid columnals and calyx plates predominate over bryozoans and brachiopods.

Microfacies E2. Grain-supported to pressure-welded crinoidal calcarenite with association of calcite cement and calcilutite matrix. Common bryozoan and brachiopod bioclasts predominate over pelecypods, ostracods, and annelid tubes.

Microfacies E3 (Fig. 12.19,7). Pressure-welded crinoidal calcarenite with calcilutite matrix. Bryozoans and brachiopods are common constituents. Neomorphic syntaxial overgrowths around crinoids replace the matrix.

Microfacies AE. Very arenaceous crinoidal calcarenite similar to microfacies E2, but with lower frequency of carbonate components due to increased amount of detrital quartz.

Bryozoan Group

Microfacies F1 (Fig. 12.19,8). Grain-supported to pressure-welded bryozoan-brachiopodal calcarenite with abundant sparite cement. Ramose and fenestrate bryozoans predominate over crinoids, pelecypods, and trilobites.

Microfacies F2. Grain-supported bryozoan calcarenite with association of sparite cement and calcilutite matrix. Brachiopod, crinoid, and pelecypod bioclasts are common.

Microfacies F3 (Fig. 12.20,1). Grain-supported bryozoan calcarenite with argillaceous calcilutite matrix. Brachiopod bioclasts are associated with crinoids and calcispheres.

Microfacies AF. Very arenaceous bryozoan calcarenite similar to microfacies F2 but with lesser frequency of carbonate components due to increased amount of detrital quartz.

Oncoidal Group

Microfacies G (Fig. 12.20,2). Grain-supported oncoidal biocalcarenite with sparite cement. The oncoid cores are foraminifers, pellets, and bioclasts of crinoids and bryozoans, which also occur as uncoated constituents associated with brachiopods and endothyrids.

Oolitic Group

Microfacies H1 (Fig. 12.20,3). Grain-supported to pressure-welded oolitic biocalcarenite with sparite cement. Ooid cores consist of crinoid, bryozoan, and pelecypod bioclasts also occurring as uncoated constituents.

Microfacies H2 (Fig. 12.20,4). Grain-supported oolitic biocalcarenite with sparite cement and pelletoidal calcilutite matrix, or with matrix only. Ooid cores consist of crinoid and bryozoan bioclasts, which also occur as uncoated grains together with echinoids, foraminifers, and calcispheres.

Silty Calcilutite Group

Microfacies I (Fig. 12.20,5). Silty, arenaceous, and pyritic calcilutite devoid of any bioclasts, frequently pelletoidal and bioturbated. Neomorphism into pseudomicrosparite is common.

Description of Siliciclastic Microfacies

Siltstone Group

Microfacies J (Fig. 12.20,6). Laminated and bioturbated siltstone consisting of angular and poorly sorted quartz grains, muscovite flakes, and micrite pellets with interstitial argillaceous matrix and local patches of microsparite.

Sandstone Group

1. Fine- to medium-grained, pressure-welded quartz arenite with hematitic sparite cement (Fig. 12.20,7). Fairly well-sorted quartz grains are associated with oxidized bioclasts, plagioclases, and muscovite flakes.
2. Fine- to medium-grained pressure-welded quartz arenite with hematitic sparite cement

FIGURE 12.17 (*continued*) tion by mutual penetration leads to polygonal shapes. Both photomicrographs: plane-polarized light. From Lacey and Carozzi (1967). Reprinted by permission of Société Nationale des Pétroles d'Aquitaine.

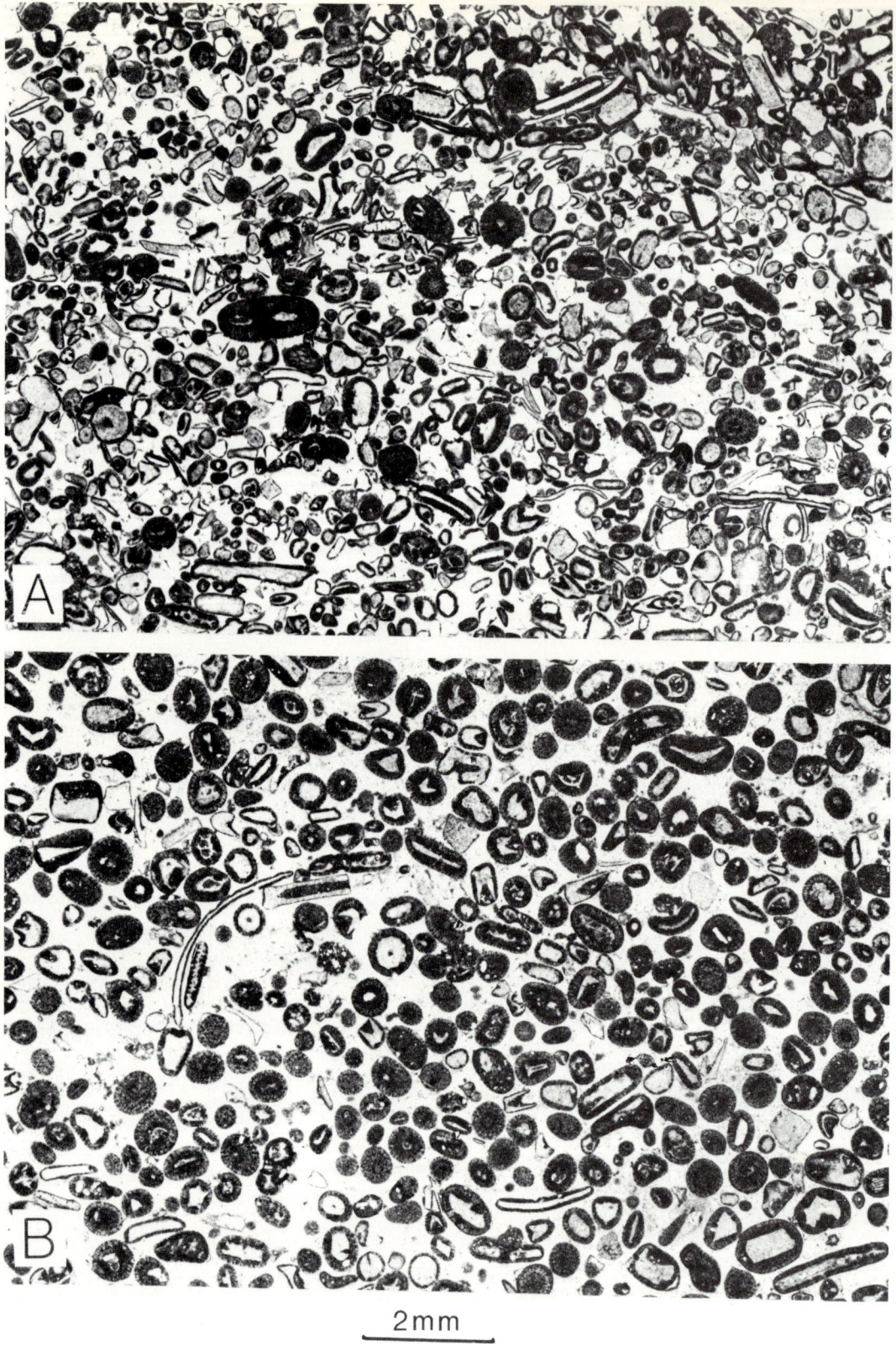

2mm

FIGURE 12.18 Ste. Genevieve Limestone (Middle Mississippian), southern part of Illinois Basin. *Allochthonous ooids*. A. The ooids with variable external diameter possess cores consisting of bioclasts of crinoids, bryozoans, brachiopods, pelecypods, and pellets. Abundant superficial ooids developed around brachiopod and pelecypod shells and numerous bioclasts. The cement is clear sparite. B. Rather well-developed ooids with cores consisting of crinoid and bryozoan bioclasts. Despite variability of size, abundant superficial ooids de-

and argillaceous matrix. It is similar to the previous microfacies but contains as much as 25% clay minerals.

3. Fine- to medium-grained quartz wacke with argillaceous matrix (Fig. 12.20,8). Quartz grains (65%) are associated with highly compacted lithoclasts of siltstone and shale, oxidized bioclasts of bryozoans and crinoids, and grains of plagioclase and zircon. Interstitial argillaceous matrix consists of mixed layer illite/smectite, discrete illite and chlorite, and rare kaolinite.

Shale Group

The shales have the same composition as the argillaceous matrix of the quartz wackes. They are usually fossiliferous with predominant bryozoan bioclasts.

The Ideal Carbonate Shallowing-Upward Sequence

The 20 carbonate microfacies of the Glen Dean Formation can be organized into a shallowing-upward sequence (Fig. 12.21) that expresses an environmental evolution divided into five major domains: slope, outer bank, interbank, inner bank, and lagoon. A typical detailed section (Figs. 12.22, 12.23) of such a shallowing-upward trend shows reciprocal behavior of microscopic parameters.

The Ideal Carbonate Depositional Model

This platform model (Figs. 12.24, 12.25, 12.26) consists of a shallow, open marine, subtidal environment in which two broad hydrodynamic banks or buildups were juxtaposed. The behavior of the various parameters allows recognition of the following subdivisions in an onshore direction: (1) lower and upper slope characterized by sponge spicules, calcispheres, endothyrids, earlandiids, archaediscids, ostracods, brachiopod spines, echinoid spines, quartz grains, pellets, and abundant bioturbation; (2) outer hydrodynamic bank consisting of three subenvironments, including foraminiferal (endothyrids and earlandiids), crinoidal, and oncoidal; (3) interbank with crinoids, bryozoans, calcispheres, and quartz grains; (4) inner hydrodynamic bank consisting of three subenvironments, including bryozoan, oncoidal, and oolitic; and (5) the lagoon, which is essentially devoid of organisms.

Although hydrodynamic buildups show distinct zonation, the distribution of many components displays a more general dispersion due to the action of currents and tides. Crinoid fragments, with clasticity and frequency peaks in microfacies E1 and E3, are distributed in all other microfacies. Echinoids, sharing the same original environment with crinoids, have their spines concentrated in microfacies D2. Bryozoans and brachiopods, which characterize microfacies F1, F2, and F3, also appear to have been redistributed seaward, brachiopods mostly as spines. The smaller benthic foraminifers (endothyrids and earlandiids), which are typical of the front of the outer bank, appear to have undergone little transportation, whereas ostracods, sponge spicules, and calcispheres eventually settled in the same general environment. Intraclasts, originating mainly from the breakdown of oncoids, were also redistributed away from the main oncolitic microfacies. Finally, ooids were dispersed oceanward from their site of formation, and detrital quartz was winnowed from the high-energy surfaces and concentrated in low areas during its seaward transport across the buildups.

Study of the microfacies of the siliciclastic group and their depositional environment indicates that the carbonate depositional model terminated landward into a siltstone wedge (microfacies J) during periods of inactive siliciclastic deposition. During periods of active siliciclastic deposition, three distinct environments were juxtaposed: delta complex, carbonate platform, and basin. Deltaic sands and muds invaded the carbonate platform and cut across it by means of channel systems which allowed them at times to reach the lower slope of the carbonate platform, accumulate in the depocenter, and even prograde as far as the proximal portion of the carbonate platform located on the opposite western margin of the basin (Feiznia and Carozzi, 1981). With respect to the geographic distribution of carbonate microfacies and their corresponding environments (Fig. 12.27), foraminiferal subbanks were present mainly on the eastern shelf (Indiana and Kentucky) and oncolitic subbanks concentrated mostly in the northern part of the same shelf (Indiana), whereas bryozoan subbanks were more extensive on the western shelf (Illinois). Dolomitization was more abundant in the eastern shelf and shall be discussed later in the carbonate diagenesis section.

FIGURE 12.18 (*continued*) veloped around brachiopod and pelecypod shells and numerous bioclasts. Both photomicrographs: plane-polarized light. From Lacey and Carozzi (1967). Reprinted by permission of Société Nationale des Pétroles d'Aquitaine.

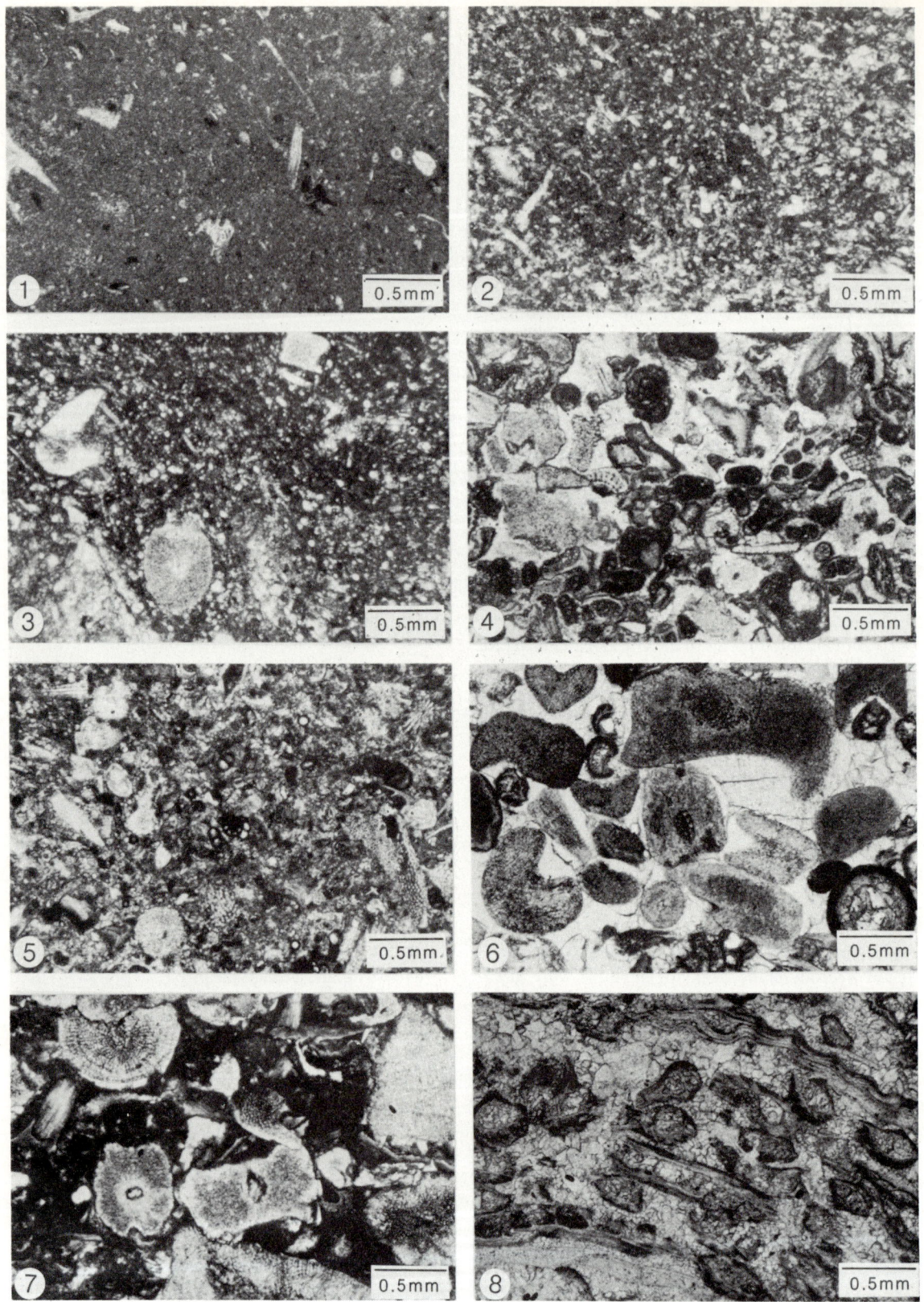

FIGURE 12.19 Glen Dean Formation (Middle Mississippian), eastern shelf of Illinois Basin. Typical microfacies. 1. Microfacies A. 2. Microfacies B. 3. Microfacies C. 4. Microfacies D1. 5. Microfacies D3. 6. Microfacies E1. 7. Microfacies E3. 8. Microfacies F1. See text for detailed descriptions. All photomicrographs: plane-polarized light. From Feiznia (1983).

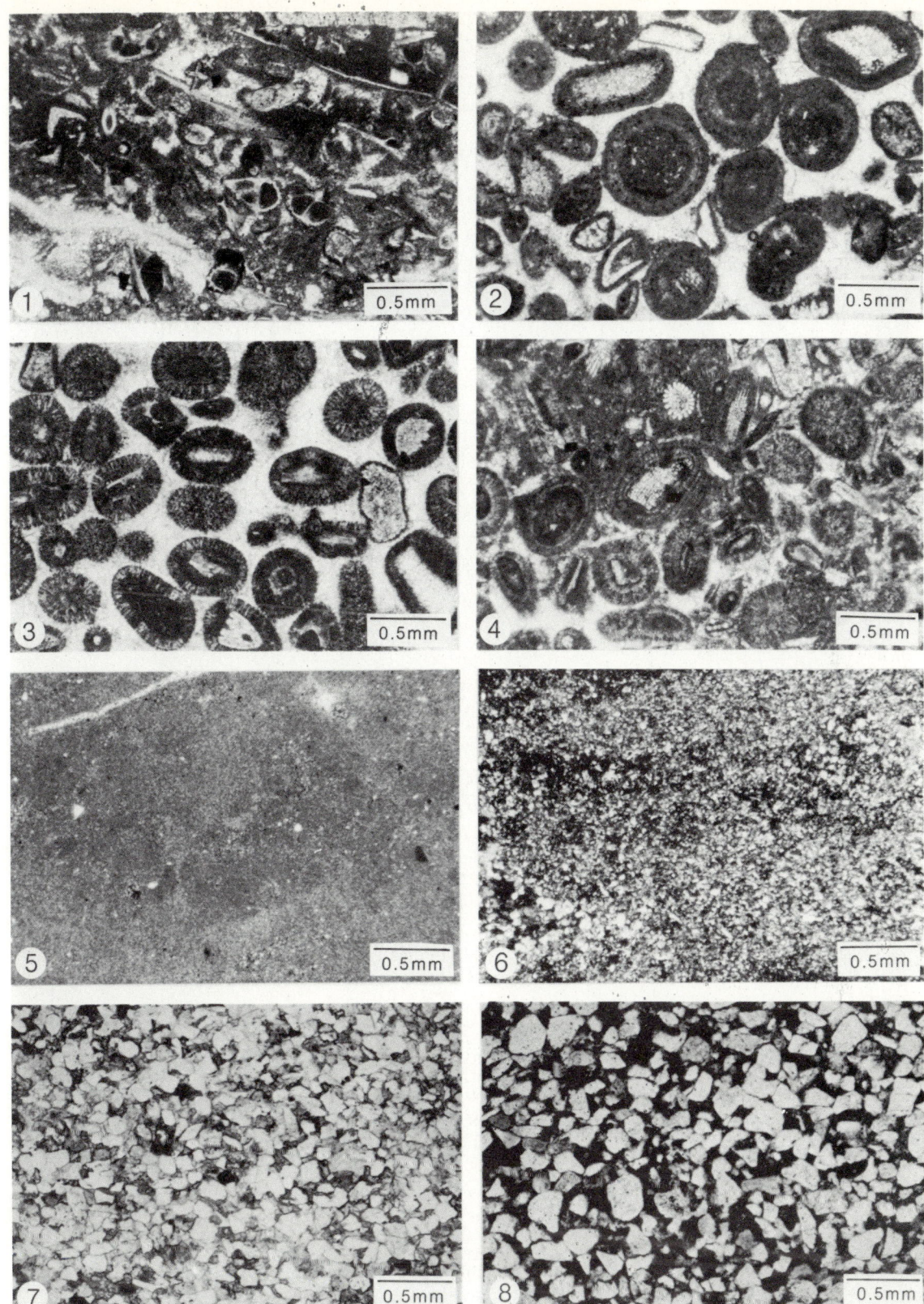

FIGURE 12.20 Glen Dean Formation (Middle Mississippian), eastern shelf of Illinois Basin. Typical microfacies (continued). 1. Microfacies F3. 2. Microfacies G. 3. Microfacies H1. 4. Microfacies H2. 5. Microfacies I. 6. Microfacies J. 7. Quartz arenite with hematitic sparite cement. 8. Quartz wacke with argillaceous matrix. See text for detailed descriptions. All photomicrographs: plane-polarized light. From Feiznia (1983).

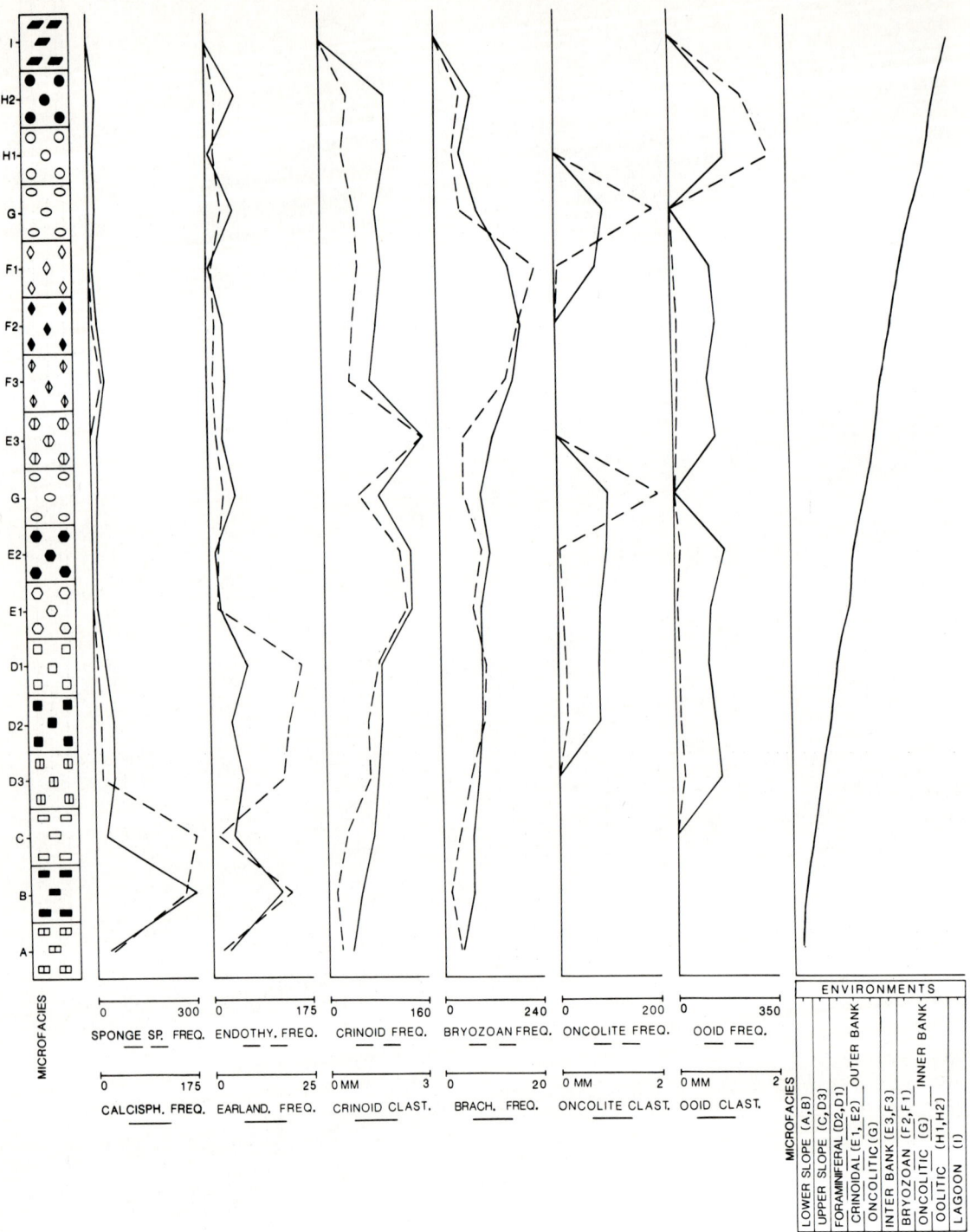

FIGURE 12.21 Glen Dean Formation (Middle Mississippian), eastern shelf of Illinois Basin. Ideal shallowing-upward sequence. From Feiznia (1983).

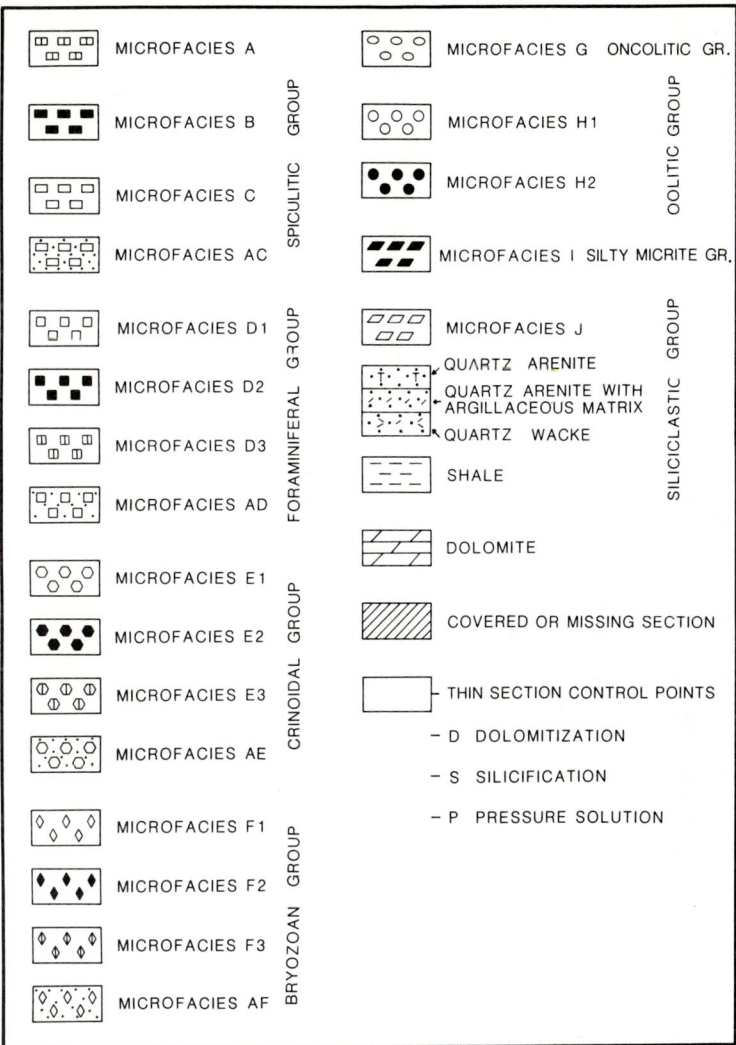

FIGURE 12.22 Glen Dean Formation (Middle Mississippian), eastern shelf of Illinois Basin. Symbols for microfacies. From Feiznia (1983).

The Ideal Siliciclastic Depositional Model

A system of streams (Michigan River) introduced siliciclastic sediments from the northeast and deposited them as a deltaic fringe behind or partially overlapping a carbonate platform (Fig. 12.28). The quartz arenites probably represent distributary channel fills and distributary mouth bars, whereas quartz arenites interbedded with shales correspond to interdistributary plain facies and shales to prodelta facies. Some of the sediments introduced by deltas were carried across the carbonate platform by means of submarine channels (recognizable on isopach maps) and deposited down the slope into the basinal environment by turbidity currents as inferred submarine fans (Fig. 12.29). In the submarine fans, quartz wackes and quartz arenites probably represent channel-axis facies, the two types of sandstones interbedded with shales are probably interchannel facies, and shales alone are most likely distal facies.

Carbonate Diagenesis

The depositional and diagenetic sequence of the Glen Dean Formation, as shown by detailed petrographic analysis, consists of two distinct sequences (I and II) that succeeded each other in time (Figs. 12.30, 12.31). Distribution and intensity of dolomitization (accompanied by silicification) show that it is essentially confined to the top of the outer bank (Fig. 12.26) and that it is geographically concentrated in the eastern shelf (Fig. 12.27). These conditions and the absence of sabkha and evaporitic facies indicate that dolomitization took place in a mixed freshwater phreatic-marine environment generated by the combination of subaerial exposure of the tops of the banks

328 Part III / Carbonate Platforms

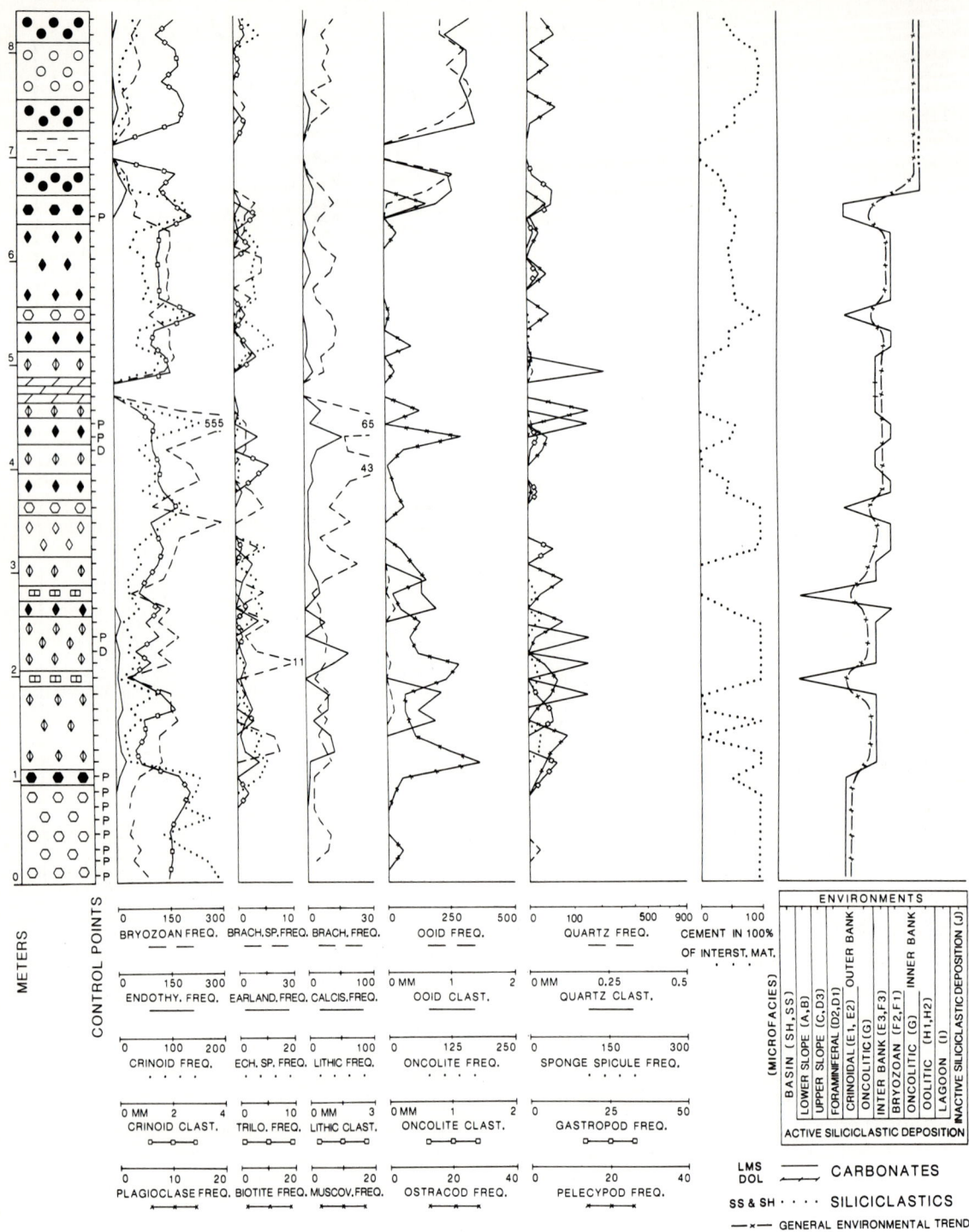

FIGURE 12.23 Glen Dean Formation (Middle Mississippian), eastern shelf of Illinois Basin. Typical example of vertical variation of microscopic parameters during a shallowing-upward sequence. From Feiznia (1983).

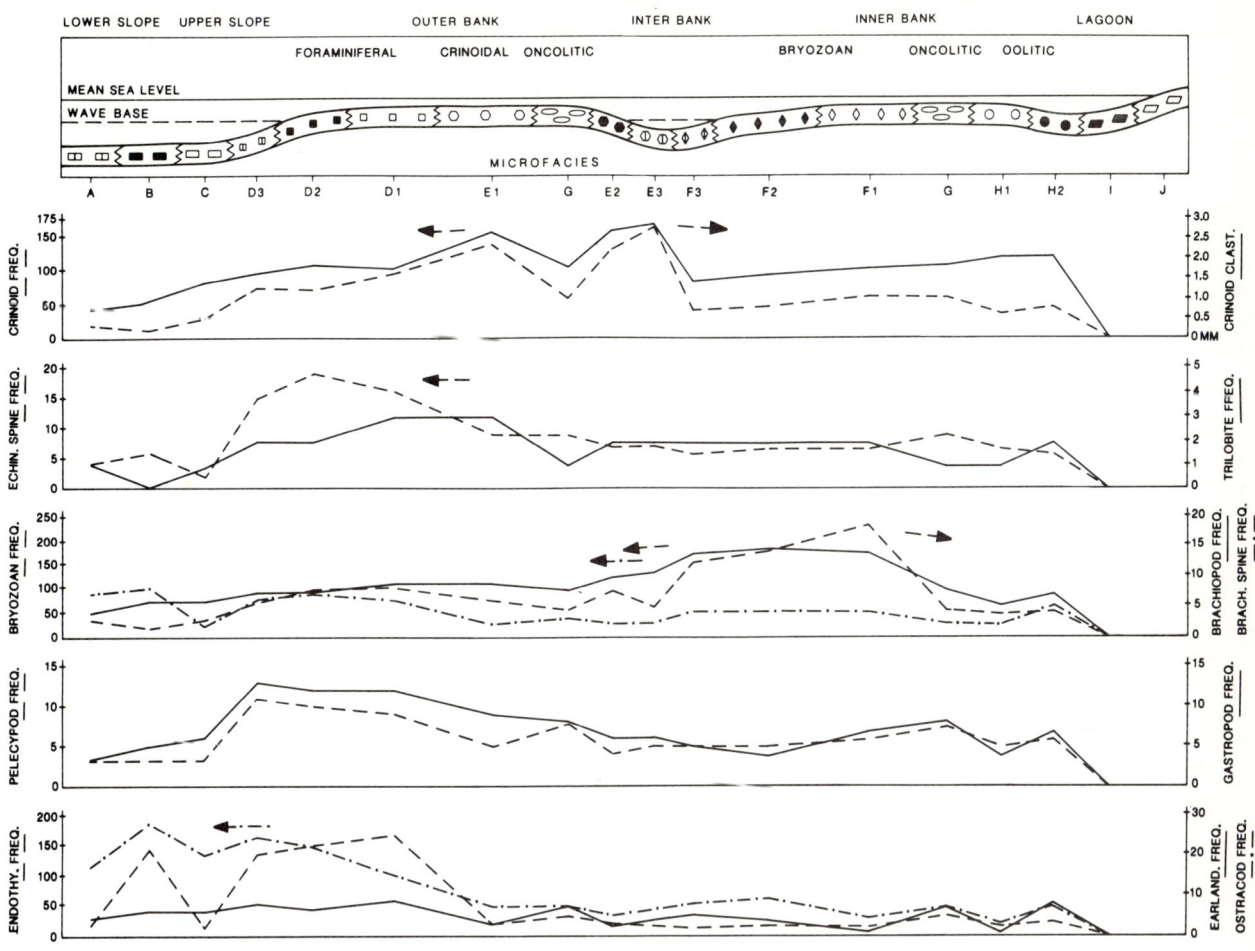

FIGURE 12.24 Glen Dean Formation (Middle Mississippian), eastern shelf of Illinois Basin. Ideal depositional model. From Feiznia (1983).

with local freshwater lenses and the additional contribution of freshwater from the deltaic complex entering the carbonate banks from the eastern shoreline.

12.4 CHARACTERISTIC CONSTITUENTS: SMALLER BENTHIC FORAMINIFERS, CRINOIDS, AND BRYOZOANS

Broad platforms under agitated conditions that develop longitudinal rows of hydrodynamic buildups (as in the previous case) may be invaded at times by prograding deltaic siliciclastic sediments. The latter cut across carbonates by means of channel systems and reach basinal areas as turbidites. Upon excessive siliciclastic sediment yield, turbidity currents onlap the opposite side of the basin dispersing sand and silt over similar carbonate platforms that lean against inactive margins.

The Glen Dean Formation (Middle Mississippian) of the western shelf of the Illinois Basin is located in southwestern Illinois. This carbonate platform displays two unusual properties: first, it was adjacent to an inactive basin margin, the Ozark shoal; second, it was at times invaded as far as its proximal portion by fine-grained distal deltaic siliciclastic sediments that originated from the opposite side of the basin (Michigan River delta) and crossed the eastern carbonate platform described in the previous example (Feiznia, 1983, Feiznia and Carozzi, 1987).

The Glen Dean Formation of the western shelf (Feiznia and Carozzi, 1981) was investigated by means of four stratigraphic field sections. The petrographic study was based on 266 samples collected at an average vertical interval of 13 cm for carbonates and sandstones, and 30 cm for shales.

Description of Microfacies

Eight distinct microfacies were recognized in this mixed carbonate-siliciclastic platform that consists of a series of hydrodynamic bioclastic buildups and intervening de-

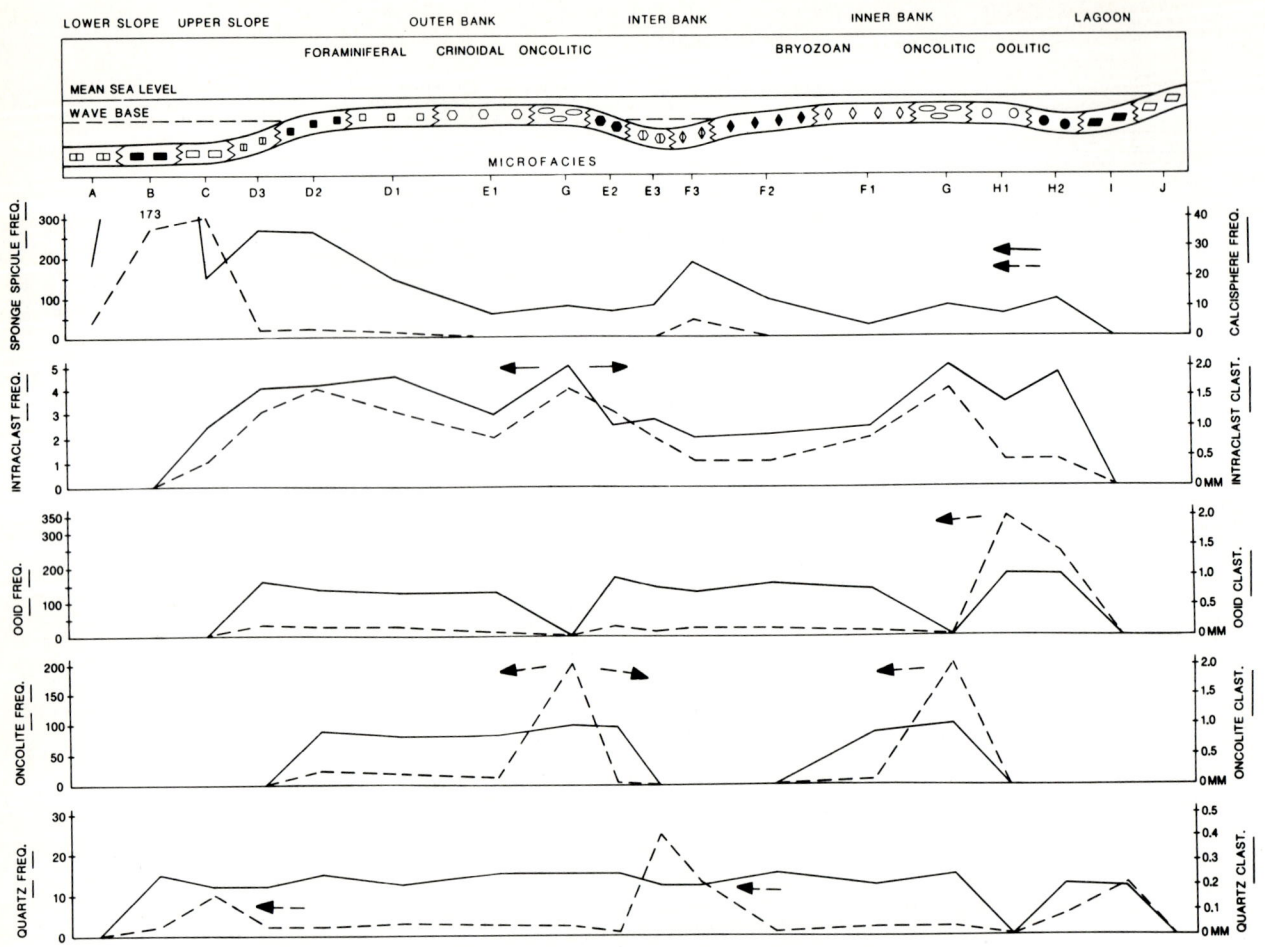

FIGURE 12.25 Glen Dean Formation (Middle Mississippian), eastern shelf of Illinois Basin. Ideal depositional model (continued). From Feiznia (1983).

pressions. The microfacies are described in a landward direction and in a general shallowing-upward order.

Microfacies 1 (Fig. 12.32A). Moderately sorted quartz feldspathic arenite with iron-oxide pigments, cryptocrystalline calcite cement, and matrix of clay minerals. It consists of angular grains of quartz, plagioclase altered to sericite, biotite, and muscovite flakes. Lithoclasts of dark calcareous siltstone are abundant and distorted by compaction. Lamination and low-angle cross-bedding are visible in the field.

Microfacies 2. Laminated calcareous shale to siltstone consisting of a groundmass of clay minerals, predominantly illite and smectite, abundant silt-size angular grains of quartz and feldspars, and scattered bioclasts of bryozoans and crinoids.

Microfacies 3 (Fig. 12.32B,C). Deeply bioturbated biocalcisiltite with abundant calcified monaxonic sponge spicules involved in spiral structures (Fig. 12.32C) together with abundant fecal pellets. Other scattered sand-size bioclasts are bryozoan zoaria, crinoids, benthic calcareous foraminifers, and ostracods. Angular and fine-grained detrital quartz is locally present.

Microfacies 4 (Fig. 12.32D,E,F). Fine grain-supported and bioturbated biocalcarenite with calcisiltite to calcilutite matrix. Bioclasts in decreasing order of importance are: *Endothyra* (Fig. 12.32E), *Earlandia* (Fig. 12.32F), fenestrate bryozoans, crinoids, ostracods, gastropods, calcispheres, and sponge spicules. Fine sand-size angular grains of detrital quartz are locally common. Bioturbation is shown by irregular to spiral arrangement of bioclasts within the matrix. Incipient dolomitization occurs as small rhombs replacing matrix.

Microfacies 5 (Fig. 12.32G,H). Coarse and well-sorted crinoidal calcarenite with pressure-solution and syntaxial rim cement. The common bryozoans are deformed and crushed by pressure-solution and abundant

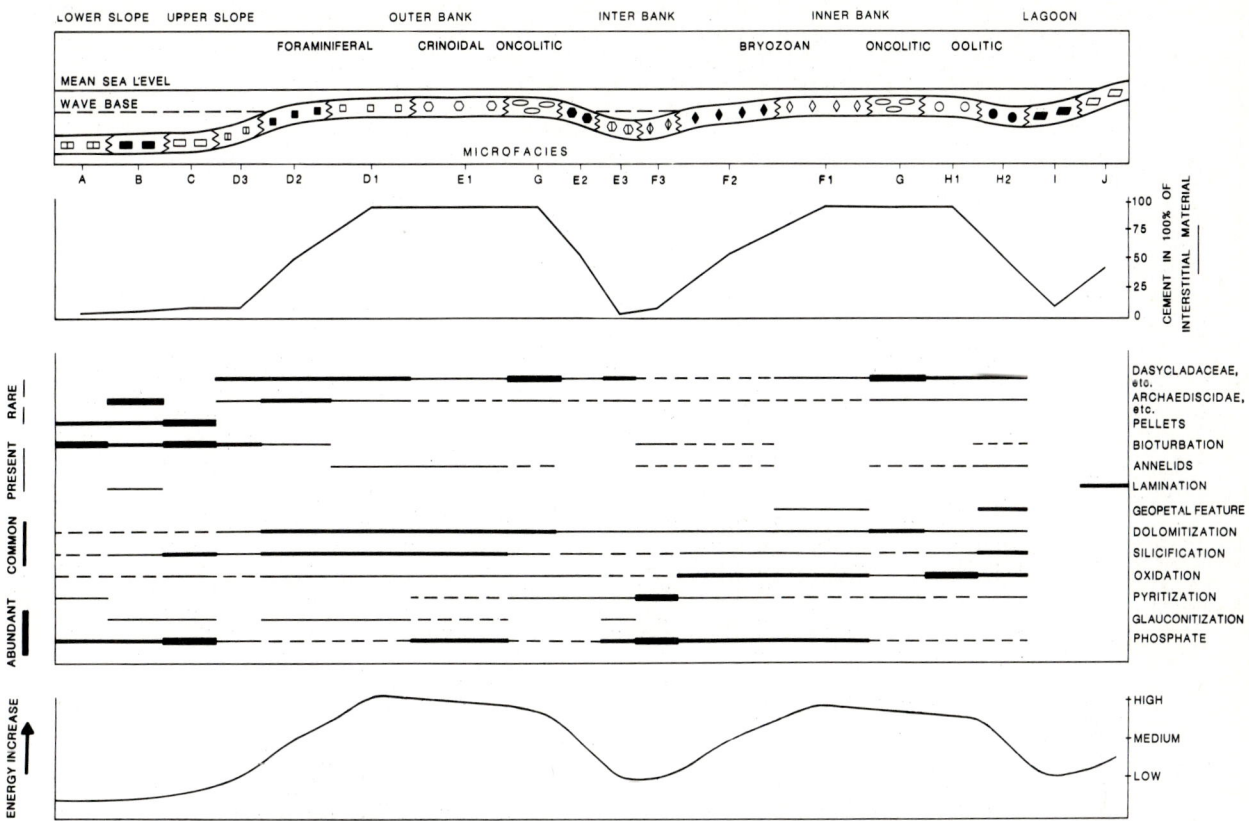

FIGURE 12.26 Glen Dean Formation (Middle Mississippian), eastern shelf of Illinois Basin. Ideal depositional model (continued). From Feiznia (1983).

cement (Fig. 12.32H). Bioclasts of pelecypods, gastropods, and a few reworked ooids are present. Scattered dolomite rhombs indicate incipient dolomitization.

Microfacies 6 (Fig. 12.33A,B,C). Grain-supported crinoid-bryozoan calcarenite with association of calcisiltite matrix and cavity-filling bladed to equant sparite cement. Bioturbation generates a more bioclastic matrix locally (Fig. 12.33B), and crinoid bioclasts often display syntaxial rims. Other common constituents are pelecypods, gastropods, ostracods, *Endothyra*, reworked ooids, and intraclasts of oxidized calcilutite. Matrix is altered locally by neomorphism to patches of pseudomicrosparite (Fig. 12.33C) and replaced commonly by dolomite rhombs.

Microfacies 7 (Fig. 12.33D,E,F). Grain-supported bryozoan-brachiopod calcarenite with abundant bladed to equant sparite cement. Association of hollow ramose bryozoans, fenestrate bryozoans, and thin-shelled brachiopods generated a very open framework that had a high primary porosity, and did not suffer much subsequent compaction (Fig. 12.33E,F). Additional bioclasts scattered within this framework are crinoids, pelecypods, and trilobites associated with some reworked ooids and pellets.

Microfacies 8 (Fig. 12.33G,H). Poorly sorted, grain-supported oolitic biocalcarenite with sparite cement and calcisiltite matrix often changed by neomorphism into pseudomicrosparite and pseudosparite. Neomorphism has often affected bioclasts making their boundaries hazy and upon completion has led to "ghost" structures (Fig. 12.33H). All ooids show abraded edges, indicating reworking from an adjacent hydrodynamic buildup and redeposition in this microfacies. Ooid cores consist predominantly of crinoid and bryozoan bioclasts, which also occur as uncoated particles. Minor constituents are brachiopods, pelecypods, gastropods, and trilobites.

The Ideal Shallowing-Upward Sequence

The eight microfacies of the Glen Dean Formation on the western shelf of the Illinois Basin can be organized into a shallowing-upward sequence (Figs. 12.34, 12.35) which, with minor oscillations, is represented by an actual field section. The latter expresses the succession in time

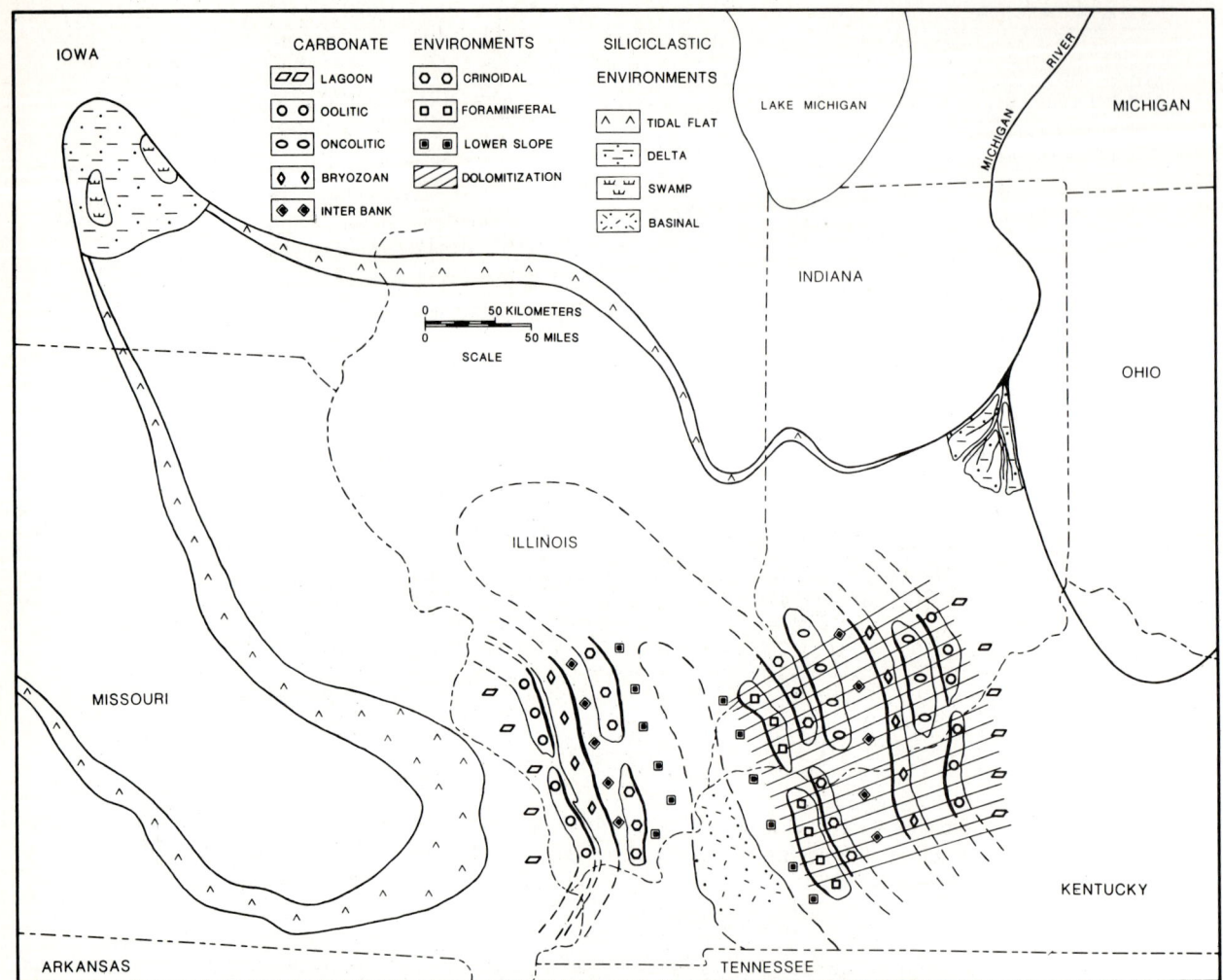

FIGURE 12.27 Glen Dean Formation (Middle Mississippian), eastern shelf of Illinois Basin. Basinwide distribution of carbonate and siliciclastic environments during inactive deltaic conditions. From Feiznia (1983).

of the following environments: distal deltaic, lower slope, upper slope, outer shoal (buildup), intershoal, inner shoal (buildup), and oolitic trough.

The Ideal Depositional Model

The carbonate platform (Figs. 12.36, 12.37) is dominated by an outer hydrodynamic buildup or shoal where crinoids were thriving and from which tidal currents were distributing bioclasts landward and seaward, and by an inner hydrodynamic buildup or shoal consisting of debris of brachiopods and hollow branching forms of bryozoans (*Callocladia*), and developed in more protected conditions. The depression between the two buildups is characterized by gastropods. Another community of the latter, together with fenestrate bryozoans (*Archimedes*), defines the upper slope, which also displays peaks of frequency of *Endothyra*, *Earlandia*, and calcispheres. The lower slope is the site of deposition of ostracods and sponge spicules transported from more internal parts of the platform.

Pelecypod bioclasts characterize the somewhat restricted environment of the oolitic trough, where reworked ooids and intraclasts are concentrated after transportation from an inferred adjacent oolitic bar and lagoon.

Dolomitization is an incipient postdepositional process which affects mainly microfacies 6, 5, and 4 and to a smaller extent microfacies 3, that is, the seaward portion of the carbonate platform. This region would be the discharge area of freshwater originating from any freshwater lens that may have developed upon subaerial

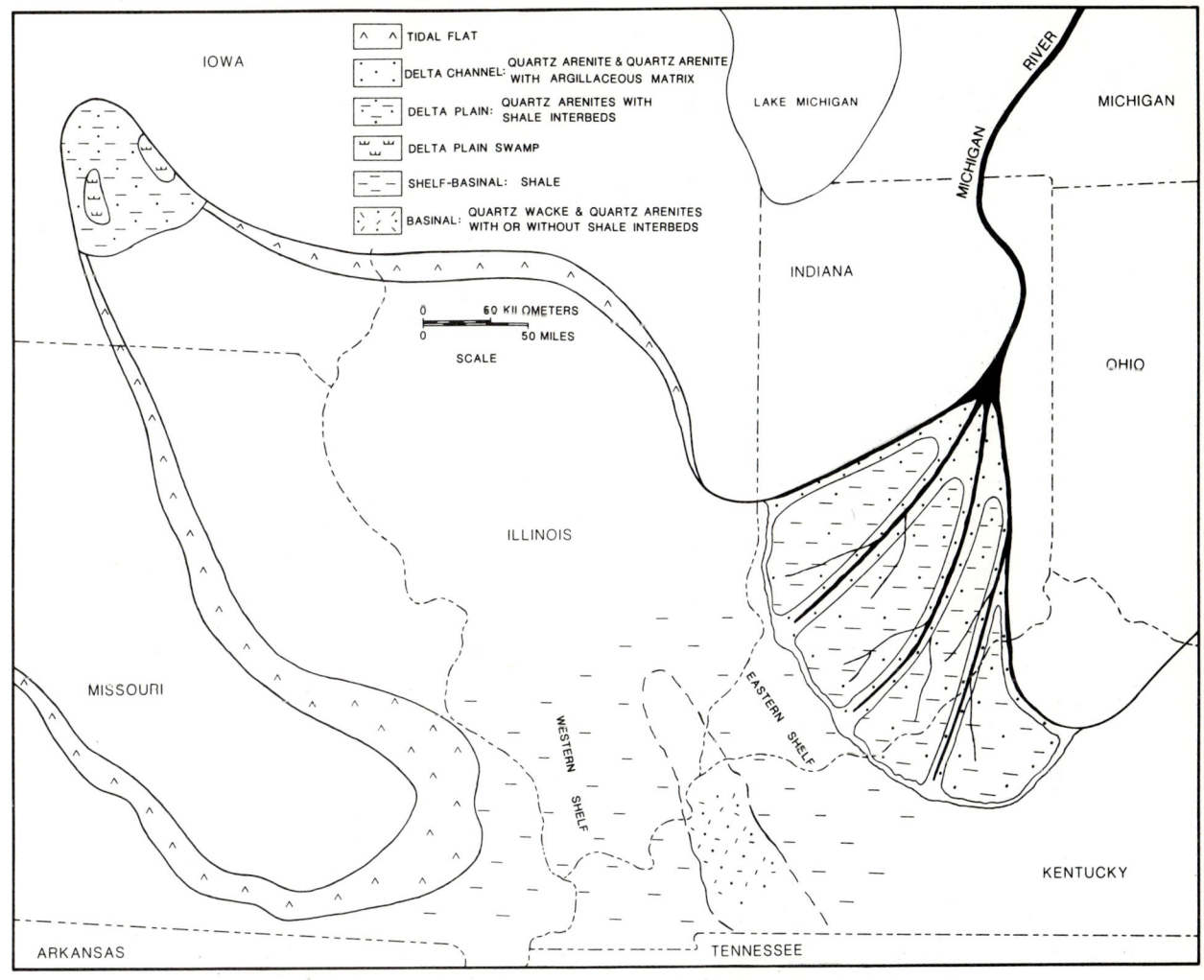

FIGURE 12.28 Glen Dean Formation (Middle Mississippian), eastern shelf of Illinois Basin. Basinwide distribution of siliciclastics during active deltaic conditions. From Feiznia (1983).

exposure of either the inner buildup or the inferred oolitic bar. Freshwater may also have come from the inactive basin margin, a pre-Mississippian carbonate terrane. Therefore, dolomitization was probably formed by a mixed freshwater-seawater process (dorag model). A similar mechanism is postulated for the distribution of silicification that is identical with that of dolomitization, reaching formation of chert nodules in microfacies 4.

Aggrading neomorphism of the calcisiltite matrix is particularly well developed in microfacies 8, 6, and 4. Although interpreted as a freshwater phreatic process, it may have been initiated earlier under the action of freshwater circulation related to the inactive basin margin or the prograding delta system (Carozzi, 1971a).

The distribution of ooids and intraclasts in the proposed depositional model indicates that the model is incomplete and that two additional environments can be predicted (Fig. 12.38) before reaching the inactive basin margin, namely, an oolitic hydrodynamic buildup or bar and a lagoon.

Sandstones and shales intercalated with the carbonates of the upper part of the Glen Dean Formation in southwestern Illinois represent the onlap of the distal deltaic system of the Michigan River after it had crossed the carbonate platform of the eastern shelf of the Illinois Basin and reached the center of the basin (Fig. 12.39). This unusual situation, which seems to represent deposition up the depositional slope of the carbonate platform, may have been due to a settling from suspension of fine-grained terrigenous material or more probably to deposition from distal deltaic turbidity currents. If from turbidity flows, it can be presumed that either the heads were thicker than the vertical rise or that they were actually flowing upslope by inertia.

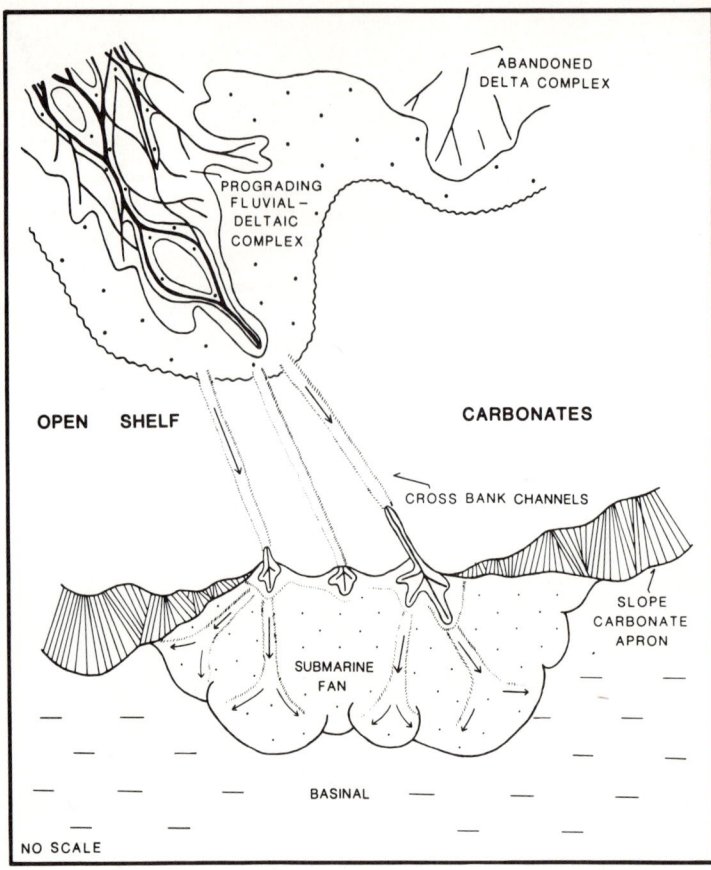

FIGURE 12.29 Glen Dean Formation (Middle Mississippian), eastern shelf of Illinois Basin. Ideal siliciclastic depositional model. From Feiznia (1983).

The General Evolution of the Depositional Environments

A composite section (Fig. 12.40) shows a vertical evolution consisting of three episodes. The first one is a shallowing from slope to oolitic trough microfacies across the entire carbonate platform. The distal deltaic environment was influential during the second episode, whereas the third episode corresponds to a return of carbonate slope deposition but still received an influx of detrital quartz.

12.5 CHARACTERISTIC CONSTITUENTS: OOIDS

Platforms with frontal oolitic hydrodynamic buildups afford generally effective protection to internal areas that consist of intertidal to supratidal lagoons or sabkhas adjacent to inactive hinterlands. However, under conditions of ephemeral and powerful siliciclastic sediment yield, strongly prograding fan deltas can entirely replace lagoons and sabkhas by fan delta plain and even fluvial deposits. A distal aspect is the generation of unusual mixed carbonate–fan delta slope sediments often accompanied by widespread slumping.

The Quintuco–Loma Montosa Formation (Lower Cretaceous) of the Neuquén Basin, Mendoza Province, Argentina (Carozzi, et al., 1981c) was investigated by using 21 wells and by detailed petrographic study of 2604 thin sections, 937 of which were made from cores and 1667 from cuttings. Average vertical spacing of thin sections from cores was 0.70 m and from cuttings 4.5 m. The total aggregated thickness of the investigated column was 8965 m, of which 606 m are represented by cores and 8359 m by cuttings.

Description of Microfacies

Association of lithologies, sedimentary structures, microfacies, and microfaunas revealed that the Quintuco–Loma Montosa Formation consists of the stratigraphic superposition of three major depositional models. The first model displays the following juxtaposition of environments in a landward direction: basinal, slope, hydrodynamic oolitic buildup, lagoon or sabkha, and fan delta. The second model represents an intense siliciclastic progradation which destroyed oolitic buildups and consists of

DIAGENETIC FEATURES \ DIAGENETIC ENVIRONMENTS	MARINE PHREATIC	MIXING MARINE FRESHWATER PHREATIC I	FRESHWATER PHREATIC I	VADOSE	FRESHWATER PHREATIC II	MIXING MARINE FRESHWATER PHREATIC II	BURIAL
MICRITIZATION	▬▬						
REORGANIZATION OF INTERNAL SEDIMENT	▬▬						
EARLY COMPACTION I (PRESSURE SOLUTION, SPALLING OF OOID CORTICAL LAYERS, CRUSHING OF MICRITE ENVELOPES)	▬▬						
ISOPACHOUS BLADED CEMENT	▬▬						
EARLY COMPACTION II (PRESSURE SOLUTION)	▬▬						
MICROCRYSTALLINE CEMENT (INTRACLASTS)	?▬▬▬▬▬?						
LEACHING OF ARAGONITIC SHELLS			▬▬				
SPARITE AND SINGLE CRYSTAL CEMENTS			▬▬				
SYNTAXIAL RIM CEMENT			▬▬				
NEOMORPHISM (MATRIX AND OOIDS)			▬▬				
DESICCATION FEATURES				▬▬			
VADOSE SILT DEPOSITION				▬▬			
SPARITE CEMENT (IN UPPER PARTS OF DESICCATION FEATURES)					▬▬		
SILICIFICATION						▬▬	
DOLOMITIZATION						▬▬	
LATE COMPACTION (STYLOLITIZATION, DEFORMATION OF BIOCLASTS, CHAIN ONCOLITES)							▬▬
LATE DOLOMITIZATION (SADDLE DOLOMITE)							▬▬
ANHYDRITIZATION							▬▬

TIME →

FIGURE 12.30 Glen Dean Formation (Middle Mississippian), eastern shelf of Illinois Basin. Generalized diagenetic sequence. From Feiznia (1983).

the following environments: basinal, mixed carbonate–fan delta slope represented by unusual mixed carbonate-siliciclastic sediments, fan delta front and plain, and fluvial. The third model corresponds to the return of carbonate sedimentation and consists of the following environments: basinal, slope, oolitic-bioclastic platform, lagoon or sabkha, and fan delta. The major microfacies of these various depositional environments are described in a general landward and shallowing order.

Basinal Environment

Microfacies 1 (Fig. 12.41A). Calcareous siltstone with minute angular grains of detrital quartz and scattered calcispheres associated with rare ostracods and agglutinated foraminifers.

Slope Environment

Microfacies 2 (Fig. 12.41B). Argillaceous calcisiltite with upper surface cemented into a phosphatic, glauconitic, and pyritic hardground displaying scouring and perforations. The overlying deposit is a grain-supported oolitic biocalcarenite with sparite cement. Main constituents are reworked ooids, bioclasts of crinoids, echinoids and pelecypods, and angular grains of detrital quartz.

Microfacies 3 (Fig. 12.41C). Bioturbated arenaceous grain-supported biocalcarenite with argillaceous micrite matrix. Major bioclasts are pelecypod shells surrounded by micrite envelopes and filled with sparite cement, ostracods, crinoids, echinoids, dasyclads associated with reworked ooids, and micritic intraclasts.

Mixed Carbonate–Fan Delta Slope Environment

Microfacies 4 (Fig. 12.41D). Very arenaceous grain-supported biocalcarenite with argillaceous micrite matrix. Predominant bioclasts are crinoids, echinoids, ostracods, pelecypods, and annelids associated with reworked ooids.

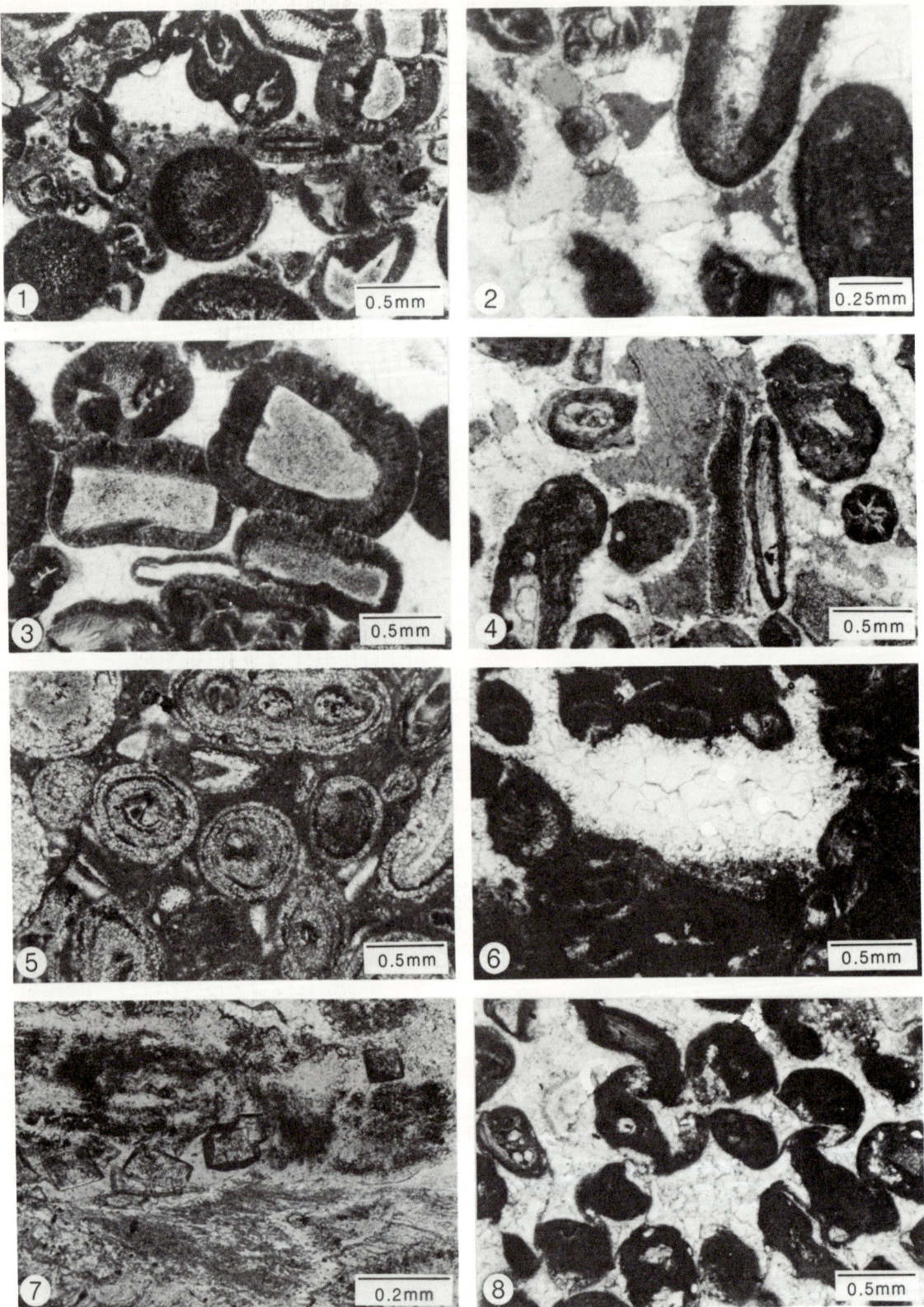

FIGURE 12.31 Glen Dean Formation (Middle Mississippian), eastern shelf of Illinois Basin. Diagenetic features. *Marine phreatic.* 1. Geopetal arrangement in graded layers of internal sediment consisting of pelletoidal calcilutite, and overlain by cavity-filling equant mosaic

Microfacies 5 (Fig. 12.41E). Coarse-grained and arenaceous grain-supported biocalcarenite with common pressure-solution and interstitial argillaceous microsparite cement. Predominant bioclasts are crinoids, echinoids, pelecypods, bryozoans, and ostracods, associated with micrite intraclasts, reworked ooids, and coarse angular grains of detrital quartz.

Microfacies 6 (Fig. 12.41F). Grain-supported oolitic biocalcarenite with slightly argillaceous cavity-filling sparite cement. Major bioclasts are pelecypods with micrite envelopes and filled with sparite cement, echinoids, crinoids, bryozoans, dasyclads, and ostracods. Reworked ooids of all sizes display abrasion features and are associated with a few micrite intraclasts.

Fan Delta Environment

Microfacies 7 (Fig. 12.41G). Polymictic orthoconglomerate with abundant cobbles of rhyolite with typical embayed quartz phenocrysts set in an argillaceous mudstone matrix.

Microfacies 8 (Fig. 12.41H). Quartz-feldspathic lithic arenite with an interstitial argillaceous microsparite cement. Dark lithic clasts are altered acidic to intermediate volcanics.

Hydrodynamic Oolitic Buildup Environment

Microfacies 9 (Fig. 12.42A). Grain-supported oolitic calcarenite with cavity-filling sparite cement. Well-developed normal to superficial ooids display cores consisting of micrite pellets or bioclasts of pelecypods, echinoids, crinoids, bryozoans, red algae, dasyclads, and ostracods. Some ooids are deformed by compaction.

Microfacies 10 (Fig. 12.42B). Grain-supported oolitic biocalcarenite with an association of micrite matrix and irregularly distributed patches of cavity-filling sparite cement. Ooids with predominant cores of pellets display concentric rings which are selectively altered by neomorphism into pseudomicrosparite aggregates. Bioclasts consist of angular and subrectangular fragments of pelecypod shells, in part displaying original structure or micrite envelopes with filling of sparite cement.

Bioclastic Platform Environment

Microfacies 11 (Fig. 12.42C). Grain-supported to pressure-welded echinoid and crinoid biocalcarenite with patches of interstitial micrite matrix. Other constituents are bioclasts of red algae, bryozoans, dasyclads, annelids, micrite-filled gastropod shells, reworked ooids, and micrite intraclasts.

Microfacies 12 (Fig. 12.42D). Pelecypod accumulated limestone with abundant interstitial pelletoidal micrite matrix weakly altered by neomorphism into pseudomicrosparite. Superposed disarticulated valves are parallel to bedding and display perfectly preserved internal structure. They are encrusted often by colonies of annelids and bryozoans.

Sabkha Environment

Microfacies 13 (Fig. 12.42E). Dolomitic siltstone with chickenwire texture due to incipient displacive micronodules of anhydrite. Concentrations of calcispheres are replaced by microcrystalline anhydrite.

Microfacies 14 (Fig. 12.42F). Stromatolite-constructed limestone with irregular and pelleted dark algal mats separated by intermat arenaceous micrite in-

FIGURE 12.31 (*continued*) of sparite cement. 2. First generation of isopachous rim cement consisting of cloudy fine blades perpendicular to oncoid margins. Remaining interstitial space filled with equant mosaic of sparite. 3. Early compaction II after first generation of isopachous rim cement and prior to any subsequent cementation shown by ooids pressure-welded and separated by peripheral band of bladed cement of uniform thickness. *Freshwater phreatic I*. 4. Second generation of cavity-filling sparite cement changing from fine to coarse mosaic (left) and from bladed to single crystal (center). 5. Neomorphism of ooid cortical layers to pseudosparite and of calcilutite matrix to pseudomicrosparite. *Freshwater vadose and freshwater phreatic II*. 6. Desiccation cavity displaying geopetal filling of vadose silt overlain by sparite cement. *Mixing marine freshwater phreatic II*. 7. Cross cutting of silicified brachiopod shell structure by dolomite rhombs indicates that silicification preceded dolomitization. *Burial*. 8. Chains of distorted oncoids with rearrangement of interstitial coarse sparite mosaic due to late compaction. All photomicrographs: plane-polarized light, except 2, crossed nicols. From Feiznia (1983).

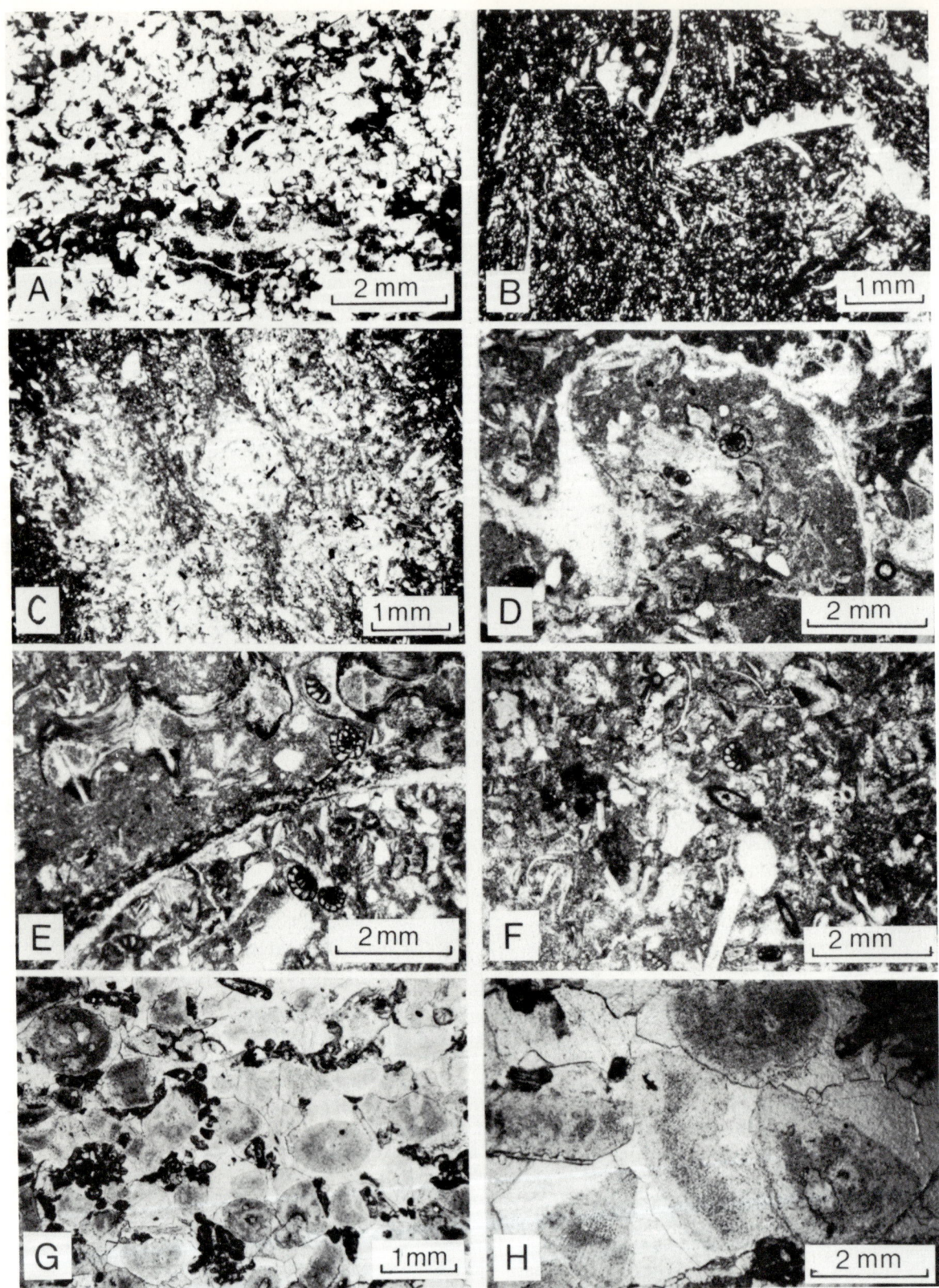

FIGURE 12.32 Glen Dean Formation (Middle Mississippian), western shelf of Illinois Basin. Typical microfacies. A. Microfacies 1. B and C. Microfacies 3. D to F. Microfacies 4. G and H. Microfacies 5. See text for detailed descriptions. All photomicrographs: plane-polarized light. From Feiznia and Carozzi (1981).

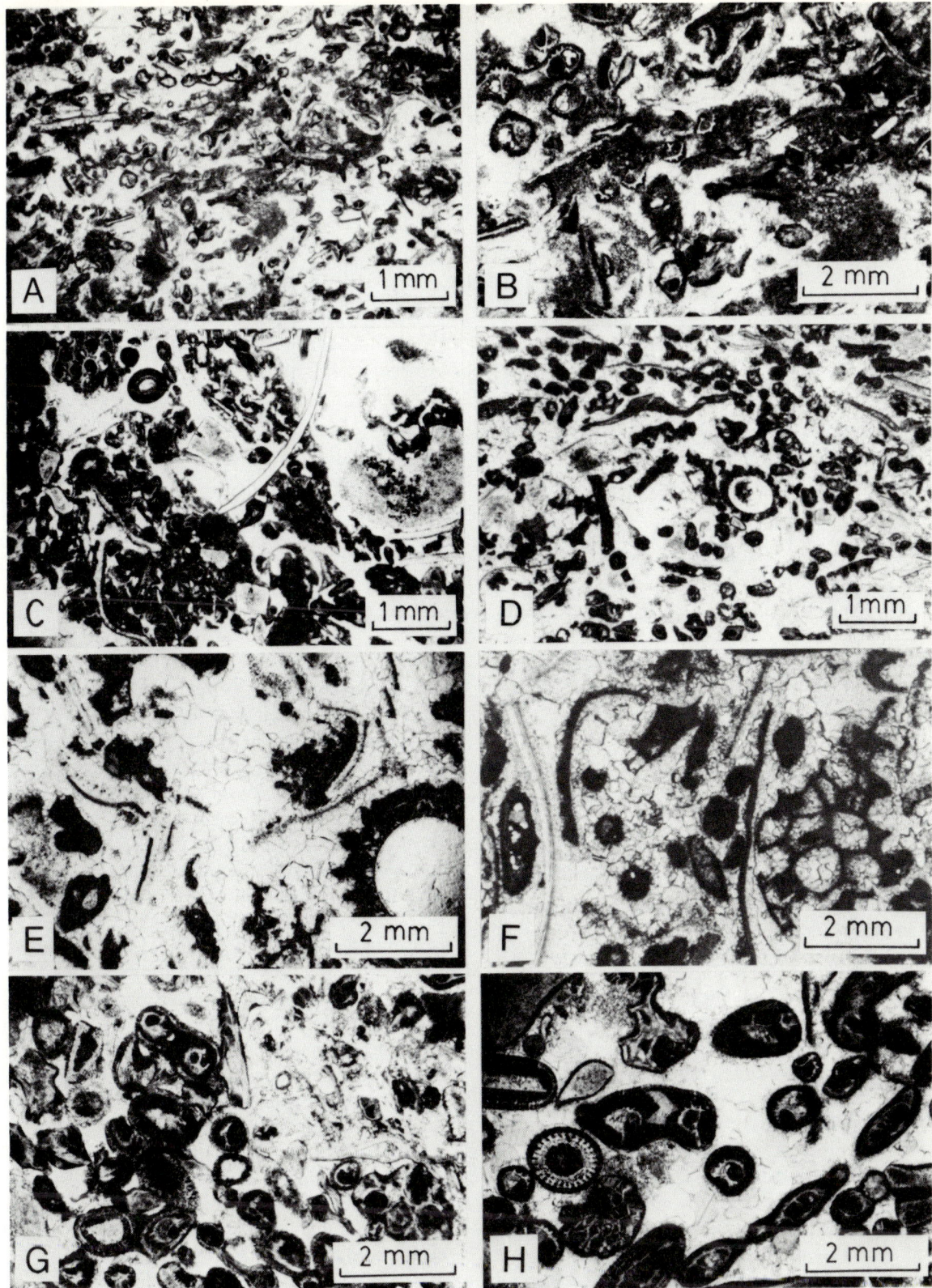

FIGURE 12.33 Glen Dean Formation (Middle Mississippian), western shelf of Illinois Basin. Typical microfacies (continued). A to C. Microfacies 6. D to F. Microfacies 7. G and H. Microfacies 8. See text for detailed descriptions. All photomicrographs: plane-polarized light. From Feiznia and Carozzi (1981).

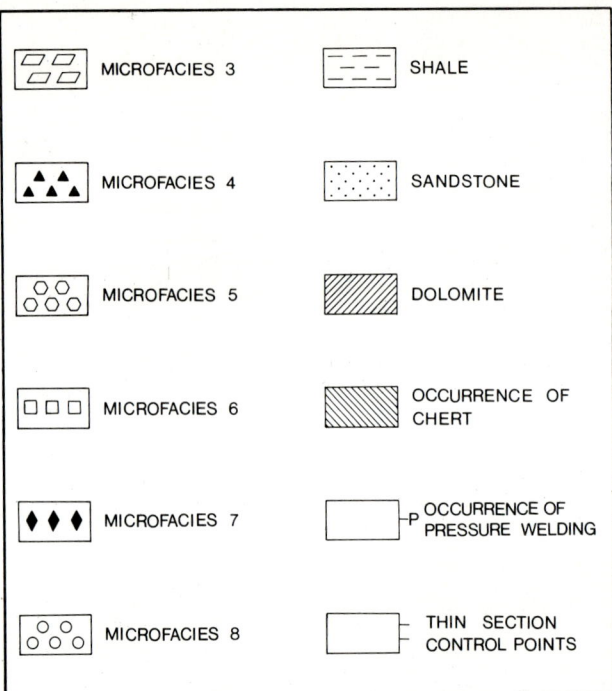

FIGURE 12.34 Glen Dean Formation (Middle Mississippian), western shelf of Illinois Basin. Symbols for microfacies. From Feiznia and Carozzi (1981).

cipiently distorted by growth of micronodules of anhydrite.

Lagoon Environment

Microfacies 15 (Fig. 12.42G). Grain-supported to pressure-welded coarse-grained arenaceous biocalcarenite with association of interstitial micrite matrix and scattered patches of microsparite cement. Major constituents are micrite-filled ostracods, sparite-filled gastropods, and reworked normal ooids.

Microfacies 16 (Fig. 12.42H). Grain-supported to pressure-welded fine-grained oolitic biocalcarenite with interstitial microsparite cement. Major bioclasts are pelecypods, echinoids, and dasyclads with scattered ostracods and large gastropods. Abundant superficial ooids with cores of micrite pellets are associated with a few angular grains of detrital quartz.

The Vertical Depositional Sequences

This petrographic study applied to oil exploration can be used as an example of graphically condensing a great amount of data down to the most critical parameters usable in regional cross sections and basinwide interpretation. The representation of well Y.P.F. RN AA x-1 (Agua Amarga) is a typical case. The first graphic expression (Fig. 12.43) displays the following data: geophysical logs, general lithologic composition, porosity from logs and measured on plugs, petrographic composition calculated as follows: siliciclastic grains + carbonate grains + normal ooids + superficial ooids + bioclasts + matrix + cement + dolomite + anhydrite-gypsum + porosity = 100%, grain size of constituents and types of bioclasts. The second graphic representation (Fig. 12.44) displays lithology simplified to clastics versus carbonates; composition of the carbonates simplified to the following constituents in percentage: siliciclastic grains, normal ooids, superficial ooids, dolomite, and anhydrite; energy level of carbonates expressed by a scale of mudstone (M), wackestone (W), packstone (P), and grainstone (G); presence of accessory minerals (glauconite, phosphate), and associations of microfacies which may be used to characterize depositional environments. The third and final graphic representation (Fig. 12.45) uses the envelope curves of the variation of parameters of the second representation, namely composition and energy level to which is added the percentage of fauna. Interpretation of the sequence in terms of environments of deposition completes the diagram. This final simplified version of the petrographic study contains diagnostic parameters which can be easily used in basinwide cross sections.

The Ideal Depositional Models

During the first carbonate episode of the evolution of the Quintuco–Loma Montosa Formation, the depositional model centers on hydrodynamic oolitic buildups which

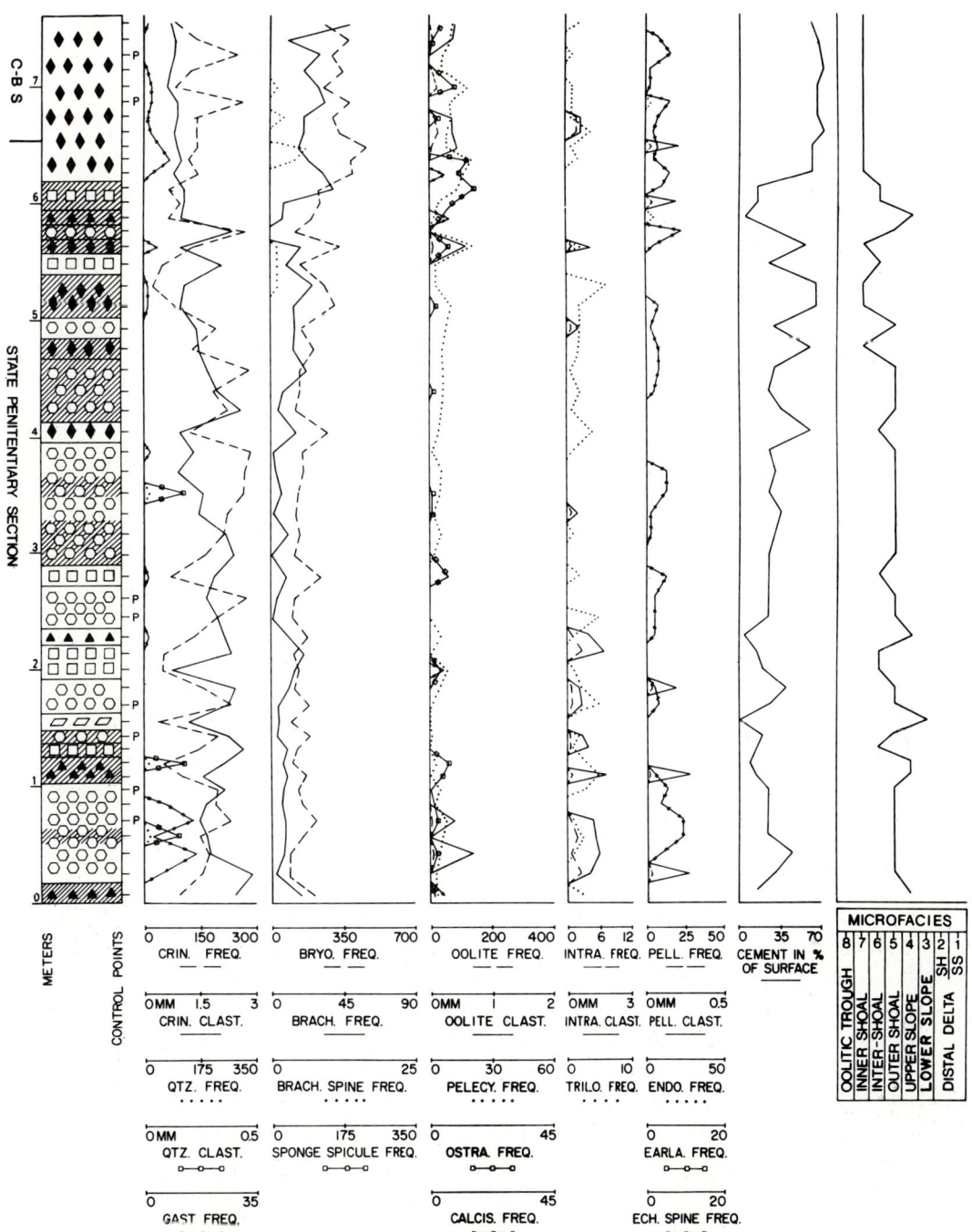

FIGURE 12.35 Glen Dean Formation (Middle Mississippian), western shelf of Illinois Basin. Typical example of vertical variation of microscopic parameters during a shallowing-upward sequence. From Feiznia and Carozzi (1981).

342 Part III / Carbonate Platforms

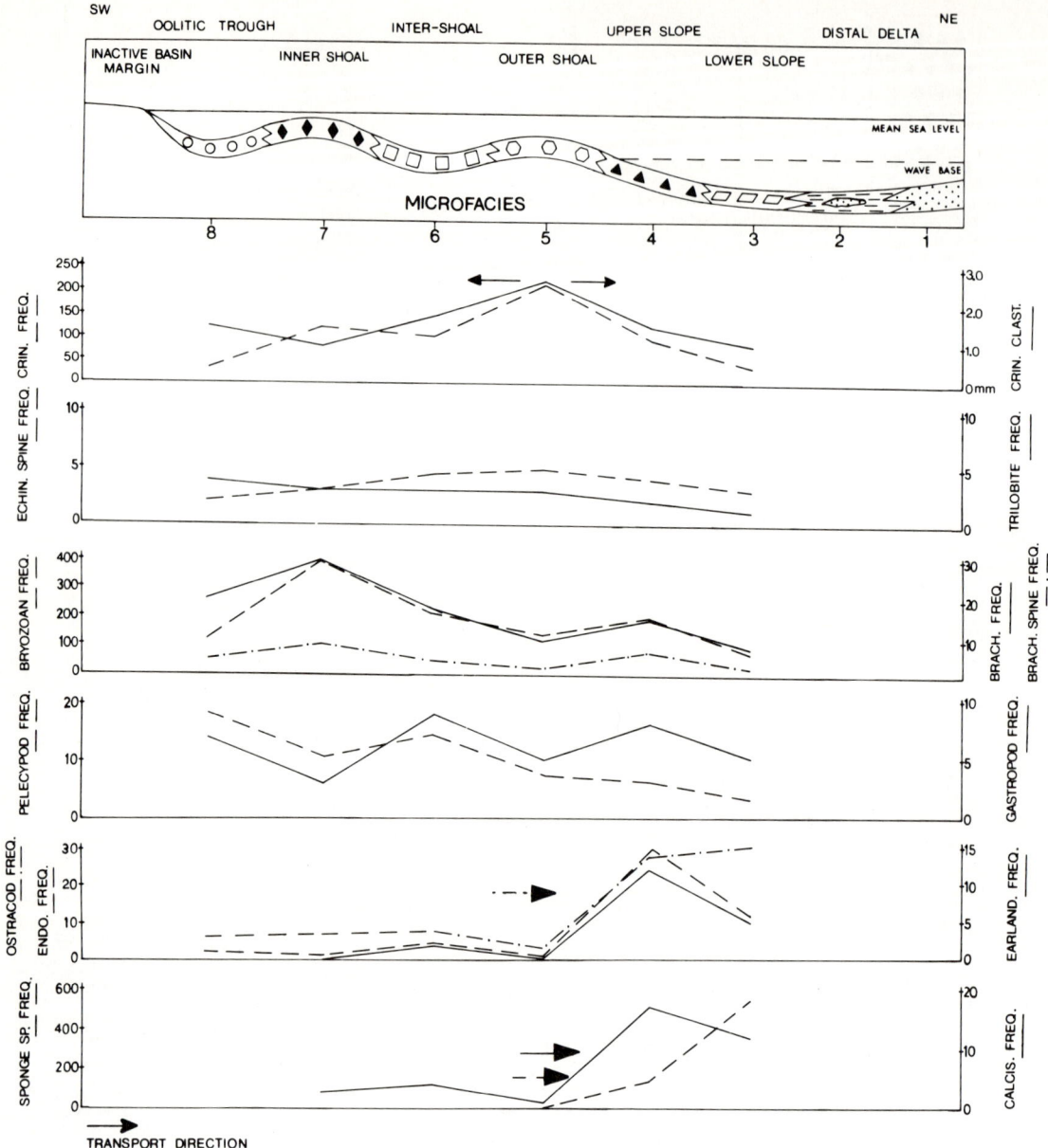

FIGURE 12.36 Glen Dean Formation (Middle Mississippian), western shelf of Illinois Basin. Ideal depositional model. From Feiznia and Carozzi (1981).

supported either a lagoon (Fig. 12.46A), or a combination lagoon-sabkha (Fig. 12.46B), or a sabkha (Fig. 12.46C). Progradation of the fan delta system is represented by its mixed carbonate–fan delta slope preceded by the fan delta front and plain and the coarser fluvial portion (Fig. 12.47D). The second carbonate episode corresponds to a depositional model centered on an oolitic-bioclastic platform which supported either a lagoon (Fig. 12.47E), or a sabkha (Fig. 12.47F), or a combination lagoon-sabkha (Fig. 12.47G).

All these various environmental situations are characterized by the behavior of diagnostic parameters such as normal and superficial ooids, reworked ooids, distribution of dolomite and anhydrite in the sabkha, in the lagoon, and as contaminating material in the fan delta plain, behavior of bioclasts, and distribution of coarse siliciclastics. It is the reciprocal behavior of all these constituents which provides the key to differentiation of the various depositional environments.

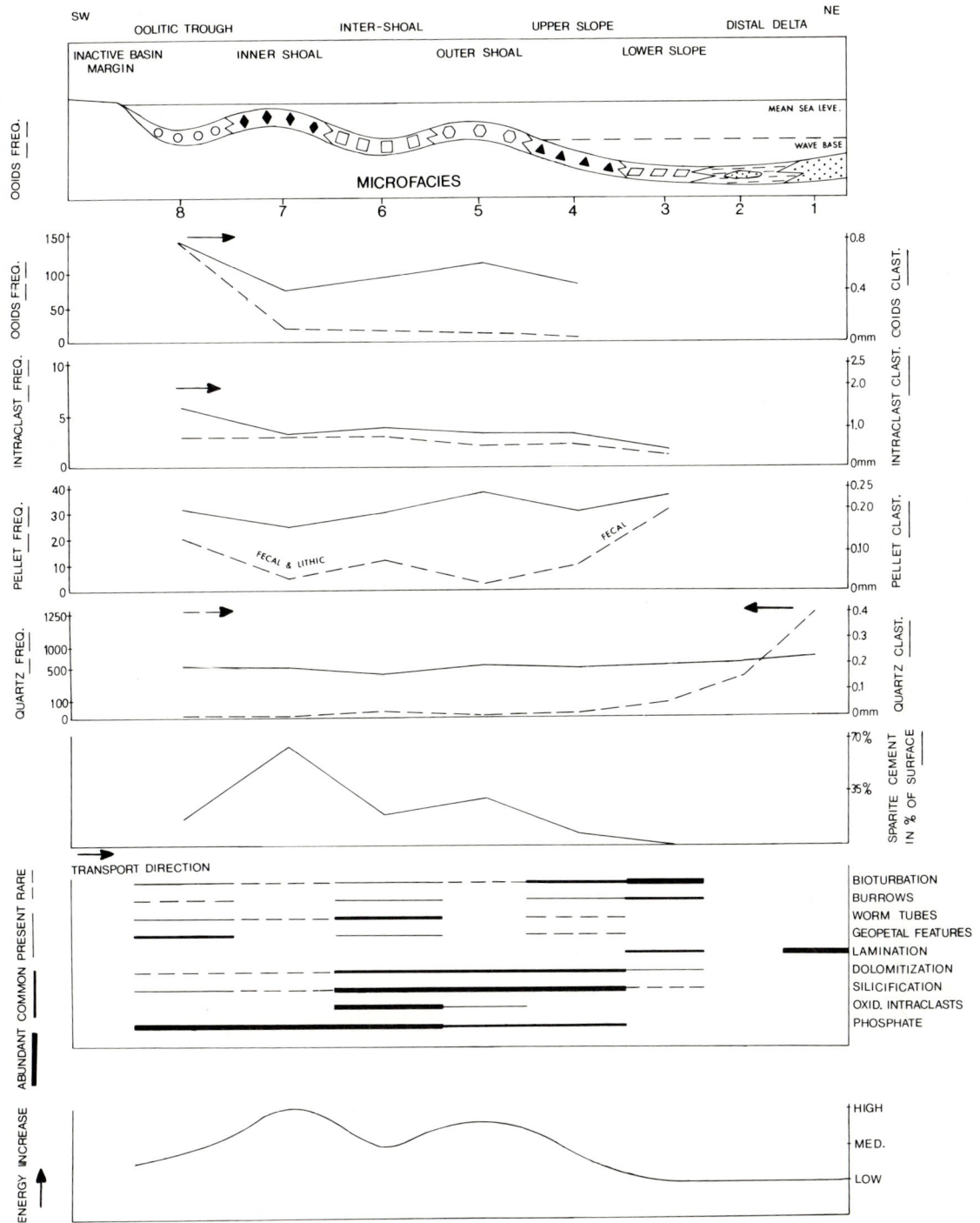

FIGURE 12.37 Glen Dean Formation (Middle Mississippian), western shelf of Illinois Basin. Ideal depositional model (continued). From Feiznia and Carozzi (1981).

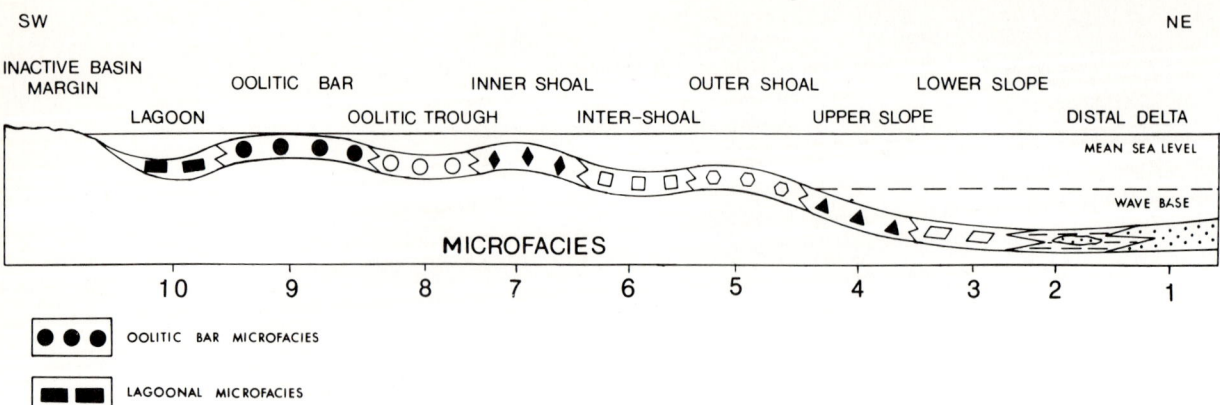

FIGURE 12.38 Glen Dean Formation (Middle Mississippian), western shelf of Illinois Basin. Predicted complete ideal depositional model. From Feiznia and Carozzi (1981).

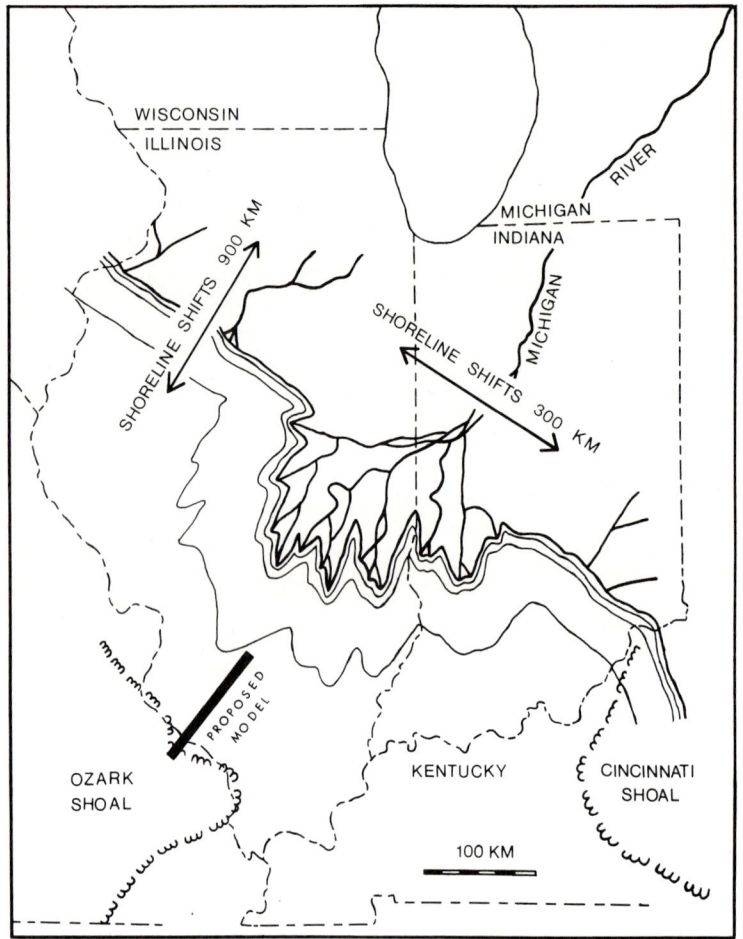

FIGURE 12.39 Glen Dean Formation (Middle Mississippian), western shelf of Illinois Basin. Variation of Michigan River delta in Chesterian times. From Feiznia and Carozzi (1981).

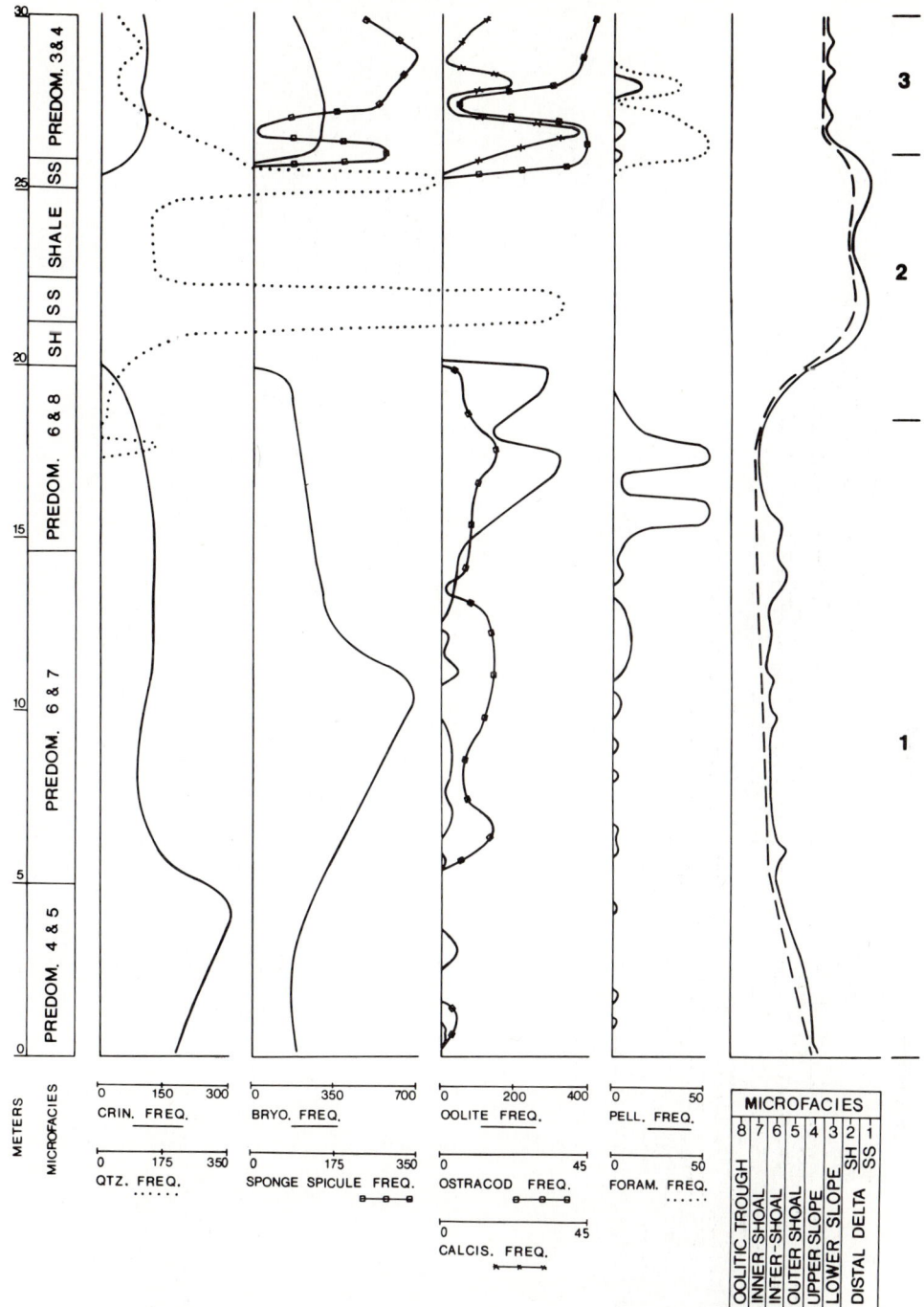

FIGURE 12.40 Glen Dean Formation (Middle Mississippian), western shelf of Illinois Basin. General vertical evolution of depositional environments. From Feiznia and Carozzi (1981).

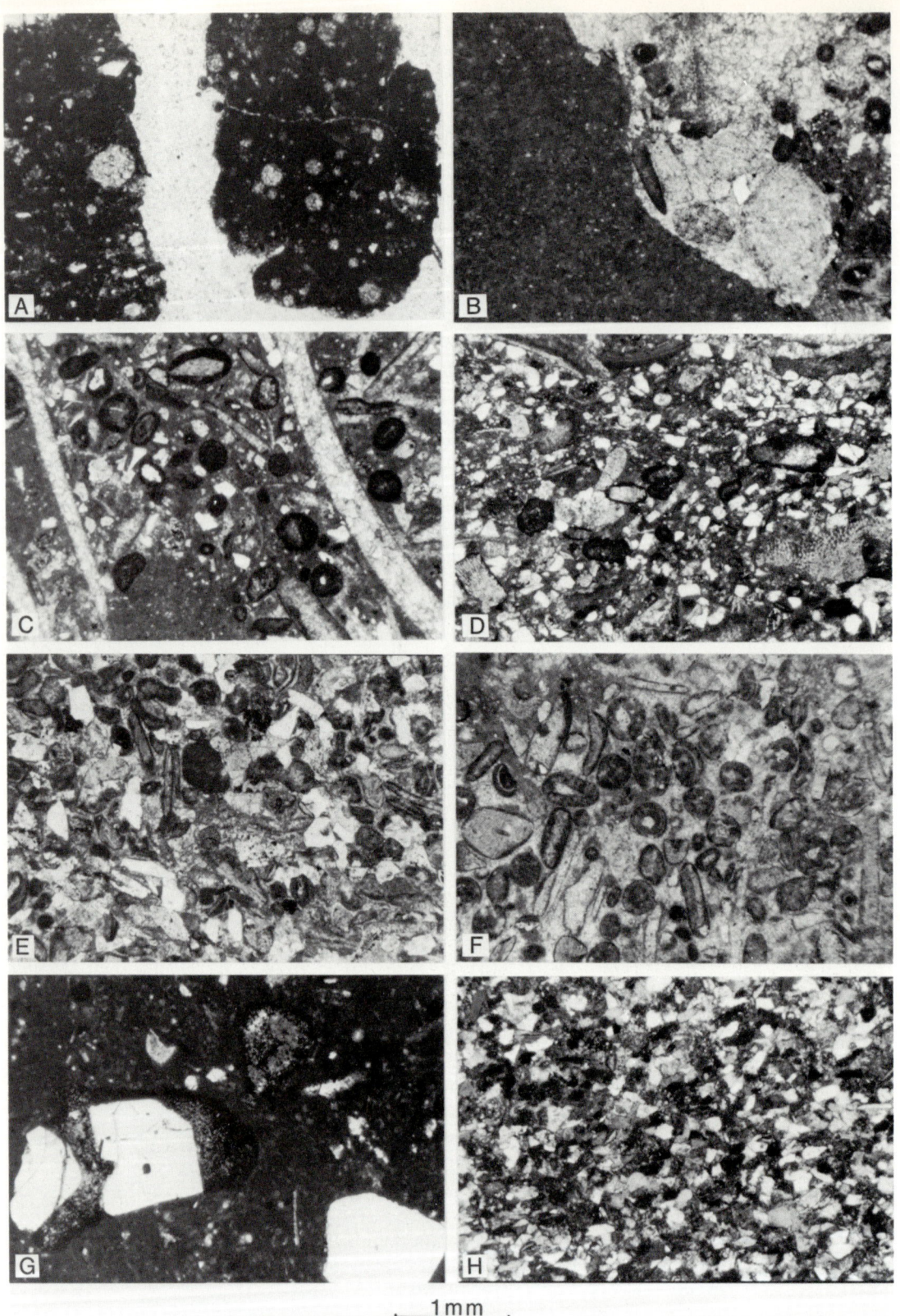

FIGURE 12.41 Quintuco–Loma Montosa Formation (Lower Cretaceous), Neuquén Basin, Argentina. Typical microfacies. A. Microfacies 1. B. Microfacies 2. C. Microfacies 3. D. Microfacies 4. E. Microfacies 5. F. Microfacies 6. G. Microfacies 7. H. Microfacies 8. See text for detailed descriptions. All photomicrographs: plane-polarized light, except H, crossed nicols. From Carozzi *et al.* (1981c). Reprinted by permission of Yacimientos Petrolíferos Fiscales (Y.P.F.).

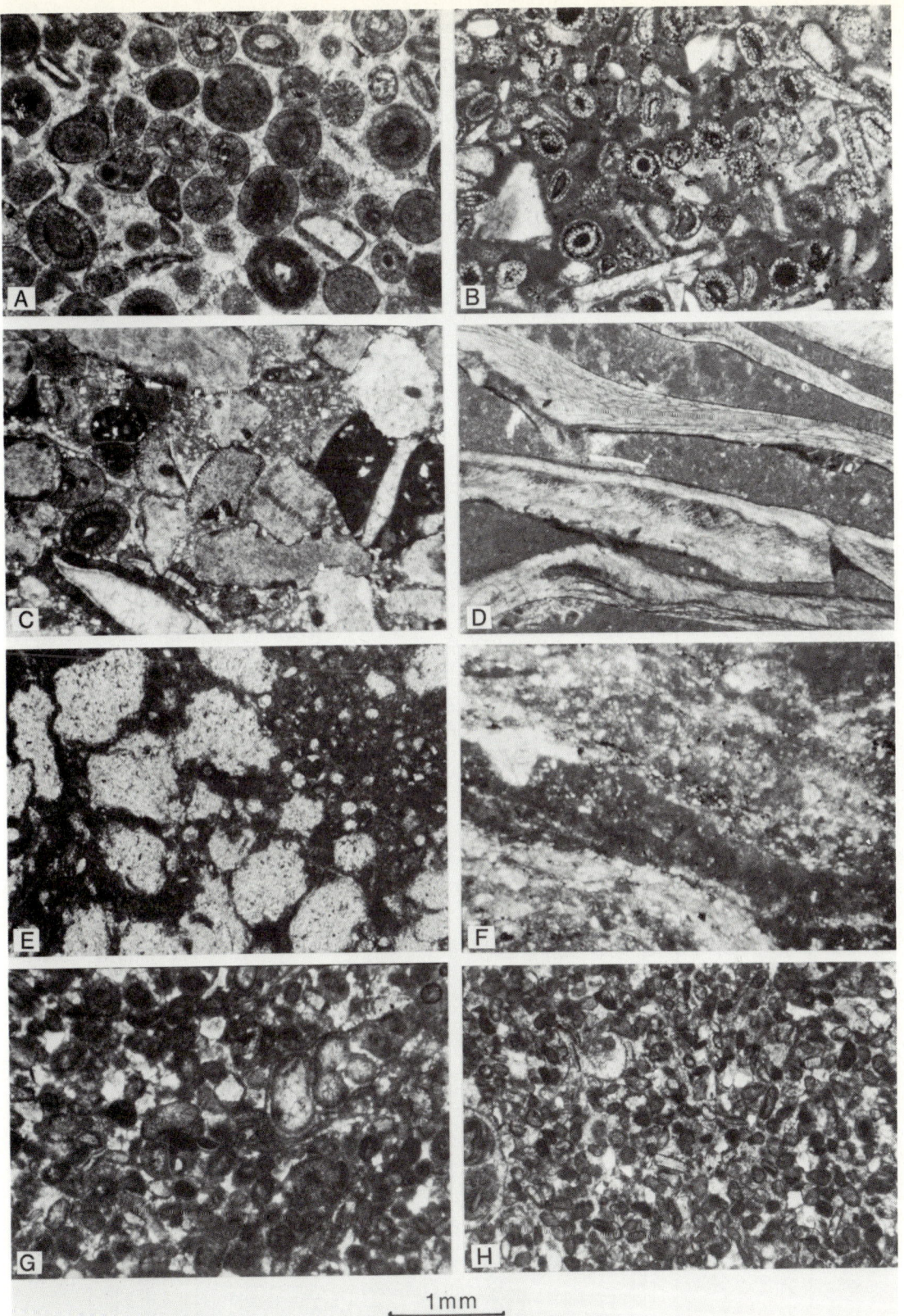

FIGURE 12.42 Quintuco–Loma Montosa Formation (Lower Cretaceous), Neuquén Basin, Argentina. Typical microfacies (continued). A. Microfacies 9. B. Microfacies 10. C. Microfacies 11. D. Microfacies 12. E. Microfacies 13. F. Microfacies 14. G. Microfacies 15. H. Microfacies 16. See text for detailed descriptions. All photomicrographs: plane-polarized light, except A and E, crossed nicols. From Carozzi *et al.* (1981c). Reprinted by permission of Yacimientos Petrolíferos Fiscales (Y.P.F.).

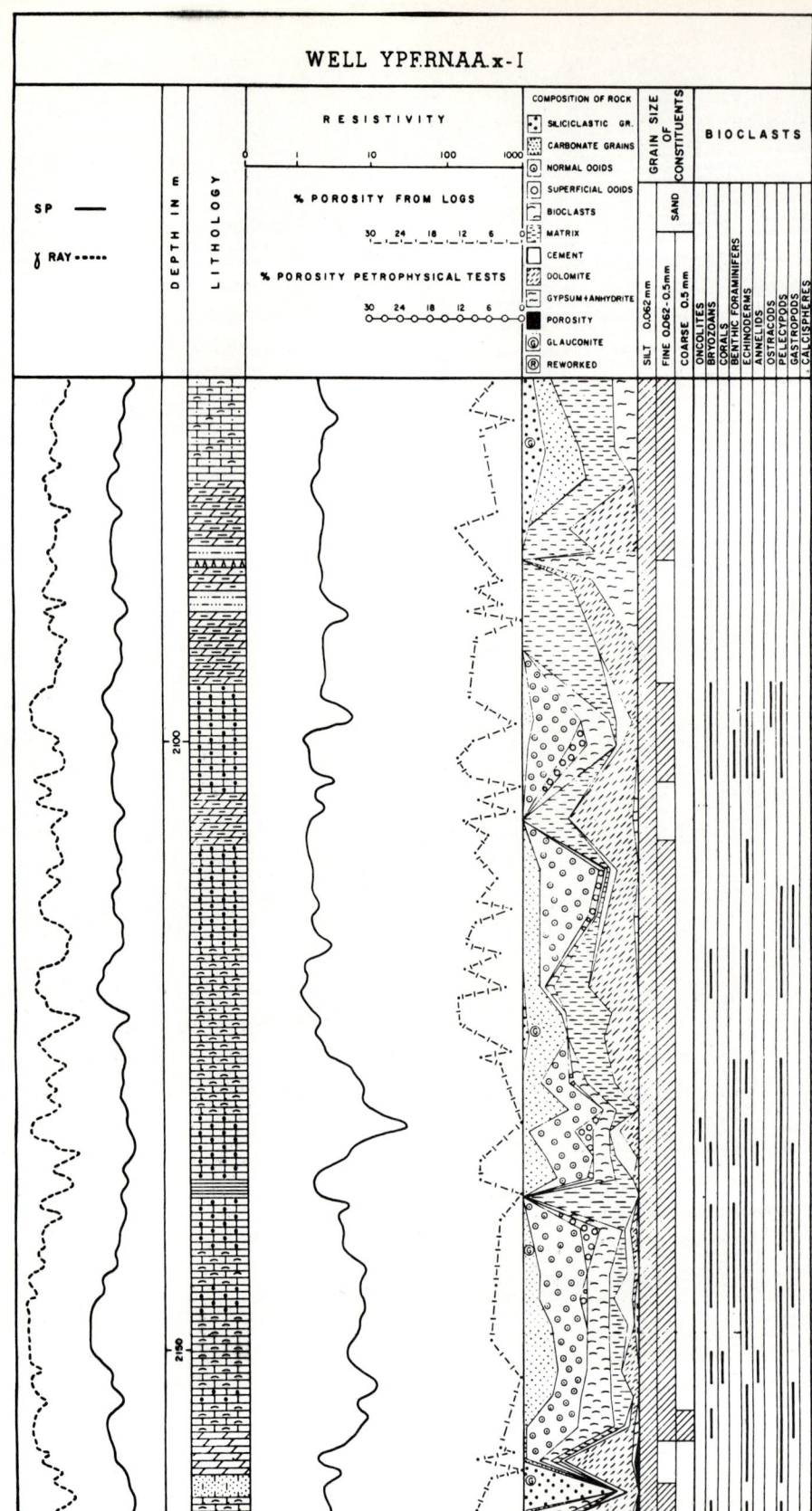

FIGURE 12.43 Quintuco–Loma Montosa Formation (Lower Cretaceous), Neuquén Basin, Argentina. Detailed graphic representation of microscopic parameters in well Y.P.F. RN AA x-1 (Agua Amarga). From Carozzi *et al.* (1981c). Reprinted by permission of Yacimientos Petrolíferos Fiscales (Y.P.F.).

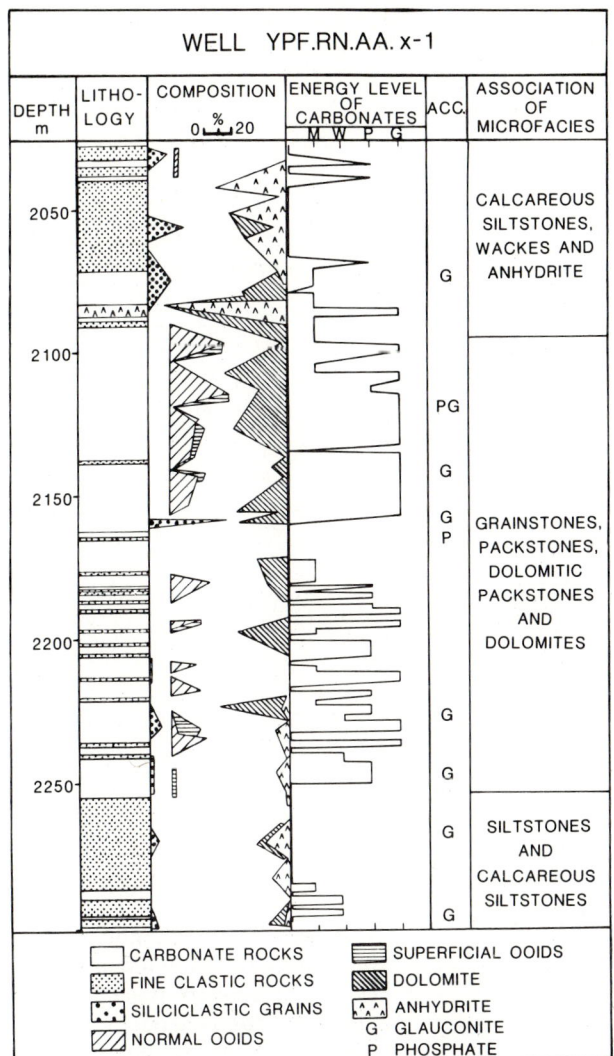

FIGURE 12.44 Quintuco–Loma Montosa Formation (Lower Cretaceous), Neuquén Basin, Argentina. Simplified graphic representation of microscopic parameters and association of microfacies in well Y.P.F. RN AA x-1 (Agua Amarga). From Carozzi et al. (1981c). Reprinted by permission of Yacimientos Petrolíferos Fiscales (Y.P.F.).

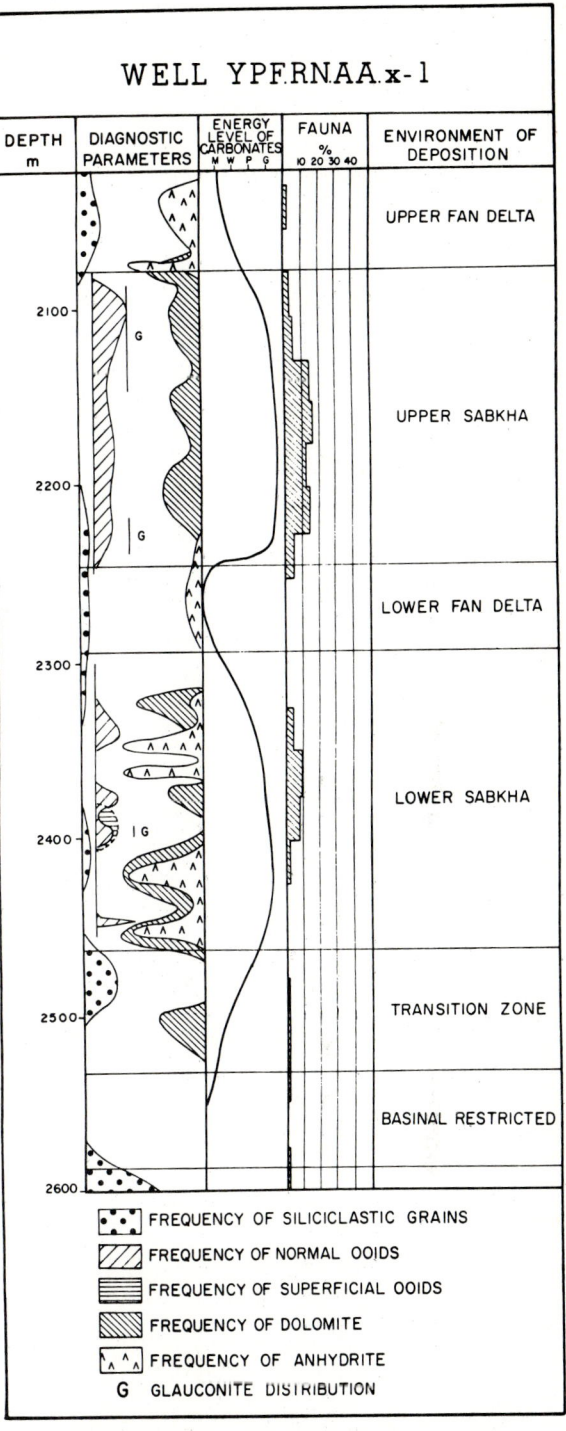

FIGURE 12.45 Quintuco–Loma Montosa Formation (Lower Cretaceous), Neuquén Basin, Argentina. Generalized graphic representation of microscopic parameters and depositional environments in well Y.P.F. RN AA x-1 (Agua Amarga). From Carozzi et al. (1981c). Reprinted by permission of Yacimientos Petrolíferos Fiscales (Y.P.F.).

350 Part III / Carbonate Platforms

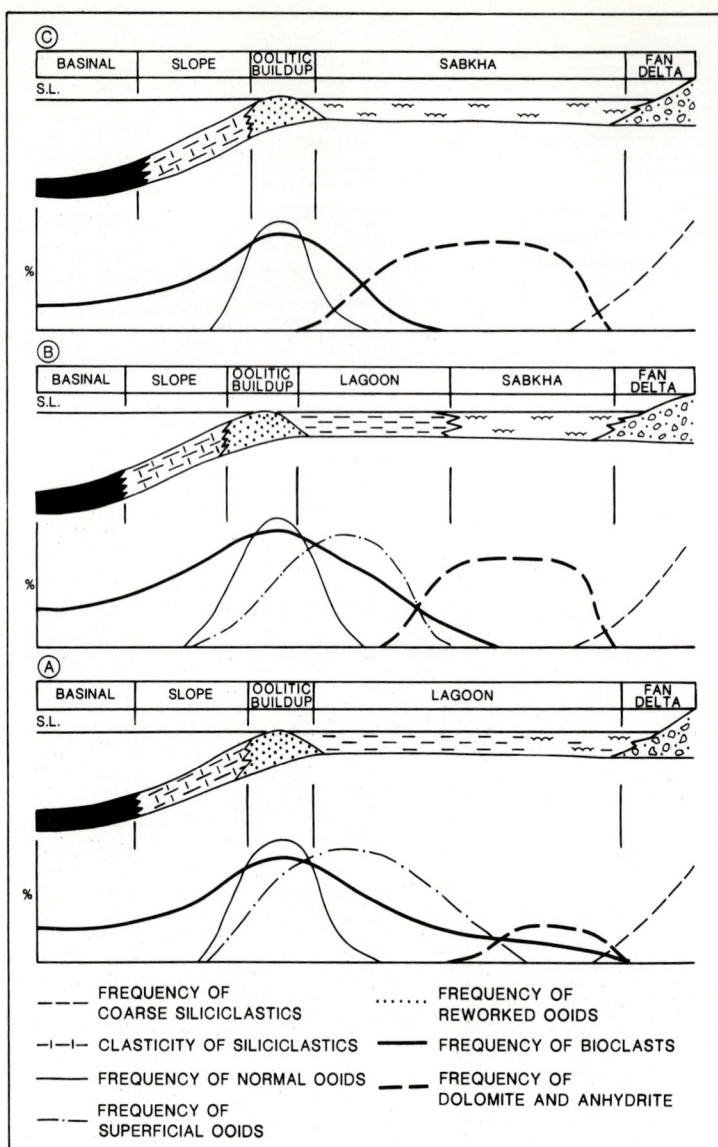

FIGURE 12.46 Quintuco–Loma Montosa Formation (Lower Cretaceous), Neuquén Basin, Argentina. Ideal depositional model. See text for discussion. From Carozzi *et al*. (1981c). Reprinted by permission of Yacimientos Petrolíferos Fiscales (Y.P.F.).

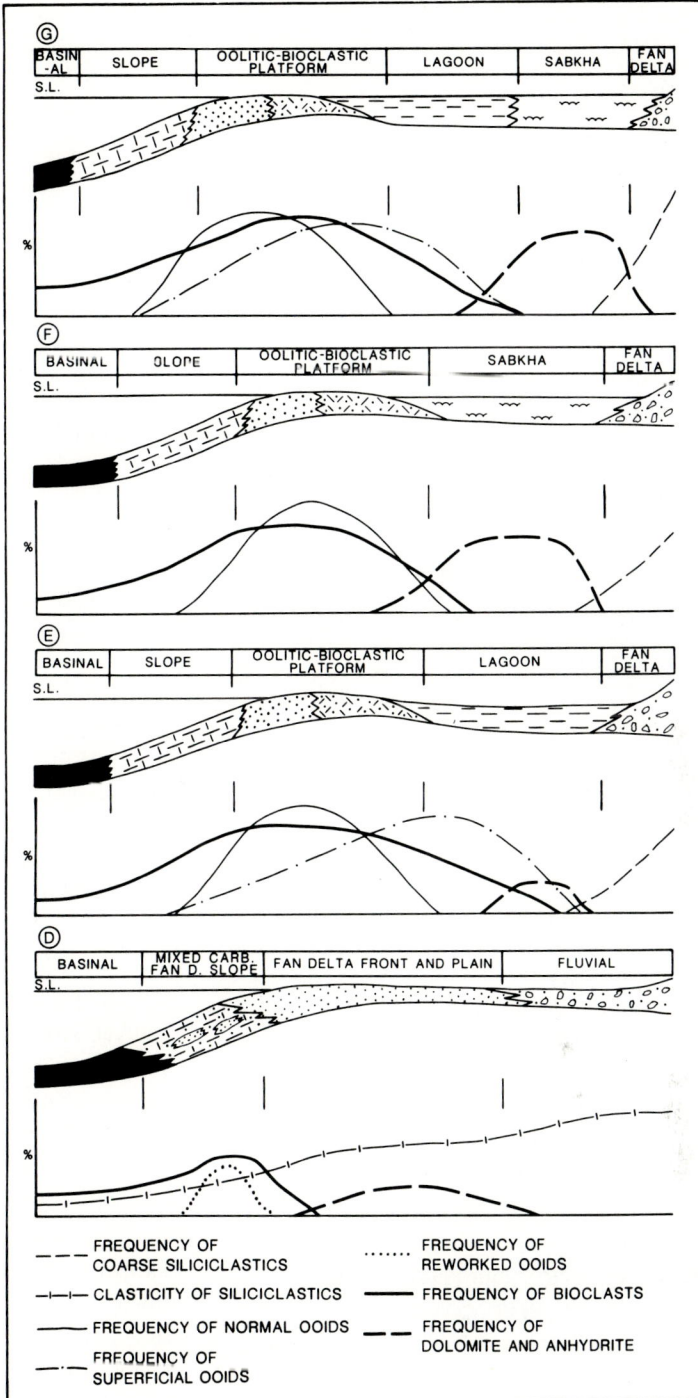

FIGURE 12.47 Quintuco–Loma Montosa Formation (Lower Cretaceous), Neuquén Basin, Argentina. Ideal depositional model (continued). See text for discussion. From Carozzi *et al.* (1981c). Reprinted by permission of Yacimientos Petrolíferos Fiscales (Y.P.F.).

13

Carbonate Platforms with Frontal Bioconstructed to Hydrodynamic Buildups

Frontal bioconstructed buildups accompanied by appreciable hydrodynamic accumulations of bioclasts, commonly as slope talus or peripheral aprons but nevertheless successful in constructing wave-resistant frameworks, belong to platforms submitted to the action of very agitated seas. Stromatoporoids, corals, red algae, and some stromatolites are the most prolific colonial organisms under these conditions. They can provide protection to associated communities of more delicate forms such as several other types of algae and *Amphipora*.

13.1 CHARACTERISTIC CONSTITUENTS: STROMATOPOROIDS, CORALS, AND *AMPHIPORA*

The Coral Zone of the basal Jeffersonville Limestone (Middle Devonian) of southeastern Indiana (Carozzi and Hulse, 1963) is a platform which consists of a relatively thin sequence of carbonates ranging from 11 to 20 ft (3.3 to 6 m). It was investigated along a transverse profile of nine stratigraphic sections by means of 171 thin sections with an average vertical spacing of 10 in. (25.4 cm).

Description of Microfacies

Microfacies are described in order of general shallowing.

Microfacies 1 (Fig. 13.1A). Calcisiltite with scattered sand-size bioclasts of crinoids, bryozoans, brachiopods, pelecypods, and ostracods. Matrix is dominantly bioclastic with pyrite pigments and minute angular grains of detrital quartz. Minute rhombs of secondary dolomite are scattered throughout the groundmass.

Microfacies 2 (Fig. 13.1B). Grain-supported biocalcarenite with bioclastic matrix containing numerous isolated horn corals and stromatoporoid colonies in growth position. Interstitial material consists of sand-size, poorly sorted bioclasts of crinoids, bryozoans, brachiopods, pelecypods, and ostracods often pressure-welded. The bioclastic matrix shows the same constituents with a fluidal texture and associated with pyrite pigments and bituminous matter.

Microfacies 3 (Fig. 13.1C). Stromatoporoid (cabbage head) and *Amphipora*-constructed limestone with a small amount of pelletoidal interstitial bioclastic matrix consisting of bioclasts of stromatoporoids, *Amphipora*, bryozoans, crinoids, and brachiopods.

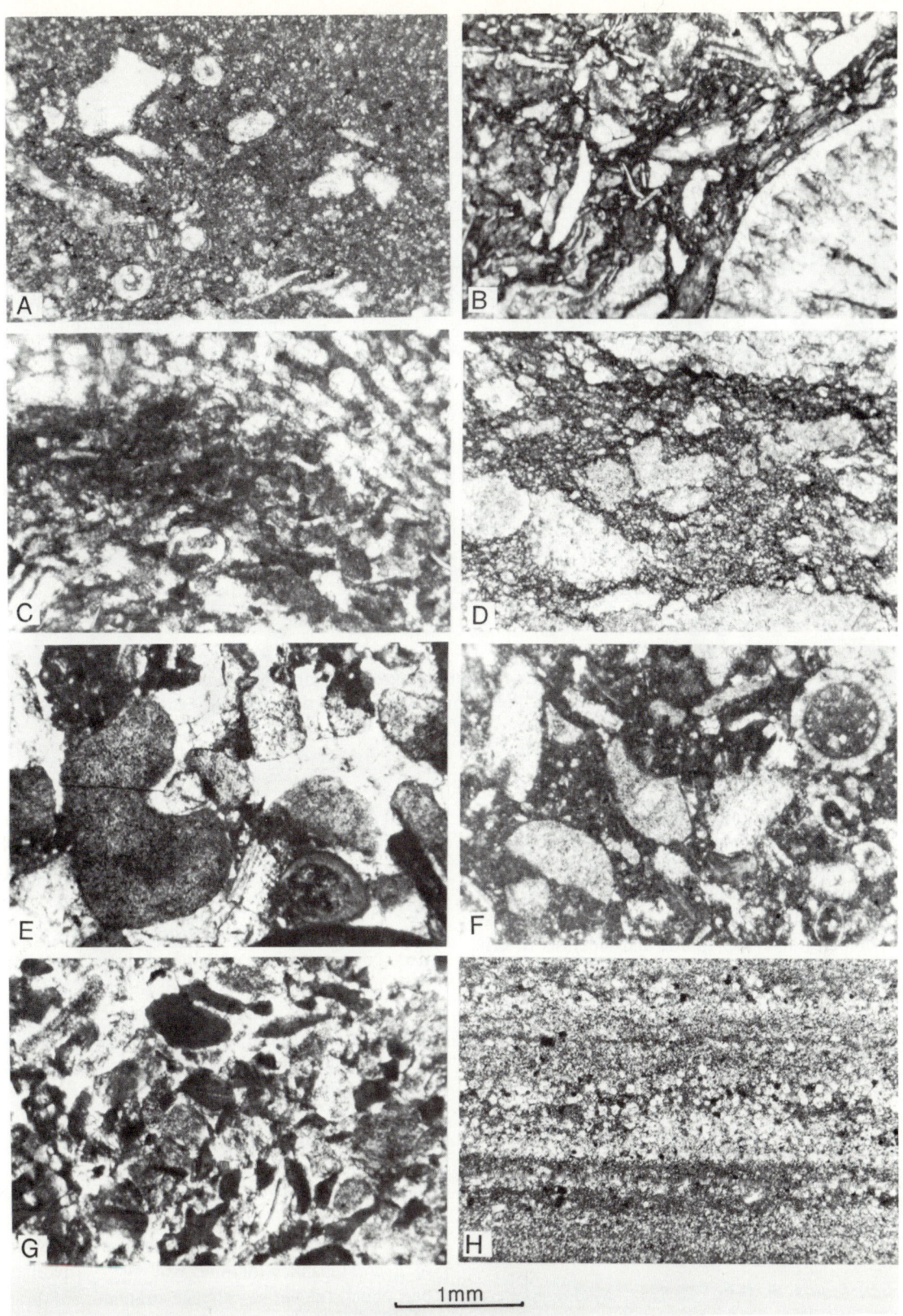

FIGURE 13.1 Coral Zone of Jeffersonville Limestone (Middle Devonian), southeast Indiana. Typical microfacies. A. Microfacies 1. B. Microfacies 2. C. Microfacies 3. D. Microfacies 4. E. Microfacies 5. F. Microfacies 6. G. Microfacies 6a. H. Laminated evaporitic dolosiltite. See text for detailed descriptions. All photomicrographs: plane-polarized light. From Carozzi and Hulse (1963).

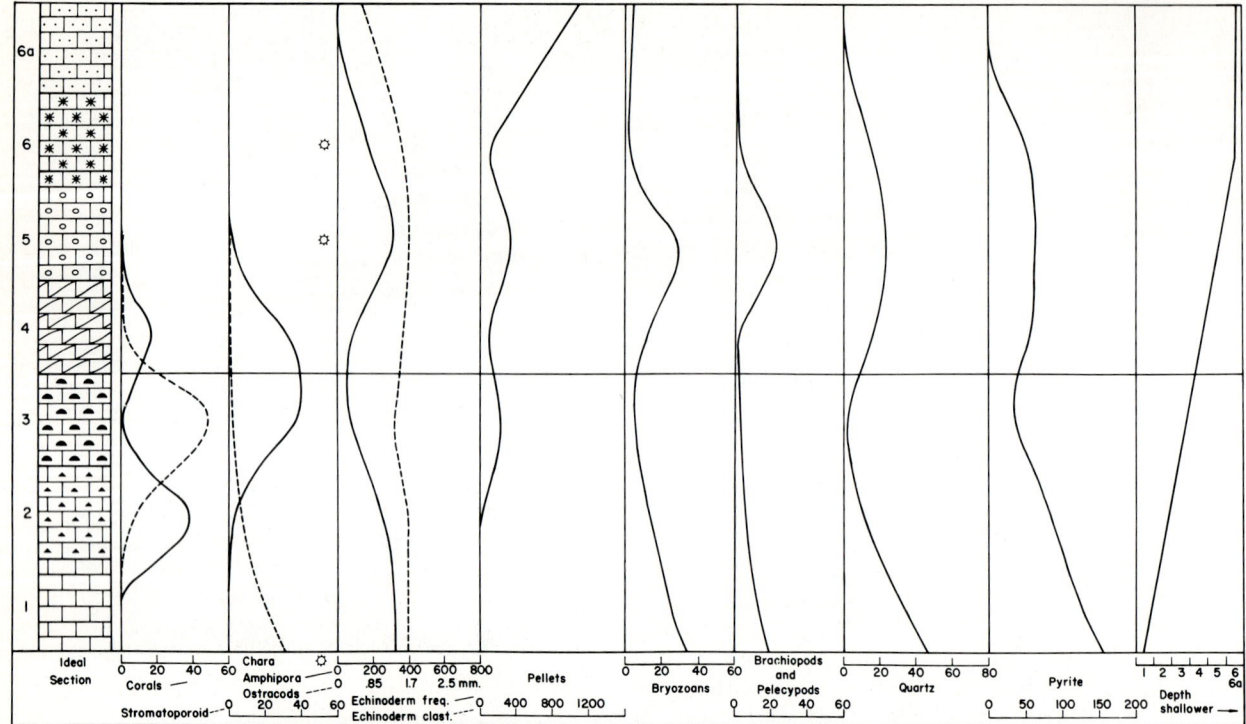

FIGURE 13.2 Coral Zone of Jeffersonville Limestone (Middle Devonian), southeast Indiana. Ideal shallowing-upward sequence. From Carozzi and Hulse (1963).

Microfacies 4 (Fig. 13.1D). Coarse-grained, irregularly stratified, grain-supported calcarenite to calcirudite with bituminous bioclastic matrix and local streaks of detrital quartz grains. Components consist mainly of broken and abraded solitary corals, fragments of stromatoporoid colonies and *Amphipora*, and bioclasts of crinoids and bryozoans. Matrix displays an accentuated fluidal texture. This microfacies results from a powerful reworking of microfacies 2 and appears to be storm influenced. It is also interbedded with matlike colonies of stromatoporoids.

Microfacies 5 (Fig. 13.1E). Grain-supported biocalcarenite consisting mainly of bioclasts of crinoids, bryozoans, and ostracods associated with a few pellets and floated *Chara* oogonia. Cementation is by interstitial sparite mosaic and syntaxial overgrowths around crinoids.

Microfacies 6 (Fig. 13.1F). Mud-supported biocalcarenite with bioclastic to calcisiltite matrix. Bioclasts are from crinoids, bryozoans, brachiopods, pelecypods, ostracods, trilobites, and floated *Chara* oogonia. Matrix consists of the same material with pyrite pigments and very abundant minute grains of detrital quartz.

Microfacies 6a (Fig. 13.1G). Grain-supported pelletoidal biocalcarenite with sparite cement. Micrite pellets are lithic and associated with bioclasts of crinoids, bryozoans, and ostracods.

Laminated Dolosiltite (Fig. 13.1H). Thin dark layers of microcrystalline anhedral dolomite with abundant pyrite pigments, small pellets, and a few grains of detrital quartz. Thicker and lighter layers consist of a mosaic of medium crystalline subhedral to euhedral dolomite often with zoned individuals. Originally a laminated stromatolitic calcisiltite.

The Ideal Shallowing-Upward Sequence

The ideal sequence (Fig. 13.2) divides itself naturally into two parts. The superposition of microfacies 1, 2, and 3 shows open marine conditions and development of a coral bank with bioclastic matrix overlain by a barrier constructed by cabbage stromatoporoids and *Amphipora*, whereas all other constituents become subordinate. The second part (microfacies 4 to 6a) begins with storm-influenced destructional conditions of the barrier, still associated with matlike stromatoporoid colonies grading into protected conditions with abundant crinoids,

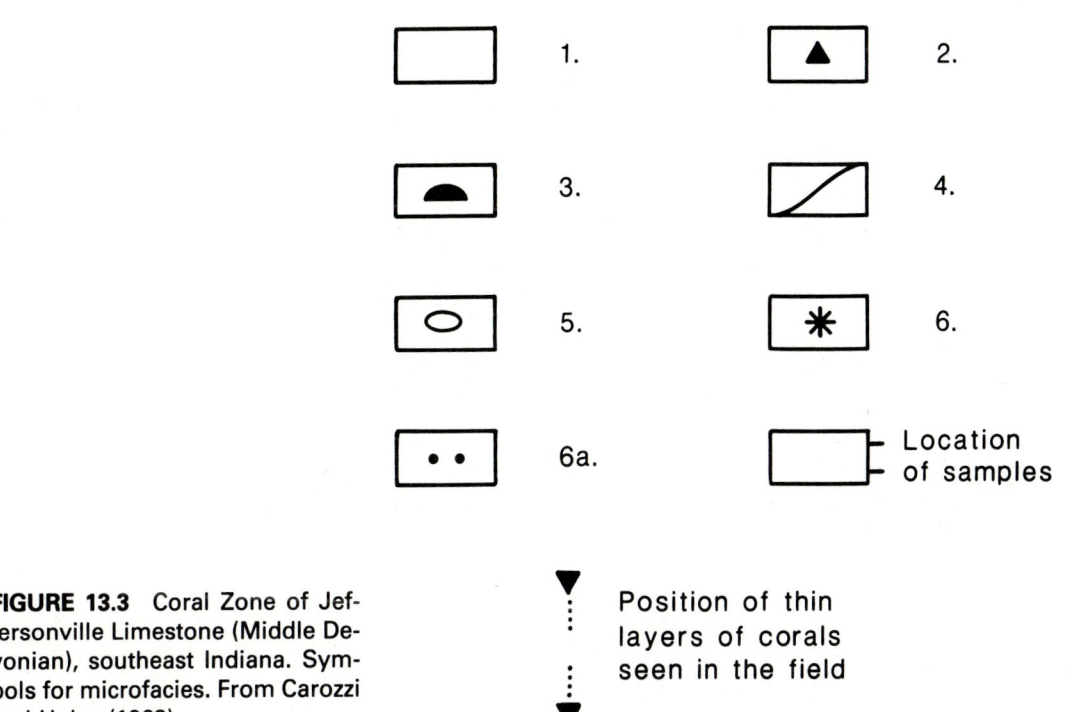

FIGURE 13.3 Coral Zone of Jeffersonville Limestone (Middle Devonian), southeast Indiana. Symbols for microfacies. From Carozzi and Hulse (1963).

bryozoans, pelecypods, and brachiopods. Eventually lagoonal conditions predominated and consisted of pelletoidal facies (microfacies 6a) with floated *Chara* oogonia. The latter indicate penecontemporaneous lacustrine conditions further inland from which fine detrital quartz also originated. Most of it bypassed the barrier and settled eventually in open marine conditions. Because detrital quartz is silt size throughout the investigated sections, only its frequency curve is given.

Examination of the various sections in a landward direction illustrates detailed variation of microscopic parameters and general evolution of the carbonate platform. Section 1 (Figs. 13.3, 13.4) is almost an ideal shallowing-upward sequence, section 4 (Fig. 13.5) corresponds to the greatest development of the storm-affected destructional microfacies 4, section 5 (Fig. 13.6) illustrates the constructed barrier itself, while section 7 (Fig. 13.7) is a thinner back-barrier to lagoonal pelletoidal environment.

The Ideal Depositional Model

The depositional model (Fig. 13.8) can be divided into four stages. The first one is a carbonate ramp with incipient development at the wave base of a bank of isolated horn corals in a bioclastic matrix. Stages 2 and 3 correspond to the growth and temporary destruction by storms of the barrier constructed by stromatoporoids and *Amphipora*. This barrier grew over the dead coral bank and generated more quiet back-barrier conditions which graded landward into a pelletoidal lagoonal environment with adjacent freshwater lakes (stage 4). Thin-laminated dolomite sequences indicate local subevaporitic conditions.

13.2 CHARACTERISTIC CONSTITUENTS: STROMATOPOROIDS AND CORALS

Bioconstructed buildups do not necessarily represent single or multiple phases of growth of framework-building communities which were separated or terminated by the effect of local environmental changes. Instances occur where discrete bodies show major episodes of deepening and shallowing by symmetrical superposition of different types of colonial organisms, although the megascopic aspect remains massive and no lateral displacement has taken place.

Another aspect of the Coral Zone of the basal Jeffersonville Limestone (Middle Devonian) near Columbus,

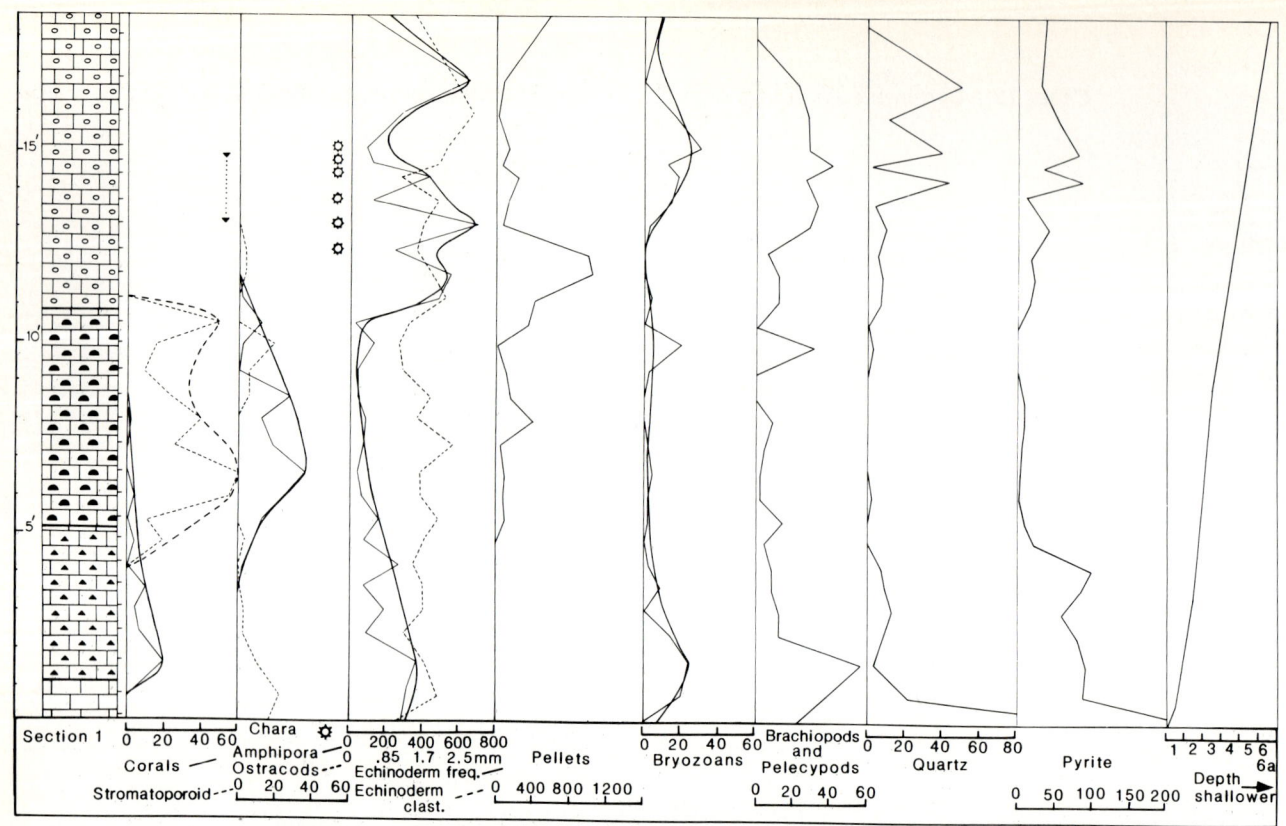

FIGURE 13.4 Coral Zone of Jeffersonville Limestone (Middle Devonian), southeast Indiana. Vertical variation of microscopic parameters in field section 1. See Figure 13.8 for location. From Carozzi and Hulse (1963).

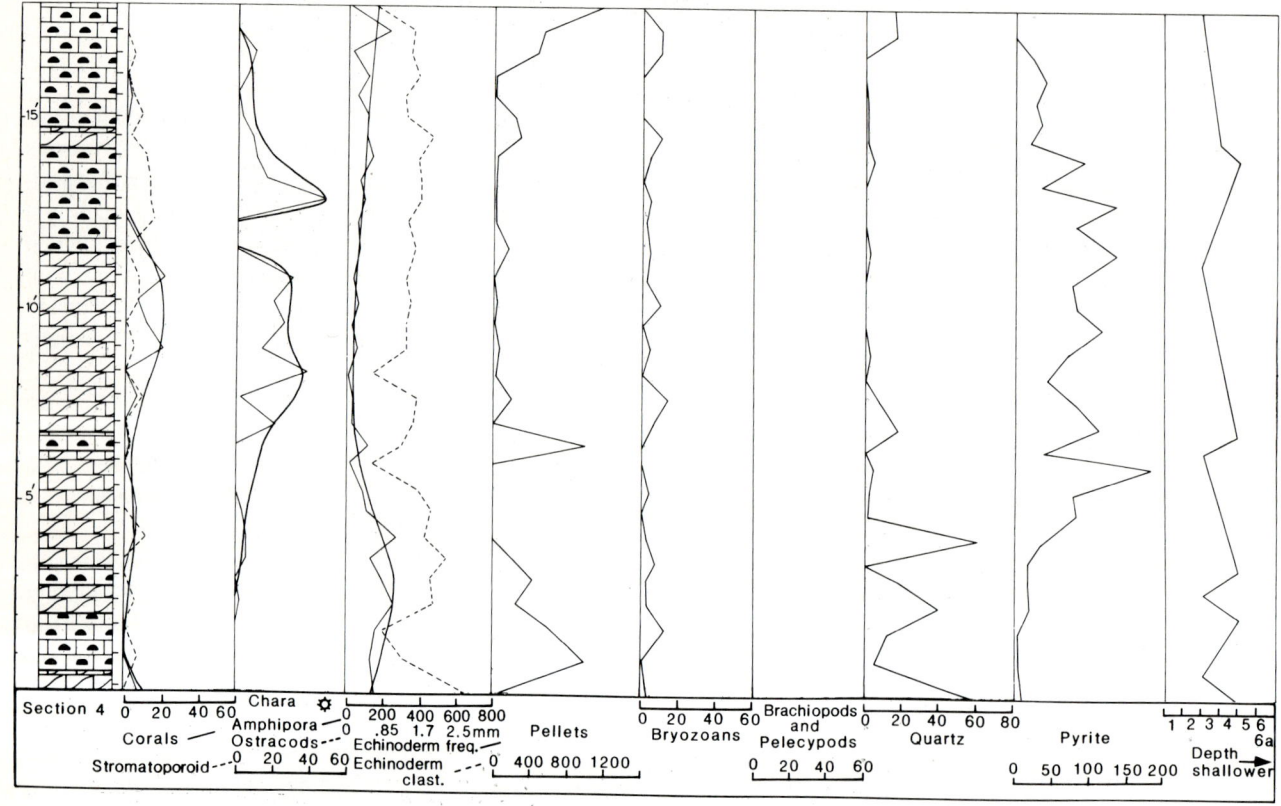

FIGURE 13.5 Coral Zone of Jeffersonville Limestone (Middle Devonian), southeast Indiana. Vertical variation of microscopic parameters in field section 4. See Figure 13.8 for location. From Carozzi and Hulse (1963).

Chap. 13 / Carbonate Platforms with Frontal Bioconstructed to Hydrodynamic Buildups 357

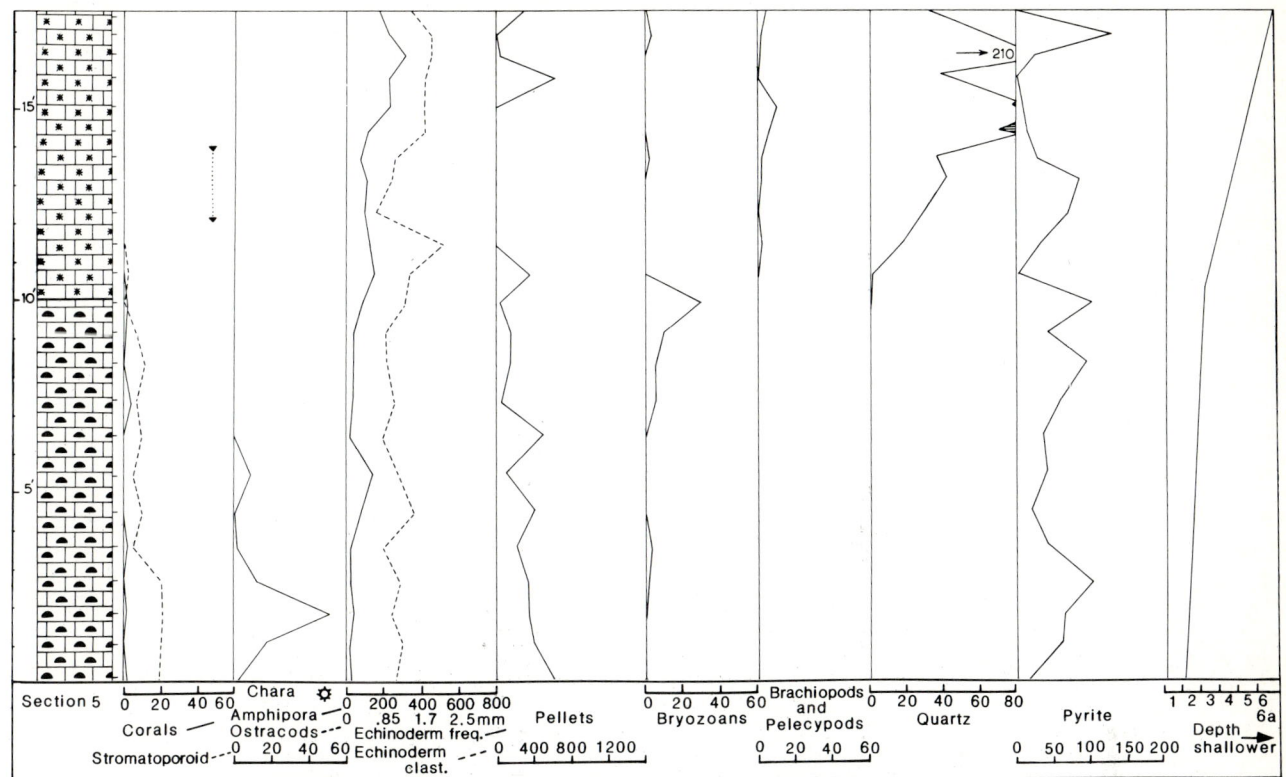

FIGURE 13.6 Coral Zone of Jeffersonville Limestone (Middle Devonian), southeast Indiana. Vertical variation of microscopic parameters in field section 5. See Figure 13.8 for location. From Carozzi and Hulse (1963).

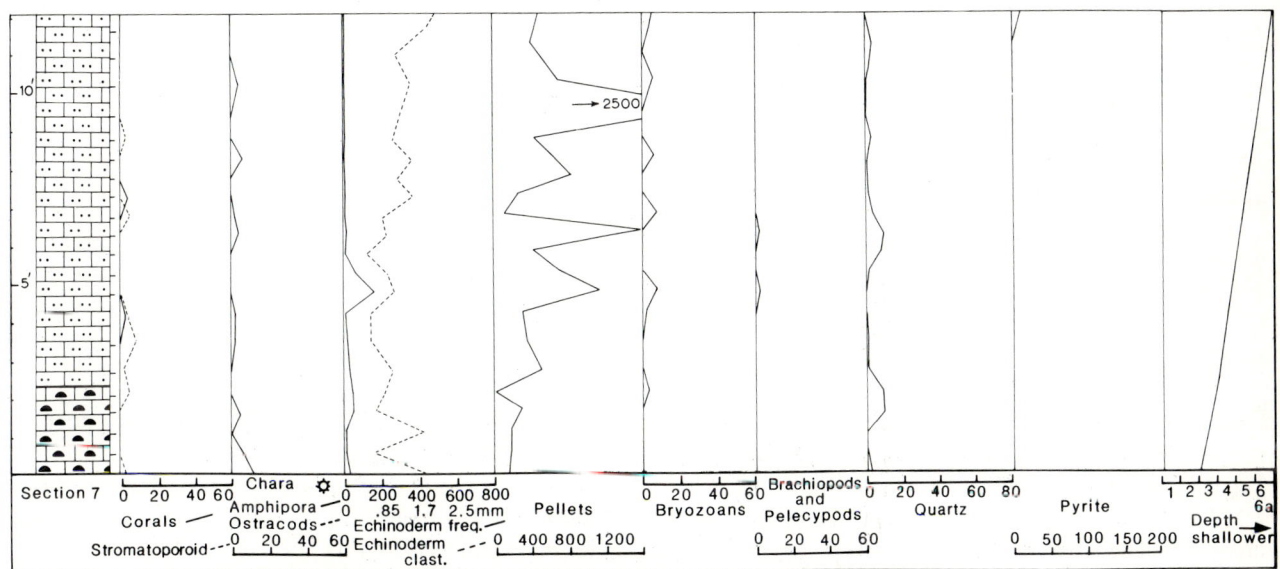

FIGURE 13.7 Coral Zone of Jeffersonville Limestone (Middle Devonian), southeast Indiana. Vertical variation of microscopic parameters in field section 7. See Figure 13.8 for location. From Carozzi and Hulse (1963).

358 Part III / Carbonate Platforms

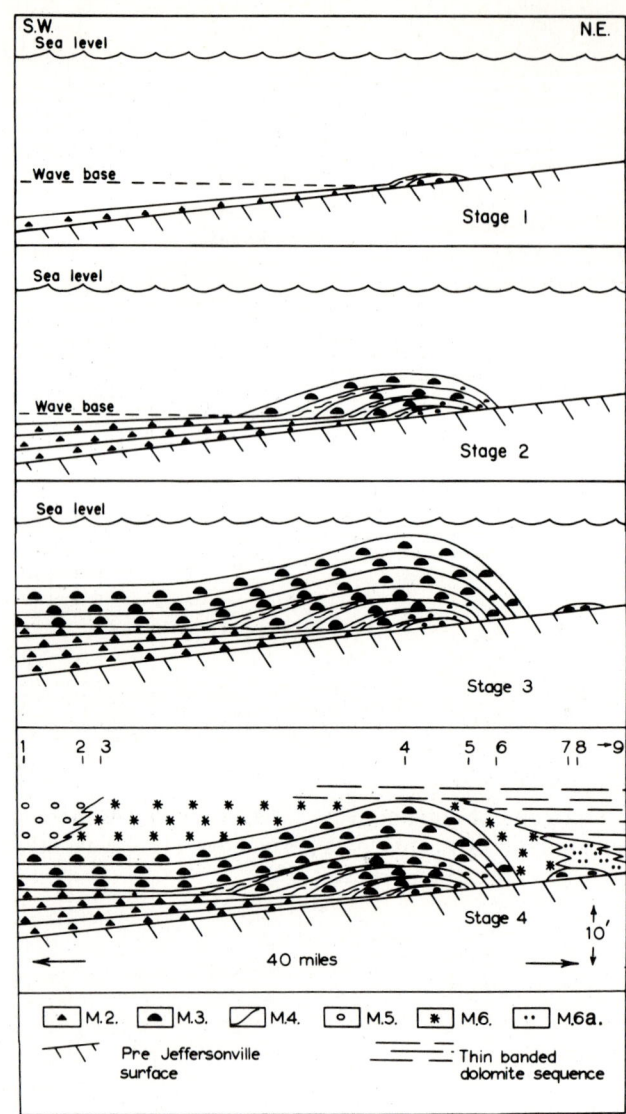

FIGURE 13.8 Coral Zone of Jeffersonville Limestone (Middle Devonian), southeast Indiana. Ideal depositional model. From Carozzi and Hulse (1963).

southeastern Indiana (Carozzi and Lundwall, 1959) consists of a single buildup 85 ft (25.5 m) long with an average thickness of 9 ft (2.74 m). It was investigated by means of 10 detailed stratigraphic sections (Fig. 13.9) with 130 thin sections at an average vertical interval of 1 ft (0.30 m). The buildup is underlain by the Geneva Dolomite and its evolution also terminates in dolomitic microfacies.

Description of Microfacies

Because the vertical evolution of the microfacies indicates a phase of deepening of the environment followed by a phase of shallowing, the microfacies are described in that order.

Microfacies of the Geneva Dolomite (Fig. 13.10A). Massive subhedral to euhedral dolomite with abundant bituminous matter. Irregularly distributed patches of coarser dolomite correspond to the "ghosts" of original bioclasts scattered in a matrix which was probably bioclastic.

Microfacies 1 (Fig. 13.10B). Bioconstructed limestone of matlike stromatoporoid colonies separated by irregular zones and pockets of a grain-supported biocalcarenite with bioclastic matrix. Bioclasts are of stromatoporoids, dasyclads, horn corals, and crinoids.

Microfacies 2 (Fig. 13.10C). Coarse, poorly sorted, grain-supported biocalcarenite with association of calcisiltite matrix and patches of sparite cement. Bioclasts are from stromatoporoids and dasyclads, associated with lithic pellets of micrite.

Microfacies 3 (Fig. 13.10D). Medium-grained, well-sorted, grain-supported, pelletoidal biocalcarenite with abundant horn corals in growth position and sparite cement. Bioclasts consist of stromatoporoids,

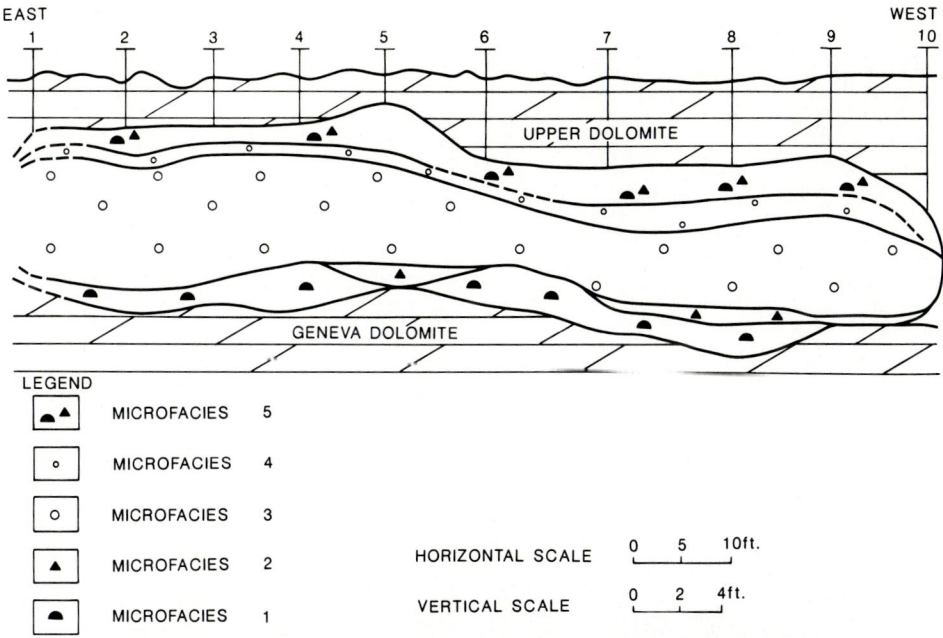

FIGURE 13.9 Buildup of Coral Zone of Jeffersonville Limestone (Middle Devonian), southeast Indiana. General cross section showing distribution of microfacies. From Carozzi and Lundwall (1959). Reprinted by permission of the Society of Economic Paleontologists and Mineralogists.

horn corals, crinoids, pelecypods, brachiopods, and textulariids. Pellets are micritic and of lithic origin.

Microfacies 4 (Fig. 13.10E). Fine to medium-grained, grain-supported, pelletoidal calcarenite with sparite cement. Lithic pellets of micrite are well rounded and associated with a few bioclasts of crinoids and ostracods.

Microfacies 5 (Fig. 13.10F). Bioconstructed limestone of matlike stromatoporoids and horn corals separated by a grain-supported biocalcarenite with bioclastic matrix. Unsorted and angular bioclasts are from stromatoporoids, horn corals, and pelecypods.

Microfacies of Upper Dolomite. Fine-grained laminated subhedral bituminous dolomite with extremely rare "ghosts" of small bioclasts, and desiccation cracks.

The Ideal Vertical Sequence

Vertical evolution of average microscopic parameters (Fig. 13.11) shows that matlike stromatoporoid colonies are much better developed in the deepening phase (microfacies 2) than in the shallowing phase (microfacies 5). Horn corals, although accompanying stromatoporoids, reach their best development (associated with textulariids) in deeper conditions than the latter (microfacies 3).

Crinoids and ostracods are most abundant just before the return of the stromatoporoid colonies (microfacies 4). Detailed variation of microscopic parameters are well shown in section 2 (Fig. 13.12), section 6 (Fig. 13.13), and section 9 (Fig. 13.14), giving an idea of the variability within the buildup.

The Ideal Depositional Model

The depositional model can be divided into six phases (Figs. 13.15, 13.16) which show clearly the symmetrical evolution in time of the buildup. When deposition of microfacies 4 (phase 4) took place during deepest conditions, differential compaction seems to have occurred in the underlying Geneva Dolomite. The terminal dolomite (phase 6) indicates lagoonal conditions and is overlain by laminated dolosiltite with desiccation cracks.

13.3 CHARACTERISTIC CONSTITUENTS: STROMATOPOROIDS AND *AMPHIPORA*

Mature platforms with frontal bioconstructed to hydrodynamic buildups display complex margins showing well-developed longitudinal zones combined with transverse channels. Zonation is controlled by the degree of exposure to very agitated conditions of colonial organisms

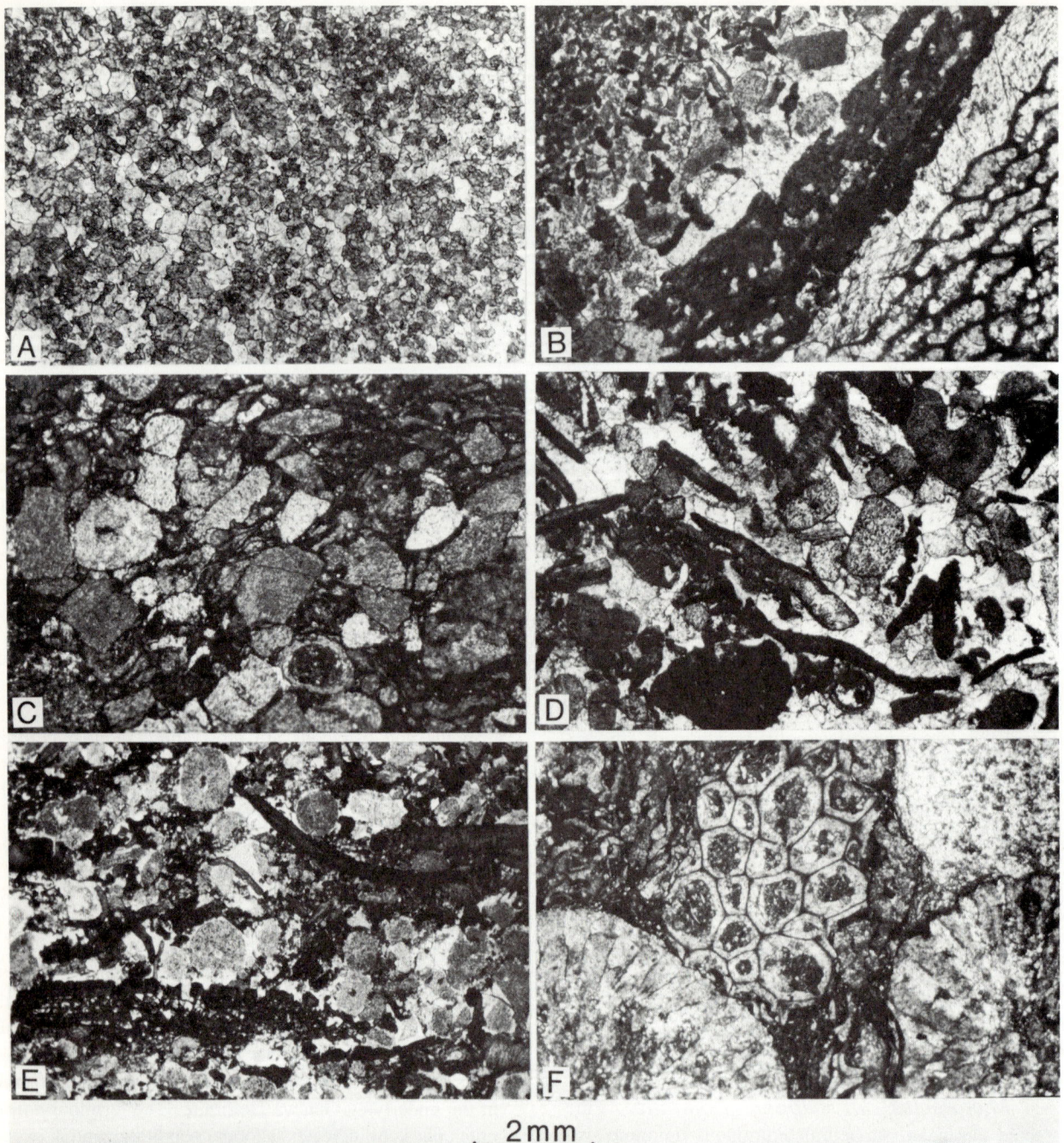

FIGURE 13.10 Buildup of Coral Zone of Jeffersonville Limestone (Middle Devonian), southeast Indiana. Typical microfacies. A. Geneva Dolomite. B. Microfacies 1. C. Microfacies 2. D. Microfacies 3. E. Microfacies 4. F. Microfacies 5. See text for detailed descriptions. All photomicrographs: plane-polarized light. From Carozzi and Textoris (1967). Reprinted by permission of E. J. Brill.

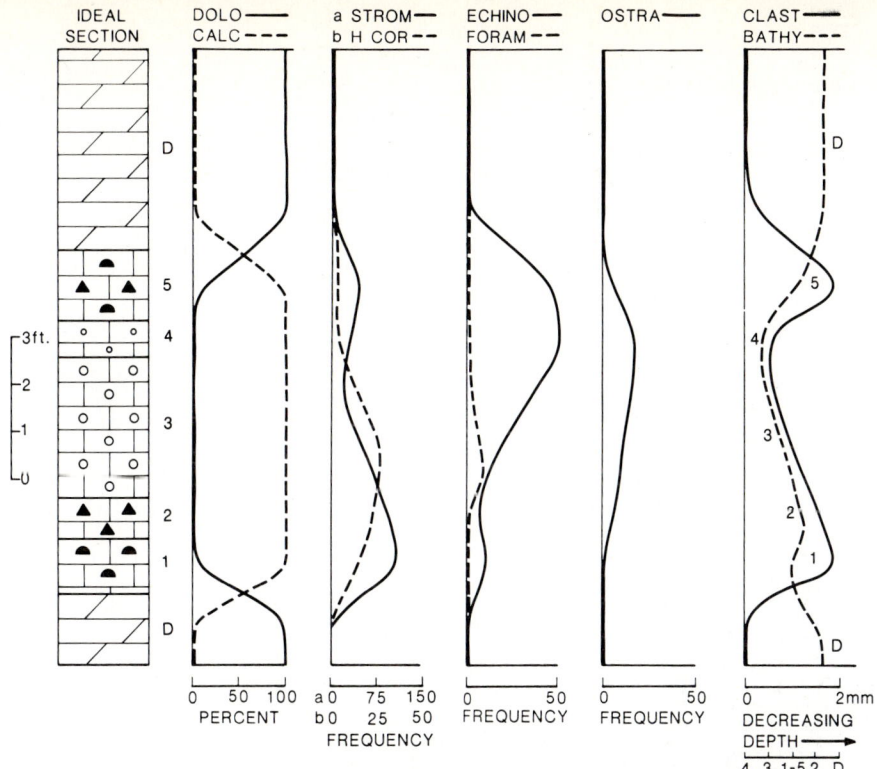

FIGURE 13.11 Buildup of Coral Zone of Jeffersonville Limestone (Middle Devonian), southeast Indiana. Ideal vertical sequence. From Carozzi and Lundwall (1959). Reprinted by permission of the Society of Economic Paleontologists and Mineralogists.

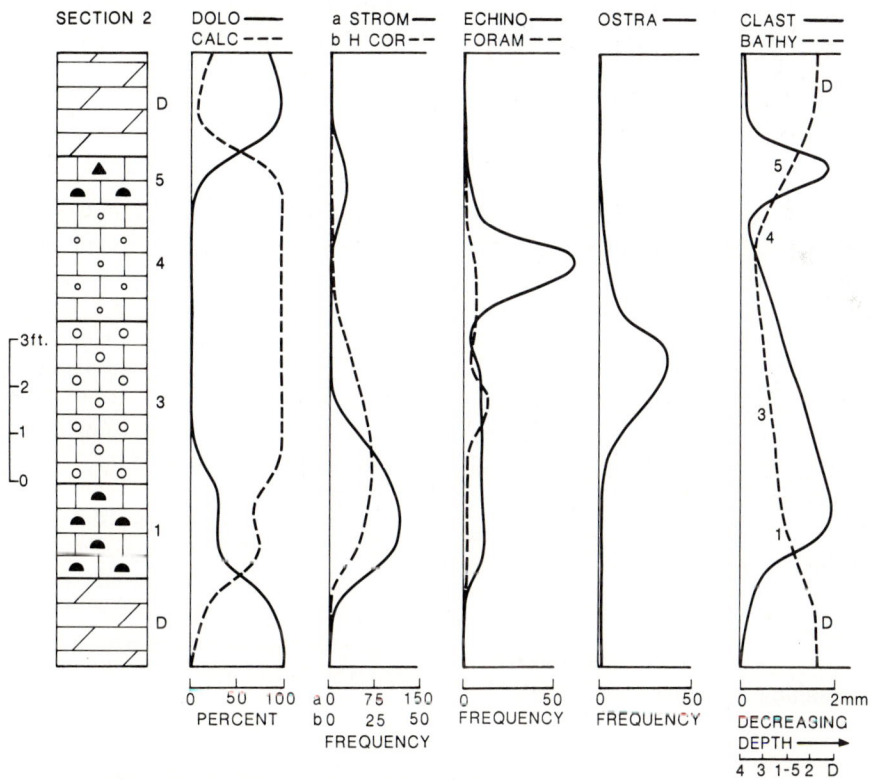

FIGURE 13.12 Buildup of Coral Zone of Jeffersonville Limestone (Middle Devonian), southeast Indiana. Vertical variation of microscopic parameters in field section 2. See Figure 13.9 for location. From Carozzi and Lundwall (1959). Reprinted by permission of the Society of Economic Paleontologists and Mineralogists.

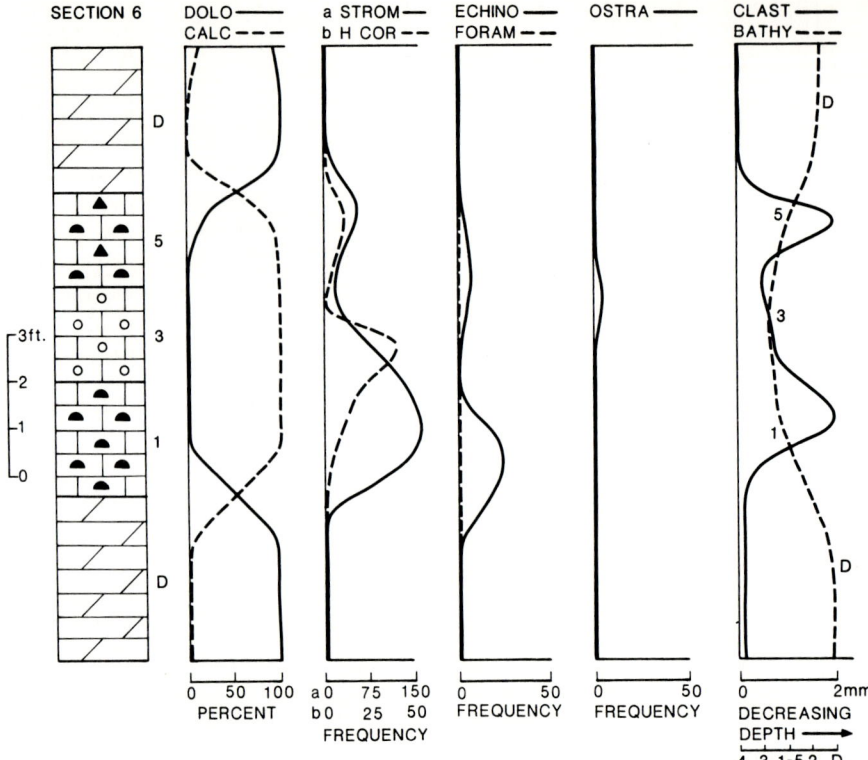

FIGURE 13.13 Buildup of Coral Zone of Jeffersonville Limestone (Middle Devonian), southeast Indiana. Vertical variation of microscopic parameters in field section 6. See Figure 13.9 for location. From Carozzi and Lundwall (1959). Reprinted by permission of the Society of Economic Paleontologists and Mineralogists.

and by the protection they afford to each other. Longitudinal rows of constructed buildups, such as various types of stromatoporoids (heads and mats) or *Amphipora*, are separated by furrows filled with concentrations of bioclasts mainly located immediately behind each type of colony.

The Swan Hills Member of the Beaverhill Lake Formation, Upper Devonian, was investigated in the Shell Swan Hills 10-17 well, Canadian Rocky Mountain foothills, northern Alberta, Canada (Carozzi, 1961a). In this particular well, a thickness of 438 ft was cored continuously and studied by more than 400 thin sections at approximately a 1-ft vertical interval. This platform is controlled predominantly by constructed buildups of stromatoporoids and *Amphipora* with associated accumulation of bioclasts resulting from their mechanical destruction.

Description of Microfacies

The investigated section shows eight microfacies which are organized by rhythmic alternations. Five of them are basic types, three are subtypes which may be missing locally. Microfacies are described in a landward direction and general shallowing-upward order.

Microfacies 1 (Fig. 13.17A). Dark-to light-colored medium-grained, grain-supported, pelletoidal biocalcarenite with sparite cement. Major bioclasts are bryozoans, brachiopods, pelecypods associated with abundant fragmented ostracods, calcispheres, rare stromatoporoids, and *Amphipora*.

Microfacies 2 (Fig. 13.17B). Dark stromatoporoid-constructed limestone. The framework consists of cabbage-type stromatoporoids with a few *Amphipora*, and interstitial matrix is a poorly sorted fine-grained mud-supported biocalcarenite with a calcisiltite matrix. Bioclasts are from stromatoporoids, *Amphipora*, gastropods, pelecypods, brachiopods, ostracods, and abundant calcispheres.

Microfacies 2a (Fig. 13.17C). Dark fine-grained, poorly sorted, grain-supported biocalcarenite with calcisiltite matrix. Angular bioclasts are predominantly from stromatoporoids and *Amphipora* with abundant calcispheres, a few ostracods, and pellets.

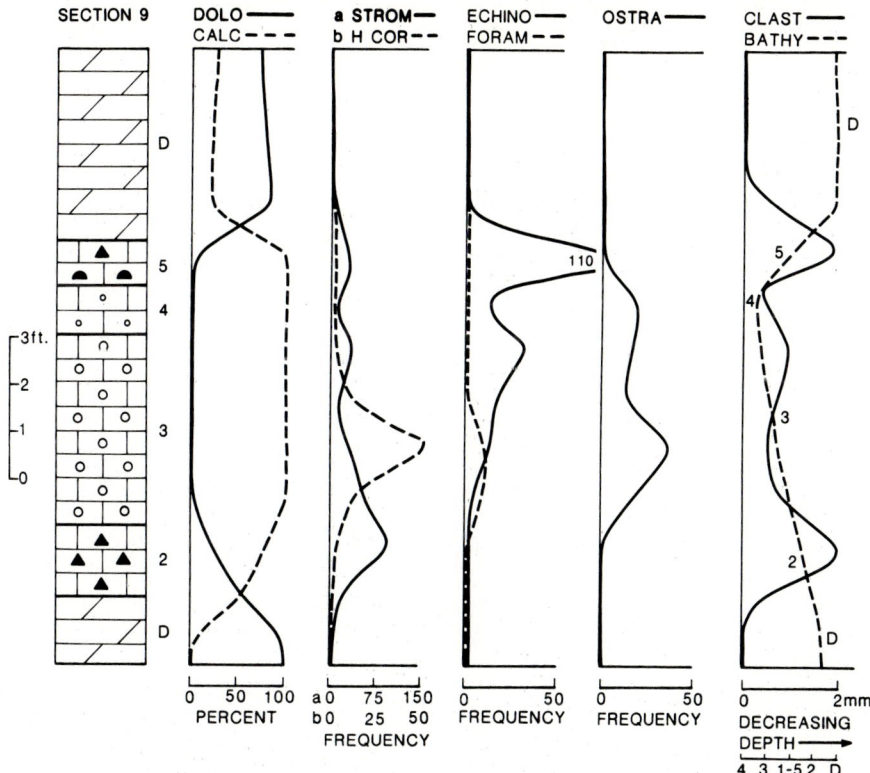

FIGURE 13.14 Buildup of Coral Zone of Jeffersonville Limestone (Middle Devonian), southeast Indiana. Vertical variation of microscopic parameters in field section 9. See Figure 13.9 for location. From Carozzi and Lundwall (1959). Reprinted by permission of the Society of Economic Paleontologists and Mineralogists.

Microfacies 3 (Fig. 13.17D). Dark-colored *Amphipora*-constructed limestone. The framework consists predominantly of branching-type *Amphipora*, either complete or broken and oriented parallel to bedding. Interstitial material is a poorly sorted, grain-supported, pelletoidal biocalcarenite with a calcisiltite matrix. Angular bioclasts are predominantly from *Amphipora* with rare stromatoporoids, ostracods, and local concentrations of calcispheres.

Microfacies 3a (Fig. 13.17E). Dark to light-colored, coarse-grained, grain-supported, pelletoidal biocalcarenite with an association of calcisiltite matrix and irregular patches of sparite cement. Scattered among pellets are larger fragments of *Amphipora*, concentrations of calcispheres, and arenaceous benthic foraminifers.

Microfacies 4 (Fig. 13.17F). Light-colored, very coarse-grained, grain-supported pelletoidal biocalcarenite with sparite cement. Bioclasts, which reach a clasticity of 1.680 mm, are mainly unsorted and consist of angular debris of *Amphipora* associated with larger fragments of branching forms. Complete ostracods and frequent calcispheres are present.

Microfacies 4a (Fig. 13.17G). Light-colored, medium-grained, grain-supported, pelletoidal biocalcarenite with calcisiltite matrix. Bioclasts are angular fragments predominantly of *Amphipora*, mat-type stromatoporoids, gastropods, articulated ostracods, and abundant calcispheres. Among these components are irregularly scattered numerous larger fragments of branching *Amphipora*.

Microfacies 5 (Fig. 13.17H). Light-colored, very fine-grained, grain-supported, pelletoidal biocalcarenite with calcisiltite matrix grading locally into a biocalcisiltite. Intense bioturbation has led to a very irregular distribution of bioclasts and to large patches of sparite cement in burrows. Larger bioclasts are from mat-type stromatoporoids, branching *Amphipora*, entire thin-shelled gastropods, concentrations of articulated smooth ostracods, sponge spicules, and calcispheres.

The Ideal Shallowing-Upward Sequence

The vertical succession of the eight microfacies just described has taken place several times, and in this case, the ideal sequence was actually deposited as such (Fig.

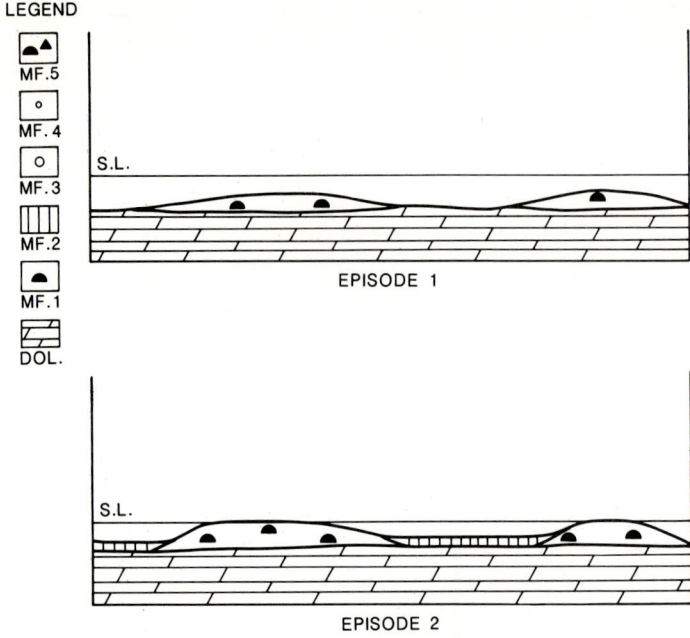

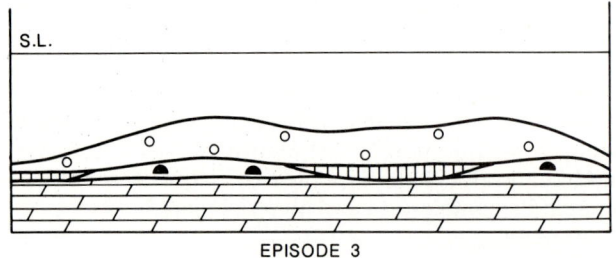

FIGURE 13.15 Buildup of Coral Zone of Jeffersonville Limestone (Middle Devonian), southeast Indiana. Ideal depositional model. See text for discussion. From Carozzi and Lundwall (1959). Reprinted by permission of the Society of Economic Paleontologists and Mineralogists.

13.18). Variation curves of microscopic parameters of the ideal sequence (Fig. 13.19) are first-order envelopes of actual variations.

The Ideal Depositional Model

The platform consists of a succession of three major constructed ridges in a landward direction (Fig. 13.20). It was probably initiated by early colonies of encrusting bryozoans that served as a support for the development of the first ridge of cabbage-type stromatoporoids and its associated destructional microfacies located immediately behind it. Similarly, under this protection and in shallower water was a ridge constructed by more delicate *Amphipora* colonies. The margin of the real platform was the highest-energy environment of mat-type stromatoporoids and *Amphipora* protecting a quiet lagoon with abundant ostracods and sponge colonies and even freshwater areas with *Chara*. Distribution of calcispheres, ostracods, and floated *Chara* stems and oogonia in a seaward direction indicates the effects of tidal channels across the platform. Two gastropod communities are clearly displayed, whereas brachiopods and pelecypods are predominant in deeper water. The proposed model encompasses the entire range of dasyclads and appears to have been within the photic zone, allowing an estimation of maximum water depth at about 150 ft (46 m).

The General Evolution of the Depositional Environments

A general bathymetric interpretation (Fig. 13.21) displays five shallowing-upward sequences during which an *Amphipora*-stromatoporoid community gradually developed a complex carbonate platform, terminated by two deepening-upward sequences corresponding to the drowning of the platform.

13.4 CHARACTERISTIC CONSTITUENTS: CORALS, STROMATOPOROIDS, AND STROMATOLITES

Mature and extensive platforms which have well-developed but narrow bioconstructed margins often support extensive internal areas where many types of biocon-

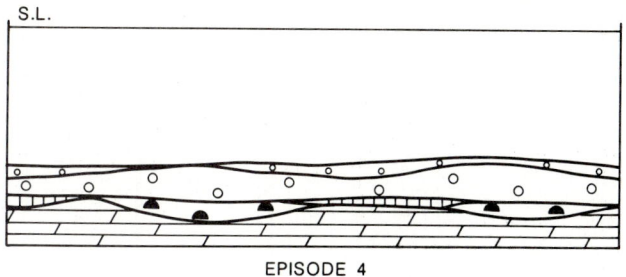

EPISODE 4

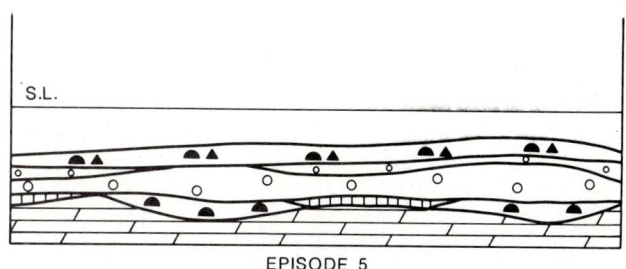

EPISODE 5

FIGURE 13.16 Buildup of Coral Zone of Jeffersonville Limestone (Middle Devonian), southeast Indiana. Ideal depositional model (continued). See text for discussion. From Carozzi and Lundwall (1959). Reprinted by permission of the Society of Economic Paleontologists and Mineralogists.

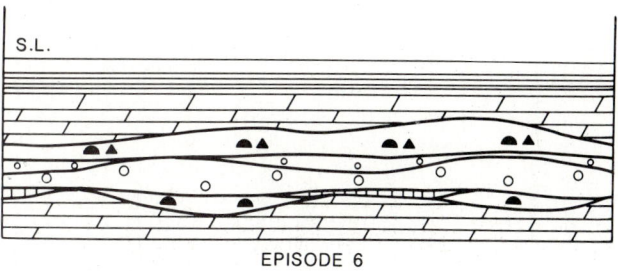

EPISODE 6

structed buildups of corals and stromatoporoids are scattered among mounds and sheets of crinoidal bioclasts. Even more internal areas consist of quiet, but not necessarily restricted lagoons, with stromatolite colonies among pelletoidal and massive micritic muds.

The Traverse Group (Givétian) of the northern part of the southern peninsula of Michigan (Roche and Carozzi, 1970) displays spectacular sections of abundant but relatively small constructed and hydrodynamic buildups with surrounding sediments in a series of active and abandoned quarries. These buildups range in length from 10 to 60 m and in height from 4 to 5 m. Adjacent carbonates designated as "interbuildup" range from open marine subtidal biocalcisiltites to biocalcarenites to restricted intertidal to supratidal calcisiltites and calcilutites which display in the most isolated situations desiccation breccias, collapse breccias, and pseudomorphs of evaporitic minerals. The petrographic study was done with 730 thin sections made from samples collected from each locality (Fig. 13.22) at an average vertical interval of 15 cm.

The paleogeography of the Traverse Group is not sufficiently known to establish the precise location of these buildups with respect to a major shelf edge barrier which is assumed to have existed in the subsurface immediately south of the investigated area. Periodic influx of clay minerals, fine detrital quartz, plant debris, and *Chara* stems indicate influences from low-lying land areas to the north. The investigated buildups appear to belong to the inner part of the barrier itself and to some more internal areas.

Description of Microfacies

The microfacies of the investigated sections fall into two megagroups. The first consists predominantly of interbuildup subtidal calcarenites which were deposited around and above the relatively small coral-stromatoporoid bioconstructed cores; all these microfacies are therefore associated under the designation of "open marine interbuildup-buildup," letter "I" followed by a numeral. The second group consists predominantly of lagoonal calcisiltites and calcilutites which were deposited around and above small stromatolitic buildups; all these microfacies are therefore associated under the designation of

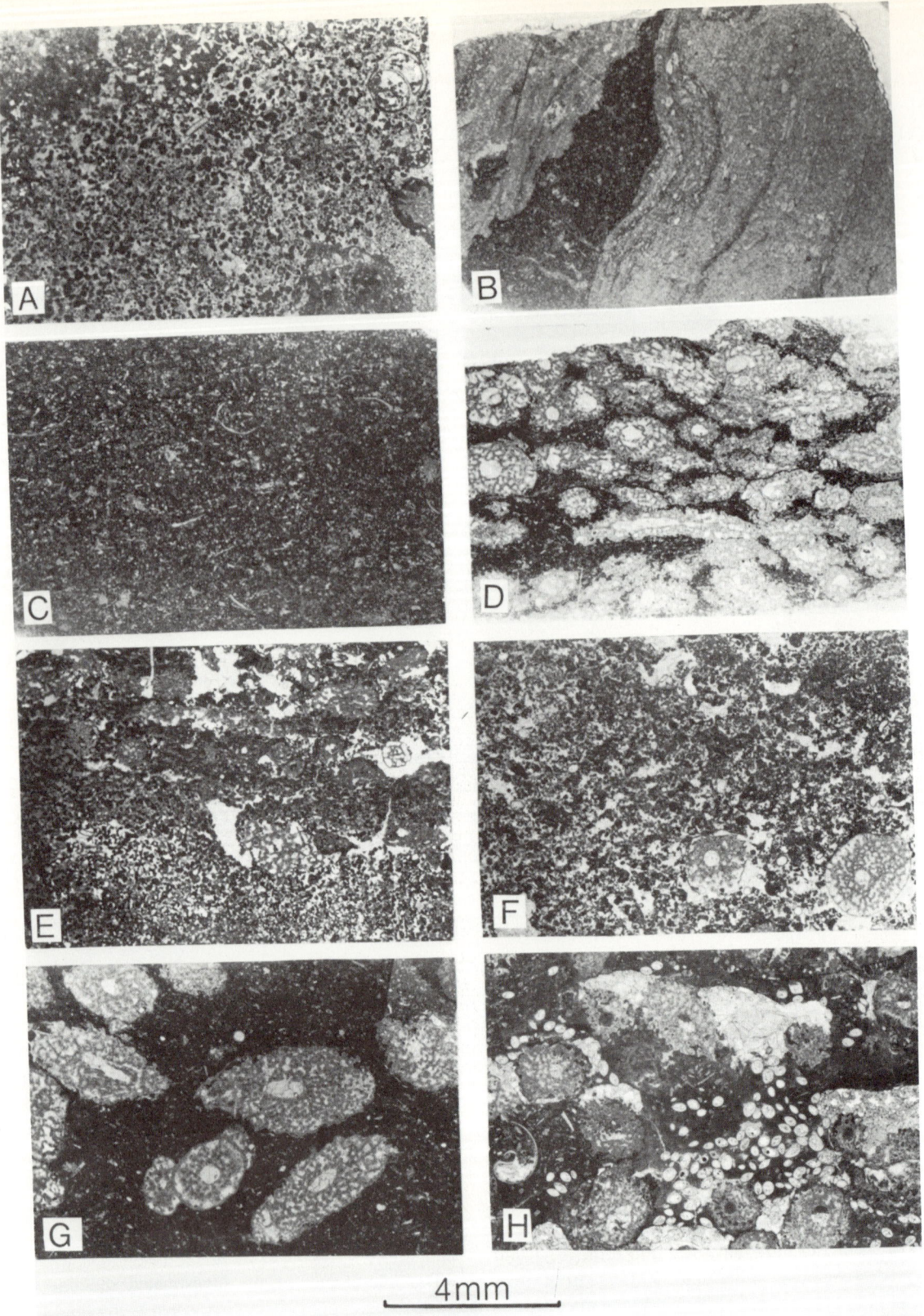

FIGURE 13.17 Beaverhill Lake Formation (Upper Devonian), Canadian Rocky Mountain foothills, north Alberta. Typical microfacies. A. Microfacies 1. B. Microfacies 2. C. Microfacies 2a. D. Microfacies 3. E. Microfacies 3a. F. Microfacies 4. G. Microfacies 4a. H. Microfacies 5. See text for detailed descriptions. All photomicrographs: plane-polarized light. From Carozzi (1961a). Reprinted by permission of the Society of Economic Paleontologists and Mineralogists.

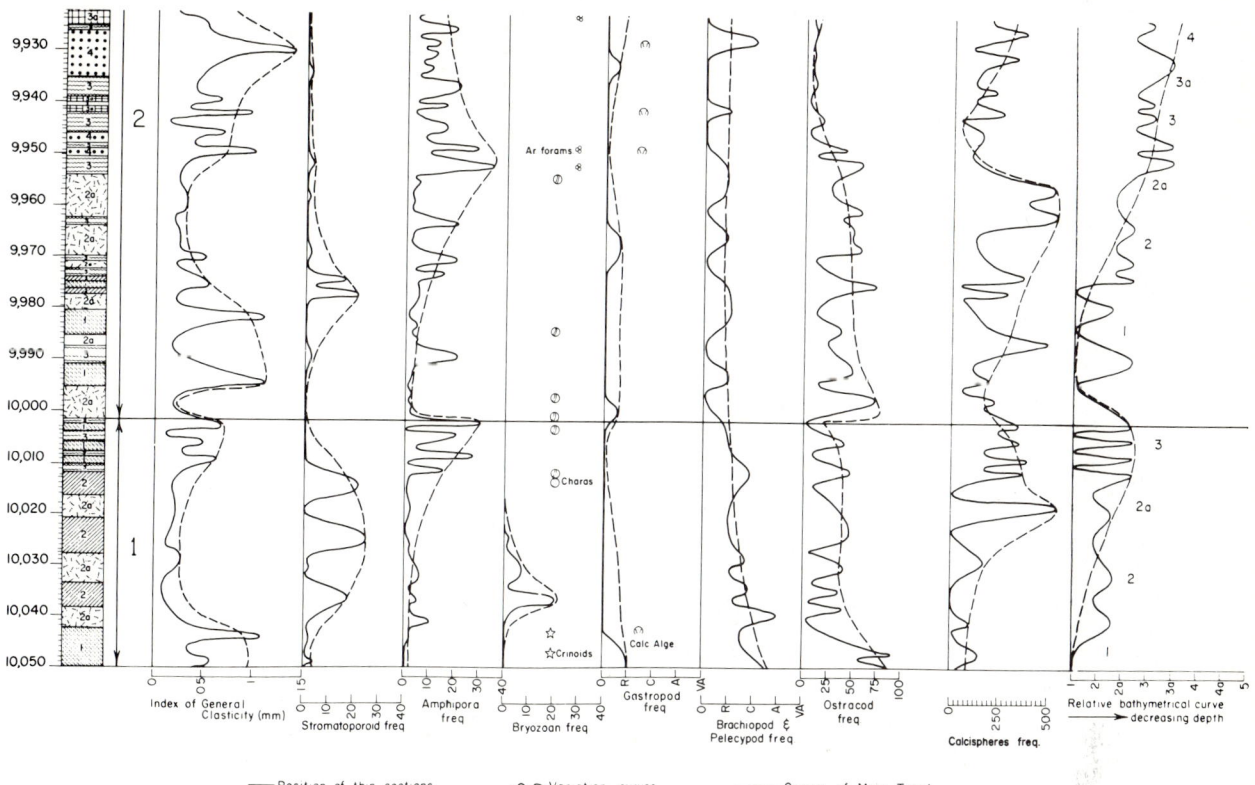

FIGURE 13.18 Beaverhill Lake Formation (Upper Devonian), Canadian Rocky Mountain foothills, north Alberta. Typical example of variation of microscopic parameters during two successive shallowing-upward sequences. From Carozzi (1961a). Reprinted by permission of the Society of Economic Paleontologists and Mineralogists.

"lagoonal interbuildup-buildup," letter "L" followed by an appropriate numeral.

Open Marine Interbuildup-Buildup Environment

The eight microfacies of this group account for 68% of all thin sections examined. They are described in order of increasing energy of the environment of deposition indicated by smaller to larger numbers.

Microfacies I-1 (Figs. 13.23A,B,C, 13.24A,B). Dark brown bituminous biocalcisiltite, often burrowed and bioturbated with up to 10% of irregularly scattered sand-size bioclasts which are predominantly large crinoid columnals associated with intact bryozoan fronds, whole brachiopods, and occasionally entire delicate branching corals (*Depasophyllum adentum*). Burrows are usually filled by coarser biocalcisiltite, often altered into pseudosparite by neomorphism (Fig. 13.23A), or in places by geopetal vadose silt overlain by sparite cement (Fig. 13.23B). Locally, this microfacies grades into a well-laminated variety with fine bioclasts of crinoids, brachiopods, and monaxonic sponge spicules (Fig. 13.24B).

Microfacies I-2 (Figs. 13.23D, 13.25B). Dark pelletoidal and bioturbated biocalcisiltite with scattered sand-size bioclasts of crinoids, brachiopods, and ostracods, associated with concentrations of calcispheres (Fig. 13.25B) and monaxonic sponge spicules (Fig. 13.23D). Pellets, predominantly fecal, are uniformly sized, but irregularly shaped small intraclasts occur also. Bioturbation is responsible for irregular patches of interparticle microsparite cement.

Microfacies I-3 (Figs. 13.23E, 13.24C, 13.25C, 13.26D). Dark and coarse-grained, mud-supported biocalcarenite with bioturbated calcisiltite matrix, often displaying a fluidal texture (Fig. 13.25C). Major bioclasts are crinoids followed in decreasing number by bryozoans, corals, brachiopods, and ostracods. Locally, the bioturbated matrix may show intense neomorphism into pseudomicrosparite (Fig. 13.23E).

Microfacies I-4 (Figs. 13.24D, 13.25D, 13.26E). Fine grain-supported to pressure-welded pelletoidal biocalcarenite with calcisiltite matrix. Intense

368 Part III / Carbonate Platforms

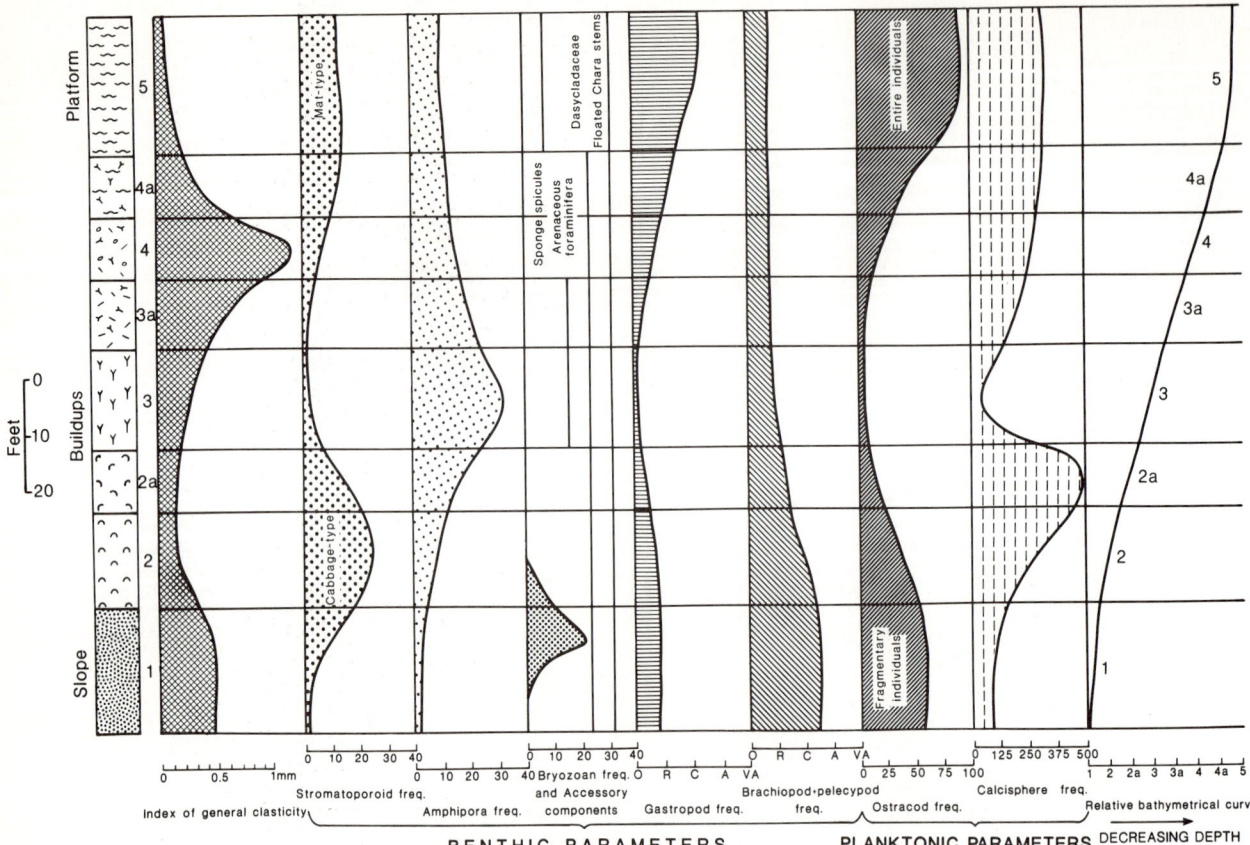

FIGURE 13.19 Beaverhill Lake Formation (Upper Devonian), Canadian Rocky Mountain foothills, north Alberta. Ideal shallowing-upward sequence. From Carozzi (1961a). Reprinted by permission of the Society of Economic Paleontologists and Mineralogists.

bioturbation is shown by the great variability in concentration and orientation of the various bioclasts. Crinoids and bryozoans predominate over ostracods and coral debris. Pellets, mostly of fecal origin, are concentrated locally in relation to bioturbation (Fig. 13.24D).

Microfacies I-5 (Figs. 13.23F,G, 13.24E, 13.25E,F, 13.26F, 13.27F, 13.28A,B,C, 13.29G,H). Coarse grain-supported to pressure-welded pelletoidal biocalcarenite with bituminous calcisiltite matrix, often altered by neomorphism into irregularly scattered patches of pseudomicrosparite. Bioclasts of bryozoans and crinoids predominate over brachiopods and corals (*Favosites* and *Thamnopora*). Fecal pellets are associated with a small amount of intraclasts, and calcispheres are rare. This microfacies contains large heads of corals (*Hexagonaria*) and of stromatoporoids which have rolled out of growth position and are frequently upside down.

Microfacies I-6 (Figs. 13.23H, 13.24F, 13.25G,H, 13.26G, 13.27G,H, 13.28D). Coarse grain-supported to pressure-welded pelletoidal biocalcarenite with association of bituminous calcisiltite matrix and interstitial patches of sparite cement. Major bioclasts are crinoids, corals (*Favosites*, *Hexagonaria*), bryozoans, and stromatoporoids. This microfacies contains reworked corals and stromatoporoids as the previous one, but also occurs frequently as interstitial material inside the framework of bioconstructed buildups.

Microfacies I-7 (Fig. 13.26H). Coarse grain-supported to pressure-welded intraclastic-algal calcarenite with sparite cement. Subrounded intraclasts of dark micrite and rounded pellets are associated with abundant bioclasts of *Nuia*, ostracods, and calcispheres. This microfacies displays lagoonal affinities but is kept in this group because of the greater energy of its environment of deposition.

Microfacies I-8 (Fig. 13.30). Framework constructed by superposed heads and mats of stromatoporoids with associated colonial and isolated corals (*Hexagonaria*, *Favosites*). Interstitial material consists of coarse fragments of the above colonial organisms set in microfa-

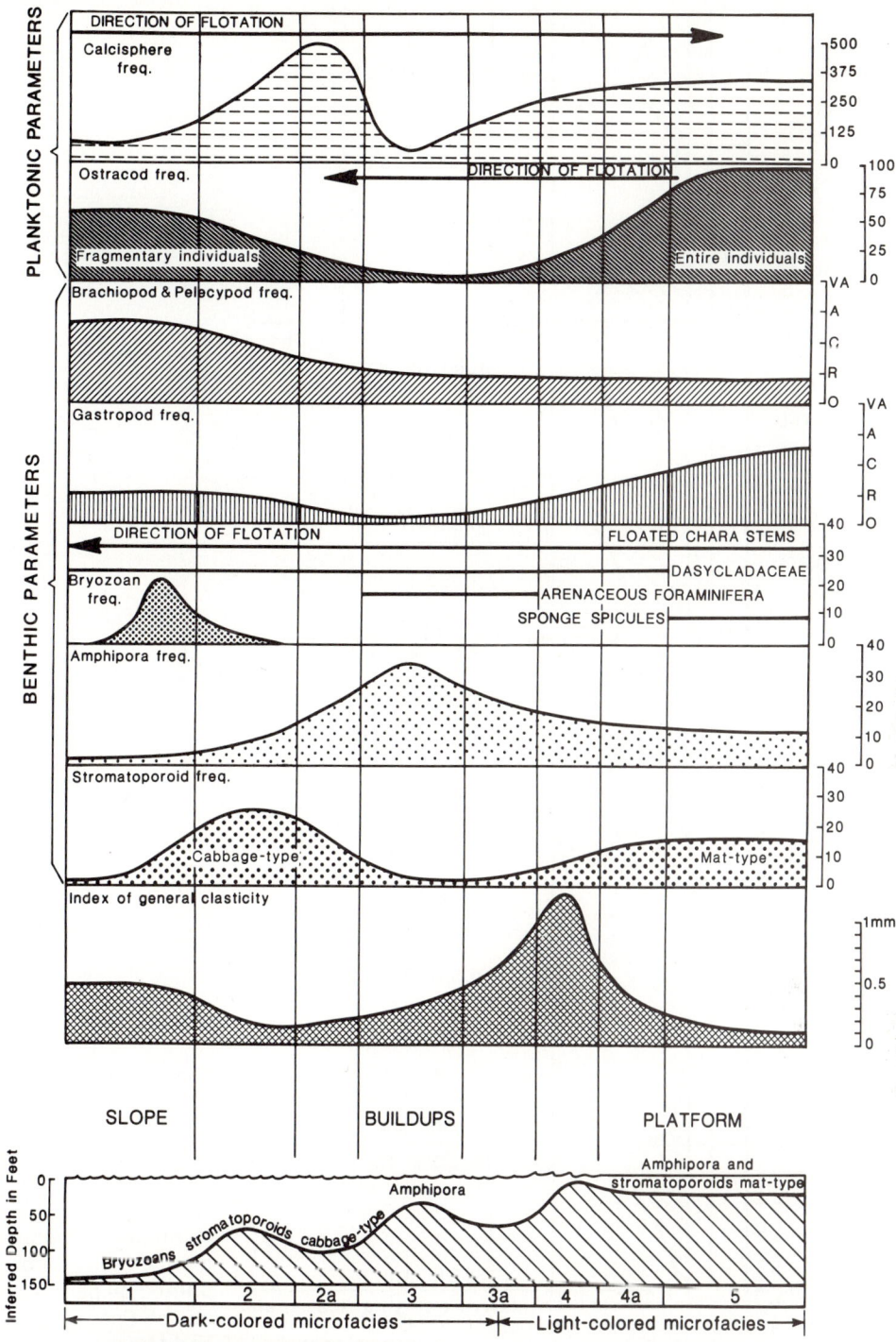

FIGURE 13.20 Beaverhill Lake Formation (Upper Devonian), Canadian Rocky Mountain foothills, north Alberta. Ideal depositional model. From Carozzi (1961a). Reprinted by permission of the Society of Economic Paleontologists and Mineralogists.

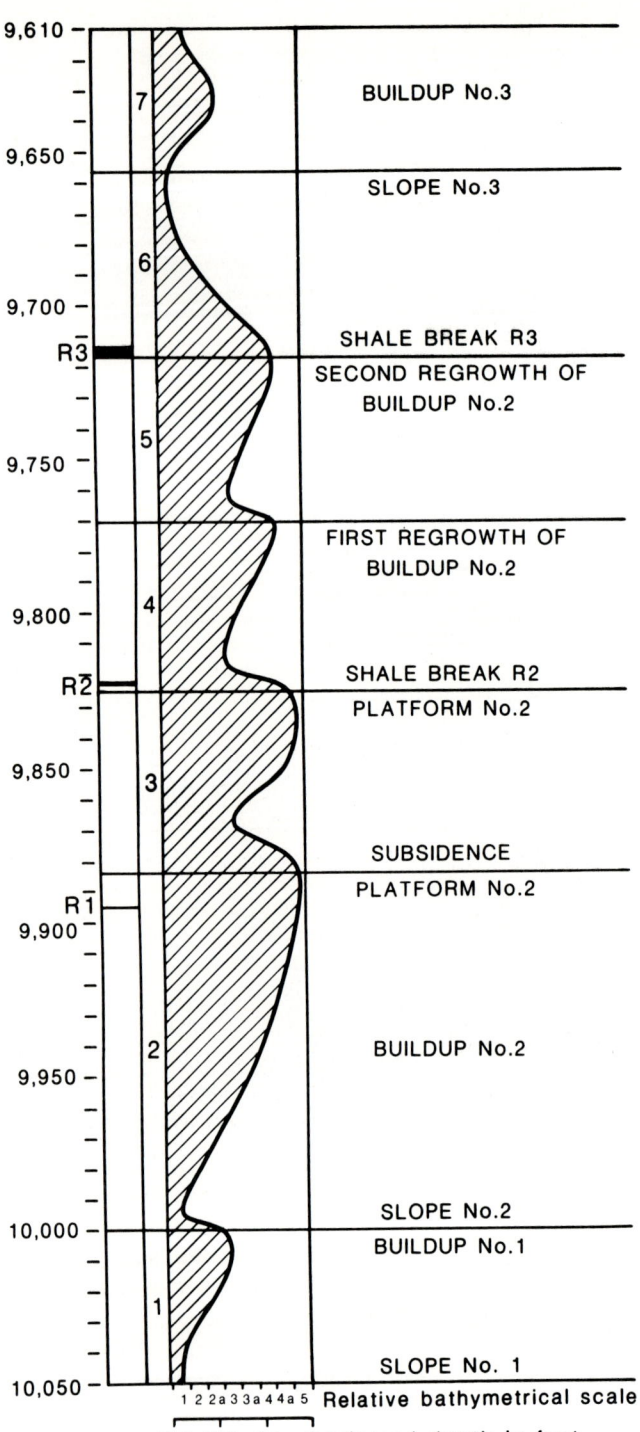

FIGURE 13.21 Beaverhill Lake Formation (Upper Devonian), Canadian Rocky Mountain foothills, north Alberta. General vertical evolution of depositional environments. From Carozzi (1961a). Reprinted by permission of the Society of Economic Paleontologists and Mineralogists.

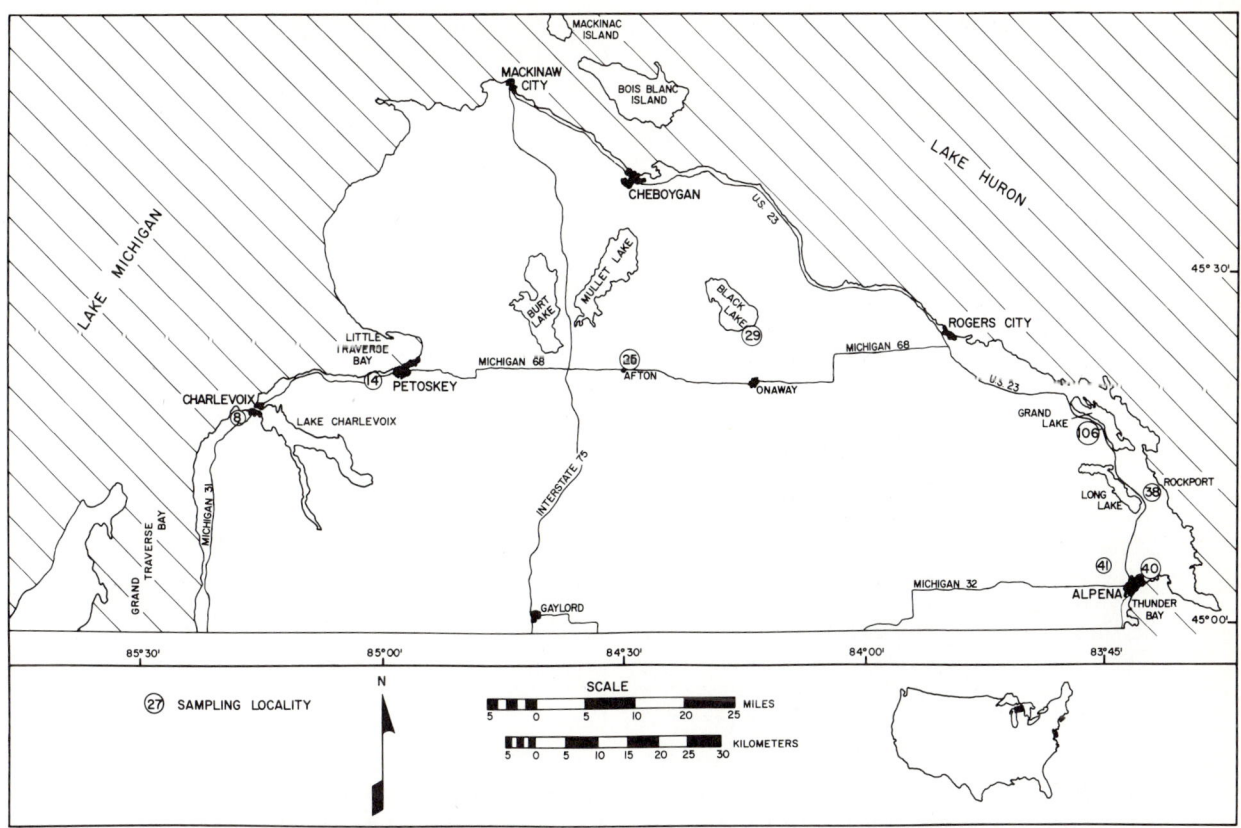

FIGURE 13.22 Traverse Group (Givétian), northern part of southern peninsula of Michigan. Location map of investigated sections. From Roche and Carozzi (1970). Reprinted by permission of Société Nationale des Pétroles d'Aquitaine.

cies I-6. Most of the constructed buildups consist of this association of two microfacies.

Lagoonal Interbuildup-Buildup Environment

Four microfacies of this group account for the remaining 32% of all thin sections examined. These microfacies are also described in order of increasing energy of the environment of deposition indicated by smaller to larger numbers with the qualification that ranges of energy are much smaller in these conditions than in the open marine environment.

Microfacies L-1 (Figs. 13.26A, 13.27A,B, 13.29A). Massive to finely pelletoidal calcilutite to calcisiltite with scattered calcispheres and ostracods and minute unidentifiable bioclasts. Common fenestral texture filled by sparite or dolosparite replacing the former (Fig. 13.27A); some fenestrae have a subrectangular shape which might indicate an origin from dissolution of anhydrite crystals (Fig. 13.26A).

Microfacies L-2 (Figs. 13.25A, 13.26B, 13.27C,D, 13.28E,G,H). Dominantly pelletoidal to fine intraclastic calcisiltite with abundant *Parathurammina* (Fig. 13.28G), calcispheres (Fig. 13.25A), ostracods belonging entirely to *Welleria aftonensis* (Fig. 13.26B), occasionally associated with a few minute crinoidal bioclasts. Fenestral texture with large development of generally horizontal cavities, combined with abundant randomly oriented smaller ones, is characteristic of this microfacies. Fenestrae are filled with microsparite to equant coarse sparite mosaic.

Microfacies L-3 (Figs. 13.26C, 13.29B,C,D). Stromatolitic mats alternating with trapped layers of pelletoidal to intraclastic calcisiltite with rare scattered ostracods and calcispheres and displaying minute interparticle patches of microsparite cement. The laminated texture is interrupted by fenestral cavities generally horizontal and well interconnected by vertical gas-escape tubules (Fig. 13.26C), subrectangular molds of anhydrite crystals, and square molds of halite crystals are very frequent (Fig. 13.29C,D).

372 Part III / Carbonate Platforms

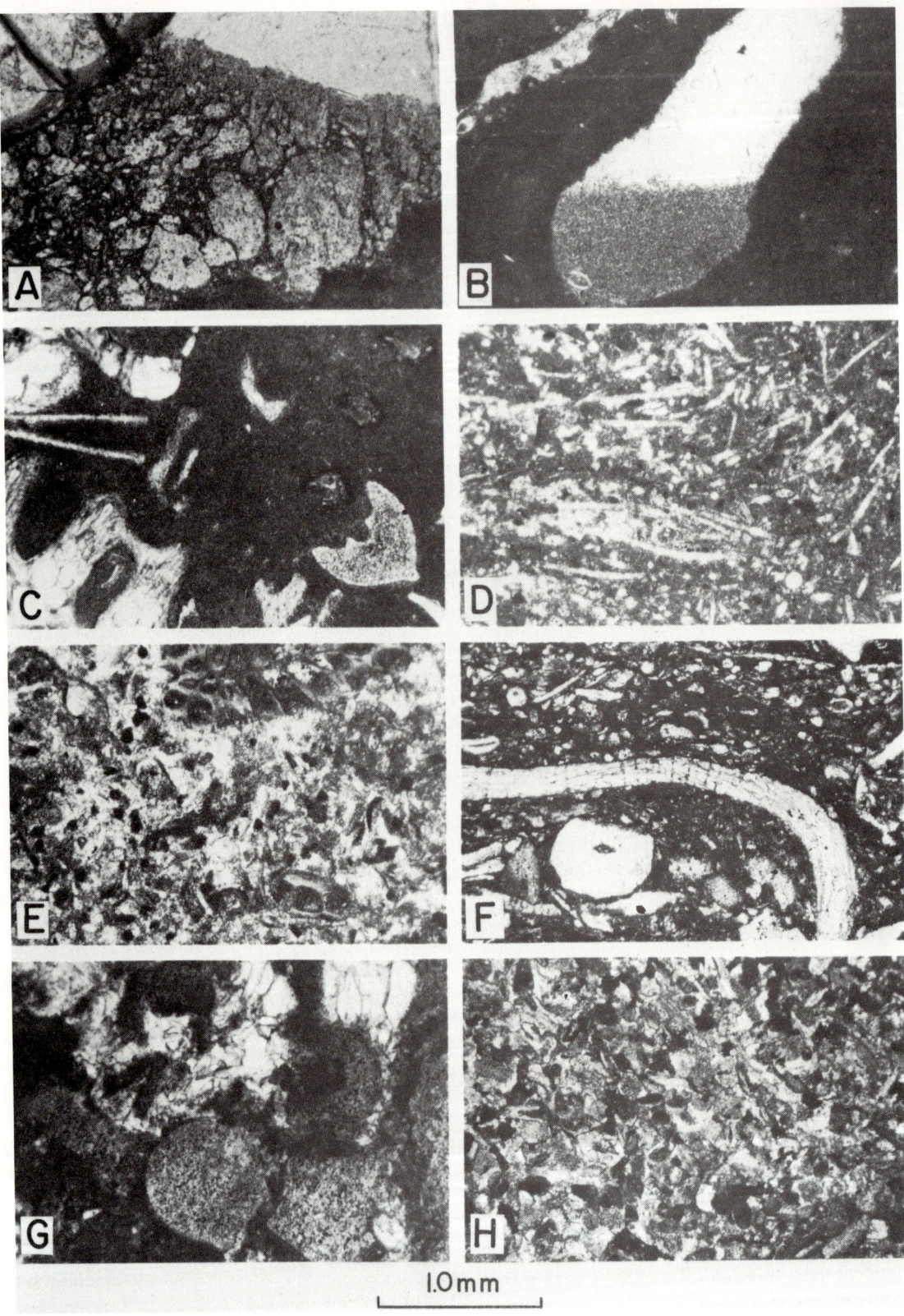

FIGURE 13.23 Traverse Group (Givétian), northern part of southern peninsula of Michigan. Typical microfacies of Four Mile Dam and Norway Point Formations. Locality 41. A. Calcisiltite with scattered organic debris (I-1), detail of burrowed calcisiltite, organically com-

Microfacies L-4 (Figs. 13.27E, 13.28F, 13.29E,F). Stromatolitic and intraclastic calcarenite to calcirudite with cavity-filling sparite cement and pelletoidal calcisiltite matrix. This microfacies results from fragmentation and redeposition of stromatolitic algal mats or even stromatoporoids (Fig. 13.27E), and of desiccated calcisiltite and calcilutite fragmented in place or transported as flat pebbles (Fig. 13.29E,F). Some textural aspects of the calcirudite varieties with graded-bedding indicate storm action (Fig. 13.28F).

The description above of "open marine interbuildup-buildup" and of "lagoonal interbuildup-buildup" environments raises the issue whether prolific growth of colonial organisms over an extensive platform can lead to a certain amount of local environmental restriction which would be even more accentuated in the more internal so-called lagoonal conditions. Geochemical studies (Carozzi, 1971b) showed that the "open marine interbuildup-buildup" environment is indeed open marine reducing (350 ppm of B) with slight restrictions in some areas between and around well-developed buildups (440 ppm of B). Concentrations of benthic organic matter up to 1.6% organic carbon on total rock and related organophilic trace metals (Ni, Cu, and Mo) are normal for such conditions.

The "lagoonal interbuildup-buildup" is essentially oxidizing and protected but rarely restricted hypersaline (500 ppm of B). Common boron contents of 230 to 300 ppm are related to a decrease of Sr content, itself due to the disappearance of the constructing and benthic organisms. Therefore, no appreciable increase of salinity took place in these lagoons because evaporation was not very active. Rare occurrences of molds of anhydrite and halite are due to short episodes of desiccation of carbonate muds. Lagoonal microfacies contain little organic matter (0.25% organic carbon on total rock) and show no enrichments in trace metals. The only exception is Locality 8, which was a deeper lagoon adjacent to buildups where some organic matter and associated trace metals were preserved.

Both open marine and lagoonal microfacies can display abundant supplies of detrital organic matter of vegetal origin (organic carbon content on total rock up to 2.6%). This influx of plant material, not accompanied by freshwater influences, indicates a long-distance transport by slow currents circulating between buildups and across lagoons. This situation confirms paleogeographical and microfacies data which indicate that an appreciable distance separated the investigated large platform domain of the Traverse Group from its surrounding continental margins. It also implies large-scale destruction of the vegetation cover of the low-lying lands surrounding the Michigan Basin.

Description of Sections and Ideal Depositional Models

Description of individual field sections shows detailed vertical variation of the megascopic and microscopic parameters, numerous and complex shallowing-upward sequences, and related ideal depositional models which in most cases are based on geometrical relationships visible on quarry walls. Localities are numbered according to a standard list maintained by the University of Michigan and the Michigan Geological Survey. They are described in stratigraphic order and from east to west (Fig. 13.22).

Rockport Quarry (Locality 38)

This section (Figs. 13.31, 13.32) grades upward from an extensive buildup constructed by stromatoporoids and subordinated corals with normal marine fauna, through a transitional zone to lagoonal conditions with

FIGURE 13.23 (*continued*) minuted crinoidal debris fill lower portion of a burrow next to the exterior wall of the coral *Depasophyllum adentum*. B. Calcisiltite with scattered organic debris (I-1), fine silt fills the bottom of a vertically oriented burrow and sparite occurs in upper part. C. Concentration of coarse crinoid and bryozoan debris in an otherwise featureless calcisiltite (I-1). D. Pelletoidal calcisiltite with numerous sponge spicules (I-2). E. Mud-supported pelletoidal calcarenite with partially recrystallized matrix (I-3). F. Grain-supported pelletoidal-crinoidal calcarenite with calcisiltite matrix (I-5) showing large fragment of punctate brachiopod. G. Coarse grain-supported crinoidal calcarenite with partially recrystallized muddy matrix (I-5). H. Pressure-welded pelletoidal-crinoidal calcarenite (I-6). See text for detailed descriptions. All photomicrographs: plane-polarized light. From Roche and Carozzi (1970). Reprinted by permission of Société Nationale des Pétroles d'Aquitaine.

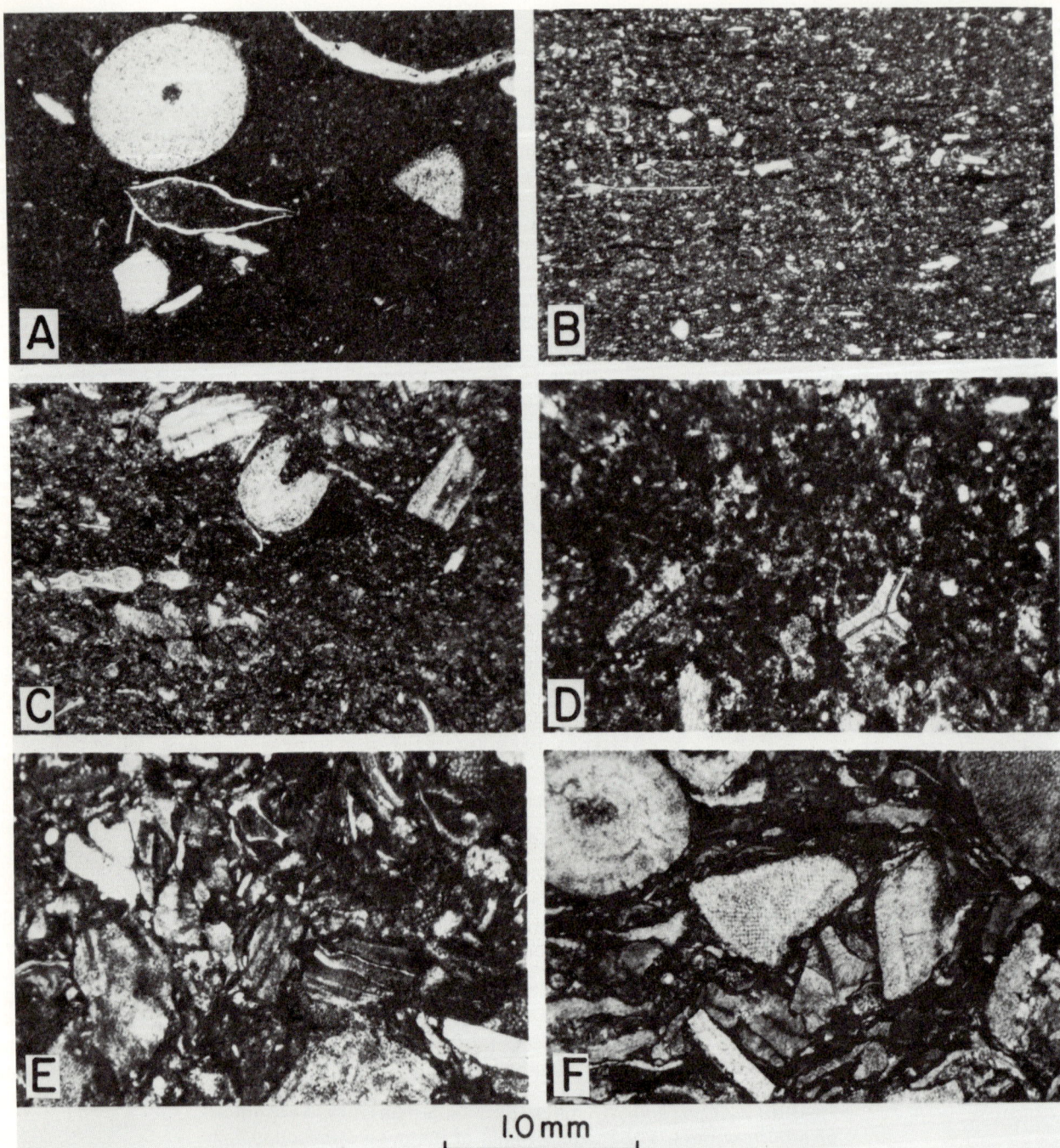

FIGURE 13.24 Traverse Group (Givétian), northern part of southern peninsula of Michigan. Typical microfacies of Gravel Point Formation. Locality 14. A. Calcisiltite with scattered debris of crinoids and brachiopods (I-1). B. Well-laminated bituminous calcisiltite with fine debris of crinoids, brachiopods, and sponge spicules (I-1). C. Mud-supported calcarenite with crinoids, brachiopods, and bryozoans (I-3). D. Fine grain-supported pelletoidal calcarenite with calcisiltite matrix (I-4). E. Coarse grain-supported bryozoan-coral calcarenite with calcisiltite matrix (I-5). F. Bituminous pressure-welded crinoid-coral-bryozoan calcarenite (I-6). See text for detailed descriptions. All photomicrographs: plane-polarized light. From Roche and Carozzi (1970). Reprinted by permission of Société Nationale des Pétroles d'Aquitaine.

abundance of ostracods, calcispheres, and intense production of fecal pellets. This shallowing-upward evolution occurs by means of a number of small oscillations.

The ideal depositional model (Fig. 13.32) shows the bioconstructed buildup consisting of interbedded microfacies I-6 and I-8 with bioclastic aprons of I-5 toward open marine conditions and I-4 toward the lagoon. In the latter, L-4 contains only stromatoporoid clasts while interbedded L-1 and L-2 display dolosparite in fenestral textures.

Grand Lake Section (Locality 106)

This section (Fig. 13.33) shows a substratum of open marine pelletoidal biocalcarenite I-4 which becomes gradually colonized upward by *Hexagonaria*. The latter prepared the establishment of a small buildup constructed by mats and tumbled heads of stromatoporoids (I-8), with interstitial bituminous calcarenite (I-6). The buildup was eventually encroached upon and buried by the same crinoid-bryozoan calcarenites over which it developed. The ideal depositional model (Fig. 13.33) shows the general geometrical relationships of the various microfacies.

Quarry of Huron Portland Cement, Alpena (Locality 40)

This section (Fig. 13.34) begins with lagoonal conditions and then displays upward several oscillations during which all the open marine microfacies I-2 to I-6 combine in the development of hydrodynamic buildups at times covered by thin buildups constructed by stromatoporoid mats and *Hexagonaria* colonies. As shown in the ideal depositional model (Fig. 13.35), the two bioclastic buildups also separated the open marine from the lagoonal environments.

Four Mile Dam Section (Locality 41)

This section (Fig. 13.36) begins with a mound of greenish bituminous calcisiltite (microfacies I-1) containing delicately branching corals of the genus *Depasophyllum adentum*. This mud-accumulated buildup is overlain by coarse-grained pelletoidal biocalcarenites (microfacies I-5) with intercalated stromatoporoid mats. Upward within the same sequence of calcarenites, energy decreases and increases again. The ideal depositional model (Fig. 13.36) shows that the various bioclastic sediments, which overlapped and eventually buried the mud buildup, are separated by short and local submarine erosional surfaces.

Black Lake Section (Locality 29)

This section (Fig. 13.37) shows prevailing lagoonal conditions with abundance of ostracods, calcispheres, pellets, and episodes of desiccations of stromatolite mats leading to flat-pebble breccias and locally graded-bedded calcirudites, indicating storm action. One thin buildup of stromatoporoid mats and coral debris in the upper part of the section indicates the proximity of open marine conditions. The ideal depositional model (Fig. 13.37) shows the geometrical relationships between the various microfacies.

Campbell Quarry, Afton (Locality 25)

The section of this quarry (Fig. 13.38) shows a vertical transition from a typical lagoonal environment with calcispheres, ostracods, pellets, and molds of evaporite crystals with interbedded stromatolitic buildups to open marine conditions where the spectrum of pelletoidal biocalcarenites (I-3 to I-6) forming a hydrodynamic buildup culminates with intraclastic-*Nuia* calcarenites (microfacies I-7). The ideal depositional model (Fig. 13.38) indicates that in this case also a hydrodynamic buildup separates open marine from lagoonal environments.

Quarry of Penn-Dixie Cement Co., Petoskey (Locality 14)

This section (Fig. 13.39) consists entirely of gentle oscillations involving almost the entire spectrum of open marine bituminous pelletoidal calcisiltites and biocalcarenites (I-1 to I-6) with interbedded thin buildups constructed by either stromatoporoids or *Hexagonaria*. The entire sequence dips generally 10° to the east and the ideal depositional model (Fig. 13.40) expresses an interpretation as an open marine slope, possibly draping over some large buried buildup.

Quarry of Medusa Portland Cement Co., Charlevoix (Locality 8)

This section (Fig. 13.41) displays complex oscillations between lagoonal conditions with large stromatolitic buildups and associated destructive intraclastic and desiccation microfacies L-4, and open marine conditions with a hydrodynamic buildup consisting of bioclastic to intraclastic-*Nuia* calcarenites (I-4 to I-7). As in previous cases, thin buildups constructed by stromatoporoid mats and *Hexagonaria* are intercalated among biocalcarenites. The ideal depositional model (Fig. 13.41) shows that the hydrodynamic buildup separates open marine from lagoonal environments and that both types of sediments display complex interfingering.

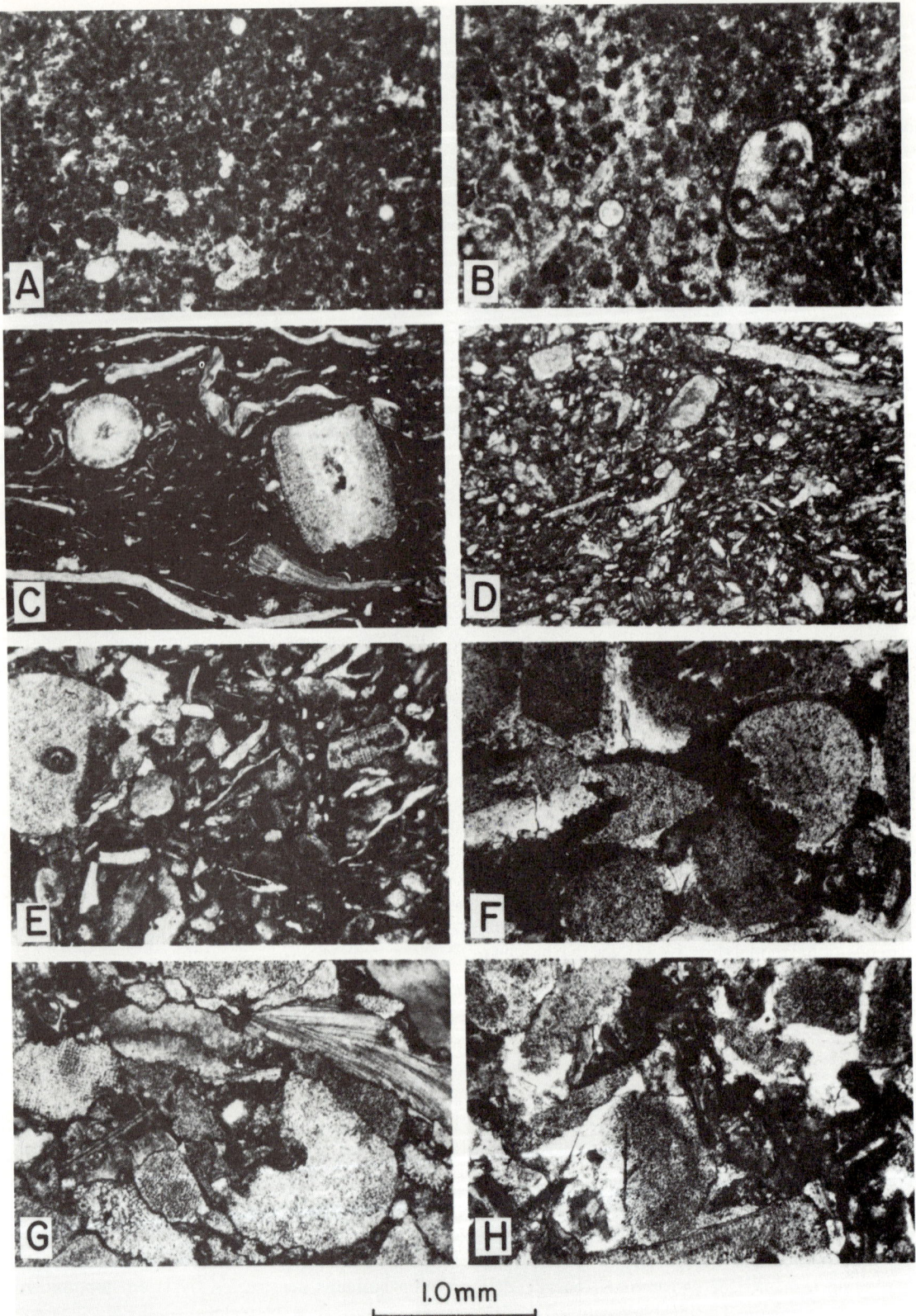

FIGURE 13.25 Traverse Group (Givétian), northern part of southern peninsula of Michigan. Typical microfacies of Newton Creek and Alpena Formations. Locality 40. A. Pelletoidal calcisiltite with calcispheres and occasional debris of crinoids (L-2). B. Pelletoidal calcisil-

13.5 CHARACTERISTIC CONSTITUENTS: *DONEZELLA* AND CRINOIDS

Platforms facing very agitated seas may be preceded oceanward by broad slopes which provide, at the wave base, favorable conditions for the development of prolific algal-constructed buildups (*Donezella*, for instance). Under the destructive action of waves, they produce important accumulations of bioclasts located as talus immediately behind the constructed mounds. Consequently, the slope is divided into a deeper outer section and a shallower inner section where algae reappear just before the true edge of the platform. The platform is characterized by greater energy crinoidal and oolitic deposits.

The Atokan Limestone (Middle Pennsylvanian) of the western margin of the Delaware Basin in Reeves County, Texas (Von Bergen, 1985; Carozzi and Von Bergen, 1987), was investigated by means of three cores from a gas-producing area. More than 375 thin sections were studied petrographically from samples taken at an average vertical interval of 1 ft (0.30 m) for a total thickness of 400 ft (120 m). Samples were taken at larger intervals (2 to 4 ft, 0.61 to 1.22 m) within uniform microfacies, and at much smaller intervals (as small as 0.1 ft, 3 cm) at boundaries between microfacies and within microfacies showing significant changes in components and textures. Porosity was measured in 350 cylindrical plugs drilled as close as possible to the corresponding thin section to permit comparison of petrographic features with porosity values. This study emphasizes mesogenetic diagenesis and, in particular, types, frequency, and amplitude of stylolites as critical factors in the generation of gas reservoirs of commercial interest.

Stylolites were carefully measured for both frequency and degree of development or amplitude. Frequency, relative to the entire thin section, is expressed as rare (1), present (2 to 5), common (6 to 15), and abundant (>15). Amplitude is the vertical distance measured from peak to adjacent valley of the largest observed stylolite per thin section and is expressed as low (<0.25 mm), medium (0.25 to 1.0 mm), and high (>1.0 mm). Stylolites were also divided into two major categories, sutured and nonsutured. Nonsutured or microstylolites display a smooth and sinuous peak-valley transition with amplitudes consistently low (<50 μm). Nonsutured stylolites are found typically in concentrations referred to as swarms. In contrast, sutured stylolites always show an angular or jagged peak-valley relationship regardless of the amplitude which may range from 50 μm to over 1.0 mm. Sutured stylolites are not found concentrated into swarms.

Description of Microfacies

Nine distinct microfacies were recognized in a depositional model which ranges from outer slope to lagoonal conditions and displays a well-defined morphological and environmental zonation according to which microfacies are described in a landward direction and general shallowing order.

Outer Slope

Microfacies 1 (Fig. 13.42A). Argillaceous, foraminiferal biocalcisiltite with common monaxonic sponge spicules, calcispheres, and crinoidal bioclasts. Other rare bioclasts are *Endothyra*, *Tuberitina*, ammodiscids, paleotextulariids, bryozoans, echinoid spines, ostracods, trilobites, fusulinids, and gastropods. The groundmass is bioturbated locally and displays rare fecal pellets and silt-size grains of detrital quartz.

FIGURE 13.25 (*continued*) tite with calcispheres, algae, and benthic debris (I-2). This section displays features which are transitional between lagoonal and interbiohermal deposition. C. Black and very bituminous mud-supported calcarenite (I-3), "so-called shale" at base of Alpena Formation, showing fluidal texture. D. Fine grain-supported to pressure-welded calcarenite of crinoid and bryozoan debris (I-4). E. Grain-supported crinoid-bryozoan calcarenite with calcisiltite matrix (I-5). F. Coarse grain-supported crinoidal calcarenite (I-5) with calcisiltite matrix partially recrystallized. G. Coarse pressure-welded calcarenite of crinoids, bryozoans, and corals (I-6); sausage-shaped element with medial canal (upper center) is a portion of a *Favosites* sp. wall. H. Coarse, pressure-welded crinoid-bryozoan calcarenite (I-6) showing secondary rims of calcite. See text for detailed descriptions. All photomicrographs: plane-polarized light. From Roche and Carozzi (1970). Reprinted by permission of Société Nationale des Pétroles d'Aquitaine.

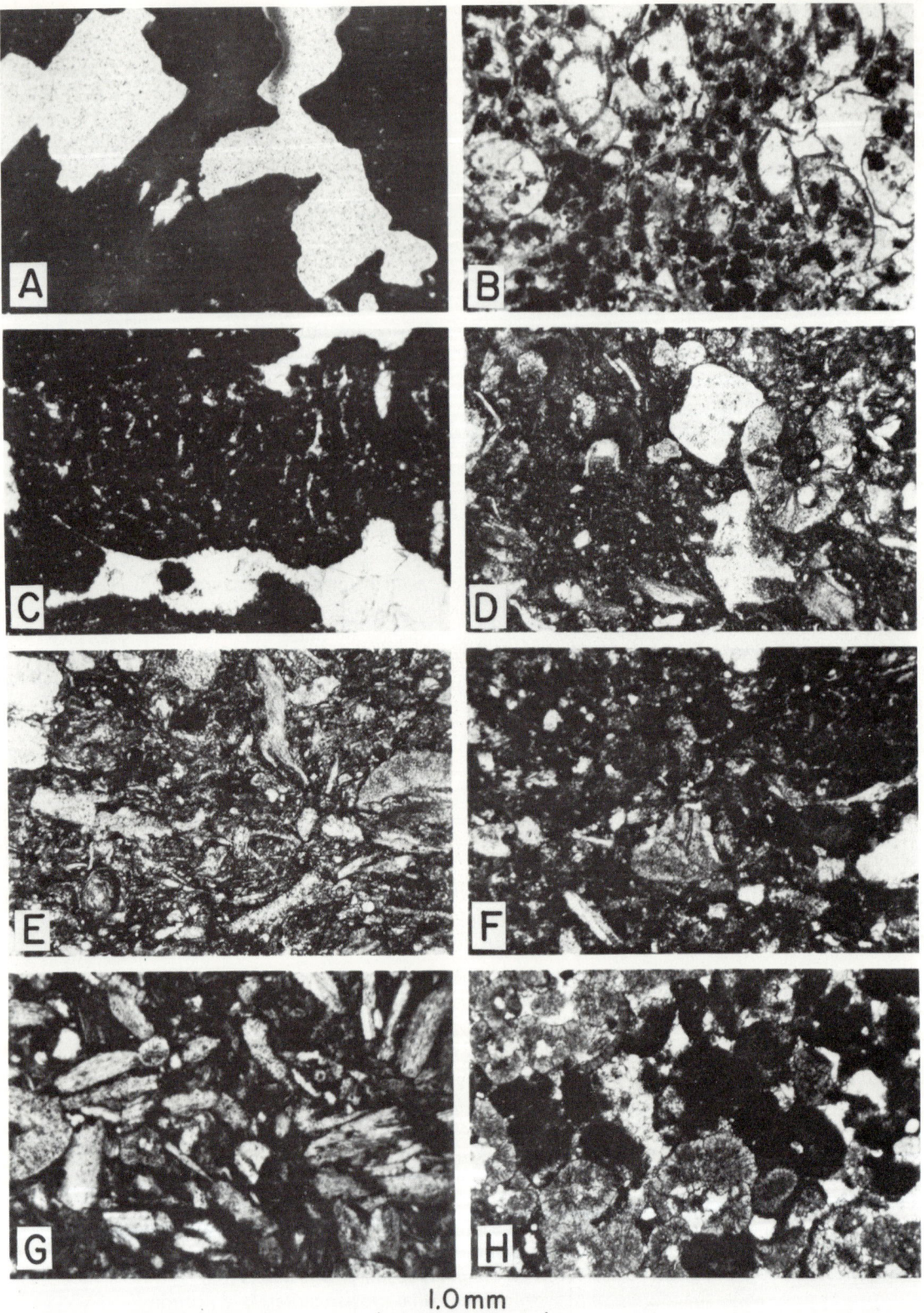

FIGURE 13.26 Traverse Group (Givétian), northern part of southern peninsula of Michigan. Typical microfacies of Koehler and Gravel Point Formations. Locality 25. A. Calcilutite, devoid of fossil debris (L-1), showing tetragonal-shaped casts, perhaps from anhydrite. B.

Donezella Mound

Microfacies 2 (Fig. 13.42B). Bioconstructed *Donezella* limestone with compacted and uncompacted colonies. In uncompacted *Donezella* colonies, their typical branching networks of tubes made of dark, cryptocrystalline to microcrystalline calcite are often surrounded by a well-developed submarine isopachous and fibrous rim cement with remaining interstitial space filled by phreatic sparite cement. A fine, pelletoidal internal sediment is often trapped between branching thalli of the algae.

Donezella colonies may be compacted, thus causing destruction of the open, branching network of the thalli. Compacted colonies display a lower cement/matrix ratio due to reduction of interstitial space. Compacted colonies lack the well-developed fibrous rim cement of uncompacted colonies but show bladed calcite crystals surrounding some filaments. This calcite cement may have replaced the earlier fibrous rim cement.

In some instances, early cemented uncompacted and compacted colonies have been reworked into intraclasts which may reach 10 mm in diameter, set in a calcisiltite matrix, and form a *Donezella* calcirudite.

Crinoidal fragments are rare to present. Other rare bioclasts are *Endothyra*, *Tuberitina*, ammodiscids, smaller arenaceous foraminifers, paleotextulariids, bryozoans, calcispheres, ostracods, fusulinids, and echinoid spines. Bioturbation is rare.

Donezella Backmound

Microfacies 3 (Fig. 13.42C). Grain-supported *Donezella* calcarenite with variable amounts of calcisiltite matrix and sparite cement. The branching growth habit observed in microfacies 2 is absent and *Donezella* algae consist of poorly to well-sorted fragmented segments of colonies. Algal surfaces display evidence of early cementation (bladed calcite cement coating thalli rims) and the interstitial space contains a patchy distribution of cavity-filling, phreatic sparite cement, and calcisiltite matrix.

Crinoid fragments are rare to common. Other rare bioclasts consist of fusulinids, ammodiscids, *Endothyra*, *Tuberitina*, paleotextulariids, smaller arenaceous foraminifers, bryozoans, ostracods, calcispheres, echinoid spines, and brachiopods. Fecal pellets and bioturbation are locally present.

Early submarine cementation and subsequent reworking resulted in the formation of intraclasts. These early cemented clasts of grain-supported *Donezella* calcarenite are set in a calcisiltite matrix and the rock displays a calcirudite texture.

Lower Inner Slope

Microfacies 4 (Fig. 13.42D). Spiculitic slightly argillaceous pelletoidal calcisiltite with scattered bioclasts of crinoids, bryozoans, ostracods, brachiopods, and paleotextulariids. The pelletoidal texture is due to abundant merged fecal pellets.

Upper Inner Slope

Microfacies 5 (Fig. 13.42E). Poorly sorted, grain-supported *Donezella*-crinoidal calcarenite with calcisiltite matrix and large patches of cavity-filling phreatic sparite cement combined with syntaxial overgrowths around large unabraded crinoidal fragments. Small *Donezella* colonies are present and isopachous rim cement occurs on some *Donezella* fragments. Other rare bioclasts are: bryozoans, brachiopods, fusulinids, echinoid spines, calcispheres, and ostracods.

FIGURE 13.26 (*continued*) Pelletoidal calcisiltite (L-2). Observe abundant intact tests of the ostracod *Welleria aftonensis*. Matrix is mostly recrystallized, although most tests are filled with pore-filling sparite. C. Detail of well-laminated merged-pelletoidal calcisiltite with fenestral cavities (L-3). The smaller, vertically oriented cavities are typical of this microfacies and may possibly represent algal filaments. D. Mud-supported crinoid-bryozoan calcarenite with calcisiltite matrix (I-3). E. Fine grain-supported bryozoan-crinoidal calcarenite with calcisiltite matrix (I-4). F. Grain-supported pelletoidal calcarenite with abundant coarse benthic debris (I-5). G. Pressure-welded crinoid-bryozoan calcarenite (I-6). H. Grain-supported algal-lithoclastic-oolitic calcarenite with recrystallized to precipitated calcite cement (I-7) showing fragment of *Chara* in upper left, and numerous specimens of *Nuia* sp. See text for detailed descriptions. All photomicrographs: plane-polarized light. From Roche and Carozzi (1970). Reprinted by permission of Société Nationale des Pétroles d'Aquitaine.

380 Part III / Carbonate Platforms

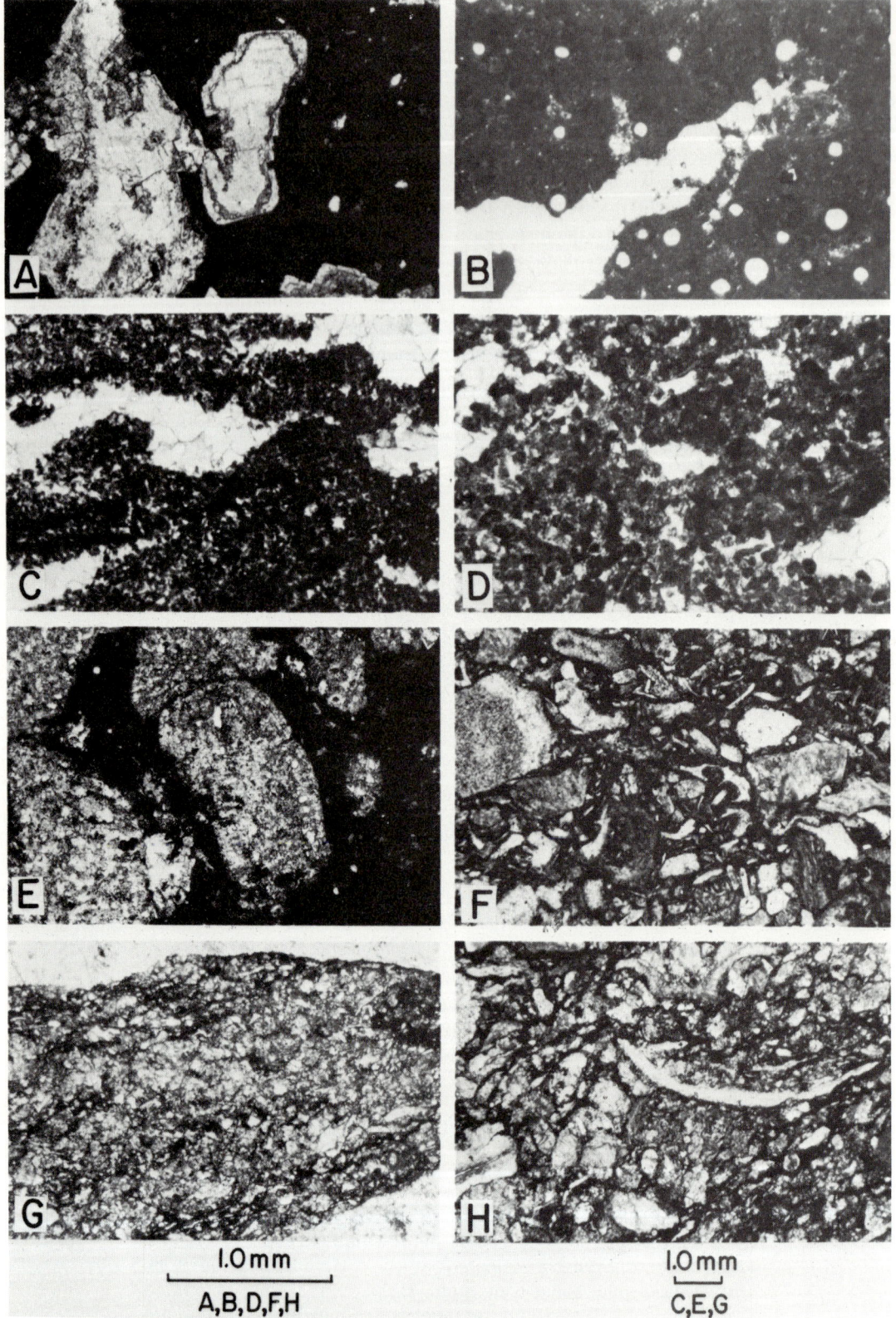

FIGURE 13.27 Traverse Group (Givétian), northern part of southern peninsula of Michigan. Typical microfacies of the Rockport Quarry and Ferron Point Formations at locality 38. A. Calcisiltite with scattered organic debris (L-1), showing replacement of dolomite by cal-

Crinoid Flat

Microfacies 6 (Fig. 13.42F). Coarse-grained, grain-supported, crinoidal calcarenite with syntaxial overgrowth cement. Other constituents are common intraclasts of *Donezella* bioconstructed limestone, and rare *Endothyra*, *Tetrataxis*, paleotextulariids, smaller arenaceous foraminifers, bryozoans, fusulinids, and brachiopods.

Oolitic Shoal

Microfacies 7 (Fig. 13.42G). Coarse-grained, well-sorted, oolitic calcarenite with minor calcisiltite matrix, fibrous rim submarine cement, and interparticle cavity-filling phreatic sparite cement. Ooid size range is 1.5 to 2.0 mm. Bioclasts commonly forming ooid nuclei include crinoids, bryozoans, ammodiscids, ostracods, fusulinids, and calcispheres along with rare echinoid spines, *Endothyra*, paleotextulariids, smaller arenaceous foraminifers, gastropods, and brachiopods.

Oolitic Backshoal

Microfacies 8 (Fig. 13.42H). Poorly sorted, grain-supported, foraminiferal, crinoidal biocalcarenite with patchy distribution of calcisiltite matrix and cavity-filling phreatic cement. Common to abundant bioclasts are crinoids, ammodiscids, bryozoans, *Endothyra*, fusulinids, ostracods, echinoid spines, brachiopods, and smaller arenaceous foraminifers. *Tuberitina*, calcispheres, and paleotextulariids are present, whereas gastropods, trilobites, *Tetrataxis*, and phylloid algae are rare. Fecal pellets, ooids, and oolitic intraclasts are locally common. Bioturbation is irregularly distributed. Minor early submarine cementation and subsequent reworking resulted in the formation of common foraminiferal calcarenitic intraclasts.

Lagoon

Microfacies 9 (Fig. 13.49A). Phylloid algal calcisiltite to mud-supported calcarenite. *Archaeolithophyllum missouriense* fragments are common to abundant, and algal fronds occasionally exhibit excellent cell structure. However, most of the algal fragments have been either filled by a coarse, mosaic calcite cement following complete solution or replaced by neomorphic pseudosparite. Other associated bioclasts consist of crinoids, ostracods, ammodiscids, fusulinids, calcispheres, bryozoans, *Endothyra*, *Tuberitina*, and smaller arenaceous foraminifers. Rare bioclasts include echinoid spines, pelecypods, *Tetrataxis*, brachiopods, paleotextulariids, *Girvanella*, and stromatolite fragments. Fecal pellets are common to abundant locally. Micritic intraclasts are rare. Bioturbation is common locally, often in the form of dissolution-enlarged, cement-lined, dolomitized burrows.

The Ideal Shallowing-Upward Sequence

The vertical superposition of microfacies 1 to 9 encompasses nine distinct depositional environments which represent a well-characterized shallowing-upward sequence (Figs. 13.43, 13.44). This trend is displayed perfectly in an interval of the "BA" Fee core (Fig. 13.45) where it begins in the *Donezella* backmound (microfacies 3) and terminates in the oolitic shoal (microfacies 7), and in an interval of the "AZ" Fee core (Fig. 13.46) where it begins in the upper inner slope (microfacies 5) and terminates in the lagoon (microfacies 9). Both sections show the detailed variations of the microscopic parameters.

The Ideal Depositional Model

The proposed model (Figs. 13.47, 13.48) shows clearly the relationships between constituents, microfacies, and inferred submarine morphology. *Donezella* is the pre-

FIGURE 13.27 (*continued*) cite in fenestral cavity. B. Calcisiltite with calcispheres (L-1), showing calcite-filled fenestral cavity. C. Pelletoldal calcisiltite with sparse organic debris (L-2), and calcite-filled fenestral cavities parallel to bedding. D. Pelletoidal calcisiltite (L-2), detail of above slide. E. Lithoclastic breccia with calcisiltite matrix (L-4) and large rounded clasts of stromatoporoid mats. F. Grain-supported calcarenite of crinoids and bryozoans with bituminous calcisiltite matrix (I-5). G. Pressure-welded bituminous calcarenite (I-6), example of calcarenite between stromatoporoid mats. H. Coarse pressure-welded bituminous calcarenite (I-7). See text for detailed descriptions. All photomicrographs: plane-polarized light. From Roche and Carozzi (1970). Reprinted by permission of Société Nationale des Pétroles d'Aquitaine.

382 Part III / Carbonate Platforms

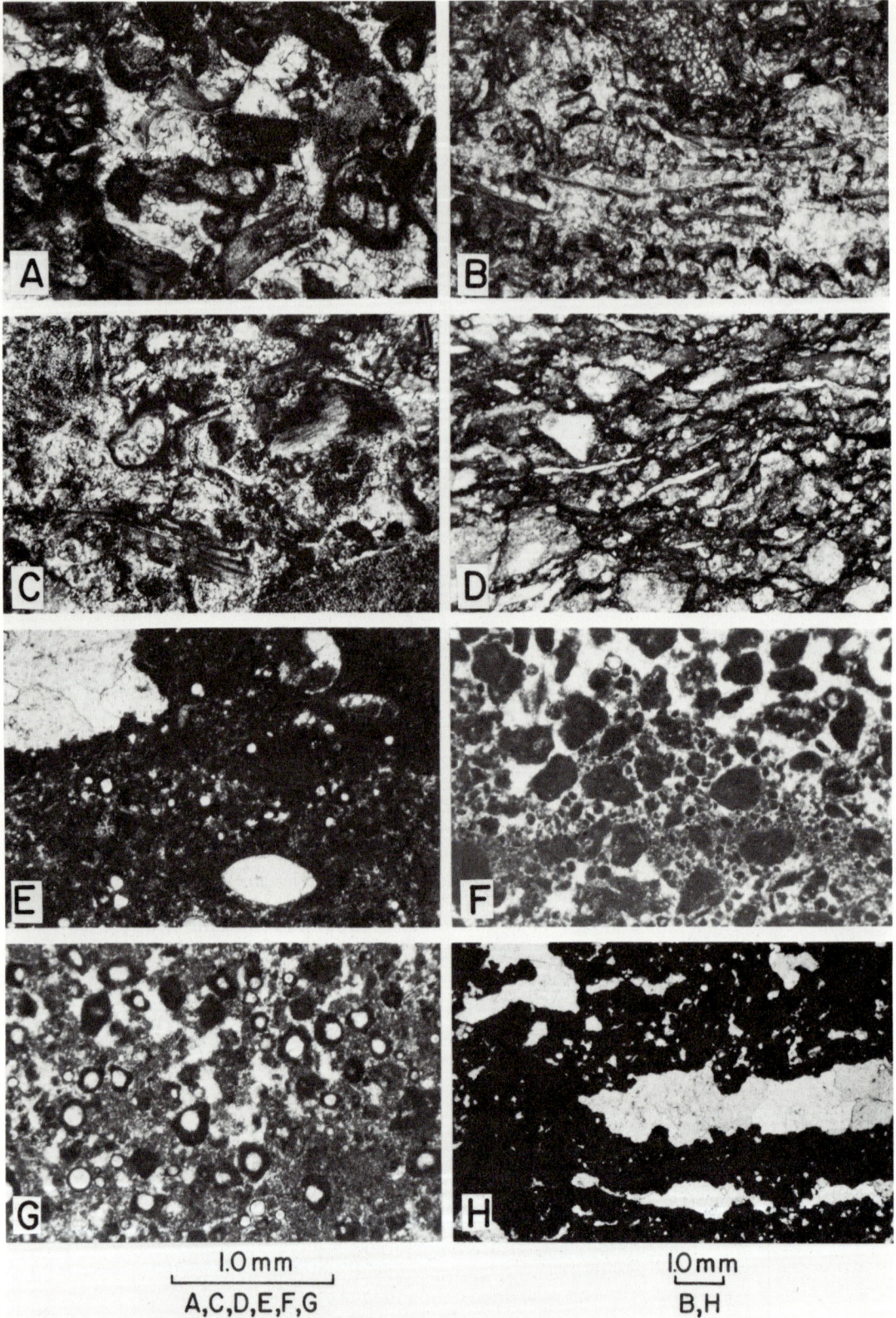

FIGURE 13.28 Traverse Group (Givétian), northern part of southern peninsula of Michigan. Typical microfacies of Rockport Quarry Formation. Locality 106. A. Grain-supported calcarenite of bryozoans and crinoids with recrystallized calcisiltite matrix (I-5). B. Same as

dominant and exclusive constituent of microfacies 2 and 3. In the former, it occurs as a bioconstructed form accompanied by the formation of large (10 mm) intraclasts resulting from early submarine cementation and subsequent fragmentation of the colonies, hence conditions at or above the wave base; in the latter, it builds a backmound calcarenitic talus. In microfacies 5, *Donezella* reappears in colonies and in fragmented tubes, hence in water conditions similar in terms of depth, clarity, and salinity to the mound of microfacies 2. *Archaeolithophyllum missouriense* occurs primarily in microfacies 9 in association with the following features, which indicate a restricted lagoonal environment: frequency peaks of fecal pellets and bioturbation, large relative percentage of calcisiltite matrix, microfauna restricted to ostracods in comparison to its abundance in microfacies 8, presence of minor amounts of *Girvanella* and stromatolite mats, and finally exposure features such as dissolution vugs and vadose silt.

Landward decrease of clasticity and frequency of crinoids away from microfacies 6 is due to the unfavorable shallow-water conditions of the oolitic shoal, conditions which are also affecting all the other organisms. This reduction is increased further by the fact that bioclasts act as ooid nuclei and are therefore counted as ooids. Final landward decrease of crinoids is attributed to restricted lagoonal conditions. The basinward decrease of clasticity and frequency is due to smaller-energy environments which are less favorable to crinoid growth, and to transportation of fragments away from the crinoid flat.

Intraclasts consist of four major types: *Donezella* bioconstructed and calcarenitic intraclasts, oolitic intraclasts, biocalcarenite intraclasts, and minor micritic intraclasts. *Donezella* bioconstructed intraclasts are the largest in microfacies 3 and demonstrate the above wave base position of the mound. They also are the most abundant in microfacies 6, where they accumulated under the greatest energy conditions of the model after their derivation from microfacies 5.

Monaxonic sponge spicules were transported in quiet, deeper, and possibly stagnant environments (microfacies 4), while ammodiscids, fusulinids, bryozoans, and ostracods reached their peaks of frequency in the oolitic backshoal (microfacies 8). Ostracods alone tolerated lagoonal conditions. Calcispheres settled in the outer slope environment, indicating a mechanical accumulation after a period of passive flotation. Fecal pellets display their maximum frequency in the lagoon and decrease rapidly seaward. Their concentration in lesser energy environments indicates infaunal activity and preservation of its products.

Authigenic feldspar (albite) shows a peak of frequency in microfacies 5 and again in microfacies 8. Landward and seaward from microfacies 8 are moderate decreases in abundance. No authigenic feldspar occurs in microfacies 6. This variation seems to correspond to changes in frequency and amplitude of sutured stylolites. Seaward from microfacies 5 is a rapid decline in authigenic feldspar with no additional increases. This reflects in part the presence of nonsutured stylolites in microfacies 1 and 4. The reason for the lack of abundant feldspar in microfacies 2 and 3 was not determined.

Stylolites occur throughout all microfacies. In general, however, nonsutured stylolites are restricted to microfacies containing argillaceous material (microfacies 1 and 4). Sutured stylolites are present to common in the remaining microfacies corresponding to pure carbonates. The nonsutured stylolites always possess low amplitudes, while the amplitude of the sutured stylolites varies. The increase of amplitude is directly related to increasing carbonate purity.

Observed trends in secondary burial porosity indicate an increase which parallels the increase in frequency and amplitude of sutured stylolites. A peak in porosity occurs in microfacies 5, which parallels a peak in high-

FIGURE 13.28 (*continued*) A. Parallelism of bryozoan fronds indicates fairly quiet conditions of deposition. C. As above (I-5) with calcisiltite matrix partially recrystallized. D. Pressure-welded bituminous calcarenite (I-6), which occurs typically between coral heads and stromatoporoid mats. Locality 29. E. Pelletoidal calcisiltite (L-2) with ostracods, calcispheres, and fenestral cavity filled with sparite. F. Lithoclastic and pelletoidal calcirudite (L-4). Lithoclasts are mainly rounded fragments of pelletoidal calcisiltite. Matrix is recrystallized to pseudosparite. G. Pelletoidal calcisiltite, partially recrystallized (L-2), showing radial canals in individual of *Parathurammina* at lower left center. H. Pelletoidal calcisiltite (L-2) with calcite-filled fenestral cavities due to burrowing. See text for detailed descriptions. All photomicrographs: plane-polarized light. From Roche and Carozzi (1970). Reprinted by permission of Société Nationale des Pétroles d'Aquitaine.

384 Part III / Carbonate Platforms

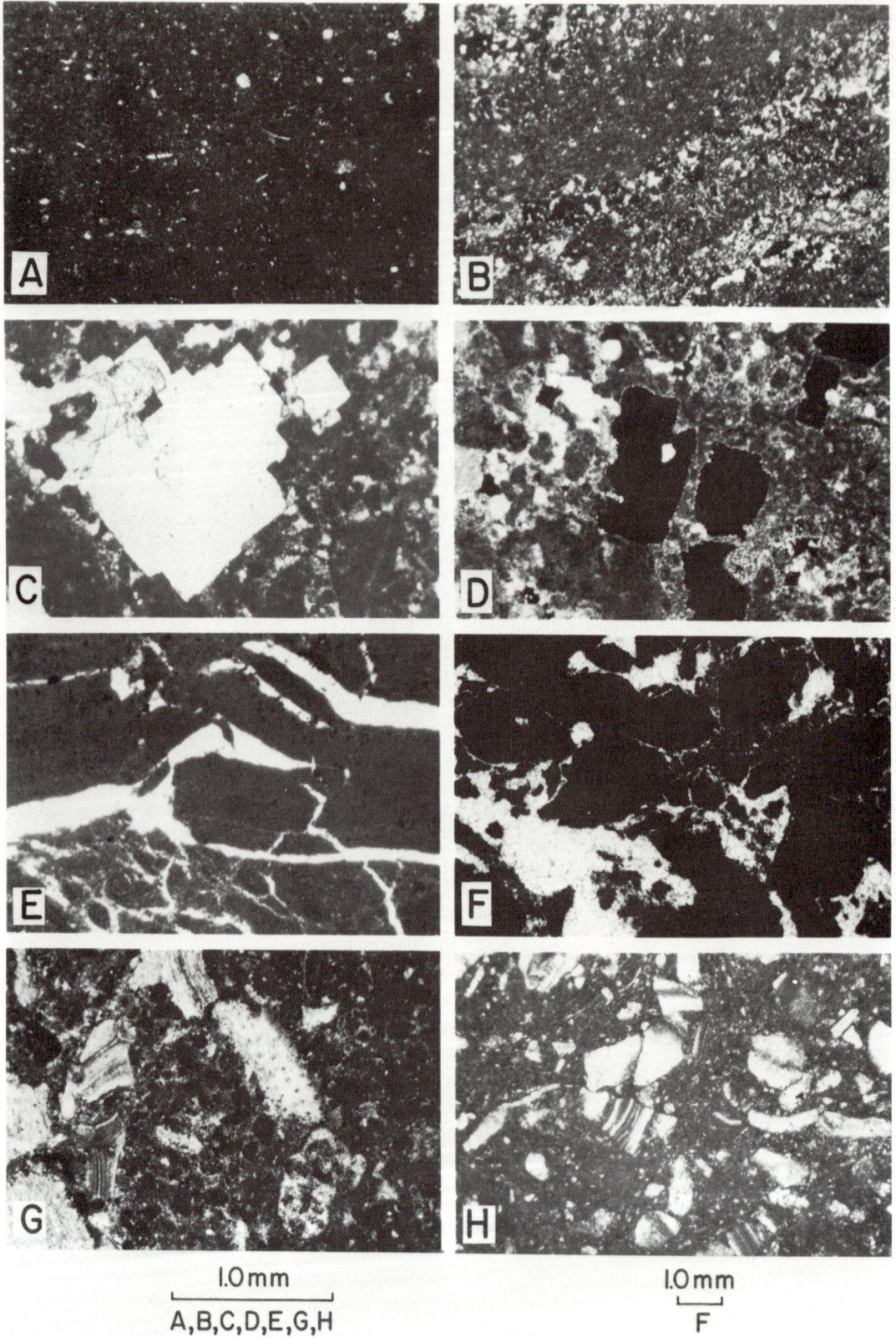

FIGURE 13.29 Traverse Group (Givétian), northern part of southern peninsula of Michigan. Typical microfacies of Gravel Point Formation. Locality 8. A. Calcisiltite to calcilutite with pellets, calcispheres, and minute unidentifiable debris (L-1). B. Well-laminated calcisiltite, probably stromatolitic (L-3). C. Laminated pelletoidal and lithoclastic

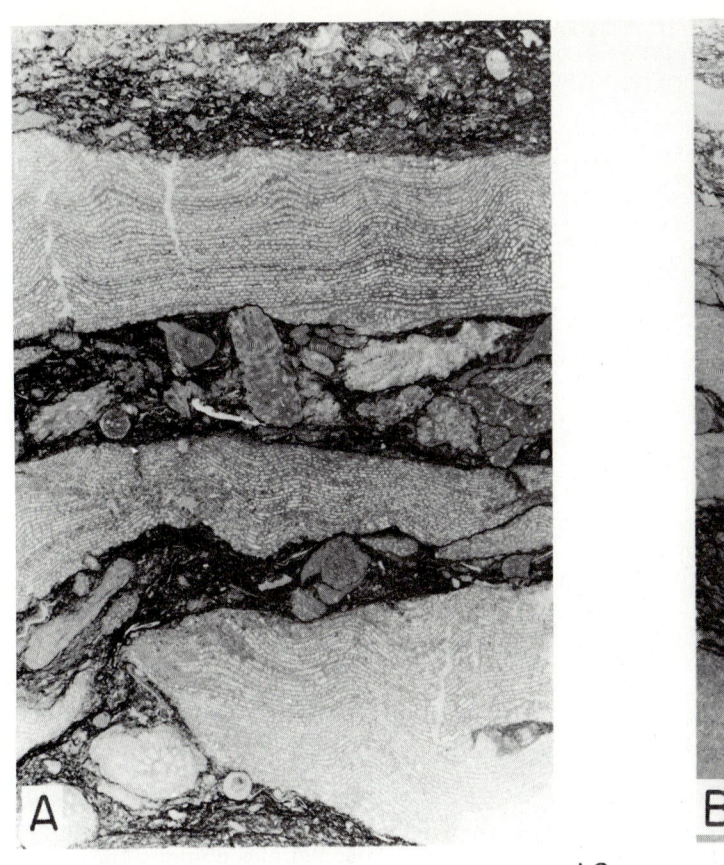
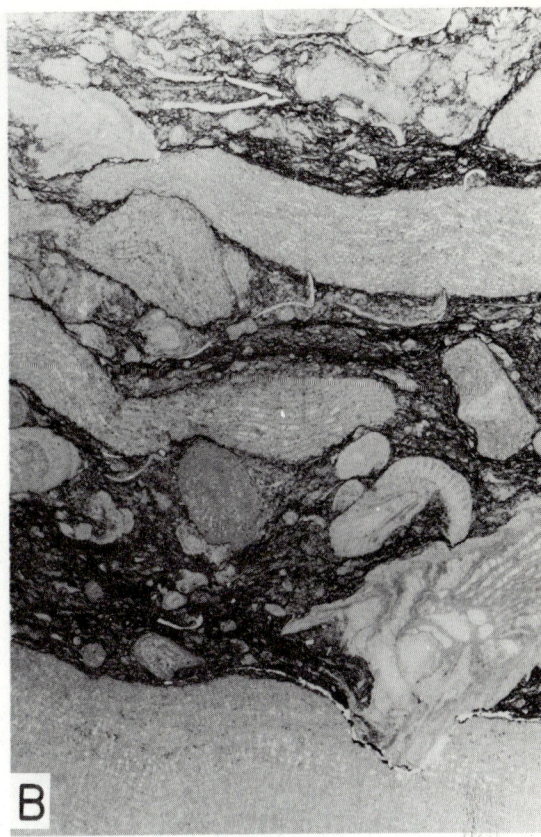

1.0 cm

FIGURE 13.30 Traverse Group (Givétian), northern part of southern peninsula of Michigan. Typical microfacies of Rockport Quarry Formation at locality 38. A. Thin stromatoporoid mats interlayered with very coarse to fine bituminous, grain-supported calcarenite. Note broken ends of mats and abundance of rudaceous stromatoporoid debris (I-6 and I-8). B. Same as A. Observe pressure-welding of solitary coral into mat at bottom (I-6 and I-8). See text for detailed descriptions. All photomicrographs: plane-polarized light. From Roche and Carozzi (1970). Reprinted by permission of Société Nationale des Pétroles d'Aquitaine.

FIGURE 13.29 (*continued*) calcarenite (L-3). Molds of cubic crystals are quite common in this part of the sequence, and are most likely remnants of halite. D. As above (L-3). Casts of tetragonal crystals, probably originally anhydrite, are found in association with cubic molds. This specimen is taken about 15 cm above that of C. Nicols crossed. E. Lithoclastic calcisiltite (L-4). Calcisiltite mud is desiccated and cracked in place. Fine laminae can be seen easily in the larger pieces. Movement by waves or currents resulted in a texture similar to that of next photomicrograph. F. Lithoclastic breccia of calcisiltite clasts (L-4), with partially recrystallized matrix. G. Grain-supported pelletoidal calcarenite with unusual amount of calcisiltite matrix (I-5). On the left are good examples of *Favosites* or *Thamnopora* wall. H. Grain-to-mud supported calcarenite (I-5). See text for detailed descriptions. All photomicrographs: plane-polarized light. From Roche and Carozzi (1970). Reprinted by permission of Société Nationale des Pétroles d'Aquitaine.

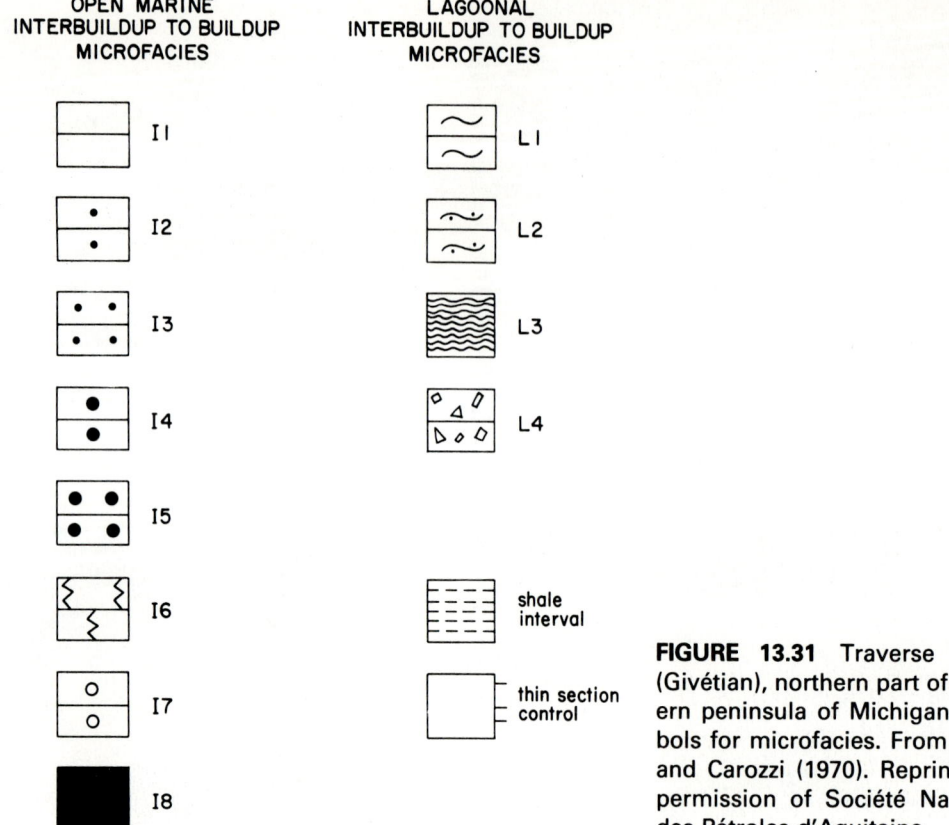

FIGURE 13.31 Traverse Group (Givétian), northern part of southern peninsula of Michigan. Symbols for microfacies. From Roche and Carozzi (1970). Reprinted by permission of Société Nationale des Pétroles d'Aquitaine.

amplitude sutured stylolites. A decrease in porosity landward corresponds to lower-amplitude stylolites in microfacies 6 to 9. Seaward, porosity decreases to minimum values in microfacies 4 and 1 corresponding to peaks in nonsutured stylolites. A moderate rise in porosity in microfacies 2 and 3 is again associated with medium-amplitude sutured stylolites. The relationship between porosity and high-amplitude sutured stylolites is well shown also in the example of shallowing-upward sequence of well "AZ" Fee (Fig. 13.46).

Dolomite is a rather minor constituent, although three peaks of frequency were identified. The two peaks associated with microfacies 5 and 7 correspond to increases in stylolite amplitude. This burial dolomite consists of rhombs about 0.05 mm in size; it replaces matrix, cement, and components and is concentrated along stylolite seams. The dolomite peak associated with microfacies 9 is due primarily to the presence of a fine-grained mosaic of small rhombs (not exceeding 0.025 mm in size) which preferentially replaced burrow- and vug-filling vadose silts and internal sediments. This early dolomitization, which is fabric selective and texture preserving, can be attributed to a small-scale dorag model. This mixing model involves phreatic meteoric waters that originated after shallowing-upward processes caused emergence. The mixing zone permeated downward through the marine sediments as freshwater gradually mixed with and displaced interstitial marine waters.

Diagenesis

Observed diagenetic features are numerous (Fig. 13.49) and range from depositional through deep burial conditions. Although many of the early diagenetic features may be grouped under a single heading and are listed in a particular order (Fig. 13.50), this ordering is somewhat arbitrary because several of these processes probably are synchronous.

Of particular interest among the burial processes is the formation of authigenic albite which consists of euhedral, rectangular laths replacing fossils, matrix, and cement (Fig. 13.49G). Some crystals contain minute inclusions of matrix and show the characteristic "four-ling" twinning. Authigenic albite crystals occur only in microfacies containing sutured stylolites and show a substantial increase in concentration along the latter. Two textural patterns were observed which depict the time relationship between stylolites and authigenic feldspar. In the first case, feldspar crystals cross-cut stylolites, maintain their euhedral morphology, and contain insoluble residues that are continuous with adjacent penetrating stylolites (Fig. 13.49G, top). In the second case, feldspar

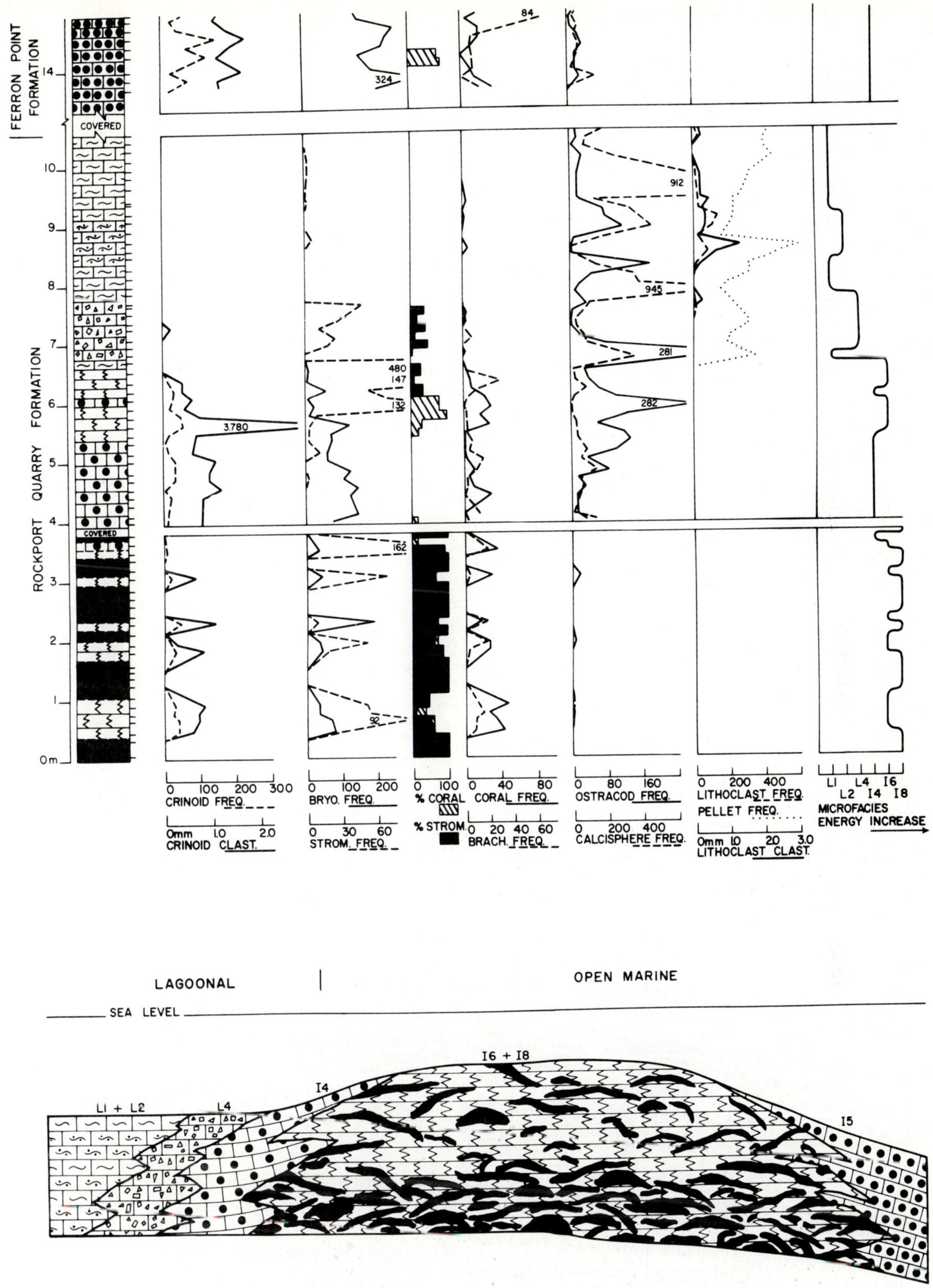

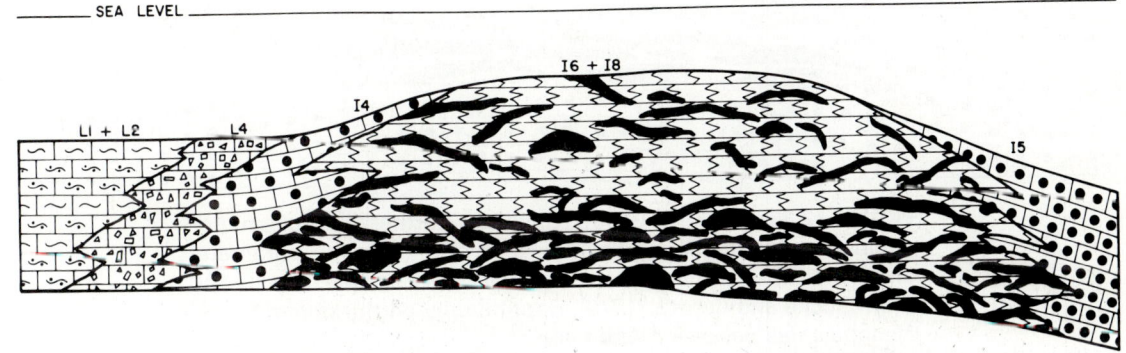

FIGURE 13.32 Traverse Group (Givétian), northern part of southern peninsula of Michigan. Vertical variation of microscopic parameters and environmental interpretation of Rockport Quarry (locality 38). From Roche and Carozzi (1970). Reprinted by permission of Société Nationale des Pétroles d'Aquitaine.

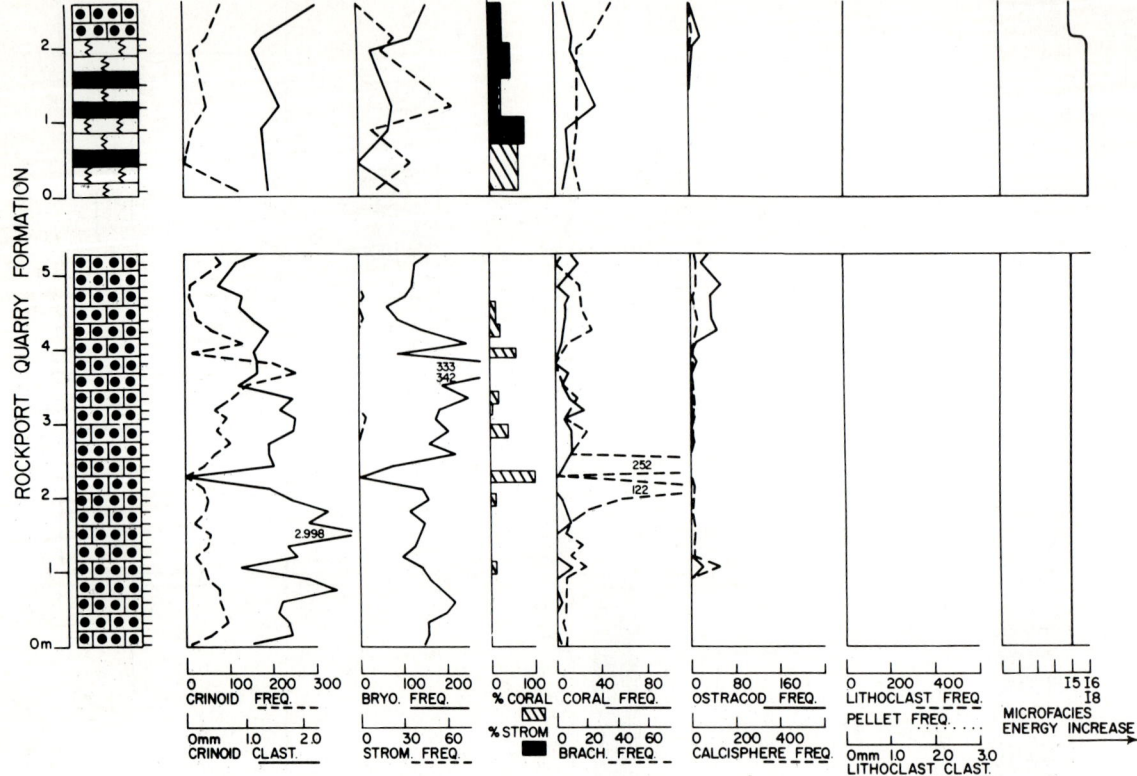

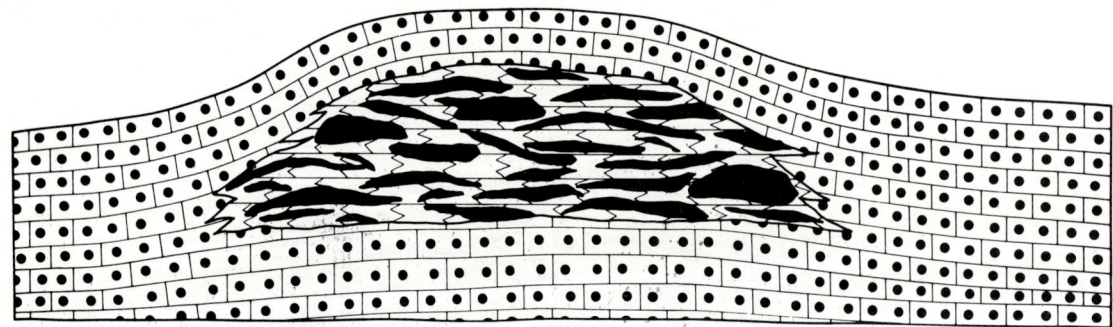

FIGURE 13.33 Traverse Group (Givétian), northern part of southern peninsula of Michigan. Vertical variation of microscopic parameters and environmental interpretation of Grand Lake section (locality 106). From Roche and Carozzi (1970). Reprinted by permission of Société Nationale des Pétroles d'Aquitaine.

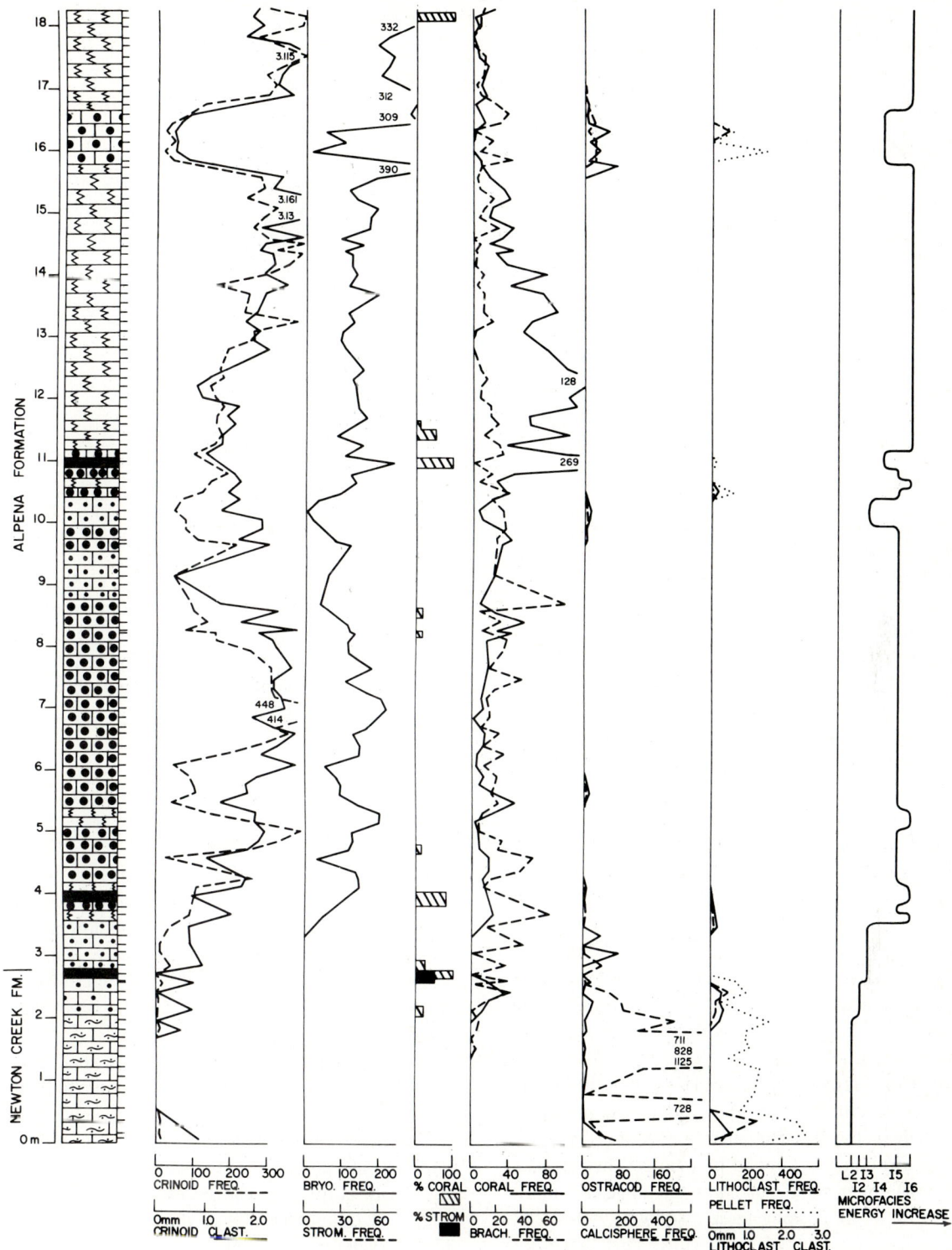

FIGURE 13.34 Traverse Group (Givétian), northern part of southern peninsula of Michigan. Vertical variation of microscopic parameters of Huron Portland Cement Quarry (locality 40). From Roche and Carozzi (1970). Reprinted by permission of Société Nationale des Pétroles d'Aquitaine.

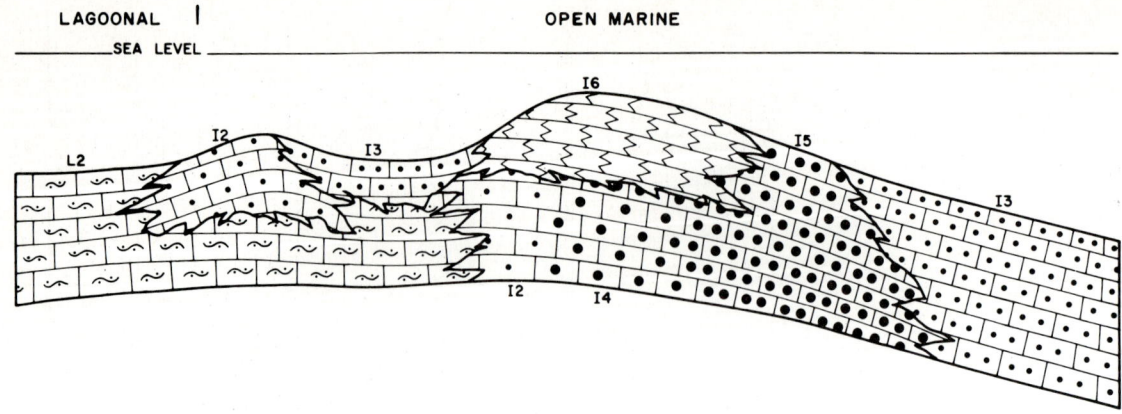

FIGURE 13.35 Traverse Group (Givétian), northern part of southern peninsula of Michigan. Environmental interpretation of Huron Portland Cement Quarry (locality 40). From Roche and Carozzi (1970). Reprinted by permission of Société Nationale des Pétroles d'Aquitaine.

crystals are no longer completely euhedral but possess dissolved or serrated contacts against stylolite seams (Fig. 13.49G left). These patterns indicate that feldspar generation followed or preceded stylolitization, respectively.

Burial dissolution was observed as oomoldic, biomoldic, and vuggy nonfabric selective porosity (Fig. 13.49E,F). The most common type of late porosity development is dissolution of ooids (Fig. 13.49E). Ooids are either entirely or partially dissolved. The late character of this dissolution process is demonstrated by concentration of oomolds along stylolite seams and their marked decrease away from stylolites. Furthermore, oomolds are never occluded by early diagenetic cements. This oomold-stylolite relationship is evidence that stylolites acted as conduits for diagenetic fluids and were critical in this phase of dissolution. Stylolites were enhancing instead of inhibiting porosity development. Some oomolds contain hydrocarbons indicating that oil migration took place sometime after oomold formation. Oomolds may be filled with baroque dolomite and/or an extremely coarse, late sparite cement.

A more convincing example of stylolite-related porosity is development of pores that follow or outline stylolite edges (Fig. 13.49H). In this case, undersaturated fluids moving along stylolite seams enlarged them considerably, resulting in significant pores. Continued nonfabric selective dissolution along these "stylolitic pores" resulted in the formation of small vugs. This "stylolitic porosity" is developed only along sutured, medium- to high-amplitude stylolites.

In summary, stylolites acted as reservoirs themselves and carried diagenetic fluids which created adjacent halos of secondary porosity (mostly oomoldic and biomoldic) and connected otherwise isolated porous zones. Whether this burial porosity was preserved depends upon the amount of subsequent cementation (Fig. 13.50).

Observed diagenetic pathways for each microfacies vary in complexity (Fig. 13.51). Some are relatively simple (microfacies 1), others are intricate (microfacies 9). Microfacies which contain sutured stylolites clearly display a more complex late diagenetic history which includes late dissolution.

The Evolution of the Depositional Environments

The vertical succession of microfacies in each of the three investigated wells shows well-developed shallowing-upward sequences followed by rapid deepening and are designated by Roman numerals. Three such sequences occur in well "BA" Fee (Fig. 13.52), four in well "AZ" Fee (Fig. 13.53), and two in the Chapman West well (Fig. 13.54).

Only sequences I, II, and III in the "AZ" Fee and "BA" Fee wells can be correlated due to their proximity, similar well depths, and relatively equivalent Atokan thicknesses. Correspondence between stylolite parameters and porosity is also well illustrated. The most significant correlation can be observed from 13,004 to 12,994 ft (microfacies 7) of the "BA" Fee core (Fig. 13.52), where a substantial and consistent amount of well-developed sutured stylolites corresponds to a rise in porosity. Dolomite and matrix in this section are negligible.

The negative (or neutral) influence of nonsutured stylolites on porosity, even when abundant, is shown in the section from 13,251 to 13,245 ft of the Chapman West core (Fig. 13.54).

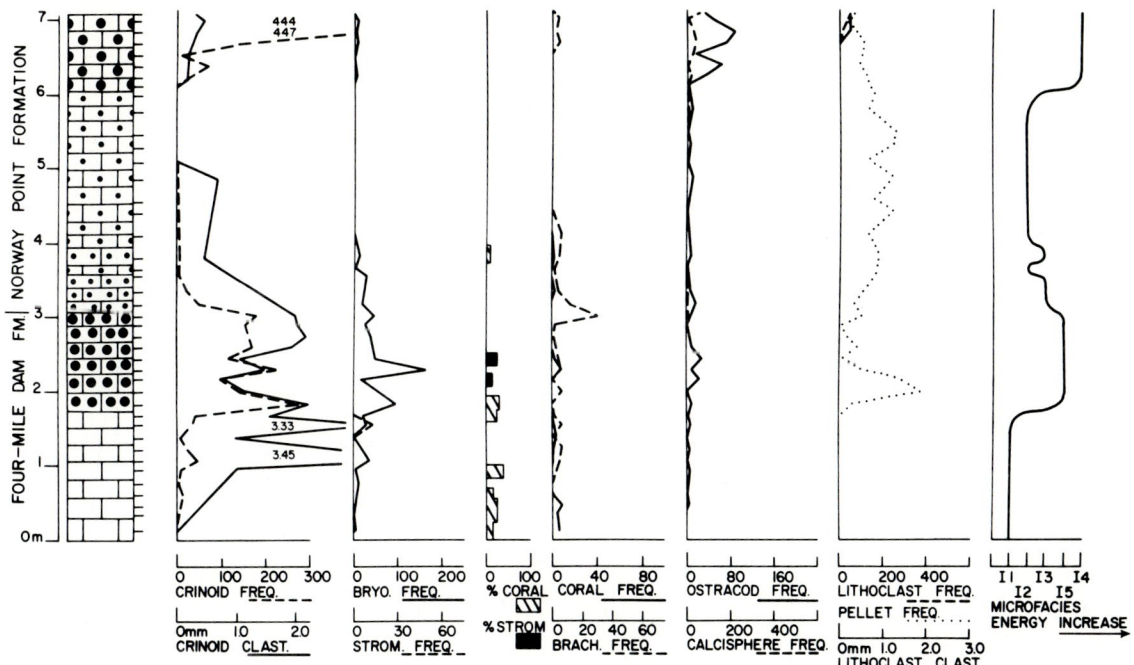

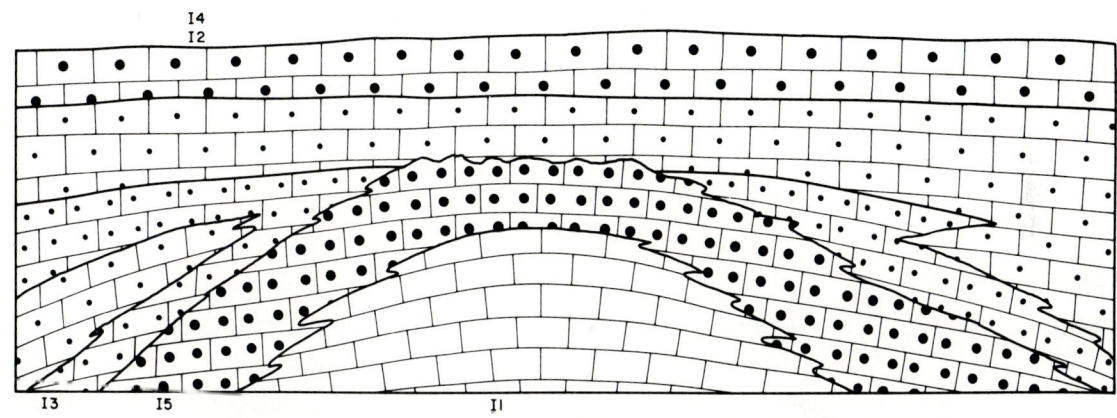

FIGURE 13.36 Traverse Group (Givétian), northern part of southern peninsula of Michigan. Vertical variation of microscopic parameters and environmental interpretation of Four Mile Dam section (locality 41). From Roche and Carozzi (1970). Reprinted by permission of Société Nationale des Pétroles d'Aquitaine.

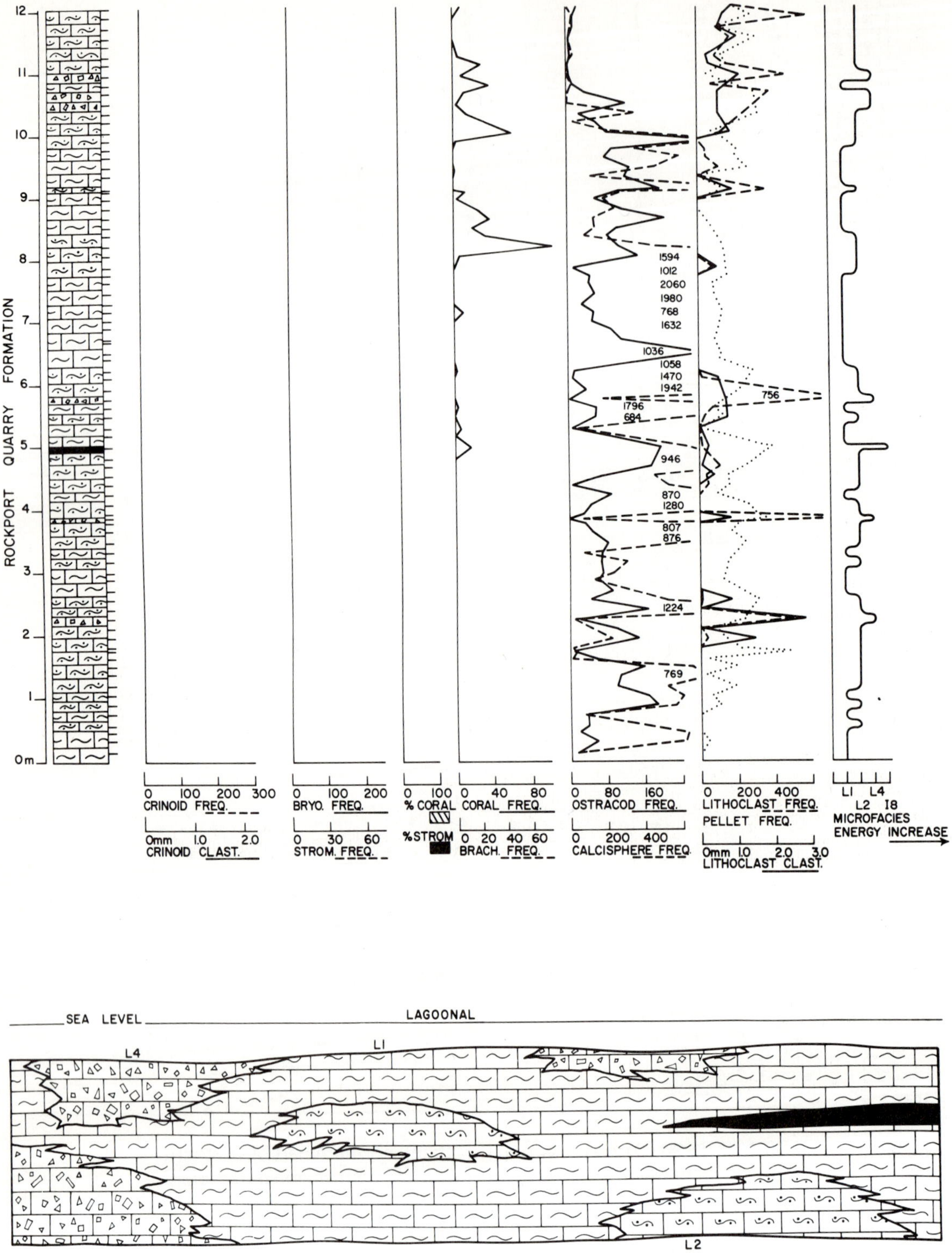

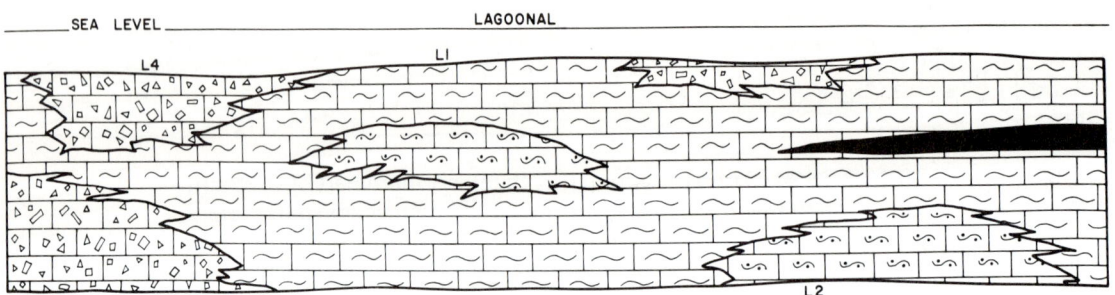

FIGURE 13.37 Traverse Group (Givétian), northern part of southern peninsula of Michigan. Vertical variation of microscopic parameters and environmental interpretation of Black Lake section (locality 29). From Roche and Carozzi (1970). Reprinted by permission of Société Nationale des Pétroles d'Aquitaine.

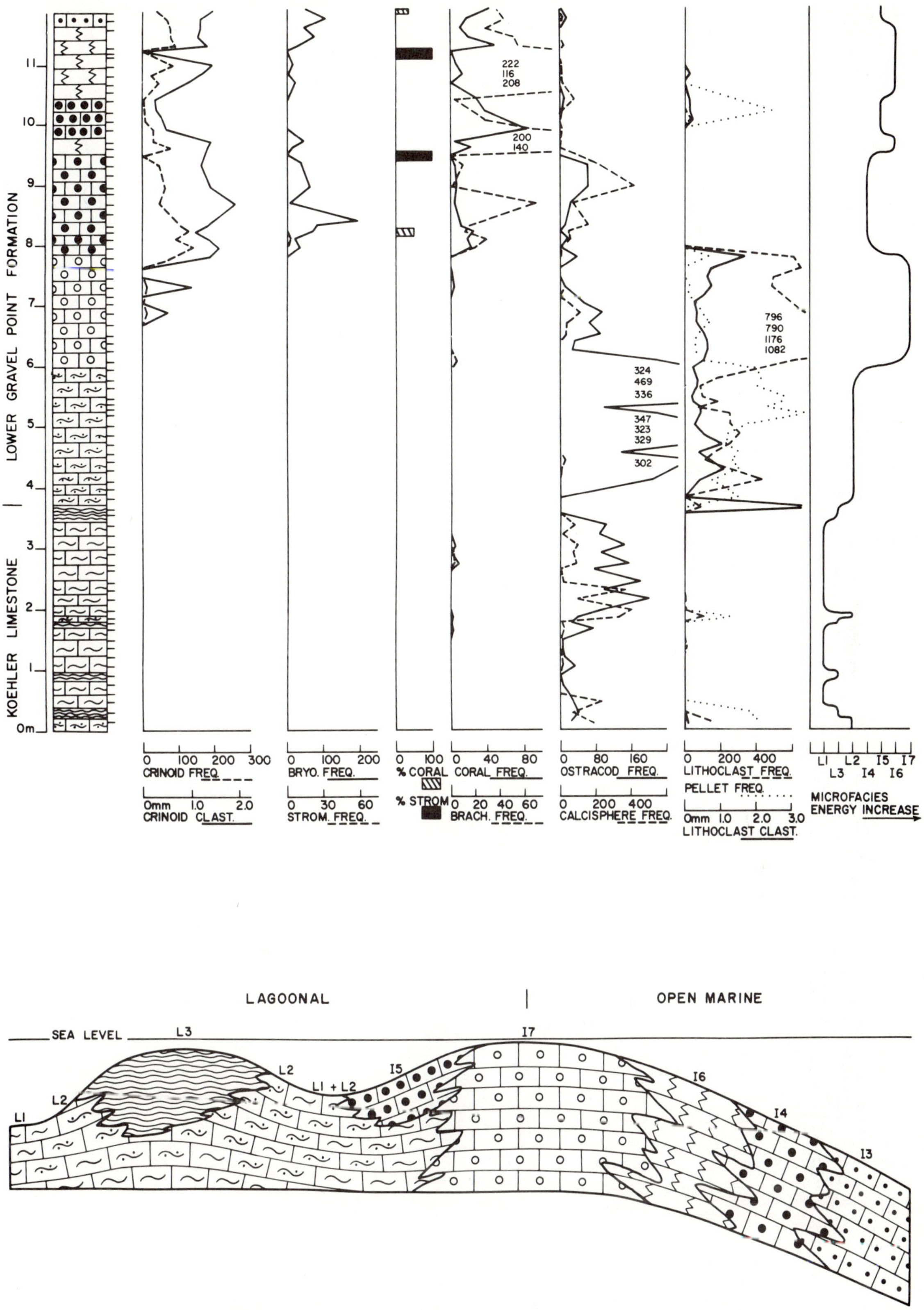

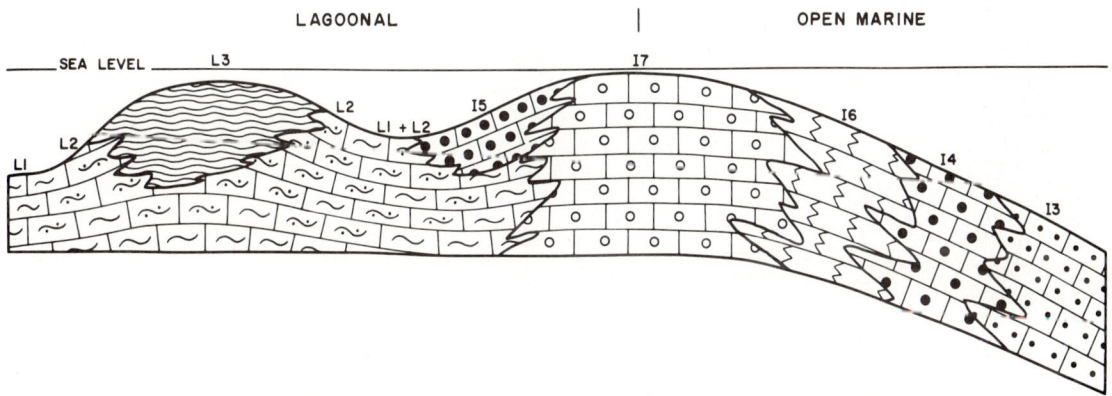

FIGURE 13.38 Traverse Group (Givétian), northern part of southern peninsula of Michigan. Vertical variation of microscopic parameters and environmental interpretation of Campbell Quarry (locality 25). From Roche and Carozzi (1970). Reprinted by permission of Société Nationale des Pétroles d'Aquitaine.

394 Part III / Carbonate Platforms

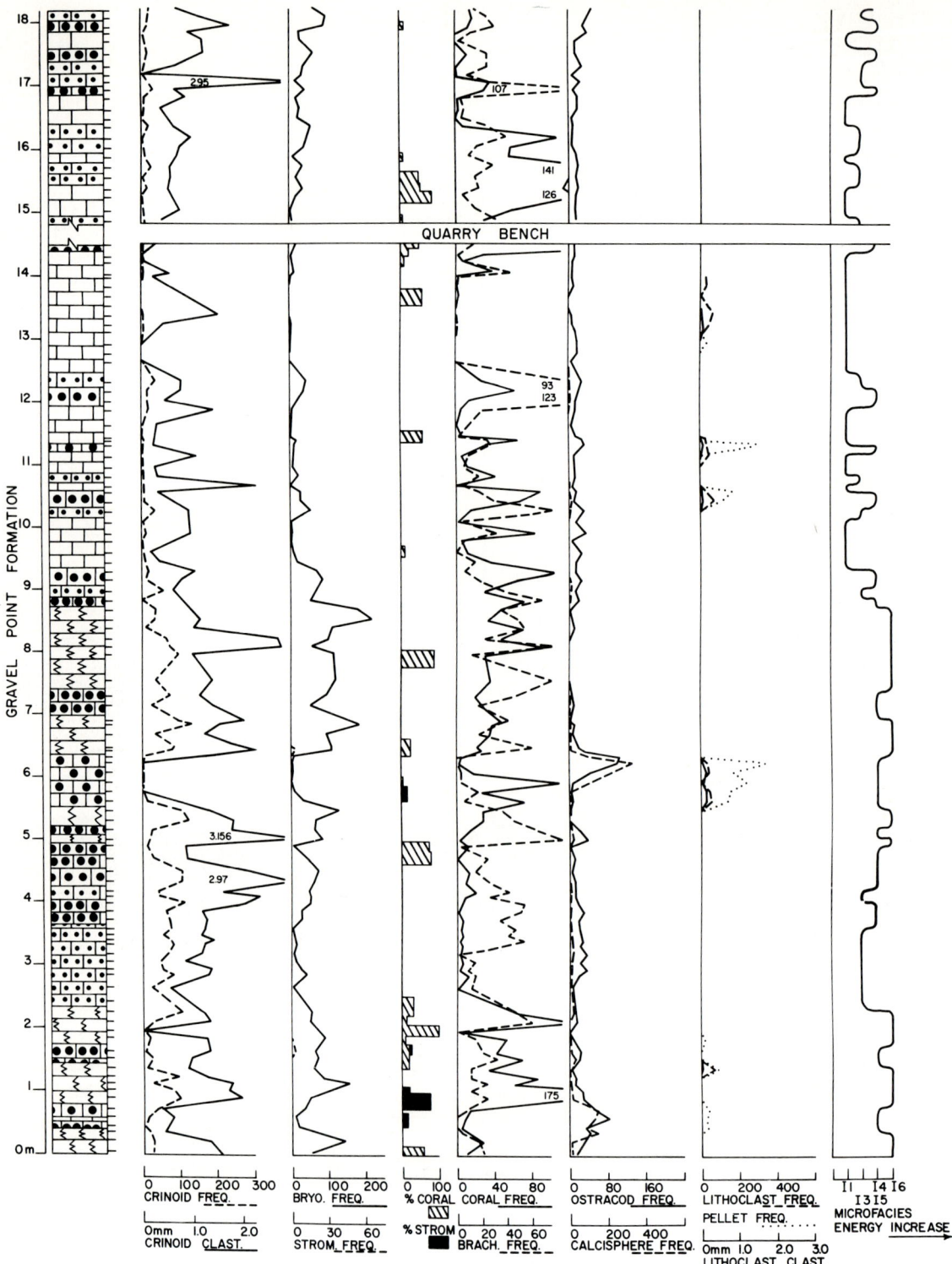

FIGURE 13.39 Traverse Group (Givétian), northern part of southern peninsula of Michigan. Vertical variation of microscopic parameters of Penn-Dixie Cement Co. Quarry (locality 14). From Roche and Carozzi (1970). Reprinted by permission of Société Nationale des Pétroles d'Aquitaine.

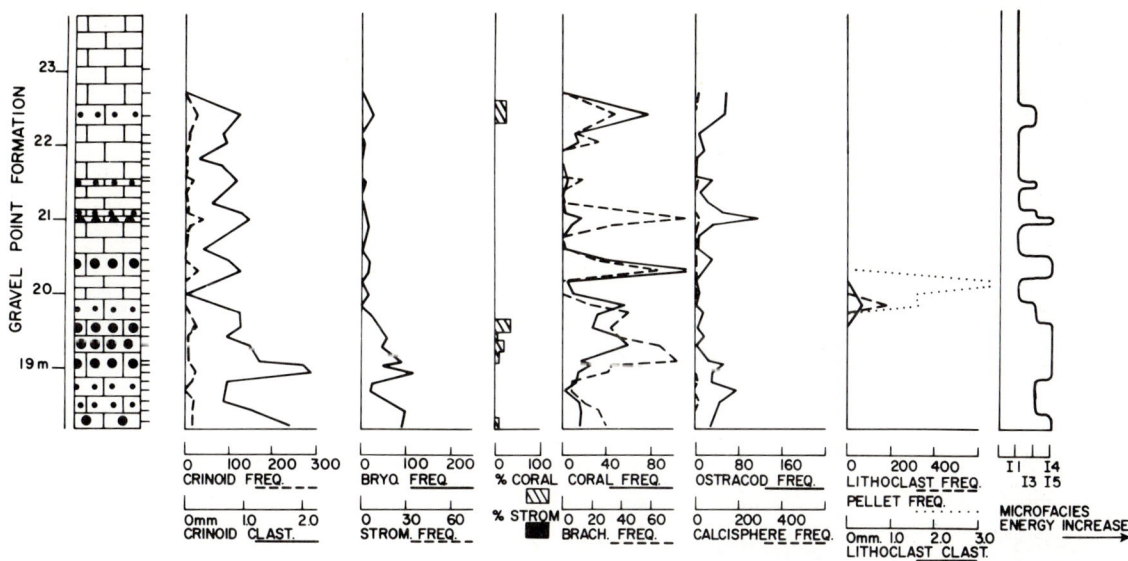

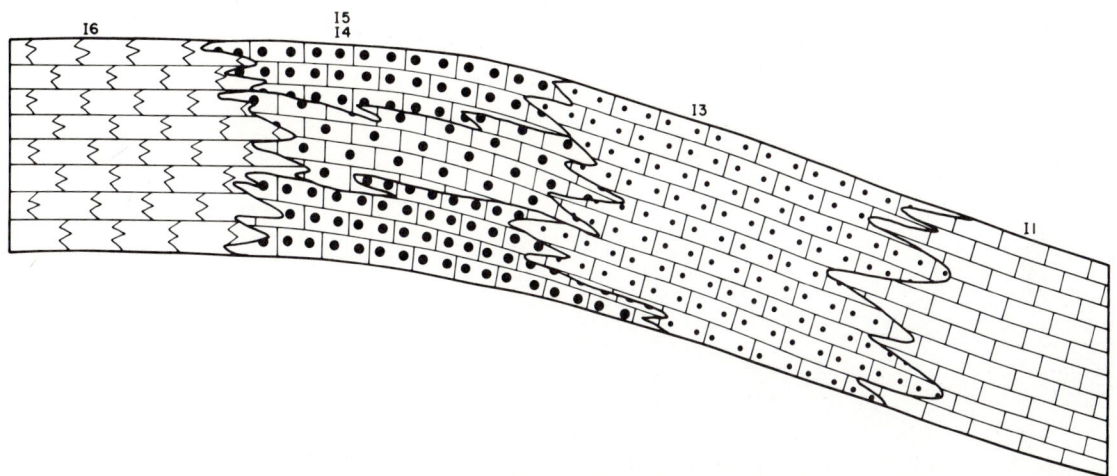

FIGURE 13.40 Traverse Group (Givétian), northern part of southern peninsula of Michigan. Vertical variation of microscopic parameters (continued) and environmental interpretation of Penn-Dixie Cement Co. Quarry (locality 14). From Roche and Carozzi (1970). Reprinted by permission of Société Nationale des Pétroles d'Aquitaine.

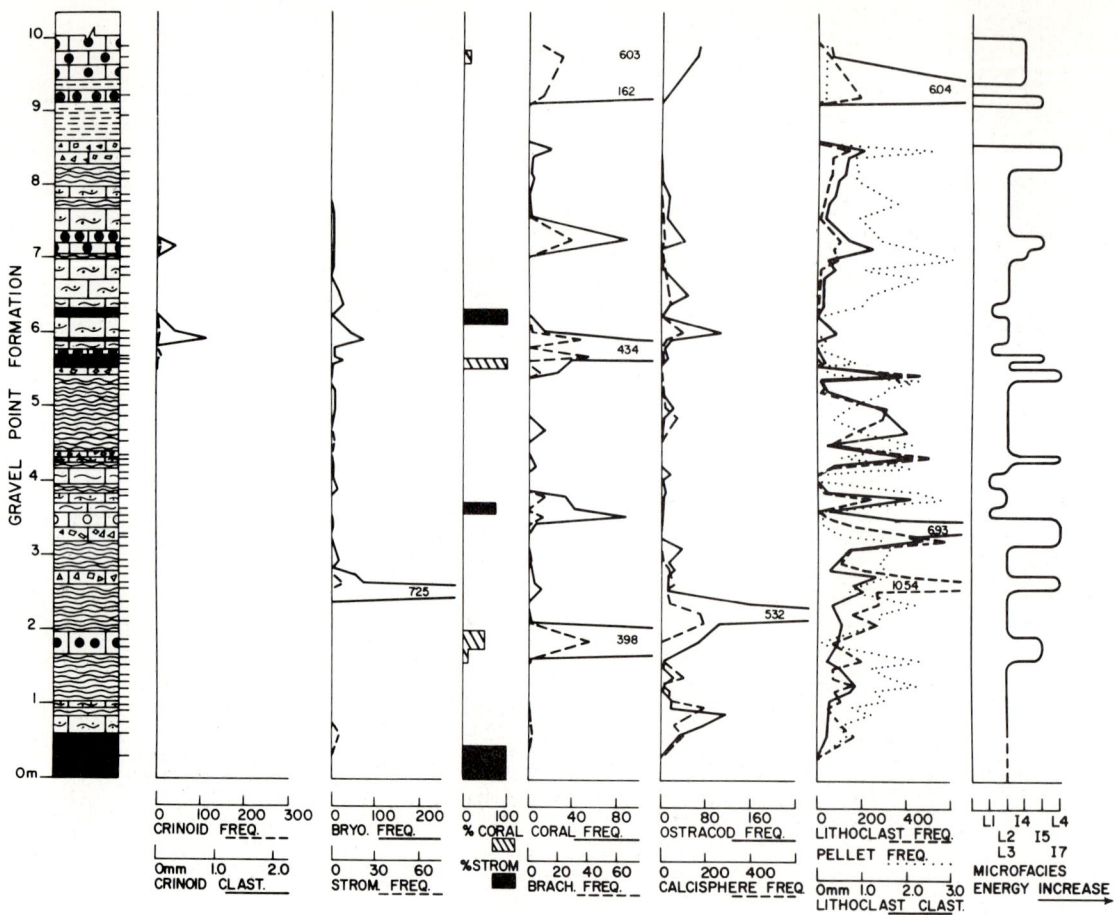
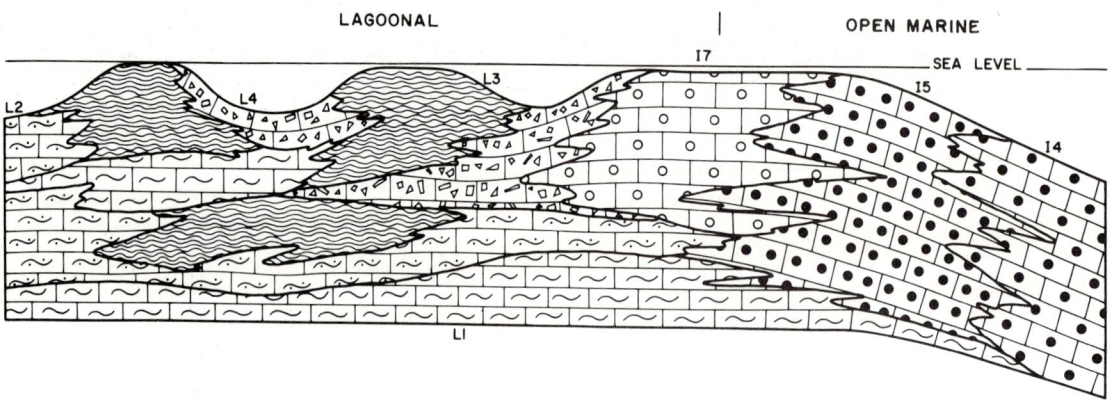

FIGURE 13.41 Traverse Group (Givétian), northern part of southern peninsula of Michigan. Vertical variation of microscopic parameters and environmental interpretation of Medusa Portland Cement Co. Quarry (locality 8). From Roche and Carozzi (1970). Reprinted by permission of Société Nationale des Pétroles d'Aquitaine.

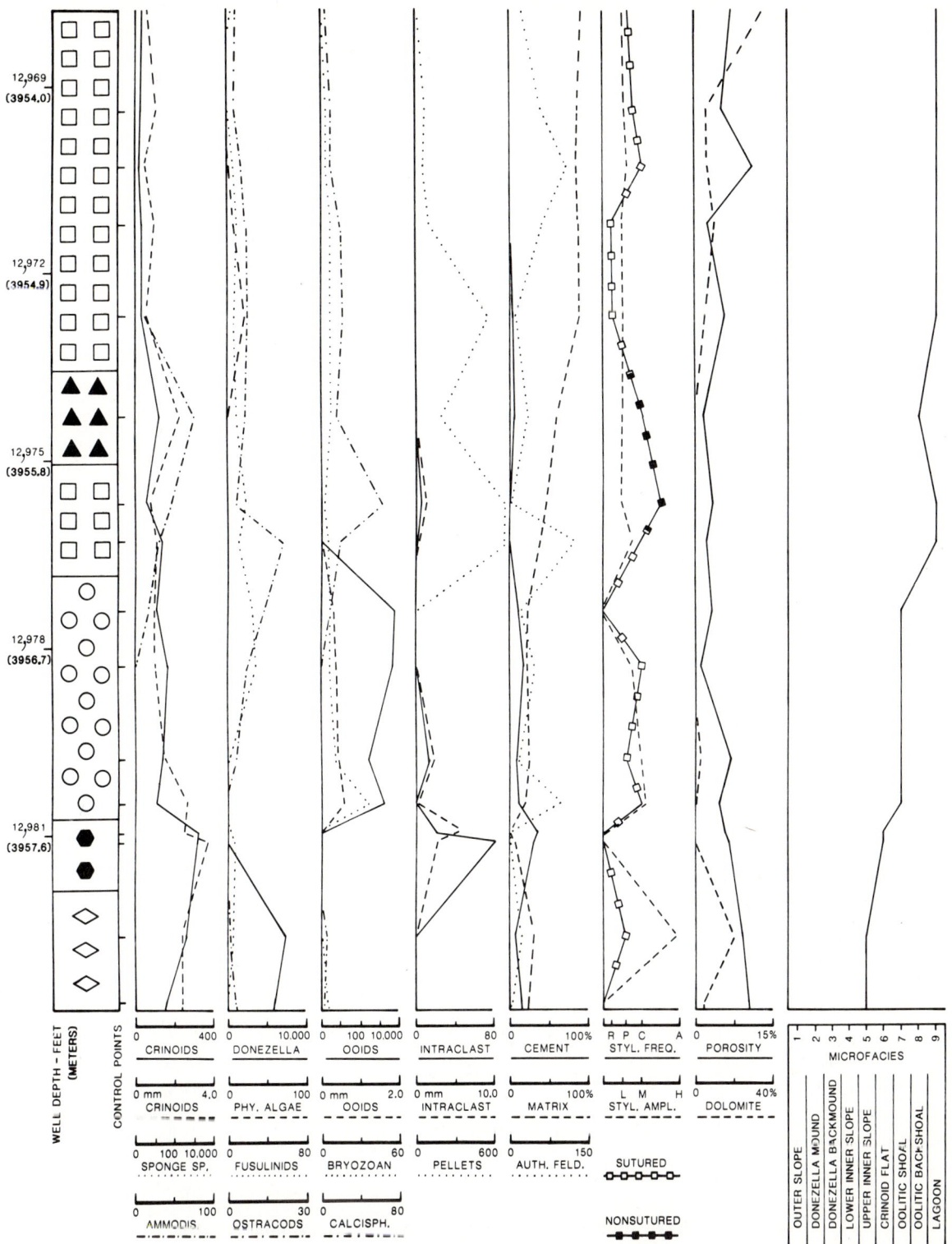

FIGURE 13.46 Atokan Limestone (Middle Pennsylvanian), west margin of Delaware Basin, Reeves County, Texas. Typical example of vertical variation of microscopic parameters during a shallowing-upward sequence of well "AZ" Fee. From Von Bergen (1985).

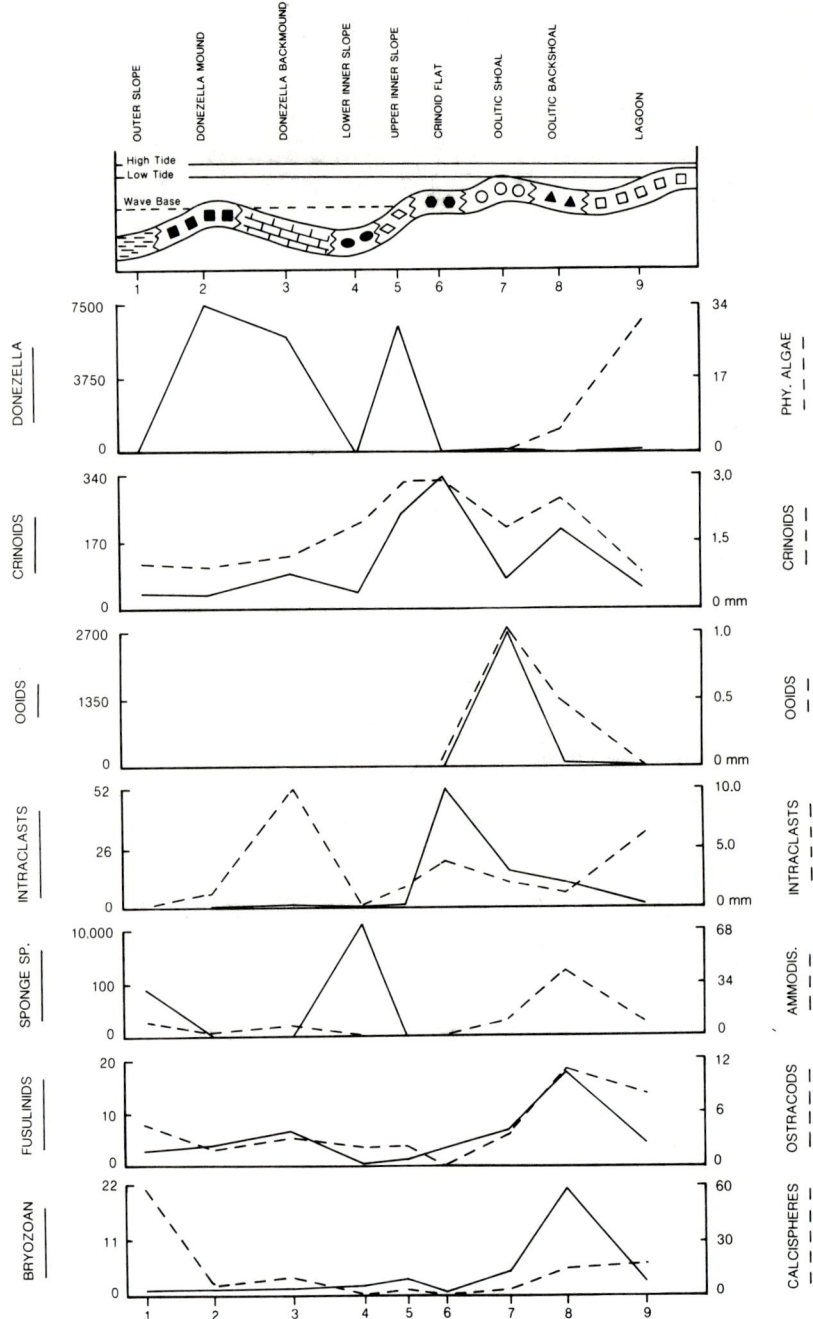

FIGURE 13.47 Atokan Limestone (Middle Pennsylvanian), west margin of Delaware Basin, Reeves County, Texas. Ideal depositional model. From Carozzi and Von Bergen (1987). Reprinted by permission of Scientific Press Ltd.

Chap. 13 / Carbonate Platforms with Frontal Bioconstructed to Hydrodynamic Buildups 403

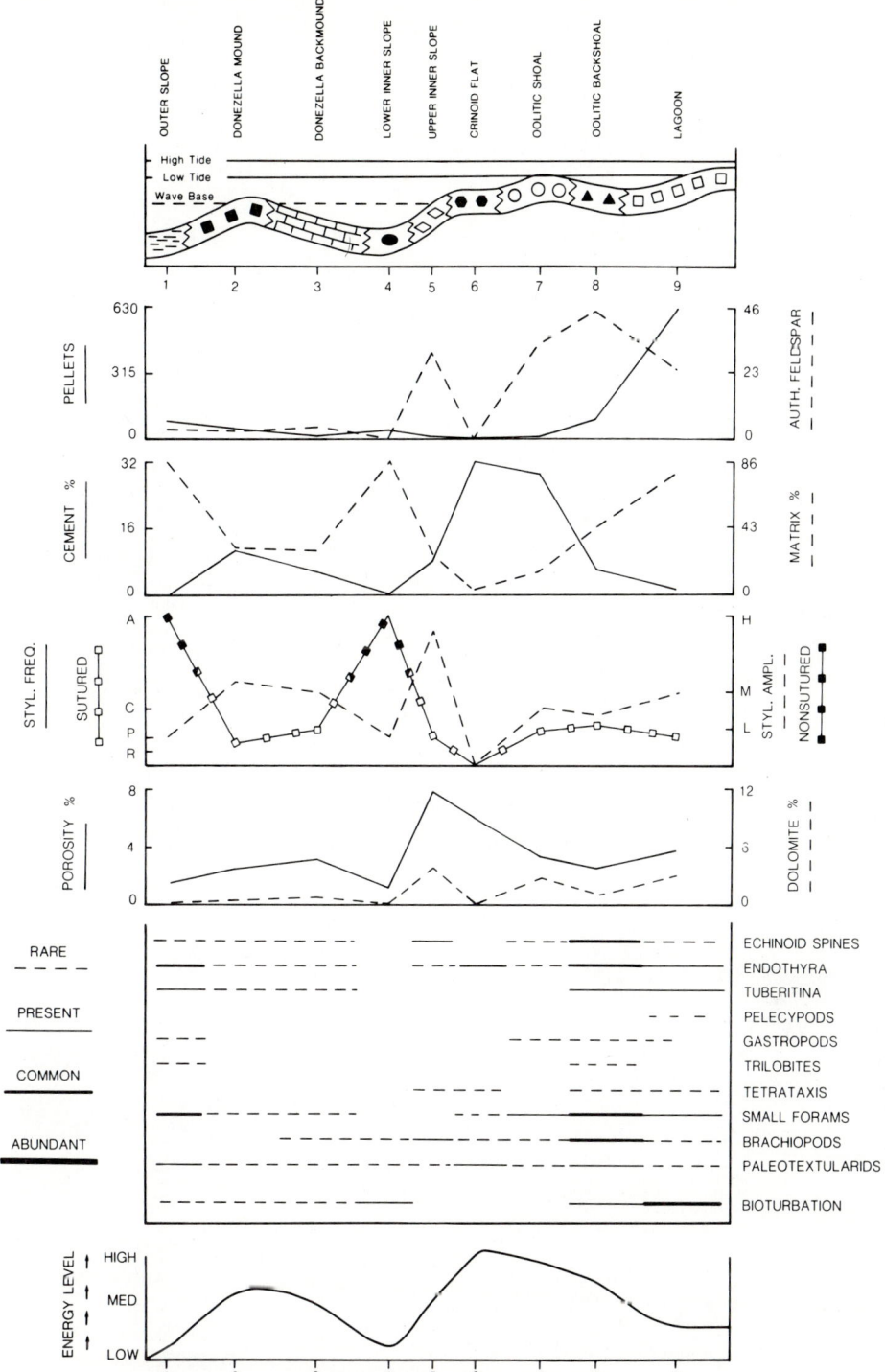

FIGURE 13.48 Atokan Limestone (Middle Pennsylvanian), west margin of Delaware Basin, Reeves County, Texas. Ideal depositional model (continued). From Carozzi and Von Bergen (1987). Reprinted by permission of Scientific Press Ltd.

404 Part III / Carbonate Platforms

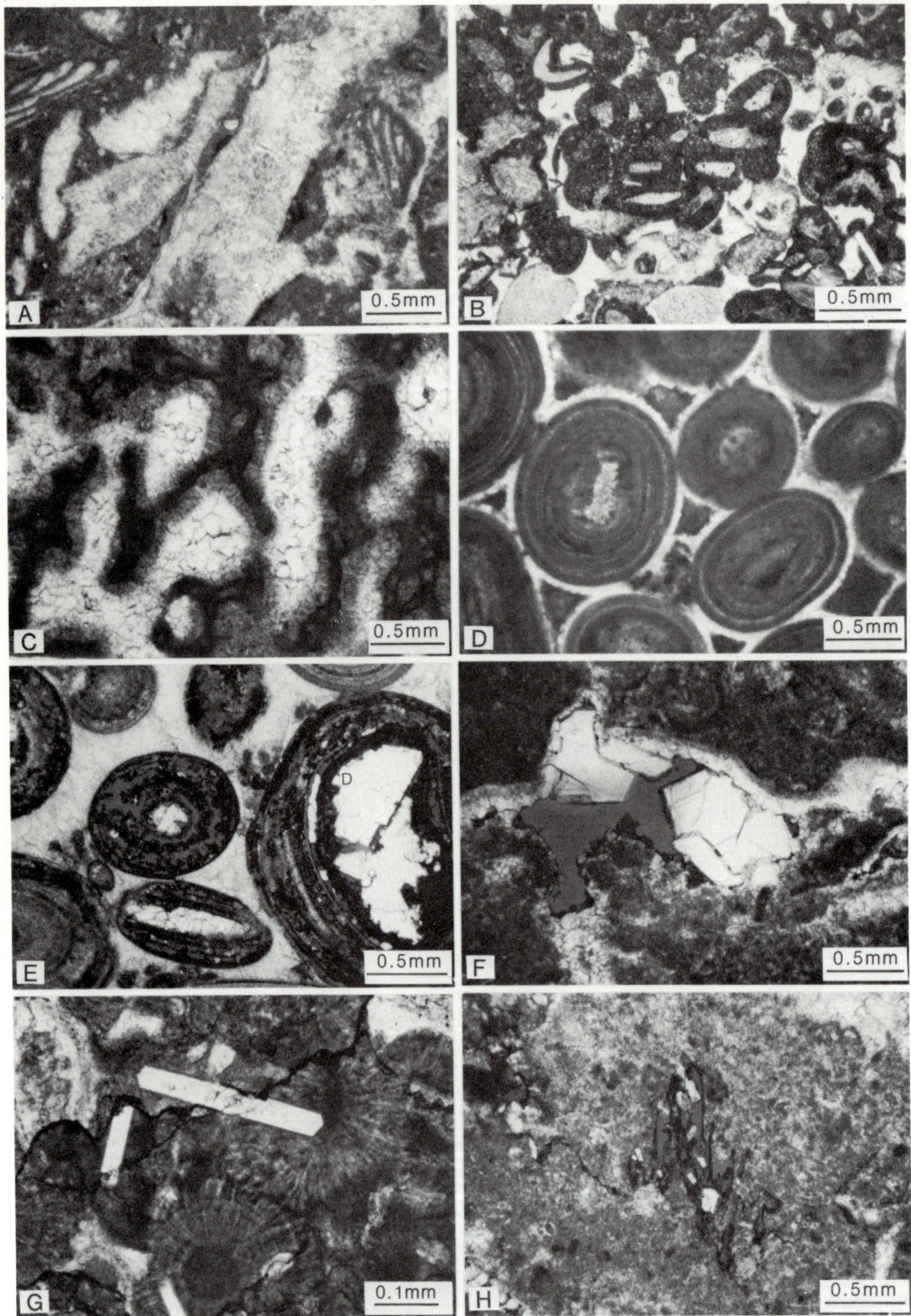

FIGURE 13.49 Atokan Limestone (Middle Pennsylvanian), west margin of Delaware Basin, Reeves County, Texas. Typical microfacies (continued) and diagenetic sequence. A. Microfacies 9. See text for

and thick pelecypods with interstitial biocalcarenite to biocalcisiltite matrix-filling intraframework spaces.

The Ideal Shallowing-Upward Sequence

The six above-described microfacies are organized in a shallowing-upward order which also expresses an increasing degree of energy of the environment of deposition culminating with the bioconstructed buildup (Fig. 13.56, upper part). The behavior of microscopic parameters allows one to distinguish three major environments which succeeded each other in time during the shallowing-upward evolution. The first environment (microfacies 1 and 2) represents interbuildup areas to lower portion of the bioclastic slope. It is characterized by concentrations of iron and silica and a predominance of planktonic ostracods over annelids and other relatively rare benthic constituents. The second environment represents the major portion of the bioclastic slope. It displays a succession of three distinct ecological peaks of frequency in decreasing depth order: annelids, dasyclads, and benthic foraminifers (miliolids and textulariids). The third environment corresponds to the constructed buildup in which the various framework-building organisms are associated with dasyclads.

A typical section (Fig. 13.56, lower part) shows detailed variation of microscopic parameters.

The Ideal Depositional Model

The internal ecological zonation of the buildups (Fig. 13.57) is very characteristic and indicates the effect of increasing shallowing and related increasing energy of the environment of deposition. This trend is also expressed by the vertical change of the intraframework material from calcilutites and calcisiltites to various types of calcarenites. The morphology of the investigated buildups (Carozzi, 1954) indicates that in spite of having generated appreciable flank deposits, they do not appear to have been destroyed very much by wave action. A bathymetric range of 5 to 15 m depth is suggested for constructed limestones. The depth of the second environment corresponding to the maximum development in increasing water depth of foraminifers (textulariids and miliolids), dasyclads, and annelids can be estimated to have been between 15 and 25 m. The third environment displays the last dasyclads in its upper part and represents the limit of the photic zone (45 m) and deeper.

13.7 CHARACTERISTIC CONSTITUENTS: RED AND GREEN ALGAE

Very agitated seas represent ideal conditions for development along platform edges of frontal buildups consisting of complex communities of many types of red and green algae. These communities with abundant encrusting forms are very resistant to wave action, produce a relatively small amount of bioclasts, and provide efficient protection for internal areas of the platforms. These internal areas consist of lagoonal intraclastic to pelletoidal carbonates with stromatolites and grade landward into coastal sabkhas.

The Bonfim Formation (Cenomanian) of the inland portion of the Barreirinhas Basin in northeast Brazil is 800 m thick and is entirely in the subsurface. It was investigated by means of 2300 thin sections made from cores and chips of 36 exploratory wells (Carozzi *et al.*, 1973). A total of 22 distinct microfacies was observed making up a complex sedimentologic evolution that con-

FIGURE 13.49 (*continued*) detailed description. *Marine phreatic.* B. Interpenetration of ooids due to early compaction. Contacts between ooids are nonsutured. C. Early isopachous fibrous marine cement coating thalli of *Donezella*. This early cement was followed by freshwater phreatic, cavity-filling sparite. D. Early marine cement coating ooids. Early compaction also can be seen which preceded the isopachous rim cement. Some calcisiltite internal sediment or micrite matrix occupies the remaining pore space. *Deep burial.* E. Large baroque dolomite rhomb (D) and late calcite cement are infilling previously moldic ooid on the right. Observe early fibrous submarine cement on ooid surfaces. F. Dissolved phylloid algal blade with biomoldic porosity (dark gray), partially occluded by euhedral crystals of late calcite. G. Stylolitized oolitic calcarenite with authigenic albite. Formation of larger crystal (center top) which includes insoluble residue followed stylolitization. Irregular ends of smaller crystal (left) indicates that its formation preceded stylolitization. H. Well-developed stylolitic porosity containing several authigenic albite crystals. All photomicrographs: plane-polarized light. From Von Bergen (1985).

DIAGENETIC PROCESSES AND FEATURES	DIAGENETIC ENVIRONMENTS						POROSITY	
	MARINE PHREATIC	FRESHWATER VADOSE	UNDERSAT. FRESHWATER PHREATIC	SATURATED FRESHWATER PHREATIC	MIXING (Marine & Freshwater Phreatic)	DEEP BURIAL	LOSS	GAIN
MICRITIZATION AND BORING (Micrite Envelopes)	—							
EARLY CEMENTATION (Fibrous Rim & Peloidal, Micritic Cements)	—							
PARTIAL LITHIFICATION OF LIME MUDS	—							
EARLY COMPACTION (Syneresis Cracks, Compacted Algae & Ooids)	—							
BORINGS IN FIRM SEDIMENT	—							
REWORKING (Intraclasts)	—							
NONSELECTIVE DISSOLUTION (Collapse Breccia, Vugs, Enlarged Burrows)		—						
VADOSE SILT		—						
ISOPACHOUS RIM CEMENT (Lining Burrows & Vugs)	—							
INTERNAL SEDIMENTS	—							
EARLY DOLOMITIZATION (Texture Preserving)					—			
SELECTIVE DISSOLUTION OF UNSTABLE GRAINS			—					
NEOMORPHISM (Stabilization of Rim Cement)				—				
MOSAIC AND SYNTAXIAL CEMENTS				—				
LATE COMPACTION (Stylolites and Chain Ooids)						—		
SILICIFICATION (Micronodules & Bioclasts)						—		
AUTHIGENIC FELDSPAR						—		
STYLOLITIC DOLOMITIZATION						—		
LATE DISSOLUTION (Stylolitic, Oomoldic & Biomoldic Porosity)						—		
LATE DOLOMITIZATION (Baroque Dolomite)						—		
LATE CALCITE I (Replacing Baroque Dolomite & Filling Late Porosity)						—		
LATE FRACTURATION						—		
LATE CALCITE II (Infilling Late Fractures)						—		

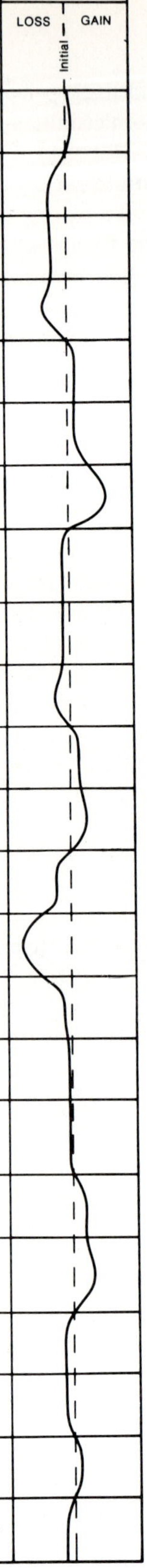

FIGURE 13.50 Atokan Limestone (Middle Pennsylvanian), west margin of Delaware Basin, Reeves County, Texas. Generalized diagenetic sequence showing inferred variations of relative porosity with changes in diagenetic environments and processes. From Carozzi and Von Bergen (1987). Reprinted by permission of Scientific Press Ltd.

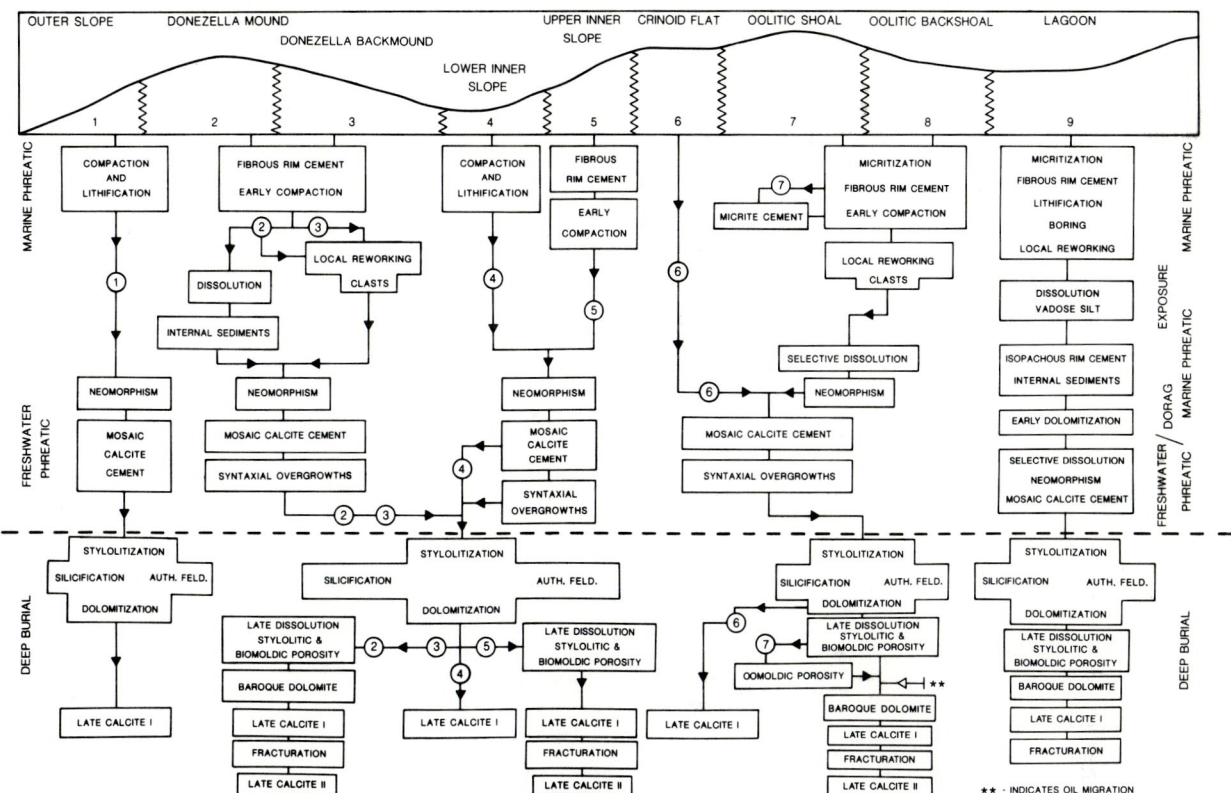

FIGURE 13.51 Atokan Limestone (Middle Pennsylvanian), west margin of Delaware Basin, Reeves County, Texas. Main diagenetic pathways for each microfacies of ideal model. The arrows in this flow diagram illustrate the probable sequence of events for each microfacies and the general diagenetic environments are indicated in the margin. The horizontal, heavy dashed line separates the eogenetic events (top) from those belonging to the mesogenetic or deep burial environment. From Carozzi and Von Bergen (1987). Reprinted by permission of Scientific Press Ltd.

sisted of the succession through time of three depositional models. Microfacies are described in a general shallowing-upward order and are subsequently integrated into their respective models.

Description of Microfacies

Basinal Environment

Microfacies 1 (Fig. 13.58A). Micrite with clotted texture, locally pelletoidal with abundant pyrite pigments and silt-size detrital quartz. *Hedbergella* is the most common planktonic foraminifer associated with calcispheres, radiolarians, and planktonic pelecypods (microfilaments). This fauna is not as abundant as is normally the case in an open sea pelagic carbonate, and some environmental restriction is implied. Rare bioclasts of echinoids are observed.

Slope Environment

Microfacies 2 (Fig. 13.58B). Grain-supported biocalcarenite with micrite matrix. The predominant and poorly sorted bioclasts include the red algae *Lithothamnium* and *Solenopora* and the green algae *Cayeuxia* and *Pycnoporidium*. They are associated with debris of echinoids and mollusks, benthic foraminifers, and ostracods. Slumping features are common, indicating transportation and redeposition.

Microfacies 3 (Fig. 13.58C). Grain-supported oolitic calcarenite with micrite matrix. Ooids display a variety of cores, including microoncoids, small gastropods, bioclasts of pelecypods, and echinoids. They show frequent abrasion and fragmentation and have been strongly affected by pressure-solution and dislocated by compaction. The interstitial micrite contains small bioclasts of red algae, pelecypods, gastropods, echinoids, benthic foraminifers, bryozoans, and intraclasts. Slump-

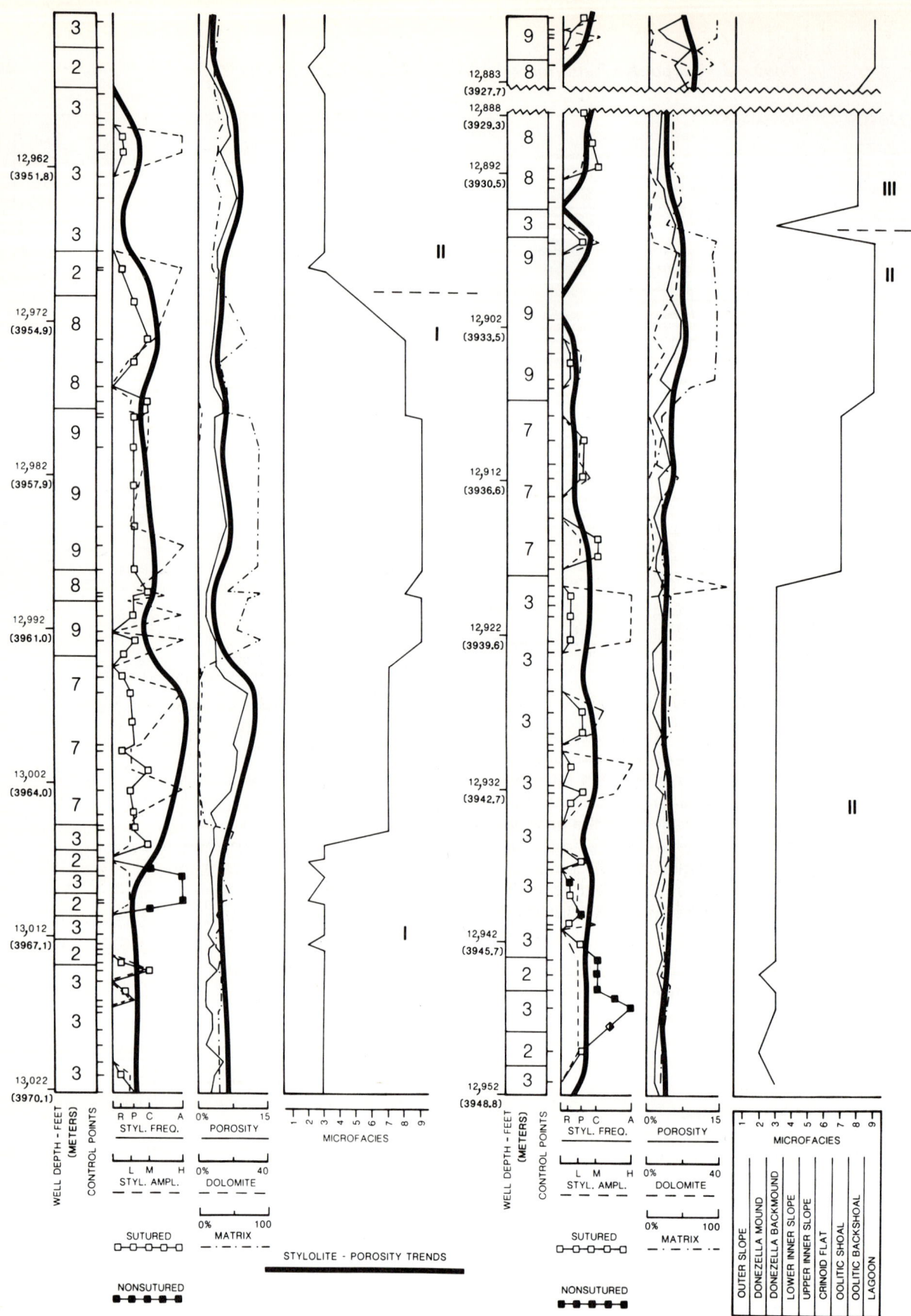

FIGURE 13.52 Atokan Limestone (Middle Pennsylvanian), west margin of Delaware Basin, Reeves County, Texas. General vertical evolution of depositional environments and stylolitic porosity trends in well "BA" Fee. From Carozzi and Von Bergen (1987). Reprinted by permission of Scientific Press Ltd.

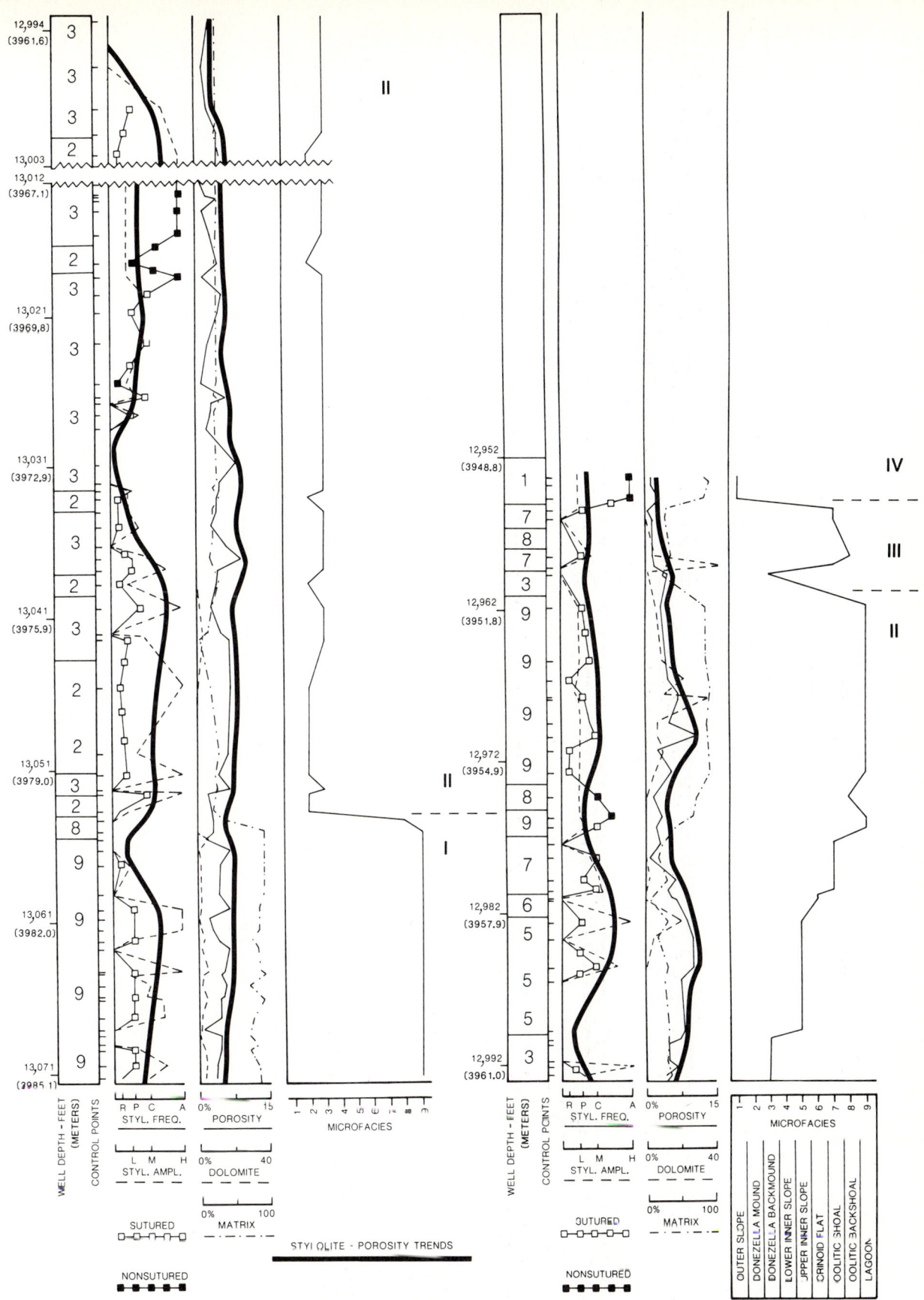

FIGURE 13.53 Atokan Limestone (Middle Pennsylvanian), west margin of Delaware Basin, Reeves County, Texas. General evolution of depositional environments and stylolitic porosity trends in well "AZ" Fee. From Von Bergen (1985).

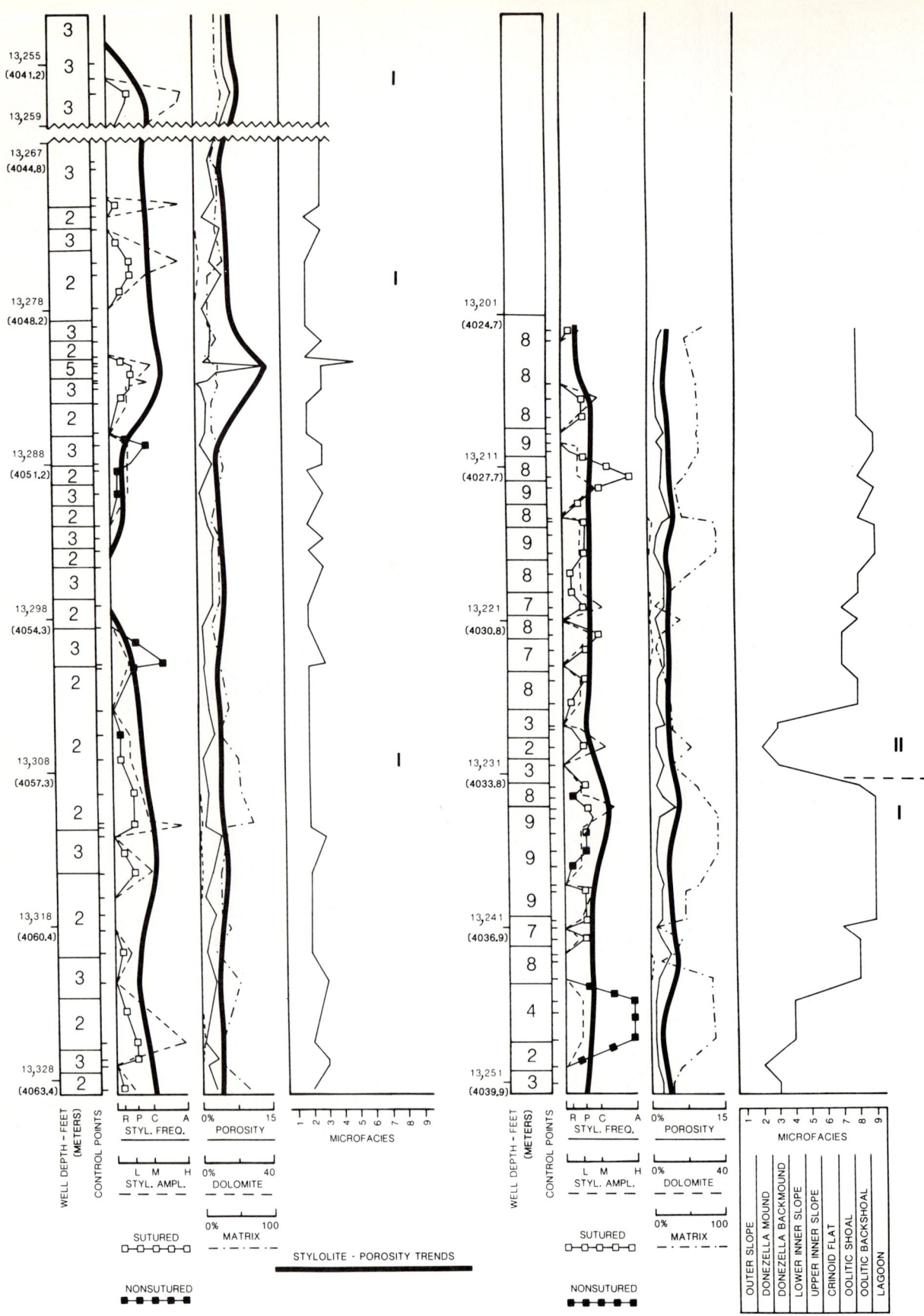

FIGURE 13.54 Atokan Limestone (Middle Pennsylvanian), west margin of Delaware Basin, Reeves County, Texas. General vertical evolution of depositional environments and stylolitic porosity trends in well Chapman West. From Von Bergen (1985).

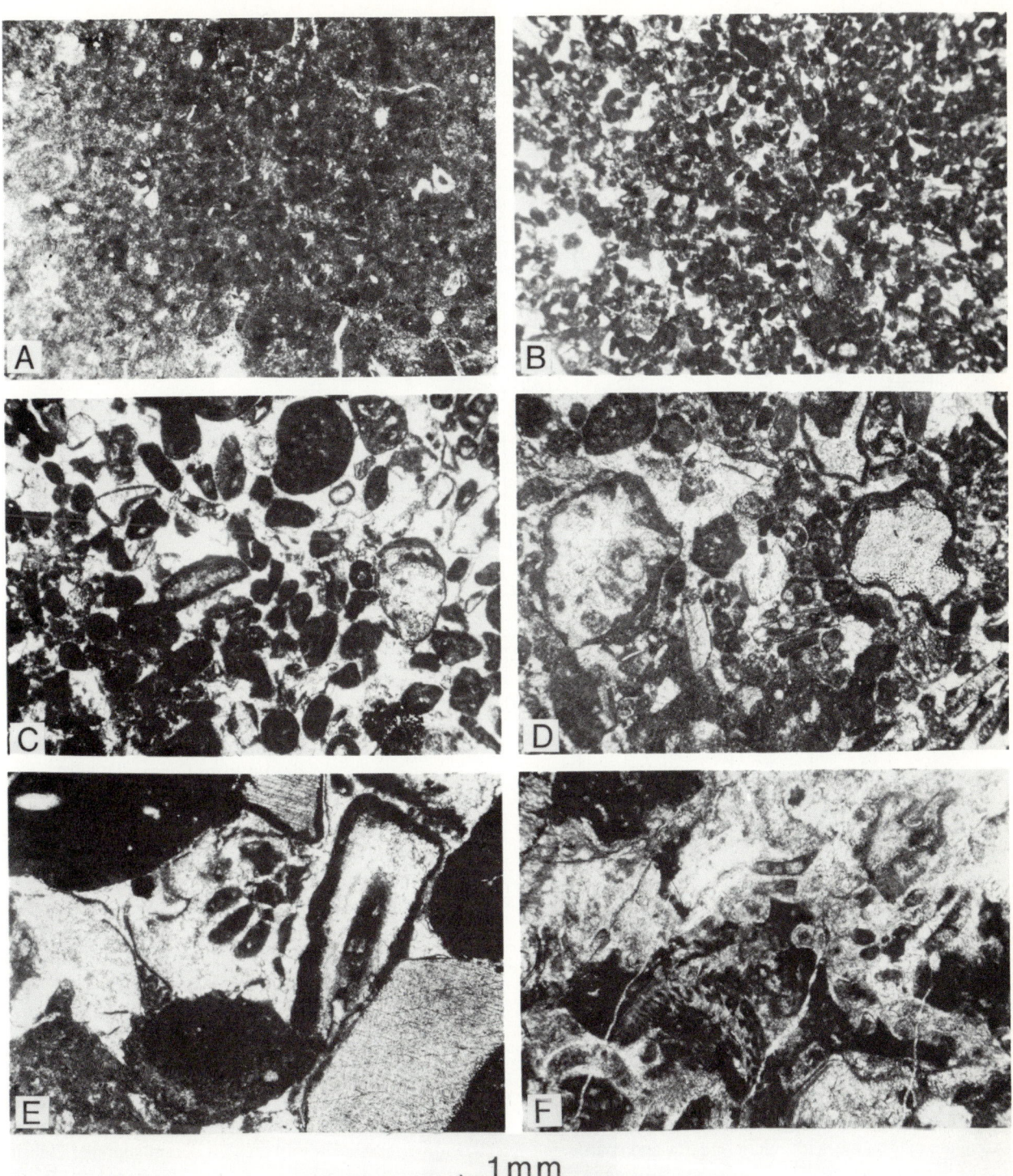

FIGURE 13.55 Kimmeridgian-Portlandian (Upper Jurassic), Salève, Haute Savoie, France. Typical microfacies. A. Microfacies 1. B. Microfacies 2. C. Microfacies 3. D. Microfacies 4. E. Microfacies 5. F. Microfacies 6. See text for detailed descriptions. All photomicrographs: plane-polarized light. From Carozzi (1955b).

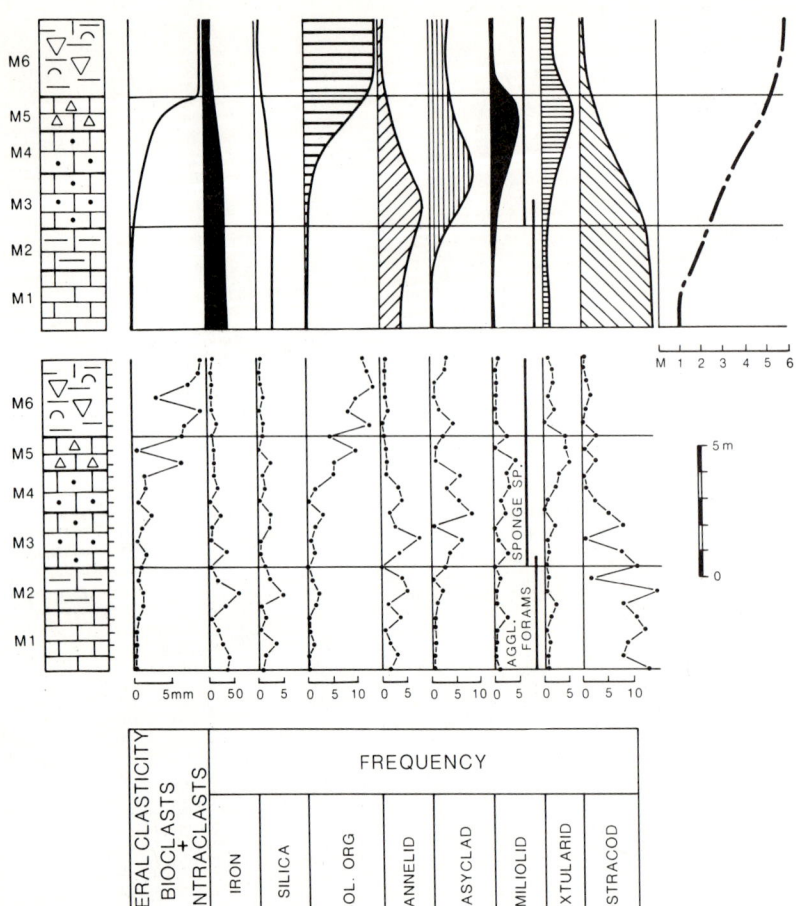

FIGURE 13.56 Kimmeridgian-Portlandian (Upper Jurassic), Salève, Haute Savoie, France. Ideal shallowing-upward sequence (upper part) and typical example of vertical variation of microscopic parameters during a shallowing-upward sequence (lower part). Redrawn from Carozzi (1955b).

ing features are common, indicating transportation and redeposition.

External Platform with Red Algae Buildups

Microfacies 4 (Fig. 13.58D). Red algae-constructed limestone consisting of complex colonies of *Lithothamnium*, *Lithophyllum*, *Solenopora*, and *Amphiroa* to which are associated the green algae *Cayeuxia* and *Pycnoporidium*. Interstitial material is micrite with minute bioclasts and scattered patches of sparite cement.

Microfacies 5 (Fig. 13.58E). Grain-supported red algae calcirudite with micrite matrix and sparite cement. Poorly sorted and subangular bioclasts are derived from *Lithothamnium*, *Solenopora*, and *Amphiroa* with smaller contribution from *Cayeuxia* and *Pycnoporidium*. Bioclasts of echinoids and mollusks are associated with miliolids and *Trocholina*. Interstitial material is micritic matrix with patches of sparite cement.

Microfacies 6 (Fig. 13.58F). Grain-supported red algae biocalcarenite with sparite cement. Predominant well-sorted and well-rounded bioclasts are derived from *Lithothamnium* and *Solenopora*, while whole colonies of *Amphiroa* and *Lithophyllum* are not so important. Other bioclasts are echinoids, bryozoans, mollusks, and green algae together with some reworked ooids. Cementation is by cavity-filling sparite and syntaxial overgrowths around echinoids; local patches of micrite are altered by neomorphism into pseudosparite.

Microfacies 7, 8, and 9 (Figs. 13.58G,H, 13.59A). These three microfacies are related closely and distinguished according to predominance of bioclasts of green algae (microfacies 7, Fig. 13.58G), echinoids (microfacies 8, Fig. 13.58H), and mollusks (microfacies 9, Fig. 13.59A). They all display a grain-supported fabric with sparite cement. In microfacies 7, bioclasts of *Bouenia*, *Arabicodium*, and *Halimeda* are the most abundant. Minor components are benthic foraminifers (sometimes agglutinated), *Lithothamnium*, and reworked ooids and intraclasts.

Microfacies 10 (Fig. 13.59B). Grain-supported oolitic calcarenite with sparite cement. Well-developed and well-sorted ooids display cores consisting of bioclasts of algae, mollusks, and bryozoans. Many

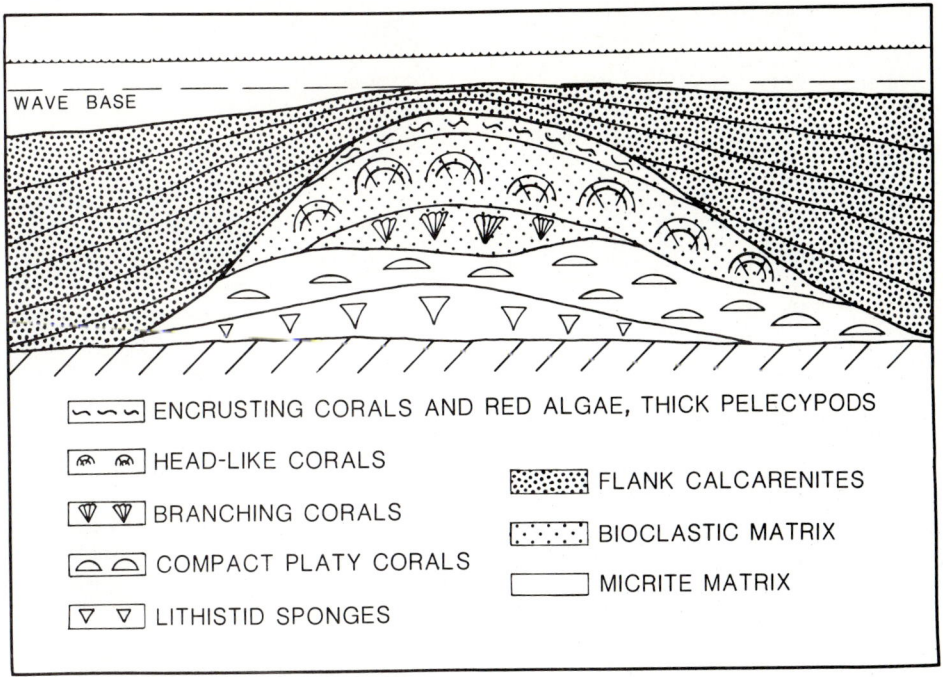

FIGURE 13.57 Kimmeridgian-Portlandian (Upper Jurassic), Salève, Haute Savoie, France. Internal ecological zonation of buildup. Modified from Wilson (1975).

ooids display extensive perforation due to endolithic algae. The effect of pressure-solution is widespread and leads to reciprocal deformation of the ooids into polygonal shapes. Associated bioclasts are derived from red algae, echinoids, bryozoans, and benthic foraminifers. The interstitial sparite cement displays local patches of micrite. Oomoldic porosity is present and secondary dolomitization occurs at all degrees of intensity.

External Ramp

The external ramp environment developed at the end of Bonfim times when the red algae buildups had disappeared and were replaced by extensive biocalcarenites. At this particular time, microfacies 6 to 9 became predominant and were bordered both seaward and landward by three lower-energy but equivalent bioclastic microfacies (11, 12, and 13).

Microfacies 11, 12, and 13 (Fig. 13.59C,D,E). These three microfacies are very closely related and distinguished according to the predominance of bioclasts of green algae (microfacies 11, Fig. 13.59C), echinoids (microfacies 12, Fig. 13.59D), and mollusks (microfacies 13, Fig. 13.59E). The matrix of these grain-supported rocks is a micrite containing smaller debris of the major constituents along with benthic foraminifers and ostracods. The micrite matrix is locally altered by neomorphism into pseudomicrosparite.

Internal Ramp

Microfacies 14 (Fig. 13.59F). Grain-supported pelletoidal calcarenite with sparite cement. Well-rounded and well-sorted pellets of dark micrite (average size, 0.12 mm) appear lithic in origin, and some of the elongate ones are oriented parallel to bedding. In places, concentrations of larger elliptical pellets of dark brown micrite (average size, 0.25 mm) appear fecal. Minor components are bioclasts of *Lithothamnium*, echinoids, mollusks, green algae, and benthic foraminifers.

Microfacies 15 (Fig. 13.59G). Grain-supported merged pelletoidal calcarenite with micrite matrix. The pellets are identical to those of microfacies 14 but heavily merged and reciprocally interpenetrated. The micrite matrix is often altered by neomorphism into pseudomicrosparite, and contains small bioclasts of echinoids, benthic foraminifers, pelecypods, and reworked ooids. Small patches of sparite are syntaxial overgrowths around echinoid debris.

Microfacies 16 (Fig. 13.59H). Mud-supported biocalcarenite with micrite matrix. The most frequent bioclasts are echinoids, pelecypods, bryozoans, oncoids, benthic foraminifers, ostracods, and calcispheres. Intraclasts, subrounded and extensively pyritized, are derived from intraformational reworking of several types of micritic muds.

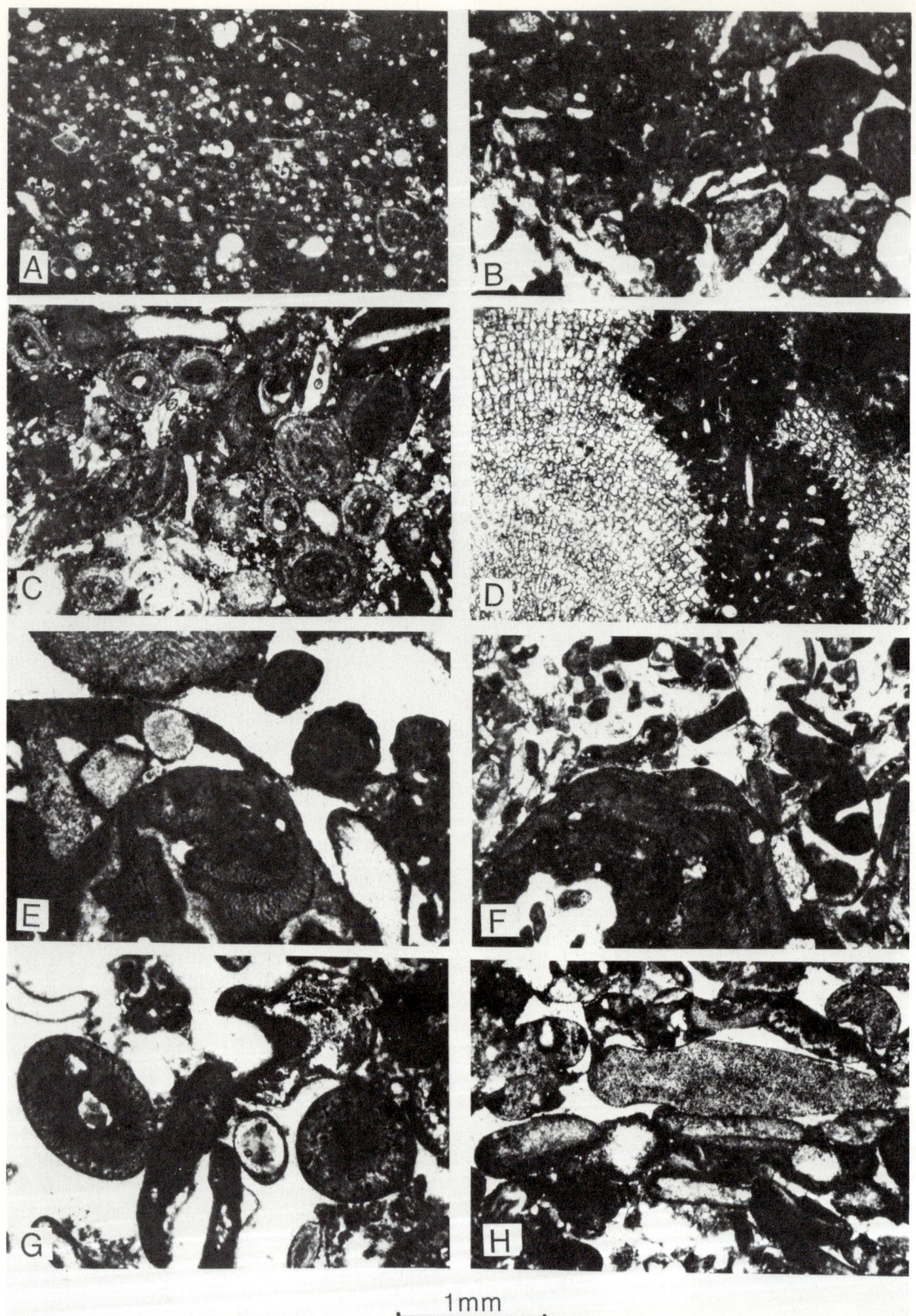

FIGURE 13.58 Bonfim Formation (Cenomanian), Barreirinhas Basin, northeast Brazil. Typical microfacies. A. Microfacies 1. B. Microfacies 2. C. Microfacies 3. D. Microfacies 4. E. Microfacies 5. F. Microfacies 6. G. Microfacies 7. H. Microfacies 8. See text for detailed descriptions. All photomicrographs: plane-polarized light. From Carozzi *et al.* (1973).

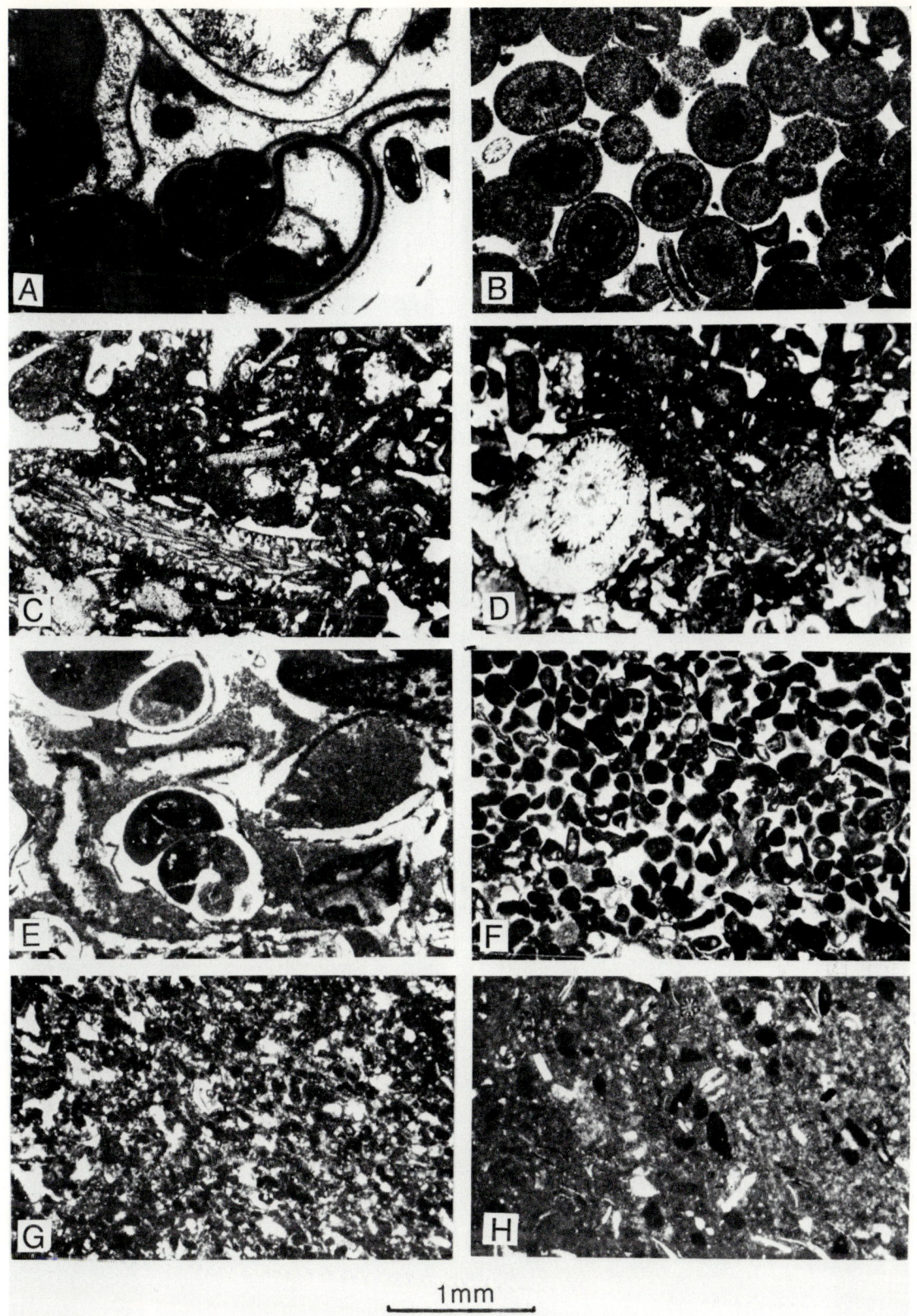

FIGURE 13.59 Bonfim Formation (Cenomanian), Barreirinhas Basin, northeast Brazil. Typical microfacies (continued). A. Microfacies 9. B. Microfacies 10. C. Microfacies 11. D. Microfacies 12. E. Microfacies 13. F. Microfacies 14. G. Microfacies 15. H. Microfacies 16. See text for detailed descriptions. All photomicrographs: plane-polarized light. From Carozzi *et al.* (1973).

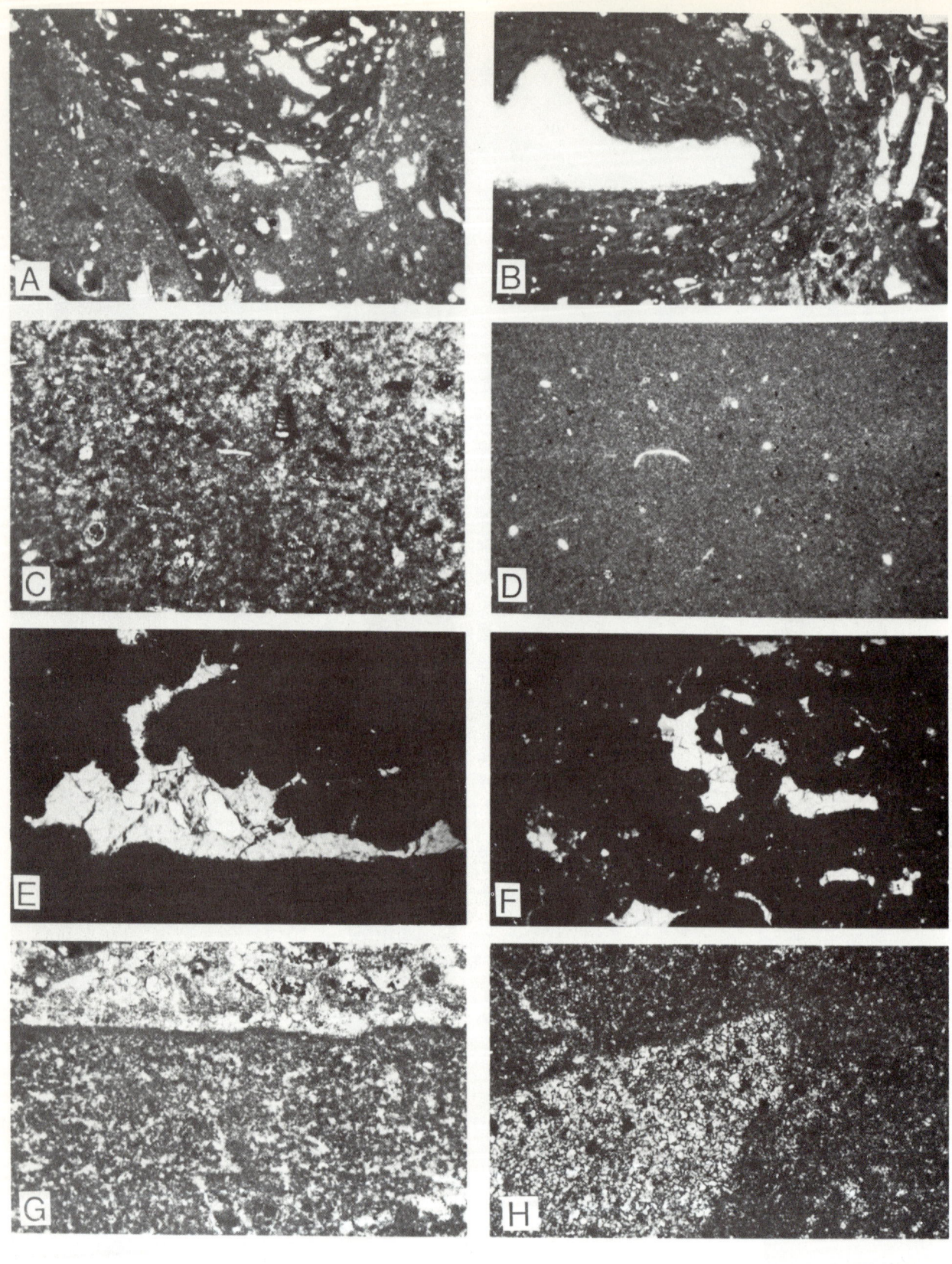

FIGURE 13.60 Bonfim Formation (Cenomanian), Barreirinhas Basin, northeast Brazil. Typical microfacies (continued). A and B. Microfacies 17. C. Microfacies 18. D. Microfacies 19. E and F. Microfacies 20. G. Microfacies 21. H. Microfacies 22. See text for detailed descriptions. All photomicrographs: plane-polarized light. From Carozzi et al. (1973).

Microfacies 17 (Fig. 13.60A,B). Micrite with scattered blue-green algal oncoids. The micrite matrix contains small bioclasts of echinoids, pelecypods, bryozoans, benthic foraminifers, and ostracods, together with a few reworked ooids and pellets. Blue-green algal oncoids range in size from a fraction of 1 mm to 3 cm. Their irregularly concentric algal mats developed around cores of bioclasts of mollusks (Fig. 13.60B), echinoids, bryozoans, corals, worm tubes, and benthic foraminifers. This microfacies is frequently dolomitized.

Microfacies 18 (Fig. 13.60C). Merged pelletoidal micrite with benthic organisms. The heterogeneously textured micrite matrix contains abundant pyrite pigments and scattered small debris of echinoids and mollusks associated with benthic foraminifers and ostracods. Occasional larger debris of red or blue-green algae may be found. Neomorphism into pseudomicrosparite is frequent and irregularly distributed.

Microfacies 19 (Fig. 13.60D). Massive micrite with pyrite pigments, fine silt-size detrital quartz, and ostracods.

Supratidal Environment

Microfacies 20 (Fig. 13.60E,F). Fenestral dolomicrite. In the homogeneous groundmass, which may contain occasional thin stromatolitic mats, are scattered pellets, small intraclasts, and a few miliolids. Fenestrae, elongated parallel to bedding, are filled with freshwater phreatic calcite mosaic. Some of the cavities appear to be enlarged molds of dissolved intraclasts or bioclasts (Fig. 13.60F).

Microfacies 21 (Fig. 13.60G). Stromatolite-constructed dolomite. Wavy algal mats consisting of dark organic-rich dolomicrite alternate with thick layers of trapped pelletoidal dolomicrosparite.

Microfacies 22 (Fig. 13.60H). Desiccation dolobreccia. Angular to subangular fragments of dolomicrosparite are set, often with an imbricate structure, in a groundmass of dolomicrite. The latter may contain windblown silt-size grains of detrital quartz, clay minerals, and rare plant remains.

The Ideal Depositional Models

Model 1 (Fig. 13.61, lower part) corresponds to the best development of the red algae bank with its associated biocalcirudites, biocalcarenites, and oolitic calcarenites which formed a well-defined barrier generating widespread supratidal lagoons and flats behind it. These flats were of sabkha type and gave rise to an effective dolomitization by seepage reflux through the red algae bank. Model 2 (Fig. 13.61, middle part) corresponds to a time when the bank began to deteriorate and buildups were smaller and less abundant. The existence of numerous transverse tidal channels strongly decreased the effectiveness of the bank as a barrier. A widespread internal platform developed and the supratidal environment was appreciably reduced. The low-energy micritic sediments of the lagoon were relatively impervious and reduced the dolomitization of the external platform by seepage reflux to a minimum. Model 3 (Fig. 13.61, upper part) represents the disappearance of the red algae bank. It consisted of a gentle seaward-sloping ramp containing a central band of biocalcarenites cemented with sparite and flanked on both sides by zones of biocalcarenites with micrite matrix. This shoal-like system followed in part the trend of the underlying dead red algae bank of which it was the last residual expression. The widespread internal part of the ramp was similar to the previous internal platform and the supratidal environment remained reduced. The seepage reflux was nonexistent, and the widespread secondary dolomitization of medium intensity observed at this level was a late dorag type related to the erosional unconformity and exposure at the end of carbonate deposition. This rare case of a platform reverting to a ramp resulted from drowning due to eustatic or tectonic causes.

The Bonfim Formation with its widespread combination of seepage reflux by sabkha brines and mixed seawater-freshwater during its final emergence possesses an appreciable porosity and hence contains potential oil and gas reservoirs onshore and offshore.

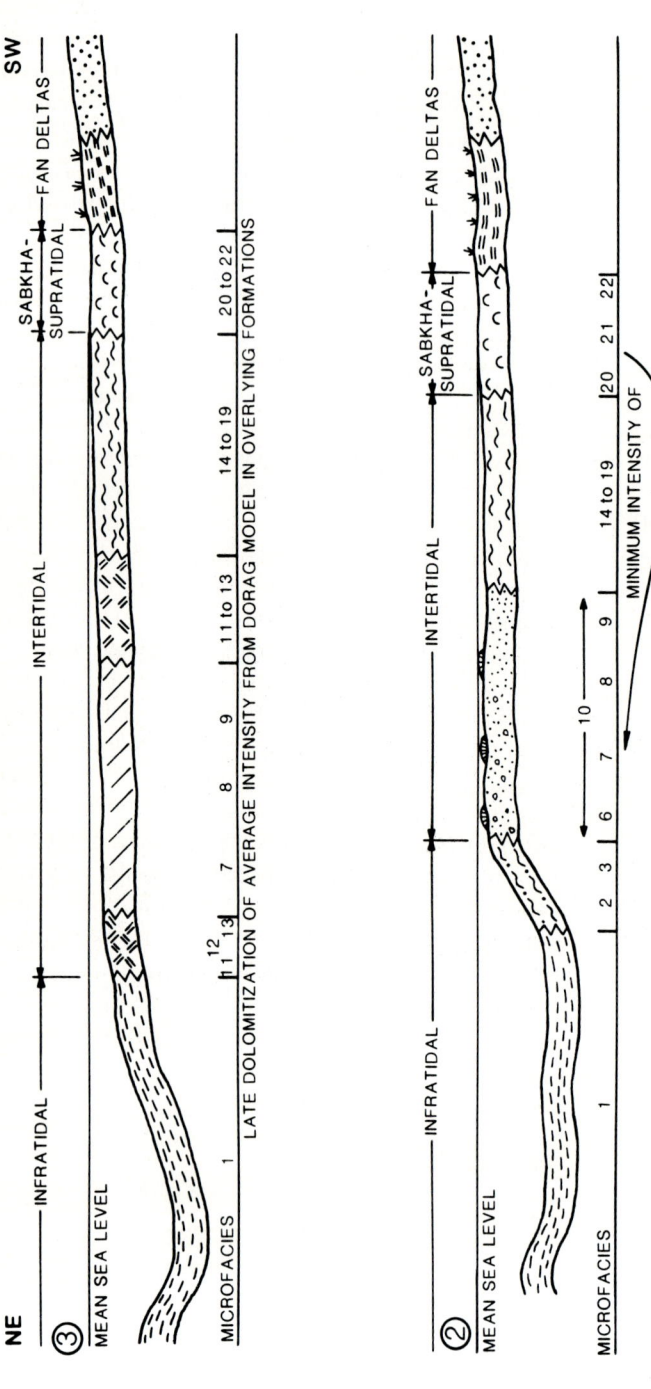
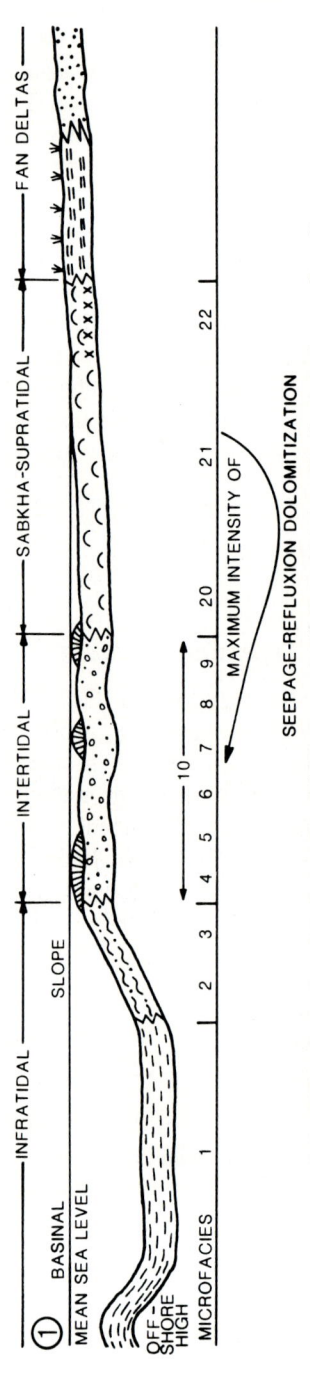

FIGURE 13.61 Bonfim Formation (Cenomanian), Barreirinhas Basin, northeast Brazil. Ideal depositional models 1, 2, and 3. From Carozzi et al. (1973).

14

Carbonate Platforms with Frontal Bioconstructed Buildups

Among efficient builders of platform edges in very agitated conditions are columnar to hemispherical stromatolite colonies. Although they generate desiccation breccias through temporary subaerial exposure, their bioclast production is relatively small. Stromatolites often occur on platforms which display widespread frontal oolitization by the effect of cold slope-ascending currents and which support extensive lagoons also with stromatolites, or coastal sabkhas.

Under rough sea conditions, similar effective constructions are built by the association of corals–red algae–encrusting foraminifers which generates large frontal aprons of calcirudites grading into deeper-water carbonate turbidites.

14.1 CHARACTERISTIC CONSTITUENTS: STROMATOLITES UNDER STORM PROCESSES

The Allentown Dolomite, Upper Cambrian, of Warren County, New Jersey (Zadnik and Carozzi, 1963), was investigated at two field sections (not overlapping) and reaching a total thickness of 2202.3 ft (661 m). Detailed petrographic study was completed on 1198 samples collected at an average vertical interval of 1.5 ft (0.45 m). The Allentown Dolomite displays numerous types of shallowing-upward sequences, sometimes with internal microsequences, and associated in large megasequences of similar nature.

Description of Microfacies

Petrographic study of both sections permitted recognition of six microfacies described in order of decreasing relative depth of deposition.

Microfacies 1 (Fig. 14.1A). Weakly laminated dololutite consisting of a uniform mosaic of anhedral and interlocked dolomite crystals with relatively abundant interstitial pyrite pigments. Grains of detrital quartz are distributed regularly throughout the groundmass of what was originally a laminated calcilutite.

Microfacies 2 (Fig. 14.1B). Grain-supported, intraclastic, and oolitic dolarenite with a dolosiltite matrix that appears to be relatively dark colored and contains interstitial cubes and pigments of pyrite distributed uniformly. Intraclasts derived from microfacies 1 are often aligned parallel to bedding, and the relatively small ooids display a poorly preserved concentric structure. Medium-size abundant grains of detrital quartz with very irregular boundaries due to marginal replacement by carbonates (Page and Carozzi, 1962), are scattered uniformly in the matrix or concentrated in streaks and small lenses. This microfacies was originally an intraclastic, oolitic, and perhaps pelletoidal calcarenite with a calcisiltite matrix.

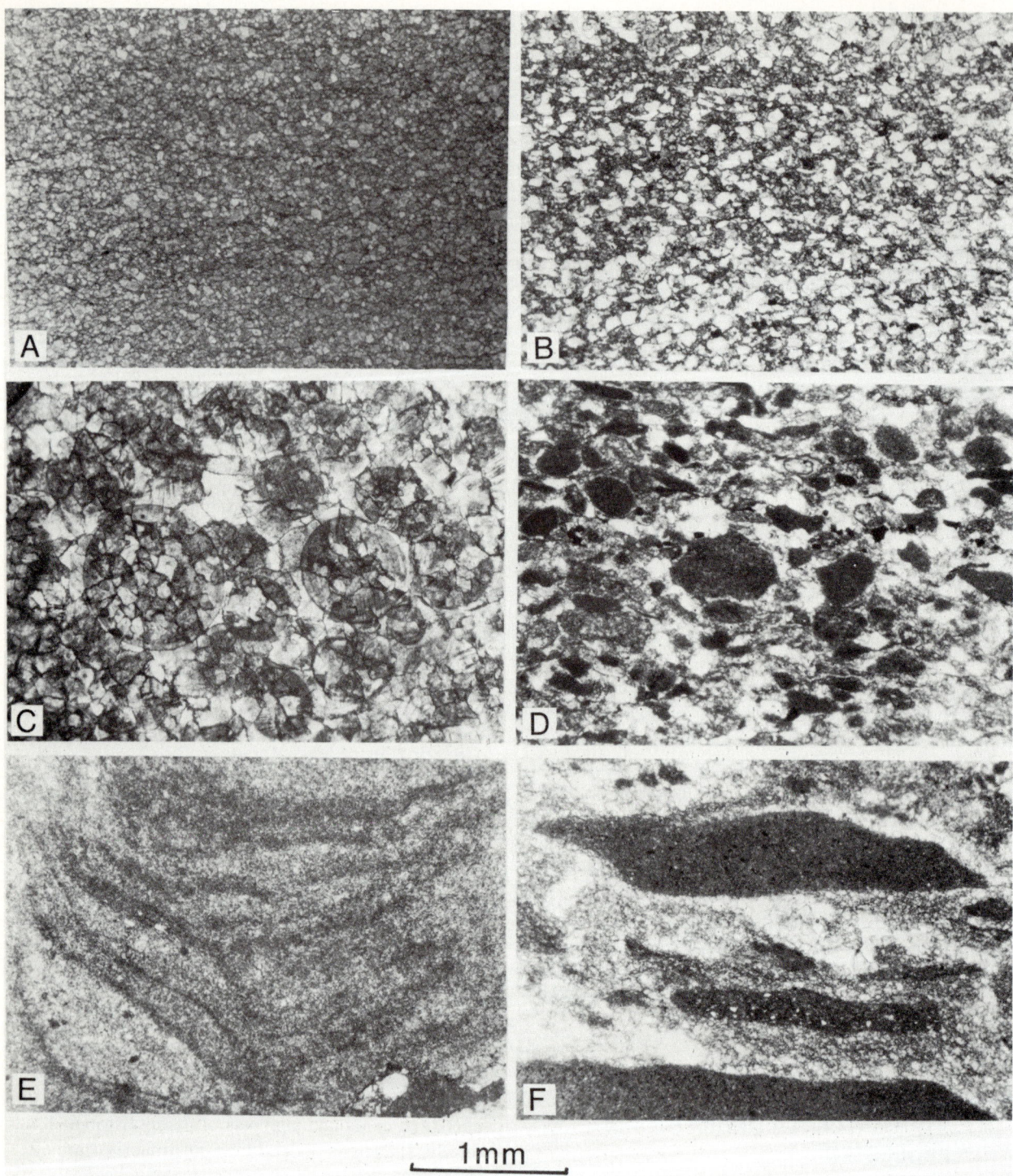

FIGURE 14.1 Allentown Dolomite (Upper Cambrian), Warren County, New Jersey. Typical microfacies. A. Microfacies 1. B. Microfacies 2. C. Microfacies 3. D. Microfacies 4. E. Microfacies 5. F. Microfacies 6. See text for detailed descriptions. All photomicrographs: plane-polarized light. From Zadnik and Carozzi (1963).

Microfacies 3 (Fig. 14.1C). Grain-supported oolitic to pseudo-oolitic dolarenite with dolosparite cement. The normal ooids are developed perfectly and have clearly visible concentric rings and rounded nuclei of dololutite (microfacies1) that also form the uncoated pseudo-ooids. Despite the fact that large subhedral to euhedral crystals of dolomite have developed throughout the rock irrespective of its original texture, ooids and pseudo-ooids stand out from the clear matrix by their darker color because of organic matter that delineates their internal structures or boundaries. Occurring often in this microfacies are several types of ooids distorted by compaction processes as well as half-moon ooids representing originally alternating carbonate-evaporite concentric layers with diagenetic dissolution of the latter (Carozzi, 1961c, 1963). Grains of detrital quartz and pyrite pigments are relatively rare. This microfacies was originally an oolitic calcarenite with sparite cement.

Microfacies 4 (Fig. 14.1D). Grain-supported intraclastic dolorudite with interstitial material consisting of intraclastic dolarenite or oolitic dolarenite with a dark dolomicrosparite cement. The subangular to rounded intraclasts are poorly sorted and display a wide range of sizes. They consist of fragments derived from the dololutite of microfacies 1, the intraclastic dolarenite of microfacies 2, and the oolitic dolarenite of microfacies 3 mixed in all proportions but often with one type predominating in a given thin section. Grains of detrital quartz are fairly abundant and a few large isolated reworked ooids may occur in places. This microfacies originally was a grain-supported intraclastic calcirudite with an interstitial intraclastic calcarenite with a calcisiltite matrix.

Microfacies 5 (Fig. 14.1E). Stromatolite-constructed dolomite displaying laterally linked hemispheroidal colonies with continuous to pelleted mats of darker color separated by lighter-colored trapped dolosiltite. Between the latter and in the V-shaped depressions separating the stromatolite heads are concentrations of very small grains of detrital quartz, pellets, and small ooids in a dolosiltite matrix.

Microfacies 6 (Fig. 14.1F). Coarse, grain-supported desiccation dolorudite. The flat pebbles, either oriented parallel to bedding or imbricated, consist of dololutite, intraclastic dolarenite, and oolitic dolarenite corresponding, respectively, to clasts of microfacies 1, 2, and 3. They are associated with crescent-shaped fragments with concentric lamellae derived from the destruction of stromatolitic heads of microfacies 5. However, one type of clast may predominate in a given thin section. The interstitial material is an intraclastic dolarenite with abundant and large grains of detrital quartz and reworked ooids set in a cement of dolosparite. Locally, quartz grains may be abundant enough to form most of the interstitial material between flat pebbles. In this case a quartzitic texture by pressure solution is generated. This microfacies was originally a desiccation calcirudite in which storm action played an important role.

The Ideal Shallowing-Upward Sequence

The six microfacies of the Allentown Dolomite (Fig. 14.2) can be organized into a shallowing-upward sequence. The curve of general clasticity pertains to intraclasts and varies in agreement with the parameters of detrital quartz by reaching a first peak in the intraclastic dolorudite of microfacies 4 and the largest peak in the desiccation dolorudite of microfacies 7 which terminates the shallowing-upward process. The two populations of ooids are clearly displayed. The index of crystallinity expresses the variation of the maximum apparent size of the dolomite crystals. Usually, it is a very useful indicator of the original grain size of carbonate sediment before dolomitization. This "blueprint" relationship displays a peak encompassing microfacies 3 and 4 and results from the combination of the peak of ooid size and of intraclast size, respectively. This relationship does not hold for the desiccation dolomite (microfacies 6), where the abundance and large size of detrital quartz grains have prevented the growth of dolomite crystals. The increase upward of pyrite indicates increasing iron oxide supply with shallowing conditions subsequently changed diagenetically into sulfide during burial.

A typical interval of field section No. 2 at Carpentersville, New Jersey (Figs. 14.3, 14.4), shows the reciprocal behavior of microscopic parameters in the superposition of several shallowing-upward sequences. The latter, which may contain smaller microsequences (for instance, 82, 83, and 84), associate themselves in shallowing-upward megasequences (such as XV) in which the superposed sequences terminate with progressively shallower microfacies.

The Ideal Depositional Model

The carbonate platform of the Allentown Dolomite may be interpreted as a depositional model under predominant storm action (Fig. 14.5). Indeed, the general juxtaposition of slope deposits, oolitic flat, stromatolite buildup, and inferred lagoon was modified almost constantly by storms moving in a landward direction as demonstrated by the presence of aprons of intraclastic calcirudites both in front and in the back of the stromatolitic buildup, but also by the cumulative reworking of microfacies in a landward direction. The detrital quartz sediment yield came from the opposite direction, where a coastal and continental sabkha, with dunes, was adjacent to the in-

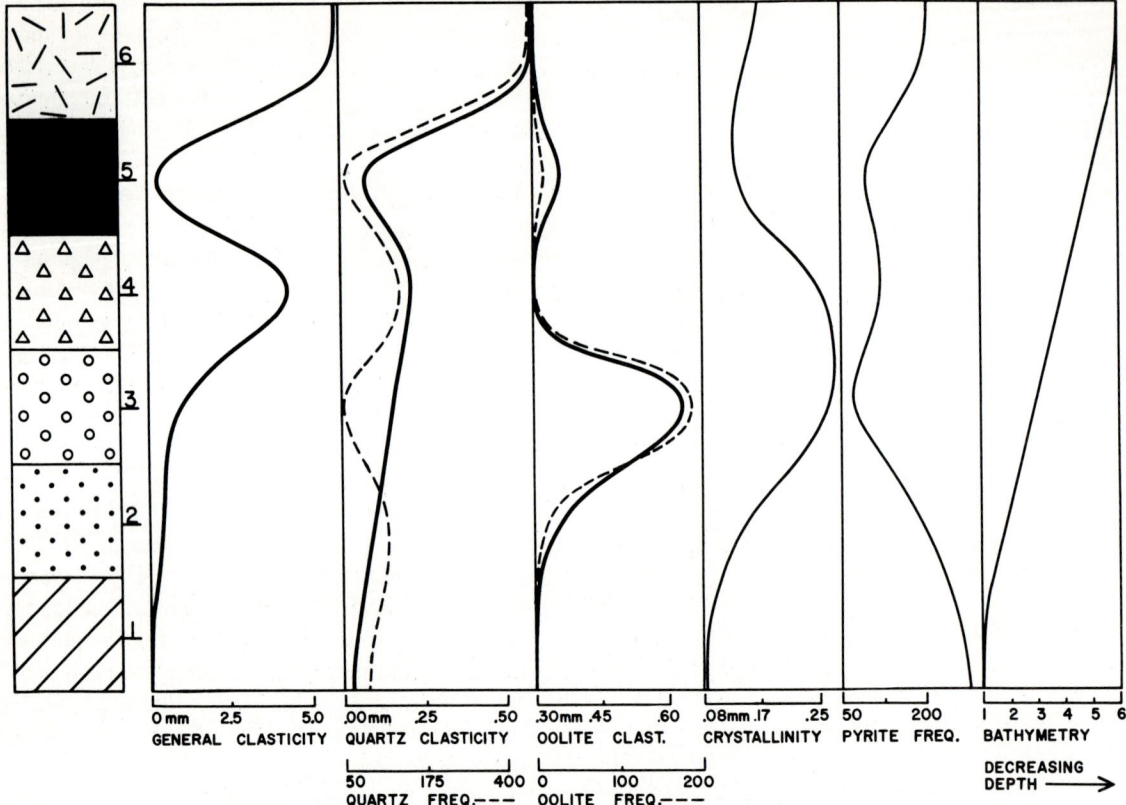

FIGURE 14.2 Allentown Dolomite (Upper Cambrian), Warren County, New Jersey. Ideal shallowing-upward sequence. From Zadnik and Carozzi (1963).

ferred lagoon. Both environments were not encountered in the investigated sections.

General Evolution of the Depositional Environments

Field section No. 2 at Carpentersville, New Jersey, can be taken as an example (Figs. 14.6, 14.7) for large-scale interpretation. It consists of 104 shallowing-upward sequences associated into 21 shallowing-upward megasequences. Curves of general trend of the microscopic parameters have been computed by dividing the section in segments of 20 ft (6.10 m) thickness, regardless of the sequence boundaries, and by calculating for each segment the mean value of each parameter.

The curves of main trend of microscopic parameters display at this scale the same general relationship recorded by the ideal shallowing-upward sequence. The curve of relative bathymetry indicates the existence of distinct periods which show particular bathymetric characteristics. Megasequences I through VII show rapid depth variations corresponding to relatively thin shallowing-upward sequences most of which terminate with the desiccation dolorudite (microfacies 6). Megasequences VIII to XI correspond to moderate oscillations producing shallowing-upward sequences of irregular thickness. Most of them are thicker than those of the preceding shallower phase, and only a small number of them terminate with the desiccation dolorudite. Megasequences XII through XIX repeat a shallower phase during which the shallowing-upward sequences are thin and display strong oscillations. Above megasequence XIX, a deeper phase begins with weak oscillations and thicker shallowing-upward sequences.

The curve of relative bathymetry, which shows these distinct periods of oscillations, allows one to carry the analysis of the sedimentation one step further by illustrating the mean values of the parameters for groups of shallowing-upward sequences (Fig. 14.8). Thus section 2 can be divided into four parts on the basis of the nature of the shallowing-upward sequences and of the microfacies in which they terminate. Sequences 1 to 45 correspond to a period of shallower depth with strong oscillations. This period is followed by the interval of sequences 46 to 68, which displays moderate oscillations and intermediate depth. Sequences 69 to 95 represent another interval of shallower depth followed by a deeper phase with weak oscillations which ends the sec-

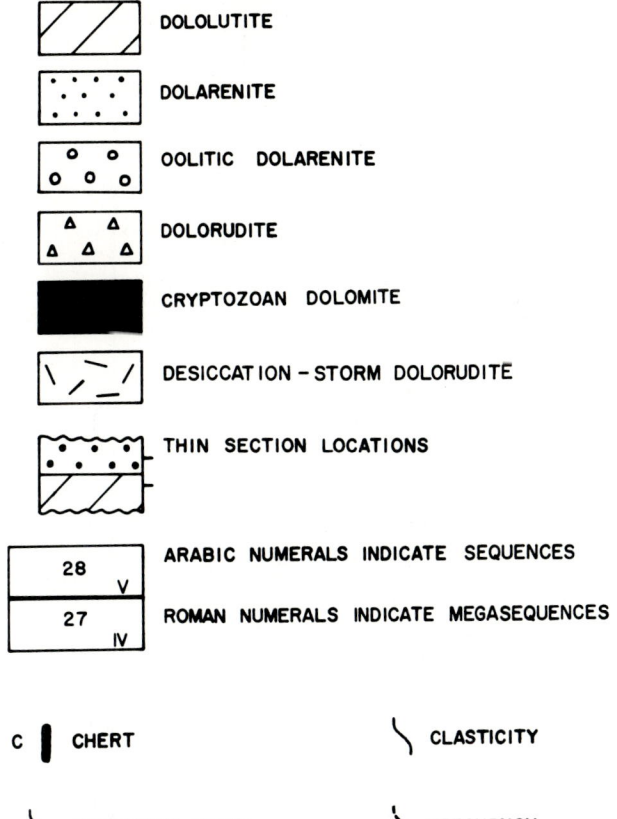

FIGURE 14.3 Allentown Dolomite (Upper Cambrian), Warren County, New Jersey. Symbols for microfacies. From Zadnik and Carozzi (1963).

tion. This evolution is summarized by the curve designated as "range of oscillations," whereas the curve expressing the average thickness of shallowing-upward sequences per group shows that the thinner sequences characterize the shallower environments and the thicker sequences correspond to deeper environments with less frequent oscillations of smaller amplitude. The behavior of the curves of microscopic parameters is the same, generally, as it was in the previous generalization and in the ideal shallowing-upward sequence.

14.2 CHARACTERISTIC CONSTITUENTS: STROMATOLITES

In addition to frontal oolitization processes, platforms with stromatolite buildups often display a reduced generation of smaller ooids between buildups and also in the adjacent part of the internal lagoon. A repetition of longitudinal rows of buildups separated with lagoons of variable width, and interconnected by tidal channels, is frequent.

A sequence about 10 ft (3.05 m) thick of stromatolitic dolomites belonging to the Shakopee Dolomite (Lower Ordovician) was studied in a quarry near Wyalusing, southwest Wisconsin (Carozzi and Davis, 1964). More than 100 thin sections were investigated petrographically and collected at 75 different locations in the quarry wall (Fig. 14.9) where several shallowing-upward sequences are displayed.

Description of Microfacies

The various microfacies are described in general shallowing order.

Microfacies 1 (Fig. 14.10A,B). Grain-supported oolitic dolarenite with clear dolosparite cement. Ooids are well developed, with pelletoidal cores corresponding to fragments of stromatolite colonies or with cores of well-rounded grains of detrital quartz. Flat pebbles of stromatolitic material occur locally. Silicification is frequent, emphasizing the cavity-filling nature of the original sparite cement. Half-moon ooids (Carozzi, 1963) are common (Fig. 14.10B).

Microfacies 2 (Fig. 14.10C,D,E). Stromatolite-constructed dolomite showing associations of mats and laterally linked hemispheroidal colonies. Numerous dark layers display a pelletoidal texture developed in place by desiccation and exposure of the mats, indicating

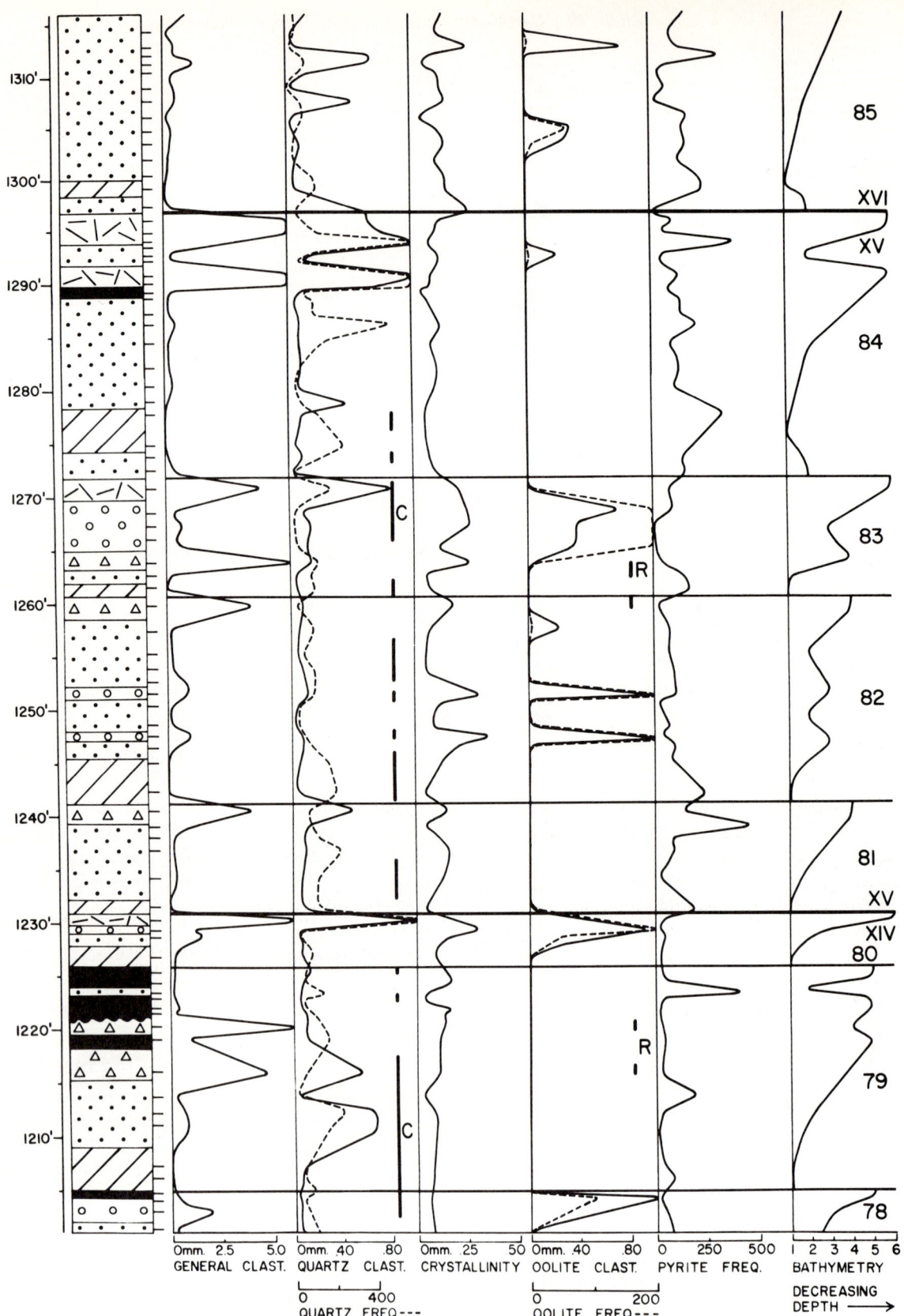

FIGURE 14.4 Allentown Dolomite (Upper Cambrian), Warren County, New Jersey. Typical example of vertical variation of microscopic parameters during shallowing-upward sequences in field section 2 at Carpentersville. From Zadnik and Carozzi (1963).

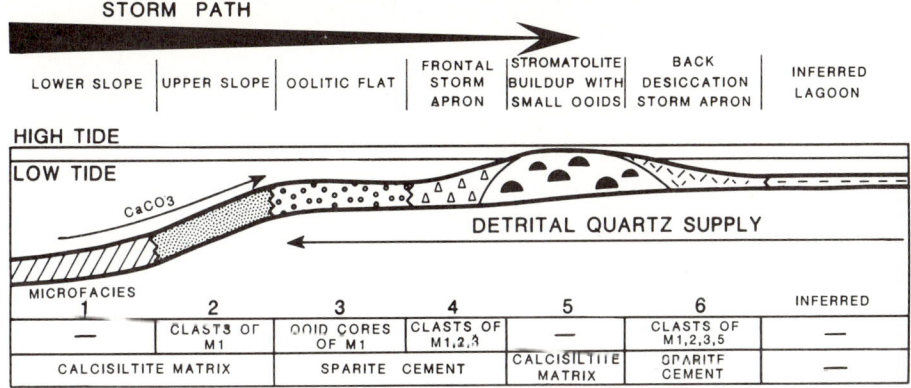

FIGURE 14.5 Allentown Dolomite (Upper Cambrian), Warren County, New Jersey. Ideal depositional model.

the intertidal nature of the environment (Fig. 14.10C). Between colonies are irregular depressions partially filled by algal clasts, ooids, and concentrations of rounded grains of detrital quartz (Fig. 14.10D,E). In many instances, the lighter-colored intermat trapped material, originally a micrite mud, is unusually thick.

Microfacies 3 (Fig. 14.10F,G). Intraformational stromatolitic dolorudite in which very angular, subrectangular, or crescent-shaped clasts of stromatolitic colonies predominate over rounded ones (Fig. 14.10G). The interstitial material consists of clasts of same origin, small reworked ooids, abundant rounded grains of detrital quartz set in a pelletoidal dolosiltite matrix. One clast of silicified microfacies 1 demonstrates the early nature of silicification which preceded reworking. Microfacies 3 is a greater-energy tempestite due to massive destruction of microfacies 2 followed by deposition on top of it, and transportation of the clasts in a landward direction.

Microfacies 4 (Fig. 14.10H). Pelletoidal subhedral to euhedral dolomite showing "ghosts" of pellets inside frequently zoned rhombs. Rounded and angular grains of detrital quartz are very abundant and are associated with reworked ooids and small stromatolitic clasts. This microfacies locally contains small scattered stromatolitic colonies.

The Ideal Shallowing-Upward Sequence

Superposition of the four microfacies (Fig. 14.11) shows that the major generation of ooids was in front of the stromatolitic colonies and that only smaller reworked ones occurred higher in the sequence. Storm-induced destruction accumulated clasts of stromatolites in minor amounts in front of the colonies, but mostly in back of them, as would be expected from such a process. The increase in the frequency of detrital quartz in microfacies 4 and the appearance of angular grains indicates eolian contribution from landward sabkha environments. Strictly speaking, the shallowing-upward trend terminates at the stromatolitic barrier (microfacies 3), and the final lagoon (microfacies 4) represents a slight deepening.

A generalized section (Fig. 14.12) shows detailed variation of microscopic parameters and the occurrence of three distinct sequences. Sequences I and II consist typically of a long, shallowing-upward trend ending in microfacies 3 and a short, deepening-upward trend ending in lagoonal conditions (microfacies 4 and 2).

The Ideal Depositional Model

The proposed model (Fig. 14.13), which is very characteristic of stromatolite-controlled carbonate platforms of the Early Paleozoic, shows that the lagoonal environment may display minor stromatolitic colonies dispersed in it as well as a transition landward to coastal and continental sabkhas. A possible lateral juxtaposition of several stromatolitic barriers and related lagoons should also be considered.

14.3 CHARACTERISTIC CONSTITUENTS: STROMATOLITES AND *NUIA*

Platforms in which stromatolites are associated with *Nuia* display advanced morphologic differentiation. A first oceanward bar controlled by *Nuia* and crinoids, and an adjacent lagoon, precede several rows of stromatolitic buildups with intervening oolitic flats with stromatolite mats and scattered evaporites. Major evaporitic flats may extend between the carbonate complex and the coastal quartzose dunes.

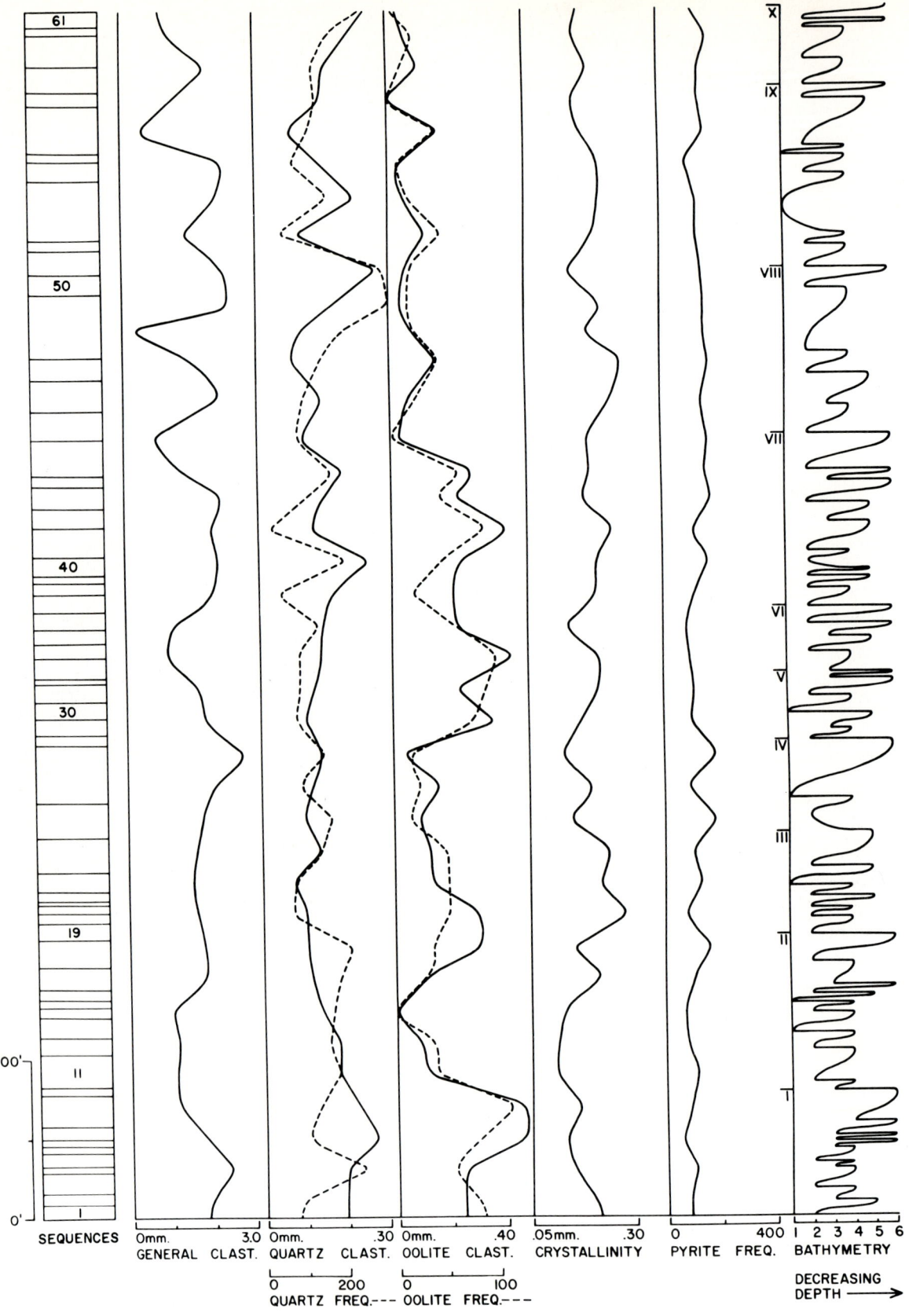

FIGURE 14.6 Allentown Dolomite (Upper Cambrian), Warren County, New Jersey. General vertical evolution of depositional environments in field section 2 at Carpentersville. From Zadnik and Carozzi (1963).

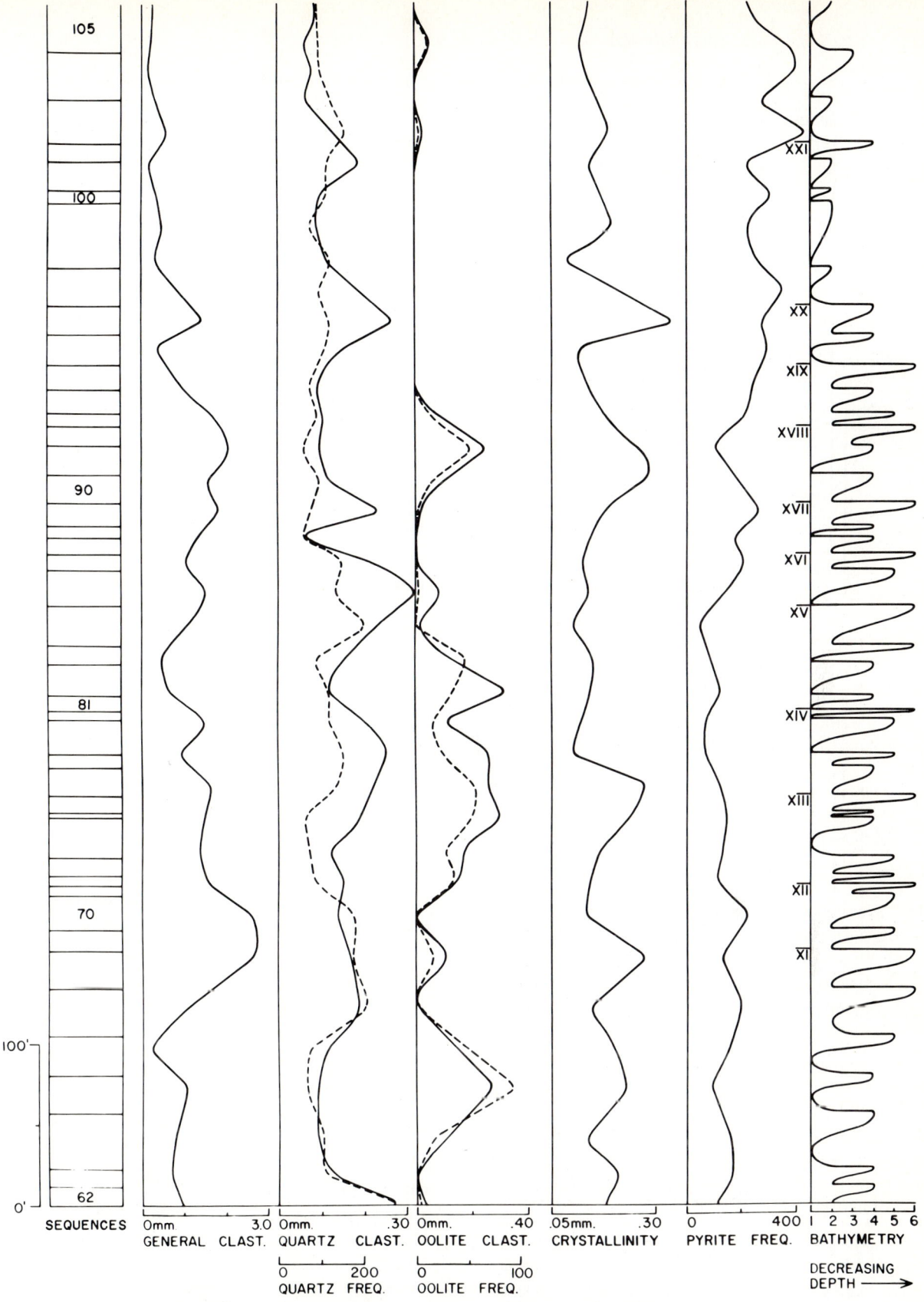

FIGURE 14.7 Allentown Dolomite (Upper Cambrian), Warren County, New Jersey. General vertical evolution of depositional environments in field section 2 at Carpentersville (continued). From Zadnik and Carozzi (1963).

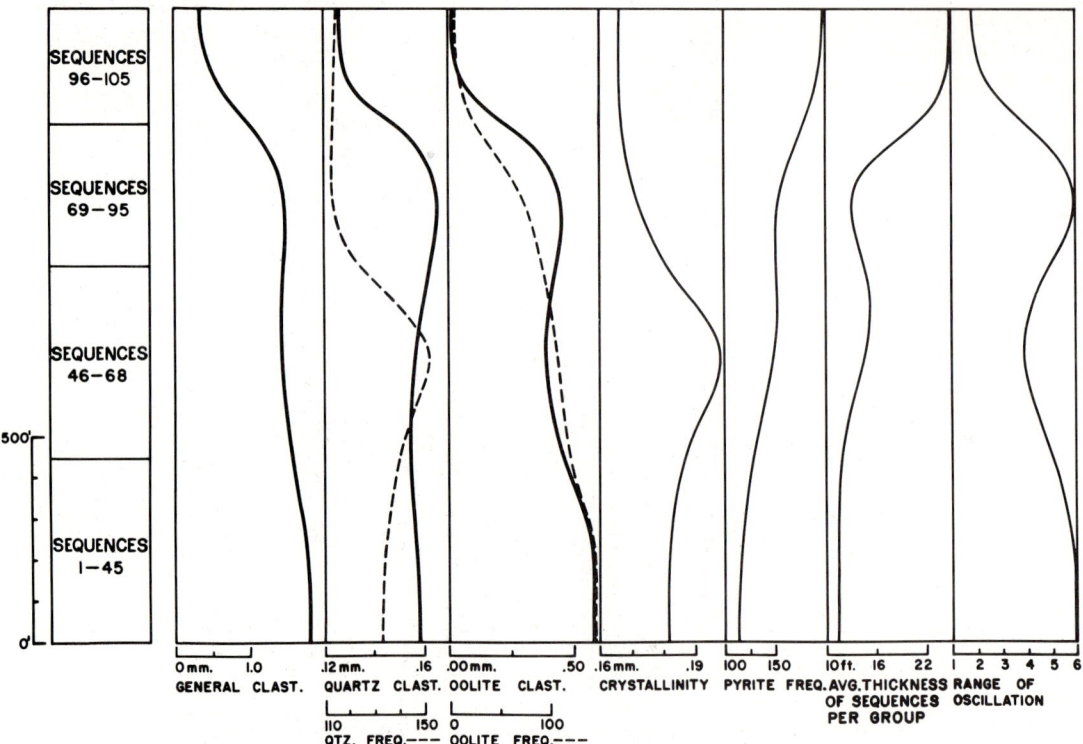

FIGURE 14.8 Allentown Dolomite (Upper Cambrian), Warren County, New Jersey. General vertical evolution of depositional environments by groups of shallowing-upward sequences in field section 2 at Carpentersville. From Zadnik and Carozzi (1963).

The Joachim Dolomite (Middle Ordovician) of the southwest margin of the Illinois Basin in southeastern Missouri and southern Illinois (Okhravi and Carozzi, 1983) was investigated with 10 field sections and a complete core from the central part of the Illinois Basin. The average thickness of the Joachim Dolomite is 60 m and a total of 1154 samples were studied petrographically. Among them 915 samples were obtained from outcrops (average vertical interval of 11 cm) and the rest from the core (average vertical interval of 30.5 cm).

Description of Microfacies

Petrographic study led to the recognition of 11 microfacies, which were secondarily dolomitized to a variable degree. They are designated as deposited originally, and described in a landward and shallowing general direction according to the depositional model discussed later.

Microfacies 1 (Fig. 14.14A,B). Dolomitized pelletoidal calcilutite to calcisiltite with up to 10% scattered sand-size bioclasts of *Leperditia,* pelmatozoans, and *Nuia* with its typical fibro-radiated structure. Extensive bioturbation is represented by burrows and borings filled by sparite cement or coarser matrix. Patchy dolomitization (average 59.3%) consists of equal-size medium crystalline rhombs replacing matrix and often partially the large ostracods (Fig. 14.14B).

Microfacies 2 (Fig. 14.14C,D). Dolomitized mud-supported intraclastic calcarenite to calcirudite with few bioclasts of *Nuia* and large ostracods. Intraclasts (tempestite) are poorly sorted, angular to subrounded, and reworked from early cemented microfacies 1. Some intraclasts are composite indicating multiple episodes of reworking (Fig. 14.14D). Dolomitization (average 53.2%) is limited to the matrix, which now appears as a mosaic of medium crystalline rhombs.

Microfacies 3 (Fig. 14.14E,F). Dolomitized grain-supported intraclastic biocalcarenite with interstitial pelletoidal calcisiltite matrix and poikilotopic sparite cement. Rounded intraclasts (tempestite), which originated from the reworking of early cemented microfacies 1 and 2, are associated with oncoids consisting of cores of large ostracods encrusted by blue-green algae (Fig. 14.14F). Analysis of cementation indicates the following succession: beachrock or submarine isopachous rim cement; dolomitization (average 28%) by finely crystalline rhombs partially replacing matrix, bioclasts, and intraclasts; and influx of vadose silt and precipitation of phreatic poikilotopic sparite.

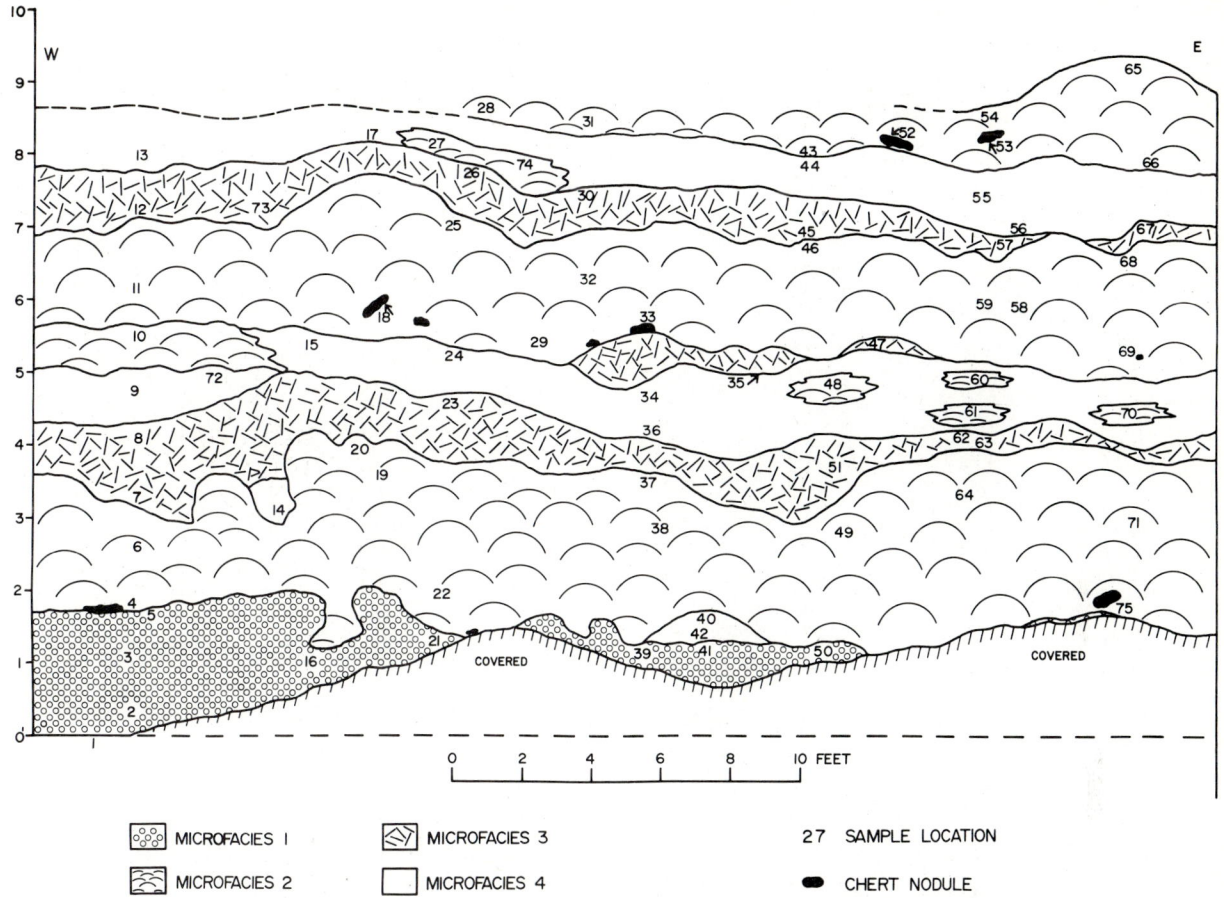

FIGURE 14.9 Shakopee Dolomite (Lower Ordovician), Wyalusing, Wisconsin. Distribution of microfacies and location of samples in quarry wall at Wyalusing. From Carozzi and Davis (1964).

Microfacies 4 (Fig. 14.14G,H). Dolomitized mud-supported to grain-supported intraclastic pelletoidal calcarenite to calcirudite with calcisiltite matrix. Intraclasts (tempestite) are poorly sorted desiccation chips of calcilutite which are also further disintegrated into pellets. The rare bioclasts are ostracods. Bimodal quartz grains of desert origin are scattered commonly in interstitial spaces otherwise filled by calcisiltite, vadose silt, and poikilotopic sparite. Desiccated layers of calcilutite are common with fractures filled by gypsum and anhydrite cements (Fig. 14.14H). Dolomitization (average 58.3%) is partially replacing matrix and intraclasts by finely to very finely crystalline rhombs.

Microfacies 5 (Fig. 14.15A,B). Dolomitized pelletoidal calcilutite to calcisiltite with up to 10% bioclasts of *Leperditia*, gastropods, and *Lingula*. Pellets are mainly lithic in origin with a few fecal ones. Bioturbation is developed strongly with vertical, horizontal, oblique, and L-shaped burrows and borings which indicate early induration (Fig. 14.15B). Internal sediments of pellets and fine detrital quartz are rare. After extensive dolomitization (average 55.8%) by very finely crystalline rhombs, the cavities were filled by vadose silt, phreatic sparite cement, or have remained empty.

Microfacies 6 (Fig. 14.15C,D). Dolomitized stromatolite-constructed limestone consisting of dark mat-type colonies or *Spongiostromata* colonies (Fig. 14.15D), separated by biocalcisiltite layers with large ostracods. Molds of gypsum and anhydrite crystals and euhedral authigenic quartz are developed inside the colonies which display, furthermore, desiccation cracks, breccias (tempestite), and fenestrae filled by poikilotopic sparite overlying vadose silt. Dolomitization (average 72.1%) is by finely crystalline rhombs.

Microfacies 7 (Fig. 14.15E,F). Dolomitized mud-supported to grain-supported oolitic intraclastic (tempestite) calcarenite with matrix of calcisiltite, interbedded with dark, thin stromatolitic mats. Superficial ooids often deformed to spastolites appear as oomoldic pores or are filled by poikilotopic sparite, secondary gypsum, or quartz cements (Fig. 14.15F). Dolomitization

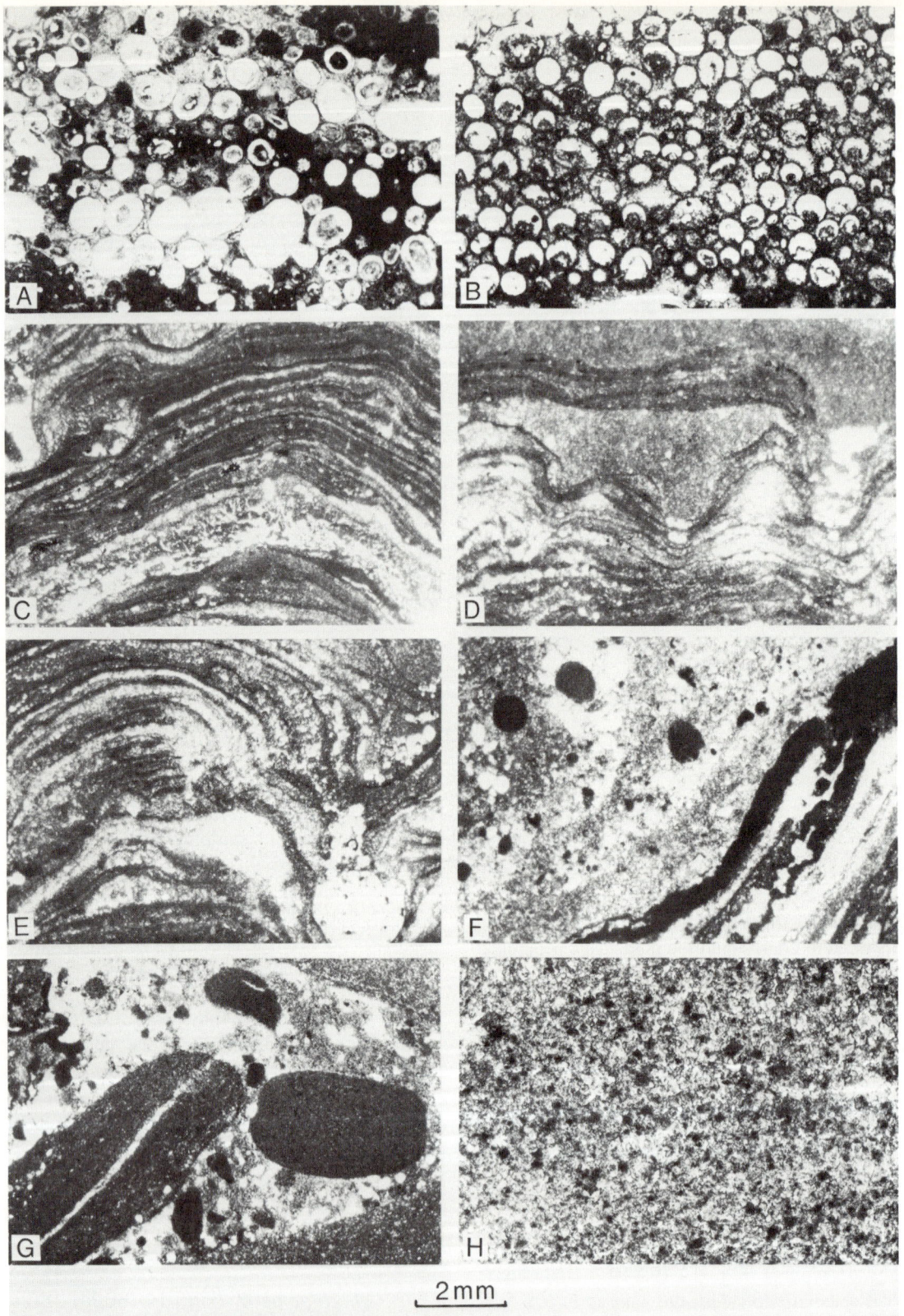

FIGURE 14.10 Shakopee Dolomite (Lower Ordovician), Wyalusing, Wisconsin. Typical microfacies. A and B. Microfacies 1. C to E. Microfacies 2. F and G. Microfacies 3. H. Microfacies 4. See text for detailed explanations. All photomicrographs: plane-polarized light. From Carozzi and Davis (1964).

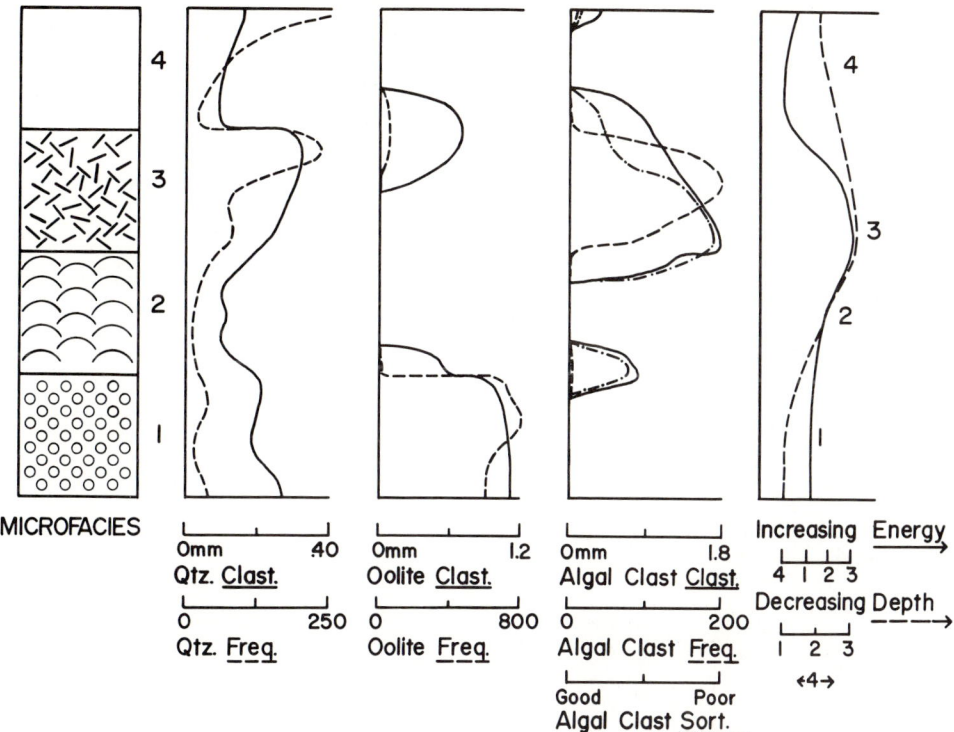

FIGURE 14.11 Shakopee Dolomite (Lower Ordovician), Wyalusing, Wisconsin. Ideal shallowing-upward sequence. From Carozzi and Davis (1964).

(average 60%) is by finely crystalline subhedral to euhedral rhombs.

Microfacies 8 (Fig. 14.15G,H). Alternating dolomitized poorly developed, dark, organic-rich stromatolitic mats and light-colored laminae of calcisiltite with bimodal grains of detrital quartz. Dolomitization (average 64%) is by finely to coarsely crystalline rhombs. Common collapse breccia texture indicates dissolution of associated halite, with cracks filled by vadose silt and poikilotopic sparite cement (Fig. 14.15H).

Microfacies 9 (Fig. 14.16A,B,C). Dolomitized pelletoidal calcisiltite with fenestral fabric, scattered bimodal quartz grains, and intercalated stromatolitic mats with desiccation fractures. Hopper-shaped halite casts are preserved as moldic porosity (Fig. 14.16B). Fenestrae are filled by vadose silt and poikilotopic sparite cement. This microfacies grades into a dolomitized flat-pebble conglomerate (tempestite) with interstitial pelletoidal calcisiltite matrix and sparite cement (Fig. 14.16C). The subangular to subrounded clasts are derived from microfacies 6 and 8. Dolomitization (average 42.7%) is by finely crystalline rhombs.

Microfacies 10 (Fig. 14.16D,E,H). Dolomitized calcilutite to nodular anhydrite with chickenwire structure. The latter has been either preserved (Fig. 14.16E) or replaced pseudomorphically by sparite (Fig. 14.16D), or dissolved and the cavities filled by vadose silt and poikilotopic sparite. Rosettes of euhedral authigenic quartz replace anhydrite and predate sparite (Fig. 14.16H). Dolomitization (average 50.5%) occurs mostly as replacement of matrix and anhydrite by finely crystalline rhombs.

Microfacies 11 (Fig. 14.16F,G). Unimodal to bimodal quartz arenite with up to 10% potassic feldspars and matrix of sericite and quartz, or quartz overgrowths (Fig. 14.16G), or poikilotopic sparite cement, or association of dolomite and anhydrite cements. These are characteristic dune and interdune desert arenites.

The Ideal Shallowing-Upward Sequence

The 11 microfacies of the Joachim Dolomite (Fig. 14.17) can be organized into a shallowing-upward sequence which expresses an environmental evolution divided into four major domains: subtidal, intertidal, supratidal coastal sabkha, and continental sabkha. Marine conditions were low energy, restricted and graded into lagoonal hypersaline and evaporitic flats, periodically exposed, subjected to storms, and invaded by dunes. A typical detailed section (Figs. 14.18, 14.19) of such a shallowing-upward trend shows the reciprocal behavior of micro-

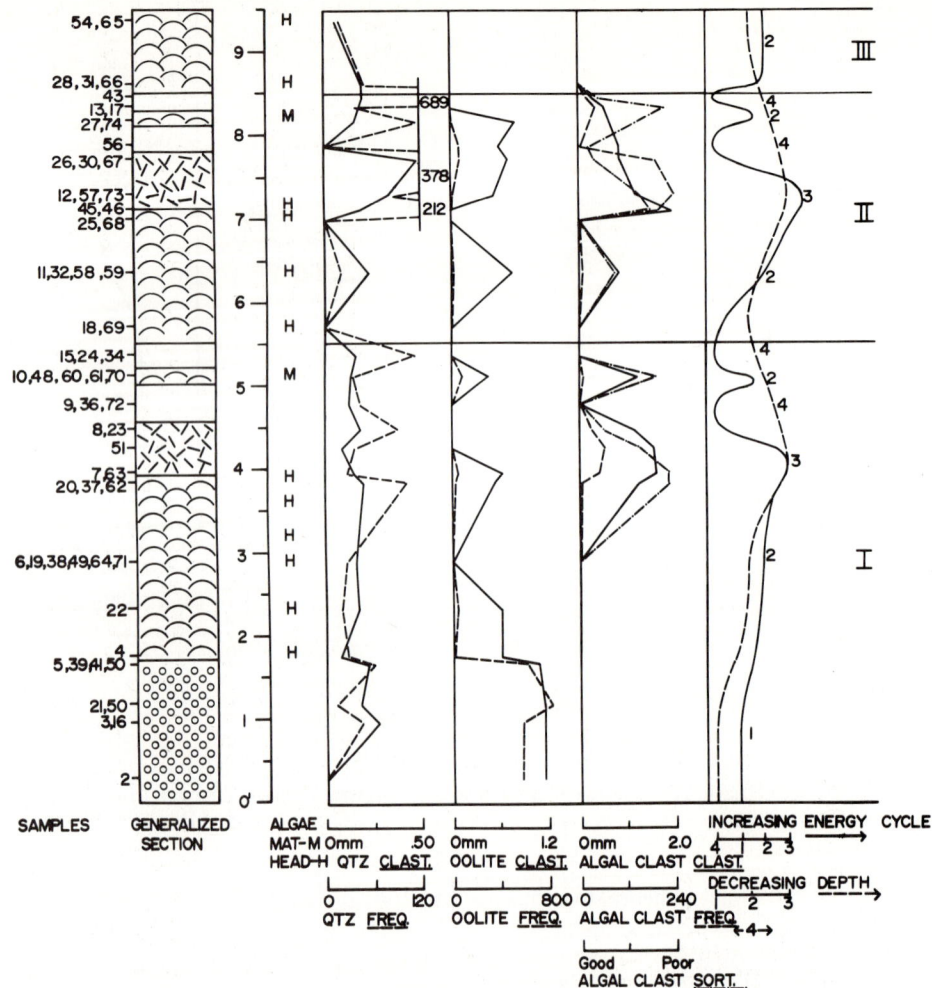

FIGURE 14.12 Shakopee Dolomite (Lower Ordovician), Wyalusing, Wisconsin. Typical example of vertical variation of microscopic parameters during three successive shallowing-upward sequences with I and II both shallowing upward and deepening upward. From Carozzi and Davis (1964).

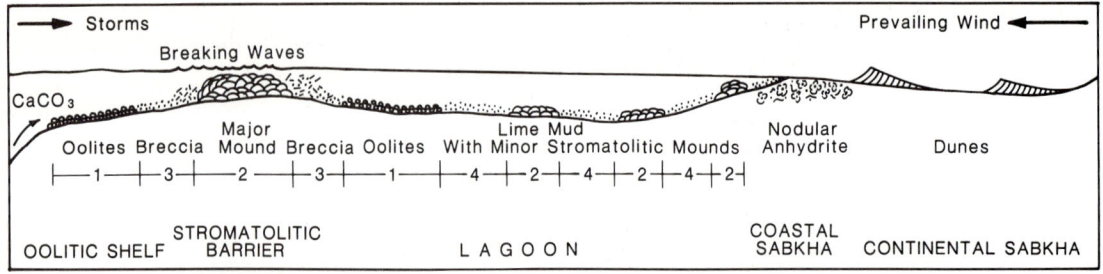

FIGURE 14.13 Shakopee Dolomite (Lower Ordovician), Wyalusing, Wisconsin. Ideal depositional model. From Carozzi and Davis (1964).

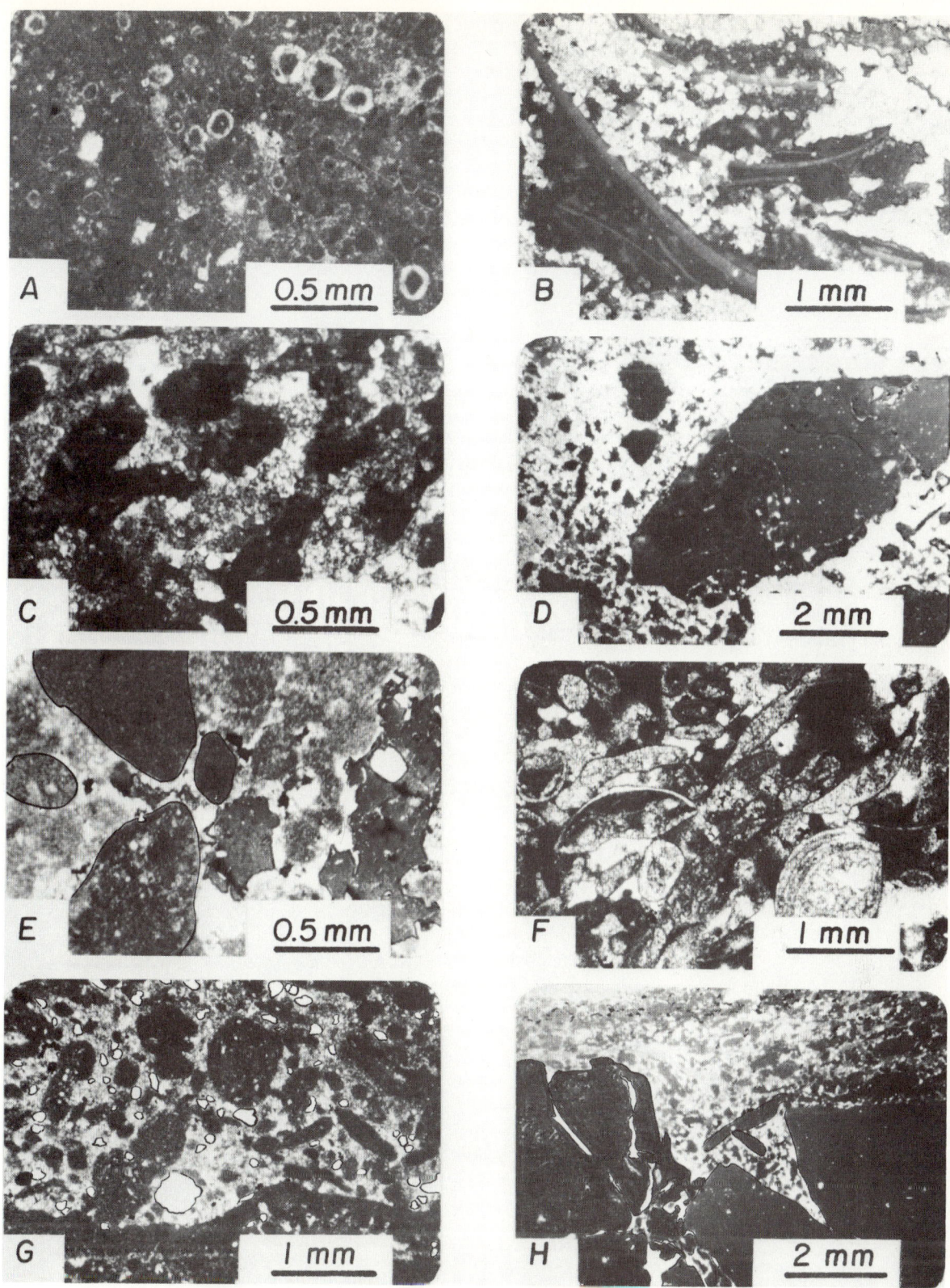

FIGURE 14.14 Joachim Dolomite (Middle Ordovician), southwest margin of Illinois Basin. Typical microfacies. A and B. Microfacies 1. C and D. Microfacies 2. E and F. Microfacies 3. G and H. Microfacies 4. See text for detailed descriptions. All photomicrographs: crossed nicols, except D and G, plane-polarized light. From Okhravi and Carozzi (1983).

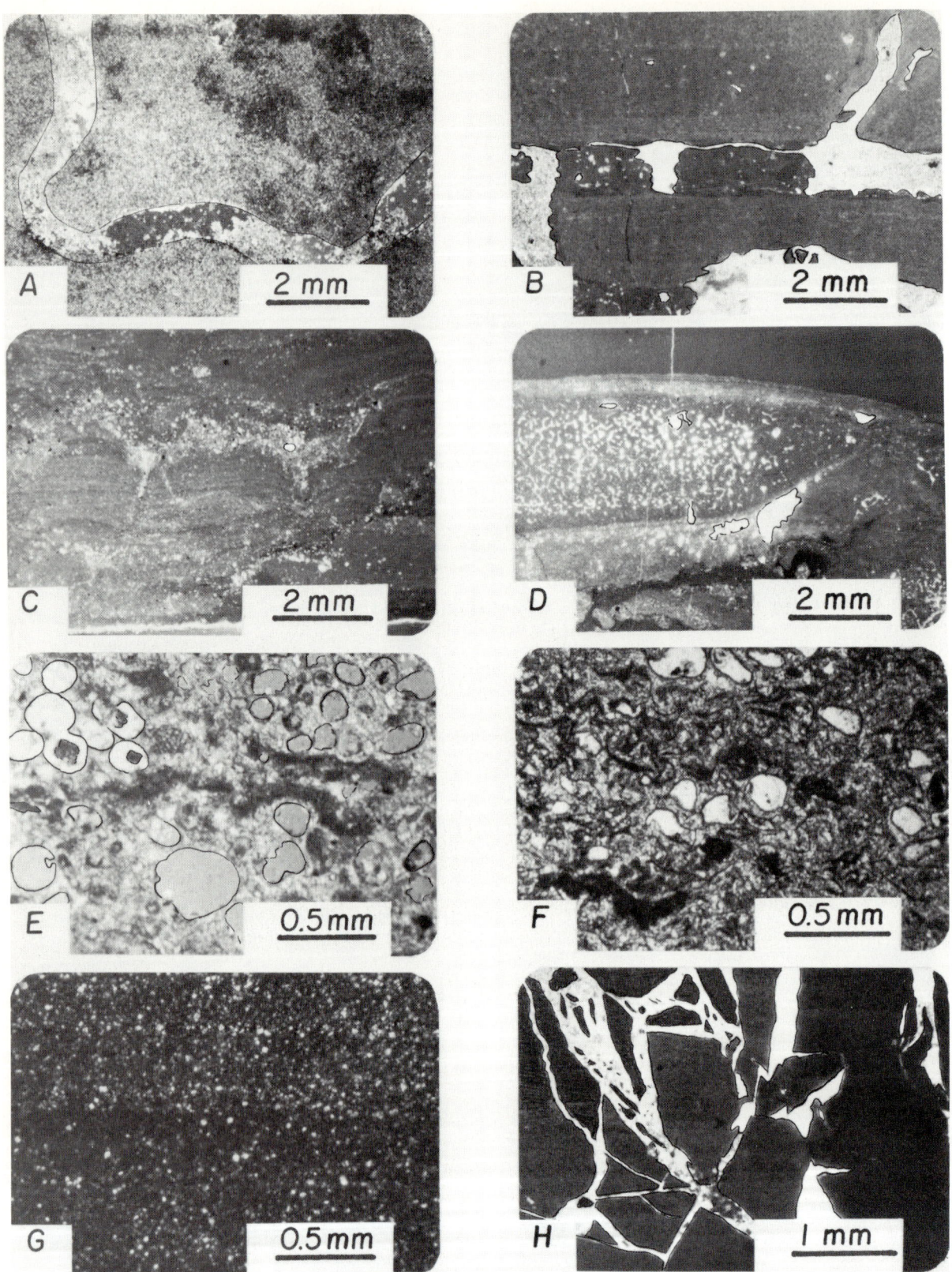

FIGURE 14.15 Joachim Dolomite (Middle Ordovician), southwest margin of Illinois Basin. Typical microfacies (continued). A and B. Microfacies 5. C and D. Microfacies 6. E and F. Microfacies 7. G and H. Microfacies 8. See text for detailed descriptions. All photomicrographs: plane-polarized light, except A, B, and E, crossed nicols. From Okhravi and Carozzi (1983).

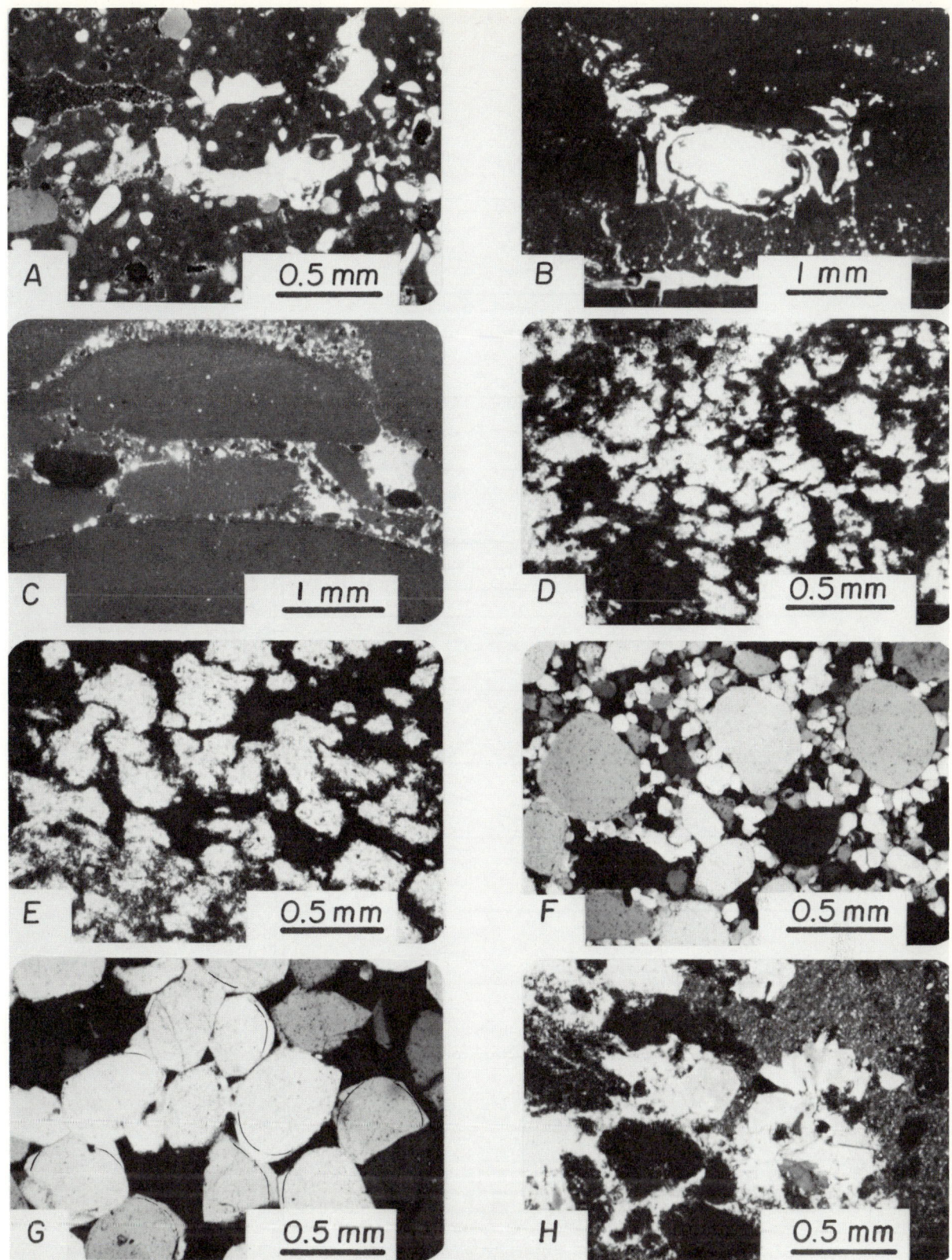

FIGURE 14.16 Joachim Dolomite (Middle Ordovician), southwest margin of Illinois Basin. Typical microfacies (continued). A to C. Microfacies 9. D and E. Microfacies 10. F and G. Microfacies 11. H. Microfacies 10. See text for detailed descriptions. All photomicrographs: crossed nicols, except C and E, plane-polarized light. From Okhravi and Carozzi (1983).

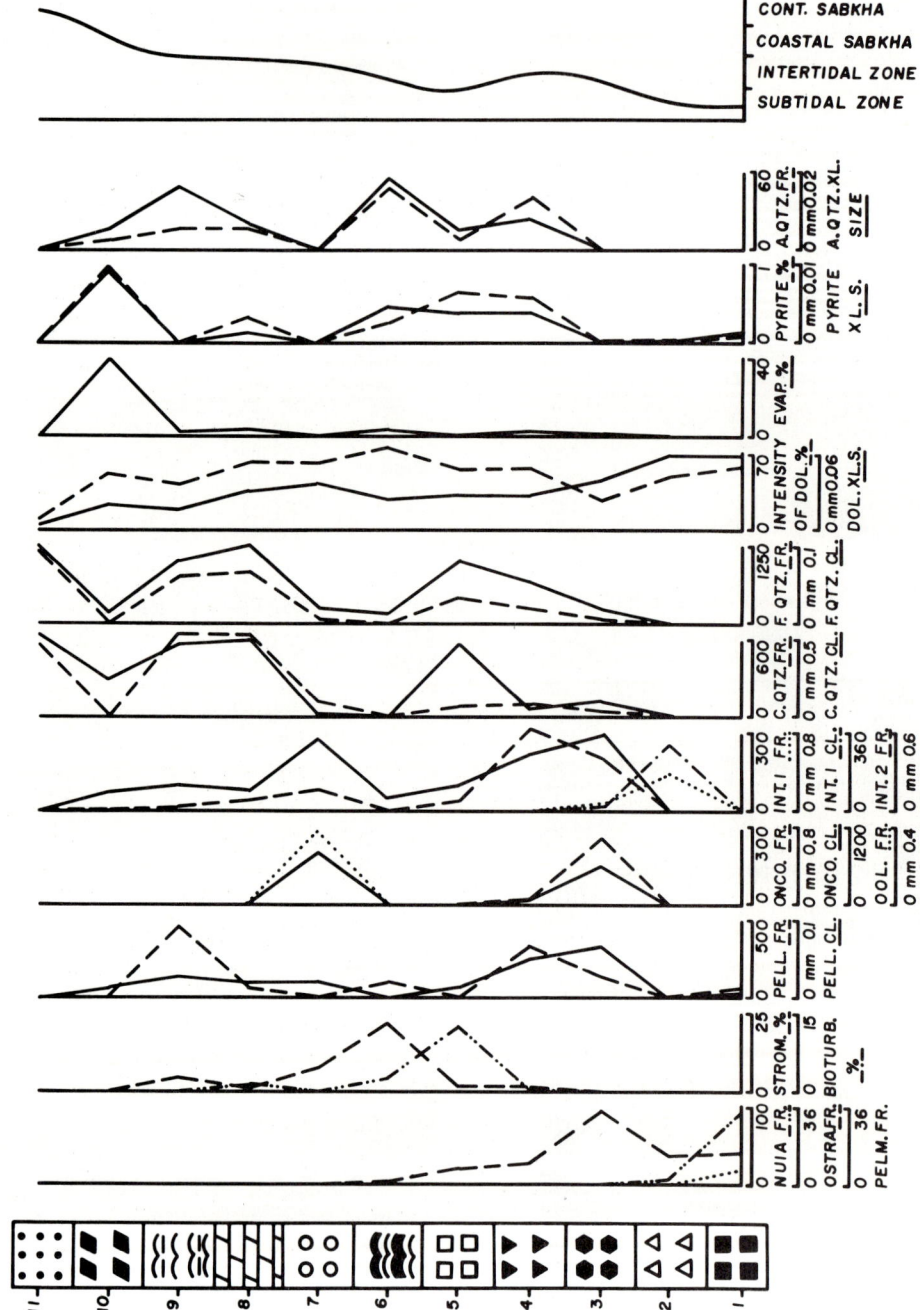

FIGURE 14.17 Joachim Dolomite (Middle Ordovician), southwest margin of Illinois Basin. Ideal shallowing-upward sequence. From Okhravi and Carozzi (1983).

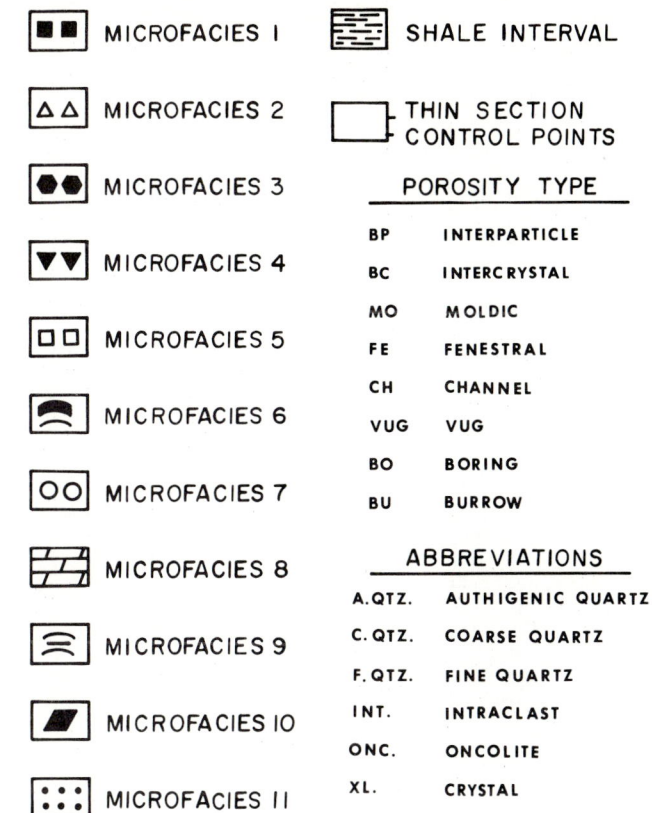

FIGURE 14.18 Joachim Dolomite (Middle Ordovician), southwest margin of Illinois Basin. Symbols for microfacies. From Okhravi and Carozzi (1983).

scopic parameters as well as a general graphic representation of variation of major mineralogical and textural components in percent surface area.

The Ideal Depositional Model

This platform model (Figs. 14.20, 14.21), which consists of two major stromatolitic ridges, is preceded oceanward by an intraclastic-bioclastic intertidal bar controlled mainly by *Nuia* and crinoids. The distribution of the major microscopic components can be summarized as follows.

The greatest frequency of pelmatozoans occurs in offshore conditions (microfacies 1), indicating that their main habitat was located seaward. The frequency of pelmatozoans decreases rapidly landward because of increasingly unfavorable living conditions.

Nuia formed a dense population in offshore conditions (microfacies 1). Its behavior is similar to that of pelmatozoans but is even more dispersed on the seaward side of the intra-bioclastic bar.

Ostracods show their frequency peak on the intra-bioclastic bar (microfacies 3). Apparently, this is the result of mechanical accumulation by passive flotation. The frequency decreases rapidly on both sides of the bar and then continues to decrease slowly landward. The gradual increase seaward indicates final deposition of ostracods offshore.

Stromatolite mats reach their maximum development in the constructed stromatolitic ridge (microfacies 6). Their rapid seaward decrease correlates with the increase of bioturbation intensity in the lagoon, indicating that the lower limits of algal mats were biologically controlled. Stromatolite mats decrease in frequency across oolitic and carbonate flats (microfacies 7 and 8) and terminate with a small peak of frequency corresponding to the inner stromatolitic ridge (microfacies 9). This landward limit of stromatolite mats coincides with the upper limit of the intertidal zone and apparently was controlled by desiccation.

Both the frequency and clasticity curves of oncoids show their peaks in coincidence with the intra-bioclastic bar (microfacies 3), which offers the required conditions of agitation for their generation by destruction of stromatolitic mats.

Bioturbation is most extensive in the lagoon (microfacies 5) and shows a small peak in shoreface conditions (microfacies 2).

Intraclasts are of two different kinds, intraclasts 1 and intraclasts 2. The former consist of partially dolomitized micrite of relatively deeper water with *Nuia* and ostracod fragments, whereas the latter consist mostly

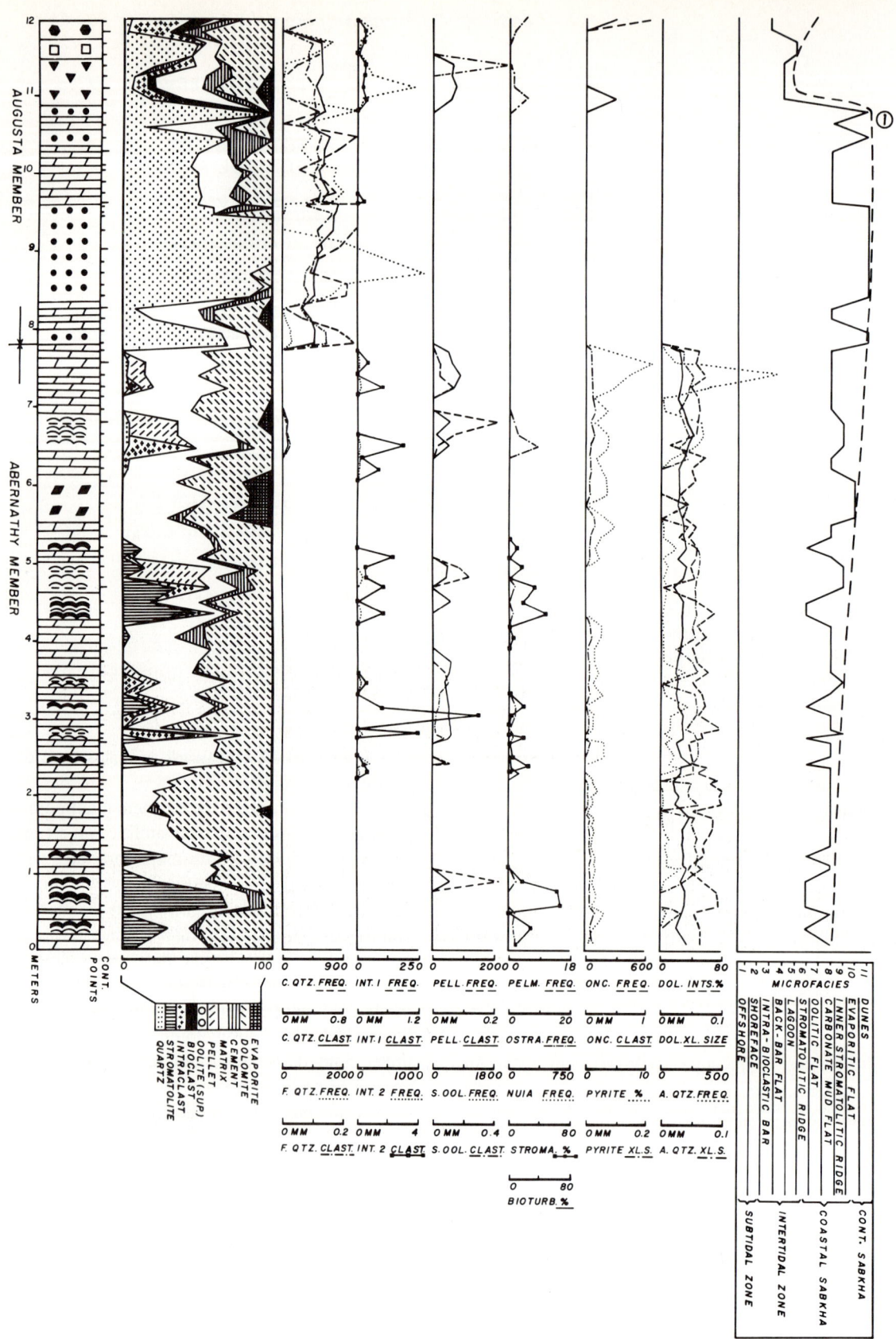

FIGURE 14.19 Joachim Dolomite (Middle Ordovician), southwest margin of Illinois Basin. Typical example of vertical variation of microscopic parameters during a shallowing-upward sequence. From Okhravi and Carozzi (1983).

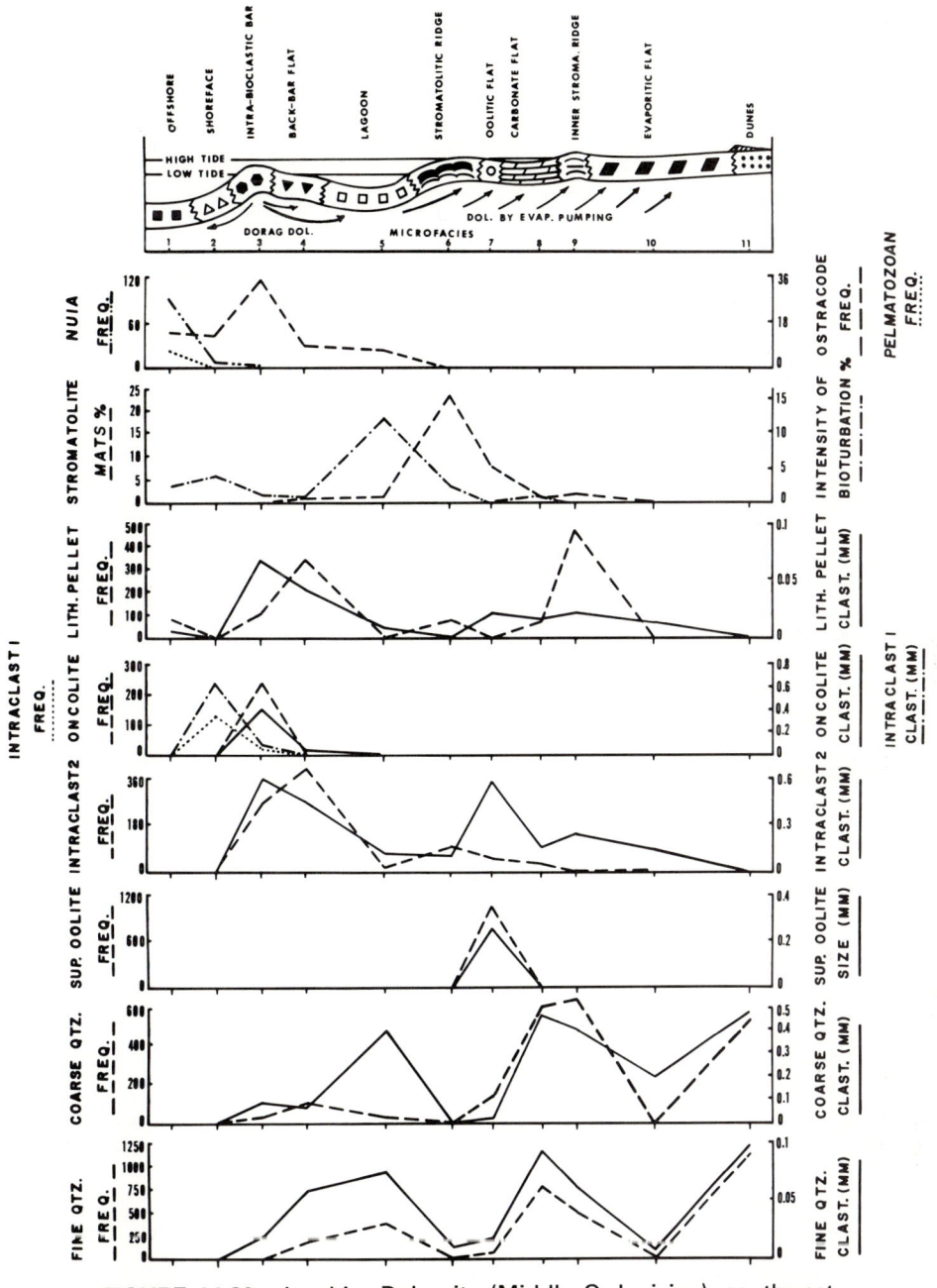

FIGURE 14.20 Joachim Dolomite (Middle Ordovician), southwest margin of Illinois Basin. Ideal depositional model. From Okhravi and Carozzi (1983).

of finely crystalline dolomitized ostracod-bearing micrite devoid of *Nuia*, and derived from the reworking of early-cemented microfacies 1 and 2. Frequency and clasticity of intraclasts 1 reach their peaks in the shoreface environment (microfacies 2) and then decrease both seaward and landward. This indicates that their maximum production and accumulation in shoreface conditions resulted, probably, from some periodic strong currents.

Frequency of intraclasts 2 is elevated on the bar and reaches its peak in the back-bar flat (microfacies 4). Clasticity of intraclasts 2 does not coincide with the frequency peak; this indicates active production by periodic exposure of the bar (microfacies 3) and distribution behind it (microfacies 4). The clasticity curve shows a second high peak in the oolitic flat because of a local relative increase in energy which reworked micritic mud.

The frequency and clasticity curves of the lithic pellets are almost parallel to those of intraclasts 2, indicating that they originated from them. The same lack of coincidence between peaks exists with the greatest clasticity on the bar (microfacies 3) and the greatest frequency of distribution behind it (microfacies 4). The frequency

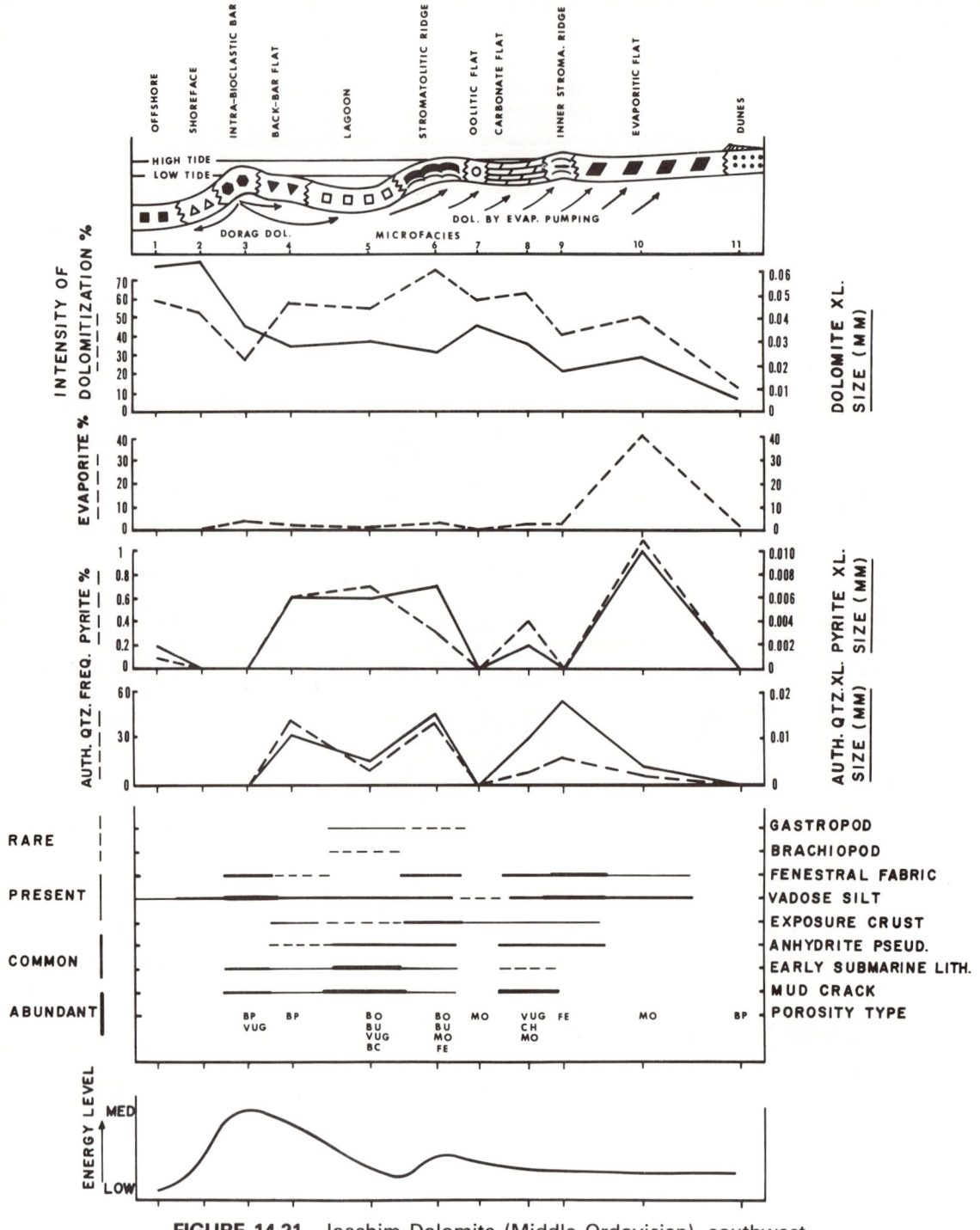

FIGURE 14.21 Joachim Dolomite (Middle Ordovician), southwest margin of Illinois Basin. Ideal depositional model (continued). From Okhravi and Carozzi (1983).

curve shows a second peak in the inner stromatolitic ridge (microfacies 9), which indicates periodic exposure of that area as well.

Superficial ooids show peaks of their frequency and clasticity in coincidence with the oolitic flat (microfacies 7) where they were generated by moderate local agitation in a supersaturated environment. Their occurrence correlates with the second maximum peak of the clasticity of intraclasts 2, which confirms the local conditions of relatively greater energy, some superficial ooids containing nuclei consisting of intraclasts 2.

The frequency and clasticity of the coarse quartz display their largest values in the dune environment (microfacies 11), and both decrease seaward with several fluctuations. This shows a continental source for the quartz with dispersion across the carbonate system. The frequency and clasticity curves are almost parallel with each other, indicating a steady sediment yield and distribution of the grains. The peaks, which coincide with the carbonate flat (microfacies 8) and lagoon (microfacies 5), result from trapping effects of the eolian sand in depressions.

The frequency and clasticity of the fine quartz show their largest values in the dune environment (microfacies 11). They are perfectly parallel with each other and with the frequency and clasticity curves of the coarse quartz, indicating similarity in transportation processes and depositional conditions.

Dolomitization is expressed by two curves: intensity in percentage of surface and crystallinity by means of the size of the rhombs. The intensity indicates the juxtaposition of two distinct processes. The first encompasses microfacies 1 through 5 symmetrically increasing on both sides of microfacies 3 and indicating a freshwater-seawater mixing (dorag model) generated by an active freshwater lens below the temporarily exposed intra-bioclastic bar. The second process involves microfacies 6 through 11 with its maximum under the stromatolitic ridge and a steplike decrease landward. It is interpreted as the result of evaporative pumping under sabkha conditions. The crystallinity of the dolomite increases in a steplike fashion in an oceanward direction across the entire carbonate system. It is an expression of the original grain size of the sediments, which is shown equally by the behavior of the general energy variation of the entire system. This "blueprint" relationship has been observed in other dolomitized carbonate models (Zadnik and Carozzi, 1963; see also previous discussion, this chapter). The evaporite percentage curve reaches its peak in the evaporitic flat (microfacies 10) where the greatest rate of evaporation would be expected.

The pyrite percentage and pyrite crystal size curves are almost parallel. They show their greatest peaks in the evaporitic flat (microfacies 10), indicating that pyrite crystals were formed as a result of subsurface sulfate reduction. The other smaller peaks in the carbonate flat (microfacies 8), during the interval back-bar flat–lagoon–stromatolitic ridge (microfacies 4–5–6), and toward offshore (microfacies 1), appear related to reducing conditions due to local restriction or deepening of the environment of deposition.

The frequency and crystal size curves of authigenic quartz are parallel. Three peaks are apparent: back-bar flat (microfacies 4), stromatolitic ridge (microfacies 6), and inner stromatolitic ridge (microfacies 9). The only common character of these three environments is the abundance of organic matter which may have promoted this authigenesis.

Each of the investigated sections of the Joachim Dolomite (Figs. 14.22, 14.23, 14.24) consists of the superposition of 11 shallowing-upward sequences, all of which display numerous internal microsequences. Further analysis shows that it is possible to associate these sequences in four large-scale episodes (designated by Roman numerals) that either correspond to shallowing-upward megasequences or to intervals of time during which the shallowing-upward sequences present strong similarities in size, amplitude, or pattern.

14.4 CHARACTERISTIC CONSTITUENTS: STROMATOLITES (*SPONGIOSTROMATA*)

Spongiostromata colonies, scattered at random across widespread carbonate platforms, often acquire bulbous pillarlike shapes. Because of the protected environment in which they grow, their production of bioclasts or intraclasts is small to nonexistent. However, along the paths of gentle tidal currents, hydrodynamic buildups develop consisting of accumulations of fossiliferous calcisiltites encrusted by thin bioconstructed layers of corals, stromatoporoids, and stromatolitic mats.

The Burnt Bluff Group (Silurian) of the western margin of the Michigan Basin in Wisconsin is an entirely dolomitized platform sequence approximately 100 ft (30.5 m) thick which displays a large number of stromatolite-constructed buildups belonging to the group *Spongiostromata* (Soderman and Carozzi, 1963). These structures are columnar with an upper margin; they are 10 to 15 ft (3.05 to 4.5 m) high and 20 to 25 ft (6.10 to 7.5 m) wide. Dispersed among them are hydrodynamic buildups, partly encrusted by stromatolites and of similar general dimensions. This study was based on 645 samples collected from buildup and interbuildup areas at an average vertical interval of 0.46 ft (15.24 cm). In some places, samples were taken continuously or nearly so (0.1-ft or 3-cm interval).

Description of Microfacies

Microfacies are described in order of increasing energy from the interbuildup areas, into the buildups themselves and the overlying deposits.

Microfacies 1 (Fig. 14.25A). Coarsely crystalline unzoned massive dolomite with abundant tests of leperditid ostracods. Originally, it was a calcilutite.

Microfacies 1A (Fig. 14.25B). Medium crystalline subhedral dolomite with poorly preserved ooids, small algal mats, and traces of cross-bedding. Originally, it was a calcisiltite resulting from the winnowing of microfacies 1.

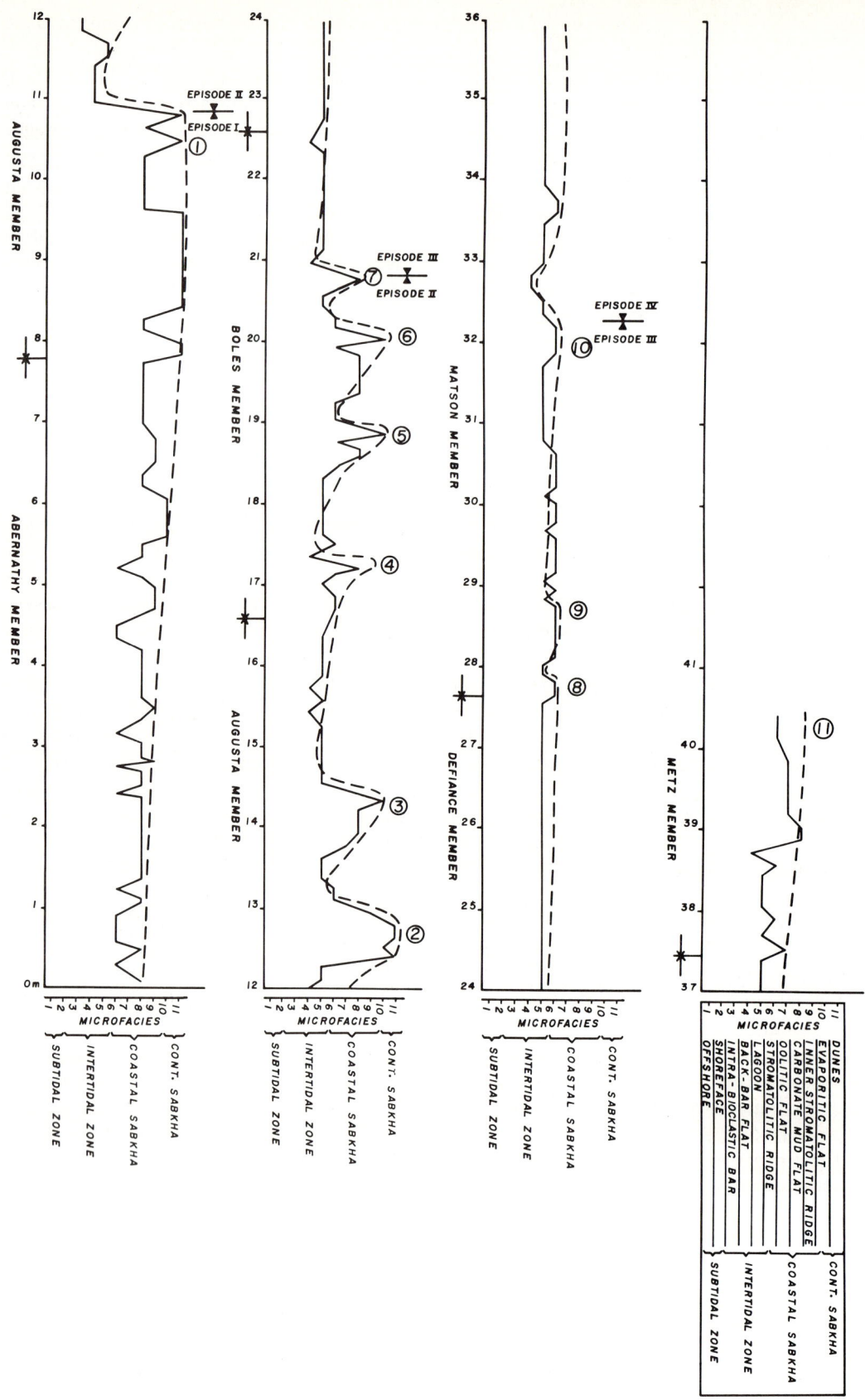

FIGURE 14.22 Joachim Dolomite (Middle Ordovician), southwest margin of Illinois Basin. General vertical evolution of depositional environments (sections A, B, C, D, and G). From Okhravi and Carozzi (1983).

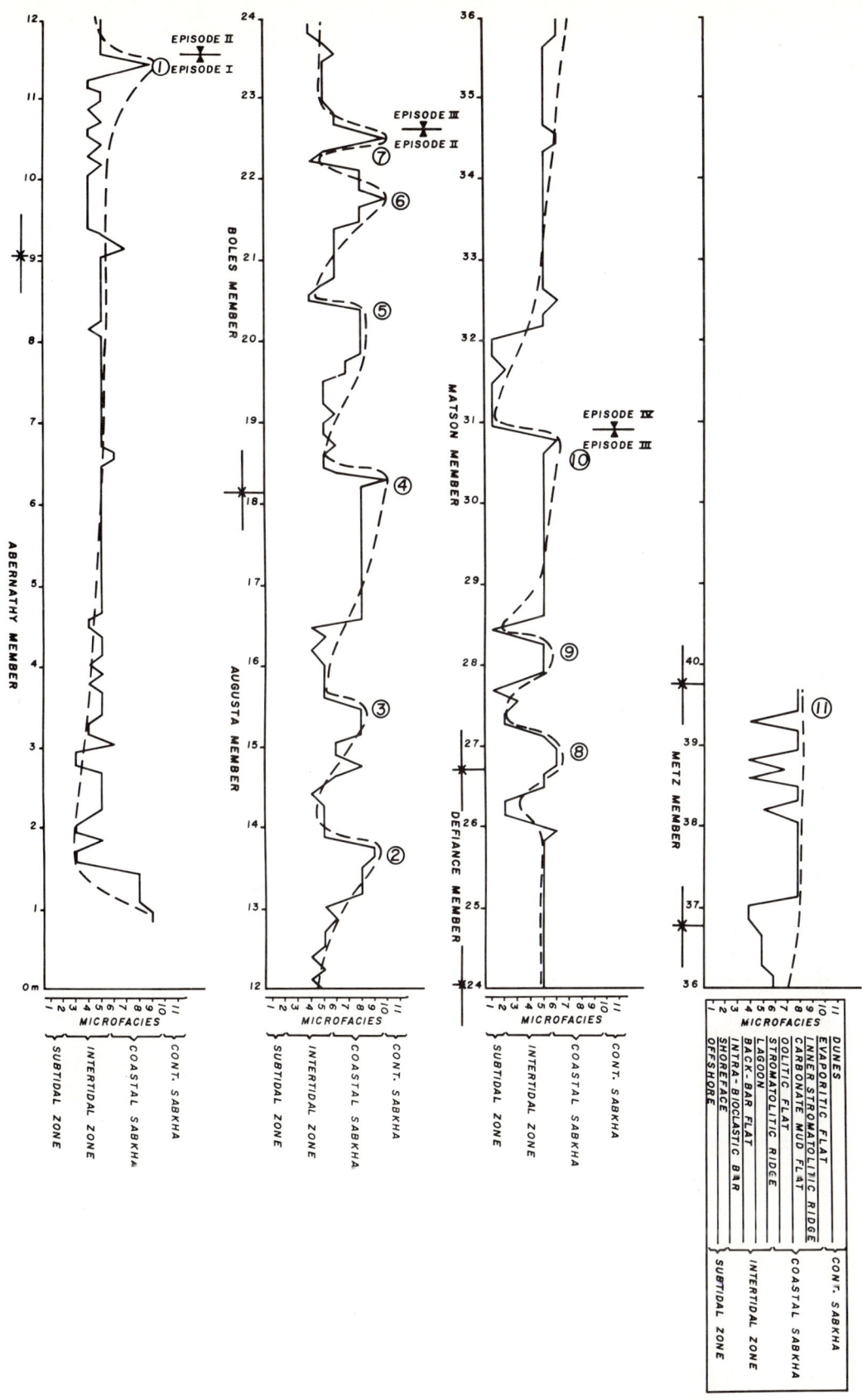

FIGURE 14.23 Joachim Dolomite (Middle Ordovician), southwest margin of Illinois Basin. General vertical evolution of depositional environments (sections F and H). From Okhravi and Carozzi (1983).

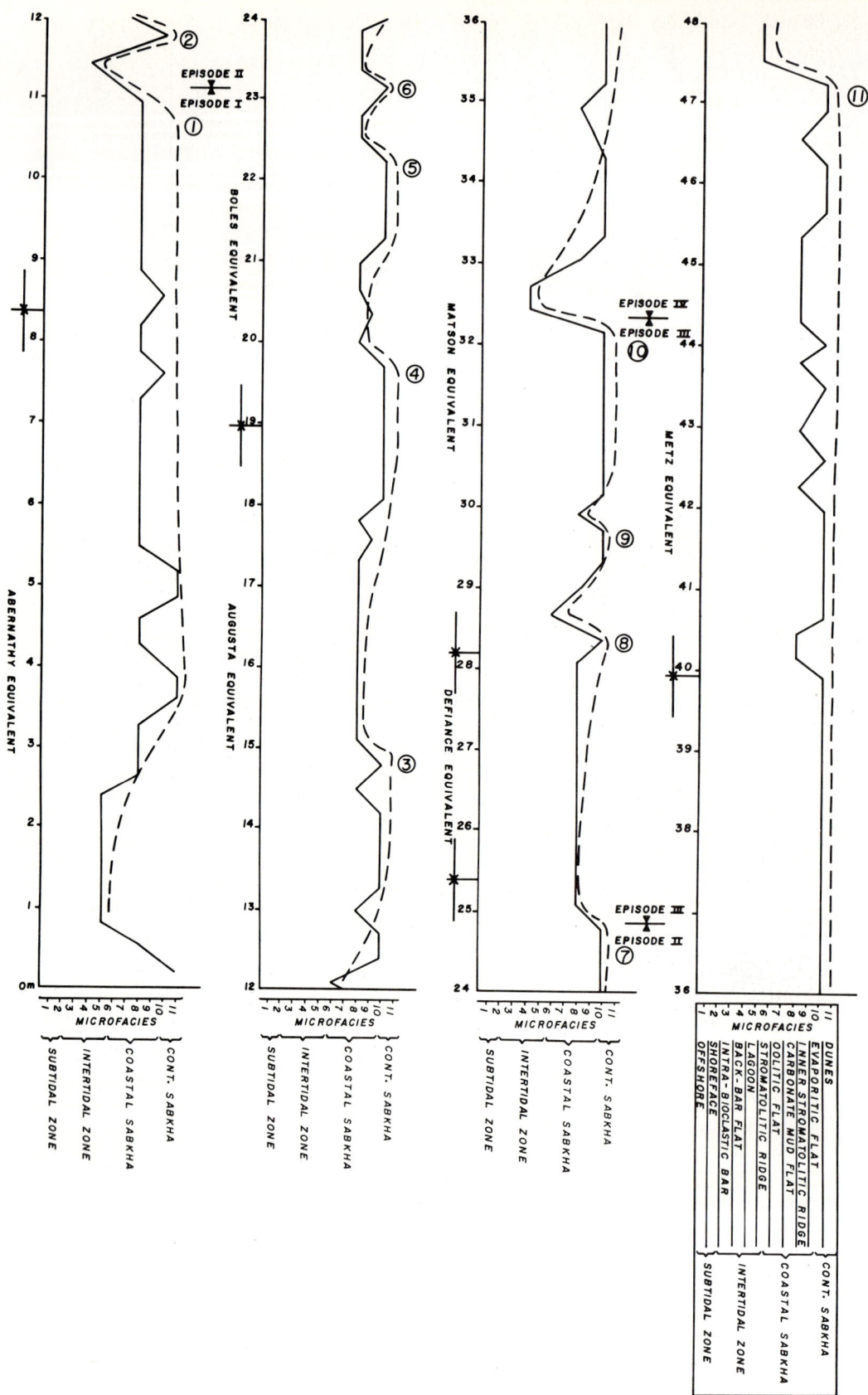

FIGURE 14.24 Joachim Dolomite (Middle Ordovician), southwest margin of Illinois Basin. General vertical evolution of depositional environments (section K). From Okhravi and Carozzi (1983).

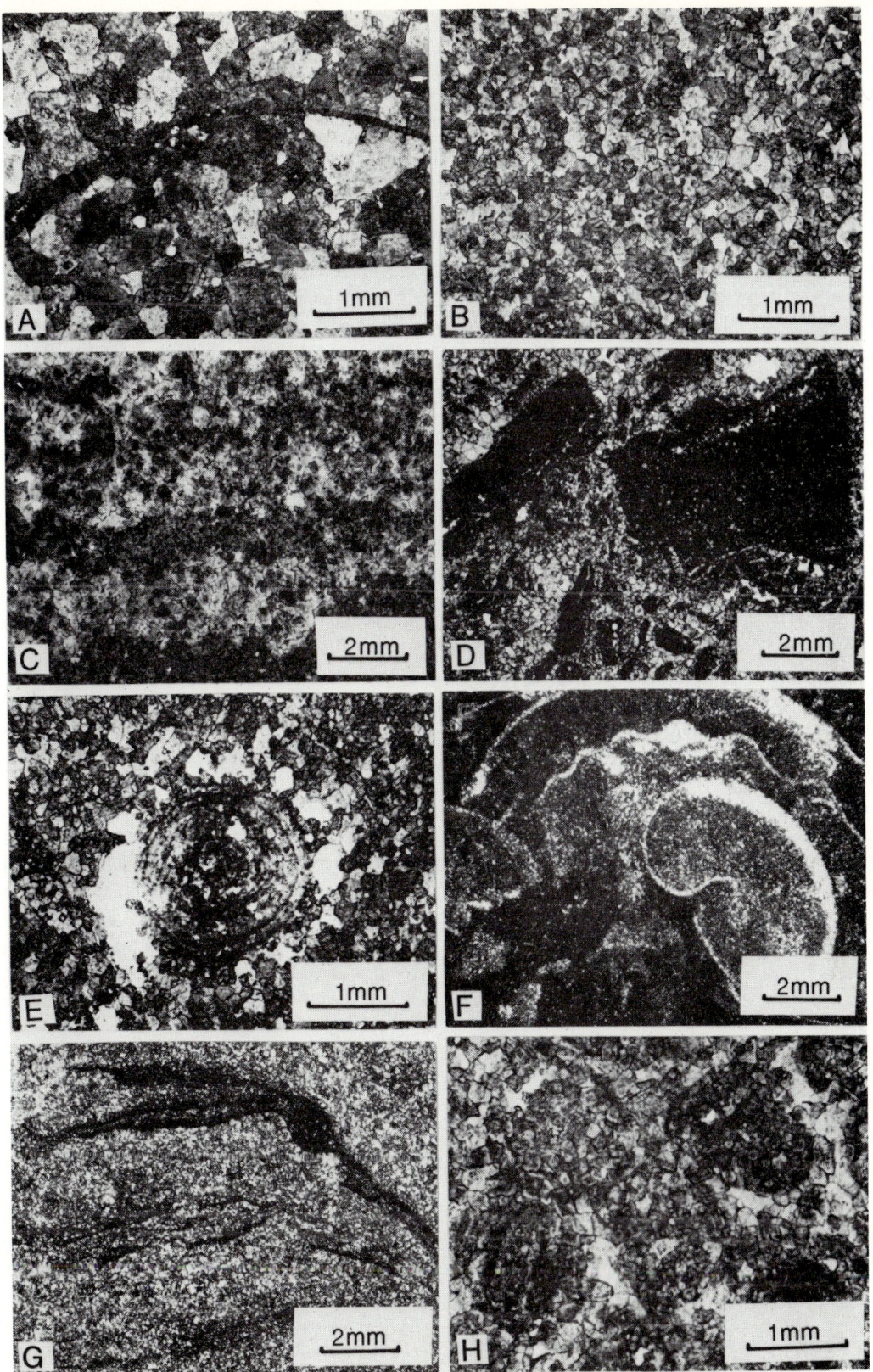

FIGURE 14.25 Burnt Bluff Group (Silurian), west margin of Michigan Basin. Typical microfacies. A. Microfacies 1. B. Microfacies 1A. C and D. Microfacies 2. E to G. Microfacies 3. H. Microfacies 4. See text for detailed descriptions. All photomicrographs: plane-polarized light. From Soderman and Carozzi (1963). Reprinted by permission of the American Association of Petroleum Geologists.

Microfacies 2 (Fig. 14.25C,D). Alternating layers of lighter-colored coarsely crystalline dolomite and darker medium crystalline dolomite. Originally, it was interbedded calcilutite and calcisiltite, with inversion of crystal size by dolomitization. Intercalations of intraformational microbreccias with finely crystalline angular dark clasts of dolomite set in coarsely crystalline dolomite matrix are frequent (Fig. 14.25D). Originally, it was a microbreccia of calcisiltite fragments in calcilutite matrix.

Microfacies 3 (Fig. 14.25E,F,G). Medium to finely crystalline dolomite with a great variety of constituents: ooids, entire leperditids, brachiopods, rugose and tabulate corals, small stromatoporoid heads, gastropods, ostracods, and cephalopods (Fig. 14.25F). Locally thin algal mats are well preserved (Fig. 14.25G). Originally, it was a fossiliferous calcisiltite with instances of preserved cross-bedding and traces of bioturbation.

Microfacies 4 (Fig. 14.25H). Clear coarsely crystalline dolomite with well-sorted and closely packed ooids with ghosts of concentric rings preserved in medium crystalline darker and zoned dolomite rhombs. Crinoid bioclasts and rounded intraclasts are rare constituents. Originally, it was an oolitic biocalcarenite with sparite cement.

Microfacies 5 (Fig. 14.26A,B). Oncoidal dolorudite with intercalated algal mats, rare bioclasts of crinoids, and smooth ostracods. Two main aspects occur: oncoids consist of a dark finely crystalline dolomite and are set in a coarse clear dolomite cement, or oncoids have been replaced by a clear coarsely crystalline anhedral dolomite and are set in a dark finely crystalline dolomite matrix (Fig. 14.26B). Originally, it was an oncoidal calcirudite with sparite cement and calcisiltite matrix, respectively.

Microfacies 6 (Fig. 14.26C,D). This microfacies forms buildups and is interpreted as constructed by *Spongiostromata*. It appears as a coarsely crystalline unzoned dolomite mosaic with pockets of large leperditid tests, or as an uneven textured, often bimodal association of large unzoned crystals surrounded by medium crystalline aggregates (Fig. 14.26D). Originally, it was an algal-bioconstructed limestone.

Microfacies 7 (Fig. 14.26E). Medium to finely crystalline dolomite with abundant bioclasts of leperditids, dark finely crystalline subrounded intraclasts, and scattered well-rounded quartz grains of St. Peter sandstone type. Originally, it was an arenaceous biocalcarenite with a calcisiltite matrix.

Microfacies 8 (Fig. 14.26F,G,H). Bioconstructed dolomite, either by stromatolites consisting of irregular algal mats of dark finely crystalline dolomite separated by thicker layers of clear coarsely crystalline dolomite (originally calcilutite), or by dark bands of *Spongiostromata* colonies with well-preserved labyrinthic structure (Fig. 14.26G), or by tabulate corals with coarsely crystalline dolomite-filling interseptal cavities (Fig. 14.26H).

Ideal Shallowing-Upward Sequence of Nasbro Buildup

The sequence of microfacies 1 through 5 corresponds to a continuous shallowing and increase of energy level, starting with calcilutites and terminating with oncoidal calcirudites. Microfacies 6, the algal bioconstructed limestone forming the buildups, is laterally equivalent to microfacies 1, 2, and 3. Therefore, microfacies 6 formed initially in relatively deep water and grew from such conditions almost to levels at which microfacies 4 developed. Microfacies 7 is interbedded with the upper portions of microfacies 3 and consequently is interpreted as a lateral equivalent of microfacies 4. The three varieties of the bioconstructed microfacies 8 replace each other locally and are assigned to a relative depth comparable to microfacies 5.

The buildup at Nasbro, Wisconsin, is the first example of zonation of these microfacies (Figs. 14.27, 14.28). The main trend curves (Figs. 14.29, 14.30) summarize and contrast the vertical evolution of the microscopic parameters for the nonbiohermal and biohermal sections and show the bathymetric relationships of the microfacies. The latter indicate an initial deepening to depths at which microfacies 1 was deposited, followed upward by a gradual shallowing in the nonbiohermal section, and by a sudden shallowing through rapid growth in the biohermal section. Section 5 (Fig. 14.31) shows detailed variation of microscopic parameters.

The Ideal Depositional Model of Nasbro Buildup

Six stages may be recognized in the development of this buildup (Fig. 14.32). Local winnowing created, apparently, a favorable location (stage 1) for the rapid growth of algal pillarlike buildups (stage 2). As sedimentation proceded around the buildups, peripheral trough structures developed around them (stages 3 and 4), probably due to mild channeling. The buildup did not contribute any debris to the surrounding sediments and was draped over eventually by oolitic biocalcarenites and overlain in turn by oncoidal calcirudites (stages 5 and 6).

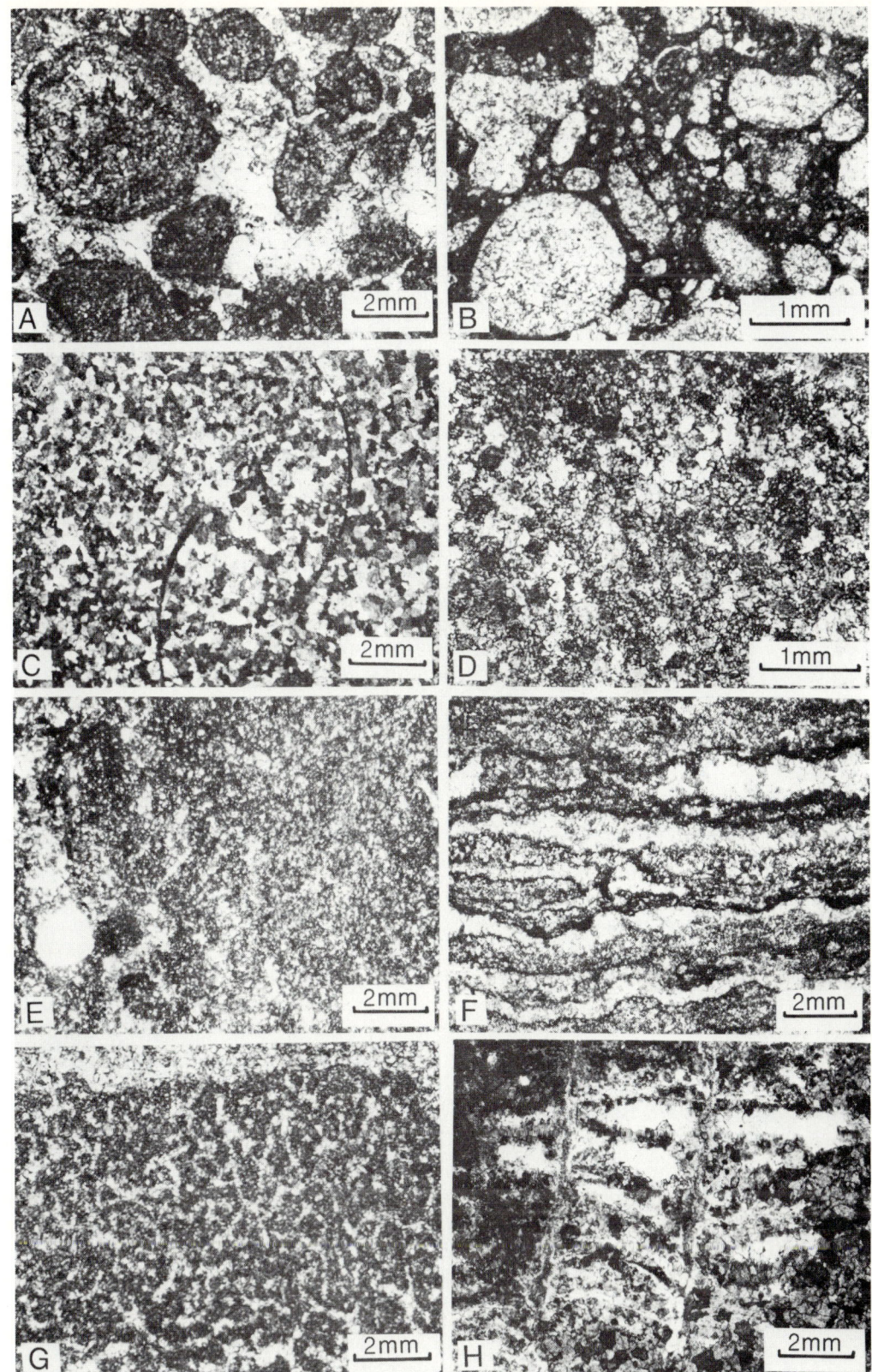

FIGURE 14.26 Burnt Bluff Group (Silurian), west margin of Michigan Basin. Typical microfacies (continued). A and B. Microfacies 5. C and D. Microfacies 6. E. Microfacies 7. F to H. Microfacies 8. See text for detailed descriptions. All photomicrographs: plane-polarized light. From Soderman and Carozzi (1963). Reprinted by permission of the American Association of Petroleum Geologists.

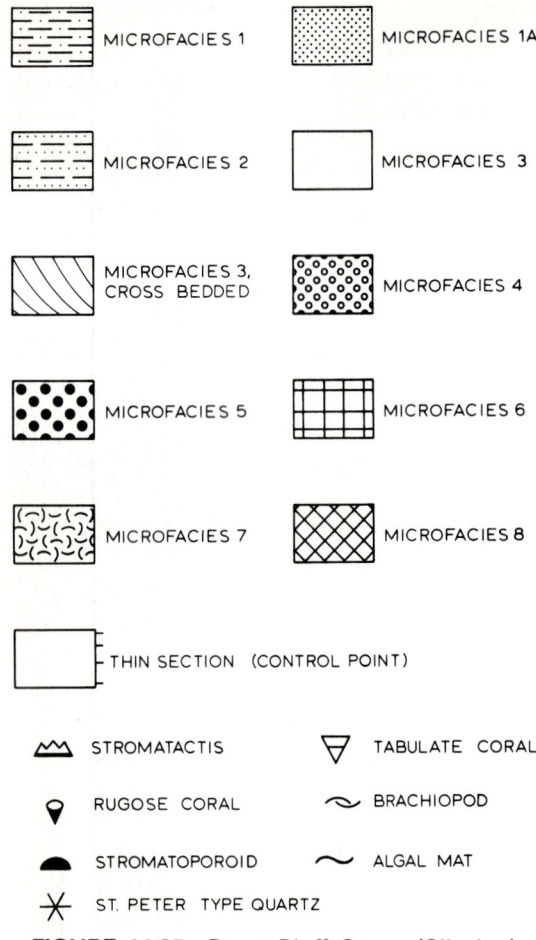

FIGURE 14.27 Burnt Bluff Group (Silurian), west margin of Michigan Basin. Symbols for microfacies. From Soderman and Carozzi (1963). Reprinted by permission of the American Association of Petroleum Geologists.

The Ideal Shallowing-Upward Sequence of Sturgeon Bay Buildup

The buildups at Sturgeon Bay, Wisconsin, display the entire spectrum of the described microfacies (Fig. 14.33). Main trend curves (Figs. 14.34, 14.35) generalize and contrast the evolution of the microscopic parameters for nonbiohermal and biohermal sections. Bathymetric curves show that after a time of relatively deep-water fluctuations, a single episode of shallowing occurred, followed by a minor shallow fluctuation. In the biohermal section, the shallowing occurred more rapidly at first, caused by the local slight elevation of the seafloor by algal growth. Section 5 (Fig. 14.36) shows detailed variation of microscopic parameters.

The Ideal Depositional Model of Sturgeon Bay Buildup

Six stages may be recognized in the development of this buildup (Fig. 14.37). The upper surface of calcilutites and calcisiltites appears of marine erosional nature (stage 1). Algae became established on the high places of this surface and started to build upward (stage 2). Algal colonies were covered and killed by the deposition of microfacies 2, which was reworked continuously. The resulting clasts of calcisiltite were transported to and deposited on the flanks of adjacent buildups (stage 3). As shallowing progressed, microfacies 2 was eroded from the top of the buildups (stage 4) and followed by the deposition of cross-bedded fossiliferous calcisiltites, arenaceous varieties, and finally, bioconstructed tabular buildups of stromatolites and tabulate corals (stages 5 and 6).

Organically Encrusted Hydrodynamic Buildup of Chilton

This crudely bedded and moundlike body (Fig. 14.38) displays numerous episodes of fossiliferous calcisiltite (microfacies 3) that contain internal erosional surfaces and are encrusted at irregular intervals by thin bioconstructed layers (microfacies 8) of corals, stromatoporoids, and stromatolites, probably developed after early submarine induration of the underlying calcisiltite. Section 3 (Fig. 14.39) shows detailed variation of microscopic parameters.

14.5 CHARACTERISTIC CONSTITUENTS: CORALS, RED ALGAE, AND ENCRUSTING FORAMINIFERS

Frontal buildups constructed by the association of corals, red algae, and encrusting foraminifers which thrive in rough seas support very diversified platforms. These wave-resistant buildups generate thick, frontal calcirudites that grade into deeper-water calcarenites (supporting pinnacle reefs) and eventually into carbonate turbidites. The protected lagoons contain numerous small bioaccumulated to hydrodynamic buildups of smaller and larger foraminifers, red algae, and finger corals strongly reworked by the action of tidal channels.

The Miocene reef systems of the Visayan Islands, central Philippines (Carozzi et al., 1976) were analyzed by both surface and subsurface investigation in a poorly known area from a sedimentologic viewpoint. The local stratigraphy of many islands had to be reevaluated by an intensive study of index and facies microfossils, followed by the microscopic study and environmental inter-

Chap. 14 / Carbonate Platforms with Frontal Bioconstructed Buildups 449

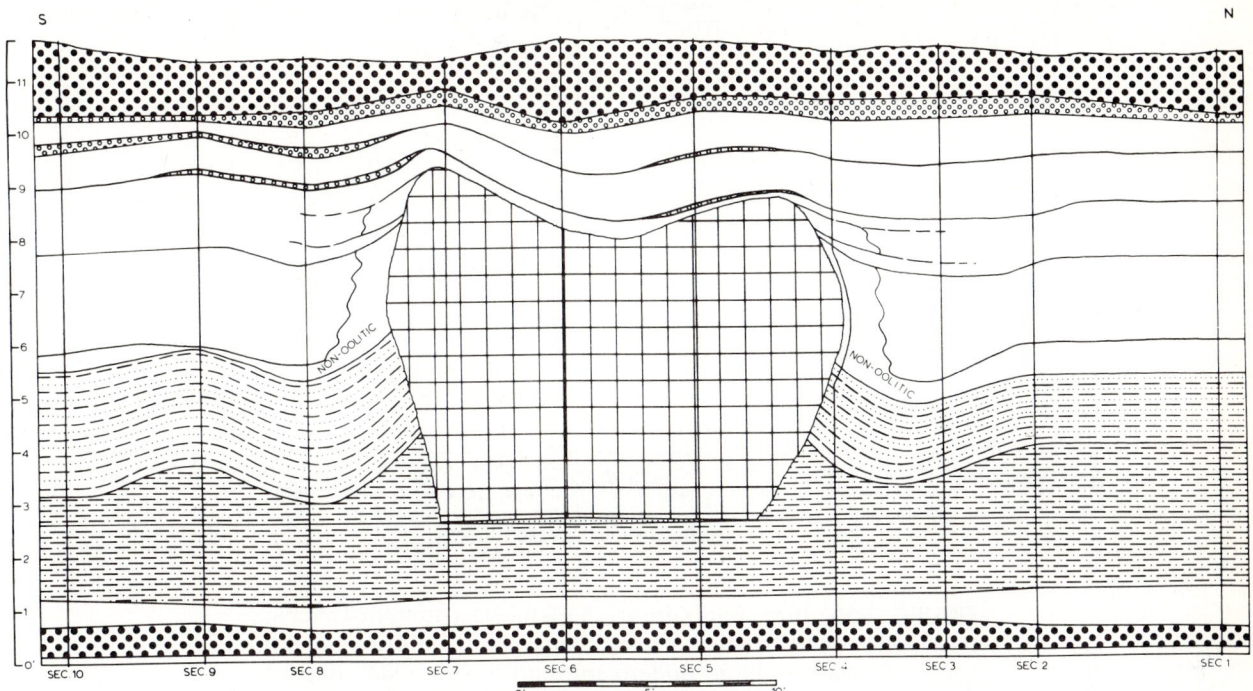

FIGURE 14.28 Burnt Bluff Group (Silurian), west margin of Michigan Basin. Microfacies zonation in stromatolite-constructed Nasbro buildup. From Soderman and Carozzi (1963). Reprinted by permission of the American Association of Petroleum Geologists.

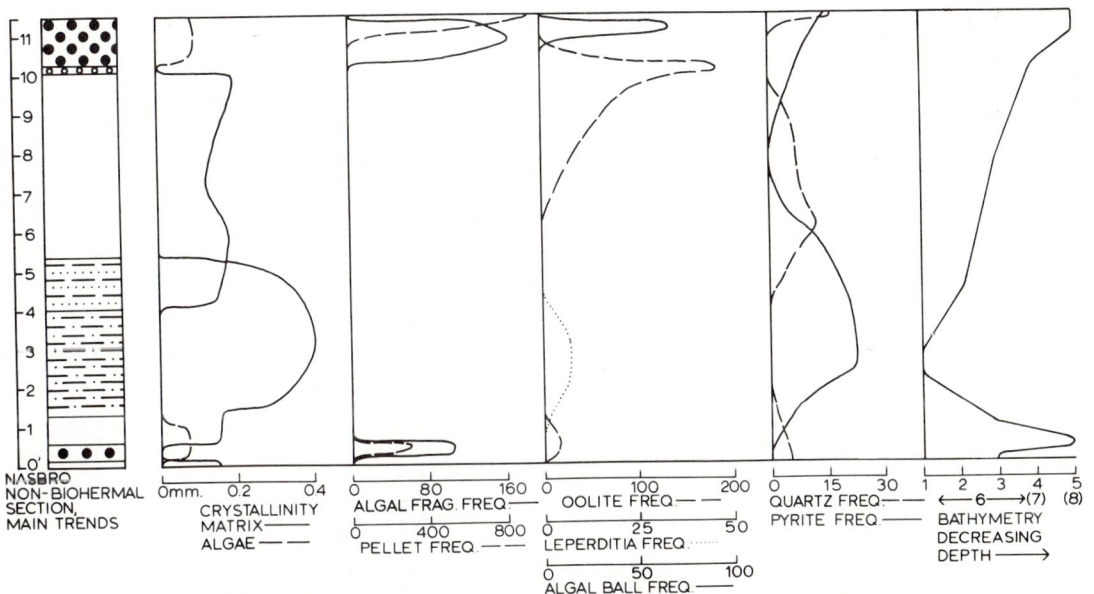

FIGURE 14.29 Burnt Bluff Group (Silurian), west margin of Michigan Basin. Main trend variation of microscopic parameters of a nonbiohermal section at Nasbro. From Soderman and Carozzi (1963). Reprinted by permission of the American Association of Petroleum Geologists.

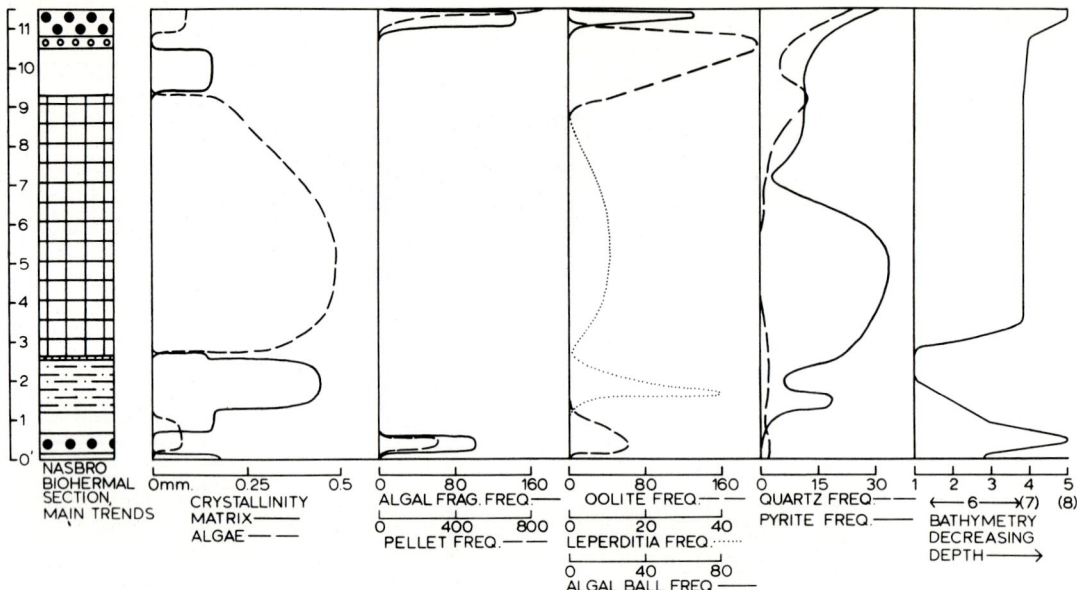

FIGURE 14.30 Burnt Bluff Group (Silurian), west margin of Michigan Basin. Main trend variation of microscopic parameters of a biohermal section at Nasbro. From Soderman and Carozzi (1963). Reprinted by permission of the American Association of Petroleum Geologists.

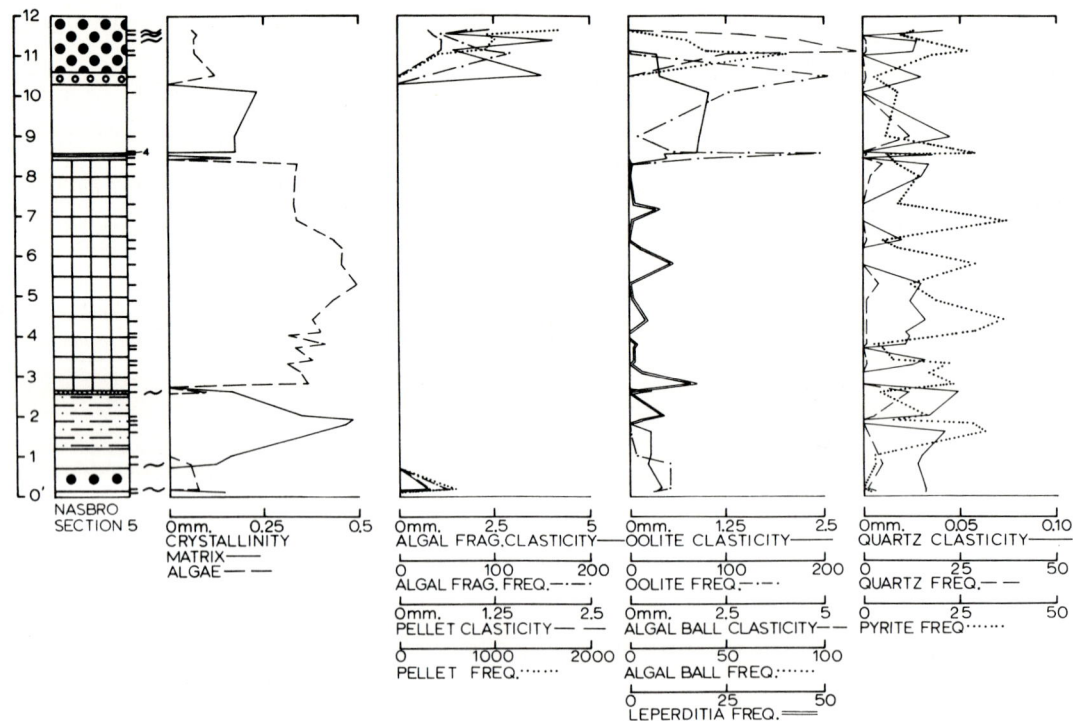

FIGURE 14.31 Burnt Bluff Group (Silurian), west margin of Michigan Basin. Typical example of vertical variation of microscopic parameters in section 5 at Nasbro. See Figure 14.28 for location. From Soderman and Carozzi (1963). Reprinted by permission of the American Association of Petroleum Geologists.

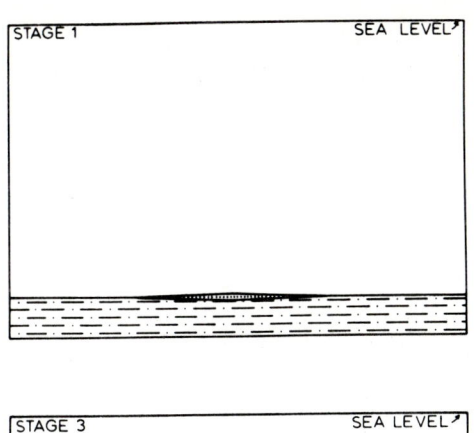
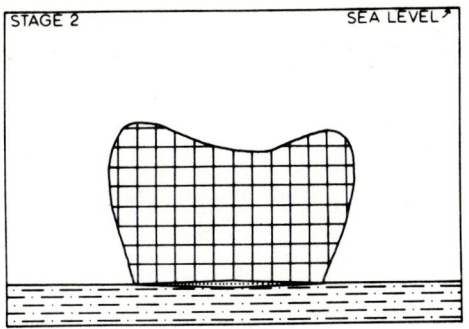
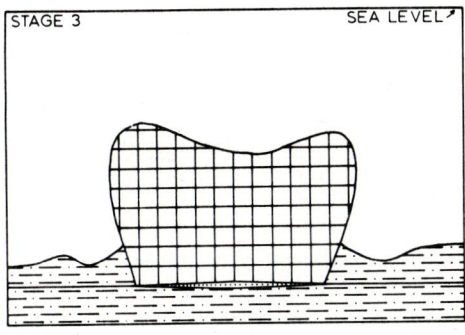
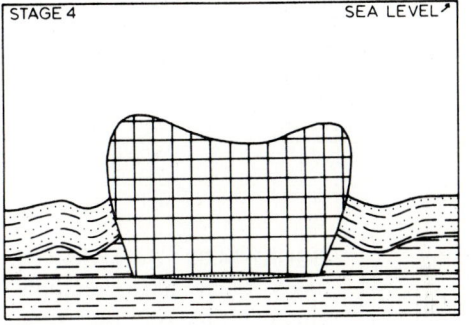
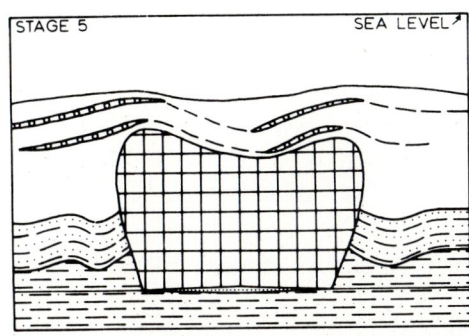
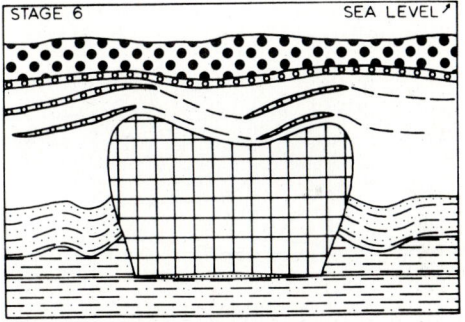

FIGURE 14.32 Burnt Bluff Group (Silurian), west margin of Michigan Basin. Ideal depositional model of stromatolite-constructed Nasbro buildup. From Soderman and Carozzi (1963). Reprinted by permission of the American Association of Petroleum Geologists.

pretation of more than 10,000 thin sections of carbonate rocks and associated clastics.

Description of Microfacies

Microfacies are described in terms of juxtaposed environments numbered consecutively from the coastline in a seaward direction.

Estuarine, Mangrove Tidal Flat, or Beach Environment

In this environment, which forms the shoreward limit, medium- to fine-grained clastics predominate over carbonates.

Microfacies 1 (Fig. 14.40A). Poorly sorted mixed biocalcarenite and graywacke that displays a great variability in composition from carbonates to clastics. The most common variety is a weakly bioturbated, arenaceous biocalcarenite with abundant grains of detrital quartz, plagioclases, hornblende, opaques, and volcanic glass, as well as lithoclasts of altered intermediate to basic volcanics associated with bioclasts of echinoids, ostracods, thin pelecypods, gastropods, miliolids, rotaliids, textulariids, relatively rare larger benthic foraminifers, and sponge spicules. Locally, pyritized plant debris are common as well as floated planktonic foraminifers. This microfacies either has a matrix of argillaceous and pyritic calcisiltite or is cemented by extensive pressure-

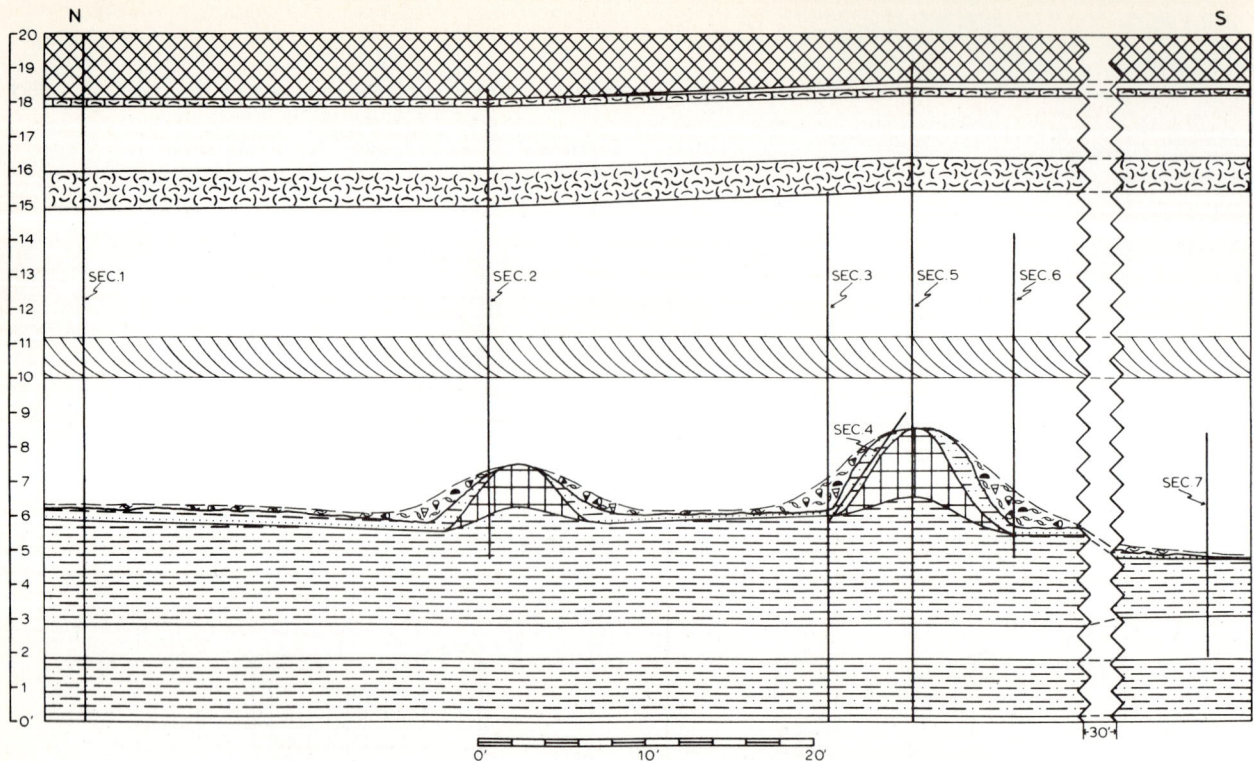

FIGURE 14.33 Burnt Bluff Group (Silurian), west margin of Michigan Basin. Microfacies zonation in stromatolite-constructed Sturgeon Bay buildup. From Soderman and Carozzi (1963). Reprinted by permission of the American Association of Petroleum Geologists.

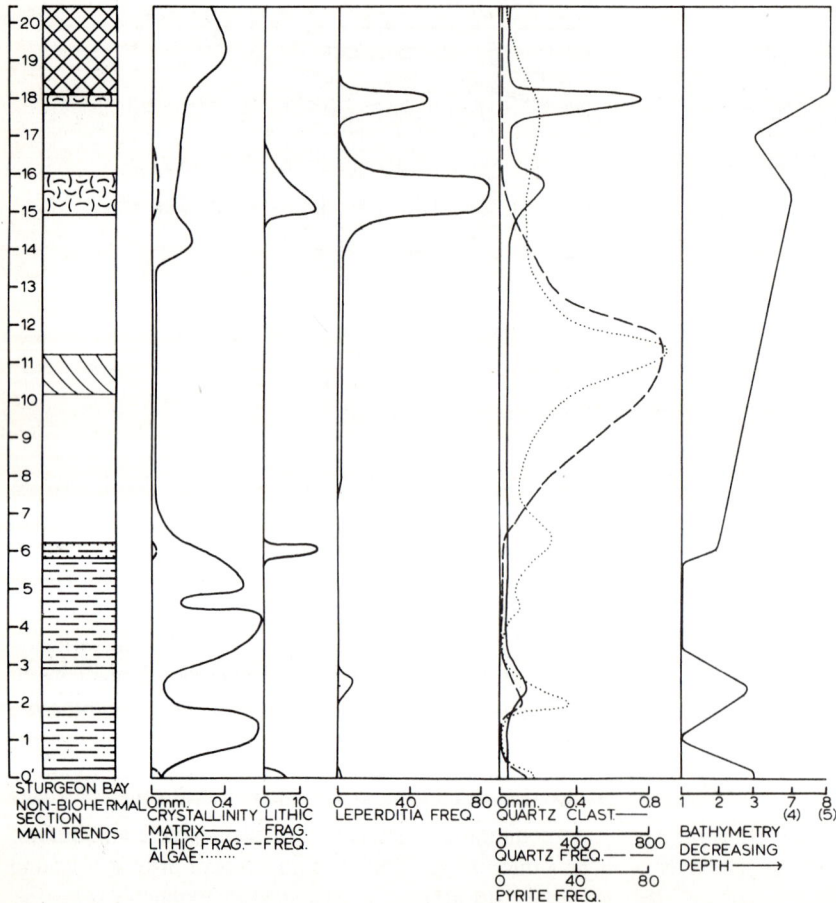

FIGURE 14.34 Burnt Bluff Group (Silurian), west margin of Michigan Basin. Main trend variation of microscopic parameters of a nonbiohermal section at Sturgeon Bay. From Soderman and Carozzi (1963). Reprinted by permission of the American Association of Petroleum Geologists.

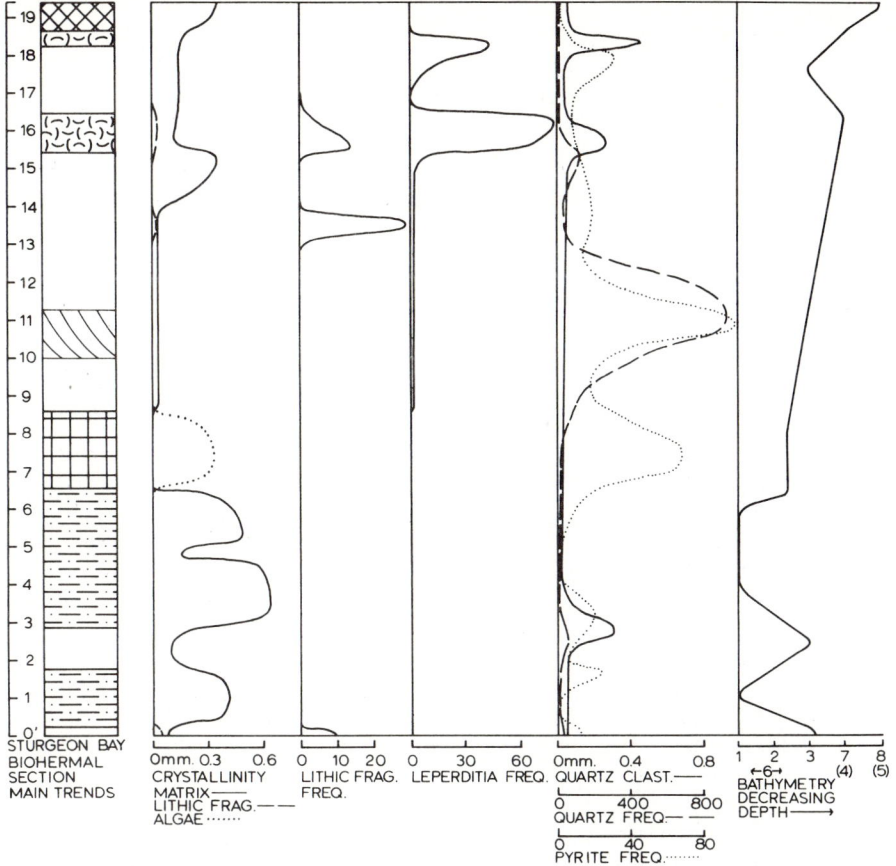

FIGURE 14.35 Burnt Bluff Group (Silurian), west margin of Michigan Basin. Main trend variation of microscopic parameters of a biohermal section at Sturgeon Bay. From Soderman and Carozzi (1963). Reprinted by permission of the American Association of Petroleum Geologists.

solution associated with overgrowths around echinoid bioclasts.

Shoreface Environment

This environment represents the major zone of transition between clastic and carbonate deposition.

Microfacies 2 (Fig. 14.40B). Poorly sorted, coarse to fine biocalcarenite with either an argillaceous calcisiltite or bioclastic matrix or a combination of pressure-solution and local patches of interstitial microsparite. Bioclasts are orbitoids, *Cycloclypeus*, *Operculina*, *Amphistegina*, rotaliids, red algae, echinoids, gastropods, pelecypods, ostracods, and smaller arenaceous foraminifers. The detrital sediment yield consists of lithoclasts of altered volcanics and grains of quartz, feldspars, and altered mafics. Bioturbation is rare.

Lagoonal Environment

This environment is by far the most complex and has an association of five microfacies. This association actually consists of a fine-grained carbonate background (microfacies 4) in which are distributed in a general seaward direction the following types of banks: smaller arenaceous foraminifer banks (microfacies 5), gastropod–red algae banks (microfacies 6), larger foraminifer banks (microfacies 3), and finally, adjacent to the backreef environment, patch reefs of finger corals designated as microfacies 4–8 so as to distinguish them from the main coral–red algae-encrusting foraminifer-constructed buildup (microfacies 8). The general juxtaposition of these banks with their particular organic communities indicates that the lagoonal environment gradually changes from restricted conditions on its landward side to more open conditions along its back-reef boundary.

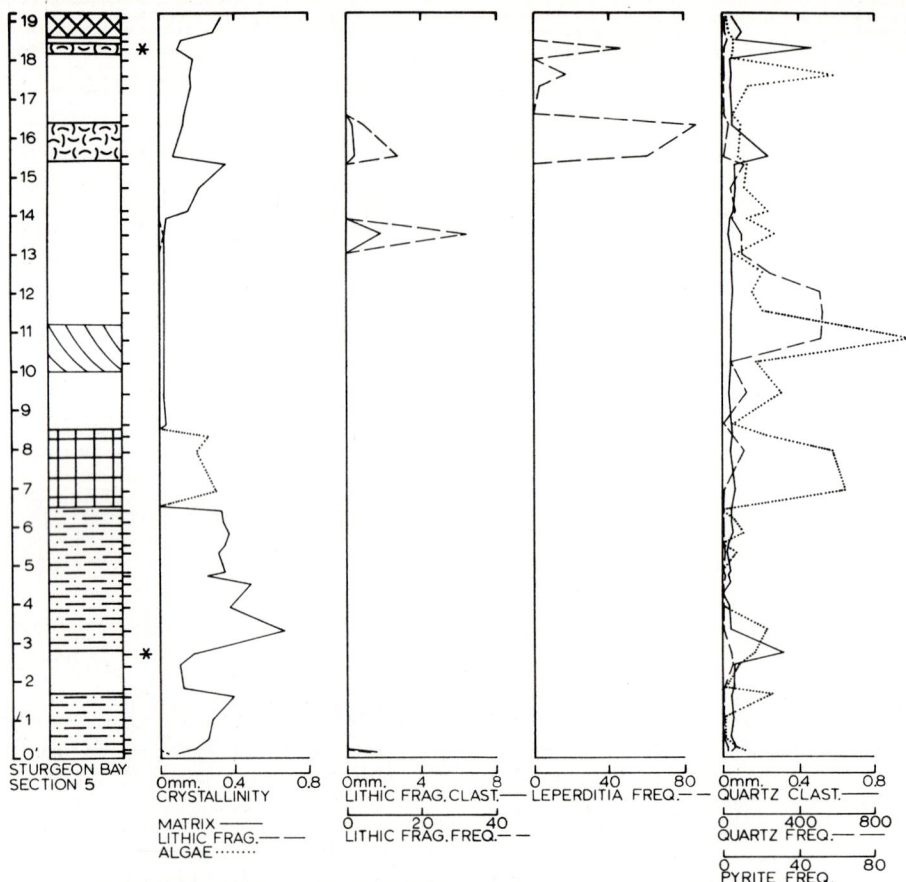

FIGURE 14.36 Burnt Bluff Group (Silurian), west margin of Michigan Basin. Typical example of vertical variation of microscopic parameters in section 5 at Sturgeon Bay. See Figure 14.33 for location. From Soderman and Carozzi (1963). Reprinted by permission of the American Association of Petroleum Geologists.

Microfacies 3 (Fig. 14.40C,D). Foraminiferal bioaccumulated limestone with interstitial calcisiltite matrix and local pressure-solution. Superposed larger benthic foraminifers such as orbitoids and *Miogypsina* are aligned parallel to bedding or are locally disturbed by bioturbation. Subordinate constituents are *Amphistegina* (Fig. 11.40D), *Halimeda*, red algae, echinoids, ostracods, and finger corals.

Microfacies 4 (Fig. 14.40E). Poorly sorted biocalcarenite with pelletoidal calcisiltite matrix to pelletoidal calcisiltite with scattered sand-size bioclasts. Abundant subrounded colonies of red algae encrusted by bryozoans are commonly associated with smaller arenaceous foraminifers, textulariids, miliolids, echinoids, ostracods, dasyclads, and rare planktonic foraminifers. Intense bioturbation is shown by imbrication or irregular distribution of larger bioclasts. Better sorting is revealed in places by pressure-welding of the bioclasts associated with small patches of interstitial microsparite cement.

Microfacies 5 (Fig. 14.40F). Poorly sorted foraminiferal biocalcarenite with bituminous calcisiltite matrix grading into a calcisiltite with scattered sand-size bioclasts. Abundant miliolids, textulariids, and rotaliids are associated with orbitoids, echinoids, ostracods. *Halimeda*, pelecypods, gastropods, and rare planktonic foraminifers. The general alignment of the components parallel to bedding is often disturbed by bioturbation.

Microfacies 6 (Fig. 14.40G). Pelecypod–red algae bioaccumulated limestone with bituminous pelletoidal calcisiltite matrix. *Operculina*, *Halimeda*, and finger corals are also common and are associated with miliolids, textulariids, echinoids, and ostracods. Pockets of intense bioturbation are common.

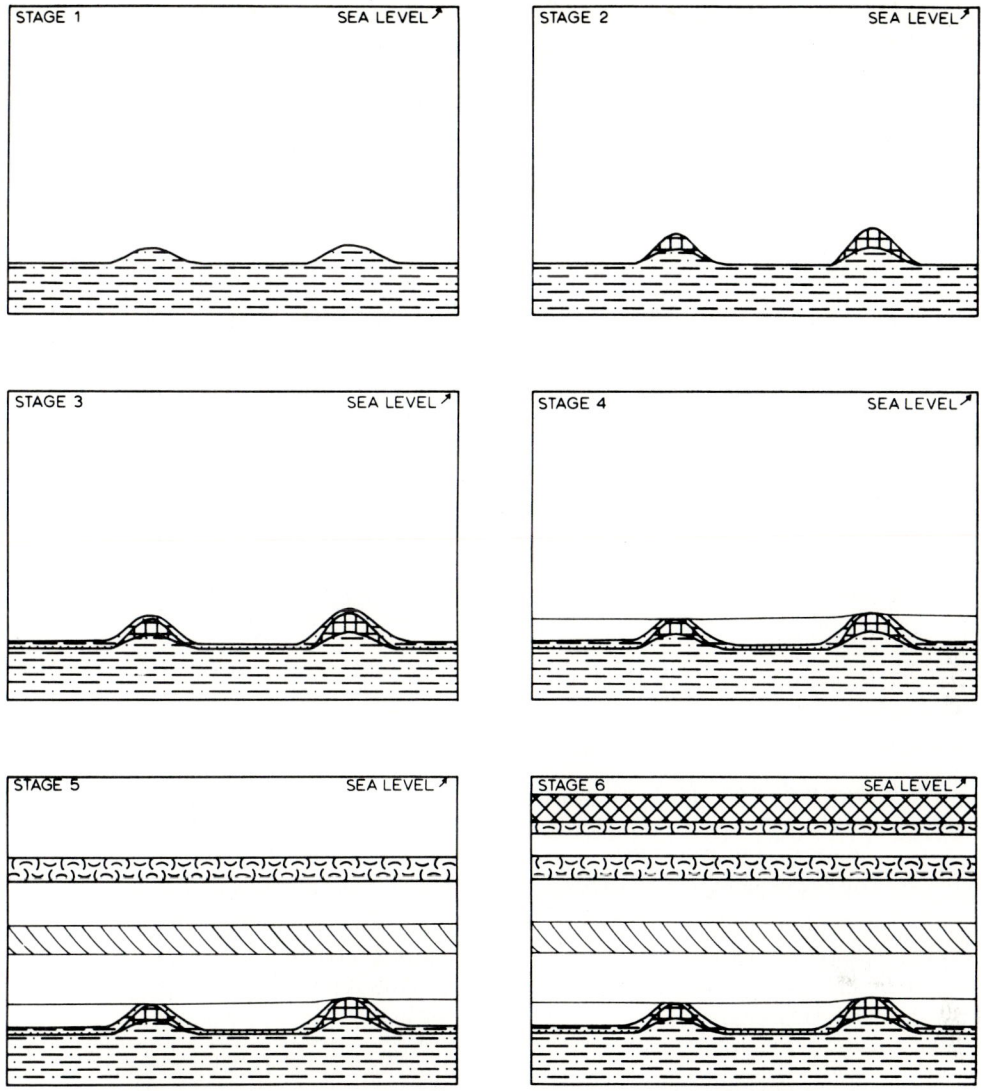

FIGURE 14.37 Burnt Bluff Group (Silurian), west margin of Michigan Basin. Ideal depositional model of stromatolite-constructed Sturgeon Bay buildup. From Soderman and Carozzi (1963). Reprinted by permission of the American Association of Petroleum Geologists.

Microfacies 4-8 (Fig. 14.40H). Finger-coral bioaccumulated limestone with an abundant matrix of dark massive calcisiltite, locally pelletoidal. Small bioclasts of sponge spicules, ostracods, rare miliolids, and *Elphidium* are scattered throughout the matrix.

Back-Reef Environment

This environment is developed mostly in tidal channels between constructed buildups, extends deep into the lagoonal environment as tongues, and eventually spreads out also as the fore-reef apron. The local occurrence of patch reefs of finger and cup corals within the back-reef environment provides a variety of bioclasts which are redistributed in this relatively greater energy environment, together with components derived from the various banks of the lagoon. Whenever continental sediment yield is active, lithoclasts and grains of volcanic and metamorphic rocks are also distributed along these main axes of communication between lagoon and open sea across the buildup barrier.

Microfacies 7 (Fig. 14.41A,B). Well-sorted bimodal biocalcarenite with abundant cavity-filling sparite cement and local pressure-solution (Fig. 14.41B). Lithoclasts of various lagoonal microfacies (particularly microfacies 4) are associated with encrusting red algae, large

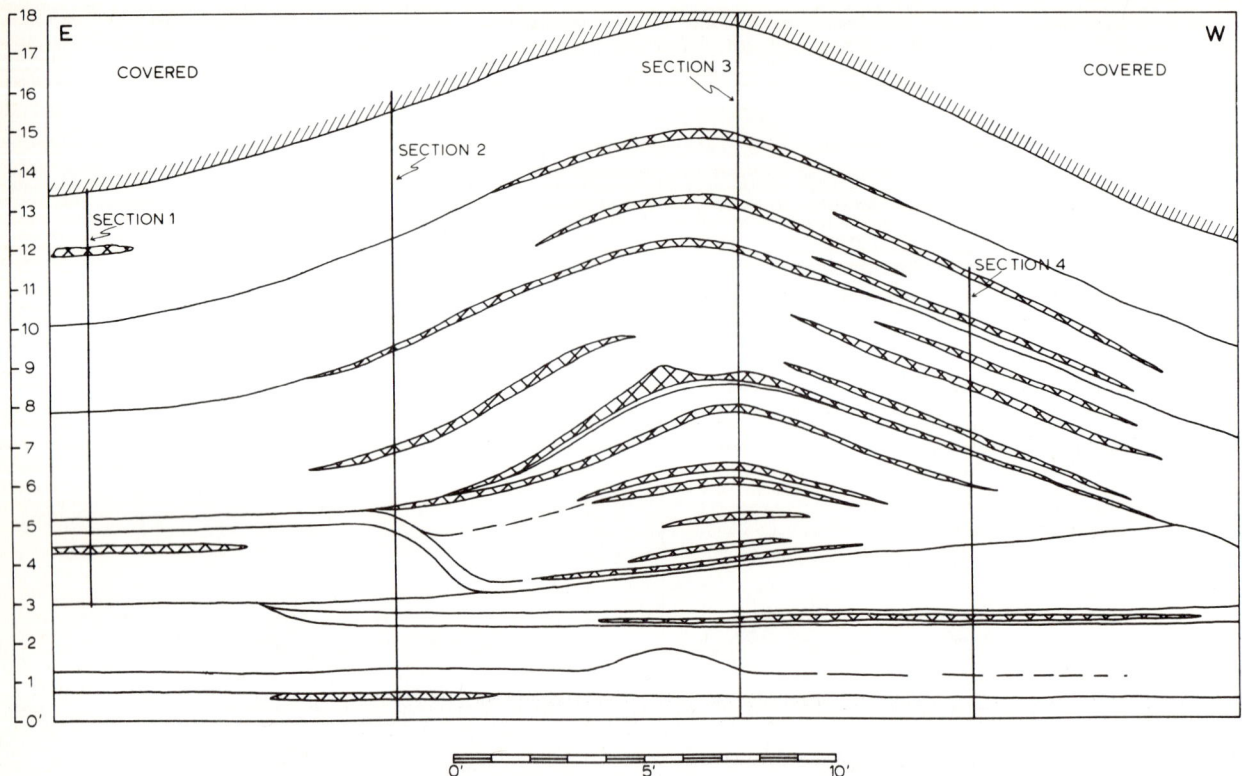

FIGURE 14.38 Burnt Bluff Group (Silurian), west margin of Michigan Basin. Microfacies zonation of Chilton organically encrusted hydrodynamic buildup. From Soderman and Carozzi (1963). Reprinted by permission of the American Association of Petroleum Geologists.

orbitoids, *Spiroclypeus, Cycloclypeus, Elphidium, Carpenteria, Miogypsina, Halimeda,* echinoid spines, miliolids, textulariids, and ostracods.

Main Coral–Red Algae-Encrusting Foraminifer Buildup Environment

Microfacies 8 (Fig. 14.41C,D,E). This constructed limestone consists of massive coral colonies and cup corals encrusted to a variable degree by red algae, bryozoans, and foraminifers and set in an abundant interstitial matrix of dark calcisiltite with bioclasts of orbitoids, *Amphistegina, Halimeda, Borelis,* echinoids, ostracods, miliolids, small gastropods, pelecypods, and rare planktonic foraminifers. The common neomorphism of the framework builders into pseudomicrosparite and pseudosparite is due to vadose and phreatic diagenesis. All the porosity is burial secondary biomoldic (Fig. 14.41D) at the expense of all the constituents (porosity 31.2%, permeability 51 md), or enlarged moldic to vuggy (Fig. 14.41E) when dissolution affects also the interstitial calcisiltite (porosity 26.1%, permeability 146 md). Some pores are obliterated by abundant concentrations of vadose silt overlain by phreatic sparite cement.

Fore-Reef Talus Environment

This environment builds a discontinuous zone along the front of the major constructed buildups facing the open ocean. These greater energy carbonates result from the accumulation, and often slumping, of the coarse products generated by the mechanical destruction of buildups by waves and tides. Associated with early cemented large subangular fragments of microfacies 8 are lithoclasts of lagoonal microfacies, particularly 4 as well as lithoclasts of volcanics, all of which demonstrate the active role played by tidal channels.

Microfacies 9 (Fig. 14.41F,G). Coarse biolithocalcirudite with poorly sorted subangular to subrounded fragments of coral colonies, red algae colonies, lithoclasts of calcisiltite (microfacies 4), abundant bioclasts of *Lepidocyclina, Miogypsina, Operculina, Amphistegina, Carpenteria,* echinoids, and planktonic foraminifers. Locally, lithoclasts of porphyritic andesites

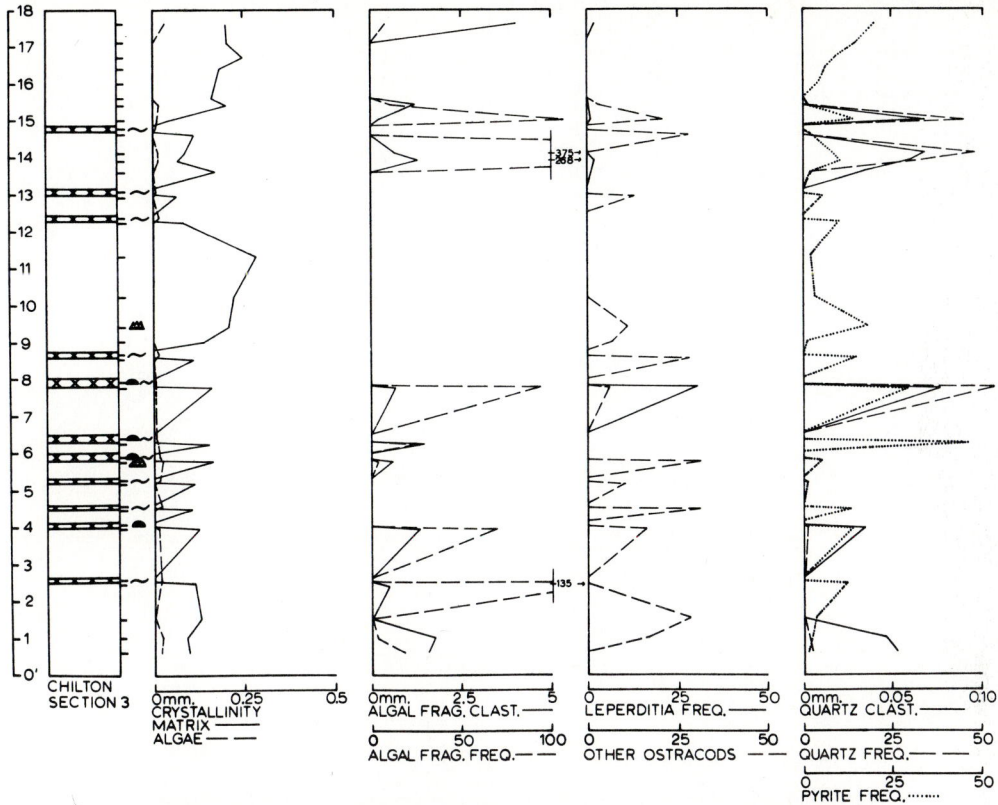

FIGURE 14.39 Burnt Bluff Group (Silurian), west margin of Michigan Basin. Typical example of vertical variation of microscopic parameters in section 3 at Chilton. See Figure 14.38 for location. From Soderman and Carozzi (1963). Reprinted by permission of the American Association of Petroleum Geologists.

(Fig. 14.41F) and isolated hornblende phenocrysts are abundant. Cementation is by widespread pressure-solution with interstitial patches of cavity-filling sparite and rare pelletoidal calcisiltite matrix.

Open Marine Environment

It consists of pelagic carbonate muds grading into shales in deeper water and interrupted locally by submarine fans of graded-bedded calcarenites formed in front of major tidal channels.

Microfacies 10 (Fig. 14.41H). Massive to clotted calcisiltite with abundant planktonic foraminifers: globigerinids, orbulinids, *Globoquadrina*, *Sphaeroidinellopsis*, *Globorotalia*, monaxonic sponge spicules, and ostracods. Pyrite pigments and rare minute grains of detrital quartz are scattered throughout the groundmass.

Microfacies 11 (Fig. 14.42). Turbidite graded sequence starting at the base with an arenaceous biocalcarenite consisting of size-sorted, broken benthic organisms, planktonic foraminifers, and grains of detrital minerals set in a sparite cement, grading upward into a finer biocalcarenite with argillaceous calcisiltite matrix, then into a mud-supported biocalcarenite, and eventually into alternating laminae of argillaceous biocalcisiltite with variable amounts of bioclasts.

Diagenesis

Diagenetic features observed in these Miocene carbonates are complex but can be interpreted as representing depositional and postdepositional changes that affected the main coral-red algal reef barrier and its adjacent microfacies.

The marine phreatic environment is represented by perforations of bioclasts, micrite envelopes, isopachous fibrous rim cement, spectacular radiaxial cement, and geopetal internal sediments. The freshwater vadose and undersaturated freshwater phreatic environments are characterized by leaching of some aragonitic

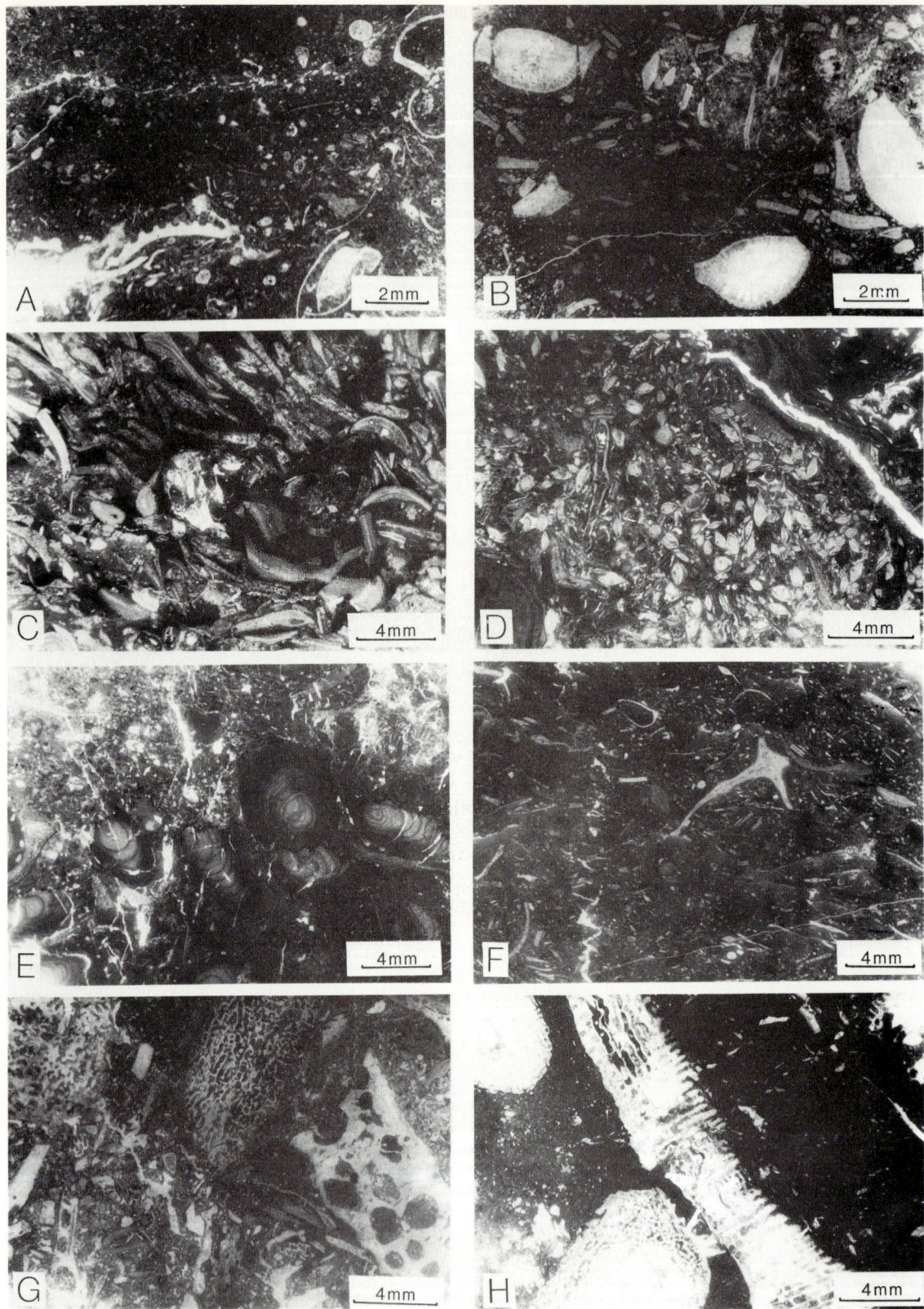

FIGURE 14.40 Miocene of the Visayan Islands, central Philippines. Typical microfacies. A. Microfacies 1. B. Microfacies 2. C and D. Microfacies 3. E. Microfacies 4. F. Microfacies 5. G. Microfacies 6. H. Microfacies 4-8. See text for detailed descriptions. All photomicrographs: plane-polarized light. From Carozzi et al. (1976).

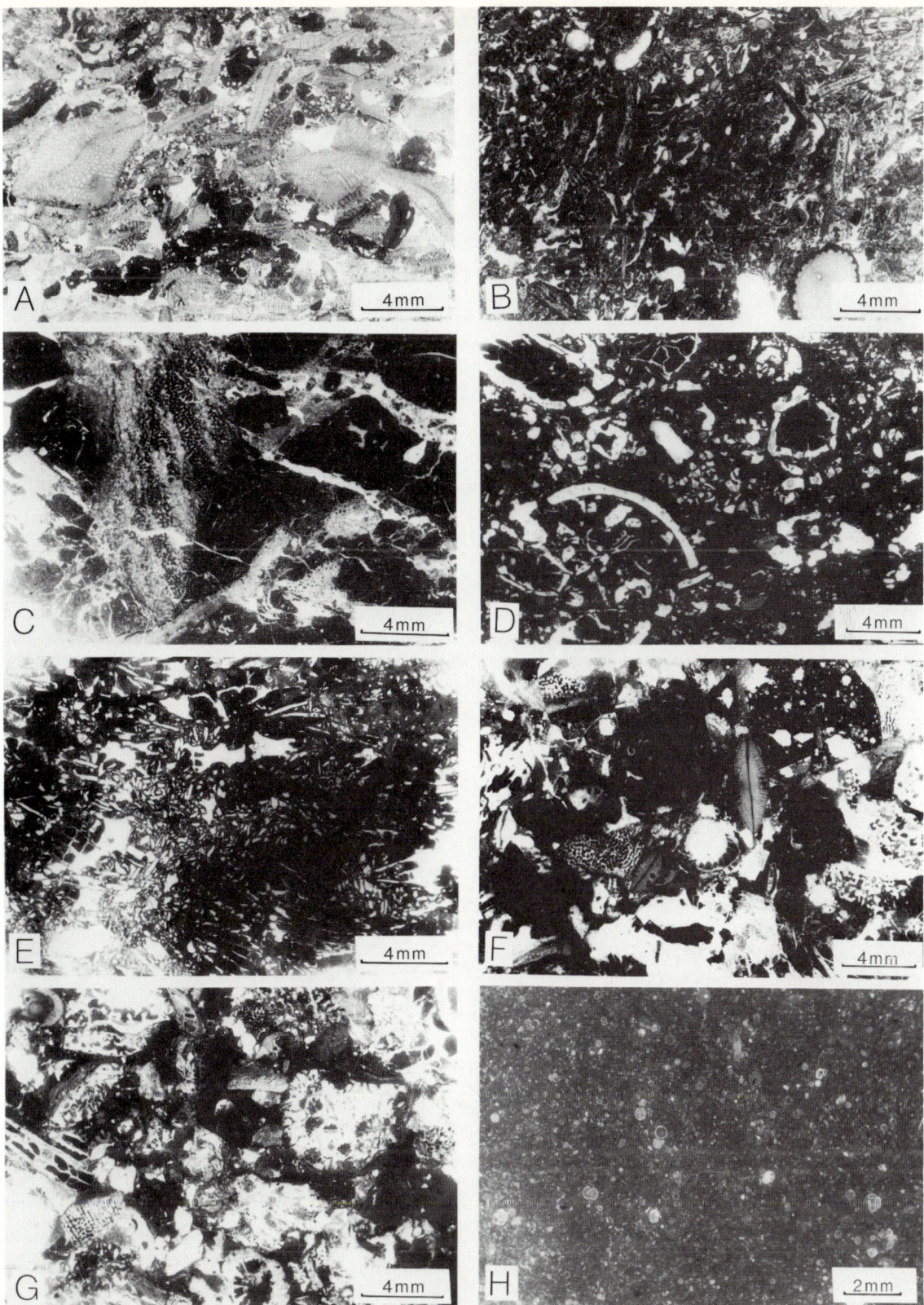

FIGURE 14.41 Miocene of the Visayan Islands, central Philippines. Typical microfacies (continued). A and B. Microfacies 7. C to E. Microfacies 8. F and G. Microfacies 9. H. Microfacies 10. See text for detailed descriptions. All photomicrographs: plane-polarized light. From Carozzi *et al.* (1976).

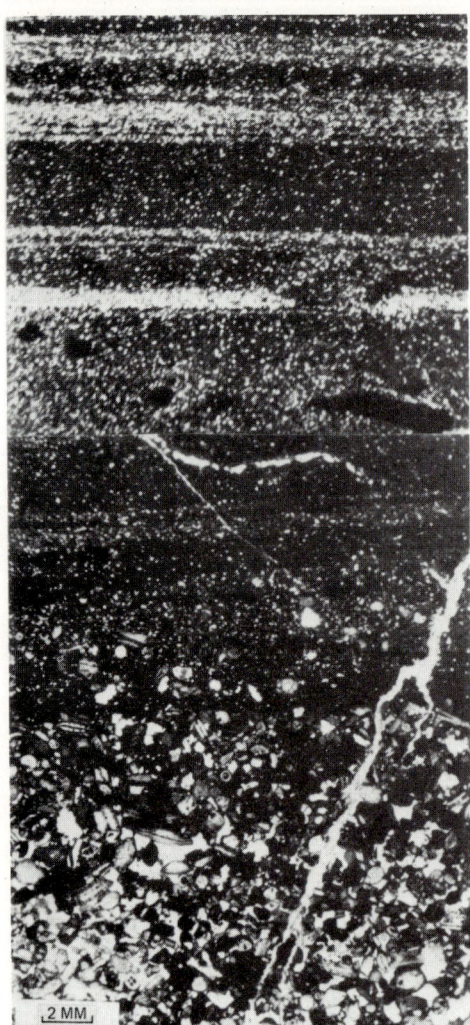

FIGURE 14.42 Miocene of the Visayan Islands, central Philippines. Typical microfacies (continued). Microfacies 11. See text for detailed description. Plane-polarized light. From Carozzi *et al.* (1976).

constituents and a small amount of nonfabric selective dissolution. This early porosity is often obliterated by deposition of vadose silt which has also taken place in fissure systems. The saturated freshwater phreatic environment is marked by intense neomorphism of corals and interstitial matrix as well as by sparite cementation. The burial environment is responsible for a highly developed fabric selective dissolution of all constituents. This biomoldic porosity frequently extends to enlarged moldic and to vuggy when involving the interstitial matrix and accounts for the observed reservoir properties.

The Ideal Depositional Model

The proposed model (Figs. 14.43, 14.44, 14.45) is valid for the entire circum-Pacific belt of Cenozoic reefs and corresponds to a context of tectonically active island arcs where vertical movements lead to the superposition of successive generations of constructed buildups separated by rapid episodes of emergence or submergence. The pervasive effect of andesitic volcanism is an important factor in providing seals for carbonate buildups that have acquired burial secondary porosity. The investigation of microfacies in a jungle terrain, as in the Philippines, can be greatly facilitated by the identification of the larger foraminifers (Fig. 14.46), which by themselves contribute to the recognition of the major environments of the proposed model.

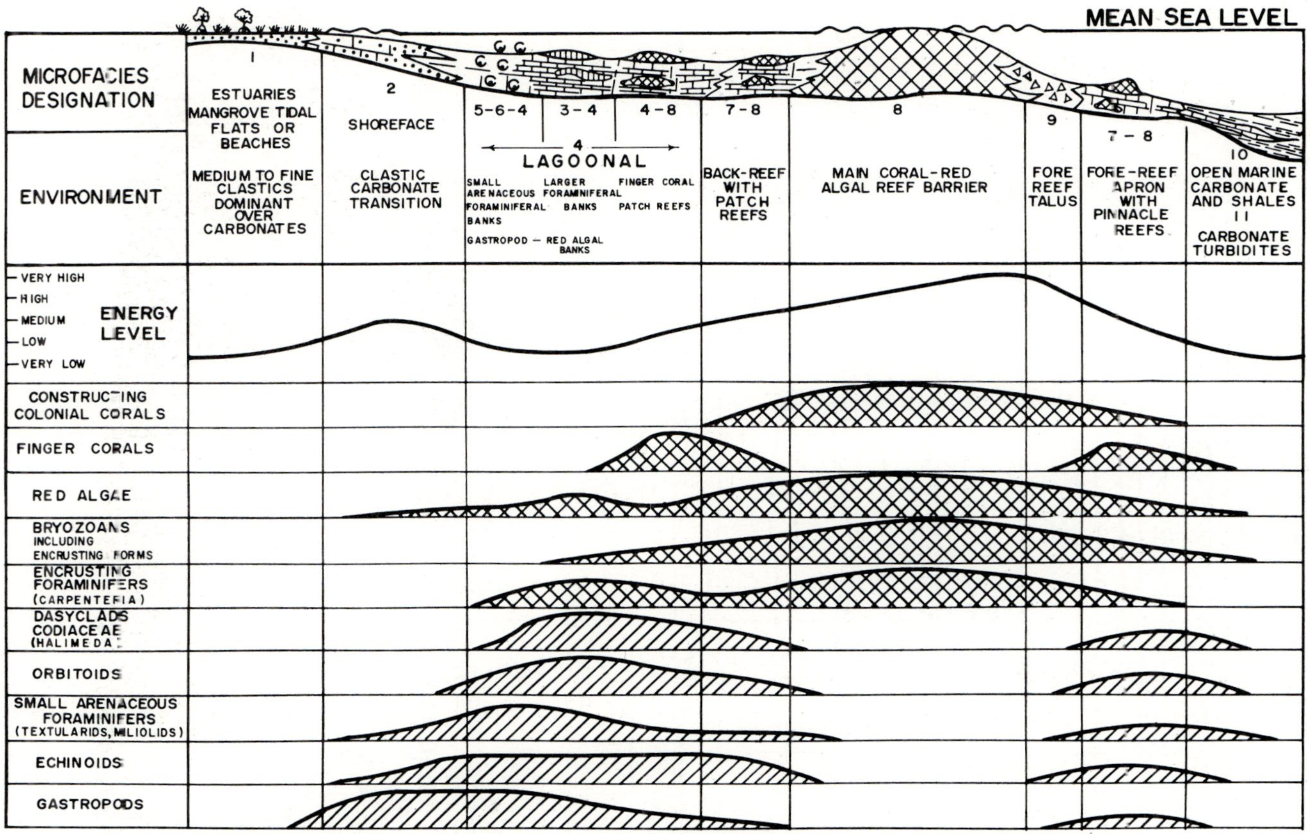

FIGURE 14.43 Miocene of the Visayan Islands, central Philippines. Ideal depositional model. Modified from Carozzi et al. (1976).

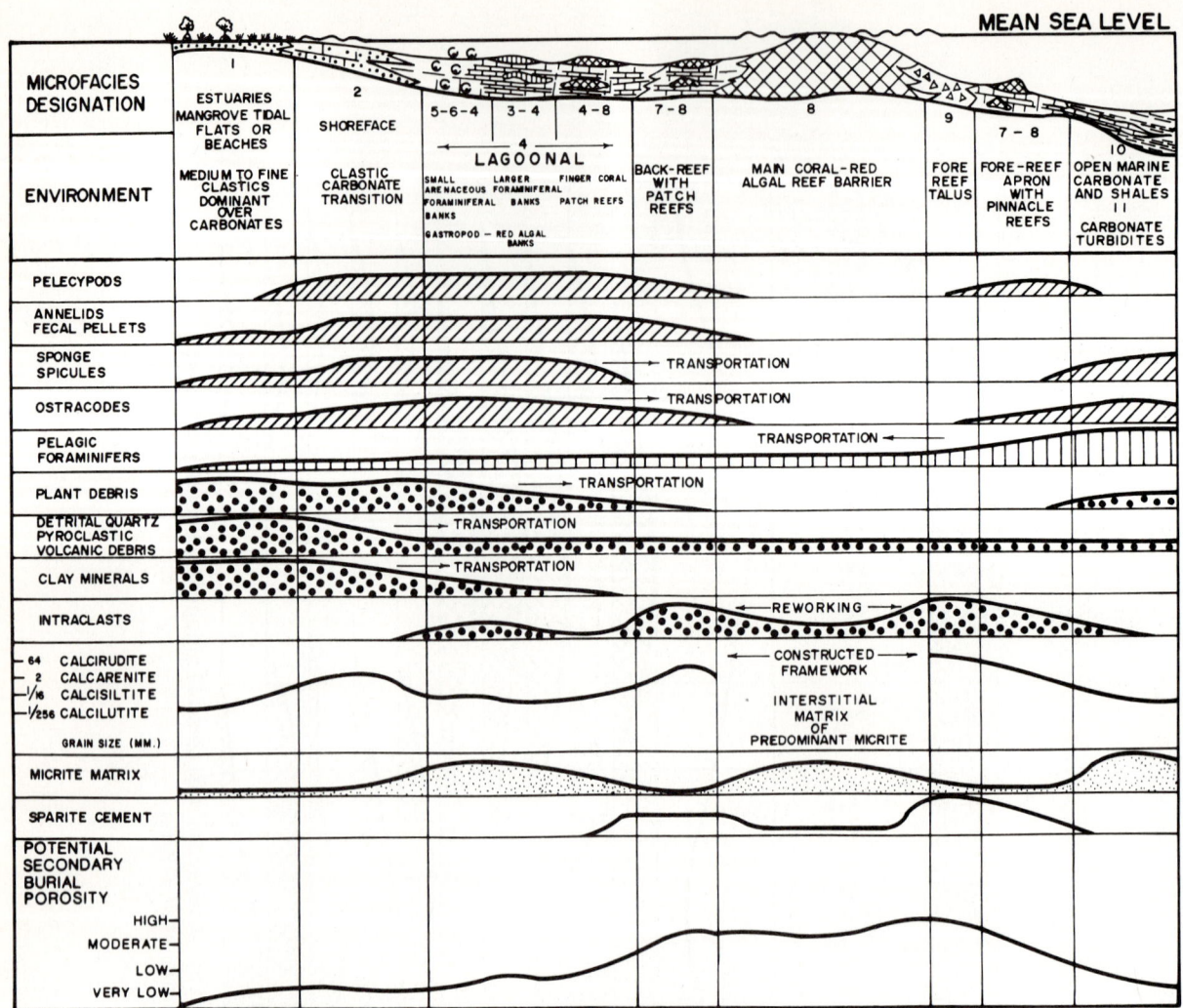

FIGURE 14.44 Miocene of the Visayan Islands, central Philippines. Ideal depositional model (continued). Modified from Carozzi *et al.* (1976).

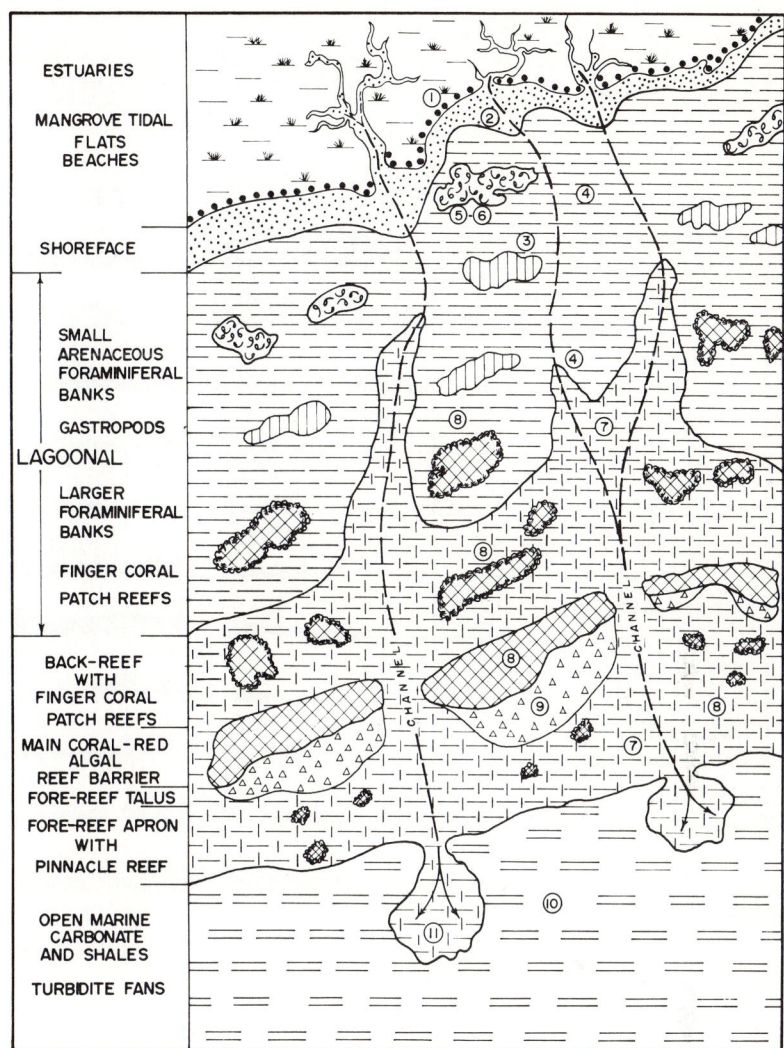

FIGURE 14.45 Miocene of the Visayan Islands, central Philippines. Schematic plan view of ideal depositional model. From Carozzi *et al.* (1976).

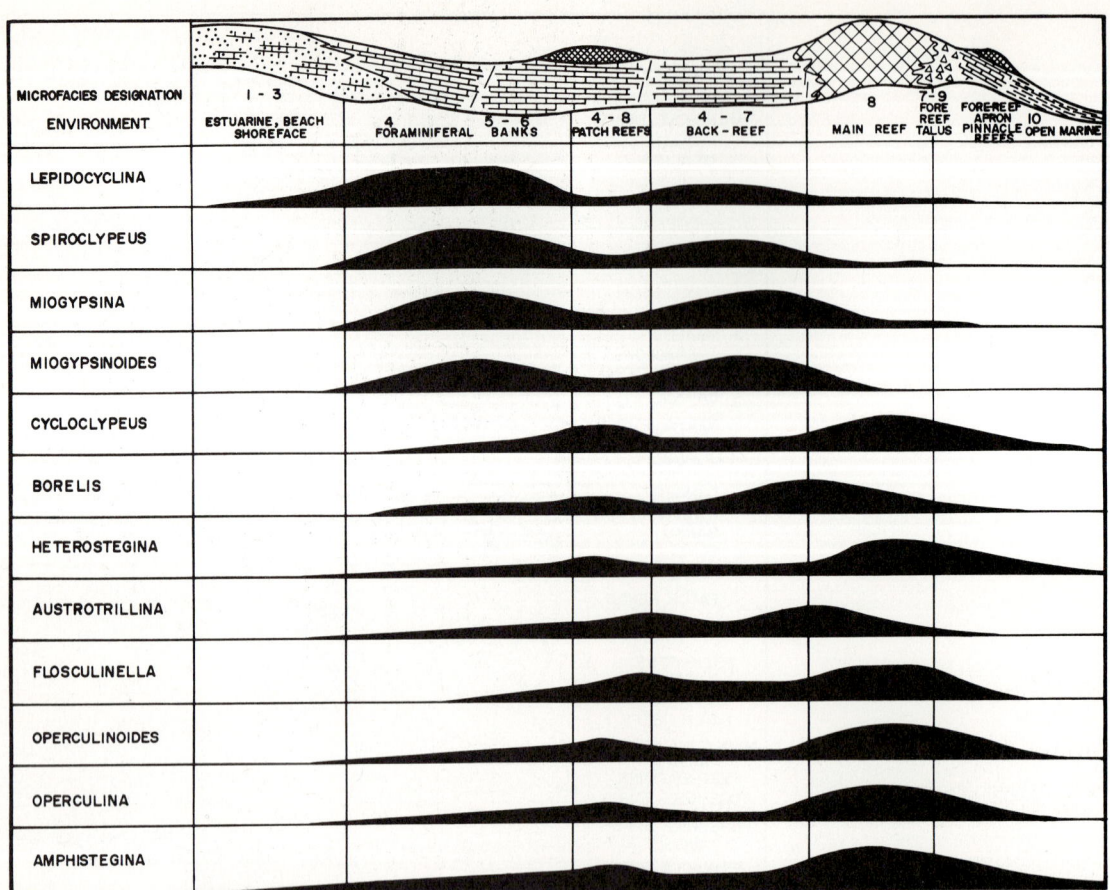

FIGURE 14.46 Miocene of the Visayan Islands, central Philippines. Ideal model of distribution of larger benthic foraminifers. From Carozzi *et al.* (1976).

15

Carbonate Turbidites in Cratonic and Orogenic Settings

Carbonate turbidites differ from their siliciclastic equivalents by a simpler composition expressing their intrabasinal origin. Indeed, they consist mainly of selected benthic bioclasts (such as crinoids, bryozoans, smaller and larger, foraminifers), intraclasts, and pellets of same age as the turbidite layer, or slightly older when reworked from margins and slopes of carbonate basins. Carbonate turbidites are often triggered along margins of buildups and platforms by slumping of excessive biogenic production, by the effect of storms raking platforms, and other causes that generate siliciclastic types.

15.1 CHARACTERISTIC CONSTITUENTS: CRINOIDS

During the final stage of evolution of the Niagaran buildups of Indiana (Textoris and Carozzi, 1964; see also Chapter 10, this volume), the association of a current-swept platform morphology and an overproduction of crinoids developed ideal conditions for generation of slumping and turbidity currents as major mechanisms for the radial distribution down the flanks of crinoidal bioclasts. A large quarry near Lapel, central Indiana, was studied because it provides a complete horizontal section from buildup to interbuildup deeper environment (Carozzi and Frost, 1966). Five sections, reaching a total thickness of 102 ft (30.6 m), were studied by means of 162 thin sections with an average vertical interval of 7.5 in. (19 cm).

Description of Microfacies

As a consequence of turbidity currents, microfacies are described in a proximal to distal order.

Microfacies 1 (Fig. 15.1A,B). Grain-supported biodolarenite consisting predominantly of moderately sorted crinoid bioclasts associated with debris of stromatoporoids, bryozoans, rugose corals, tabulate corals, and ostracods. A cement of dolosparite has replaced the original cavity-filling sparite. This microfacies shows numerous instances of complex whirl-shaped concentrations of bioclasts due to slumping.

Microfacies 2 (Fig. 15.1C,D,E). Mud-supported biodolarenite in which bioclasts "float" in a matrix of slightly argillaceous dolosiltite. Bioclasts are the same essentially as those of microfacies 1, crinoids being predominant over bryozoans but smaller in size and less sorted. Lithic and fecal pellets may occur locally. The matrix contains very angular grains of detrital quartz, muscovite flakes, and a small amount of anhedral crystals and pigments of pyrite. This microfacies displays all the textural features of turbidites consisting of crinoid-bryozoan bioclasts, namely sharp basal contact, grada-

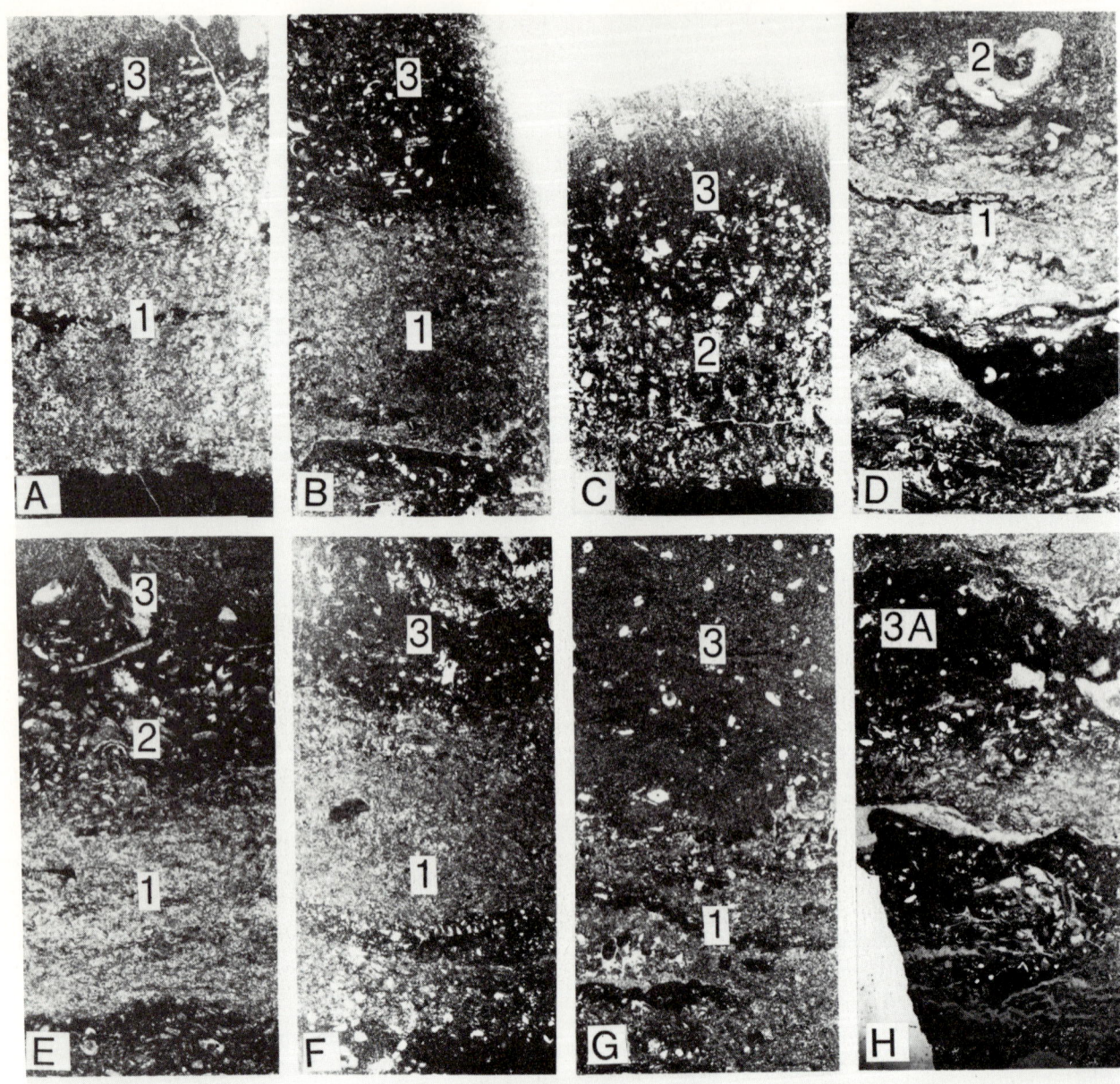

FIGURE 15.1 Niagaran turbidites (Silurian), central Indiana. Typical turbidites. A and B. Single turbidite consisting of superposition of microfacies 1 and 3. C. Single turbidite consisting of superposition of microfacies 2 and 3. D. Double turbidite consisting of superposition of microfacies 1 and 2. E. Single turbidite consisting of superposition of microfacies 1, 2, and 3. F and G. Single turbidite consisting of superposition of microfacies 1 and 3. H. Double turbidite showing development of microfacies 3A. See text for detailed descriptions. All photomicrographs: plane-polarized light. From Carozzi and Frost (1966). Reprinted by permission of the Society of Economic Paleontologists and Mineralogists.

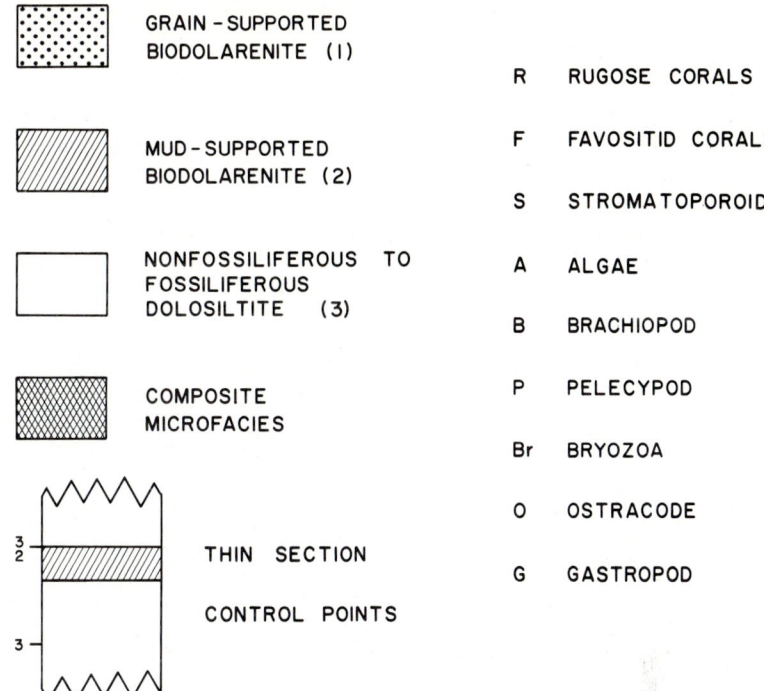

FIGURE 15.2 Niagaran turbidites (Silurian), central Indiana. Symbols for microfacies. From Carozzi and Frost (1966). Reprinted by permission of the Society of Economic Paleontologists and Mineralogists.

tional upper limits, absence of grading, or incipient graded bedding.

Microfacies 3 (Fig. 15.1E,F,G). Mud-supported fine and slightly argillaceous biodolarenite to dolosiltite with scattered sand-size bioclasts. This textural change occurs both within a single turbidite and in a distal direction. Bioclasts are predominantly well-preserved crinoids of smaller size than in the other microfacies, but locally bryozoan fronds, fragments of brachiopods, and ostracods may occur. Compared with the two other microfacies, a marked increase exists in the frequency of silt-size detrital quartz, muscovite flakes, pyrite, and illite clay minerals.

Microfacies 3A (Fig. 15.1H). Argillaceous dolosiltite, with or without scattered minute bioclasts of crinoids, with abundant silt-size detrital quartz, muscovite flakes, and pyrite concentrations. This microfacies represents the deeper and lower-energy interbuildup environment in which turbidites pinch out. It also occurs between individual turbidites in a vertical direction.

The Vertical Sequences

Examination of the individual sections in proximal to distal order shows detailed variation of microscopic parameters. Whenever microfacies forming turbidites cannot be distinguished, they are designated in the sections as composite microfacies (Fig. 15.2). Section 1 (Fig. 15.3) results predominantly from slumping and it is the most proximal. Turbidites appear well developed in median section 3 (Fig. 15.4) and in distal section 5 (Fig. 15.5).

The Ideal Depositional Model

The graphic expression of variation of average parameters from a proximal to a distal location (Fig. 15.6) shows the interference between crinoidal bioclasts brought in by slumping and turbidity currents, and the detrital (quartz, muscovite), organic (monoaxonic sponge spicules), and diagenetic (pyrite) constituents of the deeper interbuildup environment. It was noticed (Carozzi and Zadnik, 1959), that sponge spicules, the normal fauna of the deepest interbuildup environment, are also oriented parallel to bedding at their peak of frequency. As one approaches the flank beds, it is seen that sponge spicules, before disappearing, display increasingly variable orientations, expressing greater turbulence. This textural property can be used as a tool for predicting the proximity of buildups even in the absence of well-developed turbidites.

A tentative correlation (Fig. 15.7) between sections of various turbidites in a proximal to distal direction, based on textural and compositional characters of the individual turbidites, is difficult because of interference between turbidity currents and flow directions which are not necessarily radial. At any rate, the decrease of grain

468 *Part IV / Carbonate Slopes and Basins*

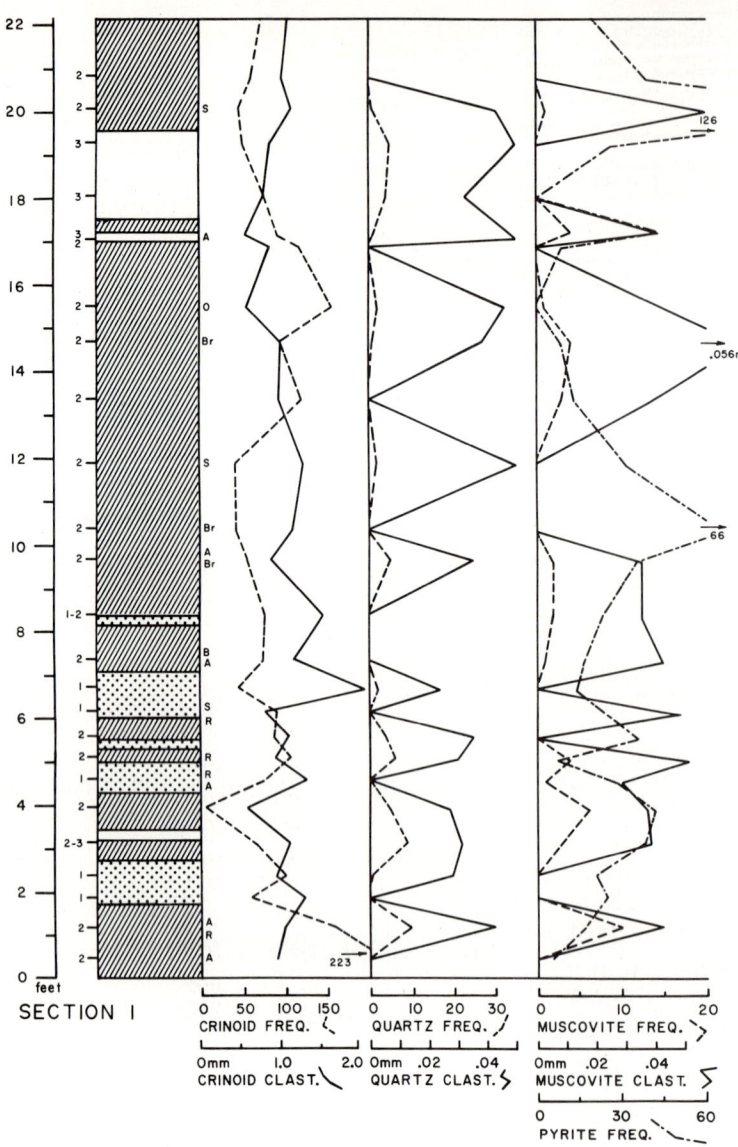

FIGURE 15.3 Niagaran turbidites (Silurian), central Indiana. Typical example of vertical variation of microscopic parameters in proximal section 1. From Carozzi and Frost (1966). Reprinted by permission of the Society of Paleontologists and Mineralogists.

size of bioclasts in a distal direction obviously is displayed.

15.2 CHARACTERISTIC CONSTITUENTS: CRINOIDS AND PELLETS

Certain carbonate turbidites contain a variety of benthic bioclasts, such as crinoids, ostracods, brachiopods, trilobites, and sponge spicules associated with pellets, intrabasinal glauconite grains, and planktonic radiolarians. Furthermore, a limited amount of extrabasinal constituents may occur, namely silt-size quartz, clay minerals, and silica. Under these circumstances, the latter is responsible for the extensive chertification of the turbidites.

This example refers to the Bailey Limestone, the basal unit of the Lower Devonian Series of the southern part of the Illinois Basin (Carozzi and Banaee, 1984). The Bailey Limestone occupies a large portion of the deepest part of the basin, in particular the Metropolis Depression of structural origin, where it reaches a maximum thickness of 120 m (Fig. 15.8). It consists of approximately 100 thin to medium-bedded couplets (10 to 15 cm thick) of limestones and cherty limestones or cherts displaying the characteristic sedimentary structures of turbidites (Bouma sequence), among which sole markings and graded bedding are prominent. The Bailey Limestone was investigated near Grand Tower, Illinois, in a section 40 m thick along the Mississippi River bluffs by means of more than 400 thin sections corresponding

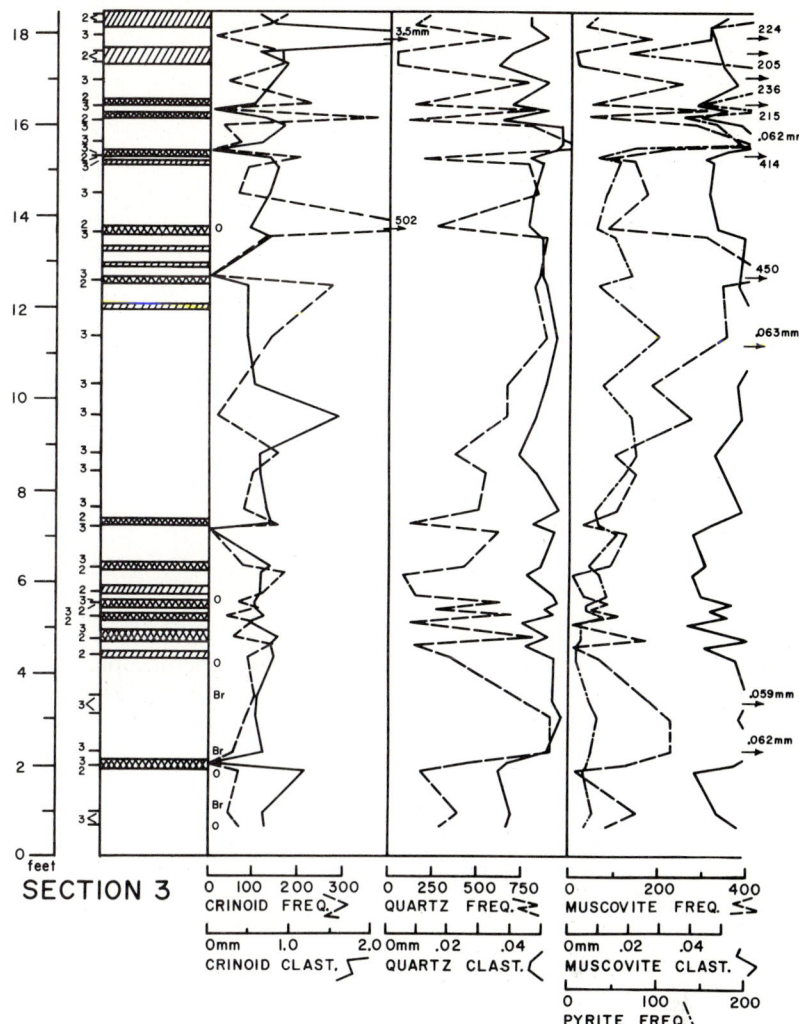

FIGURE 15.4 Niagaran turbidites (Silurian), central Indiana. Typical example of vertical variation of microscopic parameters in median section 3. From Carozzi and Frost (1966). Reprinted by permission of the Society of Economic Paleontologists and Mineralogists.

to samples collected at an average vertical interval of 10 cm, which provided an adequate representation of the style of deposition of turbidites.

Description of Microfacies

Microfacies are designated according to the Bouma sequence and described in order of decreasing grain size, decreasing proportion of benthic bioclasts, increasing proportion of planktonic bioclasts, and increasing amount of pelletoidal matrix (Fig. 15.9).

Microfacies 1 (Ta). Coarse-grained crinoidal biocalcarenite with association of syntaxial rim cement and bioclastic matrix. Texture is graded bedded to massive and bioturbation is low.

Microfacies 2 (Tb). Fine-grained crinoidal biocalcarenite with bioclastic to pelletoidal matrix. Texture is incipiently graded to parallel laminated and bioturbation is medium.

Microfacies 3 (Tc). Crinoidal and pelletoidal biocalcisiltite with scattered sponge spicules and radiolarians. Texture is parallel laminated and bioturbation is moderate to extensive.

Microfacies 4 (Td). Spiculitic, pelletoidal biocalcisiltite. Texture is massive or laminated and bioturbation is abundant.

Microfacies 5 (Te). Pelletoidal bioclastic calcisiltite. Texture is massive and bioturbation is medium.

The Ideal Depositional Sequence

In an average turbidite-interturbidite couplet (Fig. 15.10), the behavior of microscopic parameters can be summarized as follows. Frequency and clasticity of crinoids;

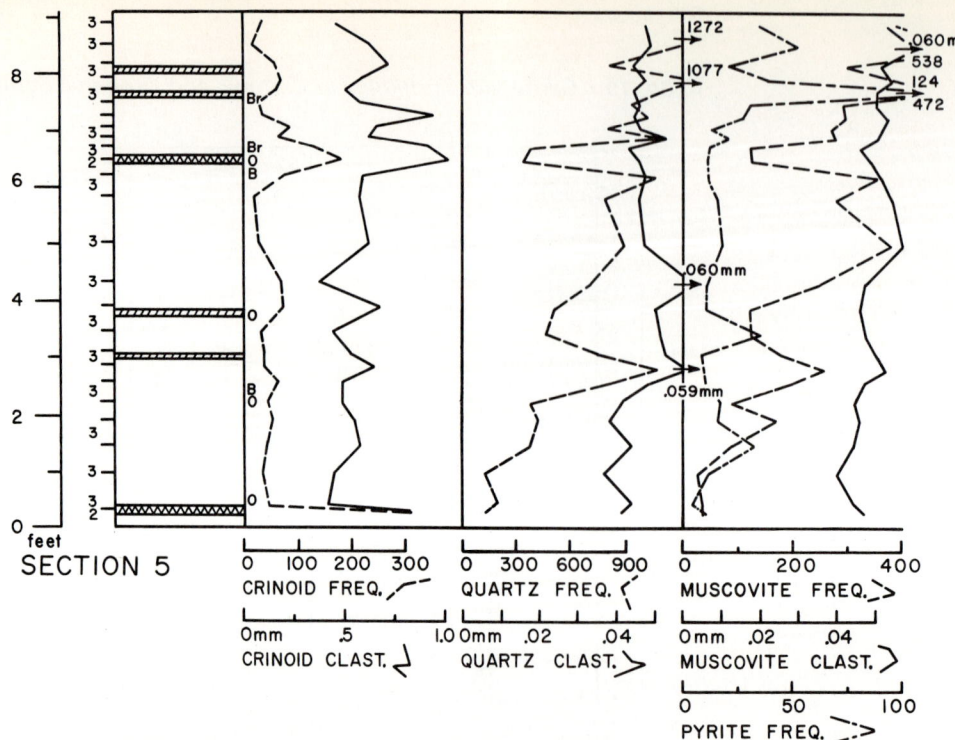

FIGURE 15.5 Niagaran turbidites (Silurian), central Indiana. Typical example of vertical variation of microscopic parameters in distal section 5. From Carozzi and Frost (1966). Reprinted by permission of the Society of Economic Paleontologists and Mineralogists.

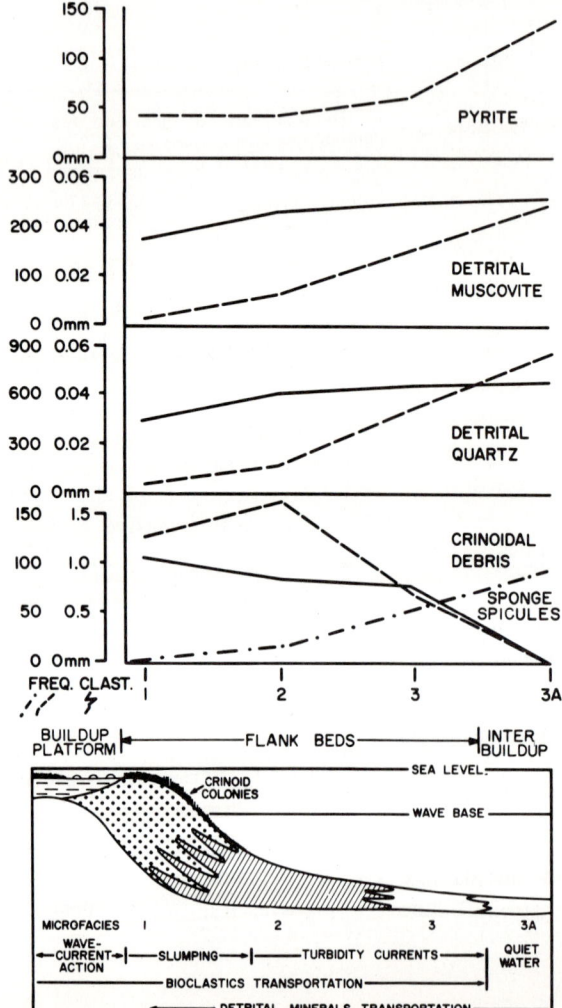

FIGURE 15.6 Niagaran turbidites (Silurian), central Indiana. Ideal depositional model. From Carozzi and Frost (1966). Reprinted by permission of the Society of Economic Paleontologists and Mineralogists.

Chap. 15 / Carbonate Turbidites in Cratonic and Orogenic Settings   471

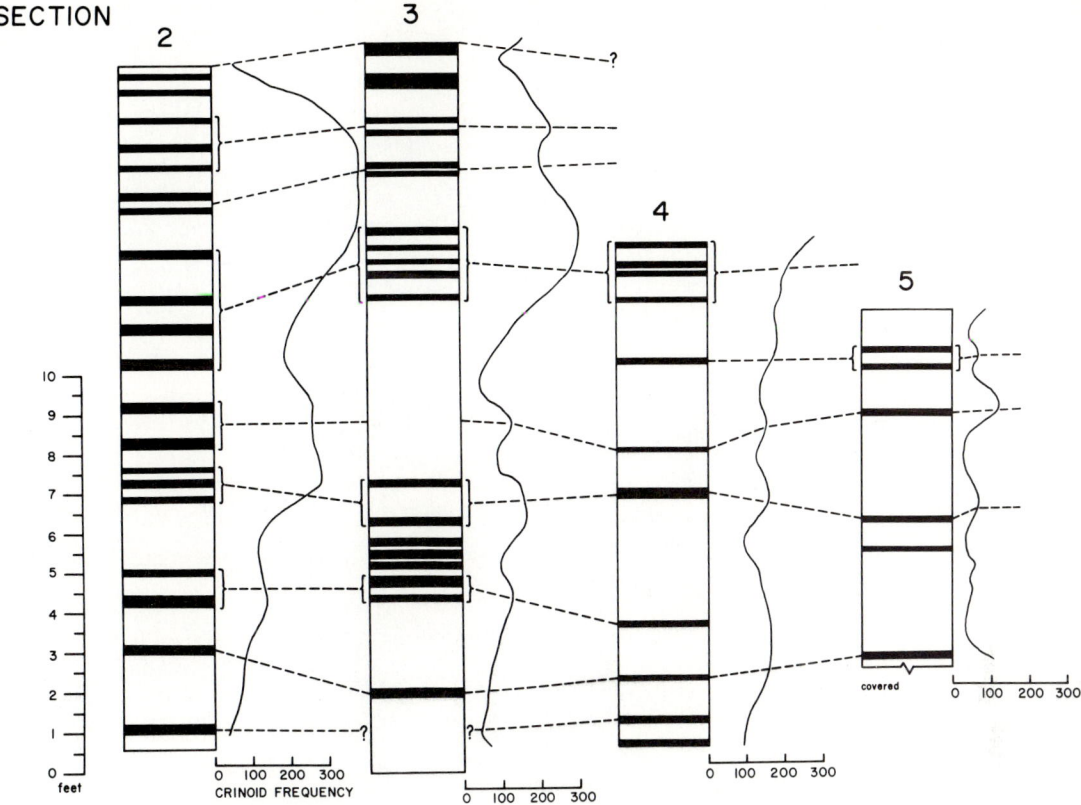

FIGURE 15.7 Niagaran turbidites (Silurian), central Indiana. Tentative correlation between investigated sections. From Carozzi and Frost (1966). Reprinted by permission of the Society of Economic Paleontologists and Mineralogists.

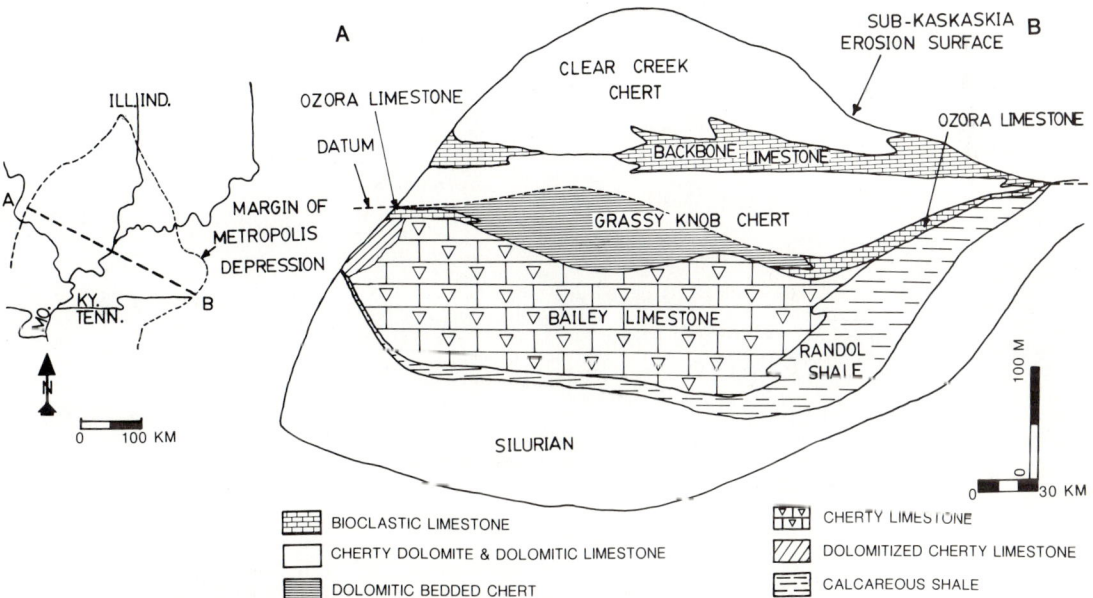

FIGURE 15.8 Bailey Limestone (Lower Devonian), southern part of Illinois Basin. Lithologic cross section of the Uppermost Silurian and Lower Devonian rocks in Metropolis Depression. From Carozzi and Banaee (1984). Reprinted by permission of the Illinois State Academy of Science.

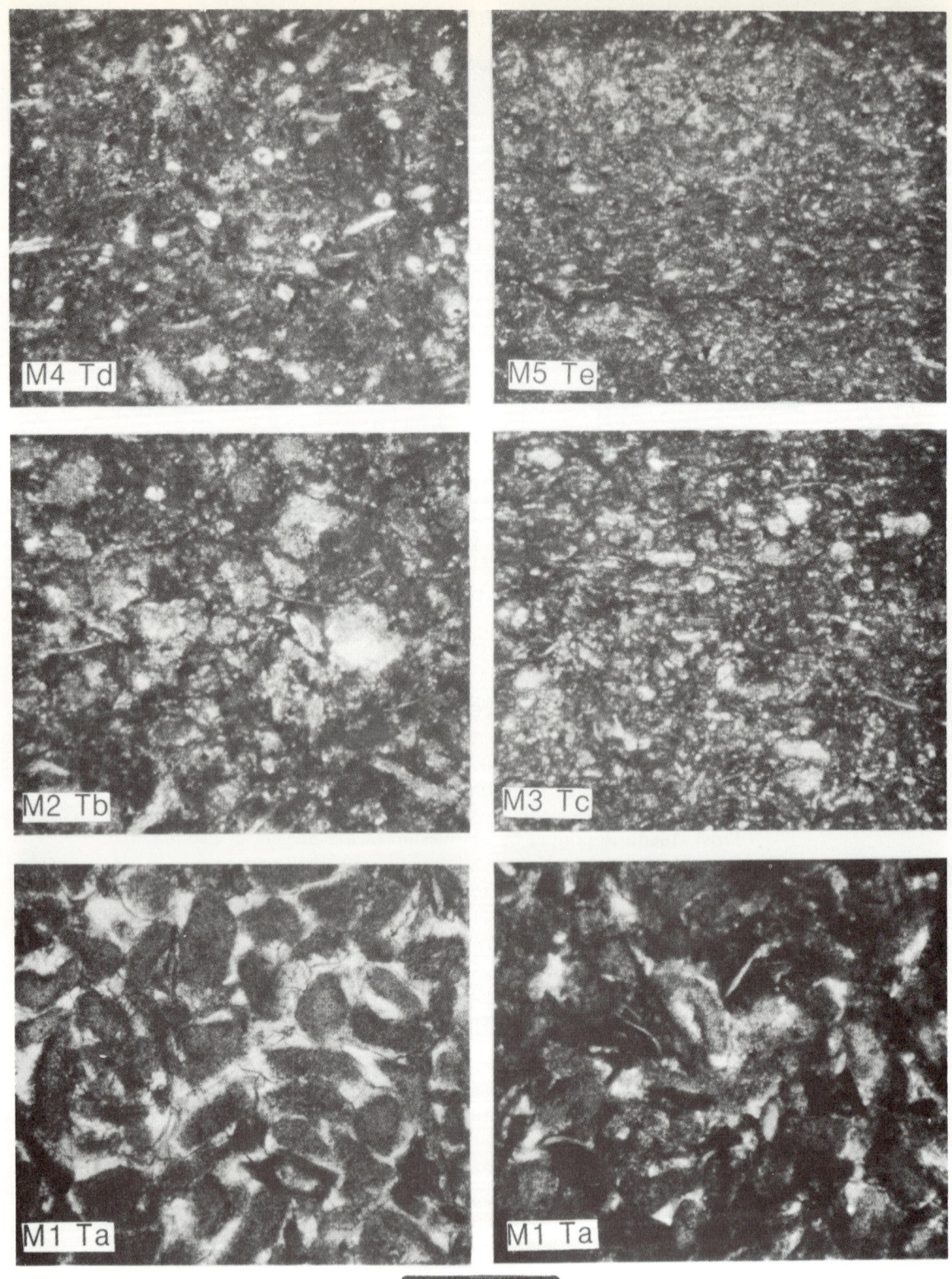

FIGURE 15.9 Bailey Limestone (Lower Devonian), southern part of Illinois Basin. Typical microfacies arranged from bottom to top and from left to right for illustrating Bouma sequence. Ta. Microfacies 1 (coarse-grained on left, medium-grained on right). Tb. Microfacies 2. Tc. Microfacies 3. Td. Microfacies 4. Te. Microfacies 5. See text for detailed descriptions. All photomicrographs: plane-polarized light. From Carozzi and Banaee (1984). Reprinted by permission of the Illinois State Academy of Science.

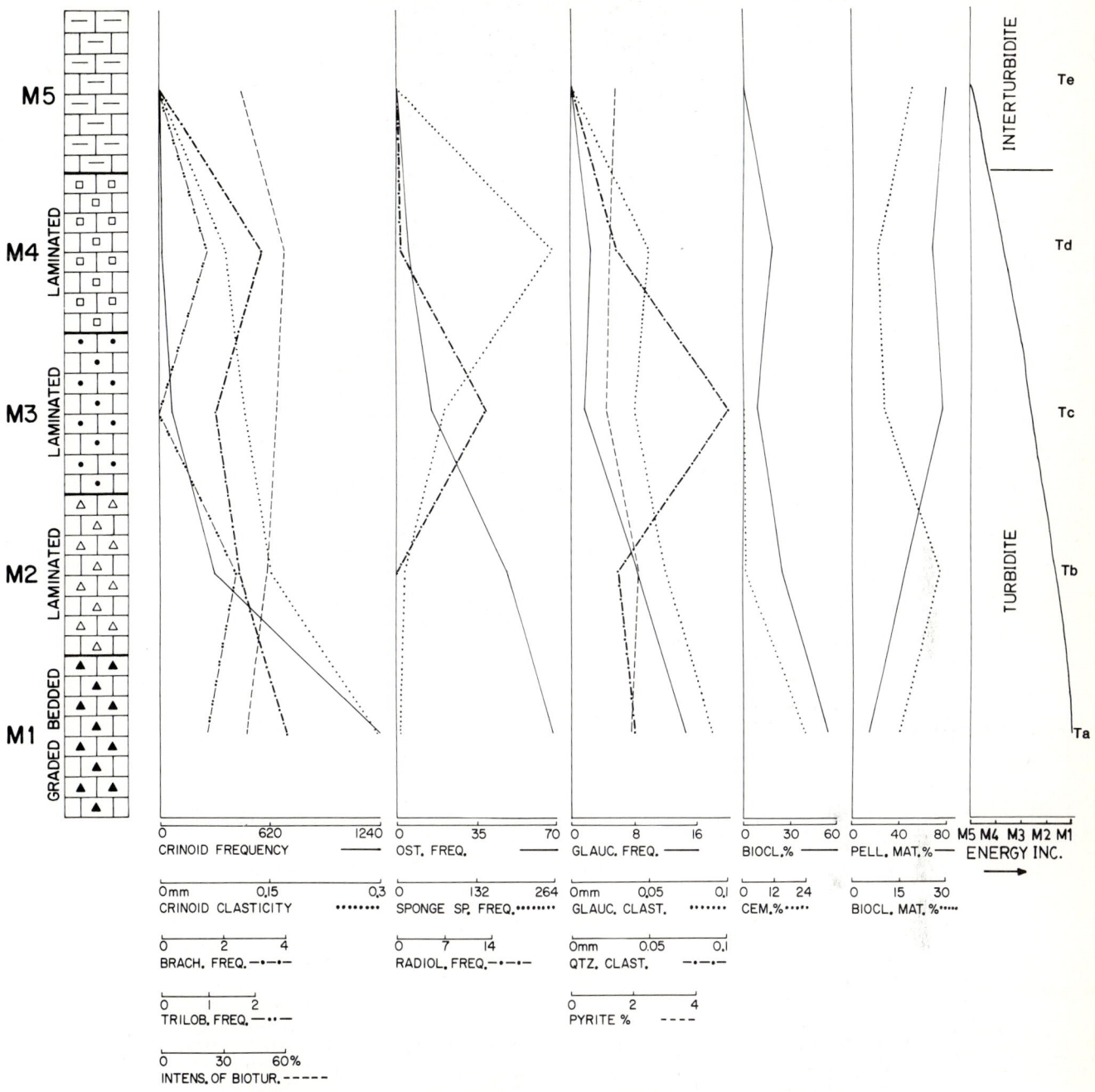

FIGURE 15.10 Bailey Limestone (Lower Devonian), southern part of Illinois Basin. Ideal depositional sequence. From Carozzi and Banaee (1984). Reprinted by permission of the Illinois State Academy of Science.

frequency of ostracods, brachiopods, and trilobites; percentage of all bioclasts and of cement; as well as frequency and clasticity of glauconite grains decrease upward. Frequency of sponge spicules, percentage of pelletoidal matrix, and intensity of bioturbation (until microfacies 4) increase upward. Clasticity of detrital quartz and frequency of radiolarians show a peak value in microfacies 3 that is size controlled. Percentage of bioclastic matrix and of pyrite pigments are random.

In summary, the ideal depositional sequence shows that relatively coarse-grained detrital biogenic components (dominated by benthic fauna) are gradually replaced upward by smaller and lighter detrital (quartz) and biogenic components (planktonic radiolarians and sponge

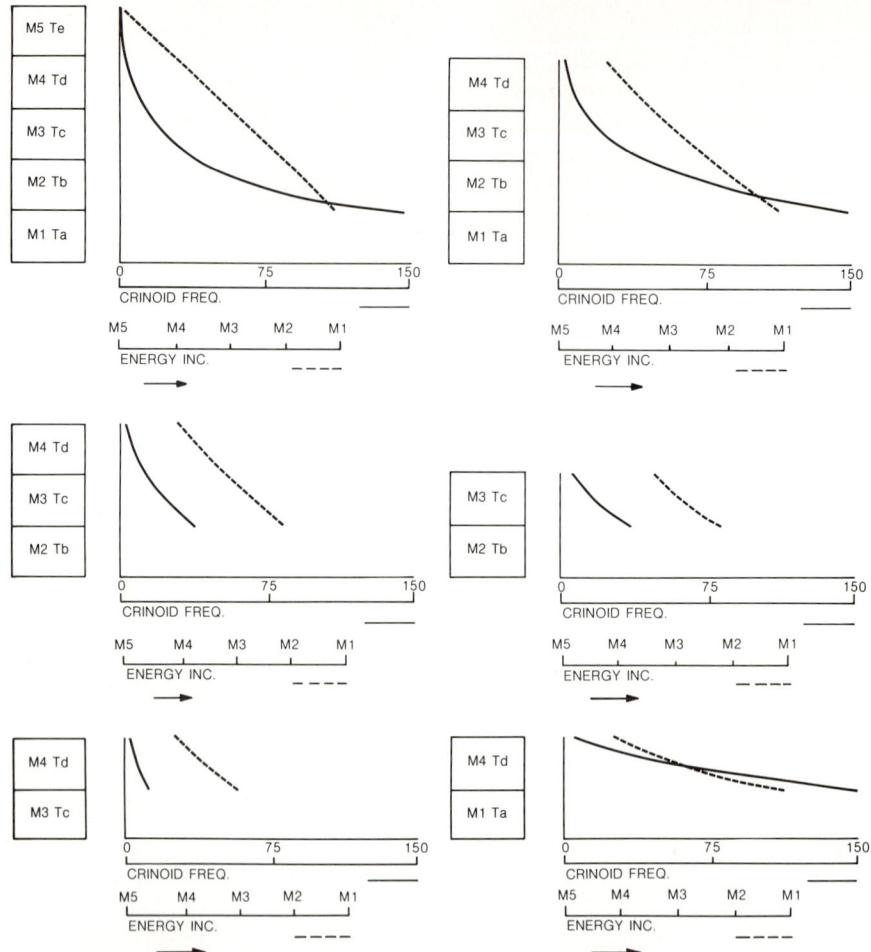

FIGURE 15.11 Bailey Limestone (Lower Devonian), southern part of Illinois Basin. Incomplete Bouma sequences. From Carozzi and Banaee (1984). Reprinted by permission of the Illinois State Academy of Science.

spicules) and by pelletoidal matrix as a typical expression of turbidites in relatively deep water.

Complete Bouma sequences as represented in the ideal sequence are, in general, relatively rare, and incomplete sequences (Tb-c-d, Tb-c, Tc-d, Ta-d) are found more commonly (Fig. 15.11).

The Origin of Components

Bioclasts including crinoids, ostracods, brachiopods, trilobites, sponge spicules, and detrital glauconite grains originated from the shelf foreslope or slope environments. Radiolarians were contemporaneous planktonic fauna. Detrital quartz grains of limited size range were probably recycled from Silurian or finely arenaceous older deposits exposed on the shelves and the shorelines of the Metropolis Depression. The same interpretation applies to clay minerals and the silica responsible for early chertification.

The Vertical Evolution of the Investigated Section

A typical example of the interval from 30 to 40 m of the investigated section (Figs. 15.12, 15.13) shows detailed variation of parameters and interpretation of the energy curve. Because the present study dealt only with one vertical column, it was not possible to establish by direct observation the evolution in time and space of the corresponding submarine fan.

Application of Walker's index of proximality $P = A + \frac{1}{2}B$ (in which A and B are the percentage of beds beginning, respectively, with units Ta and Tb within groups of turbidites based on changes in the thickness

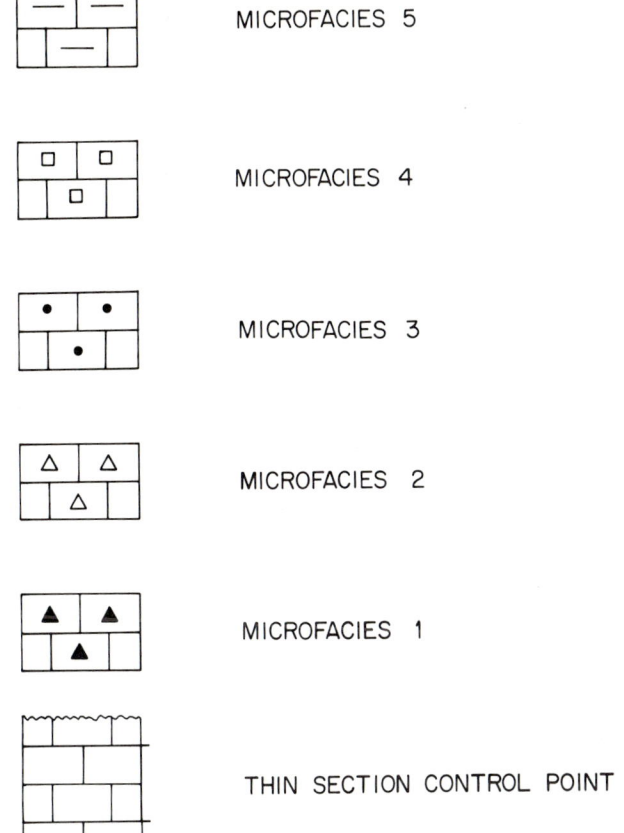

FIGURE 15.12 Bailey Limestone (Lower Devonian), southern part of Illinois Basin. Symbols for microfacies. From Carozzi and Banaee (1984). Reprinted by permission of the Illinois State Academy of Science.

and/or frequency of occurrence of individual units of the Bouma sequence) allows detection of oscillations from inner fan through mid fan and outer fan into basin plain subenvironments of an assumed submarine fan system (Fig. 15.14).

15.3 CHARACTERISTIC CONSTITUENTS: BENTHIC BIOCLASTS, PELLETS, OOIDS, AND QUARTZ

Under conditions of tectonic instability in Alpine carbonate basins, pelagic calcilutites contain numerous thin turbidites. The latter result from the long-distance transportation and redeposition of a large variety of constituents, such as lithic clasts, bioclasts, ooids, pellets, and detrital quartz. This unusual composition indicates original deposition over narrow carbonate platforms with frontal constructed buildups supporting lagoons, freshwater lakes, and fluviodeltaic systems.

The rocks forming the Morcles Nappe of the High Calcareous Alps of France and Switzerland were deposited originally in a basin 60 km wide which, during the Upper Jurassic, was characterized by a sequence of pelagic calcilutites reaching an average thickness of 150 m. These basinal carbonates contain ammonoids, tintinnoids, radiolarians, pelagic crinoids (*Saccocoma*), ostracods, and very rare minute grains of detrital quartz. They display at least nine intercalations of thin turbidites, about 20 cm thick, which originated from the northwestern margin of the basin (Carozzi, 1952c,d, 1955a, 1957). That margin was the peneplained surface of the Aiguilles Rouges massif which, during Late Jurassic, was an unstable shallow-water platform with freshwater lakes and brackish lagoons, separated from the basinal environment by a large but discontinous barrier of coral buildups.

During the Middle Oligocene main alpine orogenic phase, the Mont Blanc massif was thrusted northwest over the carbonate basin. Basin filling was pushed in the same direction, as a nappe, thus overriding the Aiguilles Rouges block. The basin is today reduced to a root zone 2 km wide, whereas the Morcles Nappe thus formed, which is part of the High Calcareous Alps of Savoy, France, shows an overthrust of 20 km.

Because most of the turbidity currents originated in the northwestern coastline of the basin, namely the southern slope of the Aiguilles Rouges massif, the effect of the orogeny made the proximal portions of the turbidites belong to the so-called autochthonous unit, while the distal portions of the same turbidites were included as overturned sequences in the arch bend and the reverse limb of the Morcles Nappe (Fig. 15.15).

Five field sections with a total thickness of 680 m were investigated petrographically by means of 600 thin sections. The average vertical interval was about 1.5 m but shorter for the thin turbidites to account adequately for their petrography.

Description of Microfacies

The turbidites are either microbreccias or microconglomerates, pressure-welded or with a calcisiltite matrix. Lithic clasts include oolitic limestones and intraclasts of pelagic micrite. Bioclasts belong to corals, rudistids, crinoids, brachiopods, bryozoans, annelids, arenaceous benthic foraminifers, dasyclads, and *Characeae* associated with reworked ooids and pellets (Fig. 15.16C,D). In many instances, angular grains of fine silt- to sand-size angular quartz are present (Fig. 15.16A,B). Basal contacts of the turbidites are sharp, with flute casts and load casts. Internal structure ranges from massive to

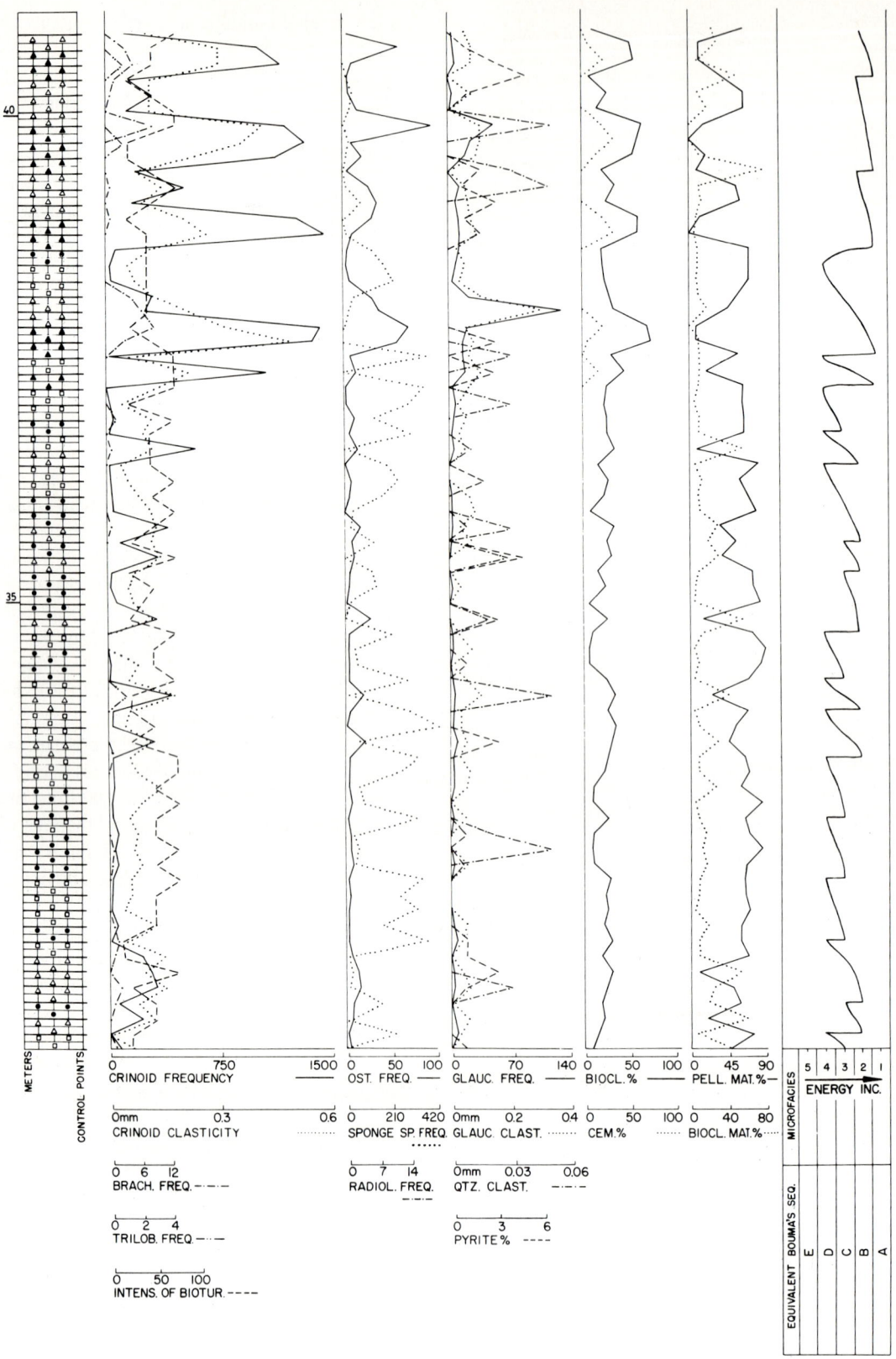

FIGURE 15.13 Bailey Limestone (Lower Devonian), southern part of Illinois Basin. Typical example of vertical variation of microscopic parameters. From Carozzi and Banaee (1984). Reprinted by permission of the Illinois State Academy of Science.

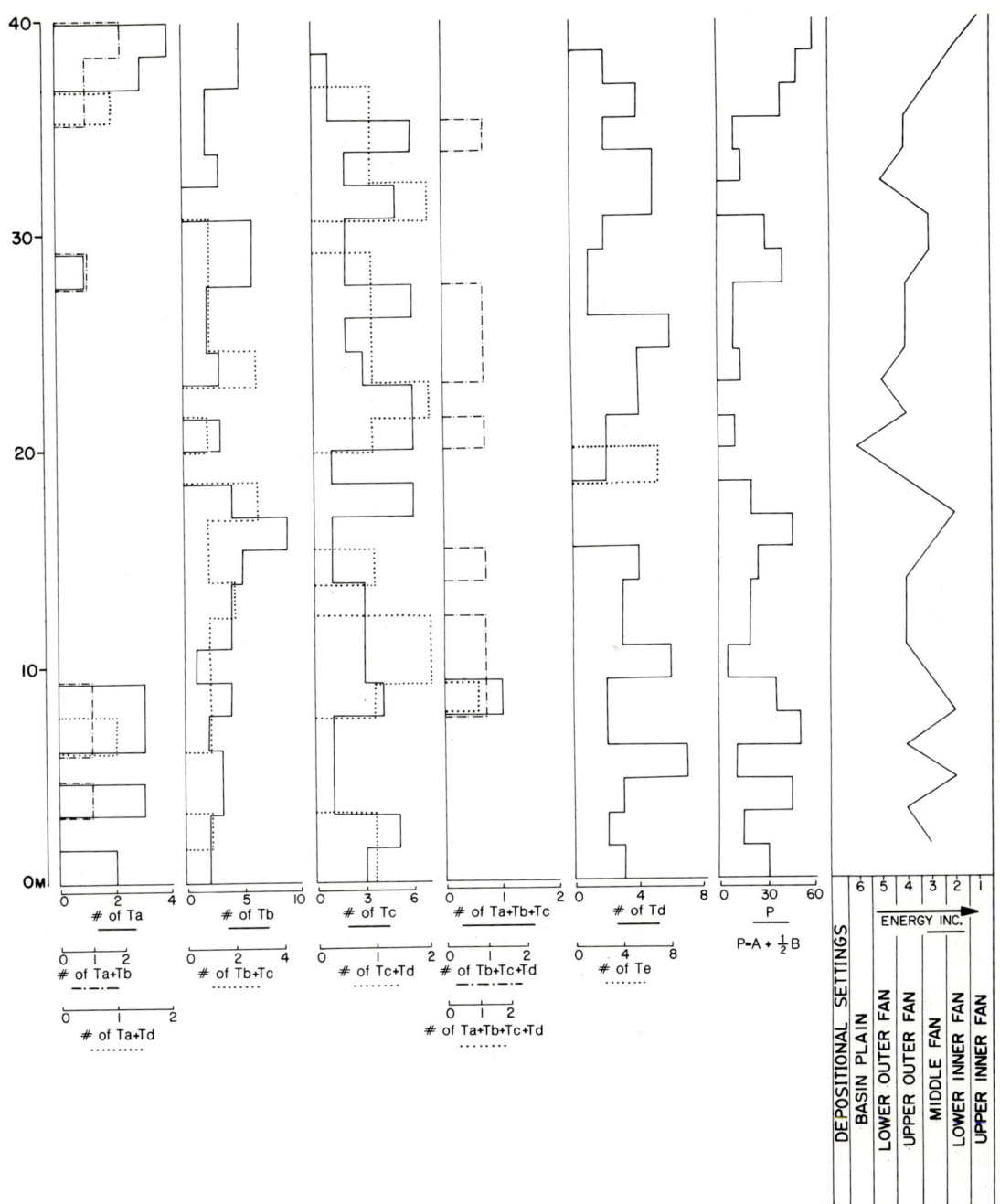

FIGURE 15.14 Bailey Limestone (Lower Devonian), southern part of Illinois Basin. Vertical variation of individual units of Bouma sequence, of complete and incomplete Bouma sequences, of P index, and environmental interpretation of investigated section. From Carozzi and Banaee (1984). Reprinted by permission of the Illinois State Academy of Science.

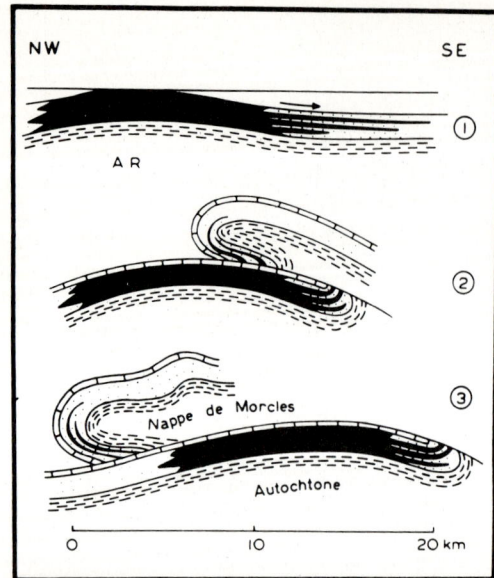

FIGURE 15.15 Upper Jurassic, High Calcareous Alps, France and Switzerland. Diagram showing the tectonic effects of overthrusting on the deposits of turbidity currents in the Upper Jurassic of Haute-Savoie. 1. Emplacement of clastic beds derived from the coral reefs of Aiguilles Rouges (A.R.). Only three beds are shown. Reefs and clastic beds in black, pelagic deposits dotted. 2. Initial stage in the development of Nappe de Morcles. This must have started after emplacement of the clastic beds. 3. Present position of Nappe de Morcles after a displacement of about 20 km over Aiguilles-Rouges and its autochthonous cover. From Carozzi (1957). Reprinted by permission of the Society of Economic Paleontologists and Mineralogists.

incipiently graded. In the latter case the bioclasts are size sorted. Upper contacts of the turbidites are gradational, usually through an increasing proportion of calcilutite matrix with associated radiolarians and tintinnoids.

The proximal section of Vogealle (Fig. 15.17) is used as a typical example of graphic representation of microscopic parameters. Turbidites are divided into three groups: a lower and an upper one which are microconglomerates (circles on black background), and a middle one which consists of microbreccias (triangles on black background).

Clasticity curves were drawn for extraclasts and intraclasts (designated as maximum clasticity in the diagrams) as well as for quartz and muscovite. Frequency curves were established for reworked ooids, reworked benthic bioclasts, planktonic components, and pyrite pigments. In an attempt to characterize the turbidites further, size and frequency were measured for crystals of authigenic feldspars.

Horizontal Distribution and Correlation of the Turbidites

Examination of the stratigraphic sections along the depositional dip from a proximal position (Fig. 15.17) through a median one (Fig. 15.18) to a distal position (Fig. 15.19) shows the gradual disappearance of some of the turbidites. The correlation from section to section was undertaken by using all textural and compositional properties (Fig. 15.20) and proves that the extent of the nine turbidites was over a distance of 40 km down depositional slope (Fig. 15.21). In summary, the first group of turbidites (1 to 4) consists of microconglomerates with reworked ooids which display repeated graded bedding within single beds resulting from several turbidites originating from different point sources and flowing in the same direction; the second group of turbidites (5, 6, and 7) consists of microbreccias with simple graded bedding; the third group of turbidites (8 and 9) consists again of microconglomerates with massive to chaotic structures as well as simple to multiple graded bedding.

Horizontal Grading

Horizontal grading can be expressed readily by the measurement of maximum clasticity index taken at the bottom of graded beds as shown over the observational distance of 40 km for several of the investigated turbidites (Fig. 15.22). The same procedure reveals beds deposited by slumping over relatively short distances of displacement and hence lacking horizontal sorting. They appear as two horizontal lines connected by a curved line generated through mixing (Fig. 15.23). It is also possible to see the mixing of two individual turbidity currents halfway down the slope where a powerful current took over the distal portion of a weaker one (Fig. 15.23).

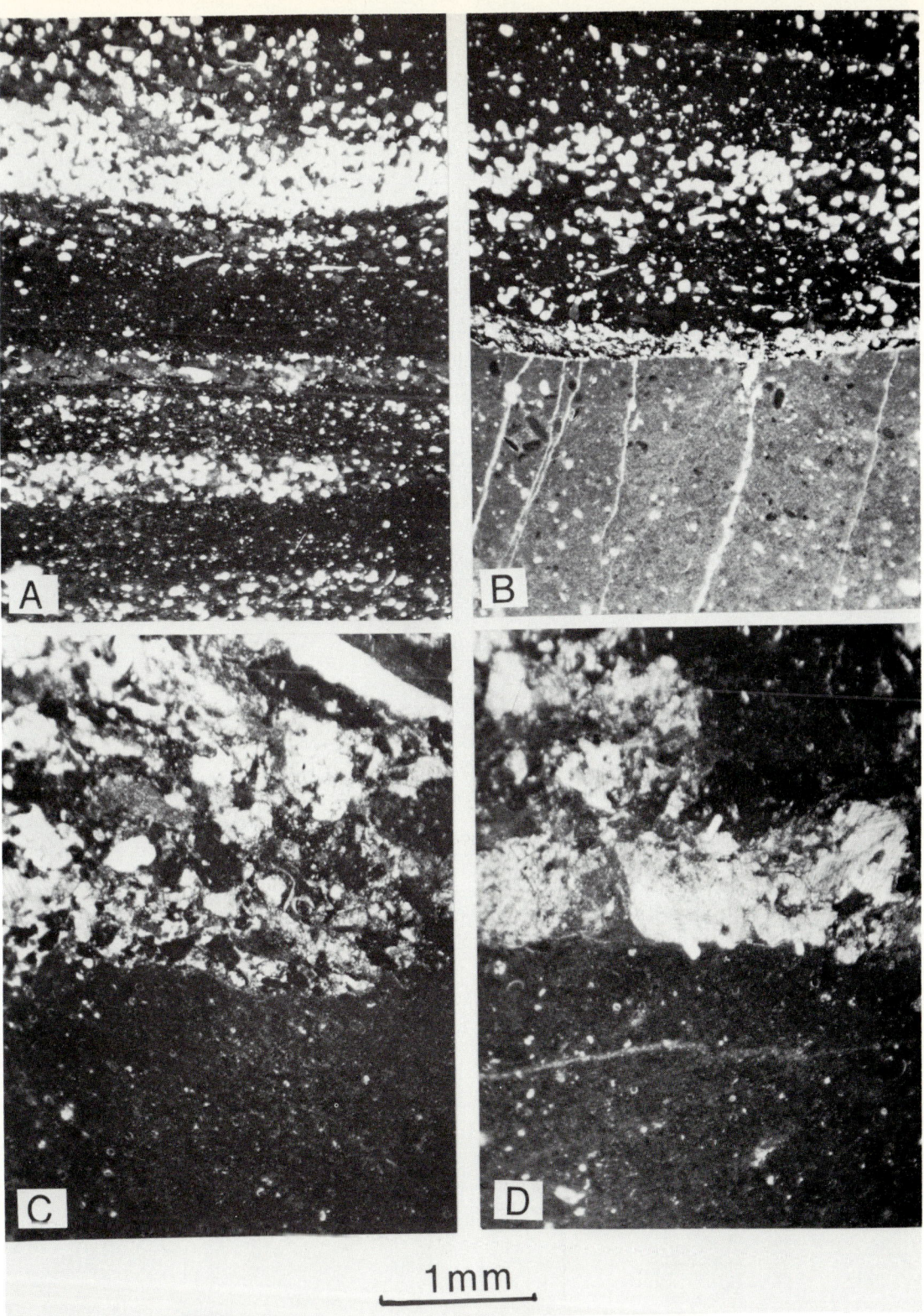

FIGURE 15.16 Upper Jurassic, High Calcareous Alps, France and Switzerland. Typical turbidite microfacies. A. Quartzose type with multiple incipient graded bedding and internal erosional surface. B. Basal contact of quartzose type on pelagic calcilutite. C. Bioclastic type with poorly sorted debris of crinoids, bryozoans, dasyclads, and arenaceous benthic foraminifers, showing incipient load cast. D. Bioclastic type with large fragments of corals and crinoids. Observe irregular basal contact with pelagic calcilutite with tintinnoids transected by authigenic feldspar crystals. See text for detailed descriptions. All photomicrographs: plane-polarized light. From Carozzi (1955a).

480 Part IV / Carbonate Slopes and Basins

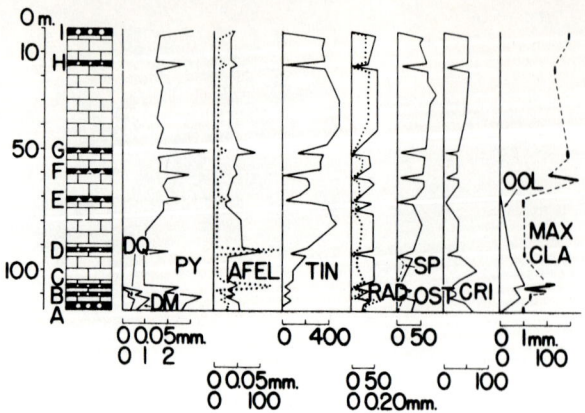

FIGURE 15.17 Upper Jurassic, High Calcareous Alps, France and Switzerland. Typical vertical variation of microscopic parameters in proximal section of Vogealle. DQ, detrital quartz; DM, detrital muscovite; PY, pyrite; AFEL, authigenic feldspars; TIN, tintinnoids; RAD, radiolarians; SP, sponge spicules; OS, ostracods; CRI, crinoids; OOL, ooids; MAC CLA, maximum clasticity. From Carozzi (1957). Reprinted by permission of the Society of Economic Paleontologists and Mineralogists.

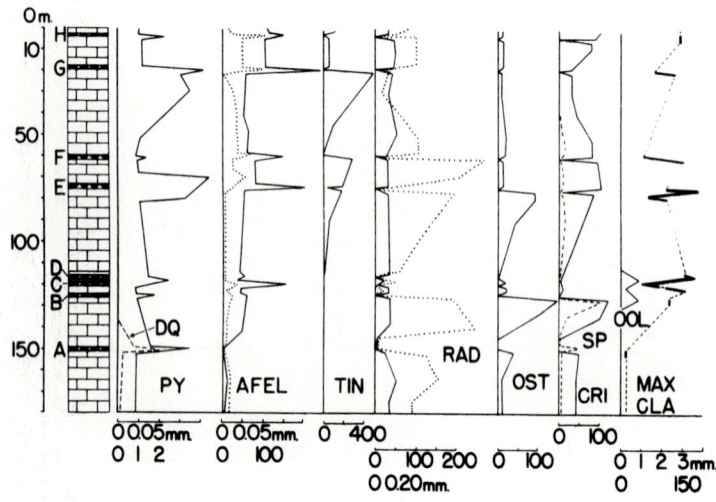

FIGURE 15.18 Upper Jurassic, High Calcareous Alps, France and Switzerland. Typical vertical variation of microscopic parameters in median section of Commune. See Figure 15.17 for abbreviation of symbols. From Carozzi (1957). Reprinted by permission of the Society of Economic Paleontologists and Mineralogists.

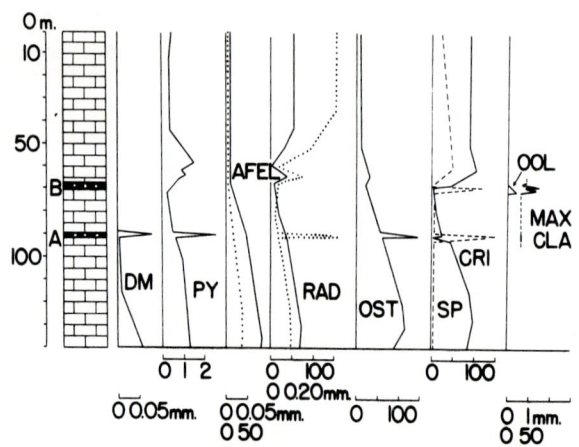

FIGURE 15.19 Upper Jurassic, High Calcareous Alps, France and Switzerland. Typical vertical variation of microscopic parameters in distal section of La Giettaz. See Figure 15.17 for abbreviation of symbols. From Carozzi (1957). Reprinted by permission of the Society of Economic Paleontologists and Mineralogists.

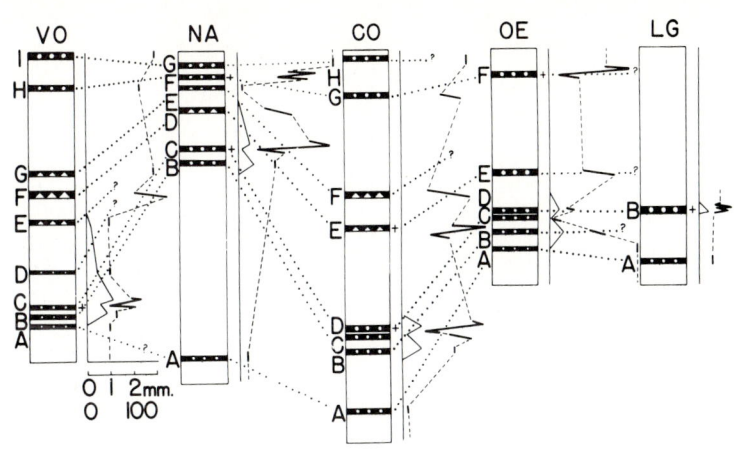

FIGURE 15.20 Upper Jurassic, High Calcareous Alps, France and Switzerland. Tentative correlation of turbidites in investigated sections. VO, Vogealle; NA, Nantbride; CO, Commune; OE, Oex-Arpenaz; LG, La Giettaz. From Carozzi (1957). Reprinted by permission of the Society of Economic Paleontologists and Mineralogists.

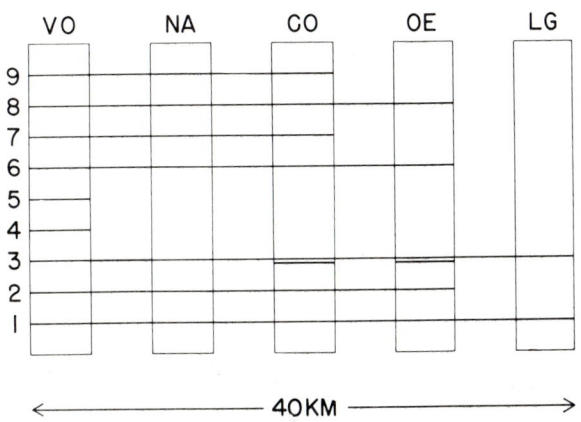

FIGURE 15.21 Upper Jurassic, High Calcareous Alps, France and Switzerland. Schematic distribution of turbidites down depositional slope. See Figure 15.20 for abbreviation of section names. From Carozzi (1957). Reprinted by permission of the Society of Economic Paleontologists and Mineralogists.

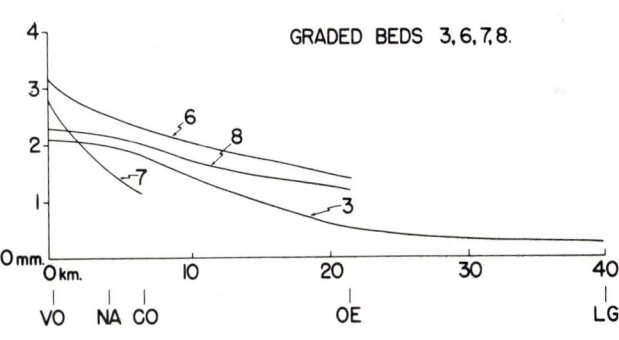

FIGURE 15.22 Upper Jurassic, High Calcareous Alps, France and Switzerland. Clasticity index (horizontal grading) in beds deposited by single turbidity currents. See Figure 15.20 for abbreviation of section names. From Carozzi (1957). Reprinted by permission of the Society of Economic Paleontologists and Mineralogists.

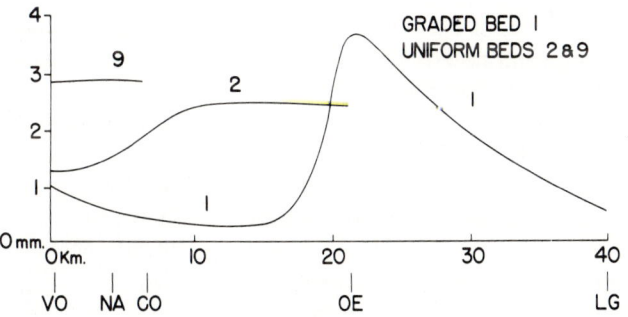

FIGURE 15.23 Upper Jurassic, High Calcareous Alps, France and Switzerland. Clasticity index in beds deposited by slumping and through interference processes between turbidity currents. See Figure 15.20 for abbreviation of section names. From Carozzi (1957). Reprinted by permission of the Society of Economic Paleontologists and Mineralogists.

16

Carbonate Slopes in Cratonic Settings

In carbonate models under arid climates, phosphatization occurs at the interface of deeper anoxic phosphate-rich waters upwelling onto slopes and platforms of any kind and beneath shallower oxygenated waters. Direct precipitation of apatite in interstitial waters is followed by penecontemporaneous replacement of carbonate precursors and intraformational reworking, which generates phosphatic intraclasts, nodules, pellets, and ooids.

16.1 CHARACTERISTIC FEATURES: PHOSPHATIZATION PROCESSES

This study focuses on the contact between the Galena (Middle Ordovician) and the Maquoketa (Upper Ordovician) Groups and its characteristic thin and widespread basal Maquoketa phosphorites of the Upper Mississippi Valley in eastern Missouri and eastern Iowa which were investigated by means of four complete cores (Black, 1985). A total of 190 samples were collected with sample intervals ranging from 4 to 60 cm over a total aggregated thickness of 24 m. The sample interval was determined by the frequency of facies changes observed during macroscopic study of the cores. Special efforts were made to sample continuously across the contacts between phosphatic and nonphosphatic facies to allow study of the relationships between microfacies, mineralogic composition, and behavior of Fe, Mg, Ca, Sr, and P. This combination of petrography and chemistry is necessary to understand the nature and evolution of phosphate formation.

The petrographic classification of marine phosphorites used in this study is modified from Slansky (1980) and is similar to that of carbonate rocks. It utilizes two basic parameters: grain size of the predominant ($> 10\%$) phosphatic figured constituents, and nature of matrix or cement. Phosphalutite (0.01 to 0.03 mm), phosphasiltite (0.03 to 0.064 mm), phospharenite (0.064 to 2 mm), and phospharudite (> 2 mm) are further qualified by prefixes such as bio-, pel-, oo-, intra-, extra-, to describe bioclasts, pellets, ooids, intraclasts, and extraclasts.

Phosphalutites and phosphasiltites may contain up to 10% sand-size components of organic or inorganic origin. Phospharenites containing 10 to 30% sand-size constituents in a matrix of phosphalutite or phosphasiltite are called matrix-supported, those with more than 30% sand-size constituents and possessing an interstitial matrix as above or a cement of precipitated collophanite (microsphatite < 0.01 mm) are called grain-supported. Nonphosphatic constituents that may form part of the matrix (quartz, glauconite, clay minerals) or part of the cement (carbonates, silica) are designated according to their mineralogy, grain or crystal size, and relative abundance, such as grain-supported oophospharenite with 10% quartz-glauconite matrix, or grain-supported pel-phospharenite with 20% sparite cement. Phospharudites contain more than 30% granule (2 to 4 mm) or pebble (4 to 64 mm)-size constituents with interstitial material of all previous phosphatic types.

Description of Microfacies

Petrographic and mineralogic study showed the existence of 16 distinct dolomitized microfacies which can be divided into a carbonate sequence (microfacies 1 to 11) and a phosphorite sequence (microfacies 12 to 16). These two sequences are associated in a carbonate-phosphorite depositional model divided, geomorphologically in a landward direction, in six different environments: basin, outer slope, bioclastic bar, lower inner slope, upper inner slope, and platform. The various microfacies are described in this general shallowing-upward order. However, it should be pointed out that microfacies 12 to 16 are approximate depositional equivalents to microfacies 8 to 11 in terms of relative position to shoreline and energy, but differ in their early diagenetic chemical environment, which corresponded to phosphatization.

Basin

Microfacies 1 (Fig. 16.1A). Dolomitized (40 to 100%) calcareous bituminous shale interbedded with calcisiltite laminae. Minute flattened ostracods, phosphatic pellets and spherulites, and occasional sparite-filled trilobites are associated with the calcisiltite stringers. Pyritized organic debris, pyrite rhombs, and euhedral rhombs of dolomite are pervasive. Average mineral composition: 5% calcite, 22% dolomite, 19% quartz, 3% potassium feldspar, 33% clay minerals, 5% apatite, and 13% pyrite. Microfacies 1A is a nondolomitized variety which contains calcite pseudomorphs after nodular anhydrite.

Microfacies 2 (Fig. 16.1B). Dolomitized (100%) calcareous mudstone with less than 10% scattered, well-rounded phosphatic intraclasts and pellets. Rare brachiopod and cephalopod fragments (replaced by apatite) are present. Scattered pyritized organic matter is common in the groundmass, which displays abundant equant dolomite rhombs. Average mineral composition: 31% dolomite, 9% quartz, 2% potassium feldspar, 44% clay minerals, 5% apatite, and 9% pyrite.

Outer Slope

Microfacies 3 (Fig. 16.1C). Dolomitized (100%) argillaceous biomicrite with less than 10% scattered bioclasts of crinoids, ostracods, cephalopods, and brachiopods, phosphatic lithic pellets, and intraclasts. Brachiopods and ostracods are commonly replaced by pyrite, but the latter occurs also as a replacement of other scattered minute organic debris. Bioturbation is common and represented by well-defined burrows characterized by larger dolomite rhombs and concentrations of pyritized and phosphatic constituents. Average mineral composition: 48% dolomite, 10% quartz, 2% potassium feldspar, 29% clay minerals, 2% apatite, and 9% pyrite.

Microfacies 4 (Fig. 16.1D). Dolomitized (100%) argillaceous biocalcisiltite with 10% scattered sand-size bioclasts of ostracods and brachiopods predominant over crinoids, cephalopods, and pelecypods. Large, thin ostracods are often silicified and cephalopod fragments replaced by apatite, whereas brachiopod and pelecypod bioclasts are replaced by dolosparite. Phosphatic pellets, spherulites, and rare intraclasts are concentrated in burrows and circular arrangements due to bioturbation. Phosphatic spherulites are defined as tiny grains (50 to 100 μm), spherical to slightly ellipsoidal in shape consisting of well-crystallized apatite, but devoid of internal structure. They are considered to be organic in origin, possibly fecal, due to their great regularity of shape, and clear distinction from phosphatic lithic pellets, which consist generally of poorly crystalline apatite. Average mineral composition: 47% dolomite, 10% quartz, 2% potassium feldspar, 26% clay minerals, 6% apatite, and 9% pyrite.

Bioclastic Bar

Microfacies 5 (Fig. 16.1E). Dolomitized (100%) matrix-supported biocalcarenite with argillaceous calcilutite to calcisiltite matrix. Bioclasts reach 20% and consist mostly of crinoids, brachiopods, and trilobites with lesser amounts of ostracods, phosphatic spherulites, and intraclasts. If not pyritized or silicified, bioclasts are replaced by dolosparite. Marginal replacement of silicified crinoid fragments by dolomite rhombs indicates silicification preceded dolomitization. Bioturbation gives a mottled texture to the matrix. Average mineral composition: 56% dolomite, 10% quartz, 1% potassium feldspar, 22% clay minerals, 4% apatite, and 7% pyrite.

Microfacies 6 (Fig. 16.1F). Dolomitized (100%) grain-supported biocalcarenite with argillaceous calcilutite to calcisiltite matrix. Trilobite and brachiopod bioclasts predominate over crinoids. All bioclasts are either pyritized or replaced by dolosparite. Poorly crystalline and organic-rich phosphatic pellets and intraclasts are present in moderate amounts. Coarser dolomite crystals are concentrated in burrows. Average mineral composition: 71% dolomite, 5% quartz, 1% potassium feldspar, 16% clay minerals, 3% apatite, and 4% pyrite.

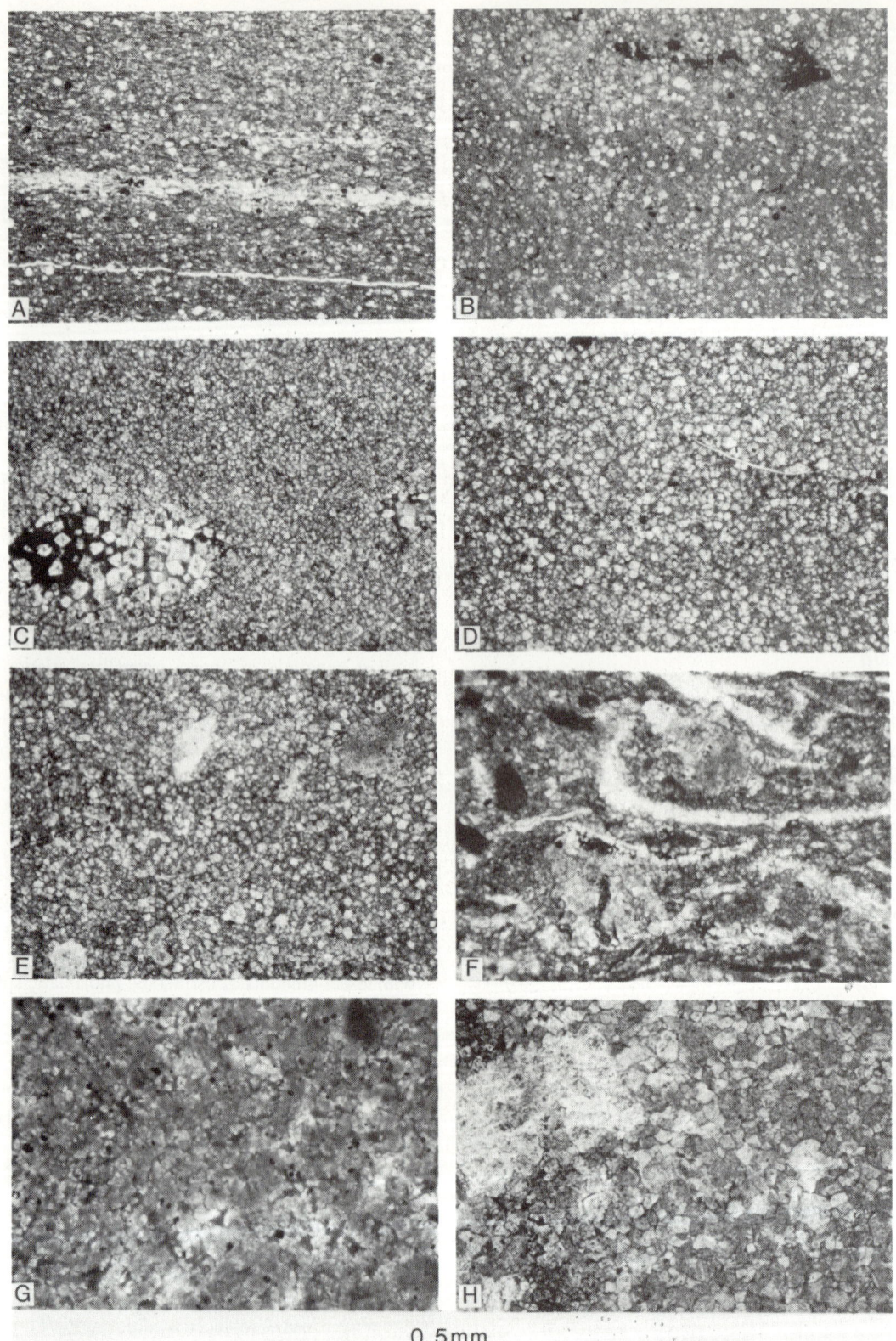

FIGURE 16.1 Maquoketa phosphorites (Upper Ordovician), Upper Mississippi Valley. Typical microfacies. A. Microfacies 1. B. Microfacies 2. C. Microfacies 3. D. Microfacies 4. E. Microfacies 5. F. Microfacies 6. G. Microfacies 7. H. Microfacies 8. See text for detailed descriptions. All photomicrographs: plane-polarized light. From Black (1985).

Lower Inner Slope

Microfacies 7 (Fig. 16.1G). Dolomitized (100%) organic matter-rich biomicrite with less than 10% sand-size bioclasts with scattered small patches of microsparite cement due to intense bioturbation. The next frequent bioclasts are micritized crinoids associated with lesser amounts of brachiopods and bone fragments. Disseminated pyrite pigments and nodules are abundant. Average mineral composition: 73% dolomite, 2% quartz, 5% clay minerals, 5% apatite, and 15% pyrite.

Microfacies 8 (Fig. 16.1H). Dolomitized (100%) biocalcisiltite with up to 10% scattered sand-size bioclasts which are mostly crinoids associated with lesser amounts of brachiopods, bone fragments, and phosphatic intraclasts. Bioturbation is extensive leading to a mottled texture. Dolomitization which displays cloudy-centered rhombs often with rims of ferroan dolomite was preceded by partial replacement of crinoids and brachiopods by silica, and a patchy replacement by phosphates related to bioturbation. Average mineral composition: 82% dolomite, 5% quartz, traces of potassium feldspar, 7% clay minerals, 3% apatite, and 3% pyrite.

Upper Inner Slope

Microfacies 9 (Fig. 16.2A). Dolomitized (100%) matrix-supported biocalcarenite with calcisiltite matrix and up to 30% sand-size bioclasts. Crinoids and brachiopods predominate over bone fragments, phosphatic intraclasts, and lithic pellets. Most bioclasts are replaced by dolosparite, except where silicification or phosphatization has occurred previously. Phosphatic constituents are concentrated within and around the frequent burrows due to bioturbation which correspond to patches of coarser and clear dolomite rhombs. Average mineral composition: 91% dolomite, 2% quartz, traces of potassium feldspar, 5% clay minerals, and 2% apatite.

Microfacies 10 (Fig. 16.2B). Dolomitized (100%) matrix- to grain-supported biocalcarenite with scattered phosphatic ooids, pellets, and intraclasts in a bioturbated calcisiltite matrix with patches of interparticle sparite cement. Bioclasts of crinoids, trilobites, and brachiopods predominate over cephalopods, replacement is either by dolosparite or apatite. Bioturbation leads to a patchy replacement of the matrix by microcrystalline apatite and concentration of the phosphatic ooids and pellets reworked from microfacies 12 to 16. Average mineral composition: 75% dolomite, 3% quartz, traces of potassium feldspar, 7% clay minerals, 9% apatite, and 6% pyrite.

Microfacies 11 (Fig. 16.2C). Dolomitized (100%) grain-supported biocalcarenite with calcisiltite matrix and patchy sparite cement. Crinoid and brachiopod bioclasts predominate over trilobites and ostracods; replacement is either by dolosparite, apatite, or silica. Bioturbation concentrated phosphatic lithic pellets and intraclasts and irregularly distributed the various bioclasts. Silicification of brachiopod and crinoid fragments preceded dolomitization. Average mineral composition: 91% dolomite, 2% quartz, 5% clay minerals, and 2% apatite.

Phosphatized Upper Inner Slope

Microfacies 12 (Fig. 16.2D). Dolomitized (100%) matrix- to grain-supported intraclastic oolitic pelphospharenite with an argillaceous calcisiltite matrix and patches of interparticle sparite cement. Bioclasts of cephalopods and brachiopods are replaced by apatite and silica, and silicified ostracods are filled by phosphalutite. Phosphatic ooids and pellets are associated with intraclasts of phosphalutite and phosphasiltite themselves containing ooids, lithic pellets, bioclasts, and minute grains of detrital quartz and feldspar. The heterogeneous distribution of the components indicates widespread bioturbation. Patchy phosphatization of the matrix by microcrystalline apatite preceded dolomitization. Average mineral composition: 37% dolomite, 7% quartz, 1% potassium feldspar, 19% clay minerals, 30% apatite, and 6% pyrite.

Microfacies 13 (Fig. 16.2E). Dolomitized (100%) grain-supported intraclastic oolitic pelphospharenite to phospharudite with argillaceous calcisiltite matrix and patches of sparite cement. Phosphatic intraclasts, ooids, and lithic pellets with abundant apatite-replaced orthocone cephalopods and brachiopods predominate over large silicified ostracods filled with phosphalutite. Partial phosphatization of the matrix developed a few nodules, while subsequent dolomitization partially replaced phosphatic intraclasts and pellets. Nodular and disseminated pyrite is ubiquitous. Average mineral composition: 26% dolomite, 5% quartz, 1% potassium feldspar, 9% clay minerals, 45% apatite, and 14% pyrite.

Microfacies 14 (Fig. 16.2F). Dolomitized (100%) grain-supported to pressure-welded intraclastic pelletoidal oophospharenite to phospharudite with argillaceous calcisiltite matrix and patches of sparite cement. Phosphatic intraclasts, ooids, and lithic pellets predominate over lesser amounts of cephalopods, brachiopods, and large ostracods. Bioclasts are either replaced by apatite or silica. Early phosphatization also generated large nodules whereas late dolomitization replaced the original

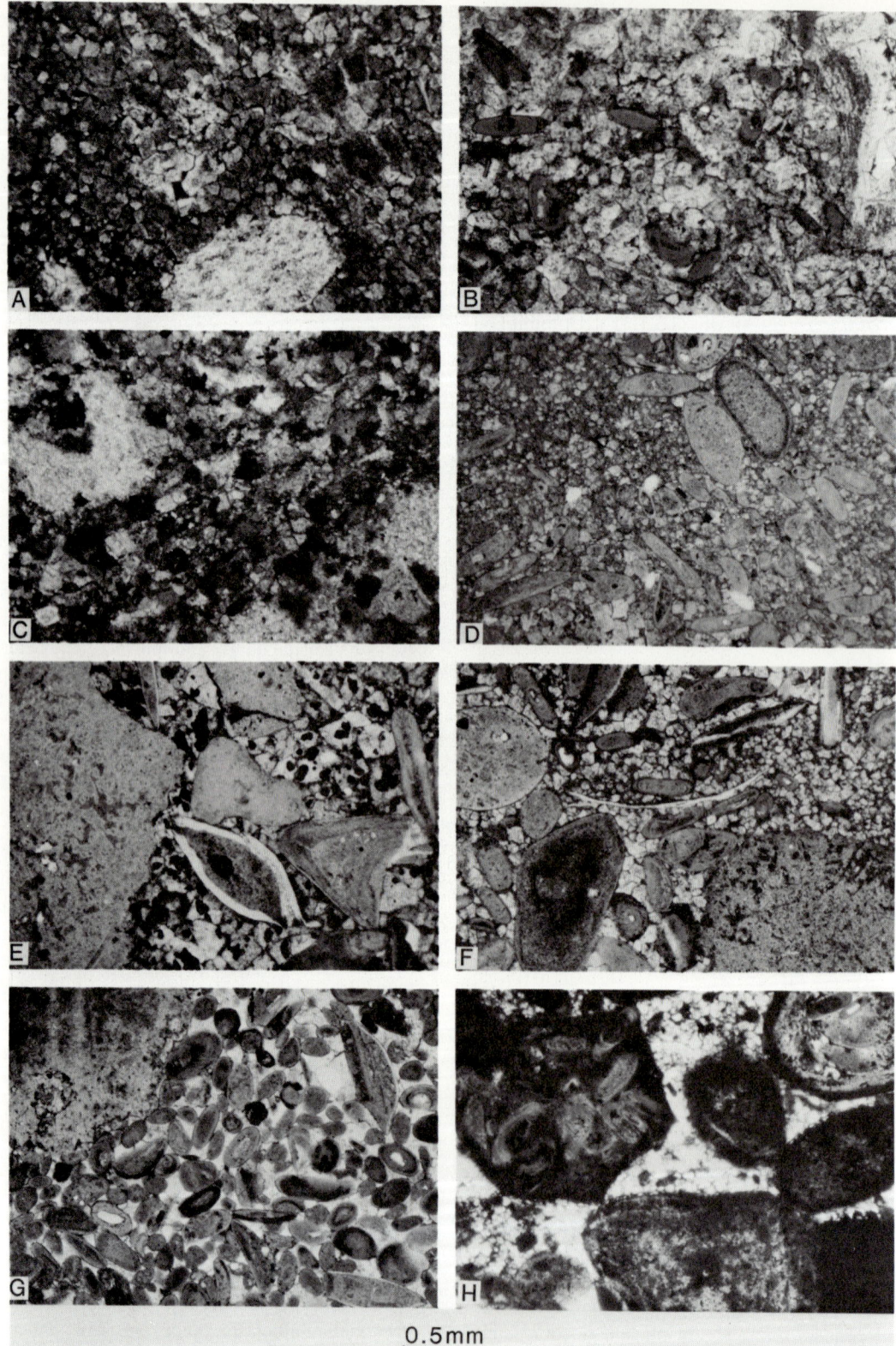

FIGURE 16.2 Maquoketa phosphorites (Upper Ordovician), Upper Mississippi Valley. Typical microfacies (continued). A. Microfacies 9. B. Microfacies 10. C. Microfacies 11. D. Microfacies 12. E. Microfacies 13. F. Microfacies 14. G. Microfacies 15. H. Microfacies 16. See text for detailed descriptions. All photomicrographs: plane-polarized light. From Black (1985).

matrix and cement and the margins of phosphatic grains and nodules. Pyrite replacement of grains and matrix is also common. Average mineral composition: 37% dolomite, 3% quartz, traces of potassium feldspar, 12% clay minerals, 44% apatite, and 4% pyrite.

Phosphatized Platform

Microfacies 15 (Fig. 16.2G). Dolomitized (100%) pressure-welded intraclastic oolitic biopharenite with sparite cement. Phosphatic ooids, pellets, and intraclasts of oolitic biophospharenite are abundant. The most frequent bioclasts are apatite-replaced cephalopods and brachiopods together with silicified ostracods filled with phosphalutite. Pyrite replacement of phosphatic grains and orthocone cephalopods is common. Subsequent dolomitization led to partial replacement of apatite grains and of original interparticle sparite cement by a mosaic of dolosparite. Average mineral composition: 19% dolomite, 1% quartz, traces of potassium feldspar, 3% clay minerals, 60% apatite, and 17% pyrite.

Microfacies 16 (Fig. 16.2H). Dolomitized (100%) nodular intraphospharudite with pelletoidal oophospharenite matrix and patchy sparite cement. Besides large nodules due to replacement of the matrix by apatite, phosphatic intraclasts predominate and consist of phosphatic ooids, pellets, brachiopods, and other bioclasts set in a phosphalutite matrix very rich in organic matter. Isolated bioclasts of brachiopods and ostracods are scattered in the interstitial material, which can be replaced locally by masses of pyrite. Subsequent dolomitization also marginally replaced all phosphatic constituents besides original matrix and cement. Average mineral composition: 23% dolomite, 4% quartz, 1% potassium feldspar, 7% clay minerals, 51% apatite, and 14% pyrite.

The Ideal Shallowing-Upward Sequence

The four cores investigated consist of complete sections that encompass the uppermost Galena, the Galena-Maquoketa contact, and the lower Maquoketa. They all display a complex environmental evolution consisting of six phases (Figs. 16.3, 16.4, 16.5).

> *Phase 1*: shallowing-upward sequence from lower inner slope to upper inner slope, that is, microfacies 7 to 11 (uppermost Galena).
>
> *Phase 2*: phosphatization of the upper inner slope and platform with formation of microfacies 12 to 16 (phosphorites of Galena-Maquoketa contact).
>
> *Phase 3*: rapid deepening down to basin and outer slope with microfacies 1 to 4 and equally rapid shallowing-upward back to upper inner slope (Lower Maquoketa).

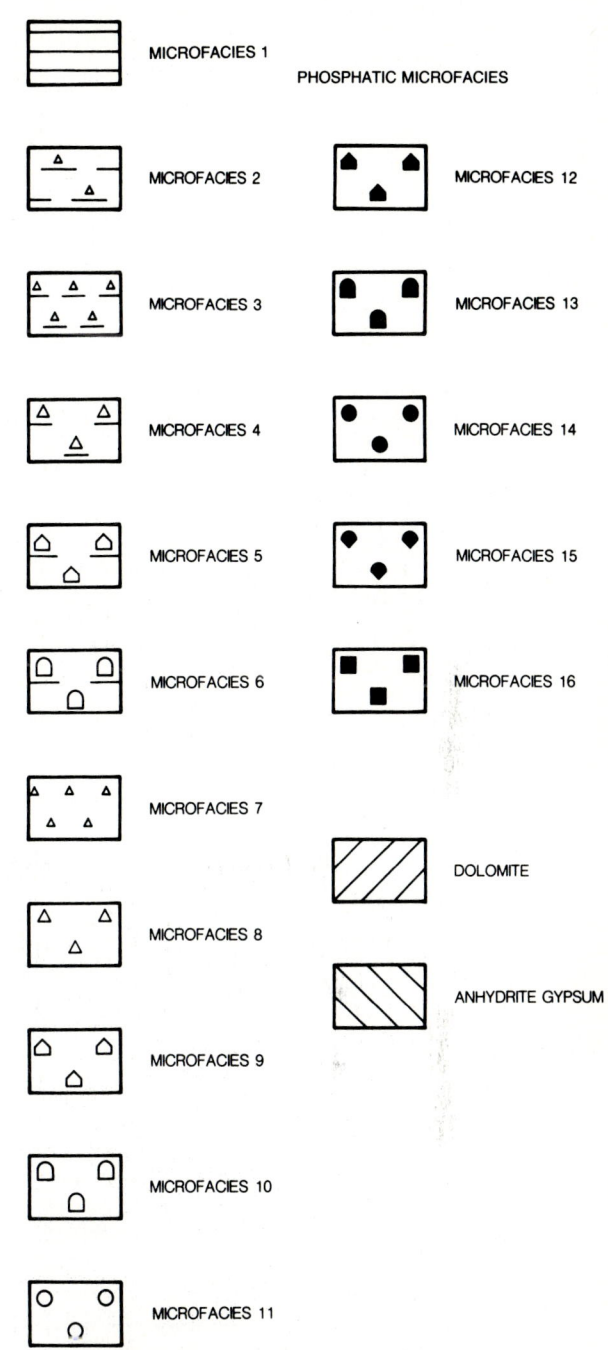

FIGURE 16.3 Maquoketa phosphorites (Upper Ordovician), Upper Mississippi Valley. Symbols for microfacies. From Black (1985).

Phase 4: phosphatization of the upper inner slope with formation of microfacies 12, 13, and 14 and concentration of phosphatic grains by reworking (Lower Maquoketa).

Phase 5: rapid deepening down to basinal conditions with microfacies 1 (Lower Maquoketa).

Phase 6: shallowing-upward sequence with several oscillations from basinal conditions to bioclastic bar, that is, microfacies 1 to 6 (Lower Maquoketa).

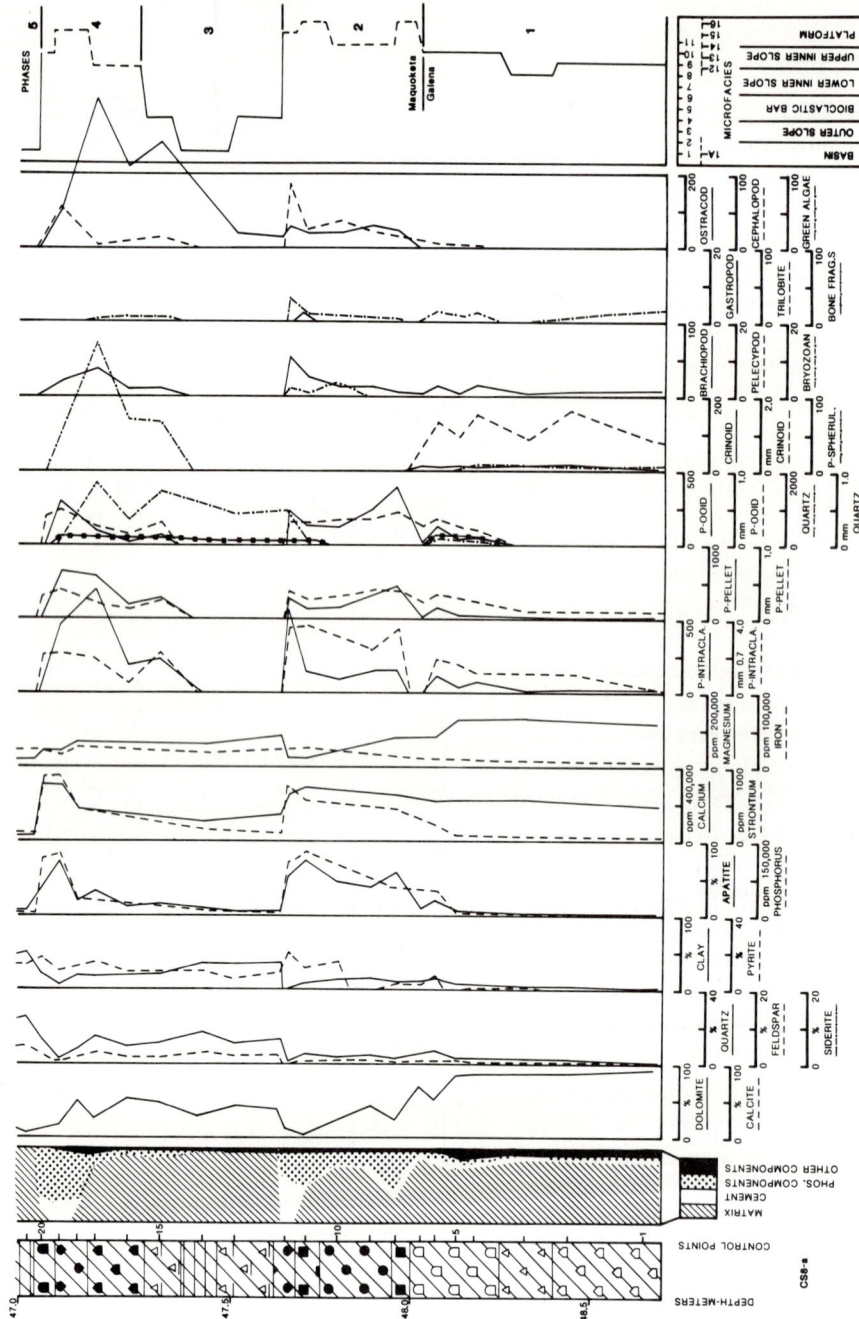

FIGURE 16.4 Maquoketa phosphorites (Upper Ordovician), Upper Mississippi Valley. Typical example of vertical variation of microscopic parameters in section CS8. From Black (1985).

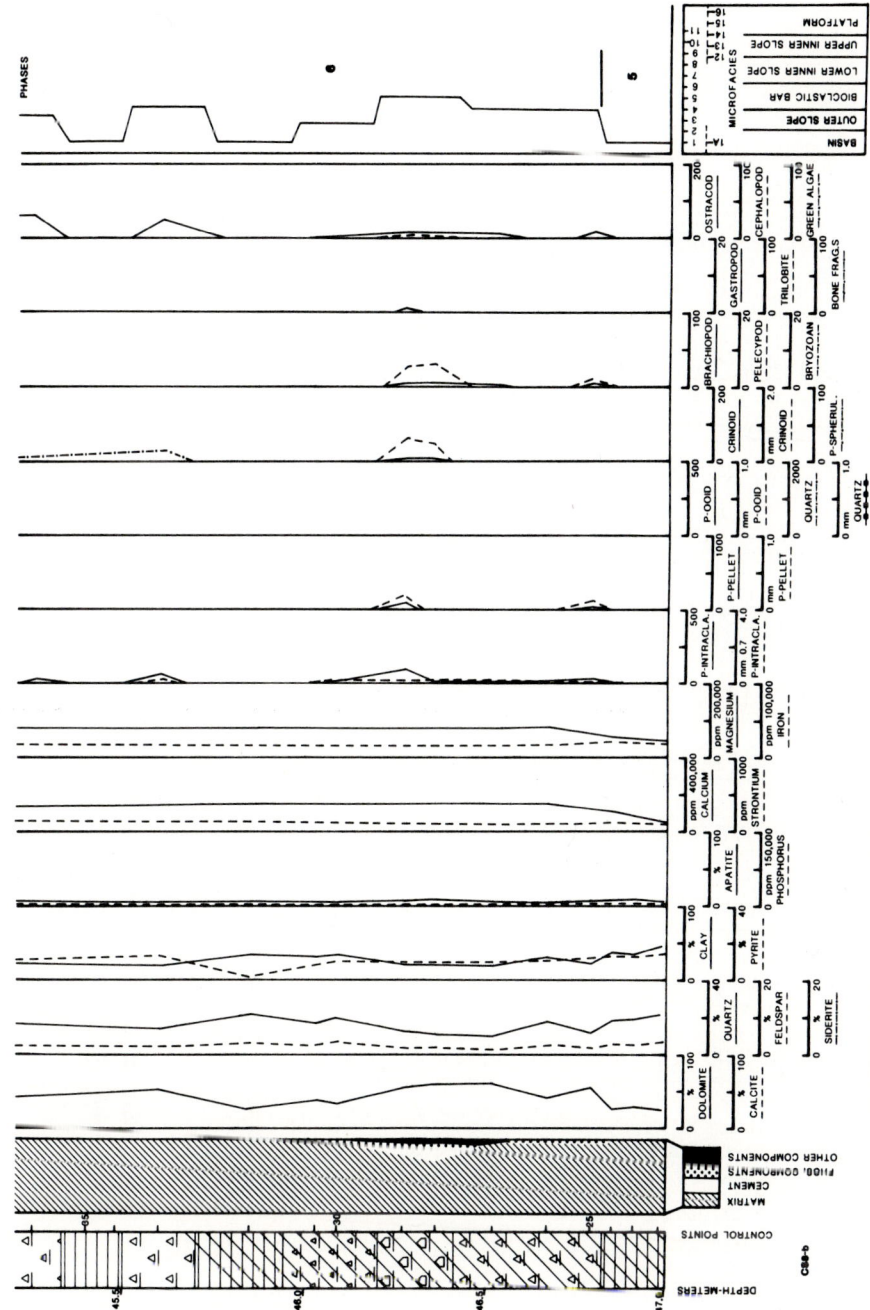

FIGURE 16.5 Maquoketa phosphorites (Upper Ordovician), Upper Mississippi Valley. Typical example of vertical variation of microscopic parameters in section CS8 (continued). From Black (1985).

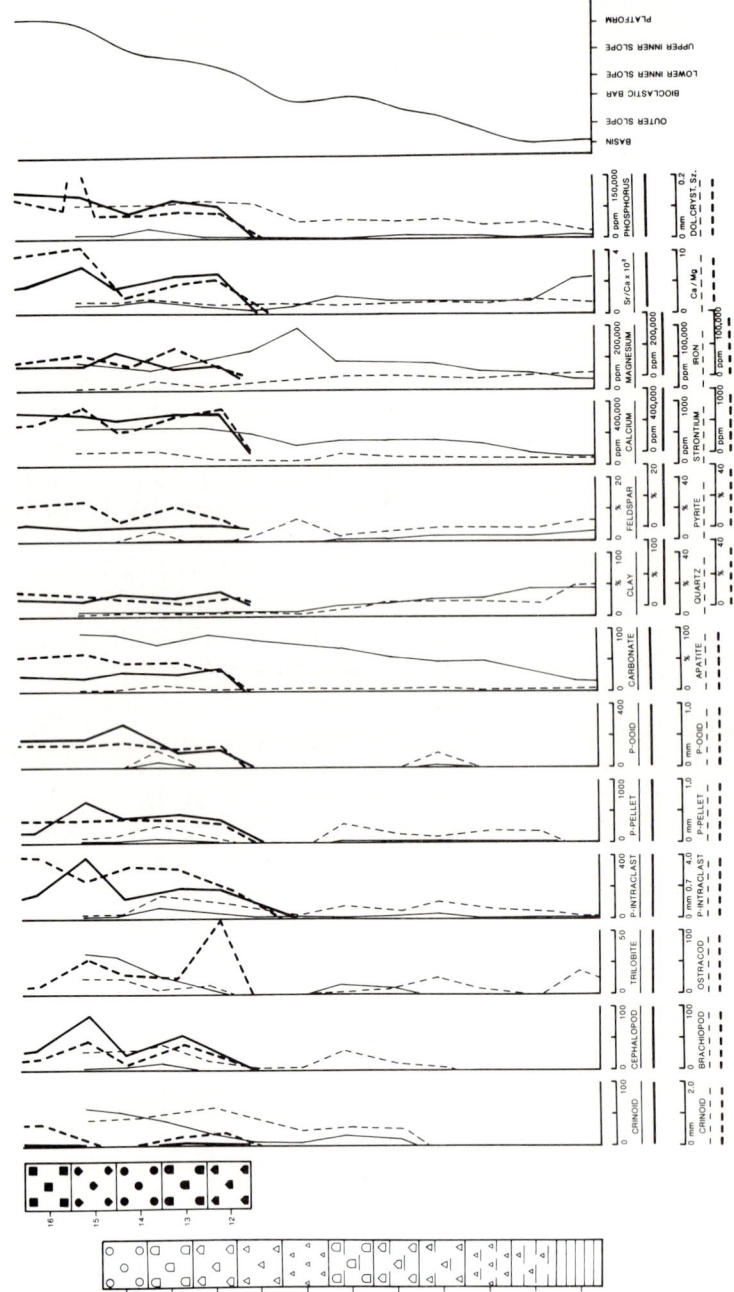

FIGURE 16.6 Maquoketa phosphorites (Upper Ordovician), Upper Mississippi Valley. Ideal shallowing-upward sequence. From Black (1985).

This environmental evolution can be represented graphically by a general shallowing-upward sequence (Fig. 16.6) consisting of the carbonate sequence of microfacies 1 to 11 partially overlapping the phosphatized sequence of microfacies 12 to 16. Variations of petrographic and chemical parameters for the latter sequence are drafted in heavier lines on Figures 16.6, 16.7, and 16.8.

The Ideal Depositional Model

In the proposed model (Figs. 16.7, 16.8), with the exception of the platform microfacies, the sediments were deposited almost *in situ* or by fairly weak bottom currents both downslope and upward. The presence of weak currents along the slope is indicated by the change in depositional conditions from microfacies 8 to 11 to microfacies 12 to 16. The largest frequency of crinoids occurs in microfacies 10 and 11, and generally decreases seaward, indicating that crinoid colonies mainly existed on the platform and that only minor colonies of smaller crinoids lived under slope conditions with a slightly larger frequency in association with the bioclastic bar. Few crinoids are present in microfacies 12 to 16, due to their apparent intolerance for unfavorable physicochemical conditions which resulted in the synsedimentary phosphatization of platform carbonates.

The frequency curve of brachiopods generally mimics that of crinoids; brachiopods participate appreciably in the generation of the bioclastic bar. They increase in frequency in microfacies 12 to 16, indicating a greater tolerance than crinoids for synsedimentary phosphatization. Orthocone cephalopods represent the predominant fraction of bioclasts in microfacies 12 to 16. This occurrence indicates that they were either autochthonous platform dwellers during phosphatization or that they were brought in by weak bottom currents and concentrated mechanically in a greater energy microfacies. A nearshore and an offshore population of ostracods is clearly displayed, and their peak of frequency in microfacies 12 on the edge of the area of synsedimentary phosphatization may represent a mechanical concentration by weak bottom currents. Trilobite frequency is greatest in microfacies 11 (platform edge) and decreases seaward with a second increase associated with the bioclastic bar. This pattern is expected of scavengers in relation to other peaks of benthic constituents such as crinoids and brachiopods.

Most of the phosphatic constituents were generated in upper inner slope to platform conditions (microfacies 12 to 16) as shown by the behavior of phosphatic intraclasts, phosphatic pellets derived from their disintegration, phosphatic ooids, and phosphatic spherulites of fecal origin. Weak bottom currents transported seaward a reduced number of the smaller phosphatic constituents, which occasionally were concentrated in the bioclastic bar.

The overall greater matrix content and the very small cement content of the basin, outer slope, and lower inner slope environments are indicative of general low energy of the proposed model. Increased cement, decreased matrix and their irregular behavior in microfacies 12 to 16 are characteristic of the higher-energy synsedimentary phosphatization processes over the platform. Similar conclusions can be reached by observing variation of curves of all other nonphosphatic components versus the phosphatic ones.

Crystallinity of dolomite (expressed by the maximum dolomite rhomb size) reflects the grain size of the original replaced carbonate as observed in other instances, and hence bears a direct relationship to the energy curve of the environment of deposition (Okhravi and Carozzi, 1983; see also Chapter 14, this volume).

Variations in mineralogic data lead to a refinement of the petrographic characteristics of microfacies determination and provide further insight into the physicochemical conditions of the environments of deposition. Carbonate percent increases steadily toward the platform, recording increased carbonate precipitation in more aerated and shallower conditions. However, carbonate percent decreases in microfacies 12 to 16 due to its replacement by apatite expressing penecontemporaneous phosphatization. The extrabasinal sediment yield of fine-grained detrital quartz, feldspar, and clay minerals gradually increases in frequency seaward, corresponding to deposition of deeper-water argillaceous microfacies which are clearly different from those of the greater energy platform, where such small grain-size constituents were winnowed. Pyrite percent increases basinward, indicating variation in reducing conditions and related accumulation of organic matter. The larger pyrite content in microfacies 12 to 16 is reflective of the conditions leading to precipitation of apatite on carbonate slope and platform: low dissolved oxygen, decrease in pH, high biological productivity, and a rapid accumulation of organic matter.

Total calcium (expressed in ppm) varies in agreement with the total carbonate content of the microfacies. The larger values in microfacies 12 to 16 are due to the large amount of Ca within the fluorapatite structure. The magnesium content (expressed in ppm) is representative of the amount of dolomite present. It shows a trend parallel to the carbonate percent, and roughly parallel to total calcium. The lesser magnesium values in microfacies 12 to 16 are due not only to the corresponding larger Ca and P values, but possibly also to an actual decrease in the chemical activity of magnesium during

492 Part IV / Carbonate Slopes and Basins

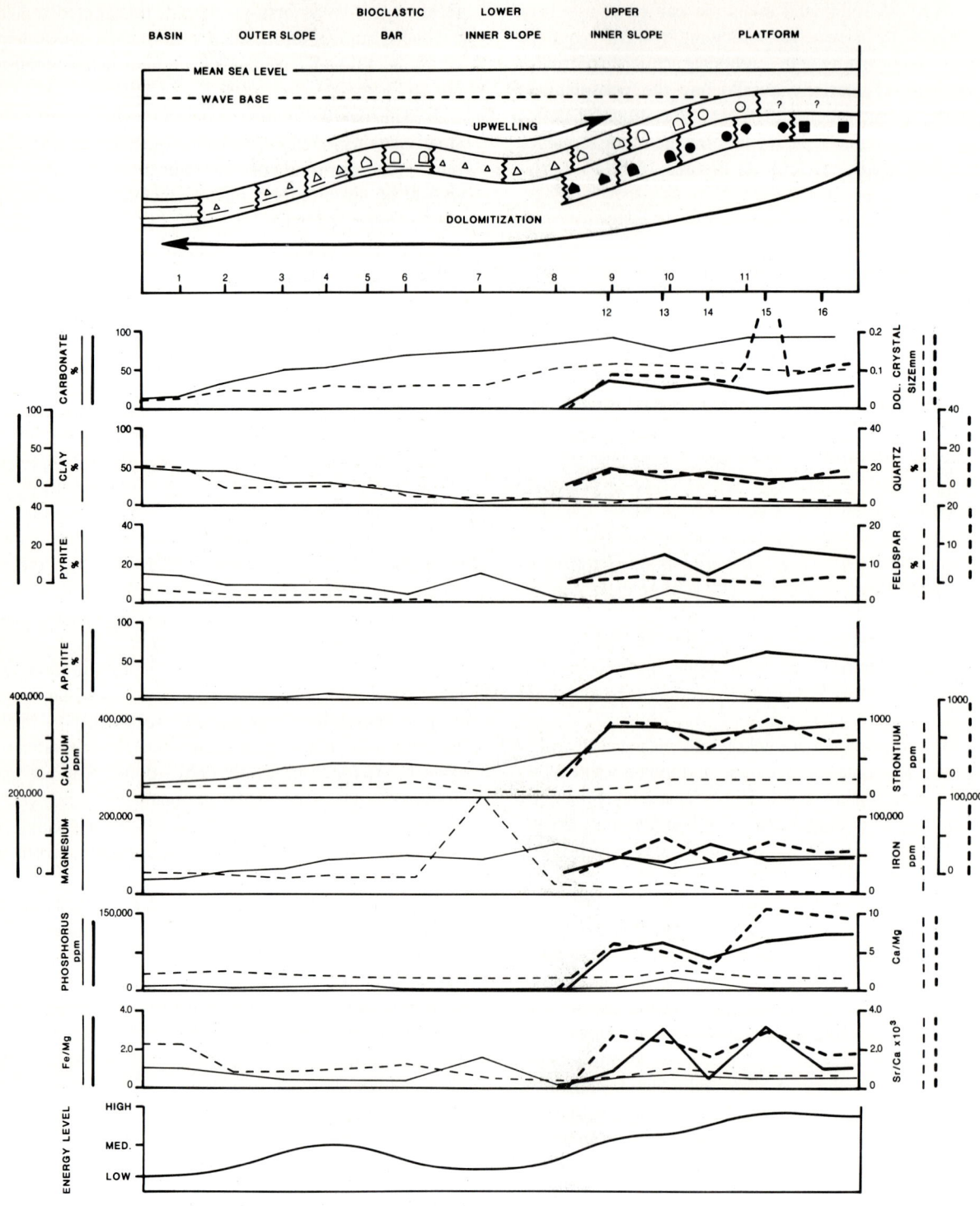

FIGURE 16.7 Maquoketa phosphorites (Upper Ordovician), Upper Mississippi Valley. Ideal depositional model. From Black (1985).

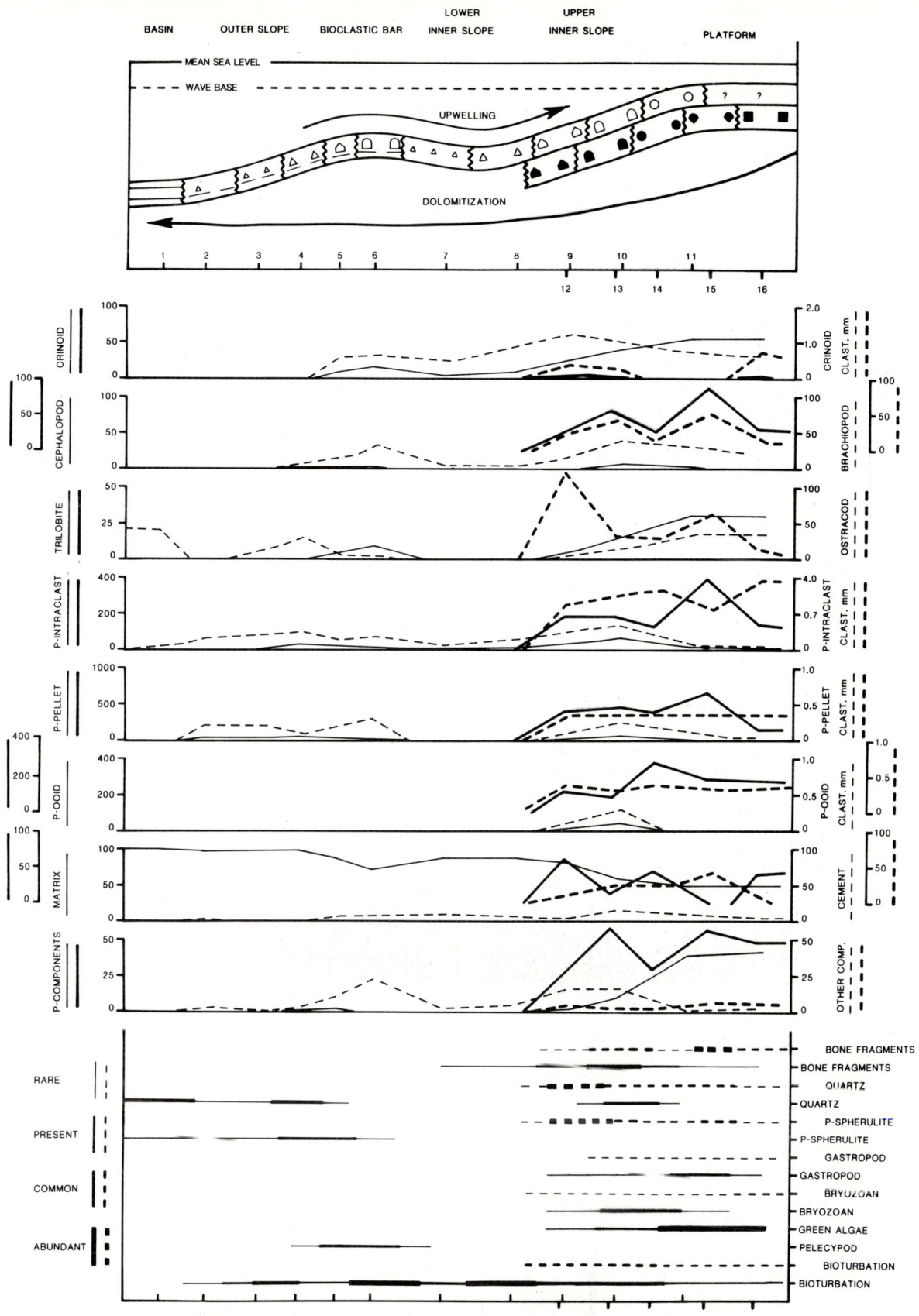

FIGURE 16.8 Maquoketa phosphorites (Upper Ordovician), Upper Mississippi Valley. Ideal depositional model (continued). From Black (1985).

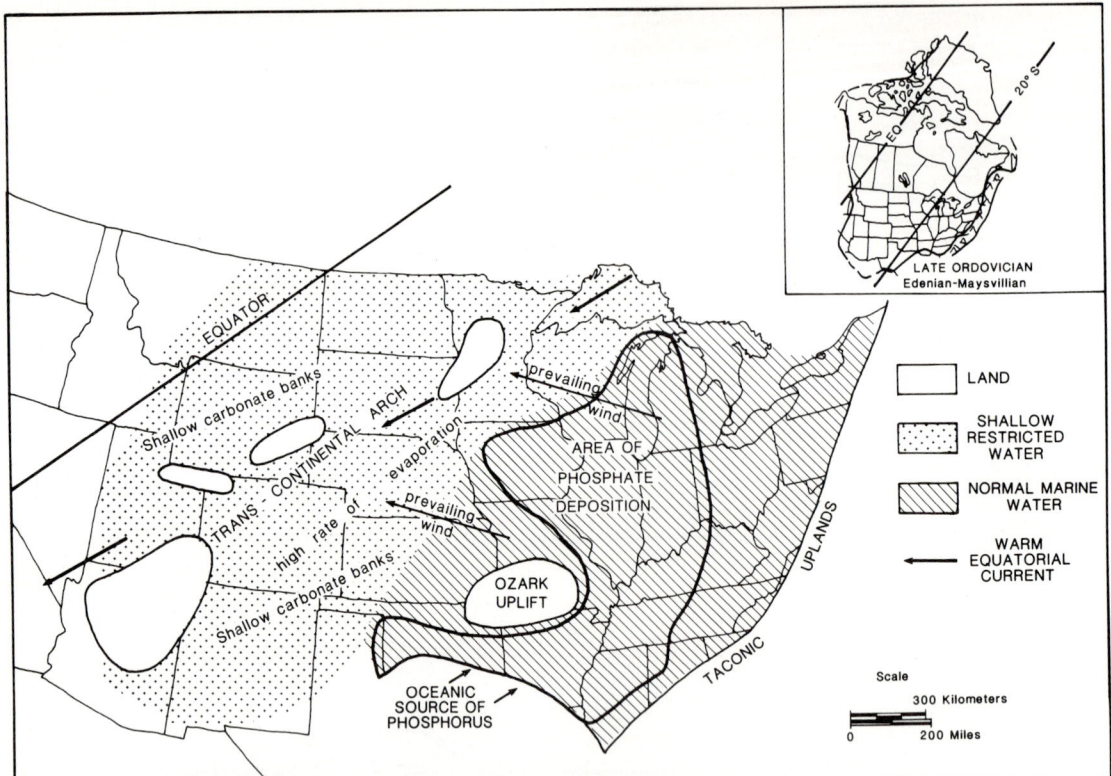

FIGURE 16.9 Maquoketa phosphorites (Upper Ordovician), Upper Mississippi Valley. Generalized paleogeographic map of the Upper Ordovician in the central United States, showing major structural features, general wind and current direction, and generalized distribution of paleoenvironments. Normal marine waters existed in the eastern half of the continent whereas more restricted shallow-water environments were present in the west. The heavy black line outlines the general region of phosphate deposition during lower Maquoketa-Maysvillian time resulting from an anticlockwise upwelling under a quasi-geostrophic open ocean model. From Black (1985).

deposition because this element is considered an inhibitor to precipitation of marine apatites. The curve of iron follows the curve of pyrite percent for all microfacies.

The strontium content (expressed in ppm) is relatively small and its depletion is known to be an effect of dolomitization. Little variation occurs from microfacies 1 to 11, but strontium generally follows the same trend as calcium due to its substitution for Ca in calcite. The concentrations of Sr in microfacies 12 to 16 are significantly larger because strontium commonly is fixed in the fluorapatite lattice. The Sr/Ca ratio follows the same trend as Sr except for a significant increase in microfacies 1, which may result from possible absorption of strontium on clay minerals. Total phosphorus is small (generally less than 10,000 ppm) in microfacies 1 to 11; it increases markedly in microfacies 12 to 16 due to precipitation of fluorapatite. The curve is therefore parallel to percent apatite. The Ca/Mg ratio is more useful as an indicator of changes in carbonate chemistry than Ca or Mg alone because it is independent of the actual percent carbonate present (calcite or dolomite). For microfacies 1 to 11, the Ca/Mg ratio is fairly constant. The increase in microfacies 10, and in 12 to 16 results from the increase in calcium associated with precipitation of fluorapatite. In addition, the presence of Mg ions in solution acts as an inhibitor to the precipitation of apatite and thus some of the negative correlation between P and Mg is probably due to a decrease in the activity of Mg in interstitial solutions.

The Fe/Mg curve follows a similar trend to that of percent pyrite and ppm iron. The large fluctuations in the Fe/Mg ratio for microfacies 12 to 16 are due to corresponding variations in Mg and Fe content which are magnified through the use of ratios and do not appear to be the result of any significant environmental fluctuations.

Diagenesis and Regional Paleogeography

Petrographic studies showed two distinct periods of diagenesis: an early or penecontemporaneous phase, and a late one. The early phase is characterized by widespread phosphatization, contemporaneous precipitation of pyrite, and minor silicification through growth of authigenic crystals and replacement of bioclasts. During Ordovician time, the area investigated (Fig. 16.9) was located at approximately 20° south latitude within an arid climatic belt which covered most of the central United States, with prevailing wind direction from the southeast. Phosphatization took place at the interface of fluctuating deeper anoxic phosphate-rich waters upwelling in an anticlockwise direction under quasi-geostrophic open ocean conditions onto the carbonate slope and platform, and beneath shallower oxygenated waters. Direct precipitation of apatite in the sediment interstitial waters was followed by penecontemporaneous replacement of carbonate precursors and intraformational reworking leading to generation of phosphatic intraclasts, pellets, and ooids. Phases 3 and 4 (Fig. 16.4) occurred very quickly, indicating that the thermocline was probably fairly shallow, and minor fluctuations caused by epeirogenic and glacio-eustatic changes of sea level (expressing the Late Ordovician glaciation) could produce the cyclicity of phosphorites, brown and black shales, and argillaceous carbonates which occurred during early Maquoketa time.

The late phase of diagenesis consists of postdepositional precipitation of anhydrite and gypsum represented today by calcite pseudomorphs after chickenwire anhydrite and selenitic gypsum. It was followed by phreatic calcite cementation and a large-scale episode of dorag dolomitization which took place at the end of Maquoketa time (Late Ordovician), as described in the study of the Galena Group in Chapter 4.

17

Lacustrine Carbonates with Bioaccumulated to Hydrodynamic Buildups

Lacustrine carbonate models are generally of lesser energy than their marine equivalents, water motion being controlled by winds and storms rather than tides. These models are also simpler ecologically because of the smaller variety of lacustrine fauna and flora. Nevertheless, phytoplankton productivity can be prolific; hence euxinic carbonate muds and shales become important source beds of hydrocarbons. Reservoirs are provided mainly by bioaccumulated to hydrodynamic buildups formed by pelecypods, ostracods, and stromatolites.

17.1 CHARACTERISTIC CONSTITUENTS: PELECYPODS, OSTRACODS, AND BASIC HYALOCLASTITES

The Lagoa Feia Formation (Lower Cretaceous) was deposited during the rift valley stage of the Campos Basin, offshore Rio de Janeiro, Brazil (Bertani and Carozzi, 1984, 1985). It consists of an association of lacustrine carbonates, siliciclastics, and volcaniclastics ranging from 200 to more than 1500 m in thickness. The burial conditions are in excess of 3000 m, and the association of siltstones and porous carbonates contains mature source rocks and potential reservoirs.

This study was based on 27 exploratory wells drilled over an area of 1600 km^2. Petrographic and petrophysical analysis was made on 780 thin sections (average vertical interval of 56 cm) cut from 47 cores representing 449 m of section.

Description of Microfacies

The petrographic study showed the existence of 17 microfacies which were divided into four major groups: terrigenous-dominated sequence, ostracod-dominated sequence, pelecypod-dominated sequence, and basic volcaniclastic-dominated sequence. Subdivisions within the four groups were made according to the grain size of terrigenous components, relative proportion of terrigenous to carbonate components, and relative proportion of matrix to cement and primary porosity.

Terrigenous-Dominated Sequence

Microfacies T1 (Fig. 17.1A). Poorly sorted, matrix-supported lithic conglomerate to conglomeratic lithic feldspathic wacke. Coarse lithic grains consist mainly of basalt either altered to clay minerals or calcitized. Matrix is composed of sand- to silt-size grains of volcanic glass altered to clay minerals and minor amounts of feldspars and quartz.

Microfacies T2 (Fig. 17.1B). Lithic feldspathic quartz arenite with calcite cement. Lithic grains of volcanic glass and feldspars are frequently calcitized. Pri-

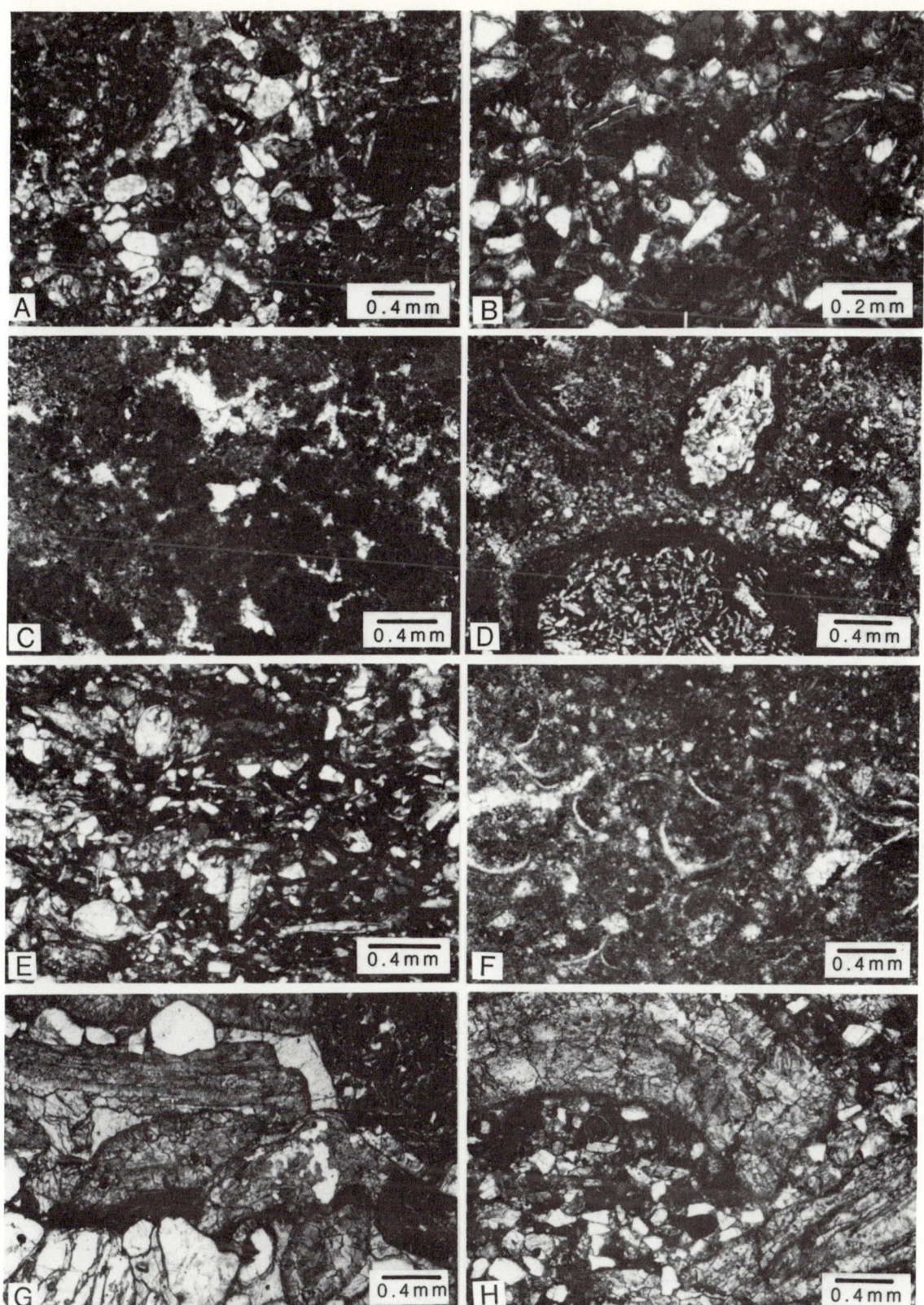

FIGURE 17.1 Lagoa Feia Formation (Lower Cretaceous), Campos Basin, offshore Rio de Janeiro, Brazil. Typical microfacies. A. Microfacies T1. B. Microfacies T2. C. Microfacies M1. D. Microfacies OT2. E. Microfacies OT3. F. Microfacies OS4. G. Microfacies PT2. H. Microfacies PT3. See text for detailed descriptions. All photomicrographs: plane-polarized light. From Bertani and Carozzi (1984).

mary interparticle porosity is observed occasionally. Lateral grading into varieties with matrix composed of fine glass debris and clay minerals is frequent. Lithic grains of volcanic glass are highly altered to trioctahedral smectites.

Microfacies T3. Argillaceous siltstone to silty shale composed of minute debris of volcanic glass, feldspars, and quartz.

Ostracod-Dominated Sequence

Microfacies M1 (Fig. 17.1C). Pelletoidal micrite with rare fragments of ostracods, commonly rich in clay minerals and silt-size grains of glass and feldspars. Disruption by intrasedimentary growths of gypsum and anhydrite replaced by calcite is frequent. Root casts and desiccation cracks are common.

Microfacies OT2 (Fig. 17.1D). Grain-supported oncoidal calcarenite with micrite matrix often dolomitized. Ostracod bioclasts are rare, basalt fragments usually form the nucleus of oncoids.

Microfacies OT3 (Fig. 17.1E). Poorly sorted, very fine-grained lithic arenite to siltstone, rich in ostracod shell fragments, with matrix composed of micrite and clay minerals. Terrigenous components are mainly basic glass, feldspars, and a small amount of quartz.

Microfacies OS4 (Fig. 17.1F). Biocalcisiltite to matrix-supported biocalcarenite. Bioclasts consist mainly of fragments of ostracod shells and to a lesser extent of pelecypods. Micrite matrix shows neomorphism or is dolomitized. Bioturbation, desiccation cracks, and calcite pseudomorphs after gypsum and anhydrite nodules are common.

Microfacies OS6. Bioaccumulated ostracod limestone consisting of thick and ornamented ostracod shells. Matrix is composed of pseudomicrosparite and silt-size ostracod shell fragments. Calcite pseudomorphs after gypsum and anhydrite crystals occur in places.

Microfacies OS7. Well-laminated, micrite-rich shale. Scattered fish debris and rare ostracod bioclasts are the only skeletal components. Geochemical analysis reveals that these sediments are rich in mature organic matter, commonly presenting 1 to 2% organic carbon content.

Pelecypod-Dominated Sequence

Microfacies PT2 (Fig. 17.1G). Well-sorted arenaceous biocalcarenite to bioclast-rich lithic feldspathic arenite. Bioclasts are reworked pelecypod shells altered by neomorphism to pseudosparite and siliciclastics consist of fine- to medium-grained lithic fragments of basalt and basic glass, feldspars, and quartz. Bladed rim and sparitic calcite cements, and locally zeolites, partially to completely fill primary intergranular pores.

Microfacies PT3 (Fig. 17.1H). Grain-supported arenaceous biocalcarenite with carbonate and terrigenous matrix. Bioclasts are reworked pelecypod shells altered by neomorphism to pseudosparite, and siliciclastics consist of fine sand-size grains of basic glass, feldspars, and quartz. Matrix is composed of a mixture of micrite, silt-size bioclasts, fine basic glass debris, and clay minerals.

Microfacies P4 (Fig. 17.2A). Grain-supported pelecypod biocalcarenite with pseudosparitic matrix. Pelecypod shell fragments show intense breakage and reworking. They display alteration by neomorphism, or are silicified, or dissolved with molds refilled by sparite cement. Calcite pseudomorphs after anhydrite crystals are common.

Microfacies P5 (Fig. 17.2B). Well-sorted pelecypod biocalcarenite with bladed rim and sparite calcite cements. Reworked pelecypod bioclasts show neomorphism, or are silicified or partially dissolved. Incompletely filled intergranular spaces are preserved commonly as reduced primary porosity (in black in Fig. 17.2B).

Microfacies P6 (Fig. 17.2C). Grain-supported bioaccumulated pelecypod limestone showing neomorphism of argillaceous micrite matrix. Shell breakage and deformation due to compaction are intense locally.

Basic Volcaniclastic-Dominated Sequence

Microfacies H (Fig. 17.2D,E,F). Hyaloclastite composed of glass globules about 1 mm in diameter set in a fine-grained glass matrix usually replaced by calcite. Glass globules may show intense internal fracturing (Fig. 17.2D) and concentric internal structure due to cooling (Fig. 17.2E), or elongate shape due to plastic deformation of glass while still hot and somewhat viscous. Diagenetic alteration of basic glass is intense; it consists of palagonitization, zeolitization, silicification, dolomitization, calcitization, and alteration to clay minerals.

Kerolitic ooids (Fig. 17.2F) are associated with the hyaloclastite, forming layers on top of the basic volcaniclastic sequences. They display a nucleus composed of basic glass altered to kerolite, and a cortex of tangentially oriented very finely fibrous kerolite. Ooids are usually well sorted and cemented by microcrystalline quartz. Compaction and pressure-welding also occur.

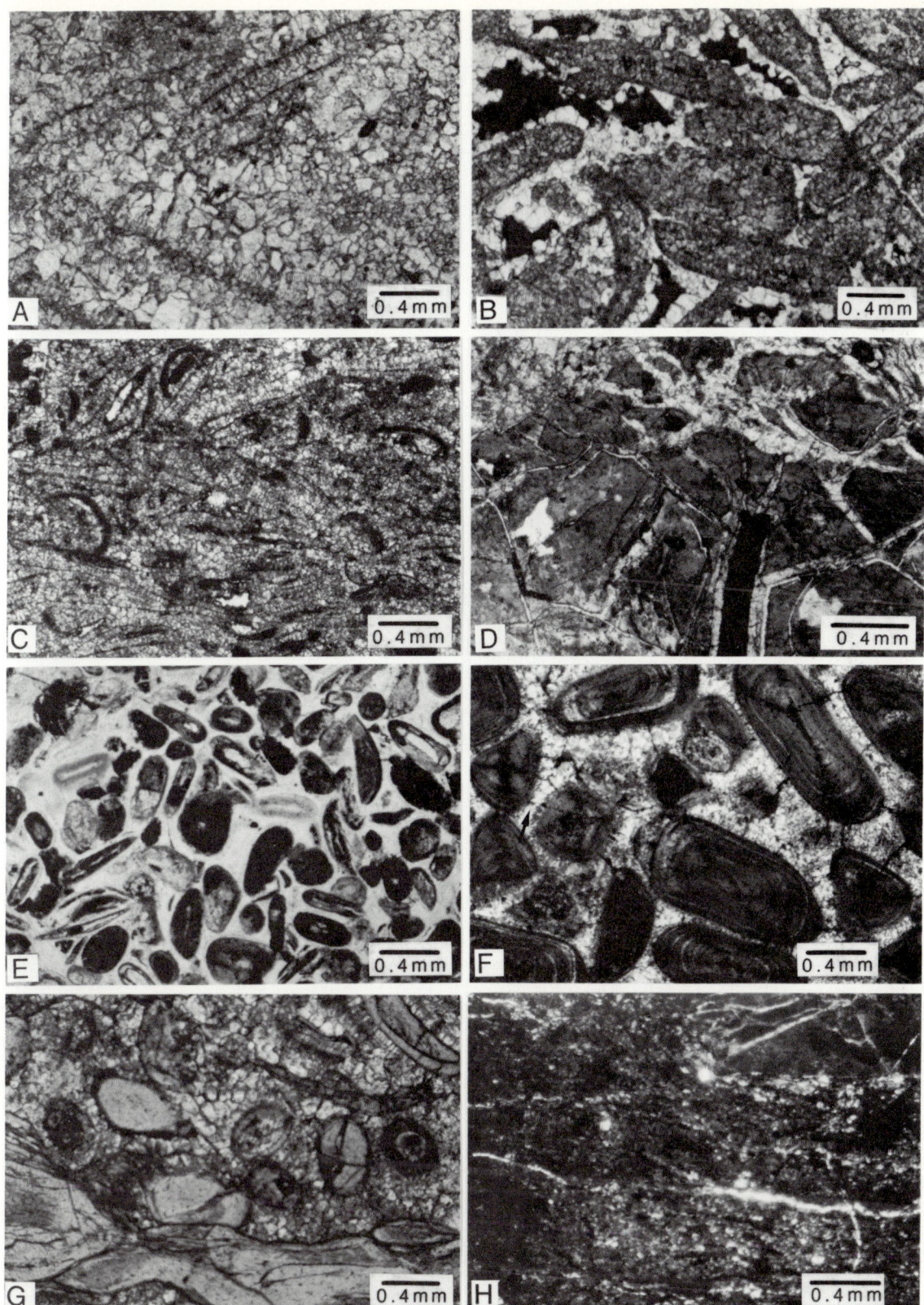

FIGURE 17.2 Lagoa Feia Formation (Lower Cretaceous), Campos Basin, offshore Rio de Janeiro, Brazil. Typical microfacies (continued). A. Microfacies P4. B. Microfacies P5. C. Microfacies P6. D to F. Microfacies H. G. Microfacies HC. H. Microfacies HT. See text for detailed descriptions. All photomicrographs: plane-polarized light, except F, crossed nicols. From Bertani and Carozzi (1984).

500 Part IV / Carbonate Slopes and Basins

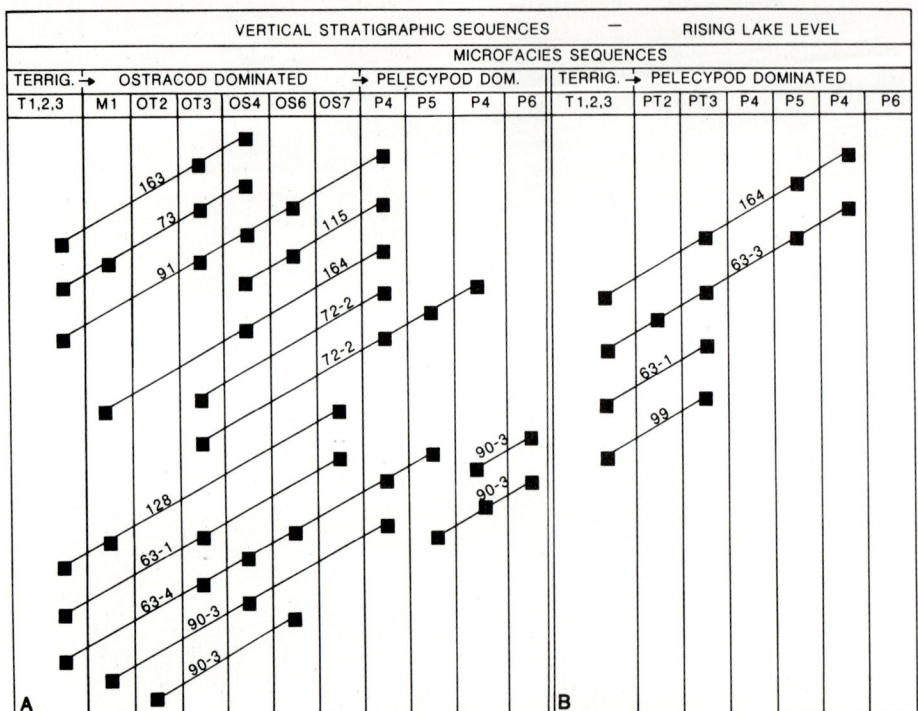

FIGURE 17.3 Lagoa Feia Formation (Lower Cretaceous), Campos Basin, offshore Rio de Janeiro, Brazil. Vertical sequences of observed microfacies for rising lake level. Numbers on oblique lines refer to wells. A. Basinward. B. Shoreward. From Bertani and Carozzi (1984).

Microfacies HC (Fig. 17.2G). Microfacies H with a carbonate instead of glass matrix. Matrix consists of micrite with ostracod and pelecypod bioclasts, usually altered by neomorphism, or are silicified, or dolomitized. Micrite matrix in direct contact with glass particles often shows neomorphism to a rim of short prismatic calcite crystals.

Microfacies HT (Fig. 17.2H). Hyalotuff composed of very fine basic glass particles, parallel laminated, usually deeply altered to trioctahedral smectites, with very rare grains of quartz and feldspars.

Vertical Sequences of Microfacies

Four types of vertical sequences of microfacies were observed (Figs. 17.3, 17.4).

- A. Terrigenous-dominated sequence–Ostracod-dominated sequence–Pelecypod-dominated sequence (except terrigenous-rich pelecypod microfacies).
- B. Terrigenous-dominated sequence–Pelecypod-dominated sequence.
- A′. Pelecypod-dominated sequence (except terrigenous-rich pelecypod microfacies)–Ostracod-dominated sequence–Terrigenous-dominated sequence.
- B′. Pelecypod-dominated sequence–Terrigenous-dominated sequence.

Two characteristics of these sequences are relevant for the definition of depositional models: the symmetry between sequences A and A′, and B and B′; the absence of terrigenous-rich pelecypod microfacies in sequences A and A′, and of ostracod-rich microfacies in sequences B and B′.

Ideal Vertical Sequences of Microfacies

Ideal models of vertical microfacies sequences based on the concept of an alternatively expanding and contracting lake were constructed (Figs. 17.5, 17.6). In this depositional setting, sediments of the lacustrine environment interfinger with terrigenous-dominated sequences deposited in terrigenous flats toward the basin margin.

During periods of expansion, or rising lake level (Fig. 17.5), a playa lake stage with ostracod-dominated sediments is followed by a pluvial lake stage with pelecypod-dominated microfacies. Toward the center of the basin (Fig. 17.5A), terrigenous sediments are overlain

Chap. 17 / Lacustrine Carbonates with Bioaccumulated to Hydrodynamic Buildups 501

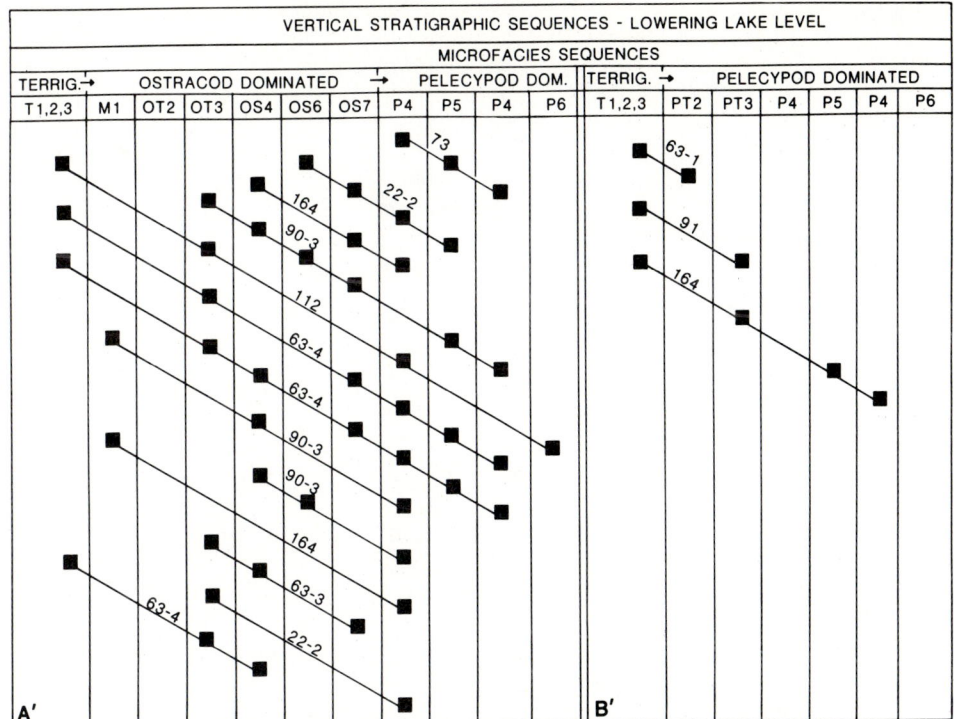

FIGURE 17.4 Lagoa Feia Formation (Lower Cretaceous), Campos Basin, offshore Rio de Janeiro, Brazil. Vertical sequences of observed microfacies for falling lake level. Numbers on oblique lines refer to wells. A'. Basinward. B'. Shoreward. From Bertani and Carozzi (1984).

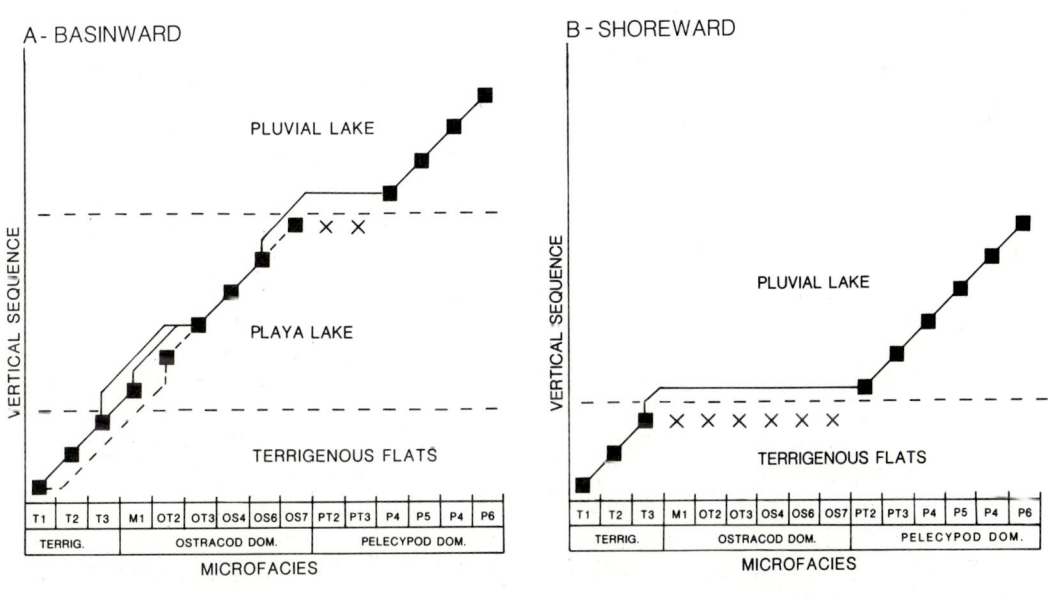

FIGURE 17.5 Lagoa Feia Formation (Lower Cretaceous), Campos Basin, offshore Rio de Janeiro, Brazil. Ideal vertical sequence of microfacies for rising lake level. From Bertani and Carozzi (1984).

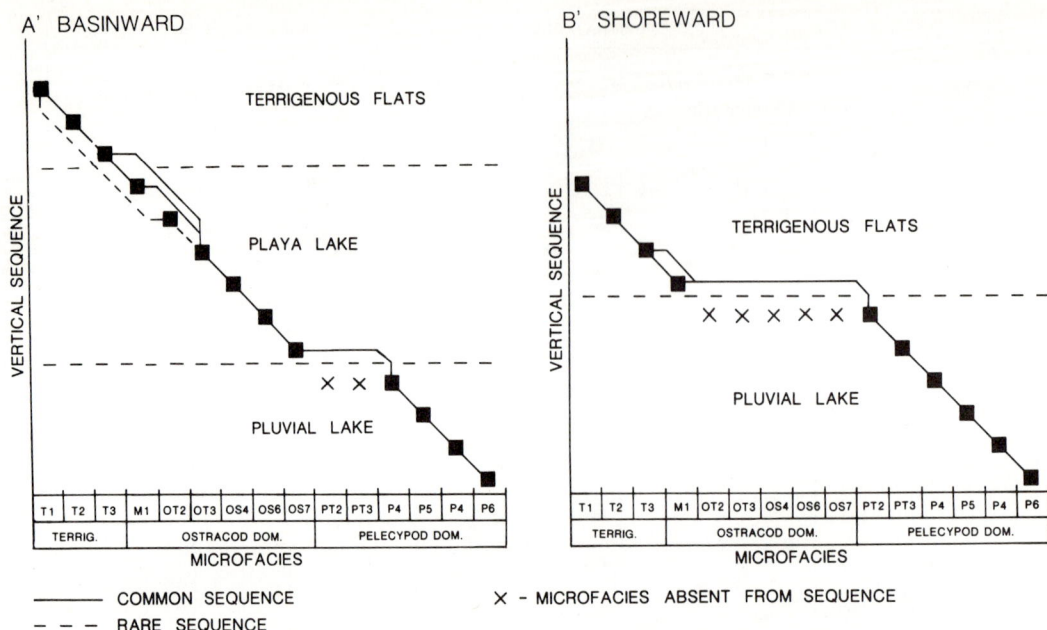

FIGURE 17.6 Lagoa Feia Formation (Lower Cretaceous), Campos Basin, offshore Rio de Janeiro, Brazil. Ideal vertical sequence of microfacies for falling lake level. From Bertani and Carozzi (1984).

by ostracod-rich microfacies and then by pelecypod-rich ones. Terrigenous-rich pelecypod microfacies are absent from this basinal sequence because at the time of the pluvial stage, the lake expanded considerably and the terrigenous sediment yield was restricted to its margins.

Near the basin margin (Fig. 17.5B), terrigenous-rich pelecypod microfacies directly overlie terrigenous sediments of the terrigenous flats. Ostracod-rich microfacies are absent from the basin margin sequence because at the time the lake had expanded to the basin margin, it was in the pluvial regime.

Symmetric evolution took place during periods of contraction, or lowering lake level, leading to deposition of vertical sequences near the center of the basin (Fig. 17.6A) and toward its margin (Fig. 17.6B) from which terrigenous-rich pelecypod microfacies and ostracod-rich ones are absent, respectively.

The marked symmetry between sequences of microfacies deposited during periods of rising and falling lake level suggests that climatic variations played an important role in the evolution of the depositional environments of the Lagoa Feia Formation. This formation can thus be interpreted as resulting from cyclic depositional episodes oscillating between two extremes: a playa lake and a pluvial lake. Basic volcaniclastic rocks may be intercalated with sediments of both depositional environments. A typical example shows detailed variation of microscopic parameters (Figs. 17.7, 17.8).

The Ideal Playa Lake Depositional Model

The playa lake model consists of the following environments: alluvial plain (alluvial fans, sand flats, and mudflats), carbonate flat, foreshore and shoreface lacustrine, offshore lacustrine, and euxinic offshore lacustrine (Fig. 17.9).

Within this depositional setting, nine microfacies are recognized, each with composition, texture, and early diagenetic minerals reflecting local subenvironments (Fig. 17.10). Carbonate flats developed in the low flat areas adjacent to the saline lake and were flooded periodically during rainy periods or by seiches caused by winds. Extended arid periods led to evaporative concentration of water and finally, to subaerial exposure.

The foreshore is characterized by oncoidal calcarenites with a dolomitized micrite matrix of low energy, whereas the shoreface environment is characterized by intense bioturbation. Under offshore conditions, shoals developed consisting of bioaccumulated ostracod shells due to *in situ* winnowing of matrix by weak currents or wind-generated waves.

The Ideal Pluvial Lake Depositional Model

The pluvial lake depositional model consists of a less saline lake which expands over the area previously occupied by the playa lake and which interfingers with the

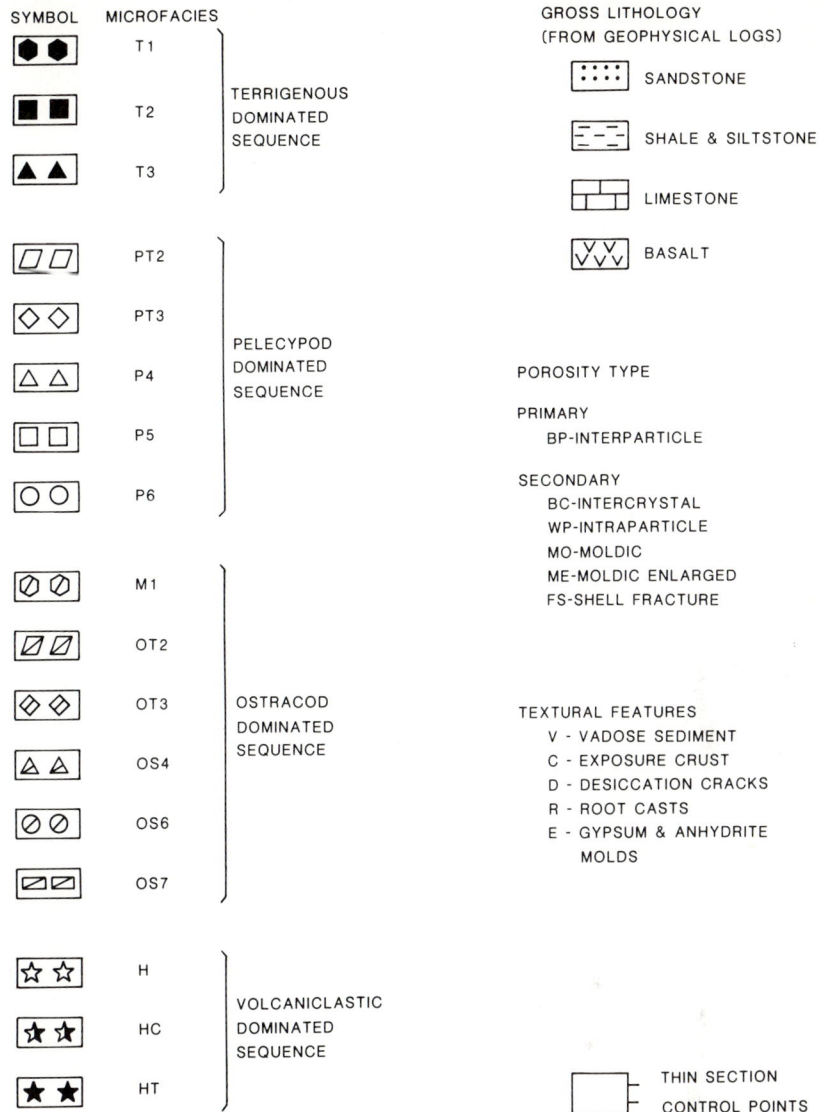

FIGURE 17.7 Lagoa Feia Formation (Lower Cretaceous), Campos Basin, offshore Rio de Janeiro, Brazil. Symbols for microfacies. From Bertani and Carozzi (1984).

alluvial plain (Fig. 17.11). However, local conditions of elevated salinity still occurred, as indicated by evaporitic minerals in sediments of this model. As in the playa lake model, the alluvial plain environment consists of alluvial fans, sand flats, and mudflats. The lacustrine environment consists of five distinct microfacies, each with composition, texture, and early diagenetic minerals reflecting local subenvironments (Fig. 17.12).

The most common types of offshore sediments are grain-supported pelecypod biocalcarenites with pseudomicrosparite matrix. These matrix-rich sediments were deposited below the wave base, but frequent intense breakage of pelecypod shells suggests *in situ* reworking by periodic storms without significant matrix winnowing. Of further interest are the offshore bioclastic bars of matrix-free, well-sorted pelecypod biocalcarenites deposited under high-energy conditions, whereas in more distal areas are located banks of bioaccumulated thin pelecypod shells with argillaceous micrite matrix.

The Ideal Subaqueous Basic Volcanism Model

Basic volcaniclastic rocks were formed by subaqueous extrusion of basaltic magma, which interacted with water and unconsolidated sediments to generate hyaloclastites, hyaloclastites with carbonate matrix, and hyalotuffs (Fig. 17.13). During late stages of volcanic episodes, kerolitic ooids were formed on top of the hyaloclastite system as a result of convective currents of heated lake water. The cortex of kerolitic ooids was formed by direct chemical precipitation in a Mg-rich and agitated environment.

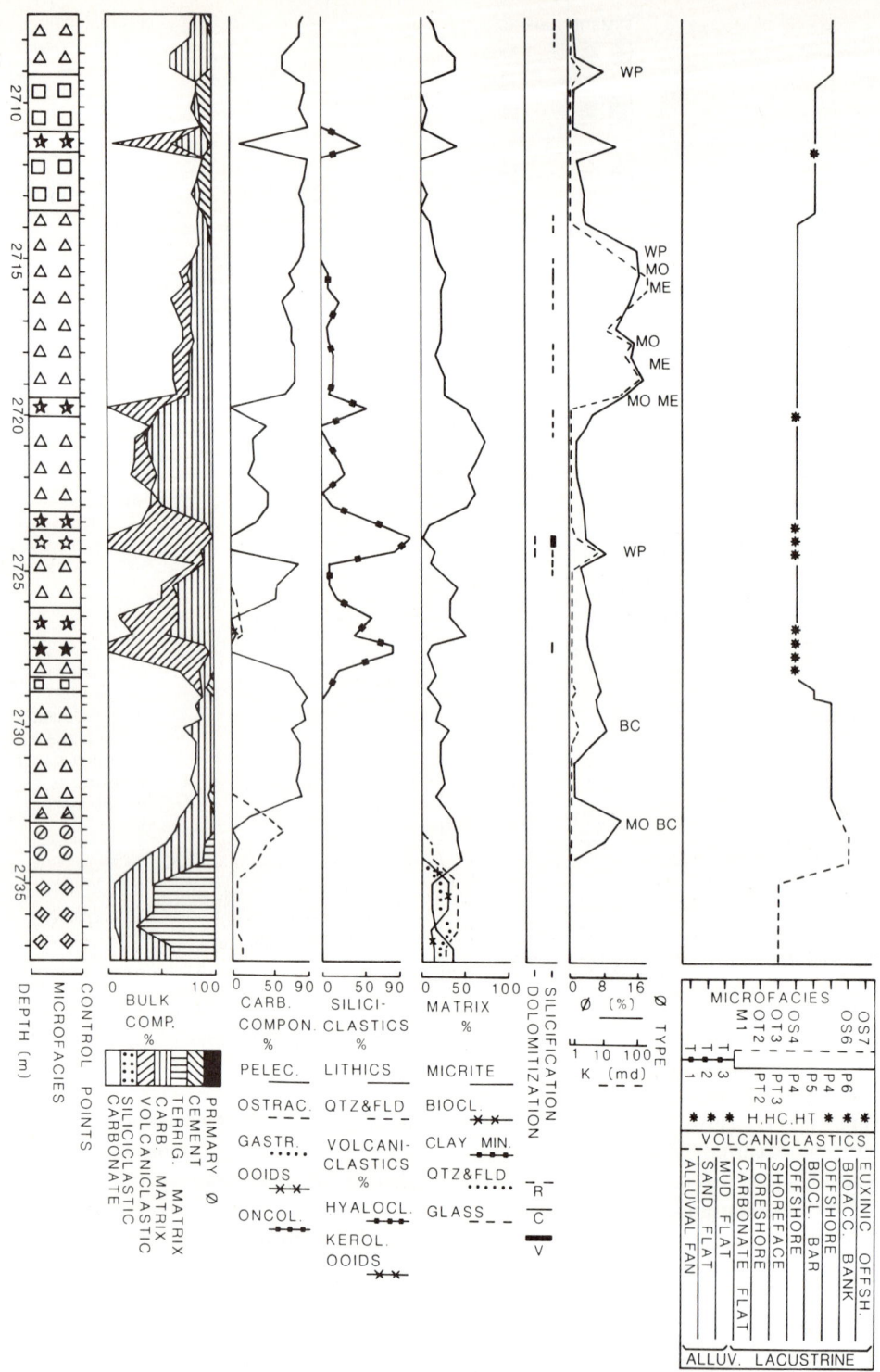

FIGURE 17.8 Lagoa Feia Formation (Lower Cretaceous), Campos Basin, offshore Rio de Janeiro, Brazil. Typical example of vertical variation of microscopic parameters in well 63-4. From Bertani and Carozzi (1984).

Chap. 17 / Lacustrine Carbonates with Bioaccumulated to Hydrodynamic Buildups

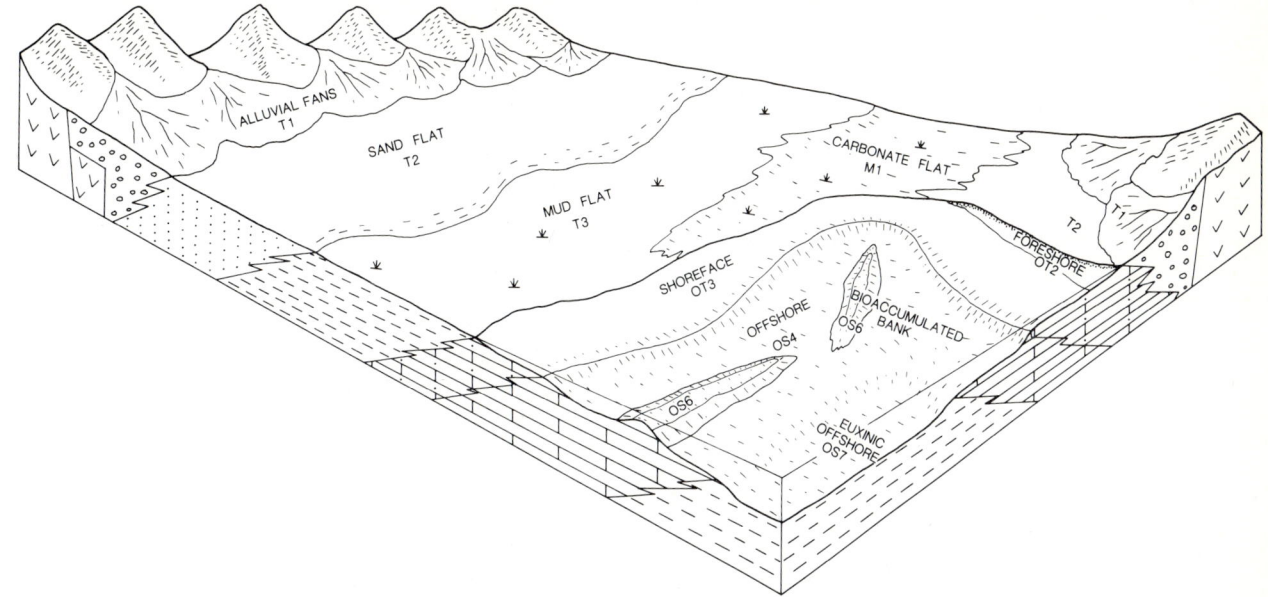

FIGURE 17.9 Lagoa Feia Formation (Lower Cretaceous), Campos Basin, offshore Rio de Janeiro, Brazil. Block diagram of ideal depositional model for playa lake system. From Bertani and Carozzi (1984).

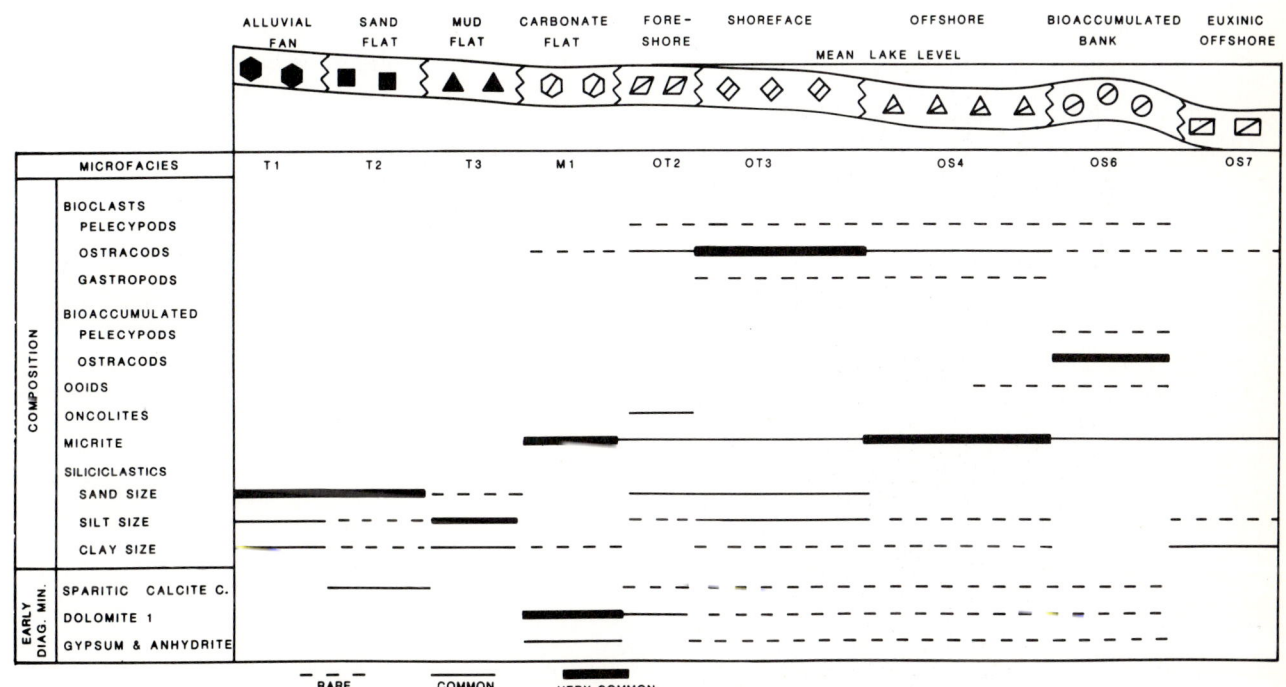

FIGURE 17.10 Lagoa Feia Formation (Lower Cretaceous), Campos Basin, offshore Rio de Janeiro, Brazil. Sediment composition and early diagenetic minerals of playa lake system. From Bertani and Carozzi (1984).

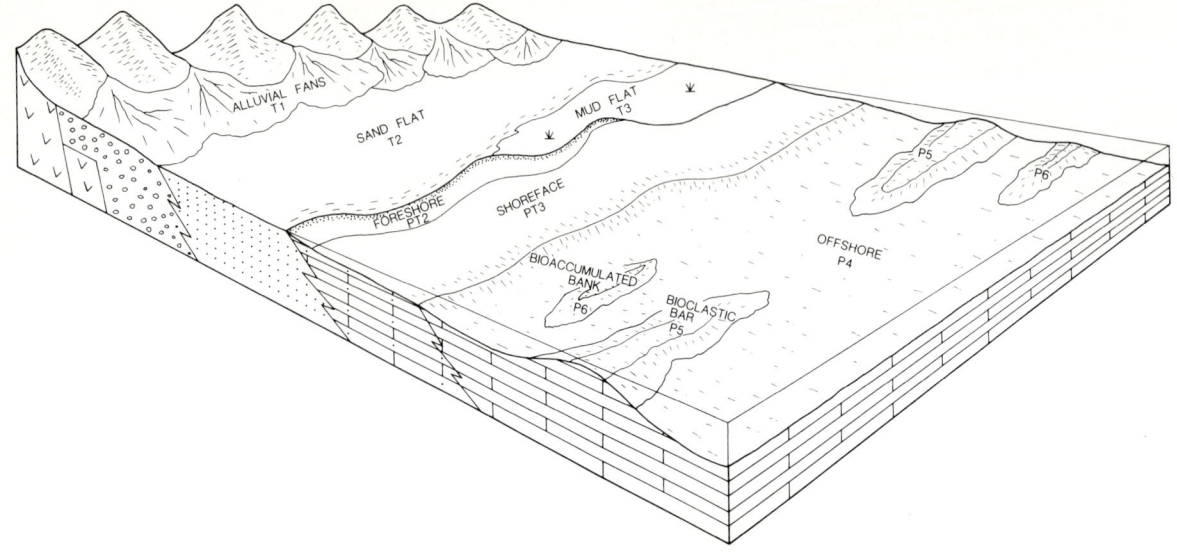

FIGURE 17.11 Lagoa Feia Formation (Lower Cretaceous), Campos Basin, offshore Rio de Janeiro, Brazil. Block diagram of ideal depositional model for pluvial lake system. From Bertani and Carozzi (1984).

Diagenesis and Porosity Evolution

Each of the four sequences of microfacies of the Lagoa Feia Formation described above presents its distinct diagenetic imprint related to pore water chemistry and mineralogic composition.

Due to climatic variations, the terrigenous sediments were submitted to alternating extensive dry periods (playa lake stage), and extensive rainy periods (pluvial stage), with corresponding diagenetic changes before burial (Fig. 17.14). Primary porosity remains unaltered during rainy period diagenesis, but significant amounts of moldic and intraparticle porosity can be generated. Diagenesis during dry periods destroys porosity effectively through calcite cementation. Little or no compaction took place during burial diagenesis in rocks previously ce-

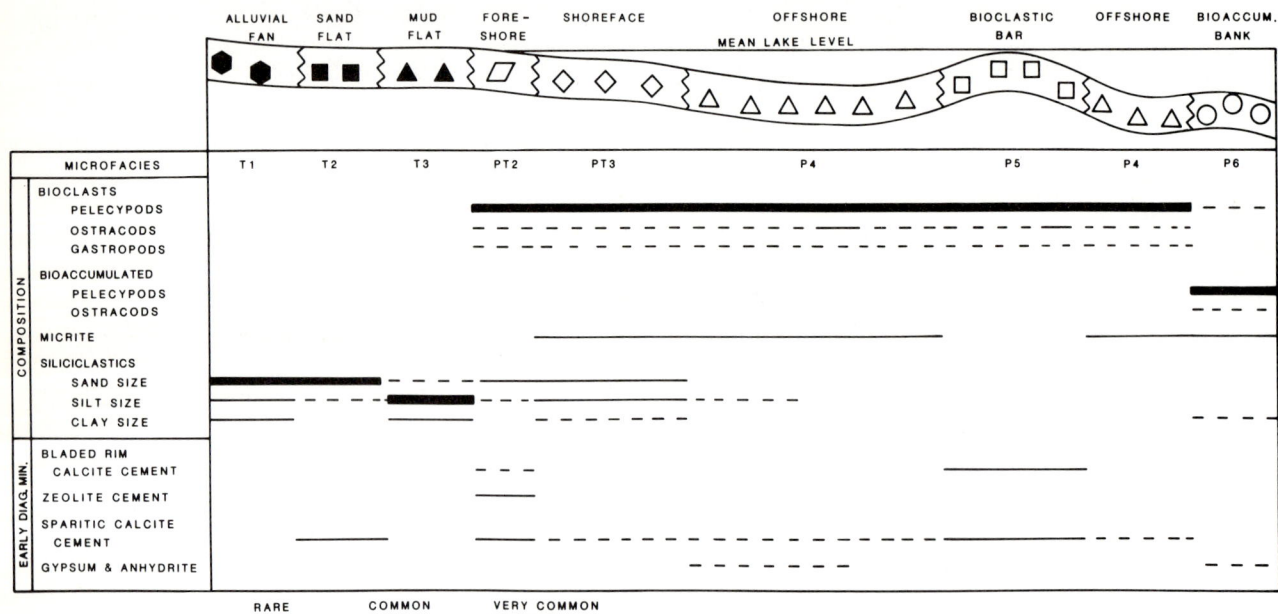

FIGURE 17.12 Lagoa Feia Formation (Lower Cretaceous), Campos Basin, offshore Rio de Janeiro, Brazil. Sediment composition and early diagenetic minerals of pluvial lake system. From Bertani and Carozzi (1984).

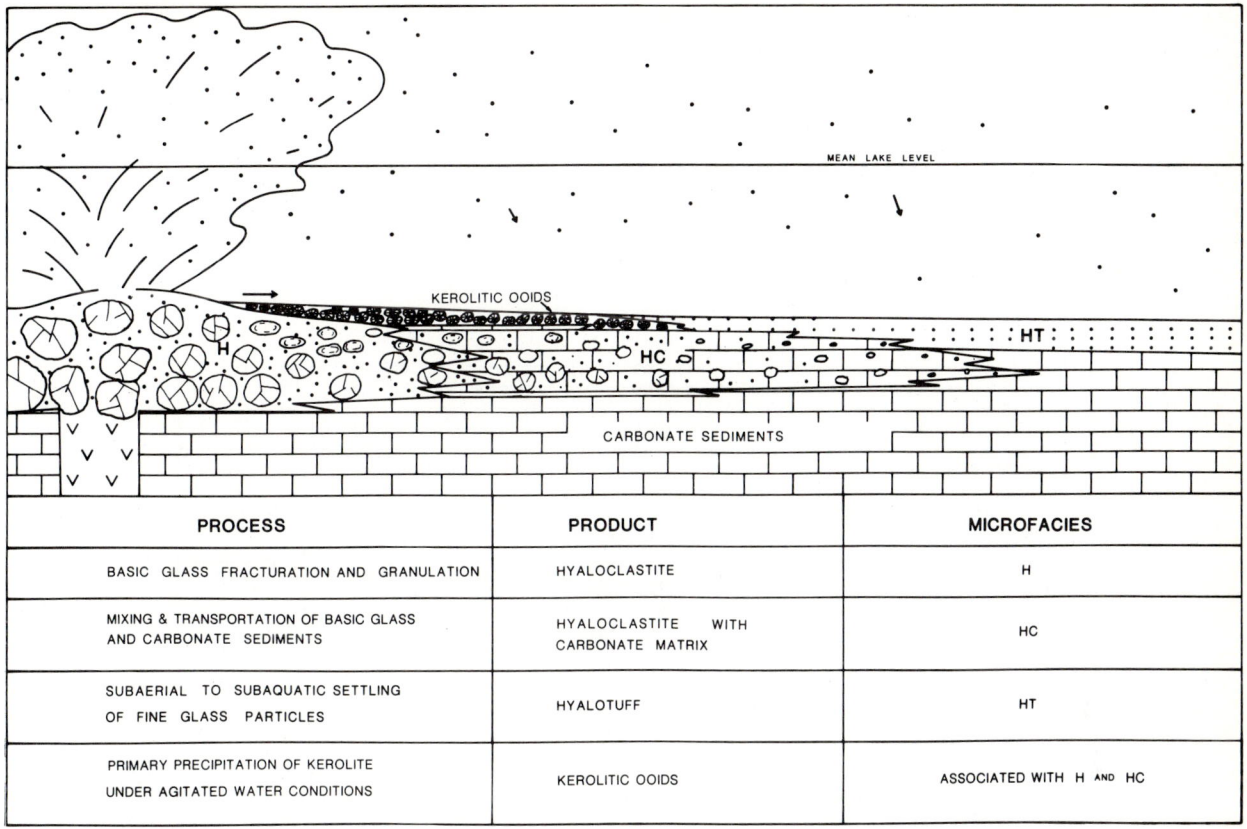

FIGURE 17.13 Lagoa Feia Formation (Lower Cretaceous), Campos Basin, offshore Rio de Janeiro, Brazil. Ideal depositional model for subaqueous basic volcanism. From Bertani and Carozzi (1984).

mented by calcite or calcitized. They represent the best potential to become reservoir rocks upon burial through calcite dissolution by subsurface brines charged with CO_2 derived from hydrocarbon maturation.

Numerous processes operated on the ostracod-rich sediments in distinct diagenetic environments. The evolution of the latter followed various pathways related to climatically controlled lake-level and water-table oscillations combined with basin subsidence (Fig. 17.15). The succession of diagenetic processes (Fig. 17.16) illustrates the complexity of the ostracod-dominated sequence, which is characterized by a larger proportion of carbonate and terrigenous matrix in all its microfacies.

The initial larger primary porosity consisted mainly of intercrystal micropores within matrix and to a lesser extent of ostracod internal cavities not filled by matrix. During diagenesis, this primary porosity was almost completely destroyed (Fig. 17.16) through compaction (in the lacustrine phreatic, vadose freshwater and burial environments) and through matrix neomorphism and cementation by sparitic calcite (in the freshwater phreatic environment).

Sediments exposed to vadose freshwater conditions developed moldic, moldic enlarged, and vuggy porosity (Fig. 17.16). At this stage, sediments with abundant more soluble components (pelecypod bioclasts, evaporites, glass fragments) may have reached improved porosity. However, most of these pores were obliterated by sparitic calcite cement precipitated during subsequent freshwater phreatic diagenesis.

Secondary pores preserved from cementation are usually isolated, connected only by secondary intercrystal microporosity developed in the matrix altered by neomorphism.

The factors controlling the evolution of the diagenetic environments of the pelecypod-dominated sequence are the same as those of the ostracod-dominated sequence (Fig. 17.17). The succession of the diagenetic processes (Fig. 17.18) illustrates the even greater complexity of the pelecypod-dominated sequence. Of particular interest are the several stages of pedogenesis developed in the freshwater vadose zone. They range from micritization of sparite (sparmicritization) by endolithic organisms to formation of calcretes with abundant rootlets and pisolitization.

Primary depositional porosity of the pelecypod microfacies was modified greatly by several diagenetic processes which led to its destruction (Fig. 17.18). Pri-

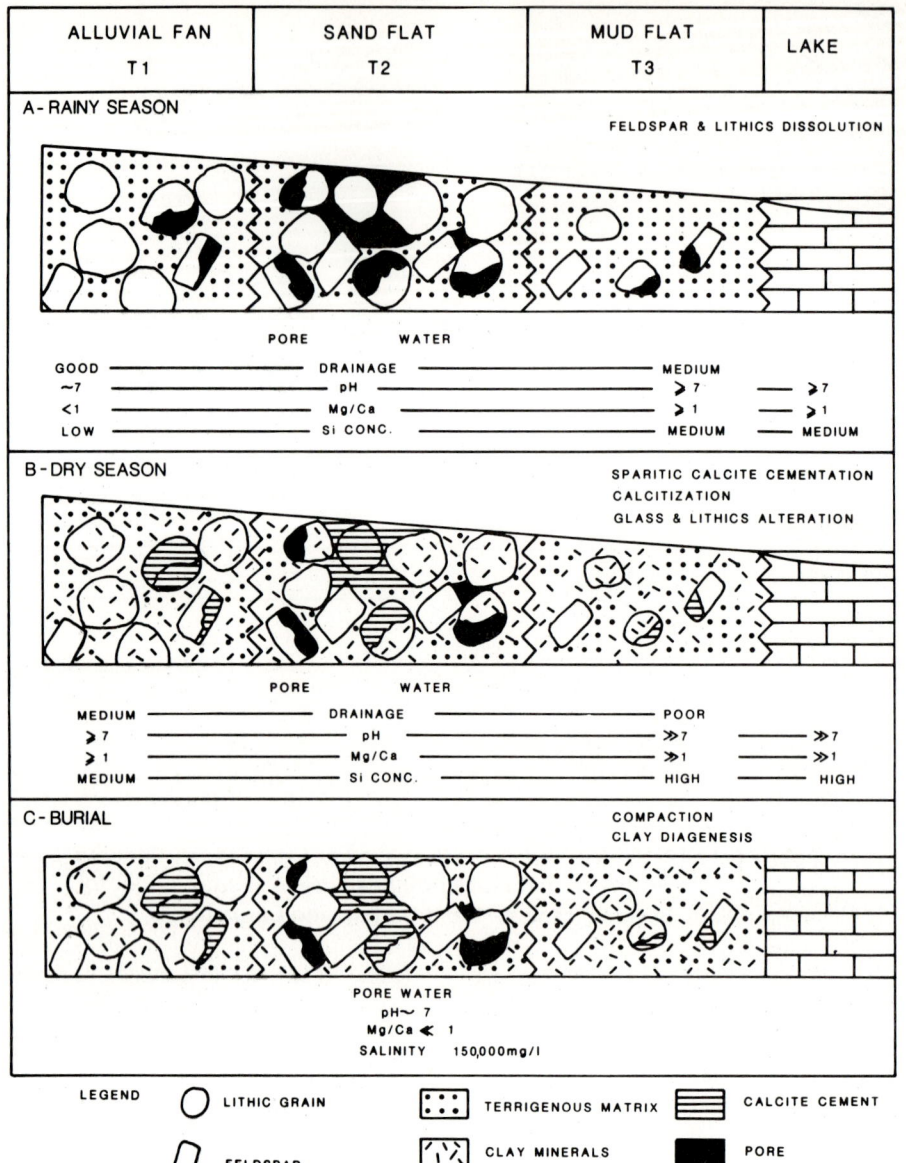

FIGURE 17.14 Lagoa Feia Formation (Lower Cretaceous), Campos Basin, offshore Rio de Janeiro, Brazil. Geochemical model for the diagenesis of the terrigenous dominated sequence. From Bertani and Carozzi (1984).

mary intercrystal porosity tended to be destroyed by early compaction, neomorphism, and late compaction. Primary interparticle porosity was less susceptible to destruction by early compaction, but cementation in the active lacustrine phreatic and freshwater phreatic environments partially or completely obliterated it.

Well-developed intraparticle, moldic, and moldic enlarged porosity was developed during vadose freshwater dissolution, but extensive cementation occluded it when sediments were submitted to freshwater phreatic conditions.

The ideal conditions for generation and preservation of reservoir rocks consisted of extensive exposure of the sediments through lowering of lake level, followed by rapid lake expansion, and sediment burial. This diagenetic pathway was followed by porous layers of pelecypod microfacies intercalated in less porous ones. It favored generation of considerable secondary porosity (moldic, moldic enlarged, and vuggy) and prevented or minimized the effects of freshwater phreatic conditions (Fig. 17.19).

To understand fully the relationships between mea-

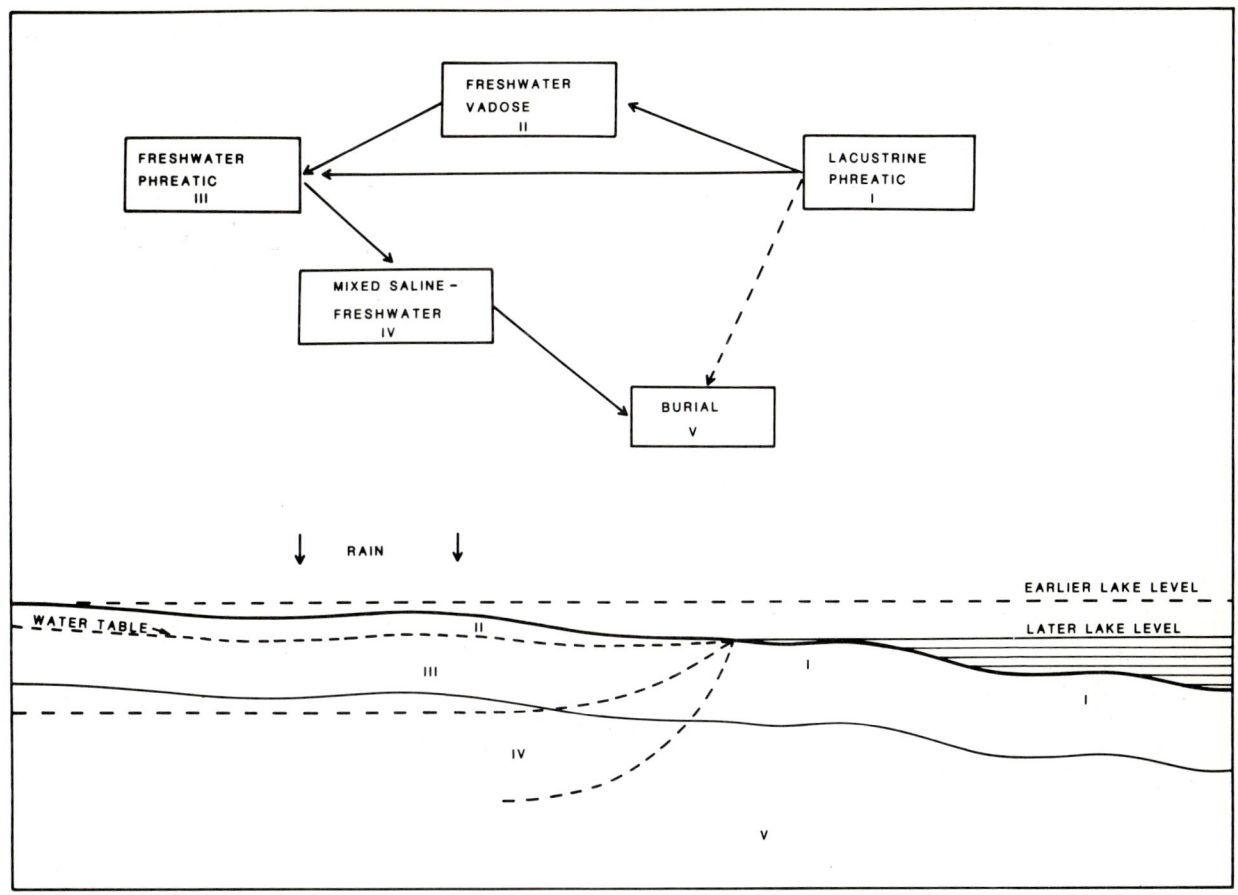

FIGURE 17.15 Lagoa Feia Formation (Lower Cretaceous), Campos Basin, offshore Rio de Janeiro, Brazil. Evolution of diagenetic environments for the ostracod dominated sequence. From Bertani and Carozzi (1984).

sured porosity and the ostracod- and pelecypod-dominated microfacies, the latter were grouped according to dominant biogenic content and proportion of matrix as follows:

Group 1—Pelecypod dominated, matrix-free: well-sorted pelecypod biocalcarenites and lithic arenites rich in pelecypod bioclasts (microfacies P5 and PT2)

Group 2—Pelecypod dominated, matrix-rich: pelecypod biocalcarenites with pseudomicrosparite matrix, and arenaceous pelecypod biocalcarenites with micrite and terrigenous matrix (microfacies P4 and PT3)

Group 3—Pelecypod (bioaccumulated) dominated, matrix-rich: bioaccumulated pelecypod limestones with abundant argillaceous micrite matrix (microfacies P6)

Group 4—Ostracod dominated, matrix-rich: micrites, biocalcisiltites, matrix-supported biocalcarenites, bioaccumulated limestones, and bioclast-rich siltstones, characterized by a high proportion of micrite matrix and ostracod shells (microfacies M1, OT3, OS4, and OS6)

The frequency distributions of measured porosities for these groups (Fig. 17.20) are similar and characterized by a peak between 2 and 6% and a skewness toward low values. Porosity values greater than 12% were observed only in pelecypod-rich microfacies of groups 1 and 2. The tendency to bimodality observed in the frequency distribution curve of group 3 is probably due to the small number of measured samples. Analysis of variance of the frequency distributions reveals no significant differences among the porosity means of the four groups of microfacies, at a 95% confidence level.

Despite the similarity in average porosities, the four groups of microfacies differ in the dominant porosity types (Fig. 17.21).

In high-energy pelecypod biocalcarenites and bioclast-rich lithic arenites (group 1, microfacies PT2 and

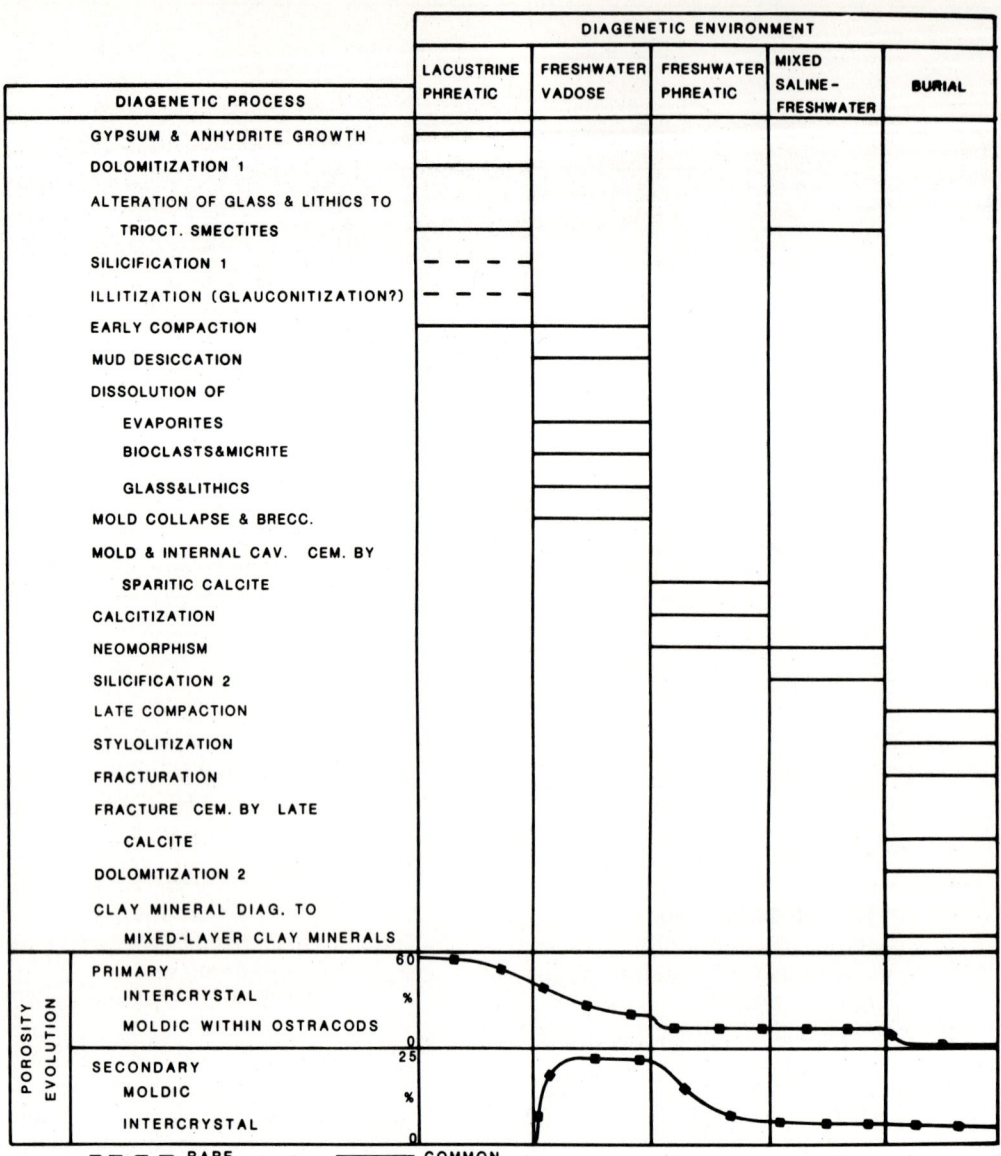

FIGURE 17.16 Lagoa Feia Formation (Lower Cretaceous), Campos Basin, offshore Rio de Janeiro, Brazil. Generalized diagenetic sequence of the ostracod dominated sequence. From Bertani and Carozzi (1984).

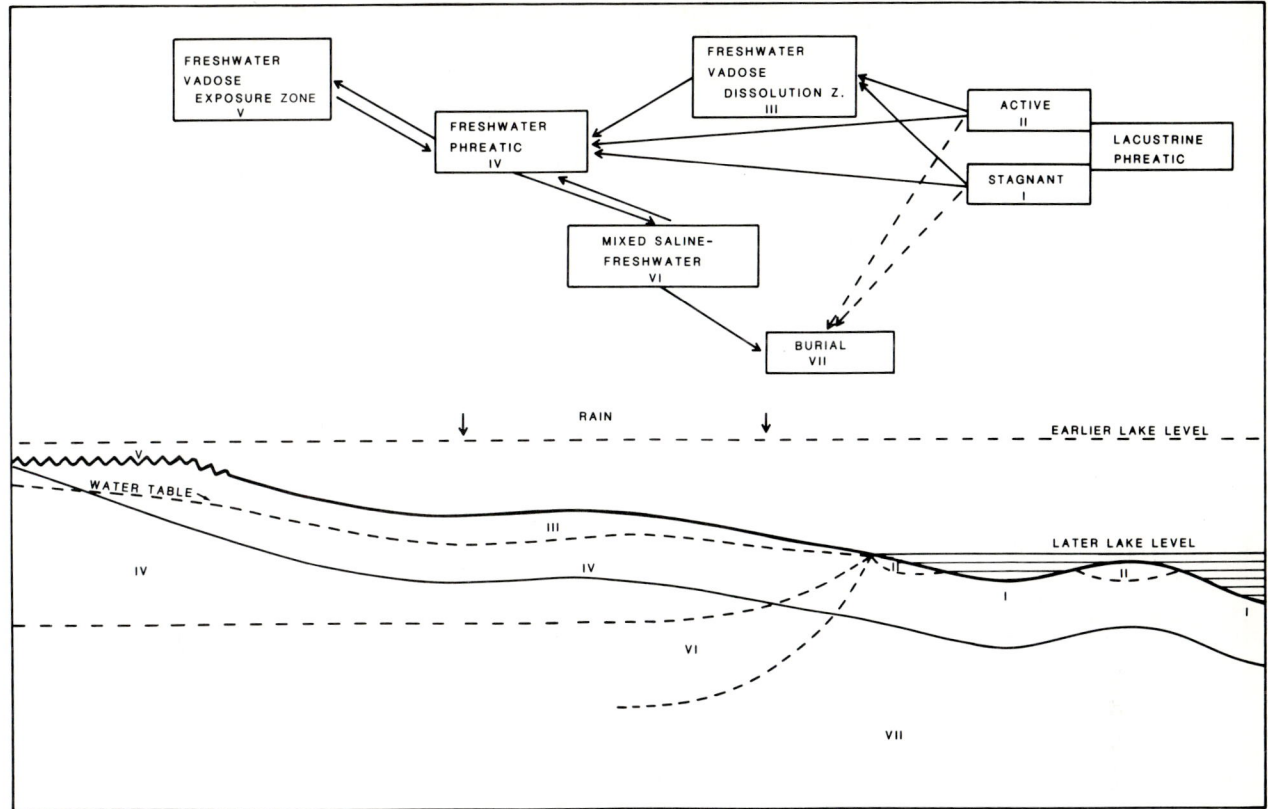

FIGURE 17.17 Lagoa Feia Formation (Lower Cretaceous), Campos Basin, offshore Rio de Janeiro, Brazil. Evolution of diagenetic environments for the pelecypod dominated sequence. From Bertani and Carozzi (1984).

P5), primary interparticle porosity is the dominant type and secondary intraparticle and moldic pores from dissolution of bioclasts and lithic grains are common.

In pelecypod-rich, matrix-rich microfacies (group 2, microfacies PT3 and P4), moldic, moldic enlarged, and intraparticle pores are the major contributors to total rock porosity. Intercrystal microporosity within carbonate matrix and micrite envelopes is common in microfacies P4.

In the bioaccumulated pelecypod limestones (group 3, microfacies P6) and in all ostracod-dominated microfacies (group 4, microfacies OS4, OS6, OT3, and M1), intercrystal microporosity within pseudomicrosparite matrix is the dominant type. In group 3, moldic and moldic enlarged pores developed by dissolution of pelecypod shells and micrite matrix are common, whereas in group 4, moldic porosity was formed mainly through dissolution of lithic grains and basic glass.

The basic volcaniclastic-dominated sequence displays four stages of diagenesis (Fig. 17.22): lava-water-sediment interaction, aquathermal diagenesis, lacustrine phreatic diagenesis by influence of normal lake water, and finally, freshwater vadose and phreatic diagenesis under the action of meteoric water. No porosity was generated during this evolution.

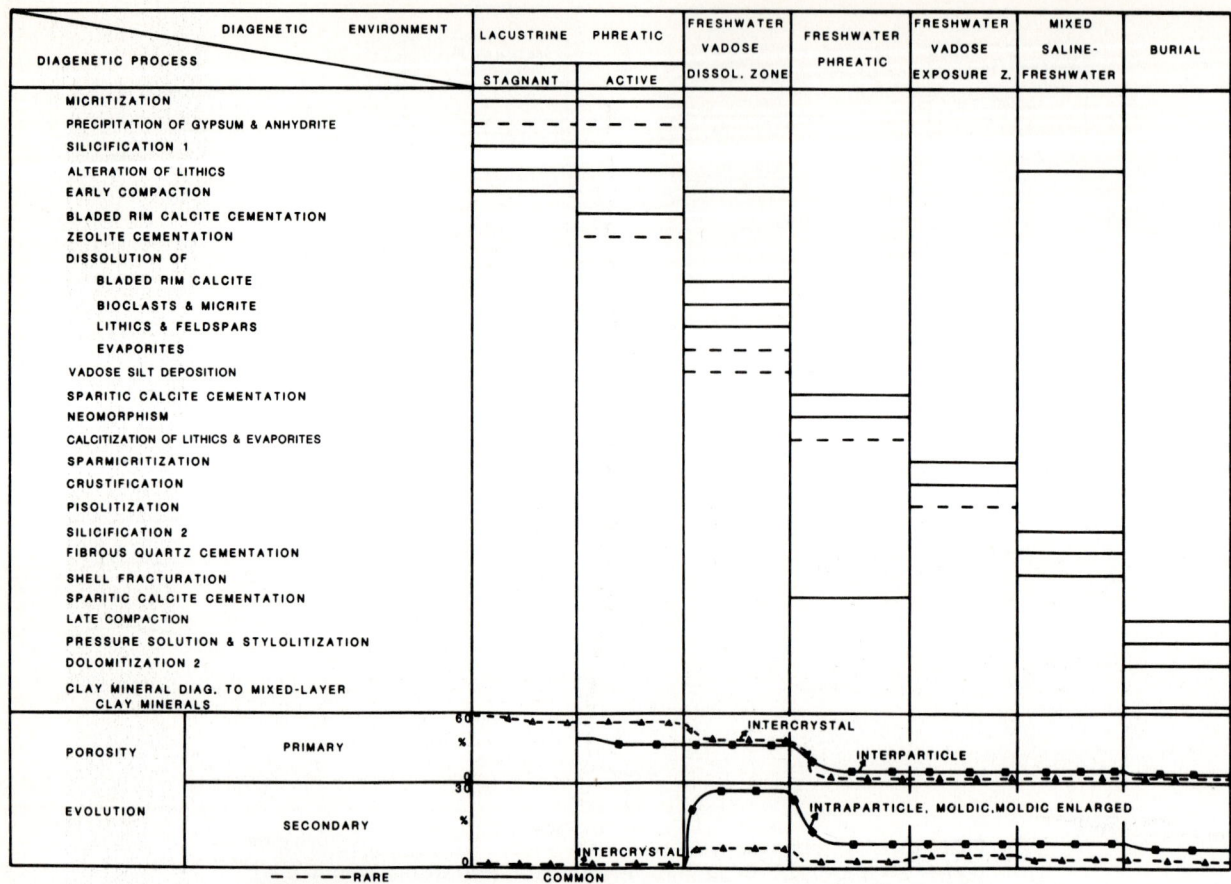

FIGURE 17.18 Lagoa Feia Formation (Lower Cretaceous), Campos Basin, offshore Rio de Janeiro, Brazil. Generalized diagenetic sequence of the pelecypod dominated sequence. From Bertani and Carozzi (1984).

FIGURE 17.19 Lagoa Feia Formation (Lower Cretaceous), Campos Basin, offshore Rio de Janeiro, Brazil. Illustration of diagenetic processes and types of porosity. A. Pelecypod microfacies. Primary interparticle porosity (in black) reduced by subhedral crystals of sparite subsequently silicified. B. Pelecypod microfacies. Early cementation by bladed rim calcite showing short prismatic habit and rhombohedral terminations (arrow). C. Pelecypod microfacies. Partial dissolution of bladed calcite cement (arrow), coarse sparite cement and neomorphosed pelecypod shells. D. Pelecypod microfacies. Primary interparticle porosity (in dark) reduced by bladed rim calcite cement. E. Lithic arenite microfacies. Primary interparticle porosity (in dark) reduced by isopachous zeolite cement (arrow). F. Pelecypod microfacies. Intraparticle porosity (in black) formed by selective dissolution of layers inside pelecypod shells. G. Pelecypod microfacies. Moldic porosity (in dark) formed by dissolution of well-rounded pelecypod bioclasts. H. Lithic arenite microfacies. Moldic porosity (in black) formed by dissolution of rounded volcaniclastic lithic grains. All photomicrographs: plane-polarized light. From Bertani and Carozzi (1984).

Chap. 17 / Lacustrine Carbonates with Bioaccumulated to Hydrodynamic Buildups 513

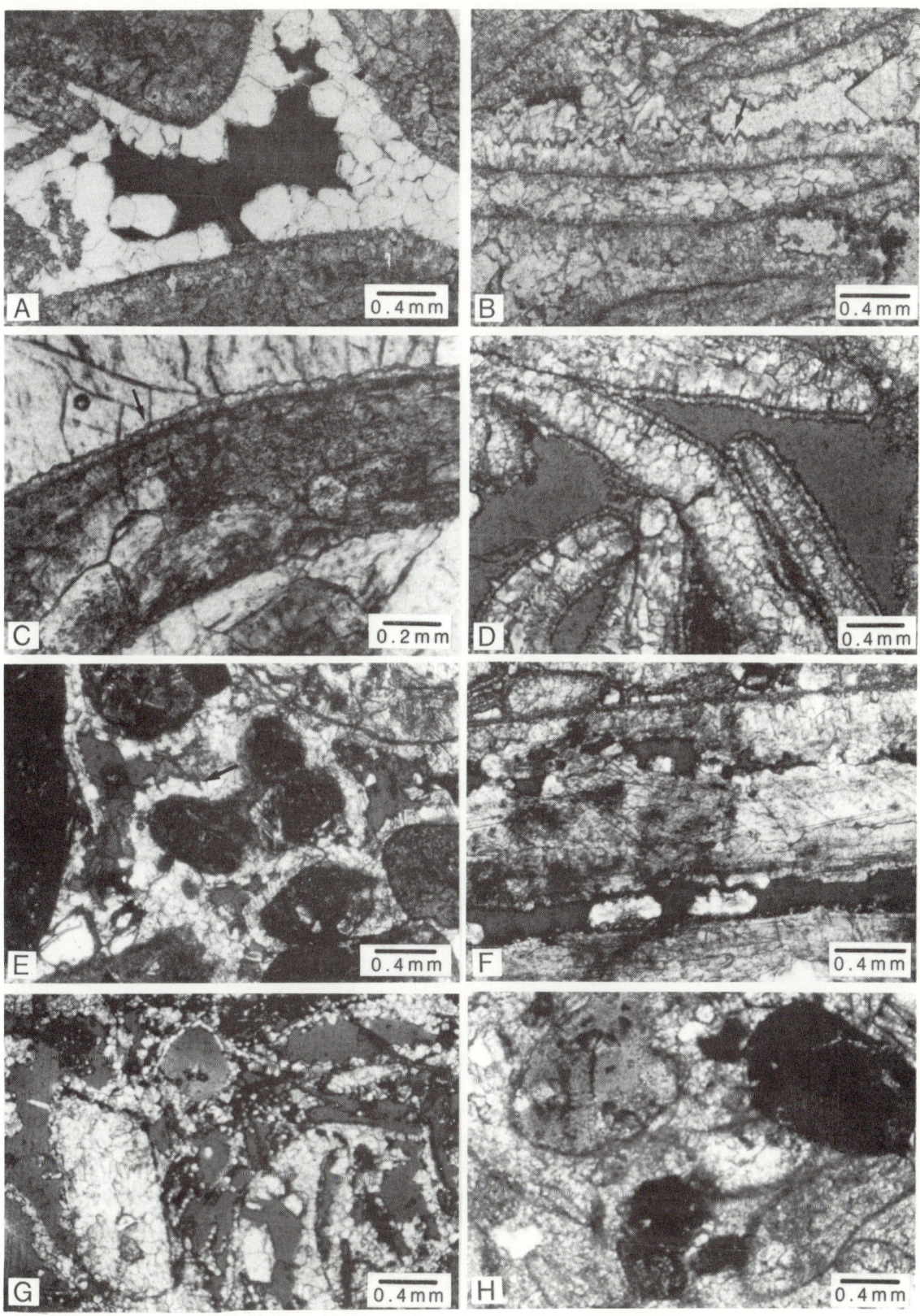

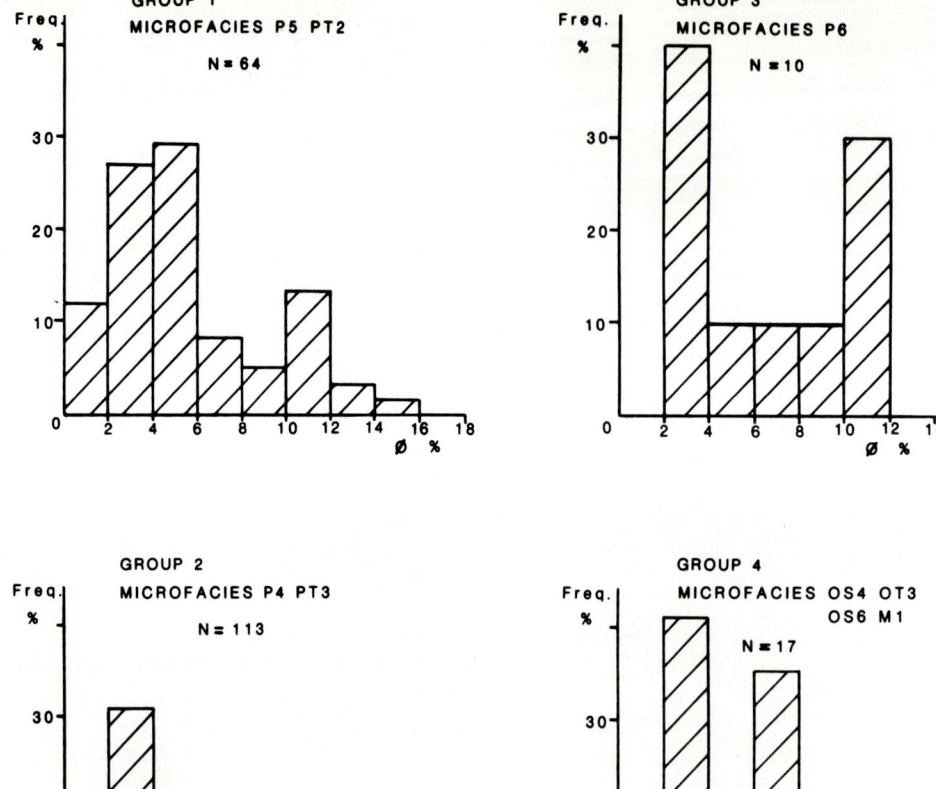

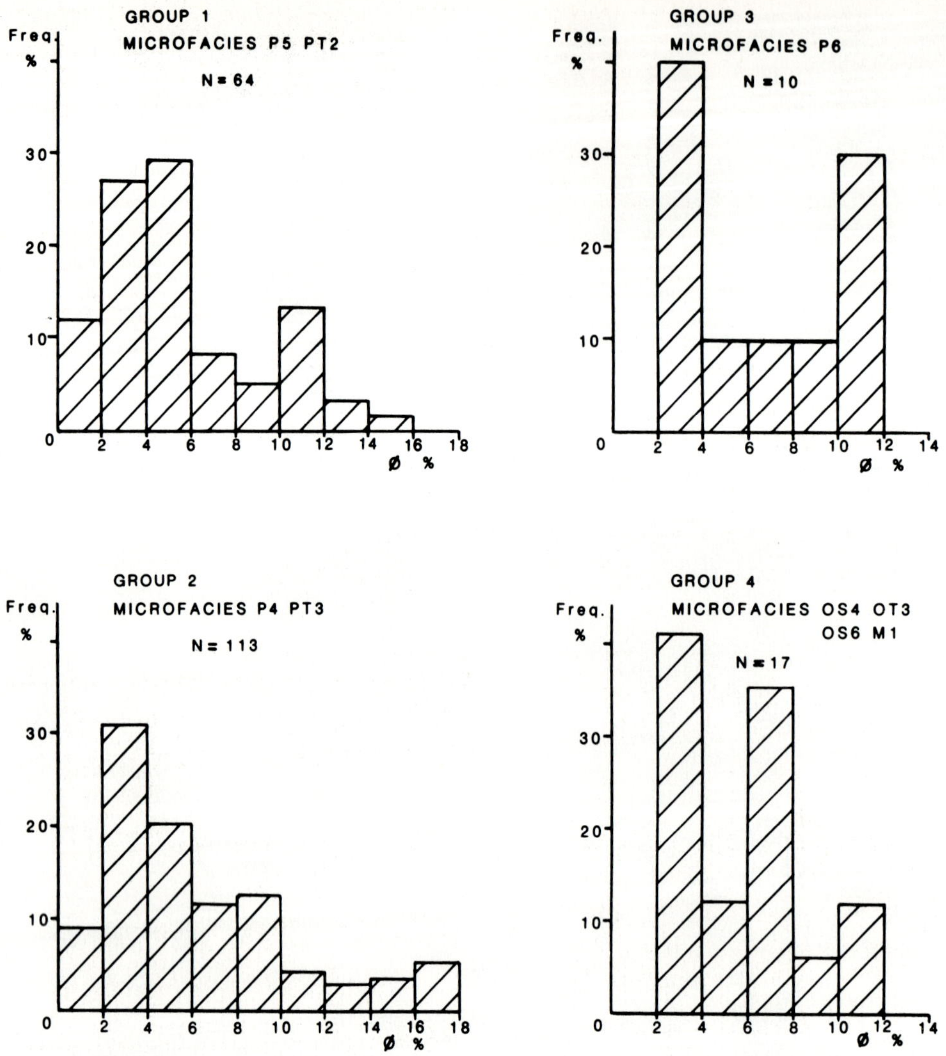

FIGURE 17.20 Lagoa Feia Formation (Lower Cretaceous), Campos Basin, offshore Rio de Janeiro, Brazil. Frequency distribution of measured porosities in ostracod and pelecypod dominated microfacies. From Bertani and Carozzi (1984).

FIGURE 17.22 Lagoa Feia Formation (Lower Cretaceous), Campos Basin, offshore Rio de Janeiro, Brazil. Evolution of diagenetic environments for the basic volcaniclastic sequence. From Bertani and Carozzi (1984).

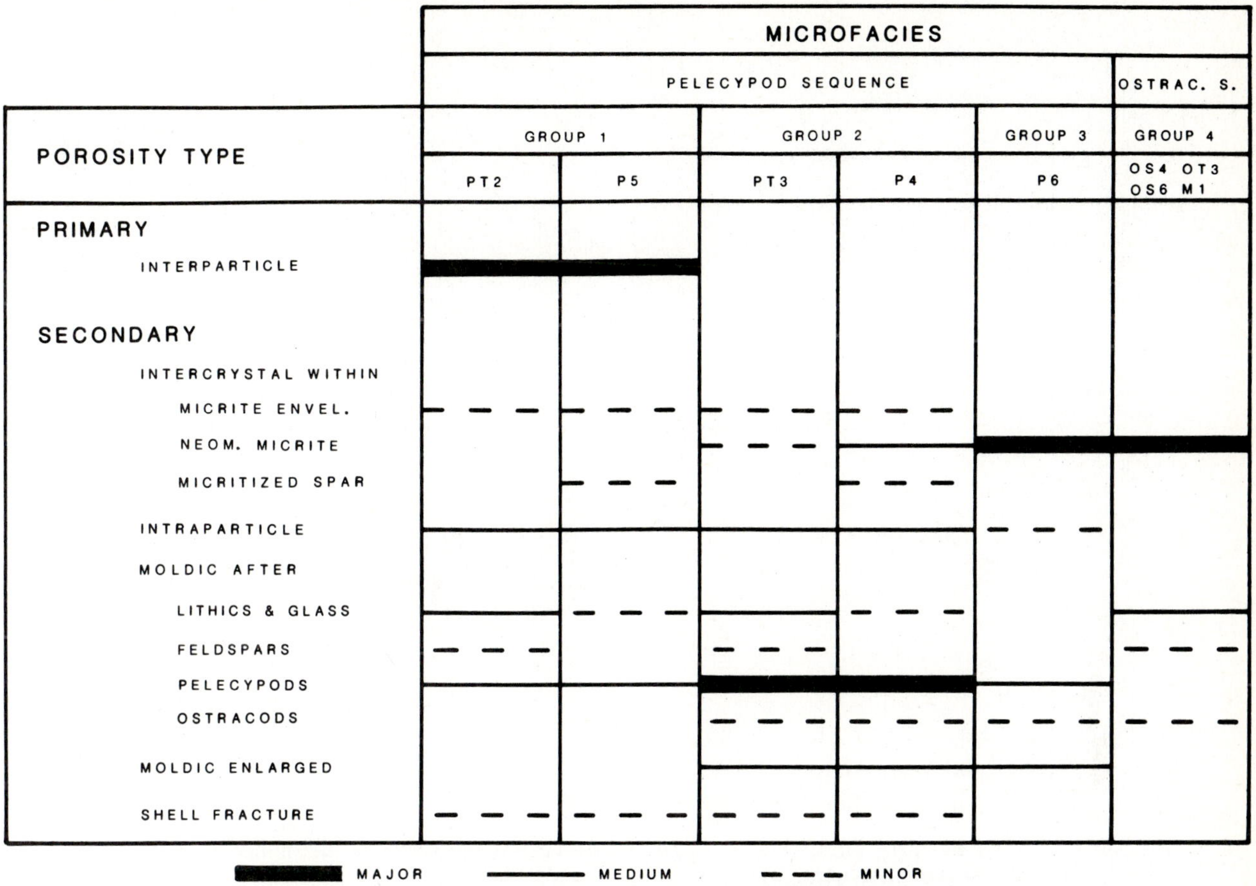

Chap. 17 / Lacustrine Carbonates with Bioaccumulated to Hydrodynamic Buildups 515

		MICROFACIES					
		PELECYPOD SEQUENCE					OSTRAC. S.
POROSITY TYPE		GROUP 1		GROUP 2		GROUP 3	GROUP 4
		PT2	P5	PT3	P4	P6	OS4 OT3 OS6 M1
PRIMARY							
	INTERPARTICLE	■■■ MAJOR	■■■ MAJOR				
SECONDARY							
	INTERCRYSTAL WITHIN MICRITE ENVEL.	---	---	---	---	■■■	
	NEOM. MICRITE			---		■■■	
	MICRITIZED SPAR		---		---		
	INTRAPARTICLE					---	
	MOLDIC AFTER						
	LITHICS & GLASS		---	---	---		
	FELDSPARS	---		---	---		---
	PELECYPODS			■■■	■■■		
	OSTRACODS			---	---	---	---
	MOLDIC ENLARGED						
	SHELL FRACTURE	---	---	---	---		

■■■ MAJOR   ——— MEDIUM   - - - MINOR

FIGURE 17.21 Lagoa Feia Formation (Lower Cretaceous), Campos Basin, offshore Rio de Janeiro, Brazil. Relative contribution of porosity types to total porosity. From Bertani and Carozzi (1984).

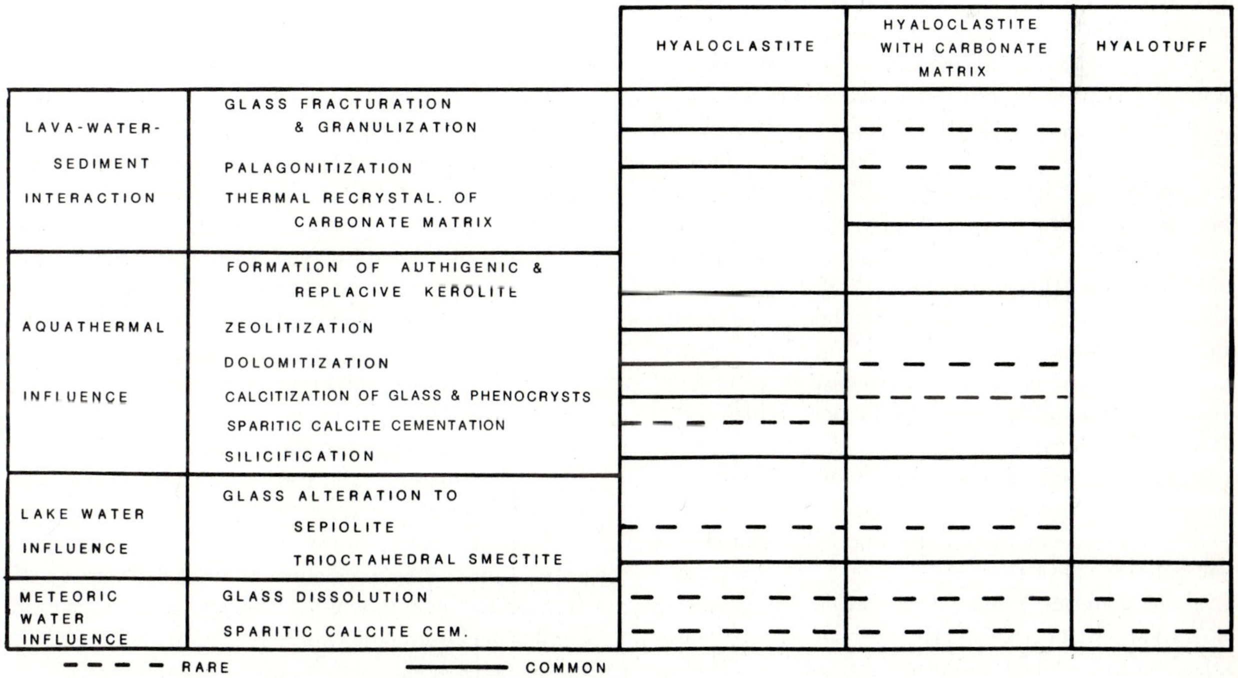

		HYALOCLASTITE	HYALOCLASTITE WITH CARBONATE MATRIX	HYALOTUFF
LAVA-WATER-SEDIMENT INTERACTION	GLASS FRACTURATION & GRANULIZATION		---	
	PALAGONITIZATION		---	
	THERMAL RECRYSTAL. OF CARBONATE MATRIX			
AQUATHERMAL INFLUENCE	FORMATION OF AUTHIGENIC & REPLACIVE KEROLITE	———		
	ZEOLITIZATION			
	DOLOMITIZATION		---	
	CALCITIZATION OF GLASS & PHENOCRYSTS		---	
	SPARITIC CALCITE CEMENTATION	---		
	SILICIFICATION			
LAKE WATER INFLUENCE	GLASS ALTERATION TO SEPIOLITE	---	---	
	TRIOCTAHEDRAL SMECTITE			
METEORIC WATER INFLUENCE	GLASS DISSOLUTION	---	---	---
	SPARITIC CALCITE CEM.		---	

- - - RARE   ——— COMMON

18

Synthesis

18.1 INTRODUCTION TO THE GENETIC CLASSIFICATION OF CARBONATE ROCK DEPOSITIONAL MODELS

The genetic classification presented here is based on a selected number of previously described case histories. It is organized in a process-oriented manner which emphasizes the role of the major depositional and diagenetic factors. Indeed, the natural evolution of carbonate depositional systems begins with various types of ramps which are modified subsequently by incipient to well-developed buildups of a bioaccumulated, hydrodynamic, or bioconstructed nature. Eventually, these buildups form the frontal structures of broad carbonate platforms which display several rows of more internal buildups, complex lagoons, and tidal flats. This evolution is characterized by the following features: an expansion of carbonate sedimentation over cratonic areas as a result of increased submergence of the craton, a general increase in the energy (bedshear) level of the depositional models which corresponds to an increasing preservational effect of tidal processes, and a general increase in the morphological complexity of the seafloor.

The first feature is a direct function of the interaction between subsidence and eustatic oscillations of sea level and is treated in more detail later in the discussion of a new eustatic model. The second feature, which expresses the gradual dominance of tidal depositional systems as the width of the inundated craton increases, parallels cratonic siliciclastic depositional systems as emphasized by Klein (1982). The third feature is intrinsic to the accumulation and construction activities of carbonate-secreting benthic faunas. In its incipient stage of hydrodynamic buildups, it could be compared to subtidal siliciclastic sand bodies.

Whether this evolution of carbonate depositional systems is completed depends on the interplay of many variables, still not clearly understood, among which the most important appear to be the rate of subsidence, eustatic oscillations of sea level, tectonism, rate of calcium carbonate productivity of benthic communities, types of organisms which as a result of Phanerozoic evolutionary processes have displayed great variations in their accumulation and construction capabilities through time (see Figs. 1.3 and 1.11), and finally, regional climate.

Depositional models in general are assumed (Walker, 1984) to fulfill four major functions, namely to serve as a norm for comparative purposes, as a framework and guide for future studies, as a predictor in new geological situations, and finally, as an integrated basis for environmental interpretation.

Words alone, however, cannot express clearly norms, frameworks, predictors, or integrated bases (hereafter called "integrations"). This synthesis is therefore based on a new graphic expression (Figs. 18.1 to 18.49). This expression integrates for each graphic representation

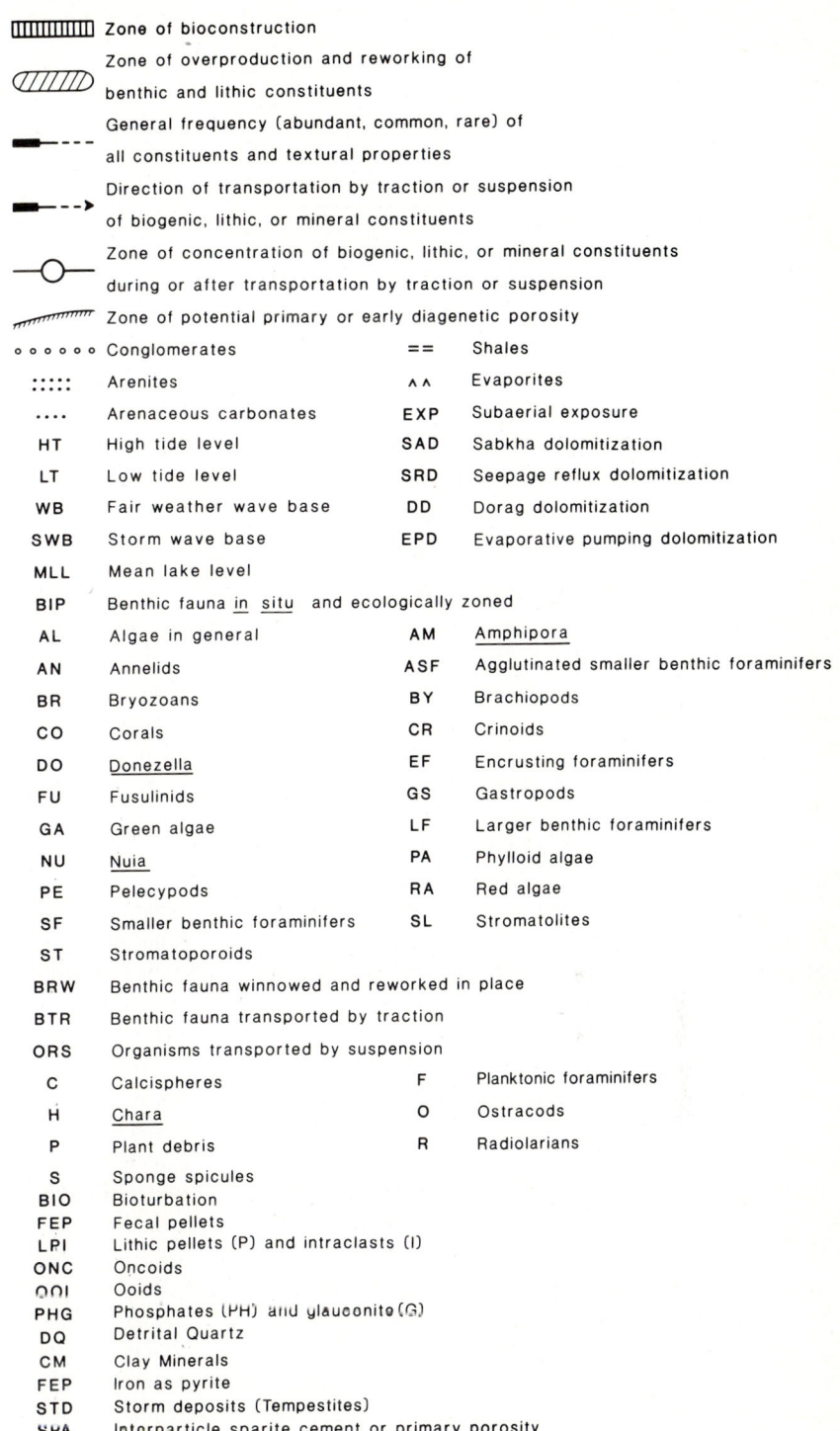

FIGURE 18.1 Explanatory symbols for integrated depositional models.

the following fundamental features from top to bottom (see Fig. 18.1 for explanatory symbols).

At the top is drafted the shape of the submarine topography, with coastline toward the right, as well as the position of high tide (HT), low tide (LT), fair weather wave base (WB), and storm wave base (SWB). Features of subaerial exposure are designated as EXP.

Below this, microfacies and their horizontal extent are indicated by the same numbers or letters used in the description of case histories and are located immediately beneath submarine topographic profiles.

Processes of dolomitization are represented by arrows beneath microfacies designations that indicate the direction of flow of dolomitizing fluids. They are divided

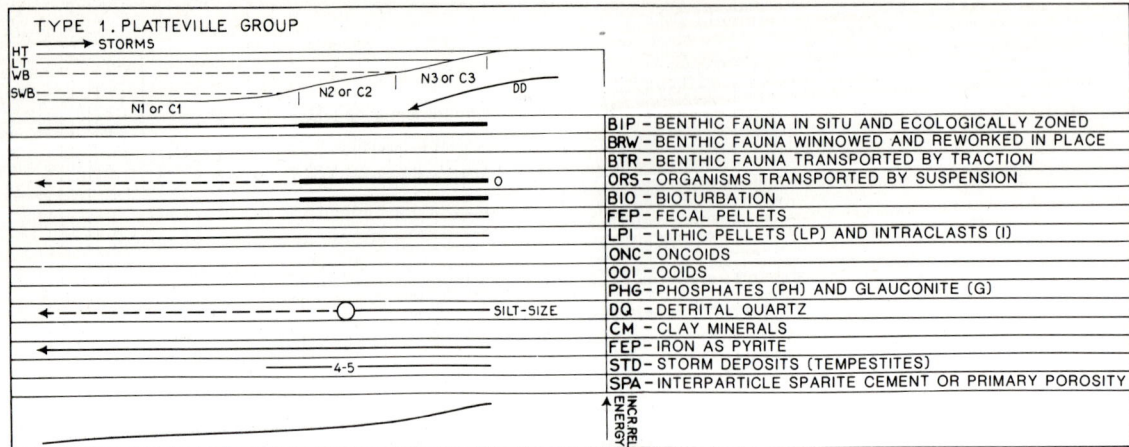

FIGURE 18.2 Integration of depositional models of simple ramps. Type 1. Platteville Group.

into four major types: sabkha dolomitization (SAD), seepage reflux dolomitization (SRD), dorag dolomitization (DD), and evaporative pumping dolomitization (EPD).

Further down in each graphic representation are *in situ* ecologically zoned benthic faunal constituents, namely, autochthonous faunas (BIP), bioturbation (BIO), fecal pellets (FEP), lithic pellets and intraclasts (LPI), oncoids (ONC), ooids (OOI), phosphates and glauconite (PHG), storm deposits (STD), and interparticle sparite cement or primary porosity (SPA). Their relative frequency (abundant, common, and rare) and lateral extent are represented by bars (see Fig. 18.1 for explanatory symbols). Areas of bioconstruction are designated by rectangular boxes with vertical ruling and abbreviations indicating the organisms involved. Benthic constituents winnowed and reworked *in situ* are labeled BRW.

Benthic constituents that have undergone transportation by traction, both landward and seaward, are named BTR, and benthic and planktonic constituents transported in suspension, ORS. Both categories are represented by arrows expressing their relative frequency (abundant, common, and rare) and direction of transportation (see Fig. 18.1 for explanatory symbols). Abbreviations indicate the organisms involved. Zones of overproduction from which distribution of transported biogenic constituents takes place are represented by ellipsoids with oblique ruling and abbreviations indicating the organisms involved. If concentrations of any constituents occurred during or after transportation by traction or suspension, such zones are designated by circles with abbreviations indicating the organisms involved.

Below the benthic and planktonic constituents are the intrabasinal and extrabasinal lithic and mineral constituents represented by arrows; these include fecal pellets, lithic pellets and intraclasts, oncoids, ooids, phosphates, glauconite, detrital quartz (DQ), clay minerals (CM), and iron as pyrite (FEP). These constituents have undergone transportation by traction or suspension and arrows express their relative frequency (abundant, common, and rare) and direction of transportation with abbreviations indicating the constituents involved (see Fig. 18.1 for explanatory symbols). Zones of intraformational reworking from which distribution takes place are represented by ellipsoids with oblique ruling, and abbreviations indicate the constituents involved. Zones of concentration during or after transportation by traction or suspension are designated by circles with abbreviations indicating the constituents involved.

In the lower part of each graphic representation are drafted variations of relative energy of the depositional environment by means of a curve with energy increasing upward. In Figure 18.2, the first example of integration, the names of all constituents are spelled out, whereas only abbreviations are used thereafter (Figs. 18.3 to 18.49).

Furthermore, particular diagrams (Figs. 18.12, 18.18, 18.22, 18.37, 18.47) show the various submarine topographic profiles superposed in a general shallowing order for each group of ramps or platforms. The resulting families of curves act as final predictors for potential reservoir conditions (zones with oblique ruling), expressed by the occurrence of primary porosity (or its interparticle sparite cement filling) combined with possible enhancement due to secondary porosity introduced by the various types of early diagenetic dolomitization. In spite of oversimplification, the diagrams represent, nevertheless, the most efficient manner of displaying the generalized interplay of depositional and early diagenetic features of carbonate models.

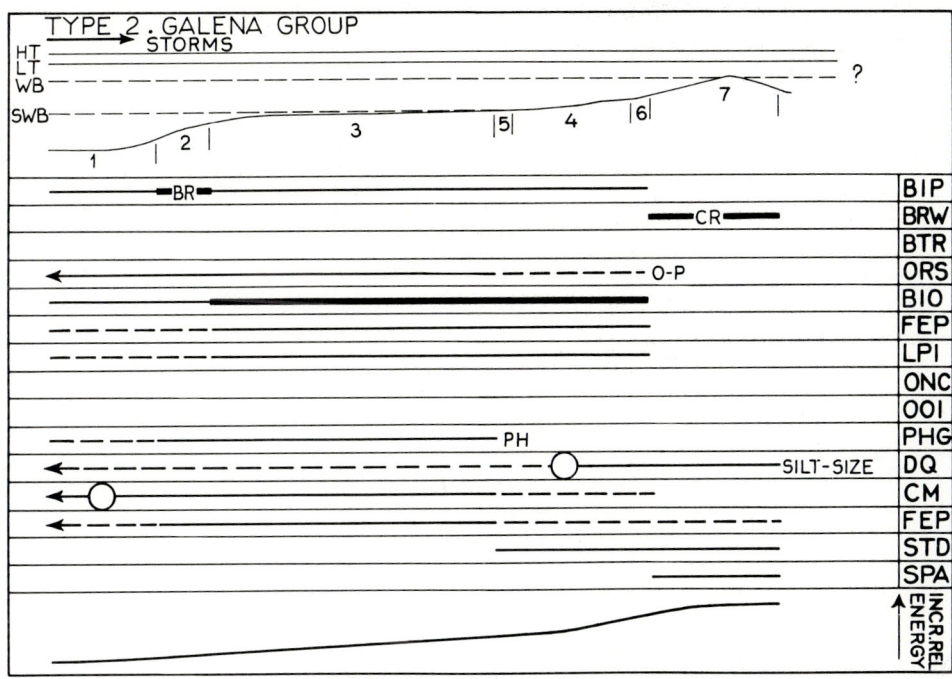

FIGURE 18.3 Integration of depositional models of simple ramps. Type 2. Galena Group.

18.2 GENETIC CLASSIFICATION OF SIMPLE RAMPS

An ideal classification of simple ramps in open marine conditions is based on the increase of general energy expressed by a decrease of relative depth of water and on the relationship between fair weather wave base and ramp morphology. An example is also given of a simple ramp under semirestricted conditions.

Type 1 (Fig. 18.2)

This type exemplifies integration of the Platteville Group (Middle Ordovician), north-central Illinois (Figs. 4.4, 4.7). It is a relatively deep, infratidal, and low-energy ramp which is well below wave base level with a benthic, *in situ*, and ecologically zoned fauna. It is also characterized by well-developed types of environmentally controlled bioturbation. Lithic pellets and intraclasts are generated *in situ* by weak bottom currents without appreciable transport.

Type 2 (Fig. 18.3)

This type exemplifies integration of the Galena Group (Middle Ordovician) of the Mississippi Valley (Figs. 4.14, 4.18, 4.19). It is a moderate to shallow infratidal ramp closer to wave base level than the previous type. Its benthic fauna is ecologically zoned except in the shallowest area, where winnowing and reworking *in situ* occur; lithic pellets and intraclasts are also generated *in situ*. Storm deposits are limited to the reworking and redeposition *in situ* of bottom materials. Shoreline microfacies may show some potential primary porosity.

Type 3 (Fig. 18.4)

This type exemplifies integration of the Menard Formation (Upper Mississippian), southwest margin of the Illinois Basin (Figs. 4.30, 4.33). It is a shallow infratidal ramp of which a large portion is located immediately below the wave base level. These conditions lead to widespread winnowing and fragmentation of some benthic constituents but no major dispersion of bioclasts. Ooids from an inferred shoreline microfacies are distributed into deeper waters by bottom currents. Shoreline microfacies possess potential primary porosity.

Type 4 (Figs. 18.5, 18.6, 18.7)

The profile of this type of gentle ramp cuts across wave base and most of its shallower microfacies show association both of *in situ* and transported benthic faunas. Microfacies with oncoids, lithic pellets, intraclasts, and ooids are well represented and display potential porosity.

A pure carbonate example (Fig. 18.5) exemplifies the integration of the Bird Spring Group (Middle Morrowan to Lower Missourian) of southeast Nevada (Figs. 4.37, 4.40, 4.41). It includes episodes of active influx

520 Part V / Toward an Explanation

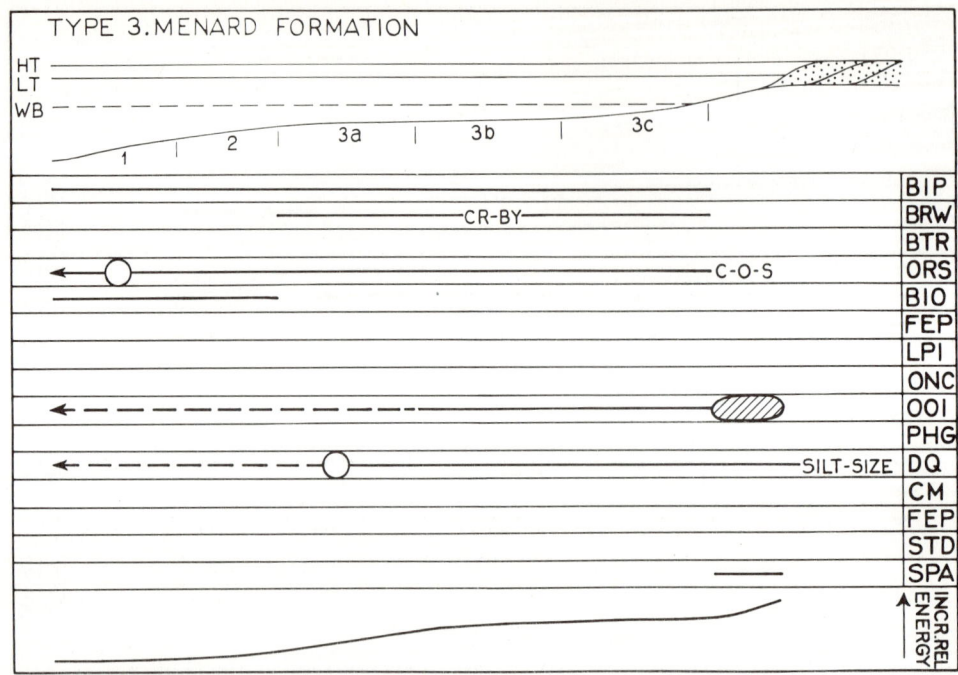

FIGURE 18.4 Integration of depositional models of simple ramps. Type 3. Menard Formation.

of detrital quartz (Fig. 18.6) with destructive effects on the benthic fauna, accompanied by disappearance of oolitic microfacies in favor of arenaceous biocalcarenites. A similar type of a ramp grading into shoreline microfacies of arenaceous biocalcarenites and pure quartz arenites (Fig. 18.7) exemplifies the integration of the simple ramp model of the Arrow Canyon Formation and Crystal Pass Limestone (Upper Devonian), southeast Nevada (Figs. 9.7, 9.9, 9.10). In this case, storm deposits consist of clasts of the entire spectrum of microfacies, indicating extensive transport landward followed by seaward transport until final deposition.

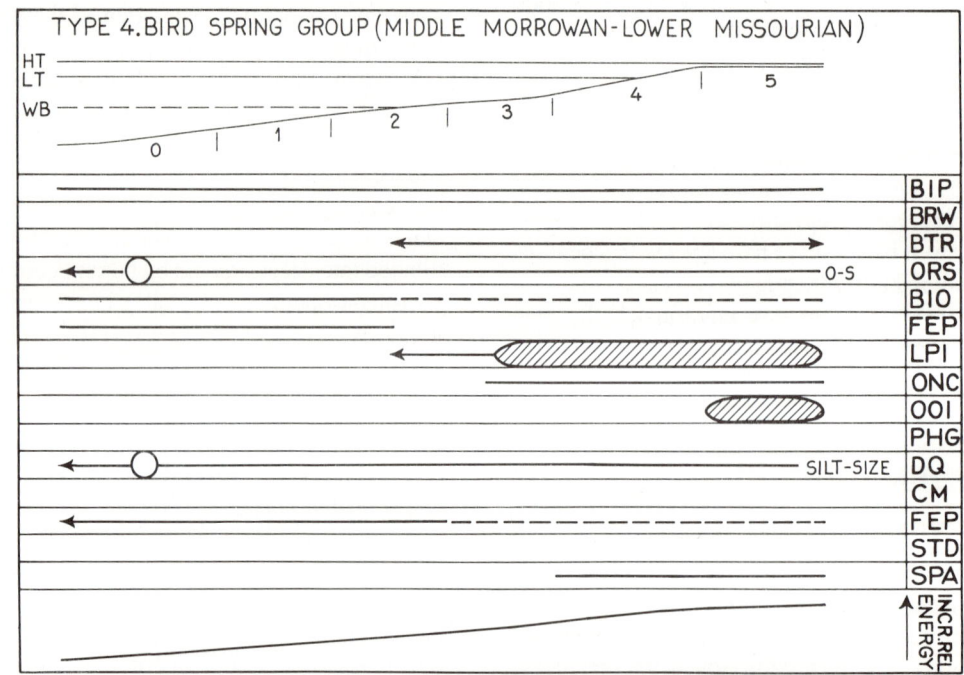

FIGURE 18.5 Integration of depositional models of simple ramps. Type 4. Bird Spring Group (Middle Morrowan–Lower Missourian).

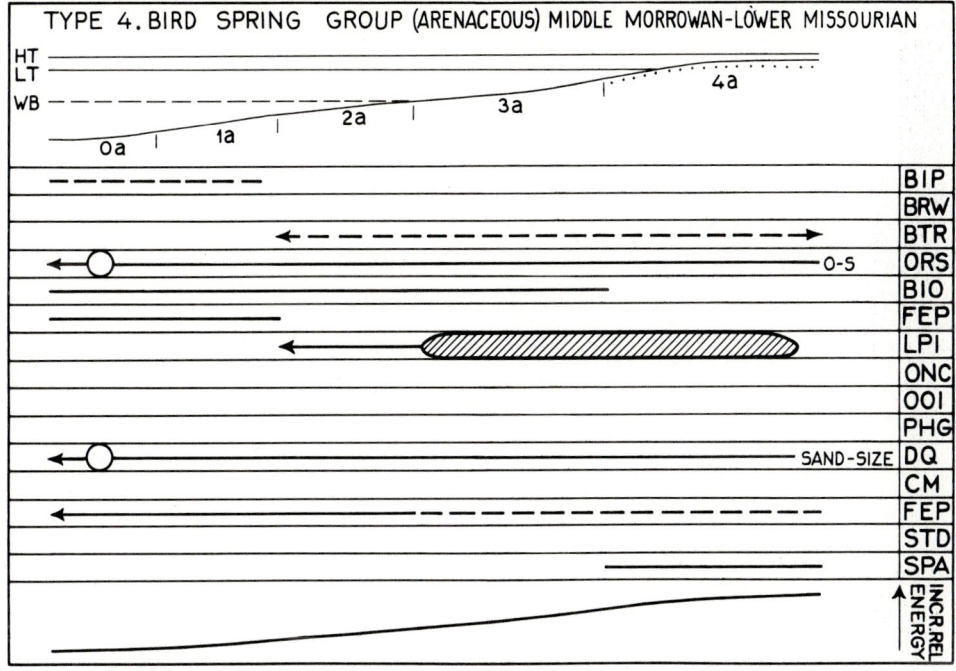

FIGURE 18.6 Integration of depositional models of simple ramps. Type 4. Bird Spring Group, arenaceous (Middle Morrowan–Lower Missourian).

Type 5 (Figs. 18.8, 18.9, 18.10)

The profile of these ramps is more accentuated than that of the previous ones and corresponds to a sharp declivity where both benthic fauna is the most prolific and wave base action the strongest. This situation leads to reworking and dispersion of bioclasts. Integration (Fig. 18.8) of model 2 of the Kinkaid Formation (Upper Mississippian), southern part of the Illinois Basin (Figs. 8.12, 8.13), shows these features together with widespread subaerial exposure and suspension transport of many organisms. Storm deposits, concentrated in shoreline ar-

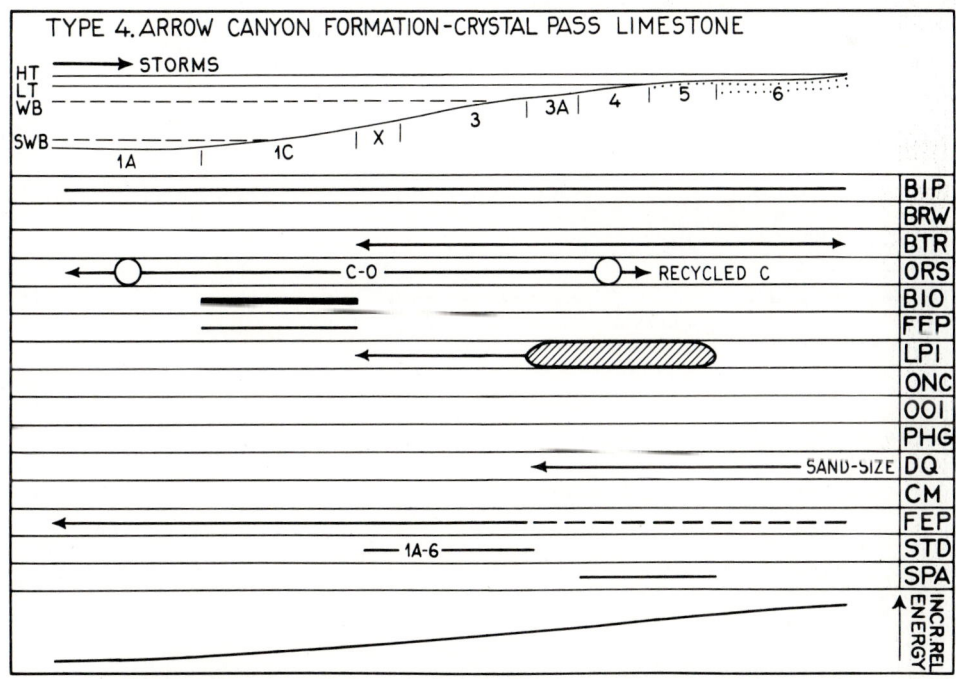

FIGURE 18.7 Integration of depositional models of simple ramps. Type 4. Arrow Canyon Formation and Crystal Pass Limestone.

522 Part V / Toward an Explanation

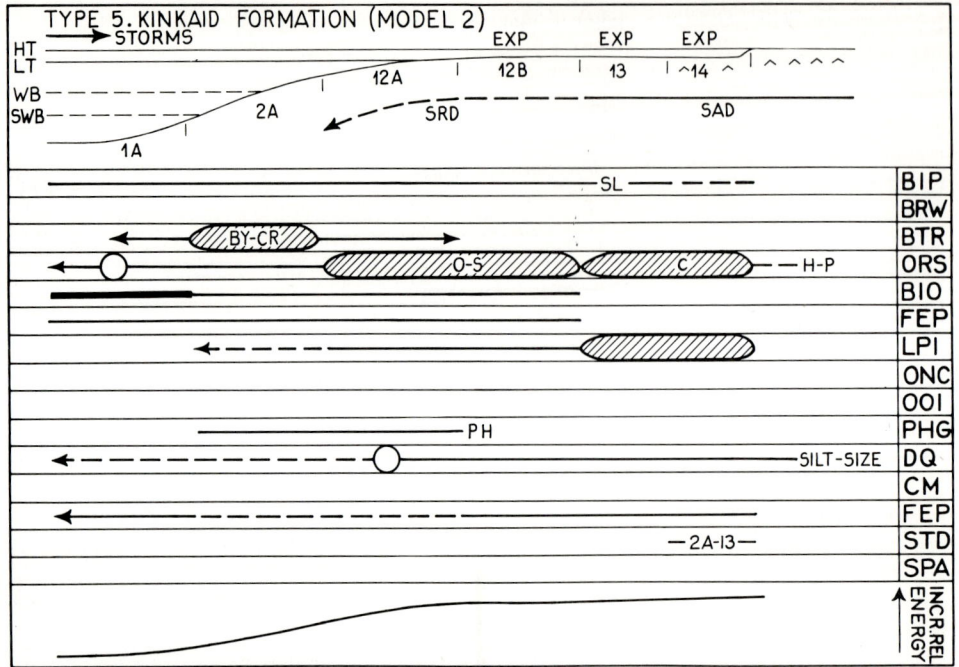

FIGURE 18.8 Integration of depositional models of simple ramps. Type 5. Kinkaid Formation (model 2).

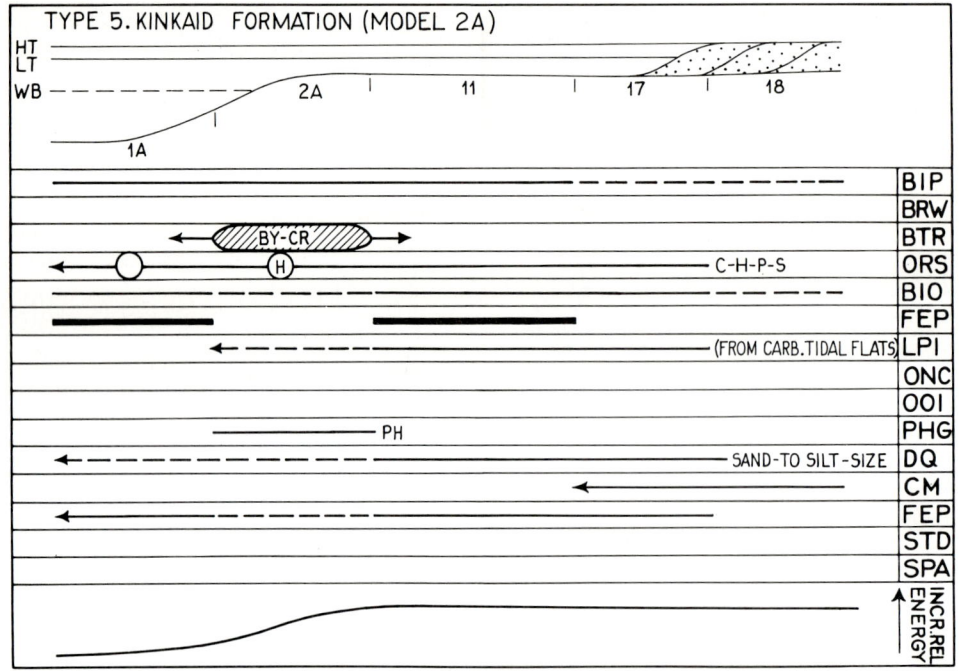

FIGURE 18.9 Integration of depositional models of simple ramps. Type 5. Kinkaid Formation (model 2A).

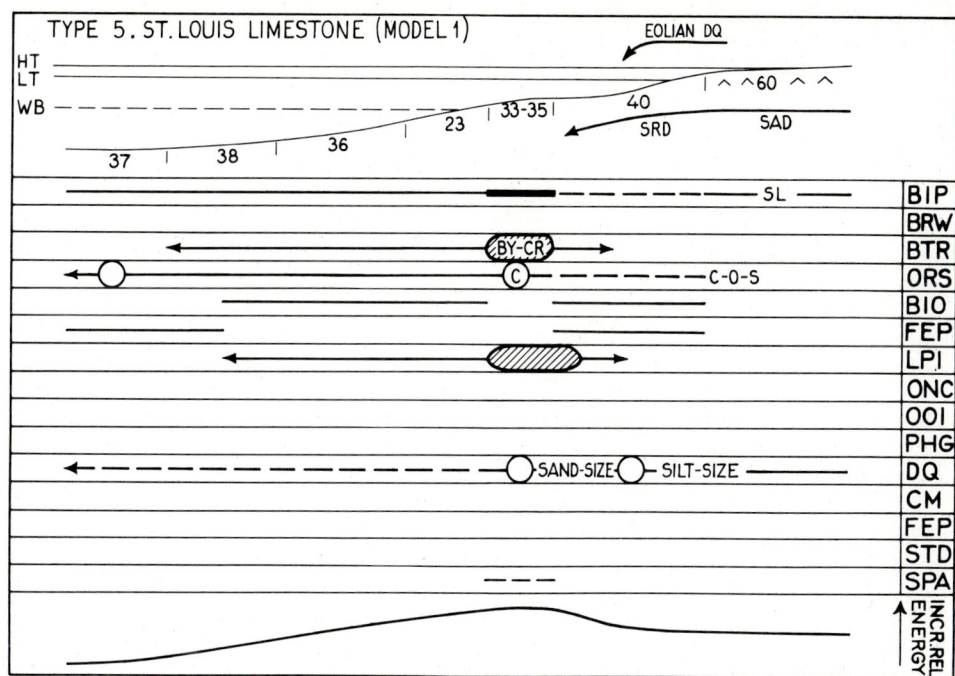

FIGURE 18.10 Integration of depositional models of simple ramps. Type 5. St. Louis Limestone (model 1).

eas, indicate only landward transport. Microfacies with potential porosity do not occur, but dolomitization by seepage reflux from adjacent sabkhas is well developed.

Integration (Fig. 18.9) of model 2A of the Kinkaid Formation (Fig. 8.14), a lateral equivalent of model 2, shows a mixed carbonate-siliciclastic situation with a prograding delta system and related destructive effect of the siliciclastic influx on benthic fauna.

Integration (Fig. 18.10) of model 1 of the St. Louis Limestone (Middle Mississippian) of the Illinois Basin (Figs. 3.6, 3.7) shows another example of the destructive effect on benthic fauna, this time by seepage reflux brines from adjacent evaporitic tidal flats. Of interest is the association of silt-size eolian quartz with sand-size quartz transported by aqueous processes.

Type 6 (Fig. 18.11)

This type of ramp under semirestricted marine conditions exemplifies integration of the Itaituba–Nova Olinda Formations (Carboniferous-Permian), of the Amazon Basin, northern Brazil (Fig. 5.4). It shows an accentuated profile like the ramps of type 5, but in a relatively low energy basin where the benthic fauna is *in situ* and ecologically zoned in all its components, from marine to restricted, with abundant agglutinated smaller benthic foraminifers. The action of wave base is limited to the generation of oncoids. Microfacies with potential primary porosity are relatively extensive, but their development is weak in spite of their combination with seepage reflux from adjacent evaporitic flats.

Distribution of Potential Primary to Early Diagenetic Reservoirs in Simple Ramps

A schematic diagram (Fig. 18.12) shows the five types of simple ramps arranged in increasing shallowing conditions and corresponding to a family of curves of submarine morphology. In general, ramps are systems unfavorable to the generation of microfacies with potential primary porosity. Such microfacies correspond essentially for all five types to the greater-energy environments of shorelines, with possible improvement introduced by early dolomitization due to seepage reflux of heavy brines from adjacent evaporitic flats (coastal sabkhas) under tropical arid climates. The accentuated profile of the ramps of type 5 introduces the possibility of another location at the point of strongest declivity where greater-energy conditions are due to wave base action.

18.3 GENETIC CLASSIFICATION OF RAMPS WITH BIOACCUMULATED TO HYDRODYNAMIC BUILDUPS

An ideal classification of these ramps, in open marine conditions, is based on the increase of general energy expressed by a general decrease of relative water depth

524 Part V / Toward an Explanation

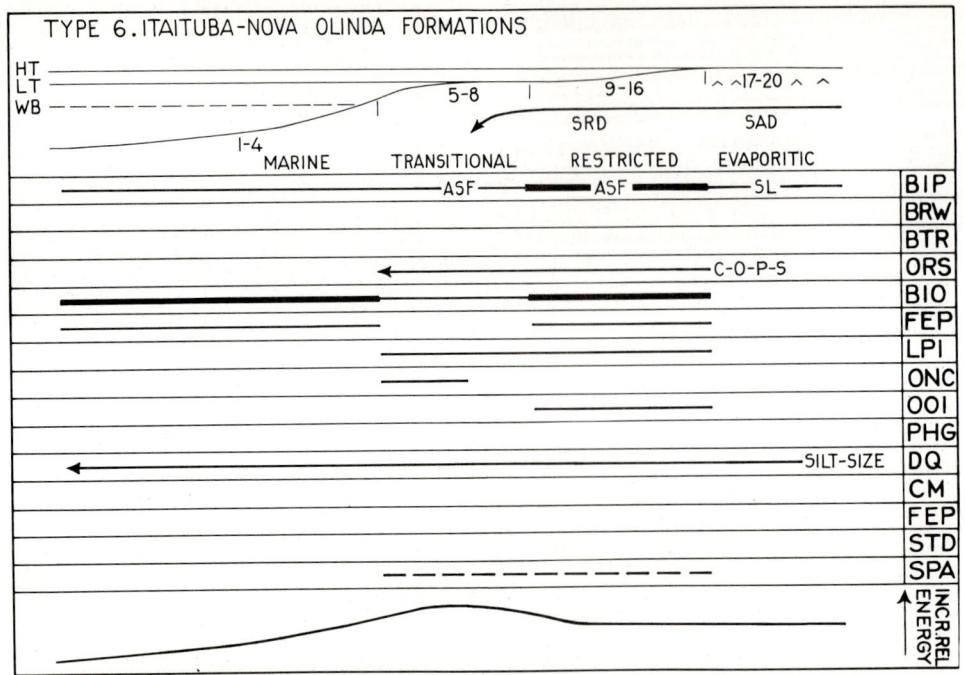

FIGURE 18.11 Integration of depositional models of simple ramps. Type 6. Itaituba–Nova Olinda Formations.

of the depositional model combined, commonly, with changes of submarine morphology introduced by the growth of hydrodynamic buildups. This growth ranges from below to above wave base and reaches finally incipient to well-developed subaerial exposure (beachrock, solution features, paleosoils). The growth of hydrodynamic buildups above the wave base introduces lagoonal conditions behind them with a variable degree of restriction. Whenever subaerial exposure is reached, dolomitization may occur under the appropriate climatic conditions by a dorag-type process when freshwater lenses develop.

Type 1 (Fig. 18.13)

It is a relatively deep and quiet ramp with a bioaccumulated to hydrodynamic buildup consisting predominantly of crinoids associated with brachiopods. Because the buildup does not reach wave base, dispersion of its bioclasts is limited. This represents integration of the hydrodynamic buildup model of the Brereton Limestone (Middle Pennsylvanian) of the southwest margin of the Illinois Basin (Figs. 9.38, 9.39) which is directly under the influence of prograding deltas. The location of this model in interdeltaic embayments leads to carbonates with detrital quartz, clay minerals, phosphates, glauconite, and pyrite. Microfacies with potential primary porosity are limited to the central part of the buildup.

Type 2 (Fig. 18.14)

This ramp is morphologically more differentiated than type 1. In it a bioaccumulated to hydrodynamic buildup, consisting mainly of crinoids and bryozoans, reaches the wave base and generates behind it a well-characterized lagoon. It exemplifies integration of the Rundle Group

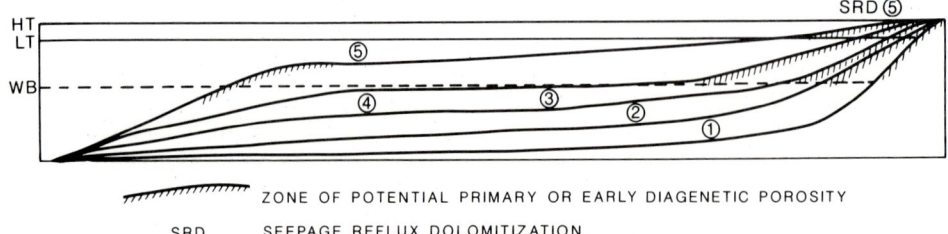

FIGURE 18.12 Distribution of potential primary reservoirs in simple ramps. Circled numbers refer to types of models.

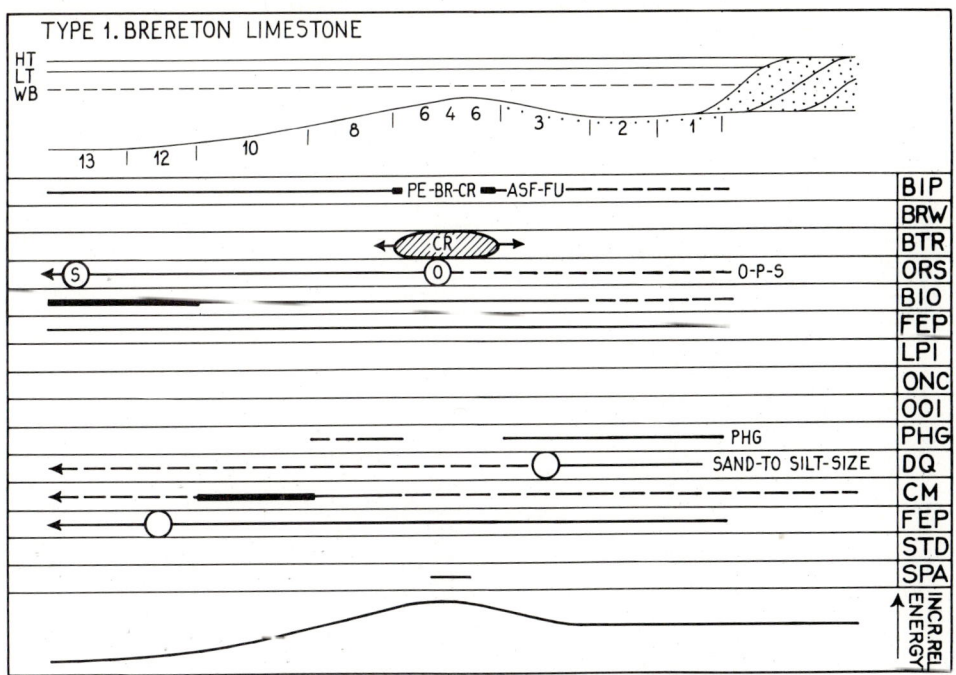

FIGURE 18.13 Integration of depositional models of ramps with bioaccumulated to hydrodynamic buildups. Type 1. Brereton Limestone.

(Mississippian) of the Canadian Rocky Mountains Front Range in Alberta (Figs. 8.19, 8.22). Its main features are a general unidirectional basinward transport of all constituents, a mechanical concentration of several sand- to silt-size components in the upper slope, where their size is in equilibrium with the local energy, a buildup with potential primary porosity, and finally, a lagoon characterized by abundant ostracods and sponges. The lagoon also displayed an active generation of fecal pellets due to bioturbation, as well as lithic pellets and intraclasts, and a shoreline oolitic environment accompanied by seepage reflux dolomitizing solutions from the adjacent sabkha.

Type 3 (Fig. 18.15)

This type of ramp integrates model 2 of the St. Louis Limestome (Middle Mississippian) of the Illinois Basin (Figs. 3.8, 3.9), where the buildup is predominantly hydrodynamic and consists of crinoids, bryozoans, and endothyrids. It reaches very close to low tide level but is not exposed subaerially. The dispersion of bioclasts and intraclasts is well developed, both landward and seaward. The buildup displays microfacies with potential primary porosity; it is an effective barrier to the basinward transportation of detrital quartz. The lagoonal microfacies contain bioaccumulations of bryozoans and brachiopods, and the more shoreward microfacies possess a very reduced fauna due to seepage reflux of brines from adjacent sabkhas.

Type 4 (Fig. 18.16)

This type of ramp exemplifies integration of the Salem Limestone (Middle Mississippian) of the southeastern margin of the Illinois Basin (Figs. 7.6, 7.7, 7.8). It displays a hydrodynamic buildup consisting of an association of crinoids, bryozoans, endothyrids, and ooids. The dispersion of constituents is well developed, both landward and seaward. The buildup shows features of temporary subaerial exposure and represents an effective obstacle to basinward transport of detrital quartz. Microfacies with potential primary porosity occur in the buildup, which shows dorag-type dolomitization on its seaward side, and in the shallower algal flat in the lagoon from which constituents are dispersed. The shoreward side of the lagoon is under the influence of dolomitization by seepage reflux from adjacent sabkhas.

Type 5 (Fig. 18.17)

This ramp shows a well-developed bioaccumulated to hydrodynamic buildup with a frontal part consisting of crinoids and bryozoans, and a rear part of ooids. This type exemplifies integration of model 1 of the Kinkaid Formation (Upper Mississippian) of the southern part

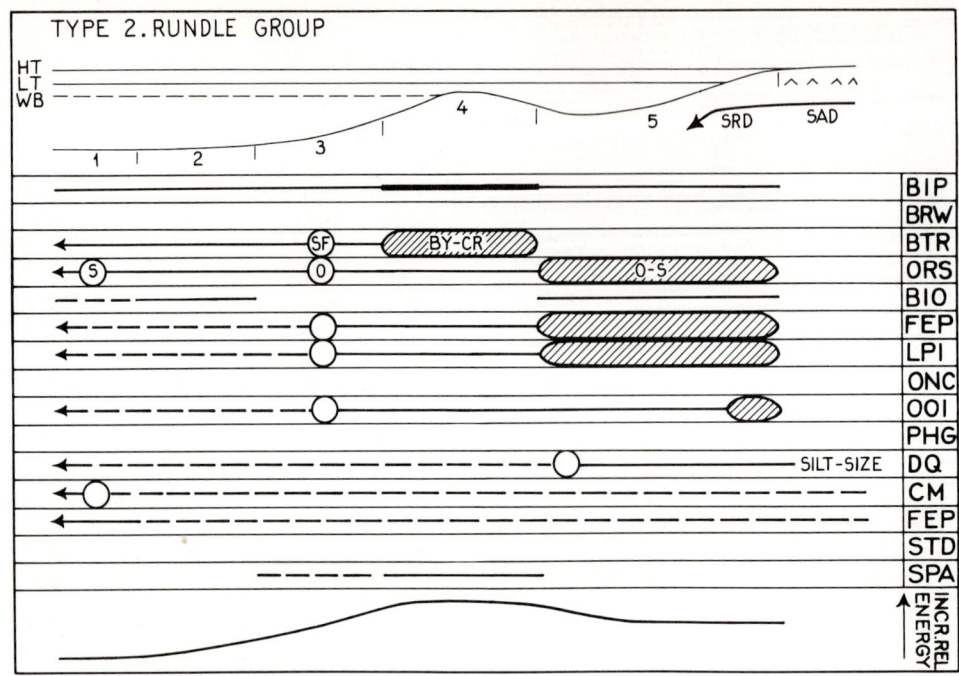

FIGURE 18.14 Integration of depositional models of ramps with bioaccumulated to hydrodynamic buildups. Type 2. Rundle Group.

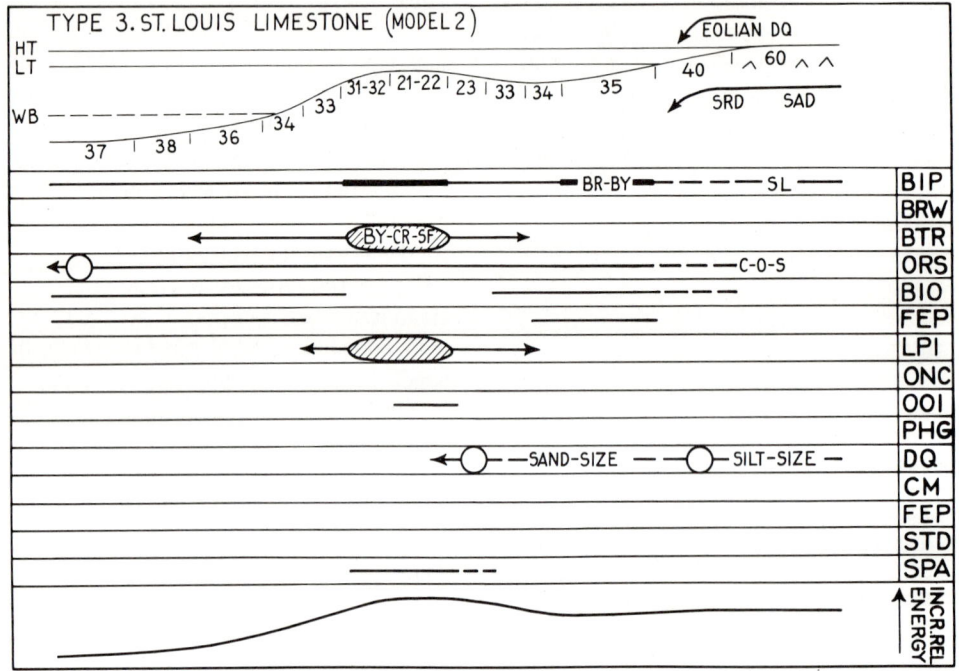

FIGURE 18.15 Integration of depositional models of ramps with bioaccumulated to hydrodynamic buildups. Type 3. St. Louis Limestone (model 2).

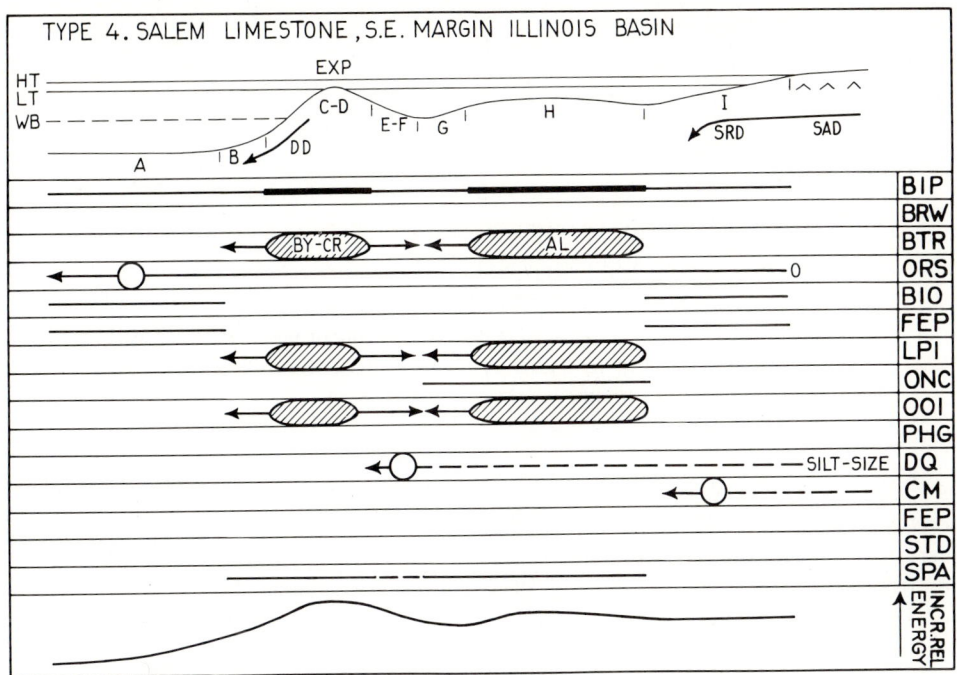

FIGURE 18.16 Integration of depositional models of ramps with bioaccumulated to hydrodynamic buildups. Type 4. Salem Limestone, southeast margin of Illinois Basin.

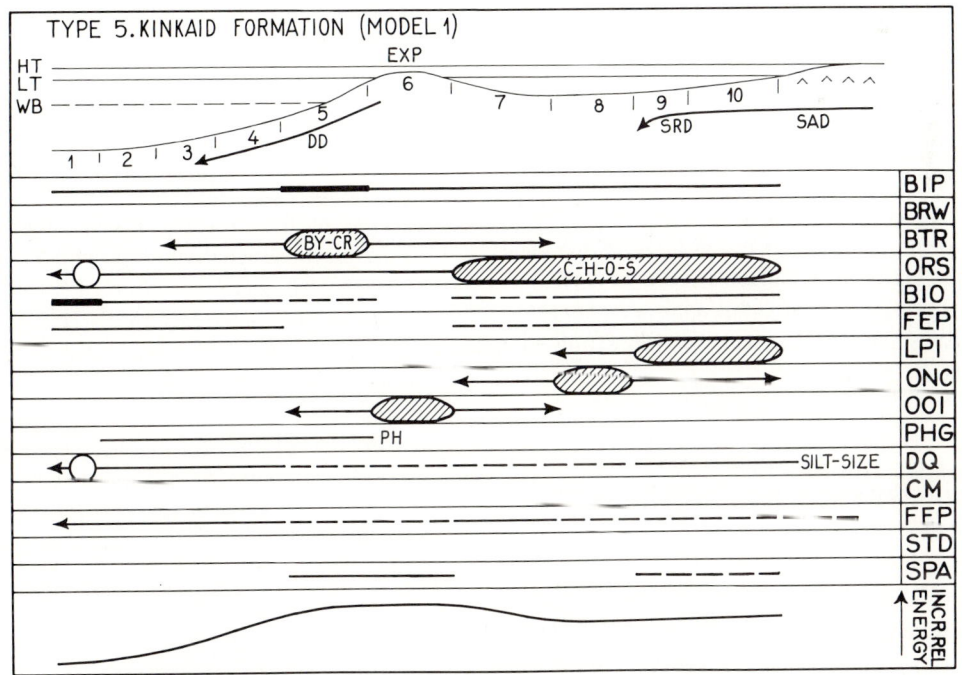

FIGURE 18.17 Integration of depositional models of ramps with bioaccumulated to hydrodynamic buildups. Type 5. Kinkaid Formation (model 1).

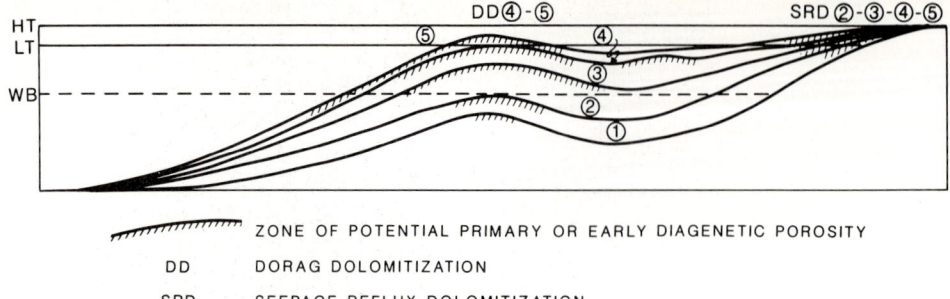

FIGURE 18.18 Distribution of potential primary reservoirs in ramps with bioaccumulated to hydrodynamic buildups. Circled numbers refer to types of models.

of the Illinois Basin (Figs. 8.7, 8.8). The buildup is exposed subaerially at low tide and its microfacies with potential primary porosity are improved by dorag-type dolomitization, which involves a large portion of the seaward slope. The lagoonal environment is very rich in calcispheres, ostracods, sponges, and *Chara*. Lithic pellets and intraclasts, together with oncoids, are widespread. The shoreline microfacies of the lagoon are dolomitized by seepage reflux from adjacent sabkhas.

Distribution of Potential Primary to Early Diagenetic Reservoirs in Ramps with Bioaccumulated to Hydrodynamic Buildups

A schematic diagram (Fig. 18.18) shows the five types of ramps with bioaccumulated to hydrodynamic buildups arranged in increasing general shallowing conditions, and corresponding to a family of curves of submarine morphology. This order leads to an increase in the importance of microfacies with potential primary porosity, which, upon subaerial exposure, are improved by dorag-type dolomitization due to freshwater lenses. Only occasionally, may lagoonal microfacies, such as algal flats, contain some primary porosity; this applies also to shoreline microfacies submitted to seepage reflux from adjacent sabkhas.

18.4 GENETIC CLASSIFICATION OF RAMPS WITH BIOCONSTRUCTED BUILDUPS

An ideal classification of these ramps, in open to restricted marine conditions, is also based on the increase of general energy expressed by a decrease of relative depth of the depositional model, associated with the growth of buildups and their relationship with the position of the wave base. Consequently, these ramps share many features with those displaying bioaccumulated to hydrodynamic buildups.

Type 1 (Fig. 18.19)

This type of ramp, in relatively deep and quiet water, corresponds to an interdeltaic embayment and displays a bioconstructed buildup of phylloid algae and bryozoans which remains entirely below the wave base. Hence no dispersion of bioclasts occurs. This exemplifies integration of the bioconstructed buildup model of the Brereton Limestone (Middle Pennsylvanian) of the southwest margin of the Illinois Basin (Figs. 9.36, 9.37). It is under the direct influence of prograding deltas, and its carbonates are rich in detrital quartz, clay minerals, phosphates, and glauconite. Due to the general lower energy of this environment, microfacies with potential primary porosity are absent.

Type 2 (Fig. 18.20)

This ramp with its incipient bioconstructed buildup of *Amphipora* and stromatoporoids exemplifies integration of the bioconstructed buildup ramp model of the Arrow Canyon Formation and Crystal Pass Limestone (Upper Devonian) of southeast Nevada (Figs. 9.7, 9.9, 9.10). In this case, which displays a shoreline of pure quartz arenites, the buildup grows up to wave base or just above it and acts as an obstacle for seaward distribution of detrital quartz; however, no dispersion of bioclasts exists from this buildup. Storm deposits consist of clasts from the entire spectrum of microfacies indicating extensive transport, first landward then seaward, until final deposition.

Type 3 (Fig. 18.21)

This integration corresponds to a portion of the restricted evaporitic basin of the Cayugan (Upper Silurian) in northern Ohio (Figs. 9.1, 9.3). The model consists of a shore-

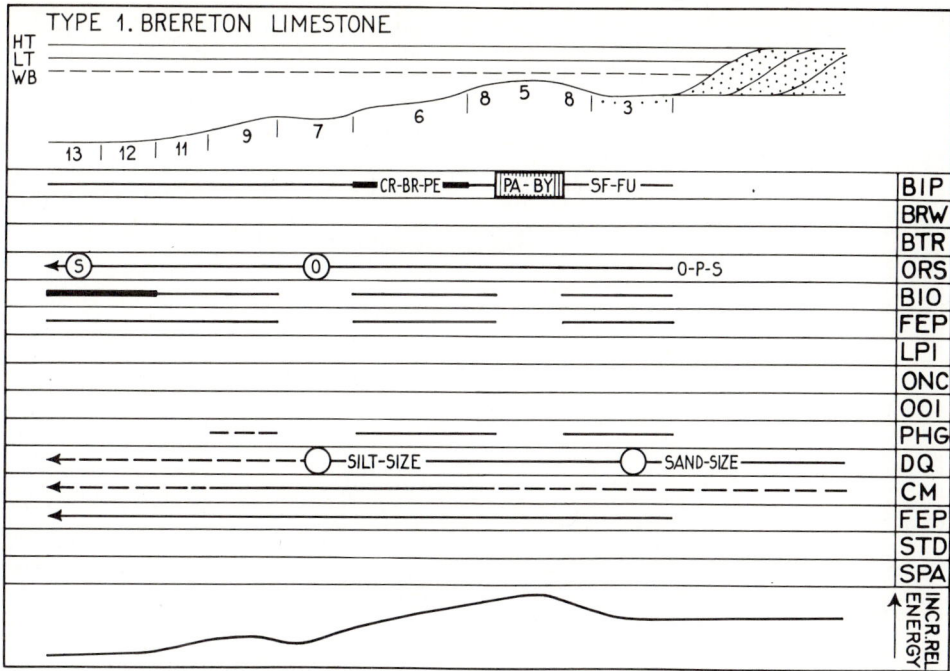

FIGURE 18.19 Integration of depositional models of ramps with bioconstructed buildups. Type 1. Brereton Limestone.

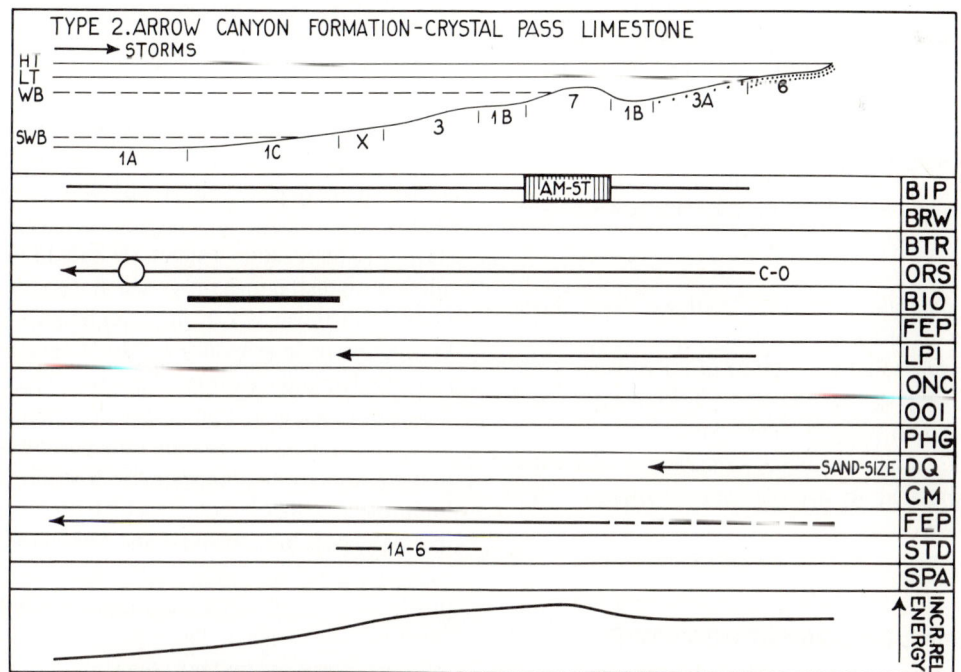

FIGURE 18.20 Integration of depositional models of ramps with bioconstructed buildups. Type 2. Arrow Canyon Formation and Crystal Pass Limestone.

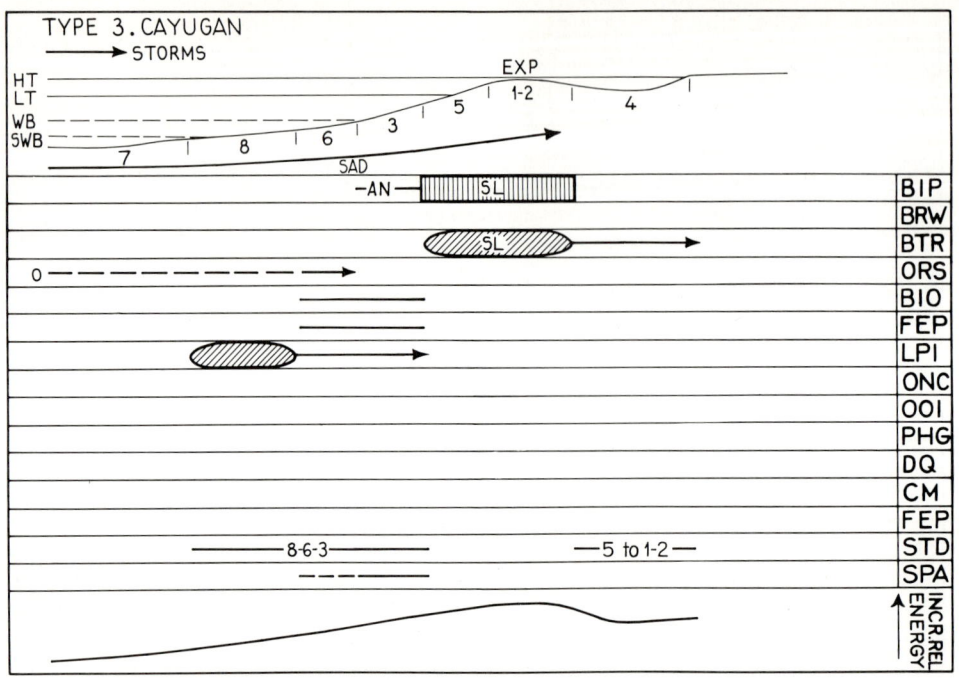

FIGURE 18.21 Integration of depositional models of ramps with bioconstructed buildups. Type 3. Cayugan.

line bioconstructed buildup of matlike and hemispherical stromatolites with features of subaerial exposure. Some storm deposits are derived from basinal microfacies and are redeposited eventually *in situ*; others result from intense destruction of stromatolitic buildups with accumulation behind these buildups. All microfacies of this model are dolomitized completely by sabkha processes, and those presenting a potential primary porosity are restricted to near-wave-base conditions.

Distribution of Potential Primary to Early Diagenetic Reservoirs in Ramps with Bioconstructed Buildups

A schematic diagram (Fig. 18.22) shows these three types of ramps in order of increasing shallowing conditions and the very small importance of microfacies with potential primary porosity. However, this representation is based on a small number of investigated cases, and a more realistic picture would be similar to that of ramps with bioaccumulated to hydrodynamic buildups (Fig. 18.18).

18.5 GENERAL SYNTHESIS OF RAMPS

Features and processes of general significance for these related types of depositional models are analyzed below.

Fair-Weather Wave Base, Tidal Currents, and Submarine Morphology

In carbonate models, the intersection of wave base with submarine topography has been a widely used criterion for environmental interpretation because it is based on easily observable textural features. It is indeed the place where relatively deeper grain-supported biocalcarenites, with a micrite or bioclastic matrix, grade into relatively shallower and similar deposits with interstitial sparite cement or primary porosity. However, microfacies studies which analyze the behavior of clasticity and frequency curves of inorganic and organic constituents and their state of preservation (intact, abraded, or fragmented) allow us to distinguish those which indicate *in situ* conditions from those which have undergone transportation by traction or in suspension. This transportation, which is caused mainly by tidal currents, represents one of the most fundamental processes in carbonate sedimentation. The effects of incoming tidal currents are expressed first by a landward transport by traction (decreasing in importance in that direction) which involves mainly benthic bioclasts, lithic pellets, and intraclasts, and second, by a simultaneous landward transport in suspension of planktonic constituents (foraminifers, Mesozoic calcispheres, radiolarians). The effects of outgoing tidal currents are expressed first by a seaward transport by traction (decreasing in importance in that direction) which involves mainly benthic bioclasts, lithic pellets, intraclasts,

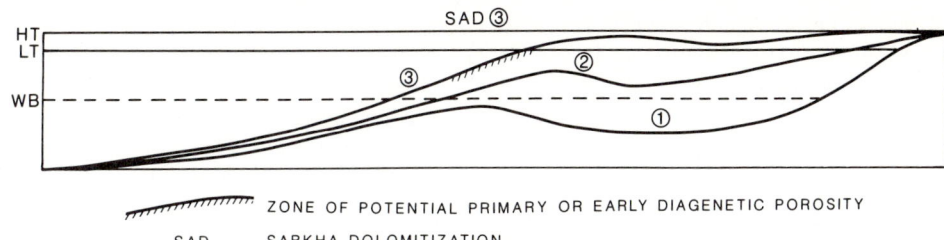

FIGURE 18.22 Distribution of potential primary reservoirs in ramps with bioconstructed buildups. Circled numbers refer to types of models.

and coarse-grained extrabasinal constituents such as quartz, and second, by a simultaneous seaward transport in suspension of delicate lagoonal constituents (ostracods, sponge spicules, Paleozoic calcispheres) and fine-grained extrabasinal constituents such as silt-size quartz, clay minerals, plant debris, stems, and oogonia of *Chara*. In short, all carbonate models display the final average results of various types of interferences between effects of wave base and tidal currents.

The effect of fair-weather wave base can be observed in single ramps with gentle slopes under relatively deeper water, where it separates lesser-energy carbonate sediments with interstitial micritic to bioclastic matrix from larger-energy carbonates with potential primary interparticle porosity limited essentially to shorelines (Figs. 18.2, 18.3). Effects of incoming tides are not detectable, those of outgoing tides are limited to the local generation of lithic pellets and small intraclasts without any major transportation by traction, and to the transport of silt-size quartz, ostracods, and sponge spicules in suspension. The weakness of these processes is shown by the benthic fauna which remains *in situ* and zoned ecologically.

In simple ramps with gentle slopes under relatively shallow water, fair-weather wave base causes a much broader effect on sediments after its intersection with submarine topography. This effect, which increases in intensity landward, consists of winnowing and abrasion *in situ* of selected benthic constituents (Fig. 18.4). An incipient effect of incoming and outgoing tidal currents might be considered here.

In other simple ramps with gentle slopes but under shallower conditions (Figs. 18.5, 18.6, 18.7), the effects of incoming and outgoing tidal currents predominate over the action of fair-weather wave base. Consequently, benthic bioclasts and lithic pellets are reworked, fragmented, and transported by traction over a limited distance both landward and seaward, whereas lagoonal constituents are carried in suspension seaward.

Whenever the slope of simple ramps presents a more accentuated profile and corresponds to a sharp declivity where the benthic fauna reaches its maximum development (up to bioaccumulation), the effects of fair-weather base are combined with those of incoming and outgoing tidal currents (Figs. 18.8, 18.9, 18.10). This combination leads to a maximum reworking of benthic communities (Figs. 18.8, 18.9, 18.10), and eventually to the concentration of their bioclasts together with lithic pellets and intraclasts into hydrodynamic buildups (Figs. 18.13 to 18.17). Further dispersion of constituents beyond the limits of buildups in a landward and seaward direction is extremely widespread, together with a seaward dispersion of benthic lagoonal organisms.

Analysis of seaward transport by traction until final deposition in deeper water shows that settling occurs at different places along the slope, according to where the grain size of a given constituent reaches equilibrium with local conditions of energy. Complications are introduced when hydrodynamic buildups, temporarily exposed or not, are themselves obstacles to such a seaward transport (Figs. 18.13 to 18.16). The same observations can be made for seaward transport in suspension of constituents which may even occur by itself. For instance, investigated examples of ramps with bioconstructed buildups, which developed below wave base or have barely reached it, show seaward transport by suspension of delicate bioclasts and minute lithic pellets, whereas the great majority of the benthic fauna is unaffected and remains ecologically zoned (Figs. 18.19, 18.20).

Potential primary porosity may occur inside hydrodynamic buildups and in immediately adjacent areas of larger energy, whereas protected lagoonal areas landward of the buildups are generally devoid of it. However, potential primary porosity may reappear locally in the shallower areas of lagoons and again along shorelines under local conditions of greater agitation.

Coexistence of *in Situ* and Transported Organic and Inorganic Constituents

Alternating tidal currents and their related seaward and landward transport of certain benthic and inorganic constituents are demonstrated indirectly by the behavior of *in situ* benthic fauna. The latter, in addition to being ecologically zoned, often displays environmentally controlled types of bioturbation which confirm its autochthonous character (Figs. 18.2, 18.3). Appreciable production of fecal pellets is also related directly to *in situ* benthic faunas. However, bioturbation and fecal pellets, although associated frequently, do not coincide in distribution necessarily (Figs. 18.4, 18.6, 18.9, 18.10) because fecal pellets can be produced by nonbioturbating infauna, and viceversa, bioturbation activity is not accompanied necessarily by the production of recognizable fecal pellets.

Distribution Pattern of Oolitic Microfacies

In simple ramps, the generation of ooids is limited to intertidal agitated shoreline areas (Figs. 18.4, 18.5, 18.11). In ramps with hydrodynamic buildups, major oolitization may develop in addition throughout the upper subtidal to intertidal parts of buildups (Figs. 18.16, 18.17) as well as in certain more agitated portions of the lagoons (Fig. 18.14). Ooids can be reworked, abraded, and redistributed by tidal currents both landward and seaward from their areas of generation. In some instances, ooids are generated in shoreline conditions and redistributed by currents across the entire carbonate system into the basinal environment (Fig. 18.14).

Distribution Pattern of Oncoids

Oncoids result from the mechanical destruction of blue-green algal mats followed by their regeneration as individual bodies under the protection of baffling systems (see Falkenhein *et al.* 1981). In simple ramps (Fig. 18.5), and in ramps displaying an accentuated profile with sharp declivity, oncoids are generated in areas where fair-weather wave base is most active (Fig. 18.11) and also in shallower conditions. In ramps with hydrodynamic buildups reaching temporary subaerial exposure, oncoids develop in agitated conditions of the lagoon (Figs. 18.16, 18.17).

Distribution Pattern of Phosphates and Glauconite

In simple ramps with gentle slopes, phosphatization in the form of synsedimentary replacement of carbonate muds and organisms below the water-sediment interface occurs only in upper slope positions, namely in temporarily oxygen-depleted waters (Fig. 18.3). In ramps with sharp declivity, phosphatization is located immediately below fair-weather wave base (Figs. 18.8, 18.9). In ramps with hydrodynamic buildups, the location of phosphates is similar to that in the previous case (Fig. 18.17). However, in some of these ramps, as in those with bioconstructed buildups with shorelines corresponding to prograding deltaic systems, phosphatization accompanied by glauconitization occurs also in protected areas between buildups and delta fronts (Figs. 18.13, 18.19).

Distribution Pattern of Extrabasinal Materials

Whenever simple ramps display coastlines consisting of evaporitic flats (sabkhas) or inactive older carbonate terrane undergoing karstic processes, extrabasinal influx is essentially nonexistent or very fine silt-size angular grains of detrital quartz (often of eolian origin) and clay minerals predominate (Figs. 18.2, 18.3, 18.5, 18.8, 18.10, 18.11). When the source area consists of an association of older carbonate and siliciclastic rocks drained by sluggish streams, narrow beaches occur which consist of mixed bioclasts and detrital quartz with local generation of ooids (Fig. 18.5). Greater influx from continental origin leads to mixed carbonate-siliciclastic models in which ramps terminate shoreward into pure quartz arenite beaches (Figs. 18.6, 18.7), estuaries, argillaceous tidal flats, prograding fan deltas, and deltas of lithic arenites to wackes (Figs. 18.4, 18.9). Regardless of this great variation in the amount of siliciclastic sediment influx, coarse sand-size grains of detrital quartz, clay minerals, and iron oxides (preserved as pyrite) are distributed across these carbonate systems in a seaward direction, probably together with colloidal organic matter in suspension. Whereas clasticity of detrital quartz decreases in the direction of transport, frequency varies as a function of the available range of grain size leading to concentrations at various places where a particular size reaches equilibrium with local bottom energy conditions (Figs. 18.2 to 18.6, 18.8). In certain cases, frequency distribution is uniform whether grain size is sand or silt; elsewhere it is possible to observe concentrations of eolian quartz as opposed to that transported by aqueous processes (Fig. 18.10).

In the examples above of simple ramps which terminate in arenaceous shorelines and in prograding deltas, benthic faunas are affected strongly by the large volume of sand-size quartz grains: they become either scarce in this unfavorable environment or display widespread effects of abrasion of bioclasts (Figs. 18.6, 18.9).

The behavior of silt-size grains of detrital quartz in ramps with hydrodynamic buildups follows the general pattern of simple ramps, but the distribution is often complicated by additional areas of trapping behind the buildups (Figs. 18.13, 18.14, 18.15). Some buildups (Figs. 18.15, 18.16) are barriers to transportation so that all terrigenous materials settle in lagoonal environments, or buildups may act as selective filters (Fig. 18.17).

Ramps with bioconstructed buildups display trapping conditions similar to those with hydrodynamic buildups. Indeed, concentration of sand-size quartz grains occurs behind buildups (Figs. 18.19, 18.20), or additional trapping of silt-size quartz may be provided downslope by any depression of the submarine topography (Fig. 18.19).

Clay minerals transported mainly in suspension, because of their fine grain size, tend to settle in relatively deep and low-energy conditions during or after seaward transport. In simple ramps, they settle in lower slope to basinal conditions (Fig. 18.3). If flocculation is rapid, such as in prodeltaic conditions, argillaceous microfacies occur in shallower areas (Fig. 18.9). In ramps with bioaccumulated to hydrodynamic buildups and in ramps where hydrodynamic buildups do not reach low tide, and hence do not represent major obstacles to basinward transport, clay minerals, as in simple ramps, settle in slope environments (Figs. 18.13, 18.14). When buildups become exposed subaerially, clay minerals may remain confined to lagoonal environments (Fig. 18.16). Ramps with bioconstructed buildups should display conditions similar to the previous type. However, the only case investigated here was a subtidal buildup and a ubiquitous distribution of clay minerals, mostly concentrated in the upper slope (Fig. 18.19).

In some of these ramps, petrographic studies are sufficient to reach an estimate of the amount of iron of extrabasinal origin, generally preserved as pyrite, and to draw some environmental conclusions. In most simple ramps, pyrite is more abundant below fair-weather wave base in slope to basinal conditions (Figs. 18.2, 18.3, 18.5, 18.6, 18.7). In simple ramps with accentuated profiles, pyrite tends to characterize the upper and lower parts of the ramp and to show a smaller frequency at the point of maximum declivity where fair-weather wave base action is the strongest (Figs. 18.8, 18.9). In ramps with bioaccumulated to hydrodynamic buildups that do not reach low tide level, pyrite concentrates again in deep and low-energy areas (Figs. 18.13, 18.14). When buildups reach subaerial exposure, pyrite occurs also in lagoonal conditions behind the buildup (Fig. 18.17). Ramps with bioconstructed buildups should display conditions similar to those of the previous type. Two of the cases above show abundant pyrite everywhere (Fig. 18.19) or concentrations below the wave base in slope environment (Fig. 18.20).

A general relationship between the distribution of clay minerals and the concentration of pyrite exists in most carbonate ramps, namely in slope and basinal environments associated with euxinic conditions which preserve organic matter. Hence carbonate source beds are likely to be generated under these conditions.

Distribution Pattern of Dolomitization

Patterns of distribution of early dolomitization processes are of interest to explain the generation of additional secondary porosity which improves primary depositional porosity. Simple ramps under conditions of hot and arid climate with intermittent rains are bordered frequently by coastal sabkhas with characteristic evaporitic flats (stromatolites, dolomite, anhydrite, and halite), and in certain cases by ephemeral lakes of variable salinity. Dolomitization of carbonate sediments of the upper ramp is produced by seepage reflux of heavy brines flowing from evaporitic flats in a seaward direction (Figs. 18.8, 18.11). The submarine outflow of these brines may in some cases cause a nefarious effect on the *in situ* benthic fauna which becomes extremely reduced (Fig. 18.10). In ramps with hydrodynamic buildups which do not undergo temporary subaerial exposure but are under similar arid climatic conditions, seepage reflux is the same as in simple ramps (Figs. 18.14, 18.15). When hydrodynamic buildups become exposed temporarily and experience intermittent rains, a freshwater lens is generated in their upper part. This system of mixed freshwater-seawater (dorag-type process) produced dolomitization of carbonate sediments on the seaward slope of the buildup, where phreatic waters return to the sea through submarine springs (Fig. 18.16). This process may involve appreciable portions of the slope sediments (Fig. 18.17). Identical conditions of dorag-type dolomitization could occur when bioconstructed buildups on ramps undergo temporary subaerial exposure, but no such example was studied. The Cayugan example (Fig. 18.21) is due to a sabkha-type system.

Distribution Pattern of Storm Deposits

Storm effects are related directly to the intersection of the storm wave base with submarine morphology. In simple ramps with very gentle slopes under relatively deep water, the effects are limited to one or several areas where, depending on the intensity of the storm, the related wave base intersects the seafloor (Figs. 18.2, 18.3). There, reworking of bottom sediments takes place locally followed by redeposition of the materials essentially *in situ*. In simple ramps with very gentle slopes, but under shallower conditions, storm deposits are developed largely around the area of intersection of storm wave base and submarine topography (Fig. 18.7). These deposits consist of an accumulation of sediments in depths ranging from storm wave base to shallower conditions. The storm process consisted, therefore, of an extensive landward transport of materials followed by a final seaward return of these materials by slumping and turbidity currents. In some simple ramps, storm deposits of sediments ranging in depth from storm wave base to shallower conditions are accumulated in high intertidal to supratidal environments (Fig. 18.8). In investigated ramps where bioconstructed buildups reach wave base (Fig. 18.20), storm deposits are located immediately above storm wave base and because they consist of clasts of all the microfacies, they indicate that extensive landward and seaward transport was also involved. In a buildup constructed by stromatolites (Fig. 18.21) in a shoreline position, storm breccias are located immediately behind it.

Distribution Pattern of Potential Primary to Early Diagenetic Reservoirs

Potential primary interparticle porosity is characteristic of larger-energy carbonates which in simple ramps (Fig. 18.12) are located for types 1 to 4 along shorelines, and for type 5 also near the point of intersection of the fair-weather wave base with submarine topography due to its sharp declivity. Porosity conditions for all these types may be improved by seepage reflux dolomitization from adjacent supratidal flats. In ramps with incipient to well-developed bioaccumulated to hydrodynamic buildups (Figs. 18.18) in types 1 to 5, potential primary porosity occurs in the buildups, in their seaward and landward slopes, as well as along shorelines, and occasionally in type 4 in shallower lagoonal areas. Improvements of porosity conditions are introduced by doragtype dolomitization when the buildups are subaerially exposed and in shorelines when seepage reflux processes are active. In ramps with bioconstructed buildups (Fig. 18.22), potential porosity exists only when buildups reach subaerial exposure, in which case it is on the front of the buildup under conditions of sabkha dolomitization. Regardless of the amount of porosity generated by these early processes, the areas that they affected can become targets for further secondary porosity generated by burial processes.

18.6 GENETIC CLASSIFICATION OF PLATFORMS WITH FRONTAL BIOACCUMULATED BUILDUPS

Platforms that develop along the margins of relatively quiet intracratonic basins consist essentially of micritic buildups which extend from the edge of platforms down to foreslope environments. Although the origin of the accumulation of the micrite mud often remains enigmatic, it is generally considered to be algal, and in the Paleozoic, it often displays stromatactis structures before being stabilized by encrusting bryozoans. The latter, in turn, provide a favorable substratum for subsequent prolific growth of crinoids which eventually may lead, as in the Niagaran (Middle Silurian) of the southern margin of the Michigan Basin, to final bioconstructed situations (Figs. 10.10, 10.11, 10.12).

Type 1 (Fig. 18.23)

In this Mesozoic type integrated from the Chachao Formation (Valanginian) of the Neuquén–South Mendoza Basin, Argentina (Figs. 10.15 to 10.19), the large-scale micritic and strongly bioturbated buildup displays a prolific *in situ* and ecologically zoned fauna of pelecypods, colonial annelids, and sponges. In this particular case, the origin of micrite mud could be entirely fecal. Of interest is the abundant fish fauna and the fact that Mesozoic calcispheres appear to be a kind of plankton from the open sea, contrary to Paleozoic calcispheres, which are of lagoonal origin and represent spores of some dasyclads.

This particular example supports only an incipient coral growth. It reaches low tide level and displays a temporary episode of subaerial exposure accompanied by localized dolomitization of dorag type. The occurrence of microfacies with potential primary porosity is very reduced in this type of micritic buildup and corresponds to ephemeral reworking preceding subaerial exposure.

18.7 GENETIC CLASSIFICATION OF PLATFORMS WITH FRONTAL BIOACCUMULATED TO HYDRODYNAMIC BUILDUPS

An ideal classification of these platforms in open marine conditions is based on the nature, shape, number, and importance of frontal buildups. The latter act as barriers under conditions of increasing energy of waves and tidal currents and undergo increasing destruction by reworking and redistribution of the materials. This process of spreading of debris invades the lagoonal environment located behind the buildups and reduces its size. Platforms in semirestricted conditions are also discussed.

Type 1 (Figs. 18.24, 18.25)

This type involves frontal bioaccumulated (bioconstructed locally) and hydrodynamic buildups formed by the association of red algae and larger benthic foraminifers. Although these buildups are well developed, they do not represent an effective barrier to the action of

Chap. 18 / Synthesis   535

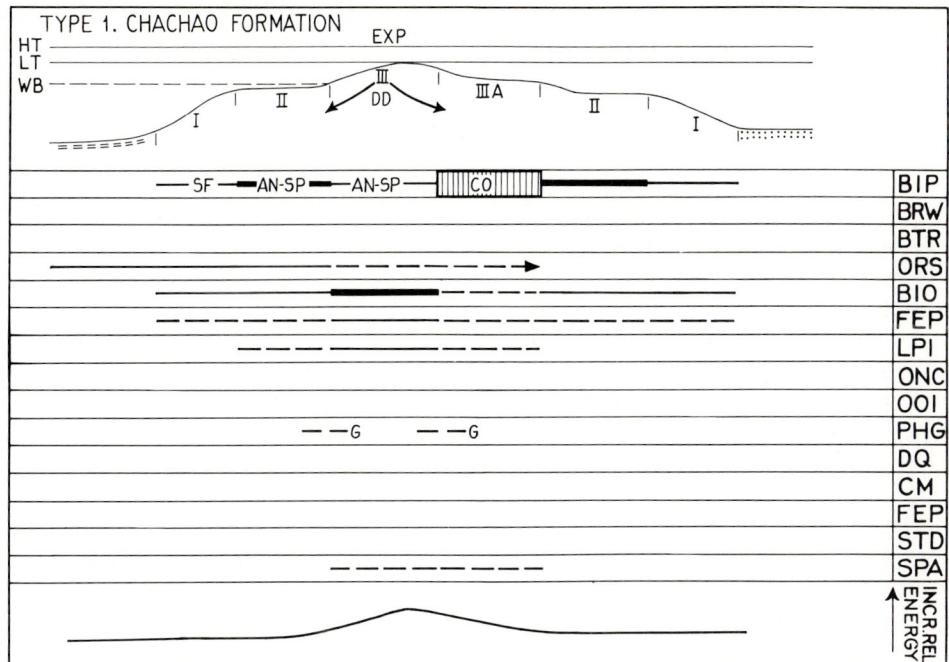

FIGURE 18.23 Integration of depositional models of platforms with frontal bioaccumulated buildups. Type 1. Chachao Formation.

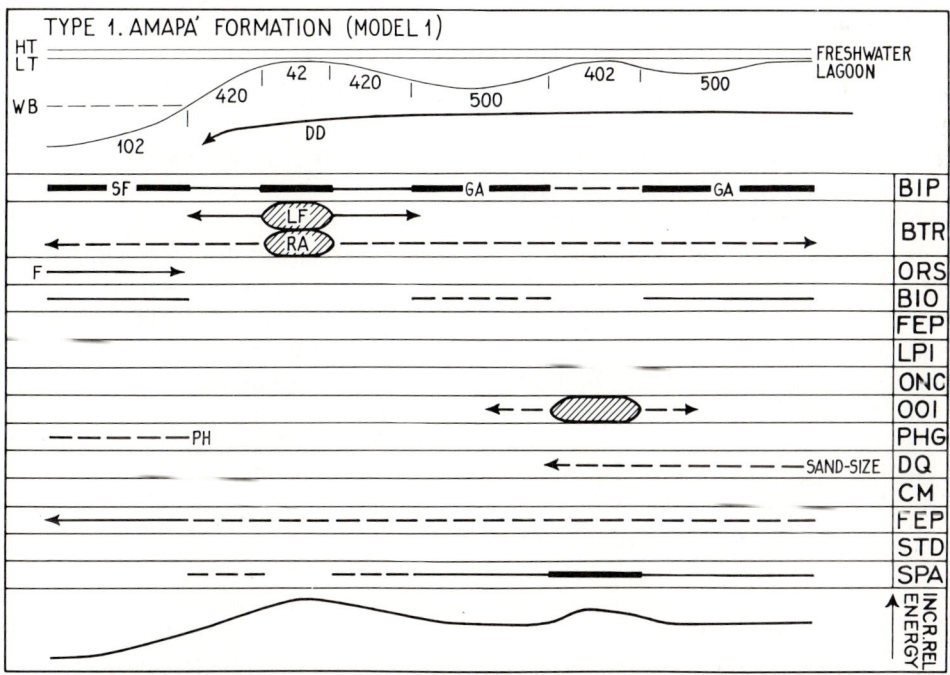

FIGURE 18.24 Integration of depositional models of platforms with frontal bioaccumulated to hydrodynamic buildups. Type 1. Amapá Formation (model 1).

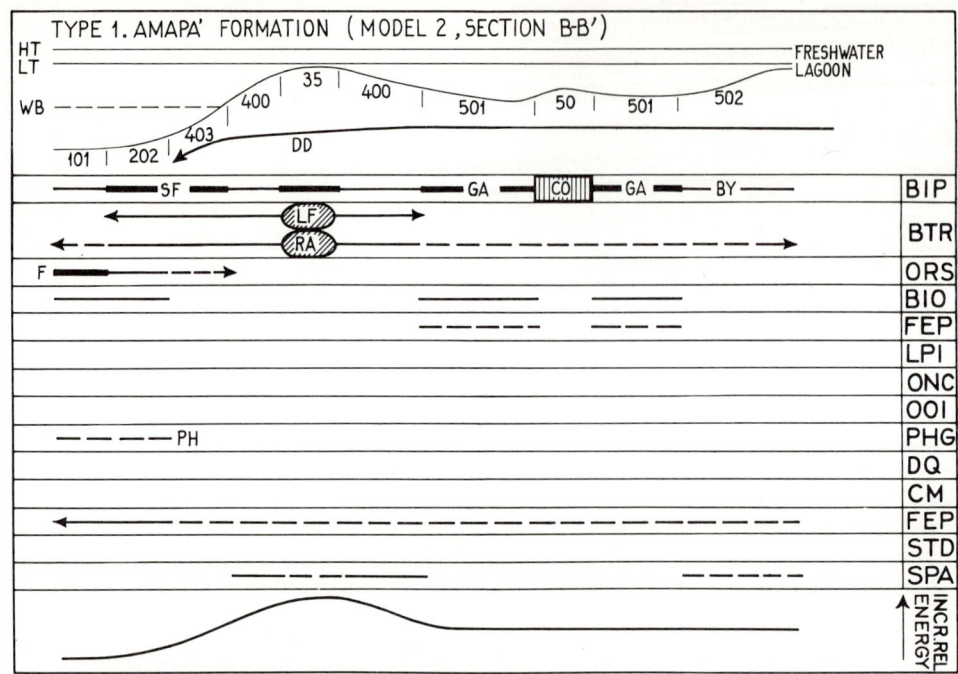

FIGURE 18.25 Integration of depositional models of platforms with frontal bioaccumulated to hydrodynamic buildups. Type 1. Amapá Formation (model 2, section B-B′).

waves and tidal currents. Furthermore, they are interrupted by large transverse channels. In one situation (Fig. 18.24), the integration of model 1 of the Amapá Formation (Paleocene–Middle Miocene) of the Foz do Amazonas Basin, Brazil (Figs. 11.69, 11.70), the lagoonal environment is relatively open and agitated and displays hydrodynamic oolitic buildups developed among sparite-cemented calcarenites with dasyclads. As a consequence, microfacies with potential primary porosity are only moderately represented in both bioclastic flanks of the frontal buildup (which itself has an interstitial micrite matrix) but are well developed in the lagoonal environment. However, an unusually generalized dorag process generated by circulation of undersaturated waters from adjacent freshwater lagoons can introduce, into the entire carbonate system, additional porosity by dissolution, mainly in the lagoonal carbonates, and by dolomitization, mainly in the frontal buildup.

In a second situation (Fig. 18.25), the frontal buildup consists of an association of red algae–larger benthic foraminifers–encrusting foraminifers. It represents integration of model 2, section B-B′ of the Amapá Formation (Figs. 11.79, 11.82), which acts as a more effective barrier. The lagoon remains relatively open with scattered small buildups constructed by finger corals among dasyclad-bryozoan calcarenites with micrite matrix. Microfacies with potential primary porosity are well represented in both bioclastic flanks of the frontal bioaccumulated-bioconstructed buildup which itself has a minor amount of interstitial sparite. The lagoonal environment is almost devoid of microfacies with potential primary porosity except along its shores. As in the previous case, an unusually generalized dorag process provides additional porosity by dissolution and dolomitization to the entire system.

Type 2 (Figs. 18.26, 18.27)

In this type, frontal buildups predominantly are hydrodynamic and consist of a concentration of bioclasts derived from reworking of the seaward slope microfacies rich in benthic fauna, and of lithic pellets and intraclasts originating from the reworking and subaerial exposure of lagoonal sediments. This clear expression of tidal current action indicates shallow lagoons with intense reworking located behind frontal buildups which afford little protection.

In one case (Fig. 18.26) corresponding to integration of the *Nuia* model of the Pogonip Group (Lower Ordovician) of southeast Nevada (Figs. 11.6, 11.7), the frontal hydrodynamic buildup consists of bioclasts of crinoids, *Nuia,* lithic pellets, and intraclasts. Most of the latter originate from the lagoon, but others are produced on the buildup itself by disintegration of beachrock, indicating temporary subaerial exposure. Silt-size detrital quartz is also trapped in the buildup, whereas some intra-

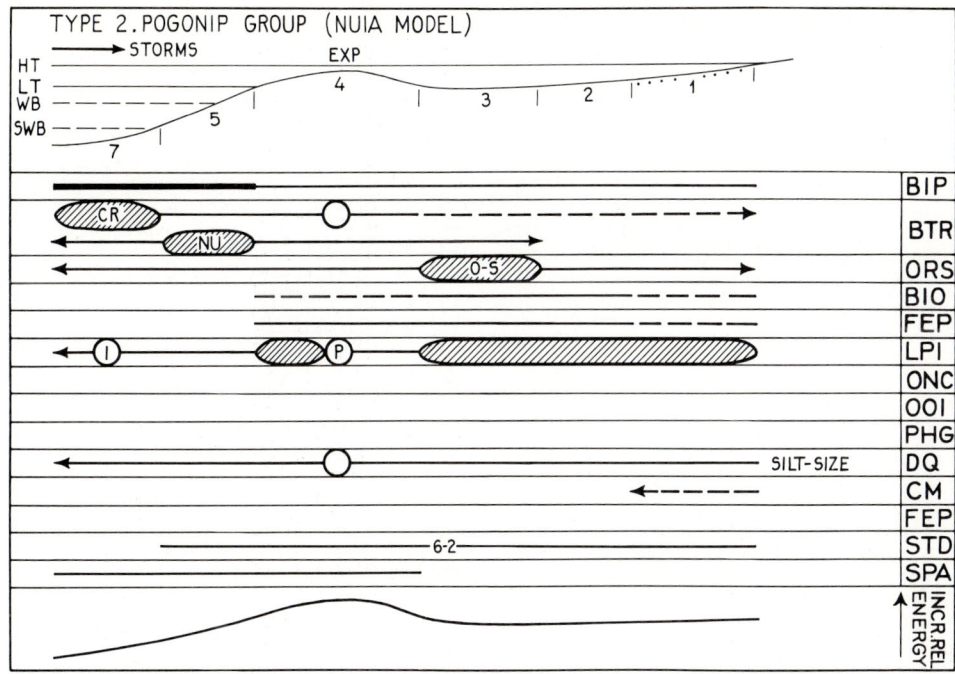

FIGURE 18.26 Integration of depositional models of platforms with frontal bioaccumulated to hydrodynamic buildups. Type 2. Pogonip Group (*Nuia* model).

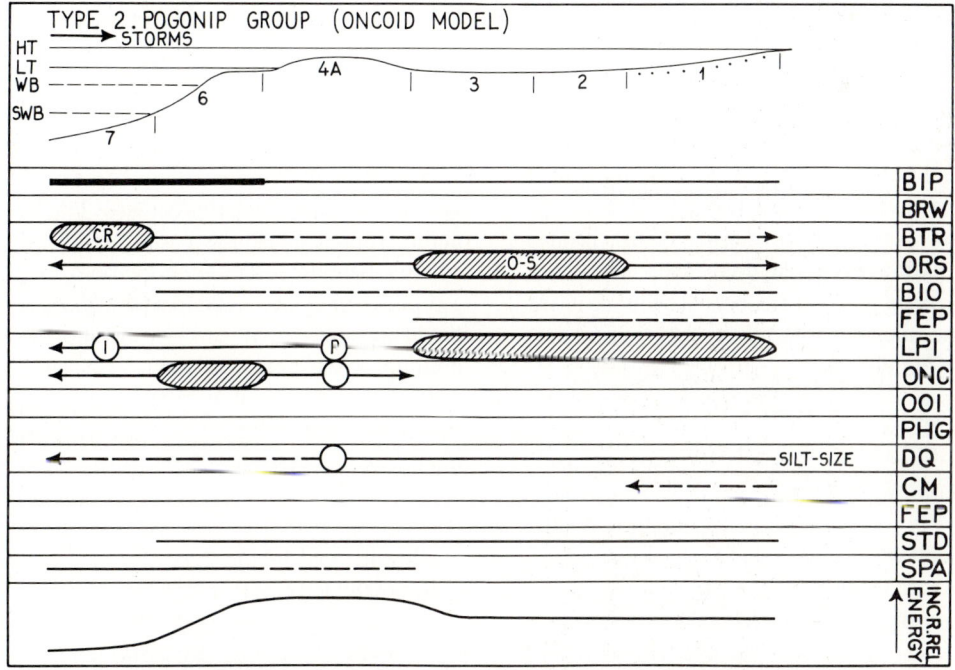

FIGURE 18.27 Integration of depositional models of platforms with frontal bioaccumulated to hydrodynamic buildups. Type 2. Pogonip Group (oncoid model).

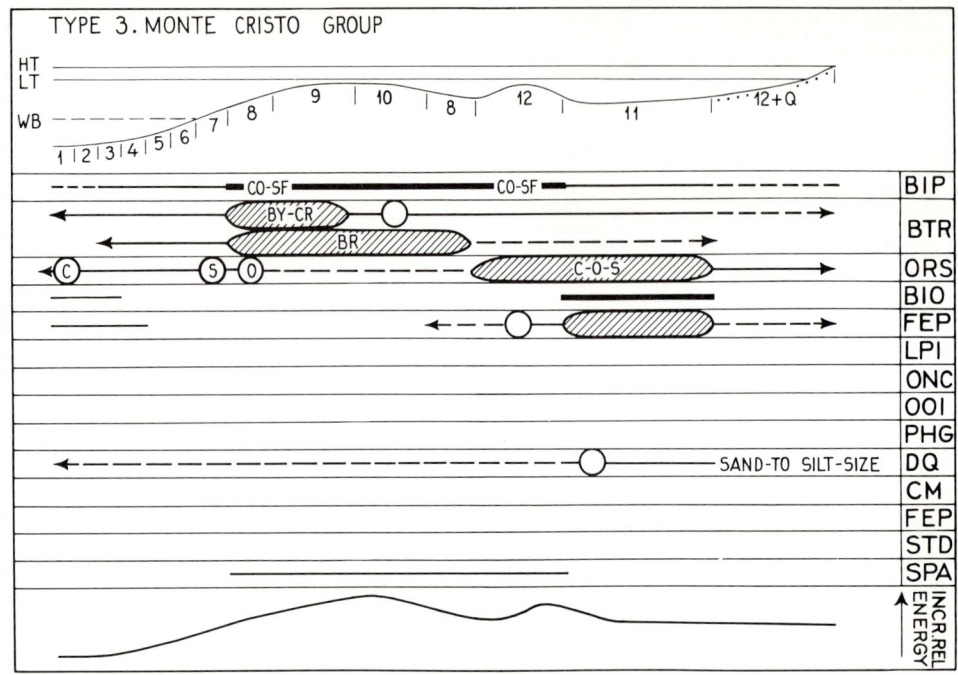

FIGURE 18.28 Integration of depositional models of platforms with frontal bioaccumulated to hydrodynamic buildups. Type 3. Monte Cristo Group.

clasts are transported all the way from the lagoon into the deepest microfacies of the slope. As a consequence of intense reworking of the seaward slope, microfacies with potential primary porosity are located there and in the hydrodynamic buildup itself. In the second case (Fig. 18.27) representing integration of the oncoid model of the same Pogonip Group (Figs. 11.9, 11.10), the frontal hydrodynamic buildup consists of bioclasts of crinoids from slope microfacies, of clasts of oncoids derived from an area immediately seaward where oncoids developed at wave base, and finally, of lithic pellets and intraclasts derived from the reworking of lagoonal sediments. As in the preceding case, silt-size detrital quartz is also trapped in the hydrodynamic buildup and microfacies with potential primary porosity are located on the seaward slope and on the oncoid flat, whereas their importance decreases in the buildup itself.

Type 3 (Figs. 18.28, 18.29)

This type of platform consists of one or several frontal buildups due to the hydrodynamic concentration of various types of bioclasts derived from local bioaccumulations. To this organic contribution may be added the generation *in situ* of ooids or the concentration of fecal pellets, lithic pellets, and intraclasts also formed *in situ*, or partially derived from reworking of the lagoonal environment. Nevertheless, the frontal buildups are still a rather inefficient barrier to the action of waves and tidal currents leading to a lagoon which undergoes appreciable reworking and becomes a source of lithic pellets and intraclasts.

In the first case (Fig. 18.28), which corresponds to the integration of the model of the Monte Cristo Group (Mississippian) of southeast Nevada (Figs. 11.40, 11.41, 11.42), the edge of the platform consists first of a broad hydrodynamic buildup that results mainly from the concentration of bioclasts of crinoids, bryozoans, and brachiopods, followed by a second buildup in which fecal pellets derived from the lagoon are associated with bioclasts of crinoids and bryozoans. The entire hydrodynamic complex displays coral growths associated with smaller benthic foraminifers at both ends, specifically *Syringopora* at the seaward and *Lithostrotionella* at the landward end. Some of the protection afforded by the buildups to the lagoon is shown by its important bioturbation, abundant generation of fecal pellets, and rich benthic fauna. The important influx of sand- to silt-size detrital quartz, which generates the shoreline arenaceous microfacies, is stopped in its seaward transport mainly at the back of the hydrodynamic buildup complex. All these conditions lead to the concentration of all microfacies with potential primary porosity in that complex.

In a second case (Fig. 18.29), which corresponds to the integration of the Bird Spring Group (Upper Missourian–Wolfcampian) of southeast Nevada (Figs. 11.49, 11.54), the frontal hydrodynamic buildup consists of bioclasts from a rich fauna dominated by crinoids, bryo-

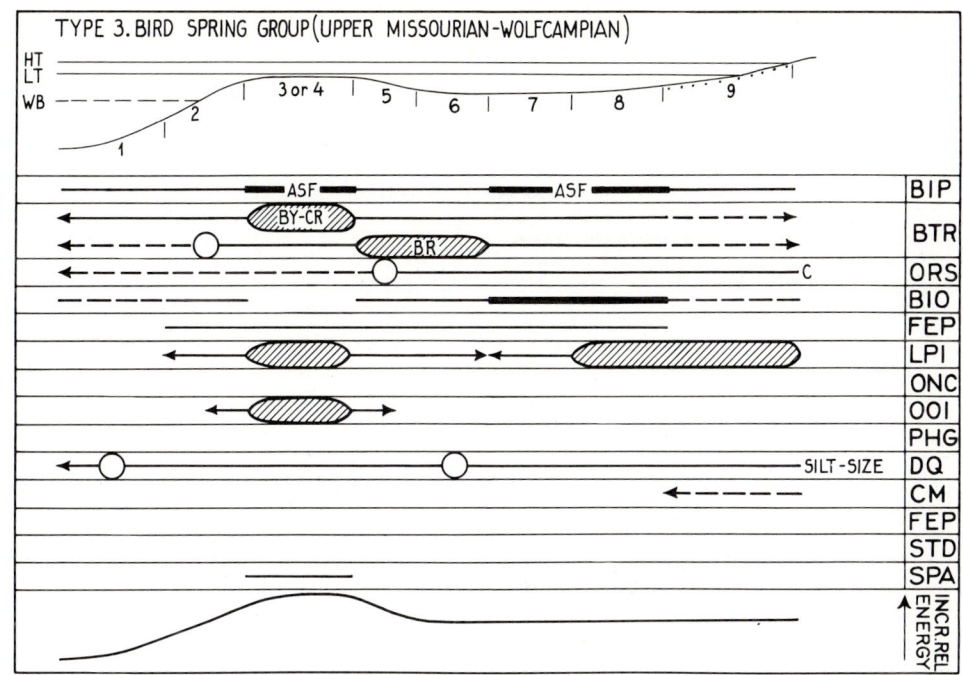

FIGURE 18.29 Integration of depositional models of platforms with frontal bioaccumulated to hydrodynamic buildups. Type 3. Bird Spring Group (Upper Missourian–Wolfcampian).

zoans, and agglutinated smaller benthic foraminifers. Oolitization processes occur on the buildup itself, where agitation also generates lithic pellets and intraclasts. The landward slope of the buildup displays microfacies with abundant brachiopods from which bioclasts are widely distributed and eventually concentrated on the seaward slope. The lagoonal environment is well developed with abundant agglutinated smaller benthic foraminifers, strong bioturbation, and reworking on the most landward side with generation of lithic pellets and intraclasts. The influx of detrital quartz which generates the shoreline arenaceous microfacies is concentrated during its seaward transport at both the back and front of the hydrodynamic buildup complex. Only the central part of the latter displays potential primary porosity.

Type 4 (Figs. 18.30, 18.31)

This type of platform consists of well-developed frontal buildups of a hydrodynamic nature resulting from the mechanical concentration of various bioclasts, ooids, lithic pellets, and intraclasts, all of local origin. All these constituents are dispersed mainly by tidal currents both in a landward and seaward direction. Hence the protected and quiet lagoon located behind the buildups receives most of its bioclasts from the buildups. The lagoon may even display bioaccumulated microfacies in addition to an important *in situ* production of fecal pellets related to strong bioturbation.

In one case (Fig. 18.30), corresponding to integration of model 3 of the St. Louis Limestone (Middle Mississippian) of the Illinois Basin (Figs. 3.10, 3.11), the broad and complex hydrodynamic buildup is an association of several types of sparite-cemented biocalcarenites of higher energy, consisting mainly of bioclasts of crinoids, bryozoans, and smaller benthic foraminifers along with abundant ooids, lithic pellets, and intraclasts. Both landward and seaward flanks of the hydrodynamic complex show similar microfacies but with interstitial micrite matrix. The low-energy, protected lagoon has bioaccumulations of bryozoans and brachiopods, whereas its most landward portion is practically devoid of organisms because of the effect of heavy brines of seepage reflux from the adjacent sabkha. Silt-size eolian quartz concentrates in that particular area, whereas sand-size grains, during their seaward transport, concentrate in some microfacies of the buildup. The higher-energy environment of these buildup microfacies also gives them primary porosity. Some of it may occur in the shoreward part of the lagoon due to seepage reflux dolomitization.

In a second case (Fig. 18.31), representing the integration of the Ste. Genevieve Limestone (Middle Mississippian) of the southern part of the Illinois Basin (Figs. 12.10, 12.13), the broad frontal hydrodynamic buildup consists entirely of sparite-cemented oolitic calcarenites. The cores of the ooids are bioclasts of crinoids, bryozoans, smaller benthic foraminifers, and lithic pellets. The dispersion of the ooids by tidal currents both

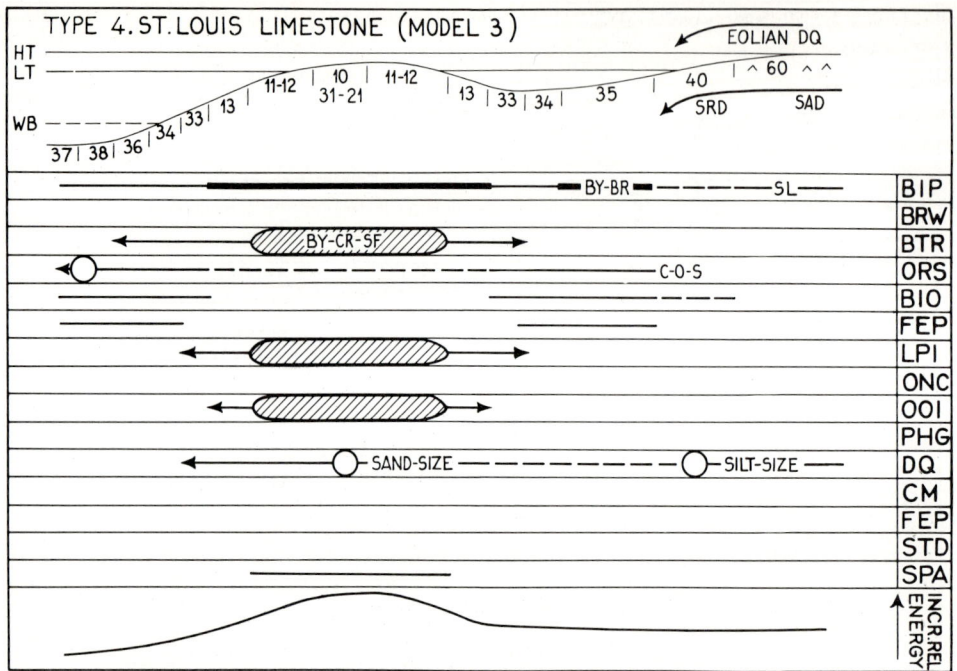

FIGURE 18.30 Integration of depositional models of platforms with frontal bioaccumulated to hydrodynamic buildups. Type 4. St. Louis Limestone (model 3).

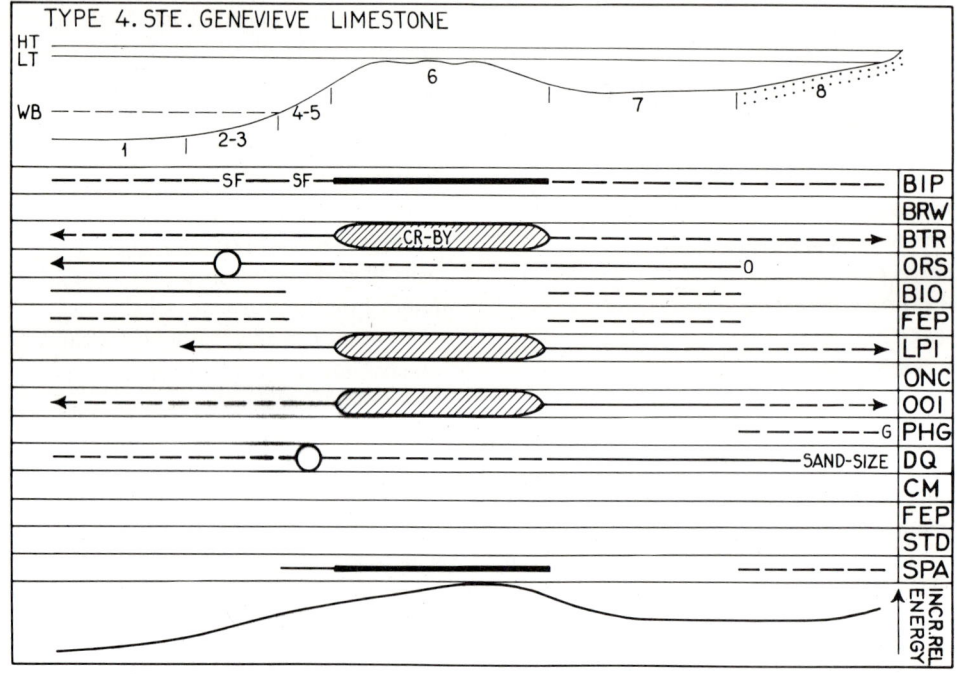

FIGURE 18.31 Integration of depositional models of platforms with frontal bioaccumulated to hydrodynamic buildups. Type 4. Ste. Genevieve Limestone.

seaward and landward from the buildup involves the entire carbonate system. The lagoon appears well protected and grades landward into estuaries with pure quartz arenites. Detrital quartz originating from the latter is transported seaward but concentrates mainly behind the buildup and to a minor extent in front. Microfacies with potential primary porosity characterize the oolitic buildup and the upper part of its frontal slope.

Type 5 (Figs. 18.32, 18.33, 18.34)

The platforms of this type are very broad and reach a greater degree of complexity. They consist of the juxtaposition of several bioaccumulated to hydrodynamic buildups (often in greater number than shown on the integrated models). Each buildup is characterized by its own particular benthic fauna whose bioclasts, along with ooids, lithic pellets, and intraclasts, are dispersed by waves and tidal currents into the lower-energy intervening topographically lower areas that represent interbuildup troughs. Lagoons, strictly speaking, are often reduced to a narrow space between the most internal buildup and the shoreline. The latter ranges from inactive carbonate terranes to sabkhas and distal deltaic environments.

In the first case (Fig. 18.32), which represents the integration of the Salem Limestone (Middle Mississippian) of the southwest margin of the Illinois Basin (Figs. 12.5, 12.6), the platform consists of two distinct hydrodynamic buildups reaching subaerial exposure. The frontal one is made of larger-energy oolitic biocalcarenites with abundant endothyrids, bioclasts of crinoids, and brachiopods together with lithic pellets, intraclasts, and oncoids. Most of these constituents are dispersed by tidal currents toward the seaward slope of the platform and toward the outer open lagoon which precedes the second hydrodynamic buildup. The latter consists of sparite-cemented crinoid-bryozoan calcarenites with a smaller degree of dispersion of bioclasts. Most of the silt-size detrital quartz of eolian origin settles in the inner restricted lagoon, whereas the sand-size grains, transported seaward, concentrate in the more internal hydrodynamic buildup. Microfacies with potential primary porosity are limited to both buildups and their frontal microfacies. Dorag dolomitization in each exposed buildup improves porosity, whereas the seepage reflux affecting the inner lagoon shows little effect.

In a second case (Fig. 18.33), which represents the integration of the Glen Dean Formation (Middle Mississippian) of the western shelf of the Illinois Basin (Figs. 12.36, 12.37, 12.38), the platform consists of two hydrodynamic buildups reaching subaerial exposure preceded seaward by a smaller one reaching wave base level. The greater energy of the system leads to a wide dispersion by tidal currents of bioclasts, ooids, lithic pellets, and intraclasts from the three buildups toward the intervening troughs and the seaward platform slope. The outer buildup is characterized by a coarse and well-sorted crinoidal calcarenite with syntaxial rim cement. The first

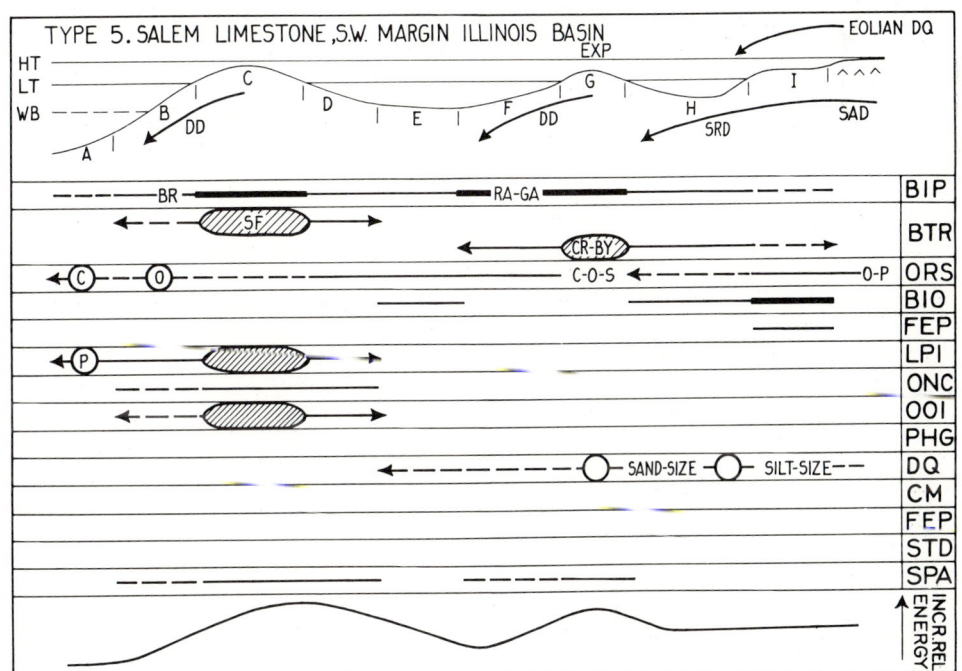

FIGURE 18.32 Integration of depositional models of platforms with frontal bioaccumulated to hydrodynamic buildups. Type 5. Salem Limestone, southwest margin of Illinois Basin.

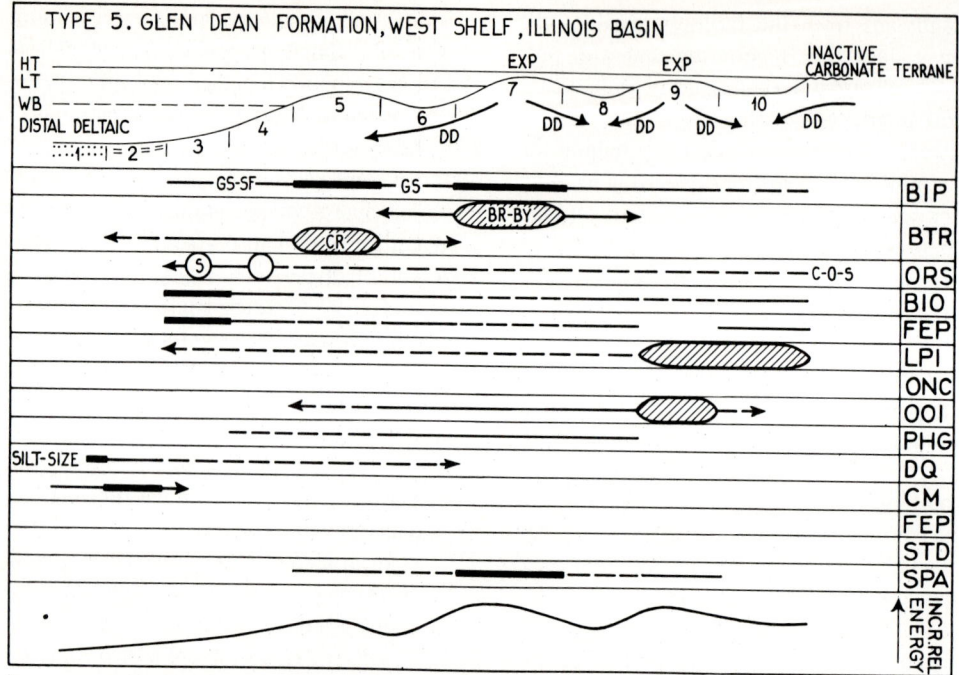

FIGURE 18.33 Integration of depositional models of platforms with frontal bioaccumulated to hydrodynamic buildups. Type 5. Glen Dean Formation, western shelf of Illinois Basin.

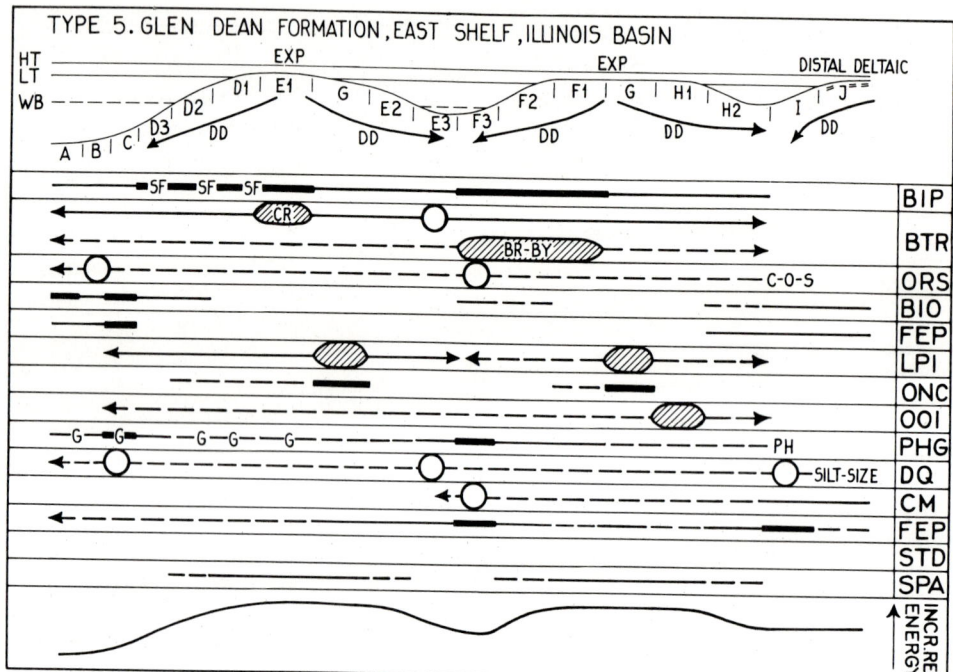

FIGURE 18.34 Integration of depositional models of platforms with frontal bioaccumulated to hydrodynamic buildups. Type 5. Glen Dean Formation, eastern shelf of Illinois Basin.

hydrodynamic buildup at the edge of the platform consists of bryozoan-brachiopod calcarenites with sparite cement, and the more internal buildup consists of oolitic calcarenites with lithic pellets, intraclasts, and sparite cement. Of interest are concentrations of gastropods on both slopes of the outer smaller buildup and the location of certain types of bryozoans, such as *Archimedes* on the outer slope and *Callocladia* on the first major buildup.

This model is adjacent to an inactive carbonate terrane and receives detrital quartz and clay minerals from the seaward side as distal deltaic products originating from the opposite shelf of the basin. Microfacies with potential primary porosity coincide with the three hydrodynamic buildups and extend also to their intervening troughs as a consequence of the relatively higher energy of this carbonate system. Potential primary porosity is improved by well-developed dorag dolomitization related to each exposed buildup and to waters that originate from the adjacent karstic terrane.

In a third case (Fig. 18.34), which corresponds to the integration of the same Glen Dean Formation (Middle Mississippian) but on the eastern shelf of the Illinois Basin (Figs. 12.24, 12.25, 12.26), the platform is extremely wide. It consists of the juxtaposition of two complex hydrodynamic buildups which are subaerially exposed and separated by a narrow intervening trough. The lagoon leading to a distal deltaic environment is also very reduced. In spite of the extremely widespread dispersal of various bioclasts, oncoids, ooids, and lithic pellets by tidal currents, each of the hydrodynamic buildups preserved its own ecological zonation. The external buildup displays, in a landward direction, the following zonation: smaller benthic foraminifers–crinoids–oncoids, with dispersion of the last two constituents. The internal buildup shows, in the same direction, the following zonation: bryozoans–brachiopods–oncoids–ooids, with an equally strong dispersion of all constituents. The silt-size detrital quartz, during its seaward transport, is trapped mainly in the depressions of the platform and on the external slope. Microfacies with potential primary porosity are located in the central part of both buildups with decreasing importance along their flanks. Potential primary porosity is improved by well-developed dorag dolomitization coinciding with the exposed portions of the buildups and from waters originating from the adjacent delta complex.

Type 6 (Figs. 18.35, 18.36)

Platforms of this type developed on the passive margin of Brazil in the proto South Atlantic during marine semirestricted conditions which followed the evaporitic phase. They are characterized by bioaccumulated to hydrodynamic buildups which consist almost entirely of oncoids and ooids and are separated by intervening troughs filled by several types of calcilutites with benthic and plank-

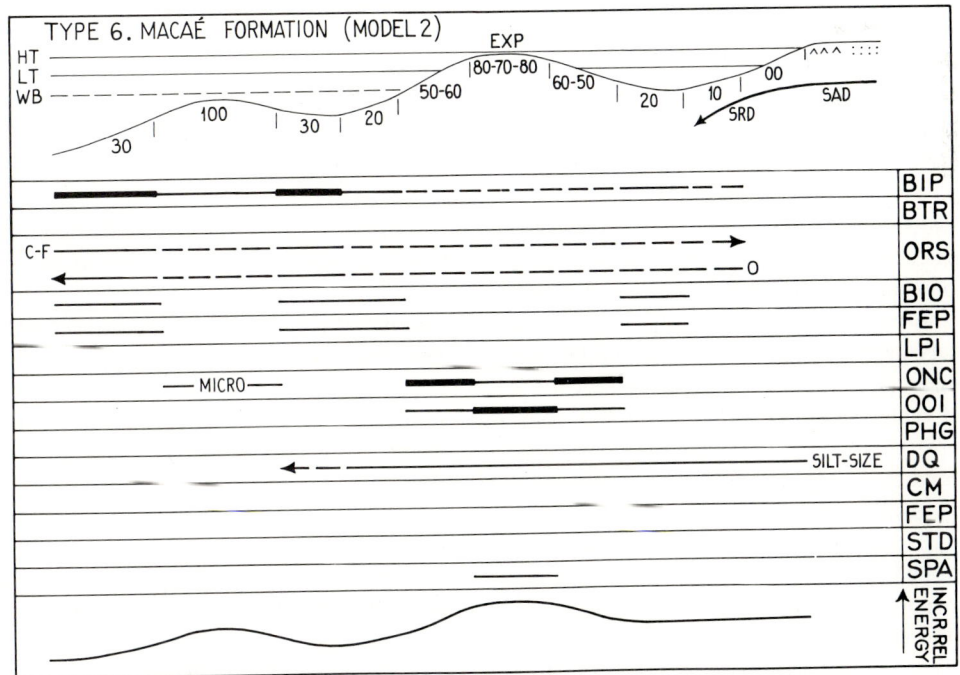

FIGURE 18.35 Integration of depositional models of platforms with frontal bioaccumulated to hydrodynamic buildups. Type 6. Macaé Formation (model 2).

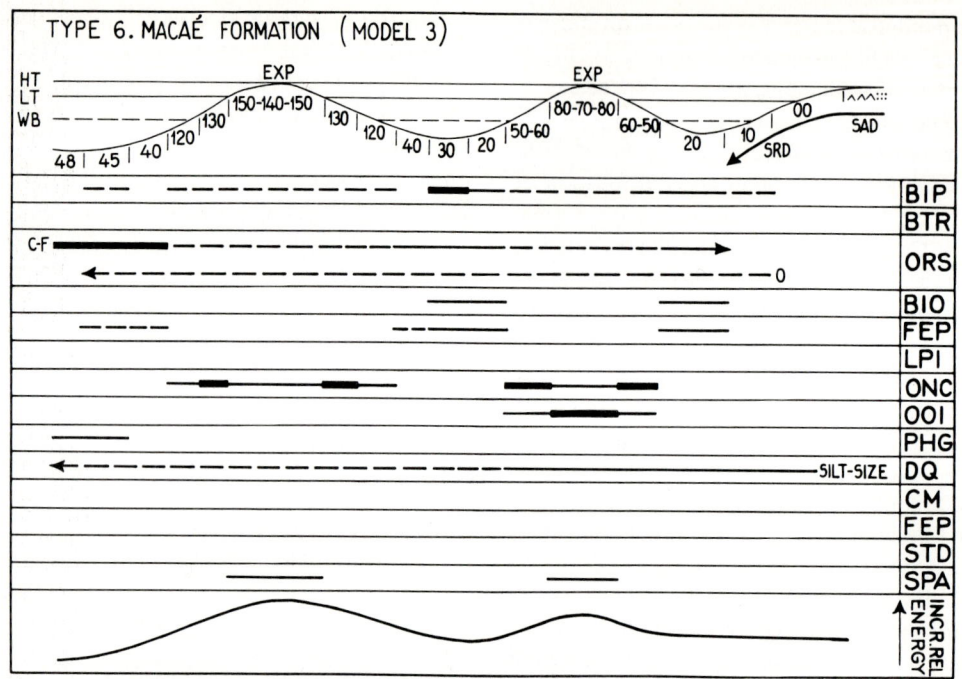

FIGURE 18.36 Integration of depositional models of platforms with frontal bioaccumulated to hydrodynamic buildups. Type 6. Macaé Formation (model 3).

tonic fauna. The energy level of these platforms is relatively small, just sufficient to generate oncoids and ooids, but incapable of dispersing any benthic constituents away from the buildups.

In one case (Fig. 18.35), which integrates model 2 of the Macaé Formation (Albian-Cenomanian) of the Campos Basin (Fig. 11.63), the platform consists of an offshore and relatively small hydrodynamic buildup of micro-oncoids which barely reaches the wave base level and is flanked by microfacies with benthic fauna, and of a larger shoreface hydrodynamic buildup of oncoids and ooids reaching subaerial exposure. The foreshore trough is relatively poor in benthic fauna. Calcispheres and planktonic foraminifers of oceanic origin penetrate the entire carbonate system. Microfacies with potential primary porosity are limited to the central part of the shoreface buildup and are enhanced by beachrock conditions and vadose dissolution. Seepage reflux dolomitization causes no appreciable effect.

In a second case (Fig. 18.36), which corresponds to the integration of model 3 of the Macaé Formation (Fig. 11.64), the better developed platform consists of two juxtaposed hydrodynamic buildups, both displaying subaerial exposure. The offshore one is only oncoidal, whereas the shoreface one is both oncoidal and oolitic. The intervening depressions are relatively poor in benthic fauna, whereas the penetration of planktonic foraminifers and calcispheres of oceanic origin is strong across the entire system. Microfacies with potential primary porosity are limited to the central parts of both buildups and enhanced by beachrock conditions and vadose dissolution. In this case also, seepage reflux dolomitization causes no appreciable effect.

18.8 GENERAL SYNTHESIS OF PLATFORMS WITH FRONTAL BIOACCUMULATED TO HYDRODYNAMIC BUILDUPS

Features and processes of general significance for these types of platforms are examined below.

Fair-Weather Wave Base, Tidal Currents, and Submarine Morphology

In these platforms, contrary to ramps, the effect of fair-weather wave base becomes limited to frontal slopes. It can involve occasionally certain interbuildup troughs (Figs. 18.34, 18.36), but as a whole, this factor becomes negligible and is replaced by the action of tidal currents. Of importance is the relationship between the shape and size of the buildups and their effectiveness as obstacles to tidal currents. Indeed, some buildups barely reach low tide level and undergo very short periods of subaerial exposure (Figs. 18.24 to 18.31), whereas others reach well above low tide level and display characteristic features reaching paleosoils (Figs. 18.32 to 18.36). Another typical consequence of these conditions is the behavior of the lagoonal environment, which ranges from fairly open (Figs. 18.24, 18.25) through conditions of variable reworking (Figs. 18.26 to 18.29) to low energy and well-protected environments (Figs. 18.30, 18.31).

These platforms are characterized as a result of tidal current action by the formation of hydrodynamic buildups and related dispersion of bioclasts, lithic pellets, and intraclasts along their flanks in deeper water, either in the case of frontal buildups down the platform slope, or in the case of more internal buildups, toward the intervening depressions. This mechanism of dispersion, in both a landward and a seaward direction, is selective, and a number of benthic organisms remain essentially *in situ*, preserving their ecological zoning.

The process of dispersion in platforms of types 1, 3, and 4 (Figs. 18.24, 18.25, 18.28 to 18.31) affects mainly the frontal buildup and the range of dispersion varies according to the type of bioclast. In all instances, transport may be accompanied by concentrations of certain bioclasts at places where their grain size is in equilibrium with local conditions of energy. This situation is well shown in platforms of type 3 (Fig. 18.28, 18.29).

In platforms consisting of several juxtaposed buildups such as type 5 (Figs. 18.32, 18.33, 18.34), each buildup, although retaining its ecologically zoned benthic fauna, undergoes its own process of dispersion of bioclasts, lithic pellets, intraclasts, and oncoids, which leads to an extremely complex pattern, particularly in the case of larger energy and broad platforms (Fig. 18.34). The unusual case in which an important transportation of bioclasts occurs from deeper to shallower conditions along a platform outer slope is seen in platforms of type 2 (Figs. 18.26, 18.27). It appears related to storm action and is discussed in more detail below.

Whenever lagoons undergo appreciable reworking, the generated lithic pellets and intraclasts together with fecal pellets produced by bioturbation are dispersed mainly seaward. They may concentrate at particular places during their transport or contribute to the generation of hydrodynamic buildups, until they reach, through tidal channels, the platform slopes. Such is the case of platforms of type 2 (Figs. 18.26, 18.27), type 3 (Figs. 18.28, 18.29), and in some of type 5 (Figs. 18.33, 18.34).

In all these platforms, the seaward transport of delicate lagoonal organisms by suspension, together with *Chara* oogonia and stems and plant debris, is extremely common. These constituents reach platform slopes, eventually, where they settle in deeper water (Figs. 18.26, 18.27, 18.30, 18.31). Settling or concentration of these constituents may occur at different places on the frontal slope or in the back of major buildups (Figs. 18.28, 18.29, 18.32, 18.33). In broad platforms with larger energy consisting of several juxtaposed buildups, some of these constituents settle in interbuildup troughs (Fig. 18.34).

Mesozoic and Cenozoic platforms display two opposite and major directions of transportation in suspension; planktonic foraminifers and calcispheres originating from the open sea penetrate the carbonate system, whereas lagoonal constituents are transported seaward (Figs. 18.24, 18.25, 18.35, 18.36).

Coexistence of *in Situ* and Transported Organic and Inorganic Constituents

Alternating tidal currents and their related seaward and landward transport of certain benthic and inorganic constituents is demonstrated indirectly by the behavior of constituents which remain *in situ* and are still zoned ecologically even in the most reworked hydrodynamic buildups. Typical cases are represented by smaller benthic foraminifers and green algae in platforms of type 1 (Figs. 18.24, 18.25); by corals, smaller benthic foraminifers, and agglutinated smaller benthic foraminifers in platforms of type 3 (Figs. 18.28, 18.29); by bryozoans and brachiopods in platforms of type 4 (Fig. 18.30); and by red and green algae, gastropods, and smaller benthic foraminifers in platforms of type 5 (Figs. 18.32, 18.33, 18.34). Bioturbation and production of fecal pellets are often, but not necessarily, associated with the same degree of frequency (Figs. 18.26, 18.27). Instances of distinct fecal pellet production occur in slope and lagoonal conditions (Figs. 18.30, 18.31) and in depressions between buildups (Fig. 18.33). Lagoonal fecal pellets are transported seaward and participate in the formation of hydrodynamic buildups (Fig. 18.28).

Distribution Pattern of Oolitic Microfacies

In platforms of type 1 (Fig. 18.24), oolitic microfacies are limited to small hydrodynamic buildups located in lagoons between green algae calcarenites. In platforms of type 3 (Fig. 18.29) and type 4 (Figs. 18.30, 18.31), oolitization processes are well developed in the frontal hydrodynamic buildups, and dispersion of ooids by tidal currents increases to the extent that when the buildups become entirely oolitic (Fig. 18.31), reworked, broken, and abraded ooids are distributed across the entire carbonate system. In platforms of type 5 consisting of several juxtaposed hydrodynamic buildups, ooids may be found in the frontal buildups (Fig. 18.32), but they generally occur in more restricted conditions associated with the most internal buildup (Figs. 18.33, 18.34). They show, nevertheless, appreciable dispersion. In semirestricted marine conditions of platforms of type 6 (Figs. 18.35, 18.36), oolitization processes characterize the central part of the more internal buildup and no dispersion occurs from this isolated area.

Distribution Pattern of Oncoids

In one platform of type 2 (Fig. 18.27), oncoids develop on a bench at wave level in front of the hydrodynamic buildup in which they concentrate mechanically. In platforms of type 5, consisting of juxtaposed hydrodynamic buildups, oncoids occur either in the frontal buildups (Fig. 18.32), or in the middle and rear part of all buildups (Fig. 18.34). In one semirestricted marine platform of type 6 (Fig. 18.35), the small offshore hydrodynamic buildup consists of micro-oncoids, and the larger shoreface buildup is oncoidal-oolitic with best-developed oncoids along its margins. A similar situation is shown by another case (Fig. 18.36) where the large offshore buildup is entirely oncolitic and the shoreface buildup oncolitic and oolitic.

Distribution Pattern of Phosphates and Glauconite

Phosphatization indicating temporary occurrence of oxygen-depleted waters occurs in lower and upper slope (forebank) microfacies of platforms of type 1 (Figs. 18.24, 18.25). Glauconite is encountered in estuarine conditions in the shoreward part of platforms of type 4 (Fig. 18.31). The most important occurrence of both phosphate and glauconite is in platforms of type 5, which are mixed carbonate-siliciclastic models (Figs. 18.33, 18.34) submitted to deltaic influence. Both minerals reach their maximum abundance in relatively low energy interbuildup troughs or in platform frontal slope areas.

Distribution Pattern of Extrabasinal Materials

As in carbonate ramps, the influx of sand- to silt-size quartz varies as a function of the geological composition, morphology, and climate of the source areas. Its distribution in platforms with frontal hydrodynamic buildups is complicated by morphologic variation of the platforms themselves. In those of type 2 (Figs. 18.26, 18.27), silt-size quartz is concentrated during its seaward transport in the buildups themselves. In the type 3 platform (Figs. 18.28, 18.29), it is concentrated immediately behind the buildups and may settle eventually in deeper slope environment as well. In a platform of type 4 (Fig. 18.30), silt-size quartz of eolian origin settles in the most shoreward portion of the lagoon, whereas sand-size quartz concentrates in the hydrodynamic buildup itself. Similar conditions are also shown by one example of a type 5 platform (Fig. 18.32) where sand-size quartz concentrates in the more internal hydrodynamic buildup and is not transported beyond the outer open lagoon. In a platform of type 5 (Fig. 18.33), which is adjacent to an inactive carbonate terrane, silt-size detrital quartz of distal deltaic origin is transported from the opposite side of the basin and penetrates the carbonate system no farther than the depression behind the most external buildup. In a type 5 platform consisting of two juxtaposed broad buildups (Fig. 18.34), silt-size quartz of deltaic origin settles during its seaward transport in three distinct low-energy environments: lagoon, interbuildup trough, and frontal slope below the wave base level.

Clay minerals generally are not frequent in higher-energy platforms with frontal bioaccumulated to hydrodynamic buildups. Whenever present (Figs. 18.26, 18.27, 18.29), they are deposited in the most shoreward portions of lagoons. In platforms receiving clay minerals of deltaic origin (Figs. 18.33, 18.34), they behave like silt-size detrital quartz and settle in low-energy and relatively deeper areas.

In these platforms, extrabasinal iron, present in form of pyrite, is rather rare and tends to behave as clay minerals concentrating in frontal slopes (Figs. 18.24, 18.25). In type 5 platforms submitted to deltaic influences (Fig. 18.34), pyrite is widespread and, as silt-size quartz and clay minerals, it concentrates in lagoon and low-energy interbuildup depressions.

Distribution Pattern of Dolomitization

In these platforms, all major processes of early dolomitization are represented because of the great variety of their hinterlands, of climate, and of frequent subaerial exposure of the hydrodynamic buildups. Platforms of type 1 (Figs. 18.24, 18.25) are dolomitized by large-scale dorag processes, being adjacent to well-developed freshwater lagoons fed by fan deltas. One platform of type 4 (Fig. 18.30) and those of type 6 (Figs. 18.35, 18.36) grade landward into sabkha evaporitic flats and seepage reflux of brines is responsible for scarcity of fauna in the landward portion of the lagoon and for early dolomitization of its sediments. In platforms of type 5, consisting of juxtaposed hydrodynamic buildups with exposure features, each buildup possesses an active freshwater lens in its upper part generating a mixed freshwater-seawater (dorag-type) dolomitization. This process may dolomitize only the seaward flank of the buildup (Fig. 18.32), or both flanks (Figs. 18.33, 18.34), and reach well into the deposits of interbuildup troughs. It may be combined with seepage reflux dolomitization from adjacent sabkhas (Fig. 18.32) or with another dorag process due to freshwater flow from an adjacent carbonate terrane (Fig. 18.33) or from waters derived from a delta (Fig. 18.34). In platforms of type 6, only sediments of their foreshore trough are dolomitized by seepage reflux of brines from adjacent sabkhas (Figs. 18.35, 18.36).

Distribution Pattern of Storm Deposits

As mentioned earlier, platforms of type 2 (Figs. 18.26, 18.27) represent the only case where the frontal hydrodynamic buildups consist mainly of bioclasts of crinoids, *Nuia*, and clasts of oncoids transported from deeper to shallower conditions. This situation, combined with the widespread and frequent presence (up to 130 beds) of intraclastic calcirudites in the *Nuia* and oncoid models, consisting of clasts of microfacies 5 to 2 and 6 to 2, demonstrates the powerful action of storms transporting materials landward across the entire carbonate system. All the other types of investigated platforms with hydrodynamic buildups do not show any recognizable storm deposits.

Distribution Pattern of Potential Primary to Early Diagenetic Reservoirs

A schematic diagram (Fig. 18.37) shows these types of platforms associated in three main groups: type 1, types 2 to 4, and types 5 and 6. These groups express the combination of increasing energy, increasing action of tidal currents, and effect of morphology of various platform types. The great extent of microfacies with potential primary porosity in type 1 results from the open character of the system to the action of waves and tidal currents. In the second group (types 2 to 4), only the main and well-developed frontal buildups display primary porosity, whereas in the third group (types 5 and 6), each of the several buildups shows favorable conditions. Further potential early diagenetic porosity by dolomitization is introduced by seepage reflux of brines from adjacent sabkhas (Figs. 18.30, 18.32, 18.35, 18.36), by mixed freshwater-seawater systems of dorag type related to subaerial exposure of hydrodynamic buildups (Figs. 18.32, 18.33, 18.34), and even by the flow of freshwater from inactive carbonate source areas (Fig. 18.33), and from adjacent deltas (Fig. 18.34).

18.9 GENETIC CLASSIFICATION OF PLATFORMS WITH FRONTAL BIOCONSTRUCTED TO HYDRODYNAMIC BUILDUPS

An ideal classification of these platforms in open marine conditions is based on the nature, shape, number, and importance of frontal bioconstructed buildups with increasing energy of the action of tidal currents. However, in contrast to platforms with bioaccumulated to hydrodynamic buildups, which show, under increasing action of tidal currents (types 1 to 5), a greater number of hydrodynamic buildups spreading their debris over the platform at the expense of the lagoon, this group of platforms displays increasingly wave- and tide-resistant frontal buildups which protect the more internal areas where various types of small buildups develop. They shed very little debris and lagoons remain wide, complex, and relatively quiet. In many instances, the precise location of what would be called the platform "edge" can be disputed.

Type 1 (Figs. 18.38, 18.39)

In this platform, one or several frontal bioconstructed buildups do not reach or barely reach the wave base. They precede the shallowest and larger one, which supports the lagoonal environment.

In the first case (Fig. 18.38), which represents integration of the Beaverhill Lake Formation (Upper Devonian) of Alberta (Figs. 13.19, 13.20), the platform displays a succession of three bioconstructed buildups of decreasing depth until a well-characterized and bioturbated lagoonal environment is reached which is associ-

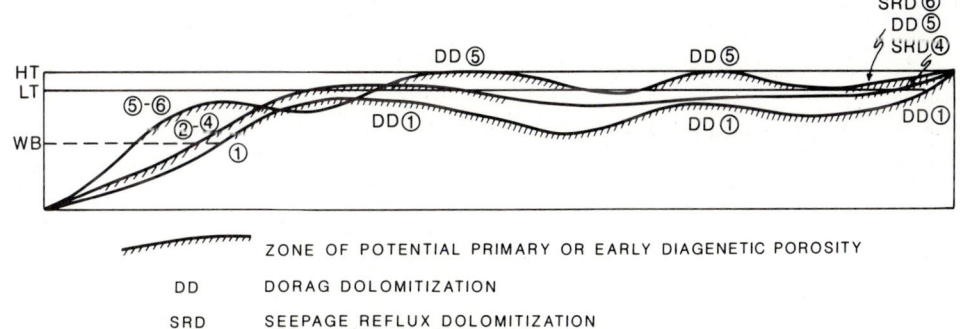

FIGURE 18.37 Distribution of potential primary reservoirs in platforms with frontal bioaccumulated to hydrodynamic buildups. Circled numbers refer to types of models.

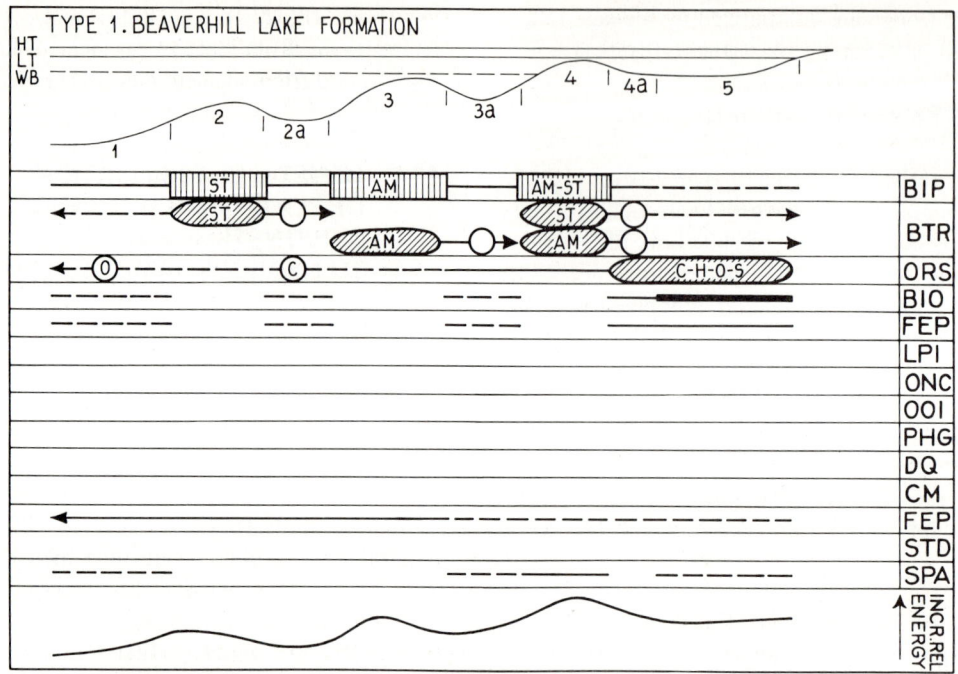

FIGURE 18.38 Integration of depositional models of platforms with frontal bioconstructed to hydrodynamic buildups. Type 1. Beaverhill Lake Formation.

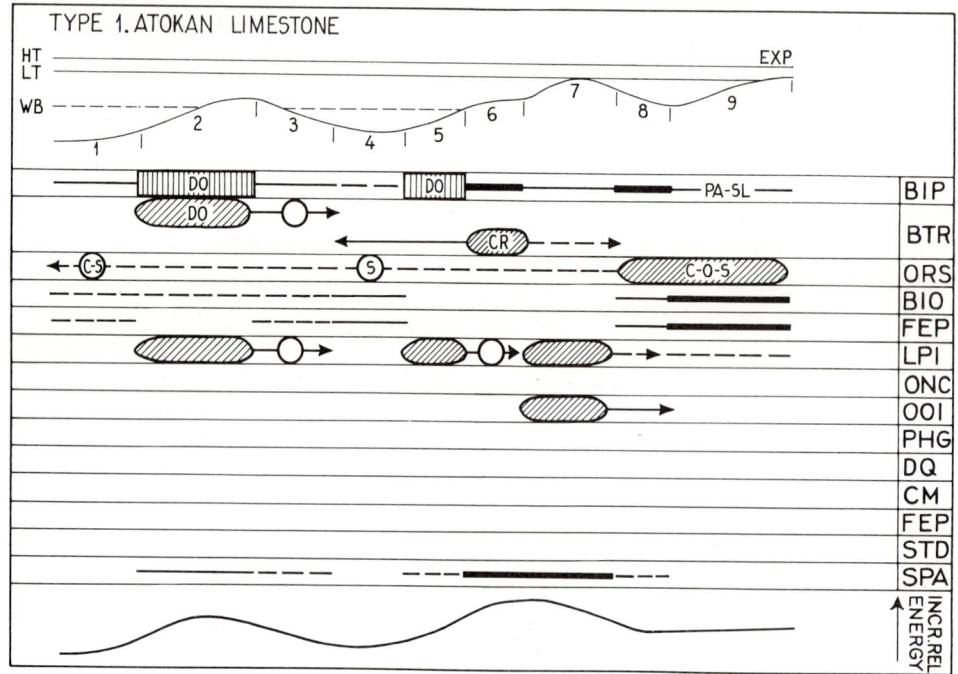

FIGURE 18.39 Integration of depositional models of platforms with frontal bioconstructed to hydrodynamic buildups. Type 1. Atokan Limestone.

ated with freshwater lakes. The first buildup is constructed by cabbage-like stromatoporoids and remains below the wave base level with products of its reworking accumulated behind it. Similarly, the second buildup constructed by *Amphipora* reached the fair-weather wave base with its destructional products behind. The third buildup is above the wave base and constructed by an association of *Amphipora* and mat-like stromatoporoids. Its reworked products are also transported landward into the lagoon. Microfacies with potential primary porosity coincide with the third bioconstructed buildup; other occurrences in slope conditions and in the lagoon are insignificant.

In the second case (Fig. 18.39), which integrates the Atokan Limestone (Middle Pennsylvanian) of the Delaware Basin (Figs. 13.44, 13.47, 13.48), the platform consists of an external buildup constructed by *Donezella* essentially at the wave base. It is intensively reworked with important accumulation of its bioclasts behind it. A trough leads to the repetition of a similar bathymetric position with another bioconstructed buildup by *Donezella* but under less agitated conditions. The latter could be considered the platform edge. Shallower conditions lead to a crinoid flat followed by an oolitic hydrodynamic buildup, and finally, to a lagoon with abundant bioturbation, phylloid algae, and stromatolites. The production of intraclasts, indicating early submarine lithification, corresponds to both *Donezella* buildups and the oolitic one with general landward transportation. Microfacies with potential primary porosity are related to the external *Donezella* bioconstructed buildup, but mainly to the crinoid flat and the oolitic hydrodynamic buildup.

Type 2 (Fig. 18.40)

This type of platform is more complex than the previous one and consists of several buildups, bioconstructed or hydrodynamic, and subaerially exposed. It exemplifies integration of the Joachim Dolomite (Middle Ordovician) of the Upper Mississippi Valley (Figs. 14.17, 14.20, 14.21). A first oncolitic-bioclastic hydrodynamic buildup, often subaerially exposed, precedes two stromatolite-bioconstructed buildups which are also exposed and form with the intervening carbonate flat the major portion of the platform. A so-called lagoon separates the two areas of buildups. It is a fairly open environment with brachiopods and gastropods. The well-developed adjacent coastal sabkha with evaporites and the continental sabkha with dunes of coarse quartz lead, respectively, to dolomitization of most of the platform by evaporative pumping, and to an important influx of detrital quartz. The latter is trapped during its seaward transport in all the deeper and quiet areas of the platform together with pyrite. The frontal intraclastic-bioclastic buildup is dolomitized by its own dorag process. Storm deposits are common and derived from each exposed buildup. Each

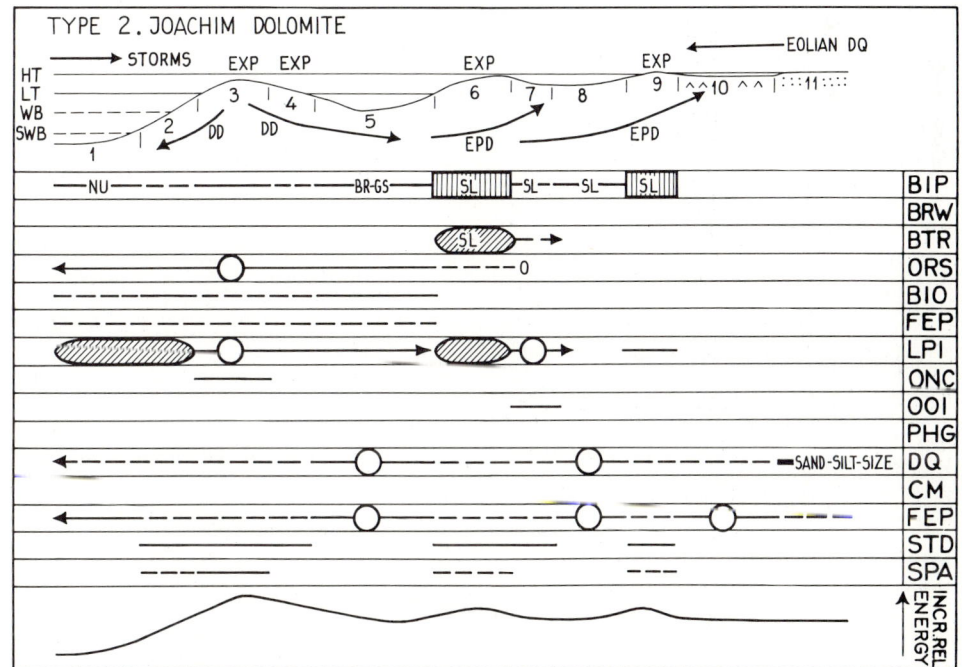

FIGURE 18.40 Integration of depositional models of platforms with frontal bioconstructed to hydrodynamic buildups. Type 2. Joachim Dolomite.

one also shows, in particular the frontal one, microfacies with potential primary porosity further improved by the two processes of early dolomitization.

Type 3 (Figs. 18.41, 18.42, 18.43)

These platforms consist of single frontal buildups constructed by various types of organisms; the buildups are relatively narrow and subaerially exposed. They are flanked by accumulations of bioclasts due to the action of waves and tidal currents. These buildups provide an efficient protection to a wide lagoonal environment in which are scattered many smaller bioconstructed, bioaccumulated, and even hydrodynamic buildups.

In the first case (Fig. 18.41), which corresponds to the integration of the Allentown Dolomite (Upper Cambrian) of New Jersey (Figs. 14.2, 14.5), the frontal bioconstructed buildup, often subaerially exposed, consists of stromatolites. It is preceded by an extensive oolitic apron and protects an inferred lagoon. Two distinct locations of oolitization occur, the major one on the frontal apron, and the minor one between the stromatolitic colonies of the buildup. This carbonate system has been subjected to extensive storm action which generated abundant intraclasts at the expense of essentially all the microfacies, and accumulated them mainly on both flanks of the buildup. As a consequence, microfacies with potential primary porosity are located in the oolitic apron and in both flanks of the buildup itself, which has an interstitial micrite matrix. In addition, all microfacies are dolomitized from a general seepage reflux process from adjacent evaporitic sabkhas.

In the second case (Fig. 18.42), which exemplifies intergration of the Shakopee Dolomite (Lower Ordovician) of southwest Wisconsin (Figs. 14.11, 14.13), the subaerially exposed frontal buildup is entirely constructed by stromatolites flanked by intraclastic microfacies derived from its reworking by waves, tidal currents, and mainly storms. Two domains of oolitization occur, the major one in the upper part of the frontal slope, the second minor one on the external edge of the lagoon; both areas lead to ample dispersion of their ooids. The lagoon with abundant bioturbation and scattered small bioconstructed buildups of stromatolites displays an important influx of sand-size detrital quartz which remains mostly concentrated in the lagoon; some of it is of eolian origin from the adjacent evaporitic sabkha. Microfacies with potential primary porosity characterize the intraclastic flanks of the frontal buildup rather than the latter itself, which probably has an interstitial micrite matrix in spite of its location at the point of maximum energy of the depositional model, a situation also encountered in subsequent bioconstructed platform types. All microfacies are entirely dolomitized by a large-scale seepage reflux process from adjacent sabkhas.

In a third example (Fig. 18.43), corresponding to the integration of the Miocene of the Visayan Islands of the Philippines (Figs. 14.43, 14.44, 14.45), the subaer-

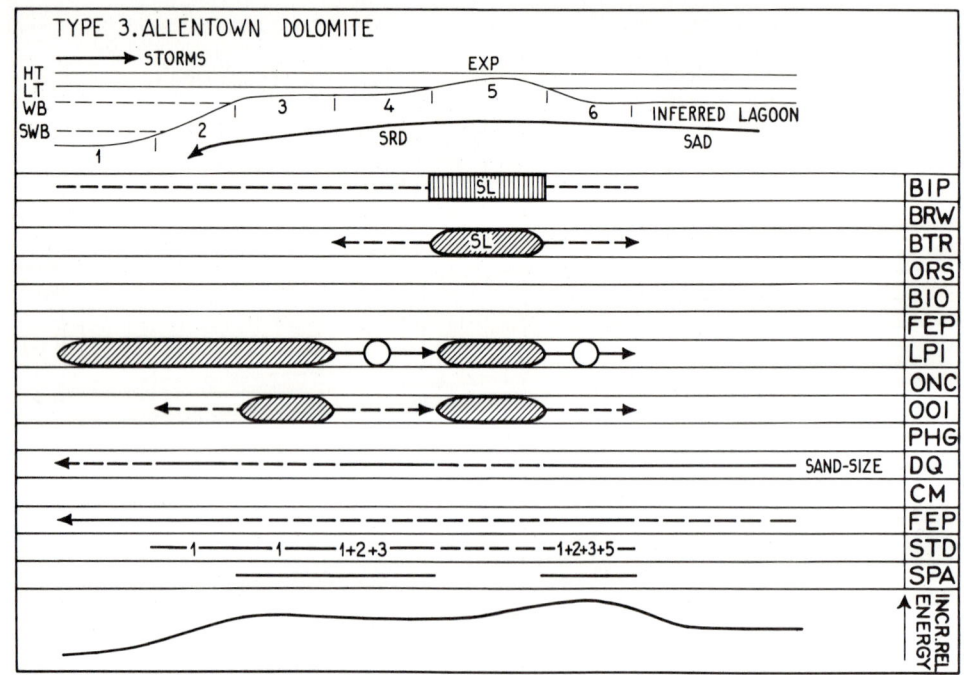

FIGURE 18.41 Integration of depositional models of platforms with frontal bioconstructed to hydrodynamic buildups. Type 3. Allentown Dolomite.

Chap. 18 / Synthesis 551

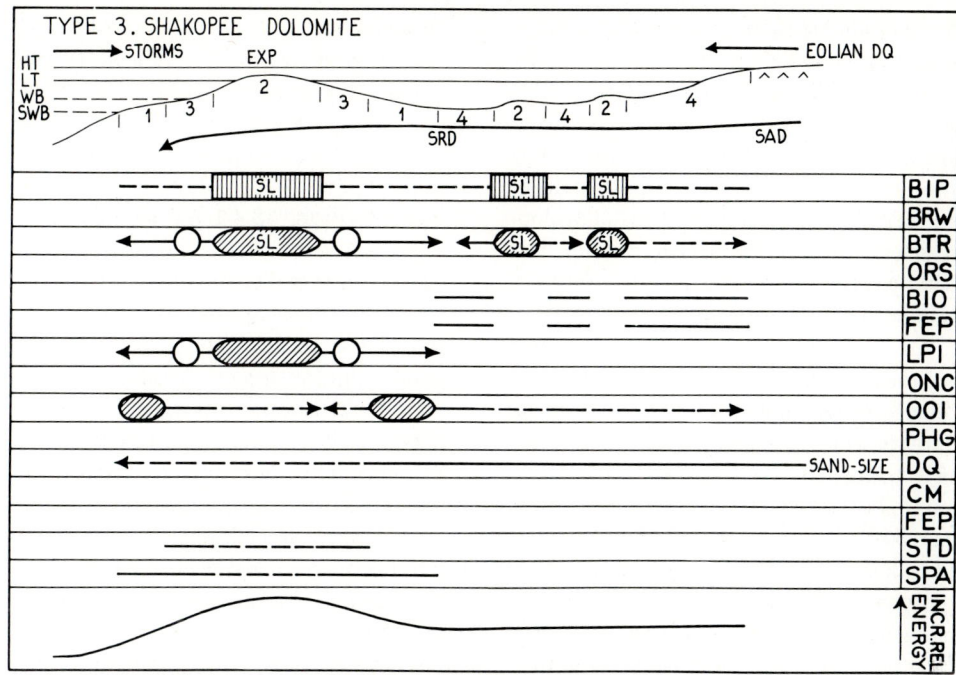

FIGURE 18.42 Integration of depositional models of platforms with frontal bioconstructed to hydrodynamic buildups. Type 3. Shakopee Dolomite.

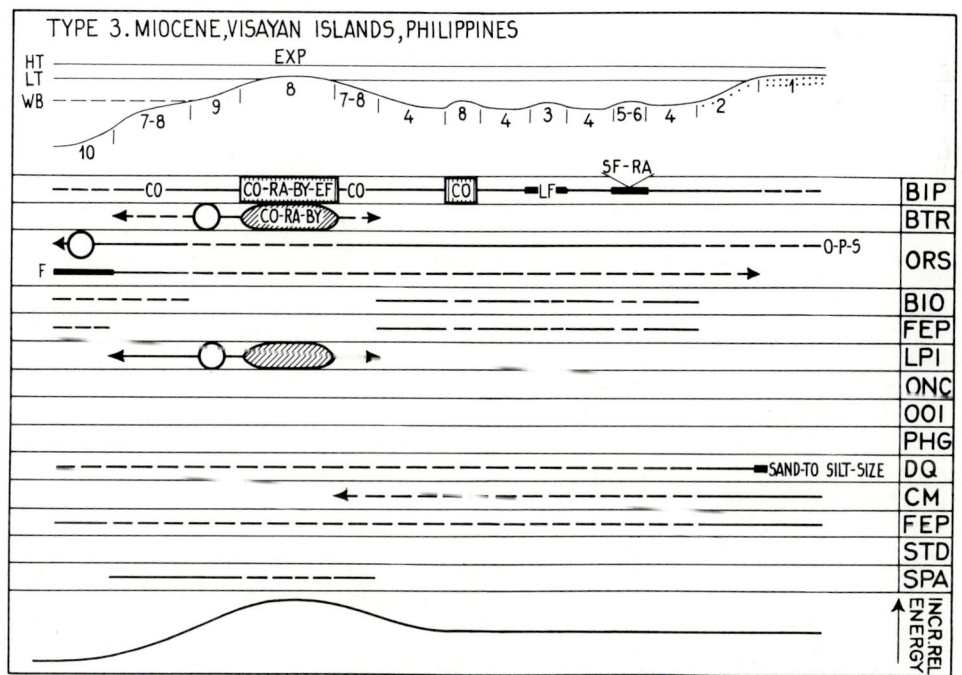

FIGURE 18.43 Integration of depositional models of platforms with frontal bioconstructed to hydrodynamic buildups. Type 3. Miocene, Visayan Islands, Philippines.

ially exposed frontal buildup facing open oceanic conditions is complex and strongly wave resistant but relatively narrow. It is constructed by the association of corals–red algae–bryozoans-encrusting foraminifers. Most of its bioclasts and intraclasts are shed as a frontal talus of rudites leading to an upper slope bioclastic environment where pinnacle-shaped coralline buildups develop. A vast lagoon extends behind the frontal buildup, and although crossed by numerous tidal channels, it displays an ecologically zoned succession of small buildups which in a landward direction include coral bioconstructed buildups, bioaccumulated buildups of larger benthic foraminifers, and finally, bioaccumulated buildups of smaller benthic foraminifers and red algae. The importance of the tidal channels is shown by their well-sorted biocalcarenites containing bioclasts and intraclasts derived from all the lagoonal buildups and grading into submarine fans as turbidities and by the behavior of the organisms transported in suspension. These are planktonic foraminifers penetrating far into the lagoon and ostracods, sponge spicules, and plant debris dispersed seaward. An important influx of detrital quartz, clay minerals, and iron originating from volcanic source areas concentrates mostly in the lagoon but is transported seaward across the entire carbonate system also. Microfacies with potential primary porosity are located mainly in the front and rear intraclastic and bioclastic deposits of the frontal buildup which itself displays an interstitial micrite matrix in spite of being the area of maximum energy. However, frequent subaerial exposure of the frontal buildup may introduce some secondary porosity by vadose and upper phreatic undersaturated dissolution.

Type 4 (Figs. 18.44, 18.45, 18.46)

The platforms belonging to this group reach a great width and include some of the largest found in the Phanerozoic record (Devonian). They consist of the juxtaposition of various bioconstructed buildups, sometimes associated with hydrodynamic ones, and separated by relatively narrow depressions that generally contain bioclasts from the adjacent buildups. The lagoons located behind them are complex, extensive, and well protected.

In one example (Fig. 18.44), which represents integration of the Bonfim Formation (Cenomanian) of the Barreirinhas Basin, northeast Brazil (Fig. 13.61), the frontal buildup consists of an upper oncoidal slope preceding two sites of bioconstruction by different types of red algae associated with green algae. These sites are separated by a depression containing calcirudites and calcarenites consisting of bioclasts and intraclasts from the adjacent buildups and oolitic calcarenites. The lagoon shows several types of stromatolitic buildups. The effects of tidal currents in the transverse channels across the

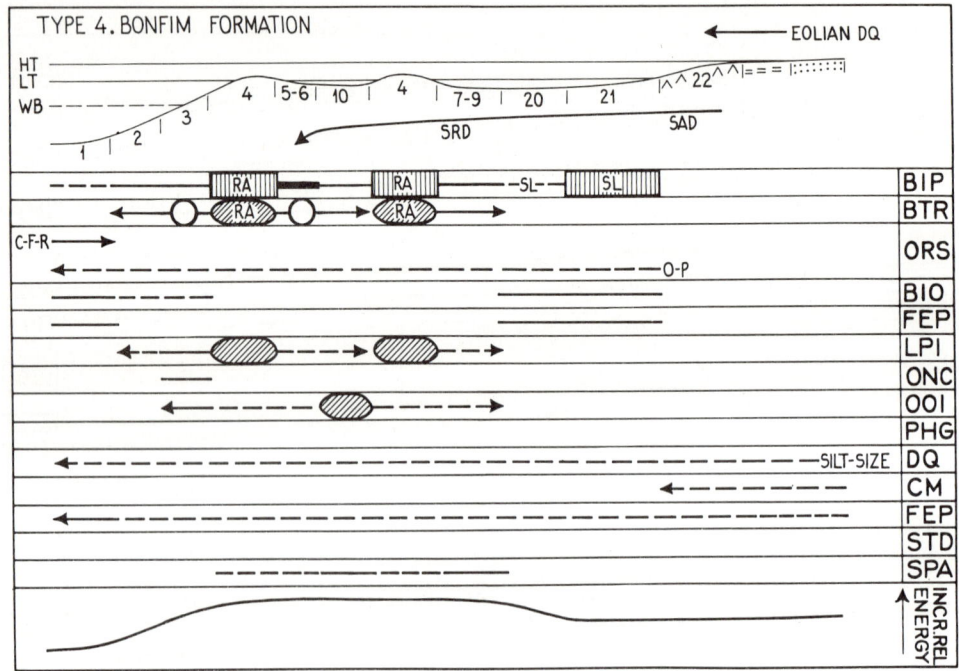

FIGURE 18.44 Integration of depositional models of platforms with frontal bioconstructed to hydrodynamic buildups. Type 4. Bonfim Formation.

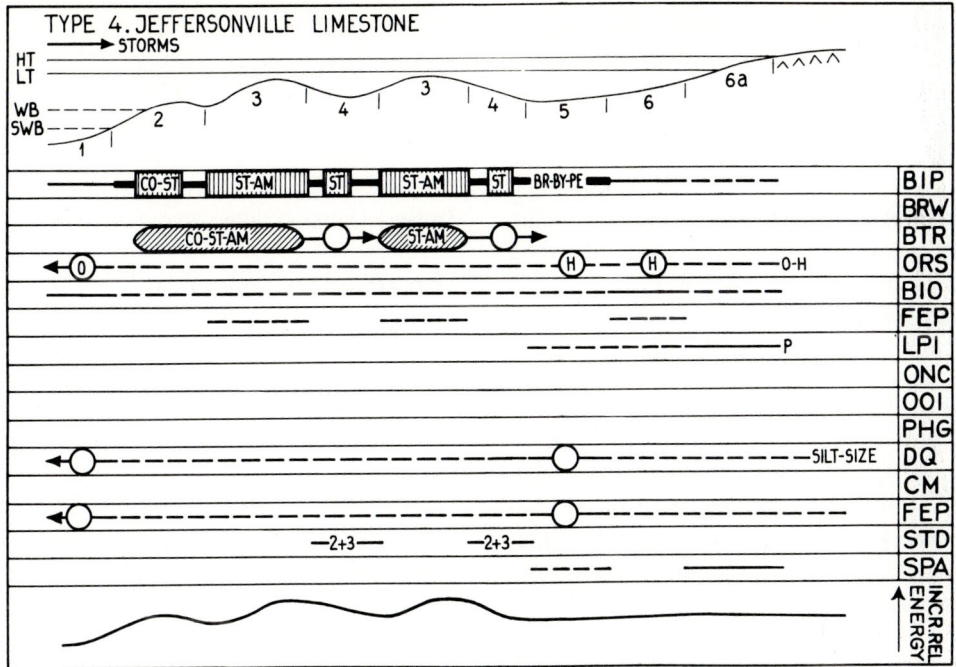

FIGURE 18.45 Integration of depositional models of platforms with frontal bioconstructed to hydrodynamic buildups. Type 4. Jeffersonville Limestone.

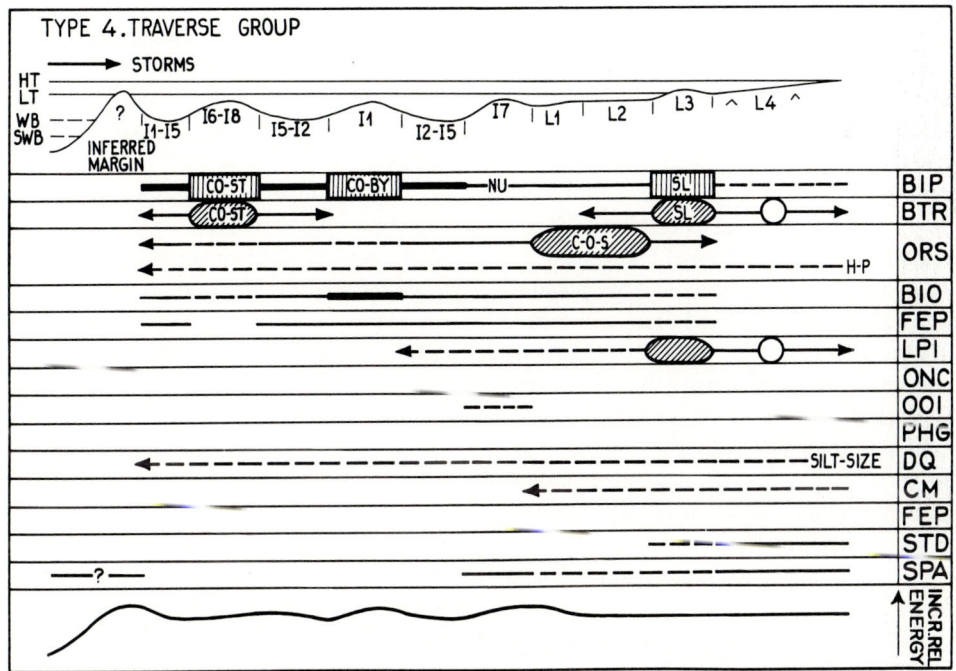

FIGURE 18.46 Integration of depositional models of platforms with frontal bioconstructed to hydrodynamic buildups. Type 4. Traverse Group.

frontal buildup do not appear very strong because planktonic foraminifers, calcispheres, and radiolarians remain in the slope environment, whereas only ostracods and plant debris are transported seaward. Microfacies with potential primary porosity correspond essentially to the entire frontal buildup and conditions are improved by seepage reflux dolomitization from adjacent evaporitic flats. The latter are associated with fan deltas, yielding detrital quartz to the carbonate system by eolian and subaqueous transport.

In a second example (Fig. 18.45), which integrates the Jeffersonville Limestone (Middle Devonian) of southeast Indiana (Figs. 13.2, 13.8), the frontal bioconstruction consists, in a landward direction, of at least three distinct buildups, the first one by corals and cabbage-like stromatoporoids, the second and the third by cabbage-like stromatoporoids and *Amphipora*. In the intervening depressions, their bioclasts accumulate and associate with mat-like stromatoporoids and storm deposits. The adjacent lagoon also displays a rich fauna with brachiopods, pelecypods, and bryozoans. It grades landward, eventually, into pelletoidal microfacies preceding an association of dolomitic-evaporitic flats and freshwater lakes from which abundant floated *Chara* originate. Microfacies with potential primary porosity are rare in this type of platform, where most of the bioconstructed buildups, although of greater energy, show interstitial micrite matrix. Occurrences of primary porosity in the lagoon are of little importance.

In a third example (Fig. 18.46), which represents integration of the huge platform of the Traverse Group (Givétian) of Michigan (Figs. 13.32 to 13.41), the frontal portion, which must have been very narrow, is unknown. Immediately behind it are several rows of buildups constructed by cabbage-like heads and mats of stromatoporoids together with numerous types of corals (*Hexagonaria*, *Favosites*, and *Thamnopora*), separated by depressions containing bituminous crinoidal biocalcarenites with abundant coral bioclasts, and often overturned colonies which have rolled downslope. In a more internal position are micrite buildups supporting bioconstructed microfacies by delicate branching corals (*Depasophyllum*) and bryozoans, and finally, hydrodynamic buildups with *Nuia*. The very quiet lagoon displays stromatolite-constructed buildups and an abundance of calcispheres, ostracods, sponge spicules which are transported seaward together with abundant *Chara*, plant debris, and colloidal organic matter. The shoreline lagoonal microfacies contain isolated occurrences of anhydrite and halite, and storm deposits. Microfacies with potential primary porosity, as in the previous example, are rare. All frontal bioconstructed buildups have a micrite matrix, and porosity is limited to the hydrodynamic buildups and the storm deposits of the lagoon shoreline.

18.10 GENERAL SYNTHESIS OF PLATFORMS WITH FRONTAL BIOCONSTRUCTED TO HYDRODYNAMIC BUILDUPS

Features and processes of general significance for these types of platforms are analyzed below.

Fair-Weather Wave Base, Tidal Currents, and Submarine Morphology

In these platforms, the effect of the fair-weather wave base is limited to frontal slopes except in cases where the edge of the platform consists of several smaller and deeper bioconstructed buildups preceding the shallower ones which support the lagoon (Figs. 18.38, 18.39). With increasing wave action and tidal current velocities, the platforms begin to display associations of several bioconstructed and hydrodynamic buildups that are frequently subaerially exposed (Fig. 18.40). Then in type 3, the platform consists of a single bioconstructed buildup, often subaerially exposed, relatively narrow, and flanked by aprons made of bioclasts derived from the buildup often associated with abundant oolitic microfacies. The lagoons are broad and contain several types of smaller buildups which are bioaccumulated, bioconstructed, and even hydrodynamic (Figs. 18.41, 18.42, 18.43). Finally, under the highest-energy conditions of type 4, extremely complex and broad frontal associations of numerous types of bioconstructed to hydrodynamic buildups are generated. They are separated by relatively narrow depressions filled with debris from the buildups and protect extensive lagoons (Figs. 18.44, 18.45, 18.46).

The importance of the dispersion by traction of bioclasts and intraclasts by tidal currents (which mechanically rework the constructed buildups) increases from type 1 through 4 as a function of the general increase of energy and complexity of the platforms. The landward and seaward dispersion is regulated to a large extent by tidal channels which, when observed, show the same sediment fill as the intervening troughs and depressions between buildups, namely, a rich *in situ* benthic fauna along with a variety of transported constituents (compare Fig. 18.40 with Figs. 18.45, 18.46). Well-protected lagoons of the Paleozoic examples display abundant calcispheres, ostracods, and sponge spicules which, during their seaward transport in suspension toward final deposition, participate in hydrodynamic buildups (Fig. 18.40), become trapped in depressions between bioconstructed buildups (Figs. 18.38, 18.39), or when frontal buildups are broad and act as effective barriers, remain blocked behind them (Figs. 18.45, 18.46). The effect of tidal currents is also very clear in the Mesozoic examples, where planktonic constituents do not penetrate the carbonate system (Fig. 18.44), and in the Cenozoic cases, where they easily reach the shores of the lagoon (Fig. 18.43). However, in both cases, lagoonal constituents reach slope environments.

Coexistence of *in Situ* and Transported Organic and Inorganic Constituents

This coexistence is demonstrated amply in the wider platforms and particularly in the depressions between adjacent bioconstructed buildups where rich benthic fauna *in situ* receives abundant bioclasts and intraclasts derived from the buildups (Figs. 18.45, 18.46).

Distribution Pattern of Oolitic Microfacies

With the exception of two cases where oolitization is well developed in the frontal areas of the platforms (Figs. 18.41, 18.42) and due to the effect of ascending cold currents, generation of ooids is localized in particular conditions such as inside a stromatolitic buildup (Fig. 18.41), as an internal hydrodynamic buildup (Figs. 18.39, 18.46), in a depression between bioconstructed buildups (Fig. 18.44), or in various locations in the lagoon (Figs. 18.40, 18.42).

Distribution Pattern of Oncoids

Oncoids are relatively rare in these types of larger-energy platforms and limited to frontal slope environments at the wave base level (Figs. 18.40, 18.44).

Distribution Pattern of Phosphates and Glauconite

Phosphates are absent in these platforms because the latter correspond to well-oxygenated conditions and the deepest portions of their frontal slopes were not observed. Glauconite is also missing because in these carbonate systems the influx of clay minerals is relatively small and rates of carbonate sedimentation are large, two factors which do not favor glauconitization.

Distribution Pattern of Extrabasinal Materials

Some of these platforms are totally devoid of detrital quartz (Figs. 18.38, 18.39), whereas others show a variable amount of sand- to silt-size quartz with an appreciable eolian contribution because of adjacent evaporitic sabkhas (Figs. 18.40, 18.42, 18.44). Quartz grains during their seaward transport may become trapped or concentrated in lagoons and in troughs between bioconstructed buildups before reaching slope environments (Figs. 18.40, 18.42, 18.43, 18.45, 18.46). Iron, in form of pyrite, displays a similar distribution (Figs. 18.40, 18.45).

Distribution Pattern of Dolomitization

For climatic reasons, only platforms adjacent to evaporitic sabkhas display large-scale seepage reflux dolomitization (Figs. 18.41, 18.42, 18.44) or even evaporative pumping dolomitization (Fig. 18.40). Furthermore, under these arid conditions, dorag-type dolomitization occurs where subaerially exposed buildups contain ephemeral freshwater lenses due to intermittent rain (Fig. 18.40).

Distribution Pattern of Storm Deposits

Storm deposits are widespread in these types of platforms. In type 2 (Fig. 18.40), the exposed tops of the hydrodynamic and bioconstructed buildups are reworked heavily into intraclasts by storms. In type 3 (Fig. 18.41), storms accumulate intraclasts in the front and the back of the frontal stromatolitic buildups with a cumulative effect indicating an important landward transport. In another case (Fig. 18.42), intraclastic storm microfacies flank both sides of the frontal stromatolitic buildup. In type 4, storm deposits either originate from reworking of the top of bioconstructed buildups (Fig. 18.45) or are limited to lagoonal conditions (Fig. 18.46).

Distribution Pattern of Potential Primary to Early Diagenetic Reservoirs

A double schematic diagram (Fig. 18.47) shows the distribution of microfacies with potential primary porosity with increasing general energy, and shallowing from type 1 to 4. In types 1 and 2, potential primary porosity appears above the wave base level coinciding with bioconstructed highs. In types 3 and 4, although general energy is the largest, the presence of interstitial micritic to bioclastic matrix instead of sparite cement inside some bioconstructed buildups tends to limit potential porous microfacies to flanks of buildups (type 3) or to coarse biocalcarenites in depressions between bioconstructed buildups (type 4).

18.11 GENETIC CLASSIFICATION OF LACUSTRINE MODELS

Only two lacustrine models have been discussed in this volume; they both belong to rift-valley stages, hence the coverage is not sufficient to account for all types of

556 Part V / Toward an Explanation

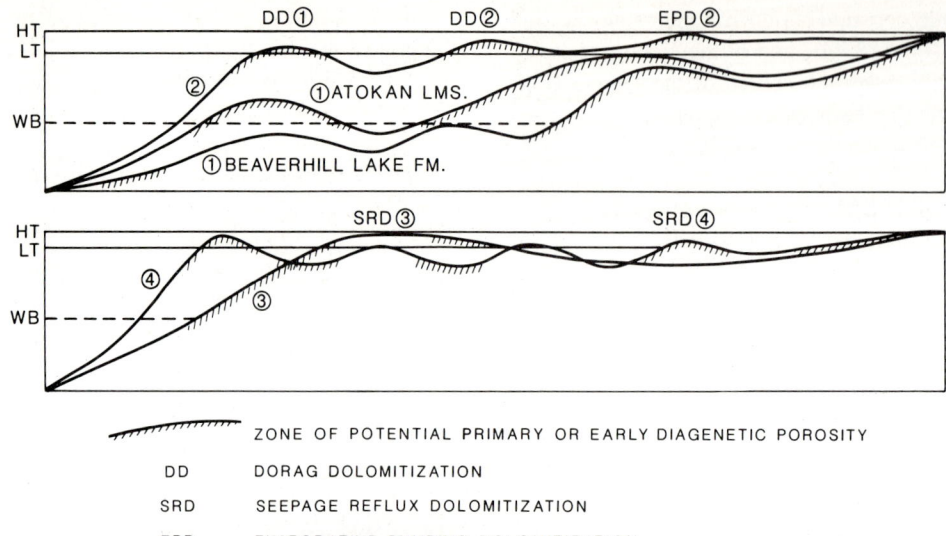

FIGURE 18.47 Distribution of potential primary reservoirs in platforms with frontal bioconstructed to hydrodynamic buildups. Circled numbers refer to types of models.

carbonate lacustrine models. However, they are ramps that are extremely sensitive to climatic conditions and controlled by bioaccumulated to hydrodynamic buildups made by ostracods, gastropods, and pelecypods and by stromatolite-bioconstructed buildups. Given these restrictions, they share many features with marine carbonate ramps.

Type 1 (Fig. 18.48)

This type corresponds to integration of the playa lake stage (ostracod model) of the Lagoa Feia Formation (Lower Cretaceous) of the Campos Basin, offshore Brazil (Figs. 17.9, 17.10). The gentle ramp displays a bioaccumulated ostracod-pelecypod buildup with a lakeward

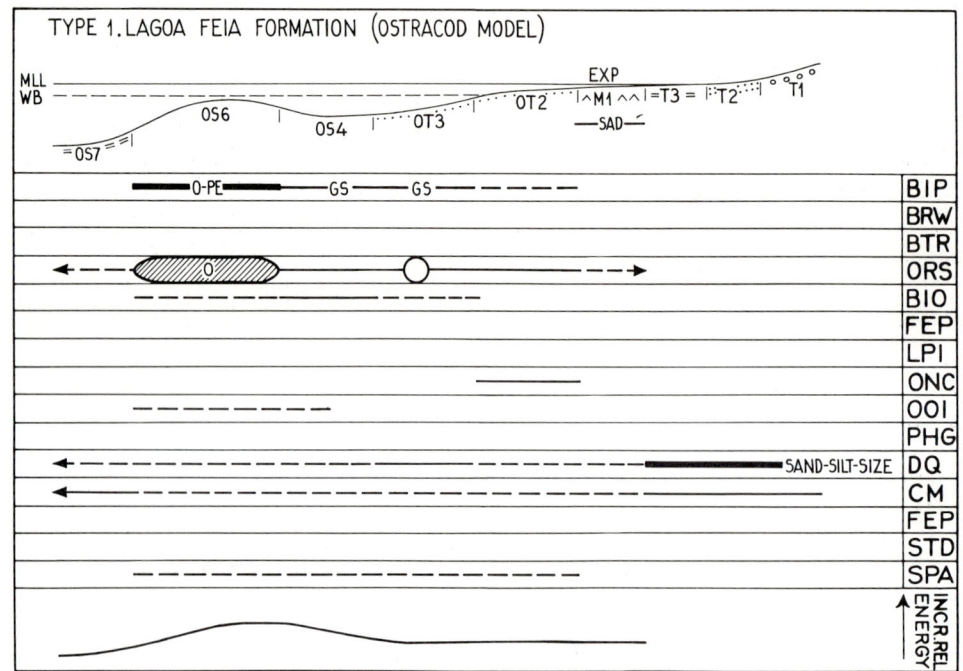

FIGURE 18.48 Integration of lacustrine depositional models. Type 1. Lagoa Feia Formation (ostracod model).

slope grading into basinal, laminated, and micritic bituminous shales rich in organic matter and a landward slope leading into a relatively low-energy depression. The deeper portion of the depression consists of bioturbated microfacies containing ostracods transported from the buildups and *in situ* gastropods. These microfacies grade shoreward into fine-grained arenaceous carbonates, lithic arenites, and siltstones, locally evaporitic, which in turn grade into terrigenous flats and alluvial fans. A small amount of ooids occur in the buildup, whereas oncoids form at the fair-weather wave base level along the shoreline. The influx of detrital quartz concentrates along the lake shores and decreases in abundance lakeward, whereas clay minerals concentrate in deeper water. Microfacies with potential primary porosity are widespread but of little importance due to the abundance of micrite matrix.

Type 2 (Fig. 18.49)

This type corresponds to the integration of the pluvial lake stage (pelecypod model) of the same Lagoa Feia Formation (Figs. 17.11, 17.12). It is also a gentle ramp leading into basinal, laminated, micrite-rich bituminous shales. The ramp is first modified by a bioaccumulated pelecypod-ostracod buildup below the fair-weather wave base level and is also modified more shoreward by a pelecypod-gastropod hydrodynamic buildup that reaches above the wave base. From this shallower buildup, bioclasts are dispersed both lakeward and landward. In the landward direction is a depression with moderately agitated waters. In its deeper portion it contains bioturbated pelecypod biocalcarenites, and shoreward under shallower conditions it consists of arenaceous pelecypod biocalcarenites grading into terrigenous flats and alluvial fans. In general, this ramp is in a higher-energy environment than type 1. Consequently, microfacies with potential primary porosity exist mainly in the hydrodynamic buildup and in some shoreline microfacies. Storm deposits occur in the two types of buildups. Their importance is moderate, and they are represented by repeated zones of intense breaking and imbrication of large pelecypod shells.

18.12 GENERAL SYNTHESIS OF LACUSTRINE MODELS

A comparison between marine and lacustrine carbonate ramps with bioaccumulated to hydrodynamic buildups reveals similarities, although lacustrine ramps are generally formed under lower-energy conditions.

Fair-Weather Wave Base, Seiches, and Sublacustrine Morphology

Generation of hydrodynamic pelecypod-gastropod buildups is due to the action of the fair-weather wave base (Fig. 18.49), possibly combined with the effect of

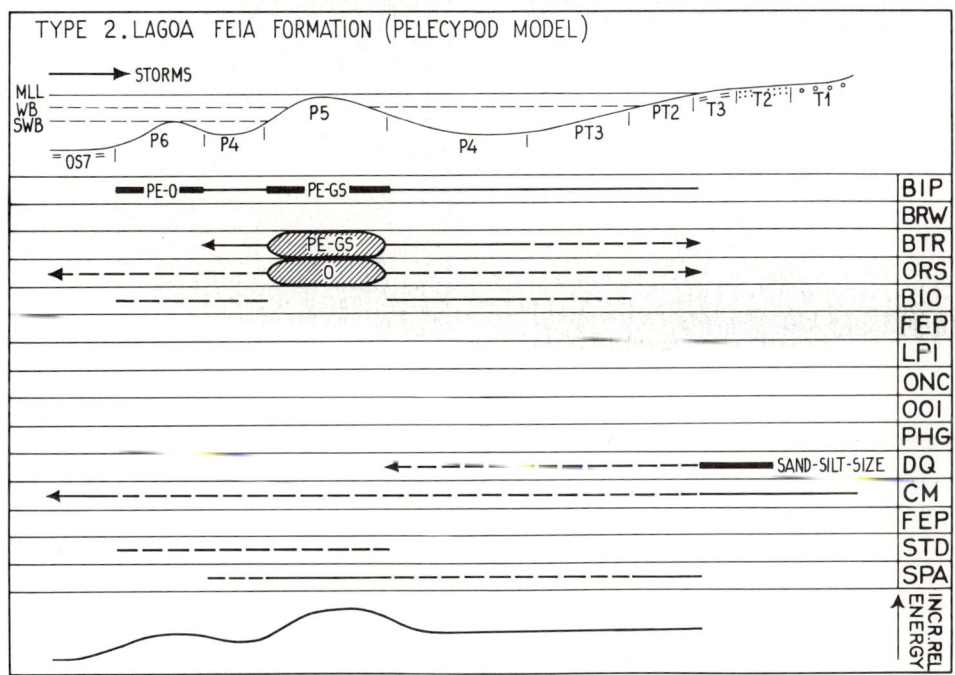

FIGURE 18.49 Integration of lacustrine depositional models. Type 2. Lagoa Feia Formation (pelecypod model).

seiches. Probably both are responsible for some of the winnowing and reworking in bioaccumulated buildups which almost reach wave base (Fig. 18.48). Transportation by traction of bioclasts takes place, probably by currents related to seiches along the flanks of the hydrodynamic buildups, both in landward and lakeward directions (Fig. 18.49). Constituents transported in suspension are limited to ostracods, which in both types of ramps proliferate in coincidence with the bioaccumulated (Fig. 18.48) and hydrodynamic buildups (Fig. 18.49). They are dispersed landward and lakeward with concentration in the depressed areas behind the buildups (Fig. 18.48). This major difference from conditions observed in marine ramps is due to the fact that real lagoons do not exist in the investigated lacustrine models where the maximum development of ostracods occurs in open waters.

Coexistence of *in Situ* and Transported Organic and Inorganic Constituents

This coexistence is clearly shown by the behavior of benthic constituents such as pelecypods and gastropods which can be determined petrographically to occur both *in situ* as entire individuals and as bioclasts.

Distribution Pattern of Oolitic Microfacies

Oolitic microfacies are very rare in these cases. Ooids occur in a small amount in the bioaccumulated buildup (Fig. 18.48), probably in relation to gentle agitation by the overlying wave base.

Distribution Pattern of Oncoids

Oncoids are observed at the wave base level in shoreline conditions of type 1 (Fig. 18.48). They imply destruction of stromatolitic mats, and their nuclei consist of basalt lithoclasts.

Distribution Pattern of Extrabasinal Materials

The influx of detrital quartz (among many other coarse-grained minerals) and of clay minerals is important in both types of lacustrine ramps because they grade landward into terrigenous flats and alluvial fans. In type 1 (Fig. 18.48), both minerals are dispersed lakeward across the entire carbonate system. In type 2 (Fig. 18.49), the pelecypod-gastropod hydrodynamic buildup acts as a barrier and limits the distribution of detrital quartz to the depression behind it.

Distribution Pattern of Dolomitization

Dolomitization occurs only as a penecontemporaneous replacement of sabkha type in the exposed carbonate flats of type 1 (Fig. 18.48), where it is associated with evaporites and zeolites.

Distribution Pattern of Storm Deposits

They occur in the bioaccumulated and hydrodynamic buildups of type 2 (Fig. 18.49) and are expressed by repeated zones of reworking, fragmentation, and imbrication *in situ* of pelecypod and gastropod shells without dispersion.

Distribution Pattern of Potential Primary to Early Diagenetic Reservoirs

Microfacies with potential primary porosity are limited to the well-sorted pelecypod-gastropod biocalcarenites which form the shallower hydrodynamic buildup of type 2 (Fig. 18.49); elsewhere their occurrence even along shorelines is unimportant.

18.13 PALEOCLIMATIC IMPLICATIONS OF CARBONATE ROCK DEPOSITIONAL MODELS

Carbonate sediments and their fauna indicate warm equatorial to tropical waters. However, only extrabasinal constituents and other associated sediments afford data on the degree of aridity or of humidity within that particular latitudinal belt. Hence depositional models investigated by microfacies techniques pose important climatic implications. These are determined by the following major criteria: influx of detrital quartz and clay minerals which leads to a variety of mixed models (Mount, 1984) in which carbonates display siliciclastic shorelines such as beaches, dunes, estuaries, deltas, and fan deltas; influx of constituents from freshwater lakes (*Chara*) or rainy forest terrane (plant debris); association with evaporites and related types of dolomitization; and occurrence of various types of storm deposits.

It is appropriate, therefore, to set these independently obtained climatic interpretations within the framework provided by the paleogeographic and paleoclimatic base maps of Scotese et al. (1979) and Ziegler et al. (1979, 1982) for the Phanerozoic. In most cases, the climatic interpretations are in agreement with that framework.

Paleogeographic Interpretations

Cambrian Paleogeography (Fig. 18.50A)

During Cambrian time Laurentia, like most other continents, was in a low-latitude position straddling the equator. The equatorial rainy belt is represented by coarse clastics derived from the erosion of the American Mid-Continent high and the Canadian shield. The Allentown Dolomite (Upper Cambrian) of New Jersey (1) indicates an arid climate and was located at about 20° south latitude. Numbers in brackets indicate on each map of Fig. 18.50 the location of carbonate models investigated.

Ordovician Paleogeography (Fig. 18.50B)

From Late Cambrian through Ordovician time, Laurentia underwent a counterclockwise rotation while remaining in low latitudes. Coarse clastics decreased in importance, and carbonates occurred in the latitudinal belt of 0 to 20° north and south, whereas evaporites were disposed symmetrically at about 20° north and south latitude. The carbonate sequences of the Mid-Continent of North America (2) are the Shakopee Dolomite (Lower Ordovician) of Wisconsin, the Joachim Dolomite (Middle Ordovician) of Iowa, the Platteville Group (Middle Ordovician) of Illinois, and the Galena Group (Middle Ordovician) of Illinois. Their association with well-developed coastal evaporitic sabkhas and continental sabkhas with dunes indicates an arid climate. Indeed, they clustered around 18° south latitude. The carbonate sequence of the Pogonip Group (Lower Ordovician) of southeast Nevada (3) displays shoreline arenaceous and argillaceous microfacies, which along with an appreciable general influx of quartz grains, indicate subhumid conditions. It was located around 12° south latitude.

Silurian Paleogeography (Fig. 18.50C)

During Silurian time, Laurentia remained essentially in its Ordovician location. The Middle Silurian was a time of broad extent of shallow seas with the consequent disappearance of the earlier sources of coarse clastics. Carbonates were widespread in low latitudes and evaporites were confined to 10 to 30° north and south latitude. The carbonate sequences of the Mid-Continent of North America (4) such as the Niagaran (Middle Silurian) of Indiana, the Burnt Bluff Group (Middle Silurian) of Wisconsin, and the Cayugan (Upper Silurian) of Indiana, surrounded the Michigan Basin and were intimately related to evaporites and an arid climate. They clustered at 18° south latitude.

Devonian Paleogeography (Fig. 18.50D)

During Devonian time, the Iapetus Ocean, which separated Laurentia and Baltica, closed and caused the Caledonian and Acadian orogenies of northern Europe and eastern North America. The collision led to the formation of the Laurussia continent which moved to the northeast. The carbonate sequences of the Mid-Continent of North America are the large bioconstructed platforms of the Jeffersonville Limestone (Middle Devonian) of Indiana (5), and of the Traverse Group (Middle Devonian) of Michigan (6). They indicate a humid climate by the abundance of continental organic matter, plant debris, and floated *Chara* from freshwater lakes located behind platforms, although traces of local and minor evaporites exist. These platforms ranged between the equator and 4° north latitude. In the Rocky Mountains, the platform of the Beaverhill Lake Formation (Upper Devonian) of Alberta (8) indicates similar humid conditions. It was located at 18° north latitude. The Arrow Canyon Formation and Crystal Pass Limestone (Upper Devonian) of southeast Nevada (7) contain pure quartz sand in shoreline sequences that indicate less humid conditions. This sequence was located at 10° north latitude.

Mississippian Paleogeography (Fig. 18.50E)

Laurussia drifted a few degrees east in Mississippian time, but maintained its latitudinal position of the Devonian. It was a period of well-differentiated climates around the world with the first coal deposition, extensive continental glaciations, and carbonate sediments limited to low latitudes and evaporites widely distributed between 5 and 30° north and south latitude. Evaporites occurred with greater abundance on the western side of large land masses or high mountains. The carbonate sequences of the Mid-Continent of North America and of the Illinois Basin (9) in particular display the effects of a climatic change through time which began with pure carbonate systems followed by carbonates with adjacent evaporitic sabkhas in arid conditions and terminated with humid conditions and associated prograding deltas containing abundant plant debris foreshadowing the advent of the Pennsylvanian coal-bearing cyclothems. These carbonates are in stratigraphic order as follows: Burlington Limestone, a pure carbonate platform, Salem Limestone with evaporitic sabkhas and rare plant debris, St. Louis Limestone with extensive evaporitic flats, Ste. Genevieve Limestone with estuaries and deltas, Glen Dean Formation with prograding deltas, Menard Formation also with prograding deltas, and Kinkaid Formation with a combination of local evaporites and prograding deltas containing abundant plant debris and floated *Chara*. All these

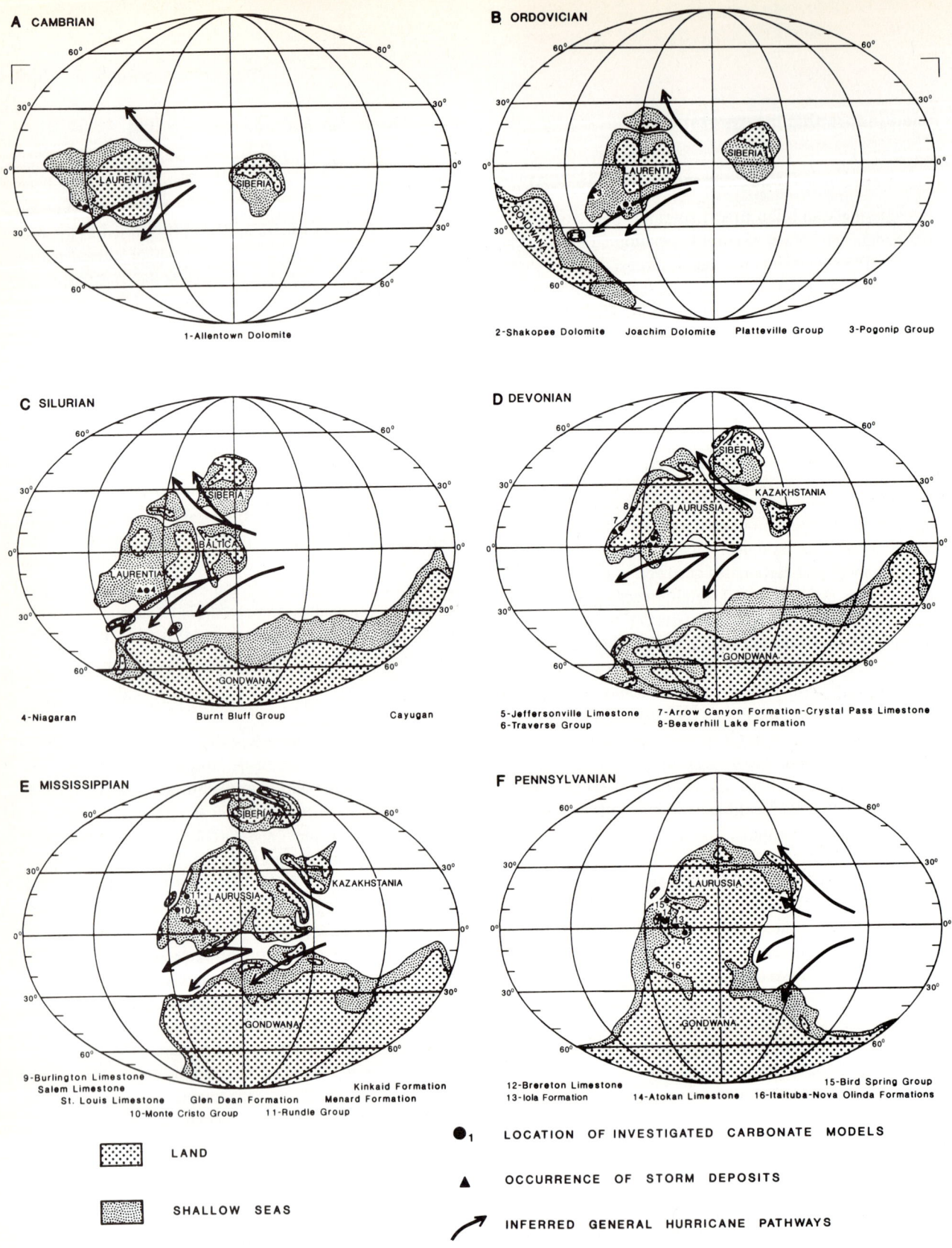

FIGURE 18.50 Cambrian to Pennsylvanian paleogeographic maps with inferred general pathways of tropical hurricanes. Only the portion of the hemisphere surrounding North America was modified from the paleogeographic base maps of Scotese et al. (1979). Inferred general hurricane pathways modified from Marsaglia and Klein (1983).

sequences are located on the appropriate paleoequator. In the Rocky Mountains, the Monte Cristo Group of southeast Nevada (10) and the Rundle Group of Alberta (11) with associated sabkha evaporites are respectively located at 10 and 20° north latitude.

Pennsylvanian Paleogeography (Fig. 18.50F)

During Pennsylvanian time, Gondwana rotated clockwise and collided with Laurussia. The rotation of the two continents increased the width of the Tethys and the collision resulted in the folding of the Ouachita, the Appalachian, and Hercynian orogenic belts. In Laurussia, the equatorial rainy zone was represented by a belt of coal swamps extending from the middle of the present-day United States to Europe. Carbonate sequences include the Brereton Limestone of Illinois (12), which consists of coal-bearing cyclothems that display incipient phylloid algal buildups developed in an interdeltaic embayment with abundant influx of detrital quartz, clay minerals, and bituminous material. This limestone was located at 2° south latitude. The Iola Formation of southeast Kansas (13), which consists of marine cyclothems and is associated with prograding deltas and an influx of plant debris, was located at 2° north latitude. The Atokan Limestone of the Delaware Basin (14) had no influx of detrital quartz, indicating a dry climate (but not evaporitic), and was located at 5° north latitude. In the Rocky Mountains of southeast Nevada, the Bird Spring Group (15), which includes an upper portion of Permian age, shows semihumid conditions with a moderate influx of detrital quartz, and was located at 15° north latitude.

In South America, the Itaituba–Nova Olinda Formations of the Amazon Basin (16) were deposited in a semirestricted marine embayment. These carbonates were associated with extensive evaporitic flats, but received some clastic sediment and plant debris from surrounding lowlands. This arid climate affected the entire embayment, which extended from 10 to 35° south latitude.

Mesozoic Paleogeography

During Aptian-Cenomanian time, carbonates developed on both passive margins upon opening of the proto-South Atlantic following the rift valley stage, which involved deposition of lacustrine carbonates and evaporites (Lagoa Feia Formation of the Campos Basin). Marine carbonates developed adjacent to a combination of evaporitic lagoons and active fan delta systems. From south to north they are the Macaé Formation of the Campos Basin, the Barra Nova Formation of the Espirito Santo Basin, and the Bonfim Formation of the Barreirinhas Basin. These formations extended from 35° south latitude to the paleoequator. The first two sequences consist of platforms with only oncoidal-oolitic hydrodynamic buildups, which indicate that there was still some restriction in the proto-South Atlantic, whereas the Bonfim Formation is a major platform with bioaccumulated to hydrodynamic buildups of red and green algae, which indicate open equatorial waters.

Farther south, in the intracratonic Neuquén-Mendoza Basin of Argentina, the oolitic carbonates of the Quintuco-Loma Montosa Formation (Lower Cretaceous) are associated with extensive evaporitic sabkhas and fan deltas under arid conditions. They were located at 35° south latitude. Still farther south, the bioaccumulated buildups of micrite with pelecypods of the Chachao Formation (Valanginian) graded landward into abundant coarse deltaic clastics that indicate a subhumid climate. These buildups were located at 40° south latitude.

Cenozoic Paleogeography

The major platform built by red algae and larger benthic foraminifers of the Amapá Formation (Paleocene–Middle Miocene) of the Foz do Amazonas Basin is adjacent to extensive freshwater lagoons and active fan delta systems that existed at all times behind the platform and drained across it by means of huge canyons. A humid equatorial climate is amply demonstrated in these carbonates, which were located at the paleoequator.

Interpretation of Storm Deposits

Storm deposits described in the various case histories are intraclastic to bioclastic calcirudites to calcarenites that display the following characteristic features: basal erosional surface (truncation); random orientation to imbrication of elongate intraclasts and large bioclasts; "umbrella" geopetal textures; random general fabric to incipient graded bedding; vertical gradation of interstitial material from sparite cement to bioclastic and micritic matrix; intense bioturbation that is interrupted or indicates an attempted escape upward; and periodicity independent of carbonate cyclicity. Hummocky stratification, a controversial bedform considered typical of siliciclastic tempestites (Dott and Bourgeois, 1982), has been clearly demonstrated in some carbonate equivalents (Kreisa, 1981; Aigner, 1982; Handford, 1986) but not in the case histories above.

A paleogeographic interpretation of storm deposits can be attempted by applying the paleo-storm model proposed by Marsaglia and Klein (1983) to the same base maps used for the previous paleoclimatic interpretation. This model assumes that conditions for the generation of ancient hurricanes and winter storms are similar to those of the present, that meteorologic phenomena have not changed appreciably through geologic time, and that the location of continents is the main factor regulating atmosphere circulation patterns, and hence hurricane generation patterns. Present-day hurricanes originate in latitudes of 5 to 10° N and S over open ocean and generally impinge on the eastern side of land masses between latitudes 20 to 35° N and S, up to a maximum of 45°. Generation of hurricanes in the eastern side of oceans is considered insignificant. A relatively unaffected equatorial belt extends between hurricane zones of the northern and southern hemispheres. Winter storms range from 25° N and S latitude to their zone of maximum intensity at about 45° N and S latitude poleward, hence they partially overlap the tropical hurricane zones.

To establish inferred pathways of ancient tropical hurricanes (the only ones to be considered for carbonate models), it should be kept in mind that appreciable deviations from an average direction indicated by arrows (Fig. 18.50A to F) can be expected as in present hurricane systems in the North Atlantic. With these reservations in mind, the paleo-storm model of Marsaglia and Klein (1983), with a few modifications, can account for the origin of most of the storm deposits.

For instance, in the Cambrian paleogeography (Fig. 18.50A), the Allentown Dolomite of New Jersey (1) displays extensive storm deposits that could be related peripherally to the proposed general hurricane pathways affecting the eastern and southeastern coast of Laurentia. In the Ordovician paleogeography (Fig. 18.50B), the various carbonate models of the Mid-Continent of North America (2), namely the Shakopee Dolomite, Joachim Dolomite, Platteville Group, and Galena Group, show extensive storm deposits that are in agreement with the proposed general hurricane pathways along the southeastern coast of Laurentia. However, the Pogonip Group of southeast Nevada (3), which was located on the west coast of Laurentia, shows an important storm record that includes more than 130 tempestite beds of calcirudites in a section 400 m thick. These beds are repeated with a periodicity independent from the cyclicity of the interbedded carbonates. Thus this case cannot be attributed to hurricane-induced swells or other long-distance effects as proposed by Marsaglia and Klein (1983) and remains a notable exception. These tempestites perhaps indicate that the concept according to which generation of hurricanes in the eastern portions of oceans is insignificant should be reconsidered, or that they were formed by a different kind of storm that was not a hurricane.

The only example studied in the Silurian paleogeography (Fig. 18.50C) is the Cayugan of the Michigan Basin (4), which could be peripherally related to the proposed general hurricane pathways affecting the southeast coast of Laurentia. In the Devonian paleogeography (Fig. 18.50D), moderate storm deposits of the Jeffersonville Limestone of Indiana (5) and of the Traverse Group of Michigan (6) could also be peripherally related to the proposed general hurricane pathways affecting the southern coast of Laurussia. Storm deposits of the Arrow Canyon Formation and Crystal Pass Limestone of southeast Nevada (7), which were located on the west coast of Laurussia, are much less important than those of the underlying Pogonip Group, but again appear to be an exception and raise the same question as previously mentioned.

In the Mississippian paleogeography (Fig. 18.50E), among the numerous investigated carbonate systems of the Mid-Continent (9), only the Burlington Limestone displays a tornado deposit, and the Kinkaid Formation contains small to moderate storm deposits. They could be peripherally related to the proposed general hurricane pathways affecting the southern coast of Laurussia. The Monte Cristo Group (10) and the Rundle Group (11) located along the west coast of Laurussia show no storm deposits. In the Pennsylvanian paleogeography (Fig. 18.50F), all investigated carbonate models (12 to 16) were located along the west coast of the Laurussia-Gondwana block and show no storm deposits except some non-hurricane-related storm action in the Iola Formation (13).

18.14 CYCLICITY OF CARBONATE SEDIMENTATION

Analysis of the numerous examples of carbonate environments in this volume shows that the fundamental expression of carbonate sedimentation, regardless of microfacies type, geologic age, and geotectonic setting, is the so-called "ideal shallowing-upward sequence." It is, in fact, a *small-scale asymmetric cycle* ranging in thickness from 1 to 5 m and representing a duration of tens of thousands of years. This cycle consists of a relatively slow shallowing-upward sequence of distinct microfacies that ends with or without subaerial exposure and is followed by a relatively rapid deepening represented either by a hiatus or surface of nondeposition, a single microfacies, or a very thin sequence of microfacies without any well-defined sequential order. In the examples reviewed herein, the hiatus or surface of nondeposition is extremely widespread and by far the most characteristic feature (see, for example, the Upper Cambrian Allentown Dolomite, Figs. 14.4, 14.6, 14.7, 18.51). However, a single microfacies may occur, as in the Lower Ordovician

Shakopee Dolomite (Fig. 14.11, microfacies 4), or a relatively thin sequence of microfacies with no clear sequential order, as in the Middle Mississippian Salem Limestone (Fig. 12.7, episode III). In rare instances, a definite sequence of microfacies indicates deepening, as in the Valanginian Chachao Formation (Fig. 10.15).

The limits between successive small-scale asymmetric cycles which correspond to a cessation of sedimentation or to a rapid and important deepening of environment indicate geologically instantaneous events of allogenic nature. These limits represent surfaces that can be traced over tens of kilometers and they may represent isochronous surfaces that extend at least basinwide. Small-scale asymmetric cycles thus can be interpreted as thin time-stratigraphic units. Within them, lateral changes of microfacies may occur corresponding to the gradation between adjacent synchronous depositional environments. These changes are also recorded by the behavior of the clasticity curves of detrital quartz and other constituents transported by tidal currents (Carozzi, 1951c).

These fundamental small-scale asymmetric cycles consist of a number of *asymmetric microcycles* that display short shallowing-upward phases followed by minor hiatuses or shorter deepening episodes. Again, the Upper Cambrian Allentown Dolomite is a typical example (see Fig. 14.4, cycles 79, 82, 83, and 84 and Fig. 18.51, cycles 11, 12, and 16).

Furthermore, the fundamental small-scale asymmetric cycles are associated most often in *asymmetric megacycles*, ranging in thickness from 50 to 300 m, and representing a duration of hundreds of thousands of years. A typical asymmetric megacycle consists of a relatively slow shallowing-upward sequence of distinct superposed small-scale asymmetric cycles which begin and terminate in gradually shallowing conditions. This sequence may or may not terminate with subaerial exposure but is followed by a relatively rapid deepening represented, as in small-scale asymmetric cycles, either by a hiatus or surface of nondeposition, or by a thin sequence of microfacies without any well-defined sequential order. Again, the Upper Cambrian Allentown Dolomite can be used as a typical example (see Fig. 18.51, megacycle III). Other characteristic examples are the Middle Ordovician Joachim Dolomite (Figs. 14.22, 14.23, 14.24), the Mississippian Rundle Group (Fig. 8.21), the Middle Mississippian St. Louis Limestone (Figs. 3.4, 3.13), and the Atokan to Missourian portion of the Bird Spring Group (Fig. 4.43).

Small-scale asymmetric cycles may show other less frequent associations such as superpositions without any specific trend indicating periods of stability, as in the case of the interval Missourian to Wolfcampian of the Bird Spring Group (Fig. 11.55), or asymmetric megacycles indicating a gradual deepening trend, as in the case of the Morrowan portion of the Bird Spring Group (Fig. 4.42). Study of certain sequences reveals a succession in time of these variants of general trend. Such is the case of the Upper Cambrian Allentown Dolomite (Figs. 14.6, 14.7), the Lower Ordovician Pogonip Group (Figs. 11.12, 11.13, 11.14), the Mississippian Monte Cristo Group (Fig. 11.43), and the Upper Mississippian Kinkaid Group (Fig. 8.17). More attention should be given in future studies to the poorly understood nature of asymmetric megacycles that indicate a deepening trend, and to the results of computer modeling of carbonate cycles in general (Read *et al.*, 1986).

In summary, associations of carbonate microfacies predominantly display a hierarchy of three types of asymmetric shallowing-upward sequences: microcycles, cycles, and megacycles (Fig. 18.52). Among them, asymmetric cycles represent the fundamental expression in carbonate environments of episodic sedimentation under the effects of allogenic causes. Although controversy over the identification of allogenic causes remains, the repetition of small-scale asymmetric cycles has been explained by two major concepts, the *eustatic model* and the *autocyclic model*.

In the eustatic model (Wilkinson, 1982; James, 1984), the rate of carbonate sedimentation is considered constant, whereas the rate of subsidence or the absolute position of sea level change periodically. During periods of stability or slowly rising sea level, carbonate sedimentation accretes or progrades, generating a typical shallowing-upward sequence. This pattern is interrupted by a sudden and rapid period of deeper water with reduced deposition or interrupted sedimentation. Sea level remains relatively stationary in this new position, and progradation of the carbonates again begins to deposit a new shallowing-upward sequence on top of the preceding one.

A more sophisticated version of the eustatic model has been presented by Goodwin and Anderson (1985) under the designation of punctuated aggradational cycles (abbreviated as PAC). This general hypothesis of episodic sedimentation, documented mainly by field work on carbonates, assumes that the stratigraphic record consists of small-scale (1 to 5 m thick), basinwide, shallowing-upward cycles separated by surfaces indicating abrupt changes to deeper facies. This pattern, pervasive in time and environments, is assumed to have been produced by relatively long periods (tens of thousands of years) of sea level stability, separated by geologically instantaneous sea level rises (punctuation events), both essentially unaffected except by a few episodes of the fall of sea level. This hypothesis recognizes the hierarchy of the three types of carbonate organization mentioned above, microcycles, cycles, and megacycles. It also con-

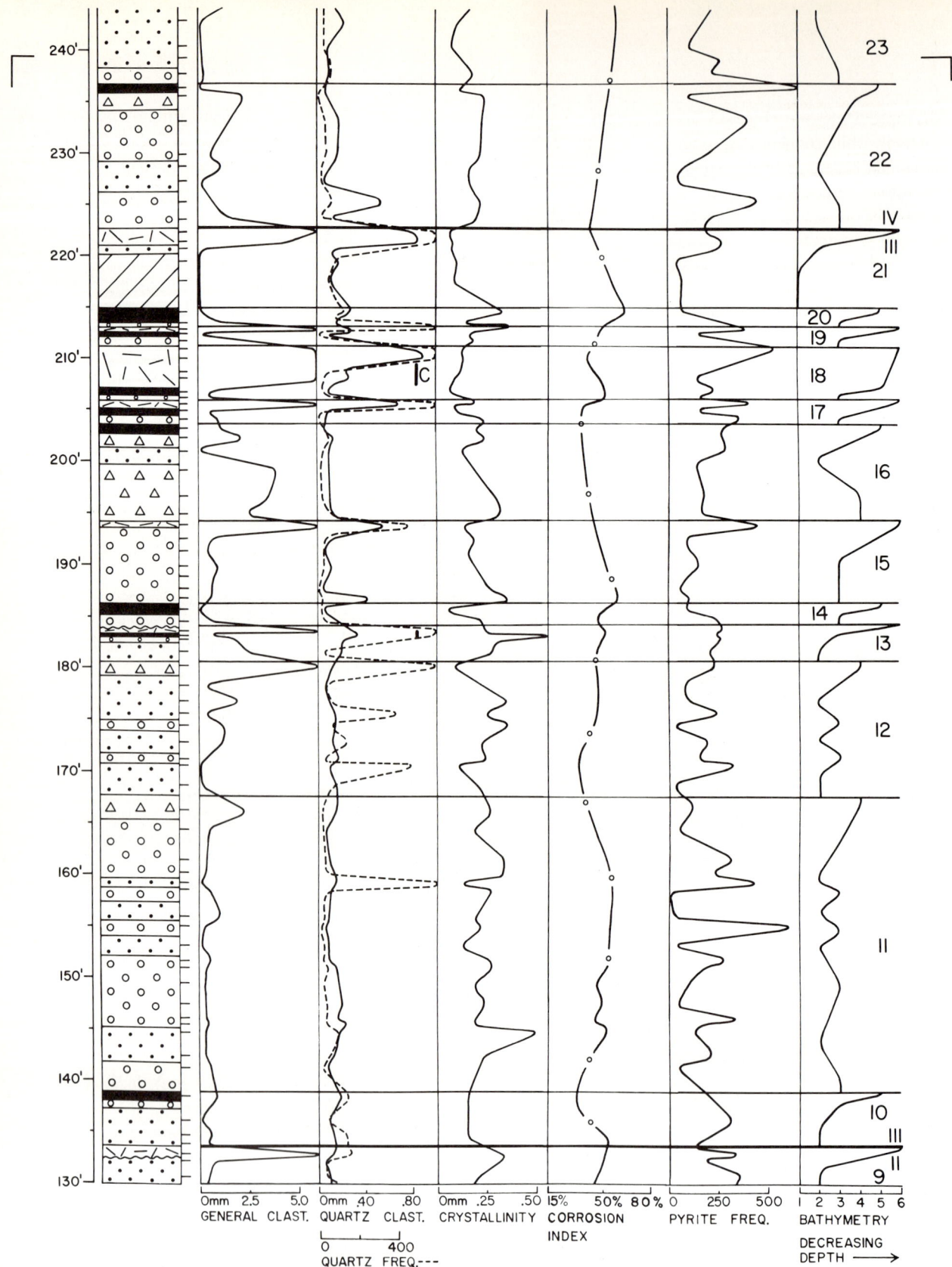

FIGURE 18.51 Allentown Dolomite (Upper Cambrian), Warren County, New Jersey. Typical example of vertical variation of microscopic parameters showing asymmetric cycles (10 through 21) with several displaying microcycles (11, 12, and 16), and associated in an asymmetric megacycle (III). Field section 1 at Riegelsville. From Zadnik and Carozzi (1963).

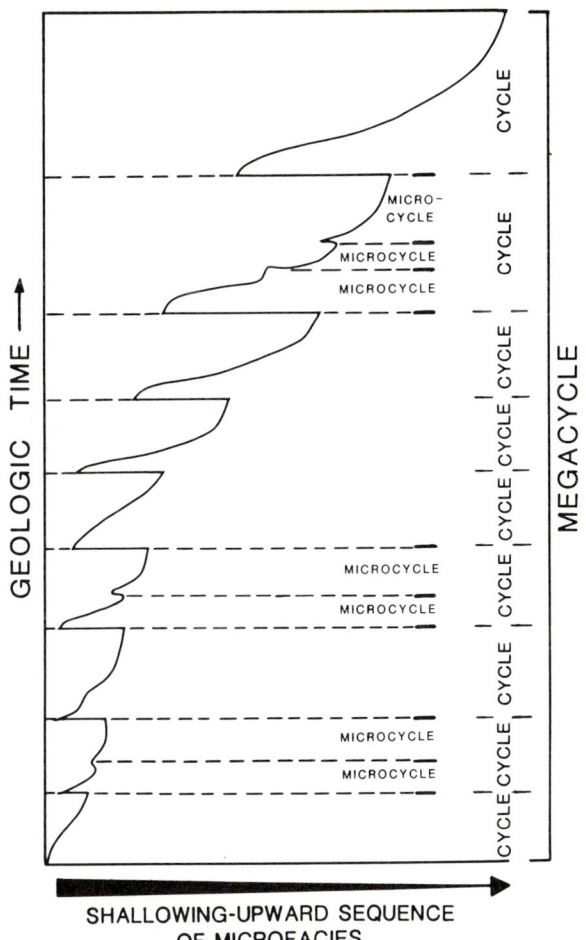

FIGURE 18.52 Theoretical relationships between asymmetric microcycles, cycles, and megacycle. From Carozzi (1986).

siders the cycle as the fundamental basinwide lithologic time-stratigraphic unit which should be used in all aspects of stratigraphic analysis. Among the several possible allogenic mechanisms for explaining punctuated aggradational cycles, glacial eustasy driven by orbital perturbations is preferred by these authors.

In the *autocyclic model* (Ginsburg, 1971), as in the eustatic one, carbonate deposition is assumed to take place on a gently inclined ramp or on a platform under conditions of gradual subsidence or slowly rising sea level, or some combination of both. The control is within the model and consists of a variable rate of carbonate sedimentation controlled by the extent of the subtidal source area. The latter produces carbonate sediments which move shoreward by wind-driven, tidal, or estuarine-like circulation and a seaward progradation of the carbonate wedge takes place. This progradation gradually reduces the surface of the sediment-producing subtidal area until carbonate production no longer exceeds subsidence. Because the relative rise of sea level continues, the entire platform becomes again subtidal and deep enough to resume sediment production and a new progradation phase begins.

However, both the eustatic model, which implies a constant rate of carbonate production, and the autocyclic model, which assumes a variable rate of carbonate production, take into account only either a stable sea level or repeated episodes of slow or rapid rate of sea level rise (Fig. 18.53). Furthermore, the autocyclic model based on the Florida Bay lagoon, the tidal flats of the Bahamas, and the Persian Gulf suffers from its uniformitarian character, restricted depositional environment, and from the fact that the landward movement of carbonate sediments is not a common mechanism of deposition in ancient carbonate models. In fact, the numerous examples of ancient carbonate sediments presented in this book indicate that they are either formed and deposited almost *in situ* or display complex patterns of relatively short-distance seaward and landward transportation, the latter being important only under storm influence.

In short, the proponents of eustatic and autocyclic models, although admitting at least in the case of the concept of punctuated aggradational cycles, the existence of a hierarchy of microcycles, cycles, and megacycles, do not seem to recognize that these cycles are similar in significance and in implications to the paracycles, cycles, and supercycles of eustatic changes of sea level (Fig. 18.54) described by Vail *et al.* (1977).

At present, eustatic changes of sea level are known to have taken place during at least the Phanerozoic on a worldwide basis. Eustatic oscillations of sea level follow a well-established pattern of repeated episodes of slow rise or "transgression," stillstand, and rapid fall or "regression." This pattern displayed by siliciclastic sedimentation must also be recorded in carbonate sediments which not only share many basins but which also belong to the same worldwide oceanic realm.

An *apparent* dilemma (Fig. 18.55) exists, nevertheless, when carbonate cycles, which consist of a slow shallowing-upward phase (apparent regression), followed by a rapid deepening phase (apparent transgression) are compared with worldwide eustatic cycles which display a pattern of slow sea level rising (transgression), followed by a rapid fall (regression). Wilkinson (1982) took the unrealistic position of considering that the shape of worldwide eustatic cycles are simply wrong and that carbonate cyclicity can be explained only by the eustatic model based on sudden and repeated sea level rises. James (1984) tried to solve the apparent dilemma of large-scale carbonate cycles by assuming a constant rate of subsi-

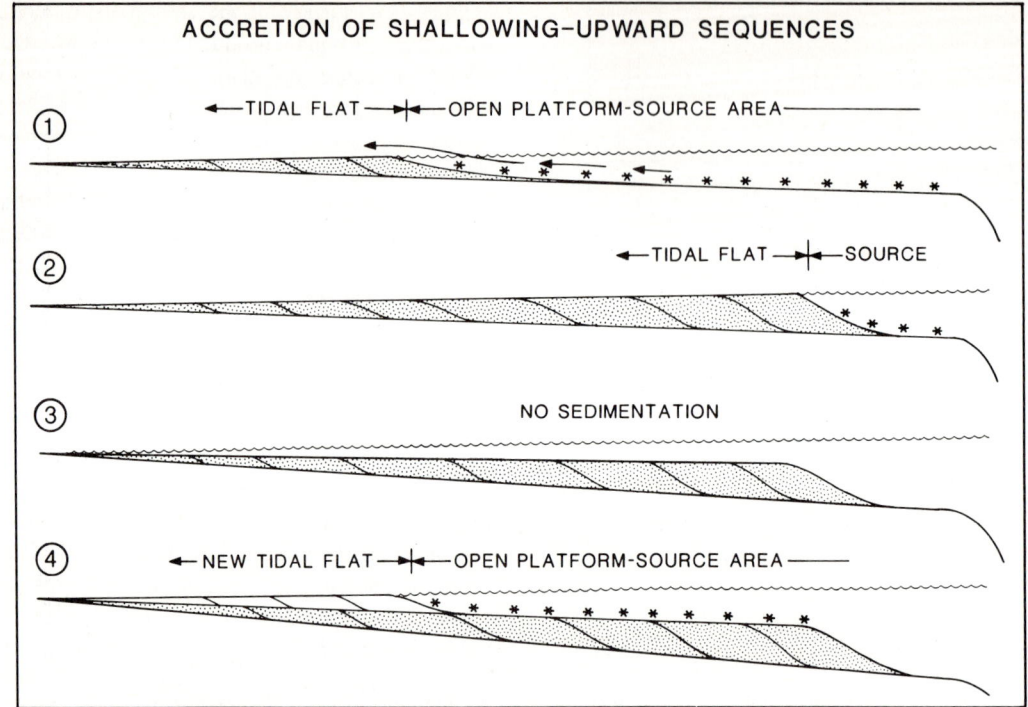

FIGURE 18.53 Diagram illustrating how two shallowing-upward sequences can be produced by progradation of a tidal flat wedge. These general conditions apply to both eustatic and autocyclic models. From James (1984). Reprinted by permission of the Geological Association of Canada.

dence and by combining it with a slightly modified Vail curve which consists of a more or less symmetric worldwide rise and fall of sea level (Fig. 18.56). According to this hypothesis, the rising sea level initially floods the platform and carbonate accretion can outpace sea level rise so that shallowing-upward sequences develop. During the long period of relatively rapid sea level rise, and depending on the rate between this process and subsidence, either subtidal conditions are maintained or a few thick shallowing-upward sequences occur. When the rate of sea level rise decreases, several shallowing-upward sequences are formed. When the sea level begins to fall and subsidence continues, the net result is a stillstand generating numerous thin shallowing-upward se-

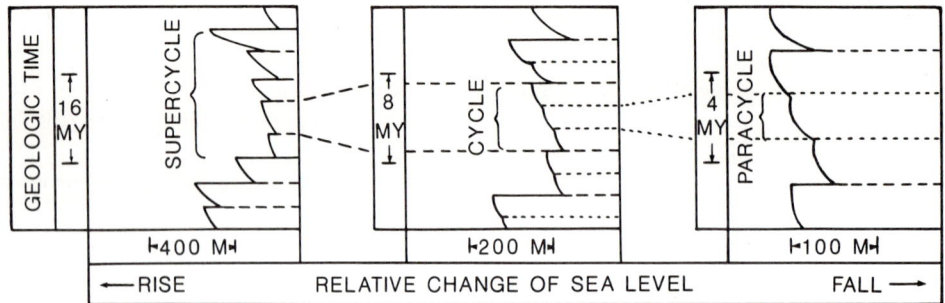

FIGURE 18.54 Chart of relative changes of sea level. Cycles consist of relative rises and falls of sea level, commonly containing several paracycles, which are smaller pulses of relative rises to stillstands. Several cycles usually form a higher-order cycle (supercycle) with patterns of successive rises between major falls. Observe asymmetry of gradual rises and abrupt falls at each scale. Modified from Vail et al. (1977).

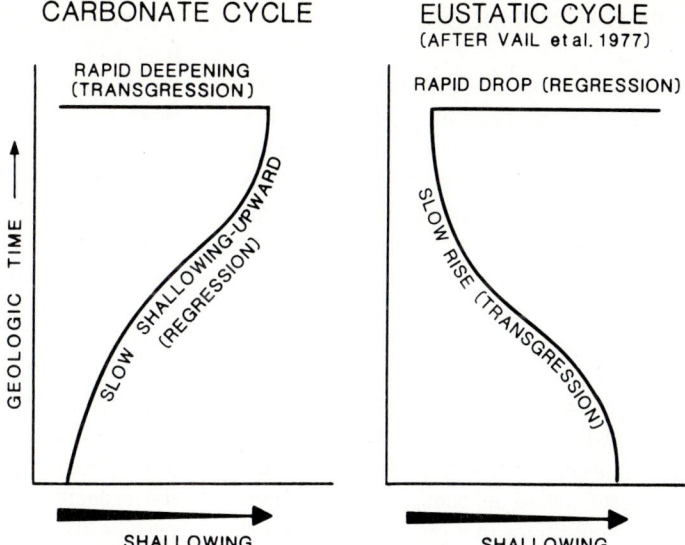

FIGURE 18.55 Theoretical illustration of apparent dilemma between carbonate cycle and Vail's eustatic cycle. From Carozzi (1986).

quences separated by long episodes of subaerial exposure. The final rapid fall of sea level outpaces subsidence and the carbonate platform is entirely exposed. The main point in James' hypothesis is to show that a large-scale asymmetric carbonate cycle may be produced by a uniform eustatic rise and fall of sea level. It is certainly a bold idea, although not in agreement with the asymmetric pattern of eustatic oscillations of sea level.

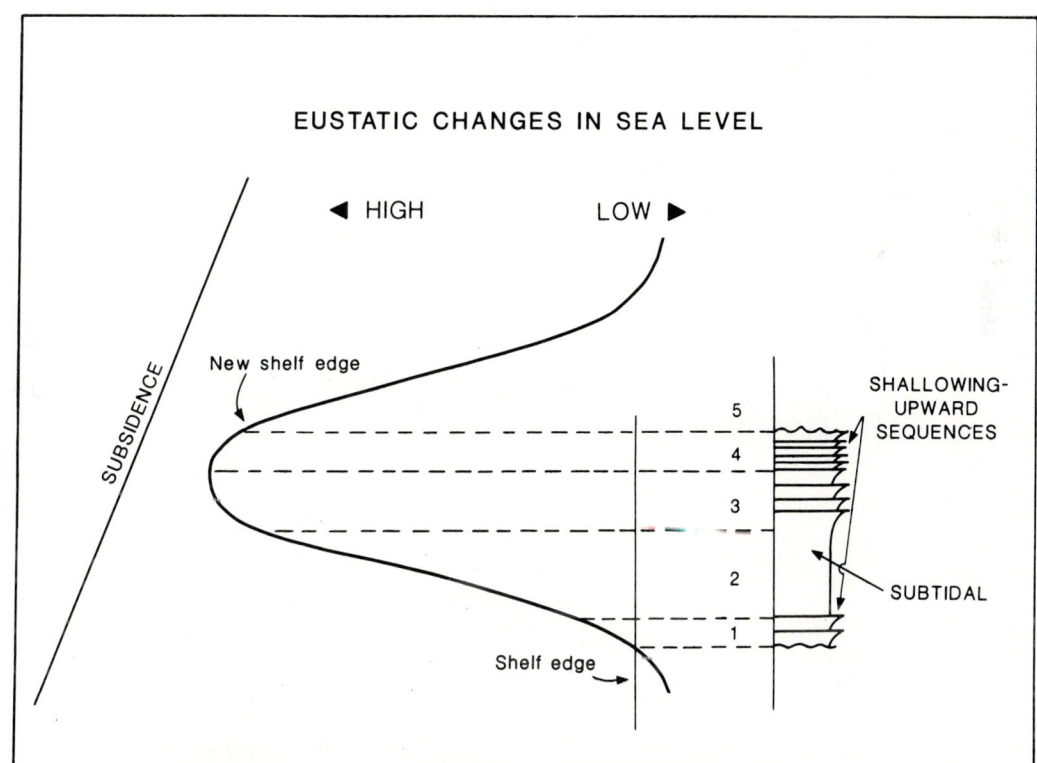

FIGURE 18.56 Diagram illustrating how a large-scale shallowing-upward sequence is produced under conditions of slow platform subsidence and a uniform eustatic rise and fall in sea level. Modified from James (1984).

18.15 A NEW EUSTATIC MODEL

Unquestionably, the origin of carbonate cycles (Carozzi, 1986) is a complex interplay of eustatic sea level oscillations, subsidence (tectonism), and the fundamental property of carbonate ramps and platforms which consists mainly of *in situ* carbonate sediment production by benthic and planktonic organisms.

The rate of carbonate productivity has been largely underestimated at both the scale of lagoons and platforms. Indeed, Neumann and Land (1975) have shown in their study of the Bight of Abaco, Bahamas, that at present lagoonal calcareous green algae produce, by their skeletal breakdown, 1.5 to 3 times the mass of aragonite mud now in the lagoon. This means that the excess is exported as suspended sediment toward off-bank and basinal environments, where it contributes to a great extent to flank accretion and to the so-called "pelagic" deposition. Furthermore, according to Schlager (1981, p. 204), "a platform with a sediment-covered flat top that simultaneously builds upward and progrades basinward must produce more sediments than it needs to match the relative rise of sea level. Consequently, its growth potential must be greater than its vertical growth rate." The latter can be estimated by the study of pre-Holocene prograding platforms which reveals accumulation rates (not corrected for compaction) ranging from 30 to 500 μm per year and average growth potential probably on the order of a thousand microns per year (Schlager, 1981). Taking into account figures provided by Schlager (1981), such as a maximum accumulation rate of 500 μm per year and a basin subsidence averaging 10 to 100 μm per year, it is safe to say that a carbonate platform shows the capacity to build upward at a rate exceeding any eustatic rise of sea level due to increases in the rate of seafloor spreading, changes in subduction and spreading

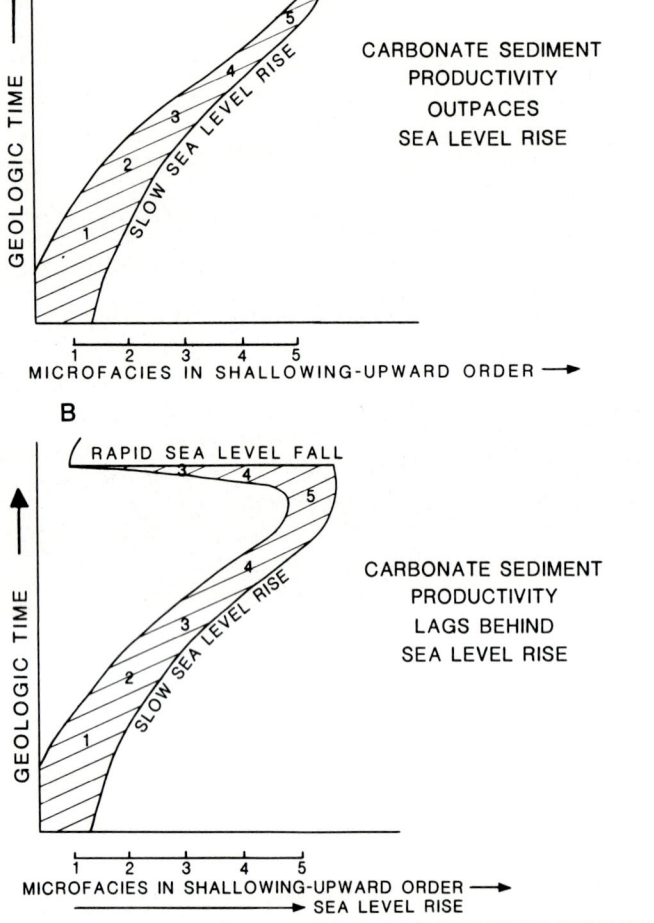

FIGURE 18.57 Theoretical illustration of how the two major types of asymmetric carbonate cycles are generated during a eustatic slow sea level rise followed by a rapid sea level fall. From Carozzi (1986).

pattern, or other long-term geologic processes which is estimated to range between 10 and 90 μm per year.

In view of these data, and under the most conservative conditions, the origin of carbonate cycles can be approached as an interplay between global slow rises and rapid falls of sea level and a rate of carbonate productivity resulting in vertical accretion and horizontal progradation that can easily be greater than sea level rise. Assuming a constant rate of subsidence, we can consider two possibilities. First (Fig. 18.57A), the rate of carbonate productivity increases with a slow rise of sea level and eventually outpaces it. A shallowing-upward sequence is generated that terminates with subaerial exposure followed by a hiatus during sea level fall. Second (Fig. 18.57B), the rate of carbonate productivity increases with a slow rise of sea level but lags behind it at all times. A shallowing-upward sequence is generated that does not reach subaerial exposure but is overlain during rapid sea level fall by a thin sequence of nonsequential carbonate sediments or a complex lag deposit.

This hypothesis resolves the previously mentioned dilemma (see Fig. 18.55). It shows that the shallowing-upward sequence is an illusion of interpretation. In reality, *the shallowing-upward sequence takes place during a slow rise of sea level, that is, during a transgression, as a reaction of carbonate productivity versus eustasy.* Furthermore, the existence of a hierarchy of three types of carbonate cycles similar to those of eustatic sea level changes contributes to the agreement of carbonate sedimentation with global processes.

A comparison of this hypothesis with the results of Ross and Ross (1985) shows that all the Lower Carboniferous examples of shallowing-upward sequences of the Mississippi Valley presented in this book, namely Burlington, Salem, St. Louis, Ste. Genevieve, Glen Dean, and Kinkaid (Grove Church), correspond indeed to major rises of sea level or transgressions (Fig. 18.58). Ross and Ross (1985) described more than 50 transgressive-regressive depositional sequences in Carboniferous and Permian shallow marine successions over stable cratonic shelves all over the world. Stratigraphic evidence shows the large areal extent of these shoreline displacements and suggests slow rises of sea level followed by rapid falls. Faunal correlations indicate that each transgressive-regressive sequence is synchronous worldwide and hence represents eustatic changes in sea level of 100 to 200 m, lasting from 1.2 to 4 m.y. with an average of 2 m.y.

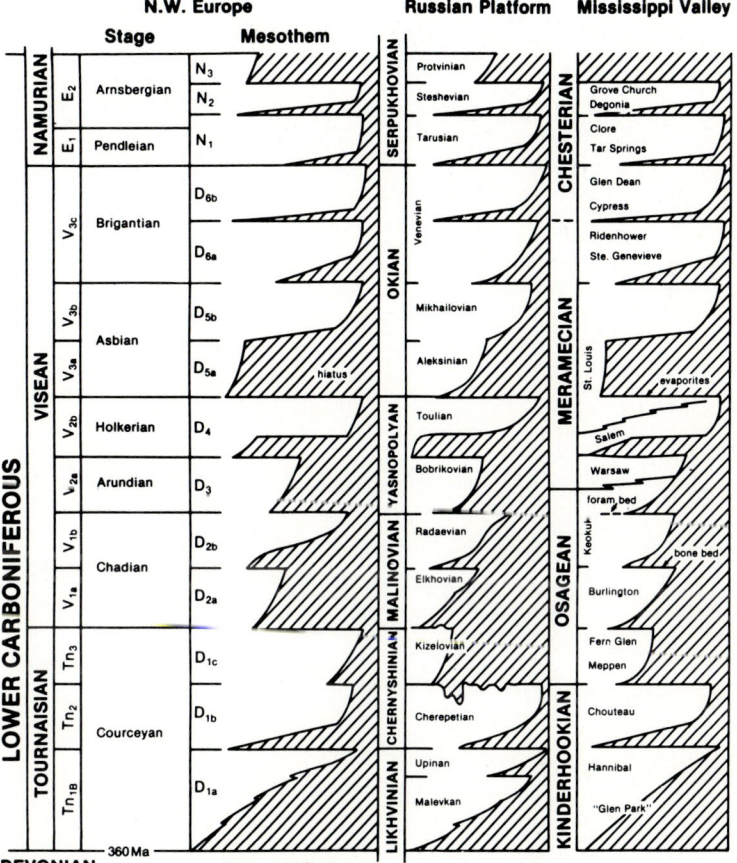

FIGURE 18.58 Correlation of lower Carboniferous transgressive-regressive sequences based on fossil zonations. From Ross and Ross (1985). Reprinted by permission of the authors and of the Geological Society of America.

What lies ahead is, in essence, the testing of Vail's curve by means of a combination of detailed microfacies analysis and biostratigraphic studies of the well-exposed sedimentary record of cratons around the world. Such a combined approach can unravel the pattern of sea level fluctuations expressed by the hierarchy of microcycles, cycles, and megacycles. In particular, a comparison of cycles on a basinwide to worldwide basis could establish their value as time-stratigraphic units for basin analysis of the future.

References

AHR, W. M., 1973. The carbonate ramp: an alternative to the shelf model. Trans. Gulf Coast Assoc. Geol. Soc., 23:221–225.

AIGNER, T., 1982. Calcareous tempestites: storm-dominated stratification in Upper Muschelkalk limestone (Middle Trias, SW-Germany). In: G. Einsele and A. Seilacher (Eds.), Cyclic and Event Stratification. Springer-Verlag, Heidelberg, pp. 180–198.

ALLING, H. L., 1945. Use of microlithologies as illustrated by some New York sedimentary rocks. Bull. Geol. Soc. Am., 56,6:737–756.

BADIOZAMANI, K., 1973. The Dorag dolomitization model—application to the Middle Ordovician of Wisconsin. Jour. Sed. Petrology, 43,4:965–984.

BAKUSH, S. H., and CAROZZI, A.V., 1986. Subtidal storm-influenced carbonate ramp model: Galena Group (Middle Ordovician) along Mississippi River (Iowa, Wisconsin, Illinois, and Missouri), U.S.A. Archives Sciences, Genève, 39,2:141–183.

BATHURST, R. G. C., 1975. Carbonate sediments and their diagenesis. Second enlarged edition. Develop. Sedimentology 12, Elsevier, Amsterdam, 658 pp.

BERTANI, R. T., and CAROZZI, A. V., 1984. Microfacies, depositional models, and diagenesis of Lagoa Feia Formation (Lower Cretaceous), Campos Basin, offshore Brazil. Petrobrás, Cenpes, Ciência Técnica Petróleo, No. 14, 104 pp.

BERTANI, R. T., and CAROZZI, A. V., 1985. Lagoa Feia Formation (Lower Cretaceous), Campos Basin, offshore Brazil: rift-valley stage lacustrine carbonate reservoirs. Jour. Petroleum Geology, part I, 8,1:37–58; part II, 8,2:199–220.

BLACK, N. R., 1985. Petrography and diagenesis of the Galena (Middle Ordovician)–Maquoketa (Late Ordovician) contact and the basal Maquoketa phosphorites in eastern Missouri and eastern Iowa, U.S.A. Unpublished MS thesis, University of Illinois at Urbana-Champaign, 133 pp.

BROWN, J. S., 1943. Suggested use of the word microfacies. Econ. Geology, 38,4:325.

CALKINS, F. C., 1943. The word "Microfacies." Econ. Geology, 38,7:608–609.

CAMPBELL, I., 1944. The word "Microfacies." Econ. Geology, 39,1:70–71.

CAROZZI, A. V., 1948. Étude stratigraphique et micrographique du Purbeckien du Jura Suisse. Archives Sciences, Genève, 1,1:1–175.

CAROZZI, A. V., 1949. Interprétation des séries sédimentaires. Le Berriasien et le Valanginien schisteux de la Giettaz (Nappe de Morcles–Aravis). Archives Sciences, Genève, 2,2:350–354.

CAROZZI, A. V., 1950. Contribution à l'étude des rythmes de sédimentation. Archives Sciences, Genève, 3,1–2:1–76.

CAROZZI, A. V., 1951a. Sédimentation rythmique dans la Nappe de Morcles–Aravis (Chaînes subalpines de Haute-Savoie, France). Proc. 3rd Internat. Congress Sedimentology, The Netherlands, pp. 81–89.

CAROZZI, A. V., 1951b. Rythmes de sédimentation dans le Crétacé helvétique. Geol. Rundschau, 39,1:177–195.

CAROZZI, A. V., 1951c. La notion de synchronisme en géologie. Revue Générale des Sciences Pures et Appliquées, Paris, 58,7–8:230–236.

CAROZZI, A. V., 1952a. Méthodes d'analyse de la sédimentation rythmique. C. R. 19th Internat. Geol. Congress, Alger, Sections XIII, XIV, pp. 71–80 (1954).

CAROZZI, A. V., 1952b. Les phénomènes de courants de turbidité dans la sédimentation alpine: une hypothèse de travail. Archives Sciences, Genève, 5,1:35–39.

CAROZZI, A. V., 1952c. Microfaune déplacée dans les niveaux "remaniés" du Malm supérieur de la Nappe de Morcles (Haute-Savoie). Archives Sciences, Genève, 5,1:39–42.

CAROZZI, A. V., 1952d. Tectonique, courants de turbidité et sédimentation. Application au Jurassique supérieur des chaînes subalpines de Haute-Savoie. Revue Générale des Sciences Pures et Appliquées, Paris, 59,7–8:229–245.

CAROZZI, A. V., 1953. Pétrographie des roches sédimentaires. Rouge & Cie., Lausanne, 250 pp.

CAROZZI, A. V., 1954. Sédimentation rythmique en milieu corallien: le Jurassique supérieur du Grand-Salève. Archives Sciences, Genève, 7,2:65–93.

CAROZZI, A. V., 1955a. Nouvelles observations microscopiques sur les dépôts de courants de turbidité du Malm de la Nappe de Morcles en Haute-Savoie. Inst. Nat. Genevois Bull., 57:1–31.

CAROZZI, A. V., 1955b. Sédimentation récifale rythmique dans le Jurassique supérieur du Grand-Salève (Haute-Savoie, France). Geol. Rundschau, 43,2:433–446.

CAROZZI, A. V., 1956. Problèmes de sédimentation et de corrélation dans le Groupe de Platteville (Ordovicien moyen) de l'Iowa, Illinois et Indiana, U.S.A. Archives Sciences, Genève, 9,3:283–302.

CAROZZI, A. V., 1957. Tracing turbidity current deposits down the slope of an Alpine Basin. Jour. Sed. Petrology, 27,3:271–281.

CAROZZI, A. V., 1958. Micro-mechanisms of sedimentation in the epicontinental environment. Jour. Sed. Petrology, 28, 2:133–150.

CAROZZI, A. V., 1960. Microscopic Sedimentary Petrography. John Wiley, New York, 485 pp.

CAROZZI, A. V., 1961a. Reef petrography in the Beaverhill Lake Formation, Upper Devonian, Swan Hills area, Alberta, Canada. Jour. Sed. Petrology, 31,4:497–513.

CAROZZI, A. V., 1961b. Oolithes remaniées, brisées et régénérées dans le Mississippien des chaînes frontales, Alberta central, Canada. Archives Sciences, Genève, 14,2:281–296.

CAROZZI, A. V., 1961c. Distorted oolites and pseudoolites. Jour. Sed. Petrology, 31,2:262–274.

CAROZZI, A. V., 1962. Cerebroid oolites. Trans. Illinois State Acad. Science, 55,3–4:239–249.

CAROZZI, A. V., 1963. Half-moon oolites. Jour. Sed. Petrology, 33,3:633–645.

CAROZZI, A. V., 1971a. Apparent relationship between development of neomorphic calcite and environmental energy level in Chesterian (Viséan-Namurian) biocalcarenites, type area in southwestern Illinois. In: O. P. Bricker (Ed.), Carbonate Cements. The Johns Hopkins University Studies in Geology, No. 19, Johns Hopkins University Press, Baltimore, pp. 205–208.

CAROZZI, A. V., 1971b. Geochemical data on back-reef carbonates: Traverse Group (Givétian) of the northern part of the Southern Peninsula of Michigan. Bull. Centre Rech. Pau, Soc. Nat. Pétroles Aquitaine, 5,2:213–222.

CAROZZI, A. V., 1972. Microscopic Sedimentary Petrography. Corrected edition. R. E. Krieger, Huntington, N.Y., 485 pp.

CAROZZI, A. V., 1981. Porosity models and oil exploration of Amapá carbonates, Paleogene, Foz do Amazonas Basin, offshore NW Brazil, Jour. Petroleum Geology, 4,1:3–34.

CAROZZI, A. V., 1983. Modelos Deposicionales Carbonaticos. Asociación Geológica Argentina, Buenos Aires, Serie B. Didáctica y Complementaria, No. 11, vol. 1, 106 pp.; vol. 2, 197 pp.

CAROZZI, A. V., 1986. New eustatic model for the origin of carbonate cyclic sedimentation. Archives Sciences, Genève, 39,1:53–66.

CAROZZI, A. V., and BANAEE, J., 1984. Bailey Limestone (Lower Devonian) of southwestern Illinois: a carbonate turbidite. Trans. Illinois Acad. Science, 77,3–4:271–282.

CAROZZI, A. V., and DAVIS, R. A., JR., 1964. Pétrographie et paléoécologie d'une série de dolomies à stromatolithes de l'Ordovicien inférieur du Wisconsin, U.S.A. Archives Sciences, Genève, 17,1:47–63.

CAROZZI, A. V., and DIABY, I., 1981. Microfacies and depositional model of the Salem Limestone (Middle Mississippian), southwestern Illinois, U.S.A. Actas VIII Congreso Geol. Argentino, II:435–457.

CAROZZI, A. V., and FALKENHEIN, F. U. H., 1985. Depositional and diagenetic evolution of Cretaceous oncolitic packstone reservoirs, Macaé Formation, Campos Basin, offshore Brazil. In: P. O. Roehl and P. W. Choquette (Eds.), Carbonate Petroleum Reservoirs. Springer-Verlag, New York, pp. 473–484.

CAROZZI, A. V., and FROST, S. H., 1966. Turbidites in dolomitized flank beds of Niagaran (Silurian) reef, Lapel, Indiana. Jour. Sed. Petrology, 36,2:563–575.

CAROZZI, A. V., and GERBER, M. S., 1978. Synsedimentary chert breccia: a Mississippian tempestite. Jour. Sed. Petrology, 48,3:705–708.

CAROZZI, A. V., and GERBER, M. S., 1979. Crinoid arenite banks and crinoid wacke inertia flows: a depositional model for the Burlington Limestone (Middle Mississippian), Illinois, Iowa, and Missouri, U.S.A. C. R. Neuvième Congr. Internat. Stratigraphie et Géologie du Carbonifère, Washington et Urbana-Champaign, 3:452–460 (1984).

CAROZZI, A. V., and HULSE, J. A., 1963. Variations latérales de microfaciès dans un banc à coraux et stromatopores du Dévonien moyen de l'Indiana, U.S.A. Archives Sciences, Genève, 16,2:309–337.

CAROZZI, A. V., and LUNDWALL, W. R., JR., 1959. Microfacies study of a Middle Devonian bioherm, Columbus, Indiana. Jour. Sed. Petrology, 29,3:343–353.

CAROZZI, A. V., and REICHELDERFER, J. L., 1987. Reservoir controls in carbonate offshore bars, Salem Limestone (Middle Mississippian), southeastern Illinois. Trans. Illinois State Acad. Science, 80:1–2:71–82.

Carozzi, A. V., and Roche, J. E., 1968. Petrography of selected Chesterian carbonates (Viséan-Namurian) from the type area in southwestern Illinois. Trans. Illinois State Acad. Science, 61,2:182–200.

Carozzi, A. V., and Soderman, J. G. W., 1962. Petrography of Mississippian (Borden) crinoidal limestones at Stobo, Indiana. Jour. Sed. Petrology, 32,3:397–414.

Carozzi, A. V., and Textoris, D. A., 1963. Les *Stromatactis* des récifs siluriens de l'Indiana sont des bryozoaires. Archives Sciences, Genève, 16,1:188–192.

Carozzi, A. V., and Textoris, D. A., 1967. Paleozoic carbonate microfacies of the Eastern Stable Interior, U.S.A. Internat. Sed. Petro. Series, XI, E. J. Brill, Leiden, 41 pp., 100 plates.

Carozzi, A. V., and Von Bergen, D., 1987. Stylolitic porosity in carbonates: a critical factor for deep hydrocarbon production. Jour. Petroleum Geology, 10,3:267–282.

Carozzi, A. V., and Zadnik, V. E., 1959. Microfacies of Wabash Reef, Wabash, Indiana. Jour. Sed. Petrology, 29, 2:164–171.

Carozzi, A. V., Bouroullec, J., Deloffre, R., and Rumeau, J. L., 1972. Microfaciès du Jurassique d'Aquitaine–Pétrographie–Diagenèse–Géochimie–Pétrophysique. Bull. Centre Rech. Pau, Soc. Nat. Pétroles Aquitaine, Volume Spécial No. 1, 594 pp. (bilingual text French-English).

Carozzi, A. V., Tibana, P., and Tessari, E., 1973. Estudo das microfácies da Formação Bonfim (Cenomaniano) da Bacia de Barreirinhas, Brasil. Petrobrás, Cenpes, Ciência Técnica Petróleo, No. 6, 86 pp.

Carozzi, A. V., Alves, R. J., and de Castro, J. C., 1974. Contrôle tectônico sinsedimentar dos carbonatos Permo-Carboníferos das formações Itaituba e Nova Olinda da Bacia do Amazonas, Brasil. Anais XXVI Congresso Brasileiro de Geologia, Belém, 3:47–64.

Carozzi, A. V., Reyes, M. V., and Ocampo, V. P., 1976. Microfacies and microfossils of the Miocene reef carbonates of the Philippines. Philippine Oil Development Co., Manila, Special Publ. No. 1, 80 pp., 20 plates.

Carozzi, A. V., Bercowski, F., Rodriguez Schelotto, M. L., Sánchez M., and Vonesch, T., 1981a. Estudio de microfacies de la Formación Chachao (Valanginiano), Provincia de Mendoza. Actas VIII Congreso Geol. Argentino, II:545–565.

Carozzi, A. V., de Castro, J. C., Beltrami, C. V., Spadini, A. R., and Wolff, B., 1981b. Microfacies, depositional model, and porosity control of the Amapá carbonates, Paleogene, Foz do Amazonas Basin, offshore NW Brazil. Actas VIII Congreso Geol. Argentino, II:651–693.

Carozzi, A. V., Orchuela, I. A., Rodriguez Schelotto, M. L., Baliña, M. M., Blanco Ibañez, S., Ferraresi, P. L., and Ambrosis, M. E., 1981c. Microfacies, paleoenvironments, and petroleum potential of the Quintuco–Loma Montosa Formation, Lower Cretaceous, Neuquén Basin, Argentina. Yacimentos Petrolíferos Fiscales (Y.P.F.), Gerencia de Planificación Geológica. Unpublished Internal Report, 89 pp.

Carozzi, A. V., Baliña, M. M., Blanco Ibañez, S., Corbari, S., Constanzo, J., Fernandez, S., Gustavino, L., Orchuela, I. A., Pedrazzini, M., and Valenzuela, M., 1982. Analisis sedimentologico y diagenetico de los carbonatos del Miembro Superior de la Fm. Quintuco en el yacimiento Loma La Lata, Provincia del Neuquén. Trab. Tecn. Primer Congr. Nac. Hidrocarburos, Petróleo y Gas. Exploración. Instituto Argentino del Petróleo, Buenos Aires, pp. 65–84.

Carozzi, A. V., Falkenhein, F. U. H., and Franke, M. R., 1983. Depositional environment, diagenesis, and reservoir properties of oncolitic packstones, Macaé Formation (Albian-Cenomanian), Campos Basin, offshore Rio de Janeiro, Brazil. In: T. M. Peryt (Ed.), Coated Grains. Springer-Verlag, Heidelberg, pp. 330–343.

Carss, B. W., and Carozzi, A. V., 1965. Petrology of Upper Devonian pelletoidal limestones, Arrow Canyon Range, Clark County, Nevada. Sedimentology, 4,3:197–224.

Carvalho, M. D., Paumer, M. L., Coelho de Lima, C., Babinski, N. A., Fanha, A. B., and Carozzi, A. V., 1982a. Sistemas deposicionais, evolução diagenética e geoquímica dos carbonatos e evaporitos da área da plataforma de São Mateus (E.S.). Unpublished Internal Report Petrobrás, Cenpes, No. 340, 102 pp.

Carvalho, M. D., Paumer, M. L., Coelho de Lima, C., Babinski, N. A., Fanha, A. B., and Carozzi, A. V., 1982b. Sistemas deposicionais e evolução diagenética dos carbonatos e evaporitos da plataforma de São Mateus (E.S.). Anais XXXII Congresso Brasileiro de Geologia, Salvador, Bahia, 5:2328–2335.

Cayeux, L., 1935. Les roches sédimentaires de France—Roches carbonatées (calcaires et dolomies). Masson, Paris, 436 pp. See also L. Cayeux, 1970. Sedimentary Rocks of France. Carbonate rocks (limestones and dolomites), translated and updated by A. V. Carozzi, Hafner, Darien, Conn., 438 pp.

Choquette, P. W., and Pray, L. C., 1970. Geologic nomenclature and classification of porosity in sedimentary carbonates. Am. Assoc. Petroleum Geologists Bull., 54,2:207–250.

Cook, H. E., and Enos, P. (Eds.), 1977. Deep-water carbonate environments. Soc. Econ. Paleontologists and Mineralogists, Special Publ. No. 25, 336 pp.

Cook, H. E., McDaniel, P. N., Mountjoy, E. W., and Pray, L. C., 1972. Allochthonous carbonate debris flows at Devonian bank ('reef') margins, Alberta, Canada. Canadian Petroleum Geology Bull., 20,3:439–497.

Cook, H. E., Hine, A. C., and Mullins, H. T. (Eds.), 1983. Platform margin and deep water carbonates. S.E.P.M. lecture notes for short course No. 12, 573 pp.

Crevello, P. D., and Harris, P. M. (Eds.), 1985. Deepwater carbonates: buildups, turbidites, debris flows and chalk, a core workshop. Soc. Econ. Paleontologists and Mineralogists, Core Workshop No. 6, 527 pp.

Cuvillier, J., 1952. La notion de "microfaciès" et ses applications. VIII Congr. Naz. Metano e Petrolio, Sect. I:1–7.

Cuvillier, J., 1958. Micropaléontologie moderne. Revue de Micropaléontologie, 1,1:5–8.

Cuvillier, J., 1961. Étude et utilisation rationnelle de microfaciès. Revue de Micropaléontologie, 4,1:3–6.

CUVILLIER, J., and SACAL, V., 1956. Stratigraphic correlations by microfacies in western Aquitaine. Internat. Sed. Petro. Series, II, E. J. Brill, Leiden, 33 pp., 100 plates.

DAVIS, RICHARD A., JR., 1983. Depositional systems. A genetic approach to sedimentary geology. Prentice-Hall, Englewood Cliffs, N.J., 669 pp.

DAWSON, W. C., and CAROZZI, A. V., 1986. Anatomy of a phylloid algal buildup, Raytown Limestone, Iola Formation, Pennsylvanian, southeast Kansas, U.S.A. Sed. Geology, 47,3/4:221–261.

DEMIRMEN, F., 1969. Multivariate procedures and FORTRAN IV program for evaluation and improvement of classifications. Kansas State Geol. Survey, Computer Contrib. 31, 51 pp.

DIABY, I., and CAROZZI, A. V., 1984. The St. Louis Limestone (Middle Mississippian) of Illinois Basin, U.S.A.—a carbonate ramp-bar-platform model. Archives Sciences, Genève, 37, 3:123–169.

DOTT, R. H., JR., and BOURGEOIS, J., 1982. Hummocky stratification: significance of its variable bedding sequences. Geol. Society America Bull., 93,8:663–680.

DUNHAM, R. J., 1962. Classification of carbonate rocks according to depositional texture. In: W. E. Ham (Ed.), Classification of carbonate rocks. Am. Assoc. Petroleum Geologists, Memoir 1, pp. 108–121.

ELF-AQUITAINE, 1975. Essai de caractérisation sédimentologique des dépôts carbonatés. 1. Éléments d'analyse. Boussens et Pau, 173 pp.

ELF-AQUITAINE, 1977. Essai de caractérisation sédimentologique des dépôts carbonatés. 2. Éléments d'interprétation. Boussens et Pau, 231 pp.

ELF-AQUITAINE, REECKMANN, A., and FRIEDMAN, G. M., 1982. Exploration for Carbonate Petroleum Reservoirs. Wiley-Interscience, New York, 213 pp.

FAIRBRIDGE, R. W., 1954. Stratigraphic correlation by microfacies. Am. Jour. Science, 252,11:683–694.

FALKENHEIN, F. U. H., FRANKE, M. R., and CAROZZI, A. V., 1981. Petroleum geology of the Macaé Formation (Albian-Cenomanian), Campos Basin, Brazil (carbonate microfacies-depositional and diagenetic models—natural and experimental porosity). Petrobrás, Cenpes, Ciência Técnica Petróleo, No. 11, 140 pp.

FEIZNIA, S., 1983. Depositional and diagenetic environments of carbonate-siliciclastic rocks of the Glen Dean Formation (Upper Mississippian), Illinois Basin, U.S.A. Unpublished Ph.D. thesis, University of Illinois at Urbana-Champaign, 210 pp.

FEIZNIA, S., and CAROZZI, A. V., 1981. The Glen Dean Formation (Upper Mississippian), southwestern Illinois, U.S.A.: a mixed distal deltaic-carbonate platform depositional model. Actas VIII Congreso Geol. Argentino, II:459–487.

FEIZNIA, S., and CAROZZI, A. V., 1987. Tidal and deltaic controls on carbonate platforms: Glen Dean Formation (Upper Mississippian) of Illinois Basin, U.S.A.: Sedimentary Geology, 54,3/4,201–243.

FISHER, J. H. (Ed.), 1977. Reefs and evaporites—concepts and depositional models. Am. Assoc. Petroleum Geologists, Studies in Geology No. 5, 196 pp.

FLÜGEL, E., 1972. Mikrofazielle Untersuchungen in der alpinen Trias. Methoden und Probleme. Mitt. Ges. Geol. Bergbaustud. Österreich, Innsbruck, 21,2:7–64.

FLÜGEL, E., 1982. Microfacies Analysis of Limestones. Springer-Verlag, Heidelberg, 633 pp.

FOLK, R. L., 1962. Spectral classification of limestone types. In: W. E. Ham (Ed.), Classification of carbonate rocks. Am. Assoc. Petroleum Geologists, Memoir 1, pp. 62–84.

GERBER, M. S., 1978. Carbonate microfacies of the Burlington crinoidal limestone (Middle Mississippian), western Illinois, southeastern Iowa, and northeastern Missouri, U.S.A. Unpublished M.S. thesis, University of Illinois at Urbana-Champaign, 78 pp.

GINSBURG, R. N., 1971. Landward movement of carbonate mud: new model for regressive cycles in carbonates (abstract). Am. Assoc. Petroleum Geologists Bull., 55,2:340.

GOODWIN, P. W., and ANDERSON, E. J., 1985. Punctuated aggradational cycles: a general hypothesis of episodic stratigraphic accumulation. Jour. Geology, 93,5:515–533.

GRESSLY, A., 1838. Observations géologiques sur le Jura soleurois. Nouv. Mém. Soc. Helv. Sc. Nat. Neuchâtel, 2, 349 pp., 14 plates.

HALLEY, R. B., and LOUCKS, R. G. (Eds.), 1980. Carbonate reservoir rocks—a core workshop. Soc. Econ. Paleontologists and Mineralogists, Core Workshop No. 1, 183 pp.

HANDFORD, C. R., 1986. Facies and bedding sequences in shelf-storm-deposited carbonates—Fayetteville Shale and Pitkin Limestone (Mississippian), Arkansas. Jour. Sed. Petrology, 56,1:123–137.

HANSEN, M. W., and CAROZZI, A. V., 1974. Carbonate microfacies of the Monte Cristo Group (Mississippian), Arrow Canyon Range, Nevada. Earth Science Bull. Wyoming Geol. Assoc., 7,4:13–54.

HARRIS, P. M. (Ed.), 1983. Carbonate buildups—a core workshop. Soc. Econ. Paleontologists and Mineralogists, Core Workshop No. 4, 593 pp.

HARRIS, P. M., 1984. Carbonate sands—a core workshop. Soc. Econ. Paleontologists and Mineralogists, Core Workshop No. 5, 464 pp.

HEATH, C. P., LUMSDEN, D. N., and CAROZZI, A. V., 1967. Petrography of a carbonate transgressive-regressive sequence: the Bird Spring Group (Pennsylvanian), Arrow Canyon Range, Clark County, Nevada. Jour. Sed. Petrology, 37,2:377–400.

HECKEL, P. H., 1974. Carbonate buildups in the geological record: a review. In: L. F. Laporte (Ed.), Reefs in time and space. Soc. Econ. Paleontologists and Mineralogists, Special Publ. No. 18, pp. 90–154.

HOVELACQUE, M., and KILIAN, C. C. W., 1900. Album de microphotographies des roches sédimentaires. Gauthier-Villars, Paris, 14 pp., 69 plates.

IRWIN, M. L., 1965. General theory of epeiric clear water sedimentation. Am. Assoc. Petroleum Geologists Bull., 49, 4:445–459.

JAMES, N. P., 1983. Reef environment. In: P. A. Scholle, D. G. Bebout, and C. H. Moore (Eds.), Carbonate depositional environments. Am. Assoc. Petroleum Geologists, Memoir 33, pp. 345–440.

JAMES, N. P., 1984. Shallowing-upward sequences in carbonates. In: R. G. Walker, (Ed.), Facies Models. Geoscience Canada, Reprint Series 1, 2nd ed., pp. 213–228.

JAMES, N. P., and MOUNTJOY, E. W., 1983. Shelf-slope break in fossil carbonate platforms: an overview. In: D. S. Stanley and G. T. Moore (Eds.), The shelfbreak: critical interface on continental margins. Soc. Econ. Paleontologists and Mineralogists, Special Publ. No. 33, pp. 189–206.

KHAWLIE, M. R., and CAROZZI, A. V., 1976. Microfacies and geochemistry of the Brereton Limestone (Middle Pennsylvanian) of southwestern Illinois. Archives Sciences, Genève, 29,1:67–110.

KLEIN, G. DEV., 1982. Probable sequential arrangement of depositional systems on cratons. Geology, 10,1:17–22.

KREISA, R. D., 1981. Storm-generated sedimentary structures in subtidal marine facies with examples from the middle and upper Ordovician of southwestern Virginia. Jour. Sed. Petrology, 51,3:823–848.

KUENEN, P. H., and CAROZZI, A. V., 1953. Turbidity currents and sliding in geosynclinal basis of the Alps. Jour. Geology, 61,4:363–373.

KUHNHENN, G. L., and CAROZZI, A. V., 1977. Carbonate microfacies of the Platteville Group (Middle Ordovician), Lee and Lasalle Counties, Illinois. Archives Sciences, Genève, 30, 2:179–212.

LACEY, J. E., and CAROZZI, A. V., 1967. Critères de distinction entre oolithes autochtones et allochtones. Application au calcaire de Sainte-Geneviève (Viséen) de l'Illinois, U.S.A. Bull. Centre Rech, Pau, Soc. Nat. Pétroles Aquitaine, 1,2:279–313.

LASEMI, Y., and CAROZZI, A. V., 1981. Carbonate microfacies and depositional environments of the Kinkaid Formation (Upper Mississippian) of the Illinois Basin, U.S.A. Actas VIII Congreso Geol. Argentino, II:357–384.

LONGMAN, M. W., 1980. Carbonate diagenetic textures from nearsurface diagenetic environments. Am. Assoc. Petroleum Geologists Bull., 64,4:461–487.

LONGMAN, M. W., 1981. A process approach to recognizing facies of reef complexes. In: D. F. Toomey (Ed.), European fossil reef models. Soc. Econ. Paleontologists and Mineralogists, Special Publ. No. 30, pp. 9–40.

LUMSDEN, D. N., 1965. Microfacies of the Middle Bird Spring Group (Pennsylvanian-Permian), Arrow Canyon Range, Clark County, Nevada, U.S.A. Unpublished Ph.D thesis, University of Illinois at Urbana-Champaign, 104 pp.

MARSAGLIA, K. M., and KLEIN, G. DEV., 1983. The paleogeography of Paleozoic and Mesozoic storm depositional systems. Jour. Geology, 91,2: 117–142.

MATTES, B. W., and MOUNTJOY, E. W., 1980. Burial dolomitization of the Upper Miette buildup, Jasper National Park, Alberta. In: D. H. Zenger, J. B. Dunham, and R. L. Ethington (Eds.), Concepts and models of dolomitization. Soc. Econ. Paleontologists and Mineralogists, Special Publ. No. 28, pp. 259–297.

MIDDLETON, G. V., 1973. Johannes Walther's Law of the correlation of facies. Geol. Soc. Am. Bull. 84,3:979–988.

MOUNT, J. F., 1984. Mixing of siliciclastic and carbonate sediments in shallow shelf environments. Geology, 12,7:432–435.

MOUNTJOY, E. W., COOK, H. E., PRAY, L. C., and MCDANIEL, P. N., 1972. Allochthonous carbonate debris flows—worldwide indicators of reef complexes, banks or shelf margins. Proc. 24th Internat. Geol. Congress, Montreal, Section 6, pp. 172–189.

NELSON, C. S., 1978. Temperate shelf carbonate sediments in the Cenozoic of New Zealand. Sedimentology, 25,6:737–771.

NEUMANN, A. C., and LAND, L. S., 1975. Lime mud deposition and calcareous algae in the Bight of Abaco, Bahamas: a budget. Jour. Sed. Petrology, 45,4:763–786.

NOWAK, F. J., and CAROZZI, A. V., 1972. Microfacies of the Upper Bird Spring Group (Pennsylvanian-Permian), Arrow Canyon Range, Clark County, Nevada, U.S.A. Archives Sciences, Genève, 25,3:343–382.

OKHRAVI, R., and CAROZZI, A. V., 1983. The Joachim Dolomite: a Middle Ordovician sabkha of southeast Missouri and southern Illinois, U.S.A. Archives Sciences, Genève, 36, 3:373–424.

PAGE, N. J., and CAROZZI, A. V., 1962. Étude du remplacement diagénétique du quartz détritique par les carbonates dans des dolomies cambriennes. Archives Sciences, Genève, 14,3:461–491.

PARÉJAS, E., and LILLIE, A., 1935a. Données microscopiques sur le Crétacé supérieur de Châtelard-en-Bauges (Savoie). C. R. Soc. Phys. Hist. Nat., Genève, 52,3:272–275.

PARÉJAS, E., and LILLIE, A., 1935b. Données microscopiques sur le Crétacé supérieur de Vormy (Aravis, Haute-Savoie). C. R. Soc. Phys. Hist. Nat. Genève, 52,3:275–277.

PERYT, T. (Ed.), 1983. Coated Grains. Springer-Verlag, Heidelberg, 655 pp.

PURDY, E. G., 1963. Recent calcium carbonate facies of the Great Bahama Bank, Parts 1 and 2. Jour. Geology, 71,3:334–355, 4:472–497.

RAO, C. P., and CAROZZI, A. V., 1971. Application of computer techniques to the petrographic study of oolitic environments, Ste. Genevieve Limestone (Middle Mississippian), southern Illinois and eastern Missouri. Archives Sciences, Genève, 24,1:17–55.

READ, J. F., 1982. Carbonate platforms of passive (extensional) continental margins: types, characteristics, and evolution. Tectonophysics, 81,3/4:195–212.

READ, J. F., 1985. Carbonate platform facies models. Am. Assoc. Petroleum Geologists Bull., 69,1:1–21.

READ, J. F., GROTZINGER, J. P., BOVA, J. A., and KOERSCHNER, W. F., 1986. Models for generation of carbonate cycles. Geology, 14,2:107–110.

READING, H. G. (Ed.), 1986. Sedimentary Environments and Facies, 2nd ed. Elsevier, New York, 615 pp.

REICHELDERFER, J. L., 1985. Microfacies, diagenesis, and porosity development in the Salem Limestone (Middle Mississippian) southern Illinois, U.S.A. Unpublished MS thesis, University of Illinois at Urbana-Champaign, 85 pp.

RICH, D. W., and CAROZZI, A. V., 1981. Natural porosity in oolitic limestones: an attempt at a genetic classification. Actas VIII Congreso Geol. Argentino, II:593–635.

ROCHE, J. E., and CAROZZI, A. V., 1970. Petrography of back-reef carbonates: Traverse Group (Givétian) of the northern part of the Southern Peninsula of Michigan. Bull. Centre Rech. Pau., Soc. Nat. Pétroles Aquitaine, 4,1:137–189.

RODRIGUEZ SCHELOTTO, M. L., and CAROZZI, A. V., 1981. Tratamiento de cuttings como técnica complementaria para el estudio de microfacies. Actas VIII Congreso Geol. Argentina, 2:351–356.

RODRIGUEZ SCHELOTTO, M. L., ORCHUELA, I. A., BALIÑA, M. M., BLANCO IBAÑEZ, S., FERRARESI, P., AMBROSIS, M. E., and CAROZZI, A. V., 1981. Medios depositacionales y microfacies de la Formación Quintuco (Berriasiano-Valanginiano) en el yacimiento Loma La Lata, Provincia del Neuquén. Actas VIII Congreso Geol. Argentino, II:503–520.

ROEHL, P. O., and CHOQUETTE, P. W. (Eds.), 1985. Carbonate Petroleum Reservoirs. Springer-Verlag, New York, 480 pp.

ROSS, C. A., and ROSS, J. R. P., 1985. Late Paleozoic depositional sequences are synchronous and worldwide. Geology, 13,3:194–197.

SCHLAGER, W., 1981. The paradox of drowned reefs and carbonate platforms. Geol. Soc. Am. Bull., Part I, 92,4:197–211.

SCHNEIDERMANN, N., and HARRIS, P. M. (Eds.), 1985. Carbonate cements. Soc. Econ. Paleontologists and Mineralogists, Special Publ. No. 36, 408 pp.

SCHOLLE, P. A., BEBOUT, D. G., and MOORE, C. H. (Eds.), 1983. Carbonate depositional environments. Am. Assoc. Petroleum Geologists, Memoir 33, 708 pp.

SCOTESE, C. R., BAMBACH, R. K., BARTON C., VAN DER VOO, R., and ZIEGLER, A. M., 1979. Paleozoic base maps. Jour. Geology, 87,3:217–277.

SHAVER, R. H., 1977. Silurian reef geometry—new dimensions to explore. Jour. Sed. Petrology, 47,4:1409–1424.

SLANSKY, M., 1980. Géologie des phosphates sédimentaires. Bureau Rech. Géol. Min., Mémoire No. 114, 92 pp.

SODERMAN, J. W., and CAROZZI, A. V., 1963. Petrography of algal bioherms in Burnt Bluff Group (Silurian), Wisconsin. Am. Assoc. Petroleum Geologists Bull., 47,9:1682–1708.

SOUPAC, 1976. Soupac program description, University of Illinois at Urbana-Champaign. C.S.O. vol. 1, 175 pp.

STRICKER, G. D., and CAROZZI, A. V., 1973. Carbonate microfacies of the Pogonip Group (Lower Ordovician), Arrow Canyon Range, Clark County, Nevada, U.S.A. Bull. Centre Rech. Pau, Soc. Nat. Pétroles Aquitaine, 7,2:499–451.

STRICKER, G. D., and CAROZZI, A. V., 1974. Mathematical evidence for storm deposits in Lower Ordovician carbonates, Pogonip Group, Arrow Canyon Range, Clark County, Nevada, U.S.A. C. R. Soc. Phys. Hist. Nat. Genève, N.S. 9,1–3:74–80.

TEXTORIS, D. A., 1971. Grain-size measurement in thin section. In: R. E. Carver (Ed.), Procedures in Sedimentary Petrology. Wiley-Interscience, New York, pp. 95–107.

TEXTORIS, D. A., and CAROZZI, A. V., 1964. Petrography and evolution of Niagaran (Silurian) reefs, Indiana. Am. Assoc. Petroleum Geologists Bull., 48,4:397–426.

TEXTORIS, D. A., and CAROZZI, A. V., 1966. Petrography of a Cayugan (Silurian) stromatolite mound and associated facies, Ohio. Am. Assoc. Petroleum Geologists Bull., 50,7:1375–1388.

TOOMEY, D. F. (Ed.), 1981. European fossil reef models. Soc. Econ. Paleontologists and Mineralogists, Special Publ. No. 30, 546 pp.

VAIL, P. R., MITCHUM, R. M., and THOMPSON, S. III, 1977. Seismic stratigraphy and global changes of sea level. Part 3. Relative changes of sea level from coastal onlap, Part 4. Global cycles of relative changes of sea level. In: C. E. Payton (Ed.), Seismic Stratigraphy—Application to hydrocarbon exploration. Am. Assoc. Petroleum Geologists, Memoir 26, pp. 63–81, 83–97.

VON BERGEN, D., 1985. Microfacies, depositional environments, and diagenesis of Atokan carbonates, Delaware Basin, Reeves County, Texas, U.S.A. Unpublished MS thesis, University of Illinois at Urbana-Champaign, 101 pp.

WALKER, R. G. (Ed.), 1984. Facies models. Geoscience Canada, Reprint Series 1, revised 2nd ed., 317 pp.

WALPOLE, R. L., and CAROZZI, A. V., 1961. Microfacies study of Rundle Group (Mississippian) of Front Ranges, Central Alberta, Canada. Am. Assoc. Petroleum Geologists Bull., 45,11:1810–1846.

WANLESS, H. R., ZIEBELL, W. G., ZIEMBA, E. A., and CAROZZI, A. V., 1956. Limestone texture as a key to interpreting depth of deposition. C. R. 20th Internat. Geol. Congress, Mexico City, Section V, 1:65–82 (1957).

WILKINSON, B. H., 1982. Cyclic cratonic carbonates and Phanerozoic calcite seas. Jour. Geol. Education, 30,4:189–203.

WILSON, J. L., 1970. Depositional facies across carbonate shelf margins. Trans. Gulf Coast Assoc. Geol. Soc. 20:229–233.

WILSON, J. L., 1974. Characteristics of carbonate-platform margins. Am. Assoc. Petroleum Geologists Bull., 58,4:810–824.

WILSON, J. L., 1975. Carbonate Facies in Geologic History. Springer-Verlag, Heidelberg, 471 pp.

WOLFF, B., and CAROZZI, A. V., 1984. Microfacies, depositional environments, and diagenesis of the Amapá carbonates (Paleocene-Middle Miocene), Foz do Amazonas Basin, offshore NE Brazil. Petrobrás, Cenpes, Ciência Técnica Petróleo, No. 13, 103 pp.

YANG, CHENGYUN, and CAROZZI, A. V., 1987. Practical classification and microfacies analysis of carbonate rocks. Peking University Press, Beijing, 295 pp. (bilingual text Chinese-English).

ZADNIK, V. E., and CAROZZI, A. V., 1963. Sédimentation cyclique dans les dolomies du Cambrien supérieur de Warren County, New Jersey, U.S.A. Inst. Nat. Genevois Bull., 62:1–55.

ZENGER, D. H., DUNHAM, J. B., and ETHINGTON, R. L. (Eds.), 1980. Concepts and models of dolomitization. Soc. Econ. Paleontologists and Mineralogists, Special Publ. No. 28, 320 pp.

ZIEGLER, A. M., SCOTESE, C. R., MCKERROW, W. S., JOHNSON, M. E., and BAMBACH, R. K., 1979. Paleozoic paleogeography. Ann. Rev. Earth Planet. Sci., 7:473–502.

ZIEGLER, A. M., SCOTESE, C. R., and BARRETT, S. F., 1982. Mesozoic and Cenozoic paleogeographic maps. In: P. Brosche and J. Sündermann (Eds.), Tidal Friction and the Earth's Rotation II. Springer-Verlag, Heidelberg, pp. 240–252.

Index

A

Abott Formation, 99
Agglutinated benthic foraminifers:
   Itaituba-Nova Olinda Formations, 103, 106–7
Ahr, W.M., 6
Aigner, T., 561
Algal dust, Edwardsville Formation, 233, 235
Allentown Dolomite:
   crystallinity index, 421
   detrital quartz, 419, 421
      marginal replacement by carbonates, 419
   exposure features, 421
   integration, 550
   intraclasts, 419, 421
   lithic pellets, 419, 421
   megacycles, 563
   microcycles, 563
   microfacies description, 419, 421
   model, 421–22
   oöids, 419, 421
      distorted, 421
      half-moon, 421
   paleoclimatology, 559
   paleogeography, 559
   pseudo-ooids, 421
   pyrite, 419, 421
   sampling techniques, 419
   shallowing-upward megasequence, 422–23
   shallowing-upward sequence, 421
   small-scale cycle, 562
   storm deposits, 421
   stromatolites, 421
   superposition of environments, 422–23
   typical sequence, 421
Alling, H.L., 24
Amapá Formation:
   eustatic sea level variations, 297
   exposure features, 297
   model 1, 278
      bryozoans, 275
      codiaceans, 275

Amapá Formation (*cont.*)
   model 1 (*cont.*)
      computer evaluation of porosity, 280
      dasyclads, 275
      diagenesis, 278–79
      dolomitization, dorag, 280, 296–97
      echinoids, 275
      Halimeda, 275
      integration, 534, 536
      intercrystalline porosity, 280
      larger benthic foraminifers, 275
      microfacies description, 275
      mollusks, 275
      nummulitids, 275
      ooids, 275
      ostracods, 275
      phosphates, 275
      planktonic foraminifers, 275
      porosity evolution, 279–80
      primary porosity, 279–80
      Ranikothalia, 275
      shallowing-upward sequence, 278
      smaller benthic foraminifers, 275
   model 2, 287
      Amphistegina, 286
      bryozoans, 280–82, 286–87
      channel porosity, 287
      computer evaluation of porosity, 287, 296
      corals, 280–82, 286–87
      dasyclads, 287
      diagenesis, 287
      discocyclinids, 281–82
      dolomitization, dorag, 287, 296–97
      echinoids, 280–82, 286–87
      encrusting foraminifers, 281–82, 286–87
      Halimeda, 287
      integration, 536
      intercrystalline porosity, 287
      larger benthic foraminifers, 280–82, 286–87
      lepidocyclinids, 281–82
      microfacies description, 280–82, 286–87
      moldic porosity, 287

Amapá Formation (cont.)
   model 2 (cont.)
      olistoliths, 287
      olistostromes, 280, 287
      nummulitids, 281–82
      planktonic foraminifers, 280–81
      Porites, 286
      porosity evolution, 287, 296
      red algae, 280–82, 286–87
      reservoir generation, 287, 296
      shale channels, 287
      shallowing-upward sequence, 287
      smaller benthic foraminifers, 280–82, 286–87
      typical sequence, 287
      vuggy porosity, 287
   paleoenvironmental maps, 296
   porosity distribution, 296–97
   reservoir generation, model 1, 279–80
   sampling techniques, 275
   superposition of environments, 296–97
Ammonoids, Jurassic, Morcles Nappe, Alps, 475
Amphipora:
   Arrow Canyon Formation-Crystal Pass Limestone, 161
   Beaverhill Lake Formation, 362–64
   Jeffersonville Limestone, coral zone, 352, 354–55
Amphiroa, Bonfim Formation, 412
Amphistegina:
   Amapá Formation, model 2, 286
   Miocene, Visayan Islands, 453–54, 456
Anderson, E.J., 563
Andesitic volcanism, Miocene, Visayan Islands, 460
Anhydrite:
   chickenwire texture:
      Joachim Dolomite, 431
      Quintuco-Loma Montosa Formation, 337, 340, 342
Anhydritization:
   Iola Formation, 167
   Itaituba-Nova Olinda Formations, 103, 107
   Kinkaid Formation, models 2–2A, 143
   Lagoa Feia Formation, 498
   Macaé Formation, 273
   St. Louis Limestone, 56
   Salem Limestone:
      SE Illinois, 119, 121, 126, 128
      SW Illinois, 305
   Traverse Group, 371
Annelids:
   Bonfim Formation, 417
   Brereton Limestone:
      crinoid-brachiopod buildup model, 178
      phylloid algae buildup model, 177
   Chachao Formation, 195, 202
   Jurassic, Morcles Nappe, Alps, 475
   Kimmeridgian-Portlandian, Salève, 398–99, 405
   Quintuco-Loma Montosa Formation, 337
Annelid tubes, Glen Dean Formation, E shelf Illinois Basin, 321
Arabicodium, Bonfim Formation, 412
Archaeolithophyllum, Iola Formation, 167
Archaeolithophyllum missouriense, Atokan, 381, 383
Archimedes:
   Glen Dean Formation, SW shelf Illinois Basin, 332
   Kinkaid Formation:
      model 1, 132
      models 2–2A, 135
   Menard Formation, 85
Arenite:
   Itaituba-Nova Olinda Formation, 106
      Joachim Dolomite, 431
      Niagaran, N Indiana, 193
      Ste. Genevieve Limestone, 310, 313
   quartz:
      Arrow Canyon Formation-Crystal Pass Limestone, 161
      Glen Dean Formation, E shelf Illinois Basin, 321, 323
   quartz feldspathic:
      Glen Dean Formation, SW shelf Illinois Basin, 330
      Macaé Formation, 259
   quartz feldspathic lithic:
      Lagoa Feia Formation, 496, 498
      Quintuco-Loma Montosa Formation, 337
   Arrow Canyon Formation-Crystal Pass Limestone:
      Amphipora, 161

Arrow Canyon Formation-Crystal Pass Limestone (cont.)
   bioturbation, 159
   bryozoans, 161
   buildup ramp model, 161
   calcispheres, 161
   component variation, 161
   detrital quartz, 161
   integration buildup ramp model, 528
   integration ramp model, 520
   intraclasts, 161
   lithic pellets, 161
   microfacies:
      description, 159, 161
      filiation, 161
   ostracods, 161
   paleoclimatology, 559
   paleogeography, 559
   pyrite, 161
   quartz arenites, 161
   ramp model, 161
   reworking processes, 161
   sampling techniques, 159
   shallowing-upward sequence, 161
   storm deposits, 161
   stromatoporoids, 161
   typical sequence, 161
Atokan:
   Archaeolithophyllum missouriense, 381, 383
   authigenic feldspar, relationship to stylolites, 383, 386, 390
   bioturbation, 377, 379, 381
   brachiopods, 379, 381
   bryozoans, 377, 379, 381
   burial porosity, relationship to stylolites, 383, 386, 390, 398
   calcispheres, 379, 381, 383
   crinoids, 377, 379, 381, 383
   detrital quartz, 377
   diagenesis, 386, 390
   dolomitization:
      burial, 390
      dorag, 386
      stylolitic, 386
   Donezella, 379, 381
   echinoids, 377, 379, 381, 383
   Endothyra, 377, 379, 381
   exposure features, 383
   fecal pellets, 377, 379, 381, 383
   fusulinids, 377, 379, 381
   gastropods, 377, 381
   Girvanella, 381
   integration, 549
   intraclasts, 379, 383
   microfacies description, 377, 379, 381
   model, 381, 383, 386
   ooids, 381, 383
   ostracods, 377, 379, 381, 383
   paleoclimatology, 561
   paleogeography, 561
   phylloid algae, 381
   porosity:
      biomoldic, 390
      measurements, 377
      oomoldic, 390
      stylolitic, 390
      vuggy, 390
   reservoir generation, 390
   sampling techniques, 377
   shallowing-upward sequence, 381
   smaller benthic foraminifers, 377, 379, 381
   sponge spicules, 377, 383
   stromatolites, 381, 383
   stylolites:
      distribution, 383
      non-sutured, 377
      relationships to authigenic feldspars, 383, 386, 390
      relationships to porosity, 383, 386, 390, 398
      sutured, 377
   superposition of environments, 390, 398
   Tetrataxis, 381
   trilobites, 377
   Tuberitina, 377, 379, 381

Atokan (cont.)
  typical sequence, 381
  vadose silt, 383

B

Badiozamani, K., 12
Bahama syndrome, 4
Bahama tidal flats, cyclicity, 565
Bailey Limestone:
  bioturbation, 469, 473
  brachiopods, 473–74
  clay minerals, 474
  crinoids, 469, 473–74
  depositional sequence, 474–75
  detrital glauconite, 473–74
  detrital quartz, 473–74
  graded bedding, 468
  lithic pellets, 469, 473
  microfacies description, 469
  origin of constituents, 474
  ostracods, 473–74
  pyrite, 469, 473
  radiolarians, 469, 473–74
  sampling techniques, 468–69
  silicification, submarine, 474
  sponge spicules, 469, 473–74
  trilobites, 473
  turbidites:
    Bouma sequence, 474
    proximality index, 474–75
    setting, 468
    textural features, 468, 474
  typical sequence, 474–75
  vertical sequence, 474–75
Bakush, S.H., 61
Banaee, J., 468
Barra Nova Formation:
  paleoclimatology, 561
  paleogeography, 561
  sampling techniques, 108
  stratigraphic division, 108
Basic volcanic glass, Lagoa Feia Formation, 496, 498, 500
Basic volcanic lithoclasts, Lagoa Feia Formation, 496, 498
Bathurst, R.G.C., 4
Beachrock:
  Joachim Dolomite, 428
  Macaé Formation, 262, 264, 270–71
  Pogonip Group, 210
  Salem Limestone, SW Illinois, 307
Beaverhill Lake Formation:
  Amphipora, 362–64
  bioturbation, 363
  brachiopods, 362
  bryozoans, 362
  calcispheres, 362–64
  Chara, 364
  dasyclads, 364
  gastropods, 362–64
  integration, 547, 549
  lithic pellets, 362–63
  microfacies description, 362–63
  model, 364
  ostracods, 362–64
  paleobathymetry, 364
  paleoclimatology, 559
  paleogeography, 559
  pelecypods, 362
  sampling techniques, 362
  shallowing-upward sequence, 363–64
  smaller benthic foraminifers, 363
  sponge spicules, 363–64
  stromatoporoids, 362–64
  superposition of environments, 364
  typical sequence, 363–64
Benthic fauna:
  lacustrine models, 558
  models, 518
  platforms with frontal bioaccumulated to hydrodynamic buildups, 545
  platforms with frontal bioconstructed to hydrodynamic buildups, 555
  ramps, 531–32
Bertani, R.T., 2, 496

Bight of Abaco, Bahamas, carbonate productivity, 568
Biotite, detrital:
  Lower Regencia Member, 108, 110
  Middle Regencia Member, 110, 115
Bioturbation:
  Arrow Canyon Formation-Crystal Pass Limestone, 159
  Atokan, 377, 379, 381
  Bailey Limestone, 469, 473
  Beaverhill Lake Formation, 363
  Bird Spring Group, Upper Missourian-Wolfcampian, 253
  Brereton Limestone, 175–76
    crinoid-brachiopod buildup model, 177
    phylloid algae buildup model, 177
  Burnt Bluff Group, 446
  Chachao Formation, 195–202
  Galena Group, 61, 64
  Glen Dean Formation:
    E shelf Illinois Basin, 319, 321, 323
    SW shelf Illinois Basin, 330–31
  Iola Formation, 167
  Itaituba-Nova Olinda Formation, 103
  Joachim Dolomite, 428–29, 437
  Kimmeridgian-Portlandian, Salève, 398–99
  Kinkaid Formation:
    model 1, 131–32, 135
    models 2–2A, 136–37, 143
    model 3, 144
  Lagoa Feia Formation, 498, 502
  Lower Regencia Member, 108, 110
  Macaé Formation, 259
  Maquoketa phosphorites, 483, 485, 487
  Middle Regencia Member, 115
  Miocene, Visayan Islands, 453–56
  Monte Cristo Group, 238–39, 241, 244
  Pogonip Group, 209–10
  platforms with frontal bioaccumulated to hydrodynamic buildups, 545
  Platteville Group, 57
  ramps, 532
  Rundle Group, 153
  Salem Limestone, SE Illinois, 121, 123
  Traverse Group, 367–68
Bird Spring Group, Atokan-Lower Missourian:
  brachiopods, 96
  bryozoans, 96
  calcispheres, 96
  component variation, 96
  crinoids, 96
  detrital quartz, 96
  fecal pellets, 96
  lithic pellets, 96
  megacycles, 563
  model, 96, 99
  ooids, 96
  ostracods, 96
  shallowing-upward sequence, 96
  smaller benthic foraminifers, 96
Bird Spring Group, Morrowan:
  brachiopods, 92
  bryozoans, 92, 96
  component variation, 92, 96
  crinoids, 92, 96
  detrital quartz, 92, 96
  fecal pellets, 92
  lithic pellets, 92, 96
  model, 96, 99
  oncoids, 92
  ooids, 92, 96
  ostracods, 92
  shallowing-upward sequence, 92, 96
  smaller benthic foraminifers, 92, 96
  typical sequence, 96
Bird Spring Group, Morrowan-Lower Missourian:
  comparison with Illinois cyclothems, 99
  integration, 519–20
  microfacies description, 92
  model, 96, 99
  paleoclimatology, 561
  paleogeography, 561
  sampling techniques, 92
  superposition of environments, 96, 99

Bird Spring Group, Upper Missourian-Wolfcampian:
  bioturbation, 253
  brachiopods, 253
  bryozoans, 253
  calcispheres, 253
  component variation, 253
  crinoids, 253
  detrital quartz, 253
  Endothyra, 253
  exposure features, 253
  fecal pellets, 253
  fusulinids, 253
  integration, 538–39
  lithic pellets, 253
  lithoclasts, 253
  microfacies description, 251, 253
  model, 255
  ooids, 253
  paleoclimatology, 561
  paleogeography, 561
  sampling techniques, 251
  shallowing-upward sequence, 253
  smaller benthic foraminifers, 253
  superposition of environments, 255–56
  typical sequence, 253
Black, N.R., 482
Bone fragments, Maquoketa phosphorites, 485
Bonfim Formation:
  Amphiroa, 412
  annelids, 417
  Arabicodium, 412
  Bouenia, 412
  calcispheres, 407, 413
  Cayeuxia, 407, 412
  clay minerals, 417
  detrital quartz, 417
  dolomitization:
    dorag, 417
    sabkha, 417
  echinoids, 407, 412–13, 417
  exposure features, 417
  fecal pellets, 407, 412–13, 417
  fenestral textures, 417
  gastropods, 407, 412–13, 417
  green algae, 407, 412–413, 417
  Halimeda, 412
  Hedbergella, 407
  integration, 552, 554
  intraclasts, 407, 412–13, 417
  lithic pellets, 407, 412–13, 417
  Lithophyllum, 412
  Lithothamnium, 407, 412–13
  microfacies description, 407, 412–13, 417
  microfilaments, 407
  microoncoids, 407
  models, 417
  oncoids, 417
  ooids, 412–13, 417
  paleoclimatology, 561
  paleogeography, 561
  pelecypods, 407, 412–13, 417
  planktonic foraminifers, 407
  plant debris, 417
  porosity, oomoldic, 413
  Pycnoporidium, 407, 412
  radiolarians, 407
  red algae, 407, 412–13, 417
  reservoir generation, 417
  sampling techniques, 405, 407
  slumping, 407, 412
  smaller benthic foraminifers, 407, 412–13, 417
  Solenopora, 407, 412
  stromatolites, 417
  superposition of models, 417
  Trocholina, 412
Borelis, Miocene, Visayan Islands, 456
Boron, Traverse Group, 373
Bouenia, Bonfim Formation, 412
Bourgeois, J., 561
Brachiopods:
  Atokan, 379, 381

Brachiopods (*cont.*)
  Bailey Limestone, 473–74
  Beaverhill Lake Formation, 362
  Bird Spring Group:
    Atokan-Lower Missourian, 96
    Morrowan, 92
    Upper Missourian-Wolfcampian, 253
  Brereton Limestone:
    crinoid-brachiopod buildup model, 177
    phylloid algae buildup model, 177
  Burlington Limestone:
    arenite suite, 221
    wacke suite, 226
  Burnt Bluff Group, 446
  Edwardsville Formation, 233
  Galena Group, 69
  Glen Dean Formation:
    E shelf Illinois Basin, 319, 321, 323
    SW shelf Illinois Basin, 330–31
  Iola Formation, 167
  Itaituba-Nova Olinda Formation, 101, 103, 106
  Jeffersonville Limestone:
    buildup, 359
    coral zone, 352, 354
  Jurassic, Morcles Nappe, Alps, 475
  Kinkaid Formation, models 2–2A, 143
  Maquoketa phosphorites, 483, 485, 487, 491
  Monte Cristo Group, 248–49
  Niagaran:
    central Indiana, 467
    N Indiana, 191, 193
  Platteville Group, 59
  Pogonip Group, oncoid model, 215–16
  Rundle Group, 153
  St. Louis Limestone:
    model 1, 39, 41
    model 2, 41
    model 3, 43
  Salem Limestone:
    SE Illinois, 119, 121, 123
    SW Illinois, 303, 305
  Traverse Group, 367–68
Brecciation, St. Louis Limestone, 35
Brereton Limestone:
  bioturbation, 175–76
  crinoid-brachiopod buildup model:
    annelids, 178
    bioturbation, 177
    brachiopods, 177
    bryozoans, 178
    component variation, 177–78
    crinoids, 177
    detrital quartz, 177–78
    fusulinids, 177
    integration, 524
    ostracods, 177–78
    pelecypods, 177
    phylloid algae, 178
    smaller benthic foraminifers, 177
    sponge spicules, 177
    trilobites, 177
  fecal pellets, 175–76
  glauconite, 175–76
  Ivanovia, 175
  microfacies description, 175–76
  paleoclimatology, 561
  paleoenvironmental maps, 176
  paleogeography, 561
  phosphorites, 175–76
  phylloid algae buildup model:
    annelids, 177
    bioturbation, 177
    brachiopods, 177
    bryozoans, 177
    crinoids, 177
    detrital quartz, 177
    fusulinids, 177
    integration, 528
    ostracods, 177
    pelecypods, 177
    phylloid algae, 177

Brereton Limestone (cont.)
  phylloid algae buildup model (cont.)
    smaller benthic foraminifers, 177
    sponge spicules, 177
    trilobites, 177
  sampling techniques, 173
Brown, J.S., 24
Bryozoans:
  Amapá Formation:
    model 1, 275
    model 2, 281–82, 286–87
  Arrow Canyon Formation-Crystal Pass Limestone, 161
  Atokan, 377, 379, 381
  Beaverhill Lake Formation, 362
  Bird Spring Group:
    Atokan-Lower Missourian, 96
    Morrowan, 92, 96
    Upper Missourian-Wolfcampian, 253
  Brereton Limestone:
    crinoid-brachiopod buildup model, 178
    phylloid algae buildup model, 177
  Burlington Limestone:
    arenite suite, 221–23
    wacke suite, 223, 226
  Chachao Formation, 195, 202
  Edwardsville Formation, 231, 233
  Galena Group, 69
  Glen Dean Formation, SW shelf Illinois Basin, 330–32
  Iola Formation, 167, 171
  Itaituba Nova Olinda Formations, 101, 103, 106
  Jeffersonville Limestone, coral zone, 352, 354
  Jurassic, Morcles Nappe, Alps, 475
  Kinkaid Formation:
    model 1, 132, 135
    models 2–2A, 143
    model 3, 144
  Macaé Formation, 262
  Menard Formation, 86
  Miocene, Visayan Islands, 454–56
  Monte Cristo Group, 244, 248–49
  Niagaran:
    central Indiana, 465, 467
    N Indiana, 191, 193
  Platteville Group, 59
  Quintuco-Loma Montosa Formation, 337
  Rundle Group, 153
  Ste. Genevieve Limestone, 307, 309–10, 313
  St. Louis Limestone:
    model 1, 39, 41
    model 2, 41
    model 3, 43
  Salem Limestone:
    SE Illinois, 119, 121, 123
    SW Illinois, 303, 305
  Traverse Group, 367–68
Buildups:
  bioaccumulated:
    lacustrine models, 558
    platforms, 534, 536, 538–39, 541, 543–47
    ramps, 523–25, 528
  bioconstructed:
    platforms, 547, 549–50, 552, 554–55
    ramps, 528, 530
  classification, 4
  hydrodynamic:
    lacustrine models, 558
    platforms, 534, 536, 538–39, 541, 543–47, 549–50, 552, 554–55
    ramps, 523–25, 528
  temporal sequence, 9
Burlington Limestone:
  arenite suite:
    brachiopods, 221
    bryozoans, 221–23
    corals, 221
    crinoidal banks, 222–23, 227, 229
    crinoids, 221–23
    environmental interpretation, 222–23
    fish plates, 221
    fish scales, 221
    glauconite, 221
    ostracods, 221

Burlington Limestone (cont.)
  arenite suite (cont.)
    sedimentary structures, 221–22
    silicification, burial, 221
    smaller benthic foraminifers, 221
    trilobites, 221
  chert brecciation, synsedimentary, 229–30
  microfacies description, 220–23, 226
  model, 227
  sampling techniques, 219–20
  silicification, submarine, 229–30
  storm deposits, 226, 229–30
  superposition of environments, 227, 229
  typical sequence, 226
  vertical sequence, 226
  wacke suite:
    brachiopods, 226
    bryozoans, 223, 226
    crinoidal interbanks, 226, 227, 229
    crinoids, 223, 226
    dolomitization, burial, 226
    environmental interpretation, 226
    graded bedding, 226
    inertia flows, 226
    sedimentary structures, 223, 226
Burnt Bluff Group:
  bioturbation, 446
  brachiopods, 446
  cephalopods, 446
  corals, 446, 448
  crinoids, 446
  detrital quartz, 446
  gastropods, 446
  intraclasts, 446
  microfacies description, 441, 446
  model:
    Nasbro buildup, 446
    Sturgeon Bay buildup, 448
  oncoids, 446
  ooids, 446
  ostracods, 441, 446
  paleoclimatology, 559
  paleogeography, 559
  reworking processes, 446
  sampling techniques, 441
  shallowing-upward sequences:
    Nasbro buildup, 446
    Sturgeon Bay buildup, 448
  Spongiostromata, 441, 446
  stromatolites, 441, 446, 448
  stromatoporoids, 446, 448
  typical sequence:
    Chilton buildup, 448
    Nasbro buildup, 446
    Sturgeon Bay buildup, 448

C

Calcispheres:
  Arrow Canyon Formation-Crystal Pass Limestone, 161
  Atokan, 379, 381, 383
  Beaverhill Lake Formation, 362–64
  Bird Spring Group:
    Atokan-Lower Missourian, 96
    Upper Missourian-Wolfcampian, 253
  Bonfim Formation, 407, 413
  Chachao Formation, 201–2
  Glen Dean Formation:
    E shelf Illinois Basin, 219, 321, 323
    SW shelf Illinois Basin, 330–32
  Itaituba-Nova Olinda Formations, 103, 107
  Kinkaid Formation:
    model 1, 135
    models 2–2A, 143
    model 3, 144
  Macaé Formation, 259, 263
  Menard Formation, 86
  Monte Cristo Group, 249
  Quintuco-Loma Montosa Formation, 335
  St. Louis Limestone:
    model 1, 41

Calcispheres (*cont.*)
   St. Louis Limestone (*cont.*)
      model 2, 41
      model 3, 43
   Salem Limestone, SW Illinois, 303, 305
Calcrete:
   Chachao Formation, 195
   Lagoa Feia Formation, 507
Calkins, F.C., 24
Callocladia: Glen Dean Formation, SW shelf Illinois Basin, 332
Cambrian:
   paleoclimatology, 559
   paleogeography, 559
   storm deposits, 562
Campbell, I., 24
Carbonate platform, edges, 9
Carbonate productivity:
   Bight of Abaco, Bahamas, 568
   characteristics, 565–67, 568–70
   models, 516
   relationship to eustatism, 568–70
Carbonate ramp model, 6
Carbonate ramp to platform evolution, 9
Carbonate rocks:
   basinal, 9
   classification, 4
   debris flow, 9
   models, 9
   submarine fans, 9
   terminology, 4
   transgression-regression sequence, 569–70
   turbidites, 9
Carbondale Formation, 99
Carpenteria, Miocene, Visayan Islands, 456
Carrs, B.W., 159
Carvalho, M.D., 108
Cayeux, L., 4
Cayeuxia, Bonfim Formation, 407, 412
Cayugan, N Ohio:
   dolomitization, sabkha, 156
   evaporites, 159
   exposure features, 156
   integration, 528, 530
   intraclasts, 157, 159
   lithic pellets, 157, 159
   microfacies description, 156–57, 159
   model, 159
   ostracods, 159
   paleoclimatology, 559
   paleogeography, 559
   reworking processes, 159
   sampling techniques, 156
   Spongiostromata, 156
   storm deposits, 157, 159
   stromatolites, 156–57
   worm tubes, 156–57, 159
Cenozoic:
   paleoclimatology, 561
   paleogeography, 561
Cephalopods:
   Burnt Bluff Group, 446
   Maquoketa phosphorites, 483, 485, 487, 491
Chachao Formation:
   annelids, 195, 202
   bioturbation, 195, 202
   bryozoans, 195, 202
   calcispheres, 201–2
   calcrete crusts, 195
   collapse breccias, 195
   corals, 195, 202
   deepening-upward sequence, 202
   dolomitization, dorag, 195, 202
   echinoids, 195, 201–2
   Eriphyla, 201
   evaporites, 195
   exposure features, 195, 202
   fish coprolites, 195, 201
   fish scales, 195, 201–2
   gastropods, 195, 201–2
   glauconite, 195
   integration, 534

Chachao Formation (*cont.*)
   Lenticulina, 195, 201
   micrite mounds, 195
   microfacies description, 195, 201
   model, 202
   Myoconcha, 201
   oysters, 195, 201–2
   paleoclimatology, 561
   paleoecological zonation, 202
   paleogeography, 561
   Panopea, 195
   pelecypods, 195, 201–2
   porosity, fracture, 202
   Ptychomia, 201
   reservoir generation, 202
   sampling techniques, 195
   shallowing-upward sequence, 201–2
   small-scale cycles, 563
   smaller benthic foraminifers, 195, 201–2
   sponge spicules, 195, 202
   typical sequence, 201–2
   vadose silt, 195
Chara:
   Beaverhill Lake Formation, 364
   Jeffersonville Limestone, coral zone, 354–55
   Jurassic, Morcles Nappe, Alps, 475
   Kinkaid Formation:
      model 1, 135
      models 2–2A, 143
   Traverse Group, 365
Chert brecciation, synsedimentary:
   Burlington Limestone, 229–30
   Galena Group, 77, 79, 81
Chickenwire anhydrite, Maquoketa phosphorites, 495
Choquette, P.W., 12
Clastic dikes, Niagaran, N Indiana, 193
Clasticity index:
   definition, 28
   first use, 1
   significance, 28
Clathrodictyon, Niagaran, N Indiana, 193
Clay minerals:
   Bailey Limestone, 474
   Bonfim Formation, 417
   Galena Group, 71
   Glen Dean Formation, E shelf Illinois Basin, 323
   lacustrine models, 558
   Lagoa Feia Formation, 496, 498
   Maquoketa phosphorites, 491
   Miocene, Visayan Islands, 451
   platforms with frontal bioaccumulated to hydrodynamic buildups, 546
   ramps, 533
   Traverse Group, 365
Climacammina, Kinkaid Formation, model 1, 132
Codiaceans, Amapá Formation, model 1, 275
Collapse breccia:
   Chachao Formation, 195
   Iola Formation, 167
   Kinkaid Formation, models 2–2A, 143
   Macaé Formation, 257
   Traverse Group, 365
Compaction:
   Quintuco Formation, 13
   St. Louis Limestone, 46
Composita, Iola Formation, 167
Computer modeling, cyclicity, 563
Conchydium, Niagaran, N Indiana, 193
Cook, H.E., 9
Corals:
   Amapá Formation, model 2, 280–82, 286–87
   Burlington Limestone, arenite suite, 221
   Burnt Bluff Group, 446, 448
   Chachao Formation, 195, 202
   Itaituba-Nova Olinda Formations, 106
   Jeffersonville Limestone:
      buildup, 358–59
      coral zone, 352, 354–55
   Jurassic, Morcles Nappe, Alps, 475
   Kimmeridgian-Portlandian, Salève, 399, 405
   Miocene, Visayan Islands, 453–54, 456

Corals (cont.)
   Monte Cristo Group, 248
   Niagaran:
      central Indiana, 465
      N Indiana, 193
   Salem Limestone, SE Illinois, 119, 121, 123
   Traverse Group, 367–68, 375
Crevello, P.D., 9
Crinoids:
   Atokan, 377, 379, 381, 383
   Bailey Limestone, 469, 473–74
   Bird Spring Group:
      Atokan-Lower Missourian, 96
      Morrowan, 92, 96
      Upper Missourian-Wolfcampian, 253
   Brereton Limestone:
      crinoid-brachiopod buildup model, 177
      phylloid algae buildup model, 177
   Burlington Limestone:
      arenite suite, 221–23
      wacke suite, 223, 226
   Burnt Bluff Group, 446
   Edwardsville Formation, 231, 233
   Galena Group, 69
   Glen Dean Formation, SW shelf Illinois Basin, 330–31
   Itaituba-Nova Olinda Formations, 101, 103, 106
   Jeffersonville Limestone:
      buildup, 358–59
      coral zone, 352, 354
   Jurassic, Morcles Nappe, Alps, 475
   Maquoketa phosphorites, 483, 485, 487, 491
   Menard Formation, 86
   Monte Cristo Group, 244, 248–49
   Niagaran:
      central Indiana, 465, 467
      N Indiana, 191, 193
   Pogonic Group:
      Nuia model, 213
      oncoid model, 213
   Quintuco-Loma Montosa Formation, 335, 337
   Rundle Group, 153
   Ste. Genevieve Limestone, 307, 309–10, 313
   St. Louis Limestone:
      model 1, 39
      model 2, 41
      model 3, 43
   Salem Limestone:
      SE Illinois, 119, 121, 123
      SW Illinois, 303, 305
   Traverse Group, 367–68
Crustacean coprolites, Middle Regencia Member, 115
Crystallinity index:
   Allentown Dolomite, 421
   definition, 28
   Joachim Dolomite, 441
   significance, 28
Cuvillier, J., 24
Cycles:
   autocyclic model, 565
   new eustatic model, 568–70
   punctuated aggradational, PAC, 563
   small-scale:
      Allentown Dolomite, 562
      Chachao Formation, 563
      characteristics, 562–63
      Salem Limestone, SW Illinois, 563
      Shakopee Dolomite, 563
Cyclicity:
   characteristics, 562–63
   computer modeling, 563
   eustatic model, 563, 565
   relationship to carbonate productivity, 565–67
   relationship to eustatism, 565–67
Cycloclypeus, Miocene, Visayan Islands, 453, 456

D

Dasyclads:
   Amapá Formation:
      model 1, 275
      model 2, 287
   Beaverhill Lake Formation, 364

Dasyclads (cont.)
   Jeffersonville Limestone, buildup, 358
   Jurassic, Morcles Nappe, Alps, 475
   Kimmeridgian-Portlandian, Salève, 399, 405
   Kinkaid Formation, model 1, 135
   Miocene, Visayan Islands, 454, 456
   Quintuco-Loma Montosa Formation, 337, 340
Davis, R.A., Jr., 9, 423
Dawson, W.C., 165
Dedolomitization, Galena Group, 81
Deepening-upward sequence:
   Chachao Formation, 202
   Kinkaid Formation, model 1, 132
   Monte Cristo Group, 249, 251
Demirmen, F., 30
Depasophyllum adentum, Traverse Group, 367, 375
Devonian:
   paleoclimatology, 559
   paleogeography, 559
   storm deposits, 562
Diaby, I., 34, 119, 303
Diagenesis:
   Amapá Formation:
      model 1, 278–79
      model 2, 287
   Atokan, 386, 390
   burial, 12
   environments, 9, 12–13
   freshwater phreatic, 12
   freshwater vadose, 12
   Galena Group, 77, 79, 81
   Glen Dean Formation, E shelf Illinois Basin, 327–29
   Iola Formation, 171, 173
   Lagoa Feia Formation, 506–8, 511
      basic volcaniclastics, 511
      playa lake model, 506–8
      pluvial lake model, 506–8
   Lower Regencia Member, 110
   Macaé Formation, 269–71, 273
   Maquoketa phosphorites, 495
   marine phreatic, 12
   Middle Regencia Member, 115, 117
   Miocene, Visayan Islands, 457, 460
   mixed freshwater-seawater phreatic, 12
   pathways, 9, 12–13
   Quintuco Formation, 12–13
   St. Louis Limestone, 45–46, 48, 50, 56
   Salem Limestone, SE Illinois, 125–26, 128
Diamictites, Macaé Formation, 259
Discocyclinids, Amapá Formation, model 2, 281–82
Dolomite crystallinity:
   Joachim Dolomite, 441
   Maquoketa phosphorites, 491
Dolomitization:
   burial, 12
   Atokan, 390
   Burlington Limestone, wacke suite, 226
   Macaé Formation, 273
   Monte Cristo Group, 238–39
   Niagaran, N Indiana, 191, 193
   Quintuco Formation, 13
   St. Louis Limestone, 56
   Salem Limestone, SE Illinois, 119, 121, 123, 126, 128
   dorag, 12
   Amapá Formation, 296–97
      model 1, 280
      model 2, 287
   Atokan, 386
   Bonfim Formation, 417
   Chachao Formation, 195, 202
   Galena Group, 81
   Glen Dean Formation:
      E shelf Illinois Basin, 323, 327–29
      SW shelf Illinois Basin, 332–33
   Joachim Dolomite, 441
   Kinkaid Formation:
      model 1, 135
      model 3, 144
   Macaé Formation, 268
   Maquoketa phosphorites, 495
   Platteville Group, 57, 59

Dolomitization (*cont.*)
  dorag (*cont.*)
    Quintuco Formation, 13
    Salem Limestone:
      SE Illinois, 124
      SW Illinois, 307
  lacustrine models, 558
  models, 517–18
  platforms with frontal bioaccumulated to hydrodynamic buildups, 546–47
  platforms with frontal bioconstructed to hydrodynamic buildups, 555
  ramps, 533
  sabkha:
    Bonfim Formation, 417
    Cayugan, N Ohio, 156
    Itaituba-Nova Olinda Formations, 103, 107
    Jeffersonville Limestone:
      buildup, 359
      coral zone, 355
    Joachim Dolomite, 441
    Kinkaid Formation:
      model 1, 135
      models 2–2A, 143
    Macaé Formation, 268
    Quintuco-Loma Montosa Formation, 337, 342
    St. Louis Limestone, 46, 48
      model 1, 39
      model 2, 41
      model 3, 43
    Salem Limestone, SW Illinois, 307
    Shakopee Dolomite, 425
  stylolitic, Atokan, 386
  submarine:
    Galena Group, 77
    Macaé Formation, 257
Donezella, Atokan, 379, 381
Dott, R.H., Jr., 561
Dunham, R.J., 4

E

Earlandia, Glen Dean Formation, SW shelf Illinois Basin, 330, 332
Echinoids:
  Amapá Formation:
    model 1, 275
    model 2, 280–82, 286–87
  Atokan, 377, 379, 381, 383
  Bonfim Formation, 407, 412–13, 417
  Chachao Formation, 195, 201–2
  Glen Dean Formation, E shelf Illinois Basin, 319, 321, 323
  Itaituba-Nova Olinda Formations, 101, 103, 106
  Kimmeridgian-Portlandian, Salève, 399
  Macaé Formation, 259
  Middle Regencia Member, 115
  Miocene, Visayan Islands, 451, 453–57
  Quintuco-Loma Montosa Formation, 335, 337, 340
  Rundle Group, 153
  St. Louis Limestone, model 1, 41
  Salem Limestone, SW Illinois, 305
Edwardsville Formation:
  algal dust, 233, 235
  brachiopods, 233
  bryozoans, 231, 233
  component variation, 233
  crinoid baffling effect, 235
  crinoids, 231, 233
  detrital quartz, 233
  fish plates, 233
  microfacies description, 231, 232
  model, 233, 235
  ostracods, 233
  phytoplankton, 235
  sampling techniques, 231
  shallowing-upward sequence, 233
  siltstones, 233, 235
  sponge spicules, 233
  superposition of environments, 233
  typical section, 233
ELF-AQUITAINE, 9
Elphidium, Miocene, Visayan Islands, 455
Endothyra:
  Atokan, 377, 379, 381
  Bird Spring Group, Upper Missourian-Wolfcampian, 253

Endothyra (*cont.*)
  Glen Dean Formation, SW shelf Illinois Basin, 330–32
  Itaituba-Nova Olinda Formations, 106
  Rundle Group, 153
  St. Louis Limestone, model 2, 41
  Salem Limestone:
    SE Illinois, 119, 121, 123–24, 128
    SW Illinois, 303, 305
Energy (bedshear):
  concept, 4
  models, 518
Enos, P., 9
Eriphyla, Chachao Formation, 201
Eustatism:
  cycles, 565
  models, 516
  paracycles, 565
  relationship to carbonate productivity, 568–70
  relationship to cyclicity, 565–67
  supercycles, 565
Evaporites:
  Cayugan, N Ohio, 159
  Chachao Formation, 195
  Macaé Formation, 257
  Maquoketa phosphorites, 495
  sabkha, 9
    Rundle Group, 154
  St. Louis Limestone, 44
  Traverse Group, 365, 371
Evolution, models, 516
Exposure features:
  Allentown Dolomite, 421
  Amapá Formation, 297
  Atokan, 383
  Bird Spring Group, Upper Missourian-Wolfcampian, 253
  Bonfim Formation, 417
  Cayugan, N Ohio, 156
  Chachao Formation, 195, 202
  Glen Dean Formation, E shelf Illinois Basin, 327–29
  Iola Formation, 171
  Jeffersonville Limestone, buildup, 359
  Joachim Dolomite, 429
  Kinkaid Formation:
    model 1, 135
    models 2–2A, 137, 143
  Lagoa Feia Formation, 498, 502
  Lower Regencia Member, 110
  Niagaran, N Indiana, 193
  Pogonip Group, 212
  Traverse Group, 365, 373
Extrabasinal constituents:
  lacustrine models, 558
  models, 518
  platforms with frontal bioaccumulated to hydrodynamic buildups, 546
  platforms with frontal bioconstructed to hydrodynamic buildups, 555
  ramps, 532

F

Fairbridge, R.W., 24
Falkenhein, F.U.H., 2, 256, 269
Fan delta, Quintuco-Loma Montosa Formation, 334
Favosites, Traverse Group, 368
Feiznia, S., 319, 323, 329
Feldspar:
  authigenic:
    Jurassic, Morcles Nappe, Alps, 478
    relationship to stylolites, 383, 386, 390
  detrital:
    Glen Dean Formation, E shelf Illinois Basin, 321, 323
    Lagoa Feia Formation, 496, 498, 500
    Macaé Formation, 259, 262
    Maquoketa phosphorites, 483, 485, 487, 491
    Miocene, Visayan Islands, 451, 457
    Ste. Genevieve Limestone, 313
Fenestella, Itaituba-Nova Olinda Formations, 106
Fenestral textures:
  Bonfim Formation, 417
  Joachim Dolomite, 429
  Kinkaid Formation, models 2–2A, 137, 143
  Lower Regencia Member, 110
  Macaé Formation, 257

Fenestral textures (cont.)
  Monte Cristo Group, 244
  Salem Limestone, SW Illinois, 305
  Traverse Group, 371
Fish coprolites, Chachao Formation, 195, 201
Fisher, J.H., 9
Fish plates:
  Burlington Limestone, arenite suite, 221
  Edwardsville Formation, 233
Fish scales:
  Burlington Limestone, arenite suite, 221
  Chachao Formation, 195, 201–2
  Lagoa Feia Formation, 498
Fistulipora, Niagaran, N Indiana, 191, 193
Florida Bay, cyclicity, 565
Flügel, E., 25, 30, 33
Folk, R.L., 4
Foraminifers:
  encrusting:
    Amapá Formation, model 2, 281–82, 286–87
    Iola Formation, 167
    Miocene, Visayan Islands, 453, 456
  larger benthic:
    Amapá Formation:
      model 1, 275
      model 2, 280–82, 286–87
    Miocene, Visayan Islands, 451, 453–57, 460
  smaller benthic:
    Amapá Formation:
      model 1, 275
      model 2, 280–82, 286–87
  Atokan, 377, 379, 381
  Beaverhill Lake Formation, 363
  Bird Spring Group:
    Atokan-Lower Missourian, 96
    Morrowan, 92, 96
    Upper Missourian-Wolfcampian, 253
  Bonfim Formation, 407, 412–13, 417
  Brereton Limestone:
    crinoid-brachiopod buildup model, 177
    phylloid algae buildup model, 177
  Burlington Limestone, arenite suite, 221
  Chachao Formation, 195, 201–2
  Glen Dean Formation:
    E shelf Illinois Basin, 319, 321, 323
    SW shelf Illinois Basin, 330–31
  Itaituba-Nova Olinda Formations, 101, 103, 106
  Jeffersonville Limestone, buildup, 359
  Jurassic, Morcles Nappe, Alps, 475
  Kimmeridgian-Portlandian, Salève, 309, 405
  Kinkaid Formation:
    model 1, 135
    models 2–2A, 143
    model 3, 144
  Lower Regencia Member, 110
  Macaé Formation, 259, 262–64
  Menard Formation, 86
  Middle Regencia Member, 115
  Miocene, Visayan Islands, 451, 453–57
  Monte Cristo Group, 249
  Quintuco-Loma Montosa Formation, 335
  Rundle Group, 153
  Ste. Geneviève Formation, 313
  Salem Limestone:
    SE Illinois, 119, 121, 123
    SW Illinois, 305
Fracturation:
  St. Louis Limestone, 48, 50, 56
  Salem Limestone, SE Illinois, 126, 128
Frequency index:
  definition, 28, 30
  first use, 1
  significance, 30
Frequency percentage, visual estimates, 30
Frost, S.H., 465
Fusulinids:
  Atokan, 377, 379, 381
  Bird Spring Group, Upper Missourian-Wolfcampian, 253
  Brereton Limestone:
    crinoid-brachiopod buildup model, 177
    phylloid algae buildup model, 177

Fusulinids (cont.)
  Iola Formation, 167
  Itaituba-Nova Olinda Formations, 106

G

Galena Group:
  bioturbation, 61, 64
  brachiopods, 69
  bryozoans, 69
  chert brecciation, synsedimentary, 77, 79, 81
  clay minerals, 71
  component variation, 66, 69, 71, 76
  crinoids, 69
  dedolomitization, 81
  detrital quartz, 69, 71
  diagenesis, 77, 79, 81
  dolomite textures, 81
  dolomitization:
    dorag, 81
    submarine, 77
  fecal pellets, 69
  gastropods, 69
  hardgrounds, 81
  integration, 519
  intraclasts, 69
  lithic pellets, 69
  microfacies description, 61, 64
  Mississippi-type ores, 81
  model, 66, 69, 71, 76
  paleoclimatology, 559
  paleogeography, 559
  pelecypods, 69
  pyrite mineralization, 81
  Receptaculites, 61, 64, 69
  relative bathymetry, 69
  sampling techniques, 61
  shallowing upward sequence, 64, 66
  silicification:
    late, 81
    submarine, 77, 79, 81
  storm deposits, 64, 71, 76–77, 79, 81
  typical sequence, 66
  vertical evolution of environments, 76–77
Gastropods:
  Atokan, 377, 381
  Beaverhill Lake Formation, 362–64
  Bonfim Formation, 407, 412–13, 417
  Burnt Bluff Group, 446
  Chachao Formation, 195, 201–2
  Galena Group, 69
  Glen Dean Formation, SW shelf Illinois Basin, 330–32
  Itaituba-Nova Olinda Formations, 103, 107
  Joachim Dolomite, 429
  Kinkaid Formation:
    models 2–2A, 143
    model 3, 144
  Lower Regencia Member, 110
  Macaé Formation, 259
  Middle Regencia Member, 115
  Miocene, Visayan Islands, 451, 453–57
  Platteville Group, 59
  Pogonip Group, oncoid model, 216
  Quintuco-Loma Montosa Formation, 337, 340
  Salem Limestone, SW Illinois, 305
Gerber, M.S., 81, 219
Ginsburg, R.N., 565
Girvanella:
  Atokan, 381
  Kinkaid Formation, model 1, 132
  Macaé Formation, 262
  Pogonip Group, 209, 212
  Salem Limestone, SE Illinois, 123
Glass globules, Lagoa Feia Formation, 498, 500
Glauconite:
  Bailey Limestone, 473–74
  Brereton Limestone, 175–76
  Burlington Limestone, arenite suite, 221
  Chachao Formation, 195
  Macaé Formation, 259
  platforms with frontal bioaccumulated to hydrodynamic buildups, 546
  platforms with frontal bioconstructed to hydrodynamic buildups, 555

Glauconite (cont.)
 Quintuco-Loma Montosa Formation, 335
 ramps, 532
 Ste. Genevieve Limestone, 313
Glen Dean Formation, E shelf Illinois Basin:
 annelid tubes, 321
 bioturbation, 319, 321, 323
 brachiopods, 319, 321, 323
 calcispheres, 319, 321, 323
 carbonate environments distribution, 323
 clay minerals, 323
 detrital feldspar, 321, 323
 detrital muscovite, 321, 323
 detrital quartz, 319, 321, 323
 diagenesis, 327–29
 dolomitization, dorag, 323, 327–29
 echinoids, 319, 321, 323
 exposure features, 327–29
 integration, 543
 intraclasts, 323
 lithic pellets, 319, 321, 323
 Michigan River delta, 322, 327
 microfacies description, 319, 321
 model, 323
 oncoids, 321, 323
 ooids, 321, 323
 ostracods, 319, 321, 323
 paleoclimatology, 559, 561
 paleogeography, 559, 561
 pelecypods, 321
 quartz arenites, 321, 323
 quartz wackes, 323
 sampling techniques, 319
 shales, 323
 shallowing-upward sequence, 323
 siliciclastic environments distribution, 323, 327
 siliciclastic model, 327
 silicification, 327
 siltstones, 321, 323
 smaller benthic foraminifers, 319, 321, 323
 sponge spicules, 319, 321, 323
 submarine channels, 327
 submarine fans, 327
 trilobites, 321
 typical sequence, 323
Glen Dean Formation, SW shelf Illinois Basin:
 aggrading neomorphism, 333
 Archimedes, 332
 bioturbation, 330–31
 brachiopods, 330–31
 bryozoans, 330–32
 calcareous shale, 330
 calcispheres, 330–32
 Calocladia, 332
 crinoids, 330–31
 dolomitization, dorag, 332–33
 Earlandia, 330, 332
 Endothyra, 330–32
 fecal pellets, 330–31
 gastropods, 330–32
 integration, 541, 543
 intraclasts, 332–33
 lithoclasts, 330
 Michigan River delta, 329, 333
 microfacies description, 329–31
 model, 332–33
 ooids, 331–33
 ostracods, 330–32
 paleoclimatology, 559, 561
 paleogeography, 559, 561
 pelecypods, 330–31
 quartz feldspathic arenite, 330
 sampling techniques, 329
 shallowing-upward sequence, 331–32
 silicification, 333
 smaller benthic foraminifers, 330–31
 sponge spicules, 330–32
 superposition of environments, 334
 turbidites, 333
 typical sequence, 331
Globigerinids, Miocene, Visayan Islands, 457

Globoquadrina, Miocene, Visayan Islands, 457
Globorotalia, Miocene, Visayan Islands, 457
Goodwin, P.W., 563
Graded bedding:
 Bailey Limestone, 468
 Burlington Limestone, wacke suite, 226
 Jurassic, Morcles Nappe, Alps, 478
 Niagaran, central Indiana, 467
Graywackes, Miocene, Visayan Islands, 451
Great Bahama Bank, comparison with Ste. Genevieve Limestone, 313–14
Green algae, Bonfim Formation, 407, 412–13, 417
Gressly, A., 24
Gümbelina, 1
Gypsification, Lagoa Feia Formation, 498
Gypsum, Joachim Dolomite, 431

H

Half-moon ooids, Iola Formation, 167
Halimeda:
 Amapá Formation:
  model 1, 275
  model 2, 287
 Bonfim Formation, 412
 Miocene, Visayan Islands, 454, 456
Halite:
 Joachim Dolomite, 431
 Traverse Group, 371
Halley, R.B., 9
Handford, C.R., 561
Hansen, M.W., 235
Hardgrounds:
 Galena Group, 81
 Quintuco-Loma Montosa Formation, 335
Harris, P.M., 4, 9
Heath, C.P., 92
Heckel, P.H., 4
Hedbergella:
 Bonfim Formation, 407
Hedraites, Iola Formation, 167
Hexagonaria, Traverse Group, 368, 375
Hornblende, Miocene, Visayan Islands, 451, 457
Hovelacque, M., 24
Hulse, J.A., 352
Hummocky stratification, storm deposits, 561
Hurricane model, characteristics, 561–62
Hyaloclastites, Lagoa Feia Formation, 498, 500, 503
 alteration to clay minerals, 498
 calcitization, 498
 dolomitization, 498
 palagonitization, 498
 silicification, 498
 zeolitization, 498
Hyalotuff, Lagoa Feia Formation, 500, 503

I

Illinois cyclothems, comparison with Bird Spring Group, 99
Inoceramus, 1
Integration:
 Allentown Dolomite, 550
 Amapá Formation:
  model 1, 534, 536
  model 2, 536
 Arrow Canyon Formation-Crystal Pass Limestone:
  buildup ramp model, 528
  ramp model, 520
 Atokan, 549
 Beaverhill Lake Formation, 547, 549
 Bird Spring Group:
  Morrowan to Lower Missourian, 519–20
  Upper Missourian-Wolfcampian, 538–39
 Bonfim Formation, 552, 554
 Brereton Limestone:
  crinoid-brachiopod buildup model, 524
  phylloid algae buildup model, 528
 Cayugan, N Ohio, 528, 530
 Chachao Formation, 534
 Galena Group, 519
 Glen Dean Formation:
  E shelf Illinois Basin, 543
  SW shelf Illinois Basin, 541, 543
 Itaituba-Nova Olinda Formations, 523

Integration (*cont.*)
    Jeffersonville Limestone, coral zone, 554
    Joachim Dolomite, 549–50
    Kinkaid Formation:
        model 1, 525, 528
        models 2–2A, 521, 523
    Lagoa Feia Formation:
        playa lake model, 556–57
        pluvial lake model, 557
    Macaé Formation:
        model 2, 543–44
        model 3, 543–44
    Menard Formation, 519
    Miocene, Visayan Islands, 550, 552
    models, 516–18
    Monte Cristo Group, 538
    Platteville Group, 519
    Pogonip Group:
        Nuia model, 536, 538
        oncoid model, 538
    Rundle Group, 524–25
    Ste. Genevieve Limestone, 539, 541
    St. Louis Limestone:
        model 1, 523
        model 2, 525
        model 3, 539
    Salem Limestone:
        SE Illinois, 525
        SW Illinois, 541
    Shakopee Dolomite, 550
    Traverse Group, 554
Intraclasts:
    Allentown Dolomite, 419–21
    Arrow Canyon Formation-Crystal Pass Limestone, 161
    Atokan, 379, 383
    Bonfim Formation, 407, 412–13, 417
    Burnt Bluff Group, 446
    Cayugan, N Ohio, 157, 159
    Galena Group, 69
    Glen Dean Formation:
        E shelf Illinois Basin, 323
        SW shelf Illinois Basin, 332–33
    Iola Formation, 171
    Itaituba-Nova Olinda Formations, 103
    Joachim Dolomite, 428–29, 431, 437, 439
    Jurassic, Morcles Nappe, Alps, 475
    Kimmeridgian-Portlandian, Salève, 399
    Kinkaid Formation:
        model 1, 135
        models 2–2A, 143
    Lower Regencia Member, 110
    Middle Regencia Member, 115
    Miocene, Visayan Islands, 455–56
    platforms with frontal bioaccumulated to hydrodynamic buildups, 545
    platforms with frontal bioconstructed to hydrodynamic buildups, 555
    Quintuco-Loma Montosa Formation, 335, 337, 340
    Ste. Genevieve Limestone, 309
    St. Louis Limestone:
        model 1, 39
        model 2, 41
        model 3, 43
    Salem Limestone:
        SE Illinois, 121, 123
        SW Illinois, 303, 305
    Traverse Group, 367–68, 371, 373
Iola Formation:
    anhydritization, 167
    Archaeolithophyllum, 167
    bioturbation, 167
    brachiopods, 167
    bryozoans, 167, 171
    collapse breccia, 167
    Composita, 167
    diagenesis, 171, 173
    encrusting foraminifers, 167
    exposure features, 171
    fusulinids, 167
    half-moon ooids, 167
    Hedraites, 167
    intraclasts, 171
    microfacies description, 165, 167, 171

Iola Formation (*cont.*)
    model, 171
    oncoids, 167
    ooids, 167, 171
    paleoclimatology, 561
    paleogeography, 561
    pelmatozoans, 167, 171
    phosphorites, 167
    phylloid algae, 167, 171
    phylloid algal buildups, 165
    plant materials, 171
    plant root molds, 171
    Polypora, 167
    porosity:
        biomoldic, 173
        fracture, 173
    red algae, 167
    sampling techniques, 165
    shallowing-upward sequence, 171
    siltstones, 171
    sponge spicules, 167
    storm deposits, 171
    superposition of models, 171
    Tuberitina, 167
    typical section, 171
    vadose silt, 171
Irwin, M.L., 6
Itaituba-Nova Olinda Formations:
    agglutinated benthic foraminifers, 103, 106–7
    anhydritization, 103, 107
    bioturbation, 103
    brachiopods, 101, 103, 106
    bryozoans, 101, 103, 106
    calcispheres, 103, 107
    corals, 106
    crinoids, 101, 103, 106
    depositional environment, 101
    detrital quartz, 103, 105–6
    dolomitization, sabkha, 103, 107
    echinoids, 101, 103, 106
    Endothyra, 106
    fecal pellets, 103
    Fenestella, 106
    fusulinids, 106
    gastropods, 103, 107
    integration, 523
    intraclasts, 103
    microfacies description, 101, 103, 106
    Millerella, 106
    model, 106–7
    oncoids, 103
    ooids, 103
    ostracods, 103, 107
    paleoclimatology, 561
    paleogeography, 561
    Paleotextularia, 106
    pelecypods, 101, 103, 107
    plants debris, 103
    Plectogyra, 106
    Polypora, 106
    quartz arenites, 106
    sampling techniques, 101
    silicification, 103
    smaller benthic foraminifers, 101, 103, 106
    sponge spicules, 103, 107
    stromatolites, 103, 107
    Textrataxis, 106
    trilobites, 101, 103
Ivanovia, Brereton Limestone, 175

J

James, N.P., 4, 9, 31, 563
Jeffersonville Limestone, buildup:
    brachiopods, 359
    corals, 358–59
    crinoids, 358–59
    dasyclads, 358
    dolomitization, sabkha, 359
    exposure features, 359
    Geneva Dolomite facies, 358
    lithic pellets, 358–59

Jeffersonville Limestone, buildup (*cont.*)
  microfacies description, 358–59
  model, 359
  paleoclimatology, 559
  paleogeography, 559
  pelecypods, 359
  sampling techniques, 358
  smaller benthic foraminifers, 359
  stromatoporoids, 358–59
  typical sequences, 359
  vertical sequence, 359
Jeffersonville Limestone, coral zone:
  Amphipora, 352, 354–55
  brachiopods, 352, 354
  bryozoans, 352, 354
  Chara, 354–55
  corals, 352, 354–55
  crinoids, 352, 354
  detrital quartz, 352, 354
  dolomitization, sabkha, 355
  integration, 554
  lithic pellets, 354
  microfacies description, 352, 354
  model, 355
  ostracods, 352, 354
  paleoclimatology, 559
  paleogeography, 559
  pelecypods, 352, 354
  sampling techniques, 352
  shallowing-upward sequence, 354–55
  storm deposits, 354–55
  stromatolites, 354
  stromatoporoids, 354–55
  trilobites, 354
  typical sequences, 355
Joachim Dolomite:
  anhydrite, chickenwire, 431
  authigenic quartz, 429, 431, 441
  beachrock, 428
  bioturbation, 428–29, 437
  detrital quartz, 429, 431, 440–41
  dolomite, crystallinity, 441
  dolomitization:
    dorag, 441
    sabkha, 441
  exposure features, 429
  fecal pellets, 429
  fenestral textures, 429
  gastropods, 429
  gypsum, 431
  halite, 431
  integration, 549–50
  intraclasts, 428–29, 431, 437, 439
  Leperditia, 428–29
  Lingula, 429
  lithic pellets, 429, 431, 439–40
  megacycles, 563
  microfacies description, 428–29, 431
  model, 437, 439, 441
  Nuia, 428, 439
  oncoids, 428, 437
  ooids, superficial, 429, 440
  ostracods, 428–29, 437
  paleoclimatology, 559
  paleogeography, 559
  pelmatozoans, 428, 437
  pyrite, 441
  quartz arenite, 431
  sampling techniques, 428
  shallowing-upward sequence, 431, 437
  spastolites, 429
  Spongiostromata, 429
  storm deposits, 429, 431
  stromatolites, 429, 431, 437
  superposition of environments, 441
  typical sequence, 431, 437
  vadose silt, 428–29, 431
Jurassic, Morcles Nappe, Alps:
  ammonoids, 475
  annelids, 475
  authigenic feldspars, 478

Jurassic (*cont.*)
  brachiopods, 475
  bryozoans, 475
  Chara, 475
  corals, 475
  crinoids, 475
  dasyclads, 475
  detrital muscovite, 478
  detrital quartz, 475, 478
  intraclasts, 475
  lithic pellets, 475
  lithoclasts, 475
  microfacies description, 475, 478
  ooids, 475
  ostracods, 475
  pelagic crinoids, 475
  pyrite, 478
  radiolarians, 475
  rudistids, 475
  Saccocoma, 475
  sampling techniques, 475
  smaller benthic foraminifers, 475
  tectonic setting, 475
  tintinnoids, 475
  turbidites:
    correlation, 478
    graded bedding, 478
    horizontal distribution, 478
    horizontal grading, 478
    setting, 475
    slumping textures, 478
    textural features, 475, 478
    typical sequence, 478

K

Khawlie, M.R., 173
Kilian, C.C.W., 24
Kimmeridgian-Portlandian, Salève:
  annelids, 398–99, 405
  bioturbation, 398–99
  corals, 399, 405
  dasyclads, 399, 405
  echinoids, 399
  intraclasts, 399
  lithic pellets, 398–99
  Lithoporella, 399
  microfacies description, 398–99, 405
  model, 405
  ostracods, 398–99, 405
  paleobathymetry, 405
  pelecypods, 399, 405
  red algae, 399
  sampling techniques, 398
  shallowing-upward sequence, 405
  silicification, 405
  smaller benthic foraminifers, 399, 405
  sponges, 399
  stromatoporoids, 399, 405
  typical sequence, 405
Kinkaid Formation:
  model 1:
    Archimedes, 132
    bioturbation, 131–32, 135
    bryozoans, 132, 135
    calcispheres, 135
    Chara, 135
    Climacammina, 132
    dasyclads, 135
    deepening-upward sequence, 132
    dolomitization:
      dorag, 135
      sabkha, 135
    exposure crusts, 135
    fecal pellets, 135
    Girvanella, 132
    integration, 525, 528
    intraclasts, 135
    lithic pellets, 135
    microfacies description, 131–32
    oncoids, 135
    ooids, 135

Kinkaid Formation *(cont.)*
  model 1 *(cont.)*
    ostracods, 135
    pelecypods, 132, 135
    pelmatozoans, 132, 135
    shallowing-upward sequence, 132
    smaller benthic foraminifers, 135
    sponge spicules, 135
    stromatolites, 132
    trilobites, 132
    typical section, 132
  models 2-2A:
    anhydritization, 143
    Archimedes, 135
    bioturbation, 136-37, 143
    brachiopods, 143
    bryozoans, 143
    calcispheres, 143
    Chara, 143
    collapse breccia, 143
    dolomitization, sabkha, 143
    exposure features, 137, 143
    fecal pellets, 143
    fenestral fabric, 137, 143
    gastropods, 143
    integration, 521, 523
    intraclasts, 143
    lithic pellets, 143
    Michigan River delta, 135
    microfacies description, 135-37, 143
    ostracods, 143
    pelmatozoans, 143
    plant root molds, 143
    smaller benthic foraminifers, 143
    sponge spicules, 143
    storm deposits, 143
    stromatolites, 137, 143
  model 3:
    bioturbation, 144
    bryozoans, 144
    calcispheres, 144
    dolomitization, dorag, 144
    fecal pellets, 144
    gastropods, 144
    lithic pellets, 144
    microfacies description, 143
    ostracods, 144
    pelmatozoans, 144
    smaller benthic foraminifers, 144
    sponge spicules, 144
    vadose silt, 144
  paleoclimatology, 559, 561
  paleogeography, 559, 561
  sampling techniques, 131
  superposition of models, 131, 145
Klein, G. de V., 516, 561
Koninckopora:
  Salem Limestone:
    SE Illinois, 121, 123
    SW Illinois, 305
Kreisa, R.D., 461
Kuenen, P.H., 1
Kuhnhenn, G.L., 57

L

Lacey, J.E., 314
Lacustrine models:
  benthic fauna, 558
  clay minerals, 558
  detrital quartz, 558
  dolomitization, 558
  extrabasinal constituents, 558
  fair weather wave base, 557-58
  general synthesis, 557-58
  genetic classification, 555-57
  oncoids, 558
  ooids, 558
  porosity types, 558
  reservoir generation, 558
  seiches, 558
  storm deposits, 558

Lacustrine models *(cont.)*
  sublacustrine morphology, 557-58
  type 1, 556-57
  type 2, 557
Lagena, 1
Lagoa Feia Formation:
  anhydrization, 498
  basic volcanic glass, 496, 498, 500
  basic volcaniclastic diagenesis, 511
  basic volcanic lithoclasts, 496, 498
  bioturbation, 498, 502
  calcrete, 507
  clay minerals, 496, 498
  detrital feldspar, 496, 498, 500
  detrital quartz, 496, 498, 500
  diagenesis, 506-8, 511
    playa lake model, 506-8
    pluvial lake model, 506-8
  exposure features, 498, 502
  fish debris, 498
  glass globules, 498, 500
  gypsification, 498
  hyaloclastites, 498, 500, 503
    alteration to clay minerals, 498
    calcitization, 498
    dolomitization, 498
    palagonitization, 498
    silicification, 498
    zeolitization, 498
  hyalotuffs, 500, 503
  kerolitic ooids, 498, 503
  lithic feldspathic quartz arenite, 496, 498
  lithic feldspathic wacke, 496
  microfacies description, 496, 498, 500
  oncoids, 498
  organic carbon, 498
  ostracods, 498, 500, 502
  paleoclimatology, 561
  paleogeography, 561
  pedogenesis, 507
  pelecypods, 498, 500, 502-3
  pisolitization, 507
  plant root molds, 498
  playa lake model, 502
    integration, 556-57
  pluvial lake model, 502-3
    integration, 557
  porosity:
    evolution, 506-8
    intercrystalline, 506-8, 511
    intraparticle, 506-8, 511
    moldic, 506-7, 511
    primary interparticle, 496, 498, 506-7
    primary reduced, 496, 498
    relationship to microfacies, 508-9, 511
    vuggy, 506-7, 511
  reservoir generation, 507-8
  sampling techniques, 496
  seiches, 502
  silicification, 498, 500
  siltstone, 498
  sparmicritization, 507
  storm deposits, 503
  subaqueous basic volcanism model, 503
  typical sequences, 500, 502
  zeolitization, 498, 500
Land, L.S., 568
Lasemi, Y., 131
Lenticulina, Chachao Formation, 195, 201
Leperditia, Joachim Dolomite, 428-29
Lepidocyclina, Miocene, Visayan Islands, 456
Lepidocyclinids, Amapá Formation, model 2, 281-82
Lillie, A., 1
Lingula, Joachim Dolomite, 429
Lithoclasts:
  Bird Spring Group, Upper Missourian-Wolfcampian, 253
  Glen Dean Formation, SW shelf Illinois Basin, 330
  Jurassic, Morcles Nappe, Alps, 475
  Pogonip Group, *Nuia* model, 213
Lithophyllym, Bonfim Formation, 412
Lithoporella, Kimmeridgian-Portlandian, Salève, 399

Lithostrotionella, Monte Cristo Group, 244, 248–49
Lithothamnium, Bonfim Formation, 407, 412–13
Longman, M.W., 4, 12
Loucks, R.G., 9
Lower Regencia Member:
    bioturbation, 108, 110
    detrital biotite, 108, 110
    detrital quartz, 108, 110
    diagenesis, burial 110
    exposure features, 110
    fecal pellets, 110
    fenestral textures, 110
    gastropods, 110
    intraclasts, 110
    lithic pellets, 110
    micritization, 110
    microfacies description, 108, 110
    model, 110
    ooids, 110
    ostracods, 108, 110
    pelecypods, 110
    plant debris, 108, 110
    porosity, oomoldic, 110
    shallowing-upward sequence, 110
    smaller benthic foraminifers, 110
    stromatolites, 108, 110
    typical section, 110
Lumsden, D.N., 99
Lundwall, W.R., Jr., 358

M

Macaé Formation:
    anhydritization, 273
    baffling seagrass communities, 268
    beachrock, 262, 264, 270–71
    bioturbation, 259
    bryozoans, 262
    calcispheres, 259, 263
    collapse breccia, 257
    component variation, 264, 268–69
    detrital feldspar, 259, 262
    detrital muscovite, 259
    detrital quartz, 257, 259, 262–64
    diagenesis, 269–71, 273
    diamictites, 259
    dolomitization:
        burial, 273
        dorag, 268
        sabkha, 268
        submarine, 257
    echinoids, 259
    evaporites, 257
    fenestral textures, 257
    gastropods, 259
    Girvanella, 262
    glauconite, 259
    microfacies description, 257, 259, 262–64
    microoncoidal shoals, 268
    microoncoids, 263
    model 1, 268
    model 2, 268
    model 3, 268–69
    neomorphism, 273
    oil migration, 273
    oncoid "flour," 270
    oncoidal shoals, 268–69
    oncoids, 262–64
    ooids, 262–64
    oolitic-oncoidal shoals, 268
    ostracods, 259, 262–64
    paleoclimatology, 561
    paleoecological indexes, 264
    paleogeography, 561
    pelecypods, 259, 262–64
    phosphates, 259
    planktonic foraminifers, 259, 263
    plant remains, 257, 259
    porosity:
        channel, 270
        interparticle, 264, 271

Macaé Formation (*cont.*)
    porosity (*cont.*)
        intraparticle, 273
        vuggy, 264, 270–71
    pyrite, 259
    quartz feldspathic arenites, 259
    radiolarians, 259
    red algae, 259, 262–63
    reservoir generation, 264, 268–71, 273
    rudistids, 263
    sampling techniques, 256–57
    shallowing-upward sequence, 264
    silicification, burial, 273
    smaller benthic foraminifers, 259, 262–64
    turbidites, 259
    worm tubes, 259, 262–63
Maquoketa phosphorites:
    bioturbation, 483, 485, 487
    bone fragments, 485
    brachiopods, 483, 485, 487, 491
    cephalopods, 483, 485, 487, 491
    chickenwire anhydrite, 495
    clay minerals, 491
    crinoids, 483, 485, 487, 491
    detrital feldspar, 483, 485, 487, 491
    detrital quartz, 483, 485, 487, 491
    diagenesis, 495
    dolomite crystallinity, 491
    dolomitization, dorag, 495
    evaporites, 495
    microfacies description, 483, 485, 487
    model, 491, 494
    nodular anhydrite, 483
    ostracods, 483, 485, 487, 491
    paleoclimatology, 495
    paleogeography, 495
    pelecypods, 483, 485, 487
    petrographic classification, 482–83
    phosphatic intraclasts, 483, 485, 487
    phosphatic nodules, 485, 487
    phosphatic ooids, 485, 487, 491
    phosphatic pellets, 483, 485, 487, 491
    phosphatic spherulites, 483, 491
    phosphatization, 495
    pyritization, 483, 485, 487, 491
    sampling techniques, 482
    selenitic gypsum, 495
    shallowing-upward sequence, 487, 491
    silicification, 483, 485, 487, 495
    trace elements:
        identification, 482
        variation, 491, 494
    trilobites, 483, 485, 491
Marsaglia, K.M., 561
Mattes, B.W., 12
Megacycles:
    Allentown Dolomite, 563
    Bird Spring Group, Atokan to Lower Missourian, 563
    characteristics, 563
    Joachim Dolomite, 563
    Rundle Group, 563
    St. Louis Limestone, 563
Menard Formation:
    Archimedes, 85
    bryozoans, 86
    calcispheres, 86
    channel sandstones, 90
    component variation, 86
    crinoids, 86
    detrital quartz, 86
    energy level, 90
    integration, 519
    microfacies description, 83, 85–86
    model, 90
    neomorphism, 85–86, 90
    ooids, 86
    ostracods, 86
    paleoclimatology, 559, 561
    paleogeography, 559, 561
    reworking processes, 81
    sampling techniques, 83

Menard Formation (*cont.*)
  shallowing-upward sequence, 86
  smaller benthic foraminifers, 86
  sponge spicules, 86
  typical sequence, 86, 90
Mesozoic:
  paleoclimatology, 561
  paleogeography, 561
Michigan River delta:
  Glen Dean Formation:
    E shelf Illinois Basin, 323, 327
    SW shelf Illinois Basin, 329, 333
  Kinkaid Formation, models 2–2A, 135
Micrite matrix:
  St. Louis Limestone:
    model 1, 41
    model 2, 41
    model 3, 43
Micrite mounds:
  Chachao Formation, 195
  Niagaran, N Indiana, 191
Micritization:
  Lower Regencia Member, 110
  Middle Regencia Member, 115
  Quintuco Formation, 12
  St. Louis Limestone, 46
Microcycles:
  Allentown Dolomite, 563
  characteristics, 563
Microfacies:
  cements, 27
  correlation coefficients, 31–32
  cuttings, 27
  data sheet, 28
  definition, 24
  distribution in models, 517
  dolomite thin sections, 27
  energy (bedshear) level, 27
  field sampling, 25
  filiation diagram, 33
  final classification, 31–32
  final shallowing-upward order, 31–32
  final types, 31–32
  grain-supported texture, 27
  graphic representation, 25, 30, 32–33
  history, 24
  inorganic constituents, 27
  interpretation:
    environmental, 32
    relative bathymetrical, 32
    relative energy, 32
    shallowing-upward, 32–33
  limestone thin sections, 25, 27
  matrix, 27
  mud-supported texture, 27
  numbering technique, 32
  organic constituents, 27
  preliminary shallowing-upward order, 27–28
  preliminary types, 27
  quantitative analysis, 25, 28, 30
  relative bathymetry, 27–28
  standard types, 33
  statistical evaluation, 30–31
  vertical interval, 25
Microfilaments, Bonfim Formation, 407
Microoncoids:
  Bonfim Formation, 407
  Macaé Formation, 263
Microporosity, Middle Regencia Member, 117
Middle Regencia Member:
  bioturbation, 115
  crustacean coprolites, 115
  detrital biotite, 110, 115
  detrital quartz, 110, 115
  diagenesis, burial, 115, 117
  echinoids, 115
  fecal pellets, 115
  gastropods, 115
  intraclasts, 115
  lithic pellets, 115
  micritization, 115

Middle Regencia Member (*cont.*)
  microfacies description, 108, 110, 115
  microporosity, 117
  model, 115
  oncoids, 115
  ooids, 115
  ostracods, 115
  pelecypods, 115
  plant debris, 110, 115
  porosity:
    biomoldic, 117
    fracture, 117
    interparticle, 117
    oomoldic, 117
    stylolitic, 117
    vuggy, 117
  shallowing-upward sequence, 115
  smaller benthic foraminifers, 115
  typical section, 115
Middleton, G.V., 6, 25, 33
Millerella, Itaituba-Nova Olinda Formation, 106
Mineral constituents, models, 518
Miocene, Visayan Islands:
  Amphistegina, 453–54, 456
  andesitic volcanism, 460
  bioturbation, 453–56
  Borelis, 456
  bryozoans, 454–56
  Carpenteria, 456
  clay minerals, 451
  corals, 453–54, 456
  Cycloclypeus, 453, 456
  dasyclads, 454, 456
  detrital feldspar, 451, 457
  detrital quartz, 451, 457
  diagenesis, 457, 460
  echinoids, 451, 453–57
  Elphidium, 455
  encrusting foraminifers, 453, 456
  gastropods, 451, 453–57
  globigerinids, 457
  Globoquadrina, 457
  Globorotalia, 457
  graywackes, 451
  Halimeda, 454, 456
  hornblende, 451, 457
  integration, 550, 552
  intraclasts, 455–56
  larger benthic foraminifers, 451, 453–57, 460
  Lepidocyclina, 456
  microfacies description, 451, 453–57
  Miogypsina, 454, 456
  model, 460
  Operculina, 453–54, 456
  orbitoids, 453–57
  orbulinids, 457
  ostracods, 451, 453–57
  planktonic foraminifers, 451, 453–57
  plant debris, 451
  pelecypods, 451, 453–57
  porosity:
    biomoldic, 456, 460
    burial, 456, 460
    vuggy, 456, 460
  pyrite, 457
  red algae, 453–56
  reservoir generation, 460
  sampling techniques, 448, 451
  smaller benthic foraminifers, 451, 453–57
  Sphaeroidinellopsis, 457
  Spiroclypeus, 456
  sponge spicules, 451, 453–57
  turbidites, 457
  vadose silt, 456, 460
  volcanic lithoclasts, 451, 455–56
Miogypsina, Miocene, Visayan Islands, 454, 456
Mississippi-type ores, Galena Group, 81
Mississippian:
  paleoclimatology, 559, 561
  paleogeography, 559, 561
  storm deposits, 562

Model:
    Allentown Dolomite, 421–22
    Amapá Formation:
        model 1, 278
        model 2, 287
    Arrow Canyon Formation-Crystal Pass Limestone:
        buildup ramp model, 161
        ramp model, 161
    Atokan, 381, 383, 386
    Beaverhill Lake Formation, 364
    Bird Spring Group:
        Morrowan-Lower Missourian, 96
        Upper Missourian-Wolfcampian, 255
    Bonfim Formation, 417
    Brereton Limestone:
        crinoid-brachiopod buildup model, 177
        phylloid algae buildup model, 177
    Burlington Limestone, 227
    Burnt Bluff Group:
        Nasbro buildup, 446
        Sturgeon Bay buildup, 448
    Cayugan, N Ohio, 159
    Chachao Formation, 202
    Edwardsville Formation, 233, 235
    Galena Group, 66, 69, 71, 76
    Glen Dean Formation:
        E shelf Illinois Basin, 323
        siliciclastics, E shelf Illinois Basin, 327
        SW shelf Illinois Basin, 332–33
    Iola Formation, 171
    Itaituba-Nova Olinda Formations, 106–7
    Jeffersonville Limestone:
        buildup, 359
        coral zone, 355
    Joachim Dolomite, 437, 439, 441
    Kimmeridgian-Portlandian, Salève, 405
    Kinkaid Formation:
        model 1, 132, 135
        model 2–2A, 143
        model 3, 144
    Lagoa Feia Formation:
        playa lake, 502
        pluvial lake, 502–3
    Lower Regencia Member, 110
    Macaé Formation:
        model 1, 268
        model 2, 268
        model 3, 268–69
    Maquoketa phosphorites, 491, 494
    Menard Formation, 90
    Middle Regencia Member, 115
    Miocene, Visayan Islands, 460
    Monte Cristo Group, 244, 248–49
    Niagaran:
        central Indiana, 467–68
        N Indiana, 193
    Platteville Group, 61
    Pogonip Group:
        Nuia model, 213
        oncoid model, 213
    Quintuco-Loma Montosa Formation, 340, 342
    Ste. Genevieve Limestone, 313–14
    St. Louis Limestone:
        model 1, 39, 41
        model 2, 41, 43
        model 3, 43
    Salem Limestone:
        SE Illinois, 124–25
        SW Illinois, 305, 307
    Shakopee Dolomite, 425
    Traverse Group, 373, 375
Models in general:
    benthic constituents, 518
    carbonate productivity, 516
    characteristics, 33
    dolomitization, 517–18
    eustatism, 516
    evolution, 516
    extrabasinal constituents, 518
    functions, 516
    generic classification, 516

Models in general (*cont.*)
    ideal type, 25, 33
    integration, 516–18
    intrabasinal constituents, 518
    lacustrine, 555–58
    microfacies distribution, 517
    mineral constituents, 518
    paleoclimatology, 558–59, 561–62
    paleogeography, 558–59, 561–62
    planktonic constituents, 518
    porosity types, 518
    prediction capabilities, 33
    relative energy, 518
    reservoir generation, 518
    submarine topography, 517–18
    subsidence, 516
    tectonism, 516
Modesto Formation, 99
Monte Cristo Group:
    bioturbation, 238–39, 241, 244
    brachiopods, 248–49
    bryozoans, 244, 248–49
    calcispheres, 249
    corals, 248
    crinoids, 244, 248–49
    deepening-upward sequences, 249, 251
    detrital quartz, 249
    dolomitization, burial, 238–39
    fecal pellets, 249
    fenestral textures, 244
    integration, 538
    Lithostrotionella, 244, 248–49
    microfacies:
        description, 238–39, 241, 244
        filiation, 244
    model, 244, 248–49
    paleoclimatology, 561
    paleogeography, 561
    sampling techniques, 235, 238
    shallowing-upward sequence, 244
    smaller benthic foraminifers, 249
    sponge spicules, 249
    stromatolites, 244
    superposition of environments, 249, 251
    Syringopora, 241, 249
Mount, J.F., 558
Mountjoy, E.W., 9, 12
Muscovite, detrital:
    Glen Dean Formation, E shelf, Illinois Basin, 321, 323
    Jurassic, Morcles Nappe, Alps, 478
    Macaé Formation, 259
    Niagaran, central Indiana, 465, 467
    Ste. Genevieve Limestone, 309
Myoconcha, Chachao Formation, 201

N

Nelson, C.S., 6
Neomorphism:
    Glen Dean Formation, SW shelf Illinois Basin, 333
    Macaé Formation, 273
    Menard Formation, 85–86, 90
    Quintuco Formation, 13
Neumann, A.C., 568
Niagaran, central Indiana:
    brachiopods, 467
    bryozoans, 465, 467
    corals, 465
    crinoids, 465, 467
    detrital muscovite, 465, 467
    detrital quartz, 465, 467
    fecal pellets, 465
    lithic pellets, 465
    microfacies description, 465, 467
    model, 467–68
    ostracods, 465, 467
    pyrite, 465, 467
    sampling techniques, 465
    slumping textures, 467
    stromatoporoids, 465
    turbidites:
        graded bedding, 467

Niagaran (*cont.*)
  turbidites (*cont.*)
    proximal to distal correlation, 467–68
    setting, 465
    textural features, 465, 467
    typical distal sequence, 467
    typical median sequence, 467
    typical proximal sequence, 467
Niagaran, N Indiana:
  brachiopods, 191, 193
  bryozoans, 191, 193
  buildup evolution, 191
  clastic dikes, 193
  Clathrodictyon, 193
  Conchydium, 193
  corals, 193
  crinoids, 191, 193
  dolomitization, burial, 191, 193
  exposure features, 193
  Fistulipora, 191, 193
  lithic pellets, 191, 193
  micrite mounds, 191
  microfacies:
    description, 191, 193
    filiation, 193
  model, 193
  ostracods, 191, 193
  paleoclimatology, 559
  paleogeography, 559
  quartz arenites, 193
  sampling techniques, 191
  shallowing-upward sequence, 193
  sponge spicules, 191, 193
  stromatactis, 193
  stromatoporoids, 193
  trilobites, 191, 193
  typical sequence, 193
  vadose ooids, 193
  vadose pisooids, 193
Nodular anhydrite, Maquoketa phosphorites, 483
Nowak, F.J., 25, 251
Nuia:
  Joachim Dolomite, 428, 439
  Pogonip Group, *Nuia* model, 213
  Traverse Group, 368, 375
Nuia sibirica, Pogonip Group, 209–10, 212
Nummulitids:
  Amapá Formation:
    model 1, 275
    model 2, 281–82

O

Okhravi, R., 28, 428, 491
Olistoliths, Amapá Formation, model 2, 287
Olistostromes, Amapá Formation, model 2, 280, 287
Oncoids:
  Bird Spring Group, Morrowan, 92
  Bonfim Formation, 417
  Burnt Bluff Group, 446
  Glen Dean Formation, E shelf Illinois Basin, 321, 323
  Iola Formation, 167
  Itaituba-Nova Olinda Formations, 103
  Joachim Dolomite, 428, 437
  Kinkaid Formation, model 1, 135
  lacustrine models, 558
  Lagoa Feia Formation, 498
  Macaé Formation, 262–64
  Middle Regencia Member, 115
  platforms with frontal bioaccumulated to hydrodynamic buildups, 546
  platforms with frontal bioconstructed to hydrodynamic buildups, 555
  Pogonip Group, 212
    oncoid model, 213
  ramps, 532
  Salem Limestone, SW Illinois, 303
Ooids:
  Allentown Dolomite, 419, 421
  allochthonous, Ste. Genevieve Limestone, 314–15, 318
  Amapá Formation, model 1, 275
  Atokan, 381, 383
  autochthonous, Ste. Genevieve Limestone, 314–15, 318

Ooids (*cont.*)
  Bird Spring Group:
    Atokan-Lower Missourian, 96
    Morrowan, 92, 96
    Upper Missourian-Wolfcampian, 253
  Bonfim Formation, 412–13, 417
  broken, Ste. Genevieve Limestone, 309
  Burnt Bluff Group, 446
  cerebroid, Ste. Genevieve Limestone, 310
  complete, Ste. Genevieve Limestone, 309–10, 313
  distorted, Allentown Dolomite, 421
  Glen Dean Formation:
    E shelf Illinois Basin, 321, 323
    SW shelf Illinois Basin, 331–33
  half-moon:
    Allentown Dolomite, 421
    Shakopee Dolomite, 423
  Iola Formation, 167, 171
  Itaituba-Nova Olinda Formations, 103
  Jurassic, Morcles Nappe, Alps, 475
  kerolitic, Lagoa Feia Formation, 498, 503
  Kinkaid Formation, model 1, 135
  lacustrine models, 558
  Lower Regencia Member, 110
  Macaé Formation, 262–64
  Menard Formation, 86
  Middle Regencia Member, 115
  normal:
    Quintuco-Loma Montosa Formation, 337, 340, 342
    Shakopee Dolomite, 423, 425
  platforms with frontal bioaccumulated to hydrodynamic buildups, 545–46
  platforms with frontal bioconstructed to hydrodynamic buildups, 555
  ramps, 532
  reworked, Quintuco-Loma Montosa Formation, 335, 337, 340
  Rundle Group, 153
  St. Louis Limestone, model 3, 43
  Salem Limestone:
    SE Illinois, 121, 123
    SW Illinois, 303, 305
  superficial:
    Joachim Dolomite, 429, 440
    Quintuco-Loma Montosa Formation, 340, 342
Operculina, Miocene, Visayan Islands, 453–54, 456
Orbitoids, Miocene, Visayan Islands, 453–57
Orbulinids, Miocene, Visayan Islands, 457
Ordovician:
  paleoclimatology, 559
  paleogeography, 559
  storm deposits, 562
Organic carbon, Lagoa Feia Formation, 498
Organic matter, Traverse Group, 373
Orthonella, Salem Limestone, SE Illinois, 123
Ostracods:
  Amapá Formation, model 1, 275
  Arrow Canyon Formation-Crystal Pass Limestone, 161
  Atokan, 377, 379, 381, 383
  Bailey Limestone, 473–74
  Beaverhill Lake Formation, 362–64
  Bird Spring Group:
    Atokan-Lower Missourian, 96
    Morrowan, 92
  Brereton Limestone:
    crinoid-brachiopod buildup model, 177–78
    phylloid algae buildup model, 177
  Burlington Limestone, arenite suite, 221
  Burnt Bluff Group, 441, 446
  Cayugan, N Ohio, 159
  Edwardsville Formation, 233
  Glen Dean Formation:
    E shelf Illinois Basin, 319, 321, 323
    SW shelf Illinois Basin, 330–32
  Itaituba-Nova Olinda Formations, 103, 107
  Jeffersonville Limestone, coral zone, 352, 354
  Joachim Dolomite, 428–29, 437
  Jurassic, Morcles Nappe, Alps, 475
  Kimmeridgian-Portlandian, Salève, 398–99, 405
  Kinkaid Formation:
    model 1, 135
    models 2–2A, 143
    model 3, 144
  Lagoa Feia Formation, 498, 500, 502

Ostracods (*cont.*)
 Lower Regencia Member, 108, 110
 Macaé Formation, 259, 262–64
 Maquoketa phosphorites, 483, 485, 487, 491
 Menard Formation, 86
 Middle Regencia Member, 115
 Miocene, Visayan Islands, 451, 453–57
 Niagaran:
  central Indiana, 465–467
  N Indiana, 191, 193
 Platteville Group, 59
 Pogonip Group, oncoid model, 215
 Quintuco-Loma Montosa Formation, 335, 337, 340
 Rundle Group, 153
 Ste. Genevieve Limestone, 309–10, 313
 St. Louis Limestone:
  model 1, 41
  model 2, 41
  model 3, 43
 Salem Limestone:
  SE Illinois, 119, 121, 123
  SW Illinois, 303, 305
 Traverse Group, 367–68, 371
Oysters, Chachao Formation, 195, 201–2

P

Page, N.J., 419
Paleoclimatology:
 Allentown Dolomite, 559
 Arrow Canyon Formation-Crystal Pass Limestone, 559
 Atokan, 561
 Barra Nova Formation, 561
 Beaverhill Lake Formation, 559
 Bird Spring Group:
  Morrowan to Lower Missourian, 561
  Upper Missourian-Wolfcampian, 561
 Bonfim Formation, 561
 Brereton Limestone, 561
 Burnt Bluff Group, 559
 Cambrian, 559
 Cayugan, N Ohio, 559
 Cenozoic, 561
 Chachao Formation, 561
 Devonian, 559
 Galena Group, 559
 Glen Dean Formation:
  E shelf Illinois Basin, 559, 561
  SW shelf Illinois Basin, 559, 561
 Iola Formation, 561
 Itaituba-Nova Olinda Formations, 561
 Jeffersonville Limestone:
  buildup, 559
  coral zone, 559
 Joachim Dolomite, 559
 Kinkaid Formation, 559, 561
 Lagoa Feia Formation, 561
 Macaé Formation, 561
 Maquoketa phosphorites, 495
 Menard Formation, 559, 561
 Mesozoic, 561
 Mississippian, 559, 561
 models, 558–59, 561–62
 Monte Cristo Group, 561
 Niagaran, N Indiana, 559
 Ordovician, 559
 Pennsylvanian, 561
 Platteville Group, 559
 Pogonip Group, 559
 Quintuco-Loma Montosa Formation, 561
 Rundle Group, 561
 Ste. Genevieve Limestone, 559, 561
 St. Louis Limestone, 559, 561
 Salem Limestone:
  SE Illinois, 559, 561
  SW Illinois, 559, 561
 Shakopee Dolomite, 559
 Silurian, 559
 Traverse Group, 559
Paleoecological community, 4
Paleoecological indexes, Macaé Formation, 264

Paleogeography:
 Allentown Dolomite, 559
 Arrow Canyon Formation-Crystal Pass Limestone, 559
 Atokan, 561
 Barra Nova Formation, 561
 Beaverhill Lake Formation, 559
 Bird Spring Group:
  Morrowan to Lower Missourian, 561
  Upper Missourian-Wolfcampian, 561
 Bonfim Formation, 561
 Brereton Limestone, 561
 Burnt Bluff Group, 559
 Cambrian, 559
 Cayugan, N Ohio, 559
 Cenozoic, 561
 Chachao Formation, 561
 Devonian, 559
 Galena Group, 559
 Glen Dean Formation:
  E shelf Illinois Basin, 559, 561
  SW shelf Illinois Basin, 559, 561
 Iola Formation, 561
 Itaituba-Nova Olinda Formations, 561
 Jeffersonville Limestone:
  buildup, 559
  coral zone, 559
 Joachim Dolomite, 559
 Kinkaid Formation, 559, 561
 Lagoa Feia Formation, 561
 Macaé Formation, 561
 Maquoketa phosphorites, 495
 Menard Formation, 559, 561
 Mesozoic, 561
 Mississippian, 559, 561
 models, 558–59, 561–62
 Monte Cristo Group, 561
 Niagaran, N Indiana, 559
 Ordovician, 559
 Pennsylvanian, 561
 Platteville Group, 559
 Pogonip Group, 559
 Quintuco-Loma Montosa Formation, 561
 Rundle Group, 561
 Ste. Genevieve Limestone, 559, 561
 St. Louis Limestone, 559, 561
 Salem Limestone:
  SE Illinois, 559, 561
  SW Illinois, 559, 561
 Shakopee Dolomite, 559
 Silurian, 559
 Traverse Group, 559
Paleotextularia, Itaituba-Nova Olinda Formations, 106
Panopea, Chachao Formation, 195
Parathurammina, Traverse Group, 371
Paréjas, E., 1
Pedogenesis, Lagoa Feia Formation, 507
Pelagic crinoids, Jurassic, Morcles Nappe, Alps, 475
Pelecypods:
 Beaverhill Lake Formation, 362
 Bonfim Formation, 407, 412–13, 417
 Brereton Limestone:
  crinoid-brachiopod buildup model, 177
  phylloid algae buildup model, 177
 Chachao Formation, 195, 201–2
 Galena Group, 69
 Glen Dean Formation:
  E shelf Illinois Basin, 321
  SW shelf Illinois Basin, 330–31
 Itaituba-Nova Olinda Formations, 101, 103, 107
 Jeffersonville Limestone:
  buildup, 359
  coral zone, 352, 354
 Kimmeridgian-Portlandian, Salève, 399, 405
 Kinkaid Formation, model 1, 132, 135
 Lagoa Feia Formation, 498, 500, 502–3
 Lower Regencia Member, 110
 Macaé Formation, 259, 262–64
 Maquoketa phosphorites, 483, 485, 487
 Middle Regencia Member, 115
 Miocene, Visayan Islands, 451, 453–57
 Platteville Group, 59

Pelecypods (cont.)
    Pogonip Group, oncoid model, 216
    Quintuco-Loma Montosa Formation, 335, 337, 340
    Rundle Group, 153
    Salem Limestone:
        SE Illinois, 119, 121, 123
        SW Illinois, 303, 305
Pellets, fecal:
    Atokan, 377, 379, 381, 383
    Bird Spring Group:
        Atokan-Lower Missourian, 96
        Morrowan, 92
        Upper Missourian-Wolfcampian, 253
    Bonfim, 407, 412–13, 417
    Brereton Limestone, 175–76
    Galena Group, 69
    Glen Dean Formation, SW shelf Illinois Basin, 330–31
    Itaituba-Nova Olinda Formation, 103
    Joachim Dolomite, 429
    Kinkaid Formation:
        model 1, 135
        models 2–2A, 143
        model 3, 144
    Lower Regencia Member, 110
    Middle Regencia Member, 115
    Monte Cristo Group, 249
    Niagaran, central Indiana, 465
    platforms with frontal bioaccumulated to hydrodynamic buildups, 545
    ramps, 532
    St. Louis Limestone:
        model 1, 39
        model 2, 41
        model 3, 43
    Traverse Group, 367–68, 371, 373
Pellets, lithic:
    Allentown Dolomite, 419, 421
    Arrow Canyon Formation-Crystal Pass Limestone, 161
    Bailey Limestone, 469, 473
    Beaverhill Lake Formation, 362–63
    Bird Spring Group:
        Atokan-Lower Missourian, 96
        Morrowan, 92, 96
        Upper Missourian-Wolfcampian, 253
    Bonfim Formation, 407, 412–13, 417
    Cayugan, N Ohio, 157, 159
    Galena Group, 69
    Glen Dean Formation, E shelf Illinois Basin, 319, 321, 323
    Jeffersonville Limestone:
        buildup, 358–59
        coral zone, 354
    Joachim Dolomite, 429, 431, 439–40
    Jurassic, Morcles Nappe, Alps, 475
    Kimmeridgian-Portlandian, Salève, 398–99
    Kinkaid Formation:
        model 1, 135
        models 2–2A, 143
        model 3, 144
    Lower Regencia Member, 110
    Middle Regencia Member, 115
    Niagaran:
        central Indiana, 465
        N Indiana, 191, 193
    platforms with frontal bioaccumulated to hydrodynamic buildups, 545
    Pogonip Group:
        Nuia model, 213
        oncoid model, 213
    Quintuco-Loma Montosa Formation, 337, 340
    Rundle Group, 153
    Ste. Genevieve Limestone, 309–10, 313
    St. Louis Limestone:
        model 1, 39
        model 2, 41
        model 3, 43
    Salem Limestone:
        SE Illinois, 121, 123
        SW Illinois, 303, 305
    Shakopee Dolomite, 423, 425
    Traverse Group, 367–68, 371, 373
Pelmatozoans:
    Iola Formation, 167, 171
    Joachim Dolomite, 428, 437

Pelmatozoans (cont.)
    Kinkaid Formation:
        model 1, 132, 135
        models 2–2A, 143
        model 3, 144
    Platteville Group, 59
    See also Crinoids
Pennsylvanian:
    paleoclimatology, 561
    paleogeography, 561
    storm deposits, 562
Permeability, Salem Limestone, SE Illinois, 119, 121, 123, 128
Persian Gulf, cyclicity, 565
Peryt, T., 4
Phosphates:
    Amapá Formation, model 1, 275
    Iola Formation, 167
    Macaé Formation, 259
    platform with frontal bioaccumulated to hydrodynamic buildups, 546
    platform with frontal bioconstructed to hydrodynamic buildups, 555
    Quintuco-Loma Montosa Formation, 335
    ramps, 532
Phosphatic intraclasts, Maquoketa phosphorites, 483, 485, 487
Phosphatic nodules, Maquoketa phosphorites, 485, 487
Phosphatic ooids, Maquoketa phosphorites, 485, 487, 491
Phosphatic pellets, Maquoketa phosphorites, 483, 485, 487, 491
Phosphatic spherulites, Maquoketa phosphorites, 483, 491
Phosphorites:
    Brereton Limestone, 175–76
    Maquoketa, 482–95
Phylloid algae:
    Atokan, 381
    Brereton Limestone:
        crinoid-brachiopod buildup model, 178
        phylloid algae buildup model, 177
    Iola Formation, 167, 171
Phytoplankton, Edwardsville Formation, 235
Pisoids, Rundle Group, 153
Pisolitization, Lagoa Feia Formation, 507
Planktonic constituents, models, 518
Planktonic foraminifers:
    Amapá Formation:
        model 1, 275
        model 2, 280–81
    Bonfim Formation, 407
    Macaé Formation, 259, 263
    Miocene, Visayan Islands, 451, 453–57
Plant debris:
    Bonfim Formation, 417
    Iola Formation, 171
    Itaituba-Nova Olinda Formations, 103
    Lower Regencia Member, 108, 110
    Macaé Formation, 257, 259
    Middle Regencia Member, 110, 115
    Miocene, Visayan Islands, 451
    Salem Limestone, SW Illinois, 305
    Traverse Group, 365, 373
Plant root molds:
    Iola Formation, 171
    Kinkaid Formation, models, 2–2A, 143
    Lagoa Feia Formation, 498
Platforms with frontal bioaccumulated buildups:
    genetic classification, 534
    type 1, 534
Platforms with frontal bioaccumulated to hydrodynamic buildups:
    benthic fauna, 545
    bioturbation, 545
    clay minerals, 546
    detrital quartz, 546
    dolomitization, 546–47
    extrabasinal constituents, 546
    fair weather wave base, 544
    fecal pellets, 545
    general synthesis, 544–47
    genetic classification, 534, 536, 538–39, 541, 543–44
    glauconite, 546
    intraclasts, 545
    lithic pellets, 545
    oncoids, 546
    ooids, 545–46
    phosphates, 546

Platforms with frontal bioaccumulated to hydrodynamic buildups (cont.)
  porosity types, 547
  pyrite, 546
  reservoir generation, 547
  storm deposits, 547
  submarine morphology, 545
  tidal currents, 545
  type 1, 534, 536
  type 2, 536, 538
  type 3, 538–39
  type 4, 539, 541
  type 5, 541, 543
  type 6, 543–44

Platforms with frontal bioconstructed to hydrodynamic buildups:
  benthic fauna, 555
  detrital quartz, 555
  dolomitization, 555
  extrabasinal constituents, 555
  fair weather wave base, 554
  general synthesis, 554–57
  genetic classification, 547, 549–50, 552, 554
  glauconite, 555
  intraclasts, 555
  oncoids, 555
  ooids, 555
  phosphates, 555
  porosity types, 555
  pyrite, 555
  reservoir generation, 555
  storm deposits, 555
  submarine morphology, 554
  tidal currents, 554–55
  type 1, 547, 549
  type 2, 549–50
  type 3, 550, 552
  type 4, 552–554

Platteville Group:
  bioturbation textures, 57
  brachiopods, 59
  bryozoans, 59
  component variation, 59, 61
  detrital quartz, 57, 59
  dolomitization, dorag, 57, 59
  gastropods, 59
  integration, 519
  microfacies description, 59, 61
  models, 61
  ostracods, 59
  paleoclimatology, 559
  paleogeography, 559
  pelecypods, 59
  pelmatozoans, 59
  sampling techniques, 57
  statistical techniques, 59, 61
  storm deposits, 59, 562
  trilobites, 59
  typical sequence, 61
  Vermiporella, 59

Plectogyra, Itaituba-Nova Olinda Formations, 106

Pogonip Group:
  beachrock, 210
  bioturbation, 209–10
  exposure features, 212
  Girvanella, 209, 212
  microfacies description, 209–10, 212
  Nuia model:
    component variation, 213
    crinoids, 213
    detrital quartz, 213
    integration, 536, 538
    lithic pellets, 213
    lithoclasts, 213
    Nuia, 213
    shallowing-upward sequence, 213
  Nuia sibirica, 209–10, 212
  oncoid model:
    brachiopods, 215–16
    component variation, 213, 215–16
    crinoids, 213
    detrital quartz, 215

Pogonip Group (cont.)
  oncoid model (cont.)
    gastropods, 216
    integration, 538
    lithic pellets, 213
    oncoids, 213
    ostracods, 215
    pelecypods, 216
    shallowing-upward sequence, 213
    sponge spicules, 215
  oncoids, 212
  paleoclimatology, 559
  paleogeography, 559
  reworking processes, 210, 212
  sampling techniques, 209
  storm deposits, 212, 213, 216–18
  storm deposits, Fourier analysis, 216–17
  superposition of environments, 216–18
  typical section, 213

Polypora:
  Iola Formation, 167
  Itaituba-Nova Olinda Formations, 106

Porites, Amapá Formation, model 2, 286

Porosity:
  biomoldic:
    Amapá Formation, model 2, 287
    Atokan, 390
    Iola Formation, 173
    Lagoa Feia Formation, 506–7, 511
    Middle Regencia Member, 117
    Miocene, Visayan Islands, 456, 460
    Salem Limestone, SE Illinois, 121, 123
  burial:
    relation to stylolites, Atokan, 383, 386, 390, 398
    Miocene, Visayan Islands, 456, 460
  channel:
    Amapá Formation, model 2, 287
    Macaé Formation, 270
  classification, 12
  computer evaluation:
    Amapá Formation:
      model 1, 280
      model 2, 287, 296
  evolution:
    Amapá Formation:
      model 1, 279–80
      model 2, 287, 296
    Lagoa Feia Formation, 506–8
  fracture:
    Chachao Formation, 202
    Iola Formation, 173
    Middle Regencia Member, 117
    Salem Limestone, SE Illinois, 121
  intercrystalline:
    Amapá Formation:
      model 1, 280
      model 2, 287
    Lagoa Feia Formation, 506–8, 511
  interparticle:
    Amapá Formation, model 1, 279–80
    Lagoa Feia Formation, 496, 498, 506–7
    Macaé Formation, 264, 271
    Middle Regencia Member, 117
    Salem Limestone, SE Illinois, 121, 123
  intraparticle:
    Lagoa Feia Formation, 506–8, 511
    Salem Limestone, SE Illinois, 121, 123
  oomoldic:
    Atokan, 390
    Bonfim, 413
    Lower Regencia Member, 110
    Middle Regencia Member, 117
    Salem Limestone, SE Illinois, 121
  relationship to microfacies, Lagoa Feia Formation, 508–9, 511
  stylolitic:
    Atokan, 390
    Middle Regencia Member, 117
    Salem Limestone, SE Illinois, 121
  terminology, 12
  vuggy:
    Amapá Formation, model 2, 287

Porosity (cont.)
  vuggy (cont.)
    Atokan, 390
    Lagoa Feia Formation, 506–7, 511
    Macaé Formation, 264, 270–71
    Middle Regencia Member, 117
    Miocene, Visayan Islands, 456, 460
    Salem Limestone, SE Illinois, 121, 123
Porosity types:
  lacustrine models, 558
  Macaé Formation, 264, 270–71, 273
  models, 518
  platforms with frontal bioaccumulated to hydrodynamic buildups, 547
  platforms with frontal bioconstructed to hydrodynamic buildups, 555
  Quintuco Formation, 13
  ramps, 523, 531, 534
  ramps with bioaccumulated to hydrodynamic buildups, 528
  ramps with bioconstructed buildups, 530
  Ste. Genevieve Limestone, 314–15, 318
  Salem Limestone, SE Illinois, 119, 121, 123, 128
Porosity-permeability-microfacies relationships, Salem Limestone, SE Illinois, 124–25
Pray, L.C., 12
Proximality index, turbidites, Bailey Limestone, 474–75
Pseudo-ooids, Allentown Dolomite, 421
Ptychomla, Chachao Formation, 201
Purdy, E.G., 314
Pycnoporidium, Bonfim Formation, 497, 412
Pyrite:
  Allentown Dolomite, 419, 421
  Arrow Canyon Formation-Crystal Pass Limestone, 161
  Bailey Limestone, 469, 473
  Galena Group, 81
  Joachim Dolomite, 441
  Jurassic, Morcles Nappe, Alps, 478
  Macaé Formation, 259
  Maquoketa phosphorites, 483, 485, 487, 491
  Miocene, Visayan Islands, 457
  Niagaran, central Indiana, 465, 467
  platforms with frontal bioaccumulated to hydrodynamic buildups, 546
  platforms with frontal bioconstructed to hydrodynamic buildups, 555
  Quintuco-Loma Montosa Formation, 335
  ramps, 533

Q

Quartz, authigenic:
  Joachim Dolomite, 429, 431, 441
  See also Silicification
Quartz, detrital:
  Allentown Dolomite, 419, 421
  Arrow Canyon Formation-Crystal Pass Limestone, 161
  Atokan, 377
  Bailey Limestone, 473–74
  Bird Spring Group:
    Atokan-Lower Missourian, 96
    Morrowan, 92, 96
    Upper Missourian-Wolfcampian, 253
  Bonfim Formation, 417
  Brereton Limestone:
    crinoid-brachiopod buildup model, 177–78
    phylloid algae buildup model, 177
  Burnt Bluff Group, 446
  Edwardsville Formation, 233
  Galena Group, 69, 71
  Glen Dean Formation, E shelf Illinois Basin, 319, 321, 323
  Itaituba-Nova Olinda Formations, 103, 105–6
  Jeffersonville Limestone, coral zone, 352, 354
  Joachim Dolomite, 429, 431, 440–41
  Jurassic, Morcles Nappe, Alps, 475, 478
  lacustrine models, 558
  Lagoa Feia Formation, 496, 498, 500
  Lower Regencia Member, 108, 110
  Macaé Formation, 257, 259, 262–64
  Maquoketa phosphorites, 483, 485, 487, 491
  marginal replacement by carbonates, 419
  Menard Formation, 86
  Middle Regencia Member, 110, 115
  Miocene, Visayan Islands, 451, 457
  Monte Cristo Group, 249
  Niagaran, central Indiana, 465, 467
  platforms with frontal bioaccumulated to hydrodynamic buildups, 546

Quartz, detrital (cont.)
  platforms with frontal bioconstructed to hydrodynamic buildups, 555
  Platteville Group, 57, 59
  Pogonip Group:
    Nuia model, 213
    oncoid model, 215
  Quintuco-Loma Montosa Formation, 335, 337, 340
  ramps, 532–33
  Rundle Group, 153
  Ste. Genevieve Limestone, 307, 309–10, 313
  St. Louis Limestone:
    model 1, 41
    model 2, 41
    model 3, 41
  Salem Limestone:
    SE Illinois, 123
    SW Illinois, 305, 307
  Shakopee Dolomite, 423, 425
  Traverse Group, 365
Quintuco Formation:
  compaction, 13
  diagenesis, 12–13
  dolomitization:
    burial, 13
    dorag, 13
  micritization, 12
  neomorphism, 13
  porosity, 13
Quintuco-Loma Montosa Formation:
  anhydrite, chickenwire texture, 337, 340, 342
  annelids, 337
  bryozoans, 337
  calcispheres, 335
  crinoids, 335, 337
  dasyclads, 337, 340
  detrital quartz, 335, 337, 340
  dolomitization, sabkha, 337, 342
  echinoids, 335, 337, 340
  fan delta, 334
  gastropods, 337, 340
  glauconite, 335
  hardgrounds, 335
  intraclasts, 335, 337, 340
  lithic pellets, 337, 340
  microfacies description, 334–35, 337, 340
  mixed carbonate-fan delta slope, 335
  models, 340, 342
  ooids:
    normal, 337, 340, 342
    reworked, 335, 337, 340
    superficial, 340, 342
  ostracods, 335, 337, 340
  paleoclimatology, 561
  paleogeography, 561
  pelecypods, 335, 337, 340
  phosphates, 335
  polymictic conglomerate, 337
  pyrite, 335
  quartz feldspathic lithic arenite, 337
  red algae, 337
  sampling techniques, 334
  smaller benthic foraminifers, 335
  stromatolites, 337, 340
  superposition of environments, 340, 342
  typical sequence, 340

R

Radiolarians:
  Bailey Limestone, 469, 473–74
  Bonfim Formation, 407
  Jurassic, Morcles Nappe, Alps, 475
  Macaé Formation, 259
Ramps:
  benthic fauna, 531–32
  with bioaccumulated to hydrodynamic buildups:
    genetic classification, 523–25, 528
    porosity types, 528
    reservoir generation, 528
    type 1, 524
    type 2, 524–25
    type 3, 525

Ramps (*cont.*)
  with bioaccumulated to hydrodynamic buildups (*cont.*)
    type 4, 525
    type 5, 525, 528
  with bioconstructed buildups:
    genetic classification, 528, 530
    porosity types, 530
    reservoir generation, 530
    type 1, 528
    type 2, 528
    type 3, 528, 530
  bioturbation, 532
  clay minerals, 533
  detrital quartz, 532–33
  dolomitization, 533
  extrabasinal constituents, 532
  fair weather wave base, 530–31
  fecal pellets, 532
  genetic synthesis, 530–34
  glauconite, 532
  oncoids, 532
  ooids, 532
  phosphates, 352
  porosity types, 531, 534
  pyrite, 533
  reservoir generation, 534
  simple:
    genetic classification, 519–21, 523
    porosity types, 523
    reservoir generation, 523
    type 1, 519
    type 2, 519
    type 3, 519
    type 4, 519–20
    type 5, 521, 523
    type 6, 523
  storm deposits, 533–34
  submarine morphology, 530–31
  tidal currents, 530–31
Ranikothalia, Amapá Formation, model 1, 275
Rao, C.P., 307
Read, J.F., 6, 563
Reading, H.G., 9
Receptaculites, Galena Group, 61, 64, 69
Red algae:
  Amapá Formation, model 2, 280–82, 286–87
  Bonfim Formation, 407, 412–13, 417
  Iola Formation, 167
  Kimmeridgian-Portlandian, Salève, 399
  Macaé Formation, 259, 262–63
  Miocene, Visayan Islands, 453–56
  Quintuco-Loma Montosa Formation, 337
Reefs, 4
Reichelberger, J.L., 119
Relative bathymetry curve, first use, 1
Reservoir generation:
  Amapá Formation:
    model 1, 279–80
    model 2, 287, 296
  Atokan, 390
  Bonfim Formation, 417
  lacustrine models, 558
  Lagoa Feia Formation, 507–8
  Macaé Formation, 264, 268–71, 273
  Miocene, Visayan Islands, 460
  models, 518
  platforms with frontal bioaccumulated to hydrodynamic buildups, 547
  platforms with frontal bioconstructed to hydrodynamic buildups, 555
  ramps, 534
    with bioaccumulated to hydrodynamic buildup, 528
    with bioconstructed buildups, 530
    simple, 523
  Ste. Genevieve Limestone, 314–15, 318
  Salem Limestone, SE Illinois, 128
Reworking processes:
  Arrow Canyon Formation-Crystal Pass Limestone, 161
  Burnt Bluff Group, 446
  Cayugan, N Ohio, 159
  Menard Formation, 81
  Pogonip Group, 210, 212
Rich, D.W., 318

Roche, J.E., 83, 365
Rodriguez Schelotto, M.L., 12, 27
Roehl, P.O., 12
Ross, C.A., 569
Ross, J.R.P., 569
Rudistids:
  Jurassic, Morcles Nappe, Alps, 475
  Macaé Formation, 263
Rundle Group:
  bioturbation, 153
  brachiopods, 153
  bryozoans, 153
  crinoids, 153
  detrital quartz, 153
  echinoids, 153
  Endothyra, 153
  evaporites, sabkha, 154
  integration, 524–25
  lithic pellets, 153
  megacycles, 563
  microfacies description, 151, 153
  model, 154–55
  ooids, 153
  ostracods, 153
  paleoclimatology, 561
  paleogeography, 561
  pelecypods, 153
  pisoids, 153
  sampling techniques, 151
  shallowing-upward sequence, 153–54
  smaller benthic foraminifers, 153
  sponge spicules, 153
  superposition of environments, 155
  typical sequence, 153

S

Sacal, V., 24
Saccocoma, Jurassic, Morcles Nappe, Alps, 475
Ste. Genevieve Limestone:
  bioclastic sequence, 313
  bryozoans, 307, 309–10, 313
  comparison with Great Bahama Bank, 313–14
  crinoids, 307, 309–10, 313
  detrital feldspar, 313
  detrital muscovite, 309
  detrital quartz, 307, 309–10, 313
  glauconite, 313
  integration, 539, 541
  intraclasts, 309
  lithic pellets, 309–10, 313
  microfacies description, 307, 309–10, 313
  microfacies filiation, 313
  model, 313–14
  ooids:
    allochthonous, 314–15, 318
    autochthonous, 314–15, 318
    broken, 309, 313
    cerebroid, 310
    complete, 309–10, 313
  ostracods, 309–10, 313
  paleoclimatology, 559, 561
  paleogeography, 559, 561
  pelletoidal sequence, 313
  porosity types, 314–15, 318
  quartz arenites, 310, 313
  reservoir generation, 314–15, 318
  sampling techniques, 307
  shallowing-upward sequences, 313
  smaller benthic foraminifers, 313
  typical sequence, 313
St. Louis Limestone:
  anhydritization, 56
  basinal evolution, 44–45
  brecciation, 35
  compaction, 46
  diagenesis, 45–46, 48, 50, 56
  dolomitization:
    burial, 56
    sabkha, 46, 48
  evaporite distribution, 44
  fracturation, 48, 50, 56

St. Louis Limestone (cont.)
 inorganic components, 34
 megacycles, 563
 micritization, 46
 microfacies association, 39
 microfacies description, 35, 37, 39
 models, 39
 model 1:
  brachiopods, 39, 41
  bryozoans, 39, 41
  calcispheres, 41
  characteristics, 39
  component variation, 39, 41
  crinoids, 39
  detrital quartz, 41
  dolomitization, sabkha, 39
  echinoid spines, 41
  fecal pellets, 39
  integration, 523
  intraclasts, 39
  lithic pellets, 39
  micrite matrix, 41
  ostracods, 41
  relative energy, 41
  shallowing-upward sequence, 39
  sparite cement, 41
  sponge spicules, 41
 model 2:
  brachiopods, 41
  bryozoans, 41
  calcispheres, 41
  characteristics, 41
  component variation, 41, 43
  crinoids, 41
  detrital quartz, 41
  dolomitization, sabkha, 41
  Endothyra, 41
  fecal pellets, 41
  integration, 525
  intraclasts, 41
  lithic pellets, 41
  micrite matrix, 41
  ostracods, 41
  shallowing-upward sequence, 39
  sparite cement, 41
  sponge spicules, 41
 model 3:
  brachiopods, 43
  bryozoans, 43
  calcispheres, 43
  characteristics, 43
  component variation, 43
  crinoids, 43
  detrital quartz, 43
  dolomitization, sabkha, 43
  fecal pellets, 43
  integration, 539
  intraclasts, 43
  lithic pellets, 43
  micrite matrix, 43
  ooids, 43
  ostracods, 43
  shallowing-upward sequence, 39
  sparite cement, 43
  sponge spicules, 43
  stromatolites, 43
 organic components, 34
 paleoclimatology, 559, 561
 paleogeography, 559, 561
 petrographic techniques, 34–35
 sampling techniques, 34
 shallowing-upward sequence, 39
 silicification, 50, 56
 Stachyodes, 39
 statistical techniques, 35
 stylolitization, 56
 superposition of models, 43–44
Salem Limestone, SE Illinois:
 anhydritization, 119, 121, 126, 128
 bioclastic bar model, 124
 bioturbation, 121, 123

Salem Limestone, SE Illinois (cont.)
 brachiopods, 119, 121, 123
 bryozoans, 119, 121, 123
 corals, 119, 121, 123
 crinoids, 119, 121, 123
 detrital quartz, 123
 diagenesis, 125–26, 128
 dolomitization:
  burial, 119, 121, 123, 126, 128
  dorag, 124
 Endothyra, 119, 121, 123–24, 128
 fracturation, 126, 128
 Girvanella, 123
 integration, 525
 intraclasts, 121, 123
 Koninckopora, 121, 123
 lithic pellets, 121, 123
 microfacies:
  description, 119, 121, 123
  permeability, 119, 121, 123, 128
  porosity, 119, 121, 123, 128
 model, general, 125
 ooids, 121, 123
 oolitic bar model, 124–25
 Orthonella, 123
 ostracods, 119, 121, 123
 paleoclimatology, 559, 561
 paleogeography, 559, 561
 pelecypods, 119, 121, 123
 porosity:
  biomoldic, 121, 123
  fracture, 121
  interparticle, 121, 123
  intraparticle, 121, 123
  oomoldic, 121
  stylolitic, 121
  vuggy, 121, 123
 porosity-permeability-microfacies relationships, 124–25
 reservoir generation, 128
 sampling techniques, 119
 shallowing-upward sequence, 123–24
 silicification, 126, 128
 smaller benthic foraminifers, 119, 121, 123
 small-scale cycles, 563
 Stachyodes, 123
 stylolitization, 126, 128
 trilobites, 123
Salem Limestone, SW Illinois:
 anhydritization, 305
 beachrock, 307
 brachiopods, 303, 305
 bryozoans, 303, 305
 calcispheres, 303, 305
 crinoids, 303, 305
 detrital quartz, 305, 307
 dolomitization:
  dorag, 307
  sabkha, 307
 echinoids, 305
 Endothyra, 303, 305
 fenestral textures, 305
 gastropods, 305
 integration, 541
 intraclasts, 303, 305
 Koninckopora, 305
 lithic pellets, 303, 305
 microfacies description, 303, 305
 model, 305, 307
 oncoids, 303
 ooids, 303, 305
 ostracods, 303, 305
 paleoclimatology, 559, 561
 paleogeography, 559, 561
 pelecypods, 303, 305
 plant debris, 305
 sampling techniques, 303
 shallowing-upward sequence, 305
 smaller benthic foraminifers, 305
 sponge spicules, 303–5
 superposition of environments, 307

Salem Limestone, SW Illinois (*cont.*)
  trilobites, 305
  vadose silt, 305
Schlager, W., 568
Schneidermann, N., 4
Scholle, P.A., 9
Scotese, C.R., 558
Seiches:
  lacustrine models, 558
  Lagoa Feia Formation, 502
Selenitic gypsum, Maquoketa phosphorites, 495
Shakopee Dolomite:
  detrital quartz, 423, 425
  dolomitization, sabkha, 425
  integration, 550
  lithic pellets, 423, 425
  microfacies description, 423, 425
  model, 425
  ooids:
    half-moon, 423
    normal, 423, 425
  paleoclimatology, 559
  paleogeography, 559
  sampling techniques, 423
  shallowing-upward sequence, 425
  silicification, submarine, 425
  small-scale cycles, 563
  storm deposits, 425
  stromatolites, 423, 425
  typical sequence, 425
Shales, Glen Dean Formation, SW shelf Illinois Basin, 330
Shallowing-upward sequence:
  Allentown Dolomite, 421, 422–23
  Amapá Formation:
    model 1, 278
    model 2, 287
  Arrow Canyon Formation-Crystal Pass Limestone, 161
  Atokan, 381
  Beaverhill Lake Formation, 363–64
  Bird Spring Group:
    Atokan-Lower Missourian, 96
    Morrowan, 92, 96
    Upper Missourian-Wolfcampian, 253
  Burnt Bluff Group:
    Nasbro buildup, 446
    Sturgeon Bay buildup, 448
  Chachao Formation, 201–2
  characteristics, 6, 562–63
  Edwardsville Formation, 233
  Galena Group, 64, 66
  general interpretation, 568–70
  Glen Dean Formation:
    E shelf Illinois Basin, 323
    SW shelf Illinois Basin, 331–32
  ideal type, 25
  Iola Formation, 171
  Jeffersonville Limestone, coral zone, 354–55
  Joachim Dolomite, 431, 437
  Kimmeridgian-Portlandian, Salève, 405
  Kinkaid Formation, model 1, 132
  Lower Regencia Member, 110
  Macaé Formation, 264
  Maquoketa phosphorites, 487, 491
  Menard Formation, 86
  Middle Regencia Member, 115
  Monte Cristo Group, 244
  Niagaran, N Indiana, 193
  Pogonip Group:
    Nuia model, 213
    oncoid model, 213
  Rundle Group, 153–54
  Ste. Genevieve Limestone, 313
  St. Louis Limestone:
    model 1, 39
    model 2, 39
    model 3, 39
  Salem Limestone:
    SE Illinois, 123–24
    SW Illinois, 305
  Shakopee Dolomite, 425
Shaver, R.H., 193

Silicification:
  burial:
    Burlington Limestone, arenite suite, 221
    Galena Group, 81
    Macaé Formation, 273
  Glen Dean Formation:
    E shelf Illinois Basin, 327
    SW shelf Illinois Basin, 333
  Itaituba-Nova Olinda Formations, 103
  Kimmeridgian-Portlandian, Salève, 405
  Lagoa Feia Formation, 498, 500
  Maquoketa phosphorites, 483, 485, 487, 495
  St. Louis Limestone, 50, 56
  Salem Limestone, SE Illinois, 126, 128
  submarine:
    Bailey Limestone, 474
    Burlington Limestone, 229–30
    Galena Group, 77, 79, 81
    Shakopee Dolomite, 425
Siltstones:
  Edwardsville Formation, 233, 235
  Iola Formation, 171
  Lagoa Feia Formation, 498
Silurian:
  paleoclimatology, 559
  paleogeography, 559
  storm deposits, 562
Slansky, M., 482
Slumping:
  Bonfim Formation, 407, 412
  Niagaran, central Indiana, 467
Soderman, J.G.W., 231, 441
Solenopora, Bonfim Formation, 407, 412
SOUPAC, 31, 35
Sparite cement, St. Louis Limestone:
  model 1, 41
  model 2, 41
  model 3, 43
Sparmicritization, Lagoa Feia Formation, 507
Spastolites, Joachim Dolomite, 429
Sphaeroidinellopsis, Miocene, Visayan Islands, 457
Spiroclypeus, Miocene, Visayan Islands, 456
Sponges, Kimmeridgian-Portlandian, Salève, 399
Sponge spicules:
  Atokan, 377, 383
  Bailey Limestone, 469, 473–74
  Beaverhill Lake Formation, 363–64
  Brereton Limestone:
    crinoid-brachiopod buildup model, 177
    phylloid algae buildup model, 177
  Chachao Formation, 195, 202
  Edwardsville Formation, 232
  Glen Dean Formation, E shelf Illinois Basin, 319, 321, 323
  Glen Dean Formation, SW shelf Illinois Basin, 330–32
  Iola Formation, 167
  Itaituba-Nova Olinda Formations, 103, 107
  Kinkaid Formation:
    model 1, 135
    models 2–2A, 143
    model 3, 144
  Menard Formation, 86
  Miocene, Visayan Islands, 451, 453–57
  Monte Cristo Group, 249
  Niagaran, N Indiana, 191, 193
  Pogonip Group:
    Nuia model, 215
    oncoid model, 215
  Rundle Group, 153
  St. Louis Limestone:
    model 1, 41
    model 2, 41
    model 3, 43
  Salem Limestone, SW Illinois, 303, 305
  Traverse Group, 367
Spongiostromata:
  Burnt Bluff Group, 441, 446
  Cayugan, N Ohio, 156
  Joachim Dolomite, 429
Spoon Formation, 99

Stachyodes:
  St. Louis Limestone, 39
  Salem Limestone, SE Illinois, 123
Storm deposits:
  Allentown Dolomite, 421
  Arrow Canyon Formation-Crystal Pass Limestone, 161
  Burlington Limestone, 226, 229–30
  Cambrian, 562
  Cayugan, N Ohio, 157, 159
  characteristics, 561
  Devonian, 562
  Fourier analysis, 216–17
  Galena Group, 64, 71, 76–77, 79, 81
  hummocky stratification, 561
  hurricane model, 561–62
  Iola Formation, 171
  Jeffersonville Limestone, coral zone, 354–55
  Joachim Dolomite, 429, 431
  Kinkaid Formation, models 2–2A, 143
  lacustrine models, 558
  Lagoa Feia Formation, 503
  Mississippian, 562
  Ordovician, 562
  Pennsylvanian, 562
  platforms with frontal bioaccumulated to hydrodynamic buildups, 547
  platforms with frontal bioconstructed to hydrodynamic buildups, 555
  Platteville Group, 59, 562
  Pogonip Group, 212, 213, 216–18
  ramps, 533–34
  Shakopee Dolomite, 425
  Silurian, 562
  Traverse Group, 373
Stricker, G.D., 209
Stromatactis, Niagaran, N Indiana, 193
Stromatolites:
  Allentown Dolomite, 421
  Atokan, 381, 383
  Bonfim Formation, 417
  Burnt Bluff Group, 441, 446, 448
  Cayugan, N Ohio, 156–57
  Itaituba-Nova Olinda Formations, 103, 107
  Jeffersonville Limestone, coral zone, 354
  Joachim Dolomite, 429, 431, 437
  Kinkaid Formation:
    model 1, 132
    models 2–2A, 137, 143
  Lower Regencia Member, 108, 110
  Monte Cristo Group, 244
  Quintuco-Loma Montosa Formation, 337, 340
  St. Louis Limestone, model 3, 43
  Shakopee Dolomite, 423–25
  Traverse Group, 371, 373, 375
Stromatoporoids:
  Arrow Canyon Formation-Crystal Pass Limestone, 161
  Beaverhill Lake Formation, 362–64
  Burnt Bluff Group, 446, 448
  Jeffersonville Limestone:
    buildup, 358–59
    coral zone, 354–55
  Kimmeridgian-Portlandian, Salève, 399, 405
  Niagaran:
    central Indiana, 465
    N Indiana, 193
  Traverse Group, 368, 375
Stylolites:
  Atokan:
    distribution, 383
    nonsutured, 377
    relations to authigenic feldspars, 383, 386, 390
    relations to porosity, 383, 386, 390, 398
    sutured, 377
  St. Louis Limestone, 56
  Salem Limestone, SE Illinois, 126, 128
Subaqueous basic volcanism model, Lagoa Feia Formation, 503
Sublacustrine morphology, lacustrine models, 557–58
Submarine morphology:
  marine models, 517–18
  platforms with frontal bioaccumulated to hydrodynamic buildups, 545
  platforms with frontal bioconstructed to hydrodynamic buildups, 554
  ramps, 530–31

Subsidence, carbonate rock models, 516
Syringopora, Monte Cristo Group, 241, 249

T

Tectonism, models, 516
Tetrataxis:
  Atokan, 381
  Itaituba-Nova Olinda Formations, 106
Textoris, D.A., 24, 33, 156, 191, 193, 465
Thamnopora, Traverse Group, 368
Tidal currents:
  platforms with frontal bioaccumulated to hydrodynamic buildups, 545
  platforms with frontal bioconstructed to hydrodynamic buildups, 554–55
  ramps, 530–31
Tintinnoids, Jurassic, Morcles Nappe, Alps, 475
Toomey, D.F., 4
Trace-elements, Maquoketa phosphorites, 491, 494
Trace-metals, Traverse Group, 373
Traverse Group:
  anhydrite, 371
  bioturbation, 367–68
  boron, 373
  brachiopods, 367–68
  bryozoans, 367–68
  carbonate geochemistry, 373
  Chara, 365
  clay minerals, 365
  collapse breccias, 365
  corals, 367–68, 375
  crinoids, 367–68
  Depasophyllum adentum, 367, 375
  detrital quartz, 365
  evaporites, 365, 371
  exposure features, 365, 373
  Favosites, 368
  fecal pellets, 367–68, 371, 373
  fenestral textures, 371
  halite, 371
  Hexagonaria, 368, 375
  integration, 554
  intraclasts, 367–68, 371, 373
  lithic pellets, 367–68, 371, 373
  microfacies description, 365, 367–68, 371, 373
  models, 373, 375
  Nuia, 368, 375
  organic matter, 373
  ostracods, 367–68, 371
  paleoclimatology, 559
  paleogeography, 559
  paleosalinity, 373
  Parathurammina, 371
  plant debris, 365, 373
  sampling techniques, 365
  sponge spicules, 367
  storm deposits, 373
  stromatolites, 371, 373, 375
  stromatoporoids, 368, 375
  Thamnopora, 368
  trace-metals, 373
  typical sections, 373, 375
  Welleria aftonensis, 371
Trilobites:
  Atokan, 377
  Bailey Limestone, 473
  Brereton Limestone:
    crinoid-brachiopod buildup model, 177
    phylloid algae buildup model, 177
  Burlington Limestone, arenite suite, 221
  Glen Dean Formation, E shelf Illinois Basin, 321
  Itaituba-Nova Olinda Formations, 101, 103
  Jeffersonville Limestone, coral zone, 354
  Kinkaid Formation, model 1, 132
  Maquoketa phosphorites, 483, 485, 491
  Niagaran, N Indiana, 191, 193
  Platteville Group, 59
  Salem Limestone, SE Illinois, 123
  Salem Limestone, SW Illinois, 305
Trocholina, Bonfim Formation, 412

Tuberitina:
  Atokan, 377, 379, 381
  Iola Formation, 167
Turbidites:
  Bailey Limestone:
    Bouma sequence, 474
    setting, 468
    textural features, 468, 474
  Glen Dean Formation, SW shelf Illinois Basin, 333
  Jurassic, Morcles Nappe, Alps:
    correlation, 478
    horizontal distribution, 478
    horizontal grading, 478
    setting, 475
    slumping textures, 478
    textural features, 475, 478
    typical sequence, 478
  Macaé Formation, 259
  Miocene, Visayan Islands, 457
  Niagaran, central Indiana:
    proximal to distal correlation, 467–68
    setting, 465
    textural features, 465, 467
    typical distal sequence, 467
    typical median sequence, 467
    typical proximal sequence, 467

V

Vadose ooids, Niagaran, N Indiana, 193
Vadose pisoids, Niagaran, N Indiana, 193
Vadose silt:
  Atokan, 383
  Chachao Formation, 195
  Iola Formation, 171
  Joachim Dolomite, 428–29, 431
  Kinkaid Formation, model 3, 144

Vadose silt (*cont.*)
  Miocene, Visayan Islands, 456, 460
  Salem Limestone, SW Illinois, 305
Vail, P.R., 565
Vermiporella, Platteville Group, 59
Volcanic lithoclasts, Miocene, Visayan Islands, 451, 455–56
Von Bergen, D., 377

W

Wacke:
  lithic feldspathic, Lagoa Feia Formation, 496
  quartz, Glen Dean Formation, E shelf Illinois Basin, 323
Walker, R.G., 9, 516
Walpole, R.L., 145
Walther's law, 6, 25, 33
Wanless, H.R., 1
Wave base, fair weather:
  lacustrine models, 557–58
  platforms with frontal bioaccumulated to hydrodynamic buildups, 544
  platforms with frontal bioconstructed to hydrodynamic buildups, 554
  ramps, 530–31
Welleria aftonensis, Traverse Group, 371
Wilkinson, B.H., 563
Wilson, J.L., 9, 33
Wolff, B., 2, 275
Worm tubes:
  Cayugan, N Ohio, 156–57, 159
  Macaé Formation, 259, 262–63

Y

Yang, C., 2

Z

Zadnik, V.E., 28, 81, 419, 441, 467
Zenger, D.H., 9, 12
Zeolitization, Lagoa Feia Formation, 498, 500
Ziegler, A.M., 558